Lecture Notes in Computer Science 4051

Commenced Publication in 1973
Founding and Former Series Editors:
Gerhard Goos, Juris Hartmanis, and Jan van Leeuwen

Michele Bugliesi   Bart Preneel
Vladimiro Sassone   Ingo Wegener (Eds.)

Automata, Languages and Programming

33rd International Colloquium, ICALP 2006
Venice, Italy, July 10-14, 2006
Proceedings, Part I

 Springer

Volume Editors

Michele Bugliesi
Università Ca'Foscari
Dipartimento di Informatica
Via Torino 155, 30172 Venezia-Mestre, Italy
E-mail: bugliesi@dsi.unive.it

Bart Preneel
Katholieke Universiteit Leuven Department of Electrical Engineering-ESAT/COSIC
Kasteelpark Arenberg 10, 3001 Leuven-Heverlee, Belgium
E-mail: Bart.Preneel@esat.kuleuven.be

Vladimiro Sassone
University of Southampton
School of Electronics and Computer Science
SO17 1BJ, UK
E-mail: vs@ecs.soton.ac.uk

Ingo Wegener
Universität Dortmund
FB Informatik, LS2
Otto-Hahn-Str. 14, 44221 Dortmund, Germany
E-mail: ingo.wegener@uni-dortmund.de

Library of Congress Control Number: 2006928089

CR Subject Classification (1998): F, D, C.2-3, G.1-2, I.3, E.1-2

LNCS Sublibrary: SL 1 – Theoretical Computer Science and General Issues

ISSN	0302-9743
ISBN-10	3-540-35904-4 Springer Berlin Heidelberg New York
ISBN-13	978-3-540-35904-3 Springer Berlin Heidelberg New York

Springer is a part of Springer Science+Business Media

springer.com

© Springer-Verlag Berlin Heidelberg 2006
Printed in Germany

Typesetting: Camera-ready by author, data conversion by Scientific Publishing Services, Chennai, India
Printed on acid-free paper   SPIN: 11786986   06/3142   5 4 3 2 1 0

Preface

ICALP 2006, the 33rd edition of the International Colloquium on Automata, Languages and Programming, was held in Venice, Italy, July 10–14, 2006. ICALP is a series of annual conferences of the European Association for Theoretical Computer Science (EATCS) which first took place in 1972. This year, the ICALP program consisted of the established track A (focusing on algorithms, automata, complexity and games) and track B (focusing on logic, semantics and theory of programming), and of the recently introduced track C (focusing on security and cryptography foundation).

In response to the call for papers, the Program Committee received 407 submissions, 230 for track A, 96 for track B and 81 for track C. Out of these, 109 papers were selected for inclusion in the scientific program: 61 papers for Track A, 24 for Track B and 24 for Track C. The selection was made by the Program Committee based on originality, quality, and relevance to theoretical computer science. The quality of the manuscripts was very high indeed, and several deserving papers had to be rejected.

ICALP 2006 consisted of four invited lectures and the contributed papers. This volume of the proceedings contains all contributed papers presented at the conference in Track A, together with the paper by the invited speaker Noga Alon (Tel Aviv University, Israel). A companion volume contains all contributed papers presented in Track B and Track C together with the papers by the invited speakers Cynthia Dwork (Microsoft Research, USA) and Prakash Panangaden (Mc Gill University, Canada). The program had an additional invited lecture by Simon Peyton Jones (Microsoft Research, UK), which does not appear in the proceedings.

ICALP 2006 was held in conjunction with the Annual ACM International Symposium on Principles and Practice of Declarative Programming (PPDP 2006) and with the Annual Symposium on Logic-Based Program Synthesis and Transformation (LOPSTR 2006). Additionally, the following workshops were held as satellite events of ICALP 2006: ALGOSENSORS 2006 - International Workshop on Algorithmic Aspects of Wireless Sensor Networks; CHR 2006 - Third Workshop on Constraint Handling Rules; CL&C 2006 - Classical Logic and Computation; DCM 2006 - 2nd International Workshop on Developments in Computational Models; FCC 2006 - Formal and Computational Cryptography; iETA 2006 - Improving Exponential-Time Algorithms: Strategies and Limitations; MeCBIC 2006 - Membrane Computing and Biologically Inspired Process Calculi; SecReT 2006 - 1st Int. Workshop on Security and Rewriting Techniques; WCAN 2006 - 2nd Workshop on Cryptography for Ad Hoc Networks.

We wish to thank all authors who submitted extended abstracts for consideration, the Program Committee for their scholarly effort, and all referees who assisted the Program Committees in the evaluation process.

Thanks to the sponsors for their support, to the Venice International University and to the Province of Venice for hosting ICALP 2006 in beautiful S. Servolo. We are also grateful to all members of the Organizing Committee in the Department of Computer Science and to the Center for Technical Support Services and Telecommunications (CSITA) of the University of Venice. Thanks to Andrei Voronkov for his support with the conference management software EasyChair. It was great in handling the submissions and the electronic PC meeting, as well as in assisting in the assembly of the proceedings.

April 2006

Michele Bugliesi
Bart Preneel
Vladimiro Sassone
Ingo Wegener

Organization

Program Committee

Track A

Harry Buhrman, University of Amsterdam, The Netherlands
Mark de Berg, TU Eindhoven, The Netherlands
Uriel Feige, Weizmann Institute, Isreal
Anna Gal, University of Texas at Austin, USA
Johan Hastad, KTH Stockholm, Sweden
Edith Hemaspaandra, Rochester Institute of Technology, USA
Kazuo Iwama, Kyoto University, Japan
Mark Jerrum, University of Edinburgh, UK
Stefano Leonardi, Università di Roma, Italy
Friedhelm Meyer auf der Heide, Universität Paderborn, Germany
Ian Munro, University of Waterloo, Canada
Sotiris Nikoletseas, Patras University, Greece
Rasmus Pagh, IT Univerisy of Copenhagen, Denmark
Tim Roughgarden, Stanford University, USA
Jacques Sakarovitch, CRNS Paris, France
Jiri Sgall, Academy of Sciences, Prague, Czech Republic
Hans Ulrich Simon, Ruhr-Universität Bochum, Germany
Alistair Sinclair, University of Berkeley, USA
Angelika Steger, ETH Zürich, Switzerland
Denis Thérien, McGill University, Canada
Ingo Wegener, Universität Dortmund, Germany (Chair)
Emo Welzl, ETH Zurich, Switzerland

Track B

Roberto Amadio, Université Paris 7, France
Lars Birkedal, IT University of Copenhagen, Denmark
Roberto Bruni, Università di Pisa, Italy
Mariangiola Dezani-Ciancaglini, Università di Torino, Italy
Volker Diekert, University of Stuttgart, Germany
Abbas Edalat, Imperial College, UK
Jan Friso Groote, Eindhoven University of Technology, The Nederlands
Tom Henzinger, EPFL, Switzerland
Madhavan Mukund, Chennai Mathematical Institute, India
Jean-Éric Pin, LIAFA, France
Julian Rathke, University of Sussex, UK
Jakob Rehof, Microsoft Research, Redmont, USA

Vladimiro Sassone, University of Southampton, UK (Chair)
Don Sannella, University of Edinburgh, UK
Nicole Schweikardt, Humboldt-Universität zu Berlin, Germany
Helmut Seidl, Technische Universität München, Germany
Peter Selinger, Dalhousie University, Canada
Jerzy Tiuryn, Warsaw University, Poland
Victor Vianu, U. C. San Diego, USA
David Walker, Princeton University, USA
Igor Walukiewicz, Labri, Université Bordeaux, France

Track C

Martín Abadi, University of California at Santa Cruz, USA
Christian Cachin, IBM Research, Switzerland
Ronald Cramer, CWI and Leiden University, The Netherlands
Ivan Damgrd, University of Aarhus, Denmark
Giovanni Di Crescenzo, Telcordia, USA
Marc Fischlin, ETH Zürich, Switzerland
Dieter Gollmann, University of Hamburg-Harburg, Germany
Andrew D. Gordon, Microsoft Research, UK
Aggelos Kiayias, University of Connecticut, USA
Joe Kilian, Rutgers University, USA
Cathy Meadows, Naval Research Laboratory, USA
John Mitchell, Stanford University, USA
Mats Näslund, Ericsson, Sweden
Tatsuaki Okamoto, Kyoto University, Japan
Rafael Ostrovksy, University of California at Los Angeles, USA
Pascal Paillier, Gemplus, France
Giuseppe Persiano, University of Salerno, Italy
Benny Pinkas, HP Labs, Israel
Bart Preneel, Katholieke Universiteit Leuven, Belgium (Chair)
Vitaly Shmatikov, University of Texas at Austin, USA
Victor Shoup, New York University, USA
Jessica Staddon, PARC, USA
Frederik Vercauteren, Katholieke Universiteit Leuven, Belgium

Organizing Committee

Michele Bugliesi, University of Venice (Conference Chair)
Andrea Pietracaprina, University of Padova (Workshop Co-chair)
Francesco Ranzato, University of Padova (Workshop Co-chair)
Sabina Rossi, University of Venice (Workshop Co-chair)
Annalisa Bossi, University of Venice
Damiano Macedonio, University of Venice

Referees

Martín Abadi
Masayuki Abe
Zoe Abrams
Gagan Aggarwal
Mustaq Ahmed
Cagri Aksay
Tatsuya Akutsu
Susanne Albers
Eric Allender
Jean-Paul Allouche
Jesus Almansa
Helmut Alt
Joel Alwen
Andris Ambainis
Elena Andreeva
Spyros Angelopoulos
Elliot Anshelevich
Lars Arge
Stefan Arnborg
Sanjeev Arora
Vincenzo Auletta
Per Austrin
David Avis
Moshe Babaioff
Michael Backes
Evripides Bampis
Nikhil Bansal
Jeremy Barbay
Paulo Baretto
David Barrington
Roman Bartak
Adam Barth
Paul Beame
Luca Becchetti
Amos Beimel
Elizabeth Berg
Robert Berke
Piotr Berman
Thorsten Bernholt
Ivona Bezakova
Laurent Bienvenu
Mathieu Blanchette
Yvonne Bleischwitz

Isabelle Bloch
Avrim Blum
Liad Blumrosen
Alexander Bockmayr
Alexandra Boldyreva
Beate Bollig
Giuseppe Battista
Nicolas Bonichon
Vincenzo Bonifaci
Joan Boyar
An Braeken
Justin Brickell
Patrick Briest
Gerth Brodal
Gerth Stlting Brodal
Gerth S. Brodal
Peter Buergisser
Jonathan Buss
Gruia Calinescu
Christophe De Cannire
Flavio d'Alessandro
Alberto Caprara
Iliano Cervesato
Timothy M. Chan
Pandu Rangan
  Chandrasekaran
Krishnendu Chatterjee
Arkadev Chattopadhyay
Avik Chaudhuri
Kamalika Chaudhuri
Chandra Chekuri
Zhi-Zhong Chen
Joseph Cheriyan
Benoit Chevallier-Mames
Markus Chimani
Christian Choffrut
Marek Chrobak
Fabian Chudak
David Clarke
Andrea Clementi
Bruno Codenotti
Richard Cole
Scott Contini

Ricardo Corin
Graham Cormode
Jose Correa
Veronique Cortier
Stefano Crespi
Maxime Crochemore
Mary Cryan
Felipe Cucker
Artur Czumaj
Sanjoy Dasgupta
Ccile Delerable
Xiaotie Deng
Jonathan Derryberry
Jean-Louis Dessalles
Nikhil Devanur
Luc Devroye
Florian Diedrich
Martin Dietzfelbinger
Jintai Ding
Irit Dinur
Benjamin Doerr
Eleni Drinea
Petros Drineas
Stefan Droste
Laszlo Egri
Friedrich Eisenbrand
Michael Elkin
Leah Epstein
Kimmo Eriksson
Thomas Erlebach
Peter Gacs
Rolf Fagerberg
Ulrich Faigle
Piotr Faliszewski
Pooya Farshim
Arash Farzan
Serge Fehr
Sandor Fekete
Rainer Feldmann
Stefan Felsner
Coby Fernadess
Paolo Ferragina
Jiri Fiala

Faith Fich
Matthias Fitzi
Abraham Flaxman
Lisa K. Fleischer
Rudolf Fleischer
Fedor Fomin
Lance Fortnow
Pierre Fraignaud
Paolo Franciosa
Matt Franklin
Eiichiro Fujisaki
Satoshi Fujita
Toshihiro Fujito
Stanley Fung
Martin Furer
Jun Furukawa
Martin Frer
Bernd Gaertner
Martin Gairing
Steven Galbraith
Clemente Galdi
G. Ganapathy
Juan Garay
Naveen Garg
Ricard Gavalda
Dmitry Gavinsky
Joachim Gehweiler
Stefanie Gerke
Abhrajit Ghosh
Oliver Giel
Reza Dorri Giv
Andreas Goerdt
Eu-Jin Goh
Leslie Goldberg
Mikael Goldmann
Aline Gouget
Navin Goyal
Gregor Gramlich
Robert Granger
Alexander Grigoriev
Martin Grohe
Andre Gronemeier
Jiong Guo
Ankur Gupta
Anupam Gupta

Venkatesan Guruswami
Inge Li Grtz
Robbert de Haan
Torben Hagerup
Mohammad Taghi Haji-
aghayi
Michael Hallett
Dan Halperin
Christophe Hancart
Goichiro Hanaoka
Dan Witzner Hansen
Sariel Har-Peled
Thomas Hayes
Meng He
Lane Hemaspaandra
Javier Herranz
Jan van den Heuvel
Martin Hirt
Michael Hoffmann
Dennis Hofheinz
Thomas Hofmeister
Christopher M. Homan
Hendrik Jan Hoogeboom
Juraj Hromkovic
Shunsuke Inenaga
Piotr Indyk
Yuval Ishai
Toshimasa Ishii
Hiro Ito
Toshiya Itoh
Gabor Ivanyos
Riko Jacob
Jens Jaegerskuepper
Sanjay Jain
Kamal Jain
Klaus Jansen
Thomas Jansen
Jesper Jansson
Stanislaw Jarecki
Wojciech Jawor
David Johnson
Tibor Jordan
Philippe Jorrand
Jan Juerjens
Valentine Kabanets

Jesse Kamp
Sampath Kannan
Haim Kaplan
Sarah Kappes
Bruce Kapron
Juha Karkkainen
Julia Kempe
Johan Karlander
Howard Karloff
Jonahtan Katz
Akinori Kawachi
Claire Kenion
Krishnaram Kenthapadi
Rohit Khandekar
Subhash A. Khot
Samir Khuller
Eike Kiltz
Guy Kindler
Valerie King
Lefteris Kirousis
Daniel Kirsten
Bobby Kleinberg
Adam Klivans
Johannes Koebler
Jochen Koenemann
Petr Kolman
Guy Kortsarz
Michal Koucky
Elias Koutsoupias
Dexter Kozen
Darek Kowalski
Matthias Krause
Hugo Krawczyk
Klaus Kriegel
Alexander Kroeller
Piotr Krysta
Ludek Kucera
Noboru Kunihiro
Eyal Kushilevitz
Minseok Kwon
Shankar Ram Lakshmi-
narayanan
Joseph Lano
Sophie Laplante
Christian Lavault

Ron Lavi
Thierry Lecroq
Troy Lee
Hanno Lefmann
Francois Lemieux
Asaf Levin
Benoit Libert
Christian Liebchen
Andrzej Lingas
Helger Lipmaa
Sylvain Lombardy
Alex Lopez-Ortiz
Zvi Lotker
Laci Lovasz
Chris Luhrs
Rune Bang Lyngs
Peter Mahlmann
John Malone-Lee
Heikki Mannila
Alberto
    Marchetti-Spaccamela
Martin Marciniszyn
Gitta Marchand
Stuart Margolis
Martin Mares
Russell Martin
Toshimitsu Masuzawa
Jiri Matousek
Giancarlo Mauri
Alexander May
Elvira Mayordomo
Pierre McKenzie
Kurt Mehlhorn
Aranyak Mehta
Nele Mentens
Mark Mercer
Ron van der Meyden
Ulrich Meyer
Peter Bro Miltersen
Dieter Mitsche
Shuichi Miyazaki
Burkhard Monien
Cris Moore
Thomas Moscibroda
Mitsuo Motoki

Ahuva Mu'alem Kamesh
Munagala
Ian Munro
Kazuo Murota
Petra Mutzel
Hiroshi Nagamochi
Shin-ichi Nakano
Seffi Naor
Gonzalo Navarro
Frank Neumann
Antonio Nicolosi
Rolf Niedermeier
Jesper Buus Nielsen
Svetla Nikova
Karl Norrman
Dirk Nowotka
Robin Nunkesser
Regina O'Dell
Hirotaka ONO
Mitsunori Ogihara
Kazuo Ohta
Chihiro Ohyama
Yusuke Okada
Yoshio Okamoto
Christopher Okasaki
Ole Østerby
Janos Pach
Anna stlin Pagh
Anna Palbom
Konstantinos
    Panagiotou
Leon Peeters
Derek Phillips
Toniann Pitassi
Giovanni Pighizzini
Wolf Polak
Pawel Pralat
Pavel Pudlak
Prashant Puniya
Yuri Rabinovich
Jaikumar
    Radhakrishnan
Stanislaw Radziszowski
Harald Raecke
Prabhakar Ragde

Zia Rahman
S. Raj Rajagopalan
V. Ramachandran
Dana Randall
S. Srinivasa Rao
Ran Raz
Alexander Razborov
Andreas Razen
Ken Regan
Ben Reichardt
Christophe Reutenauer
Eleanor Rieffel
Romeo Rizzi
Martin Roetteler
Phillip Rogaway
Amir Ronen
Dominique Rossin
Peter Rossmanith
Joerg Rothe
Arnab Roy
Leo Ruest
Milan Ruzic
Kunihiko Sadakane
Cenk Sahinalp
Kai Salomaa
Louis Salvail
Peter Sanders
Mark Sandler
Rahul Santhanam
Palash Sarkar
Martin Sauerhoff
Daniel Sawitzki
Nicolas Schabanel
Christian Schaffner
Michael Schapira
Dominik Scheder
Christian Scheideler
Christian Schindelhauer
Katja Schmidt-Samoa
Georg Schnitger
Henning Schnoor
Uwe Schoening
Gunnar Schomaker
Eva Schuberth
Andreas Schulz

Nathan Segerlind
Meinolf Sellmann
Pranab Sen
Hadas Shachnai
Ronen Shaltiel
Abhi Shelat
Bruce Sheppard
Oleg M. Sheyner
David Shmoys
Detlef Sieling
Jiri Sima
Mohit Singh
Naveen Sivadasan
Matthew Skala
Steve Skiena
Michiel Smid
Adam Smith
Shakhar Smorodinsky
Christian Sohler
Alexander Souza
Paul Spirakis
Michael Spriggs
Reto Sphel
Rob van Stee
Stamatis Stefanakos
Daniel Stefankovic
Cliff Stein
Bernhard von Stengel
Tobias Storch
Madhu Sudan
Dirk Sudholt
Mukund Sundararajan
Koutarou Suzuki
Maxim Sviridenko
Tibor Szabo

Zoltan Szigeti
Troels Bjerre Sørensen
Asano Takao
Hisao Tamaki
Akihisa Tamura
Eva Tardos
Sebastiaan Terwijn
Pascal Tesson
Prasad Tetali
Ralf Thoele
Karsten Tiemann
Yuuki Tokunaga
Takeshi Tokuyama
Eric Torng
Patrick Traxler
Luca Trevisan
Tatsuie Tsukiji
Gyorgy Turan
Pim Tuyls
Ryuhei Uehara
Chris Umans
Falk Unger
Takeaki Uno
Pavel Valtr
Sergei Vassilvitskii
Ingrid Verbauwhede
Kolia Vereshchagin
Damien Vergnaud
Adrian Vetta
Ivan Visconti
Berthold Voecking
Heribert Vollmer
Sergey Vorobyov
Sven de Vries
Stephan Waack

Uli Wagner
Michael Waidner
Bogdan Warinschi
Osamu Watanabe
Brent Waters
Kevin Wayne
Stephanie Wehner
Ralf-Philipp Weinmann
Andreas Weissl
Tom Wexler
Erik Winfree
Peter Winkler
Kai Wirt
Carsten Witt
Ronald de Wolf
Stefan Wolf
David Woodruff
Mutsunori Yagiura
Hiroaki Yamamoto
Go Yamamoto
Shigeru Yamashita
Takenaga Yasuhiko
Yiqun Lisa Yin
Filip Zagorski
Guochuan Zhang
Yuliang Zheng
Ming Zhong
Hong-Sheng Zhou
Xiao Zhou
Wieslaw Zielonka
Eckart Zitzler
David Zuckerman
Philipp Zumstein
Uri Zwick

Sponsoring Institutions

IBM Italy
Venis S.P.A - Venezia Informatica e Sistemi
Dipartimento di Informatica, Università Ca' Foscari
CVR - Consorzio Venezia Ricerche

Table of Contents – Part I

Formal Languages

Approximation Algorithms I

Approximation Algorithms II

Graph Algorithms I

Algorithms I

Complexity I

Data Structures and Linear Algebra

Graphs

Complexity II

Game Theory I

Algorithms II

Game Theory II

Networks, Circuits and Regular Expressions

Fixed Parameter Complexity and Approximation Algorithms

Graph Algorithms II

Table of Contents – Part II

Cryptographic Primitives

Bounded Storage and Quantum Models

Foundations

Multi-party Protocols

Games

Semantics

Automata I

Models

Equations

Logics

Automata II

Additive Approximation for Edge-Deletion Problems (Abstract)

Noga Alon[1], Asaf Shapira[1], and Benny Sudakov[2]

[1] Tel Aviv University, Tel Aviv, Israel 69978
nogaa,asafico@post.tau.ac.il
[2] Princeton University and IAS, Princeton, NJ 08540, USA
bsudakov@math.princeton.edu

A graph property is *monotone* if it is closed under removal of vertices and edges. We consider the following algorithmic problem, called the *edge-deletion problem*; given a monotone property P and a graph G, compute the smallest number of edge deletions that are needed in order to turn G into a graph satisfying P. We denote this quantity by $E'_P(G)$. Our first result states that the edge-deletion problem can be efficiently approximated for any monotone property.

- For any fixed $\epsilon > 0$ and any monotone property P, there is a deterministic algorithm, which given a graph $G = (V, E)$ of size n, approximates $E'_P(G)$ in linear time $O(|V| + |E|)$ to within an additive error of ϵn^2.

The proof is based on a strong version of Szemerédi's Regularity Lemma proved in [4], following the proof of the original lemma in [12] (see also [10]), as well as on the algorithmic versions of this lemma, developed in [2], [9]. An alternative, related approach can be developed using the techniques of [7]. The approximation algorithm applies in many settings in which the methods of [6], [8] and [3] do not suffice, and is related to the results of [5] on testing monotone properties.

Given the above, a natural question is for which monotone properties one can obtain better additive approximations of E'_P. Our second main result essentially resolves this problem by giving a precise characterization of the monotone graph properties for which such approximations exist.

1. If there is a bipartite graph that does not satisfy P, then there is a $\delta > 0$ for which it is possible to approximate E'_P to within an additive error of $n^{2-\delta}$ in polynomial time.
2. On the other hand, if all bipartite graphs satisfy P, then for any $\delta > 0$ it is NP-hard to approximate E'_P to within an additive error of $n^{2-\delta}$.

While the proof of (1) is a relatively simple consequence of the classical result of [11], the proof of (2) requires several new ideas and involves tools and new results from Extremal Graph Theory together with spectral techniques, as well as the approach of [1] which transforms a hard sparse instance to a hard dense one by blowing it up and combining it with a random looking instance.

Interestingly, prior to this work it was not even known that computing E'_P *precisely* for the properties in (2) is NP-hard. We thus answer (in a strong form)

M. Bugliesi et al. (Eds.): ICALP 2006, Part I, LNCS 4051, pp. 1–2, 2006.

a question of Yannakakis, who asked in 1981 if it is possible to find a large and natural family of graph properties for which computing E'_P is NP-hard.

References

1. N. Alon, Ranking Tournaments, SIAM J. Discrete Math. 20 (2006), 137-142.
2. N. Alon, R. A. Duke, H. Lefmann, V. Rödl and R. Yuster, The algorithmic aspects of the Regularity Lemma, Proc. 33rd IEEE FOCS, Pittsburgh, IEEE (1992), 473-481. Also: J. of Algorithms 16 (1994), 80-109.
3. N. Alon, W. Fernandez de la Vega, R. Kannan and M. Karpinski, Random Sampling and Approximation of MAX-CSP Problems, Proc. of 34th ACM STOC, ACM Press (2002), 232-239. Also: JCSS 67 (2003), 212-243.
4. N. Alon, E. Fischer, M. Krivelevich and M. Szegedy, Efficient testing of large graphs, Proc. of 40th FOCS, New York, NY, IEEE (1999), 656-666. Also: Combinatorica 20 (2000), 451-476.
5. N. Alon and A. Shapira, Every monotone graph property is testable, Proc. of 37th STOC 2005, 128-137.
6. S. Arora, D. Karger and M. Karpinski, Polynomial time approximation schemes for dense instances of graph problems, Proc. of 28th STOC (1995). Also: JCSS 58 (1999), 193-210.
7. C. Borgs, J. Chayes, L. Lovász, V. T. Sós, B. Szegedy and K. Vesztergombi, Graph limits and parameter testing, to appear in STOC 2006.
8. A. Frieze and R. Kannan, The regularity lemma and approximation schemes for dense problems, Proc. of 37th FOCS, 1996, 12-20.
9. Y. Kohayakawa, V. Rödl and L. Thoma, An optimal algorithm for checking regularity, SIAM J. on Computing 32 (2003), no. 5, 1210-1235.
10. J. Komlós and M. Simonovits, Szemerédi's Regularity Lemma and its applications in graph theory. In: *Combinatorics, Paul Erdös is Eighty*, Vol II (D. Miklós, V. T. Sós, T. Szönyi eds.), János Bolyai Math. Soc., Budapest (1996), 295–352.
11. T. Kövari, V.T. Sós and P. Turán, On a problem of K. Zarankiewicz, *Colloquium Math.* 3 (1954), 50-57.
12. E. Szemerédi, Regular partitions of graphs, In: *Proc. Colloque Inter. CNRS* (J. C. Bermond, J. C. Fournier, M. Las Vergnas and D. Sotteau, eds.), 1978, 399–401.
13. M. Yannakakis, Edge-deletion problems, SIAM J. Comput. 10 (1981), 297-309.

Testing Graph Isomorphism in Parallel by Playing a Game

Martin Grohe and Oleg Verbitsky[*]

Institut für Informatik
Humboldt Universität zu Berlin, D-10099 Berlin, Germany
{grohe, verbitsk}@informatik.hu-berlin.de

Abstract. Our starting point is the observation that if graphs in a class C have low descriptive complexity, then the isomorphism problem for C is solvable by a fast parallel algorithm. More precisely, we prove that if every graph in C is definable in a finite-variable first order logic with counting quantifiers within logarithmic quantifier depth, then Graph Isomorphism for C is in $\mathrm{TC}^1 \subseteq \mathrm{NC}^2$. If no counting quantifiers are needed, then Graph Isomorphism for C is even in AC^1. The definability conditions can be checked by designing a winning strategy for suitable Ehrenfeucht-Fraïssé games with a logarithmic number of rounds. The parallel isomorphism algorithm this approach yields is a simple combinatorial algorithm known as the Weisfeiler-Lehman (WL) algorithm.

Using this approach, we prove that isomorphism of graphs of bounded treewidth is testable in TC^1, answering an open question from [9]. Furthermore, we obtain an AC^1 algorithm for testing isomorphism of rotation systems (combinatorial specifications of graph embeddings). The AC^1 upper bound was known before, but the fact that this bound can be achieved by the simple WL algorithm is new. Combined with other known results, it also yields a new AC^1 isomorphism algorithm for planar graphs.

1 Introduction

1.1 The Graph Isomorphism Problem

An isomorphism between two graphs G and H is a 1-to-1 correspondence between their vertex sets $V(G)$ and $V(H)$ that relates edges to edges and non-edges to non-edges. Two graphs are isomorphic if there exists an isomorphism between them. *Graph Isomorphism (GI)* is the problem of recognizing if two given graphs are isomorphic. The problem plays a prominent role in complexity theory as one of the few natural problems in NP that are neither known to be NP-complete nor known to be in polynomial time. There are good reasons to believe that GI is not NP-complete; most strikingly, this would imply a collapse of the polynomial hierarchy [7,34]. The best known graph isomorphism algorithm due to Babai, Luks, and Zemplyachenko [1,4] takes time $O(2^{\sqrt{n \log n}})$, where n denotes the

---

[*] Supported by an Alexander von Humboldt fellowship.

M. Bugliesi et al. (Eds.): ICALP 2006, Part I, LNCS 4051, pp. 3–14, 2006.
© Springer-Verlag Berlin Heidelberg 2006

number of vertices in the input graphs. The strongest known hardness result [36] says that GI is hard for DET, which is a subclass of NC^2. The complexity status of GI is determined precisely only if the problem is restricted to trees: For trees GI is LOGSPACE-complete [24,21].

However, there are many natural classes of graphs such that the restriction of GI to input graphs from these classes is in polynomial time. These include planar graphs [18,19], graphs of bounded genus [13,26], graphs of bounded treewidth [5], graphs with excluded minors [31], graphs of bounded degree [25], and graphs of bounded eigenvalue multiplicity [3]. Here we are interested in classes of graphs for which the isomorphism problem is solvable by a fast (i.e., polylogarithmic) parallel algorithm. Recall the class NC and its refinements: $NC = \bigcup_i NC^i$ and $NC^i \subseteq AC^i \subseteq TC^i \subseteq NC^{i+1}$, where NC^i consists of functions computable by circuits of polynomial size and depth $O(\log^i n)$, AC^i is an analog for circuits with unbounded fan-in, and TC^i is an extension of AC^i allowing threshold gates. As well known [22], AC^i consists of exactly those functions computable by a CRCW PRAM with polynomially many processors in time $O(\log^i n)$. Miller and Reif [27] design an AC^1 algorithm for planar graph isomorphism and isomorphism of rotation systems, which are combinatorial specifications of graph embeddings (see [28, Sect. 3.2]). Chandrasekharan [9] (see also [10]) designs an AC^2 isomorphism algorithm for k-trees, a proper subclass of graphs of treewidth k, and asks if there is an NC algorithm for the whole class of graphs with treewidth k.

We answer this question in affirmative by showing that isomorphism of graphs with bounded treewidth is in TC^1 (see Corollary 9). Furthermore, we obtain a new AC^1-algorithm for testing isomorphism of rotation systems (see Corollary 11), which by techniques due to Miller and Reif [27] also yields a new AC^1 isomorphism algorithm for planar graphs (see Corollary 12).

Remarkably, the algorithm we employ for both graphs of bounded treewidth and rotation systems is a simple combinatorial algorithm that is actually known since the late 1960s from the work of Weisfeiler and Lehman. This is what we believe makes our result on rotation systems worthwhile, even though in this case the AC^1 upper bound was known before.

1.2 The Multidimensional Weisfeiler-Lehman Algorithm

For the history of this approach to GI we refer the reader to [2,8,11,12]. We will abbreviate *k-dimensional Weisfeiler-Lehman algorithm* by *k-dim WL*. The 1-dim WL is commonly known as *canonical labeling* or *color refinement algorithm*. It proceeds in rounds; in each round a coloring of the vertices of the input graphs G and H is defined, which refines the coloring of the previous round. The initial coloring C^0 is uniform, say, $C^0(v) = 1$ for all vertices $v \in V(G) \cup V(H)$. In the $(i+1)$st round, the color $C^{i+1}(v)$ is defined to be a pair consisting of the preceding color $C^{i-1}(v)$ and the multiset of colors $C^{i-1}(u)$ for all u adjacent to v. For example, $C^1(v) = C^1(w)$ iff v and w have the same degree. To keep the color encoding short, after each round the colors are renamed (we never need more than $2n$ color names). As the coloring is refined in each round, it stabilizes after at most $2n$ rounds, that is, no further refinement occurs. The algorithm stops as soon as this happens. If the multiset of colors of the vertices of G is distinct from the multiset of colors of the vertices of H, the algorithms reports that the graphs are not isomorphic; otherwise, it declares them to be isomorphic. Clearly, this algorithm is not correct. It may report false positives, for example, if both input graphs are regular with the same vertex degree.

Following the same idea, the k-dimensional version iteratively refines a coloring of $V(G)^k \cup V(H)^k$. The initial coloring of a k-tuple $\bar{v}$ is the isomorphism type of the subgraph induced by the vertices in $\bar{v}$ (viewed as a labeled graph where each vertex is labeled by the positions in the tuple where it occurs). The refinement step takes into account the colors of all neighbors of $\bar{v}$ in the Hamming metric (see details in Sect. 3). Color stabilization is now reached in $r < 2n^k$ rounds. The k-dim WL is polynomial-time for each constant k. In 1990, Cai, Fürer, and Immerman [8] proved a striking negative result: For any sublinear dimension $k = o(n)$, the k-dim WL does not work correctly even on graphs of vertex degree 3. Nevertheless, later it was realized that a constant-dimensional WL is still applicable to particular classes of graphs, including planar graphs [14], graphs of bounded genus [15], and graphs of bounded treewidth [16].

We show that the k-dim WL admits a natural parallelization such that the number of parallel processors and the running time are closely related to n^k and r, respectively, where r denotes the number of rounds performed by the algorithm. Previous work never used any better bound on r than the trivial $r < 2n^k$, which was good enough to keep the running time polynomially bounded. In view of a possibility that r can be much smaller, we show that the r-round k-dim WL can be implemented on a logspace uniform family of TC circuits of depth $O(r)$ and size $O(r \cdot n^{3k})$. It follows that if for a class of graphs C there is a constant k such that for all $G, H \in C$ the k-dim WL in $O(\log n)$ rounds correctly decides if G and H are isomorphic or not, then there is a TC^1 algorithm deciding GI on C. We also prove a version of these results for a related algorithm we call the *count-free WL algorithm* that places GI on suitable classes C into AC^1.

1.3 Descriptive Complexity of Graphs

To prove that the k-dim WL correctly decides isomorphism of graphs from a certain class C in a logarithmic number of rounds, we exploit a close relationship between the WL algorithm and the descriptive complexity of graphs, which was discovered in [8]: The r-round k-dim WL correctly decides if two graphs G and H are isomorphic in at most r rounds if and only if G and H are distinguishable in the $(k + 1)$-variable first order logic with counting quantifiers in the language of graphs by a sentence of quantifier depth r. (In)distinguishability of two graphs in various logics can be characterized in terms of so-called Ehrenfeucht-Fraïssé games. The appropriate game here is the counting version of the r-round k-pebble game (see Sect. 5). The equivalence between correctness of the r-round k-dim WL, logical indistinguishability, and its game characterization reduces the design of a TC^1 isomorphism algorithm on C to the design of winning strategies in the $O(\log n)$-round k-pebble counting game on graphs from the class C, for a constant k. Similarly, the design of an AC^1 iso-morphism algorithm on C can be reduced to the design of winning strategies in the $O(\log n)$-round k-pebble game (without counting) on graphs from the class C.

Our results on the descriptive complexity of graphs are actually slightly stronger than it is needed for algorithmic applications: They give $O(\log n)$ upper bounds on the quantifier depth of a k-variable first-order sentence (with or without counting) required to distinguish a graph G from all other graphs. For graphs of treewidth at most k, we obtain an $O(k\cdot\log n)$ upper bound in the $(4k+4)$-variable first-order logic with counting (see Theorem 8). For rotation systems, we obtain an $O(\log n)$ upper bound in the 5-variable first-order logic without counting (see Theorem 10). The proofs are based on an analysis of Ehrenfeucht-Fraïssé games.

Various aspects of descriptive complexity of graphs have recently been inves-tigated in [6,23,29,30,37] with focus on the minimum quantifier depth of a first order sentence defining a graph. In particular, a comprehensive analysis of the definability of trees in first order logic is carried out in [6,29,37]. Here we extend it to the definability of graphs with bounded treewidth in first order logic with counting. Notice a fact that makes our results on descriptive complexity poten-tially stronger (and harder to prove): We are constrained by the condition that a defining sentence must be in a finite-variable logic.

The rest of the paper is organized as follows. In Sect. 2 we give relevant defi-nitions from descriptive complexity of graphs. The Weisfeiler-Lehman algorithm is treated in Sect. 3. Section 4 contains some graph-theoretic preliminaries. Sec-tion 5 is devoted to the Ehrenfeucht-Fraïssé game. We outline our results about graphs of bounded treewidth in Sect. 6 and about rotation systems in Sect. 7.

2   Logical Depth of a Graph

Let Φ be a first order sentence about a graph in the language of the adjacency and the equality relations. We say that Φ *distinguishes* a graph G from a graph H if Φ is true on G but false on H. We say that Φ *defines* G if Φ is true on G and false on any graph non-isomorphic to G. The quantifier rank of Φ is the maximum number of nested quantifiers in Φ. The *logical depth* of a graph G, denoted by $D(G)$, is the minimum quantifier depth of Φ defining G.

The *k-variable logic* is the fragment of first order logic where usage of only k variables is allowed. If we restrict defining sentences to the k-variable logic, this variant of the logical depth of G is denoted by $D^k(G)$. We have

$$D^k(G) = \max\left\{ D^k(G, H) : H \not\cong G \right\}, \qquad (1)$$

where $D^k(G, H)$ denotes the minimum quantifier depth of a k-variable sentence distinguishing G from H. This equality easily follows from the fact that, for each r, there are only finitely many pairwise inequivalent first order sentences about graphs of quantifier depth at most r. It is assumed that $D^k(G) = \infty$

(resp. $D^k(G, H) = \infty$) if the k-variable logic is too weak to define G (resp. to distinguish G from H).

Furthermore, let $_cD^k(G)$ (resp. $_cD^k(G, H)$) denote the variant of $D^k(G)$ (resp. $D^k(G, H)$) for the first order logic with *counting quantifiers* where we allow expressions of the type $\exists^m \Psi$ to say that there are at least m vertices with property Ψ (such a quantifier contributes 1 in the quantifier depth irrespective of m). Similarly to (1) we have

$$_cD^k(G) = \max \left\{ \, _cD^k(G, H) : H \not\cong G \right\}. \tag{2}$$

3 The k-Dim WL as a Parallel Algorithm

Let $k \geq 2$. Given an ordered k-tuple of vertices $\bar{u} = (u_1, \ldots, u_k) \in V(G)^k$, we define the *isomorphism type* of $\bar{u}$ to be the pair

$$\mathrm{tp}(\bar{u}) = \Big(\, \big\{ (i,j) \in [k]^2 : u_i = u_j \big\}, \big\{ (i,j) \in [k]^2 : \{u_i, u_j\} \in E(G) \big\} \Big),$$

where $[k]$ denotes the set $\{1, \ldots, k\}$. If $w \in V(G)$ and $i \leq k$, we let $\bar{u}^{i,w}$ denote the result of substituting w in place of u_i in $\bar{u}$.

The *r-round k-dimensional Weisfeiler-Lehman algorithm (r-round k-dim WL)* takes as an input two graphs G and H and purports to decide if $G \cong H$. The algorithm performs the following operations with the set $V(G)^k \cup V(H)^k$.

Initial coloring. The algorithm assigns each $\bar{u} \in V(G)^k \cup V(H)^k$ color $W^{k,0}(\bar{u})$ $= \mathrm{tp}(\bar{u})$ (in a suitable encoding).

Color refinement step. In the i-th round each $\bar{u} \in V(G)^k$ is assigned color

$$W^{k,i}(\bar{u}) = \Big(W^{k,i-1}(\bar{u}), \big\{\!\big\{ (W^{k,i-1}(\bar{u}^{1,w}), \ldots, W^{k,i-1}(\bar{u}^{k,w})) : w \in V(G) \big\}\!\big\} \Big)$$

and similarly with each $\bar{u} \in V(H)^k$.

Here $\{\!\{\ldots\}\!\}$ denotes a multiset. In a variant of the algorithm, which will be referred to as the *count-free version*, this is a set.

Computing an output. The algorithm reports that $G \not\cong H$ if

$$\big\{\!\big\{ W^{k,r}(\bar{u}) : \bar{u} \in V(G)^k \big\}\!\big\} \neq \big\{\!\big\{ W^{k,r}(\bar{u}) : \bar{u} \in V(H)^k \big\}\!\big\}. \tag{3}$$

and that $G \cong H$ otherwise.

In the above description we skipped an important implementation detail. To prevent increasing the length of $W^{k,i}(\bar{u})$ at the exponential rate, before every refinement step we arrange colors of all k-tuples of $V(G)^k \cup V(H)^k$ in the lexicographic order and replace each color with its number.

As easily seen, if ϕ is an isomorphism from G to H, then for all k, i, and $\bar{u} \in V(G)^k$ we have $W^{k,i}(\bar{u}) = W^{k,i}(\phi(\bar{u}))$. This shows that for the isomorphic input graphs the output is always correct. We say that the r-round k-dim WL *works correctly for a graph G* if its output is correct on all input pairs (G, H).

Proposition 1 (Cai-Fürer-Immerman [8]).

1. *The r-round k-dim WL works correctly for a graph G iff $r \geq {}_c D^{k+1}(G)$.*
2. *The count-free r-round k-dim WL works correctly for a graph G iff $r \geq D^{k+1}(G)$.*

Theorem 2. *Let $k \geq 2$ be a constant and $r = r(n)$ a function, where n denotes the order of the input graphs.*

1. *The r-round k-dim WL can be implemented by a logspace uniform family of TC circuits of depth $O(r)$ and size $O(r \cdot n^{3k})$.*
2. *The count-free r-round k-dim WL can be implemented by a logspace uniform family of AC circuits of depth $O(r)$ and size $O(r \cdot n^{3k})$.*

The proof is omitted due to space limitation and can be found in a full version of this paper [17]. The following corollary states the most important application of the previous theorem for us:

Corollary 3. *Let $k \geq 2$ be a constant.*

1. *Let C be a class of graphs G with ${}_c D^{k+1}(G) = O(\log n)$. Then Graph Isomorphism for C is in TC^1.*
2. *Let C be a class of graphs G with $D^{k+1}(G) = O(\log n)$. Then Graph Isomorphism for C is in AC^1.*

Remark 4. The Weisfeiler-Lehman algorithm naturally generalizes from graphs to an arbitrary class of structures over a fixed vocabulary. It costs no extra efforts to extend Theorem 2 as well as Corollary 3 in the general situation.

4   Graph-Theoretic Preliminaries

The distance between vertices u and v in a graph G is denoted by $d(u,v)$. If u and v are in different connected components, we set $d(u,v) = \infty$. The diameter of G is defined by $diam\,(G) = \max\{d(u,v) : u,v \in V(G)\}$. Let $X \subset V(G)$. The subgraph induced by G on X is denoted by $G[X]$. We denote $G \setminus X = G[V(G) \setminus X]$, which is the result of removal of all vertices in X from G. We call the vertex set of a connected component of $G \setminus X$ a *flap of $G \setminus X$*. We call X a *separator of G* if every flap of $G \setminus X$ has at most $|V(G)|/2$ vertices.

A *tree decomposition* of a graph G is a tree T and a family $\{X_i\}_{i \in V(T)}$ of sets $X_i \subseteq V(G)$, called *bags*, such that the union of all bags covers all $V(G)$, every edge of G is contained in at least one bag, and we have $X_i \cap X_j \subseteq X_l$ whenever l lies on the path from i to j in T.

Proposition 5 (see Robertson-Seymour [33]). *In any tree decomposition of a graph G there is a bag that is a separator of G.*

The *width* of the decomposition is $\max |X_i| - 1$. The *treewidth* of G is the minimum width of a tree decomposition of G.

Now we introduce a non-standard notation specific to our purposes. It will be convenient to regard it as a notation for two binary operations over sets of vertices. Let $A \subset V(G)$ and $v \in V(G) \setminus A$. Then $A \odot v$ denotes the union of A and the flap of $G \setminus A$ containing v. Furthermore, let $A, C \subset V(G)$ be nonempty and disjoint. Then $A \ominus C$ is the union of A, C, and the set of all those vertices $x \in V(G) \setminus (A \cup C)$ such that there are a path from x to A in $G \setminus C$ and a path from x to C in $G \setminus A$.

5   Ehrenfeucht-Fraïssé Game

Let G and H be graphs with disjoint vertex sets. The *r-round k-pebble Ehren-feucht-Fraïssé game on G and H*, denoted by $\mathrm{EHR}_r^k(G, H)$, is played by two players, Spoiler and Duplicator, with k pairwise distinct pebbles $p_1, \ldots, p_k$, each given in duplicate. Spoiler starts the game. A *round* consists of a move of Spoiler followed by a move of Duplicator. At each move Spoiler takes a pebble, say p_i, selects one of the graphs G or H, and places p_i on a vertex of this graph. In response Duplicator should place the other copy of p_i on a vertex of the other graph. It is allowed to move previously placed pebbles to other vertices and place more than one pebble on the same vertex.

After each round of the game, for $1 \le i \le k$ let x_i (resp. y_i) denote the vertex of G (resp. H) occupied by p_i, irrespectively of who of the players placed the pebble on this vertex. If p_i is off the board at this moment, x_i and y_i are undefined. If after every of r rounds the component-wise correspondence $(x_1, \ldots, x_k)$ to $(y_1, \ldots, y_k)$ is a partial isomorphism from G to H, this is a win for Duplicator; Otherwise the winner is Spoiler.

In the *counting version* of the game, the rules of $\mathrm{EHR}_r^k(G, H)$ are modified as follows. A round now consists of two acts. First, Spoiler specifies a set of vertices A in one of the graphs. Duplicator responds with a set of vertices B in the other graph so that $|B| = |A|$. Second, Spoiler places a pebble p_i on a vertex $b \in B$. In response Duplicator has to place the other copy of p_i on a vertex $a \in A$.

Proposition 6. (Immerman, Poizat, see [20, Theorem 6.10])

1. $D^k(G, H)$ equals the minimum r such that Spoiler has a winning strategy in $\mathrm{EHR}_r^k(G, H)$.
2. $cD^k(G, H)$ equals the minimum r such that Spoiler has a winning strategy in the counting version of $\mathrm{EHR}_r^k(G, H)$.

All the above definitions and statements have a perfect sense for any kind of structures, in particular, for colored graphs (i.e., graphs with unary predicates) or even more complicated structures considered in Sect. 7. The following lemma provides us with a basic primitive on which our strategy will be built.

Lemma 7. *Consider the game on graphs G and G'. Let $u, v \in V(G)$, $u', v' \in V(G')$ and suppose that u, u' and as well v, v' are under the same pebbles. Suppose also that $d(u, v) \ne d(u', v')$ and $d(u, v) \ne \infty$ (in particular, it is possible that $d(u', v') = \infty$). Then Spoiler is able to win with 3 pebbles in $\lceil \log d(u, v) \rceil$ moves.*

Proof. Spoiler uses the *halving strategy* (see [35, Chap. 2] for a detailed account).

Saying that Spoiler has a *fast win* in the game on graphs G and G', we will mean that he is able to win in the next $\log n + O(1)$ moves using constantly many pebbles irrespective of Duplicator's strategy, where n denotes the order of G. Consider the following configuration: A set of vertices A and two vertices $v \notin A$ and u are pebbled in G, while a set A' and vertices v' and u' are pebbled in G' correspondingly. Let $u \in A \odot v$ but $u' \notin A' \odot v'$. Applying Lemma 7 to graphs $G \setminus A$ and $G' \setminus A'$, we see that Spoiler has a fast win (operating with 3 pebbles but keeping all the pebbles on A and A').

Let now $u \notin A \odot v$ but $u' \in A' \odot v'$. The symmetric argument only shows that Spoiler wins in less than $\log diam\,(G') + 1$ moves, whereas $diam\,(G')$ may be much larger than n. However, Lemma 7 obviously applies in the case that $diam\,(G) \neq diam\,(G')$ and Spoiler wins fast anyway.

Assume that $diam\,(G) = diam\,(G')$. It follows that, if such A, v, A', v' are pebbled and Spoiler decides to move only inside $(A \odot v) \cup (A' \odot v')$, then Duplicator cannot move outside for else Spoiler wins fast. In this situation we say that Spoiler *forces play in* $(A \odot v) \cup (A' \odot v')$ or *restricts the game to* $G[A \odot v]$ and $G'[A' \odot v']$. Similarly, if at some moment of the game we have two disjoint sets A and C of vertices pebbled in G, then Spoiler can force further play in $(A \ominus C) \cup (A' \ominus C')$, where A', C' are the corresponding sets in G'.

6   Graphs of Bounded Treewidth

Theorem 8. *If a graph G on n vertices has treewidth k, then $_cD^{4k+4}(G) < 2(k+1)\log n + 8k + 9$.*

On the account of Corollary 3.1 this has a consequence for the computational complexity of Graph Isomorphism.

Corollary 9. *Let k be a constant. The isomorphism problem for the class of graphs with treewidth at most k is in TC^1.*

The proof of Theorem 8 is based on Equation (2) and Proposition 6.2. Let $G' \not\cong G$. We have to design a strategy for Spoiler in the Ehrenfeucht-Fraïssé game on G and G' allowing him to win with only $4k + 4$ pebbles in less than $2(k+1)\log n + 8k + 9$ moves, whatever Duplicator's strategy. Fix $(T, \{X_s\}_{s \in V(T)})$, a width-$k$ tree decomposition of G.

It is not hard to see that Spoiler can force play on K and K', some non-isomorphic components of G and G'. We hence can assume from the very beginning that G and G' are connected. Moreover, we will assume that $diam\,(G) = diam\,(G')$ because otherwise Spoiler has a fast win as discussed in Sect. 5.

We here give only a high level description of the strategy (see [17] for full details). The strategy splits the game into phases. Each phase can be of two types, Type AB or Type ABC. Whenever $X \subset V(G)$ consists of vertices pebbled in some moment of the game, by default X' will denote the set of vertices pebbled correspondingly in G' and vice versa. Saying that G is *colored according to the pebbling*, we mean that every vertex which is currently pebbled by p_j receives color j.

Phase i of type AB. Spoiler aims to ensure pebbling sets of vertices $A \subset V(G)$, $A' \subset V(G')$, vertices $v \in V(G) \setminus A$, $v' \in V(G') \setminus A'$, and perhaps sets of vertices $B \subset V(G)$, $B' \subset V(G')$ so that the following conditions are met.

AB1. Let $G_i = G[A_i \odot v_i]$ be colored according to the pebbling and G'_i be defined similarly. Then $G_i \not\cong G'_i$.

AB2. $|V(G_i)| \leq |V(G_{i-1})|/2 + k + 1$ (we set $G_0 = G$).

AB3. Both G_i and G'_i are connected.

AB4. A set B_i is pebbled if $|V(G_i)| > 2k + 2$, otherwise play comes to an endgame. B_i is a separator of G_i and $B'_i \subset V(G'_i)$.

AB5. There are distinct $r, t \in V(T)$ such that $A_i \subseteq X_r$ and $B_i \subseteq X_t$.

Phase i of type ABC. Spoiler aims to ensure pebbling sets of vertices $A, C \subset V(G)$, $A', C' \subset V(G')$ so that $A \cap C = \emptyset$, and perhaps sets $B \subset V(G)$, $B' \subset V(G')$ so that the following conditions are met.

ABC1. Let $G_i = G[A_i \ominus C_i]$ be colored according to the pebbling and G'_i be defined similarly. Then $G_i \not\cong G'_i$.

ABC2, ABC3, and *ABC4* are the same as, respectively, AB2, AB3, and AB4.

ABC5. There are pairwise distinct $r, s, t \in V(T)$ such that $s \in (\{r\} \odot t) \cap (\{t\} \odot r)$ and $A_i \subseteq X_r$, $B_i \subseteq X_s$, $C_i \subseteq X_t$.

A choice of B_i is granted to Spoiler by Proposition 5. In the next Phase $i + 1$ Spoiler restricts the game to G_i and G'_i, keeping pebbles on $A_i \cup \{v_i\}$ (or $A_i \cup C_i$) until the new A_{i+1}, v_{i+1} (or A_{i+1}, C_{i+1}) are pebbled. As soon as this is done, the pebbles on $A_i \cup \{v_i\} \setminus (A_{i+1} \cup \{v_{i+1}\})$ (or $A_i \cup C_i \setminus (A_{i+1} \cup C_{i+1})$) can be released and reused by Spoiler in further play.

Endgame. Suppose it begins after Phase l. We have $G_l \not\cong G'_l$ and the former graph has at most $2k + 2$ vertices. Spoiler restricts the game to G_l and G'_l. If Duplicator agrees, Spoiler obviously wins in no more than $2k + 2$ moves. Once Duplicator moves outside, Spoiler has a fast win as explained in Sect. 5, where *fast* means less than $\log diam(G_l) + 2 \leq \log(k + 1) + 3$ moves.

7   Graph Embeddings in Orientable Surfaces

We here consider cellular embeddings of connected graphs in orientable surfaces of arbitrary genus using for them a standard combinatorial representation, see [28, Sect. 3.2]. A *rotation system* $R = \langle G, T \rangle$ is a structure consisting of a graph G and a ternary relation T on $V(G)$ satisfying the following conditions:

(1) If $T(x, y, z)$, then y and z are in $\Gamma(x)$, the neighborhood of x in G.

(2) For every x of degree at least 2, the binary relation $T_x(y, z) = T(x, y, z)$ is a directed cycle on $\Gamma(x)$ (i.e., for every $y \in \Gamma(x)$ there is exactly one z such that $T_x(y, z)$, for every $z \in \Gamma(x)$ there is exactly one y such that $T_x(y, z)$, and the digraph T_x is connected).

Geometrically, T_x describes the circular order in which the edges of G incident to x occur in the embedding if we go around x clockwise.

Theorem 10. *Let $R = \langle G, T \rangle$ be a rotation system for a connected graph G with n vertices. We have $D^5(R) < 3 \log n + 8$.*

On the account of Corollary 3.2 this implies earlier results of Miller and Reif [27].

Corollary 11. *The isomorphism problem for rotation systems is in AC^1.*

Miller and Reif give also a reduction of the planar graph isomorphism to the isomorphism problem for rotation systems which is an AC^1 reduction provided 3-connected planar graphs are embeddable in plane in AC^1. The latter is shown by Ramachandran and Reif [32].

Corollary 12. *The isomorphism problem for planar graphs is in AC^1.*

In the rest of the section we prove Theorem 10. The proof is based on Equation (1) and Proposition 6.1. Let $R = \langle G, T \rangle$ be a rotation system with n vertices and $R' = \langle G', T' \rangle$ be a non-isomorphic structure of the same signature. We have to design a strategy for Spoiler in the Ehrenfeucht-Fraïssé game on R and R' allowing him to win with only 5 pebbles in less than $3 \log n + 8$ moves, whatever Duplicator's strategy.

The main idea of the proof is to show that a rotation system admits a natural coordinatization and that Duplicator must respect vertex coordinates. A coordinate system on $R = \langle G, T \rangle$ is determined by fixing its origins, namely, an ordered edge of G. We first define *local coordinates* on the neighborhood of a vertex x. Fix $y \in \Gamma(x)$ and let z be any vertex in $\Gamma(x)$. Then $c_{xy}(z)$ is defined to be the number of z in the order of T_x if we start counting from $c_{xy}(y) = 0$. In the global system of coordinates specified by an ordered pair of adjacent $a, b \in V(G)$, each vertex $v \in V(G)$ receives coordinates $C_{ab}(v)$ as follows. Given a path $P = a_0 a_1 a_2 \ldots a_l$ from $a_0 = a$ to $a_l = v$, let $C_{ab}(v; P) = (c_1, \ldots, c_l)$ be a sequence of integers with $c_1 = c_{ab}(a_1)$ and $c_i = c_{a_{i-1} a_{i-2}}(a_i)$ for $i \geq 2$. We define $C_{ab}(v)$ to be the lexicographically minimum $C_{ab}(v; P)$ over all P. Note that $C_{ab}(v)$ has length $d(a, v)$. By P_v we will denote the path for which $C_{ab}(v) = C_{ab}(v; P_v)$. One can say that P_v is *the extreme left shortest path from a to v*. Note that P_v is reconstructible from $C_{ab}(v)$ and hence different vertices receive different coordinates. The following observation enables a kind of the halving strategy.

Lemma 13. *Let $a, b, v \in V(G)$ and $a', b', v' \in V(G')$, where a and b as well as a' and b' are adjacent. Assume that $d(a, v) = d(a', v')$ but $C_{ab}(v) \neq C_{a'b'}(v')$. Furthermore, let u and u' lie on P_v and $P_{v'}$ at the same distance from a and a' respectively. Assume that $C_{ab}(u) = C_{a'b'}(u')$. Finally, let w and w' be predecessors of u and u' on P_v and $P_{v'}$ respectively. Then $C_{uw}(v) \neq C_{u'w'}(v')$.*

Proof. By definition, $C_{ab}(v) = C_{ab}(u)C_{uw}(v)$ and $C_{a'b'}(v') = C_{a'b'}(u')C_{u'w'}(v')$.

Lemma 14. *Suppose that $a, b, v \in V(G)$ and $a', b', v' \in V(G')$ are pebbled coherently to the notation. Assume that a and b as well as a' and b' are adjacent and that $C_{ab}(v) \neq C_{a'b'}(v')$. Then Spoiler is able to win with 5 pebbles in less than $3 \log n + 3$ moves.*

The proof of the lemma can be found in [17]. Now we are ready to describe Spoiler's strategy in the game on R and R'. In the first two rounds he pebbles a and b, arbitrary adjacent vertices in G. Let Duplicator respond with adjacent a' and b' in G'. If G contains a vertex v with coordinates $C_{ab}(v)$ different from every $C_{a'b'}(v')$ in G' or if G' contains a vertex with coordinates absent in G, then Spoiler pebbles it and wins by Lemma 14. Suppose therefore that the coordinatization determines a matching between $V(G)$ and $V(G')$. Given $x \in V(G)$, let $f(x)$ denote the vertex $x' \in V(G')$ with $C_{a'b'}(x') = C_{ab}(x)$. If f is not an isomorphism from G to G', then Spoiler pebbles two vertices $u, v \in V(G)$ such that the pairs u, v and $f(u), f(v)$ have different adjacency. Not to lose immediately, Duplicator responds with a vertex having different coordinates and again Lemma 14 applies. If f is an isomorphism between G and G', then this map does not respect the relations T and T' and Spoiler demonstrates this similarly. The proof of Theorem 10 is complete.

References

1. Babai, L.: Moderately exponential bound for graph isomorphism. In: Gécseg, F. (ed.): Fundamentals of Computation Theory. Lecture Notes in Computer Science, Vol. 117. Springer-Verlag (1981) 34–50
2. Babai, L.: Automorphism groups, isomorphism, reconstruction. In: Graham, R.L., Grötschel, M., Lovász, L. (eds.): Handbook of Combinatorics, Chap. 27. Elsevier Publ. (1995) 1447–1540
3. Babai, L., Grigoryev, D.Yu., Mount, D.M.: Isomorphism of graphs with bounded eigenvalue multiplicity. In: Proc. of the 14th ACM Symp. on Theory of Computing (1982) 310–324
4. Babai, L., Luks, E.M.: Canonical labeling of graphs. In: Proc. of the 15th ACM Symposium on Theory of Computing (1983) 171–183
5. Bodlaender, H.L.: Polynomial algorithms for Graph Isomorphism and Chromatic Index on partial k-trees. J. Algorithms **11** (1990) 631–643
6. Bohman, T., Frieze, A., Łuczak, T., Pikhurko, O., Smyth, C., Spencer, J., Verbitsky, O.: The first order definability of trees and sparse random graphs. E-print (2005) *http://arxiv.org/abs/math.CO/0506288*
7. Boppana, R.B., Håstad, J., Zachos, S.: Does co-NP have short interactive proofs? Inf. Process. Lett. **25** (1987) 127–132
8. Cai, J.-Y., Fürer, M., Immerman, N.: An optimal lower bound on the number of variables for graph identification. Combinatorica **12** (1992) 389–410
9. Chandrasekharan, N.: Isomorphism testing of k-trees is in NC, for fixed k. Inf. Process. Lett. **34** (1990) 283–287
10. Del Greco, J.G., Sekharan, C.N., Sridhar, R.: Fast parallel reordering and isomorphism testing of k-trees. Algorithmica **32** (2002) 61–72
11. Evdokimov, S., Karpinski, M., Ponomarenko, I.: On a new high dimensional Weisfeiler-Lehman algorithm. J. Algebraic Combinatorics **10** (1999) 29–45
12. Evdokimov, S., Ponomarenko, I.: On highly closed cellular algebras and highly closed isomorphism. Electronic J. Combinatorics **6** (1999) #R18
13. Filotti, I.S., Mayer, J.N.: A polynomial-time algorithm for determining the isomorphism of graphs of fixed genus. In: Proc. of the 12th ACM Symp. on Theory of Computing (1980) 236–243

14. Grohe, M.: Fixed-point logics on planar graphs. In: Proc. of the Ann. Conf. on Logic in Computer Science (1998) 6–15
15. Grohe, M.: Isomorphism testing for embeddable graphs through definability. In: Proc. of the 32nd ACM Ann. Symp. on Theory of Computing (2000) 63–72
16. Grohe, M., Marino, J.: Definability and descriptive complexity on databases of bounded tree-width. In: Proc. of the 7th Int. Conf. on Database Theory. Lecture Notes in Computer Science, Vol. 1540. Springer-Verlag (1999) 70–82
17. Grohe, M., Verbitsky, O.: Testing graph isomorphism in parallel by playing a game. E-print (2006) *http://arxiv.org/abs/cs.CC/0603054*
18. Hopcroft, J.E., Tarjan, R.E.: Isomorphism of planar graphs (working paper). In: Miller, R.E., Thatcher, J.W. (eds.): Complexity of computer computations. Plenum Press, New York-London (1972) 131–152
19. Hopcroft, J.E., Wong, J.K.: Linear time algorithm for isomorphism of planar graphs. In: Proc. of the 6th ACM Symp. on Theory of Computing (1974) 172–184
20. Immerman, N.: Descriptive complexity. Springer-Verlag (1999)
21. Jenner, B., Köbler, J., McKenzie, P., Torán, J.: Completeness Results for Graph Isomorphism. J. Comput. Syst. Sci. **66** (2003) 549–566
22. Karp, R.M., Ramachandran, V.: Parallel algorithms for shared-memory machines. In: van Leeuwen, J. (ed.): Algorithms and complexity. Handbook of theoretical computer science, Vol. A. Elsevier Publ., Amsterdam (1990) 869–941
23. Kim, J.-H., Pikhurko, O., Spencer, J., Verbitsky, O.: How complex are random graphs in first order logic? Random Structures and Algorithms **26** (2005) 119–145
24. Lindell, S.: A logspace algorithm for tree canonization. In: Proc. of the 24th Ann. ACM Symp. on Theory of Computing (1992) 400–404
25. Luks, E.M.: Isomorphism of graphs of bounded valence can be tested in polynomial time. J. Comput. Syst. Sci. **25** (1982) 42–65
26. Miller, G.L.: Isomorphism testing for graphs of bounded genus. In: Proc. of the 12th ACM Symp. on Theory of Computing (1980) 225–235
27. Miller, G.L., Reif, J.H.: Parallel tree contraction. Part 2: further applications. SIAM J. Comput. **20** 1128–1147
28. Mohar, B., Thomassen, C.: Graphs on surfaces. The John Hopkins University Press (2001)
29. Pikhurko, O., Spencer, J., Verbitsky, O.: Succinct definitions in first order graph theory. Annals of Pure and Applied Logic **139** (2006) 74–109
30. Pikhurko, O., Veith, H., Verbitsky, O.: First order definability of graphs: tight bounds on quantifier rank. Discrete Applied Mathematics (to appear)
31. Ponomarenko, I.N.: The isomorphism problem for classes of graphs that are invariant with respect to contraction. In: Computational Complexity Theory 3. Zap. Nauchn. Sem. Leningrad. Otdel. Mat. Inst. Steklov, Vol. 174 (1988) 147–177.
32. Ramachandran, V., Reif, J.: Planarity testing in parallel. J. Comput. Syst. Sci. **49** (1994) 517–561
33. Robertson, N., Seymour, P.D.: Graph minors II. Algorithmic aspects of tree-width. J. Algorithms **7** (1986) 309–322
34. Schöning, U.: Graph isomorphism is in the low hierarchy. J. Comput. Syst. Sci. **37** (1988) 312–323
35. Spencer, J.: The strange logic of random graphs. Springer Verlag (2001)
36. Torán, J.: On the hardness of graph isomorphism. SIAM J. Comput. **33** (2004) 1093–1108
37. Verbitsky, O.: The first order definability of graphs with separators via the Ehrenfeucht game. Theor. Comput. Sci. **343** (2005) 158–176

The Spectral Gap of Random Graphs with Given Expected Degrees

Amin Coja-Oghlan[1] and André Lanka[2]

[1] Humboldt-Universität zu Berlin, Institut für Informatik
Unter den Linden 6, 10099 Berlin, Germany
coja@informatik.hu-berlin.de
[2] Technische Universität Chemnitz, Fakultät für Informatik
Straße der Nationen 62, 09107 Chemnitz, Germany
lanka@informatik.tu-chemnitz.de

Abstract. We investigate the Laplacian eigenvalues of a random graph $G(n, d)$ with a given expected degree distribution d. The main result is that w.h.p. $G(n, d)$ has a large subgraph $\mathrm{core}(G(n, d))$ such that the spectral gap of the normalized Laplacian of $\mathrm{core}(G(n, d))$ is $\geq 1 - c_0 \bar{d}_{\min}^{-1/2}$ with high probability; here $c_0 > 0$ is a constant, and $\bar{d}_{\min}$ signifies the minimum expected degree. This result is of interest in order to extend known spectral heuristics for random regular graphs to graphs with irregular degree distributions, e.g., power laws. The present paper complements the work of Chung, Lu, and Vu [Internet Mathematics **1**, 2003].

1 Introduction

Numerous heuristics for graph partitioning problems are based on *spectral methods*: the heuristic sets up a matrix that represents the input graph and reads information on the *global structure* of the graph out of the eigenvalues and eigenvectors of the matrix. Spectral techniques are very popular in areas such as VLSI design, parallel computing, and scientific simulation [16,17].

Though in many cases there are worst-case examples known showing that certain spectral heuristics perform badly on general instances, they seem to perform well on many "practical" inputs. Therefore, in order to gain a better theoretical understanding, quite a few papers deal with rigorous analyses of spectral heuristics on *random* graphs, e.g., Alon, Kahale [2], Alon, Krivelevich, Sudakov [3], or McSherry [14].

However, a crucial problem with most known spectral methods is that their use is limited to essentially *regular* graphs, where all vertices have approximately the same degree (two exceptions are [6,8], cf. below). For the spectra of the matrices that are most frequently used to represent graphs (e.g., the adjacency matrix) are quite susceptible to fluctuations of the vertex degrees. In effect, in the case of irregular graphs their eigenvalues fail to mirror *global* graph properties but merely reflect the tails of the degree distribution, cf. Mihail and Papadimitriou [15].

M. Bugliesi et al. (Eds.): ICALP 2006, Part I, LNCS 4051, pp. 15–26, 2006.
© Springer-Verlag Berlin Heidelberg 2006

Nonetheless, in the past decade it turned out that many interesting types of graphs actually are extremely irregular. For example, the degree distribution of the Internet domain graph follows a *power law* [9]; that is, the number of vertices of degree d is proportional to $d^{-\gamma}$ for a constant $\gamma > 1$. Similar degree distributions occur in further networks arising, e.g., in biology. In addition, these graphs are usually *sparse* [1]. Since a power law distribution has a heavy *upper tail* (i.e., there are plenty of vertices whose degrees by far exceed the average degree), most of the known spectral methods do not apply.

In the present paper we investigate how spectral methods can be extended to irregular graphs, and in particular, to *sparse* irregular graphs. As in the regular case, random graphs turn out to be a rather useful tool to analyze spectral techniques rigorously. The random graph model we shall work with is the following: let $V = \{1, \ldots, n\}$, and let $\boldsymbol{d} = (\bar{d}(v))_{v \in V}$, where each $\bar{d}(v)$ is a positive real. Moreover, set $\bar{d} = \frac{1}{n} \sum_{v \in V} \bar{d}(v)$ and suppose that $\bar{d}(v)\bar{d}(w) \leq \bar{d}n$ for all $v, w \in V$. Then the random graph $G(n, \boldsymbol{d})$ has the vertex set V, and for any two distinct vertices $v, w \in V$ the edge $\{v, w\}$ is present with probability $p_{vw} = \bar{d}(v)\bar{d}(w)/(n\bar{d})$ independently of all others.

Hence, the *expected* degree of each vertex $v \in V$ is $\sum_{w \neq v} p_{vw} \sim \bar{d}(v)$, and the expected average degree is $\bar{d}$. In other words, $G(n, \boldsymbol{d})$ is a random graph with a given expected degree sequence $\boldsymbol{d}$. We say that $G(n, \boldsymbol{d})$ has some property $\mathcal{E}$ *with high probability (w.h.p.)* if the probability that $\mathcal{E}$ holds tends to one as $n \rightarrow \infty$. Possibly $G(n, \boldsymbol{d})$ is the simplest model of a random irregular graph; its advantage is that it can model graphs with very general degree distributions, including but not limited to power laws. There are, however, also relevant generative models of power law graphs (cf. [4] for details).

There are essentially two previous papers [6,8] that successfully apply spectral methods to the $G(n, \boldsymbol{d})$ model. Chung, Lu, and Vu [6] studied the eigenvalue distribution of the *normalized Laplacian matrix* of $G(n, \boldsymbol{d})$. They showed that its spectrum does reflect global properties, provided that $\min_{v \in V} \bar{d}(v) \geq \ln^2 n$, i.e., the graph is *dense enough*. (By contrast, in the irregular case the spectrum of the *adjacency matrix* does not mirror global properties [15].) Furthermore, the normalized Laplacian was also used by Dasgupta, Hopcroft, and McSherry [8] to devise a heuristic for partitioning sufficiently dense random irregular graphs (with average degree $\gg \ln^6 n$); their model is closely related to $G(n, \boldsymbol{d})$.

We complement the work of Chung, Lu, and Vu [6] by studying the normalized Laplacian of *sparse* random graphs $G(n, \boldsymbol{d})$, e.g., with average degree independent of n. We believe that this extension is significant, because sparse graphs are the most appropriate to model real networks [1,9]. In comparison with the dense case, dealing with sparse graphs requires new techniques, as both the proofs and the results of [6] depend crucially on the assumption that the minimum expected degree is $\gg \ln^2 n$. In addition, we indicate a few algorithmic applications of our main result, which show how spectral algorithms for random regular graphs can be extended to the irregular case.

Before discussing related work in Section 3, we state the results. Finally, in Section 4 we sketch the proof of the main theorem.

2   Results

Let us recall the definition of the normalized Laplacian $\mathcal{L}(G)$ of a graph $G = (V, E)$. Letting $d_G(v)$ denote the degree of v in G, for $v, w \in V$ we define $\ell_{vw} = 1$ if $v = w$ and $d_G(v) > 0$, $\ell_{vw} = -1/\sqrt{d_G(v)d_G(w)}$ if $\{v, w\} \in E$, and $\ell_{vw} = 0$ otherwise. Then $\mathcal{L}(G) = (\ell_{vw})_{v,w \in V}$. This matrix is singular and positive semidefinite, and its largest eigenvalue is ≤ 2 (cf. [5]). Letting $0 = \lambda_1(\mathcal{L}(G)) \leq \cdots \leq \lambda_{\#V}(\mathcal{L}(G))$ denote its eigenvalues, we call $\min\{\lambda_2(\mathcal{L}(G)), 2 - \lambda_{\#V}(\mathcal{L}(G))\}$ the *spectral gap* of $\mathcal{L}(G)$.

The spectral gap is a most interesting parameter, because it is directly related to combinatorial graph properties. To see this, let $G = (V, E)$ be an arbitrary graph. Then we say that G has (α, β)-*low discrepancy* if for *any* two disjoint sets $X, Y \subset V$ the following holds. Letting $e_G(X, Y)$ denote the number of X-Y-edges in G and setting $d(X, Y) = \sum_{(v,w) \in X \times Y} d_G(v)d_G(w)$, we have

$$|e(X,Y) - d(X,Y)/(2\#E)| \leq (1 - \alpha)\sqrt{d(X,Y)} + \beta \qquad \text{and} \qquad (1)$$

$$|2e(X,X) - d(X,X)/(2\#E)| \leq (1 - \alpha)\sqrt{d(X,X)} + \beta. \qquad (2)$$

An easy computation shows that $d(X,Y)/(2\#E)$ is the number of X-Y-edges that we would *expect* if G were a random graph with expected degree sequence $(d_G(v))_{v \in V}$; similarly, $d(X,X)/(4\#E)$ is the expected number of edges inside of X. Thus, the closer $\alpha < 1$ is to 1 and the smaller $\beta \geq 0$, the more G "looks like" a random graph if (1) and (2) hold.

Now, if the spectral gap of $\mathcal{L}(G)$ is $\geq \gamma$, then G has $(\gamma, 0)$-low discrepancy [5]. Hence, the larger the spectral gap, the more G resembles a random graph. Therefore, if the normalized Laplacian provides a reasonable way to represent graphs with degree sequence $\boldsymbol{d}$, then the spectral gap of $G(n, \boldsymbol{d})$ should be large.

Chung, Lu, and Vu [6] proved that this is indeed the case, provided that the minimum expected degree $\bar{d}_{\min} = \min_{v \in V} \bar{d}(v)$ satisfies $\bar{d}_{\min} \gg \ln^2 n$. More precisely, they proved

$$\text{spectral gap of } \mathcal{L}(G(n, \boldsymbol{d})) \geq 1 - (1 + o(1))4\bar{d}^{-1/2} - \bar{d}_{\min}^{-1} \ln^2 n \text{ w.h.p.} \qquad (3)$$

As for general graphs with average degree $\bar{d}$ the spectral gap is at most $1 - 4\bar{d}^{-1/2}$, the bound (3) is very strong and in general best possible.

However, (3) is obviously void if $\bar{d}_{\min} \leq \ln^2 n$ (because the r.h.s. is negative). In fact, the following proposition shows that if $\bar{d}$ is small, then the spectral gap of $\mathcal{L}(G(n, \boldsymbol{d}))$ is just 0, even if the expected degrees of all vertices coincide.

Proposition 1. *Let $d > 0$ be arbitrary but constant, set $d_v = d$ for all $v \in V$, and let $\boldsymbol{d} = (d_v)_{v \in V}$. Let λ_* denote the smallest positive eigenvalue of $\mathcal{L}(G(n, \boldsymbol{d}))$, and let λ^* be the largest. Then for any constant $\varepsilon > 0$ we have $\lambda_* < \varepsilon$ and $\lambda^* = 2$ w.h.p.*

Nonetheless, our main result is that even in the sparse case w.h.p. $G(n, \boldsymbol{d})$ has a *large subgraph* core(G) on which a similar statement as (3) holds.

Theorem 2. *There are constants $c_0, d_0 > 0$ such that the following holds. Suppose that $\boldsymbol{d} = (\bar{d}(v))_{v \in V}$ satisfies*

$$d_0 \leq \bar{d}_{\min} = \min_{v \in V} \bar{d}(v) \leq \max_{v \in V} \bar{d}(v) \leq n^{0.99}. \tag{4}$$

Then w.h.p. the random graph $G = G(n, \boldsymbol{d})$ has an induced subgraph $\text{core}(G)$ that enjoys the following properties.

1. *We have $\sum_{v \in G - \text{core}(G)} d_G(v) \leq n \exp(-\bar{d}_{\min}/c_0)$.*
2. *The spectral gap of $\mathcal{L}(\text{core}(G))$ is $\geq 1 - c_0 \bar{d}_{\min}^{-1/2}$.*

Thus, the spectral gap of the core is close to 1 if $\bar{d}_{\min}$ is not too small. It is instructive to compare Theorem 2 with (3), cf. Remark 8 below for details. Further, in Remark 7 we point out that the bound on the spectral gap given in Theorem 2 is best possible up to the precise value of the constant c_0.

Theorem 2 has a few interesting algorithmic implications. Namely, we can extend a couple of algorithmic results for random graphs in which all expected degrees are equal to the irregular case.

Corollary 3. *There is a polynomial time algorithm LowDisc that satisfies the following two conditions.*

Correctness. *For any input graph G, LowDisc outputs two numbers α, β such that G has (α, β)-low discrepancy.*
Completeness. *If $G = G(n, \boldsymbol{d})$ is a random graph such that $\boldsymbol{d}$ satisfies the assumption (4) of Theorem 2, then $\alpha \geq 1 - c_0 \bar{d}_{\min}^{-1/2}$ and $\beta \leq 2n \exp(-\bar{d}_{\min}/c_0)$ w.h.p.*

LowDisc relies on the fact that for a given graph G the subgraph $\text{core}(G)$ can be computed efficiently. Then, LowDisc computes the spectral gap of $\mathcal{L}(\text{core}(G))$ to bound the discrepancy of G. Hence, Corollary 3 shows that spectral techniques do yield information on the global structure of random irregular graphs $G(n, \boldsymbol{d})$, even in the sparse case.

One could object that we might as well derive by probabilistic techniques such as the "first moment method" that $G(n, \boldsymbol{d})$ has low discrepancy w.h.p. However, such arguments just show that "most" graphs $G(n, \boldsymbol{d})$ have low discrepancy. By contrast, the statement of Corollary 3 is much stronger: for a *given outcome* $G = G(n, \boldsymbol{d})$ of the random experiment we can find a *proof* that G has low discrepancy *in polynomial time*.

Since the discrepancy of a graph is closely related to quite a few prominent graph invariants that are (in the worst case) NP-hard to compute, we can apply Corollary 3 to obtain further algorithmic results on random graphs $G(n, \boldsymbol{d})$. For instance, we can bound the *independence number* $\alpha(G(n, \boldsymbol{d}))$ efficiently. In addition, we have the following result on the chromatic number $\chi(G(n, \boldsymbol{d}))$.

Corollary 4. *There exists a polynomial time algorithm BoundChi that satisfies the following conditions.*

Correctness. *For any input graph G BoundChi outputs a lower bound $\chi \leq \chi(G)$ on the chromatic number.*

Completeness. *If $G = G(n, \boldsymbol{d})$ is a random graph such that $\boldsymbol{d}$ satisfies (4), then $\chi \geq c_1 \bar{d}^{1/2}$ w.h.p., for a certain constant $c_1 > 0$.*

Corollaries 3 and 4 extends results from [10,13] for $G(n,p)$ to the $G(n,\boldsymbol{d})$ model. (Actually Corollary 4 relies on a somewhat modified construction of the subgraph $\mathrm{core}(G)$; we omit the details.)

3 Related Work

A large number of authors have studied the Erdős-Rényi model $G(n,p)$ of random graphs, where $0 \leq p \leq 1$ is the expected density of the graph. The $G(n,p)$ model is the same as $G(n,\boldsymbol{d})$ with $\bar{d}(v) = np$ for all v. With respect to the eigenvalues $\lambda_1(A) \leq \cdots \leq \lambda_n(A)$ of the adjacency matrix $A = A(G(n,p))$, Füredi and Komlós [12] proved that if $np(1 - p) \gg \ln^6 n$, then $\max\{-\lambda_1(A), \lambda_{n-1}(A)\} \leq (2 + o(1))(np(1 - p))^{1/2}$ and $\lambda_n(A) \sim np$. Furthermore, Feige and Ofek [10] showed that $\max\{-\lambda_1(A), \lambda_{n-1}(A)\} \leq O(np)^{1/2}$ and $\lambda_n(A) = \Theta(np)$ also holds w.h.p. under the weaker assumption $np \geq \ln n$.

By contrast, in the sparse case (say, $\bar{d} = np = O(1)$ as $n \to \infty$), neither $\lambda_n(A) = \Theta(\bar{d})$ nor $\max\{-\lambda_2(A), \lambda_{n-1}(A)\} \leq O(\bar{d})^{1/2}$ is true w.h.p. For if $\bar{d} = O(1)$, then the vertex degrees of $G = G(n,p)$ have (asymptotically) a Poisson distribution with mean $\bar{d}$. Consequently, the degree distribution features a fairly heavy upper tail. Indeed, the maximum degree is $\Omega(\ln n / \ln \ln n)$ w.h.p., and the high degree vertices induce both positive and negative eigenvalues as large as $\Omega(\ln n / \ln \ln n)^{1/2}$ in absolute value. Nonetheless, following an idea of Alon and Kahale [2] and building on the work of Kahn and Szemerédi [11], Feige and Ofek [10] showed that the graph $G' = (V', E')$ obtained by removing all vertices of degree, say, $> 2\bar{d}$ from G w.h.p. satisfies $\max\{-\lambda_1(A(G')), \lambda_{\#V'-1}(A(G'))\} = O(\bar{d}^{1/2})$ and $\lambda_{\#V(G')}(A(G')) = \Theta(\bar{d})$. The articles [10,12] are the basis of several papers dealing with rigorous analyses of spectral heuristics on random graphs (e.g., [3,14]). Further, the first author [7] used [10,12] to investigate the Laplacian of $G(n,p)$.

The graphs we are considering in this paper may have a significantly more general (i.e., irregular) degree distribution than even the sparse random graph $G(n,p)$. In fact, such irregular degree distributions occur in real-world networks, cf. Section 1. While such networks are frequently modeled best by *sparse* graphs, (i.e., $\bar{d} = O(1)$ as $n \to \infty$) the *maximum* degree may very well be as large as $n^{\Omega(1)}$, i.e., not only logarithmic but even polynomial in n. As a consequence, the eigenvalues of the adjacency matrix are determined by the upper tail of the degree distribution rather than by global graph properties (cf. [15]). Furthermore, the idea of Feige and Ofek [10] of just deleting the vertices of degree $\gg \bar{d}$ is not feasible, because the high degree vertices constitute a significant share of the graph. Thus, the adjacency matrix is simply not appropriate to represent power law graphs.

As already mentioned in Section 1, Chung, Lu, and Vu [6] were the first to obtain rigorous results on the normalized Laplacian. In addition to the afore-mentioned estimate (3), they also proved that the global distribution of the eigenvalues follows the semicircle law. Their proofs rely on the "trace method" of Wigner [18], i.e., Chung, Lu, and Vu (basically) compute the trace of $\mathcal{L}(G(n, \boldsymbol{d}))^k$ for a large even number k. Since this equals the sum of the k'th powers of the eigenvalues of $\mathcal{L}(G(n, \boldsymbol{d}))$, they can thus infer information on the distribution of the eigenvalues.

However, the proofs in [6] hinge upon the assumption that $\bar{d}_{\min} \gg \ln^2 n$, and indeed there seems to be no easy way to extend the trace method to the sparse case (even if $\bar{d}(v) = \bar{d}$ for all $v \in V$). Therefore, in the present paper instead of relying on the trace method we extend a technique developed by Kahn and Szemerédi [11] to analyze the spectral gap of random *regular* graphs to the irregular case. In addition, we also need to extend some methods from the first author's analysis [7] of the Laplacian of $G(n, p)$ to irregular degree distributions.

4   The Spectral Gap of the Laplacian

In this section we sketch the proof of Theorem 2. We state the definition of the core exactly in Section 4.1. Then, in Section 4.2 we outline how to prove that the Laplacian of $\text{core}(G(n, \boldsymbol{d}))$ has a large spectral gap w.h.p. In the remaining subsections we sketch some of the proof details. *We always assume that (4) holds, that $c_0, d_0 > 0$ are sufficiently large constants, and that n is large enough.* No attempt has been made to optimize the constants involved in the analysis.

Let us briefly introduce some notation. Throughout, we let $V = \{1, \ldots, n\}$. Moreover, if $G = (V, E)$ is a graph and $X, Y \subset V$, then $e_G(X, Y)$ signifies the number of X-Y-edges in G. We denote the degree of $v \in V$ by $d_G(v)$. Further, by $\mu(X, Y)$ we denote the *expected* number of X-Y-edges in a random graph $G = G(n, \boldsymbol{d})$. If $X = Y$, then we abbreviate $e_G(X) = e_G(X, X)$ and $\mu(Y) = \mu(Y, Y)$. Additionally, let $\text{Vol}(X) = \sum_{v \in X} \bar{d}(v)$.

If $M = (m_{vw})_{v, w \in V}$ is a matrix and $X, Y \subset V$, then $M_{X \times Y}$ denotes the matrix with entries $(m'_{vw})_{v, w \in V}$, where $m'_{vw} = m_{vw}$ if $(v, w) \in X \times Y$, and $m'_{vw} = 0$ otherwise. If $X = Y$, we briefly write $M_X = M_{X \times X}$.

4.1   The Definition of $\text{Core}(G(n, d))$

To motivate the definition of the core, let us discuss why $\mathcal{L}(G(n, \boldsymbol{d}))$ may have positive eigenvalues much smaller than one. Basically the reason is the existence of vertices of small degree. Indeed, consider $v \in V$ such that $\bar{d}(v) \leq 2\bar{d}$, say. Then the actual degree $d_G(v)$ is a sum of independent Bernoulli variables. Therefore, if $\bar{d}_{\max} = o(\sqrt{n})$ and $\bar{d} = O(1)$ as $n \to \infty$, then an easy computation shows that $d_G(v)$ is much smaller than $\bar{d}(v)$ (or even $d_G(v) = 0$) with probability bounded away from 0. Therefore, w.h.p. for each $0 \leq d < \bar{d}_{\min}$ there are $\Omega(n)$ vertices with degree d in $G(n, \boldsymbol{d})$. Furthermore, these small degree vertices can easily cause the spectral gap to be tiny. To see this, we call a vertex v of $G(n, \boldsymbol{d})$ a (d, d)-*star* if v has degree d, its neighbors $v_1, \ldots, v_d$ have degree d as well, and $\{v_1, \ldots, v_d\}$

is an independent set. Then, using standard random graph arguments, we can prove the following.

Lemma 5. *Let $d \leq \bar{d} = O(1) \leq \bar{d}_{\max} = o(\sqrt{n})$. Then $G(n, \boldsymbol{d})$ has a (d, d)-star w.h.p.*

Now, a (d, d)-star induces the eigenvalues $1 \pm (1 - o(1))d^{-1/2}$: define a vector $\xi = (\xi_w)_{w \in V}$ by letting $\xi_v = d^{1/2}$, $\xi_{v_i} = 1$ for $1 \leq i \leq d$, and $\xi_w = 0$ for all other vertices w. Then $\|\xi\|^{-2} \langle \mathcal{L}(G)\xi, \xi \rangle = 1 - d^{-1/2}$; similarly, one could define a vector η such that $\|\eta\|^{-2} \langle \mathcal{L}(G)\eta, \eta \rangle = 1 + d^{-1/2}$. Hence, Lemma 5 entails that w.h.p. the spectral gap is much smaller than $1 - \bar{d}_{\min}^{-1/2}$.

Thus, to construct a subgraph of $G = G(n, \boldsymbol{d})$ with a large spectral gap, we need to get rid of the small degree vertices. To this end, we consider the following process.

CR1. Initially let $H = G - \{v : d_G(v) \leq 0.01\bar{d}_{\min}\}$.
CR2. While there is a vertex $v \in H$ that has $\geq \max\{c_0, \exp(-\bar{d}_{\min}/c_0)d_G(v)\}$ neighbors in $G - H$, remove v from H.

Hence, in the first step CR1 we just remove all vertices of degree much smaller than $\bar{d}_{\min}$. This is, however, not yet sufficient; for the deletion of these vertices might create new vertices of small degree. Therefore, CR2 iteratively removes vertices that have plenty of neighbors that were removed before. The final outcome H of the process is $\mathrm{core}(G)$. Observe that by construction all vertices $v \in \mathrm{core}(G)$ satisfy

$$d_{\mathrm{core}(G)}(v) \geq \frac{\bar{d}_{\min}}{200}, \quad e(v, G - \mathrm{core}(G)) < \max\{c_0, \exp(-c_0^{-1}\bar{d}_{\min})d_G(v)\}. \quad (5)$$

Additionally, in the analysis of the spectral gap of $\mathcal{L}(\mathrm{core}(G))$ in Section 4.2, we will need to consider the following subgraph $\mathcal{S}$, which is defined by a "more picky" version of CR1–CR2:

S1. Initially, let $\mathcal{S} = \mathrm{core}(G) - \{v \in V : |d_H(v) - \bar{d}(v)| \geq 0.01\bar{d}(v)\}$.
S2. While there is a $v \in \mathcal{S}$ so that $e_G(v, G - \mathcal{S}) \geq \max\{c_0, d_G(v) \exp(-\bar{d}_{\min}/c_0)\}$, remove v from $\mathcal{S}$.

Then by (4) after the process S1–S2 has terminated, every vertex $v \in \mathcal{S}$ satisfies $|d_\mathcal{S}(v) - \bar{d}(v)| \leq \frac{\bar{d}(v)}{50}$.

An important property of $\mathrm{core}(G)$ is that given just $\bar{d}_{\min}$, G (and c_0), we can compute $\mathrm{core}(G)$ efficiently (without any further information about $\boldsymbol{d}$). This fact is the basis of the algorithmic applications (Corollaries 3 and 4). By contrast, while $\mathcal{S}$ will be useful in the *analysis* of $\mathcal{L}(\mathrm{core}(G))$, it cannot be computed without explicit knowledge of $\boldsymbol{d}$.

Let us finally point out that w.h.p. $\mathcal{S}$ and thus $\mathrm{core}(G)$ constitutes a huge fraction of $G = G(n, \boldsymbol{d})$. We sketch the proof of the following proposition in Section 4.3.

Proposition 6. *W.h.p.* $\mathrm{Vol}(V \setminus \mathrm{core}(G)) \leq \mathrm{Vol}(V \setminus \mathcal{S}) \leq \exp(-100\bar{d}_{\min}/c_0)n$.

Remark 7. Letting $d = \bar{d}_{\min}$ and assuming that $\bar{d} = O(1)$ as $n \to \infty$, one can derive that w.h.p. $\mathrm{core}(G(n,p))$ contains a (d,d)-star. Hence, a similar argument as above shows that the spectral gap of $\mathcal{L}(\mathrm{core}(G(n,\boldsymbol{d})))$ is at most $1 - \bar{d}_{\min}^{-1/2}$. Thus, Theorem 2 essentially best possible (up to the precise value of c_0).

Remark 8. While the result (3) of Chung, Lu, and Vu [6] is void if $\bar{d}_{\min} \leq \ln^2 n$, in the case $\bar{d}_{\min} \gg \ln^2 n$ its dependence on $\bar{d}_{\min}$ is better than the estimate provided by Theorem 2. In the light of Remark 7, this shows that in the dense case $\bar{d}_{\min} \gg \ln^2 n$ "bad" local structures such as $(\bar{d}_{\min}, \bar{d}_{\min})$-stars just do not occur w.h.p.

4.2   Proof of Theorem 2: Outline

We let $G = (V,E) = G(n,\boldsymbol{d})$, $H = \mathrm{core}(G)$, and $\omega = (d_G(v)^{1/2})_{v \in V}$. Then $\mathcal{L}(H)\omega = 0$, so that our task is to estimate the spectral radius of $M = \boldsymbol{E} - \mathcal{L}(H)$ restricted to the orthogonal complement of ω (where $\boldsymbol{E}$ signifies the identity matrix). A crucial issue is that the entries of M are mutually dependent random variables. For if two vertices $v, w \in H$ are adjacent, then the vw'th entry of M is $(d_H(v)d_H(w))^{-1/2}$, and of course $d_H(v), d_H(w)$ are neither mutually independent nor independent of the presence of the edge $\{v,w\}$. A further source of dependence is that we restrict ourselves to the core H of G.

Therefore, instead of M it would be much easier to deal with the matrix $\mathcal{M} = (m_{uv})_{u,v \in V}$ with entries $m_{uv} = (\bar{d}(u)\bar{d}(v))^{-1/2}$ if $\{u,v\} \in E$, and $m_{uv} = 0$ otherwise; that is, in $\mathcal{M}$ the entries are normalized by the *expected* degrees rather than by the *actual* degrees. Hence, the entries of $\mathcal{M}$ are mutually independent.

To relate M and $\mathcal{M}$, we decompose M into four blocks $M = M_{\mathcal{S}} + M_{H-\mathcal{S}} + M_{(H-\mathcal{S}) \times \mathcal{S}} + M_{\mathcal{S} \times (H-\mathcal{S})}$, where $\mathcal{S}$ is the set constructed in the process S1–S2 (cf. Section 4.1). Then $M_{\mathcal{S}}$ should be "similar" to $\mathcal{M}_{\mathcal{S}}$, because for all $v \in \mathcal{S}$ the degree $d_H(v)$ is close to its mean $\bar{d}(v)$. Thus, to analyze $M_{\mathcal{S}}$, we investigate the norm of $\mathcal{M}_{\mathcal{S}}$ on the orthogonal complement of the vector $\bar{\omega} = (\bar{d}(v)^{1/2})_{v \in H}$.

Lemma 9. *We have* $\sup_{0 \neq \chi, \xi \perp \bar{\omega}} \frac{|\langle \mathcal{M}_{\mathcal{S}} \xi, \chi \rangle|}{\|\xi\| \cdot \|\chi\|} \leq c_1 \bar{d}_{\min}^{-\frac{1}{2}}$ *for some constant* $c_1 > 0$ *w.h.p.*

The proof of Lemma 9 builds on a powerful technique developed by Kahn and Szemerédi [11] to investigate the spectral gap of random *regular* graphs. The generalization of this method to the irregular case is somewhat involved; therefore, in Section 4.5, we just give a brief sketch of the proof of Lemma 9, omitting most of the details.

Using Lemma 9, we can bound the norm of $M_{\mathcal{S}}$ on the orthogonal complement of ω. To this end, we basically need to investigate how much the actual degree distribution on $\mathcal{S}$ differs from the expected degree distribution.

Corollary 10. *There is a constant* $c_2 > 0$ *such that* $\sup_{0 \neq \xi \perp \omega} \frac{\|M_{\mathcal{S}}\xi\|}{\|\xi\|} \leq c_2 \bar{d}_{\min}^{-\frac{1}{2}}$ *w.h.p.*

To bound $\|M_{H-\mathcal{S}}\|$, we show that $H - \mathcal{S}$ is "tree-like". More precisely, we can decompose the vertex set of $H - \mathcal{S}$ into classes $Z_1, \ldots, Z_K$ such that every vertex $v \in Z_j$ has only few neighbors in the classes Z_i with indices $i \geq j$.

Lemma 11. *W.h.p. $H - \mathcal{S}$ has a decomposition $V(H - \mathcal{S}) = \bigcup_{j=1}^{K} Z_j$ such that for all j and all $v \in Z_j$ we have $e(v, \bigcup_{i=j}^{K} Z_i) \leq \max\{c_0, \exp(-\bar{d}_{\min}/c_0)d_G(v)\}$.*

Using Proposition 6 and Lemma 11, in Section 4.4 we bound $\|M_{H-\mathcal{S}}\|$.

Proposition 12. *W.h.p. $\|M_{H-\mathcal{S}}\| \leq 21\bar{d}_{\min}^{-1/2}$.*

Furthermore, similar computations as in the proof of Proposition 12 yield the following bound on $\|M_{(H-\mathcal{S})\times\mathcal{S}}\|$.

Proposition 13. *We have $\|M_{\mathcal{S}\times(H-\mathcal{S})}\| = \|M_{(H-\mathcal{S})\times\mathcal{S}}\| \leq 2c_0^{1/2}\bar{d}_{\min}^{-1/2}$.*

Finally, combining Corollary 10 with Propositions 12 and 13, we conclude that there is a constant $c_3 > 0$ such that $\sup_{0\neq\xi\perp\omega} \frac{\|M\xi\|}{\|\xi\|} \leq c_3\bar{d}_{\min}^{-\frac{1}{2}}$ w.h.p. Since $\mathcal{L}(H)\omega = 0$, this implies the assertion on the spectral gap of $\mathcal{L}(H) = \boldsymbol{E} - M$ in Theorem 2. Moreover, the first part of Theorem 2 follows directly from Proposition 6.

4.3 Proof of Proposition 6

To bound $\mathrm{Vol}(V \setminus \mathcal{S})$, we first estimate the volume of the set of vertices removed by S1.

Lemma 14. *There is a constant $k_1 > 0$ such that the set $R = \{v \in V : |d_G(v) - \bar{d}(v)| \geq 0.01\bar{d}(v)\}$ has volume $\mathrm{Vol}(R) \leq n\exp(-k_1\bar{d}_{\min})$ w.h.p.*

The proof relies on Chernoff bounds and Azuma's inequality. As a second step, we analyze the volume of the set of vertices removed during the iterative procedure in step S2.

Lemma 15. *The set T removed by S2 satisfies $\mathrm{Vol}(T) \leq n\exp(-101\bar{d}_{\min}/c_0)$ w.h.p.*

The proof of Lemma 15 relies on the fact that w.h.p. in $G(n, \boldsymbol{d})$ there do not occur sets $U, U' \subset V$ such that $e(U, U')$ exceeds its mean $\mu(U, U')$ "too much".

Lemma 16. *$G = G(n, \boldsymbol{d})$ enjoys the following property w.h.p.*

Let $U, U' \subset V$ be subsets of size $u = \#U \leq u' = \#U' \leq \frac{n}{2}$. Then at least one of the following conditions holds.

1. $e_G(U, U') \leq 300\mu(U, U')$. (6)
2. $e_G(U, U') \ln(e_G(U, U')/\mu(U, U')) \leq 300u' \ln(n/u')$.

Proof of Lemma 15 (sketch). By Lemma 14 we have $\mathrm{Vol}(R) \leq \exp(-k_1 \bar{d}_{\min})n$. Suppose S2 removes the vertices $T = \{z_1, \ldots, z_k\}$, and assume that the volume of T is "large" – say, $\mathrm{Vol}(T) > \exp(k_1 \bar{d}_{\min}/10)\mathrm{Vol}(R)$. Then we can choose $j^* < k$ as large as possible so that $\mathrm{Vol}(R \cup \{z_1, \ldots, z_{j^*}\}) \leq \exp(k_1 \bar{d}_{\min}/10)\mathrm{Vol}(R)$. Set $Z = \{z_1, \ldots, z_{j^*+1}\}$. Then trite computations show that $U = U' = R \cup Z$ violate (6), i.e., $R \cup Z$ is an "atypically dense" set of "small" volume. Hence, the assertion follows from Lemma 16. $\qquad\square$

4.4   Proof of Proposition 12

Let $Z_1, \ldots, Z_K$ be a decomposition of $H - \mathcal{S}$ as in Lemma 11. We set $Z_{\geq j} = \bigcup_{i=j}^{K} Z_i$ and define $Z_{<j}, Z_{>j}$ analogously. Let $\xi = (\xi_v)_{v \in H}$ be a unit vector, and set $\eta = (\eta_w)_{w \in H} = M_{H-\mathcal{S}}\xi$. Our objective is to bound $\|\eta\|$. For $v \in Z_j$ we set

$$\rho_v = \sum_{w \in N_H(v) \cap Z_{\geq j}} \frac{\xi_w}{(d_H(v)d_H(w))^{1/2}}, \quad \sigma_v = \sum_{w \in N_H(v) \cap Z_{<j}} \frac{\xi_w}{(d_H(v)d_H(w))^{1/2}}.$$

Then the entries of η are $\eta_v = \rho_v + \sigma_v$ if $v \in H - \mathcal{S}$, and $\eta_v = 0$ for $v \in \mathcal{S}$. Setting $\alpha_j = \sum_{v \in Z_j} \rho_v^2$ and $\beta_j = \sum_{v \in Z_j} \sigma_v^2$, we have $\|\eta\|^2 \leq 2\sum_{j=1}^{K} \alpha_j + \beta_j$.

To bound $\sum_{j=1}^{K} \alpha_j$, we apply the Cauchy-Schwarz inequality, which yields

$$\alpha_j \leq \sum_{v \in Z_j} \sum_{w \in N_H(v) \cap Z_{\geq j}} \frac{e(v, Z_{\geq j})\xi_w^2}{d_H(v)d_H(w)}. \tag{7}$$

As by Lemma 11 $e(v, Z_{\geq j}) \leq \max\{c_0, \exp(-\bar{d}_{\min}/c_0)d_H(v)\}$ for all $v \in Z_j$,

$$\alpha_j \overset{(7)}{\leq} \left(\frac{c_0}{\min_{v \in H} d_H(v)} + \exp(-\bar{d}_{\min}/c_0)\right) \sum_{v \in Z_j} \sum_{w \in N_H(v) \cap Z_{\geq j}} \frac{\xi_w^2}{d_H(w)}. \tag{8}$$

Further, as $d_H(v) \geq \frac{1}{2}d_G(v) \geq \bar{d}_{\min}/200$ for all $v \in H$ by (5), (8) implies

$$\sum_{j=1}^{K} \alpha_j \leq \frac{201}{\bar{d}_{\min}} \sum_{j=1}^{K} \sum_{w \in Z_j} \frac{e(w, Z_{\leq j})\xi_w^2}{d_H(w)} \leq \frac{201}{\bar{d}_{\min}} \sum_{w \in H-\mathcal{S}} \xi_w^2 \leq \frac{201}{\bar{d}_{\min}} \|\xi\|^2 \leq \frac{201}{\bar{d}_{\min}}.$$

Similar computations yield $\sum_{j=1}^{K} \beta_j \leq \frac{201}{\bar{d}_{\min}}$, so that $\|\eta\| \leq 21\bar{d}_{\min}^{-1/2}$, as desired.

4.5   Proof of Lemma 9

Let Q be the set of all $x \in \mathbf{R}^V$ such that $x_v = 0$ for all $v \in V \setminus \mathcal{S}$, let S consist of all $\bar{\omega} \perp x \in \mathbf{R}^V$ of norm $\|x\| \leq 1$, and set $S' = S \cap Q$. Then our objective is to prove that $\max\{|\langle \mathcal{M}x, y\rangle| : x, y \in S'\} \leq c_1 \bar{d}_{\min}^{-\frac{1}{2}}$ for a certain constant $c_1 > 0$ w.h.p. To this end, we shall replace the infinite set S' by a finite set T' such that

$$\max_{x,y \in S'} |\langle \mathcal{M}x, y\rangle| \leq 5 \max_{x,y \in T'} |\langle \mathcal{M}x, y\rangle| + 4. \tag{9}$$

Then, it suffices to prove that $\max_{x,y\in T'} |\langle \mathcal{M}x, y\rangle| \leq c_2 \bar{d}_{\min}^{-1/2}$ w.h.p. for some constant $c_2 > 0$.

Let T be the set of all lattice points $x \in (0.01n^{-1/2}\mathbf{Z})^n$ of norm $\|x\| \leq 1$ such that $|\langle \bar{\omega}, x\rangle| \leq \bar{d}^{1/2}n^{-1/2}$. Then, set $T' = T \cap Q$.

Lemma 17. *The set T' satisfies (9), and $\#T \leq c_3^n$ for some constant $c_3 > 0$.*

Given vectors $x = (x_u)_{u\in V}, y = (y_v)_{v\in V} \in \mathbf{R}^V$, we define

$$B(x,y) = \left\{ (u,v) \in V^2 : n^2\bar{d}_{\min} |x_u y_v|^2 < \bar{d}(u)\bar{d}(v) \right\}, \quad X_{x,y} = \sum_{(u,v)\in B(x,y)} m_{uv}x_u y_v.$$

Our goal is to prove that there exist constants $c_4, c_5 > 0$ such that w.h.p. $\sigma = \max_{x,y\in T} |X_{x,y}| \leq c_4\bar{d}_{\min}^{-\frac{1}{2}}$ and $\tau = \max_{x,y\in T'} \sum_{(u,v)\notin B(x,y)} |m_{uv}x_u y_v| \leq c_5\bar{d}_{\min}^{-\frac{1}{2}}$. In order to estimate σ, we first bound the expectation of $X_{x,y}$.

Lemma 18. *There is a constant $c_6 > 0$ such that $|\mathrm{E}(X_{x,y})| \leq c_6\bar{d}_{\min}^{-1/2}$ for all $x, y \in T$.*

Lemma 18 follows fairly easily from the definition of $B(x,y)$. Secondly, we can bound the probability that $X_{x,y}$ deviates from its expectation significantly.

Lemma 19. *Let $x, y \in \mathbf{R}^n$, $\|x\|, \|y\| \leq 1$. Then for any constant $C > 0$ there exists a constant $K > 0$ such that $\mathrm{P}\left[|X_{x,y} - \mathrm{E}(X_{x,y})| > K\bar{d}_{\min}^{-1/2}\right] \leq C^{-n}$.*

Proof. We shall prove below that $\mathrm{E}(\exp(n\bar{d}_{\min}^{1/2}X_{x,y})) \leq \exp((c_6 + 8)n)$. Then Markov's inequality implies that $\mathrm{P}(X_{x,y} \geq K\bar{d}_{\min}^{-1/2}) \leq \exp[(c_6 + 8 - K)n]$. Hence, choosing K large enough, we can ensure that the r.h.s. is $\leq \frac{1}{2}\exp(-Cn)$. As a similar estimate holds for $-X_{x,y} = X_{-x,y}$, we obtain the desired estimate.

To bound $\mathrm{E}(\exp(n\bar{d}_{\min}^{1/2}X_{x,y}))$, we set $\lambda = n\bar{d}_{\min}^{1/2}$, and we let α_{uv} signify the possible contribution of the edge $\{u,v\}$ to $X_{x,y}$ $(u,v \in V)$. Thus, if, e.g., $(u,v) \in B(x,y)$ and $(v,u) \notin B(x,y)$, then $\alpha_{uv} = (\bar{d}(u)\bar{d}(v))^{-1/2}x_u y_v$. Moreover, let $X_{x,y}(u,v) = \alpha_{uv}$ if $\{u,v\} \in G$, and $X_{x,y}(u,v) = 0$ otherwise. Finally, let $\mathcal{E} = \{\{u,v\} : u,v \in V\}$, so that $X_{x,y} = \sum_{\{u,v\}\in\mathcal{E}} X_{x,y}(u,v)$.

Then $\mathrm{E}(\exp(\lambda X_{x,y})) = \prod_{\{u,v\}\in\mathcal{E}} [p_{uv}(\exp(\lambda\alpha_{uv}) - 1) + 1]$, because the random variables $X_{x,y}(u,v)$, $\{u,v\} \in \mathcal{E}$, are mutually independent. Moreover, by the definition of $B(x,y)$ for all $(u,v) \in B(x,y)$ we have $\lambda\alpha_{uv} \leq 2$. Thus,

$$\mathrm{E}(\exp(\lambda X_{x,y})) \quad \leq \quad \prod_{\{u,v\}\in\mathcal{E}} \left[1 + p_{uv}\lambda\alpha_{uv} + 2p_{uv}\lambda^2\alpha_{uv}^2\right]$$

$$\overset{\text{Lemma 18}}{\leq} \quad \exp\left(c_6 n + 2\lambda^2 \sum_{\{u,v\}\in\mathcal{E}} p_{uv}\alpha_{uv}^2\right). \tag{10}$$

Furthermore, a straight computation yields $\lambda^2 \sum_{\{u,v\}\in\mathcal{E}} p_{uv}\alpha_{uv}^2 \leq 4n$. Plugging this estimate into (10), we conclude that $\mathrm{E}(\exp(\lambda X_{x,y})) \leq \exp((c_6 + 8)n)$. $\qquad \square$

Combining Lemmas 18 and 19, we conclude that there is a constant $c_4 > 0$ such that $P(|X_{x,y}| > c_4 \bar{d}_{\min}^{-1/2}) \leq (2c_3^2)^{-n}$ for any two points $x, y \in T$. Therefore, invoking Lemma 17, we get $P(\max_{x,y \in T} |X_{x,y}| > c_4 \bar{d}_{\min}^{-1/2}) \leq \#T \cdot (2c_3^2)^{-n} \leq 2^{-n}$, thereby proving that $\sigma \leq c_4 \bar{d}_{\min}^{-1/2}$ w.h.p. Finally, the estimate of τ relies on Lemma 16 (details omitted).

Acknowledgement. We thank Andreas Goerdt for helpful discussions.

References

1. Albert, R., Barabási, A.L.: Statistical mechanics of complex networks. Reviews of modern physics **74** (2002) 47–97
2. Alon, N., Kahale, N.: A spectral technique for coloring random 3-colorable graphs. SIAM J. Comput. **26** (1997) 1733–1748
3. Alon, N., Krivelevich, M., Sudakov, B.: Finding a large hidden clique in a random graph. Random Structures and Algorithms **13** (1998) 457–466
4. Bollobás, B., Riordan, O.: Mathematical results on scale-free random graphs. In Bornholdt, S., Schuster, H.G. (eds.): Handbook of graphs and networks: from the genome to the Internet, Wiley 2003, 1–34
5. Chung, F.: Spectral Graph Theory. American Mathematical Society 1997
6. Chung, F., Lu, L., Vu, V.: The spectra of random graphs with given expected degrees. Internet Mathematics **1** (2003) 257–275
7. Coja-Oghlan, A.: On the Laplacian eigenvalues of $G(n, p)$. Preprint (2005)
8. Dasgupta, A., Hopcroft, J.E., McSherry, F.: Spectral Partitioning of Random Graphs. Proc. 45th FOCS (2004) 529–537
9. Faloutsos, M., Faloutsos, P., Faloutsos, C.: On powerlaw relationships of the internet topology. Proc. of ACM-SIGCOMM (1999) 251–262
10. Feige, U., Ofek, E.: Spectral techniques applied to sparse random graphs. Random Structures and Algorithms **27** (2005) 251–275
11. Friedman, J., Kahn, J., Szemeredi, E.: On the second eigenvalue in random regular graphs. Proc. 21st STOC (1989) 587–598
12. Füredi, Z., Komlos, J.: The eigenvalues of random symmetric matrices. Combinatorica **1** (1981) 233–241
13. Krivelevich, M., Vu, V.H.: Approximating the independence number and the chromatic number in expected polynomial time. J. of Combinatorial Optimization **6** (2002) 143–155
14. McSherry, F.: Spectral partitioning of random graphs. Proc. 42nd FOCS (2001) 529–537
15. Mihail, M., Papadimitriou, C.H.: On the eigenvalue power law. Proc. 6th RANDOM (2002) 254–262
16. Pothen, A., Simon, H.D., Liou, K.-P.: Partitioning sparse matrices with eigenvectors of graphs. SIAM J. Matrix Anal. Appl. **11** (1990) 430–452
17. Schloegel, K., Karypis, G., Kumar, V.: Graph partitioning for high performance scientific simulations. in: Dongarra, J., Foster, I., Fox, G., Kennedy, K., White, A. (eds.): CRPC parallel computation handbook. Morgan Kaufmann (2000)
18. Wigner, E.P.: On the distribution of the roots of certain symmetric matrices. Annals of Mathematics **67** (1958) 325–327

Embedding Bounded Bandwidth Graphs into ℓ_1

Douglas E. Carroll[1,*], Ashish Goel[2,**], and Adam Meyerson[1,***]

[1] UCLA
[2] Stanford University

Abstract. We introduce the first embedding of graphs of low *bandwidth* into ℓ_1, with distortion depending only upon the bandwidth. We extend this result to a new graph parameter called *tree-bandwidth*, which is very similar to (but more restrictive than) treewidth. This represents the first constant distortion embedding of a non-planar class of graphs into ℓ_1. Our results make use of a new technique that we call *iterative embedding* in which we define coordinates for a small number of points at a time.

1 Introduction

Our main result is a technique for embedding graph metrics into ℓ_1, with distortion depending only upon the *bandwidth* of the original graph. A graph has bandwidth k if there exists some ordering of the vertices such that any two vertices with an edge between them are at most k apart in the ordering. While this ordering could be viewed as an embedding into one-dimensional ℓ_1 with bounded *expansion* (any two vertices connected by an edge must be close in the ordering), the *contraction* of such an embedding is unbounded (there may be two vertices which are close in the ordering but not in the original metric). Obtaining an embedding with bounded distortion (in terms of *both* expansion and contraction) turns out to be non-trivial.

In fact, our results can be extended to a new graph parameter that we call *tree-bandwidth*. We observe that metrics based on trees are easy to embed into ℓ_1 isometrically, despite the fact that even a binary tree can have large bandwidth. The *tree-bandwidth* parameter is a natural extension of bandwidth, where vertices are placed along a tree instead of being ordered linearly. We prove that the shortest path metric of an unweighted graph can be embedded into ℓ_1 with distortion depending only upon the *tree-bandwidth* of the graph (thus independent of the number of vertices).

We achieve these results by introducing a novel technique for *iterative embedding* of graph metrics into ℓ_1. The idea is to partition the graph into small sets

* Department of Computer Science, University of California, Los Angeles. Email: dcarroll@cs.ucla.edu.

** Departments of Management Science and Engineering and (by courtesy) Computer Science, Stanford University. Research supported in part by NSF CAREER award 0133968, and by an Alfred P. Sloan Fellowship. Email: ashishg@stanford.edu.

*** Department of Computer Science, University of California, Los Angeles. Email: awm@cs.ucla.edu.

and embed each set separately. The coordinates of each specific point are determined when the set containing that point is embedded. Two embeddings will be computed for each set of points. One is generated via some local embedding technique, and maintains accurate distances between the members of the same set. The other embedding copies a set of "parent" points; the goal is to maintain small distances between points and their parents. These two sets of coordinates will be carefully combined to generate the final coordinates for the new set of points. We then proceed to the next set in the ordering.

For ease of exposition we use a very simple local embedding technique in this paper. However, we have also proven a more general result in which we show that with iterative embedding, any reasonable local embedding technique suffices for embedding into ℓ_1 with distortion dependent only upon the *tree-bandwidth* (proof omitted). This leaves open the possibility that the dependence on the tree-bandwidth could be improved with a different local embedding technique.

The motivation for our work is a conjecture (stated by Gupta *et al.* [9] and others) that excluded-minor graph families can be embedded into ℓ_1 with distortion dependent only upon the set of excluded minors. This is one of the major conjectures in metric embedding, and many previous results have resolved special cases of this conjecture. However, all previous ℓ_1 embedding results either yield distortion dependent upon the number of points in the metric [4,15], or apply only to a subset of the planar graphs [13,9,6]. While our results do not resolve the conjecture, we are able to embed a well-studied subclass of graphs (bandwidth-k graphs) with distortion independent of the number of points in the metric. This is the first such result for a non-planar graph class. In addition, our definition of *tree-bandwidth* is similar to (although possibly weaker than) *treewidth*. While we conjecture that there exist families of graphs with low treewidth but unbounded tree-bandwidth, it is interesting to note that weighted treewidth-k graphs can be embedded with constant distortion into *weighted* tree-bandwidth-$O(k)$ graphs.

We note that at each step, our embedding technique requires the existence of a previously embedded "parent" set such that each point of the new set is close to one of the parents, but no point in the new set is close to any other previously embedded set. This property implies the existence of a hierarchy of small node separators (small sets of nodes which partition the graph), which is exactly the requirement for a graph of low *treewidth*. However, we also need each point to be close to *some* member of the parent set, which motivates our definition of the *tree-bandwidth* parameter.

1.1   Related Work

A great deal of recent work has concentrated on achieving tight distortion bounds for ℓ_1 embedding of restricted classes of metrics. For general metrics with n points, the result of Bourgain[4] showed that embedding into ℓ_1 with $O(\log n)$ distortion is possible. A matching lower bound (using expander graphs) was introduced by LLR [11]. It has been conjectured by Gupta *et al.* [9], and Indyk [10] that the shortest-path metrics of planar graphs can be embedded into ℓ_1 with constant distortion. Gupta *et al.* [9] also conjecture that excluded-minor graph families can be embedded into ℓ_1 with distortion that depends only on the excluded minors. In particular, this would mean that for any k the family of treewidth-k graphs could be embedded with distortion $f(k)$ independent of the number of nodes in the graph[1]. Such results would be the best possible for very general and natural classes of graphs.

Since Okamura and Seymour [13] showed that outerplanar graphs can be embedded isometrically into ℓ_1, there has been significant progress towards resolving several special cases of the aforementioned conjecture. Gupta *et al.* [9] showed that treewidth-2 graphs can be embedded into ℓ_1 with constant distortion. Chekuri *et al.* [6] then followed this by proving that k-outerplanar graphs can be embedded into ℓ_1 with constant distortion. Note that all these graph classes not only have low treewidth, but are planar. We give the first constant distortion embedding for a non-planar subclass of the bounded treewidth graphs.

Rao [15] proved that any minor excluded family can be embedded into ℓ_1 with distortion $O(\sqrt{\log n})$. This is the strongest general result for minor-excluded families. Rabinovich [14] introduced the idea of *average distortion* and showed that any minor excluded family can be embedded into ℓ_1 with constant *average distortion*.

Graphs of low treewidth have been the subject of a great deal of study. For a survey of definitions and results on graphs of bounded treewidth, see Bodlaender [2]. More restrictive graph parameters include domino treewidth [3] and bandwidth [7], [8].

2   Definitions and Preliminaries

Given two metric spaces (G, ν) and (H, μ) and an embedding $\Phi : G \to H$, we say that the *distortion* of the embedding is $\|\Phi\| \cdot \|\Phi^{-1}\|$ where

$$\|\Phi\| = \max_{x,y \in G} \frac{\mu(\Phi(x), \Phi(y))}{\nu(x, y)}, \qquad \|\Phi^{-1}\| = \max_{x,y \in G} \frac{\nu(x, y)}{\mu(\Phi(x), \Phi(y))}$$

Parameter $\|\Phi\|$ will be called the *expansion* of the embedding and parameter $\|\Phi^{-1}\|$ is called the *contraction*. We will define bandwidth and then present our definition of the generalization tree-bandwidth.

Definition 1. *Given graph $G = (V, E)$ and linear ordering $f : V \to \{1, 2, ..., |V|\}$ the* bandwidth *of f is $max\{|f(v) - f(w)| \, |(v, w) \in E\}$. The* bandwidth *of G is the minimum bandwidth over all linear orderings f.*

Definition 2. *Given a graph $G = (V, E)$, we say that it has* tree-bandwidth k *if there is a rooted tree $T = (I, F)$ and a collection of sets $\{S_i \subset V | i \in I\}$ such that:*

1. $\forall i, |S_i| \leq k$
2. $V = \bigcup S_i$

---

[1]   There is a lower bound of $\Omega(\log k)$ arising from expander graphs.

3. *the S_i are disjoint*
4. $\forall (u,v) \in E$, *u and v lie in the same set S_i or $u \in S_i$ and $v \in S_j$ and* $(i,j) \in F$.
5. *if c has parent p in T, then $\forall v \in S_c, \exists u \in S_p$ such that $d(u,v) \leq k$.*

We claimed that *tree-bandwidth* was a generalization of *bandwidth*. Intuitively, we can divide a graph of low bandwidth into sets of size k (the first k points in the ordering, the next k points in the ordering, and so forth). We then connect these sets into a path. This gives us all the properties required for tree-bandwidth except for the fifth property – there may be some node which is not close to any node which appeared prior to it in the linear ordering. We can fix this problem by defining a new linear ordering of comparable bandwidth. The proof of this fact has been deferred until the full version of the paper.

Lemma 1. *Graph $G = (V, E)$ with bandwidth b has tree-bandwidth at most 2b.*

We will now define *treewidth* and show the close relationship between the definitions of treewidth and tree-bandwidth.

Definition 3. *(i) Given a connected graph $G = (V, E)$, a DFS-tree is a rooted spanning subtree $T = (V, F \subset E)$ such that for each edge $(u, v) \in E$, v is an ancestor of u or u is an ancestor of v in T.*

(ii) The value *of DFS-tree T is the maximum over all $v \in V$ of the number of ancestors that are adjacent to v or a descendent of v.*

(iii) The edge stretch *of DFS-tree T is the the maximum over all $v, w \in V$ of the distance $d(v, w)$ where w is an ancestor of v and w is adjacent to v or a descendent of v.*

We use the following definition of treewidth due to T. Kloks and related in a paper of Bodlaender [2]:

Definition 4. *Given a connected graph $G = (V, E)$, the* treewidth *of G is the minimum value of a DFS-tree of a supergraph $G' = (V, E')$ of G where $E \subset E'$.*

The following proposition follows immediately from the definition of tree-bandwidth:

Proposition 1. *Given a connected graph $G = (V, E)$, the* tree-bandwidth *of G is the minimum edge stretch of a DFS-tree of G.*

Thus, treewidth and tree-bandwidth appear to be related in much the same way that cutwidth and bandwidth are related (see [2] for instance). The close relationship between treewidth and tree-bandwidth is cemented by the following observation (the proof is deferred until the full version of the paper):

Lemma 2. *Any metric supported on a weighted graph $G = (V, E)$ of treewidth-k can be embedded with distortion 4 into a weighted graph with tree-bandwidth-$O(k)$.*

Thus, a technique for embedding *weighted* tree-bandwidth-k graphs into ℓ_1 with $O(f(k))$ distortion would immediately result in constant distortion ℓ_1-embeddings of weighted treewidth-k graphs.

2.1 Bounded Bandwidth Example

To see that previous constant distortion embedding techniques do not handle bounded bandwidth graphs consider the following example. Construct a graph G by connecting k points in an arbitrary way, then adding k new points connected to each other and the previous k points in an arbitrary way, and repeat many times.

Clearly the graph G generated in this way has bandwidth $\leq 2k-1$. However, note that if $k \geq 3$ and some set of $2k$ consecutively added points contains $K_{3,3}$ or K_5 then G is not planar and thus previous constant distortion ℓ_1-embedding techniques cannot be applied [13,9,6]. G does have bounded treewidth, so Rao's algorithm [15] can be applied but it only guarantees $O(\sqrt{\log n})$ distortion.

2.2 Bounded Tree-Bandwidth Example

To show that bounded tree-bandwidth graphs form a broader class than the bounded bandwidth graphs consider the following example. Let $G = (V', E')$ consist of k copies of an arbitrary tree $T = (V, E)$. Construct G' from G as follows:

1. For $x \in V$, let $\{x_1, ..., x_k\}$ be the k copies of x in V'.
2. For each $x \in V$, connect $\{x_1, ..., x_k\}$ in an arbitrary way.

While the resulting graph G' clearly has tree-bandwidth k, a complete binary tree of depth d has bandwidth $\Omega(d)$ [7], thus G' may have bandwidth $\Omega(\log n)$.

Note again that if $k \geq 5$ and G' contains $K_{3,3}$ or K_5 then G' is not planar and thus previous constant distortion ℓ_1-embedding techniques cannot be applied.

Also note that there are trees T with $|V| = n$ such that any ℓ_2-embedding of T has distortion $\Omega(\sqrt{\log \log n})$ [5]. Since Rao's technique embeds first into ℓ_2 this gives a lower bound of $\Omega(\sqrt{\log \log n})$ on the distortion achievable using Rao's technique to embed G' into ℓ_1. The technique presented in this paper embeds these examples into ℓ_1 with distortion depending only on k.

Apart from being interesting from a technical viewpoint, bounded tree-bandwidth graphs may also be a good model for phylogenentic networks with limited introgression/reticulation [12]. This is a fruitful connection to explore, though it is outside the scope of this paper.

3 Algorithm

Given a graph G of tree-bandwidth k, it must have a tree-bandwidth-k decomposition $(T, \{X_i\})$. We will embed the sets X_i one set at a time according to a DFS ordering of T. When set X_i is embedded, all members of that set will be assigned values for each coordinate. Note that once a point is embedded, *its coordinates will never change - all subsequently defined coordinates will be assigned value zero for these points.* Note that when new coordinates are introduced, these are considered to be coordinates that were never used at any previous point in the algorithm.

For each set we will obtain two embeddings: one derived by extending the embedding of the parent of X_i in T and one local embedding using a simple deterministic embedding technique. We prove the existence of a method for combining these two embeddings to provide an acceptable embedding of the set X_i.

At stage i, our algorithm will compute a weight for each partition S of X_i. We would like these weights to look like $w_M(S)$ - the distance between the closest pair of points separated by S. The embedded distance between two points x, y in X_i will be the sum of weights over partitions separating x from y. The weights suggested above will guarantee no contraction and bounded expansion within X_i. We can transform weighted partitions into coordinates by introducing $w_M(S)$ coordinates for each partition S, such that the coordinate has value 1 for each $x \in S$ and value -1 for each $x \in X_i - S$.

This approach will create entirely new coordinates for each point. Since points in X_i are supposed to be close to points in $X_{p(i)}$, this can create large distortion between sets. Instead of introducing all new coordinates, we would like to "reuse" existing coordinates by forcing points in X_i to take on values similar to those taken on by points in $X_{p(i)}$.

To reuse existing coordinates we will choose a "parent" in $X_{p(i)}$ for each point $x \in X_i$ and identify x with its parent $p(x)$. The critical observation here is that each point in X_i is within distance k of some point in $X_{p(i)}$. Therefore, the partition weights (and hence distances) established by these coordinates are good approximations of the target values we would like to assign.

More precisely, for each point $x \in X_i$ there is at least one closest point in $X_{p(i)}$. Choose an arbitrary such point to be the parent of x. After identifying points in this way, each parent coordinate induces a partition S on X_i between points whose parents have values 1 and -1 in that coordinate. We can define $w_P(S)$ to be the number of parent coordinates inducing partition S. If $|w_P(S) - w_M(S)|$ is always small then the independent local weightings agree and we get a good global embedding.

Unfortunately, there are cases in which $w_P(S) - w_M(S)$ can be large. However, we can successfully combine the two metrics by using the following weighting: $w_F(S) = \max(w_M(S), w_P(S) - \mu)$. The key property of this weighting is that we do not activate too many new coordinates (since $w_P(S)$ not much less than $w_F(S)$) nor do we deactivate too many existing coordinates ($w_P(S)$ not much more than $w_F(S)$). In addition, we can show that $w_F(S)$ does not contract nor greatly expand distances between points of X_i.

3.1 MIN-SEPARATOR Embedding

We can prove that any reasonable local embedding technique suffices to obtain $O(f(k))$ distortion. However, that proof is quite involved and is omitted from this abstract. Instead, for ease of exposition, we will employ a simple local embedding technique which we call a MIN-SEPARATOR embedding and which is described below. The MIN-SEPARATOR embedding returns similar embeddings for independently embedded metrics with similar distances. This is a very useful property and greatly simplifies our overall algorithm and analysis[2].

---

[2] It is conceivable that a different local embedding technique might result in a better dependence on k.

MIN-SEPARATOR embedding: Given metric (G, d), we assign a weight for each of the distinct partitions of G. To each partition S we assign weight $w_{M(G)}(S) = d(S, G - S) = \min\{d(x, y) | x \in S, y \in G - S\}$. Note that when the source metric is clear we will denote these weights as $w_M(S)$. We then transform these weighted partitions into coordinates by introducing $w_M(S)$ coordinates for each partition S such that the coordinate has value 1 for each $x \in S$ and value -1 for each $x \in G - S$. The distances in this embedding become

$$d_{M(G)}(x, y) = \sum_{S \in 2^G : x \in S, y \in G - S} w_{M(G)}(S) = \sum_{S \in 2^G : x \in S, y \in G - S} d(S, G - S)$$

Lemma 3. *The MIN-SEPARATOR embedding does not contract distances and does not expand distances by more than 2^k.*

Proof. First we show that MIN-SEPARATOR does not contract the distance between x and y. The proof is by induction on the number of points in the metric (G, d).

If $|G| = 2$, then there is only one non-trivial partition and it has weight $d(x, y)$. For larger graphs, there must be some point z other than x, y. Let $B = G - \{z\}$; by the inductive hypothesis the claim holds on set B. However, we observe that the embedded distance $d_{M(B)}(x, y)$ is at most the embedded distance $d_{M(G)}(x, y)$. For any partition of B, we can consider two new partitions of G (one with z on each side) and observe that the total weight MIN-SEPARATOR places on these partitions must be at least the weight MIN-SEPARATOR placed on the original partition of B (this because of triangle inequality).

We now show that MIN-SEPARATOR does not expand distances by more than 2^k. For each partition S which separates x, y, $w_{M(G)}(S) \leq d(x, y)$ and since there are $< 2^k$ partitions which separate x, y, $d_{M(G)}(x, y) \leq 2^k d(x, y)$.

3.2 Combining the Local Embeddings

The algorithm EMBED-BAND relies on three critical properties of the tree-bandwidth decomposition:

1. Each node in X_i is within distance k of a node in the parent of X_i.
2. The nodes of X_i are not adjacent to any previously embedded nodes except those in the parent of X_i.
3. The number of points in X_i is at most k.

The first property enables us to prove that

$$w_P(S) - \mu \leq w_F(S) \leq w_P(S) + 2k \tag{1}$$

This is key in bounding the distortion between sets, since it indicates that we never introduce or "zero-out" too many coordinates for any partition S of X_i.

The second property means that we don't need to bound expansion between too many pairs of points. As long as we can prove that distances between points in X_i and $X_{p(i)}$ don't expand too much, the triangle inequality will allow us to bound expansion between all pairs of points.

The third property allows us to bound the distortion of the local embedding (MIN-SEPARATOR) as well as to bound the total number of coordinates introduced or zeroed out, since there are only 2^k partitions of set X_i with k points.

3.3 Example: Embedding a Cycle

It is instructive to observe what happens when embedding a cycle (see figure 1). It is clear that the first two points in the cycle (X_1) can be embedded acceptably. As we embed subsequent sets we embed the descendents of these two points. Because the pairs of points in consecutive sets diverge, each new point inherits the values of all of the coordinates of its parent. Additionally, new coordinates are added to separate the pairs of points. The union of these coordinates is enough to establish the distances between these pairs of points as they diverge.

After embedding half the points in the cycle, the pairs of points in subsequent sets begin to converge. Whenever the distance induced by the parent points exceeds the target distance of the current points (represented by the MIN-SEPARATOR distance), we set the values of μ coordinates establishing that distance to zero *for the new points*. Because points in consecutive sets are within distance k of their parents, the distances between consecutive pairs of points cannot decrease by more than $2k$ per step. Thus, zeroing μ coordinates at each step is more than sufficient to compensate for the decreasing distances.

It might appear that zeroing μ coordinates at each step would contract distances between points and their ancestors, but recall that we also define β new coordinates at each step to separate the current points from all previously embedded points and prevent such contractions.

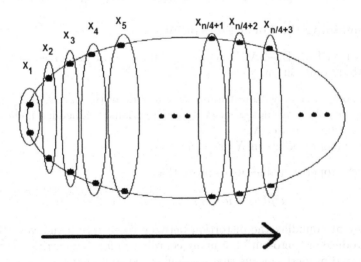

Fig. 1. Embedding a Cycle

4   Analysis

The central result of this paper follows directly from the lemmas below:

Theorem 1. *Algorithm* EMBED-BAND *embeds tree-bandwidth-k graphs into* ℓ_1 *with distortion* $\leq 2\beta = 4 \cdot 2^k \mu = 16k \cdot 2^{2k}$.

Lemma 4. *The distances between points embedded simultaneously are not contracted.*

Input: Assume $G = (V, H)$ has tree-bandwidth decomposition $(T = (I, F), \{X_i | i \in I\})$. Let $p(i)$ be the parent of $i \in T$. Assume that $p(i)$ appears before i in the ordering of the nodes of I. X_1 is the root of T.

1. $\mu \leftarrow 4k2^k$
2. for each of the $2^{k-1} - 1$ non-trivial partitions S of X_1:
   (a) $w_M(S) \leftarrow \min\{d(x, y) | x \in S, y \in X_1 - S\}$
   (b) define $w_M(S)$ new coordinates
   (c) for each new coordinate c set:
      $x_c \leftarrow 1$ if $x \in S$,
      $x_c \leftarrow -1$ if $x \in X_1 - S$
3. FOR $i \leftarrow 2$ TO $|I|$
   (a) for each $x \in X_i$, let $p(x)$ be the parent of x (closest node to x) in $X_{p(i)}$.
      (By identifying nodes x with their parents $p(x)$, each existing coordinate induces a partition on the points of X_i.)
   (b) for each of the $2^{k-1} - 1$ non-trivial partitions S of X_i:
      i. $w_M(S) \leftarrow \min\{d(x, y) | x \in S, y \in X_i - S\}$
      ii. $w_P(S) \leftarrow \#$ of existing coordinates which induce S via $X_{p(i)}$
      iii. $w_F(S) \leftarrow \max(w_M(S), w_P(S) - \mu)$
      iv. if $w_F(S) > w_P(S)$ then:
         A. for all the $w_P(S)$ coordinates that induce partition S set $x_c \leftarrow p(x)_c$ for all $x \in X_i$
         B. define $w_F(S) - w_P(S)$ new coordinates
         C. for each new coordinate c set:
            $x_c \leftarrow 1$ if $x \in S$,
            $x_c \leftarrow -1$ if $x \in X_i - S$
            ($x_c \leftarrow 0$ for all previously embedded points)
      v. If $w_F(S) \leq w_P(S)$ then:
         A. for $w_P(S) - w_F(S)$ of the coordinates that induce partition S set $x_c \leftarrow 0$ for all $x \in X_i$
         B. for the $w_F(S)$ remaining coordinates that induce partition S set $x_c \leftarrow p(x)_c$ for all $x \in X_i$
   (c) $x_c \leftarrow p(x)_c$ for all coordinates c which do not partition X_i
   (d) define an additional $\beta = 2 \cdot 2^k \mu$ coordinates and set $x_c \leftarrow 1$ for all $x \in X_i$
4. NEXT i

Fig. 2. Algorithm EMBED-BAND

Proof. If x, y are in the same tree node X_i, then the distance $d_E(x, y)$ is at least as large as the distance $d_{M(X_i)}(x, y)$ returned by MIN-SEPARATOR. This is because for every partition we use $w_F(S) = \max\{w_M(S), w_P(S) - \mu\} \geq w_M(S)$.

Lemma 5. *The distances between points embedded simultaneously are expanded by at most a factor of 2^k.*

Proof. Recall that for each partition S of X_i, we compute three weights: a local weight, a "parent" weight, and the final weight which we use to embed the current tree node.

$$w_M(S) = min\{d(x,y)|x \in S, y \in X_i - S\}$$
$$w_P(S) = \# \text{ of existing coordinates that induce } S \text{ via } X_{p(i)}$$
$$w_F(S) = max(w_M(S), w_P(S) - \mu)$$

If for all partitions S separating x and y, we have $w_F(S) = w_M(S)$, then the embedded distance will be the same as that from MIN-SEPARATOR, which is at most $2^k d(x, y)$.

Otherwise, at least one partition separating x and y has $w_F(S) = w_P(S) - \mu$. Note that by the triangle equality, $d(x, y) \leq d(p(x), p(y)) + 2k$ for all x, y. Thus every such partition has $w_F(S) \leq w_P(S) + 2k$, so by summing and observing that there are only 2^k possible partitions of k points, we have $d_E(x, y) \leq d_E(p(x), p(y)) - \mu + 2k2^k$. Applying our inductive hypothesis to points in the parent set and using $\mu = 4k2^k$ gives the desired bound.

Lemma 6. *The distances between points in different sets are expanded by at most $2\beta = 4 \cdot 2^k \mu$ where $\mu = 4k2^k$.*

Proof. Consider $x \in X_i$ and $y \in X_j$. X_i and X_j are connected by a unique path Q in T. Assume WLOG that $X_{p(i)}$ is in Q. Our proof will be by induction on the length of Q.

If $length(Q) = 1$, this means $X_j = X_{p(i)}$ and by triangle inequality we have $d_E(y, x) \leq d_E(y, p(x)) + d_E(p(x), x)$. The distortion of the first quantity is bounded because these points are in the same tree node. The second quantity is bounded by β plus the sum of differences in partition weights since we re-use coordinates when possible. Combining these, and observing that $p(x)$ is closer to x than y is, we obtain $d_E(y, x) \leq 2\beta d(y, x)$. If $length(Q) > 1$, there must be a point $z \in X_{p(i)}$ such that z lies on a shortest path between x and y in G. By the induction hypothesis, $d_E(x, z) \leq 2\beta d(x, z)$ and $d_E(z, y) \leq 2\beta d(z, y)$. Thus, $d_E(x, y) \leq d_E(x, z) + d_E(z, y) \leq 2\beta d(x, z) + 2\beta d(z, y) = 2\beta d(x, y)$ since z is on the shortest path between x and y.

Lemma 7. *The distances between points in different sets are not contracted.*

Proof. Consider $x \in X_i$ and $y \in X_j$. X_i and X_j are connected by a unique path Q in T. Assume WLOG that $X_{p(i)}$ is in path Q. x has a closest ancestor z in X_j which is at distance $d_E(z, y)$ from y. Consider the path from z to x that lies in Q. Intuitively, we activate at least β coordinates at each step and deactivate at most $2^k\mu$, so distances increase as $\approx (\beta - 2^k\mu)|Q|$. So

$$d_E(x,y) \geq max((d_E(z,y) - 2^k\mu|Q|), 0) + \beta|Q| \geq d_E(z,y) - 2^k\mu|Q| + \beta|Q|$$
$$\geq d(z,y) - 2^k\mu|Q| + \beta|Q| \geq d(x,y) - 2k|Q| - 2^k\mu|Q| + \beta|Q|$$
$$= d(x,y) + (\beta - 2k - 2^k\mu)|Q| \geq d(x,y).$$

References

1. M. Badoiu, K. Dhamdhere, A. Gupta, Y. Rabinovich, H. Raecke, R. Ravi, and A. Sidiropoulos, "Approximation Algorithms for Low-Distortion Embeddings Into Low-Dimensional Spaces", In *Proceedings of the 16th Annual ACM-SIAM Symposium on Discrete Algorithms*, 2005, pp. 119-128.
2. H. Bodlaender, "A partial k-arboretum of graphs with bounded treewidth", *Theoretical Computer Science*, 209 (1998), pp. 1-45.
3. H. Bodlaender, "A Note On Domino Treewidth", *Discrete Mathematics and Theoretical Computer Science*, 3 (1999), pp. 144-150.
4. J. Bourgain, "On Lipschitz Embeddings of Finite Metric Spaces in Hilbert Space.", *Israel Journal of Mathematics*, 52 (1985), pp. 46-52.
5. J. Bourgain, "The Metrical Interpretation of Superreflexivity in Banach Spaces.", *Israel Journal of Mathematics*, 56 (1986), pp. 222-230.
6. C. Chekuri, A. Gupta, I. Newman, Y. Rabinovich, and A. Sinclair, "Embedding k-Outerplanar Graphs into ℓ_1", In *Proceedings of the 14th Annual ACM-SIAM Symposium on Discrete Algorithms*, 2003, pp. 527-536.
7. F. Chung, P. Seymour, "Graphs with Small Bandwidth and Cutwidth", *Discrete Mathematics*, 75 (1989), pp. 113-119.
8. U. Feige, "Approximating the bandwidth via volume respecting embeddings", In *Proceedings of the 30th Annual ACM Symposium on Theory of Computing*, 1998, pp. 90-99.
9. A. Gupta, I. Newman, Y. Rabinovich, and A. Sinclair, "Cuts, trees and ℓ_1-embeddings.", In *Proceedings of the 40th Annual IEEE Symposium on Foundations of Computer Science*, 1999, pp. 399-408.
10. P. Indyk, "Algorithmic Aspects of Geometric Embeddings.", In *Proceedings of the 42th Annual IEEE Symposium on Foundations of Computer Science*, 2001, pp. 10-33.
11. N. Linial, E. London, and Y. Rabinovich, "The geometry of graphs and some of its algorithmic applications", *Combinatorica*, 15 (1995), pp. 215-245.
12. L. Nakhleh, T. Warnow, C. R. Linder, and K. St. John, "Reconstructing reticulate evolution in species - theory and practice.", *Journal of Computational Biology*, (To Appear).
13. H. Okamura, P. Seymour, "Multicommodity Flows in Planar Graphs", *Journal of Combinatorial Theory Series B*, 31 (1981), pp. 75-81.
14. Y. Rabinovich, "On Average Distortion of Embedding Metrics into the Line and into ℓ_1", In *Proceedings of the 35th Annual ACM Symposium on Theory of Computing*, 2003, pp. 456-462.
15. S. Rao, "Small distortion and volume preserving embeddings for Planar and Euclidean metrics", In *Proceedings of the 15th Annual Symposium on Computational Geometry*, 1999, pp. 300-306.
16. N. Robertson, P. Seymour, "Graph Minors II. Algorithmic Aspects of Tree-Width", *Journal of Algorithms*, 7 (1986), pp. 309-322.

On Counting Homomorphisms to Directed Acyclic Graphs[*]

Martin Dyer[1], Leslie Ann Goldberg[2], and Mike Paterson[2]

[1] School of Computing, University of Leeds, Leeds LS2 9JT, UK
[2] Dept. of Computer Science, University of Warwick, Coventry CV4 7AL, UK

Abstract. We give a dichotomy theorem for the problem of counting homomorphisms to directed acyclic graphs. H is a fixed directed acyclic graph. The problem is, given an input digraph G, to determine how many homomorphisms there are from G to H. We give a graph-theoretic classification, showing that for some digraphs H, the problem is in P and for the rest of the digraphs H the problem is #P-complete. An interesting feature of the dichotomy, absent from related dichotomy results, is the rich supply of tractable graphs H with complex structure.

1 Introduction

Our result is a dichotomy theorem for the problem of counting homomorphisms to directed acyclic graphs. A *homomorphism* from a (directed) graph $G = (V, E)$ to a (directed) graph $H = (\mathcal{V}, \mathcal{E})$ is a function from V to $\mathcal{V}$ that preserves (directed) edges. That is, the function maps every edge of G to an edge of H.

Hell and Nešetřil [6] gave a dichotomy theorem for the *decision* problem for undirected graphs H. In this case, H is an undirected graph (possibly with self-loops). The input, G, is an undirected simple graph. The question is "Is there a homomorphism from G to H?". Hell and Nešetřil [6] showed that the decision problem is in P if the fixed graph H has a loop, or is bipartite. Otherwise, it is NP-complete. Dyer and Greenhill [3] established a dichotomy theorem for the corresponding *counting* problem in which the question is "How many homomorphisms are there from G to H?". They showed that the problem is in P if every component of H is either a complete graph with all loops present or a complete bipartite graph with no loops present[1]. Otherwise, it is #P-complete. Bulatov and Grohe [2] extended the counting dichotomy theorem to the case in which H is an undirected *multigraph*. Their result will be discussed in more detail below.

In this paper, we study the corresponding counting problem for *directed* graphs. First, consider the decision problem. H is a fixed digraph and, given an input digraph G, we ask "Is there a homomorphism from G to H?". It is conjectured [7, 5.12] that there is a *dichotomy theorem* for this problem, in the sense that, for every H, the problem is either polynomial-time solvable or NP-complete. Currently, there is no graph-theoretic conjecture stating what the two classes of digraphs will look like. Obtaining such a

---

[*] Partially supported by the EPSRC grant *Discontinuous Behaviour in the Complexity of Randomized Algorithms*. Some of the work was done while the authors were visiting the Mathematical Sciences Research Institute in Berkeley.
[1] The graph with a singleton isolated vertex is taken to be a (degenerate) complete bipartite graph with no loops.

M. Bugliesi et al. (Eds.): ICALP 2006, Part I, LNCS 4051, pp. 38–49, 2006.
© Springer-Verlag Berlin Heidelberg 2006

dichotomy may be difficult. Indeed, Feder and Vardi [5, Theorem 13] have shown that the resolution of the dichotomy conjecture for *layered* (or *balanced*) digraphs, which are a small subset of *directed acyclic graphs*, would resolve their long-standing dichotomy conjecture for all *constraint satisfaction problems*. There are some known dichotomy classifications for restricted classes of digraphs. However, the problem is open even when H is restricted to oriented trees [7], which are a small subset of layered digraphs. The corresponding dichotomy is also open for the *counting* problem in general digraphs, although some partial results exist [1,2]. Note that, even if the dichotomy question for the existence problem were resolved, this would not necessarily imply a dichotomy for counting, since the reductions for the existence question may not be parsimonious.

In this paper, we give a dichotomy theorem for the counting problem in which H can be any directed acyclic graph. An interesting feature of this problem, which is different from previous dichotomy theorems for counting, is that there is a rich supply of tractable graphs H with complex structure.

The formal statement of our dichotomy is given below. Here is an informal description. First, the problem is #P-complete unless H is a layered digraph, meaning that the vertices of H can be arranged in levels, with edges going from one level to the next. We show (see Theorem 4 for a precise statement) that the problem is in P for a layered digraph H if the following condition is true (otherwise it is #P-complete). The condition is that, for every pair of vertices x and x' on level i and every pair of vertices y and y' on level $j > i$, the product of the graphs $H_{x,y}$ and $H_{x',y'}$ is isomorphic to the product of the graphs $H_{x,y'}$ and $H_{x',y}$. The precise definition of $H_{x,y}$ is given below, but the reader can think of it as the subgraph between vertex x and vertex y. The details of the product that we use (from [4]) are given below. The notion of isomorphism is the usual (graph-theoretic) one, except that certain short components are dropped, as described below. Some fairly complex graphs H satisfy this condition (see, for example, Figure 2), so for these graphs H the counting problem is in P.

Our algorithm for counting graph homomorphisms for tractable digraphs H is based on *factoring*. A difficulty is that the relevant algebra lacks unique factorisation. We deal with this by introducing "preconditioners". (See Section 6.)

Before giving precise definitions and proving our dichotomy theorem, we note that our proof relies on two fundamental results of Bulatov and Grohe [2] and Lovász [9]. These will be introduced in Section 3. Many technical lemmas are stated without proof in this extended abstract. A full version, including all proofs, can be found at http://eccc.hpi-web.de/eccc-reports/2005/TR05-121/index.html.

2 Notation and Definitions

Let $\mathbb{N}_0 = \{0, 1, 2, 3, \ldots\}$. For $m, n \in \mathbb{N}_0$, we will write $[m, n] = \{m, m+1, \ldots, n-1, n\}$ and $[n] = [1, n]$. We will generally let $H = (\mathcal{V}, \mathcal{E})$ denote a fixed "colouring" digraph, and $G = (V, E)$ an "input" digraph. We denote the *empty digraph* $(\emptyset, \emptyset)$ by **0**.

2.1 Homomorphisms

Let $G = (V, E)$, $H = (\mathcal{V}, \mathcal{E})$. If $f : V \to \mathcal{V}$, and $e = (v, v') \in E$, we write $f(e) = (f(v), f(v'))$. Then f is a *homomorphism* from G to H (or an *H-colouring* of G) if

$f(E) \subseteq \mathcal{E}$. We will denote the number of distinct homomorphisms from G to H by $\#H(G)$. Note that $\#H(\mathbf{0}) = 1$ for all H.

Let f be a homomorphism from $H_1 = (\mathcal{V}_1, \mathcal{E}_1)$ to $H_2 = (\mathcal{V}_2, \mathcal{E}_2)$. If f is also injective, it is a *monomorphism*. Then $|\mathcal{E}_1| = |f(\mathcal{E}_1)| \leq |\mathcal{E}_2|$. If there exist monomorphisms f from H_1 to H_2 and f' from H_2 to H_1, then f is an *isomorphism* from H_1 to H_2. Then $|\mathcal{E}_1| = |\mathcal{E}_2|$, so $f(\mathcal{E}_1) = \mathcal{E}_2$. If there is an isomorphism from H_1 to H_2, we write $H_1 \cong H_2$ and say H_1 is *isomorphic* to H_2. The relation $\cong$ is easily seen to be an equivalence. Usually $H_1, \ldots$ will denote equivalence classes of isomorphic graphs, and we write $H_1 = H_2$ rather than $H_1 \cong H_2$.

In this paper, we consider the particular case where $H = (\mathcal{V}, \mathcal{E})$ is a *directed acyclic graph* (DAG). In particular, H has no self-loops, and $\#H(G) = 0$ if G is not a DAG.

2.2 Layered Graphs

A DAG $H = (\mathcal{V}, \mathcal{E})$ is a *layered digraph*[2] with ℓ layers if $\mathcal{V}$ is partitioned into $(\ell + 1)$ *levels* $\mathcal{V}_i$ ($i \in [0, \ell]$) such that $(u, u') \in \mathcal{E}$ only if $u \in \mathcal{V}_{i-1}, u' \in \mathcal{V}_i$ for some $i \in [\ell]$. We will allow $\mathcal{V}_i = \emptyset$. We will call $\mathcal{V}_0$ the *top* and $\mathcal{V}_\ell$ the *bottom*. Nodes in $\mathcal{V}_0$ are called *sources* and nodes in $\mathcal{V}_\ell$ are called *sinks*. (Note that the usage of the words *source* and *sink* varies. In this paper a vertex is called a source only if it is in $\mathcal{V}_0$, a vertex in $\mathcal{V}_i$ for some $i \neq 0$ is not called a source even if it has in-degree 0, and similarly for sinks.) Layer i is the edge set $\mathcal{E}_i \subseteq \mathcal{E}$ of the subgraph $H^{[i-1,i]}$ induced by $\mathcal{V}_{i-1} \cup \mathcal{V}_i$. More generally we will write $H^{[i,j]}$ for the subgraph induced by $\bigcup_{k=i}^{j} \mathcal{V}_k$.

Let $\mathcal{G}_\ell$ be the class of all layered digraphs with ℓ layers and let $\mathcal{C}_\ell$ be the subclass of $\mathcal{G}_\ell$ in which every connected component spans all $\ell + 1$ levels. If $H \in \mathcal{C}_\ell$ and $G = (V, E) \in \mathcal{C}_\ell$, with V_i denoting level i ($i \in [0, \ell]$) and E_i denoting layer i ($i \in [\ell]$), then any homomorphism from G to H is a sequence of functions $f_i : V_i \to \mathcal{V}_i$ ($i \in [0, \ell]$) which induce a mapping from E_i into $\mathcal{E}_i$ ($i \in [\ell]$).

We use $\mathcal{C}_\ell$ to define an equivalence relation on $\mathcal{G}_\ell$. In particular, for $H_1, H_2 \in \mathcal{G}_\ell$, $H_1 \equiv H_2$ if and only if $\widehat{H_1} = \widehat{H_2}$, where $\widehat{H_i} \in \mathcal{C}_\ell$ is obtained from H_i by deleting every connected component that spans fewer than $\ell + 1$ levels.

2.3 Sums and Products

If $H_1 = (\mathcal{V}_1, \mathcal{E}_1)$, $H_2 = (\mathcal{V}_2, \mathcal{E}_2)$ are disjoint digraphs, the *union* $H_1 + H_2$ is the digraph $H = (\mathcal{V}_1 \cup \mathcal{V}_2, \mathcal{E}_1 \cup \mathcal{E}_2)$. Clearly $\mathbf{0}$ is the additive identity and $H_1 + H_2 = H_2 + H_1$. If G is connected then $\#(H_1 + H_2)(G) = \#H_1(G) + \#H_2(G)$, and if $G = G_1 + G_2$ then $\#H(G) = \#H(G_1)\#H(G_2)$.

The *layered cross product* [4] $H = H_1 \times H_2$ of layered digraphs $H_1 = (\mathcal{V}_1, \mathcal{E}_1)$, $H_2 = (\mathcal{V}_2, \mathcal{E}_2) \in \mathcal{G}_\ell$ is the layered digraph $H = (\mathcal{V}, \mathcal{E}) \in \mathcal{G}_\ell$ such that $\mathcal{V}_i = \mathcal{V}_{1i} \times \mathcal{V}_{2i}$ ($i \in [0, \ell]$), and we have $((u_1, u_2), (u_1', u_2')) \in \mathcal{E}$ if and only if $(u_1, u_1') \in \mathcal{E}_1$ and $(u_2, u_2') \in \mathcal{E}_2$. We will usually write $H_1 \times H_2$ simply as $H_1 H_2$. It is clear that $H_1 H_2$ is connected only if both H_1 and H_2 are connected. The converse is not necessarily true. An example appears in the full paper. Nevertheless, we have the following lemma.

---

[2] This is called a *balanced digraph* in [5,7]. However, "balanced" has other meanings in the study of digraphs.

Lemma 1. *If $H_1, H_2 \in \mathcal{C}_\ell$ and both of these graphs contain a directed path from every source to every sink then exactly one component of $H_1 H_2$ spans all $\ell + 1$ levels. In each other component either level 0 or level ℓ is empty.*

Note that $H_1 H_2 = H_2 H_1$, using the isomorphism $(u_1, u_2) \mapsto (u_2, u_1)$. If $G, H_1, H_2 \in \mathcal{C}_\ell$ then any homomorphism $f : G \to H_1 H_2$ can be written as a product $f_1 \times f_2$ of homomorphisms $f_1 : G \to H_1$ and $f_2 : G \to H_2$, and any such product is a homomorphism. Thus $\#H_1 H_2(G) = \#H_1(G) \, \#H_2(G)$. Observe that the directed path P_ℓ of length ℓ gives the multiplicative identity $\mathbf{1}$ and that $\mathbf{0}H = H\mathbf{0} = \mathbf{0}$ for all H. It also follows easily that $H(H_1 + H_2) = HH_1 + HH_2$, so $\times$ distributes over $+$. The algebra $\mathcal{A} = (\mathcal{G}_\ell, +, \times, \mathbf{0}, \mathbf{1})$ is a *commutative semiring*. The $+$ operation is clearly cancellative[3]. We will show in Lemma 2 that $\times$ is also cancellative, at least for $\mathcal{C}_\ell$. In many respects, this algebra resembles arithmetic on $\mathbb{N}_0$, but there is an important difference. In $\mathcal{A}$ we do not have *unique factorisation into primes*. A *prime* is any $H \in \mathcal{G}_\ell$ which has only the *trivial factorisation* $H = \mathbf{1}H$. Here we may have $H = H_1 H_2 = H_1' H_2'$ with H_1, H_2, H_1', H_2' prime and no pair equal, even if all the graphs are connected. An example is given in the full version. The layered cross product was defined in [4] in the context of interconnection networks. It is similar to the (non-layered) *direct product* [8], which also lacks unique factorisation, but they are not identical. In general, they have different numbers of vertices and edges.

3 Fundamentals

Our proof relies on two fundamental results of Bulatov and Grohe [2] and Lovász [9]. First we give the basic result of Lovász [9]. (See also [7, Theorem 2.11].) The following is essentially a special case of Lovász [9, Theorem 3.6], though stated rather differently.

Theorem 1 (Lovász). *If $\#H_1(G) = \#H_2(G)$ for all G, then $H_1 = H_2$.*

The following variant of Theorem 1 restricts H_1, H_2 and G to $\mathcal{C}_\ell$. This theorem and the subsequent lemmas and corollaries are proved in the full version.

Theorem 2. *If $H_1, H_2 \in \mathcal{C}_\ell$ and $\#H_1(G) = \#H_2(G)$ for all $G \in \mathcal{C}_\ell$, then $H_1 = H_2$.*

Corollary 1. *Suppose $H_1 = (\mathcal{V}_1, \mathcal{E}_1), H_2 = (\mathcal{V}_2, \mathcal{E}_2) \in \mathcal{C}_\ell$. If there is any $G \in \mathcal{C}_\ell$ with $\#H_1(G) \neq \#H_2(G)$, then there is such a G with $0 < |V| \leq \max_{k=1,2} |\mathcal{V}_k|$.*

Lemma 2. *If $H_1 H = H_2 H$ for $H_1, H_2, H \in \mathcal{C}_\ell$, then $H_1 = H_2$.*

Recall that $\equiv$ denotes the equivalence relation on $\mathcal{G}_\ell$ which ignores "short" components.

Lemma 3. *If $H_1, H_2, H \in \mathcal{C}_\ell$ and each of these contains a directed path from every source to every sink and $H_1 H \equiv H_2 H$ then $H_1 = H_2$.*

The second fundamental result is a theorem of Bulatov and Grohe [2, Theorem 1], which provides a powerful generalisation of a theorem of Dyer and Greenhill [3]. Let $A = (A_{ij})$ be a $k \times k$ matrix of non-negative rationals. We view A as a weighted

---

[3] This means that $H + H_1 = H + H_2$ implies $H_1 = H_2$. Similarly for $\times$.

digraph such that there is an edge (i, j) with weight A_{ij} if $A_{ij} > 0$. Given a digraph $G = (V, E)$, EVAL(A) is the problem of computing the *partition function*

$$Z_A(G) = \sum_{\sigma:V \to \{1,\dots,k\}} \prod_{(u,v) \in E} A_{\sigma(u)\sigma(v)}. \tag{1}$$

In particular, if A is the adjacency matrix of a digraph H, $Z_A(G) = \#H(G)$. Thus EVAL(A) has at least the same complexity as $\#H$. If A is *symmetric*, corresponding to a weighted *undirected* graph, the next theorem characterises the complexity of EVAL(A).

Theorem 3 (Bulatov and Grohe). *Let A be a non-negative rational symmetric matrix.*

(1) If A is connected and not bipartite, then EVAL(A) *is in polynomial time if the row rank of A is at most 1; otherwise* EVAL(A) *is #P-complete.*

(2) If A is connected and bipartite, then EVAL(A) *is in polynomial time if the row rank of A is at most 2; otherwise* EVAL(A) *is #P-complete.*

(3) If A is not connected, then EVAL(A) *is in polynomial time if each of its connected components satisfies the corresponding condition stated in (1) or (2); otherwise* EVAL(A) *is #P-complete.*

4   Reduction from Acyclic H to Layered H

Let $H = (\mathcal{V}, \mathcal{E})$ be a DAG. Clearly $\#H$ is in #P. We will call H *easy* if $\#H$ is in P and *hard* if $\#H$ is #P-complete. We will show that H is hard unless it can be represented as a *layered* digraph. Essentially, we do this using a "gadget" consisting of two opposing directed k-paths to simulate the edges of an undirected graph and then apply Theorem 3. To this end, let $N_k(u, u')$ be the number of paths of length k from u to u' in H. Say that vertices $u, u' \in \mathcal{V}$ are *k-compatible* if, for some vertex w, there is a length-k path from u to w and from u' to w. We say that H is *k-good* if, for every k-compatible pair (u, u'), there is a rational number λ such that $N_k(u, v) = \lambda N_k(u', v)$ $(\forall v \in \mathcal{V})$. In the full version, we prove the following.

Lemma 4. *If there is a k such that H is not k-good then $\#H$ is #P-complete.*

Remark 1. The statement of Lemma 4 is not symmetrical with respect to the direction of edges in H. However, if the digraph H^R is obtained from H by reversing every edge, then $\#H^R$ and $\#H$ have the same complexity, since $\#H^R(G^R) = \#H(G)$ for all G.

The main result of this section, which is proved in the full version, is the following.

Lemma 5. *If H is a DAG, but it cannot be represented as a layered digraph, then $\#H$ is #P-hard.*

5   A Structural Condition for Hardness

We can now formulate a sufficient condition for hardness of a layered digraph $H = (\mathcal{V}, \mathcal{E}) \in \mathcal{G}_\ell$. Suppose $s \in \mathcal{V}_i$ and $t \in \mathcal{V}_j$ for $i < j$. If there is a directed path in H from s to t, we let H_{st} be the subgraph of H induced by s, t, and all components of $H^{[i+1,j-1]}$ to which both s and t are incident. Otherwise, we let $H_{st} = \mathbf{0}$.

Lemma 6. *If there exist $x, x' \in \mathcal{V}_0$, $y, y' \in \mathcal{V}_\ell$ such that $H_{xy}H_{x'y'} \not\equiv H_{xy'}H_{x'y}$, and at most one of $H_{xy}, H_{xy'}, H_{x'y}, H_{x'y'}$ is $\mathbf{0}$, then $\#H$ is $\#P$-complete.*

Clearly checking the condition of the Lemma requires only constant time (since the size of H is a constant). Note that if x, x', y, y' are not all in the same component of H then at least two of $H_{xy}, H_{x'y'}, H_{xy'}, H_{x'y}$ are $\mathbf{0}$, so Lemma 6 has no content. We may generalise Lemma 6 as follows.

Lemma 7. *If there exist $x, x' \in \mathcal{V}_i$, $y, y' \in \mathcal{V}_j$ ($0 \leq i < j \leq \ell$) such that $H_{xy}H_{x'y'} \not\equiv H_{xy'}H_{x'y}$, and at most one of $H_{xy}, H_{xy'}, H_{x'y}, H_{x'y'}$ is $\mathbf{0}$, then $\#H$ is $\#P$-complete.*

Note that the "N" of Bulatov and Dalmau [1] is the special case of Lemma 7 in which $j = i + 1$, $H_{xy} = H_{xy'} = H_{x'y'} = \mathbf{1}$, and $H_{x'y} = \mathbf{0}$. More generally, any structure with $H_{xy}, H_{xy'}, H_{x'y'} \neq \mathbf{0}$ and $H_{x'y} = \mathbf{0}$ is a special case of Lemma 7, so is sufficient to prove $\#P$-completeness. Using this idea, we can prove the following.

Lemma 8. *If $H \in \mathcal{C}_\ell$ is connected and not hard, then there exists a directed path from every source to every sink.*

Lemma 8 cannot be generalised by replacing "source" with "node (at any level) with indegree 0" and replacing "sink" similarly, as the graph in Figure 1 illustrates. We call four vertices x, x', y, y' in H, with $x, x' \in \mathcal{V}_i$ and $y, y' \in \mathcal{V}_j$ ($0 \leq i < j \leq \ell$), a *Lovász violation*[4] if at most one of $H_{xy}, H_{x'y'}, H_{xy'}, H_{x'y}$ is $\mathbf{0}$ and $H_{xy}H_{x'y'} \not\equiv H_{xy'}H_{x'y}$. A graph H with no Lovász violation will be called *Lovász-good*. We show next that this property is preserved under the layered cross product.

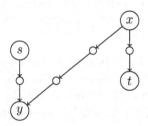

Fig. 1. An easy H with no st path

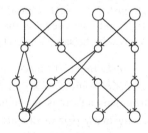

Fig. 2. A Lovász-good H

Lemma 9. *If $H, H_1, H_2 \in \mathcal{C}_\ell$ and $H = H_1 H_2$ then H is Lovász-good if and only if both H_1 and H_2 are Lovász-good.*

The requirement of H being Lovász-good is essentially a "rank 1" condition in the algebra $\mathcal{A}$ of Section 2.3, and therefore resembles the conditions of [2,3]. However, since $\mathcal{A}$ lacks unique factorisation, difficulties arise which are not present in the analyses of [2,3]. But a more important difference is that, whereas the conditions of [2,3] permit only trivial easy graphs, Lovász-good graphs can have complex structure. See Figure 2 for a small example.

---

[4] The name derives from the isomorphism theorem (Theorem 1) of Lovász.

6   Main Theorem

We can now state the dichotomy theorem for counting homomorphisms to directed acyclic graphs.

Theorem 4. *Let H be a directed acyclic graph. Then $\#H$ is in P if H is layered and Lovász-good. Otherwise $\#H$ is $\#P$-complete.*

The proof of Theorem 4 will use the following lemma, which we prove later.

Lemma 10. *Suppose $H \in \mathcal{C}_\ell$ is connected, with a single source and sink, and is Lovász-good. There is a polynomial-time algorithm for the following problem. Given a connected $G \in \mathcal{C}_\ell$ with a single source and sink, compute $\#H(G)$.*

Proof (of Theorem 4). We have already shown in Lemma 5 that any non-layered H is hard. We have also shown in Lemma 7 that H is hard if it is not Lovász-good. Suppose $H \in \mathcal{G}_\ell$ is Lovász-good. We will show how to compute $\#H(G)$.

First, we may assume that G is connected since, as noted in Section 2.3, if $G = G_1 + G_2$ then $\#H(G) = \#H(G_1)\#H(G_2)$. We can also assume that H is connected since, for connected G, $\#(H_1 + H_2)(G) = \#H_1(G) + \#H_2(G)$, but H_1 and H_2 are Lovász-good if $H_1 + H_2$ is.

So we can now assume that $H \in \mathcal{C}_\ell$ is connected and G is connected. If G has more than $\ell + 1$ non-empty levels then $\#H(G) = 0$. If G has fewer than ℓ non-empty levels then decompose H into component subgraphs $H_1, H_2, \dots$ as in the proof of Lemma 7 in the full version, and proceed with each component separately. So we can assume without loss of generality that both H and G are connected and in $\mathcal{C}_\ell$.

Now we just add a new level at the top of H with a single vertex, adjacent to all sources of H and a new level at the bottom of H with a single vertex, adjacent to all sinks of H. We do the same to G. Then we use Lemma 10.

Before proving Lemma 10 we need some definitions. Let H be a connected graph in $\mathcal{C}_\ell$. For a subset S of sources of H, let $H_S^{[0,j]}$ be the subgraph of $H^{[0,j]}$ induced by those vertices from which there is an (undirected) path to S in $H^{[0,j]}$. We say that H is *top-j disjoint* if, for every pair of distinct sources s, s', $H_{\{s\}}^{[0,j]}$ and $H_{\{s'\}}^{[0,j]}$ are disjoint, and that H is *bottom-j disjoint* if the reversed graph H^R from Remark 1 is top-j disjoint. Finally, H is *fully disjoint* if it is top-$(\ell - 1)$ disjoint and bottom-$(\ell - 1)$ disjoint.

We will say that (Q, U, D) is a *good factorisation* of H if Q, U and D are connected Lovász-good graphs in $\mathcal{C}_\ell$ such that $QH \equiv UD$, Q has a single source and sink, U has a single sink, and D has a single source.

Remark 2. The presence of the "preconditioner" Q in the definition of a good factorisation is due to the absence of unique factorisation in the algebra $\mathcal{A}$. Our algorithm for computing homomorphisms to a Lovász-good H works by factorisation. However, a non-trivial Lovász-good H can be prime. An example is given in the full version.

We can now state our main structural lemma.

Lemma 11. *If $H \in \mathcal{C}_\ell$ is connected, and Lovász-good, then it has a good factorisation (Q, U, D).*

We prove Lemma 11 below in Section 7. In the course of the proof, we give an algorithm for constructing (Q, U, D). We now describe how we use Lemma 11 (and the algorithm) to prove Lemma 10.

Proof (of Lemma 10). The proof is by induction on ℓ. The base case is $\ell = 2$. (Note that calculating $\#H(G)$ is easy in this case.) For the inductive step, suppose $\ell > 2$. Let H' denote the part of H excluding levels 0 and ℓ and let G' denote the part of G excluding levels 0 and ℓ. Using reasoning similar to that in the proof of Theorem 4, we can assume that G' is connected and then that H' is connected. Since H is Lovász-good, so is H'. Now by Lemma 11 there is a good factorisation (Q', U', D') of H'.

Let $S \subseteq \mathcal{V}_1$ be the nodes in level 1 of H that are adjacent to the source and $T \subseteq \mathcal{V}_{\ell-1}$ be the nodes in level $\ell - 1$ of H that are adjacent to the sink. Note that $\mathcal{V}_1$ is the top level of U' and $\mathcal{V}_{\ell-1}$ is the bottom level of D'.

Construct Q from Q' by adding a new top and bottom level with a new source and sink. Connect the new source and sink to the old ones. Construct D from D' by adding a new top and bottom level with a new source and sink. Connect the new source to the old one and the new sink to T. Finally, construct U from U' by adding a new top and bottom level with a new source and sink. Connect the new source to S and the new sink to the old one. See Figure 3. Note that (Q, U, D) is a good factorisation of H. To see that $QH \equiv UD$, consider the component of $Q'H'$ that includes sources and sinks. (There is just one of these. Since H' is Lovász-good, it has a directed path from every source to every sink by Lemma 8. So does Q'. Then use Lemma 1.) This is isomorphic to the corresponding component in $D'U'$ since (Q', U', D') is a good factorisation of H'. The isomorphism maps S in H' to a corresponding S in U' and now note that the new top level is appropriate in QH and DU. Similarly, the new bottom level is appropriate.

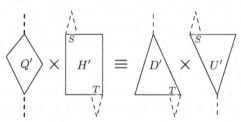

Fig. 3. Modifying H

Now let's consider how to compute $\#Q(G)$. In any homomorphism from G to Q, every node in level 1 of G gets mapped to the singleton in level 1 of Q. Thus, we can collapse all level 1 nodes of G into a single vertex without changing the problem. At this point, the top level of G and Q are not doing anything, so they can be removed, and we have a sub-problem with fewer levels. So $\#Q(G)$ can be computed recursively. The same is true for $\#D(G)$ and $\#U(G)$.

Since $G \in \mathcal{C}_\ell$, $\#QH(G) = \#Q(G)\#H(G)$. Also, since components without sources and sinks cannot be used to colour G (which has the full ℓ layers), this is equal to $\#D(G)\#U(G)$. Thus, we can output $\#H(G) = \#D(G)\#U(G)/\#Q(G)$.

That concludes the proof of Theorem 10, so it only remains to prove Lemma 11. The proof will be by induction, for which we will need the following technical lemmas, which are proved in the full version. Our proofs use the following operations on a Lovász-good connected digraph $H \in \mathcal{C}_\ell$.

Local Multiplication: Suppose that U is a connected Lovász-good single-sink graph in $\mathcal{C}_j$ on levels $0, \ldots, j$ for $j \leq \ell$. Let C be a Lovász-good connected component in $H^{[0,j]}$ with no empty levels. Then $\mathrm{Mul}(H, C, U)$ is the graph constructed from H by replacing C with the full component of UC. (Note that there is only one full component, by Lemmas 8 and 1.)

Local Division: Suppose $S \subseteq \mathcal{V}_0$, and that (Q, U, D) is a good factorisation of $H_S^{[0,j]}$. Then $\mathrm{Div}(H, Q, U, D)$ is the graph constructed from H by replacing $H_S^{[0,j]}$ with D.

Lemma 12. *Suppose that H is Lovász-good, top-$(j-1)$ disjoint, $S \subseteq \mathcal{V}_0$ and $H_S^{[0,j]}$ is connected. If (Q, U, D) is a good factorisation of $H_S^{[0,j]}$, then $\mathrm{Div}(H, Q, U, D)$ is Lovász-good.*

Lemma 13. *If H and U are Lovász-good, then $\mathrm{Mul}(H, C, U)$ is Lovász-good.*

7   Proof of Lemma 11

A *top-dangler* is a component in $H^{[1,\ell-1]}$ that is incident to a source but not to a sink. Similarly, a *bottom-dangler* is a component in $H^{[1,\ell-1]}$ that is incident to a sink but not to a source. (Note that a bottom-dangler in H is a top-dangler in H^R.)

The proof of Lemma 11 will be by induction. The base case will be $\ell = 1$, where it is easy to see that a connected Lovász-good H must be a complete bipartite graph. The ordering for the induction will be lexicographic on the following criteria (in order): (1) the number of levels, (2) the number of sources, (3) the number of top-danglers, (4) the number of sinks, and (5) the number of bottom-danglers. Thus, for example, if H' has fewer levels than H then H' precedes H in the induction. If H' and H have the same number of levels, the same number of sources and the same number of top-danglers but H' has fewer sinks then H' precedes H in the induction.

The inductive step will be broken into five cases. The cases are exhaustive but not mutually exclusive – given an H we will apply the first applicable case.

7.1   Case 1: *H is top-(j-1) disjoint and has a top-dangler with depth at most j-1.*

Let R be a top-dangler with depth j' where $j' < j$ (meaning that it is a component in $H^{[1,\ldots,\ell-1]}$ that is incident to a source but not to a sink, and that levels $j'+1, \ldots, \ell-1$ are empty and level j' is non-empty). Note that since H is top-$(j-1)$ disjoint, R must be adjacent to a single source, v, in H. This follows from the definition of top-$(j-1)$ disjoint, and from the fact that R has depth at most $j-1$.

Construct H' from H by removing R. Note that H' is connected. By construction (from H), H' is Lovász-good and has no empty levels. It precedes H in the induction order since it has the same number of levels, the same number of sources and one fewer top-dangler. By induction, it has a good factorisation (Q', U', D') so $Q'H' \equiv U'D'$.

Construct $\widehat{D}$ as follows. On layers $1, \ldots, j'$, $\widehat{D}$ is identical to D'. On layers $(j' + 2), \ldots, \ell$, $\widehat{D}$ is a path. Every node in level j' is connected to the singleton vertex in level $j' + 1$. Then clearly $\widehat{D}Q'H' \equiv \widehat{U}D'$ where $\widehat{U}$ is the single full component of $\widehat{D}U'$. (There is just one of these. Since $\widehat{D}$ is Lovász-good, it has a directed path from every source to every sink by Lemma 8. So does U'. Then use Lemma 1.) Note that $\widehat{U}$ has a single sink.

Let R' be the graph obtained from R by adding the source v. Let R'' be $Q'^{[0,j']}R'$. Form U'' from R'' and $\widehat{U}$ by identifying v with the appropriate source of $\widehat{U}$. (Note that $\widehat{U}$ has the same sources as H.) Then $\widehat{D}Q'H \equiv U''D'$.

Thus, we have a good factorisation (Q, U, D') of H by taking Q to be the full component of $\widehat{D}Q'$ and U to be the full component of U''. To see that it is a good factorisation, use Lemma 9 to show that Q and U are Lovász-good.

7.2 Case 2: For $j < \ell$, H is top-$(j-1)$ disjoint, but not top-j disjoint, and has no top-dangler with depth at most $j - 1$.

Partition the sources of H into equivalence classes $S_1, \ldots, S_k$ so that the graphs $H_{S_i}^{[0,j]}$ are connected and pairwise disjoint. See Figure 4. Since H is not top-j disjoint some equivalence class, say S_1, contains more than one source. Let $\widehat{H}$ denote $H_{S_1}^{[0,j]}$. $\widehat{H}$ is

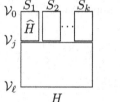

Fig. 4. H **Fig. 5.** H'

shorter than H, so it comes before H in the induction order. It is connected by construction of the equivalence classes, and it is Lovász-good by virtue of being a subgraph of H. By induction we can construct a good factorisation $(\widehat{Q}, \widehat{U}, \widehat{D})$ of $\widehat{H}$. Let $H' = \mathrm{Div}(H, \widehat{Q}, \widehat{U}, \widehat{D})$. See Figure 5. H' comes before H in the induction order because it has the same number of levels, but fewer sources. To see that H' is connected, note that H is connected and $\widehat{H}$ is connected. Since $(\widehat{Q}, \widehat{U}, \widehat{D})$ is a good factorisation), we know $\widehat{D}$ is connected, so H' is connected. By Lemma 12, H' is Lovász-good. By induction, we can construct a good factorisation (Q', U', D') of H'.

Let s be the (single) source of $\widehat{D}$. By construction, the sources of U' are $\{s\} \cup S_2 \cup \cdots \cup S_k$. See Figure 6. Let $C_1, \ldots, C_z$ be the connected components of $U'^{[0,j]}$. Let C_1

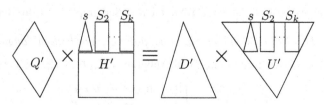

Fig. 6. $Q'H' \equiv D'U'$

be the component containing s. Since the $H_{S_i}^{[0,j]}$ are connected and pairwise disjoint, and Q' has a single source, there is a single connected component of $(Q'H')^{[0,j]}$ containing all of S_i (and no other sources) so (see Figure 6) there is a single connected component of $U'^{[0,j]}$ containing all of S_i (and no other sources). For convenience, call this C_i.

If $z > k$ then components $C_{k+1}, \ldots, C_z$ do not contain any sources. (They are due to danglers in U', which in this case are nodes that are not descendants of a source.)

Now consider $U_1 = \text{Mul}(U', C_1, \widehat{U})$, where U_1 is constructed from U' by replacing C_1 with the full component of $C_1\widehat{U}$. For $i \in \{2, \ldots, z\}$, let $U_i = \text{Mul}(U_{i-1}, C_i, \widehat{Q})$. U_z is the graph constructed from U' by replacing C_1 with the full component of $C_1\widehat{U}$ and replacing every other C_i with the full component of $C_i\widehat{Q}$.

Let $\widetilde{Q}$ extend $\widehat{Q}$ down to level ℓ with a single path. We claim that

$$Q'\widetilde{Q}H \equiv U_zD'. \tag{2}$$

To establish Equation (2), note that on levels $[j, \ldots, \ell]$ the left side is $(Q'\widetilde{Q}H)^{[j,\ell]} \equiv (Q'H')^{[j,\ell]}$. Any components of $Q'H'$ that differ from $U'D'$ do not include level ℓ, so this is equivalent to $(U'D')^{[j,\ell]} \equiv (U_zD')^{[j,\ell]}$, which is the right-hand side. So focus on levels $0, \ldots, j$. From the left-hand side, look at the component $(Q'\widetilde{Q}H)_{S_1}^{[0,j]}$. Note that it is connected. It is

$$(Q'^{[0,j]}\widehat{Q}\widehat{H})_{S_1}^{[0,j]} \equiv (Q'^{[0,j]}\widehat{U}\widehat{D})_{S_1}^{[0,j]} \equiv ((D'^{[0,j]}U'^{[0,j]})_{\{s\}}\widehat{U})_{S_1}^{[0,j]} \equiv (D'U_z)_{S_1}^{[0,j]},$$

which is the right-hand side. Then look at the component S_2.

$$(Q'\widetilde{Q}H)_{S_2}^{[0,j]} \equiv (Q'^{[0,j]}\widehat{Q}H^{[0,j]})_{S_2} \equiv ((D'^{[0,j]}U'^{[0,j]})_{S_2}\widehat{Q})_{S_2}^{[0,j]} \equiv (D'U_z)_{S_2}^{[0,j]}.$$

The other components containing sources (the only relevant components) are similar.

Having established (2), we observe that (Q, U, D') is a good factorisation of H where Q is the full component of $Q'\widetilde{Q}$ and U is the full component of U_z. Use Lemma 9 to show Q is Lovász-good and Lemma 13 to show that U_z is.

7.3   Case 3: H is top-$(\ell - 1)$ disjoint and has no top-dangler and is bottom-$(j - 1)$ disjoint and has a bottom-dangler with height at most $(j - 1)$.

We apply an analysis similar to Case 7.1 to the reversed graph H^R from Remark 1.

7.4   Case 4: For $j < \ell$, H is top-$(\ell - 1)$ disjoint and has no top-dangler and is bottom-$(j - 1)$ disjoint, but not bottom-j disjoint, and has no bottom-dangler with height at most $(j - 1)$.

We apply an analysis similar to Case 7.2 to the reversed graph H^R from Remark 1.

7.5   Case 5: H is fully disjoint. and has no top-danglers or bottom-danglers.

In the fully disjoint case (see Figure 7), the subgraphs H_{st} ($s \in \mathcal{V}_0, t \in \mathcal{V}_\ell$) satisfy

$$H_{st} \cap H_{s't'} = \begin{cases} \{s\}, & \text{if } s = s', \ t \neq t'; \\ \{t\}, & \text{if } s \neq s', \ t = t'; \\ \emptyset, & \text{if } s \neq s', \ t \neq t', \end{cases} \tag{3}$$

$H_{st} \neq \mathbf{0}$ and $H_{st}H_{s't'} \equiv H_{st'}H_{s't}$, since H is Lovász-good. We assume without loss that $|\mathcal{V}_0| > 1$ and $|\mathcal{V}_\ell| > 1$, since otherwise $(\mathbf{1}, H, \mathbf{1})$ or $(\mathbf{1}, \mathbf{1}, H)$ is a good factorisation. Choose any $s^* \in \mathcal{V}_0, t^* \in \mathcal{V}_\ell$, and let $Q = H_{s^* t^*}$. Note that Q is connected with a

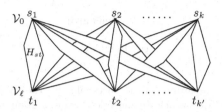

Fig. 7. Fully disjoint case

single source and sink, and is Lovász-good because H is Lovász-good. Let D be the subgraph $\bigcup_{t \in \mathcal{V}_\ell} H_{s^* t}$ of H, and let U be the subgraph $\bigcup_{s \in \mathcal{V}_0} H_{st^*}$ of H. These are both connected and Lovász-good since H is. Clearly D has a single source and U has a single sink. Also $QH \equiv DU$ follows from (3) and from the fact that there are no top-danglers or bottom-danglers and

$$(DU)_{s^* s, tt^*} = D_{s^* t}U_{st^*} \equiv H_{s^* t}H_{st^*} = H_{s^* t^*}H_{st} = QH_{st} \quad (s \in \mathcal{V}_0; \ t \in \mathcal{V}_\ell),$$
$$(4)$$

since H is Lovász-good. Thus (Q, U, D) is a good factorisation of H.

References

1. A. Bulatov and V. Dalmau, Towards a dichotomy theorem for the counting constraint satisfaction problem, in *Proc. 44th IEEE Symposium on Foundations of Computer Science*, IEEE, pp. 562–572, 2003.
2. A. Bulatov and M. Grohe, The complexity of partition functions, in *Automata, Languages & Programming: 31st International Colloquium*, Lecture Notes in Computer Science **3142**, pp. 294–306, 2004.
3. M. Dyer and C. Greenhill, The complexity of counting graph homomorphisms, *Random Structures & Algorithms* **17** (2000), 260–289.
4. S. Even and A. Litman, Layered cross product: a technique to construct interconnection networks. *Networks* **29** (1997), 219–223.
5. T. Feder and M. Vardi, The computational structure of monotone monadic SNP and constraint satisfaction: a study through Datalog and group theory, *SIAM J. Comput.* (28) (1998) 57–104.
6. P. Hell and J. Nešetřil, On the complexity of H-coloring, *Journal of Combinatorial Theory Series B* **48** (1990), 92–110.
7. P. Hell and J. Nešetřil, *Graphs and homomorphisms*, Oxford University Press, 2004.
8. W. Imrich and S. Klavžar, *Product graphs: structure and recognition*, Wiley, New York, 2000.
9. L. Lovász, Operations with structures, *Acta. Math. Acad. Sci. Hung.*, **18** (1967), 321–328.

Fault-Tolerance Threshold for a Distance-Three Quantum Code

Ben W. Reichardt

UC Berkeley

Abstract. The quantum error threshold is the highest (model-dependent) noise rate which we can tolerate and still quantum-compute to arbitrary accuracy. Although noise thresholds are frequently estimated for the Steane seven-qubit, distance-three quantum code, there has been no proof that a constant threshold even exists for distance-three codes. We prove the existence of a constant threshold. The proven threshold is well below estimates, based on simulations and analytic models, of the true threshold, but at least it is now known to be positive.

1   Introduction

Quantum operations are inherently noisy, so the development of fault-tolerance techniques is an essential part of progress toward a quantum computer. A quantum circuit with N gates can only a priori tolerate $O(1/N)$ error per gate. In 1996, Shor showed how to tolerate $O(1/\operatorname{poly}(\log N))$ error by encoding each qubit into a $\operatorname{poly}(\log N)$-sized quantum error-correcting code, then implementing each gate of the desired quantum circuit directly on the encoded qubits, alternating computation and error-correction steps [1]. Even though the corrections themselves are imperfect, noise overall remains under control – the scheme is "fault-tolerant" (Fig. 1).

Several groups [2,3,4] independently realized that by instead using a constant-sized quantum error-correcting code repeatedly concatenated on top of itself – and correcting lower levels more frequently than higher levels – a constant amount of error is tolerable, again with only polylogarithmic overhead. The tolerable noise rate, which Aharonov and Ben-Or proved to be positive (with no explicit lower bound) [2], is known as the fault-tolerance threshold. Intuitively, small, constant-sized codes can be more efficient to use because encoding into the quantum code (which is necessary at the beginning of the computation and also during error correction, in certain schemes) is a threshold bottleneck. However, the threshold proof of Aharonov and Ben-Or only applies for concatenating codes of distance five or higher. In this paper, we prove a constant noise threshold for the concatenated distance-three, seven-qubit Steane/Hamming code. (A threshold for the distance-three five-qubit code follows by the same structure of arguments.)

The attainable threshold value, and the overhead required to attain it, are together of considerable experimental interest. Thus, while work has continued on proving the existence of constant thresholds in different settings – e.g., under

M. Bugliesi et al. (Eds.): ICALP 2006, Part I, LNCS 4051, pp. 50–61, 2006.

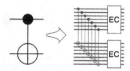

Fig. 1. Fault-tolerance overview: In fault-tolerant computing, qubits are encoded into an error-correcting code, and the ideal circuit's gates are compiled into gates acting directly on the encoded data. For example, an encoded CNOT gate can often be implemented as transversal physical CNOT gates. To prevent errors from spreading and accumulating, error-correction modules (Fig. 2) are placed between encoded gates. With a distance $d = 2t + 1$ code, one intuitively expects the "effective/logical error rate" to be $O(p^{t+1})$ if p is the physical error rate, because $t+1$ errors might be corrected in the wrong direction. One can't directly use large codes, because then even the initial encoding would fail. Instead, use a small code, and repeatedly concatenate the entire compilation procedure on itself, reducing the error rate at each level as long as the initial error rate is beneath a threshold. However, because of an inefficiency in the analysis, the classic threshold proof only gives a quadratic error-rate reduction with distance-five codes, and does not apply to distance-three codes.

physical locality constraints [5], or with non-Markovian noise [6] – a substantial amount of attention has been devoted to estimating the fault-tolerance threshold, using simulations and analytic modeling. Most of these threshold estimates have used the seven-qubit code, from basic estimates [2, 3, 7, 8] to estimates using optimized fault-tolerance schemes [9, 10, 11], to a threshold estimate with a two-dimensional locality constraint [12]. One reason the seven-qubit code has been so popular is no doubt its elegant simplicity, and its small size allows for easy, efficient simulations. However, there had been no proof that a threshold even existed for the simulated fault-tolerance schemes.

Currently, the highest error threshold estimate is due to Knill, who has estimated a threshold perhaps as high as 5% by using a very efficient distance-two code with a novel fault-tolerance scheme [13]. Being of distance two, the code only allows for error detection, not correction, so the scheme uses extensive rejection testing. This leads to an enormous overhead at high error rates, limiting the practicality of operating a quantum computer in this regime. Still, a major open problem remains to prove the existence of a threshold for a distance-two code.

We prove an error threshold lower bound of 6.75×10^{-6} in a certain error model. Our analysis is prioritized for proof simplicity and ease of presentation, not for a high threshold (although we discuss optimizations in Sect. 4). Also, it is not surprising that unproven threshold estimates should be significantly higher than proven threshold lower bounds – although actually the author is unaware of any published rigorous lower bounds besides the current work and Ref. [14] (except in the erasure error model [15]). But such a large gap between what we can prove, and what our models and simulations indicate is embarrassing. A second major open problem is to close the gap between proofs and estimates.

Our proof is based on giving a recursive characterization of the probability distribution of errors in blocks of the concatenated code. Intuitively, with a

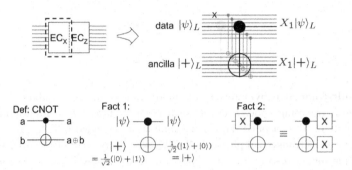

Fig. 2. Quantum error-correction overview: So-called Steane-type error correction consists of X (bit flip) error correction, followed by Z (phase flip, the dual) error correction; $X|a\rangle = |1 \oplus a\rangle$, $Z|a\rangle = (-1)^a |a\rangle$ for $a \in \{0, 1\}$. To correct X errors, prepare an ancilla in the encoded state $|+\rangle_L = \frac{1}{\sqrt{2}}(|0\rangle_L + |1\rangle_L)$. Then apply transversal CNOTs into the ancilla, measure the ancilla, and apply any necessary correction to the data. Facts 1 and 2, which each follow immediately from the definition of the CNOT gate, imply firstly that there is no logical effect and secondly that any X errors in the data are copied onto the ancilla. Thus it is safe to measure the ancilla without affecting the data's logical state, and then interpret the code's syndromes with a classical computer to determine the correction to apply.

distance-three code, two errors in a code block is a bad event, so the block error rate should drop roughly like cp^2, with p the bit error rate. After two levels of concatenation, the error rate should be like $c(cp^2)^2 = (cp)^{2^2}/c$, and so on. The threshold for improvement is at $p = 1/c$. The difficulty lies in formalizing and making rigorous this intuition.

The classic threshold proof of Aharonov and Ben-Or [2] can be reformulated to rely on a key definition of 1-goodness. Roughly, define a code block to be 1-good if it has at most one subblock which is not itself 1-good. Maintaining this definition as an inductive invariant through the logical circuit – i.e., proving that the outputs of a logical gate are 1-good if the inputs are, with high probability – allows provable thresholds for concatenating codes of distance five or higher. But this definition does not suffice for concatenating a distance-three code. For take a 1-good block, with the allowed one erroneous subblock, and apply a logical gate to it. If a single subblock failure occurs while applying the logical gate, there can be two bad subblocks total, enough to flip the state of the whole block (since the distance is only three). Therefore, the block failure rate is first-order in the subblock failure rate. Logical behavior is not necessarily improved by encoding, and the basic premise of fault tolerance, controlling errors even with imperfect controls, is violated.

Essentially, a stronger inductive assumption is required for the proof to go through. With 1-goodness, we are assuming that the block entering a computation step has no more than one bad subblock. Intuitively, though, most of the time there should be no bad subblocks at all. We capture this intuition in the stronger definition of "1-wellness." In a 1-well block, not only is there at most one bad subblock, but also the *probability* of a bad subblock is small. With this definition, the problem sketched above does not occur because the probability of there being an erroneous subblock in the input is already first-order, so a logical failure is still a second-order event. For the argument to go through, though, the definition must be carefully stated, and we need to carefully define what is required by each logical gate and how the logical gates will be implemented. Controlling the probability distribution of errors is the main technical tool and new contribution of this paper.

Very recently, Aliferis, Gottesman, and Preskill independently completed a threshold proof for distance-three codes [14], based instead on formalizing the "overlapping steps" threshold argument of Knill, Laflamme and Zurek [3]. Our probabilistic definitions may be more difficult to extend to different error models. However, the probabilistic structure is also a potential strength, in that it may make this proof more extensible toward provable thresholds for postselection-based, error-detection fault-tolerance schemes like that of Knill [13].

Sections 2 and 3 contain the necessary definitions, and the proof of a threshold for quantum stabilizer operations (meaning preparation of fresh qubits as $|0\rangle$, measurement in the computational basis, and application of Clifford group unitaries like the CNOT gate). Stabilizer operations are easy to work with because Pauli errors (bit flips and dual phase flips) propagate through linearly. The full paper extends the proof to give a threshold for full universal quantum computation. In fact, the threshold itself is unaffected by this extension – the bottleneck, as in most threshold estimates, is in achieving stabilizer operation fault-tolerance.

2 Definitions

2.1 Concatenated Steane Code

In concatenated coding, qubits are arranged into level-one code blocks of n, which are in turn arranged into level-two blocks of n, and so on. We call a single qubit a $block_0$, n grouped qubits a $block_1$, and n^k grouped qubits a $block_k$ (but often extraneous subscripts will be omitted).

In this section, we will use for simple examples the classical three-bit repetition code: $|0\rangle_L = |000\rangle, |1\rangle_L = |111\rangle$. (A level-$k$ concatenated encoding of $|0\rangle$ is $|0^{3^k}\rangle$.) This code has distance three against bit flip X errors – $X = \left(\begin{smallmatrix} 0 & 1 \\ 1 & 0 \end{smallmatrix}\right)$ so $X|0\rangle = |1\rangle, X|1\rangle = |0\rangle$ – but of course has no protection against phase flip $Z = \left(\begin{smallmatrix} 1 & 0 \\ 0 & -1 \end{smallmatrix}\right)$ errors (known as dual errors because of their behavior on the dual basis states $|\pm\rangle \equiv \frac{1}{\sqrt{2}}(|0\rangle \pm |1\rangle)$). Codewords satisfy the parity checks 110 and 011 ensuring pairwise equality of bits – equivalently, the codespace is the simultaneous +1 eigenspace of the operators $Z \otimes Z \otimes I$ and $I \otimes Z \otimes Z$. To apply a logical bit flip on an encoded state, apply bit flips transversally, to each bit: $X_L = XXX$ (tensor signs implied). A logical phase flip can be applied by a phase flip to any single bit: $Z_L = ZII$ or IZI or IIZ. The CNOT gate is defined by $\text{CNOT}|a, b\rangle = |a, a \oplus b\rangle$ for $a, b \in \{0, 1\}$. Logical CNOT is just transversal CNOT.

The smallest distance-three quantum code – protecting against X and Z errors – uses five qubits, but we will present the slightly simpler distance-three Steane code on $n = 7$ qubits. The codespace is the simultaneous $+1$ eigenspace of the six operators

$$IIIZZZZ, IZZIIZZ, ZIZIZIZ,$$
$$IIIXXXX, IXXIIXX, XIXIXIX.$$

The first three "stabilizers" are exactly the classical $[7, 4, d = 3]$ Hamming code's parity checks, so the Steane code too has distance three against bit flip errors. The last three stabilizers are the same except in the dual basis, implying that the code separately has distance three against dual phase flip errors. With this code, encoded, or logical, X and Z operators are transversal X and Z operators, respectively: i.e., $X_L = X^{\otimes 7}, Z_L = Z^{\otimes 7}$. The CNOT gate (and other so-called "Clifford group" operations) can be applied transversally.

2.2   Error Model

For ease of exposition, we consider a very simple error model, defined to start only for so-called stabilizer operations. (These operations do not form a universal quantum gate set; we will extend to universality in Sect. 5.)

Definition 1 (Base error model). *Assume each CNOT gate fails with probability p, independently of the others and earlier or simultaneous measurement outcomes, resulting in one of the sixteen Pauli products $I \otimes I, I \otimes X, \ldots, Z \otimes Z$ being applied to the involved qubits after a perfect CNOT gate. Assume that single-qubit Clifford group operations are perfect, that single-qubit preparation and measurement is perfect, and that there is no memory error.*

Several of these assumptions are not essential. There being no memory error is essential only as it implies arbitrary control parallelism, which is essential in threshold schemes. The independence assumption of the CNOT failures can be relaxed as long as the as the conditional probability of failure (regardless of earlier or simultaneous events) remains at most p. The probabilistic nature of the failures is however essential for the proof in its current form. Probabilistic failures of other operations can be straightforwardly incorporated.

As is standard, additionally assume perfect classical control with feedback based on measurements. All the required classical computations are efficient. It is often assumed that classical computations are instantaneous, although this assumption doesn't matter with no memory error and even in general it can carefully be removed.

2.3   Error States

It is important for the analysis to be able to say whether or not a given bit is in error at a given time. This is not immediate for quantum errors, because different errors can have an equivalent effect; for example if we want $\frac{1}{\sqrt{2}}(|00\rangle + |11\rangle)$ but we get $\frac{1}{\sqrt{2}}(|01\rangle + |10\rangle)$, either bit could equally well be in X error.

To avoid ambiguities, X, Z or $Y = iXZ$ errors are tracked through the circuit from their introduction by the commutation rules of Fig. 3. Some errors may have trivial effect – e.g., $Z|0\rangle = |0\rangle$ – but we still record them (we do not reduce errors modulo the stabilizer). When we later extend the proof to a universal gate set, top-level logical Pauli errors can no longer be traced through the circuit, but we will make certain that these errors happen with vanishing probability.

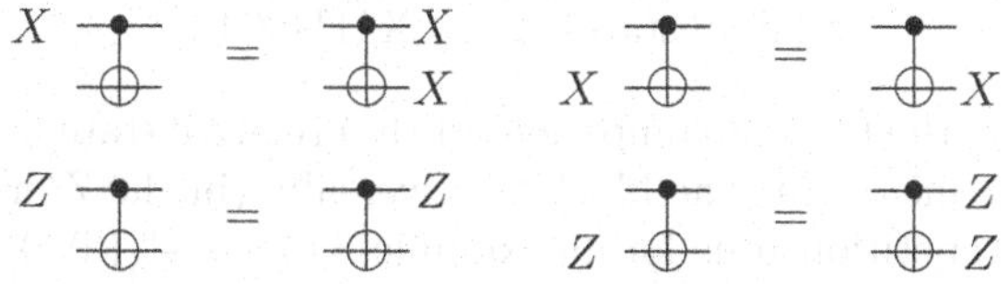

Fig. 3. Propagation of X and Z Pauli errors through a CNOT gate; X errors are copied forward and Z errors copied backward

Similarly, we want to say if a given code block is in error or not. Error-free (perfect), bottom-up decoding of a $block_k$ is defined recursively by first decoding its $subblocks_{k-1}$, to interpret their $states_{k-1}$, then decoding the block. Note that this recursive procedure is not the same as correcting to the closest codeword, but it is easier to analyze.

Definition 2 (State). *The $state_0$ of a qubit is either I, X, Y, or Z, depending on what we have tracked onto that bit. The $state_k$ of a $block_k$ is I, X, Y, or Z, determined by error-free decoding of the $states_{k-1}$ of its subblocks.*

The state of a block is determined by the states of its subblocks. We want to define the *relative* states of the subblocks, because a probability distribution over subblock errors is most naturally defined keeping in mind (i.e., relative to) the state of the enclosing block. If a block is in error, then necessarily some of its subblocks will be in error.

As a simple example, consider the classical three-bit repetition code. If the states of three bits are XII – the first bit is in error – then the block's state decodes to be I. The first bit is also in relative error. If the states of three bits are IXX, then the block's state decodes to be X. The first bit is said to be in relative error (although it is not in error). Making this definition precise, particularly in the quantum case, requires some care because different errors can be equivalent.

Definition 3 (Relative $syndrome_k$). *The relative $syndrome_k$ of a $block_k$ consist of the syndromes of the $n - 1$ code stabilizer generators on the $states_{k-1}$ of the $subblocks_{k-1}$.*

Definition 4 (Relative $state_{k-1}$). *The relative $states_{k-1}$ of $subblocks_{k-1}$ of a $block_k$ are given by the minimum weight error, counting Y errors as two, generating the relative $syndrome_k$ of the $block_k$.*

There is a unique minimum weight error, so this notion is well-defined, since every error syndrome can be achieved with at most one X and one Z error (or with one Y error).

For example if the first two subblocks are in X error, then the block's state is X error, with the third subblock in relative X error (since logical X is equivalent to XXXIIII). Here is another example with a state$_k$ of X:

$$\text{states}_{k-1} : \quad \text{I IIIIY X}$$
$$\text{relative states}_{k-1} : \quad \text{XIIIIZ I} \,.$$

Unlike its state, the (X or Z component of the) relative state of a subblock can be determined by measuring the block transversally (in the Z or X eigenbases). Again using the repetition code for an example, $|+\rangle_L = \frac{1}{\sqrt{2}}(|000\rangle + |111\rangle)$. Since $(XII)|+\rangle_L = (IXX)|+\rangle_L \propto |100\rangle + |011\rangle$, one can't measure the state of the first qubit. However, measuring in the 0/1 computational basis (Z eigenbasis) gives 100 or 011 with equal probabilities, telling us in either case that qubit 1 was in relative error (before the destructive measurement).

2.4   Logical Error Model

Definition 5 (Logical error model). *The implementation of a logical operation U_k on one or more blocks$_k$ is said to have had the correct logical effect if the following diagram commutes:*

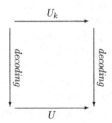

Here the vertical arrows indicate perfect recursive decoding of the involved blocks, and the lower horizontal arrow represents a perfect U on the decoded blocks.

U_k has had an incorrect logical effect if the same diagram commutes but with $P \circ U$ on the bottom arrow, where P is a Pauli operator or Pauli product on the involved blocks.

In our error model, with our implementations, every logical operation will have either the correct logical effect or an incorrect logical effect with some P probabilistically. For example, in error correction of a block, U is the identity. Error correction has the correct logical effect (no logical effect) if the state of the system following a perfect recursive decoding on the corrected blocks is the same as if we had just perfectly decoded the input blocks (and then applied the identity). In particular, this implies that the state$_k$ of the output is the same as the state$_k$ of the input, but, more than that, also logical entanglement is preserved.

2.5   Goodness and Wellness

A key problem in proving a threshold is in establishing the proper definitions for inductively controlling the errors. Once the correct definitions have been stated carefully, proving the relationships among them needed for a threshold result is fairly straightforward.

The classic proof of a threshold in this setting, due to Aharonov and Ben-Or [2], can be framed as relying on the definition:

Definition 6 (r-good$_k$). *A block$_k$ is r-good$_k$ (and not r-bad$_k$) if it has $\leq r$ subblocks$_{k-1}$ which are either in relative error or not r-good$_{k-1}$. A block$_0$ (single qubit) is r-good$_0$ if it is not in relative error.*

So in a 1-good block, we have control over errors in $n-1$ of the subblocks (they are 1-good themselves and not in relative error), but potentially no control over the state of one of the subblocks.

Definition 6 does not suffice for proving a threshold for a distance-three code because there is no room for errors in blocks which interact. We can't maintain the inductive assumption of each block being 1-good because as soon as two blocks interact, they will then each have two subblocks with uncontrolled errors (with a priori constant probability, not second-order probability as we desire). Aliferis et al. manage to use a similar definition, but change the method of induction proof to involve "overlapping steps." We instead will use a similar inductive proof to Aharonov and Ben-Or, but with different definitions to look at probability distributions of errors.

To give a threshold argument with a $d = 3$ code, we will use a definition for probability distributions over relative errors.

Definition 7 (well$_k$). *A block$_k$ is well$_k(p_1, \ldots, p_k)$ if, conditioned on its state and on the errors in all other blocks, it has at most one subblock either in relative error or not well$_{k-1}(p_1, \ldots, p_{k-1})$, and*

$$\mathbf{P}(\textit{such an uncontrolled subblock}) \leq p_k.$$

This definition conditions on the errors in all other blocks as a measure of independence. Note that we don't assert anything about the distribution of errors within a subblock in relative error. This is important because errors within relatively erroneous subblocks are typically less well controlled. (For example, consider a perfect 7-qubit block$_1$, and introduce bitwise independent errors. When the block as a whole is in error, one relative bit error is more likely than none, since two bit errors are more likely than three.)

For example, again using the three-qubit repetition code, the ensemble

$$III \text{ w/ prob. } 1 - p$$
$$IXX \quad \text{w/ prob. } p$$

is *not* well$_1(p)$. Even though the probability of a relative error is $\leq p$, conditioning on a logical state of X there is a relative error with probability one.

Definition 7 can be generalized to r-well$_k$, a requirement on probability distributions with up to r relative errors, but for a concatenated distance-three code threshold, $r = 1$ is sufficient and so I have omitted any prefix.

3   Fault-Tolerance for Stabilizer Operations

Our proof of fault-tolerance for stabilizer operations will rely on three indexed Claims A_k, B_k and C_k for, respectively, encoded ancilla preparation, error correction and encoded CNOT, at code-concatenation level k, with the following inductive dependencies:

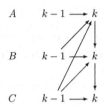

That is, a level-k encoded CNOT, or CNOT$_k$, will use CNOTs$_{k-1}$ and error corrections$_k$ – so the proof of Claim C_k will rely on Claims C_{k-1} and B_k. Error correction$_k$ will use corrections$_{k-1}$ and CNOTs$_{k-1}$, as well as ancillas$_k$. Finally, the proof of Claim A_k (ancilla$_k$ preparation) will rely on each of Claims A_{k-1}, B_{k-1} and C_{k-1}.

Each level-k operation will fail with probability A_k, B_k or C_k (failure parameters are italicized unlike the names of the claims to which they correspond). Failure parameters will drop quadratically at each level, giving a threshold as sketched in Sect. 1. That is,

$$\max\{A_k, B_k, C_k\} = O\big((\max\{A_{k-1}, B_{k-1}, C_{k-1}\})^2\big).$$

Splitting out separate error parameters in this way lets us easily track where errors are coming from, and lets us find the threshold bottlenecks for optimization. We will also define two wellness parameters a_k and b_k (since a CNOT$_k$ ends with error corrections$_k$, there is no need for a separate wellness parameter c_k).

Relaxing some of the assumptions in our error model would just require modifying these claims, and possibly adding new ones. For example, we have assumed perfect measurements, but faulty measurements would only require a fourth indexed claim dependent only on itself (a level-k measurement outcome is the decoding of n level-$(k-1)$ measurement outcomes).

Preparation of a single-qubit ancilla in state $|0\rangle_0 = |0\rangle$ or $|+\rangle_0 = |+\rangle \equiv \frac{1}{\sqrt{2}}(|0\rangle + |1\rangle)$ we assume to be perfect, so define $A_0 \equiv 0$. (Alternatively, set $A_0 > 0$ to remove this assumption.) For $k \geq 1$, we need:

Claim A_k (Ancilla$_k$). *Except with failure probability at most A_k, we can prepare a level k ancilla $|0\rangle_k$ which has a state$_k$ of I (no error) and is well$_k$ ($b_1, \ldots, b_{k-1}, a_k$).*

(The different parameters will be set explicitly within the proofs.)

Error correction$_k$ is only defined for $k \geq 1$ levels of encoding, so we may take $B_0 \equiv 0$.

Claim B_k (Correction$_k$). *With probability at least $1 - B_k$, the output is well$_k$ $(b_1, \ldots, b_k)$ and, if the input is well$_k$ $(b_1, \ldots, b_k)$, there is no logical effect.*

Additionally, if all but one of the input subblocks$_{k-1}$ are well$_{k-1}$ $(b_1, \ldots, b_{k-1})$ and not in relative error, then with probability at least $1 - B'_k$ there is no logical effect and the output is well$_k$ $(b_1, \ldots, b_k)$.

Note how powerful a successful correction$_k$ is. Even if there is no control whatsoever on the errors in the input block, the output is still well$_k$. This property is essential for getting errors fixed in a fully recursive manner, because it means that to fix an erroneous subblock$_{k-1}$, we need only apply a single-qubit correction transversally on that subblock and there is no need to worry about bit errors within that subblock.

CNOT$_0$ is simply a physical CNOT gate. $C_0 \equiv p$ is the probability of failure of a physical CNOT gate (for the oft-used simultaneous depolarization error model, p is $\frac{15}{16}$ times the depolarization rate).

Claim C_k (CNOT$_k$). *With probability at least $1 - C_k$, the output blocks$_k$ are well$_k$ $(b_1, \ldots, b_k)$ and, if the input blocks$_k$ are well$_k$ $(b_1, \ldots, b_k)$, then a logical CNOT, the correct logical effect, is applied.*

Proofs are in the full paper; there, for example, a rough Correction$_k$ failure upper bound is determined to be

$$
\begin{aligned}
B_k \equiv{} & 4A_k + \left(\tbinom{4n}{2} \right) C_{k-1}^2 + \left(\tbinom{4}{2} \right) a_k^2 + \left(\tbinom{2n}{2} \right) B_{k-1}^2 \\
& + b_k (4nC_{k-1} + 4a_k + 2nB_{k-1}) \\
& + 4nC_{k-1}(4a_k + 2nB_{k-1}) + 8na_k B_{k-1}.
\end{aligned}
$$

4 Threshold Lower Bounds

The results in Sect. 3 give the claimed positive constant threshold for stabilizer operations, because the error parameters each drop quadratically at each level of concatenation. Our goal was to complete a rigorous proof of a constant threshold, not to estimate the true threshold. Still, it is interesting how high a threshold these techniques give us. We have numerically iterated the equations of Sect. 3, taking $n = 7$. We found that the error rates converged to zero for $p = C_0 < 6.75 \times 10^{-6}$. Of course this is not a *proof* that the equations converge in this range, but the proof is clearly doable with careful numerics.

This threshold does not compare directly to the 4.18×10^{-5} rigorous threshold lower bound of Aliferis et al. [14] because their error model allows faults in single-qubit preparation and measurement as well as just CNOT gates. Also, our recursion equations were highly conservative, typically bounding the probability of a level k failure by the probability of any two level $k - 1$ failures. Aliferis et al., however, used a computer to count exactly which pairs of level $k - 1$ faults could cause a level k failure.

There are a number of optimizations that can be carried out to improve the threshold, and in the full paper we sketch some of these, increasing the threshold lower bound to 1.46×10^{-5}, and speculate on others. The bottlenecks in our equations in fact appear to be ancilla preparation subblock failure a_k and CNOT failure C_k.

These threshold estimates for stabilizer operations remain applicable for full quantum universality, discussed in Sect. 5. The techniques we use to gain universality can tolerate higher error rates than those for stabilizer operations, so achieving universality is not a bottleneck.

5   Fault-Tolerant Universality

To achieve universality, we use the technique of "magic states distillation" [16,17, 13]. Operating beneath the threshold for stabilizer operations, we can assume the error rate is arbitrarily small, and therefore condition on no stabilizer operation errors at all. Using perfect (encoded) stabilizer operations, we distill faulty copies of the "magic," single-qubit pure states $|H\rangle$ or $|T\rangle$ to perfect copies, which then give quantum universality. Details are given in the full paper.

6   Conclusion

We have proved:

Theorem 1. *For the error model specified in Def. 1, arbitrarily accurate, efficient, universal quantum computation is achievable, via a scheme based on concatenation of the $[\![7, 1, 3]\!]$ Steane code, as long as the error rate is beneath a positive constant threshold.*

It should be emphasized that preparation of reliable encoded states is a major threshold bottleneck.

A major open question in quantum fault-tolerance is to prove rigorously higher thresholds. As the highest threshold estimates rely heavily on postselection, a good understanding of the error probability distribution seems to be necessary to prove results about these schemes. Our arguments comprise a first, minor step in this direction, but to go further one needs to characterize the error distribution even within blocks which are in relative error. Dependencies between blocks are also a concern – for full independence is impossible to maintain (or even converge toward), and small deviations can grow exponentially when postselections discard large fractions of the probability mass.

There are many other open problems. How high can a threshold be proved for schemes not based on postselection? We are still orders of magnitude below the highest estimates. One promising approach is via a rigorous, computer-aided analysis of the lower levels, then plugging in to a conservative analysis once errors have dropped. Running Monte Carlo simulations, and fitting to failure and wellness parameters, would give threshold confidence intervals.

Efficiency of the scheme itself (not the analysis) is a major practical concern; even constants are very important. All constant-threshold fault-tolerance schemes should ultimately have the same efficiency in big-O notation – after startup levels, one can switch to the most efficient scheme. By fitting models to simulations, Steane has investigated concatenating on larger codes once a smaller code like the $[7, 1, 3]$ code has reduced the error rate sufficiently, and found it to be an apparently useful technique [18].

Can this threshold proof be extended to more general error models (not just probabilistic Pauli errors)?

The present work, with its improved analysis efficiency, merely sheds light on a fact which had long been assumed – a threshold for a distance-three code – but not proved. Hopefully a more solid foundation will help us as we try to address the more ambitious open questions in this field.

Research supported in part by NSF ITR Grant CCR-0121555, and ARO Grant DAAD 19-03-1-0082.

References

1. Peter W. Shor. In *Proc. Symp. Foundations of Computer Science (FOCS)*, 1996.
2. D. Aharonov and M. Ben-Or. In *Proc. Symp. Theory of Computing (STOC)*, 1997.
3. E. Knill, R. Laflamme, and W. H. Zurek. *Proc. R. Soc. Lond. A*, 454:365, 1998.
4. A. Yu Kitaev. *Russian Math. Surveys*, 52:1191–1249, 1997.
5. D. Gottesman. *J. Mod. Opt.*, 47:333–345, 2000.
6. B. M. Terhal and G. Burkard. *Phys. Rev. A*, 71:012336, 2005.
7. D. Gottesman. PhD thesis, California Institute of Technology, 1997.
8. J. Preskill. In *Introduction to Quantum Computation*, edited by H.-K. Lo, S. Popescu, and T. P. Spiller, World Scientific, Singapore, 1999.
9. C. Zalka. quant-ph/9612028, 1996.
10. B. W. Reichardt. quant-ph/0406025, 2004.
11. K. M. Svore, A. W. Cross, I. L. Chuang, and A. V. Aho. quant-ph/0508176, 2005.
12. K. M. Svore, B. M. Terhal, and D. P. DiVincenzo. *Phys. Rev. A*, 72:022317, 2005.
13. E. Knill. *Nature*, 434:39–44, 2005.
14. P. Aliferis, D. Gottesman, and J. Preskill. *Quant. Inf. Comput.*, 6:97–165, 2006.
15. E. Knill. quant-ph/0312190, 2003.
16. S. Bravyi and A. Kitaev. *Phys. Rev. A*, 71:022316, 2005.
17. B. W. Reichardt *Quant. Inf. Proc.*, 4:251–264, 2005.
18. A. M. Steane. *Phys. Rev. A*, 68:042322, 2003.

Lower Bounds on Matrix Rigidity Via a Quantum Argument

Ronald de Wolf*

CWI, Kruislaan 413, 1098 SJ, Amsterdam, The Netherlands
rdewolf@cwi.nl

Abstract. The *rigidity* of a matrix measures how many of its entries need to be changed in order to reduce its rank to some value. Good lower bounds on the rigidity of an explicit matrix would imply good lower bounds for arithmetic circuits as well as for communication complexity. Here we reprove the best known bounds on the rigidity of Hadamard matrices, due to Kashin and Razborov, using tools from *quantum* computing. Our proofs are somewhat simpler than earlier ones (at least for those familiar with quantum) and give slightly better constants. More importantly, they give a new approach to attack this longstanding open problem.

1 Introduction

1.1 Rigidity

Suppose we have some $n \times n$ matrix M whose rank we want to reduce. The *rigidity* of M measures the minimal number R of entries we need to change in order to reduce its rank to r. Formally:

$$R_M(r) = \min\{\text{weight}(M - \widetilde{M}) \mid \text{rank}(\widetilde{M}) \le r\},$$

where "weight" counts the number of non-zero entries. Here the rank could be taken over any field of interest; in this paper we consider the complex field. Roughly speaking, high rigidity means that M's rank is robust against changes: changes in few entries won't change the rank much.

Rigidity was defined by Valiant [1, Section 6] in the 1970s with a view to proving circuit lower bounds. In particular, he showed that an explicit $n \times n$ matrix M with $R_M(\varepsilon n) \ge n^{1+\delta}$ for $\varepsilon, \delta > 0$ would imply that log-depth arithmetic circuits that compute the linear map $M : \mathbb{R}^n \to \mathbb{R}^n$ need superlinear circuit size. Clearly, $R_M(r) \ge n - r$ for every full-rank matrix, since reducing the rank by 1 requires changing at least 1 entry. This bound is optimal for the identity matrix, but usually far from tight. Valiant showed that most matrices

* Supported by a Veni grant from the Netherlands Organization for Scientific Research (NWO) and also partially supported by the European Commission under the Integrated Projects RESQ, IST-2001-37559 and Qubit Applications (QAP) funded by the IST directorate as Contract Number 015848.

M. Bugliesi et al. (Eds.): ICALP 2006, Part I, LNCS 4051, pp. 62–71, 2006.

have rigidity $(n-r)^2$, but finding an *explicit* matrix with high rigidity has been open for decades.

A very natural and widely studied candidate for such a high-rigidity matrix are the Hadamard matrices. A Hadamard matrix is an orthogonal $n \times n$ matrix H with entries $+1$ and -1. Such matrices exist whenever n is a power of 2, but are conjectured to exist whenever n is a multiple of 4. Suppose we have a matrix $\widetilde{H}$ differing from H in R positions such that $\mathrm{rank}(\widetilde{H}) \leq r$. The goal in proving high rigidity is to lower bound R in terms of n and r. Alon [2] proved $R = \Omega(n^2/r^2)$, which was reproved by Lokam [3] using spectral methods. Kashin and Razborov [4] improved this to $R = \Omega(n^2/r)$. This is currently the best known for Hadamard matrices.

In view of the difficulty in proving strong lower bounds on rigidity proper, Lokam [3] also introduced a relaxed notion of rigidity. This limits the *size* of each change in entries to some parameter $\theta > 0$. Formally

$$R_M(r,\theta) = \min\{\mathrm{weight}(M - \widetilde{M}) \mid \mathrm{rank}(\widetilde{M}) \leq r, \parallel M - \widetilde{M} \parallel_\infty \leq \theta\},$$

where $\parallel \cdot \parallel_\infty$ measures the largest entry (in absolute value) of its argument. For Hadamard matrices, Lokam proved the bound $R_H(r,\theta) = \Omega(n^2/\theta)$ if $\theta \leq n/r$ and $R_H(r,\theta) = \Omega(n^2/\theta^2)$ if $\theta > r/n$. In particular, if entries can change at most by a constant then the rigidity is $\Omega(n^2)$. For the case $\theta > r/n$, Kashin and Razborov [4] improved the bound to $R_H(r,\theta) = \Omega(n^3/r\theta^2)$. Study of this relaxed notion of rigidity is further motivated by the fact that stronger lower bounds would separate the communication complexity versions of the classes PH and PSPACE [3].

Apart from Hadamard matrices, the rigidity of some other explicit matrices has been studied as well, sometimes giving slightly better bounds $R_M(r) = \Omega(n^2 \log(n/r)/r)$, for instance for Discrete Fourier Transform matrices [5,6,7]. Very recently, Lokam [8] showed a near-optimal rigidity bound $R_P(n/17) = \Omega(n^2)$ for the matrix P whose entries are the square roots of distinct primes, and proved an $\Omega(n^2/\log n)$ arithmetic circuit lower bound for the induced linear map $P : \mathbb{R}^n \to \mathbb{R}^n$. This matrix P, however, is "less explicit" than Hadamard matrices and the rigidity bound has no consequences for communication complexity because P is not a Boolean matrix. Moreover, the same circuit lower bound was already shown by Lickteig [9] (see also [10, Exercise 9.5]) without the use of rigidity.

1.2   Our Contribution

In this paper we give new proofs of the best known bounds on the rigidity of Hadamard matrices, both the standard rigidity and the relaxed one:

- if $r \leq n/2$, then $R_H(r) \geq \dfrac{n^2}{4r}$
- $R_H(r,\theta) \geq \dfrac{n^2(n-r)}{2\theta n + r(\theta^2 + 2\theta)}$

Our constant in the former bound is a bit better than the one of Kashin and Razborov [4] (their proof gives $n^2/256r$), while in the latter bound it is essentially the same. However, we feel our proof technique is more interesting than our precise result. As detailed in Section 2, the proof relies on interpreting an approximation $\widetilde{H}$ of the Hadamard matrix H as a *quantum communication system*, and then using quantum information theory bounds from [11] to relate the rank of $\widetilde{H}$ to the quality of its approximation.[1] Actually our bounds hold for all so-called *generalized* Hadamard matrices; these are the orthogonal matrices where all entries have the same magnitude. However, for definiteness we will state the results for Hadamard matrices only.

This paper fits in a recent but fast-growing line of research where results about *classical* objects are proved or reproved using *quantum* computational techniques. Other examples of this are lower bounds for locally decodable codes and private information retrieval [13,14], classical proof systems for lattice problems derived from earlier quantum proof systems [15,16], strong limitations on classical algorithms for local search [17] inspired by an earlier quantum computation proof, a proof that the complexity class PP is closed under intersection [18], formula size lower bounds from quantum lower bounds [19], and a new approach to proving lower bounds for classical circuit depth using quantum communication complexity [20].

It should be noted that the use of quantum computing is not strictly necessary for either of our results. The first is proved in two steps: (1) using the quantum approach we show that every $a \times b$ submatrix of H has rank at least ab/n and (2) using a non-quantum argument we show that an approximation $\widetilde{H}$ with small R contains a large submatrix of H and hence by (1) must have high rank. The result of (1) was already proved by Lokam [3, Corollary 2.2] using spectral analysis, so one may obtain the same result classically using Lokam's proof for (1) and our argument for (2). Either way, we feel the proof is significantly simpler than that of Kashin and Razborov [4], who show that a random $a \times a$ submatrix of H has rank $\Omega(a)$ with high probability. In contrast, the quantum aspects of our proof for the bound on $R_H(r, \theta)$ cannot easily be replaced by a classical argument, but that proof is not significantly simpler than the one of Kashin and Razborov (which uses the Hoffman-Wielandt inequality) and the constant is essentially the same.

Despite this, we feel our quantum approach has merit for two reasons. First, it unifies the two results, both of which are now proved from the same quantum information theoretic idea. And second, using quantum computational tools gives a whole new perspective on the rigidity issue, and might just be the new approach we need to solve this longstanding open problem. Our hope is that these techniques not only reprove the best known bounds, but will also push them further. In Section 5 we discuss two non-quantum approaches to the rigidity issue that followed a first version of the present paper, and point out ways in which our approach is stronger.

---

[1] The connection between the Hadamard matrix and quantum communication was also exploited in the lower bound for the communication complexity of inner product by Cleve et al. [12].

2   Relation to Quantum Communication

Very briefly, an r-dimensional *quantum state* is a unit vector of complex amplitudes, written $|\phi\rangle = \sum_{i=1}^{r} \alpha_i |i\rangle \in \mathbb{C}^r$. Here $|i\rangle$ is the r-dimensional vector that has a 1 in its ith coordinate and 0s elsewhere. The inner product between $|\phi\rangle$ and $|\psi\rangle = \sum_{i=1}^{r} \beta_i |i\rangle$ is $\langle \phi|\psi\rangle = \sum_i \alpha_i^* \beta_i$. A *measurement* is described by a set of positive semidefinite operators $\{E_i\}$ that sum to identity. If this measurement is applied to some state $|\phi\rangle$, the probability of obtaining outcome i is given by $\langle \phi|E_i|\phi\rangle$. If $\{|v_i\rangle\}$ is an orthonormal basis, then a measurement in this basis corresponds to the projectors $E_i = |v_i\rangle\langle v_i|$. In this case the probability of outcome i is $|\langle v_i|\phi\rangle|^2$. We refer to [21] for more details about quantum computing. We use $\| E \|$ to denote the operator norm (largest singular value) of a matrix E, and $\text{Tr}(E)$ for its trace (sum of diagonal entries).

Our proofs are instantiations of the following general idea, which relates (approximations of) the Hadamard matrix to quantum communication. Let H be an $n \times n$ Hadamard matrix. Its rows, after normalization by a factor $1/\sqrt{n}$, form an orthonormal set known as the *Hadamard basis*. If Alice sends Bob the n-dimensional quantum state $|H_i\rangle$ corresponding to the normalized ith row of H, and Bob measures the received state in the Hadamard basis, then he learns i with probability 1.

Now suppose that instead of H we have some rank-r $n \times n$ matrix $\widetilde{H}$ that approximates H in some way or other. Then we can still use the quantum states $|\widetilde{H}_i\rangle$ corresponding to its normalized rows for quantum communication. Alice now sends the state $|\widetilde{H}_i\rangle$. Crucially, she can do this by means of an r-dimensional quantum state. Let $|v_1\rangle, \ldots, |v_r\rangle$ be an orthonormal basis for the row space of $\widetilde{H}$. In order to send $|\widetilde{H}_i\rangle = \sum_{j=1}^{r} \alpha_j |v_j\rangle$, Alice sends $\sum_{j=1}^{r} \alpha_j |j\rangle$ and Bob applies the unitary map $|j\rangle \mapsto |v_j\rangle$ to obtain $|\widetilde{H}_i\rangle$. He measures this in the Hadamard basis, and now his probability of getting the correct outcome i is

$$p_i = |\langle H_i|\widetilde{H}_i\rangle|^2.$$

The "quality" of these p_i's correlates with the "quality" of $\widetilde{H}$: the closer the ith row of $\widetilde{H}$ is to the ith row of H, the closer p_i will be to 1.

Accordingly, Alice can communicate a random element $i \in [n]$ via an r-dimensional quantum system, with average success probability $p = \sum_{i=1}^{n} p_i/n$. But now we can apply the following upper bound on the average success probability, due to Nayak [11, Theorem 2.4.2]:[2]

$$p \leq \frac{r}{n}.$$

Intuitively, the "quality" of the approximation $\widetilde{H}$, as measured by the average success probability p, gives a lower bound on the required rank r of $\widetilde{H}$. In the next sections we instantiate this idea in two different ways to get our two bounds.

---

[2] NB: this is not the well-known and quite non-trivial random access code lower bound from the same paper, but a much simpler statement about average decoding probabilities.

We end this section with a simple proof of Nayak's bound due to Oded Regev. In general, let $|\phi_1\rangle, \ldots, |\phi_n\rangle$ be the r-dimensional states encoding $1, \ldots, n$, respectively, and $E_1, \ldots, E_n$ be the measurement operators applied for decoding. Then, using that the eigenvalues of E_i are nonnegative reals and that the trace of a matrix is the sum of its eigenvalues:

$$p_i = \langle \phi_i | E_i | \phi_i \rangle \leq \| E_i \| \leq \text{Tr}(E_i)$$

and

$$\sum_{i=1}^{n} p_i \leq \sum_{i=1}^{n} \text{Tr}(E_i) = \text{Tr}\left(\sum_{i=1}^{n} E_i\right) = \text{Tr}(I) = r.$$

3   Bound on $R_H(r)$

The next theorem was proved by Lokam [3, Corollary 2.7] using some spectral analysis. We reprove it here using a quantum argument.

Theorem 1 (Lokam). *Every $a \times b$ submatrix A of H has rank $r \geq ab/n$.*

Proof. Obtain rank-r matrix $\widetilde{H}$ from H by setting all entries outside of A to 0. Consider the a quantum states $|\widetilde{H}_i\rangle$ corresponding to the nonempty rows; they have normalization factor $1/\sqrt{b}$. For each such i, Bob's success probability is

$$p_i = |\langle H_i | \widetilde{H}_i \rangle|^2 = \left| \frac{b}{\sqrt{bn}} \right|^2 = \frac{b}{n}.$$

But we're communicating one of a possibilities using r dimensions, so Nayak's bound implies

$$\frac{1}{n} \sum_{i=1}^{n} p_i = p \leq \frac{r}{a}.$$

Combining both bounds gives the theorem.   □

Surprisingly, Lokam's result allows us quite easily to derive Kashin and Razborov's [4] bound on rigidity, which is significantly stronger than Lokam's (and Alon's). We also obtain a slightly better constant than [4]: their proof gives $1/256$ instead of our $1/4$. This is the best bound known on the rigidity of Hadamard matrices.

Theorem 2. *If $r \leq n/2$, then $R_H(r) \geq n^2/4r$.*

Proof. Consider some rank-r matrix $\widetilde{H}$ with at most $R = R_H(r)$ "errors" compared to H. By averaging, there exists a set of $a = 2r$ rows of $\widetilde{H}$ with at most aR/n errors. Now consider the submatrix A of $\widetilde{H}$ consisting of those a rows and the $b \geq n - aR/n$ columns that have no errors in those a rows. If $b = 0$

then $R \geq n^2/2r$ and we are done, so we can assume A is nonempty. This A is errorfree, hence a submatrix of H itself, and the previous theorem implies

$$r = \operatorname{rank}(\widetilde{H}) \geq \operatorname{rank}(A) \geq \frac{ab}{n} \geq \frac{a(n - aR/n)}{n}.$$

Rearranging gives the theorem. □

The condition $r \leq n/2$ is important here. If H is symmetric then its eigenvalues are all $\pm\sqrt{n}$ (because $H^T H = nI$), so we can reduce the rank to $n/2$ by adding or subtracting the diagonal matrix $\sqrt{n}I$. This shows that $R_H(n/2) \leq n$.

4   Bound on $R_H(r, \theta)$

We now consider the case where the maximal change in entries of H is bounded by θ.

Theorem 3. $R_H(r, \theta) \geq \dfrac{n^2(n - r)}{2\theta n + r(\theta^2 + 2\theta)}.$

Proof. Consider some rank-r matrix $\widetilde{H}$ with at most $R = R_H(r, \theta)$ errors, and $\| H - \widetilde{H} \|_\infty \leq \theta$. As before, define the quantum states corresponding to its rows:

$$|\widetilde{H}_i\rangle = c_i \sum_{j=1}^{n} \widetilde{H}_{ij}|j\rangle,$$

where $c_i = 1/\sqrt{\sum_j \widetilde{H}_{ij}^2}$ is a normalizing constant. Note that $\sum_j \widetilde{H}_{ij}^2 \leq (n - \Delta(H_i, \widetilde{H}_i)) + \Delta(H_i, \widetilde{H}_i)(1+\theta)^2 = n + \Delta(H_i, \widetilde{H}_i)(\theta^2 + 2\theta)$, where $\Delta(\cdot, \cdot)$ measures Hamming distance. Bob's success probability p_i is now

$$\begin{aligned}
p_i &= |\langle H_i|\widetilde{H}_i\rangle|^2 \\
&\geq \frac{c_i^2}{n}(n - \theta\Delta(H_i, \widetilde{H}_i))^2 \\
&\geq c_i^2(n - 2\theta\Delta(H_i, \widetilde{H}_i)) \\
&\geq \frac{n - 2\theta\Delta(H_i, \widetilde{H}_i)}{n + \Delta(H_i, \widetilde{H}_i)(\theta^2 + 2\theta)}.
\end{aligned}$$

Since p_i is a convex function of Hamming distance and the average $\Delta(H_i, \widetilde{H}_i)$ is R/n, we also get a lower bound for the average success probability:

$$p \geq \frac{n - 2\theta R/n}{n + R(\theta^2 + 2\theta)/n}.$$

Nayak's bound implies $p \leq r/n$. Rearranging gives the theorem. □

For $\theta \geq n/r$ we obtain the second result of Kashin and Razborov [4]:

$$R_H(r, \theta) = \Omega(n^2(n - r)/r\theta^2).$$

If $\theta \leq n/r$ we get an earlier result of Lokam [3]:

$$R_H(r, \theta) = \Omega(n(n - r)/\theta).$$

5   Non-quantum Proofs

Of course, quantum mechanical arguments like the above can always be stripped of their quantum aspects by translating to the underlying linear algebra language, thus giving a non-quantum proof. In this section we discuss the relation between our proof and two recent non-quantum approaches to rigidity. Both are significantly simpler than the Kashin-Razborov proofs [4].

5.1   Midrijanis's Proof

After reading a first version of this paper, Midrijanis [22] published a very simple argument giving the same bound $R_H(r) \geq n^2/4r$ for the special class of Hadamard matrices defined by k-fold tensor product of the basic 2×2 matrix (so $n = 2^k$)

$$H_{2^k} = \begin{pmatrix} 1 & 1 \\ 1 & -1 \end{pmatrix}^{\otimes k}.$$

Let $r \leq n/2$ be a power of 2. This H_{2^k} consists of $(n/2r)^2$ disjoint copies of $\pm H_{2r}$ and each of those has full rank $2r$. Each of those copies needs at least r errors to reduce its rank to r, so we need at least $(n/2r)^2 r = n^2/4r$ errors to reduce the rank of H_{2^k} to r. Notice, however, that this approach only obtains bounds for the case where H is defined in the above manner[3].

5.2   The Referee's Proof

An anonymous referee of an earlier version of this paper suggested that the quantum aspects were essentially redundant and could be replaced by the following spectral argument. Suppose for simplicity that the Hadamard matrix H and its rank-r approximation $\widetilde{H}$ have normalized rows, and as before let $|H_i\rangle$ and $|\widetilde{H}_i\rangle$ denote their rows. The Frobenius norm of a matrix A is $\| A \|_F = \sqrt{\sum_{i,j} A_{ij}^2}$. We can factor $\widetilde{H}^* = DE$, where D is an $n \times r$ matrix with orthonormal columns and E is an $r \times n$ matrix with $\| E \|_F = \| \widetilde{H} \|_F$. Using the Cauchy-Schwarz inequality, we bound

$$\sum_{i=1}^{n} \langle H_i | \widetilde{H}_i \rangle = \mathrm{Tr}(H\widetilde{H}^*) = \mathrm{Tr}(HDE)$$

$$\leq \| HD \|_F \cdot \| E \|_F$$

$$= \| D \|_F \cdot \| E \|_F$$

$$= \sqrt{r} \cdot \| \widetilde{H} \|_F.$$

This approach is quite interesting. It gives the same bounds when applied to the two cases of this paper (where $\sum_i \langle H_i | \widetilde{H}_i \rangle$ and $\| \widetilde{H} \|_F$ are easy to bound), with less effort than the Kashin-Razborov proofs [4]. However, it is not an unrolling

---

[3] It's not clear how new this proof is, see the comments at Lance Fortnow's weblog http://weblog.fortnow.com/2005/07/matrix-rigidity.html

of the quantum proof, since the latter upper bounds the sum of *squares* of the inner products:

$$\sum_{i=1}^{n} |\langle H_i | \widetilde{H}_i \rangle|^2 \leq r.$$

An upper bound on the sum of squares implies a bound on the sum of inner products via the Cauchy-Schwarz inequality, but not vice versa. Thus, even though the two bounds yield the same results in the two cases treated here, the quantum approach is potentially stronger than the referee's.

6   Discussion

As mentioned in the introduction, this paper is the next in a recent line of papers about classical theorems with quantum proofs. So far, these results are somewhat *ad hoc* and it is hard to see what unifies them other than the use of some quantum mechanical apparatus. A "quantum method" in analogy to the "probabilistic method" [23] is not yet in sight but would be a very intriguing possibility. Using quantum methods as a mathematical proof tool shows the usefulness of the study of quantum computers, quantum communication protocols, etc., irrespective of whether a large quantum computer will ever be built in the lab. Using the methods introduced here to prove stronger rigidity lower bounds would enhance this further.

Most lower bounds proofs for the rigidity of a matrix M in the literature (including ours) work in two steps: (1) show that all or most submatrices of M have fairly large rank, and (2) show that if the number of errors R is small, there is some (or many) big submatrix of $\widetilde{M}$ that is uncorrupted. Such an uncorrupted submatrix of $\widetilde{M}$ is a submatrix of M and hence by (1) will have fairly large rank. As Lokam [7] observes, this approach will not yield much stronger bounds on rigidity than we already have: it is easy to show that a random set of $R = O(\frac{\max(a,b)n^2}{ab} \log(n/\max(a,b)))$ positions hits every $a \times b$ submatrix of an $n \times n$ matrix. Lokam's [8] recent $\Omega(n^2)$ rigidity bound for a matrix consisting of the roots of distinct primes indeed does something quite different, but unfortunately this technique will not work for matrices over $\{+1, -1\}$ like Hadamard matrices.

To end this paper, let me describe two vague directions for improvements. First, the approach mentioned above finds a submatrix of rank at least r in $\widetilde{M}$ and concludes from this that $\widetilde{M}$ has rank at least r. However, the approach usually shows that *most* submatrices of $\widetilde{M}$ of a certain size have rank at least r. If we can somehow piece these lower bounds for many submatrices together, we could get a higher rank bound for the matrix $\widetilde{M}$ as a whole and hence obtain stronger lower bounds on rigidity.

A second idea that might give a stronger lower bound for $R_H(r)$ is the following. We used the result that every $a \times b$ submatrix of H has rank at least ab/n. This bound is tight for some submatrices but too weak for others. We conjecture (or rather, hope) that submatrices for which this bound *is* more or less tight, are very "redundant" in the sense that each or most of its rows are spanned by many sets of rows of the submatrix. Such a submatrix can tolerate a number of errors without losing much of its rank, so then we don't need to find an uncorrupted submatrix of $\tilde{H}$ (as in the current proof), but could settle for a submatrix with little corruption.

Acknowledgments

Thanks to Satya Lokam for sending me a draft of [8] and for some helpful explanations, to Oded Regev and Gatis Midrijanis for useful discussions, to Falk Unger for proofreading, and to the anonymous STACS'06 referee for the proof in Section 5.2.

References

1. Valiant, L.: Graph-theoretic arguments in low-level complexity. In: Proceedings of 6th MFCS. Volume 53 of Lecture Notes in Computer Science., Springer (1977) 162–176
2. Alon, N.: On the rigidity of an Hadamard matrix. Manuscript. His proof may be found in [24, Section 15.1.2] (1990)
3. Lokam, S.: Spectral methods for matrix rigidity with applications to size-depth trade-offs and communication complexity. Journal of Computer and Systems Sciences **63**(3) (2001) 449–473 Earlier version in FOCS'95.
4. Kashin, B., Razborov, A.: Improved lower bounds on the rigidity of Hadamard matrices. Matematicheskie Zametki **63**(4) (1998) 535–540 In Russian. English translation available at Razborov's homepage.
5. Friedman, J.: A note on matrix rigidity. Combinatorica **13**(2) (1993) 235–239
6. Shokrollahi, M.A., Spielman, D., Stemann, V.: A remark on matrix rigidity. Information Processing Letters **64**(6) (1997) 283–285
7. Lokam, S.: On the rigidity of Vandermonde matrices. Theoretical Computer Science **237**(1–2) (2000) 477–483
8. Lokam, S.: A quadratic lower bound on rigidity. (April 2005) Manuscript.
9. Lickteig, T.: Ein elementarer Beweis für eine geometrische Gradschanke für die Zahl der Operationen bei der Berechnung von Polynomen. Master's thesis, Diplomarbeit, Univ. Konstanz (1980)
10. Bürgisser, P., Clausen, M., Shokrollahi, M.A.: Algebraic Complexity Theory. Volume 315 of Grundlehren der mathematischen Wissenschaften. Springer (1997)
11. Nayak, A.: Optimal lower bounds for quantum automata and random access codes. In: Proceedings of 40th IEEE FOCS. (1999) 369–376 quant-ph/9904093.
12. Cleve, R., Dam, W. van, Nielsen, M., Tapp, A.: Quantum entanglement and the communication complexity of the inner product function. In: Proceedings of 1st NASA QCQC conference. Volume 1509 of Lecture Notes in Computer Science., Springer (1998) 61–74 quant-ph/9708019.
13. Kerenidis, I., Wolf, R. de: Exponential lower bound for 2-query locally decodable codes via a quantum argument. In: Proceedings of 35th ACM STOC. (2003) 106–115 quant-ph/0208062.
14. Wehner, S., Wolf, R. de: Improved lower bounds for locally decodable codes and private information retrieval. In: Proceedings of 32nd ICALP. Volume 3580 of Lecture Notes in Computer Science. (2005) 1424–1436 quant-ph/0403140.

15. Aharonov, D., Regev, O.: A lattice problem in quantum NP. In: Proceedings of 44th IEEE FOCS. (2003) 210–219 quant-ph/0307220.
16. Aharonov, D., Regev, O.: Lattice problems in NP∩coNP. In: Proceedings of 45th IEEE FOCS. (2004) 362–371
17. Aaronson, S.: Lower bounds for local search by quantum arguments. In: Proceedings of 35th ACM STOC. (2003) 465–474 quant-ph/0307149.
18. Aaronson, S.: Quantum computing, postselection, and probabilistic polynomial-time. quant-ph/0412187 (23 Dec 2004)
19. Laplante, S., Lee, T., Szegedy, M.: The quantum adversary method and classical formula size lower bounds. In: Proceedings of 20th IEEE Conference on Computational Complexity. (2005) quant-ph/0501057.
20. Kerenidis, I.: Quantum multiparty communication complexity and circuit lower bounds (Apr 12, 2005) quant-ph/0504087.
21. Nielsen, M.A., Chuang, I.L.: Quantum Computation and Quantum Information. Cambridge University Press (2000)
22. Midrijanis, G.: Three lines proof of the lower bound for the matrix rigidity. cs.CC/0506081 (20 Jun 2005)
23. Alon, N., Spencer, J.H.: The Probabilistic Method. second edn. Wiley-Interscience (2000)
24. Jukna, S.: Extremal Combinatorics. EATCS Series. Springer (2001)

Self-testing of Quantum Circuits*

Frédéric Magniez[1], Dominic Mayers[2], Michele Mosca[3,4], and Harold Ollivier[4]

[1] CNRS–LRI, University Paris-Sud, France
[2] Institute for Quantum Information, Caltech, USA
[3] Institute for Quantum Computing, University of Waterloo, Canada
[4] Perimeter Institute for Theoretical Physics, Waterloo, Canada

Abstract. We prove that a quantum circuit together with measurement apparatuses and EPR sources can be *self-tested*, i.e. fully verified without any reference to some trusted set of quantum devices.

To achieve our goal we define the notions of simulation and equivalence. Using these two concepts, we construct sets of simulation conditions which imply that the physical device of interest is equivalent to the one it is supposed to implement. Another benefit of our formalism is that our statements can be proved to be robust.

Finally, we design a test for quantum circuits whose complexity is polynomial in the number of gates and qubits, and the required precision.

1 Introduction

The purpose of this paper is to address the issue of deciding whether an implementation of a quantum circuit follows its specification. The precise setting in which we ask this question is that of *self-testing*. In such setting, the sources, the gates as well as the measurement apparatuses that are used, are considered as black-boxes. Moreover, none of them will be trusted to implement the quantum operator it is supposed to implement. As a consequence, the tests cannot make reference to another set of trusted and already characterized quantum devices. Such notion of self-testing follows quite closely the one defined initially for classical programs [1,2], and is indeed based on its extension to quantum devices [3,4] and to quantum testers of logical properties [5,6].

The task of self-testing a set of quantum devices has been the focus of attention of two papers [3,4], each of which considers a very particular set of assumptions. The work by Mayers and Yao [3] focuses on testing entangled EPR states shared between two distinguishable locations, A and B. The main assumptions they exploit are (1) locality, in the sense that the measurements at A commute with the measurements at B; and that (2) one can perform independent repetitions of the same experiments, in order to gather statistics (i.e., apparatuses have no memory of previous runs of the experiments). However, they do not assess the robustness of their results, i.e. if a state satisfies only approximately the required statistics then it is still close to an EPR state. Robustness is nonetheless an important property very much worth studying for practical reasons: first, one

---

* Supported by QAIP, AlgoQP, NSERC, ARDA, ORDCF, CFI, CIAR, ResQuant.

can never learn any statistics with infinite precision by sampling only; second, by their very nature, physical implementations are only approximate.

The work of Van Dam, Magniez, Mosca and Santha [4] focuses instead on testing gates. They make a number of assumptions, in addition to the above, (3) the ability to use the same gate in different places of the same experiment; (4) the ability to prepare and measure '0' and '1'; and (5) the dimension of the physical qubits (i.e., 2-level systems). Of these assumptions, the last one is certainly the most unrealistic one, but also the most crucial one. Relaxing it allows for "conspiracies" that can spoof the test.

Our work improves upon the results of [3] by making them robust. We also improve upon the paper [4] by removing the need for assumptions (3), (4) and (5). Let us detail the assumptions that we make. We assume that, (H1) the physical system we are working with consists of several identifiable sub-systems; that (H2) two subsystems interact only if we are applying a gate that has both those subsystems as input; (H3) each gate will behave identically in each experiment it is used in; and (H4) classical computation and control can be trusted.

First (Section 2), we define a precise mathematical framework for testing quantum devices. This is done by introducing the concept of *simulation* which amounts to producing the expected probability distribution for the outcomes of the measurements that are performed at the end of the computation. This alone will not be sufficient to propose efficient tests of quantum circuits. For this purpose, we introduce the concept of *equivalence* which relates the action of the devices on physical quantum systems used in the implementation, to the action of the unitary operators specifying the circuit on logical qubits.

Second (Section 3), we characterize unitary gates and circuits in terms of simulation. We explain how simulation implies equivalence, and how by composing equivalences one can derive the correctness of a physical implementation of a circuit. The main tool used in this section is the Mayers-Yao test of an EPR pair, which provides the most simple example in which simulation implies equivalence. We will then show that this test can be generalized and yields trusted input states to be used in conjunction with self-testable quantum circuits.

Last (Section 4), we prove the robustness of our characterization. In particular, we show that the EPR test of [3] can tolerate ε inaccuracy in the statistics and still yields states and measurements that are within $O(\varepsilon^{1/4})$ of their specification. Using the concepts of simulation and equivalence, such proofs are not so difficult although the robustness of the EPR test had been left open. The crucial point is to realize that the robustness of our characterization needs only to be stated on a rather small subspace in order for it to be of practical interest.

The important consequence of our study is the design of an efficient self-tester (Section 5) for quantum circuits with some specific input. Contrary to tomography which requires trusted measurement devices and an exponential number of statistics to be checked, our test has a complexity linear in the number of qubits and gates involved in the circuit, and polynomial in the required precision. We describe our tester in a general context and illustrate it with an example.

2   Testing Concepts

Notation. Set $\mathcal{H}_N = \mathbb{C}^N$, whose computational basis is $(|i\rangle)_{0 \leq i < N}$. For $\alpha \in \mathbb{R}$, let $|\alpha\rangle = \cos\alpha|0\rangle + \sin\alpha|1\rangle$. In particular $|\frac{\pi}{2}\rangle = |1\rangle$. Denote by $|\phi^+\rangle$ the EPR state $\frac{1}{\sqrt{2}}(|0\rangle \otimes |0\rangle + |1\rangle \otimes |1\rangle)$, and by $|\Phi_n^+\rangle$ the tensor product of n EPR states: $|\Phi_n^+\rangle = \frac{1}{\sqrt{2^n}}\sum_{x \in \{0,1\}^n}|x\rangle \otimes |x\rangle$. Let $\mathcal{U}(H)$ be the set of unitary transformations on H, and $\mathcal{I}(H, H')$ the set of isomorphisms between H and H' which preserve the inner product. When $H = \mathcal{H}_N$ we let $\mathcal{U}(N) = \mathcal{U}(\mathcal{H}_N)$. In case of transformations over real spaces, we use the notations $\mathcal{O}(N)$ and $\mathcal{O}(H)$ instead of $\mathcal{U}(N)$ and $\mathcal{U}(H)$. For transformations M and M' on H, and $S \subseteq H$, the notation $M =_S M'$ means that the equality holds when restricted to S. When M is a linear transformation on A, we extend M on any tensor product $A \otimes B$ by $M \otimes \mathrm{Id}_B$.

Simulation. Two states *simulate* one another when they produce the same probability distributions of outcomes for two families of projectors. Here, the projectors are used in the same way measurement devices are used in a laboratory: they act as reference systems against which systems are tested.

More precisely, we are given a family of projectors $(P^w)_{w \in \mathcal{W}}$ acting on a physical space H and a state $|\psi\rangle \in H$, whose purpose is to implement some given and fixed projectors $|w\rangle\langle w|$ on the logical space $\mathcal{H}_N$ and a state $|\phi\rangle \in \mathcal{H}_N$.

Definition 1. *A quantum state $|\psi\rangle \in H$ simulates the quantum state $|\phi\rangle \in \mathcal{H}_N$ (with respect to $(P^w)_{w \in \mathcal{W}}$), if $\|P^w|\psi\rangle\|^2 = |\langle w|\phi\rangle|^2$, for every $w \in \mathcal{W}$.*

The notion of simulation can be rephrased for the whole space H. Assume we are given a family of states $(|\psi_i\rangle)_i$ of H that respectively simulate the basis states $(|i\rangle)_i$ (with respect to fixed set of projectors $(P^w)_{w \in \mathcal{W}}$). Then we say that H *simulates* $\mathcal{H}_N$.

We now extend the simulation notion to gates.

Definition 2. *Assume that H simulates $\mathcal{H}_N$: $(|\psi_i\rangle)_i$ simulates $(|i\rangle)_i$ (with respect to $(P^w)_{w \in \mathcal{W}}$). A unitary transformation $G \in \mathcal{U}(H)$ simulates the unitary transformation $T \in \mathcal{U}(\mathcal{H}_N)$ (with respect to $(|\psi_i\rangle)_i$ and $(P^w)_{w \in \mathcal{W}}$), if $G|\psi_i\rangle$ simulates $T|i\rangle$ (with respect to $(P^w)_{w \in \mathcal{W}}$), for every i.*

Equivalence. Testing a circuit as a single unitary operation is not an option. Indeed, this would require checking a simulation condition by sampling a probability distribution with a number of realizations exponential in the number of qubits involved in the circuit. Rather, we would like to test each of the physical devices that constitute the circuit individually in order to conclude that their composition simulates the whole circuit. Unfortunately, statements about simulation cannot be composed. This is the reason for the introduction of another concept, the concept of equivalence.

The equivalence notion we introduce is motivated by results of Mayers and Yao [3], but was not explicitly stated in their work. It is a mathematical notion based on the possibility of transferring states which lie within a physical space into a logical system.

For a Hilbert space H, that will describe our physical system, we set a *logical space* $H_c = \mathcal{H}_N$ for some given integer N, and define $\bar{H} = H_c \otimes H$. We now identify H with $|0\rangle \otimes H$, and consider H as a subspace of $\bar{H}$.

First, we define the equivalence between a subspace of H and the logical system H_c with respect to a set of projectors. As for the notion of simulation, these projectors act as reference systems.

Definition 3. *Let $U \in \mathcal{U}(\bar{H})$. A subspace S of H is U-equivalent to H_c (with respect to $(P^w)_{w \in \mathcal{W}}$), if for every $w \in \mathcal{W}$, $P^w =_S U^\dagger(|w\rangle\langle w| \otimes \mathrm{Id}_H)U$.*

The above definition is equivalent to the commutative diagram:

$$\begin{array}{ccc} S & \xrightarrow{P^w} & S \\ U \downarrow & & \uparrow U^\dagger \\ \bar{H} & \xrightarrow{|w\rangle\langle w| \otimes \mathrm{Id}_H} & \bar{H} \end{array}$$

. Intuitively, the unitary transformation U ensures that the correspondence between the physical system H and the logical system H_c is well defined on S. As a consequence, the projectors P^w satisfy $P^w(S) \subseteq S$.

Define now the U-equivalence for states and gates that implies the simulation.

Definition 4. *Let S be a subspace of H. A state $|\psi\rangle \in S$ is U-equivalent to $|\phi\rangle \in H_c$ on S (with respect to $(P^w)_{w \in \mathcal{W}}$), if*
1. *S is U-equivalent to H_c,*
2. *$|\psi\rangle = U^\dagger(|\phi\rangle \otimes |\chi\rangle)$, for some $|\chi\rangle \in H$.*

Definition 5. *Let S be a subspace of H. A unitary transformation $G \in \mathcal{U}(H)$ is (U, V)-equivalent to $T \in \mathcal{U}(H_c)$ on S (with respect to $(P^w)_{w \in \mathcal{W}}$), if*
1. *S is U-equivalent to H_c,*
2. *$S' = G(S)$ is V-equivalent to H_c,*
3. *$G =_S V^\dagger(T \otimes W)U$, for some $W \in \mathcal{U}(H)$.*

This equivalence can be summarized by the following commutative diagram:

$$\begin{array}{ccccccc} S & \xleftarrow{P^w} & S & \xrightarrow{G} & S' & \xrightarrow{P^w} & S' \\ U^\dagger \uparrow & & \downarrow U & & V^\dagger \uparrow\downarrow V & & \uparrow V^\dagger \\ \bar{H} & \xleftarrow{|w\rangle\langle w| \otimes \mathrm{Id}_H} & \bar{H} & \xrightarrow{T \otimes W} & \bar{H} & \xrightarrow{|w\rangle\langle w| \otimes \mathrm{Id}_H} & \bar{H} \end{array}$$

.

Proposition 1. *Assume that $\{0, 1, \ldots, N-1\} \subseteq \mathcal{W}$. Let $(|\psi_i\rangle)_{0 \le i < N}$ be a unit vector of $P^i(S)$. If $G \in \mathcal{U}(H)$ is equivalent to $T \in \mathcal{U}(H_c)$ on S, then G simulates T with respect to $(|\psi_i\rangle)_i$.*

When $H = \bigotimes_{i=1}^n H^i$, and $P^w = \bigotimes_{i=1}^n P_{H^i}^{w^i}$, where $w = (w^1, w^2, \ldots, w^n) \in \mathcal{W}^1 \times \mathcal{W}^2 \ldots \mathcal{W}^n$, we will often use the equivalence for matrices U that can be tensor product decomposed as $U = \bigotimes_i U^i$, for some $U^i \in \mathcal{U}(\bar{H}^i)$. In that case, we will say that G is *tensor equivalent* to T. Notice that $|\chi\rangle$ and W are not required to be also tensor product decomposable. This is because we want to encompass situations where the physical implementation G of the gate creates or destroys entanglement in the hidden degrees of freedom of the quantum register. Finally, note that the tensor equivalence on H implies the equivalence for each factor H^i, if the projectors $P_{H^i}^{w^i}$ are *complete*, namely if they linearly generate the identity in H^i. This will be the case in the rest of the paper.

Norm and approximation. We consider the ℓ_2 norm $\|\cdot\|$ for states, and the corresponding operator $\|\cdot\|$ norm for linear transformations. These norms are stable by tensor product composition in the following sense: $\|u \otimes v\| = \|u\| \times \|v\|$, if u and v denote either vectors or linear transformations. We note $|\psi\rangle =^\varepsilon |\psi'\rangle$ when two vectors $|\psi\rangle, |\psi'\rangle$ are such that $\||\psi\rangle - |\psi'\rangle\| \le \varepsilon$. We extend the ℓ_2-operator norm for restrictions of linear transformations on H. Namely if M is a linear transformation on H, and S is a subspace of H we define by $\|M\|_S = \sup(\|M|\psi\rangle\| : |\psi\rangle \in S$ and $\||\psi\rangle\| = 1)$. Similarly to states, we will write $M =^\varepsilon_S N$ when $\|M - N\|_S \le \varepsilon$. We introduce the notion of ε-*simulation* by extending the notion of simulation where statistics equalities are only approximately valid up to some additive term $\le \varepsilon$. The notions of equivalence can be similarly extended to ε-*equivalence*, by replacing each equality $=_S$ by $=^\varepsilon_S$.

3 Building a Test from Simulation

We consider a *test* as a set of simulation conditions, each of which can be checked through sampling. We show how to design efficient tests for quantum circuits by studying elementary tests that characterize sources and gates, and proving that the elementary tests are enough to characterize the whole circuit.

3.1 EPR State Testing

We rephrase Mayers and Yao [3] in our framework of quantum testing we just introduced. This is the simplest situation in which simulation implies equivalence. Their main result will be stated in an extended form that is most convenient for testing several registers successively. We will then use this result as a building block for finding other situations in which simulation implies equivalence.

From now and until the end of the paper, let $\mathcal{A}_0 = \{0, \frac{\pi}{8}, \frac{\pi}{4}\}$, $\mathcal{A}_1 = \{a + \frac{\pi}{2} : a \in \mathcal{A}_0\}$, and $\mathcal{A} = \mathcal{A}_0 \cup \mathcal{A}_1$. We fix orthogonal measurements $(P_A^a, P_A^{a+\pi/2})_{a \in \mathcal{A}_0}$ and $(P_B^b, P_B^{b+\pi/2})_{b \in \mathcal{A}_0}$ respectively on two Hilbert spaces A and B. Namely, we assume that $P_A^a + P_A^{a+\pi/2} = \mathrm{Id}_A$ and $P_B^a + P_B^{a+\pi/2} = \mathrm{Id}_B$, for every $a \in \mathcal{A}_0$.

Theorem 1. *Let $H = A \otimes B \otimes C$, and $|\psi\rangle \in H$ that simulates $|\phi^+\rangle$ with respect to $(P_A^a \otimes P_B^b \otimes \mathrm{Id}_C)_{a,b \in \mathcal{A}}$. Then there exist two unitary transformations $U_{\bar{A}} \in \mathcal{U}(\bar{A})$ and $U_{\bar{B}} \in \mathcal{U}(\bar{B})$ such that $|\psi\rangle$ is $(U_{\bar{A}} \otimes U_{\bar{B}})$-equivalent to $|\phi^+\rangle$ on $S = \mathrm{span}\{P_A^a \otimes P_B^b \otimes \mathrm{Id}_C |\psi\rangle : a, b \in \mathcal{A}\}$. Moreover the dimension of S is 4.*

In [3], the theorem was initially extended from S to the supports of $|\psi\rangle$ on each side. Nonetheless our results will be stated on S since this is sufficient for our purpose, and because their respective robustness can only be stated on S.

From Theorem 1 we derive by induction over n our main tool for testing n-qubit registers. Let $A = \bigotimes_{i=1}^n A^i$ and $B = \bigotimes_{i=1}^n B^i$. We now fix $(P_{A^i}^{a^i}, P_{A^i}^{a^i+\pi/2})_{a^i \in \mathcal{A}_0}$ and $(P_{B^i}^{b^i}, P_{B^i}^{b^i+\pi/2})_{b^i \in \mathcal{A}_0}$ to be orthogonal measurements on A^i and B^i respectively for every i. We denote $P_A^a = \bigotimes_{i=1}^n P_{A^i}^{a^i}$, with $a = (a^i)_{i=1}^n$ and $P_B^b = \bigotimes_{i=1}^n P_{B^i}^{b^i}$ with $b = (b^i)_{i=1}^n$.

Corollary 1. *Let $H = A \otimes B \otimes C$, and $|\Psi\rangle \in H$ that simulates $|\phi^+\rangle$ with respect to $(P_{A^i}^{a^i} \otimes P_{B^i}^{b^i} \otimes \mathrm{Id}_C)_{a^i,b^i \in \mathcal{A}}$ for every $i = 1, 2, \ldots, n$. Then there exist two unitary transformations $U_{\bar{A}} \in \bigotimes_i \mathcal{U}(\bar{A}^i)$ and $U_{\bar{B}} \in \bigotimes_i \mathcal{U}(\bar{B}^i)$ such that $|\Psi\rangle$ is $(U_{\bar{A}} \otimes U_{\bar{B}})$-equivalent to $|\Phi_n^+\rangle$ on $S = \mathrm{span}\{P_A^a \otimes P_B^b|\psi\rangle : a, b \in \mathcal{A}^n\}$. Moreover the dimension of S is 4^n.*

Therefore, testing a $2n$-qubit EPR state can be done by checking the probabilities of $O(n)$ outcomes, whereas there are $2^{O(n)}$ possible joint measurement outcomes.

3.2 Gate Testing

One-qubit Gate Testing. As a first attempt, we state how to check that a gate is equivalent to the identity.

Proposition 2. *Let $H = A \otimes B$ and $G \in \mathcal{U}(A)$. Let $|\psi\rangle \in H$ be such that $|\psi\rangle$ and $G|\psi\rangle$ simulate $|\phi^+\rangle$ with respect to some projectors $(P_A^a)_{a \in \mathcal{A}}$ and $(P_B^b)_{b \in \mathcal{A}}$. Then, $G \otimes \mathrm{Id}_B$ is tensor equivalent to $\mathrm{Id}_{A_c} \otimes \mathrm{Id}_{B_c}$ on $S = \mathrm{span}\{P_A^a \otimes P_B^b|\psi\rangle : a, b \in \mathcal{A}\}$.*

Stating the above result allows us to exhibit simple characteristics of the general method used for proving that gates can be self-tested. First, any gate testing requires two EPR tests. These are used to ensure that the input and output states together with the measurements act properly before and after the gate. These are *conspiracy tests*. Second, the fundamental properties of EPR states is used in order to show that the gate G and the measurements commute on the input state. This allows to perform tomography of the gate G. These tests will be referred to as *tomography tests*.

We can now state the general result concerning any 1-qubit real gate. We use the fact that any real gate on one qubit of the EPR state $|\phi^+\rangle$ can be undone by doing the same real gate on the other qubit.

Theorem 2. *Let $T \in \mathcal{O}(2)$. Let $H = A \otimes B$, $G_A \in \mathcal{U}(A)$, and $G_B \in \mathcal{U}(B)$. Let $|\psi\rangle \in H$ be such that $|\psi\rangle$ and $G_A G_B|\psi\rangle$ simulate $|\phi^+\rangle$, and such that $G_A|\psi\rangle$ simulates $(T \otimes \mathrm{Id}_2)|\phi^+\rangle$. Then, G_A is tensor equivalent to T on $S = \mathrm{span}\{P_A^a \otimes P_B^b|\psi\rangle : a, b \in \mathcal{A}\}$.*

Proof. The proof proceeds in two steps. First, we show that S and $G_A(S)$ are resp. $(U_{\bar{A}} \otimes U_{\bar{B}})$- and $(V_{\bar{A}} \otimes U_{\bar{B}})$-equivalent to $A_c \otimes B_c$. Second, we prove that there exists $W \in \mathcal{U}(A)$ such that $G_A \otimes \mathrm{Id}_B =_S (V_{\bar{A}}^\dagger \otimes U_{\bar{B}}^\dagger)(T \otimes W \otimes \mathrm{Id}_B)(U_{\bar{A}} \otimes U_{\bar{B}})$.

Theorem 1 applied to $|\psi\rangle$ and $G_A G_B|\psi\rangle$ gives $U_{\bar{A}}, V_{\bar{A}} \in \mathcal{U}(\bar{A})$ and $U_{\bar{B}}, V_{\bar{B}} \in \mathcal{U}(\bar{B})$ such that S and $(G_A \otimes G_B)(S)$ are respectively $(U_{\bar{A}} \otimes U_{\bar{B}})$- and $(V_{\bar{A}} \otimes V_{\bar{B}})$-equivalent to $A_c \otimes B_c$. This implies that $(G_A \otimes \mathrm{Id}_B)(S)$ is $(V_{\bar{A}} \otimes U_{\bar{B}})$-equivalent to $A_c \otimes B_c$. That is, we have the required tensor equivalences for S and $G_A(S)$. If we define $|\chi\rangle_{AB}$ as $U_{\bar{A}} \otimes U_{\bar{B}}|\psi\rangle = |\phi^+\rangle_{A_c B_c} \otimes |\chi\rangle_{AB}$, we then have $S = U_A^\dagger \otimes U_B^\dagger(A_c \otimes B_c \otimes |\chi\rangle_{AB})$.

The simulation of $T|\phi^+\rangle$ by $G_A|\psi\rangle$ can be rewritten within the density matrix formalism as: $\mathrm{tr}\left((P_A^a \otimes P_B^b)G_A|\psi\rangle\langle\psi|G_A^\dagger\right) = \mathrm{tr}\left((|a\rangle\langle a| \otimes |b\rangle\langle b|)(T \otimes \mathrm{Id}_2)|\phi^+\rangle\langle\phi^+|(T^\dagger \otimes \mathrm{Id}_2)\right)$. Using the commutativity of the trace operator and $(\mathrm{Id}_2 \otimes |b\rangle\langle b|)|\phi^+\rangle\langle\phi^+| = \frac{1}{2}|b\rangle\langle b| \otimes |b\rangle\langle b|$, we get: $\mathrm{tr}\left((G_A^\dagger P_A^a G_A \otimes P_B^b)|\psi\rangle\langle\psi|\right) = \frac{1}{2}\mathrm{tr}\left(T^\dagger|a\rangle\langle a|T|b\rangle\langle b|\right)$.

Define the positive semi-definite operator $R_{\bar{A}\bar{B}}^a = (U_{\bar{A}} \otimes U_{\bar{B}})G_A^\dagger P_A^a G_A (U_{\bar{A}}^\dagger \otimes U_{\bar{B}}^\dagger)$. Since $|\psi\rangle$ is tensor equivalent to $|\phi^+\rangle$, we have: $\mathrm{tr}\left(R_{\bar{A}\bar{B}}^a(|b\rangle\langle b|_{A_c} \otimes |b\rangle\langle b|_{B_c} \otimes |\chi\rangle\langle\chi|_{AB})\right) = \mathrm{tr}\left(T^\dagger|a\rangle\langle a|T|b\rangle\langle b|\right)$.

Observe that the operators $U_{\bar{B}}$ and $U_{\bar{B}}^\dagger$ can be removed from the definition of $R_{\bar{A}\bar{B}}^a$ without modifying it. Therefore the previous equation can be extended for all $b, b' \in \mathcal{A}$ to $\mathrm{tr}\left(R_{\bar{A}\bar{B}}^a(|b\rangle\langle b|_{A_c} \otimes |b'\rangle\langle b'|_{B_c} \otimes |\chi\rangle\langle\chi|_{AB})\right) = \mathrm{tr}\left(T^\dagger|a\rangle\langle a|T\right)$, since the value of the left hand side does not depend on b'.

Now applying standard techniques of tomography, we get that $_{AB}\langle\chi|_{B_c}\langle b'|R_{\bar{A}\bar{B}}^a|b'\rangle_{B_c}|\chi\rangle_{AB} = (T^\dagger|a\rangle\langle a|T)$, for every $b' \in \mathcal{A}$. Since $R_{\bar{A}\bar{B}}^a$ is a semi-definite operator, the above conclusion can be rewritten as

$$R_{\bar{A}\bar{B}}^a =_{A_c \otimes B_c \otimes |\chi\rangle_{AB}} (T^\dagger|a\rangle\langle a|T) \otimes \mathrm{Id}_{A \otimes \bar{B}}. \tag{1}$$

The tensor-equivalence of $G_A(S)$ with $A_c \otimes B_c$ also gives $P_A^a =_{G_A(S)} (V_{\bar{A}}^\dagger \otimes U_{\bar{B}}^\dagger)(|a\rangle\langle a| \otimes \mathrm{Id}_{A \otimes \bar{B}})(V_{\bar{A}} \otimes U_{\bar{B}})$. Since $S = U_{\bar{A}}^\dagger \otimes U_{\bar{B}}^\dagger(A_c \otimes B_c \otimes |\chi\rangle)$, this can be used to replace P_A^a inside Equation (1). We obtain $(|a\rangle\langle a| \otimes \mathrm{Id}_{A \otimes \bar{B}})(V_{\bar{A}} \otimes U_{\bar{B}})G_A(U_{\bar{A}}^\dagger \otimes U_{\bar{B}}^\dagger)(T^\dagger \otimes \mathrm{Id}_{A \otimes \bar{B}}) =_{A_c \otimes B_c \otimes |\chi\rangle} (V_{\bar{A}} \otimes U_{\bar{B}})G_A(U_{\bar{A}}^\dagger \otimes U_{\bar{B}}^\dagger)(T^\dagger \otimes \mathrm{Id}_{A \otimes \bar{B}})(|a\rangle\langle a| \otimes \mathrm{Id}_{A \otimes \bar{B}})$. Then, we can conclude using standard linear algebra techniques that there exists $W \in \mathcal{U}(A)$ such that $G_A =_S (V_{\bar{A}}^\dagger \otimes U_{\bar{B}}^\dagger)(T \otimes W \otimes \mathrm{Id}_{\bar{B}})(U_{\bar{A}} \otimes U_{\bar{B}})$. $\quad\square$

Many-qubit Gate Testing. We now consider n-qubit real gates. We present our main result for testing gates using a slightly different formulation than in Theorem 2, which will be useful for the proof of Theorem 4. The proof is omitted since it is similar to the second step of the proof of Theorem 2.

Note that we will use that any real gate on one register of the state $|\Phi_n^+\rangle$ can be undone by doing the same real gate on the other register.

Theorem 3. Let $T \in \mathcal{O}(2^n)$. Let $H = A \otimes B \otimes C$, where $A = \bigotimes_i A^i$ and $B = \bigotimes_i B^i$. Let $G_A \in \mathcal{U}(A)$ and $G_B \in \mathcal{U}(B)$. Let $|\Psi\rangle \in H$ and $U_{\bar{A}}, V_{\bar{A}} \in \bigotimes_i \mathcal{U}(\bar{A}_i)$ and $U_{\bar{B}}, V_{\bar{B}} \in \bigotimes_i \mathcal{U}(\bar{B}_i)$ be such that:

1. $|\Psi\rangle$ is $(U_{\bar{A}} \otimes U_{\bar{B}})$-equivalent to $|\Phi_n^+\rangle$ on S with respect to $(P_A^a \otimes P_B^b)_{a,b \in \mathcal{A}^n}$,
2. $G_A G_B|\Psi\rangle$ is $(V_{\bar{A}} \otimes V_{\bar{B}})$-equivalent to $|\Phi_n^+\rangle$ on $(G_A \otimes G_B)(S)$ with respect to $(P_A^a \otimes P_B^b)_{a,b \in \mathcal{A}^n}$,
3. $G_A|\Psi\rangle$ simulates $(T \otimes \mathrm{Id}_{2^n})|\Phi_n^+\rangle$ with respect to $(P_A^a \otimes P_B^b \otimes \mathrm{Id}_C)_{a,b \in \mathcal{A}^n}$,

where $S = \mathrm{span}\{P_A^a \otimes P_B^b|\psi\rangle : a,b \in \mathcal{A}^n\}$. Then G_A is $(U_{\bar{A}} \otimes U_{\bar{B}}, V_{\bar{A}} \otimes U_{\bar{B}})$-equivalent to T on S.

As Theorems 2 & 3 exemplify, there is one restriction to the class of gates we are able to test. The ideal gates must have real-valued coefficients. Note that we are not making any assumptions about the physical implementation of gates, but rather on the ideal gates they are supposed to simulate. The problem lies in the fact that any complex gate of dimension d can be simulated using real gates and appropriate measurement devices on a $2d$-dimensional Hilbert space, in a rather standard way [7]. On the positive side, this remark means that our restriction is not a limitation, as any quantum computation can be performed with real gates and real gates can be tested.

3.3 Circuit Testing

Now we state our main theorem and its corollary which relates elementary tests of sources and gates with the simulation of a whole circuit. They derive from Corollary 1 and Theorem 3, in the sense that (i) under certain conditions simulation implies equivalence, (ii) equivalence statements can be composed and (iii) that equivalence implies simulation.

Assume that some Hilbert space H has a tensor product decomposition $H = \bigotimes_{i=1}^{n} A^i \bigotimes B^i$. For any subset $I \subseteq \{1, 2, \ldots, n\}$, let H^I denote the Hilbert space $\bigotimes_{i \in I} A^i \bigotimes_{i \in I} B^i$, and $|\Phi^+\rangle_I$ the EPR state $|\Phi_{|I|}^+\rangle$ over $\bigotimes_{i \in I} A_c^i \bigotimes_{i \in I} B_c^i$.

Theorem 4. *Let* $H = A \otimes B$, *where* $A = \bigotimes_i A^i$ *and* $B = \bigotimes_i B^i$. *Let* $I^1, I^2, \ldots, I^t \subseteq \{1, 2, \ldots, n\}$. *Let* $G_A^j \in \mathcal{U}(A^{I^j})$, $G_B^j \in \mathcal{U}(B^{I^j})$ *and* $T^j \in \mathcal{O}(A_c^{I^j})$. *Let* $|\Psi\rangle \in A \otimes B$. *Define inductively* $|\Psi'^j\rangle = (G_A^j \otimes \mathrm{Id}_B)|\Psi^{j-1}\rangle$ *and* $|\Psi^j\rangle = (G_A^j \otimes G_B^j)|\Psi^{j-1}\rangle$, *where* $|\Psi^0\rangle = |\Psi'^0\rangle = |\Psi\rangle$. *Assume:*

1. $|\Psi\rangle$ *simulates* $|\phi^+\rangle$ *with respect to* $(P_{A^i}^{a^i} \otimes P_{B^i}^{b^i})_{a^i, b^i \in \mathcal{A}}$, *for every* $i = 1, 2, \ldots, n$.
2. *For* $j = 1, \ldots, t$: $|\Psi^j\rangle$ *simulates* $|\phi^+\rangle$ *with respect to* $(P_{A^i}^{a^i} \otimes P_{B^i}^{b^i})_{a^i, b^i \in \mathcal{A}}$, *for every* $i \in I^j$.
3. *For* $j = 1, \ldots, t$: $|\Psi'^j\rangle$ *simulates* $T^j|\Phi^+\rangle_{I^j}$ *w.r.t.* $(P_{A^{I^j}}^a \otimes P_{B^{I^j}}^b)_{a, b \in \mathcal{A}^{I^j}}$.

Then $G_A^t G_A^{t-1} \cdots G_A^1$ *is tensor equivalent to* $T^t T^{t-1} \cdots T^1$ *on* $S = \mathrm{span}(P_A^a \otimes P_B^b|\Psi\rangle : a, b \in \mathcal{A}^n)$.

Corollary 2. *Let* $|\Psi\rangle \in H$ *satisfy the hypothesis of Theorem 4 for some decomposition of* $G_A \in \mathcal{U}(A)$ *and* $T \in \mathcal{U}(A_c)$ *into* t *gates acting only on a constant number of qubits. Then, for every* $x \in \{0, 1\}^n$, *the state* $\sqrt{2^n} \, \mathrm{tr}_B(P_B^x|\Psi\rangle)$ *simulates* $|x\rangle_{A_c}$ *with respect to* $(P_A^w)_{w \in \mathcal{A}^n}$. *Moreover* G_A *simulates* T *with respect to the above identification, and the number of statistics to be checked is in* $O(t)$.

4 Robustness of Testing

Until now, our interest has been focused on the possibility of self-testing a quantum circuit when outcome probabilities are known with perfect accuracy. To be of practical interest, our results must be extended to the situation of finite accuracy. We show below that it is possible and that the relevant results for testing are robust in the following way: if the statistics are close to the ideal ones, then the states, the measurements and the gates are also close to ones that are equivalent to the ideal ones. This notion of robustness follows the ones of [8,9] for classical computing and of [4] for quantum computing.

The proofs of this section follow the structure of the exact case, and are omitted due to the lack of space. They will be in the full version of the paper. We first state the robustness of Theorem 1.

Theorem 5. *Let $H = A \otimes B \otimes C$, and $|\psi\rangle \in H$ that ε-simulates $|\phi^+\rangle$ with respect to $(P_A^a \otimes P_B^b \otimes \mathrm{Id}_C)_{a,b\in\mathcal{A}}$. Then there exist $U_{\bar{A}} \in \mathcal{U}(\bar{A})$ and $U_{\bar{B}} \in \mathcal{U}(\bar{B})$ such that $|\psi\rangle$ is $(O(\varepsilon^{1/4}), (U_{\bar{A}} \otimes U_{\bar{B}}))$-equivalent to $|\phi^+\rangle$ on S.*

This result can be generalized to the case of a source producing a state $|\Psi\rangle$ that simulates n EPR pairs. In such case equivalence holds within $O(4^n\varepsilon^{1/4})$.

Corollary 3. *Let $H = A \otimes B \otimes C$, where $A = \bigotimes_i A^i$ and $B = \bigotimes_i B^i$. Let $|\Psi\rangle \in H$ be a state that ε-simulates $|\phi^+\rangle$ with respect to $(P_{A^i}^{a^i} \otimes P_{B^i}^{b^i})_{a^i,b^i\in\mathcal{A}}$, for every $i = 1, 2, \ldots, n$. Then, $|\Psi\rangle$ is $O(4^n\varepsilon^{1/4})$-equivalent to $|\Phi_n^+\rangle$.*

Another corollary we will use in the context of circuit testing concerns the case of n sources of EPR pairs that are tested simultaneously. This is qualitatively different from the previous situation as the state $|\Psi\rangle$ that is tested is assumed to be separable across the tensor product decomposition of H into $H^i = A^i \otimes B^i$.

Corollary 4. *Let $H = A \otimes B \otimes C$, where $A = \bigotimes_i A^i$ and $B = \bigotimes_i B^i$. Let $|\Psi\rangle \in H$ be a separable state across the tensor product decomposition of H into $A_i \otimes B_i$, and such that it ε-simulates $|\phi^+\rangle$ with respect to $(P_{A^i}^{a^i} \otimes P_{B^i}^{b^i})_{a^i,b^i\in\mathcal{A}}$, for every $i = 1, 2, \ldots, n$. Then, $|\Psi\rangle$ is $O(n\varepsilon^{1/4})$-equivalent to $|\Phi_n^+\rangle$.*

Now we concentrate on the robustness of Theorem 3. Note that the exponential dependency in the number n of qubits is not a problem, since we will use this theorem for constant n only (typically $n \leq 3$).

Theorem 6. *Let $T \in \mathcal{O}(2^n)$. Let $H = A \otimes B \otimes C$, where $A = \bigotimes_i A^i$ and $B = \bigotimes_i B^i$. Let $G_A \in \mathcal{U}(A)$ and $G_B \in \mathcal{U}(B)$. Let $|\Psi\rangle \in H$ and $U_{\bar{A}}, V_{\bar{A}} \in \bigotimes_i \mathcal{U}(\bar{A}_i)$ and $U_{\bar{B}}, V_{\bar{B}} \in \bigotimes_i \mathcal{U}(\bar{B}_i)$ be such that:*

1. *$|\Psi\rangle$ is $(\varepsilon, (U_{\bar{A}}\otimes U_{\bar{B}}))$-equivalent to $|\Phi_n^+\rangle$ on S with respect to $(P_A^a \otimes P_B^b)_{a,b\in\mathcal{A}^n}$,*
2. *$G_A\otimes G_B|\Psi\rangle$ is $(\varepsilon, (V_{\bar{A}}\otimes V_{\bar{B}}))$-equivalent to $|\Phi_n^+\rangle$ on $(G_A\otimes G_B)(S)$ with respect to $(P_A^a \otimes P_B^b)_{a,b\in\mathcal{A}^n}$,*
3. *$G_A|\Psi\rangle$ ε-simulates $(T \otimes \mathrm{Id}_{2^n})|\Phi_n^+\rangle$ with respect to $(P_A^a \otimes P_B^b \otimes \mathrm{Id}_C)_{a,b\in\mathcal{A}^n}$.*

Then $G_A \otimes \mathrm{Id}_B$ is $(2^{O(n)}\sqrt{\varepsilon}, (U_{\bar{A}} \otimes U_{\bar{B}}, V_{\bar{A}} \otimes U_{\bar{B}}))$-equivalent to $T \otimes \mathrm{Id}_{\bar{B}_c}$ on S.

5   Testing a Circuit on a Specific Input

We have seen in Section 3 how to test the implementation of a circuit on a whole subspace S of the input space. Surprisingly, this is much easier than to test a circuit on a particular input. In fact, using EPR pairs allows for the simultaneous testing of all possible inputs, while making the selection of a particular one difficult. The obvious choice would be to post-select the outcome of the B-side measurements of the EPR pairs. Unfortunately, the selected input state would then be prepared with exponentially small probability.

We circumvent the aforementioned difficulty using the fact that our circuits can have classically controlled feedback that decides which gates need to be applied based on some measurement results. Given a circuit for a unitary transformation T and an input x, we first measure the B-side of the (alleged) EPR states. This yields a classical state y on the A-side. Second, we design a circuit $T_{x,y}$ whose purpose is to flip the corresponding bits of y in order to get the input x, and to apply the original circuit for T. Third, we run the modified circuit on the state y that was prepared on the A-side. Finally, we test that this modified circuit implemented the correct computation. This includes verifying the gates and the preparation of all input states $|x'\rangle$—and in particular the preparation of $|x\rangle$—obtained by measuring $|\Psi\rangle$ on the B-side. See Figure 1 for an example.

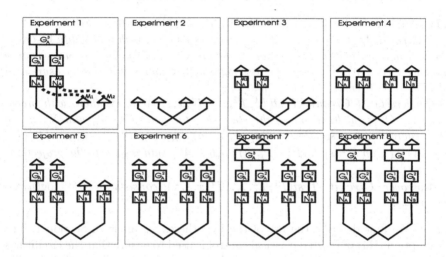

Fig. 1. The experiments to test the circuit consisting of gates $G_A^3 G_A^2 G_A^1$ on input $|00\rangle$. We first run the computation (Experiment 1) once on the modified circuit, where the intermediate measurements on the B-side yield the outcomes M_1, M_2. We now wish to check that the output of the circuit is correct. We carry on implementing Experiments 2 through 8 each a number of times in $\log(n/\gamma)/\varepsilon^8$, where ε is the required precision and γ is some confidence parameter.

The parameters of our test is a circuit for $T \in \mathcal{U}(2^n)$, that is a gate decomposition $T^t T^{t-1} \cdots T^1 = T$; a binary string $x \in \{0,1\}^n$; a precision $\varepsilon > 0$; and a confidence $\gamma > 0$. We assume that each gate T^i acts on a constant number of qubits (say ≤ 3). The input is a source of quantum states $|\Psi\rangle$ spread over n pairs of quantum registers; gates G_A^j and G_B^j acting on the same register numbers as T^j, for every j; auxiliary gates N_A^i acting on the i-th register of A; and orthogonal measurements $(P_{A^i}^a, P_{A^i}^{a+\pi/2})_{a \in \mathcal{A}_0}$ and $(P_{B^i}^b, P_{B^i}^{b+\pi/2})_{b \in \mathcal{A}_0}$. The goal is to test that, firstly, $\sqrt{2^n} \operatorname{tr}_B(P_B^b|\Psi\rangle)$ simulates $|b\rangle$ and that, secondly, the implemented circuit G_A simulates T.

Circuit Test $(T^1, T^2, \ldots, T^t \in \mathcal{U}(2^n), x \in \{0,1\}^n, \varepsilon > 0, \gamma > 0)$
1. Prepare a state $|\Psi\rangle$ of n EPR states into n pairs on $A^1 \otimes B^1, \ldots, A^n \otimes B^n$
2. Measure the B-side of $|\Psi\rangle$ using $(P_B^b)_{b \in \{0, \pi/2\}^n}$ and let y be the outcome

3. Let $T_{x,y}$ be the circuit that changes the input $|y\rangle$ into $|x\rangle$ and applies T
4. Prepare on the A-side the circuit G_A implementing $T_{x,y}$ using the t gates G_A^j and at most n gates N_A^i. Let $t' \leq t + n$ be the total number of gates
5. Run the circuit on the A-side and measure using $(P_A^a)_{a \in \{0, \pi/2\}^n}$
6. Approximate all the following statistics by repeating $O(\frac{\log(n/\gamma)}{\varepsilon})$ times the following measurements (where we use the notation of Theorem 4):
   (a) Measure $|\Psi\rangle$ using $(P_{A^i}^{a^i} \otimes P_{B^i}^{b^i})_{a^i, b^i \in \mathcal{A}_0}$, for every $i = 1, 2, \ldots, n$
   (b) For $j = 1, \ldots, t'$: Measure $|\Psi^j\rangle$ using $(P_{A^i}^{a^i} \otimes P_{B^i}^{b^i})_{a^i, b^i \in \mathcal{A}}$, for every $i \in I^j$
   (c) For $j = 1, \ldots, t'$: Measure $|\Psi'^j\rangle$ using $(P_{A^{Ij}}^a \otimes P_{B^{Ij}}^b)_{a, b \in \mathcal{A}_0^{Ij}}$
7. Accept if all the statistics are correct up to an additive error ε

Theorem 7. *Let* $T^1, T^2, \ldots, T^t \in \mathcal{U}(2^n), x \in \{0,1\}^n, \varepsilon > 0, \gamma > 0$.

*If **Circuit Test**$(T^1, T^2, \ldots, T^t, x, \varepsilon, \gamma)$ accepts then, with probability $1 - O(\gamma)$, the outcome probability distribution of the circuit (in step 5) is at total variance distance $O((t + n)\varepsilon^{1/8})$ from the distribution that comes from the measurement of $T^t T^{t-1} \cdots T^1 |x\rangle$ by $(|a\rangle\langle a|)_{a \in \{0, \pi/2\}^n}$.*

*Conversely, if **Circuit Test**$(T^1, T^2, \ldots, T^t, x, \varepsilon, \gamma)$ rejects then, with probability $1 - O(\gamma)$, at least one of the state $|\Psi\rangle$, the gates G_A^i, G_B^i and N_A^i is not $O(\varepsilon)$-equivalent to respectively either $|\Phi_n^+\rangle$, $(|a\rangle\langle a|_{A_c^i})_{a \in \mathcal{A}}$, $(|b\rangle\langle b|_{B_c^i})_{b \in \mathcal{A}})$, $T^i, {}^t(T^i)$ and NOT$_{A_c^i}$, on $S = \text{span}(P_A^a \otimes P_B^b |\Psi\rangle : a, b \in \mathcal{A}^n)$ with respect to the projections $(P_A^a \otimes P_B^b)_{a, b \in \mathcal{A}^n}$.*

*Moreover **Circuit Test**$(T^1, T^2, \ldots, T^t, x, \varepsilon, \gamma)$ consists of $O(\frac{tn}{\varepsilon} \log(n/\gamma))$ samplings.*

Proof. We first describe the use of the hypotheses we made in Section 1. The assumption (H4) of trusted classical control is used to ensure that the circuit has the same behavior on $P_B^y |\Psi\rangle$ as it would have on $|\Psi\rangle$. Hypothesis (H3) implies that we can repeat several times the same experiment, and hypotheses (H1) and (H2) allow us to state which parts of our system are separated from the others.

First, using the Chernoff-Hoeffding bound, we know that the expectation of any bounded random variable can be approximated within precision $O(\varepsilon)$ with probability $1 - O(\gamma)$ by $\frac{\log(1/\gamma)}{\varepsilon^2}$ independent samplings. Moreover if the expectation is lower bounded by a constant, then $\frac{\log(1/\gamma)}{\varepsilon}$ independent samplings are enough. In our case, the random variable is the two possible outcomes of a measurement. Call them 0 or 1. Since we can count both 0 and 1 outcomes, one of the corresponding probabilities is necessarily at least $1/2$. Therefore we get that each statistics we have from **Circuit Test** are approximated within precision $O(\varepsilon)$ with probability $1 - O(\gamma)$. From now on, we assume that all statistics are given within this precision.

First, we prove the robustness of **Circuit Test**. We derive the correct simulation of the implemented circuit using the approximate version of Corollary 2, that we get using Theorems 5 and 6. More precisely, using Corollary 4 for the initial source we get that $|\Psi\rangle$ is $O(n\varepsilon^{1/4})$-equivalent to $|\Phi_n^+\rangle$ on S. For other steps, due to the application of the j-th gate, the state $|\Psi^j\rangle$ is not necessarily a separable state across the n-registers. So we apply Corollary 3 on the registers where the j-th gate is applied, that is on a constant number of register, which gives the required $O(\varepsilon^{1/4})$-equivalence on the corresponding registers. Then, Theorem 6 concludes that the j-th gate is $O(j\varepsilon^{1/8})$-equivalent to the expected one, similarly for the intermediate states of the circuit and for the measurements. Note the error propagation is controlled by two properties: the stability of the ℓ_2 operator-norm by tensor product composition, and the triangle inequality.

Then, we focus on the run of $T_{x,y}$ in Step 5. We have to justify that the (normalized) outcome state $\sqrt{2^n}P_B^y|\Psi\rangle \in S$ of the measurement $(P_B^b)_{b\in\{0,\pi/2\}^n}$ is $O(n\varepsilon^{1/4})$-equivalent to $|y\rangle$ with respect to $(P_A^a)_{a\in\{0,\pi/2\}^n}$ on $P_B^y(S)$. Recall that by assumption the initial state $|\Psi\rangle$ is separable across the n pairs of registers, namely $|\Psi\rangle = \bigotimes_i|\psi^i\rangle$ with $|\psi^i\rangle \in A^i \otimes B^i$. For each pair of registers $A^i \otimes B^i$, using Theorem 5 we get that $|\psi^i\rangle$ is $O(\varepsilon^{1/4})$-equivalent to $|\phi^+\rangle$ with respect to $(P_{A^i}^{a^i} \otimes P_{B^i}^{b^i})_{a^i,b^i\in\mathcal{A}}$ on $S^i = \text{span}(P_{A^i}^{a^i} \otimes P_{B^i}^{b^i}|\psi^i\rangle : a^i, b^i \in \mathcal{A})$. In particular the projections $P_{A^i}^{a^i} \otimes P_{B^i}^{b^i}$ are also $O(\varepsilon^{1/4})$-equivalent to $|a^i\rangle\langle a^i| \otimes |b^i\rangle\langle b^i|$ on S^i. Therefore the normalized outcome state $\sqrt{2}P_B^{y^i}|\psi^i\rangle$ (which is in S^i) is $O(\varepsilon^{1/4})$-equivalent to $|y^i\rangle$ with respect to $(P_{A^i}^{a^i})_{a^i\in\{0,\pi/2\}}$ on $P_{B^i}^{y^i}(S^i)$. We then get our equivalence for the whole outcome state using those intermediate equivalences together with the stability of the ℓ_2 operator-norm by tensor product composition, and the triangle inequality of the norm. Finally, we combine the above approximate equivalences, one for the circuit and one for the input, and get that the outcome distribution is at total variation distance at most $O((t + n)\varepsilon^{1/8})$ from the expected one.

The second part of the theorem is the soundness of **Circuit Test**. Since ℓ_2-distance between states bounds the statistics bias of their measures, the proof of the contraposition directly follows: if our objects are ε-equivalent to the specification, then their statistics have a bias which is upper bounded by $O(\varepsilon)$.   □

References

1. Blum, M., Kannan, S.: Designing programs that check their work. J. ACM **42** (1995) 269–291
2. Blum, M., Luby, M., Rubinfeld, R.: Self-testing/correcting with applications to numerical problems. J. Computer and System Sciences **47** (1993) 549–595
3. Mayers, D., Yao, A.: Quantum cryptography with imperfect apparatus. In: Proceedings of 39th IEEE FOCS. (1998) 503–509
4. Dam, W., Magniez, F., Mosca, M., Santha, M.: Self-testing of universal and fault-tolerant sets of quantum gates. In: Proc. of 32nd ACM STOC. (2000) 688–696
5. Buhrman, H., Fortnow, L., Newman, I., Röhrig, H.: Quantum property testing. In: Proc. of 14th ACM-SIAM SODA. (2003) 480–488
6. Friedl, K., Magniez, F., Santha, M., Sen, P.: Quantum testers for hidden group properties. In: Proc. of the 28th MFCS. (2003) 419–428
7. Rudolph, T., Grover, L.: A 2–rebit gate universal for quantum computing (2002)
8. Rubinfeld, R., Sudan, M.: Robust characterizations of polynomials with applications to program testing. SIAM J. Computing **25** (1996) 23–32
9. Rubinfeld, R.: On the robustness of functional equations. SIAM J. Computing **28** (1999) 1972–1997

Deterministic Extractors
for Independent-Symbol Sources

Chia-Jung Lee[1], Chi-Jen Lu[2], and Shi-Chun Tsai[1]

[1] Department of Computer Science, National Chiao-Tung University, Hsinchu,
Taiwan
{leecj, sctsai}@csie.nctu.edu.tw
[2] Institute of Information Science, Academia Sinica, Taipei, Taiwan
cjlu@iis.sinica.edu.tw

Abstract. In this paper, we consider the task of deterministically extracting randomness from sources consisting of a sequence of n independent symbols from $\{0,1\}^d$. The only randomness guarantee on such a source is that the whole source has min-entropy k. We give an explicit deterministic extractor which can extract $\Omega(\log k - \log d - \log\log(1/\varepsilon))$ bits with error ε, for any $n, d, k \in \mathbb{N}$ and $\varepsilon \in (0,1)$. For sources with a larger min-entropy, we can extract even more randomness. When $k \geq n^{1/2+\gamma}$, for any constant $\gamma \in (0, 1/2)$, we can extract $m = k - O(d\log(1/\varepsilon))$ bits with any error $\varepsilon \geq 2^{-\Omega(n^\gamma)}$. When $k \geq \log^c n$, for some constant $c > 0$, we can extract $m = k - d(1/\varepsilon)^{O(1)}$ bits with any error $\varepsilon \geq k^{-\Omega(1)}$. Our results generalize those of Kamp & Zuckerman and Gabizon et al. which only work for bit-fixing sources (with $d = 1$ and each bit of the source being either fixed or perfectly random). Moreover, we show the existence of a non-explicit deterministic extractor which can extract $m = k - O(\log(1/\varepsilon))$ bits whenever $k = \omega(d + \log(n/\varepsilon))$. Finally, we show that even to extract from bit-fixing sources, any extractor, seeded or not, must suffer an entropy loss $k - m = \Omega(\log(1/\varepsilon))$. This generalizes a lower bound of Radhakrishnan & Ta-Shma with respect to general sources.

1 Introduction

Randomness has become a useful tool in computer science. However, when using randomness in designing algorithms or protocols, people usually assume the randomness being perfect, and the performance guarantees are based on this assumption. In reality, the random sources we (or computers) have access to are typically not so perfect at all, but only contain some crude randomness. One approach to solve this problem is to construct so-called *extractors*, which can extract almost perfect randomness from weakly random sources [32,20]. Extractors turn out to have close connections to other fundamental objects and have found a wide range of applications (e.g. [20,33,34,31,29,28,16,30]). A nice survey can be found in [25].

We measure the amount of randomness in a source by its *min-entropy*; a source is said to have min-entropy k if every element occurs with probability at most 2^{-k}. Given sources with enough min-entropy, one would like to construct an

M. Bugliesi et al. (Eds.): ICALP 2006, Part I, LNCS 4051, pp. 84–95, 2006.

extractor which extracts a string with distribution close to uniform. However, it is well known that one cannot deterministically extract even one bit from an n-bit source with min-entropy $n - 1$ [6]. In contrast, it becomes possible if we are allowed a few random bits, called a seed, to aid the extraction. Such a procedure is called a *seeded extractor*. During the past decades, a long line of research has worked on using a shorter seed to extract more randomness (e.g. [20,19,22,10,24,29,27,26]), and recently an optimal (up to constant factors) construction has been given [17].

The problem with a seeded extractor is again to get a seed which is perfectly (or almost) random. For some applications, this issue can be taken care of (for example, by enumerating all possible seed values when the seed is short), but for others, we are back to the same problem which extractors are originally asked to solve. This motivates one to consider the possibility of more restricted sources from which randomness can be extracted in a deterministic (seedless) way.

One line of research studies the case with multiple independent sources. The goal is to have a small number of independent sources with a low min-entropy requirement on sources, while still being able to extract randomness from them. With two independent sources, the requirement on the min-entropy rate (average min-entropy per bit) stayed slightly above $1/2$ for a long time [6,8], but this barrier has been broken by a recent construction which pushes the requirement slightly below $1/2$ [5]. The requirement on min-entropy rate can be lowered to any constant when there are a constant number of independent sources [3], and the number of sources has recently been reduced to three [4].

The other line of research considers the case of bit-fixing sources. In an oblivious bit-fixing source, each bit is either fixed (containing no randomness) or perfectly random, and is independent of other bits. From such a source of length n with min-entropy $n^{1/2+\gamma}$, for any constant $\gamma \in (0, 1/2)$, Kamp and Zuckerman [13] gave a seedless extractor which can extract $\Omega(n^{2\gamma})$ bits of randomness. Building on this result together with some new idea, Gabizon et al. [9] were able to extract even more randomness. In particular, when the source has min-entropy $k > n^{1/2+\gamma}$, they can extract $k - n^{1/2+\gamma}$ bits and when $k > \log^c n$ for some constant c, they can extract $k - k^{\Omega(1)}$ bits.

Note that the two lines of research discussed above can be seen as belonging to two extremes of a spectrum in the following sense. Sources in both cases consist of multiple parts which are mutually independent. In the first case, one usually has in mind sources with relatively few parts while each part is long and contains a substantial amount of randomness. In the second case, a bit-fixing source consists of many parts, while each part is only a single bit which is either random or fixed. We would like to put both cases in the same framework and study sources that lie in between these two extremes.

We consider the following more general class of sources, characterized by the parameters $n, d, k \in \mathbb{N}$. Each source in the class consists of n mutually independent parts, each of length d, and the whole source has min-entropy k. For small n and large d, this covers sources of the first type, while for large n and $d = 1$, this covers sources of the second type. For other ranges of n and d, very little is known, and the main focus of our paper is to extract randomness from such sources.

Previously, [15,14] were able to extract randomness from such a source with the condition that there are two parts in it with a combined min-entropy slightly above d. Independent of our work, Kamp et al. [12] recently also considered the same class of sources as ours and obtained some similar results. Furthermore, they showed that extractors for such sources also work for a more general class of sources which can be generated in small space.

Note that for deterministic extractors, the goal is to maximize the number m of extracted bits (or equivalently to minimize the entropy loss $k - m$) and to minimize the distance ε, which we call error, of its output distribution to the uniform one.

Our results. Our first result gives an explicit extractor which works for any min-entropy k but extracts only about $\log k$ random bits. More precisely, for any $n, d, k \in \mathbb{N}$ and $\varepsilon \in (0, 1)$, our extractor can extract $\Omega(\log k - \log d - \log \log(1/\varepsilon))$ bits with error ε. This can be seen as a generalization of the extractor of Kamp and Zuckerman [13], but note that theirs only works for bit-fixing sources and does not seem to work for the case that allows each bit having arbitrary bias. In fact, our extractor works for sources in which randomness could be distributed very non-uniformly among the n parts (e.g., some may have no min-entropy at all, but we do not know which ones), while previous constructions such as [3,4,21] do not seem to work for such sources. Independent of our work, Kamp et al. [12] also gave the same construction but used a different analysis.

To extract more randomness, we borrow the technique of Gabizon et al. [9]. We have two constructions, both built on our first construction mentioned above. First, when $k \geq n^{1/2+\gamma}$, for any constant $\gamma \in (0, 1/2)$, we can extract $m = k - O(d \log(1/\varepsilon))$ random bits with any error $\varepsilon \geq 2^{-\Omega(n^\gamma)}$. Second, when $k \geq \log^c n$, for some constant $c > 0$, we can extract $m = k - d(1/\varepsilon)^{O(1)}$ bits with error $\varepsilon \geq k^{-\Omega(1)}$. That is, when the min-entropy k is high, we can have a small entropy loss and a small error, but when k is small, the loss and error become larger. Note that the two main results in [9] follow from our two with $d = 1$ (that is, for bit-fixing source). On the other hand, we cover a large range of d and ε, and capture the tradeoff between error and entropy loss. For example, for constant d and ε, we show that the entropy loss can be lowered to a constant.

One may wonder if the entropy loss can be further reduced. We show that this is indeed possible, by proving the existence of a seedless extractor which can extract $m = k - O(\log(1/\varepsilon))$ random bits whenever $k = \omega(d + \log(n/\varepsilon))$. However, the existence is not shown in an explicit way; we only know such an extractor exists but we do not know how to construct it. Still, this shows that better explicit constructions than ours may be possible. Only for the case with $d = O(1)$, $k \geq n^{1/2+\gamma}$, and $\varepsilon \geq 2^{-\Omega(n^\gamma)}$ do we have an explicit construction matching this bound.

On the other hand, one may also wonder whether this existential upper bound on entropy loss is tight. Our final result shows that this is indeed the case by giving a matching lower bound. In fact, we show that even for the case of bit-fixing sources and even allowing a seed of length s, any extractor can only extract $k + s - \Omega(\log(1/\varepsilon))$ random bits. That is, even to extract from bit-fixing sources, any extractor, seeded or not, must suffer an entropy loss of $\Omega(\log(1/\varepsilon))$. This generalizes the result of Radhakrishnan and Ta-Shma [23], which has the same bound for seeded extractors on *general* sources.

Our techniques. Our first extractor, which extracts about $\log k$ bits, was inspired by that of Kamp and Zuckerman [13], but our approach is quite different. Instead of taking a random walk on an odd cycle, we walk on the group $\mathbb{Z}_M$ for a prime M. More precisely, given a source $\mathcal{X} = (\mathcal{X}_1, \ldots, X_n)$, we see each $\mathcal{X}_i$ as an element of $\mathbb{Z}_M$ and outputs $\mathcal{X}_1 + \cdots + \mathcal{X}_n$ over $\mathbb{Z}_M$. As in [13], we will show that each step of our walk brings the distribution closer to uniform when the symbol from the source contains some randomness. However, even for the case of $d = 1$, we cannot use the analysis from [13], which is based on bounding the second eigenvalue of the transition matrix for a perfectly random step on a cycle. This is because we may walk in a highly biased way as each bit of our source can have an arbitrary bias. Our proof is very different and elementary, and has the following interesting point. The recent breakthrough construction of multi-source extractors [3] and its subsequent works all relied on using both sums and products to increase entropy. Our analysis shows that in fact even doing sums alone can increase entropy. The increase, however, is slower, so we need a larger number of sources (as opposed to a constant number in [3]).

Then we apply the technique of [9] to extract more randomness. Our two extractors generalize the corresponding ones in [9]. The only difference is that we deal with a more general classes of sources, do a more careful analysis, and use our first extractor instead of that in [13] as a building block.

Our existential upper bound on entropy loss is proved via a probabilistic argument. That is, we generate a seedless extractor randomly, and show that it works for all of our sources with a positive probability. For each source, we can show that it fails with a small probability. However, the number of all possible sources is in fact infinite. Nevertheless, we show that it suffices to consider only a small set of sources, since any source is close to a convex combination of them. Sources in this set are those with the property that their distributions in each dimension are "almost flat" and have only a small number of possible min-entropy values.

Our lower bound proof of entropy loss follows the outline of that in [23]. Namely, given any function EXT : $\{0,1\}^n \times \{0,1\}^s \to \{0,1\}^m$ with $m \geq k + s - o(\log(1/\varepsilon))$, we show the existence of a bit-fixing source with min-entropy k on which the error of EXT exceeds ε, again using a probabilistic argument. We generate a source by randomly picking $n - k$ bits from the source and fixing them to some random values; the remaining k bits are left free and given a uniform distribution. The difficult part is to show that any such EXT fails on such a randomly chosen source with a positive probability. This probability turns out to be related to the size of some "almost" t-wise independent space, whose distribution is close to random on most sets of t dimensions. This can be seen as a relaxation of the standard notion of approximate t-wise independent space, in which the close-to-randomness property is required on *every* set of t dimensions. We prove a size lower bound on such a sample space, which seems to have an interest of its own. In particular, it immediately implies a size lower bound on any approximate t-wise independent space.

2   Preliminaries

For $n \in \mathbb{N}$, let $[n]$ denote the set $\{1, \ldots, n\}$. For $x \in \{0,1\}^n$, $i \in [n]$, and $I \subseteq [n]$, let x_i denote the bit in the i'th dimension of x and x_I denote the projection of x onto those dimensions in I. For $S \subseteq \{0,1\}^n$ and $I \subseteq [n]$, let S_I denote the set $\{x_I : x \in S\}$. For a set S and $t \in \mathbb{N}$, let $P(S, t)$ denote the collection of t-element subsets of S. All the logarithms in this paper will have base two.

When we sample from a finite set, the default distribution is the uniform one. For $n \in \mathbb{N}$, let $\mathcal{U}_n$ denote the uniform distribution over $\{0,1\}^n$. For a distribution $\mathcal{X}$ over a set S and an element $x \in S$, let $\mathcal{X}(x)$ denote the probability measure of x in the distribution $\mathcal{X}$. We say that a distribution $\mathcal{X}$ is a convex combination of distributions $\mathcal{X}^1, \ldots, \mathcal{X}^t$ over a set S, if there exist numbers $\alpha_1, \ldots, \alpha_t \geq 0$ with $\sum_{i \in [t]} \alpha_i = 1$ such that for every $x \in S$, $\mathcal{X}(x) = \sum_{i \in [t]} \alpha_i \mathcal{X}^i(x)$. We will mainly measure the distance between two distributions $\mathcal{X}, \mathcal{X}'$ over S by their L_1-distance, defined as $\|\mathcal{X} - \mathcal{X}'\|_1 = \sum_{x \in S} |\mathcal{X}(x) - \mathcal{X}'(x)|$. Another distance measure that will be used sometimes is the L_2-distance, defined as $\|\mathcal{X} - \mathcal{X}'\|_2 = \sqrt{\sum_{x \in S} (\mathcal{X}(x) - \mathcal{X}'(x))^2}$. Call a distribution ε-random if its L_1-distance to the uniform distribution is at most ε. We will measure the amount of randomness in a distribution $\mathcal{X}$ over S by its min-entropy, defined as $\mathrm{H}_\infty(\mathcal{X}) = \min_{x \in S} \log(1/\mathcal{X}(x))$. In this paper, we will focus on a special kind of sources called independent-symbol sources, which consist of n independent symbols over some set $[D]$.

Definition 1. *A distribution $\mathcal{X} = (\mathcal{X}_1, \ldots, \mathcal{X}_n)$ over the set $[D]^n$ is called an (n, D)-source if the n symbols $\mathcal{X}_1, \ldots, \mathcal{X}_n$ are distributed independently from each other. An (n, D)-source with min-entropy k is called an (n, D, k)-source. A bit-fixing source is an $(n, 2)$-source with the additional condition that each bit of the source has min-entropy either 0 or 1.*

The task of this paper is to extract randomness from such (n, D, k)-sources.

Definition 2. *For $n, D, k, s, m \in \mathbb{N}$ and $\varepsilon \in [0, 1]$, a function $\mathrm{EXT} : [D]^n \times \{0,1\}^s \to \{0,1\}^m$ is called an (n, D, k, ε)-extractor if for any (n, D, k)-source $\mathcal{X}$, $\|\mathrm{EXT}(\mathcal{X}, \mathcal{U}_s) - \mathcal{U}_m\|_1 \leq \varepsilon$.*

The second input, of s-bit long, to an extractor is called its seed. We allow the case of $s = 0$ (i.e. without a seed) and we call such an extractor a *seedless* (or *deterministic*) extractor. The entropy loss of an extractor is the value $k + s - m$. Minimizing this entropy loss is one of the main goals of extractor construction. Moreover, one usually prefers constructions which are *explicit*, in the sense that given any input, one can compute the output in polynomial time.

3   Extractor from Random Walk

In this section, we give an explicit seedless extractor for independent-symbol sources, which works for any min-entropy k but only extracts about $\log k$ bits.

Theorem 1. *For any $n, k, D \in \mathbb{N}$ and any prime number $M \geq D$, there is an explicit (n, D, k, ε)-extractor $\mathrm{EXT}_0 : [D]^n \to [M]$, with $\varepsilon \leq \sqrt{M} \cdot e^{-k/(8M^2 \log D)}$.*

Proof. We will work on the group $\mathbb{Z}_M$, for a prime M, and see any symbol $\mathcal{X}_i \in [D]$ of the source as an element in $\mathbb{Z}_M$. Throughout this section, operation $+$ or $-$ on elements in $\mathbb{Z}_M$ is understood as an operation over the group $\mathbb{Z}_M$. Our extractor $\mathrm{EXT}_0 : [D]^n \to [M]$ is then defined as $\mathrm{EXT}_0(\mathcal{X}) = \sum_i \mathcal{X}_i$, which can be seen as taking an n-step walk on the group $\mathbb{Z}_M$, using the n symbols from the source in the following way. Each time when we are at some state $v \in \mathbb{Z}_M$ (initially at $0 \in \mathbb{Z}_M$) and read a symbol a from the source, we go to the state $v + a \in \mathbb{Z}_M$. The extractor of Kamp and Zuckerman [13] for bit-fixing sources can be seen as a special case of ours, with $D = 2$ and $\mathcal{X}_i \in \{-1, 1\}$.

As in [13], we will show that each step of the walk brings the distribution closer to uniform if the symbol read from the source contains some randomness, but our analysis is totally different. See a distribution over $\mathbb{Z}_M$ as an M-dimensional vector in the natural way. Suppose the current distribution is $\mathcal{P} = (\mathcal{P}_1, \ldots, \mathcal{P}_M)$ and the next symbol in the source has a distribution $\beta = (\beta_1, \ldots, \beta_M)$ (let $\beta_i = 0$ for $D + 1 \leq i \leq M$). Then the next distribution is $\bar{\mathcal{P}} = (\bar{\mathcal{P}}_1, \ldots, \bar{\mathcal{P}}_M)$ with $\bar{\mathcal{P}}_i = \sum_{j \in \mathbb{Z}_M} \beta_j \mathcal{P}_{i-j}$ for $i \in \mathbb{Z}_M$. Let $\mathcal{U}$ denote the uniform distribution over $\mathbb{Z}_M$. Let $\delta = \mathcal{P} - \mathcal{U}$ and $\bar{\delta} = \bar{\mathcal{P}} - \mathcal{U}$, i.e., $\delta_i = \mathcal{P}_i - 1/M$ and $\bar{\delta}_i = \bar{\mathcal{P}}_i - 1/M$ for $i \in \mathbb{Z}_M$. The following lemma, to be proved in Section 3.1, shows the progress we can make after each step.

Lemma 1. $\|\bar{\delta}\|_2^2 \leq \|\delta\|_2^2 \cdot (1 - \mathrm{H}_\infty(\beta)/(4M^2 \log D))$.

This implies that $\|\mathrm{EXT}_0(\mathcal{X}) - \mathcal{U}\|_2^2 \leq \prod_{t \in [n]} (1 - \mathrm{H}_\infty(\mathcal{X}_t)/(4M^2 \log D)) \leq e^{-\sum_{t \in [n]} \mathrm{H}_\infty(\mathcal{X}_t)/(4M^2 \log D)}$. Since the n symbols of the source are independent of each other, we have $\sum_{t \in [n]} \mathrm{H}_\infty(\mathcal{X}_t) = \mathrm{H}_\infty(\mathcal{X}) = k$, so the bound above becomes $e^{-k/(4M^2 \log D)}$. Then by Cauchy-Schwartz inequality, we have the theorem. $\square$

3.1   Proof of Lemma 1

Note that for $i \in \mathbb{Z}_M$, $\bar{\delta}_i = \sum_{j \in \mathbb{Z}_M} \beta_j \delta_{i-j}$. So $\|\bar{\delta}\|_2^2 = \sum_i (\sum_j \beta_j \delta_{i-j})^2 = \sum_i \sum_j \beta_j^2 \delta_{i-j}^2 + \sum_i \sum_{j \neq \ell} \beta_j \beta_\ell \delta_{i-j} \delta_{i-\ell}$ which, using the equality $ab = (a^2 + b^2 - (a - b)^2)/2$ on the second term, equals

$$\sum_j \beta_j^2 \sum_i \delta_{i-j}^2 + \sum_{j \neq \ell} \beta_j \beta_\ell \sum_i (\delta_{i-j}^2 + \delta_{i-\ell}^2 - (\delta_{i-j} - \delta_{i-\ell})^2)/2$$

$$= \sum_j \beta_j^2 \|\delta\|_2^2 + \sum_{j \neq \ell} \beta_j \beta_\ell \|\delta\|_2^2 - \sum_{j \neq \ell} \beta_j \beta_\ell \sum_i (\delta_{i-j} - \delta_{i-\ell})^2/2$$

$$= \|\delta\|_2^2 - \sum_{j \neq \ell} \beta_j \beta_\ell \sum_i (\delta_i - \delta_{i+j-\ell})^2/2,$$

where the last line follows from the fact that $\sum_j \beta_j^2 + \sum_{j \neq \ell} \beta_j \beta_\ell = (\sum_j \beta_j)^2 = 1$. The lemma then follows easily from the following two claims: (1) for any nonzero $s \in \mathbb{Z}_M$, $\sum_{i \in \mathbb{Z}_M} (\delta_i - \delta_{i+s})^2 \geq \|\delta\|_2^2/M^2$ and (2) $\sum_{j \neq \ell} \beta_j \beta_\ell \geq H_\infty(\beta)/(2 \log D)$.

Now we prove the first claim. By an average argument, there exists an $i_0 \in \mathbb{Z}_M$ such that $\delta_{i_0}^2 \geq \|\delta\|_2^2/M$. Next, since $\sum_i \delta_i = 0$, there exists an $i_1 \in \mathbb{Z}_M$ such that δ_{i_1} and δ_{i_0} have different signs, so $|\delta_{i_0} - \delta_{i_1}|^2 \geq \delta_{i_0}^2 \geq \|\delta\|_2^2/M$. As M and s are relatively prime, there exists an integer $t \in [1, M-1]$ such that $i_1 = i_0 + ts$ over $\mathbb{Z}_M$. By a triangle inequality, $\sum_{1 \leq j \leq t} |\delta_{i_0+(j-1)s} - \delta_{i_0+js}| \geq |\delta_{i_0} - \delta_{i_0+ts}| = |\delta_{i_0} - \delta_{i_1}|$. Finally, $\sum_{i \in \mathbb{Z}_M} (\delta_i - \delta_{i+s})^2 \geq \sum_{1 \leq j \leq t} (\delta_{i_0+(j-1)s} - \delta_{i_0+js})^2$ which by Cauchy-Schwartz inequality is at least $(\sum_{1 \leq j \leq t} |\delta_{i_0+(j-1)s} - \delta_{i_0+js}|)^2/t \geq |\delta_{i_0} - \delta_{i_1}|^2/t \geq \|\delta\|_2^2/M^2$.

Next, we prove the second claim. Let $\hat{\beta} = \max\{\beta_i : i \in [M]\}$, so $H_\infty(\beta) = \log(1/\hat{\beta})$. Then $\sum_{j \neq \ell} \beta_j \beta_\ell = \sum_j \beta_j \sum_{\ell \neq j} \beta_\ell \geq \sum_j \beta_j (1 - \hat{\beta}) = 1 - \hat{\beta}$. Note that β is a distribution over $[D]$, so $\hat{\beta} \in [1/D, 1]$. For $\hat{\beta}$ in this range, we have $1 - \hat{\beta} \geq (\log(1/\hat{\beta}))(1 - 1/D)/\log D \geq H_\infty(\beta)/(2 \log D)$. This proves the second claim and completes the proof of Lemma 1.

4   Extracting More Randomness

Building on the extractor in the previous section, we have the following two extractors, which generalize the corresponding ones in [9].

Theorem 2. *For any constant $\gamma \in (0, 1/2)$, for any $D = 2^d \in \mathbb{N}$, there exists $n_0 \in \mathbb{N}$ such that for any $n \geq n_0$, $k \geq n^{1/2+\gamma}$, and $\varepsilon \geq 2^{-cn^\gamma}$, there exists an explicit seedless (n, D, k, ε)-extractor $\mathrm{EXT} : [D]^n \to \{0,1\}^m$ with $m \geq k - O(d \log(1/\varepsilon))$.*

Theorem 3. *For any $D = 2^d \in \mathbb{N}$, there exists $n_0 \in \mathbb{N}$ and constants $c_1 > 0$, $c_2 \in (0, 1)$, $c_3 \in (0, 1/c_2)$ such that for $n \geq n_0$, $k \geq \log^{c_1} n$, and $\varepsilon \geq k^{-c_2}$, there exists an explicit seedless (n, D, k, ε)-extractor $\mathrm{EXT} : [D]^n \to \{0,1\}^m$ with $m \geq k - O(d(1/\varepsilon)^{c_3})$.*

The first one works for the case of large min-entropy and can achieve a smaller error and a smaller entropy loss, while the second can work for the case of smaller min-entropy but has a larger error and a larger entropy loss. The proofs of the two theorems are very similar to the corresponding ones in [9]. The main difference is that we consider independent-symbol sources, so we cannot build on the extractor of [13] as [9] did, and instead, we build on our extractor in Theorem 1. Furthermore, we do a more careful analysis in order to cover a wider range of parameters and identify the tradeoff between error and entropy loss. For example, our theorems show that when d is small and a large ε is allowed, the entropy loss can become very small. We omit the proofs here due to space constraint.

5   Existential Upper Bound on Entropy Loss

One may wonder if it is possible to extract more randomness than our two extractors in the previous section. In this section, we show the existence of a (non-explicit) seedless extractor for independent-symbol source with entropy loss $O(\log(1/\varepsilon))$.

Theorem 4. *Suppose $k \geq c\log(Dn/\varepsilon)$ for a large enough constant c. Then there exists an (n, D, k, ε)-extractor $\mathrm{EXT} : [D]^n \to \{0,1\}^m$ with $m \geq k - O(\log(1/\varepsilon))$.*

The proof is somewhat standard, and due to the space limitation, we only sketch the idea here. The existence of such an extractor is guaranteed by a probabilistic argument: we show that a randomly chosen function is an (n, D, k, ε)-extractor with a positive probability. Our first step is to show that for any given (n, D, k)-source, a randomly chosen function fails on it with a small probability. However, the number of such sources is infinite. Our next step is to show that it suffices to consider a much smaller class of sources, namely, those (n, D, k)-sources $\mathcal{X}$ with the property that for each $i \in [D]$, $\mathcal{X}_i$ is an "almost flat" distribution and $H_\infty(\mathcal{X}_i)$ is a multiple of some number α. This is guaranteed by the fact, which is our main technical contribution in this section, that any (n, D, k, ε)-source is close an (n, D, k, ε)-source which can be expressed as a convex combination of sources with this property. The number of such sources is small, and the theorem then follows from a union bound.

6   Lower Bound on Entropy Loss

In this section, we show that the existential upper bound on the entropy loss in Section 5 is tight by giving a matching lower bound. In fact, we show that even for bit-fixing sources and even allowing a seed, any extractor must suffer an entropy loss of $\Omega(\log(1/\varepsilon))$.

Theorem 5. *Let $\mathrm{EXT} : \{0,1\}^n \times \{0,1\}^s \to \{0,1\}^m$ be an $(n, 2, k, \varepsilon)$-extractor for bit-fixing sources, with $n, s, m \in \mathbb{N}$, $k = \omega(1)$, and $\varepsilon \in (0,1)$. Then $m \leq k + s - \Omega(\log(1/\varepsilon))$.*

We will basically follow the proof idea in [23]. Briefly speaking, given any $\mathrm{EXT} : \{0,1\}^n \times \{0,1\}^s \to \{0,1\}^m$ with m exceeding the bound, we will show the existence of a bit-fixing source of min-entropy k on which EXT fails, using a probabilistic argument. Before giving the proof, let us first state some definitions and lemmas which will be needed. For any $z \in \{0,1\}^m$, let $S^{(z)}$ denote the set $\{x \in \{0,1\}^n : \exists y \in \{0,1\}^s \text{ s.t. } z = \mathrm{EXT}(x, y)\}$, and we say that z is δ-missed by $X \subseteq \{0,1\}^n$ if $|\Pr_{x \in S^{(z)}}[x \in X] - \Pr_{x \in \mathcal{U}_n}[x \in X]| \geq \delta$. We will rely on the following lemma from [23].[1]

Lemma 2. *Suppose $\mathcal{X}$ is the uniform distribution over a set $X \subseteq \{0,1\}^n$ with $|X| = 2^k$, and $\|\mathrm{EXT}(\mathcal{X}, \mathcal{U}_s) - \mathcal{U}_m\|_1 \leq \varepsilon$. Then at most $4\sqrt{\varepsilon}$ fraction of $z \in \{0,1\}^m$ can be $(2^{-(n-k)}\sqrt{\varepsilon})$-missed by X.*

---

[1] Note that this lemma does not appear explicitly in [23] but corresponds to Claim 2.7 there, which is stated in a graph-theoretical term and says that any extractor gives rise to some kind of "slice-extractor".

For $n, t \in \mathbb{N}$, $S \subseteq \{0,1\}^n$, $I \in P([n], t)$, $u \in \{0,1\}^t$, and $\beta \in (0,1)$, we say that u is β-biased in S_I if $|\Pr_{x \in S}[x_I = u] - 2^{-t}| > \beta$. Our key lemma is the following.

Lemma 3. *Suppose $n, t \in \mathbb{N}$ with $n - t = \omega(1)$, $0 < \delta \leq 1/c$ for some large enough constant c, and $S \subseteq \{0,1\}^n$ satisfies the property that over random $I \in P([n], t)$ and $u \in \{0,1\}^t$, u is $(2^{-t}\delta)$-biased in S_I with probability at most 8δ. Then $|S| \geq 2^t (1/\delta)^{\Omega(1)}$.*

Note that a set S satisfying the property in Lemma 3 can be seen as an "almost" t-wise independent space, in the sense that the uniform distribution over S looks random on most sets of t dimensions. This can be seen as a relaxation of the standard notion of approximate t-wise independent space. Lemma 3 gives a size lower bound on such a set, which seems to have an interest of its own. We will prove the lemma in Section 6.1. With this lemma, we can now prove Theorem 5.

Proof. (of Theorem 5) Assume for the sake of contradiction that $m \geq k + s - o(\log(1/\varepsilon))$. We will show that in this case EXT fails on some bit-fixing source of min-entropy k. Following [23], the existence of such a source will be shown using a probabilistic argument. The difference is that [23] had the luxury of having all possible sources of min-entropy k to search through, while we are limited to the much smaller class of bit-fixing sources, which makes our task much harder. We randomly generate such a bit-fixing source in the following way:

– Randomly pick a set $I \in P([n], n - k)$ and a string $u \in \{0,1\}^{n-k}$. Generate the source $\mathcal{X}_I^u$ which is uniform over the set $X_I^u = \{x \in \{0,1\}^n : x_I = u\}$.

Next, we will show that EXT fails with a positive probability over such a randomly generated $\mathcal{X}_I^u$. As in [23], the idea is to show that when m is large, most z's in $\{0,1\}^m$ can only have a small set $S^{(z)}$, and such z's are $(2^{-(n-k)}\sqrt{\varepsilon})$-missed by X_I^u with a non-negligible probability. As we will see next, this probability is guaranteed by Lemma 3, by observing that the condition that z is $(2^{-(n-k)}\sqrt{\varepsilon})$-missed by X_I^u is exactly the condition that u is $(2^{-(n-k)}\sqrt{\varepsilon})$-biased in $S_I^{(z)}$.

Let $t = n - k$ and $\delta = \varepsilon^{1/2}$, and note that $\mathbb{E}_z[|S^{(z)}|] = 2^{n+s}/2^m = 2^t(1/\delta)^{o(1)}$. Call z *heavy* if $|S^{(z)}| \geq 2^t(1/\delta)^{\Omega(1)}$ and call z *light* otherwise. By Markov inequality, at most $1/2$ fraction of z's are heavy. From Lemma 3, for any light z, with $|S^{(z)}| < 2^t(1/\delta)^{\Omega(1)}$, the probability over $I \in P([n], t)$ and $u \in \{0,1\}^t$ that z is $(2^{-t}\delta)$-missed by X_I^u is more than 8δ. By an average argument, there must exist $I \in P([n], t)$ and $u \in \{0,1\}^t$ such that more than 8δ fraction of light z's are $(2^{-t}\sqrt{\varepsilon})$-missed by X_I^u. Thus, for this I and u, more than $(1/2)8\delta = 4\sqrt{\varepsilon}$ fraction of all possible $z \in \{0,1\}^m$ are $(2^{-t}\sqrt{\varepsilon})$-missed by X_I^u. From Lemma 2, this implies that $\|\text{EXT}(\mathcal{X}_I^u) - \mathcal{U}_m\|_1 > \varepsilon$, a contradiction. Therefore, one must have $m \leq k + s - \Omega(\log(1/\varepsilon))$, which proves the theorem. □

6.1   Proof of Lemma 3

Consider any set S satisfying the property stated in the lemma. Our goal is to show a lower bound on the size of such a set. For r-wise independent spaces, a tight lower bound on their size is known [2,7], and we would like to apply it to get our bound. However, there are two difficulties in front of us. One is that S only guarantees some randomness property on most, instead of all, collections of t dimensions. The other is that the randomness property only guarantees being close to random instead of perfectly random. We get around these by showing that for some properly chosen $r < t$, S embeds many disjoint copies of r-wise independent spaces.

We say that S is $\sqrt{\delta}$-uniform on $I \in P([n], t)$ if the fraction of $u \in \{0, 1\}^t$ being $(2^{-t}\delta)$-biased in S_I is at most $\sqrt{\delta}$. From the property of S, a Markov inequality shows that S is not $\sqrt{\delta}$-uniform on at most $8\sqrt{\delta}$ fraction of $I \in P([n], t)$. By an average argument, there must exist some $J \in P([n], t - r)$, for some r to be determined later, such that S is not $\sqrt{\delta}$-uniform on $J \cup T$ for at most $8\sqrt{\delta}$ fraction of $T \in P([n] \setminus J, r)$. Fix one such set J. We will partition S into subsets $S_{J,v} = \{x \in S : x_J = v\}$, for $v \in \{0, 1\}^{t-r}$, and show that many of them embed an r-wise independent space.

Let us focus on the set $\bar{J} = [n] \setminus J$ and those subsets $T \in P(\bar{J}, r)$. Let $k = n - t$, so $|\bar{J}| = k + r$. Call $T \in P(\bar{J}, r)$ nice if S is $\sqrt{\delta}$-uniform on $J \cup T$. Call $v \in \{0, 1\}^{t-r}$ bad for T if (v, w) is $(2^{-t}\delta)$-biased in $S_{J \cup T}$ for some $w \in \{0, 1\}^r$. Then for any nice T, the fraction of $v \in \{0, 1\}^{t-r}$ bad for T cannot exceed $2^r\sqrt{\delta}$. Thus, the fraction of $v \in \{0, 1\}^{t-r}$ bad for at least $2^{r+1}\sqrt{\delta}$ fraction of nice T's is at most $1/2$. Fix any $v \in \{0, 1\}^{t-r}$ which is bad for at most $2^{r+1}\sqrt{\delta}$ fraction of nice T's, and thus is bad for at most $\alpha = 2^{r+1}\sqrt{\delta} + 8\sqrt{\delta}$ fraction of all possible T's in $P(\bar{J}, r)$. Next, we show that $|S_{J,v}| \geq 2^r(1/\delta)^{\Omega(1)}$.

Assume without loss of generality that $|S_{J,v}| < 2^r/(6\delta)$ (otherwise, we are done), which means that $2^{-r}(6\delta) < 1/|S_{J,v}|$. Then we have the following.

Claim. Suppose v is not bad for $T \in P(\bar{J}, r)$. Then $\forall w \in \{0, 1\}^r$, $\Pr_{x \in S_{J,v}}[x_T = w] = 2^{-r}$.

Proof. Suppose v is not bad for T, so $\forall w \in \{0, 1\}^r$, $|\Pr_{x \in S}[(x_J, x_T) = (v, w)] - 2^{-t}| \leq 2^{-t}\delta$. This implies that $|\Pr_{x \in S}[x_J = v] - 2^{-(t-r)}| \leq 2^{-(t-r)}\delta$. Then for any $w \in \{0, 1\}^r$, $\Pr_{x \in S_{J,v}}[x_T = w] = \Pr_{x \in S}[(x_J, x_T) = (v, w)] / \Pr_{x \in S}[x_J = v]$ is at most $2^{-r}(1 + \delta)/(1 - \delta) \leq 2^{-r}(1 + 3\delta)$ and at least $2^{-r}(1 - \delta)/(1 + \delta) \geq 2^{-r}(1 - 2\delta)$. That is,

$$\forall w \in \{0, 1\}^r, \quad \left| \Pr_{x \in S_{J,v}}[x_T = w] - 2^{-r} \right| \leq 2^{-r}(3\delta) < 1/(2|S_{J,v}|). \tag{1}$$

Consider the 2^r probabilities $\Pr_{x \in S_{J,v}}[x_T = w]$, for $w \in \{0, 1\}^r$, which are all multiples of $1/|S_{J,v}|$. If they were not all equal to 2^{-r}, there must exist $w, w' \in \{0, 1\}^r$ such that $\Pr_{x \in S_{J,v}}[x_T = w] < 2^{-r} < \Pr_{x \in S_{J,v}}[x_T = w']$. Then the difference between 2^{-r} and one of these two probabilities must be at least half their gap, which is at least $1/(2|S_{J,v}|)$, a contradiction to condition (1) above. Therefore, these 2^r probabilities must all be equal to 2^{-r}. □

From the claim, we next show that $S_{J,v}$ embeds an r-wise independent space, which then implies a lower bound on $|S_{J,v}|$. We differentiate two cases according to the range of δ. In the first case, when $\delta < 1/(2k)^8$, we choose

$r = \lceil (\log(1/\delta))/(4\log(2k)) \rceil \geq 2$. Note that now $\alpha = o(\sqrt[4]{\delta})$, and $|P(\bar{J}, r)| = \binom{k+r}{r} < (e(k+r)/r)^r \leq (2k)^r \leq \sqrt[4]{1/\delta}$. Since $\alpha \cdot |P(\bar{J}, r)| < 1$, v is not bad for any $T \in P(\bar{J}, r)$. This means that the set $S_{J,v}$ projected to dimensions in $\bar{J}$ is an r-wise independent space. From [2,7], such a set must have size at least $|\bar{J}|^{\Omega(r)} = (k+r)^{\Omega(r)} = 2^r(1/\delta)^{\Omega(1)}$.

In the second case, when $\delta \geq 1/(2k)^8$, we choose $r = 2$. Then $\alpha = O(\sqrt{\delta})$, and the following implies that the set $S_{J,v}$ projected to dimensions in A gives a pair-wise independent space, so by [2,7] we have $|S_{J,v}| \geq |A|^{\Omega(1)} \geq (1/\delta)^{\Omega(1)} = 2^r(1/\delta)^{\Omega(1)}$.

Claim. There exists a subset $A \subseteq \bar{J}$ of size $(1/\delta)^{\Omega(1)}$ such that v is not bad for any $T \in P(A, 2)$.

Proof. Consider the undirected graph G with vertex set $V = \bar{J}$ and edge set $E = \{T \in P(\bar{J}, 2) : v \text{ is not bad for } T\}$. Note that $|E| \geq (1-\alpha)\binom{|V|}{2}$. When $\alpha \geq 1/|V|$, one can show that $|E| > (1-2\alpha)|V|^2/2$. By the well-known Turan's theorem in graph theory (e.g., see Theorem 4.7 in [11]), G must contain a clique of size $\Omega(1/\alpha) = (1/\delta)^{\Omega(1)}$. When $\alpha < 1/|V|$, $|E| \geq (1-1/|V|)\binom{|V|}{2} > (1-2/|V|)|V|^2/2$, and Turan's theorem implies that G contains a clique of size $\Omega(|V|) = \Omega(k+r) \geq (1/\delta)^{\Omega(1)}$. Let A be the vertex set of the largest clique in G, and we have the claim. □

We have shown in both cases that $|S_{J,v}| \geq 2^r(1/\delta)^{\Omega(1)}$, for any v which is bad for at most α fraction of $T \in P(\bar{J}, r)$. Since there are at least $(1/2)2^{t-r}$ such v's, and the corresponding sets $S_{J,v}$'s are all disjoint subsets of S, we conclude that $|S| = 2^t(1/\delta)^{\Omega(1)}$.

References

1. N. Alon, O. Goldreich, J. Håstad, and R. Peralta. Simple constructions of almost k-wise independent random variables. *FOCS'90*, pp. 544–553.
2. N. Alon, L. Babai, and A. Itai. A fast and simple randomized parallel algorithm for the maximal independent set problem. *J. Algorithms*, 7(4), pp. 567–583, 1986.
3. B. Barak, R. Impagliazzo, and A. Wigderson. Extracting randomness using few independent sources. *FOCS'04*, pp. 384–393.
4. B. Barak, G. Kindler, R. Shaltiel, B. Sudakov, and A. Wigderson. Simulating Independence: New constructions of condensers, Ramsey graphs, dispersers, and extractors. *STOC'05*, pp. 1–10.
5. J. Bourgain. More on the sum-product phenomenon in prime fields and its applications. *International Journal of Number Theory*, 1(1), pp. 1–32, 2005.
6. B. Chor and O. Goldreich. Unbiased bits from sources of weak randomness and probabilistic communication complexity. *SIAM J. Comput.*, 17(2), pp. 230–261, 1988.
7. B. Chor, O. Goldreich, J. Håstad, J. Friedman, S. Rudich, and Roman Smolensky. The bit extraction problem of t-resilient functions. *FOCS'85*, pp. 396–407.
8. Y. Dodis, A. Elbaz, R. Oliveira, and R. Raz. Improved randomness extraction from two independent sources. *RANDOM'04*, pp. 334–344.

9. A. Gabizon, R. Raz, and R. Shaltiel. Deterministic extractors for bit-fixing sources by obtaining an independent seed. *FOCS'04*, pp. 394–403.
10. R. Impagliazzo, R. Shaltiel, and A. Wigderson. Extractors and pseudo-random generators with optimal seed length. *STOC'00*, pp. 1–10.
11. S. Jukna. *Extremal Combinatorics*. Springer-Verlag, 2001.
12. J. Kamp, A. Rao, S. Vahan, and D. Zuckerman. Deterministic extractors for small-space sources. *STOC'06*, to appear.
13. J. Kamp and D. Zuckerman. Deterministic extractors for bit-fixing sources and exposure-resilient cryptography. *FOCS'03*, pp. 92–101.
14. R. König and U. Maurer. Generalized strong extractors and deterministic privacy amplification. In *Proc. Cryptography and Coding*, pp. 322–339, 2005.
15. C.-J. Lee, C.-J. Lu, S.-C. Tsai, and W.-G. Tzeng. Extracting randomness from multiple independent sources. *IEEE Transactions on Information Theory*, 51(6), pp. 2224–2227, 2005.
16. C.-J. Lu. Encryption against storage-bounded adversaries from on-line strong extractors. *J. Cryptology*, 17(1), pp. 27–42, 2004.
17. C.-J. Lu, O. Reingold, S. Vadhan, and A. Wigderson. Extractors: Optimal up to constant factors. *STOC'03*, pp. 602–611.
18. J. Naor and M. Naor. Small-bias probability spaces: efficient constructions and applications. *SIAM J. Comput.*, 22(4), pp. 838–856, 1993.
19. N. Nisan and A. Ta-Shma. Extracting randomness: A survey and new constructions. *J. Comput. Syst. Sci.*, 58(1), pp. 148–173, 1999.
20. N. Nisan and D. Zuckerman. Randomness is linear in space. *J. Comput. Syst. Sci.*, 52(1), pp. 43–52, 1996.
21. R. Raz. Extractors with weak random seeds. *STOC'05*, pp. 11–20.
22. R. Raz, O. Reingold, and S. Vadhan. Extracting all the randomness and reducing the error in Trevisan's extractors. *STOC'99*, pp. 149–158.
23. J. Radhakrishnan and A. Ta-Shma. Bounds for dispersers, extractors, and depth-two superconcentrators. *SIAM J. Discrete Math.*, 13(1), pp. 2–24, 2000.
24. O. Reingold, R. Shaltiel, and A. Wigderson. Extracting randomness via repeated condensing. *FOCS'00*, pp. 12–14.
25. R. Shaltiel. Recent developments in explicit constructions of extractors. *Bulletin of the European Association for Theoretical Computer Science*, 77, pp. 67–95, 2002.
26. R. Shaltiel and C. Umans. Simple extractors for all min-entropies and a new pseudo-random generator. *FOCS'01*, pp. 648–657.
27. A. Ta-Shma, C. Umans, and David Zuckerman. Loss-less condensers, unbalanced expanders, and extractors. *STOC'01*, pp. 143–152.
28. A. Ta-Shma and D. Zuckerman. Extractor codes. *STOC'01*, pp. 193–199.
29. L. Trevisan. Extractors and pseudorandom generators. *JACM*, 48(4), pp. 860–879, 2001.
30. S. Vadhan. Constructing locally computable extractors and cryptosystems in the bounded-storage model. *J. Cryptology*, 17(1), pp. 43–77, 2004.
31. A. Wigderson and D. Zuckerman. Expanders that beat the eigenvalue bound: Explicit construction and applications. *Combinatorica*, 19(1), pp. 125–138, 1999.
32. D. Zuckerman. General weak random sources. *FOCS'90*, pp. 534–543.
33. D. Zuckerman. Simulating BPP using a general weak random source. *Algorithmica*, 16(4/5), pp. 367–391, 1996.
34. D. Zuckerman. Randomness-optimal oblivious sampling. *Random Structures and Algorithms*, 11, pp. 345–367, 1997.

Gap Amplification in PCPs Using Lazy Random Walks

Jaikumar Radhakrishnan[1,2]

[1] School of Technology and Computer Science,
Tata Institute of Fundamental Research,
Homi Bhabha Road, Mumbai 400005, India
[2] Toyota Technological Institute at Chicago,
1427 E 60th Street, Chicago, IL 60637, USA
jaikumar@tti-c.org

Abstract. We show an alternative implementation of the gap amplification step in Dinur's [4] recent proof of the PCP theorem. We construct a product G^t of a constraint graph G, so that if every assignment in G leaves an ϵ-fraction of the edges unsatisfied, then in G^t every assignment leaves an $\Omega(t\epsilon)$-fraction of the edges unsatisfied, that is, it amplifies the gap by a factor $\Omega(t)$. The corresponding result in [4] showed that one could amplify the gap by a factor $\Omega(\sqrt{t})$. More than this small quantitative improvement, the main contribution of this work is in the analysis. Our construction uses random walks on expander graphs with exponentially distributed length. By this we ensure that some random variables arising in the proof are automatically independent, and avoid some technical difficulties.

1 Introduction

Probabilistic checkable proofs occupy a central place in complexity theory today, especially in the study of the class NP and the hardness of approximation for combinatorial problems. The cornerstone of this area is the amazing PCP theorem of Arora, Lund, Motwani, Sudan and Szegedy [1], which states, e.g., that there is a constant $\epsilon > 0$, such that it is NP-hard to distinguish between satisfiable 3CNF expressions and those where only a fraction $1 - \epsilon$ of the clauses can be simultaneously satisfied. The original proof of this theorem was algebraic, and built on a long line of research that made several deep and subtle contributions [3,5,2].

Recently, Irit Dinur [4] presented a remarkable and essentially combinatorial proof the PCP theorem. This proof uses the notion of gap amplification in constraint graphs. There are several carefully chosen steps in this proof. The key new insight, however, is in the product construction based on random walks on expander graphs. In this paper, we suggest a modification of Dinur's product construction, which yields a slightly better amplification and avoids some of the complications in the original proof.

In order to describe Dinur's proof of the PCP theorem and our modification, we need some definitions.

M. Bugliesi et al. (Eds.): ICALP 2006, Part I, LNCS 4051, pp. 96–107, 2006.
© Springer-Verlag Berlin Heidelberg 2006

Definition 1.1 (Constraint graph, assignment). *A constraint graph G is a tuple $\langle V, E, \Sigma, \mathcal{C} \rangle$, where (V, E) is an undirected graph (we allow multiple edges and self-loops), Σ is a finite set called the alphabet of G, and $\mathcal{C}$ is a collection of constraints, $\langle c_e : e \in E \rangle$, where each c_e is a function from $\Sigma \times \Sigma$ to $\{0, 1\}$. An assignment is a function $A : V \to \Sigma$. We say that the assignment A satisfies an edge e of the form (u, v), if $c_e(A(u), A(v)) = 1$. We say that the assignment A satisfies G, if A satisfies all edges in G. If there is an assignment that satisfies G, then we say that G is satisfiable. We say that G is ϵ-far from satisfiable if every assignment leaves at least a fraction ϵ of the edges of G unsatisfied. Let*

$$\text{UNSAT}(G) \;=\; \max\{\epsilon : G \text{ is } \epsilon\text{-unsatisfiable}\} \;=\; \min_A \frac{|\{e : A \text{ does not satisfy } e\}|}{|E|}.$$

Constraint graphs arise naturally in computational problems. For example, consider the graph 3-coloring problem. This can be modeled as a constraint graph, where the alphabet is $\{R, G, B\}$ and the constraints on all edges are inequality functions. It is thus NP-complete to determine if a given constraint graph is satisfiable. The PCP theorem is a considerable strengthening of this assertion. Using the terminology of constraint graphs, we can state it as follows.

Theorem 1.1 (The PCP theorem). *There is a constant $\epsilon_0 > 0$, such that for for every language L in NP, there is a polynomial-time reduction f from L to the satisfaction problem for constraint graphs with alphabet $\{0, 1\}^2$, such that*

- *if $x \in L$, then $f(x)$ is satisfiable;*
- *if $x \notin L$, then $f(x)$ is ϵ_0-unsatisfiable.*

The PCP theorem implies that if we could in polynomial time approximate UNSAT(G) sufficiently closely, then P=NP.

Dinur's recent proof of the PCP theorem works with constraint graphs and uses gap amplification. The broad idea is as follows. We consider the *gap version* of the *constraint graph satisfaction problems* in which one has to distinguish satisfiable constraint graphs from those that are ϵ-far from satisfiable. We refer to ϵ as the gap of this problem. For example, the easy reduction mentioned above, reduces the 3-coloring problem to the constraint satisfaction problem with gap $\frac{1}{n^2}$, where n is the number of vertices in the original graph. Dinur shows that this gap can be amplified: she presents a general procedure that, roughly speaking, transforms a constraint graph G to another graph G' so that (1) if G is satisfiable, then G' is satisfiable, and (2) if G is ϵ-far from satisfiable, then G' is (2ϵ)-far from satisfiable; furthermore, and crucially, the size of G' is at most a constant times the size of G. By composing this with the original reduction, we reduce the 3-coloring problem to the constraint graph satisfaction problem with twice the original gap. The idea, then, is to apply this procedure approximately $2 \log n$ times starting with the original constraint graph. If the original graph is 3-colorable, then the final graph is satisfiable. On the other hand, if the original graph is not 3-colorable, then the final graph is is ϵ_0-far from satisfiable (for some constant $\epsilon_0 > 0$). Since each iteration increases the size of the graph by only a constant factor, the size of the final constraint graph is a polynomial in the size of the original input instance.

In the remainder of this section, we will present a brief overview of the various steps involved in Dinur's proof and describe our contribution.

1.1   Overview of Dinur's Proof

Dinur's proof has three steps. In the first step, we transform the input constraint graph into a constant degree expander.

Theorem 1.2 (Step 1: constant degree expander). *There are constants d, $\lambda < d$, $C_1 \geq 1$ and $D_1 \geq 1$, and a polynomial time transformation f_1 on constraint graphs such that*

- $|V(f_1(G))|, |E(f_1(G))| \leq D_1 \cdot |E(G)|$;
- *if G is satisfiable, then $f_1(G)$ is satisfiable;*
- *if G is ϵ-far from satisfiable, then $f_1(G)$ is $\left(\frac{\epsilon}{C_1}\right)$-far from satisfiable;*
- *$f_1(G)$ is a d-regular expander, with $\lambda(f_1(G)) \leq \lambda$;*
- *the alphabet of $f_1(G)$ is the same as the alphabet of G.*

Since the constant C_1 is actually greater than 1, this step reduces the gap instead of amplifying it, but it prepares the graph for Step 2, which amplifies the gap substantially and makes up for the loss suffered in the first step. This second step is the most novel ingredient in Dinur's proof, and also the subject of this paper.

Theorem 1.3 (Step 2: gap amplification). *Let $G = \langle V, E, \Sigma, C \rangle$ be a constraint graph, such that (V, E) is a d-regular expander, with second largest eigenvalue λ (in absolute value). Then, for all $t \geq 1$, we can in time polynomial in the size of the output produce a constraint graph G' such that*

- *$V(G') = V(G)$;*
- *$|E(G')| \leq d^t |V(G)|$;*
- *$\Sigma(G') = \Sigma^{d^{t/2}}$;*
- *if G is satisfiable, then G' is satisfiable;*
- *$\text{UNSAT}(G') \geq \frac{\sqrt{t}}{C_2} \min\left\{\text{UNSAT}(G), \frac{1}{t}\right\}$, where $C_2 = O\left(\left(\frac{d}{d-\lambda}\right)|\Sigma|^4\right)$.*

This gives a transformation that amplifies the gap by a factor of about $\sqrt{t}$ but at the cost of increasing the alphabet size exponentially. Repeating the first two steps with a value for t much larger than $(C_1 C_2)^2$ will no doubt amplify the gap in our constraint graphs, but this gain will be accompanied by an unaffordable increase in the size of the alphabet. Thus, before we iterate we need to somehow ensure that the alphabet is small. This is achieved in Step 3.

Theorem 1.4 (Step 3: alphabet reduction). *There is a constant $C_3 > 0$ and a polynomial-time computable transformation f_3 on constraint graphs such that for every constraint graph $G = \langle V, E, \Sigma, C \rangle$, the constraint graph $G' = f_3(G)$ satisfies the following:*

- $|V(G')|, |E(G')| \leq D_3 \cdot |E(G)|$, D_3 depends only on $\Sigma(G)$;
- $\Sigma(G') = \{0, 1\}^2$;
- if G is satisfiable, then G' is satisfiable;
- if G' is ϵ-far from satisfiable, then G is $\left(\frac{\epsilon}{C_3}\right)$-far from satisfiable.

These three steps can now be applied repeatedly; we choose $t \overset{\triangle}{=} \lceil (2C_1C_2C_3)^2 \rceil$ to ensure that the gap doubles in each iteration of this three step procedure. Since, C_1, C_2 and C_3 are absolute constants (independent of the size of the graph), we can apply this procedure until the gap becomes at least $\epsilon_0 = \frac{1}{t}$; this will take $O(\log n)$ iterations. At the start of each iteration, we have $\Sigma = \{0, 1\}^2$, so the size increases by at most a fixed multiplicative factor in each iteration, and we can afford to perform the required $O(\log n)$ iterations and still get polynomial size constraint graphs in the end.

Remark: Dinur's paper is available at the ECCC archive [4]. The lecture notes of a course given by Ryan O'Donnell and Venkat Guruswami [6] present a version of Dinur's proof that uses the lazy random walk suggested in this paper. We refer the reader to these sources for a detailed description of Steps 1 and 3, and the proofs of Theorems 1.2 and 1.4. We concentrate on the Step 2 in the remainder of this paper.

1.2   Our Contribution

The main contribution of this paper is the proof of an alternative implementation of the gap amplification step. We first state the result, and then compare it with the original version (Theorem 1.3).

Theorem 1.5 (Revised step 2: gap amplification). *Let $G = \langle V, E, \Sigma, C \rangle$ be a constraint graph, such that (V, E) is a d-regular expander with second largest eigenvalue λ (in absolute value). Then, for all $t \geq 1$, one can in time polynomial in the size of the output produce a constraint graph G' such that*

- $V(G') = V(G)$;
- $|E(G')| \leq (td)^{O(t \log |\Sigma|)} |V(G)|$;
- $\Sigma(G') = \Sigma^{d^{t+1}}$;
- *if G is satisfiable, then G' is satisfiable;*
- $\text{UNSAT}(G') \geq \frac{t}{C_2'} \min \left\{ \text{UNSAT}(G), \frac{1}{t} \right\}$, *where $C_2' = O\left(\frac{d}{d-\lambda}\right)$.*

This version differs from Dinur's original version in two respects. First, the parameters in Theorem 1.5 are better than in the original version (Theorem 1.3). For a comparable increase in the size of the alphabet, the amplification in this version is proportional to t, whereas the amplification in the original version is proportional $\sqrt{t}$. Also, the constant does not depend on $|\Sigma(G)|$. However, when this lemma is used in the proof of the PCP Theorem, t is a constant and $|\Sigma| = 4$, so this improvement is inconsequential. Dinur's proof uses random walks of a fixed length on the constraint graph, while our proof uses random walks of geometrically distributed length. This modification and its analysis are the main contribution of this paper.

To understand the differences in the two approaches, we first briefly describe Dinur's construction. Dinur's proof of Theorem 1.3 is based on a product construction. The alphabet of the new graph is enlarged so that the assignment for each vertex specifies the values for all vertices of the original graph within distance $\frac{t}{2}$. The edges of the new graph correspond to walks of length t in the original graph. The constraints on these edge are defined as follows. Consider a walk of t steps starting at vertex a and ending at vertex b. Now, some of the edges of this walk are within distance $\frac{t}{2}$ of both a and b. The constraint corresponding to this walk requires that we check that all such edge constraints are satisfied by the assignments given for a and b. Using a careful combinatorial argument, Dinur shows that on an average about $\sqrt{t}$ of the old constraints are verified by a single constraint of the new graph. Using the fact that the graph is an expander, she then established that this translates into a $\sqrt{t}$ factor amplification.

Our construction is similar, but we consider walks whose lengths are geometrically distributed, with expectation t. The advantage of this is that when we consider walks passing through a fixed edge (u, v), the starting and ending vertices of these walks are independent random variables. In Dinur's proof, such independence was needed, but could be enforced approximately only for the edges that appeared in a section of length $\sqrt{t}$ near middle of the walk. Our choice of walk length avoids this difficulty, and allows us to estimate the probability of rejection more directly. We discuss this in the next section.

2   Gap Amplification: Proof of Theorem 1.5

2.1   Preliminaries

Since we wish to use random walks on the constraint graph G, it will be convenient to work with its directed version. We replace each undirected edge connecting distinct vertices u and v, by two directed edges, one of the form (u, v) and the other of the form (v, u); we replace self-loops by a single directed edge. The adjacency matrix of this graph is symmetric. Note that if for some assignment a fraction ϵ of the edges were unsatisfied in the undirected graph, then in the directed version at least a fraction $\frac{\epsilon}{2}$ of the edges are unsatisfied. We will also assume that the number of (directed) edges in G is at least $2t$. Otherwise, we make sufficient copies of all edges; this modification scales all the eigenvalues by the same factor. Since the original graph was d-regular, the outdegree of each vertex in the directed version is exactly d, and thus the largest eigenvalue of the its adjacency matrix is d. We denote the second largest eigenvalue of G (in absolute value) by $\lambda(G)$.

We now turn to the proof of Theorem 1.5. As stated earlier, the proof is based on a product construction. It will be convenient to prove the theorem in two steps. In the first step, we describe a PCP in which the verifier reads two locations from the proof, and simultaneously checks the constraints for about t of the edges of the original graph. In the second step, we transform this PCP into a constraint graph meeting the requirements of Theorem 1.5.

2.2 The Product PCP

Fix $t \geq 1$. The PCP has the following features. There is a proof, which is supposedly derived from an assignment to the original graph. There is a verifier, who probes this proof randomly at two locations, and based on the values read, decides to accept or reject. If the original constraint graph is satisfiable, then there exists a proof that the verifier accepts with probability 1. On the other hand, if the original constraint graph is ϵ-far from satisfiable, then the verifier rejects every proof with probability at least $\epsilon \cdot \Omega(t)$. We now describe the PCP and present our analysis.

The proof: For each vertex $v \in V(G)$, the proof now provides an assignment for all vertices that are within a distance t from v. That is, the proof is a function $\mathcal{A} : V(G) \to \Sigma^{d^{t+1}}$, where $\mathcal{A}(v)$ denotes this (partial) assignment provided at vertex v. We use $\mathcal{A}(v)[w]$ to refer to the value $\mathcal{A}(v)$ assigns to vertex w, and think of $\mathcal{A}(v)[w]$ as vertex v's *opinion* for the value that should be assigned to w in order to satisfy G. Thus, every vertex within distance t of w has an opinion for w; there is no guarantee, however, that these opinions agree with each other. Vertices w that don't appear within distance t of v are not explicitly assigned a value in $\mathcal{A}(v)$; for such a vertices w, we say that $\mathcal{A}(v)[w]$ is null. Let A_1 and A_2 be two partial assignments, and let $e = (u, v)$ be an edge of G. We say that A_1 and A_2 pass the test at e, if at least one of the following conditions holds: (i) one of $A_1(u), A_1(v), A_2(u)$, and $A_2(v)$ is null; (ii) A_1 and A_2 agree on $\{u, v\}$ and $c_e(A_1(u), A_2(v)) = 1$.

The verifier: The verifier picks two random vertices, **a** and **b**, of the graph and performs a test on the values stored there.

The random walk: The two vertices **a** and **b** are generated using a random walk, as follows.
  I. Let $\mathbf{v}_0$ be a random vertex chosen uniformly from $V(G)$. Repeat Step II until some condition for stopping is met.
  II. Having chosen $\mathbf{v}_0, \mathbf{v}_1, \ldots, \mathbf{v}_{i-1}$, let $\mathbf{e}_i$ be a random edge leaving $\mathbf{v}_{i-1}$, chosen uniformly among the d possibilities. Let $\mathbf{v}_i$ be the other end point of $\mathbf{e}_i$. With probability $\frac{1}{t}$, STOP and set $\mathbf{T} = i$.

The test: Suppose the random walk visits the vertices $\mathbf{a} = \mathbf{v}_0, \mathbf{v}_1, \ldots, \mathbf{v}_T = \mathbf{b}$ using the sequence of edges, $\mathbf{e}_1, \mathbf{e}_2, \ldots, \mathbf{e}_T$. If $\mathcal{A}(\mathbf{a})$ and $\mathcal{A}(\mathbf{b})$ fail (i.e. don't pass) the test at some $\mathbf{e}_i$, the verifier rejects; otherwise, she accepts. When **a** and **b** are clear from the context, we say that *the test at $\mathbf{e}_i$ fails,* when we mean that $\mathcal{A}(\mathbf{a})$ and $\mathcal{A}(\mathbf{b})$ fail the test at $\mathbf{e}_i$.

Lemma 2.1. *Suppose G is a d-regular constraint graph with $|\lambda(G)| < d$.*

(a) If G is satisfiable, then there is a proof that the verifier accepts with probability 1.

(b) If G is ϵ-far from satisfiable, then the verifier rejects every proof with probability at least

$$\left(\frac{1}{256C}\right) \cdot t \cdot \min\left\{\epsilon, \frac{1}{t}\right\},$$

where $C = 2 + \frac{d}{d-|\lambda(G)|}$.

Proof. Part (a) is straightforward. Given a satisfying assignment A for G, let the proof be the assignment $\mathcal{A}$ such that $\mathcal{A}(v)[w] = A(w)$.

The idea for part (b) is the following. Fix an assignment $\mathcal{A}$. We will argue that for such an assignment $\mathcal{A}$ to succeed in convincing the verifier, the opinions of different vertices must be generally quite consistent. This suggests that a good fraction of $\mathcal{A}$ is consistent with a fixed underlying assignment A for G. Now, since G is ϵ-far from satisfiable, A must violate at least a fraction ϵ of the constraints in G. Since the verifier examines t edges on an average, the expected number of unsatisfied edges she encounters is $t\epsilon$. Most of the work will go into showing that when she does encounter these edges, she rejects with a sufficient probability and that these rejections are not concentrated on just a few of her walks. In our analysis we will use the following fact about the verifier's random walk. (A formal proof appears in [6].)

Lemma 2.2 (Fact about the random walk). *Let $e \in E(G)$ be of the form (u, v). Consider the verifier's walks conditioned on the event that the edge e appears exactly k times (for some $k \geq 1$) in the walk, that is, the number of i's for which $\mathbf{e}_i = e$ (in particular, $v_{i-1} = u$ and $v_i = v$) is exactly k. Conditioned on this event, consider the starting vertex, $\mathbf{a}$, and the ending vertex, $\mathbf{b}$. We claim that $\mathbf{a}$ and $\mathbf{b}$ are independent random variables. Furthermore, $\mathbf{a}$ has the same distribution as the random vertex obtained by the following random process.*

> *Start the random walk at u, but stop with probability $\frac{1}{t}$ before making each move (so we stop at u itself with probability $\frac{1}{t}$). Output the final vertex.*

Similarly, we claim that $\mathbf{b}$ can be generated using a random walk starting from v and stopping with probability $\frac{1}{t}$ before each step.

Now, fix a proof $\mathcal{A}$. Let us "decode" $\mathcal{A}$ and try to obtain an assignment A for G. The idea is to define $A(u)$ to be most popular opinion available in $\mathcal{A}$ for u, but motivated by Lemma 2.2, the popularity of an opinion will be determined by considering a random walk.

The new assignment A for G: To obtain $A(u)$, we perform a random walk starting from u mentioned in Lemma 2.2 (stopping with probability $\frac{1}{t}$ before each step). Restrict attention to those walks that stop within $t-1$ steps. Let the vertex where the walk stops be $\mathbf{b}_u$. This generates a distribution on the vertices of G. For each letter σ in the alphabet, determine the probability (under this distribution) that $\mathbf{b}_u$'s opinion for u is σ. Then, let $A(u)$ be the letter that has the highest probability. Formally,

$$A(u) \overset{\Delta}{=} \arg\max_{\sigma \in \Sigma} \, \Pr[A(\mathbf{b}_u)[u] = \sigma \text{ and } T \leq t - 1].$$

We now relate the verifier's probability of rejection to the fraction of edges of G left unsatisfied by A. Since G is ϵ-far from satisfiable, a fraction ϵ of the edges of G are left unsatisfied by A. We wish to argue that whenever the verifier encounters one of these edges in her walk, she is likely to reject the walk. Let F be a subset of these unsatisfied edges of the largest size such that $\frac{|F|}{|E|} \leq \frac{1}{t}$. Then, because we assume that $|E| \geq 2t$),

$$\min\left\{\epsilon, \frac{1}{2t}\right\} \leq \frac{|F|}{|E|} \leq \frac{1}{t}. \tag{1}$$

Now, consider the edges used by the verifier in her walk: $\mathbf{e}_1, \mathbf{e}_2, \dots, \mathbf{e}_\mathbf{T}$.

Definition 2.1 (Faulty edge). *We say that the i-th edge of the verifier's walk is faulty if*

- $\mathbf{e}_i \in F$ *and*
- $\mathcal{A}(\mathbf{a})$ *and* $\mathcal{A}(\mathbf{b})$ *fail the test at* $\mathbf{e}_i$.

Let $\mathbf{N}$ be the random variable denoting the number of faulty edges on the verifier's walk.

Since the verifier rejects whenever she encounters a faulty edge on her walk, it is enough to show that $\mathbf{N} > 0$ with high enough probability. We prove the following two claims below.

Claim. [1] (a) $\mathbf{E}[\mathbf{N}] \geq t\frac{|F|}{8|E|}$ and (b) $\mathbf{E}[\mathbf{N}^2] \leq Dt\frac{|F|}{|E|}$, where $D = 2\left(2 + \frac{d}{d-|\lambda|}\right)$.

Let us now assume that these claims hold, and complete the proof of Lemma 2.1:

$$\Pr[\textit{verifier rejects}] \geq \Pr[\mathbf{N} > 0] \geq \frac{\mathbf{E}[N]^2}{\mathbf{E}[N^2]} \geq \left(\frac{1}{64D}\right) \cdot t \cdot \left(\frac{|F|}{|E|}\right)$$

$$\geq \left(\frac{1}{128D}\right) \cdot t \min\left\{\epsilon, \frac{1}{t}\right\}.$$

For the second inequality, we used the fact that for any non-negative random variable $\mathbf{X}$, $\Pr[\mathbf{X} > 0] \geq \frac{\mathbf{E}[X]^2}{\mathbf{E}[X^2]}$ (by the Chebyshev-Cantelli inequality). For the last inequality we used (1). $\qquad \square$

---

[1] In a previous version of this paper, the right hand side of part (b) had a $|\Sigma|^2$ in the denominator; Greg Plaxton suggested this stronger version.

2.3 Proofs of the Claims

Proof (of Claim (a)). We will estimate the expected number of faulty occurrences for each edge in F. Fix one such edge $e = (u,v)$, and let $\mathbf{N}_e$ denote the number of *faulty* occurrences of e in the verifier's walk. Let $\#e$ denote the number of occurrences (not necessarily faulty) of e in the walk. Condition on the event $\#e = k$, and consider the starting vertex $\mathbf{a}$ and the ending vertex $\mathbf{b}$. By Lemma 2.2, $\mathbf{a}$ and $\mathbf{b}$ can be generated using independent lazy random walks starting at u and v respectively. The probability that the walk to generate $\mathbf{a}$ traverses t or more edges is $\left(1 - \frac{1}{t}\right)^t \leq \exp(-1)$. Thus, with probability at least $\alpha \overset{\Delta}{=} 1 - \exp(-1)$ the starting vertex $\mathbf{a}$ is at a distance at most $t-1$ from u, and hence at most t from v. Let p_u be the probability that the $\mathbf{a}$ is at a distance at most $t-1$ from u and $\mathcal{A}(\mathbf{a})[u] = A(u)$; similarly, let p_v be the probability that $\mathbf{b}$ is at a distance most $t-1$ from v and $\mathcal{A}(\mathbf{b})[v] = A(v)$. Now, the test at e fails if $\mathcal{A}(\mathbf{a})[u] \neq \mathcal{A}(\mathbf{b})[u]$ (and are both not null). This happens with probability at least $\alpha(\alpha - p_u)$. Similarly, by considering v, we conclude that the test at e fails with probability at least $\alpha(\alpha - p_v)$. Furthermore, with probability at least $p_u p_v$, we have

$$c_e(\mathcal{A}(\mathbf{a})[u], \mathcal{A}(\mathbf{b})[v]) = c_e(A(u), A(v)) = 0,$$

in which case the test at e fails. Thus, overall,

$$\Pr[\text{the test at } e \text{ fails} \mid \#e = k] \geq \max\{\alpha(\alpha - p_u), \alpha(\alpha - p_v), p_u p_v\}$$

$$\geq \alpha^2 \left(\frac{\sqrt{5}-1}{2}\right)^2 > \frac{1}{8}.$$

If $\mathcal{A}(\mathbf{a})$ and $\mathcal{A}(\mathbf{b})$ fail the test at e, then all the k occurrences of e in the walk are considered faulty. Thus,

$$\mathbf{E}[\mathbf{N}_e] = \sum_{k>0} k \cdot \Pr[\mathbf{N}_e = k]$$

$$= \sum_{k>0} k \cdot \Pr[\#e = k \text{ and the test at } e \text{ fails}]$$

$$\geq \sum_{k>0} k \cdot \Pr[\#e = k] \cdot \Pr[\text{the test at } e \text{ fails} \mid \#e = k]$$

$$\geq \sum_{k>0} k \cdot \Pr[\#e = k] \cdot \left(\frac{1}{8}\right)$$

$$= \left(\frac{1}{8}\right) \mathbf{E}[\#e]$$

$$= t\left(\frac{1}{8|E|}\right).$$

Finally, by summing over all $e \in F$, we obtain $\mathbf{E}[\mathbf{N}] = \sum_e \mathbf{E}[\mathbf{N}_e] \geq \left(\frac{|F|}{8|E|}\right) t.$ □

We have shown that the expected number of faulty edges on the verifier's walk is large. However, this does not automatically imply that the number of faulty edges is positive with reasonable probability, for it could be that faulty edges appear on the verifier's walk in bursts, and just a few walks account for the large expectation. This is where we use the fact that our underlying graph is an expander. Intuitively, one expects that a random walk in an expander graph is not likely to visit the small set of edges, F, too many times. The following proposition quantifies this intuition by showing that in the random walk the events of the form "$e_i \in F$" are approximately pairwise independent.

Proposition 2.1 (similar to [4, Proposition 2.4]). *For $j > i$,*

$$\Pr[e_j \in F \mid e_i \in F] \leq \left(1 - \frac{1}{t}\right)^{j-i} \left(\frac{|F|}{|E|} + \left(\frac{|\lambda(G)|}{d}\right)^{j-i-1}\right).$$

Proof (of Claim (b)). Let χ_i be the indicator random variable for the event "$e_i \in F$"; then, $\Pr[\chi_i = 1] = \frac{|F|}{|E|}\left(1 - \frac{1}{t}\right)^{i-1}$. We then have

$$\mathbf{E}[\mathbf{N}^2] \leq 2 \sum_{1 \leq i \leq j < \infty} \mathbf{E}[\chi_i \chi_j]$$

$$\leq 2 \sum_{i=0}^{\infty} \Pr[\chi_i = 1] \sum_{j \geq i} \Pr[\chi_j = 1 \mid \chi_i = 1]$$

$$\leq 2 \sum_{i=1}^{\infty} \Pr[\chi_i = 1] \left[1 + \sum_{\ell \geq 1} \left(1 - \frac{1}{t}\right)^{\ell} \left(\frac{|F|}{|E|} + \left(\frac{|\lambda(G)|}{d}\right)^{\ell-1}\right)\right] \quad (2)$$

$$\leq 2t \frac{|F|}{|E|} \left(1 + t\frac{|F|}{|E|} + \frac{d}{d - |\lambda(G)|}\right),$$

where we used Proposition 2.1 in (2). The claim follows because we have assumed (see (1) above) that $\frac{|F|}{|E|} \leq \frac{1}{t}$.   □

2.4   The Product Constraint Graph

It is relatively straightforward to model the PCP described above as a constraint graph. There is one technicality that we need to take care of: the verifier's walks are not bounded in length, and a naive translation would lead to a graph with infinitely many edges. We now observe that we can truncate the verifier's walk without losing much in the rejection probability.

Verifier with truncated walks: We will show that a version of Lemma 2.1 holds even when the verifier's walks are truncated at $T^* = 5t$, and she just accepts if her walk has not stopped within these many steps.

Lemma 2.3 (Truncated walks). *Suppose G is a d-regular constraint graph with alphabet Σ and $|\lambda(G)| < d$. Consider the verifier with truncated walks.*

(a) *If G is satisfiable, then there is a proof that the verifier accepts with probability 1.*

(b) *If G is ϵ-far from satisfiable, then the verifier rejects every proof with probability at least*

$$\left(\frac{1}{512C}\right) \cdot t \cdot \min\left\{\epsilon, \frac{1}{t}\right\},$$

where $C = 2 + \frac{d}{d - |\lambda(G)|}$.

Proof. We only show how the previous proof is to be modified in order to justify this lemma. If the verifier's walk is truncated before stopping, then no edge on the walk is declared faulty. Under this definition, let $\mathbf{N}'$ be the number of faulty edges in the verifier's random walk ($\mathbf{N}' = 0$ whenever the walk is truncated). Let us redo Claim (a). Let $\mathcal{I}\{\mathbf{T} \geq T^* + 1\}$ be the indicator random variable for the event $T \geq T^* + 1$, that is, $\mathcal{I}\{\mathbf{T} \geq T^* + 1\} = 1$ if $T \geq T^* + 1$ and 0 otherwise. Then, $\mathbf{N}' = \mathbf{N} - \mathbf{N} \cdot \mathcal{I}\{\mathbf{T} \geq T^* + 1\}$, and

$$\mathbf{E}[\mathbf{N}'] = \mathbf{E}[\mathbf{N}] - \mathbf{E}[\mathbf{N} \cdot \mathcal{I}\{\mathbf{T} \geq T^* + 1\}].$$

We already have a lower bound for $\mathbf{E}[\mathbf{N}]$ in Claim (a), so it is sufficient to obtain an upper bound for the second term on the right, which accounts for the contribution to $\mathbf{E}[\mathbf{N}]$ from long walks.

The contribution to $\mathbf{E}[\mathbf{N}]$ from walks of length ℓ is at most $\ell|F|/|E|$ times the probability that $\mathbf{T} = \ell$. So, the contribution to $\mathbf{E}[\mathbf{N}]$ from walks of length at least $T^* + 1$ is

$$\Pr[\mathbf{T} \geq T^* + 1] \cdot \mathbf{E}[\mathbf{T} \mid \mathbf{T} \geq T^* + 1] \cdot \frac{|F|}{|E|}.$$

The first factor is at most $(1 - \frac{1}{t})^{T^*}$, the second is $T^* + t$. So,

$$\mathbf{E}[\mathbf{N} \cdot \mathcal{I}\{\mathbf{T} \geq T^* + 1\}] \leq \exp\left(-\frac{T^*}{t}\right)(T^* + t) \cdot \frac{|F|}{|E|} \leq t \cdot \frac{|F|}{|E|} \cdot \frac{1}{16}.$$

Thus,

$$\mathbf{E}[\mathbf{N}'] \geq t \cdot \frac{|F|}{|E|} \cdot \frac{1}{8} - t \cdot \frac{|F|}{|E|} \cdot \frac{1}{16} \geq t \cdot \frac{|F|}{|E|} \cdot \frac{1}{16}.$$

Note that $\mathbf{N}' \leq \mathbf{N}$, so the upper bound in Claim (b) applies to $\mathbf{E}[\mathbf{N}'^2]$ as well. Now, Lemma 2.3 follows from the inequality $\mathbf{E}[\mathbf{N}' > 0] \geq \mathbf{E}[\mathbf{N}']^2/\mathbf{E}[\mathbf{N}'^2]$. $\square$

Definition 2.2 (The product constraint graph). *Let G be a d-regular constraint graph with alphabet Σ. The product graph G^t is defined as follows.*

- *The vertex set of the graph G^t is $V(G)$.*
- *The alphabet for G^t is $\Sigma^{d^{t+1}}$.*
- *The edges and their constraints correspond to the verifier's actions outlined above. We imagine that after picking the starting vertex a, the verifier's moves are described by a random string of length T^* over the set $[d] \times [t]$. The*

first component determines which outgoing edge the verifier takes, and she stops after that step if the second component is 1 (say). For each vertex a and each sequence τ, we have a directed edge labeled τ leaving a, corresponding to the walk starting from vertex a determined by τ.

- *If the walk does not terminate at the end of τ, then the ending vertex of this edge is also a (it is a self-loop), and the constraint on that edge the constant 1.*
- *If the walk does terminate, and the final vertex is b, then the edge labeled τ connects a to b, and its constraint is the conjunction of all constraints of G checked by the verifier along this walk.*

Thus, every vertex has $(dt)^{T^*}$ edges leaving it, and the total number of edges in G^t is exactly $|V(G)| \cdot (dt)^{T^*}$.

The following theorem is now an immediately consequence of Lemma 2.3.

Theorem 2.1. *Let $G = \langle V, E, \Sigma, \mathcal{C} \rangle$ be a d-regular constraint graph. Then,*

$$\text{UNSAT}(G^t) \geq \frac{t}{C_2'} \cdot \min\left\{\text{UNSAT}(G), \frac{1}{t}\right\}, \text{ where } C_2' = O\left(\frac{d}{d-|\lambda(G)|}\right).$$

Acknowledgments

Thanks to Eli Ben-Sasson, Irit Dinur, Prahladh Harsha, Adam Kalai, Nanda Raghunathan and Aravind Srinivasan for their comments. I am grateful to the referees for their suggestions. In a previous version of this paper, the constant C_2' in Theorem 2.1 had a bound of the form $O\left(|\Sigma|^4 \left(\frac{d}{d-|\lambda(G)|}\right)\right)$; I thank Greg Plaxton for suggesting the tighter analysis presented in this version.

References

1. S. Arora, C. Lund, R. Motwani, M. Sudan, and M. Szegedy: Proof verification and intractability of approximation problems. *J. ACM,* 45(3):501–555, 1998.
2. S. Arora and S. Safra: Probabilistic checking of proofs: A new characterization of NP. *J. ACM,* 45(1):70–122, 1998.
3. L. Babai, L. Fortnow, L.A. Levin, M. Szegedy: Checking Computations in Polylogarithmic Time. STOC 1991: 21-31.
4. I. Dinur: The PCP Theorem by Gap Amplification. ECCC TR05-046. http://eccc.uni-trier.de/eccc-reports/2005/TR05-046
5. U. Feige, S. Goldwasser, L. Lovász, S. Safra, and M. Szegedy: Approximating the clique is almost NP-complete. *J. ACM,* 43(2):268–292, 1996.
6. R. O'Donnell and V. Guruswami. Course notes of CSE 533: The PCP Theorem and Hardness of Approximation. http://www.cs.washington.edu/education/courses/533/05au/

Stopping Times, Metrics
and Approximate Counting

Magnus Bordewich[1], Martin Dyer[2], and Marek Karpinski[3]

[1] Durham University, Durham DH1 3LE, UK
m.j.r.bordewich@durham.ac.uk
[2] Leeds University, Leeds LS2 9JT, UK
dyer@comp.leeds.ac.uk
[3] University of Bonn, 53117 Bonn, Germany
marek@cs.uni-bonn.de

Abstract. In this paper we examine the importance of the choice of metric in path coupling, and its relationship to *stopping time analysis*. We give strong evidence that stopping time analysis is no more powerful than standard path coupling. In particular, we prove a stronger theorem for path coupling with stopping times, using a metric which allows us to analyse a one-step path coupling. This approach provides insight for the design of better metrics for specific problems. We give illustrative applications to hypergraph independent sets and SAT instances, hypergraph colourings and colourings of bipartite graphs, obtaining improved results for all these problems.

1 Introduction

Markov chain algorithms are an important tool in approximate counting [16]. Coupling has a long history in the theory of Markov chains [8], and can be used to obtain quantitative estimates of convergence times [1]. The idea is to arrange the joint evolution of two arbitrary copies of the chain so that they quickly occupy the same state. For all pairs of states, the coupling must specify a distribution on pairs of states so that both marginals give precisely the transition probabilities of the chain. Good couplings are usually not easy to design, but path coupling [6] has recently proved a useful technique for constructing and analysing them. The idea here is to restrict the design of the coupling to pairs of states which are close in some suitable *metric* on the state space, and then (implicitly) obtain the full coupling by composition of these pairs. For example, for independent sets in a graph or hypergraph, the pairs of interest might be independent sets which differ in one vertex (the *change vertex*) and the metric might be Hamming distance.

The limitations of path coupling analysis are always caused by certain "bad" pairs of states. But these pairs may be very unlikely to occur in a typical realisation of the coupling. Consequently, path coupling has been augmented by other techniques, such as *stopping time* analysis. The stopping time approach is applicable when the bad pairs have a reasonable probability of becoming less bad as time proceeds. As an illustration, consider the bad pairs for the Glauber dynamics on hypergraph independent sets [3]. These involve almost fully occupied

M. Bugliesi et al. (Eds.): ICALP 2006, Part I, LNCS 4051, pp. 108–119, 2006.

edges containing the change vertex. However, it seems likely that the number of occupied vertices in these edges will be reduced before we must either increase or decrease the distance between the coupled chains. This observation allows a greatly improved analysis [3]. See [3,11,14,18] for some other applications of this technique. General theorems for applying stopping times appear in [3,14].

The stopping time approach is a multistep analysis, and appears to give a powerful extension of path coupling. However, in this paper we provide strong evidence that the stopping time approach is no more powerful than single-step path coupling. We observe that, in cases where stopping times can be employed to advantage, equally good or better results can be achieved by using a suitably tailored *metric* in the one-step analysis. The intuition behind the choice of metric is precisely that used in the stopping time approach. We will illustrate this with several examples.

In fact, our first example is a proof of a theorem for path coupling using stopping times, relying on a particular choice of metric which enables us to work with the standard one-step path coupling. The resulting theorem is stronger than those in [3,14]. The proof implies that all results obtained using stopping times can just as well be obtained using standard path coupling and the right choice of metric. This does not immediately imply that we can abandon the analysis of stopping times. Determining the metric used in our proof involves bounding the expected distance at a stopping time. But it does suggest that it may be better to do a one-step analysis using a metric indicated by the stopping time.

With this insight, we revisit the Glauber dynamics for hypergraph independent sets. Equivalently, these are satisfying assignments of monotone SAT formulas, and this relationship is discussed in the full paper [4]. We also revisit hypergraph colourings, analysed in [3] using stopping times. We find that we are able to obtain stronger results than those obtained in [3], using metrics suggested by stopping time considerations but then optimised. The technical advantage arises mainly from the possibility of using simple linearity of expectation where stopping time analysis uses concentration inequalities and union bounds.

We note that this paper does not contain the first uses of "clever" metrics with path coupling. See [7,17] for examples. But we do give the first general approach to designing a good metric. While there have been instances in the literature of optimising the *chain* [13,20], the only previous analysis of which we are aware which uses optimisation of the *metric* appeared in [17].

The organisation of the paper is as follows. In section 2 we prove a better stopping time theorem than was previously known, using only standard path coupling. In section 3 we give our improved results for sampling independent sets in hypergraphs. In section 4 we give improved results for sampling colourings of 3-uniform hypergraphs. Finally, in section 5 we give a completely new application, to the "scan" chain for sampling colourings of bipartite graphs. For even relatively small values of Δ, our results improve Vigoda's [20] celebrated $11\Delta/6$ bound on the number of colours required for rapid mixing.

2   Path Coupling and Stopping Times

Let $\mathcal{M}$ be a Markov chain on state space Ω. Let d be an integer valued metric on $\Omega \times \Omega$, and let (X_t, Y_t) be a path coupling for $\mathcal{M}$, i.e. a coupling defined on a path-generating set $S \subseteq \Omega \times \Omega$. See, for example, [12]. We define T_t, a stopping time for the pair $(X_t, Y_t) \in S$, to be the smallest $t' > t$ such that $d(X_{t'}, Y_{t'}) \neq d(X_t, Y_t)$. We will define a new metric d′ such that contraction in d at T_t implies contraction in d′ at every t' with positive probability $T_t = t'$.

Let $\alpha > 0$ be a constant such that $\mathbb{E}[d(X_{T_t}, Y_{T_t})] \leq \alpha d(X_t, Y_t)$ for all $(X_t, Y_t) \in S$. If $\alpha < 1$, then for any $(X_t, Y_t) \in S$, we define d′ as follows.

$$d'(X_t, Y_t) = (1 - \alpha)d(X_t, Y_t) + \mathbb{E}[d(X_{T_t}, Y_{T_t})] \leq d(X_t, Y_t). \tag{1}$$

The metric is extended in the usual way to pairs $(X_t, Y_t) \notin S$, using shortest paths. See [12]. We will apply path coupling with the metric d′ and the original coupling. First we show a contraction property for this metric.

Lemma 1. *If* $\mathbb{E}[d(X_{T_t}, Y_{T_t})] \leq \alpha d(X_t, Y_t) < d(X_t, Y_t)$ *for all* $(X_t, Y_t) \in S$, *then*

$$\mathbb{E}[d'(X_k, Y_k) \,|\, X_0, Y_0] \leq \big(1 - (1 - \alpha)\Pr(T_0 \leq k)\big)d'(X_0, Y_0).$$

Proof. We prove this by induction on k. It obviously holds for $k = 0$, since $T_0 > 0$. Using $\mathbb{1}_\mathcal{A}$ to denote the 0/1 indicator of event $\mathcal{A}$, we may write (1) as

$$d'(X_0, Y_0) = (1-\alpha)d(X_0, Y_0) + \mathbb{E}[d(X_{T_k}, Y_{T_k})\mathbb{1}_{T_0 > k}] + \mathbb{E}[d(X_{T_0}, Y_{T_0})\mathbb{1}_{T_0 \leq k}], \tag{2}$$

since if $T_0 > k$ then $T_k = T_0$. Similarly, we have that $\mathbb{E}[d'(X_k, Y_k)]$

$$
\begin{aligned}
&= \mathbb{E}[d'(X_k, Y_k)\mathbb{1}_{T_0 > k}] + \mathbb{E}[d'(X_k, Y_k)\mathbb{1}_{T_0 \leq k}] \\
&= (1 - \alpha)\mathbb{E}[d(X_k, Y_k)\mathbb{1}_{T_0 > k}] + \mathbb{E}[d(X_{T_k}, Y_{T_k})\mathbb{1}_{T_0 > k}] + \mathbb{E}[d'(X_k, Y_k)\mathbb{1}_{T_0 \leq k}]. \\
&= (1 - \alpha)\mathbb{E}[d(X_0, Y_0)\mathbb{1}_{T_0 > k}] + \mathbb{E}[d(X_{T_k}, Y_{T_k})\mathbb{1}_{T_0 > k}] + \mathbb{E}[d'(X_k, Y_k)\mathbb{1}_{T_0 \leq k}]. \tag{3}
\end{aligned}
$$

Subtracting (2) from (3), we have that $\mathbb{E}[d'(X_k, Y_k)] - d'(X_0, Y_0)$

$$= -(1 - \alpha)\mathbb{E}[d(X_0, Y_0)\mathbb{1}_{T_0 \leq k}] + \mathbb{E}[(d'(X_k, Y_k) - d(X_{T_0}, Y_{T_0}))\mathbb{1}_{T_0 \leq k}].$$

For $T_0 \leq k$, since $k - T_0 \leq k - 1$ the inductive hypothesis implies $\mathbb{E}[d'(X_k, Y_k) \,|\, X_{T_0}, Y_{T_0}] \leq d'(X_{T_0}, Y_{T_0}) \leq d(X_{T_0}, Y_{T_0})$, (if $(X_k, Y_k) \notin S$ this is implied by linearity). Hence

$$\mathbb{E}[d'(X_k, Y_k)] - d'(X_0, Y_0) \leq -(1 - \alpha)\mathbb{E}[d(X_0, Y_0)\mathbb{1}_{T_0 \leq k}],$$

But now $\mathbb{E}[d(X_0, Y_0)\mathbb{1}_{T_0 \leq k}] = \Pr(T_0 \leq k)d(X_0, Y_0) \geq \Pr(T_0 \leq k)d'(X_0, Y_0).$

We may now prove the first version of our main result.

Theorem 1. *Let* $\mathcal{M}$ *be a Markov chain on state space* Ω. *Let* d *be an integer valued metric on* Ω, *and let* (X_t, Y_t) *be a path coupling for* $\mathcal{M}$. *Let* T_t *be the above stopping times. Suppose for all* $(X_0, Y_0) \in S$ *and for some integer* k *and* $p > 0$, *that* *(i)* $\Pr[T_0 \leq k] \geq p$,

(ii) $\mathbb{E}[d(X_{T_0}, Y_{T_0})/d(X_0, Y_0)] \leq \alpha < 1$.

Then the mixing time $\tau(\varepsilon)$ of $\mathcal{M}$ satisfies $\tau(\varepsilon) \leq \frac{k}{p(1-\alpha)} \ln\left(\frac{eD}{\varepsilon(1-\alpha)}\right)$, where $D = \max\{d(X, Y) : X, Y \in \Omega\}$.

Proof. From Lemma 1, d' contracts by a factor $1 - (1-\alpha)p \leq e^{-(1-\alpha)p}$ for every k steps of $\mathcal{M}$. Note also that $d' \leq D$. It follows that, at time $\tau(\varepsilon)$, we have

$$\Pr(X_\tau \neq Y_\tau) \leq \mathbb{E}[d(X_\tau, Y_\tau)] \leq \frac{\mathbb{E}[d'(X_\tau, Y_\tau)]}{1-\alpha} \leq \frac{D e^{-(1-\alpha)p\tau/k}}{1-\alpha} \leq \varepsilon,$$

from which the theorem follows.   □

If $1 - \alpha$ is small compared to ε, it is possible to do better than this. A proof of the following appears in the full paper [4].

Theorem 2. *If $\mathcal{M}$ satisfies the conditions of Theorem 1, the mixing time $\tau(\varepsilon)$ of $\mathcal{M}$ satisfies $\tau(\varepsilon) \leq \frac{k(2-\alpha)}{p(1-\alpha)} \ln\left(\frac{2eD}{\varepsilon}\right)$, where $D = \max\{d(X, Y) : X, Y \in \Omega\}$.*

Remark 1. One of the most interesting features of these theorems is that their proofs employ only standard path coupling, but with a metric which has some useful properties. Thus, for any problem to which stopping times might be applied, there exists a metric from which the same result could be obtained using one-step path coupling.

Remark 2. We may compare this stopping time theorem with those in [3,14]. The main result of [14] (Theorem 3) concerns bounded stopping times, where $T_0 \leq M$ for all $(X_0, Y_0) \in S$, and gives a mixing time of $O(M(1-\alpha)^{-1} \log D)$. By setting $k = M$ and $p = 1$ in Theorem 2, we obtain the same mixing time up to minor changes in constants, but with a proof that does not involve defining a multistep coupling. For unbounded mixing times, [14, Corollary 4] gives a bound $O(\mathbb{E}[T](1-\alpha)^{-2}W \log D)$ by truncating the stopping times, where W denotes the maximum of $d(X_t, Y_t)$ over all $(X_0, Y_0) \in S$ and $t \leq T$. In most applications $\mathbb{E}[T] \leq k/p$, so we obtain an improvement of order $W(1-\alpha)^{-1}$. By comparison with [3], we obtain a more modest improvement, of order $\log W \log(D(1-\alpha)^{-1})/\log D$.

Remark 3. Further improvements to Theorem 2 seem unlikely, other than in constants. The term k/p must be present, since it bounds a single stopping time. A term $1/(1-\alpha) \log(D/\varepsilon) = \Theta(\log_\alpha(D/\varepsilon))$ also seems essential, since it bounds the number of stopping times required.

3 Hypergraph Independent Sets

We now turn our attention to hypergraph independent sets. These were previously studied in [3]. Let $\mathcal{H} = (\mathcal{V}, \mathcal{E})$ be a hypergraph of maximum degree Δ and minimum edge size m. A subset $S \subseteq V$ of the vertices is *independent* if no edge is a subset of S. Let $\Omega(\mathcal{H})$ be the set of all independent sets of $\mathcal{H}$. We define the Markov chain $\mathcal{M}(\mathcal{H})$ with state space $\Omega(\mathcal{H})$ by the following transition process (*Glauber dynamics*). If the state of $\mathcal{M}$ at time t is X_t, the state at $t + 1$ is determined by the following procedure.

1. Select a vertex $v \in V$ uniformly at random,
2. (i) if $v \in X_t$ let $X_{t+1} = X_t \backslash \{v\}$ with probability $1/2$,
   (ii) if $v \notin X_t$ and $X_t \cup \{v\}$ is independent, let $X_{t+1} = X_t \cup \{v\}$ with probability $1/2$,
   (iii) otherwise let $X_{t+1} = X_t$.

This chain is easily shown to be ergodic with uniform stationary distribution. The natural coupling for this chain is the "identity" coupling, the same transition is attempted in both copies of the chain. If we try to apply standard path coupling to this chain, we immediately run into difficulties. The change in the expected *Hamming distance* between X_t and Y_t after one step could be as high as $\frac{\Delta}{2n} - \frac{1}{n}$, and we obtain rapid mixing only in the case $\Delta = 2$.

For $(\sigma, \sigma \cup \{w\}) \in S$, let $E_i(w, \sigma)$ be the set of edges containing w which have i occupied vertices in σ. Using a result like Theorem 1 above, it is shown in [3] that, for the stopping time T given by the first epoch at which the Hamming distance between the coupled chains changes,

$$\mathbb{E}[d_{\text{Ham}}(X_T, Y_T | X_0 = \sigma, Y_0 = \sigma \cup \{w\})] \leq 2 \sum_{i=0}^{m-2} p_i |E_i| \leq 2p_1 \Delta, \qquad (4)$$

where the p_i is the probability that $d(X_T, Y_T) = 2$ if w is in a single edge with i occupied vertices. Since $p_1 < 1/(m-1)$, we obtain rapid mixing when $2\Delta/(m-1) \leq 1$, i.e. when $m \geq 2\Delta + 1$. See [3] for details.

The approach of section 2 would lead us to define a metric for which the distance between σ and $\sigma \cup \{w\}$ is $(1 - 2p_1\Delta) + 2\sum_{i=0}^{m-2} p_i |E_i|$. By Lemma 1, we know that this metric contracts in expectation. However, prompted by the form of this metric, but retaining the freedom to optimise constants, we will instead define the new metric d to be $d(\sigma, \sigma \cup \{w\}) = \sum_{i=0}^{m-2} c_i |E_i|$, where $0 < c_i \leq 1$ ($0 \leq i \leq m-2$) are a nondecreasing sequence of constants to be determined. Using this metric, we obtain the following theorem.

Theorem 3. *Let Δ be fixed, and let $\mathcal{H}$ be a hypergraph such that $m \geq \Delta + 2 \geq 5$, or $\Delta = 3$ and $m \geq 2$. Then the Markov chain $\mathcal{M}(\mathcal{H})$ has mixing time $O(n \log n)$.*

Proof. Without loss of generality, we take $c_{m-2} = 1$ and we will define $c_{-1} = c_0, c_{m-1} \geq \Delta + 1$. Note that c_{-1} has no real role in the analysis, and is chosen only for convenience, but c_{m-1} is chosen so that $c_{m-1} - c_{m-2} \geq \Delta \geq d(\sigma, \sigma')$ for any pair $(\sigma, \sigma') \in S$. We require $c_i > 0$ for all i so that we will always have $d(\sigma, \sigma') > 0$ if $\sigma \neq \sigma'$.

Now consider the expected change in distance between σ and $\sigma \cup \{w\}$ after one step of the chain.

If w is chosen, then the distance decreases by $\sum_{i=0}^{m-2} c_i|E_i|$. The contribution to the expected change in distance is $-\frac{2}{2n}\sum_{i=0}^{m-2} c_i|E_i|$.

If we insert a vertex v in an edge containing w, then we increase the distance by $(c_{i+1}-c_i) \geq 0$ for each edge in E_i containing v. This holds for $i = 0, \ldots, m-2$, by the choice of $c_{m-1} = \Delta + 1$. Let U be the set of unoccupied neighbours of w, and $\nu_i(v)$ be the number of edges with i occupants containing w and v. Then

$$\sum_{v\in U} \nu_i(v) = \sum_{v\in U}\sum_{e\in E_i} \mathbb{1}_{v\in e} = \sum_{e\in E_i}\sum_{v\in e\cap U} 1 = \sum_{e\in E_i}(m-i-1) = (m-i-1)|E_i|.$$

implies that $$\sum_{v\in U}\frac{1}{2n}\sum_{i=0}^{m-2} \nu_i(v)(c_{i+1}-c_i) = \frac{1}{2n}\sum_{i=0}^{m-2}(c_{i+1}-c_i)(m-i-1)|E_i|.$$

If we delete a vertex v in an edge containing w, then we decrease the distance by $(c_i - c_{i-1})$ for each edge in E_i containing v. This holds for $i = 0, \ldots, m-2$, by the choice of c_{-1}. Let O be the set of occupied neighbours of w, and $\nu_i(v)$ be the number of edges with i occupants containing w and v. Then a similar argument gives the contribution as

$$-\sum_{v\in O}\frac{1}{2n}\sum_{i=0}^{m-2}\nu_i(v)(c_i - c_{i-1}) = -\frac{1}{2n}\sum_{i=0}^{m-2}(c_i - c_{i-1})i|E_i|.$$

Let $d_0 = d(\sigma, \sigma \cup \{w\})$, and let d_1 be the distance after one step of the chain. The change in expected distance $E' = \mathbb{E}[d_1 - d_0]$ satisfies

$$2nE' \leq -2\sum_{i=0}^{m-2} c_i|E_i| + \sum_{i=0}^{m-2}(c_{i+1}-c_i)(m-i-1)|E_i| - \sum_{i=0}^{m-2}(c_i-c_{i-1})i|E_i|$$

$$= \sum_{i=0}^{m-2}\left(ic_{i-1} - (m+1)c_i + (m-i-1)c_{i+1}\right)|E_i|.$$

We require $\mathbb{E}[d_1-d_0] \leq -\gamma$, for some $\gamma \geq 0$, which holds for all possible choices of E_i if and only if $(m-i-1)c_{i+1}-(m+1)c_i+ic_{i-1} \leq -\gamma$ for all $i = 0,1,\ldots,m-2$. Thus we need a solution to

$$ic_{i-1} - (m+1)c_i + (m-i-1)c_{i+1} \leq -\gamma \quad (i = 0,\ldots,m-2), \quad (5)$$

$$0 = c_{-1} < c_0 \leq c_1 \leq \cdots \leq c_{m-3} \leq c_{m-2} = 1, \quad c_{m-1} \geq \Delta+1, \quad \gamma \geq 0,$$

with $\gamma > 0$ if possible. Solving for the optimal solution gives

$$c_i = \frac{\gamma\sum_{j=0}^{i}\binom{m-1}{j} - \frac{m-\Delta-2+\gamma}{m}\sum_{j=0}^{i}\binom{m}{j}}{\binom{m-1}{i}} \quad (i = 0,\ldots,m-2),$$

$$\gamma = \frac{2^m - 1 - m}{(m-2)2^{m-1}+1}\left(m - \Delta - 2 + \frac{m(m-1)}{2^m - 1 - m}\right).$$

Let $f(m) = m - 2 + \frac{m(m-1)}{2^m-1-m}$, then we can have $\gamma \geq 0$ if and only if $f(m) \geq \Delta$, and $\gamma > 0$ if and only if $f(m) > \Delta$.

If $m \geq 5$ then $m(m-1)/(2^m - 1 - m) < 1$, so we will have $f(m) > \Delta$ exactly when $m \geq \Delta + 2$. For smaller values of m, $f(2) = 2$, $f(3) = 2\frac{1}{2}$ and $f(3) = 3\frac{1}{11}$.

The new case here is $\Delta = 3, m \geq 4$. In any case for which $f(m) > \Delta$, standard path coupling arguments yield the mixing times claimed since we have contraction in the metric and the minimum distance is at least c_0. Mixing for $\Delta = 3, m \leq 3$ was shown in [13]), so we have mixing for $\Delta = 3$ and every m. ☐

Remark 4. The independent set problem here has a natural *dual,* that of sampling an *edge cover* from a hypergraph with edge size Δ and degree m. An edge cover is a subset of $\mathcal{E}$ whose union contains V. For the graph case of this sampling problem, with arbitrary m, see [5]. By duality this gives the case $\Delta = 2$ of the independent set problem here.

4   Colouring 3-Uniform Hypergraphs

In our second application, also from [3], we consider proper colourings of 3-uniform hypergraphs. We again use Glauber dynamics. Our hypergraph $\mathcal{H}$ will have maximum degree Δ, uniform edge size 3, and we will have a set of q colours. For a discussion of the easier problem of colouring hypergraphs with larger edge size see [3]. A colouring of the vertices of $\mathcal{H}$ is proper if no edge is monochromatic. Let $\Omega'(\mathcal{H})$ be the set of all proper q-colourings of $\mathcal{H}$. We define the Markov chain $\mathcal{C}(\mathcal{H})$ with state space $\Omega'(\mathcal{H})$ by the following transition process. If the state of $\mathcal{C}$ at time t is X_t, the state at $t+1$ is determined by

1. selecting a vertex $v \in V$ and a colour $k \in \{1, 2, \ldots, q\}$ uniformly at random,
2. let X_t' be the colouring obtained by recolouring v colour k
3. if X_t' is a proper colouring let $X_{t+1} = X_t'$
   otherwise let $X_{t+1} = X_t$.

This chain is easily shown to be ergodic with the uniform stationary distribution. For some large enough constant Δ_0, it was shown in [3] to be rapidly mixing for $q > 1.65\Delta$ and $\Delta > \Delta_0$, using a stopping times analysis. Here we improve this result, and simplify the proof, by using a carefully chosen metric which is prompted by the new insight into stopping times analyses. If w is the change vertex, the intuition in [3] was that edges which contain both colours of w are initially "dangerous" but tend to become less so after a time. Thus our metric will be a function of the numbers of edges containing w with various relevant colourings.

Theorem 4. *Let Δ be fixed, and let $\mathcal{H}$ be a 3-uniform hypergraph of maximum degree Δ. Then if $q \geq \lceil \frac{3}{2}\Delta + 1 \rceil$, the Markov chain $\mathcal{C}(\mathcal{H})$ has mixing time $O(n \log n)$.*

Proof. Consider two proper colourings X and Y differing in a single vertex w. Without loss of generality let the change vertex w be coloured 1 in X and 2 in Y. We will partition the edges $e \in \mathcal{E}$ containing w into four classes E_1, E_2, E_3, E_4, determined by the colouring of $e \setminus \{w\}$, as follows:

$$E_1 : \{1,2\}, \quad E_2 : \bigcup_{i>2}\{1,i\} \cup \{2,i\}, \quad E_3 : \bigcup_{i>2}\{i,i\}, \quad E_4 : \bigcup_{2<i<j}\{i,j\}.$$

Instead of Hamming distance, we define a metric d by $d(X,Y) = \sum_{i=1}^{4} c_i |E_i|$, where $1 = c_1 \geq c_2 \geq c_3 \geq c_4 > 0$, and for convenience $c_0 = \Delta + 1$. Note that $d(X,Y) \leq \Delta$ if X, Y have Hamming distance 1. The diameter is therefore at most Δn in the metric d. Arguing as in Section 3, we have

$$
\begin{aligned}
nq\mathbb{E}[d_1 - d_0] \leq\ & -(q - |E_3|)(c_1|E_1| + c_2|E_2| + c_3|E_3| + c_4|E_4|) \\
& +|E_1|\big(-2(q - \Delta - 1)(c_1 - c_2) + 2(c_0 - c_1)\big) \\
& +|E_2|\big(-(q - \Delta - 2)(c_2 - c_4) - (c_2 - c_3) + (c_0 - c_2) + (c_1 - c_2)\big) \quad (6) \\
& +|E_3|\big(-2(q - \Delta - 2)(c_3 - c_4) + 4(c_2 - c_3)\big) \\
& +|E_4|\big(2(c_3 - c_4) + 4(c_2 - c_4)\big).
\end{aligned}
$$

If we set $c_1 = 1$,

$$c_2 = \frac{2q - 2\Delta + 1}{2q - \Delta + 1}, \quad c_3 = c_4 = \frac{2q - 3\Delta + 1}{2q - \Delta + 1}, \quad \gamma = \frac{2q^2 - q(3\Delta - 1) - 4\Delta}{2q - \Delta + 1}, \quad (7)$$

then (6) yields

$$\mathbb{E}[d_1] \leq\ d_0 - \frac{\gamma\Delta}{nq} \leq \left(1 - \frac{\gamma}{nq}\right)d_0. \quad (8)$$

The condition $\gamma \geq 0$ is equivalent to

$$q \geq \tfrac{3\Delta-1}{4}\left(1 + \sqrt{1 + \tfrac{32\Delta}{(3\Delta-1)^2}}\right), \quad \text{i.e.}\quad q \geq \lceil \tfrac{3}{2}\Delta \rceil + 1. \quad (9)$$

Note that we have $c_i > 0$ $(i = 1,\dots,4)$ under this condition. Note also that $\gamma > 0$ and hence, using (8), the mixing time satisfies

$$\tau(\varepsilon) \leq \frac{2q^2 - q\Delta + q}{2q^2 - q(3\Delta - 1) - 4\Delta}\, n \ln\left(\frac{\Delta n}{\varepsilon}\right).$$

5   Colouring Bipartite Graphs

Let $G = (V, E)$ be a bipartite graph with bipartition V_1, V_2, and maximum degree Δ. For $v \in V$, let $\mathcal{N}(v) = \{w : \{v, w\} \in E\}$ denote the neighbourhood of v, and let $d(v) = |\mathcal{N}(v)|$ be its degree. Let $Q = [q]$ be a colour set, and $X : V \to Q$ be a colouring of G, not necessarily proper. Let $C_X(v) = \{X(w) : w \in \mathcal{N}(v)\}$ be the set of colours occurring in the neighbourhood of v, and $c_X(v)$ denote the size of $C_X(v)$. We consider the Markov chain MULTICOLOUR on colourings of G, which in each step picks one side of the bipartition at random, and then recolours every vertex on that side, followed by recolouring every vertex in the other half of the bipartition. If the state of MULTICOLOUR at time t is X_t, the state at time $t+1$ is given by

MULTICOLOUR

1. choosing $r \in \{1, 2\}$ uniformly at random,
2. for each vertex $v \in V_r$,

(i) choosing a colour $q(v) \in Q \backslash C_{X_t}(v)$ uniformly at random,
(ii) setting $X_{t+1}(v) = q(v)$. (Heat bath recolouring)
3. for each vertex $v \in V \backslash V_r$,
   (i) choosing a colour $q(v) \in Q \backslash C_{X_{t+1}}(v)$ uniformly at random,
   (ii) setting $X_{t+1}(v) = q(v)$.

Note that the order in which the vertices are processed in steps 2 and 3 is immaterial, and that in step 3, $C_{X_{t+1}}(v)$ is well defined since all of v's neighbours have been recoloured in step 2. We prove the following theorem.

Theorem 5. *The mixing time of* MULTICOLOUR *is* $O(\log(n))$ *for* $q > f(\Delta)$, *where* f *is a function such that*

(1) $f(\Delta) \to \beta \Delta$, *as* $\Delta \to \infty$, *where* β *satisfies* $\beta e^\beta = 1$,
(3) $f(\Delta) < \lceil 11\Delta/6 \rceil$ *for* $\Delta \geq 31$.
(2) $f(\Delta) \leq \lceil 11\Delta/6 \rceil$ *for* $\Delta \geq 14$.

This chain is a single-site dynamics intermediate between Glauber and SCAN (which uses the same vertex update procedure as Glauber, but choses the vertices in a deterministic order). It is easy to see that it is ergodic if $q > \Delta + 1$, and has equilibrium distribution uniform on all proper colourings of G. Observe also that it uses many fewer random bits than Glauber. Indeed the following easy Corollary of Theorem 5 is proved in the full paper [4].

Corollary 1. *The mixing time for* SCAN *is at most that for* MULTICOLOUR.

To prove Theorem 5 we need the following lemmas, whose proofs are given in [4].

Lemma 2. *For* $1 \leq i \leq \Delta$ *let* S_i *be a subset of* $(Q - q_0)$ *such that* $m_i = |S_i| \geq q - \Delta$. *Let* s_i *be selected uniformly at random from* S_i, *independently for each* i. *Finally let* $C = \{s_i : 1 \leq i \leq \Delta\}$ *and* $c = |C|$. *Then*

$$\mathbb{E}[q - c \mid s_1 = q_1] \geq 1 + (q-2)\left(1 - \frac{1}{q-\Delta}\right)^{\frac{(\Delta-1)(q-\Delta)}{q-2}} = \alpha.$$

Lemma 3. *For* $1 \leq i \leq \Delta$ *let* S_i *be a subset of* $(Q - q_0)$ *such that* $m_i = |S_i| \geq q - \Delta$. *Let* s_i *be selected uniformly at random from* S_i, *independently for each* i. *Finally let* $C = \{s_i : 1 \leq i \leq \Delta\}$ *and* $c = |C|$. *Then*

$$\mathbb{E}\left[\frac{1}{q-c} \mid s_1 = q_1\right] \leq \frac{1}{\alpha}\left(1 + \frac{(q-\alpha-1)(\alpha-1)}{(q-\Delta)(q-2)\alpha}\right) = \alpha'.$$

Proof (Proof of Theorem 5). In the path coupling setting, we will take S to be the set of pairs of colourings which differ at exactly one vertex. Let v be the change vertex for some pair $(X, Y) \in S$, and assume without loss that $v \in V_1$. The distance between X and Y is defined to be $d(X, Y) = \sum_{w \in \mathcal{N}(v)} \frac{1}{q - c_{X,Y}(w)}$, where $c_{X,Y}(w)$ is taken to be $\min\{c_X(w), c_Y(w)\}$ in the case that they differ. We couple as follows (the usual path coupling for Glauber dynamics). If we are recolouring a vertex which is not a neighbour of v, then the sets of available colours in X and Y are the same, and we use the same colour in both copies of the chain. If we are recolouring a vertex $w \in \mathcal{N}(v)$ then there are three cases:

1. $|\{X(v), Y(v)\} \cap \{X(z) : z \in \mathcal{N}(w)\backslash\{v\}\}| = 2.$
   Colours $X(v)$ and $Y(v)$ are not available for w in either X or Y, the sets of available colours are the same, and we use the same colour in both X, Y.
2. $|\{X(v), Y(v)\} \cap \{X(z) : z \in \mathcal{N}(w)\backslash\{v\}\}| = 1.$
   Without loss assume $X(v)$ is not available to w in either X or Y, and $Y(v)$ is only available in X. For any colour other than $Y(v)$, we couple the same colour for w in X and Y. For $Y(v)$, we couple recolouring w with $Y(v)$ in X by uniformly recolouring w from the available colours in Y.
3. $|\{X(v), Y(v)\} \cap \{X(z) : z \in \mathcal{N}(w)\backslash\{v\}\}| = 0.$
   Here colour $Y(v)$ is only available in chain X, and $X(v)$ in only available in Y. We couple these colours together, and for each other colour available to both X, Y, we recolour w with the same colour.

In case 1, there is no probability of w being coloured differently in the two chains. In the other cases, the probability of disagreement at w is $\frac{1}{q - c_{X,Y}(w)}$.

Let X', Y' be the colourings after recolouring V_r (half a step of MULTI-COLOUR) and X'', Y'' be the colourings after the full step of MULTICOLOUR. We use primes and double primes to denote the quantities in X' and X'' respectively, corresponding to those in X. If we randomly select V_1 to be recoloured first, then the two copies of the chain have coupled in X' and Y' since the vertices in V_1 have the same set of available colours in each chain.

So suppose that we select V_2 to be recoloured first. The only vertices in V_2 that have different sets of available colours are the neighbours of v. Let $\mathcal{N}(v) = \{w_1, \ldots, w_k\}$ and consider the path $W_0, W_1, \ldots, W_{k+1}$ from X' to Y', where for $1 \leq i \leq k$, W_i agrees with X' on all vertices except $w_1, \ldots, w_i$ which are coloured as in Y', and $W_0 = X'$ and $W_{k+1} = Y'$. Then for $i \leq k$ we have

$$d(W_{i-1}, W_i) = \mathbb{1}_{w_i} \sum_{z \in \mathcal{N}(w_i)} \frac{1}{q - c_{W_{i-1}, W_i}(z)} \leq \mathbb{1}_{w_i} \sum_{z \in \mathcal{N}(w_i)} \frac{1}{q - c_{W_i}(z)}, \quad (10)$$

where $\mathbb{1}_{w_i}$ indicates whether X' and Y' differ on w_i.

Note that $\Pr[\mathbb{1}_{w_i} = 1] \leq \frac{1}{q - c_{X,Y}(w_i)}$. Furthermore, by the construction of the coupling either conditioning on $\mathbb{1}_{w_i} = 1$ is the same as conditioning that $W_{i-1}(w_i) = q_1$, or that $W_i(w_i) = q_1$, for some q_1. We assume without loss that this is W_i. Then for each $z \in \mathcal{N}(w_i) - v$ the selection of colours in $C_{W_i}(z)$ satisfies the conditions of Lemma 3, since we may take $q_0 = X(z)$ and q_1 as above. For v, there is no colour q_0 which is necessarily unavailable for all its neighbours, since some are coloured as in X' and some as in Y'. Hence we use a slightly weaker bound on α and α', given by

$$\alpha_v = (q-1)\left(1 - \frac{1}{q-\Delta}\right)^{\frac{(\Delta-1)(q-\Delta)}{q-1}} \quad \text{and} \quad \alpha'_v = \frac{1}{\alpha_v}\left(1 + \frac{(q - \alpha_v)(\alpha_v)}{(q - \Delta)(q-1)\alpha_v}\right).$$

Hence for $i \leq k$, $\mathbb{E}[d(W_{i-1}, W_i)] \leq \frac{1}{q - c_{X,Y}(w_i)}((\Delta - 1)\alpha' + \alpha'_v)$. The value of $d(W_k, W_{k+1})$ is still $d(X, Y)$ since the vertices in V_1 have not yet been recoloured.

Now we consider the vertices in V_1. We apply the same analysis as above to each path segment W_{i-1}, W_i, but augment the analysis using the fact that at the time a vertex $z \in V_1$ is recoloured, its neighbours (in V_2) will already have been randomly recoloured. Let the neighbours of w_i be $z_1, z_2, \ldots z_l$, and consider the path $Z_0, Z_1, \ldots Z_{l+1}$, where for $1 \leq j \leq l$, Z_j agrees with W_{i-1} on all vertices except $z_1, \ldots, z_j$ which are coloured as in W_i, and $Z_0 = W_{i-1}$ and $Z_{l+1} = W_i$. Arguing as above, for $j \leq l$ we have

$$d(Z_{j-1}, Z_j) = \mathbb{1}_{z_j} \sum_{w \in \mathcal{N}(z_j)} \frac{1}{q - c_{Z_{i-1}, Z_i}(w)}.$$

But now $\Pr[\mathbb{1}_{z_j} = 1 \mid W_{i-1}, W_i] \leq \frac{1}{q - c_{W_{i-1}, W_i}(z_j)} \mathbb{1}_{w_i}$. This is similar to equation (10), and the same argument gives $\mathbb{E}[\mathbb{1}_{z_j} = 1] \leq \frac{1}{q - c_{X,Y}(w_i)} \alpha'$, for $z_j \neq v$ and $\mathbb{E}[\mathbb{1}_{z_j} = 1] \leq \frac{1}{q - c_{X,Y}(w_i)} \alpha'_v$ if $z_j = v$. Also, since it depends only on the colouring of V_2, we have $d(Z_l, Z_{l+1}) = d(W_{i-1}, W_i)$. So

$$\mathbb{E}[\sum_{j=1}^{l+1} d(Z_{j-1}, Z_j)] \leq \frac{1}{q - c_{X,Y}(w_i)}((\Delta - 1)\alpha' + \alpha'_v)(((\Delta - 1)\alpha' + \alpha'_v) + 1).$$

Finally note that W_k and W_{k+1} differ only in V_1, so after recolouring V_1 they have coupled. Hence

$$\mathbb{E}[d(X'', Y'')] = \frac{1}{2} \sum_{i=1}^{k} \sum_{j=1}^{l+1} \mathbb{E}[d(Z_{j-1}, Z_j)]$$

$$\leq \sum_{i=1}^{k} \frac{(\Delta - 1)\alpha' + \alpha'_v)(((\Delta - 1)\alpha' + \alpha'_v) + 1}{2(q - c_{X,Y}(w_i))}$$

$$= d(X, Y)((\Delta - 1)\alpha' + \alpha'_v) \frac{(((\Delta - 1)\alpha' + \alpha'_v) + 1)}{2}.$$

This gives contraction if $((\Delta - 1)\alpha' + \alpha'_v) < 1$. For large Δ, α' and α'_v both approach $\frac{1}{q} e^{\Delta/q}$. Hence we have contraction when $\frac{\Delta}{q} e^{\Delta/q} < 1$. For small Δ, we can compute the smallest integral q giving contraction (see table). If we have contraction, standard path coupling gives the mixing time bounds claimed.   □

Δ	q	$\lceil 11\Delta/6 \rceil$	q/Δ	Δ	q	$\lceil 11\Delta/6 \rceil$	q/Δ
22	40	41	1.82	35	63	65	1.80
23	42	43	1.83	40	72	74	1.80
25	46	46	1.84	50	90	92	1.80
30	55	55	1.83	10000	17634	18334	1.76

Minimum values of q for contraction.

Remark 5. Our analysis shows that one-step analysis of a single-site chain on graph colourings need not break down at $q = 2\Delta$ [15,19]. This apparent boundary seems merely to be an artefact of using Hamming distance.

References

1. D. Aldous, Random walks on finite groups and rapidly mixing Markov chains, in *Séminaire de Probabilités XVII*, Springer Verlag, Berlin, 1983, pp. 243–297.
2. P. Berman, M. Karpinski & A. D. Scott, Approximation hardness of short symmetric instances of MAX-3SAT, *Elec. Coll. on Comp. Compl.*, ECCC TR03-049, 2003.
3. M. Bordewich, M. Dyer & M. Karpinski, Path coupling using stopping times, *Proc. 15th Int. Symp. on Fundamentals of Computation Theory*, Springer Lecture Notes in Computer Science **3623**, pp. 19–31, 2005. (A full version of the paper appears as http://arxiv.org/abs/math.PR/0501081.)
4. M. Bordewich, M. Dyer and M. Karpinski, Metric construction, stopping times and path coupling, http://arxiv.org/abs/math.PR/0511202, 2005.
5. R. Bubley & M. Dyer, Graph orientations with no sink and an approximation for a hard case of #SAT, in *Proc. 8th ACM-SIAM Symp. on Discrete Algorithms*, SIAM, 1997, pp. 248–257.
6. R. Bubley & M. Dyer, Path coupling: A technique for proving rapid mixing in Markov chains, in *Proc. 38th IEEE Symp. on Foundations of Computer Science*, IEEE, 1997, pp. 223–231.
7. R. Bubley & M. Dyer, Faster random generation of linear extensions, in *Proc. 9th ACM-SIAM Symp. on Discrete Algorithms*, ACM-SIAM, pp. 350–354, 1998.
8. W. Doeblin, Exposé de la théorie des chaînes simples constantes de Markoff à un nombre fini d'états, *Revue Mathématique de l'Union Interbalkanique* **2** (1938), 77–105.
9. M. Dyer & A. Frieze, Randomly colouring graphs with lower bounds on girth and maximum degree, *Random Structures and Algorithms* **23** (2003), 167–179.
10. M. Dyer, A. Frieze, T. Hayes & E. Vigoda, Randomly coloring constant degree graphs, in *Proc. 45th IEEE Symp. on Foundations of Computer Science*, IEEE, 2004, pp. 582–589.
11. M. Dyer, L. Goldberg, C. Greenhill, M. Jerrum & M. Mitzenmacher, An extension of path coupling and its application to the Glauber dynamics for graph colorings, *SIAM Journal on Computing* **30** (2001), 1962–1975.
12. M. Dyer & C. Greenhill, Random walks on combinatorial objects, in *Surveys in Combinatorics* (J. D. Lamb & D. A. Preece, Eds.), London Math. Soc. Lecture Note Series **267**, Cambridge University Press, Cambridge, 1999, pp. 101–136.
13. M. Dyer & C. Greenhill, On Markov chains for independent sets, *Journal of Algorithms* **35** (2000), 17–49.
14. T. Hayes & E. Vigoda, Variable length path coupling, in *Proc. 15th ACM-SIAM Symp. on Discrete Algorithms*, ACM-SIAM, 2004, pp. 103–110.
15. M. Jerrum, A very simple algorithm for estimating the number of k-colorings of a low-degree graph, *Random Structure & Algorithms* **7** (1995), 157–165.
16. M. Jerrum, Counting, sampling and integrating: algorithms and complexity, ETH Zürich Lectures in Mathematics, Birkhäuser, Basel, 2003.
17. M. Luby & E. Vigoda, Fast convergence of the Glauber dynamics for sampling independent sets, *Random Structures & Algorithms* **15** (1999), 229–241.
18. M. Mitzenmacher & E. Niklova, Path coupling as a branching process, 2002.
19. J. Salas & A. Sokal, Absence of phase transition for anti-ferromagnetic Potts models via the Dobrushin uniqueness theorem, *J. Stat. Phys.* **86** (1997), 551–579.
20. E. Vigoda, Improved bounds for sampling colorings, *J. Math. Phys.* **41** (1999), 1555–1569.

Algebraic Characterization
of the Finite Power Property

Michal Kunc*

Department of Mathematics, University of Turku, and
Turku Centre for Computer Science, FIN-20014 Turku, Finland
kunc@math.muni.cz
http://www.math.muni.cz/~kunc/

Abstract. We give a transparent characterization, by means of a certain syntactic semigroup, of regular languages possessing the finite power property. Then we use this characterization to obtain a short elementary proof for the uniform decidability of the finite power property for rational languages in all monoids defined by a confluent regular system of deletion rules. This result in particular covers the case of free groups solved earlier by d'Alessandro and Sakarovitch by means of an involved reduction to the boundedness problem for distance automata.

1   Introduction

A language L is said to have the finite power property if its iteration L^+ is a union of finitely many powers of L. The problem to algorithmically determine whether a given regular language possesses the finite power property is one of the most prominent questions in the theory of regular languages. It was formulated by Brzozowski during the SWAT conference in 1966, and solved independently by Hashiguchi [4] and Simon [13] more than ten years later. Results on this problem were the starting point of a fruitful and still active research, leading in particular to Hashiguchi's solution of the star-height problem [6]. The approach of Simon initiated the development of the theory of automata with multiplicities over the tropical semiring, which is now a standard method of dealing with problems related to the product operation on regular languages (see [14] for a survey).

On the other hand, the approach of Hashiguchi is combinatorial: he works directly with an automaton for the given language and uses an argument based on the pigeon hole principle. Our solution of the problem can be viewed as uncovering the algebraic background of Hashiguchi's arguments. First steps in this direction were already performed by Kirsten [8]. Here we present a fully algebraic treatment of this technique, and we formulate a simple and easily verifiable algebraic condition on a certain syntactic semigroup, which is equivalent to the finite power property. This approach also allows us to slightly generalize the result to all monoids where length of elements can be well defined and where every two factorizations of any element have a common refinement. These two properties

* Supported by the Academy of Finland under grant 208414.

M. Bugliesi et al. (Eds.): ICALP 2006, Part I, LNCS 4051, pp. 120–131, 2006.

are sufficient for the two main arguments of the proof: induction on the length of elements and localization of the problem to regular $\mathcal{J}$-classes, respectively.

Then we show that deciding the finite power property for rational languages in finitely generated monoids where the word problem is solved by a confluent regular system of deletions, can be uniformly reduced to monoids where the first result can be applied. More precisely, for every rational language in such a monoid we construct a different monoid according to the behaviour of deletions with respect to this language. Note that free groups can be defined by a confluent finite rewriting system consisting of the deletion rules $aa^{-1} \to \varepsilon$ and $a^{-1}a \to \varepsilon$, for each of the free generators a. Therefore, our result generalizes the decidability result for free groups of d'Alessandro and Sakarovitch [3], who follow the usual approach and reduce the problem to testing whether a distance automaton is bounded, which is decidable due to a difficult result of Hashiguchi [5].

Basic concepts employed in this paper are recalled in the following section. For a more comprehensive introduction to semigroup theory, formal languages, rational transductions and rewriting we refer the reader to [7], [11], [1] and [2], respectively.

2   Preliminaries

The sets of positive and non-negative integers are denoted by $\mathbb{N}$ and $\mathbb{N}_0$, respectively. For any set S, the notation $\wp(S)$ stands for the set of all subsets of S. As usual, we denote by A^+ the semigroup of all non-empty finite words over a finite alphabet A, and by A^* the monoid obtained by adding the empty word ε to A^+. The length of a word $w \in A^*$ is written as $|w|$.

Let $\mathfrak{M}$ be a monoid with identity element 1. Any subset $L \subseteq \mathfrak{M}$ is called a *language* in $\mathfrak{M}$. The *product* of two languages K and L in $\mathfrak{M}$ is defined as $KL = \{\, st \mid s \in K,\ t \in L \,\}$. The subsemigroup of $\mathfrak{M}$ generated by a language L, which is equal to $\bigcup_{n \in \mathbb{N}} L^n$, is denoted by L^+ and called the *iteration* of L. The submonoid of $\mathfrak{M}$ generated by L is $L^* = L^+ \cup \{1\}$. Further, for $n \in \mathbb{N}$ we write $L^{\leq n} = L \cup L^2 \cup \cdots \cup L^n$, and we say that a language L possesses the *finite power property* (FPP) if there exists $n \in \mathbb{N}$ such that $L^+ = L^{\leq n}$.

A language L in a monoid $\mathfrak{M}$ is *recognizable* if there exists a homomorphism $\sigma \colon \mathfrak{M} \to \mathfrak{S}$ to a finite semigroup $\mathfrak{S}$ satisfying $L = \sigma^{-1}\sigma(L)$, i.e. such that the membership of elements of $\mathfrak{M}$ in L depends only on their σ-images. The *syntactic homomorphism* of L is the projection homomorphism $\sigma \colon \mathfrak{M} \to \mathfrak{M}/\!\equiv$, where the congruence $\equiv$ of $\mathfrak{M}$ is defined by the condition

$$v \equiv w \iff (\forall x, y \in \mathfrak{M})(xvy \in L \iff xwy \in L).$$

The factor monoid $\mathfrak{M}/\!\equiv$ is called the *syntactic monoid* of L; it is the smallest monoid recognizing L.

A language L in $\mathfrak{M}$ is *rational* if it belongs to the smallest family of languages in $\mathfrak{M}$ containing all finite languages and closed under the *rational operations*: union, product and iteration. Kleene's theorem states that a language in a free monoid A^* is rational if and only if it is recognizable; such a language is then called *regular*. If the monoid $\mathfrak{M}$ is generated by a finite set A, i.e. there is an onto homomorphism $\gamma \colon A^* \twoheadrightarrow \mathfrak{M}$, then a language in $\mathfrak{M}$ is rational if and only if it is of the form $\gamma(L)$ for some regular language L in A^*.

Let $\mathfrak{S}$ be an arbitrary semigroup. An *ideal* of $\mathfrak{S}$ is a non-empty subset $I \subseteq \mathfrak{S}$ such that for all $s \in I$ and $t \in \mathfrak{S}$, we have $st \in I$ and $ts \in I$. For any ideal I of $\mathfrak{S}$, the *Rees factor semigroup* $\mathfrak{S}/I$ is defined on the set $(\mathfrak{S} \setminus I) \cup \{0\}$, where 0 is a new zero element, and elements $s, t \in \mathfrak{S} \setminus I$ are multiplied as in $\mathfrak{S}$, except for $st = 0$ when $st \in I$ holds in $\mathfrak{S}$.

The ideal of a semigroup $\mathfrak{S}$ generated by a given element $s \in \mathfrak{S}$ is equal to $\mathfrak{S}^1 s \mathfrak{S}^1$, where $\mathfrak{S}^1$ denotes the monoid obtained from $\mathfrak{S}$ by adding a new identity element 1. The quasi-order $\leq_{\mathcal{J}_\mathfrak{S}}$ on $\mathfrak{S}$ is defined, for any $s, t \in \mathfrak{S}$, by the rule $s \leq_{\mathcal{J}_\mathfrak{S}} t \iff s \in \mathfrak{S}^1 t \mathfrak{S}^1$. The Green relation $\mathcal{J}_\mathfrak{S}$ of the semigroup $\mathfrak{S}$ is the equivalence relation on $\mathfrak{S}$ associated with the quasi-order $\leq_{\mathcal{J}_\mathfrak{S}}$, i.e. two elements of $\mathfrak{S}$ are $\mathcal{J}$-equivalent if they generate the same ideal. Consequently, the quasi-order $\leq_{\mathcal{J}_\mathfrak{S}}$ determines a partial order of $\mathcal{J}$-classes of $\mathfrak{S}$.

A straightforward application of the pigeon hole principle to a $\mathcal{J}$-class of a finite semigroup gives the following useful lemma.

Lemma 1 (Kirsten [8]). *Let J be a $\mathcal{J}$-class of a finite semigroup $\mathfrak{S}$. Let $n \in \mathbb{N}$ and let $s_1, \ldots, s_n$ be a sequence of elements of $\mathfrak{S}$ satisfying $s_1 \cdots s_n \in J$. Let $N \subseteq \{1, \ldots, n\}$ with $|N| > |J|$ be such that $s_i \in J$ for all $i \in N$. Then there exist $k, l \in N$, $k < l$, such that $s_k \cdots s_l = s_k$.*

Recall that an element s of a semigroup which satisfies $ss = s$ is called an *idempotent*. A $\mathcal{J}$-class J of a finite semigroup $\mathfrak{S}$ is *regular* if it contains an idempotent, or equivalently, if there exist elements $s, t \in J$ such that their product st belongs to J too.

Further, we recall one of the basic constructions of semigroups, which will be used here to encode deletions. Assume we have a semigroup $\mathfrak{S}$ and an element $0 \notin \mathfrak{S}$. Let L and R be arbitrary finite sets, and let $P \colon R \times L \to \mathfrak{S} \cup \{0\}$ be any mapping (this mapping can be understood as an $(R \times L)$-matrix with entries belonging to $\mathfrak{S}$ or equal to 0). The *Rees matrix semigroup* $\mathcal{M}^0(\mathfrak{S}; L, R; P)$ over $\mathfrak{S}$ is defined on the set $(L \times \mathfrak{S} \times R) \cup \{0\}$, where 0 is a new zero element, by the multiplication formula

$$(l, s, r) \cdot (l', s', r') = \begin{cases} (l, s \cdot P(r, l') \cdot s', r') & \text{if } P(r, l') \neq 0, \\ 0 & \text{if } P(r, l') = 0. \end{cases}$$

A function from A^* to B^* is *rational* if it can be realized by a rational transducer, i.e. a finite automaton with output. According to Sakarovitch [12], a semigroup $\mathfrak{S}$ is called *rational* if there exist a finite set A, an onto homomorphism $\alpha \colon A^+ \twoheadrightarrow \mathfrak{S}$ and a rational function $\beta \colon A^+ \to A^+$ satisfying $\alpha\beta = \alpha$ and $\ker(\beta) = \ker(\alpha)$. This means that $\mathfrak{S}$ is isomorphic to the semigroup defined on the set $\beta(A^+)$ by the rule $u \cdot v = \beta(uv)$. An important property of rational monoids is that they satisfy Kleene's theorem:

Proposition 1 (Sakarovitch [12]). *In every rational monoid, the family of rational sets is equal to the family of recognizable sets.*

When dealing with rational monoids, we will call rational sets regular as in free monoids. Because rational functions algorithmically preserve regularity, a language L in a rational monoid is regular if and only if the language $\beta\alpha^{-1}(L)$ in A^* is regular. Moreover, rational operations on regular languages in a rational monoid can be performed algorithmically as we can calculate with the corresponding subsets $\beta\alpha^{-1}(L)$ of $\beta(A^+)$ using the obvious rules, e.g. $\beta\alpha^{-1}(K \cdot L) = \beta(\beta\alpha^{-1}(K) \cdot \beta\alpha^{-1}(L))$.

The class of rational semigroups possesses several useful closure properties with respect to basic semigroup constructions. In our considerations, the following two constructions will be employed.

Proposition 2 (Sakarovitch [12]). *The class of rational semigroups is algorithmically closed under taking Rees factors by regular ideals.*

Proposition 3. *Let $\mathfrak{M}$ be a rational monoid with identity element 1. Let L and R be finite sets and $P\colon R \times L \to \mathfrak{M} \cup \{0\}$ a mapping such that $P(\rho, \lambda) = 1$ for certain $\lambda \in L$ and $\rho \in R$. Then the Rees matrix semigroup $\mathcal{M}^0(\mathfrak{M}; L, R; P)$ over $\mathfrak{M}$ is rational and can be algorithmically constructed.*

3 Free Monoids

In this section we consider the FPP in a monoid $\mathfrak{M}$ with identity element 1 and zero element 0, and satisfying the following conditions:

1. There is a mapping $\ell\colon \mathfrak{M} \setminus \{0\} \to \mathbb{N}_0$ assigning to non-zero elements of $\mathfrak{M}$ their *length*, and satisfying $\ell(xy) \geq \ell(x) + \ell(y)$ for $x, y \in \mathfrak{M}$ such that $xy \neq 0$.
2. For every $u, v, w, t \in \mathfrak{M}$ satisfying $uv = wt \neq 0$, there exists $x \in \mathfrak{M}$ such that either $ux = w$ and $xt = v$ or $wx = u$ and $xv = t$.
3. The languages $\{0\}$ and $\{1\}$ in $\mathfrak{M}$ are recognizable.

Remark 1. Any free monoid with the length of a word defined in the usual way and with a zero element added satisfies the above conditions.

Lemma 2. *Let $\mathfrak{M}$ be a monoid satisfying condition 2. Let $m, n \in \mathbb{N}$, $m \leq n$, and $w, v_1, \ldots, v_m \in \mathfrak{M}$ be such that $w^n = v_1 \cdots v_m \neq 0$. Then there exist $k \in \mathbb{N}_0$, $k < m$, and elements $x, y \in \mathfrak{M}$ satisfying $v_1 \cdots v_k x = w^k$, $y v_{k+2} \cdots v_m = w^{n-k-1}$ and $v_{k+1} = xwy$.*

Let $L \subseteq \ell^{-1}(\mathbb{N})$ be an arbitrary recognizable language in $\mathfrak{M}$ consisting of elements of non-zero length, and such that its iteration L^+ is also recognizable. Let $\sigma\colon \mathfrak{M} \twoheadrightarrow \mathfrak{S}$ be a homomorphism onto a finite semigroup recognizing the languages L, L^+, $\{0\}$ and $\{1\}$. Consider the mapping $\tau\colon \mathfrak{M} \to \wp(\mathfrak{S}^3) \cup \{0\}$ defined by the rules $\tau(0) = 0$ and for $w \in \mathfrak{M} \setminus \{0\}$:

$$\tau(w) = \{\, (\sigma(x), \sigma(y), \sigma(z)) \mid x, y, z \in \mathfrak{M},\ w = xyz \,\}.$$

Lemma 3. *For every monoid $\mathfrak{M}$ satisfying condition 2, the kernel of τ is a congruence of $\mathfrak{M}$.*

Proof. Let $v, w \in \mathfrak{M} \setminus \{0\}$. Since σ recognizes $\{0\}$, we have $vw = 0$ if and only if $\sigma(vw) = 0$, which holds exactly when $\zeta\eta\vartheta\kappa\lambda\mu = 0$ for every $(\zeta, \eta, \vartheta) \in \tau(v)$ and $(\kappa, \lambda, \mu) \in \tau(w)$. And from property 2 it immediately follows that if $vw \neq 0$, then the triple $(\alpha, \beta, \gamma) \in \mathfrak{S}^3$ belongs to $\tau(vw)$ if and only if there exist $(\zeta, \eta, \vartheta) \in \tau(v)$ and $(\kappa, \lambda, \mu) \in \tau(w)$ satisfying either $\alpha = \zeta\eta\vartheta\kappa$, $\beta = \lambda$ and $\gamma = \mu$, or $\alpha = \zeta$, $\beta = \eta\vartheta\kappa\lambda$ and $\gamma = \mu$, or $\alpha = \zeta$, $\beta = \eta$ and $\gamma = \vartheta\kappa\lambda\mu$. $\qquad\square$

By Lemma 3 there exists a unique semigroup operation on $\tau(\mathfrak{M})$ such that τ is a homomorphism. Let us denote by $\mathfrak{T}$ the subsemigroup $\tau(L^+)$ of $\tau(\mathfrak{M})$.

Remark 2. Note that if $\tau(v) = \tau(w)$, then in particular $\sigma(v) = \sigma(w)$. Therefore, the homomorphism τ recognizes all languages recognized by σ, and it also means that $\tau(v) \; \mathcal{J}_{\mathfrak{T}} \; \tau(w)$ implies $\sigma(v) \; \mathcal{J}_{\mathfrak{S}} \; \sigma(w)$. Further observe that the identity element $\tau(1) \in \tau(\mathfrak{M})$ does not belong to $\mathfrak{T}$ since τ recognizes $\{1\}$.

Theorem 1. *Let $\mathfrak{M}$ be an arbitrary monoid satisfying properties 1 through 3. Then for any recognizable language $L \subseteq \ell^{-1}(\mathbb{N})$ in $\mathfrak{M}$ such that L^+ is also recognizable, and for σ, $\mathfrak{S}$, τ and $\mathfrak{T}$ defined above, the following conditions are equivalent:*

1. *L possesses the FPP.*
2. *For all $w \in L^+$, there exists $n \in \mathbb{N}$ such that $w^n \in L^{\leq n}$.*
3. *Every non-zero regular $\mathcal{J}$-class of $\mathfrak{T}$ contains some element of $\tau(L)$.*
4. *For all $w \in L^+ \setminus \{0\}$ such that $\tau(w)$ belongs to a regular $\mathcal{J}$-class of $\mathfrak{T}$, there exist $y \in L$ and $x, z \in L^*$ satisfying $w = xyz$ and $\sigma(y) \; \mathcal{J}_{\mathfrak{S}} \; \sigma(w)$.*
5. *$L^+ = L^{\leq (j+1)^h}$, where j is the maximal size of a $\mathcal{J}$-class of $\mathfrak{S}$ and h is the length of the longest chain of $\mathcal{J}$-classes in $\mathfrak{T}$.*

Proof. $1 \Longrightarrow 2$ is trivial.

$2 \Longrightarrow 3$. Let J be a non-zero regular $\mathcal{J}$-class of $\mathfrak{T}$. Then there exists an element $w \in L^+ \setminus \{0\}$ such that $\tau(w)$ is an idempotent belonging to J. Let $n \in \mathbb{N}$ be such that $w^n \in L^{\leq n}$. Then by Lemma 2 one can find $k \in \mathbb{N}_0$ and elements $x, y \in \mathfrak{M}$ and $u, v \in L^*$ satisfying $ux = w^k$, $yv = w^{n-k-1}$ and $xwy \in L$. Therefore we have

$$\tau(u)\tau(xwy)\tau(v) = \tau(w^n) = \tau(w) \quad \text{and}$$
$$\tau(xwyv)\tau(w)\tau(uxwy) = \tau(xw^{n+2}y) = \tau(xwy).$$

This shows that $\tau(xwy) \in J$ as required.

$3 \Longrightarrow 4$. Let $w \in L^+ \setminus \{0\}$ be such that $\tau(w)$ belongs to a regular $\mathcal{J}$-class of $\mathfrak{T}$. Then there exists $u \in L$ satisfying $\tau(u) \; \mathcal{J}_{\mathfrak{T}} \; \tau(w)$. Therefore $\tau(w) = \tau(tuv)$ for certain $t, v \in L^*$. By the definition of τ, one can find elements $x, y, z \in \mathfrak{M}$ for which $w = xyz$, $\sigma(x) = \sigma(t)$, $\sigma(y) = \sigma(u)$ and $\sigma(z) = \sigma(v)$. Because σ recognizes both L and L^*, this in particular means that $y \in L$ and $x, z \in L^*$. Finally, we have also $\sigma(y) = \sigma(u) \; \mathcal{J}_{\mathfrak{S}} \; \sigma(w)$ due to Remark 2.

$4 \Longrightarrow 5$. Let us prove that $L^+ \cap \tau^{-1}(J) \subseteq L^{\leq (j+1)^{h_J}-1}$ for every $\mathcal{J}$-class J of $\mathfrak{T}$, where h_J denotes the length of the longest chain of $\mathcal{J}$-classes in $\mathfrak{T}$ greater or equal to J. We proceed by induction on h_J. Let $w \in L^+ \cap \tau^{-1}(J)$.

If J is non-regular, consider the longest prefix $x \in L^* \setminus \tau^{-1}(J)$ of w such that there exist $y \in L$ and $z \in L^*$ satisfying $w = xyz$. Then $\tau(xy) \in J$ and therefore $\tau(z) \notin J$, as J is not regular. Hence, by the induction hypothesis, we obtain

$$w \in L^{\leq 2((j+1)^{h_J-1}-1)+1} \subseteq L^{\leq (j+1)^{h_J}-1}.$$

If $w = 0$ then $w = xz$ for some $x, z \in L^+ \setminus \{0\}$ and we get $w \in L^{\leq (j+1)^{h_J}-1}$ as in the previous case.

If J is regular and non-zero, denote the $\mathcal{J}$-class of $\sigma(w)$ in $\mathfrak{S}$ by I and consider a decomposition $w = w_0 v_1 w_1 v_2 \cdots v_n w_n$, where $w_0, w_1, \ldots, w_n \in L^*$ and $v_1, v_2, \ldots, v_n \in L \cap \sigma^{-1}(I)$, such that $\ell(v_1) + \ell(v_2) + \cdots + \ell(v_n)$ is maximal (note that this number is bounded by $\ell(w)$), and among such decompositions the number n is minimal.

If $n > |I|$ then Lemma 1 implies that there exist $k, l \in \{1, \ldots, n\}$, $k < l$, such that $\sigma(v_k w_k \cdots w_{l-1} v_l) = \sigma(v_k)$. Because σ recognizes L, this in particular means that $v_k w_k \cdots w_{l-1} v_l \in L \cap \sigma^{-1}(I)$, contradicting the choice of the decomposition of w. Thus, we have $n \leq |I|$.

Assuming $\tau(w_i) \in J$ for some $i \in \{0, \ldots, n\}$, by condition 4 we obtain certain elements $y \in L$ and $x, z \in L^*$ satisfying $w_i = xyz$ and $\sigma(y) \, \mathcal{J}_{\mathfrak{S}} \, \sigma(w_i)$. Since $\sigma(w_i) \in I$ can be derived using Remark 2, this means that $\sigma(y) \in I$, which contradicts the maximality of the decomposition of w. Therefore $\tau(w_i) \notin J$ for all $i \in \{0, \ldots, n\}$, and so the induction hypothesis gives

$$w \in L^{\leq (|I|+1)((j+1)^{h_J-1}-1)+|I|} \subseteq L^{\leq (j+1)^{h_J}-1}.$$

$5 \Longrightarrow 1$ is trivial. $\square$

Remark 3. Condition 2 of Theorem 1 was conjectured to be equivalent to the FPP by Linna [10]; later this was proved true for free monoids by Hashiguchi [4].

Let us now present examples demonstrating that both ingredients of the construction of the semigroup $\mathfrak{T}$ (i.e. the decomposition of words to triples and the restriction to $\tau(L^+)$) are essential.

Example 1. Let us take the language $L = \{a\} \cup bA^*$ over the alphabet $A = \{a, b\}$. This language clearly does not possess the FPP. Because L satisfies $L^+ = A^+$, the syntactic homomorphism σ of L recognizes all the languages L, L^+ and $\{\varepsilon\}$. The syntactic monoid of L has four elements α, β, γ and δ, which correspond to the languages $\{\varepsilon\} = \sigma^{-1}(\alpha)$, $\{a\} = \sigma^{-1}(\beta)$, $aA^+ = \sigma^{-1}(\gamma)$ and $bA^* = \sigma^{-1}(\delta)$. The only regular $\mathcal{J}$-class of $\sigma(A^+)$ is $\{\gamma, \delta\}$, and it contains the element $\delta \in \sigma(L)$ (note that the same is true also for the subsemigroup $\sigma(L^+)$).

But in the semigroup $\mathfrak{T} = \tau(L^+)$, there is really a regular $\mathcal{J}$-class which does not contain any element of $\tau(L)$, namely the $\mathcal{J}$-class of the idempotent

$$\tau(a^6) = \{\, (\zeta, \eta, \vartheta) \mid \gamma \in \{\zeta, \eta, \vartheta\} \subseteq \{\alpha, \beta, \gamma\} \,\}.$$

In order to verify this, let us take an arbitrary element of $\mathfrak{T}$ which is $\mathcal{J}$-equivalent to $\tau(a^6)$, and assume that it belongs to $\tau(L)$. Such an element must be of the form $\tau(va^6w)$, where $v, w \in L^*$. Because τ recognizes L, we have $va^6w \in L$, and therefore $va^6w \in bA^*$. Since $\tau(va^6w) \; \mathcal{J}_{\mathfrak{T}} \; \tau(a^6)$, there exist $x, y \in L^*$ satisfying $\tau(xva^6wy) = \tau(a^6)$. Then $\tau(xva^6wy)$ contains the triple $(\sigma(x), \sigma(va^6w), \sigma(y)) = (\sigma(x), \delta, \sigma(y))$, which cannot belong to $\tau(a^6)$, contradicting the previous equality.

Example 2. The language $L = \{ab\} \cup \{ab, ba\}^* ba \{ab, ba\}^*$ over the alphabet $A = \{a, b\}$ also does not have the FPP. Again, the syntactic homomorphism σ of L recognizes the languages L, L^+ and $\{\varepsilon\}$. We are going to verify that every regular $\mathcal{J}$-class of $\tau(A^+)$ containing some element of $\tau(L^+)$ contains also an element of $\tau(L)$. Because $\tau(L^+) = \tau(\{ab, ba\}^+) = \tau(L) \cup \tau(ab(ab)^+)$, it is enough to deal with the $\mathcal{J}$-classes of $\tau(A^+)$ containing $\tau((ab)^n)$ for $n \geq 2$.

For $n \leq 5$, we have $\tau^{-1}\tau((ab)^n) = \{(ab)^n\}$, since the τ-image of each of these words is characterized by the presence of some triples formed from the two different elements $\sigma(ab)$ and $\sigma(ab(ab)^+)$. Therefore for $n \leq 5$, the element $\tau((ab)^n)$ forms a non-regular singleton $\mathcal{J}$-class of $\tau(A^+)$. Further, one can calculate that $\tau((ab)^6)$ is an idempotent, and $\tau((ab)^n) = \tau((ab)^6)$ for every $n \geq 6$. In this case, one gets $\tau((ab)^6) \; \mathcal{J}_{\tau(A^+)} \; \tau((ba)^7) \in \tau(L)$.

Corollary 1. *The FPP is uniformly decidable for regular languages consisting of elements of non-zero length in rational monoids satisfying conditions 1 and 2.*

Proof. First, note that condition 3 holds for every rational monoid, and so Theorem 1 can be applied. By condition 5 of Theorem 1, it is enough to construct a semigroup $\mathfrak{S}$ recognizing L, L^+, $\{0\}$ and $\{1\}$, and test whether $L^+ = L^{\leq m}$, for $m = n^{2^{n^3}}$, where n is the cardinality of $\mathfrak{S}$. $\qquad\square$

Based on the results of Kirsten [8], the author [9] observed that each language recognized by a given finite semigroup $\mathfrak{S}$ has the FPP if and only if $\mathfrak{S}$ is a chain of simple semigroups, i.e. for all $s, t \in \mathfrak{S}$, either $st \; \mathcal{J}_{\mathfrak{S}} \; s$ or $st \; \mathcal{J}_{\mathfrak{S}} \; t$. Let us now show how one can derive this fact using Theorem 1 instead of Kirsten's results.

Corollary 2. *Let $\mathfrak{M}$ be a monoid satisfying properties 1 through 3, and let $\mathfrak{S}$ be a finite semigroup which is a chain of simple semigroups. Then every language $L \subseteq \ell^{-1}(\mathbb{N})$ in $\mathfrak{M}$ recognized by some homomorphism $\rho: \mathfrak{M} \to \mathfrak{S}$, and such that L^+ is also recognizable, has the FPP.*

Proof. We are going to verify condition 3 of Theorem 1. Consider $n \in \mathbb{N}$ and arbitrary elements $w_1, \ldots, w_n \in L$ such that $\tau(w_1 \cdots w_n)$ is an idempotent, and choose any $i \in \{1, \ldots, n\}$ for which $\rho(w_i)$ belongs to the smallest of the $\mathcal{J}$-classes of $\mathfrak{S}$ determined by the elements $\rho(w_1), \ldots, \rho(w_n)$. Because $\mathfrak{S}$ is a chain of simple semigroups, we have $\rho(w_i) \; \mathcal{J}_{\mathfrak{S}} \; \rho(w_1 \cdots w_n)$. Let m be the cardinality of the $\mathcal{J}$-class of $\rho(w_i)$ in $\mathfrak{S}$. Applying Lemma 1 to the sequence resulting from concatenating $2m + 1$ copies of the sequence $\rho(w_1), \ldots, \rho(w_n)$, and to the set $N = \{i, 2n + i, \ldots, 2mn + i\}$ (i.e. N consists of all odd occurrences of $\rho(w_i)$), we obtain a positive integer k such that $\rho(w_i) = \rho(w_i \cdots w_n (w_1 \cdots w_n)^k w_1 \cdots w_i)$. This shows that $w_i \cdots w_n (w_1 \cdots w_n)^k w_1 \cdots w_i \in L$. On the other hand, we have $\tau(w_1 \cdots w_n) = \tau((w_1 \cdots w_n)^{k+2})$ since $\tau(w_1 \cdots w_n)$ is an idempotent, and therefore the element $\tau(w_i \cdots w_n (w_1 \cdots w_n)^k w_1 \cdots w_i) \in \tau(L)$ belongs to the same $\mathcal{J}$-class of $\mathfrak{T}$ as $\tau(w_1 \cdots w_n)$. Hence, Theorem 1 implies that L has the FPP. $\quad\square$

4   Monoids Defined by Deletions

Let $\mathfrak{G}$ be a monoid generated by a finite set A whose word problem can be solved by a confluent regular system of deletion rules $\mathcal{R} = \{ w \to \varepsilon \mid w \in R \}$, where $R \subseteq A^+$ is a regular language. In other words, we have an onto homomorphism $\gamma \colon A^* \twoheadrightarrow \mathfrak{G}$ such that for every $v, w \in A^*$,

$$\gamma(v) = \gamma(w) \iff \mathrm{norm}(v) = \mathrm{norm}(w) \,,$$

where $\mathrm{norm}(w)$ denotes the normal form of w with respect to $\mathcal{R}$. We will also use the notation $\mathrm{norm}(L) = \{ \mathrm{norm}(w) \mid w \in L \}$ for a language $L \subseteq A^*$.

Lemma 4. *The language* $D = \{ w \in A^* \mid \mathrm{norm}(w) = \varepsilon \}$ *is context-free and algorithmically computable from* R. *For every regular language* $L \subseteq A^*$, *the language* $\mathrm{norm}(L)$ *is regular and can be algorithmically computed using* R *and* L.

Proof. Let $(A, Q, q_0, Q_\mathrm{f}, \delta)$ be a deterministic finite automaton recognizing R. The language D can be defined by a context-free grammar with the set of non-terminals $Q \cup \{S\}$ (where S is the initial symbol) and the derivation rules $S \to q_0$, $q \to aS\delta(q, a)$ for every $q \in Q$ and $a \in A$, $q \to \varepsilon$ for $q \in Q_\mathrm{f}$, $S \to SS$ and $S \to \varepsilon$.

Let $d \notin A$ be a new symbol and consider the context-free substitution φ from $(A \cup \{d\})^*$ to A^* defined by the rule $\varphi(d) = D$ and identical otherwise. Let $\psi \colon (A \cup \{d\})^* \to A^*$ be the homomorphism sending d to ε and leaving other symbols unchanged. Then $\mathrm{norm}(L) = \psi(\varphi^{-1}(L)) \setminus A^* R A^*$ and since both inverse context-free substitution and homomorphism effectively preserve regularity, the language $\mathrm{norm}(L)$ is regular and can be algorithmically computed. $\qquad\square$

Let $\gamma(L) \subseteq \mathfrak{G}$ be a rational language defined by a regular language $L \subseteq A^*$; by Lemma 4 we can assume that $L \subseteq \mathrm{norm}(A^*)$. Let the language L be given by a homomorphism $\sigma \colon A^* \to \mathfrak{S}$ to a finite monoid $\mathfrak{S}$ recognizing the three languages L, $\mathrm{norm}(A^*)$ and $\{\varepsilon\}$.

We are going to use σ to construct a monoid $\mathfrak{M}$ where deletions of $\mathcal{R}$ are performed symbolically, and a language K in this monoid such that $\gamma(L)$ has the FPP in $\mathfrak{G}$ if and only if K has the FPP in $\mathfrak{M}$. To achieve this, we need to avoid sequences of elements of L which are reduced to the empty word using $\mathcal{R}$, and calculate only with those which are not completely deleted. The following lemma shows that such a modification does not affect the FPP because all deleted sequences can be produced using only a bounded number of words from L.

Lemma 5. *For given regular languages* R *and* L, *one can algorithmically calculate a positive integer* m *such that*

$$\varepsilon \in \mathrm{norm}(L^+) \implies \varepsilon \in \mathrm{norm}(L^{\leq m})$$

and for all $x, y, z, u \in A^*$ *and* $w \in L^+$ *satisfying* $yz \in L$ *and* $xu \in L$, *we have*

$$\mathrm{norm}(zwx) = \varepsilon \implies \exists \bar{x}, \bar{z} \in A^*, \bar{w} \in L^{\leq m} \colon y\bar{z}, \bar{x}u \in L \ \& \ \mathrm{norm}(\bar{z}\bar{w}\bar{x}) = \varepsilon \,,$$

$$\mathrm{norm}(zw) = \varepsilon \implies \exists \bar{z} \in A^*, \bar{w} \in L^{\leq m} \colon y\bar{z} \in L \ \& \ \mathrm{norm}(\bar{z}\bar{w}) = \varepsilon \,,$$

$$\mathrm{norm}(wx) = \varepsilon \implies \exists \bar{x} \in A^*, \bar{w} \in L^{\leq m} \colon \bar{x}u \in L \ \& \ \mathrm{norm}(\bar{w}\bar{x}) = \varepsilon \,.$$

Proof. First, we calculate the context-free language $L_{r,s} = \sigma^{-1}(r)L^+\sigma^{-1}(s) \cap D$, for every $r, s \in \mathfrak{S}$. For each of these languages we test whether it is non-empty, and if $L_{r,s} \neq \emptyset$, we find any word in $L_{r,s}$, which belongs to $\sigma^{-1}(r)L^{m_{r,s}}\sigma^{-1}(s)$ for a certain $m_{r,s} \in \mathbb{N}$. We set $m = \max\{ m_{r,s} \mid r, s \in \mathfrak{S}, \ L_{r,s} \neq \emptyset \}$. Since σ recognizes both L and $\{\varepsilon\}$, we can easily verify the required properties. □

Now we construct, for the language L, a rational monoid to which we are going to apply results of the previous section. Let $\mathfrak{M}$ be defined on the set $(\mathfrak{S} \times \mathrm{norm}(A^*) \times \mathfrak{S}) \cup \{1, 0\}$, where 1 is the identity element and 0 is the zero element, by the rule

$$(p, u, q)(r, v, s) = \begin{cases} (p, uv, s) & \text{if } uv \in \mathrm{norm}(A^*) \ \& \ \varepsilon \in \mathrm{norm}(\sigma^{-1}(q)L^*\sigma^{-1}(r)), \\ 0 & \text{otherwise,} \end{cases}$$

for every $p, q, r, s \in \mathfrak{S}$ and $u, v \in \mathrm{norm}(A^*)$. Intuitively, the words u and v are factors of the resulting concatenation which are not affected by deletions when producing the normal form. And the elements q and r represent any suitable words from $\sigma^{-1}(q)$ and $\sigma^{-1}(r)$, which originated as a suffix and a prefix, respectively, of certain words from L, and which can be deleted using $\mathcal{R}$ together with several words from L between them.

We define the length of elements of $\mathfrak{M}$ as $\ell((r, v, s)) = |v|$ and $\ell(1) = 0$.

Lemma 6. *The above defined $\mathfrak{M}$ is a rational monoid, which can be algorithmically constructed from R and L and satisfies conditions 1 and 2.*

Proof. Because $A^* \setminus \mathrm{norm}(A^*) = A^*RA^*$ is an ideal of the monoid A^*, the set

$$I = (\mathfrak{S} \times A^*RA^* \times \mathfrak{S}) \cup \{0\}$$

is an ideal of the Rees matrix semigroup $\mathcal{M}^0(A^*; \mathfrak{S}, \mathfrak{S}; P)$ over A^*. Therefore $\mathfrak{M}$ is a monoid as it is isomorphic to $(\mathcal{M}^0(A^*; \mathfrak{S}, \mathfrak{S}; P)/I)^1$, where P is defined by the formula

$$P(q, r) = \begin{cases} \varepsilon & \text{if } \varepsilon \in \mathrm{norm}(\sigma^{-1}(q)L^*\sigma^{-1}(r)), \\ 0 & \text{otherwise.} \end{cases}$$

Since the finitely generated free monoid A^* is rational and P is computable and satisfies $P(1, 1) = \varepsilon$, by Proposition 3 the semigroup $\mathcal{M}^0(A^*; \mathfrak{S}, \mathfrak{S}; P)$ is rational too and can be algorithmically constructed. In order to prove that $\mathfrak{M}$ is also rational, let us consider the equivalence relation $\sim_I$ on $\mathcal{M}^0(A^*; \mathfrak{S}, \mathfrak{S}; P)$ defined by the rules

$$(p, u, q) \sim_I (r, v, s) \iff p = r \ \& \ \sigma(u) = \sigma(v) \ \& \ q = s,$$
$$0 \sim_I (r, v, s) \iff v \in A^*RA^*.$$

Because σ recognizes $\mathrm{norm}(A^*)$, the relation $\sim_I$ is easily seen to be a congruence of $\mathcal{M}^0(A^*; \mathfrak{S}, \mathfrak{S}; P)$ of finite index recognizing I. Hence, the ideal I is algorithmically regular, and by Proposition 2 the monoid $\mathfrak{M}$ is rational and can be computed from $\mathcal{M}^0(A^*; \mathfrak{S}, \mathfrak{S}; P)$.

One can easily verify that condition 1 is true even in the stronger form with $\ell(xy) = \ell(x) + \ell(y)$. Condition 2 trivially holds if one of the elements is the identity element. Otherwise, we have

$$(p, uv, s) = (p, u, q)(r, v, s) = (\bar{p}, \bar{u}, \bar{q})(\bar{r}, \bar{v}, \bar{s}) = (\bar{p}, \bar{u}\bar{v}, \bar{s})$$

for certain $p, q, r, s, \bar{p}, \bar{q}, \bar{r}, \bar{s} \in \mathfrak{S}$ and $u, v, \bar{u}, \bar{v} \in \text{norm}(A^*)$. If $|u| \leq |\bar{u}|$ then there exists $x \in \text{norm}(A^*)$ such that $ux = \bar{u}$ and $x\bar{v} = v$, and we immediately obtain $(p, u, q)(r, x, \bar{q}) = (\bar{p}, \bar{u}, \bar{q})$ and $(r, x, \bar{q})(\bar{r}, \bar{v}, \bar{s}) = (r, v, s)$ as required. The case $|u| > |\bar{u}|$ can be treated symmetrically. □

Let us consider the following language in $\mathfrak{M}$:

$$K = \{ (\sigma(x), y, \sigma(z)) \mid x, y, z \in A^*, \ y \neq \varepsilon, \ xyz \in L \}$$

Lemma 7. *The language K is regular and a congruence of $\mathfrak{M}$ of finite index recognizing K can be algorithmically constructed from R and L.*

Proof. We prove that K is recognized by the congruence $\sim$ of $\mathfrak{M}$ corresponding to $\sim_I$. This congruence has two one-element classes $\{0\}$ and $\{1\}$ and on the set $\mathfrak{S} \times \text{norm}(A^*) \times \mathfrak{S}$ it is defined as

$$(p, u, q) \sim (r, v, s) \iff p = r \ \& \ \sigma(u) = \sigma(v) \ \& \ q = s.$$

Take an element $(\sigma(x), y, \sigma(z))$ of K, where $x, y, z \in A^*$ are such that $y \neq \varepsilon$ and $xyz \in L$, and assume that $(r, v, s) \sim (\sigma(x), y, \sigma(z))$. Then $r = \sigma(x)$, $\sigma(v) = \sigma(y)$ and $s = \sigma(z)$, and consequently also $\sigma(xvz) = \sigma(xyz)$. Therefore we have $v \neq \varepsilon$ and $xvz \in L$ since σ recognizes both $\{\varepsilon\}$ and L. Thus, the element $(r, v, s) = (\sigma(x), v, \sigma(z))$ belongs to K. □

Proposition 4. *The language $\gamma(L) \subseteq \mathfrak{G}$ has the FPP if and only if the language $K \subseteq \mathfrak{M}$ has the FPP.*

Proof. Let us first assume that $\gamma(L)$ possesses the FPP, i.e. that there exists $k \in \mathbb{N}$ such that $\text{norm}(L^+) = \text{norm}(L^{\leq k})$. We are going to prove that $K^+ \setminus \{0\} = K^{\leq k+2} \setminus \{0\}$, which is sufficient to verify the FPP for K. Let (r, y, s) be an arbitrary non-zero element of K^+. Then

$$(r, y, s) = (\sigma(x_1), y_1, \sigma(z_1)) \cdots (\sigma(x_l), y_l, \sigma(z_l))$$

for some $x_i, y_i, z_i \in A^*$ satisfying $y_i \neq \varepsilon$ and $x_i y_i z_i \in L$, for $i = 1, \ldots, l$. In particular, this implies $y = y_1 \cdots y_l \in \text{norm}(A^*)$. It suffices to consider the case of $l \geq 3$. By the definition of multiplication in $\mathfrak{M}$, we can assume that for every $i = 1, \ldots, l-1$ there exists $w_i \in L^*$ satisfying $\text{norm}(z_i w_i x_{i+1}) = \varepsilon$ (note that z_i and x_i can be replaced by another elements of $\sigma^{-1}\sigma(z_i)$ and $\sigma^{-1}\sigma(x_i)$, respectively, since σ recognizes L). Now we have

$$w_1 x_2 y_2 z_2 w_2 \cdots w_{l-2} x_{l-1} y_{l-1} z_{l-1} w_{l-1} \in L^+,$$

which means that there exists $u \in L^{\leq k}$ such that

$$\text{norm}(w_1 x_2 y_2 z_2 w_2 \cdots w_{l-2} x_{l-1} y_{l-1} z_{l-1} w_{l-1}) = \text{norm}(u).$$

Because $y_2 \cdots y_{l-1} \in \text{norm}(A^*)$, we obtain

$$\text{norm}(z_1 u x_l) = \text{norm}(z_1 w_1 x_2 y_2 z_2 w_2 \cdots w_{l-2} x_{l-1} y_{l-1} z_{l-1} w_{l-1} x_l) = y_2 \cdots y_{l-1}.$$

Let us now consider one sequence of deletions using $\mathcal{R}$ producing $y_2 \cdots y_{l-1}$ from $z_1 u x_l$, and group together all neighbouring letters which are not deleted. As $L \subseteq \text{norm}(A^*)$ holds, this regrouping is of the form

$$z_1 = \bar{y}_1 \bar{z}_1, \quad u = v_0 \tilde{x}_1 \tilde{y}_1 \tilde{z}_1 v_1 \cdots v_{n-1} \tilde{x}_n \tilde{y}_n \tilde{z}_n v_n, \quad x_l = \bar{x}_l \bar{y}_l,$$

for certain $n \in \{0, \ldots, k\}$, $\bar{y}_1, \bar{z}_1, \bar{x}_l, \bar{y}_l \in A^*$, $\tilde{x}_i, \tilde{y}_i, \tilde{z}_i \in A^*$, for $i = 1, \ldots, n$, and $v_i \in L^*$, for $i = 0, \ldots, n$, which satisfy $\tilde{y}_i \neq \varepsilon$, $\tilde{x}_i \tilde{y}_i \tilde{z}_i \in L$, for $i = 1, \ldots, n$,

$$\text{norm}(\bar{z}_1 v_0 \tilde{x}_1) = \text{norm}(\tilde{z}_i v_i \tilde{x}_{i+1}) = \text{norm}(\tilde{z}_n v_n \bar{x}_l) = \varepsilon,$$

for $i = 1, \ldots, n-1$, and $\bar{y}_1 \tilde{y}_1 \cdots \tilde{y}_n \bar{y}_l = y_2 \cdots y_{l-1}$ (if n is equal to 0, then $\text{norm}(\bar{z}_1 v_0 \bar{x}_l) = \varepsilon$). This immediately gives a decomposition

$$(\sigma(x_1), y_1 \bar{y}_1, \sigma(\bar{z}_1))(\sigma(\tilde{x}_1), \tilde{y}_1, \sigma(\tilde{z}_1)) \cdots (\sigma(\tilde{x}_n), \tilde{y}_n, \sigma(\tilde{z}_n))(\sigma(\bar{x}_l), \bar{y}_l y_l, \sigma(z_l))$$

of (r, y, s), where each element belongs to the language K (note that we have $\sigma(x_1 y_1 \bar{y}_1 \bar{z}_1) = \sigma(x_1 y_1 z_1) \in \sigma(L)$), and therefore $(r, y, s) \in K^{\leq k+2}$.

In order to prove the converse, let $K^+ = K^{\leq k}$ for some $k \in \mathbb{N}$. We will verify that $\text{norm}(L^+) = \text{norm}(L^{\leq m(k+1)+k})$, where m is the number guaranteed by Lemma 5. Let u be an arbitrary word from the language L^+. We have to show $\text{norm}(u) \in \text{norm}(L^{\leq m(k+1)+k})$. The first statement of Lemma 5 allows us to assume that $\text{norm}(u) \neq \varepsilon$. Because $L \subseteq \text{norm}(A^*)$, the word u can be written in the form

$$u = v_0 x_1 y_1 z_1 v_1 \cdots v_{l-1} x_l y_l z_l v_l,$$

for a certain $l \in \mathbb{N}$, where $x_i, y_i, z_i \in A^*$ satisfy $y_i \neq \varepsilon$ and $x_i y_i z_i \in L$, for $i = 1, \ldots, l$, $v_i \in L^*$, for $i = 0, \ldots, l$, $\text{norm}(u) = y_1 \cdots y_l$ and

$$\text{norm}(v_0 x_1) = \text{norm}(z_i v_i x_{i+1}) = \text{norm}(z_l v_l) = \varepsilon,$$

for $i = 1, \ldots, l-1$. This implies that

$$(\sigma(x_1), y_1 \cdots y_l, \sigma(z_l)) = (\sigma(x_1), y_1, \sigma(z_1)) \cdots (\sigma(x_l), y_l, \sigma(z_l)) \in K^l.$$

By the assumption, there exist $n \in \mathbb{N}$, $n \leq k$, and $\tilde{x}_i, \tilde{y}_i, \tilde{z}_i \in A^*$ such that $\tilde{y}_i \neq \varepsilon$ and $\tilde{x}_i \tilde{y}_i \tilde{z}_i \in L$, for $i = 1, \ldots, n$, which satisfy

$$(\sigma(x_1), \text{norm}(u), \sigma(z_l)) = (\sigma(\tilde{x}_1), \tilde{y}_1, \sigma(\tilde{z}_1)) \cdots (\sigma(\tilde{x}_n), \tilde{y}_n, \sigma(\tilde{z}_n)).$$

According to the definition of the operation of $\mathfrak{M}$, words $\tilde{x}_i$, for $i = 2, \ldots, n$, and $\tilde{z}_i$, for $i = 1, \ldots, n-1$, can be chosen so that there exist words $w_i \in L^*$, for $i = 1, \ldots, n - 1$, satisfying $\mathrm{norm}(\tilde{z}_i w_i \tilde{x}_{i+1}) = \varepsilon$. By Lemma 5 we can find $\bar{x}_i, \bar{z}_i \in A^*$, for $i = 1, \ldots, n$, and $\bar{w}_i \in L^{\leq m} \cup \{\varepsilon\}$, for $i = 0, \ldots, n$, such that $\bar{x}_i \tilde{y}_i \bar{z}_i \in L$, for $i = 1, \ldots, n$, and

$$\mathrm{norm}(\bar{w}_0 \bar{x}_1) = \mathrm{norm}(\bar{z}_i \bar{w}_i \bar{x}_{i+1}) = \mathrm{norm}(\bar{z}_n \bar{w}_n) = \varepsilon,$$

for $i = 1, \ldots, n - 1$ (in order to get $\mathrm{norm}(\bar{w}_0 \bar{x}_1) = \varepsilon$, note that $x_1 \tilde{y}_1 \tilde{z}_1 \in L$ and $\mathrm{norm}(v_0 x_1) = \varepsilon$). Therefore we have

$$\mathrm{norm}(u) = \tilde{y}_1 \cdots \tilde{y}_n = \mathrm{norm}(\bar{w}_0 \bar{x}_1 \tilde{y}_1 \bar{z}_1 \bar{w}_1 \cdots \bar{w}_{n-1} \bar{x}_n \tilde{y}_n \bar{z}_n \bar{w}_n),$$

and so $\mathrm{norm}(u)$ belongs to $\mathrm{norm}(L^{\leq m(k+1)+k})$ as required.   □

Theorem 2. *The FPP is uniformly decidable for rational languages in finitely generated monoids whose word problem is solved by a confluent regular system of deletions.*

Proof. For given regular languages R and L, we can construct by Lemmas 6 and 7 the rational monoid $\mathfrak{M}$ and the regular language $K \subseteq \mathfrak{M}$ such that testing whether $\gamma(L)$ has the FPP in $\mathfrak{G}$ is equivalent to testing whether K has the FPP in $\mathfrak{M}$ (by Proposition 4). Because K contains only elements of positive length, this can be algorithmically decided using Corollary 1.   □

References

1. Berstel, J.: *Transductions and Context-Free Languages.* Teubner, Stuttgart (1979).
2. Book, R.V., Otto, F.: *String-Rewriting Systems.* Springer, New York (1993).
3. d'Alessandro, F., Sakarovitch, J.: The finite power property in free groups. *Theoret. Comput. Sci.* **293**(1) (2003) 55–82.
4. Hashiguchi, K.: A decision procedure for the order of regular events. *Theoret. Comput. Sci.* **8**(1) (1979) 69–72.
5. Hashiguchi, K.: Limitedness theorem on finite automata with distance functions. *J. Comput. System Sci.* **24** (1982) 233–244.
6. Hashiguchi, K.: Algorithms for determining relative star height and star height. *Inform. and Comput.* **78**(2) (1988) 124–169.
7. Howie, J.M.: *Fundamentals of Semigroup Theory.* Clarendon Press, Oxford (1995).
8. Kirsten, D.: The finite power problem revisited. *Inform. Process. Lett.* **84**(6) (2002) 291–294.
9. Kunc, M.: Regular solutions of language inequalities and well quasi-orders. *Theoret. Comput. Sci.* **348**(2-3) (2005) 277–293.
10. Linna, M.: Finite power property of regular languages. In: Nivat, M. (ed.): *Automata, Languages and Programming.* North-Holland, Amsterdam (1973) 87–98.
11. Rozenberg, G., Salomaa, A. (eds.): *Handbook of Formal Languages.* Springer, Berlin (1997).
12. Sakarovitch, J.: Easy multiplications. I. The realm of Kleene's theorem. *Inform. and Comput.* **74**(3) (1987) 173–197.
13. Simon, I.: Limited subsets of a free monoid. In: *Proc. 19th Annu. Symp. on Foundations of Computer Science.* IEEE, Piscataway, N.J. (1978) 143–150.
14. Simon, I.: Recognizable sets with multiplicities in the tropical semiring. In: Chytil, M., Janiga, L., Koubek, V. (eds.): *Mathematical Foundations of Computer Science 1988.* Lecture Notes in Comput. Sci., Vol. 324, Springer, Berlin (1988) 107–120.

P-completeness of Cellular Automaton Rule 110

Turlough Neary[1] and Damien Woods[2]

[1] TASS, Department of Computer Science,
National University of Ireland Maynooth, Ireland
tneary@cs.may.ie
[2] Department of Mathematics and Boole Centre for Research in Informatics,
University College Cork, Ireland
d.woods@bcri.ucc.ie

Abstract. We show that the problem of predicting t steps of the 1D cellular automaton Rule 110 is P-complete. The result is found by showing that Rule 110 simulates deterministic Turing machines in polynomial time. As a corollary we find that the small universal Turing machines of Mathew Cook run in polynomial time, this is an exponential improvement on their previously known simulation time overhead.

1 Introduction

In this paper we solve an open problem regarding the computational complexity of Rule 110 which is one of the simplest cellular automata. We show that the prediction problem for Rule 110 is P-complete. Rule 110 is a nearest neighbour, one dimensional, binary cellular automaton [1]. It is composed of a sequence of cells $\ldots p_{-1}p_0p_1 \ldots$ where each cell has a binary state $p_i \in \{0,1\}$. At timestep $t+1$ the value of cell $p_{i,t+1} = F(p_{i-1,t}, p_{i,t}, p_{i+1,t})$ is given by the synchronous local update function F

$$
\begin{array}{ll}
F(0,0,0) = 0 & F(1,0,0) = 0 \\
F(0,0,1) = 1 & F(1,0,1) = 1 \\
F(0,1,0) = 1 & F(1,1,0) = 1 \\
F(0,1,1) = 1 & F(1,1,1) = 0
\end{array}
$$

The problem of RULE 110 PREDICTION is defined as follows.

Definition 1 (RULE 110 PREDICTION). *Given an initial Rule 110 configuration, a cell index i and a natural number t written in unary. Is cell p_i in state 1 at time t?*

This problem is in P as a Turing machine simulates the cellular automaton in $O(t^2)$ steps by repeatedly traversing from left to right. From Matthew Cook's [2] result one infers a NC lower bound on the problem. Cook showed that Rule 110 simulates Turing machines via the following sequence of simulations

$$\text{Turing machine} \mapsto \text{2-tag system} \mapsto \text{cyclic tag system} \mapsto \text{Rule 110} \qquad (1)$$

M. Bugliesi et al. (Eds.): ICALP 2006, Part I, LNCS 4051, pp. 132–143, 2006.

where $A \mapsto B$ denotes that A is simulated by B. The universality of 2-tag systems [3] is well-known and Cook supplied the latter two simulations. Each of these simulations runs in polynomial time (that is, B runs in a number of steps that is polynomial in the number of A's steps) with the exception of the exponentially slow 2-tag system simulation of Turing machines [3]. This slowdown is due to the 2-tag system's unary encoding of Turing machine tape contents. Thus via Equation (1), Rule 110 is an exponentially slow simulator of Turing machines and so it has remained open as to whether RULE 110 PREDICTION is P-complete.

In this work we replace the tag system with a *clockwise Turing machine* to give the following chain of simulations

$$\text{Turing machine} \mapsto \text{clockwise Turing machine}$$
$$\mapsto \text{cyclic tag system} \mapsto \text{Rule 110} \tag{2}$$

Each simulation runs in polynomial time and the reduction from Turing machine to Rule 110 is computable by a logspace Turing machine. Thus our work shows that Rule 110 simulates Turing machines efficiently, giving the following result.

Theorem 1. RULE 110 PREDICTION is logspace complete for P.

Rule 110 is a very simple (2 state, nearest neighbour, one dimensional) cellular automaton, and Matthew Cook [2] gave four small universal Turing machines[1] that simulate its computation. Their size given as (number of states, number of symbols), are respectively $(2, 5), (3, 4), (4, 3)$ and $(7, 2)$. In terms of program size these machines are a significant improvement on previous small universal Turing machine results [4,5,6,7,8,9]. However in terms of time complexity Cook's machines offer no improvement over the exponentially slow machines of Rogozhin et al. [4,5,6,7]. A corollary of our work is that Matthew Cook's small universal Turing machines are polynomial time simulators of Turing machines.

The prediction problem for a number of classes of cellular automata has been shown to be P-complete. However Rule 110 is the simplest so far, in the sense that previous P-completeness results have been shown for more general cellular automata (e.g. more states, neighbours or dimensions). For example prediction of cellular automata of dimension $d \geqslant 1$ with an arbitrary number of states is known to be P-complete [10]. Lindgren and Nordahl [11] show that prediction for one dimensional nearest neighbour cellular automata is P-complete for seven states and Ollinger's result [12] improves this to six. If the update rule depends on the states of five neighbours then four states are sufficient [11,10]. Moore [13] shows that prediction of binary majority voting cellular automata is P-complete for dimension $d \geqslant 3$. On the other hand, the prediction problem for a variety

---

[1] Cook's small "universal Turing machines" deviate from the usual Turing machine definition in the following way: their blank tape consists of an infinitely repeated word to the left and another to the right. Intuitively this change of definition seems to make quite a difference to program size, especially since Cook encodes a program in one of these repeated words. This has no bearing on our P-completeness result as we require only a bounded initial configuration for RULE 110 PREDICTION.

of linear and quasilinear cellular automata is in NC [14,15]. The question of whether RULE 110 PREDICTION is P-complete has been asked, either directly or indirectly, in a number of previous works (for example [14,15,16]).

2   Clockwise Turing Machines

A clockwise Turing machine is like a standard single-tape Turing machine [17] except for the following details: (i) the tape is assumed to be circular, (ii) the tape head moves only clockwise on the tape, (iii) the machine's transition function is of the form $f : Q \times \Sigma \rightarrow (\Sigma \cup \Sigma\Sigma) \times Q$. Here Q and Σ are the machine's finite set of states and tape symbols respectively. A transition rule $t = (q_x, \sigma, v, q_y) \in f$, is executed as follows. If the write value v is an element of Σ then the tape cell containing the read symbol is overwritten by this value and the head moves clockwise to the next cell. Otherwise if $v \in \Sigma\Sigma$ then the tape cell containing the read symbol is replaced with two cells that each contain one of v's symbols and the head moves clockwise to the next cell.

It is not difficult to give a clockwise Turing machine R_M that simulates a single-tape Turing machine M with a quadratic time overhead. We can think of M's right moves as clockwise moves by R_M with $v \in \Sigma$. However if M is increasing its tape length by reading a blank symbol and moving right, then we proceed differently. In this case R_M inserts two symbols, $v = \sigma r$, where σ is M's write symbol. Then R_M moves clockwise, traversing the entire tape, until it meets the 'rightmost end of tape marker' symbol r. If M runs in time $T(n)$ then R_M simulates a right move by M in $O(T(n))$ time.

A left move (when reading a non-blank symbol) by M is simulated by a single traversal of the circular tape that leaves a marker and then shifts each symbol one step clockwise. Upon reaching the marker the left move simulation is complete. A left move by M, when reading a blank at the leftmost tape end, is simulated using a similar strategy to that above. Proof details are to be found in a previous paper [8].

Lemma 1. *Let M be a single-tape Turing machine that runs in time $T(n)$. Then there is a clockwise Turing machine R_M that simulates the computation of M in time $O(T^2(n))$.*

In the next section we prove that cyclic tag systems simulate clockwise Turing machines. In order to simplify this proof we state the result for clockwise Turing machines that have a binary tape alphabet $\Sigma = \{a, b\}$. As with standard Turing machines, using a binary alphabet causes at most a constant factor increase in the time, space and number of states.

3   Cyclic Tag Systems

Cyclic tag systems were used by Cook [2] to show that Rule 110 is universal.

Definition 2 (cyclic tag system). *A cyclic tag system $C = \alpha_0, \ldots, \alpha_{p-1}$, is a list of binary words $\alpha_m \in \{0, 1\}^*$ called appendants.*

A *configuration* of a cyclic tag system consists of (i) a *marker* that points to a single appendant α_m in C, and (ii) a word $w = w_0 \ldots w_{|w|-1} \in \{0,1\}^*$. We call w the *data word*. Intuitively the list C is a *program* with the marker pointing to instruction α_m. In the initial configuration the marker points to appendant α_0 and w is the binary input word.

Definition 3 (computation step of a cyclic tag system). *A computation step is deterministic and acts on a configuration in one of two ways:*

- *If $w_0 = 0$ then w_0 is deleted and the marker moves to appendant $\alpha_{(m+1 \bmod p)}$.*
- *If $w_0 = 1$ then w_0 is deleted, the word α_m is appended onto the right end of w, and the marker moves to appendant $\alpha_{(m+1 \bmod p)}$.*

We write $c_1 \vdash c_2$ when configuration c_2 is obtained from c_1 via a single computation step. We let $c_1 \vdash^i c_2$ denote a sequence of exactly i computation steps. A cyclic tag system completes its computation if (i) the data word is the empty word or (ii) it enters a forever repeating sequence of configurations. The complexity measures of time and space are defined in the obvious way.

Example 1. (cyclic tag system computation) Let $C = 00, 01, 11$ be a cyclic tag system with input word 011. Below we give the first four steps of the computation. In each configuration C is given on the left with the marked appendant highlighted in bold font.

$$\mathbf{00}, 01, 11 \quad 011 \quad \vdash \quad 00, \mathbf{01}, 11 \quad 11 \quad \vdash \quad 00, 01, \mathbf{11} \quad 101$$
$$\vdash \quad \mathbf{00}, 01, 11 \quad 0111 \quad \vdash \quad 00, \mathbf{01}, 11 \quad 111 \quad \vdash \quad \ldots$$

3.1 Cyclic Tag Systems Simulate Clockwise Turing Machines

Much of the proof of Theorem 1 is given by the following lemma.

Lemma 2. *Let R be a binary clockwise Turing machine with $|Q|$ states that runs in time $T(n)$. Then there is a cyclic tag system C_R that simulates the computation of R in time $O(|Q|T^2(n) \log T(n))$.*

Proof. Let $R = (Q, \{a,b\}, f, q_1, q_{|Q|})$ where $Q = \{q_1, \ldots, q_{|Q|}\}$ are the states, $\{a,b\}$ is the binary alphabet, f is the transition function, and $q_1, q_{|Q|} \in Q$ are the initial and final states respectively. In the sequel $\sigma_j \in \{a,b\}$. The bulk of the proof is concerned with simulating a single (but arbitrary) transition rule of R.

Encoding. We define the cyclic tag system (program) to be of the form $C_R = \alpha_0, \ldots, \alpha_{2z-1}$ where $z = 30|Q| + 61$. Given an initial configuration of R (consisting of current state $q_i \in Q$, read symbol σ_1, and tape contents $\sigma_1 \ldots \sigma_s \in \{a,b\}^*$) we encode this as a configuration of C_R as follows

$$\alpha_0, \ldots, \alpha_{2z-1} \quad \langle 1, q_i \rangle \langle \sigma_1 \rangle \ldots \langle \sigma_s \rangle \mu^{s'} \tag{3}$$

Here $\mu = 10^{z-1}$ and

$$s' = 2^{\lceil \log_2 s \rceil} \tag{4}$$

Table 1.1. (Stage 1. Halve counter). Every second μ is marked off by being changed to $\mu\!\!\!/$.

encoded object	encoded object length	initial marker index	index y of appendant	appendant α_y
$\langle 1, q_i \rangle = 0^{30i+20} 10^{2z-30i-21}$	$2z$	0	$30i + 20$	$\langle 1', q_i \rangle$
$\langle 1, q_i \rangle = 0^{30i+20} 10^{2z-30i-21}$	$2z$	z	$z + 30i + 20$	$0^z \langle 1', q_{i,s<s'} \rangle$
$\langle 1, q_{i,s<s'} \rangle = 0^{30i+25} 10^{2z-30i-26}$	$2z$	0	$30i + 25$	$\langle 1', q_{i,s<s'} \rangle$
$\langle 1, q_{i,s<s'} \rangle = 0^{30i+25} 10^{2z-30i-26}$	$2z$	z	$z + 30i + 25$	$0^z \langle 1', q_{i,s<s'} \rangle$
$\langle a \rangle = 010^{2z-2}$	$2z$	0	1	$\langle a \rangle$
$\langle a \rangle = 010^{2z-2}$	$2z$	z	$z + 1$	$\langle a \rangle$
$\langle b \rangle = 0^2 10^{2z-3}$	$2z$	0	2	$\langle b \rangle$
$\langle b \rangle = 0^2 10^{2z-3}$	$2z$	z	$z + 2$	$\langle b \rangle$
$\langle a\!\!\!/ \rangle = 0^3 10^{2z-4}$	$2z$	0	3	$\langle a\!\!\!/ \rangle$
$\langle a\!\!\!/ \rangle = 0^3 10^{2z-4}$	$2z$	z	$z + 3$	$\langle a\!\!\!/ \rangle$
$\langle b\!\!\!/ \rangle = 0^4 10^{2z-5}$	$2z$	0	4	$\langle b\!\!\!/ \rangle$
$\langle b\!\!\!/ \rangle = 0^4 10^{2z-5}$	$2z$	z	$z + 4$	$\langle b\!\!\!/ \rangle$
$\mu = 10^{z-1}$	z	0	0	$\mu\!\!\!/$
$\mu = 10^{z-1}$	z	z	z	μ'
$\mu' = 0^6 10^{2z-7}$	$2z$	0	6	μ'
$\mu' = 0^6 10^{2z-7}$	$2z$	z	$z + 6$	μ'
$\mu\!\!\!/ = 0^5 10^{2z-6}$	$2z$	0	5	$\mu\!\!\!/$
$\mu\!\!\!/ = 0^5 10^{2z-6}$	$2z$	z	$z + 5$	$\mu\!\!\!/$

are used for a 'tape length' counter. The values of appendants α_j are given during the proof below. States q_i and tape symbols $\{a, b\}$ of R are encoded as:

$$\langle 1, q_i \rangle = 0^{30i+20} 10^{2z-30i-21}$$
$$\langle a \rangle = 010^{2z-2}$$
$$\langle b \rangle = 0^2 10^{2z-3}$$

Our simulation algorithm consists of a number of stages. In a CTS configuration the current stage x of our algorithm is identifiable by the notation $\langle x, q_i \rangle$.

How to read the tables. We define the cyclic tag system C_R via a number of tables that specify encoded objects (e.g. encoded symbols, states) in the data word and the appendants they map to. Each table row gives an "encoded object" followed by the "encoded object length". The "initial marker index" gives the location of the program marker immediately before the encoded object is read. Each encoded object indexes an appendant α_y, where y is specified by the "index y of appendant" column and α_y is specified by the "appendant α_y" column.

To aid the reader we carefully describe the initial steps in the simulation of a transition rule. We encode a configuration that is arbitrary except for its tape length (which is 3). Initially the marker is pointing at appendant α_0 and the data word is $\langle 1, q_i \rangle \langle \sigma_1 \rangle \langle \sigma_2 \rangle \langle \sigma_3 \rangle \mu\mu\mu\mu \in \{0, 1\}^{12z}$. The leftmost $2z$ symbols in the data word encode the current state q_i. From Table 1.1 this is $\langle 1, q_i \rangle = 0^{30i+20} 10^{2z-30i-21}$. The computation begins by deleting the $30i + 20$ leftmost 0 symbols while moving the marker rightwards through the appendants, one step for each deletion. The leftmost data symbol is now 1, this is deleted and causes the appendant α_{30i+20} to be appended onto the rightmost end of the data word. From Table 1.1 we see that $\alpha_{30i+20} = \langle 1', q_i \rangle$. Then $2z - 30i - 21$ contiguous 0 symbols are deleted while moving the marker one step for each deletion. Since $|\langle 1, q_i \rangle| = 2z$ and there are exactly $2z$ appendants in C_R, the marker is once again positioned at α_0. We write these $2z$ steps as

$$\alpha_0, \ldots, \alpha_{2z-1} \quad \langle 1, q_i \rangle \langle \sigma_1 \rangle \langle \sigma_2 \rangle \langle \sigma_3 \rangle \mu\mu\mu\mu$$
$$\vdash^{2z} \alpha_0, \ldots, \alpha_{2z-1} \quad \langle \sigma_1 \rangle \langle \sigma_2 \rangle \langle \sigma_3 \rangle \mu\mu\mu\mu \langle 1', q_i \rangle$$

Algorithm overview. Our cyclic tag system algorithm has three stages. Stages 1 and 2 isolate the encoded read symbol of R which is located immediately to the right of $\langle 1, q_i \rangle$. These stages make use of the tape-length counter specified by Equations (3) and (4). In Stage 1 every second μ is marked and then in Stage 2 every second $\langle \sigma \rangle$ is marked. This process is iterated until all μ objects are marked ($1 + \log_2 s'$ iterations). The first six configurations of Fig. 1 illustrate this process. The encoded read symbol is now isolated as it is the only unmarked encoded tape symbol. The computation then enters Stage 3 which uses the encoded current state and (isolated) encoded read symbol to index an appendant that encodes the write symbol(s) and next state. In the final two configurations of Fig. 1 the new encoded current state and write value are appended and the counter is doubled to maintain the equality in Equation (4).

Stage 1. Halve counter. The counter value is specified by Equation (4) as the number of μ (or later, μ') objects. This value is halved by marking half of the μ objects (changing μ to $\not\mu$) using Table 1.1. In this table we see that $|\mu| = z$ so exactly two μ objects are read for a single traversal of the marker through all $2z$ appendants. Every second μ indexes $\not\mu$ and every other μ indexes μ'. The encoded state $\langle 1, q_i \rangle$ indexes $\langle 1', q_i \rangle$ or $\langle 1', q_{i, s < s'} \rangle$, which sends control to Table 1.2.

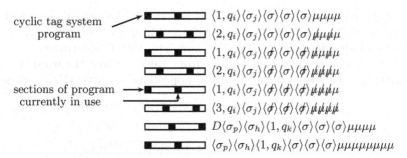

Fig. 1. CTS simulation of transition rule $(q_i, \sigma_j, \sigma_p \sigma_h, q_k)$. The CTS program is illustrated on the left. In the data word the encoded current state $\langle x, q_i \rangle$ directs the control flow by determining the sections of the CTS program that are used in Stage x.

Table 1.2. (Stage 1. Check counter value). Here $\langle 1', q_i \rangle$ or $\langle 1', q_{i,s<s'} \rangle$ is used to check if the counter is 0.

encoded object	encoded object length	initial marker index	index y of appendant	appendant α_y
$\langle 1', q_i \rangle = 0^{30i+21} 10^{2z-30i-12}$	$2z + 10$	0	$30i + 21$	$\langle 2, q_i \rangle$
$\langle 1', q_i \rangle = 0^{30i+21} 10^{2z-30i-12}$	$2z + 10$	z	$z + 30i + 21$	$\langle 3, q_i \rangle$
$\langle 1', q_{i,s<s'} \rangle = 0^{30i+26} 10^{2z-30i-17}$	$2z + 10$	0	$30i + 26$	$\langle 2, q_{i,s<s'} \rangle$
$\langle 1', q_{i,s<s'} \rangle = 0^{30i+26} 10^{2z-30i-17}$	$2z + 10$	z	$z + 30i + 26$	$\langle 3, q_{i,s<s'} \rangle$
$\langle a \rangle = 010^{2z-2}$	$2z$	10	11	$\langle a' \rangle$
$\langle a \rangle = 010^{2z-2}$	$2z$	$z + 10$	$z + 11$	$\langle a \rangle$
$\langle b \rangle = 0^2 10^{2z-3}$	$2z$	10	12	$\langle b' \rangle$
$\langle b \rangle = 0^2 10^{2z-3}$	$2z$	$z + 10$	$z + 12$	$\langle b \rangle$
$\langle \not{a} \rangle = 0^3 10^{2z-4}$	$2z$	10	13	$\langle \not{a} \rangle$
$\langle \not{a} \rangle = 0^3 10^{2z-4}$	$2z$	$z + 10$	$z + 13$	$\langle \not{a} \rangle$
$\langle \not{b} \rangle = 0^4 10^{2z-5}$	$2z$	10	14	$\langle \not{b} \rangle$
$\langle \not{b} \rangle = 0^4 10^{2z-5}$	$2z$	$z + 10$	$z + 14$	$\langle \not{b} \rangle$
$\mu' = 0^6 10^{2z-7}$	$2z$	10	16	μ'
$\mu' = 0^6 10^{2z-7}$	$2z$	$z + 10$	$z + 16$	μ'
$\not{\mu} = 0^5 10^{2z-6}$	$2z$	10	15	$\not{\mu}$
$\not{\mu} = 0^5 10^{2z-6}$	$2z$	$z + 10$	$z + 15$	$\not{\mu}$

We continue the above simulation (we later generalise to an arbitrary number of tape symbols).

$$
\begin{aligned}
&\boldsymbol{\alpha_0}, \ldots, \alpha_{2z-1} && \langle \sigma_1 \rangle \langle \sigma_2 \rangle \langle \sigma_3 \rangle \mu\mu\mu\mu \langle 1', q_i \rangle \\
\vdash^{2z} \ &\boldsymbol{\alpha_0}, \ldots, \alpha_{2z-1} && \langle \sigma_2 \rangle \langle \sigma_3 \rangle \mu\mu\mu\mu \langle 1', q_i \rangle \langle \sigma_1 \rangle \\
\vdash^{4z} \ &\boldsymbol{\alpha_0}, \ldots, \alpha_{2z-1} && \mu\mu\mu\mu \langle 1', q_i \rangle \langle \sigma_1 \rangle \langle \sigma_2 \rangle \langle \sigma_3 \rangle \\
\vdash^{z} \ &\alpha_0, \ldots, \boldsymbol{\alpha_z}, \ldots, \alpha_{2z-1} && \mu\mu\mu \langle 1', q_i \rangle \langle \sigma_1 \rangle \langle \sigma_2 \rangle \langle \sigma_3 \rangle \not{\mu} \\
\vdash^{z} \ &\boldsymbol{\alpha_0}, \ldots, \alpha_{2z-1} && \mu\mu \langle 1', q_i \rangle \langle \sigma_1 \rangle \langle \sigma_2 \rangle \langle \sigma_3 \rangle \not{\mu} \mu' \\
\vdash^{2z} \ &\boldsymbol{\alpha_0}, \ldots, \alpha_{2z-1} && \langle 1', q_i \rangle \langle \sigma_1 \rangle \langle \sigma_2 \rangle \langle \sigma_3 \rangle \not{\mu} \mu' \not{\mu} \mu'
\end{aligned}
$$

The algorithm tests if the counter is 0 by checking if exactly one unmarked μ was read. If so $\langle 3, q_i \rangle$ is appended and we enter Stage 3. Otherwise $\langle 2, q_i \rangle$ is appended and we enter Stage 2. Table 1.2 simulates this 'if' statement.

As we continue our simulation we note from Table 1.2 that the word $\langle 1', q_i \rangle$ is of length $2z + 10$. Hence the marker is at appendant α_{10} after $\langle 1', q_i \rangle$ is read:

$$
\begin{aligned}
&\boldsymbol{\alpha_0}, \ldots, \alpha_{2z-1} && \langle 1', q_i \rangle \langle \sigma_1 \rangle \langle \sigma_2 \rangle \langle \sigma_3 \rangle \not{\mu} \mu' \not{\mu} \mu' \\
\vdash^{2z+10} \ &\alpha_0, \ldots, \boldsymbol{\alpha_{10}}, \ldots, \alpha_{2z-1} && \langle \sigma_1 \rangle \langle \sigma_2 \rangle \langle \sigma_3 \rangle \not{\mu} \mu' \not{\mu} \mu' \langle 2, q_i \rangle \\
\vdash^{14z} \ &\alpha_0, \ldots, \boldsymbol{\alpha_{10}}, \ldots, \alpha_{2z-1} && \langle 2, q_i \rangle \langle \sigma_1' \rangle \langle \sigma_2' \rangle \langle \sigma_3' \rangle \not{\mu} \mu' \not{\mu} \mu'
\end{aligned}
$$

Immediately above is the first configuration of Stage 2.

Stage 2. Mark half of the encoded tape symbols. The ultimate aim of this stage is to isolate the encoded read symbol. Each iteration of this stage uses

Table 2.1. (Stage 2. Mark half of the encoded tape symbols). Rows 3 to 6 are used to mark off every second $\langle\sigma\rangle$.

encoded object	encoded object length	initial marker index	index y of appendant	appendant α_y
$\langle 2, q_i\rangle = 0^{30i+12}10^{2z-30i-3}$	$2z+10$	10	$30i+22$	$0^{2z-20}\langle 1, q_i\rangle$
$\langle 2, q_{i,s<s'}\rangle = 0^{30i+17}10^{2z-30i-8}$	$2z+10$	10	$30i+27$	$0^{2z-20}\langle 1, q_{i,s<s'}\rangle$
$\langle a'\rangle = 0^7 10^{z-8}$	z	20	27	$\langle a\rangle$
$\langle a'\rangle = 0^7 10^{z-8}$	z	$z+20$	$z+27$	$\langle\not a\rangle$
$\langle b'\rangle = 0^8 10^{z-9}$	z	20	28	$\langle b\rangle$
$\langle b'\rangle = 0^8 10^{z-9}$	z	$z+20$	$z+28$	$\langle\not b\rangle$
$\langle\not a\rangle = 0^3 10^{2z-4}$	$2z$	20	23	$\langle\not a\rangle$
$\langle\not a\rangle = 0^3 10^{2z-4}$	$2z$	$z+20$	$z+23$	$\langle\not a\rangle$
$\langle\not b\rangle = 0^4 10^{2z-5}$	$2z$	20	24	$\langle\not b\rangle$
$\langle\not b\rangle = 0^4 10^{2z-5}$	$2z$	$z+20$	$z+24$	$\langle\not b\rangle$
$\mu' = 0^6 10^{2z-7}$	$2z$	20	26	μ
$\mu' = 0^6 10^{2z-7}$	$2z$	$z+20$	$z+26$	μ
$\not\mu = 0^5 10^{2z-6}$	$2z$	20	25	$\not\mu$
$\not\mu = 0^5 10^{2z-6}$	$2z$	$z+20$	$z+25$	$\not\mu$

Table 2.1 to mark off every second (even numbered) encoded tape symbol $\langle\sigma_j\rangle$. As we continue our simulation we note from Table 2.1 that $|\langle 2, q_i\rangle| = 2z + 10$. Hence the marker is at appendant α_{20} after reading $\langle 2, q_i\rangle$.

$$
\begin{array}{ll}
\alpha_0,\ldots,\boldsymbol{\alpha_{10}},\ldots,\alpha_{2z-1} & \langle 2, q_i\rangle\langle\sigma'_1\rangle\langle\sigma'_2\rangle\langle\sigma'_3\rangle\not\mu\mu'\not\mu\mu' \\
\vdash^{2z+10}\ \alpha_0,\ldots,\boldsymbol{\alpha_{20}},\ldots,\alpha_{2z-1} & \langle\sigma'_1\rangle\langle\sigma'_2\rangle\langle\sigma'_3\rangle\not\mu\mu'\not\mu\mu'0^{2z-20}\langle 1, q_i\rangle \\
\vdash^{z}\ \alpha_0,\ldots,\boldsymbol{\alpha_{z+20}},\ldots,\alpha_{2z-1} & \langle\sigma'_2\rangle\langle\sigma'_3\rangle\not\mu\mu'\not\mu\mu'0^{2z-20}\langle 1, q_i\rangle\langle\sigma_1\rangle \\
\vdash^{z}\ \alpha_0,\ldots,\boldsymbol{\alpha_{20}},\ldots,\alpha_{2z-1} & \langle\sigma'_3\rangle\not\mu\mu'\not\mu\mu'0^{2z-20}\langle 1, q_i\rangle\langle\sigma_1\rangle\langle\not\sigma_2\rangle \\
\vdash^{9z}\ \alpha_0,\ldots,\boldsymbol{\alpha_{z+20}},\ldots,\alpha_{2z-1} & 0^{2z-20}\langle 1, q_i\rangle\langle\sigma_1\rangle\langle\not\sigma_2\rangle\langle\sigma_3\rangle\not\mu\mu\not\mu\mu \\
\vdash^{2z-20}\ \alpha_0,\ldots,\boldsymbol{\alpha_z},\ldots,\alpha_{2z-1} & \langle 1, q_i\rangle\langle\sigma_1\rangle\langle\not\sigma_2\rangle\langle\sigma_3\rangle\not\mu\mu\not\mu\mu
\end{array}
$$

If we are simulating a transition rule that has write value from $\Sigma\Sigma = \{a, b\}^2$, and the tape length is a power of 2, then we must double the counter value in order to satisfy Equation (4). This doubling occurs in Stage 3. However the tape length test happens in Stage 1 using Table 1.1 as follows.

Suppose that the encoded tape length is not a power of 2 and thus $s < s'$. Then, on some iteration, Stage 2 reads an odd number, strictly greater than 1, of unmarked encoded tape symbols. If this occurs then $\langle 1, q_i\rangle$ indexes the appendant $\langle 1', q_{i,s<s'}\rangle$. To see this, notice that in Stage 2 the tape symbols a, b are encoded as $\langle a'\rangle, \langle b'\rangle$ where $|\langle a'\rangle| = |\langle b'\rangle| = z$. If C_R reads an odd number of these then the initial marker index is at z. Suppose otherwise that the encoded tape length is a power of 2. Then $\langle 1, q_i\rangle$ always indexes the appendant $\langle 1', q_i\rangle$ in Stage 1. In

summary, if $\langle 1', q_{i,s<s'} \rangle$ is not appended before Stage 3 begins then the number of tape symbols is a power of 2 and $s = s'$.

The simulation continues as follows:

$$
\begin{array}{ll}
\alpha_0, \ldots, \boldsymbol{\alpha_z}, \ldots, \alpha_{2z-1} & \langle 1, q_i \rangle \langle \sigma_1 \rangle \langle \cancel{\sigma}_2 \rangle \langle \sigma_3 \rangle \cancel{\mu}\mu\cancel{\mu}\mu \\
\vdash^{2z} \alpha_0, \ldots, \boldsymbol{\alpha_z}, \ldots, \alpha_{2z-1} & \langle \sigma_1 \rangle \langle \cancel{\sigma}_2 \rangle \langle \sigma_3 \rangle \cancel{\mu}\mu\cancel{\mu}\mu 0^z \langle 1', q_{i,s<s'} \rangle \\
\vdash^{13z} \alpha_0, \ldots, \alpha_{2z-1} & \langle 1', q_{i,s<s'} \rangle \langle \sigma_1 \rangle \langle \cancel{\sigma}_2 \rangle \langle \sigma_3 \rangle \cancel{\mu}\mu' \cancel{\mu}\mu\cancel{\mu} \\
\vdash^{16z+10} \alpha_0, \ldots, \boldsymbol{\alpha_{10}}, \ldots, \alpha_{2z-1} & \langle 2, q_{i,s<s'} \rangle \langle \sigma'_1 \rangle \langle \cancel{\sigma}_2 \rangle \langle \sigma'_3 \rangle \cancel{\mu}\mu' \cancel{\mu}\mu\cancel{\mu} \\
\vdash^{14z+10} \alpha_0, \ldots, \boldsymbol{\alpha_{20}}, \ldots, \alpha_{2z-1} & 0^{2z-20} \langle 1, q_{i,s<s'} \rangle \langle \sigma_1 \rangle \langle \cancel{\sigma}_2 \rangle \langle \cancel{\sigma}_3 \rangle \cancel{\mu}\mu\cancel{\mu}\mu\cancel{\mu} \\
\vdash^{17z-20} \alpha_0, \ldots, \boldsymbol{\alpha_z}, \ldots, \alpha_{2z-1} & \langle 1', q_{i,s<s'} \rangle \langle \sigma_1 \rangle \langle \cancel{\sigma}_2 \rangle \langle \cancel{\sigma}_3 \rangle \cancel{\mu}\mu\cancel{\mu}\mu\cancel{\mu} \\
\vdash^{16z+10} \alpha_0, \ldots, \boldsymbol{\alpha_{z+10}}, \ldots, \alpha_{2z-1} & \langle 3, q_{i,s<s'} \rangle \langle \sigma_1 \rangle \langle \cancel{\sigma}_2 \rangle \langle \cancel{\sigma}_3 \rangle \cancel{\mu}\cancel{\mu}\cancel{\mu}\cancel{\mu}\cancel{\mu}
\end{array}
$$

Immediately above is the first the configuration of Stage 3.

Stage 3. Complete simulation of transition rule. In this stage an appendant α_y is indexed, based on the value of the encoded current state and encoded read symbol using Table 3.1. The printing of appendant α_y simulates the encoded write value, encoded next state, and the clockwise tape head movement.

Using Table 3.1 we read the encoded current state, either $\langle 3, q_i \rangle$ or $\langle 3, q_{i,s<s'} \rangle$, after which the initial marker index is either $30i + 30$ or $30i + 40$ respectively. The encoded read symbol was already isolated and uniquely retains its original value of $\langle a \rangle$ or $\langle b \rangle$; this value points at the appendant α_y (rows 3 to 10). All other (non-isolated) encoded tape symbols are of the form $\langle \cancel{a} \rangle$ or $\langle \cancel{b} \rangle$ and they point to the appendants $\langle a \rangle$ or $\langle b \rangle$ respectively.

The simulated transition rule is of the form $(q_i, \sigma_j, \sigma_p, q_k)$ or $(q_i, \sigma_j, \sigma_p \sigma_h, q_k)$, respectively encoded as the appendants $\langle \sigma_p \rangle \langle 1, q_k \rangle$ or $\langle \sigma_p \rangle \langle \sigma_h \rangle \langle 1, q_k \rangle$. In the present example we simulate the rule $(q_i, \sigma_1, \sigma_4, q_k)$:

$$
\begin{array}{ll}
\alpha_0, \ldots, \boldsymbol{\alpha_{z+10}}, \ldots, \alpha_{2z-1} & \langle 3, q_{i,s<s'} \rangle \langle \sigma_1 \rangle \langle \cancel{\sigma}_2 \rangle \langle \cancel{\sigma}_3 \rangle \cancel{\mu}\cancel{\mu}\cancel{\mu}\cancel{\mu}\cancel{\mu} \\
\vdash^{z+30i+30} \alpha_0, \ldots, \boldsymbol{\alpha_{30i+40}}, \ldots, \alpha_{2z-1} & \langle \sigma_1 \rangle \langle \cancel{\sigma}_2 \rangle \langle \cancel{\sigma}_3 \rangle \cancel{\mu}\cancel{\mu}\cancel{\mu}\cancel{\mu}\cancel{\mu} 0^{2z-30i-40} \\
\vdash^{2z} \alpha_0, \ldots, \boldsymbol{\alpha_{30i+40}}, \ldots, \alpha_{2z-1} & \langle \cancel{\sigma}_2 \rangle \langle \cancel{\sigma}_3 \rangle \cancel{\mu}\cancel{\mu}\cancel{\mu}\cancel{\mu}\cancel{\mu} 0^{2z-30i-40} \langle \sigma_4 \rangle \langle 1, q_k \rangle \\
\vdash^{12z} \alpha_0, \ldots, \boldsymbol{\alpha_{30i+40}}, \ldots, \alpha_{2z-1} & 0^{2z-30i-40} \langle \sigma_4 \rangle \langle 1, q_k \rangle \langle \sigma_2 \rangle \langle \sigma_3 \rangle \mu\mu\mu\mu \\
\vdash^{2z-30i-40} \alpha_0, \ldots, \alpha_{2z-1} & \langle \sigma_4 \rangle \langle 1, q_k \rangle \langle \sigma_2 \rangle \langle \sigma_3 \rangle \mu\mu\mu\mu
\end{array}
$$

Alternatively if the rule is of the form $(q_i, \sigma_1, \sigma_4 \sigma_5, q_k)$ then the latter configuration is instead

$$
\alpha_0, \ldots, \alpha_{2z-1} \quad \langle \sigma_4 \rangle \langle \sigma_5 \rangle \langle 1, q_k \rangle \langle \sigma_2 \rangle \langle \sigma_3 \rangle \mu\mu\mu\mu
$$

The simulation of the transition rule is now complete. The marker in C_R's program is at appendant α_0. The encoded write value is written, the new encoded state $\langle 1, q_k \rangle$ is established and the (clockwise) tape head movement is simulated.

We have given a sequence of configurations that explicitly simulate the application of a transition rule. We used arbitrary initial and next states $q_i, q_k \in Q$, and arbitrary tape symbols $\sigma_j \in \{a, b\}$.

Table 3.1. (Stage 3. Simulate transition rule). This table prints the encoded write value and establishes the new encoded current state $\langle 1, q_k \rangle$. If the counter does not need to be doubled this table completes simulation of the transition rule.

encoded object	encoded object length	initial marker index	index y of appendant	appendant α_y
$\langle 3, q_i \rangle = 0^{30i+14}10^{z+5}$	$z + 30i + 20$	$z + 10$	$z + 30i + 24$	$0^{2z-30i-30}$
$\langle 3, q_{i,s<s'} \rangle = 0^{30i+19}10^{z+10}$	$z + 30i + 30$	$z + 10$	$z + 30i + 29$	$0^{2z-30i-40}$
$\langle a \rangle = 010^{2z-2}$	$2z$	$30i + 30$	$30i + 31$	$\langle \sigma_p \rangle \langle 1, q_k \rangle$
$\langle a \rangle = 010^{2z-2}$	$2z$	$30i + 30$	$30i + 31$	$D\langle \sigma_p \rangle \langle \sigma_h \rangle \langle 1, q_k \rangle$
$\langle b \rangle = 0^2 10^{2z-3}$	$2z$	$30i + 30$	$30i + 32$	$\langle \sigma_p \rangle \langle 1, q_k \rangle$
$\langle b \rangle = 0^2 10^{2z-3}$	$2z$	$30i + 30$	$30i + 32$	$D\langle \sigma_p \rangle \langle \sigma_h \rangle \langle 1, q_k \rangle$
$\langle a \rangle = 010^{2z-2}$	$2z$	$30i + 40$	$30i + 41$	$\langle \sigma_p \rangle \langle 1, q_k \rangle$
$\langle a \rangle = 010^{2z-2}$	$2z$	$30i + 40$	$30i + 41$	$\langle \sigma_p \rangle \langle \sigma_h \rangle \langle 1, q_k \rangle$
$\langle b \rangle = 0^2 10^{2z-3}$	$2z$	$30i + 40$	$30i + 42$	$\langle \sigma_p \rangle \langle 1, q_k \rangle$
$\langle b \rangle = 0^2 10^{2z-3}$	$2z$	$30i + 40$	$30i + 42$	$\langle \sigma_p \rangle \langle \sigma_h \rangle \langle 1, q_k \rangle$
$\langle \not a \rangle = 0^3 10^{2z-4}$	$2z$	$30i + 30$	$30i + 33$	$\langle a \rangle$
$\langle \not a \rangle = 0^3 10^{2z-4}$	$2z$	$30i + 40$	$30i + 43$	$\langle a \rangle$
$\langle \not b \rangle = 0^4 10^{2z-5}$	$2z$	$30i + 30$	$30i + 34$	$\langle b \rangle$
$\langle \not b \rangle = 0^4 10^{2z-5}$	$2z$	$30i + 40$	$30i + 44$	$\langle b \rangle$
$\not\mu = 0^5 10^{2z-6}$	$2z$	$30i + 30$	$30i + 35$	μ
$\not\mu = 0^5 10^{2z-6}$	$2z$	$30i + 40$	$30i + 45$	μ

The simulation is specific in the sense that the length of the tape data is fixed. The computation of C_R remains similar for any length of tape data that is not a power of 2. If the tape length is a power of 2, and thus $s = s'$, then C_R enters Stage 3 via $\langle 3, q_i \rangle$ instead of $\langle 3, q_{i,s<s'} \rangle$. On the one hand, if the tape data does not increase in length, the remainder of the computation proceeds in a similar manner to the above simulation. On the other hand, if the tape data increases in length [i.e. we are simulating a transition rule of the form $(q_i, \sigma_j, \sigma_p\sigma_h, q_k)$] then rows 4 or 6 of Table 3.1 are executed. The appendants in these rows contain the subword D. After reading D (using Table 3.2) the marker points at α_{40} which causes each μ in the counter to index the appendant $\mu\mu$. This doubles the counter's value and completes the simulation of the transition rule.

The simulation is also specific in the sense that the encoded state is the leftmost object in the data word when we begin simulating a transition rule. This generalises to an arbitrary encoded state position. To see this notice that the encoded state directs control flow of the algorithm through Stages 1 to 3. The order of executing the stages is unaffected by the relative *position* of the encoded state in the data word.

We have shown how C_R simulates an arbitrary transition rule of R. To simulate halting C_R enters a repeating sequence of configurations. The halt state $q_{|Q|}$ is encoded in the normal way as $\langle 1, q_{|Q|} \rangle = 0^{30|Q|+20}10^{2z-30|Q|-21}$. We define the appendant at index $30|Q| + 20$ to be $\langle 1, q_{|Q|} \rangle$. Therefore $\langle 1, q_{|Q|} \rangle$ indexes a copy of itself. Also after $\langle 1, q_{|Q|} \rangle$ is read, each encoded tape symbol indexes a copy of itself. This causes C_R to enter a forever repeating sequence of configurations.

Table 3.2. (Stage 3. Double counter). Each μ indexes the appendant $\mu\mu$.

encoded object	encoded object length	initial marker index	index y of appendant	appendant α_y
$D = 0^{39}10^{2z}$	$2z + 40$	0	39	0^{2z-40}
$\langle 1, q_k \rangle = 0^{30k+20}10^{2z-30k-21}$	$2z$	40	$30k + 60$	$\langle 1, q_k \rangle$
$\langle a \rangle = 010^{2z-2}$	$2z$	40	41	$\langle a \rangle$
$\langle b \rangle = 0^210^{2z-3}$	$2z$	40	42	$\langle b \rangle$
$\mu = 10^{z-1}$	z	40	40	$\mu\mu$
$\mu = 10^{z-1}$	z	$z + 40$	$z + 40$	$\mu\mu$
$\mu' = 0^610^{2z-7}$	$2z$	40	46	$\mu\mu'$
$\mu = 0^510^{2z-6}$	$2z$	40	45	$\mu\mu'$

Space analysis. At time $T(n)$ there are $O(T(n))$ encoded objects (state and symbols) in C_R's data word; each of length $O(|Q|)$. Thus C_R uses $O(|Q|T(n))$ space.

Time analysis. Simulating a transition rule involves 3 stages. Each stage executes in $O(|Q|T(n))$ steps. To simulate a single transition rule the counter is halved $O(\log T(n))$ times, (i.e. Stages 1 and 2 are executed $O(\log T(n))$ times) and Stage 3 is executed once. Thus $O(|Q|T(n)\log(T(n))$ time is sufficient to simulate a transition rule and $O(|Q|T^2(n)\log T(n))$ time is sufficient to simulate the computation of R.   □

A consequence of the previous lemma is that Rule 110 simulates Turing machines in polynomial time. Matthew Cook's [2] universal Turing machines (see footnote on page 133) simulate Rule 110 in quadratic time, which in turn (using Cook's construction) simulates Turing machines in exponential time. We have improved this time bound to polynomial.

Corollary 1. *Matthew Cook's small universal Turing machines simulate Turing machines in polynomial time.*

Finally we show that the reduction from the GENERIC MACHINE SIMULATION PROBLEM (GMSP) [10] to RULE 110 PREDICTION is computable by a logspace transducer Turing machine. The GMSP is stated as: given a word x, an encoding $\langle M \rangle$ of a single-tape Turing machine M, and an integer t in unary, does M accept x within t steps?

Lemma 3. *The GMSP is logspace reducible to* RULE 110 PREDICTION.

Proof. From Section 2 the number of states of the binary clockwise Turing machine R_M is linear in the number of states and symbols of M. We encode these machines as words in a straightforward way such that for their lengths: $|\langle R_M \rangle| = O(|\langle M \rangle|)$. Also the input x_R to R_M is of length linear in $|x|$, the length of M's input. The conversion is clearly logspace computable.

We reduce the simulation problem for R_M to the analogous problem for cyclic tag systems. In the proof of Lemma 2 we showed how to construct C_{R_M}. We encode C_{R_M} as a word $\langle C_{R_M} \rangle$. The value z used in the proof of Lemma 2 is linear in $|Q|$, the number of states of R_M. There are $2z$ appendants, each of length $O(|Q|)$, giving an encoded program length of $O(|Q|^2)$. From Equation (3) the input $\langle x_R \rangle$ to C_{R_M} is of length $O(|Q||x_R|)$. Thus the encoded appendants and input are logspace constructable.

To show that a logspace transducer Turing machine generates a Rule 110 instance from $\langle C_{R_M} \rangle \# \langle x_R \rangle \#^t$ we examine Cook's Rule 110 simulation of cyclic tag systems [2]. The input is written directly as the states of $O(|\langle x_R \rangle|)$ contiguous cells beginning at, say, cell p_0. On the left of the input a constant word (representing Cook's 'ossifiers') is repeated $O(t)$ times. On the right the cyclic tag system program (list of appendants and 'leaders') is written $O(t)$ times. □

Since we already know that RULE 110 PREDICTION is in P, the proof of Theorem 1 is complete.

References

1. Wolfram, S.: Statistical mechanics of cellular automata. Reviews of Modern Physics **55** (1983) 601–644
2. Cook, M.: Universality in elementary cellular automata. Complex Systems **15** (2004) 1–40
3. Cocke, J., Minsky, M.: Universality of tag systems with $P = 2$. Journal of the ACM **11** (1964) 15–20
4. Rogozhin, Y.: Small universal Turing machines. TCS **168** (1996) 215–240
5. Baiocchi, C.: Three small universal Turing machines. In Margenstern, M., Rogozhin, Y., eds.: Machines, Computations, and Universality. Volume 2055 of LNCS., Chişinău, Moldova, MCU, Springer (2001) 1–10
6. Kudlek, M., Rogozhin, Y.: A universal Turing machine with 3 states and 9 symbols. In Kuich, W., Rozenberg, G., Salomaa, A., eds.: Developments in Language Theory (DLT) 2001. Volume 2295 of LNCS., Vienna, Springer (2002) 311–318
7. Minsky, M.: Size and structure of universal Turing machines using tag systems. In: Recursive Function Theory, Symp. in Pure Math. Volume 5., AMS (1962) 229–238
8. Neary, T., Woods, D.: A small fast universal Turing machine. Technical Report NUIM-CS-TR-2005-12, Dept. of Computer Science, NUI Maynooth (2005)
9. Neary, T., Woods, D.: Small fast universal Turing machines. TCS (To appear.)
10. Greenlaw, R., Hoover, H.J., Ruzzo, W.L.: Limits to parallel computation: P-completeness theory. Oxford university Press, Oxford (1995)
11. Lindgren, K., Nordahl, M.G.: Universal computation in simple one-dimensional cellular automata. Complex Systems **4** (1990) 299–318
12. Ollinger, N.: The quest for small universal cellular automata. In Widmayer, P., et al., eds.: International Colloquium on Automata, Languages and Programming (ICALP). Volume 2380 of LNCS., Malaga, Spain, Springer (2002) 318–329
13. Moore, C.: Majority-vote cellular automata, Ising dynamics and P-completeness. Journal of Statistical Physics **88** (1997) 795–805
14. Moore, C.: Quasi-linear cellular automata. Physica D **103** (1997) 100–132
15. Moore, C.: Predicting non-linear cellular automata quickly by decomposing them into linear ones. Physica D **111** (1998) 27–41
16. Aaronson, S.: Book review: A new kind of science. Quantum Information and Computation **2** (2002) 410–423
17. Hopcroft, J.E., Ullman, J.D.: Introduction to automata theory, languages, and computation. Addison-Wesley (1979)

Small Sweeping 2NFAs
Are Not Closed Under Complement

Christos A. Kapoutsis

Computer Science and Artificial Intelligence Laboratory
Massachusetts Institute of Technology
cak@mit.edu

Abstract. A two-way nondeterministic finite automaton is *sweeping*
(SNFA) if its input head can change direction only on the end-markers.
For every n, we exhibit a language that can be recognized by an n-state
SNFA but requires $2^{\Omega(n)}$ states on every SNFA recognizing its complement.

1 Introduction

Understanding the power of nondeterminism is one of the most important goals
of the theory of computation. In the past four decades, huge efforts have been
invested into problems like P vs. NP and L vs. NL, with limited success. To some,
this is creating the suspicion that essentially the same elusive idea lies at the core
of all problems of this kind, little affected by the particulars of the underlying
computational model or resource.

In this context, a possibly advantageous approach is to focus on weak models
of computation. Provided that they are also powerful enough to be relevant,
such models allow us to meaningfully study the power of nondeterministic algorithms in a much simpler setting, closer to the set-theoretic objects produced by
their computations and in some distance from our often misleading algorithmic
intuitions about how these computations may behave.

One such model is the two-way finite automaton. The question whether nondeterminism strictly increases its power, in the sense that it allows exponential
economy in the number of states, was raised by Seiferas [1] in the early 70's. Now
known as the 2D vs. 2N question, it was reduced by Sakoda and Sipser [2] to the
study of certain complete problems and remains essentially as wide open as its
famous counterparts above. The conjecture is that indeed 2D≠2N, and its more
precise variants are quite surprising—see [3] for a brief history and discussion.

Given that small two-way deterministic finite automata (2DFAs) are closed
under complement [4,5], one way to confirm the conjecture is by proving that
this closure fails in the nondeterministic case (2NFAs). In this track, Geffert,
Mereghetti and Pighizzini [5] have recently studied the special case of small
unary 2NFAs, but concluded that these *are* in fact closed under complement.

Following the same track, we study a different special case. We focus on *sweeping* 2NFAs (SNFAs), which are 2NFAs that can change the direction of their input
head only on the end-markers. We prove that small SNFAs are *not* closed under
complement—reaffirming, in a sense, the promise of the general direction.

M. Bugliesi et al. (Eds.): ICALP 2006, Part I, LNCS 4051, pp. 144–156, 2006.
© Springer-Verlag Berlin Heidelberg 2006

The sweeping restriction was originally introduced by Sipser [6], in the first major step towards the conjecture, where he showed that no small SDFA can solve *liveness*—a problem that even small one-way nondeterministic finite automata (1NFAs) can solve. Indeed, our proof has the structure of that argument: we show that *no small* SNFA *can solve the complement of liveness*. Note that this was already known for 1NFAs (by a relatively simple argument of [2]) and SDFAs (by a combination of the arguments of [6] and [4]), so our theorem can be seen as a generalization of those facts to sweeping bidirectionality and to nondeterminism, respectively. In fact, this generalization was already asked for in [6].

2   Preliminaries and Outline

We write $[n]$ for the set $\{1, 2, \ldots, n\}$. If Σ is an alphabet, Σ^* is the set of all finite strings over Σ. If z is a string, then $|z|$, z_t, and z^t are its length, t-th symbol, and t-fold concatenation with itself. A property $P \subseteq \Sigma^*$ is *infinitely right-extensible* if every string in P has a right extension in P: $(\forall y \in P)(\exists z)(|z| \neq 0 \ \& \ yz \in P)$; *infinitely left-extensible* properties are defined symmetrically.

2.1   Sets, Functions, and Relations

If U is a set, then $\overline{U}$, $|U|$, $\mathcal{P}(U)$, and U^2 denote its complement, size, powerset, and set of pairs. The following simple lemma plays a central role in our proof.

Lemma 1. *Let* $(u_i)_{i \in I}$ *and* $(v_i)_{i \in I}$ *be two sequences of subsets of a set* U, *where* I *is a set of indices totally ordered by* $<$. *If for all* $i', i \in I$ *we have*

$$i' < i \implies u_{i'} \cap v_i = \emptyset \qquad and \qquad i' = i \implies u_{i'} \cap v_i \neq \emptyset,$$

then $|I| \leq |U|$.

Proof. For each $i \in I$, let a_i be any element of the non-empty intersection $u_i \cap v_i$. If the list $(a_i)_{i \in I}$ contains a repetition, say $a_{i'} = a_i =: a$ for two indices $i' < i$, then $a = a_{i'} \in u_{i'}$ and $a = a_i \in v_i$; hence $a \in u_{i'} \cap v_i$, a contradiction. Therefore the list $(a_i)_{i \in I}$ contains $|I|$ distinct elements of U. Hence, $|I| \leq |U|$.

Let $V \subseteq \mathcal{P}(U)$ be a set of points in the lattice of subsets of U. For $u \in V$, the part of V below u is $V_u := \{u' \in V \mid u' \subseteq u\}$; the *height* $h_V(u)$ of u in V is the length of the longest chain $\emptyset \neq u_1 \subsetneq \cdots \subsetneq u_k$ in V_u. For $f : V \to V$, we say f is *monotone* if it respects inclusion: $u' \subseteq u \implies f(u') \subseteq f(u)$; we say f is an *automorphism* if its restriction to V_u is a bijection from V_u to $V_{f(u)}$, for all u. Clearly, every automorphism respects heights: $h_V(u) = h_V(f(u))$, for all u. By f^t we mean the t-fold composition of f with itself; if $t = 0$, this is the identity.

Lemma 2. *Suppose* $f : V \to V$, *where* $V \subseteq \mathcal{P}(U)$ *is a* finite *set of points from the lattice of a set* U. *If* f *is* injective *and* monotone, *then it is an automorphism.*

Proof. Pick any $u \in V$, set $v := f(u)$, and let f_u be the restriction of f to V_u. We will show f_u is a bijection from V_u to V_v. Since f is monotone, f_u has all its values in V_v: $u' \in V_u \implies u' \subseteq u \implies f(u') \subseteq f(u) \implies f_u(u') \in V_v$. Since f is injective, so is f_u. So, f_u is an injection from V_u to V_v. To show that it is a bijection, it is sufficient to show that V_v does not have more elements than V_u.

Since f is injective and V is finite, f is a permutation of V. Hence, for some $t \geq 1$, f^t is the identity. Let $f' := f^{t-1}$. Since f is injective and monotone, f' is also injective and monotone. Moreover, $u = f^t(u) = f^{t-1}(f(u)) = f'(v)$. Now the same argument as in the previous paragraph shows that the restriction f'_v of f' to V_v is an injection from V_v to V_u. Consequently, $|V_v| \leq |V_u|$.

Let $R \subseteq U^2$ be a binary relation. We write $R(\cdot)$ for the mapping of each $u \subseteq U$ to the set $R(u) := \{b \in U \mid (\exists a \in u)(aRb)\}$ of all elements related to elements of u; we usually write $R(a)$ for $R(\{a\})$. Clearly, $R(\cdot)$ is monotone. If $R' \subseteq U^2$ is also a binary relation, we write $R' \circ R$ for the composition: $a(R' \circ R)b \iff (\exists c \in U)(aR'c \ \& \ cRb)$. Clearly, $(R' \circ R)(u) = R(R'(u))$, for all u.

A total order $<$ on $\mathcal{P}(U)^2$ is *nice* if each pair "escapes" from every strictly smaller pair in at least one component: $(u', v') < (u, v) \implies u' \not\supseteq u \lor v' \not\supseteq v$. It is not hard to verify that nice orders on $\mathcal{P}(U)^2$ exist, for every finite U.

2.2 Sweeping Automata and Liveness

A *sweeping deterministic finite automaton* (SDFA, [6]) is a triple $M = (q_s, \delta, q_f)$, where δ is the *transition function*, partially mapping $Q \times (\Sigma \cup \{\square\})$ to Q, for some set Q of *states*, some *alphabet* Σ, and some *end-marker* $\square \notin \Sigma$, while q_s and q_f are the *start* and *final* states. An input $z \in \Sigma^*$ is presented to M between two copies of $\square$. The computation starts at q_s, on the symbol to the right of the left copy of $\square$, heading rightward. The next state is always derived from δ and the current state and symbol. The next position is always the adjacent one in the direction of motion; except when the current symbol is $\square$ and the next state is not q_f, in which case the next position is the adjacent one in the opposite direction. Note that the computation can either loop, or hang, or fall off the string $\square z \square$ into q_f. In this last case we say that M *accepts* z.

More generally, for any $z \in \Sigma^*$ and $p \in Q$, the *left computation of M from p on z* is the unique sequence

$$\text{LCOMP}_{M,p}(z) := (q_t)_{1 \leq t \leq m}$$

where $q_1 = p$; every next state is $q_{t+1} = \delta(q_t, z_t)$, provided that $t \leq |z|$ and the value of δ is defined; and m is the first t for which this last provision fails. If $m = |z| + 1$, the computation *exits into* q_m; otherwise, $1 \leq m \leq |z|$ and the computation *hangs at* q_m. The *right computation of M from p on z*, $\text{RCOMP}_{M,p}(z) := (q_t)_{1 \leq t \leq m}$, is defined symmetrically, with $q_{t+1} = \delta(q_t, z_{|z|+1-t})$.

If M is allowed more than one next move at each step, we say that it is *nondeterministic* (SNFA). Formally, this means that δ *totally* maps $Q \times (\Sigma \cup \{\square\})$ to the *powerset* of Q and implies that, on any $z \in \Sigma^*$, M exhibits a *set* of computations. If at least one of them falls off $\square z \square$ into q_f, then M accepts z.

Similarly, $\text{LCOMP}_{M,p}(z)$ is now a *set* of computations. To encode how states connect via left computations, we define the binary relation $\text{LVIEW}_M(z) \subseteq Q^2$

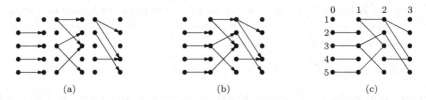

Fig. 1. (a) Three symbols in Σ_5; e.g., the third symbol is $\{(1,2),(1,4),(2,5),(4,4)\}$. (b) The string defined by them. (c) The string simplified and indexed; here $\xi = \{(3,5)\}$.

$$(p,q) \in \text{LVIEW}_M(z) \iff \big(\exists c \in \text{LCOMP}_{M,p}(z)\big)(c \text{ exits into } q),$$

and call it the *left behavior of M on z*. Then, for $u \subseteq Q$, the set $\text{LVIEW}_M(z)(u)$ of states reachable via left computations from within u is the *left view of u on z*. The *right behavior* $\text{RVIEW}_M(z)$ of M on z and the *right view* $\text{RVIEW}_M(z)(u)$ of u on z are defined similarly. Note that, if $|z| = 1$, the automaton has the same behavior in both directions: $\text{LVIEW}_M(z) = \text{RVIEW}_M(z) = \{(p,q) \mid \delta(p,z) \ni q\}$. Also, if extending z does not cause a view to include any new states, then this remains true on all identical further extensions, as described in the next lemma.

Lemma 3. *The following implications are true, for all $t \geq 1$:*

- $\text{LVIEW}_M(z)(u) \supseteq \text{LVIEW}_M(z\tilde{z})(u) \implies \text{LVIEW}_M(z)(u) \supseteq \text{LVIEW}_M(z\tilde{z}^t)(u),$
- $\text{RVIEW}_M(z)(u) \supseteq \text{RVIEW}_M(\tilde{z}z)(u) \implies \text{RVIEW}_M(z)(u) \supseteq \text{RVIEW}_M(\tilde{z}^t z)(u).$

Liveness. For $n \geq 1$, we consider the alphabet $\Sigma_n := \mathcal{P}([n]^2)$ of all directed 2-column graphs with n nodes per column and only rightward arrows (Fig. 1a). An m-long string over Σ_n is naturally viewed as a directed $(m + 1)$-column graph (Fig. 1b), in which for simplicity we often omit the direction of the arrows (Fig. 1c). We say that the string has *connectivity* $\xi \subseteq [n]^2$ if ξ correctly describes all connections between the outer columns: $(a,b) \in \xi$ iff there exists an m-long path from the a-th node of the 0-th column to the b-th node of the m-th column. We write $B_{n,\xi}$ for the set of all strings of connectivity ξ. The strings of $B_{n,\emptyset}$ are called *dead*; all other strings are called *live*. We define $B_n := \overline{B_{n,\emptyset}}$ as the collection of all live strings. So, B_n is the property of *liveness* —as defined in [2].

2.3 Outline

It is easy to see that B_n can be recognized by a SNFA (a 1NFA, actually) with only n states. Our goal is to prove that, in contrast, for the complementary language $\overline{B_n} = B_{n,\emptyset}$ a SNFA would need exponentially many states.

Theorem 1. *Every SNFA that recognizes $B_{n,\emptyset}$ has $2^{\Omega(n)}$ states.*

The rest of the article proves this fact. We fix n and a SNFA $M = (q_s, \delta, q_f)$ over a set Q of k states that recognizes $B_{n,\emptyset}$. We will prove that $k = 2^{\Omega(n)}$.

The proof is based on Lemma 1. We build two sequences $(X_\iota)_{\iota \in \mathcal{I}}$ and $(Y_\iota)_{\iota \in \mathcal{I}}$ that are related as in the lemma. The indices are all pairs of non-empty subsets of $[n]$, the universe is all sets of 1 or 2 steps of M:[1]

$$\mathcal{I} := \{(\alpha, \beta) \mid \emptyset \neq \alpha, \beta \subseteq [n]\} \qquad \mathcal{S} := \{\{s', s\} \mid s', s \in Q^2\},$$

and the total order $<$ is the restriction on $\mathcal{I}$ of some *nice* order on $\mathcal{P}([n])^2$. If we indeed construct these sequences, then the lemma says $|\mathcal{I}| \leq |\mathcal{S}|$, therefore

$$(2^n - 1)^2 \leq k^2 + \binom{k^2}{2},$$

hence $k = 2^{\Omega(n)}$. For the remainder, we fix $\mathcal{I}$ and $\mathcal{S}$ as here.

Note that from now on some subscripts in our notation are redundant. We thus drop them: e.g., $B_{n,\emptyset}$ and $\text{LVIEW}_M(z)(u)$ become $B_\emptyset$ and $\text{LVIEW}(z)(u)$.

Also, before moving on, let us prove a fact that will be useful later: In order to accept a dead string but reject a live one, M must produce on the dead string a single-state view that "escapes" the corresponding view on the live string.

Lemma 4. *Let z' be live and z dead. Then at least one of the following is true:*
- $\text{LVIEW}(z')(p) \not\supseteq \text{LVIEW}(z)(p)$ *for some $p \in Q$.*
- $\text{RVIEW}(z')(p) \not\supseteq \text{RVIEW}(z)(p)$ *for some $p \in Q$.*

Proof. Suppose $\text{LVIEW}(z')(p) \supseteq \text{LVIEW}(z)(p)$ and $\text{RVIEW}(z')(p) \supseteq \text{RVIEW}(z)(p)$, for all p. Pick any *accepting* computation c of M on z. Break c into its *traversals* $c_1, \ldots, c_m$, in the natural way: for $j < m$, each c_j starts at some state p_j next to a $\square$ and ends at some state q_j on the other $\square$; $p_1 = q_s$; $\delta(q_j, \square) \ni p_{j+1}$; and $c_m = (q_f)$. Then, for each odd (resp., even) $j < m$, we know q_j is in $\text{LVIEW}(z)(p_j)$ (resp., in $\text{RVIEW}(z)(p_j)$) and thus also in $\text{LVIEW}(z')(p_j)$ (resp., $\text{RVIEW}(z')(p_j)$); hence, some computation c_j' of M on z' starts and ends identically to c_j. If we also set $c_m' := (q_f)$ and concatenate $c_1', \ldots, c_m'$, we end up with a computation c' of M on z' which is also accepting. So, M accepts z', a contradiction. ∎

3 Hard Inputs and the Two Sequences

3.1 Generic Strings

Consider any $y \in \Sigma^*$ and the set of views produced via left computations on it:

$$\text{LVIEWS}(y) := \{\text{LVIEW}(y)(u) \mid u \subseteq Q\},$$

i.e., the range of $\text{LVIEW}(y)(\cdot)$. How does this set change if we extend y into yz?

Let $\text{LMAP}(y, z)$ be the function that for every left view produced on y returns its left view on z —i.e., $\text{LMAP}(y, z)$ simpy restricts $\text{LVIEW}(z)(\cdot)$ to $\text{LVIEWS}(y)$. It is easy to verify that $\text{LVIEWS}(yz)$ contains all values of this function, and is covered by them. In other words, $\text{LMAP}(y, z)$ is a *surjection* from $\text{LVIEWS}(y)$ to $\text{LVIEWS}(yz)$. This immediately implies that $|\text{LVIEWS}(y)| \geq |\text{LVIEWS}(yz)|$.

---

[1] A *step* of M is any $s \in Q^2$. Also, note that $\{s', s\}$ represents a singleton when $s' = s$.

The next fact encodes this conclusion, along with the obvious remark that $\mathrm{LMAP}(y, z)$ is monotone. It also shows the symmetric facts, for left extensions and right views. The set $\mathrm{RVIEWS}(y)$ consists of all views produced on y via right computations, and $\mathrm{RMAP}(z, y)$ is the restriction of $\mathrm{RVIEW}(z)(\cdot)$ on $\mathrm{RVIEWS}(y)$.

Fact 1. *For all y, z: $\mathrm{LMAP}(y, z)$ monotonically surjects $\mathrm{LVIEWS}(y)$ to $\mathrm{LVIEWS}(yz)$, so $|\mathrm{LVIEWS}(y)| \geq |\mathrm{LVIEWS}(yz)|$; symmetrically, in the other direction, $\mathrm{RMAP}(z, y)$ monotonically surjects $\mathrm{RVIEWS}(y)$ to $\mathrm{RVIEWS}(zy)$, so $|\mathrm{RVIEWS}(y)| \geq |\mathrm{RVIEWS}(zy)|$.*

Now suppose y belongs to an infinitely right-extensible property $P \subseteq \Sigma^*$. What happens to the size of $\mathrm{LVIEWS}(y)$ if we keep extending y into $yz, yzz', \ldots$ inside P? Although there are infinitely many extensions, the size of the set can decrease only finitely many times. So, at some point it must stop changing. When this happens, we have arrived at a very useful tool. We define it as follows.

Definition 1. *Let $P \subseteq \Sigma^*$. A string y is L-generic over P if $y \in P$ and*

$$(\forall yz \in P)\big[|\mathrm{LVIEWS}(y)| = |\mathrm{LVIEWS}(yz)|\big].$$

An R-generic string over P is defined symmetrically, with left-extensions and $\mathrm{RVIEWS}(\cdot)$. A string that is both L-generic and R-generic over P is called generic.

Lemma 5. *Let $P \subseteq \Sigma^*$. If P is non-empty and infinitely right-extensible (resp., left-extensible), then there exist L-generic (resp., R-generic) strings over P. If y_{L} is L-generic and y_{R} is R-generic, then every string $y_{\mathrm{L}} x y_{\mathrm{R}} \in P$ is generic.*

Proof. For the last claim, we just note that all right-extensions of an L-generic string inside P are also L-generic, and the same is true in the other direction.

Generic strings were introduced in [6] (for SDFAs and over B_n). Intuitively, they are among the *richest* strings with property P, in the sense that they exhibit a greatest subset of the "features" that M is "prepared to pay attention to". This makes them useful in building hard inputs, as described in the next lemma and in Sect. 3.2. For the lemma, we will also need the following simple fact.

Fact 2. *For all y, z: $\mathrm{LVIEWS}(yz) \subseteq \mathrm{LVIEWS}(z)$ and $\mathrm{RVIEWS}(zy) \subseteq \mathrm{RVIEWS}(z)$.*

Proof. By Fact 1, $\mathrm{LVIEWS}(yz)$ is the range of $\mathrm{LMAP}(y, z)$, which is a restriction of $\mathrm{LVIEW}(z)(\cdot)$; so, the first containment follows. Similarly in the other direction.

Lemma 6. *Suppose y is generic over $P \subseteq \Sigma^*$, and $x \in \Sigma^*$. If $yxy \in P$, then*
- *$\mathrm{LMAP}(y, xy)$ is an automorphism on $\mathrm{LVIEWS}(y)$, and*
- *$\mathrm{RMAP}(yx, y)$ is an automorphism on $\mathrm{RVIEWS}(y)$.*

Proof. Suppose $yxy \in P$. Then $|\mathrm{LVIEWS}(y)| = |\mathrm{LVIEWS}(yxy)|$ (since y is generic) and $\mathrm{LVIEWS}(yxy) \subseteq \mathrm{LVIEWS}(y)$ (by Fact 2). Hence, $\mathrm{LVIEWS}(y) = \mathrm{LVIEWS}(yxy)$. By this and Fact 1, we conclude $\mathrm{LMAP}(y, xy)$ surjects $\mathrm{LVIEWS}(y)$ onto itself, which is possible only if it is injective. Since $\text{LMAP}(y, xy)$ is also monotone, Lemma 2 implies it is an automorphism. The fact about $\text{RMAP}(yx, y)$ is proved similarly.

3.2 Constructing the Hard Inputs

Fix $\iota = (\alpha, \beta) \in \mathcal{I}$ and let $P_\iota := B_{\alpha \times \beta}$ be the property of connecting exactly every leftmost node in α to every rightmost node in β. Easily, P_ι is non-empty and infinitely extensible in both directions. So, an L-generic string y_L and an R-generic string y_R exist (Lemma 5). Then, for $\eta = [n]^2$ the complete symbol, we easily see that $y_L \eta y_R \in P_\iota$, too. Hence, this string is generic over P_ι (Lemma 5). We define $y_\iota := y_L \eta y_R$. We also define the symbol $x_\iota := \overline{\beta \times \alpha}$.

Lemma 7. *The two sequences $(y_\iota)_{\iota \in \mathcal{I}}$ and $(x_\iota)_{\iota \in \mathcal{I}}$ are such that, for all $\iota', \iota \in \mathcal{I}$:*

$$\iota' < \iota \implies y_\iota x_{\iota'} y_\iota \in P_\iota \qquad and \qquad \iota' = \iota \implies y_\iota x_{\iota'} y_\iota \in B_\emptyset.$$

Proof. Fix $\iota' = (\alpha', \beta')$ and $\iota = (\alpha, \beta)$ and let $z := y_\iota x_{\iota'} y_\iota$. Note that the connectivities of y_ι and $x_{\iota'}$ are respectively $\xi := \alpha \times \beta$ and $\xi' := \overline{\beta' \times \alpha'}$.

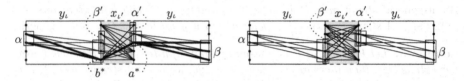

If $\iota' < \iota$ (on the left), then $\alpha' \not\supseteq \alpha$ or $\beta' \not\supseteq \beta$ (since $<$ is nice). Suppose $\beta' \not\supseteq \beta$ (if $\alpha' \not\supseteq \alpha$, use a similar argument) and fix any $b^* \in \beta \setminus \beta'$ and any $a^* \in \alpha$. For any $a, b \in [n]$, consider the a-th leftmost and b-th rightmost nodes of z. If $a \notin \alpha$ or $b \notin \beta$, then the two nodes do not connect in z, since neither can "see through" y_ι. If $a \in \alpha$ and $b \in \beta$, then $(a, b^*) \in \xi$ and $(b^*, a^*) \in \xi'$ and $(a^*, b) \in \xi$, so the two nodes connect via a path of the form $a \rightsquigarrow b^* \rightarrow a^* \rightsquigarrow b$. Overall, $z \in P_\iota$.

If $\iota' = \iota$ (on the right), then $\xi' = \overline{\beta \times \alpha}$. Suppose $z \notin B_\emptyset$. Then some path in z connects the leftmost to the rightmost column. Suppose it is of the form $a \rightsquigarrow b^* \rightarrow a^* \rightsquigarrow b$. Then $b^* \in \beta$ and $(b^*, a^*) \in \xi'$ and $a^* \in \alpha$, a contradiction.

3.3 Constructing the Two Sequences

Suppose $\iota' < \iota$. Since the extension $y_\iota x_{\iota'} y_\iota$ of y_ι preserves P_ι (Lemma 7), each of $\text{LMAP}(y_\iota, x_{\iota'} y_\iota)$ and $\text{RMAP}(y_\iota x_{\iota'}, y_\iota)$ is an automorphism (Lemma 6). Put another way, the interaction between the steps of M on $x_{\iota'}$ and its two behaviors on y_ι is such that these two mappings are automorphisms. Put formally, both

- the restriction of $\left(S_{\iota'} \circ \text{LVIEW}(y_\iota)\right)(\cdot)$ on $\text{LVIEWS}(y_\iota)$ and
- the restriction of $\left(S_{\iota'} \circ \text{RVIEW}(y_\iota)\right)(\cdot)$ on $\text{RVIEWS}(y_\iota)$

are automorphisms, for $S_{\iota'} := \{(p, q) \mid \delta(p, x_{\iota'}) \ni q\} = \text{LVIEW}(x_{\iota'}) = \text{RVIEW}(x_{\iota'})$.

What if $\iota' = \iota$? What is the status of $\text{LMAP}(y_\iota, x_\iota y_\iota)$ and $\text{RMAP}(y_\iota x_\iota, y_\iota)$? We can show that, since $y_\iota x_\iota y_\iota$ is dead (Lemma 7), we cannot have both functions be automorphisms[2]. However, something stronger is true: *we can even convince ourselves that one of the functions is not an automorphism by pointing at only 1 or 2 of the steps of M on x_ι.* The next figure shows three examples of this. In each, we sketch the left behavior of M on y_ι and all single-state views, and consider all heights to be with respect to $\text{LVIEWS}(y_\iota)$.

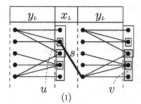 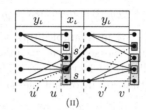 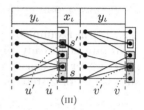

Example I shows only 1 of the steps of M on x_ι, say $s = (p, q)$ —many more may be included in S_ι. Is $\text{LMAP}(y_\iota, x_\iota y_\iota)$ an automorphism? Normally, we would need to know the entire S_ι to answer this question. Yet, in this case s is enough to answer no. To see why, note that the view v of q on y_ι has height 2, while one of the views that contain p is u, of height 1. Irrespective of the rest of S_ι, $\text{LMAP}(y_\iota, x_\iota y_\iota)$ will map u to a view that contains v and thus has height 2 or more. So, it does not respect heights, which implies it is not an automorphism.

Example II shows 2 of the steps in S_ι, say $s' = (p', q')$ and $s = (p, q)$. Is $\text{LMAP}(y_\iota, x_\iota y_\iota)$ an automorphism? Observe that neither step alone can force a negative answer: the view v' of q' on y_ι has height 1, as does the lowest view u' containing p'; similarly for s, u, v, and height 2. Hence, individually each of s' and s may very well participate in sets of steps that induce automorphisms. Yet, they cannot belong to the same such set. To see why, suppose they do. Since $u' \subseteq u$, the image of u would be $v' \cup v$ or a superset. Since $v' \not\subseteq v$, the height of that image would be greater than the height of v, and thus greater than the height of u, violating the respect to heights.

Example III also shows 2 of the steps in S_ι, say $s' = (p', q')$ and $s = (p, q)$, neither of which can disqualify $\text{LMAP}(y_\iota, x_\iota y_\iota)$ from being an automorphism. Yet, together they can. To see why, suppose both steps participate in the same automorphism. Then the image of u' must be exactly v': otherwise, it would be some strict superset of v', of height 2 or more, disrespecting the height of u'. On the other hand, u must map to a set that contains v, and thus also v'. Hence, v' must be the exact image of some $u^* \subseteq u$. But then both u^* and u' map to v', when $u^* \neq u'$ (since $u' \not\subseteq u$), a contradiction to the map being injective.

In short, each step in S_ι severely restricts the form of $\text{LMAP}(y_\iota, x_\iota y_\iota)$ and $\text{RMAP}(y_\iota x_\iota, y_\iota)$. And, either individually or in pairs, some steps can be so restrictive

---

[2] If they were, they would be bijections (because each of $\text{LVIEWS}(y_\iota)$ and $\text{RVIEWS}(y_\iota)$ has a maximum). Hence, M would not be able to distinguish between the live y_ι and the dead $y_\iota(x_\iota y_\iota)^t$, for t any exponent that turns both bijections into identities. (Note that this is true even for the n-state SNFA that solves liveness. Therefore, this observation alone can give rise to no interesting lower bound for k.)

that they cannot be part of any set of steps that induces an automorphism in both directions. To describe this formally, we introduce the next definition.

Definition 2. *A set of steps $S \subseteq Q^2$ is* compatible *with y_ι if there exists a set $\hat{S}$ such that $S \subseteq \hat{S} \subseteq Q^2$ and the following are both automorphisms:*
- *the restriction of $(\hat{S} \circ \text{LVIEW}(y_\iota))(\cdot)$ on $\text{LVIEWS}(y_\iota)$, and*
- *the restriction of $(\hat{S} \circ \text{RVIEW}(y_\iota))(\cdot)$ on $\text{RVIEWS}(y_\iota)$.*

E.g., $\{s\}$ in Example I and $\{s', s\}$ in Examples II,III are incompatible with y_ι.

We are now ready to define the sequences promised in Sect. 2.3. For each $\iota \in \mathcal{I}$, we let X_ι consist of all sets of 1 or 2 steps of M on x_ι, and Y_ι consist of all sets of 1 or 2 steps of M that are incompatible with y_ι:

$$X_\iota := \{S \in \mathcal{S} \mid S \subseteq S_\iota\}, \qquad Y_\iota := \{S \in \mathcal{S} \mid S \text{ is incompatible with } y_\iota\}.$$

We need, of course, to show that the sequences relate as in Lemma 1.

The case $\iota' < \iota$ is easy. Each $S \in X_{\iota'}$ can be extended to the set of all steps of M on $x_{\iota'}$ (i.e., $\hat{S} := S_{\iota'}$), which does induce automorphisms, so $X_{\iota'} \cap Y_\iota = \emptyset$.

The case $\iota' = \iota$ is harder. We analyze it in the next section.

4   The Main Argument

Suppose $\iota' = \iota$. Our goal is to exhibit a singleton or two-set $S \subseteq S_\iota$ that is incompatible with y_ι. First, some preparation.

The witness. Consider the strings $y_\iota(x_\iota y_\iota)^t = (y_\iota x_\iota)^t y_\iota$, for all $t \geq 1$. Since $y_\iota x_\iota y_\iota$ is dead, so are all of them. Since y_ι is live, Lemma 4 says for all $t \geq 1$:

- $\text{LVIEW}(y_\iota)(p) \not\supseteq \text{LVIEW}(y_\iota(x_\iota y_\iota)^t)(p)$ for some $p \in Q$, or
- $\text{RVIEW}(y_\iota)(p) \not\supseteq \text{RVIEW}((y_\iota x_\iota)^t y_\iota)(p)$ for some $p \in Q$.

Namely, in order to accept the extensions $y_\iota(x_\iota y_\iota)^t = (y_\iota x_\iota)^t y_\iota$ but reject the original y_ι, M must exhibit on each of them a single-state view that "escapes" its counterpart on the original. In a sense, among all $2k$ single-state views on each extension, the escaping one is a "witness" for the fact that the extension is accepted, and Lemma 4 says that *every extension has a witness*. Of course, this allows for the possibility that different extensions may have different witnesses. However, we can actually find the same witness for all extensions:

Fact 3. *At least one of the following is true:*
- $\text{LVIEW}(y_\iota)(p) \not\supseteq \text{LVIEW}(y_\iota(x_\iota y_\iota)^t)(p)$ *for some $p \in Q$ and all $t \geq 1$.*
- $\text{RVIEW}(y_\iota)(p) \not\supseteq \text{RVIEW}((y_\iota x_\iota)^t y_\iota)(p)$ *for some $p \in Q$ and all $t \geq 1$.*

Proof. Suppose neither is true. Then each of the $2k$ single-state views has an extension on which it fails to escape from its counterpart on y_ι. Namely, every p has some $t_{p,\text{L}} \geq 1$ such that $\text{LVIEW}(y_\iota)(p) \supseteq \text{LVIEW}(y_\iota(x_\iota y_\iota)^{t_{p,\text{L}}})(p)$ and some $t_{p,\text{R}} \geq 1$ such that $\text{RVIEW}(y_\iota)(p) \supseteq \text{RVIEW}((y_\iota x_\iota)^{t_{p,\text{R}}} y_\iota)(p)$. Consider the exponent

$$t^* := \left(\textstyle\prod_{p \in Q} t_{p,\text{L}}\right) \cdot \left(\textstyle\prod_{p \in Q} t_{p,\text{R}}\right)$$

and the extension $z := y_\iota(x_\iota y_\iota)^{t^*} = (y_\iota x_\iota)^{t^*} y_\iota$. Then each p has some $t \geq 1$ such that $z = y_\iota((x_\iota y_\iota)^{t_{p,L}})^t$, and thus Lemma 3 implies $\text{LVIEW}(y_\iota)(p) \supseteq \text{LVIEW}(z)(p)$; similarly, $\text{RVIEW}(y_\iota)(p) \supseteq \text{RVIEW}(z)(p)$. Overall, all single-state views on z fall within their counterparts on y_ι, contradicting Lemma 4.

We fix p to be a witness as in Fact 3. We assume p is of the first type, involving left views (otherwise, a symmetric argument applies). Moreover, among all witnesses of this type, we select p so as to minimize the height of $\text{LVIEW}(y_\iota)(p)$ in $\text{LVIEWS}(y_\iota)$. We let $V := \text{LVIEWS}(y_\iota)$, $h := h_V$, and $v_0 := \text{LVIEW}(y_\iota)(p)$.

By the selection of p, no $\tilde{p}$ with $\text{LVIEW}(y_\iota)(\tilde{p}) \subsetneq v_0$ can be a witness of the first type. Hence, for every such $\tilde{p}$ there is some $\tilde{t} \geq 1$ such that $\text{LVIEW}(y_\iota)(\tilde{p}) \supseteq \text{LVIEW}(y_\iota(x_\iota y_\iota)^{\tilde{t}})(\tilde{p})$. We fix t^* to be the product of all such $\tilde{t}$. Then:

Fact 4. *For all such $\tilde{p}$ and all $\lambda \geq 1$: $\text{LVIEW}(y_\iota)(\tilde{p}) \supseteq \text{LVIEW}(y_\iota(x_\iota y_\iota)^{\lambda t^*})(\tilde{p})$.*

Proof. Fix such a $\tilde{p}$ and the $\tilde{t}$ for which $\text{LVIEW}(y_\iota)(\tilde{p}) \supseteq \text{LVIEW}(y_\iota(x_\iota y_\iota)^{\tilde{t}})(\tilde{p})$. Fix any $\lambda \geq 1$. Then λt^* is a multiple of $\tilde{t}$ and Lemma 3 applies.

Escape computations. For all $t \geq 1$, collect into a set $\mathcal{C}_t$ all computations $c \in \text{LCOMP}_p(y_\iota(x_\iota y_\iota)^t)$ that exit into some $q \notin v_0$. These are the *escape computations* for p on the t-th extension. We also define $\mathcal{C} := \cup_{t \geq 1} \mathcal{C}_t$.

Let us see how an escape computation looks like. Pick any $c \in \mathcal{C}$ (Fig. 2a), say on the t-th extension, exiting into q. Let $s_1, \ldots, s_t$ be the steps of c on x_ι, where $s_j = (p_j, q_j) \in S_\iota$. These are the *critical steps* along c. Let $v_j := \text{LVIEW}(y_\iota)(q_j)$ be the view of the right end-point of s_j. Along with v_0, these views form the list $v_0, v_1, \ldots, v_t$ of the *major views* along c. Clearly, each of them contains the left end-point of the following critical step: $v_{j-1} \ni p_j$ (similarly, $v_t \ni q$). So, for each s_j there exist views $u \in V$ that contain its left end-point and are contained in the preceding major view: $v_{j-1} \supseteq u \ni p_j$ (similarly, $v_t \supseteq u \ni q$). Among them, let u_{j-1} be one of minimum height in V (select u_t similarly). Then the list $u_0, \ldots, u_{t-1}, u_t$ are the *minor views* along c.

We will find an incompatible S among the critical steps of such computations.

Case 1: Some $c \in \mathcal{C}$ contains some critical step s such that the singleton $\{s\}$ is incompatible with y_ι. Then we can select $S := \{s\}$, and we are done.

Case 2: For all $c \in \mathcal{C}$ and all critical steps s in c, the singleton $\{s\}$ is compatible with y_ι. In this case, we will find an incompatible two-set.

Steepness. First of all, every $c \in \mathcal{C}$ (say with t, s_j, v_j, u_j as above) has every major view at least as high as the next minor one ($h(v_j) \geq h(u_j)$, since $v_j \supseteq u_j$) and every minor view at least as high as the next major one ($h(u_j) \geq h(v_{j+1})$, otherwise $\{s_{j+1}\}$ would be incompatible, as in Example I). Hence, every $c \in \mathcal{C}$ has views of monotonically decreasing height ($h(v_0) \geq h(u_0) \geq h(v_1) \geq \cdots \geq h(u_t)$). To capture the "rate" of this decrease, we record the list of minor view heights $H_c := (h(u_j))_{0 \leq j \leq t}$, and order each $\mathcal{C}_t$ lexicographically: $c' \leq c$ iff $H_{c'} \leq_{\text{lex}} H_c$. With respect to this total order, "smaller" computation means "steeper".

Long and steepest computation. We fix t to be a multiple of t^* which is at least $|V|$, and select c to be steepest in $\mathcal{C}_t$. We let q, s_j, v_j, u_j be as usual.

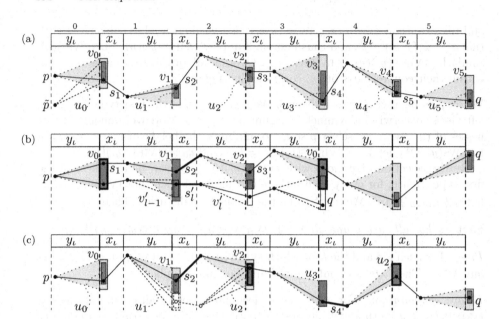

Fig. 2. (a) An escape computation $c \in \mathcal{C}_5$, exiting into q. (b) An example of Case 2A, for $j = 3$ and $l = 2$; in dashes, the new computation $c' \in \mathcal{C}_j$. (c) An example of Case 2B, for $j' = 2$ and $j = 4$; in dashes, the hypothetical case $u_{j'-1} \supseteq u_{j-1}$ and c'.

Since $t \geq |V|$, the list $u_0, \dots, u_t$ contains repetitions. Let $j' < j$ be the indices for the earliest one. Then $u_{j'} = u_j$, so $h(u_{j'}) = h(u_j)$, and thus all views in between have the same height: $h(u_{j'}) = h(v_{j'+1}) = \cdots = h(v_j) = h(u_j)$. As a result, each major view equals the next minor one: $v_{j'+1} = u_{j'+1}, \dots, v_j = u_j$.

Case 2A: $j' = 0$. Then $h(u_0) = h(v_1) = \cdots = h(v_j) = h(u_j)$, and therefore $v_1 = u_1, \dots, v_j = u_j$. In fact, we also have $h(v_0) = h(u_0)$, and therefore $v_0 = u_0$.

To see why, suppose $h(v_0) \neq h(u_0)$. Then $v_0 \supsetneq u_0$. Since $u_0 \in V$, some state $\tilde{p}$ has $\text{LVIEW}(y_\iota)(\tilde{p}) = u_0$ (Fig. 2a), and thus Fact 4 applies to it (since $u_0 \subsetneq v_0$). In particular, $\text{LVIEW}(y_\iota)(\tilde{p}) \supseteq \text{LVIEW}\big(y_\iota(x_\iota y_\iota)^t\big)(\tilde{p})$ (since t is a multiple of t^*). On the other hand, u_0 contains the left end-point of s_1, so the part of c after s_1 shows that $q \in \text{LVIEW}\big(y_\iota(x_\iota y_\iota)^t\big)(\tilde{p})$, and thus $q \in \text{LVIEW}(y_\iota)(\tilde{p}) = u_0$. Since $u_0 \subseteq v_0$, this means that c is not an escape computation, a contradiction.

So, $h(v_0) = h(u_0) = \cdots = h(v_j) = h(u_j)$ and $v_0 = u_0, \dots, v_j = u_j$ (Fig. 2b). By the selection of p, its view on the j-th extension escapes v_0. Pick any $c' \in \mathcal{C}_j$, with exit state $q' \notin v_0$, critical steps $s'_1, \dots, s'_j$, and major views $v'_0, \dots, v'_j$. Then $v'_0 = v_0$ (since both c' and c start at p) and $q' \in v'_j \setminus v_j$ (since $v_j = u_j = u_0 = v_0$ and $q' \notin v_0$). So, the respective major views start with inclusion $v'_0 \subseteq v_0$ but end with non-inclusion $v'_j \not\subseteq v_j$. So there is $1 \leq l \leq j$ so that $v'_{l-1} \subseteq v_{l-1}$ but $v'_l \not\subseteq v_l$.

We are now ready to prove that $\{s'_l, s_l\}$ is incompatible with y_ι. The argument is as in Example II. Suppose the two steps participate in a set inducing an automorphism f. Since $v'_{l-1} \subseteq v_{l-1}$, both s'_l and s_l have their left end-points in

v_{l-1}. Hence, $f(v_{l-1}) \supseteq v'_l \cup v_l$. Since $v'_l \not\subseteq v_l$, the height of $f(v_{l-1})$ is greater than that of v_l. But $h(v_{l-1}) = h(v_l)$. Therefore $h(f(v_{l-1})) > h(v_{l-1})$, a contradiction.

Case 2B: $j' \neq 0$. Then we can talk of the minor views $u_{j'-1}$ and u_{j-1} that precede the first repetition. Of course, $u_{j'-1} \neq u_{j-1}$. In fact, $u_{j'-1} \not\supseteq u_{j-1}$.

To see why, suppose $u_{j'-1} \supseteq u_{j-1}$ (Fig. 2c). Then $u_{j'-1} \supsetneq u_{j-1}$ (since $u_{j'-1} \neq u_{j-1}$) and thus $h(u_{j'-1}) > h(u_{j-1})$. Moreover, s_j has its left end-point in $v_{j'-1}$ (since $v_{j'-1} \supseteq u_{j'-1} \supseteq u_{j-1}$) while its right end-point has view $u_{j'}$ (since $v_j = u_j = u_{j'}$). Hence, by replacing $s_{j'}$ with s_j, we get a new computation c' that is also in $\mathcal{C}_t$. In addition, $H_{c'}$ differs from H_c only in that $h(u_{j'-1})$ is replaced by $h(u_{j-1})$. But then c' is strictly steeper than c, a contradiction.

We are now ready to prove that $\{s_{j'}, s_j\}$ is incompatible with y_ι. The argument is as in Example III. Suppose the two steps participate in a set inducing an automorphism f. Because of s_j, $f(u_{j-1}) \supseteq u_j$; but $h(u_{j-1}) = h(u_j)$ and f respects heights, so in fact $f(u_{j-1}) = u_j$. Because of $s_{j'}$, $f(u_{j'-1}) \supseteq u_{j'} = u_j$; so there exists $u^* \subseteq u_{j'-1}$ such that $f(u^*) = u_j$. Overall, $u^* \neq u_{j-1}$ (since exactly one is in $u_{j'-1}$) and $f(u^*) = f(u_{j-1})$. Hence f is not injective, a contradiction.

This concludes the analysis of the case $\iota' = \iota$ and thus the proof of Theorem 1.

5 Conclusion

We proved that small SNFAs are not closed under complement. In order to stay close to the combinatorial core of the problem, we used a non-standard transition function (implicit direction of motion; unusual reject and accept) and a large alphabet (exponential in n). It is not hard to show that *the lower bound remains exponential even under more standard definitions and over the binary alphabet.* In addition, by selecting the hard inputs more carefully in Sect. 3.2, we can ensure that a small 2DFA can correctly decide liveness on all of them. This way, we also have a proof that 2DFAs *can be exponentially more succinct than* SNFAs, which generalizes the analogous known relationship between 2DFAs and SDFAs [6,7,8]. More details about these claims will appear in the full version of this article.

An interesting next question concerns the exact value of our lower bound (for our definition and alphabet). The smallest known SNFA for $B_{n,\emptyset}$ is the obvious 2^n-state 1DFA. *Is this really the best* SNFA *algorithm?* If so, then nondeterminism and sweeping bidirectionality together are completely useless in this context.

Of course, the full 2D vs. 2N question remains as wide open and challenging as ever: *Is there a small* 2DFA *for liveness?*

References

1. Seiferas, J.I.: Manuscript communicated to Michael Sipser. (1973)
2. Sakoda, W.J., Sipser, M.: Nondeterminism and the size of two-way finite automata. In: Proceedings of the Symposium on the Theory of Computing. (1978) 275–286
3. Kapoutsis, C.: Deterministic moles cannot solve liveness. In: Proceedings of the Workshop on Descriptional Complexity of Formal Systems. (2005) 194–205
4. Sipser, M.: Halting space-bounded computations. Theoretical Computer Science **10** (1980) 335–338

5. Geffert, V., Mereghetti, C., Pighizzini, G.: Complementing two-way finite automata. In: Proceedings of the International Conference on Developments in Language Theory. (2005) 260–271
6. Sipser, M.: Lower bounds on the size of sweeping automata. Journal of Computer and System Sciences **21**(2) (1980) 195–202
7. Berman, P.: A note on sweeping automata. In: Proceedings of the International Colloquium on Automata, Languages, and Programming. (1980) 91–97
8. Micali, S.: Two-way deterministic finite automata are exponentially more succinct than sweeping automata. Information Processing Letters **12**(2) (1981) 103–105

Expressive Power of Pebble Automata[*]

Mikołaj Bojańczyk[1,2], Mathias Samuelides[2],
Thomas Schwentick[3], and Luc Segoufin[4]

[1] Warsaw University
[2] LIAFA, Paris 7
[3] Universität Dortmund
[4] INRIA, Paris 11

Abstract. Two variants of pebble tree-walking automata on binary trees are considered that were introduced in the literature. It is shown that for each number of pebbles, the two models have the same expressive power both in the deterministic case and in the nondeterministic case. Furthermore, nondeterministic (resp. deterministic) tree-walking automata with $n + 1$ pebbles can recognize more languages than those with n pebbles. Moreover, there is a regular tree language that is not recognized by any tree-walking automaton with pebbles. As a consequence, FO+posTC is strictly included in MSO over trees.

1 Introduction

A pebble automaton is a sort of sequential automaton which moves from node to node in a tree, along its edges. Besides a finite set of states it has a finite set $\{1, \ldots, n\}$ of pebbles which it can drop at and lift from nodes. There is a restriction though: if pebbles $i + 1, \ldots, n$ are on the tree, only pebble i can be dropped or pebble $i + 1$ can be lifted. Pebble automata were introduced in [4] as a model with intermediate expressive power between tree-walking automata [1,7] and parallel bottom-up or top-down automata. They are closely related to some aspects of XML languages. Furthermore, they are a building block of *pebble transducers* which were used to capture XML transformations (cf. [8,6]). Besides the number of pebbles, there are other parameters of pebble automata that can be varied. For example, they may be deterministic or nondeterministic, and they may have different policies of lifting a pebble: in the original model [4], a pebble can be lifted only if it is at the current node (*head position*), in the strong model, which was used to obtain a logical characterization in [5], it can be lifted everywhere. Not much is known about the relationships between the classes induced by the different models. Until recently it was even conceivable that deterministic tree-walking automata (sequential automata *without* pebbles) could recognize all regular languages. In [2,3] this has been refuted and it has been shown that nondeterministic tree-walking automata do not recognize all regular tree languages but are strictly more expressive than deterministic tree-walking automata.

---

[*] Work supported by the French-German cooperation programme PROCOPE, KBN Grant 4 T11C 042 25, and the EU-TMR network GAMES.

M. Bugliesi et al. (Eds.): ICALP 2006, Part I, LNCS 4051, pp. 157–168, 2006.

The current paper sheds some more light on the relationship between the pebble automata classes. In a nutshell, (a) whether pebbles are strong or not does not change the expressive power but (b) increasing the number of pebbles or moving from the deterministic to the nondeterministic model increases the expressive power. We next give an overview of the results of this paper. We write PA for the class of tree languages recognized by nondeterministic pebble automata. We add a subscript n for the restriction to n-pebble automata, 'D' to indicate deterministic automata and 's' for the strong model, e.g., $sDPA_n$ is the class of tree languages recognized by deterministic strong n-pebble automata. REG denotes the class of regular tree languages. The main result of this paper is that pebble automata do not recognize all regular tree languages.

Theorem 1.1. $PA \subsetneq REG$.

This result is refined by showing that the hierarchy for pebble automata based on the number of pebbles is strict for both nondeterministic and deterministic pebble automata, settling open questions raised in [4,5].

Theorem 1.2. *For each* $n \geq 0$, $PA_n \subsetneq PA_{n+1}$ *and* $DPA_n \subsetneq DPA_{n+1}$.

Furthermore, for each n, there is a language recognized by a nondeterministic tree-walking automaton but not by a deterministic n-pebble automaton. This improves the result in [2] that tree-walking automata (pebble automata with no pebbles) can not always be determinized.

Theorem 1.3. *For each* $n \geq 0$, $TWA \not\subseteq DPA_n$.

It remains open whether DPA is strictly included in PA. In [5], *strong* pebble automata were introduced as a model which corresponds to natural logics on trees. It was stated as an open question whether this model is stronger than the original one. We were surprised that this is actually not the case.

Theorem 1.4. *For each* $n \geq 0$, $sPA_n = PA_n$ *and* $sDPA_n = DPA_n$.

This proof is effective, but the state space increases n-fold exponentially. In a recent paper [9], it was shown that DPA_n is closed under complement but the closure under complement of $sDPA_n$ was left open. Nevertheless, it was shown that the complement of a language in $sDPA_n$ is in $sDPA_{3n}$. From Theorem 1.4 we get the following stronger result:

Corollary 1.5. *For each* $n \geq 0$, $sDPA_n$ *is closed under complement.*

In [5], the expressive power of strong pebble automata has been characterized in terms of logics. It was shown that FO+DTC=sDPA and FO+posTC=sPA. Here, FO+DTC is the extension of first-order logic with unary deterministic transitive closure operators and FO+posTC is the extension with positive unary transitive closure operators. By combining these results with ours and the fact that the regular tree languages are captured by monadic second-order logic (MSO), we immediately obtain the following result.

Corollary 1.6. *FO+posTC $\subsetneq$ MSO.*

Whether FO+TC $\subsetneq$ MSO and FO+DTC $\subsetneq$ FO+posTC remains open.

Section 2 gives precise definitions and develops some related terminology. In Section 3 we prove some basic facts about the behavior of pebble automata on trees, in particular we show a kind of universality of n-pebble automata: for each n-pebble automaton $\mathcal{A}$, there is an n-pebble automaton which on a tree t computes, in some sense, the complete behavior of $\mathcal{A}$ on t, for all possible contexts in which t may occur. In Section 4 we use these techniques to prove our separation results. Finally, in Section 5, we prove that strong pebbles give no additional power, thereby completing the proof of Corollary 1.6. Because of space limitation most proofs are missing and are available in the full version of this paper.

We are deeply indebted to Joost Engelfriet for carefully reading a previous draft of this paper and, in particular, pointing out a significant shortcoming in one of the proofs.

2   Definitions

We consider finite, binary trees labeled by a given finite alphabet Σ. We insist that each non-leaf node has exactly two children. A set of trees over a given alphabet is called a **tree language**. Given a tree t and a node v of t, we denote by $t|_v$ the Σ-tree corresponding to the subtree of t rooted at v. Let $*$ be a new symbol not in Σ. A **context** is a tree over $\Sigma \cup (\Sigma \times \{*\})$, where the label with $*$ occurs only once and at a leaf. This unique leaf whose label contains $*$ is called the **port** of the context. Given a context C and a tree t such that the label of the root of t is the same as the Σ-part of the label of the port of C, we denote by $C[t]$ the tree which is constructed from C and t by replacing the $*$-leaf with t. The context $C_{t,v}$ is the context resulting from t by removing all proper descendants of v and adding $*$ to the label of v.

Informally, a pebble automaton – just like a tree walking automaton – *walks* through its input tree from node to node along the edges. Additionally it has a fixed set of pebbles, numbered from 1 to n that it can place in the tree. At each time, pebbles $i, \ldots, n$ are placed on some nodes of the tree, for some i. In one step, the automaton can stay at the current node, move to its parent, to its left or to its right child, or it can lift pebble i or place pebble $i-1$ on the current node. Which of these transitions can be applied depends on the current state, the label and the type of the current node (root, left or right child — leaf or inner node), the set of pebbles at the current node and the number i.

We consider two kinds of pebble automata which differ in the way they can lift a pebble. In the standard model a pebble can be lifted only if it is on the current node. In the **strong** model this restriction does not apply.

We turn to the formal definition of pebble automata. The set types $= \{r, 0, 1\} \times \{l, i\}$ describes the possible types of a node. Here, r stands for the root, 0 for a left child, 1 for a right child, l for a leaf and i for an internal node (not a leaf). We indicate the possible kinds of moves of a pebble automaton by elements of the set $\{\epsilon, \uparrow, \diagup, \diagdown, \text{lift}, \text{drop}\}$, where informally $\uparrow$ stands for 'move to parent', ϵ stands for 'stay', $\diagup$ for 'move to left child' and $\diagdown$ for 'move to right child'. Clearly, drop refers to dropping a pebble and lift to lifting a pebble. Finally, $2^{[n]}$ denotes the powerset of $\{1, \ldots, n\}$.

Definition 2.1. An n-**pebble automaton** is a tuple $\mathcal{A} = (Q, \Sigma, I, F, \delta)$, where Q is a finite set of *states*, $I, F \subseteq Q$ are respectively the sets of *initial* and *accepting* states, and δ is the *transition relation* of the form

$$\delta \subseteq (Q \times \text{types} \times \{0, \ldots, n\} \times 2^{[n]} \times \Sigma) \times (Q \times \{\epsilon, \uparrow, \nearrow, \searrow, \text{lift}, \text{drop}\}).$$

A tuple $(q, \beta, i, S, \sigma, q', m) \in \delta$ intuitively means that if $\mathcal{A}$ is in state q with pebbles $i, \ldots, n$ on the tree, the current node has the pebbles from S, has type β and is labeled by σ then $\mathcal{A}$ can enter state q' and do a move according to m. A **pebble set** of $\mathcal{A}$ is a set $P \subseteq \{1, \ldots, n\}$. For a tree t, a P-**pebble assignment** is a function f which maps each $j \in P$ to a node in t. A P-**pebbled tree** is a tree t with an associated P-**pebble assignment**. A **pebbled tree** is a P-pebbled tree, for some P. We usually do not explicitly denote f. Analogous notions are defined for contexts. For $0 \leq i \leq n$, an i-**configuration** c is a tuple (v, q, f), where v is a node, q a state and f a $\{i+1, \ldots, n\}$-pebble assignment. We call v the **current node**, q the **current state** and f the **current pebble assignment**. We also write $(v, q, v_{i+1}, \ldots, v_n)$ if $f(j) = v_j$, for each $j \geq i+1$. We write $c \vdash_{\mathcal{A}, t} c'$ to denote that the automaton can make a (single step) transition from configuration c to c'. We denote the transitive closure of $\vdash_{\mathcal{A}, t}$ by $\vdash^+_{\mathcal{A}, t}$. The relation $\vdash_{\mathcal{A}, t}$ is basically defined in the obvious way following the intuition described above. However, there is a restriction of the lift-operation. A lift-transition can only be applied to an i-configuration (v, q, f) if $f(i+1) = v$, i.e., if pebble $i+1$ is at the current node. In Section 5 we also consider **strong** pebble automata for which this restriction does not hold. A **run** is a nonempty sequence $c_1, \ldots, c_l$ of configurations such that $c_j \vdash_{\mathcal{A}, t} c_{j+1}$ holds for each j. It is **accepting** if it starts and ends in the root of the tree with no pebble on the tree, the first state in I and the last state in F. The automaton $\mathcal{A}$ **accepts** a tree if it has an accepting run on it. A set of Σ-trees L is **recognized** by an automaton that accepts exactly the trees in L. Finally, we say that a pebble automaton is **deterministic** if δ is a function from $Q \times \text{types} \times \{1, \ldots, n\} \times 2^{[n]} \times \Sigma$ to $Q \times \{\epsilon, \uparrow, \nearrow, \searrow, \text{lift}, \text{drop}\}$.

We use PA_n (sPA_n) to denote the class of tree languages recognized by some (strong) pebble automaton using n pebbles and DPA_n (sDPA_n) for the corresponding deterministic classes. We write PA for $\bigcup_{n>0} \text{PA}_n$ and so forth.

Note that a (strong or standard) pebble automaton without pebbles is just a tree walking automaton. Thus, we also write TWA and DTWA for PA_0 and DPA_0, respectively.

An i-**run** is a run from an i-configuration to an i-configuration in which pebble $i+1$ is never lifted. An i-**loop** is an i-run from a configuration (v, p, f) to a configuration (v, q, f). Therefore, an i-loop is determined by the source i-configuration (v, p, f) and the target state q. An i-**move** is an i-run with only two i-configurations: the first and last one. It can be (a) a single transition, or (b) a *drop i* transition, followed by an $(i-1)$-loop followed by a *lift i* transition.

If the automaton is strong it can also be (c) *drop i*, followed by a (non-loop) $(i - 1)$-run, followed by *lift i*.

3   Behaviors and How to Compute Them

Let an n-pebble automaton $\mathcal{A}$ be fixed for the rest of the section. It is important in this section that we work with a *standard* pebble automaton and not with a *strong* one. Let v be a node in a tree t and $c = (v, p, v_{i+1}, \ldots, v_n)$ an i-configuration. Intuitively, whether or not there is an i-loop that starts in c clearly only depends on $t|_v$ and $C_{t,v}$ together with the pebble placement. Nevertheless, the exact relationship is not obvious: e.g., the automaton might enter $t|_v$, drop pebble i, move to $C_{t,v}$, drop pebble $i - 1$ and then enter $t|_v$ again. Thus, the behavior of $\mathcal{A}$ depends on $t|_v$ and $C_{t,v}$ in an interleaving manner.

In this section, we will formalize the intuitive notion of *behavior* of $\mathcal{A}$ through the notion of simulation. Intuitively, a tree s is simulated by a tree t if all loops in s also exist in t. The *behavior* of a tree is its simulation equivalence class (the set of trees that both simulate it, and are simulated by it). We show that, for each $\mathcal{A}$, (1) there are only finitely many different behaviors, (2) behaviors are compositional, and (3) the behavior of a tree can be computed by another pebble automaton with the same number of pebbles.

Two pebble assignments f and g are i-**compatible** if their domains partition $\{i+1, \ldots, n\}$. A pebbled tree t with assignment f is i-**compatible** with a pebbled context C with assignment g if f and g are i-compatible and the pebbles assigned to the root of t by f are exactly the pebbles assigned to the port of C by g.

Given a pebbled context C and an i-compatible pebbled tree t, let $\mathbf{loops}_i(C, t)$ denote the set of pairs (p, q) for which there is an i-loop ρ in $C[t]$ from $(v, p, f \cup g)$ to $(v, q, f \cup g)$, where v is the junction node between C and t. An i-loop is a **tree i-loop** if it involves no i-configurations outside t, it is a **context i-loop** if it involves no i-configurations outside C. By $\mathbf{tree\text{-}loops}_i(C, t)$ ($\mathbf{context\text{-}loops}_i$ (C, t)) we denote the corresponding set where ρ is a tree (context) i-loop. Clearly, $\mathbf{tree\text{-}loops}_i(C, t) \cup \mathbf{context\text{-}loops}_i(C, t) \subseteq \mathbf{loops}_i(C, t)$.

Definition 3.1. Let t, s be P-pebbled trees. We say s is i-**simulated** by t if, for every i-compatible pebbled context C, $\mathbf{tree\text{-}loops}_i(C, s) \subseteq \mathbf{tree\text{-}loops}_i(C, t)$.

We define i-simulation of pebbled contexts analogously. If s is j-simulated by t, for every $j \in \{0, \ldots, i\}$ we say that s is i^*-**simulated** by t. Two P-pebbled trees (resp. contexts) are said to be i-**equivalent** if they i-simulate each other; they are i^*-**equivalent** if they i^*-simulate each other. We will denote context equivalence classes by γ and tree equivalence classes by τ. We write $\tau_i(t)$ (resp. $\gamma_i(C)$) for the i^*-equivalence class of a pebbled tree t (resp. pebbled context C). We show next that there are only finitely many i-equivalence classes (and therefore finitely many i^*-equivalence classes.) The following technical lemma shows that the notion of i^*-simulation actually also covers context i-loops, not only tree i-loops.

Lemma 3.2. *Let $i \leq n$. Let s, t be pebbled trees and C a pebbled context, such that s and t are i-compatible with C.*

1. **context-loops$_0$**$(C, s) = $ **context-loops$_0$**(C, t)
2. *For $i > 0$, if s is $(i - 1)^*$-simulated by t, then* **context-loops$_i$**$(C, s) \subseteq$ **context-loops$_i$**(C, t).
3. *If s is i^*-simulated by t, then* **loops$_i$**$(C, s) \subseteq$ **loops$_i$**(C, t).

We associate with every tree i^*-equivalence class τ a (pebbled) tree t_τ of this class and likewise we choose a (pebbled) context C_γ, for each γ. If γ is a $(i - 1)^*$-equivalence class, then from the dual of Lemma 3.2(2) we can conclude that **tree-loops$_i$**$(C, t) = $ **tree-loops$_i$**(C_γ, t), for every context C of class γ.

Given a pebbled tree t, its **tree i-behavior** B_t^i, for $i > 0$, is a function that maps $(i - 1)^*$-equivalence class γ to the set of pairs **tree-loops$_i$**(C_γ, t). It is defined only for γ such that C_γ is i-compatible with t. For $i = 0$, B_t^i is simply the set of tree 0-loops of t. The context i-behavior B_C^i is defined analogously.

There is a natural order on i-behaviors: $B_s^i \leq B_t^i$ if $B_s^i(\gamma) \subseteq B_t^i(\gamma)$ holds for all γ. The following technical lemma shows that the i-behaviors completely determine the i-equivalence classes and their simulations:

Lemma 3.3. *Let s, t be P-pebbled trees. Then $B_s^i \leq B_t^i$ iff s is i-simulated by t.*

Thus, $B_s^i = B_t^i$ iff s and t are i-equivalent and from now on we also refer to the i-equivalence class of a tree as its tree i-behavior. A simple induction shows that, for each $i \leq n$, there are finitely many tree (resp. context) i-equivalence classes. The above construction is nonelementary, and this cannot be improved. One can easily show that the number of behaviors is at least as big as the smallest depth of an accepted tree. Using a standard construction for first-order logic, one can construct an n-pebble automaton with $O(n)$ states that only accepts trees whose depth is a tower of n exponentials.

We show next that i-behaviors behave composition- ally. For instance, the i-behavior of a tree depends only on the i-behaviors of its two subtrees and the label of the root. Let R, P_0, P_1 be a partition of $\{i + 1, \ldots, n\}$ and let a be a la- bel. For trees t_0, t_1 pebbled with P_0, P_1, respectively, we write **Compose**(a, R, t_0, t_1) for the

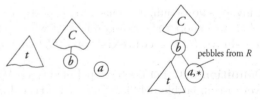

Fig. 1. The left-composed context **Compose** $(C, a, R, t, *)$

pebbled tree consisting of an a-labeled and R-pebbled root which has t_0 and t_1 as left and right subtrees, respectively. Similarly, for a P_0-pebbled tree t and a P_1-pebbled context C, **Compose**$(C, a, R, t, *)$ is the context composed from C and t as illustrated in Fig. 1. Likewise, **Compose**$(C, a, R, *, t)$ is the context where the port is the left sibling of t. Given ordered sets A, B, C, an operation $f : A \times B \to C$ is **monotone** if $a \leq a', b \leq b'$ implies $f(a, b) \leq f(a', b')$.

Lemma 3.4. *Once the label a and pebble set R are fixed, the composition operations are monotone with respect to i^*-simulation.*

In particular, i^*-equivalence is a congruence for the composition operations. Thus, it makes sense to write $\mathbf{Compose}(a, R, \tau_0, \tau_1)$ for the i^*-equivalence class of any tree with an a-labeled, R-pebbled root and subtrees of i^*-equivalence class τ_0 and τ_1. The proof of Lemma 3.4 is by induction on i and is straightforward by composing the subruns of the automaton in each of the subcomponents.

In the following lemma we assume that pebbles $i + 1, \ldots, n$ in a tree are suitably encoded by an (enlarged) alphabet. The proof is omitted in this abstract.

Lemma 3.5. *For every $i \leq n$ and tree i-behavior B^i, there is an i-pebble automaton $\mathcal{A}'$ that recognizes the pebbled trees t with $B^i_t \geq B^i$. Likewise for contexts. If $\mathcal{A}$ is deterministic, $\mathcal{A}'$ can be chosen deterministic, as well.*

Finally, we show one more closure property of pebble automata. For $i \geq 0$, the i^*-**behavior** of a tree t is defined as the sequence $B^0_t, \ldots, B^i_t$ (or, equivalently, the i^*-equivalence class of t; see the paragraph after Lemma 3.3). An i^*-behavior folding of a tree t is a tree that is obtained from t by replacing, for some nodes v of t, the subtree $t|_v$ with a single node labeled by the i^*-behavior of $t|_v$. The techniques from Lemmas 3.5 can be generalized to i^*-behavior foldings:

Lemma 3.6. *For every $i \leq n$ and tree i-behavior B^i, there is an i-pebble automaton $\mathcal{B}$ that recognizes the i^*-behavior foldings of pebbled trees t with $B^i_t \geq B^i$. Likewise for contexts. If $\mathcal{A}$ is deterministic, $\mathcal{B}$ can be chosen deterministic, too.*

4   The Pebble Automata Hierarchy

In this section we will prove Theorems 1.1, 1.2 and 1.3. In Subsection 4.1, we define the separating tree languages that we will use. In Subsection 4.2 we introduce *oracle automata*, a slight extension of tree-walking automata and show that the results (cf. Theorem 4.1 below) of [2] and [3] can be generalized to these models. Finally, in Subsection 4.3 we show the mentioned results.

4.1   The Separating Languages

In this section, we will mostly deal with trees over the alphabet $\{\mathbf{a}, \mathbf{b}\}$. Moreover we require that only leaves can be labeled by $\mathbf{a}$. We call these trees **quasi-blank trees**. An inner node of a quasi-blank tree is labeled by $\mathbf{b}$ and a leaf of a quasi-blank tree is labeled either by $\mathbf{a}$ or $\mathbf{b}$. For a quasi-blank tree t we define its **branching structure** $b(t)$. The branching structured results from t by first removing all nodes from t besides the $\mathbf{a}$-labelled leaves and their ancestors. Then, all inner nodes with only one child are removed. Thus, $b(t)$ consists only of the $\mathbf{a}$-leaves, and of deepest common ancestors of $\mathbf{a}$-leaves. Note that the descendant-relation of the nodes of $b(t)$ is inherited from t. By $\mathcal{L}_{\mathrm{branch}}$ we denote the set of quasi-blank trees t such that all the paths from root to leaf of $b(t)$ have even length. Let $\mathcal{L}_{3l}$ be the set of quasi-blank trees t such that $b(t)$ is . Thus a quasi-blank tree in $\mathcal{L}_{3l}$ has exactly three **a**-leaves whose branching structure corresponds to the tree depicted above. Likewise, $\mathcal{L}_{3r}$ is the language of trees with branching structure . Note that each quasi-blank tree with 3 **a**-leaves is either in $\mathcal{L}_{3l}$ or in $\mathcal{L}_{3r}$. We use the following result.

Theorem 4.1. $\mathcal{L}_{3l}$ *and* $\mathcal{L}_{3r}$ *are in* TWA *but not in* DTWA *[2].* $\mathcal{L}_{branch}$ *is in* REG *but not in* TWA *[3].*

Actually, in [3] a slightly stronger result was shown: for each TWA $\mathcal{A}$, there are trees $s' \in \mathcal{L}_{branch}$ and $t' \notin \mathcal{L}_{branch}$ such that each root-to-root loop of $\mathcal{A}$ in s' also exists in t'. For the construction in this section we would need yet a stronger statement, namely that s' and t' have *the same* root-to-root loops. To this end, we define another tree language $\mathcal{L}_{even}$ on top of $\mathcal{L}_{branch}$, as follows. We recall that in a finite binary tree each node can be naturally addressed by a $\{0, 1\}$-string describing the path from the root to the node where 0 corresponds to taking the left child of a node. In that spirit, a 0^*1-*node* is a right child of a node of the leftmost path. Let $\mathcal{L}_{even}$ be the set of trees t for which $b(t)$ has an even number of 0^*1-nodes v whose subtree has all branches of even length.

Proposition 4.2. *For every TWA $\mathcal{A}$, there are trees $s \in \mathcal{L}_{even}$ and $t \notin \mathcal{L}_{even}$ which have the same root-to-root loops of $\mathcal{A}$.*

Proof. Let $\mathcal{A}$ be given. Let s' and t' be as guaranteed by Theorem 4.1. We can assume that t' simulates s'. (That is, replacing t' by s' in any context gives at least as many root-to-root loops.) This can be enforced in a straightforward manner. Let m be $|Q \times Q|$, the number of pairs of states of $\mathcal{A}$, and thus the number of different tree-loops of $\mathcal{A}$. For $i \geq 0$, let t_i denote the tree which has a leftmost branch of length $m + 1$ which has s' and t' subtrees as right offspring. More precisely, a node of the form $0^j 1$ has s' as subtree if $j \leq i$ and otherwise t'. Clearly, t_i is in $\mathcal{L}_{even}$ iff i is even. Note that t_{i+1} is obtained from t_i by replacing one subtree s' with t'. It is easy to see that therefore t_{i+1} has all root-to-root loops of $\mathcal{A}$ that t_i has. Thus, the t_i, for $0 \leq i \leq m + 1$, induce a monotone sequence of $m + 2$ sets of root-to-root loops and, consequently, there must be an i such that the sets induced by t_i and t_{i+1} are identical. We can choose one of them as s and the other as t. □

We now define the languages that will be used in our separation proofs. They all consist of trees of a certain shape. A tree is n-**leveled**, for $n \geq 0$, if each of its paths from the root to a leaf is labeled by a sequence of the form $(\mathbf{cb}^*)^n(\mathbf{a} + \mathbf{b})$. Thus, in an n-leveled tree the root is labeled with c, there are n antichains labeled by **c**, some leaves have label **a** and all the other nodes are labeled by **b**. Note that

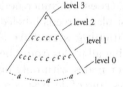

Fig. 2. A leveled tree

a 0-leveled tree consists of a single node labeled with **a** or **b**. A node is said to be **on level** i if its subtree is an i-leveled tree; it must therefore be labeled by **c**. A tree is **leveled** if it is n-leveled for some n. For a language $\mathcal{K}$ of $(n-1)$-leveled trees, the $\mathcal{K}$**-folding** of an n-leveled tree t is defined as follows. The label of the root is set to **b**. All nodes below level $n-1$ are removed. Each node v at level $n-1$ is labeled by **a** if $t|_v \in \mathcal{K}$ and by **b** otherwise. The folding of a 0-leveled tree is just the tree itself with the root label set to **b**. In the remainder of the section, we only consider leveled trees and their subtrees. Let languages $\mathcal{L}_0, \mathcal{L}_1, \ldots$ and $\mathcal{M}_0, \mathcal{M}_1, \ldots$ be defined as follows.

- $\mathcal{L}_0 = \mathcal{M}_0$ contains only the single node tree with label **a**.
- $\mathcal{L}_n$ is the set of all n-leveled trees whose $\mathcal{L}_{n-1}$-folding is in $\mathcal{L}_{\text{even}}$.
- $\mathcal{M}_n$ is the set of n-leveled trees whose $\mathcal{M}_{n-1}$-folding is in $\mathcal{L}_{3l}$.

Note that $\mathcal{L}_1 = \mathcal{L}_{\text{even}}$ and $\mathcal{M}_1 = \mathcal{L}_{3l}$.

Proposition 4.3. *For each $n \geq 1$, (a) $\mathcal{L}_n \in \text{DPA}_n - \text{PA}_{n-1}$, and (b) $\mathcal{M}_n \in \text{TWA} - \text{DPA}_{n-1}$.*

Proposition 4.3 (a) immediately implies Theorem 1.2. Likewise, Theorem 1.3 immediately follows from Proposition 4.3 (b). The lower bounds are shown in the following subsections. The upper bounds are shown by induction, the difficulty being the initial case which will be detailed in the full version.

4.2   Oracle Automata

The general idea of the lower bound proofs of Propositions 4.3 is that once an $(n-1)$-pebble automaton drops a pebble in the top level of an n-leveled tree t, with the remaining $n-2$ pebbles it cannot check whether the subtree of a node at level $n-1$ is in $\mathcal{L}_{n-1}$ (resp., $\mathcal{M}_{n-1}$). Thus, whenever the automaton uses a pebble at a node v in the top level it is *blind* with respect to the properties of the nodes at level $n-1$. But it still can check properties of v that depend on the position of v in the unlabeled version of t. In this subsection, we formalize this intuition by the notion of *oracle automata* which are an extension of tree-walking automata by *structure oracles*. Then we show that Theorem 4.1 also holds for oracle automata. A **structure oracle** $\mathcal{O}$ is a (parallel) deterministic bottom-up tree automaton [10] that is label invariant. That is, any two trees that have the same nodes get assigned the same state by $\mathcal{O}$. Therefore, a structure oracle is defined by its state space Q, an initial state $s_0 \in Q$ and a transition function $Q \times Q \to Q$. We write $t^{\mathcal{O}}$ for the state of $\mathcal{O}$ assigned to a tree t. This notation is extended to contexts: given a context C, $C^{\mathcal{O}} : Q \to Q$ is defined by $C^{\mathcal{O}}(q) = (C[t])^{\mathcal{O}}$, where t is some tree with $q = t^{\mathcal{O}}$. (All states are assumed reachable.) For a tree t, a node v of t, and a structure oracle $\mathcal{O}$, the **structural** $\mathcal{O}$-**information** about (t, v) is the pair $((C_{t,v})^{\mathcal{O}}, (t|_v)^{\mathcal{O}}) \in Q^Q \times Q$.

It should be noted that the result of any unary query expressible in monadic second-order logic which does not refer to the label predicates can be calculated based on the structural $\mathcal{O}$-information for some $\mathcal{O}$ (and vice-versa). Since the only type of oracles we use in this paper are structure oracles, we just write oracle from now on. An **oracle tree-walking automaton** is a tree-walking automaton $\mathcal{A}$ (with state set Q) extended by a structure oracle $\mathcal{O}$ (with state set P). The only difference to a usual tree-walking automaton is in the definition of the transition relation. It is of the form: $\delta \subseteq (Q \times (P^P \times P) \times \Sigma) \times (Q \times \{\epsilon, \uparrow, \swarrow, \searrow$, lift, drop$\})$. Whether a transition of $\mathcal{A}$ is allowed depends on the current state of $\mathcal{A}$, the label of the current node v and the structural $\mathcal{O}$-information about (t, v). Note that this generalizes tree-walking automata, since the structural information can include the type. The **size** of an oracle tree-walking automaton is defined as $|P| + |Q|$. The following proposition generalizes Theorem 4.1 and Proposition 4.2 to oracle automata:

Proposition 4.4. *(a) For each deterministic oracle automaton, there are trees $s \in \mathcal{L}_{3l}$, $t \notin \mathcal{L}_{3l}$ that have the same root-to-root loops.*
(b) For each oracle automaton, there are trees $s \in \mathcal{L}_{\text{even}}$, $t \notin \mathcal{L}_{\text{even}}$ that have the same root-to-root loops.

4.3   The Proof of the Lower Bounds

This subsection is devoted to the lower bound part of Proposition 4.3. To this end, let $n \geq 1$ and $\mathcal{A}$ be an $(n-1)$-pebble automaton with m states.

We will inductively construct trees s_i and t_i, $i = 1, \ldots, n$, such that, for each i, (1) s_i and t_i are i-leveled, (2) $s_i \in \mathcal{L}_i$, $t_i \notin \mathcal{L}_i$, and (3) s_i and t_i are $(i-1)^*$-equivalent. The base trees s_1 and t_1 are taken from the following lemma, which is an immediate consequence of Proposition 4.4.

Lemma 4.5. *For every k, there are 1-leveled trees $s_1 \in \mathcal{L}_1$, $t_1 \notin \mathcal{L}_1$ that have the same root-to-root loops for every nondeterministic oracle tree-walking automaton of size $\leq k$.*

Let s_1 and t_1 be the trees obtained by this lemma for k large enough, depending on $\mathcal{A}$ and n. (The exact constraints on k are stated in the proof of Lemma 4.6). For $i > 1$, s_i is obtained from s_1 by replacing every **a** leaf with s_{i-1} and every **b** leaf with t_{i-1}. The tree t_i is analogously obtained from t_1. It is immediate that s_i and t_i are i-leveled trees and that $s_i \in \mathcal{L}_i$ and $t_i \notin \mathcal{L}_i$.

The lower bound of Proposition 4.3 (a) follows directly from Lemma 3.2 and:

Lemma 4.6. *For each $i = 0, \ldots, n-1$, the trees s_{i+1} and t_{i+1} are i^*-equivalent.*

Proof. The proof is by induction on i. For the base case $i = 0$, we need to show that the trees s_1 and t_1 admit the same 0-loops, i.e. loops that do not use any pebbles. But this follows from Lemma 4.5, since it corresponds to loops of a tree-walking automaton without pebbles (we do not even need the oracle). Since Lemma 4.5 talks about root-to-root loops, and we want s_1 and t_1 to be equivalent in any context, we need k to be greater than the state space of any automaton recognizing a 0-behavior from Lemma 3.5.

Let thus $i \geq 1$. We assume that s_i and t_i are $(i-1)^*$-equivalent, we need to show that s_{i+1} and t_{i+1} are i^*-equivalent. An $(i+1)$-leveled tree where all i-leveled subtrees are either s_i or t_i is called **difficult**. Clearly both s_{i+1} and t_{i+1}

are difficult. Let τ_s and τ_t be the i^*-behaviors of s_i and t_i, respectively. Note that τ_s and τ_t may be different, our induction assumption only says that the $(i-1)^*$-behaviors of s_i, t_i are the same. The **behavior folding** $\bar{t}$ of a difficult tree t is the i^*-behavior folding of t where every occurrence of t_i is replaced by a single node labeled with τ_t, similarly for s_i. Note that the behavior foldings of s_{i+1}, t_{i+1} are essentially the trees s_1, t_1, except that **a** is replaced by τ_s and **b** is replaced by τ_t.

Let B be a j-behavior, with $j \leq i$. In order to complete the proof of the lemma, we need to show that B is the j-behavior of s_{i+1} if and only if it is the j-behavior of t_{i+1}. Let $\mathcal{C}$ be the automaton from Lemma 3.6 that accepts i^*-foldings of trees with j-behavior B. We only consider the most difficult case, when $j = i$ and $\mathcal{C}$ has i pebbles. We will show that

Claim. $\mathcal{C}$ accepts the behavior folding of s_{i+1} iff it accepts the behavior folding of t_{i+1}.

The general idea is that over behavior foldings of difficult trees, the i-pebble automaton $\mathcal{C}$ can be simulated by an oracle tree-walking automaton. That is, we will construct an oracle tree-walking automaton $\mathcal{D}$ that accepts exactly the same behavior foldings of difficult trees as $\mathcal{C}$. The size of $\mathcal{D}$ will depend only on the size of $\mathcal{C}$ (and hence in turn, on the size of $\mathcal{A}$). The result follows, as long as the k used in defining s_1 and t_1 was chosen large enough so that $\mathcal{D}$ cannot distinguish the behavior foldings of s_{i+1} and t_{i+1} (which are the same as s_1, t_1).

We now proceed to show how the simulating oracle tree-walking automaton $\mathcal{D}$ is defined. Recall that an i-run of the automaton $\mathcal{C}$ in the behavior folding of a difficult tree t (actually in any tree) can be decomposed into a sequence of i-moves each of one of the following types:

- a single transition in which pebble i is not dropped on the tree;
- a *drop pebble i* transition, followed by an $i-1$-loop, followed by *lift pebble i*.

Clearly, a single transition of the former type can be simulated by a tree-walking automaton (even without any oracle). It remains to show how to simulate an i-move of the latter type.

Claim. Let v be a node in the behavior folding $\bar{t}$ of a difficult pebbled tree. Whether or not there is an $(i-1)$-loop from a state p to a state q in v does not depend on the labels of $\bar{t}$.

The proof of this claim can be found in the full version of the paper. $\quad\square$

The proof of the lower bound of Proposition 4.3 (b) is completely analogous.

Proof (of Theorem 1.1). We will define a regular tree language $\mathcal{L}$ that is not recognized by any pebble automaton. Note that we can not use the union of all $\mathcal{L}_i$, since this language requires checking that all paths have the same number of **c** labels. The general idea though, is the same: the intersection of $\mathcal{L}$ with the set of i-leveled trees will be exactly $\mathcal{L}_i$. In particular, all the trees s_i from the previous lemma belong to $\mathcal{L}$, but none of the trees t_i does. Therefore, no pebble automaton can recognize $\mathcal{L}$. Now we define the language $\mathcal{L}$. Every path in every tree from $\mathcal{L}$ is of the form $(\mathbf{cb}^*)^*(\mathbf{a}+\mathbf{b})$. The tree with the single node $\mathbf{a}$ is in $\mathcal{L}$. Furthermore, a tree is in $\mathcal{L}$ if its $\mathcal{L}$-folding is in $\mathcal{L}_{even}$. Here, the $\mathcal{L}$-folding of a tree with paths of the form $(\mathbf{cb}^*)^*(\mathbf{a}+\mathbf{b})$ is obtained by replacing each node whose only $\mathbf{c}$ ancestor is the root by a leaf with $\mathbf{a}$ if its subtree is in $\mathcal{L}$, and by a leaf with $\mathbf{b}$ otherwise. This language clearly satisfies the desired properties. $\square$

We do not know if the language $\mathcal{M}$, analogously constructed from the $\mathcal{M}_i$, is in TWA. If it was we would get TWA $\nsubseteq$ DPA, and thus, by the result of [5], FO+DTC $\subsetneq$ FO+posTC.

5   Strong Pebbles Are Weak

The proof of Theorem 1.4 makes use of the techniques developed in Section 3. As an intermediate model it uses k-**weak** n-pebble automata in which pebbles $1, \ldots, k$ are weak (and can be lifted only when the head is on them) and pebbles $k+1, \ldots, n$ are strong (and can be lifted from anywhere). The theorem follows from the following two lemmas by induction.

Lemma 5.1. *For every* $0 \leq k < n$, *each* k-*weak* n-*pebble automaton* $\mathcal{A}$ *has an equivalent* $(k+1)$-*weak* n-*pebble automaton* $\mathcal{A}'$.

Lemma 5.2. *For every* $k < n$, *each* k-*weak pebble deterministic automaton* $\mathcal{A}$ *with* n *pebbles has an equivalent* $(k+1)$-*weak pebble deterministic automaton* $\mathcal{A}'$ *with* n *pebbles.*

References

1. A. V. Aho, J. D. Ullman Translations on a Context-Free Grammar. In *Information and Control*, 19(5): 439-475, 1971.
2. M. Bojańczyk and T. Colcombet. Tree-Walking Automata Cannot Be Determinized. TCS, to appear.
3. M. Bojańczyk and T. Colcombet. Tree-walking automata do not recognize all regular languages. In *STOC*, 2005.
4. J. Engelfriet and H.J. Hoogeboom. Tree-walking pebble automata. In *Jewels are forever*, (J. Karhumäki et al., eds.), Springer-Verlag, 72-83, 1999.
5. J. Engelfriet and H.J. Hoogeboom. Nested Pebbles and Transitive Closure. In *STACS*, 2006.
6. J. Engelfriet, S. Maneth. A comparison of pebble tree transducers with macro tree transducers. In *Acta Inf.* 39(9): 613-698, 2003.
7. J. Engelfriet, H.-J. Hoogeboom, J.-P.Van Best. Trips on Trees. In *Acta Cybern.* 14(1): 51-64, 1999.
8. T. Milo, D. Suciu and V. Vianu. Typechecking for XML transformers. In *J. Comput. Syst. Sci.*, 66(1): 66-97, 2003.
9. A. Muscholl, M. Samuelides and L. Segoufin. Complementing deterministic tree-walking automata. In *IPL*, to appear.
10. H. Comon et al. Tree Automata Techniques and Applications. Available at http://www.grappa.univ-lille3.fr/tata

Delegate and Conquer: An LP-Based Approximation Algorithm for Minimum Degree MSTs[*]

R. Ravi and Mohit Singh

Tepper School of Business, Carnegie Mellon University,
Pittsburgh PA 15213
{ravi, mohits}@andrew.cmu.edu

Abstract. In this paper, we study the minimum degree minimum spanning tree problem: Given a graph $G = (V, E)$ and a non-negative cost function c on the edges, the objective is to find a minimum cost spanning tree T under the cost function c such that the maximum degree of any node in T is minimized.

We obtain an algorithm which returns an MST of maximum degree at most $\Delta^* + k$ where Δ^* is the minimum maximum degree of any MST and k is the distinct number of costs in any MST of G. We use a lower bound given by a linear programming relaxation to the problem and strengthen known graph-theoretic results on minimum degree subgraphs [3,5] to prove our result. Previous results for the problem [1,4] used a combinatorial lower bound which is weaker than the LP bound we use.

1 Introduction

The minimum spanning tree problem is a fundamental problem in combinatorial optimization. It also has various applications, especially in network design. A favorable property of a connecting network is not only to have the lowest possible cost but also to have small load on all nodes. A natural way to formulate this problem is via the minimum degree minimum spanning tree (MDMST) problem. In an instance of the MDMST problem, we are given a graph $G = (V, E)$ and a non-negative cost function c on the edges, and the objective is to find a minimum cost spanning tree T under the cost function c such that the maximum degree of T is minimized. Here, the maximum degree of T is the maximum degree among all vertices in T.

The MDMST problem is closely related to the Hamiltonian path problem. If the maximum degree of an MST in an unweighted graph is at most 2, we get a Hamiltonian path. Since we do not assume that the costs are metric, no approximation is possible unless we relax the degree constraints [6]. Hence, for the MDMST problem, the natural criterion for approximation is the maximum degree of the minimum spanning tree.

---

[*] Tepper School of Business, Carnegie Mellon University. Supported by NSF ITR grant CCR-0122581 (The ALADDIN project) and NSF grant CCF-0430751.

M. Bugliesi et al. (Eds.): ICALP 2006, Part I, LNCS 4051, pp. 169–180, 2006.

1.1   Previous Work

For the MDMST problem, Fischer [4] gave a polynomial time algorithm which returns a minimum spanning tree with maximum degree $b\Delta^* + \log_b n$ for any $b > 1$ where Δ^* is the maximum degree of the optimal MST based on the techniques on Furer and Raghavachari [5]. A generalization of the MDMST problem is the bounded degree minimum spanning tree problem (BDMST) in which one is given degree bounds (B_v for vertex v) in an undirected graph with edge costs c and we demand a minimum cost tree satisfying the degree bounds. The BDMST problem is closely related to the well-studied Travelling Salesman Problem [8]. In particular, if we set $B_v = 2$ for each vertex v, the BDMST problem reduces to the Travelling Salesman Path Problem which has been studied by Lam and Newman [11].

For the BDMST problem, Konemann and Ravi [9,10] gave bi-criteria approximation algorithms which return a spanning tree with $O(B_v + \log n)$ bound on the degree of vertex v and cost $O(c_{opt})$. Here n is the number of vertices in the input graph and c_{opt} is the minimum cost of a spanning tree obeying the degree bounds. Chaudhuri et al [1,2] gave a quasi-polynomial time algorithm for the MDMST problem which returns a tree of maximum degree $O(\Delta^* + \frac{\log n}{\log \log n})$ and a polynomial time algorithm that returns a tree of maximum degree $O(\Delta^*)$. They also generalize both their algorithm for the BDMST problem giving algorithms with similar bounds on the degree as in the MDMST problem and cost $O(c_{OPT})$. All these results [9,1,2] for the BDMST problem are derived from results for the MDMST problem [4,1,2], thus motivating us to concentrate on the latter. Subsequent to our work, Goemans [7] has shown an algorithm for the BDMST problem which returns a tree of optimal cost and degree of vertex v at most $B_v + 2$ for each $v \in V$.

An interesting restriction of the MDMST problem arises when all costs are in $\{1, \infty\}$. Then, as every spanning tree of cost 1 edges is an MST, the MDMST problem reduces to finding a spanning tree in the undirected graph induced by the cost one edges with minimum maximum degree. Fürer and Raghavachari [5] gave an algorithm which returns a tree with maximum degree within $\Delta^* + 1$, where Δ^* is the degree of the optimal tree.

1.2   Our Work and Contributions

All previous algorithms for the MDMST problem worked with a combinatorial lower bound given by a *witness set*. The major contribution in this paper is working with a *stronger* lower bound given by a natural linear programming relaxation of the problem. Also, we strengthen the existing results of Fürer and Raghavachari [5] and Ellingham and Zha [3]. This helps us prove our main theorem below. Here, the maximum degree restriction can be generalized to specify separate bounds on individual nodes.

Theorem 1. *Given an instance of the minimum degree minimum spanning tree problem on a graph $G = (V, E)$ with a cost function c on the edges and a degree bound B_v on vertex v for each $v \in V$, there exists a polynomial time algorithm*

*which shows either that the degree upper bounds are infeasible for any minimum
spanning tree of G or returns an MST in which the degree of each vertex v is at
most $B_v + k$ where k is the number of distinct costs in any MST.*

Note that our Theorem 1 strictly generalizes the result of [5] since $k = 1$ in an
unweighted graph. We introduce the following new ideas to prove Theorem 1:

- We use linear programming relaxation as a check for infeasibility instead
  of the witness set that has been used previously [4,1]. If the degree bounds
  are feasible for a *fractional MST*, we use the optimal LP solution to divide
  the total degree bound of a vertex v into k parts, each assigned to a set of
  incident edges of a particular cost.

  Our strategy is to deal with edges of distinct costs separately. We use the
  known results for the unweighted case of the problem given by Fürer and
  Raghavachari [5] and its generalizations by Ellingham and Zha [3] to prove
  a weaker version of theorem 1 with degree guarantees of $B_v + 2k - 1$ instead
  of $B_v + k$ as claimed in the theorem. This we prove in Section 3.
- We strengthen the existing results of [5] in Theorem 4, by showing that when
  we do not find a witness for infeasibility we can obtain a solution where the
  degree bound is strictly satisfied for *one chosen vertex* while still ensuring
  that the violation of this bound is at most one for any other vertex. Similarly,
  we strengthen the results of Ellingham and Zha [3] in Theorem 5 (Section 4).
  We believe that these improvements are interesting in their own right.
- We use the strengthened guarantees in Theorem 4 and Theorem 5 to prove
  Theorem 1. We do this by applying the methods of Theorem 5 on different
  unweighted subgraphs, each naturally defined by edges of a particular cost
  that are used in an MST. This application proceeds in the top-down order
  by considering subgraph of progressively decreasing costs. At each step, we
  assign vertices to cost classes in which they can exceed their degree bound
  by at most one. We then inductively ensure that any such vertex does not
  exceed its bound in any other cost class. The resulting delegate-and-conquer
  algorithm is presented in Section 5, along with a proof of our main result.

2   Structure of MSTs

In this section, we prove some properties of MSTs. We then show the implication of these properties on the structure of the optimal solution to the linear
programming relaxation to the MDMST problem.

2.1   Forest over Forest Problem

We define a new problem which will be used later for the MDMST problem.

Given a forest F of a graph G, we call H a F-tree of G if H does not contain
any edge $e = \{u, v\}$ such that both u and v are in the same component of F and
$F \cup H$ is a spanning tree over each connected component of G. Note that for
any F-tree H of G, $|H| = $ (number of connected components in F) − (number
of connected components of G).

In an instance of the *forest over forest problem*, we are given an *unweighted* graph $G = (V, E)$ a forest F with connected components $\mathcal{C}(F) = \{C_1, \ldots, C_k\}$ and a degree bound B_v for each vertex $v \in V$. The problem is to find a F-tree H of G such that $deg_H(v) \le B_v$.

We also define a notion of *witness set* which forms the basis of the algorithms of Ellingham and Zha [3] and Fürer and Raghavachari [5]. Given a set $W \subset V$ and partition $\mathcal{P}$ of connected components of F, we say $(W, \mathcal{P})$ is a witness if each edge e with endpoints in different sets of $\mathcal{P}$ must have at least one endpoint in W. The following lemma is straightforward and proved in [3].

Lemma 1. *[3] If $(W, \mathcal{P})$ is a witness, then $\sum_{w \in W} deg_H(w) \ge |\mathcal{P}| - \kappa(G)$ for any F-tree H of G, where $\kappa(G)$ is the number of connected components of G.*

The following theorem was proved by Ellingham and Zha [3] for the forest over forest problem.

Theorem 2. *[3] There exists a polynomial time algorithm which given an instance of the forest over forest problem over a graph $G = (V, E)$ and a forest F with degree bound B_v for each vertex $v \in V$, returns a F-tree H and a witness $(W, \mathcal{P})$ such that:*

1. *If $W \ne \phi$, then the witness $(W, \mathcal{P})$ shows that $\sum_{w \in W} deg_{H'}(w) \ge (\sum_{w \in W} B_w) + 1$ for each F-tree H' of G, i.e., the degree bounds are infeasible for any F-tree of G.*
2. *If $W = \phi$, then $deg_H(v) \le B_v + 1$ for each $v \in V$.*

The MDST problem (unweighted MDMST problem) is a special case when $F = \phi$. Then the problem reduces to finding a spanning tree of G and the guarantees of the above theorem are exactly the same as those of Fürer and Raghavachari [5].

2.2   Laminar Structure of an MST

Given a graph $G = (V, E)$ with cost function c on the edges, let the cost function c take at most k different values on the edges of the MST. Without loss of generality, we can delete all edges of G of other costs since they do not occur in *any* MST. We also assume, without loss of generality, that the range of c is $\{1, \ldots, k\}$ as the particular values do not change the structure of any MST.

Let $G^{\le i}$ denote the graph over $V(G)$ with only those edges of $E(G)$ that cost at most i. We let $G^{\le 0}$ denote the graph over vertex set $V(G)$ with no edges. Let G^i denote the graph with vertex set $V(G)$ and edges in $E(G)$ which cost exactly i. The following lemma is a standard result about minimum spanning trees.

Lemma 2. *T is a minimum spanning tree of a graph G iff T^i is a $G^{\le i-1}$-tree of $G^{\le i}$ for each i.*

Hence, we also delete all edges e of cost i or higher which have both endpoint of $G^{\le i-1}$ without affecting any MST T of G. More importantly, Lemma 2 implies that we can independently select the edges of each cost class one at a time and solve the appropriate unweighted forest over forest problem to form an MST. The main issue is to manage the degree of any vertex across different cost classes.

2.3 LP Relaxation

We formulate the following integer program MST_{IP} for the problem considered in Theorem 4 which is a generalization of the MDMST problem.

$$opt_B = min \ \sum_{e \in E} c_e x_e \tag{1}$$

$$s.t. \ x(\delta(v)) \leq B_v \ \forall v \in V, \tag{2}$$

$$x \in SP_G, \tag{3}$$

$$x \ integer. \tag{4}$$

Here SP_G is the spanning tree polyhedron, i.e., a linear description of the convex hull of all spanning trees of G. It is well known that optimization over SP_G can be achieved in polynomial time [12]. We then relax the integrality conditions to obtain MST_{LP}. If the optimum value of MST_{LP} is more than the cost of an MST of G, then clearly the problem is infeasible.

Let x^* denote an optimal basic feasible solution to MST_{LP}. The following lemma follows directly from LP duality and is implicit in [9].

Lemma 3. *[9] The optimal basic feasible solution x^* to MST_{LP} can be written as a convex combination of spanning trees, i.e, there exists spanning trees $T_0, \ldots, T_n$ and constants $\lambda_0, \ldots, \lambda_n$ such that $x^* = \sum_{i=0}^{n} \lambda_i T_i$, $\sum_{i=0}^{n} \lambda_i = 1$ and $\lambda_i > 0$ for each $0 \leq i \leq n$. Here n is the number of vertices in the graph G.*

The following corollary to Lemma 3 is straightforward.

Corollary 1. *If $c(x^*) = c_{MST}$ then each of the spanning trees $T_0, \ldots, T_n$ obtained from Lemma 3 are minimum spanning trees.*

Proof. As $x^* = \sum_{i=0}^{n} \lambda_i T_i$, we have $c_{MST} = c(x^*) = \sum_{i=0}^{n} \lambda_i c(T_i) \leq \sum_{i=0}^{n} \lambda_i c_{MST} = c_{MST} \sum_{i=0}^{n} \lambda_i = c_{MST}$. Hence, each of the inequalities $c_{MST} \leq c(T_i)$ must hold at equality. □

2.4 LP Relaxation for Forest over Forest Problem

Given a forest over forest problem of constructing a F-forest of graph G with degree bound B_v for each vertex $v \in V$, we formulate the following natural IP formulation for the forest over problem which we call the $IP_{FOR}(F, G)$. Observe that this is a feasibility problem as a forest over forest problem is over an unweighted graph.

$$opt = min \ \ \ \ \ \ \ \ 0 \tag{5}$$

$$s.t. \ x(\delta(v)) \leq B_v \ \forall v \in V, \tag{6}$$

$$x \in SF(G/F), \tag{7}$$

$$x \in \{0, 1\}, \tag{8}$$

Here G/F denotes the graph formed when we shrink components of each component of F in to a single vertex in G and $SF(G)$ denote the natural linear formulation for the incidence vectors of all maximal spanning forests of G generalized from [12,9] (This is the same as the formulation for the incidence vectors of the bases of the graphic matroid of G). If we relax the integrality constraints, we get a LP relaxation which we denote by $LP_{FOR}(F,G)$. Later in Lemma 5, we show that if there exists a witness showing infeasibility of the degree bounds then the above LP relaxation is also infeasible.

2.5   Decomposing MSTs into Forests over Forests

To obtain any minimum spanning tree on an edge-weighted graph G, we need to solve the $LP_{FOR}(G^{\leq i}, G^{\leq i+1})$ for each $i = 0, \ldots, k-1$ where k is the number of distinct edge-costs in any minimum spanning tree of G. Now, we show that MST_{LP} actually solves each of these forest over forest problems with appropriate degree bounds.

Let $B_v^i = \sum_{e \in \delta(v), c(e)=i} x_e^*$. Observe that $\sum_{i=1}^{k} B_v^i \leq B_v$. Let $y_e^i = x_e^*$ if $c(e) = i$ else $y_e^i = 0$ for each $i = 1, \ldots, k$ and $e \in E$. Then we have the following lemma.

Lemma 4. *For each $1 \leq i \leq k$, y^i is a feasible solution to the linear programming relaxation of the forest over forest problem of finding a $G^{\leq i-1}$-tree of $G^{\leq i}$ with degree bound B_v^i for each vertex $v \in V$.*

Proof. Let $x^* = \sum_{j=0}^{n} \lambda_j T_j$ as in Lemma 3. Let H_j^i be the forest formed by cost-i edges in tree T_j. Clearly, each of the forests H_j^i is a valid $G^{\leq i-1}$-tree of $G^{\leq i}$ but may violate the degree bounds. By definition, $y^i = \sum_{j=0}^{n} \lambda_j H_j^i$ and hence is a valid fractional solution to $LP_{FOR}(G^{\leq i-1}, G^{\leq i})$. Also, it satisfies the degree constraints by definition as $\sum_{e \in \delta(v)} y^i(e) = \sum_{e \in \delta(v), c(e)=i} x_e = B_v^i$.   □

The following lemma shows that the LP gives a stronger notion of infeasibility than any witness.

Lemma 5. *If there exists a witness $(W, \mathcal{P})$ showing that the degree bounds are infeasible then the $LP_{FOR}(F, G)$ is infeasible.*

Proof. If $(W, \mathcal{P})$ is a witness showing that the degree bounds are infeasible then for any F-tree H, $\sum_{v \in W} deg_H(v) \geq \sum_{v \in W} B_v + 1$. Hence, the above holds for F-trees $H_0, \ldots, H_n$. For any convex combination, we get

$$\sum_{v \in W} \sum_{i=0}^{n} \alpha_i deg_{H^i}(v) \geq \sum_{i=0}^{n} \alpha_i \left(\sum_{w \in W} B_w + 1 \right) \geq \sum_{v \in W} \left(\sum_{i=0}^{n} \alpha_i B_v \right) + 1 = \sum_{v \in W} B_v + 1$$

since $\sum_{i=0}^{n} \alpha_i = 1$. Since any feasible solution to $LP_{FOR}(F, G)$ dominates a convex combination of F-trees (variant of Lemma 3), the degree constraint for at least one node in W must be violated.   □

3   Weaker Algorithm

As a warm-up, we describe an algorithm which uses the linear programming relaxation for the MDMST problem and the algorithm of Ellingham and Zha [3] as stated in Theorem 2 to obtain a weaker guarantee than claimed in Theorem 1.

Given an instance of MDMST problem over $G = (V, E)$, cost function c and degree bound B_v on vertex v, the algorithm *Alg-Weak* is as follows:

1. Find x^* the optimum solution to the linear programming relaxation to the problem. If $c(x^*) > c_{MST}$, declare the problem infeasible.
2. Define $B_v^i = \sum_{e \in \delta(v), c(e)=i} x_e^*$. For each i, we construct $G^{\leq i-1}$-tree H^i of $G^{\leq i}$ with degree bounds $\lceil B_v^i \rceil$ for each vertex $v \in V$ using the algorithm described in Theorem 2.
3. Return the MST $T = \cup_i H^i$.

Theorem 3. *Algorithm Alg-Weak for the MDMST problem on graph G with a degree bound B_v on vertex v for each $v \in V$ and a cost function c returns an MST such that the degree of any vertex v is at most $B_v + 2k - 1$ or shows that the degree bounds are infeasible for any MST of G.. Here, k is the number of distinct edge costs in any MST.*

Proof. The algorithm declares the degree bounds infeasible only if $c(x^*) > c_{MST}$. Clearly, then the degree bounds are infeasible for any MST. We only need to argue that if the $c(x^*) = c_{MST}$, then the tree returned satisfies the claimed degree bounds and is an MST.

First, observe the tree T returned is an MST as it is a union of $G^{\leq i-1}$-tree H^i of $G^{\leq i}$ for each i (see Lemma 2).

We only need to show that the algorithm in Theorem 2 returns a F-tree and not a witness set showing infeasibility of the degree bounds. However, this directly follows from Lemma 5.

Observe that the degree of any vertex v in tree T is exactly $deg_T(v) = \sum_{i=1}^{k} deg_{H^i}(v) \leq \sum_{i=1}^{k} (\lceil B_v^i \rceil + 1) < \sum_{i=1}^{k} (B_v^i + 2) \leq B_v + 2k$. Here the last inequality follows from the fact that $\sum_{i=1}^{k} B_v^i \leq B_v$. which proves the degree bound as claimed. This proves Theorem 3.   □

4   A Refined Characterization of Witnesses

In this section, we strengthen Theorem 2 which will help us obtain improved guarantees for the MDMST problem. However, to illustrate the strengthening without getting mired in notation, we first state and prove the strengthening of the result of Fürer and Raghavachari [5] for spanning trees rather than for forests over forests. Recall however that Theorem 2 is a generalization of the result of Fürer and Raghavachari [5], so the ideas in this strengthening generalize with some extra work allowing us to prove Theorem 5 in the the spirit of Theorem 4 (that is described in extended version of this paper [13]).

4.1   Improving Unweighted Minimum Degree Spanning Trees

Fürer and Raghavachari [5] present an algorithm which returns a tree of maximum degree $\Delta^* + 1$ where Δ^* is the minimum maximum degree of any tree.

We prove a stronger version of their theorem which is useful for the weighted version of the problem. The algorithm is similar to the algorithm of Fürer and Raghavachari [5] but our stopping criterion is more stringent.

Given a tree T and an edge $f \notin T$, let $Cycle(T, f)$ denote the set of vertices on the unique cycle in $T \cup f$.

Theorem 4. *Given a connected graph $G = (V, E)$, degree bound B_v for each vertex $v \in V$, there is a polynomial time algorithm which returns a spanning tree T and witness set $W \subset V$ (possibly empty) such that*

1. *Infeasibility: If $W \neq \phi$, then for any tree T', $\sum_{w \in W} deg_{T'}(w) \geq \sum_{w \in W} B_w + 1$, i.e., the degree bounds are infeasible for any spanning tree of G.*
2. *Solution: If $W = \phi$, then for each node $v \in V$, $deg_T(v) \leq B_v + 1$.*
3. *Strong Solution: If $W = \phi$, then for each node in $v \in V$, there exists a tree T_v such that $deg_{T_v}(v) \leq B_v$ and for each $u \in V \setminus \{v\}$, $deg_{T_v}(u) \leq B_u + 1$.*

While the algorithm of Fürer and Raghavachari [5] results in a tree satisfying conditions 1 and 2 only, we continue to improve the solution until we satisfy condition 3 or find a new witness for infeasibility.

Algorithm Alg-Unweighted

1. Find any spanning tree T.
2. Initialize $Ugly(T) = \{v | deg_T(v) \geq B_v + 2\}$, $Bad(T) = \{v | deg_T(v) = B_v + 1\}$, $Good(T) = \{v | deg_T(v) \leq B_v\}$, $MakeGood(u) = (u)$ for each $u \in Good(T)$. Return (T, ϕ) if $Bad(T) \cup Ugly(T) = \phi$.
3. If there exists edges $e = (u_1, u_2) \in T$ and $f = (v_1, v_2) \in E \setminus T$, such that e and f are swappable (i.e., e lies in the cycle closed by f in T), $v_1, v_2 \in Good(T)$ and either u_1 or $u_2 \notin Good(T)$, then do for each $w \in Cycle(T, f) \cap (Ugly(T) \cup Bad(T))$:
   (a) $Good(T) \leftarrow Good(T) \cup \{w\}$
   (b) $Ugly(T) \leftarrow Ugly(T) \setminus \{w\}$, $Bad(T) \leftarrow Bad(T) \setminus \{w\}$.
   (c) $Makegood(w) \leftarrow (v_1, v_2)$.
4. If any w is shifted from $Ugly(T)$ to $Good(T)$ in Step 3, then $T \leftarrow Improve(w, T)$ and Return to Step 2.
5. Return $(T, W = Ugly(T) \cup Bad(T))$.

The procedure $Improve(w, T)$ is implemented as follows:

1. If $MakeGood(w) = w$, then return T.
2. If $MakeGood(w) = (u, v)$, let T_u and T_v be the subtree containing u and v in $T \setminus W$ where $W = Bad(T) \cup Ugly(T)$. Here, $Bad(T)$ and $Ugly(T)$ are as defined by the algorithm before w is shifted to $Good(T)$. Let $T'_u = Improve(u, T_u)$ and $T'_v = Improve(v, T_v)$. Return $T' = T \cup T'_u \cup T'_v \cup \{u, v\} \setminus (T_u \cup T_v \cup e)$ where $e \in Cycle(\{u, v\}, T)$ and is incident at w.

The procedure $Improve(w)$ ensures that the degree of one ugly vertex reduces by at least 1 while no new ugly vertices are introduced in the resulting swaps. This is ensured by the following Lemma from [5]. A vertex v is called *non-blocking* in T if $deg_T(v) \leq B_v$.

Lemma 6. *[5] Suppose that $w \in Bad$ is marked Good in iteration i, when edge (u, v) is scanned in Step 3c of the algorithm. Then w can be made non-blocking by applying improvements to the components of F_i containing u and v where F_i is the subgraph of T generated by nodes marked good in iteration i.*

Now, we prove Theorem 4.

Proof. Suppose $W \neq \phi$. Let $C_1, \ldots, C_r$ be the components formed after removing W from T. Clearly, there does not exist any edge from C_i to C_j for any i, j else we would have found it in Step 3. Also number of components is at least $r \geq \sum_{w \in W} deg_T(w) - 2(|W| - 1) \geq \sum_{w \in W} (B_w + 1) - 2(|W| - 1)$, since $deg_T(w) \geq B_w + 1$ for each $w \in Bad(T)$ and $deg_T(w) > B_w + 1$ for each $w \in Ugly(T)$. Let $W = \{w_1, w_2, \ldots, w_p\}$. Then, $(W, \mathcal{P} = \{C_1, \ldots, C_r\{w_1\}, \ldots, \{w_p\}\})$ is a witness as there is no swap edge wrt to W. Hence by Lemma 1 since G is connected, there must be at least $r + |W| - 1$ edges incident at vertices in W in any tree T', i.e., $\sum_{w \in W} deg_{T'}(w) \geq \sum_{w \in W} B_w + |W| - 2(|W| - 1) + |W| - 1 = \sum_{w \in W} B_w + 1$. This proves (1) in the theorem.

Suppose now that $W = \phi$. Algorithm $Alg - Unweighted$ returns a spanning tree T and set $W = Ugly(T) \cup Bad(T) = \phi$. Hence every vertex has been marked *Good* implying that for any vertex $v \in V$, $deg_T(v) \leq B_v + 1$, proving (2).

Now, we prove (3). Assume that $W = \phi$. Take any $v \in V$. Hence, $v \in Good(T)$ where T is the final tree returned by the algorithm. Either $deg_T(v) \leq B_v$ in which case $T_v = T$ suffices. Else $deg_T(v) = B_v + 1$ and v was shifted from $Bad(T)$ to $Good(T)$ in step 3b. Then by Lemma 6, there exist a series of swaps which do not increase the degree of any vertex $u \in V$ above $B_u + 1$ and make v non-blocking. The tree T_v obtained after performing these swaps by invoking $Improve(v, T_v)$ suffices for proving (3). □

4.2   Forests over Forests Revisited

In this section, we obtain strengthening of the results of Ellingham and Zha [3] on the lines of the results in Section 4.1. We are given an instance of forest over forest problem to construct a F-tree of G satisfying the degree bounds B_v for each vertex $v \in V$. We first begin with a few definitions.

Let $\mathcal{C}(F)$ be set of connected components of F. We will refer these connected components as *supernodes*. For any vertex v, we will denote F_v to be the supernode containing v.

We present the following theorem in the spirit of Theorem 4.

Theorem 5. *Given a graph $G = (V, E)$, a forest F, a degree bound B_v for each vertex $v \in V$, there exists a polynomial time algorithm StrongForest which returns a F-tree H of G and witness set $(W, \mathcal{C}_W)$ where $W \subset V$ and $\mathcal{C}_W$ is a partition of $\mathcal{C}(F)$ such that*

1. *Infeasibility: If $W \neq \phi$, then $\sum_{w \in W} deg_{H'}(w) \geq \sum_{w \in W} B_w + 1 \; \forall \; F$-trees H', i.e, the degree bounds are infeasible for any F-tree of G.*
2. *Solution: If $W = \phi$, then for each $v \in V$, $deg_H(v) \leq B_v + 1$ and in each supernode $F_i \in \mathcal{C}(F)$ there is at most one vertex for which the above inequality is satisfied at equality.*
3. *Strong Solution: If $W = \phi$ then for each supernode $F_i \in \mathcal{C}(F)$ there exists a F-tree H_i which satisfies the condition (2) above. Moreover, for each vertex $u \in F_i$ we have $deg_{H_i}(u) \leq B_u$.*

Ellingham and Zha [3] prove the above theorem with conditions 1 and a weaker version of condition 2 and we strengthen it proving condition 3. Due to space considerations we omit the algorithm and the proof of the theorem since they are very similar to that for trees, only more notationally tedious. They are included in the technical report [13]. Note that the above theorem strictly generalizes Theorem 4, by setting $F = \emptyset$.

Another point to note here is that the strong solution guarantee can be applied to each connected component of G, i.e., we can choose one supernode from each connected component of G when obtaining the strong solution. This follows from the fact the F-tree problem over each connected component of G is independent and can be treated separately. We use this fact critically later in the algorithm for the MDMST problem.

5   Delegating Vertices Using Refined Witnesses

We now describe an algorithm, which given an MDMST problem on graph $G = (V, E)$ with cost function c and degree bounds B_v gives a better guarantee than one in Section 2. We use the algorithm *StrongForest* of Theorem 5 instead of the algorithm *forest over forest* to obtain a improved guarantee for the MDMST problem.

Algorithm Delegate-and-Conquer

1. **Step 1: Initialization**
   Solve the LP relaxation for the MDMST problem to obtain the optimal solution x^* and if $c(x^*) > c_{MST}$, we declare the instance infeasible. Else, let $B_v^i = \sum_{e \in \delta(v), c(e) = i} x_e^*$ for each $v \in V$ and for each $i = 1, \dots, k$. Observe that $\sum_{i=1}^k B_v^i \leq B_v$ for each $v \in V$. Observe that by Lemma 5, invoking the algorithm *StrongForest* of Theorem 5 to construct a $G^{\leq i-1}$-tree of $G^{\leq i}$ with degree bounds $\lceil B_v^i \rceil$ will always result in an empty witness set.
2. **Step 2: Using StrongForest**
   Find a $G^{\leq k-1}$-tree H^k of $G^{\leq k}$ with degree bounds $\lceil B_v^k \rceil$ for each vertex v using the algorithm described in Theorem 5. Let $S^k = \{v | deg_{H^k}(v) = \lceil B_v^k \rceil + 1\}$. Observe that at most one vertex of any connected component of $G^{\leq k-1}$ lies in S^k. This follows from condition (2) of Theorem 5 as each connected component of $G^{\leq k-1}$ is a supernode in the forest-over-forest problem solved. Also, let $M^k = H^k$.

3. **Step 3: Delegating the vertices to cost classes**
   For $i = k - 1$ down to 1, repeat

   (a) From each connected component of $G^{\leq i}$ there is at most one vertex in S^{i+1} (proved in Lemma 7). Apply algorithm *StrongForest* of Theorem 5 to each component G_j^i of graph $G^{\leq i}$. Apply condition (3) of Theorem 5 by selecting from each component G_j^i the supernode containing v where $v \in S^{i+1}$ to obtain $G^{\leq i-1}$-tree M^i of $G^{\leq i}$.

   (b) Define $S^i = \{v | deg_{\cup_{r=i}^k M^r}(v) = (\sum_{j=i}^k \lceil B_v^j \rceil) + 1\}$

   Return $T = \cup_{i=1}^k M^i$.

Theorem 6. *Given an instance of the MDMST problem over a graph $G = (V, E)$, cost function c and degree bound B_v for vertex $v \in V$, Algorithm Delegate-and-Conquer returns either an MST T such that $deg_T(v) \leq B_v + k$ or shows that the degree bounds are infeasible for any MST. Here k is the number of different costs in any MST.*

Proof. We declare the problem infeasible when $c(x^*) > c_{MST}$ in which case the problem is clearly infeasible. The Step 2 of the algorithm returns $G^{\leq k-1}$-tree H^k of graph $G^{\leq k}$ satisfying the conditions of Theorem 5. First we prove the following claim.

Lemma 7. *There is at most one vertex of S^i in each connected component of $G^{\leq i-1}$ for each $1 \leq i \leq k$.*

Proof. The proof of the claim is by induction for $i = k$ down to 1. Clearly, this is true for S^k from condition (2) of Theorem 5. Suppose it is true for S^{i+1} such that $2 \leq i + 1 \leq k$. We claim that it is true for S^i. Observe that the candidate vertices for S^i are vertices in S^{i+1} or vertices which exceed their corresponding degree bound $\lceil B_v^i \rceil$ in M^i.

Take any connected component G_j^{i-1} of $G^{\leq i-1}$. If there is some vertex v in G_j^{i-1} that is in S^{i+1}, then G_j^{i-1} is chosen in Step 3(a) of the algorithm as the selected supernode. Hence, no vertex in this connected component exceeds the degree bound $\lceil B_v^i \rceil$ in M^i by condition (3) of Theorem 5. Hence, v remains the only vertex that might exceed its total degree bound in $\cup_{r=i}^k M^i$. Else, if the connected component of $G^{\leq i-1}$ is such that there is no vertex of S^{i+1} in it, then we introduce at most one vertex which exceeds the degree bound in M^i by condition (2) of Theorem 5. In either case there is at most one vertex of S^i in each component of $G^{\leq i-1}$. Hence, the property holds for each $1 \leq i \leq k$. □

Hence, we obtain $deg_T(v) = \sum_{i=1}^k deg_{M^i}(v) \leq (\sum_{i=1}^k \lceil B_v^i \rceil) + 1 < \sum_{i=1}^k (B_v + 1) + 1 = B_v + k + 1$. This implies that $deg_T(v) \leq B_v + k$ for each $v \in V$. □

Observe that in the above proof, if each B_v^i were integral, then we would have obtained that $deg_T(v) \leq B_v + 1$ as we would "save" $k-1$ in rounding of fractional values.

References

1. Kamalika Chaudhuri, Satish Rao, Samantha Riesenfeld, and Kunal Talwar. What would Edmonds do? Augmenting Paths and Witnesses for degree-bounded MSTs. In *Proceedings of 8th. International Workshop on Approximation Algorithms for Combinatorial Optimization Problems*, 2005.

2. Kamalika Chaudhuri, Satish Rao, Samantha Riesenfeld, and Kunal Talwar. Push Relabel and an Improved Approximation Algorithm for the Bounded-degree MST Problem. In *To Appear in ICALP*, 2006.

3. Mark Ellingham and Xiaoya Zha. Toughness, trees and walks. *J. Graph Theory*, 33:125–137, 2000.

4. T. Fischer. Optimizing the degree of minimum weight spanning trees. *Technical report, Department of Computer Science, Cornell University*, 1993.

5. Martin Furer and Balaji Raghavachari. Approximating the minimum degree spanning tree to within one from the optimal degree. In *SODA '92: Proceedings of the third annual ACM-SIAM symposium on Discrete algorithms*, pages 317–324, Philadelphia, PA, USA, 1992. Society for Industrial and Applied Mathematics.

6. Michael R. Garey and David S. Johnson. *Computers and Intractability: A Guide to the Theory of NP-Completeness*. W. H. Freeman & Co., New York, NY, USA, 1979.

7. Michel Goemans. Personal Communication.

8. G. Gutin and A P Punnen eds. Traveling salesman problem and its variations. *Kluwer Publications*, 2002.

9. J. Könemann and R. Ravi. A matter of degree: Improved approximation algorithms for degree-bounded minimum spanning trees. In *STOC '00: Proceedings of the thirty-second annual ACM symposium on Theory of computing*, pages 537–546, New York, NY, USA, 2000. ACM Press.

10. Jochen Könemann and R. Ravi. Primal-dual meets local search: Approximating MST's with nonuniform degree bounds. In *STOC '03: Proceedings of the thirty-fifth annual ACM symposium on Theory of computing*, pages 389–395, New York, NY, USA, 2003. ACM Press.

11. Fumei Lam and Alantha Newman. Traveling salesman path perfect graphs. *Preprint*, 2006.

12. George L. Nemhauser and Laurence A. Wolsey. *Integer and combinatorial optimization*. Wiley-Interscience, New York, NY, USA, 1988.

13. R. Ravi and Mohit Singh. Delegate and Conquer: An LP-based approximation algorithm for Minimum Degree MSTs. *Technical report, Tepper School of Business, Carnegie Mellon University*, 2006.

Better Algorithms for Minimizing Average Flow-Time on Related Machines

Naveen Garg* and Amit Kumar**

Indian Institute of Technology Delhi, New Delhi, India

Abstract. We consider the problem of minimising flow time on related machines and give an $O(\log P)$-approximation algorithm for the offline case and an $O(\log^2 P)$-competitive algorithm for the online version. This improves upon the previous best bound of $O(\log^2 P \log S)$ on the competitive ratio. Here P is the ratio of the maximum to the minimum processing time of a job and S is the ratio of the maximum to the minimum speed of a machine.

1 Introduction

A well-studied setting in the scheduling literature is one where we have multiple machines of differing speeds. This is commonly referred to as the related machine scenario. All machines are equally capable so that a job can be scheduled on any machine. The only distinction is that a machine of speed twice that of another machine would take half the time to finish the same job.

The jobs arrive over time and have to be scheduled so that the total flow time is minimized. The flow time of a job is the difference between its completion and release times and is equal to the total time that the job is waiting or being processed. We will permit preemption, so that a job can be stopped even before it is completed and resumed later. However, we will not permit migration, *i.e.* the job cannot be resumed, after preemption, on another machine. Thus we are looking for a preemptive, non-migratory schedule on m related machines which minimizes the total flow time. In the three field notation of scheduling problems [7] this is denoted by $Q|r_j, pmtn| \sum_j (C_j - r_j)$.

When all machines have the same speed — the setting of parallel machines — the problem is well-studied. Leonardi and Raz [8] were the first to give an online algorithm for this problem with a bounded competitive ratio. They showed that the Shortest-Remaining-Processing-Time (SRPT) rule gives a schedule which is $O(\min(\log P, \log(n/m)))$-competitive, where P is the ratio of the maximum processing time to the minimum processing time and n is the number of jobs. They also established an $\Omega(\log P)$ lower bound on the competitiveness of any randomized online algorithm. The schedule obtained by Leonardi and Raz is migratory. Awerbuch et.al. [2] gave an online non-migratory algorithm with a

---

* Work done as part of the "Approximation Algorithms" partner group of MPI-Informatik, Germany.

** Supported by IBM Faculty Award and a Max-Planck-Society travel award.

M. Bugliesi et al. (Eds.): ICALP 2006, Part I, LNCS 4051, pp. 181–190, 2006.

competitive ratio of $O(\min(\log P, \log n)))$. This algorithm had the notion of a central queue where a job would wait till it was scheduled on one of the m machines. Chekuri et.al.[3] presented a simple $O(\min(\log P, \log n/m)))$-competitive non-migratory algorithm that scheduled jobs based on their class instead of the remaining processing time. Avrahami and Azar [1] obtained the same bounds with an algorithm without a central queue so that now a job is dispatched to the appropriate machine as soon as it is released.

For the case of related machines, the first algorithm with a bounded competitive ratio was obtained by the authors [4]; this algorithm had a $O(\log^2 P \log S)$ competitive ratio, where S is the ratio of the maximum machine-speed to the minimum speed. Prior to this Goel [5] obtained a 2-competitive algorithm for unit-sized jobs.

In this paper we develop a new linear programming approach to minimizing flow time on related machines. Our LP is a natural extension of the preemptive time-indexed formulation that Goemans [6] introduced in the context of single machine scheduling approximation algorithms. This LP is, seemingly, quite weak, since a fractional solution may schedule a job simultaneously on all machines. The flow time of a job is captured as the difference between its release time and the average time the job is scheduled in the LP solution and this quantity could be much smaller than the actual flow time of the job. In spite of all this we show a way of rounding this LP solution to obtain a non-migratory schedule whose flow time is at most $O(\log P)$ times the objective value of the LP solution. This gives an offline, $O(\log P)$-approximation algorithm for this problem.

We do not know how to solve this LP in an online manner. However, we can modify the linear program so that an optimum solution to the modified LP can be found in an online manner. Our procedure for rounding the LP solution into a non-migratory schedule continues to apply even in this online setting. Finally we show that the optimum solution to our modified LP is at most $O(\log P)$ times the flow time of the best schedule. This implies an $O(\log^2 P)$-competitive online algorithm for minimizing flow time on related machines. For the case of parallel machines, we do not need to modify the LP to solve it online and so we obtain an $O(\log P)$ competitive algorithm for parallel machines.

We remark that for both the offline and the online algorithm we obtain non-migratory schedules that are, respectively, within $O(\log P)$ and $O(\log^2 P)$ of the best offline migratory schedule. This also shows that permitting job-migration does not reduce flowtime by more than an $O(\log P)$ factor.

2   Preliminaries

We consider the scheduling problem where we are given machines with different speeds. For ease of notation, we shall work with *slowness* rather than speed. The slowness of a machine is defined as the reciprocal of its speed. In other words, a job j of size p_j will require $p_j \cdot s$ amount of processing time to get completely processed by a machine of slowness s.

There are m machines and let $s_i, s_i \geq 1$ be the slowness of machine i. The machines are ordered so that $s_1 \leq s_2 \leq \cdots \leq s_m$. There are n jobs. Job j has size p_j and, as mentioned above, takes $p_j \cdot s_i$ amount of time to finish on machine i. Let r_j be the release time of job j. We say that a job j is of *class* k if $2^{k-1} \leq p_j < 2^k$. Let the minimum size of a job be 1 and P be the maximum size of a job. We shall assume, without loss of generality, that all quantities are integers.

A schedule S specifies which job gets processed on each machine at each unit of time. Of course a job can start processing only after its release date. In this paper, all our schedules will be non-migratory and pre-emptive. However, our algorithms might construct migratory schedules on way to constructing non-migratory schedules. Also note that we can restrict ourselves to only those schedules which pre-empt jobs at integral time steps.

Let OPT denote the optimal schedule. We first consider the off-line problem where the sizes and the release dates of all the jobs are known at the beginning. Then we consider the on-line version of this problem where jobs arrive over time.

3 Off-Line Algorithm

In this section we give a polynomial time $O(\log P)$-approximation algorithm for this problem. We first give a linear programming formulation for this problem and then show how to convert an optimum LP solution to a non-migratory schedule.

3.1 Linear Programming Formulation

We formulate the problem as an integer program. In this formulation, we will even allow migratory schedules as feasible solutions. In fact we will even allow the schedule to process the same job simultaneously over multiple machines. The lower bound generated from such a formulation will be enough for our purposes. For each job j, machine i and time t, we have a variable $x_{i,j,t}$ which is 1 if machine i processes job j from time t to $t + 1$, 0 otherwise. The integer program is as follows. The variable j refers to jobs, i to machines and t to time.

$$\min \sum_j \sum_i \sum_t \quad x_{i,j,t} \cdot \left(\frac{t - r_j}{p_j \cdot s_i} + \frac{1}{2} \right) \tag{1}$$

$$\sum_j x_{i,j,t} \leq 1 \qquad \text{for all machines } i \text{ and time } t \tag{2}$$

$$\sum_i \sum_t \frac{x_{i,j,t}}{s_i} = p_j \qquad \text{for all jobs } j \tag{3}$$

$$x_{i,j,t} = 0 \qquad \text{if } t < r_j, \text{ for all jobs } j, \text{ machines } i, \text{ time } t \tag{4}$$

$$x_{i,j,t} \in \{0,1\} \qquad \text{for all jobs } j, \text{ machines } i, \text{ time } t \tag{5}$$

Constraint (2) refers to the fact that a machine can process at most one job at any point of time. Equation (3) says that job j gets completed in the schedule.

Equation (4) denotes the simple fact that we cannot process a job before its release date. It should be clear that any integral solution gives rise to a schedule where jobs can migrate across machines and may even get processed simultaneously on different machines. The only non-trivial equation in the integer program is the objective function. The following lemma shows that the optimal value of this integer program is a lower bound on the total flow-time of any non-migratory schedule. Let $\Delta_j(x)$ denote the term $\sum_i \sum_t x_{i,j,t} \cdot \left(\frac{t-r_j}{p_j \cdot s_i} + \frac{1}{2}\right)$. So the objective function is to minimize $\Delta(x) = \sum_j \Delta_j(x)$.

Let S be a non-migratory schedule. S also yields a solution to the integer program in a natural way – let x' denote this solution.

Lemma 1. *The total flow-time of S is at least $\sum_j \Delta_j(x')$.*

Proof. Fix a job j and suppose S schedules it on machine i. Let the completion time of j in S be t. So its flow-time is $t - r_j$. Notice that $\Delta_j(x')$ is maximized when j is scheduled from $t - p_j \cdot s_i$ to $t - 1$. So

$$\Delta_j(x') \le \sum_{t'=1}^{p_j \cdot s_i} \left(\frac{t - r_j - t'}{p_j \cdot s_i} + \frac{1}{2}\right) \le t - r_j.$$

Thus $\sum_j \Delta_j(x')$ is at most the total flow time of S.

Recall that OPT is the optimal non-migratory schedule. The lemma above says that the integer program yields a lower bound on the total flow-time of OPT. For the purpose of subsequent discussion, we modify the integer program slightly. We replace the objective function $\sum_j \Delta_j(x)$ by $\sum_j \Delta'_j(x)$, where

$$\Delta'_j(x) = \sum_i \sum_t x_{i,j,t} \left(\frac{t - r_j}{\lceil p_j \rceil \cdot s_i} + \frac{1}{2}\right).$$

Here $\lceil p_j \rceil$ equals p_j rounded up to the nearest power of 2. Note that $\lceil p_j \rceil$ is the same for all jobs j of the same class. Thus the terms in this new objective function can be grouped together in a convenient manner. Since $\lceil p_j \rceil \ge p_j$, the optimal value of the integer program cannot increase and so the statement of Lemma 1 still holds. Further note that $\lceil p_j \rceil \le 2 \cdot p_j$, and so the optimal value of the integer program does not change by more than a factor of 2. Let $\Delta'(x) = \sum_j \Delta'_j(x)$.

We now relax the integer program by replacing the constraints (5) by $0 \le x_{i,j,t} \le 1$. Let x^* be an optimal fractional solution this linear program.

3.2   Building the Non-migratory Schedule

We now show how to get a non-migratory schedule Q. Let y be a feasible solution to the LP corresponding to the schedule Q. The solution y is related to the optimum solution x^* in the following manner: If slot (i,t) is full in x^*, i.e. $\sum_j x^*_{i,j,t} = 1$, then we will have that $\sum_{j \in J_k} y_{i,j,t} \le \sum_{j \in J_k} x^*_{i,j,t}$, where J_k denotes jobs of class k. However, if some part of slot (i,t) is empty in x^*, i.e. $\sum_j x^*_{i,j,t} < 1$, then this empty part can be used in y to schedule jobs of any class.

For each slot (i,t) and class k we associate a variable $z_{i,k,t}$ which is initially equal to $\sum_{j \in J_k} x^*_{i,j,t}/s_i$ and denotes the total *volume* of class k jobs which have been processed in slot (i,t) in the solution x^*. In the course of our algorithm we reduce $z_{i,k,t}$ till it becomes zero.

We now describe our procedure for obtaining Q. We consider the slots in increasing order of time. At time t, consider machine i and class k. If there is a job, j, of class k which has been scheduled (in Q) on machine i and not completed yet, we schedule j in this slot to the *maximum possible extent*. This is the minimum of two quantities — the space available in this slot which is $\sum_{j \in J_k} x^*_{i,j,t}$ and the remaining processing time of this job.

If there is no job, j as above, or if we finish the job in this slot, we need to identify another job of class k to schedule in this slot. If there is a job, j, of class k which has been released but not scheduled on any machine and if $p_j \le \sum_{t' \le t} \sum_{i'=i}^m z_{i',k,t'}$ then we schedule j in this slot to the maximum possible extent. Further, we reduce the variables $z_{i',k,t'}, i \le i' \le m, t' \le t$ by a total amount equal to p_j. The variables are reduced in increasing order of (i',t').

For each slot (i,t) the above procedure is repeated for every class k. This ensures that for every class k, $\sum_{j \in J_k} y_{i,j,t} \le \sum_{j \in J_k} x^*_{i,j,t}$. If some part of this slot is empty in x^* we try to use it in solution y by using the above procedure to find a job (of any class) which can be scheduled in this slot. This job is then assigned to the maximum possible extent. The process may have to be repeated till this slot is fully occupied.

We must show that this procedure will eventually schedule all the jobs. Suppose for the sake of contradiction there is a job j which does not get assigned to a machine in our schedule. Let k denote its class. Let t be the time by which all jobs finish processing in schedule Q. Initially, the sum $\sum_i \sum_t z_{i,k,t}$ equals the total processing time of all jobs in class k and every time we schedule a job of class k we remove an amount equal to its processing time from this sum. This implies that eventually, this sum is exactly p_j and so j will get scheduled when we encounter an empty slot at time t. This yields a contradiction. Thus, our algorithm yields a feasible non-migratory schedule.

Let $P(x^*)$ denote the total time for which jobs are processed in solution x^*, i.e., $P(x^*) = \sum_i \sum_j \sum_t x^*_{i,j,t}$. We define $P(y)$ similarly.

Lemma 2. $P(y) \le P(x^*)$.

Proof. When we assign a job of class k to machine i in schedule Q, we reduce $\sum_{i'=i}^m \sum_{t' \le t} z_{i',k,t'}$ by a total amount p_j. Since $s_{i'} \ge s_i$, the reduction in $\sum_{i'=i}^m \sum_{t' \le t} s_{i'} \cdot z_{i',k,t'}$ is greater than $s_i * p_j$, the processing time of j in Q. Note that the sum $\sum_i \sum_t s_i \cdot z_{i,k,t}$ equals the total processing time of all class k jobs in the solution x^*. This implies that the total processing time of class k jobs is less in y than in x^*.

Now we try to bound the flow-time of our schedule.

Lemma 3.
$$\Delta'(y) \le \Delta'(x^*) + 2\log P \cdot P(x^*).$$

Proof. Note that

$$\Delta'(y) - \Delta'(x^*) = \frac{P(y) - P(x^*)}{2} + \sum_i \sum_j \sum_t \frac{(y_{i,j,t} - x^*_{i,j,t}) \cdot t}{s_i \cdot \lceil p_j \rceil}$$

$$\le \sum_k \sum_i \sum_{j \in J_k} \sum_t \frac{(y_{i,j,t} - x^*_{i,j,t}) \cdot t}{s_i \cdot 2^k}$$

where the last inequality follows from Lemma 2.

For a fixed class k and machine i, consider the quantity $\sum_t \sum_{j \in J_k} (y_{i,j,t} - x^*_{i,j,t}) \cdot t/s_i$. A little thought yields that this is same as the sum over all t of the difference in volume of jobs of class k which get processed on machine i after t. Summing it over all i, this is same as the sum over t of the difference in the volume of jobs of class k which get processed after t in the two schedules.

Define $V_{k,t}(x^*)$ as the volume of class k jobs which get processed by time t in x^*. Define $V_{k,t}(y)$ similarly. Since the total volume of jobs of a particular class that get processed in the two schedules is the same, we see that for any fixed k,

$$\sum_i \sum_t \sum_{j \in J_k} \frac{(y_{i,j,t} - x^*_{i,j,t}) \cdot t}{s_i \cdot 2^k} = \sum_t (V_{k,t}(x^*) - V_{k,t}(y))/2^k.$$

Claim. $V_{k,t}(x^*) - V_{k,t}(y)$ is at most $m_t \cdot 2^{k+1}$, where m_t is the number of busy machines in our schedule Q at time t.

Proof. Consider the quantity $\sum_i \sum_{t' \le t} z_{i,k,t'}$. Initially this is equal to $V_{k,t}(x^*)$. Every time we schedule a job of class k we reduce this quantity by the processing time of the job. Hence by time t, the reduction in this quantity equals the total volume of jobs that have been scheduled in Q. Let $z^t_{i,k,t'}$ denote the value of the variable $z_{i,k,t'}$ after our algorithm has processed slot (i, t). Thus $V_{k,t}(x^*) - \sum_i \sum_{t' \le t} z^t_{i,k,t'}$ equals $V_{k,t}(y)$ plus "the unprocessed volume of class k jobs that have been scheduled but have not finished by time t in Q".

To bound $\sum_i \sum_{t' \le t} z^t_{i,k,t'}$, we fix a machine i and consider the sum $\sum_{t' \le a} z^a_{i,k,t'} + p^a$, where p^a is the unprocessed volume of the class k job that is scheduled but has not finished at time a. We argue that for any a this sum can never exceed 2^k.

When machine i has an unfinished job of class k, the increase $\sum_{t' \le a+1} z^{a+1}_{i,k,t'} - \sum_{t' \le a} z^a_{i,k,t'}$ is offset by the decrease $p^a - p^{a+1}$. Thus, till i has an unfinished job, this sum cannot increase. If we schedule a new job from slot $(i, a+1)$, then p^{a+1} is balanced by the decrease $\sum_{t' \le a+1} z^{a+1}_{i,k,t'} - \sum_{t' \le a} z^a_{i,k,t'}$. This is because, when we scheduled j we reduced $z_{i,k,t'}, t' \le a+1$ by an extent of p_j.

Thus, this sum can increase beyond 2^k only if no job of class k is scheduled in slot $(i, a+1)$. If $\sum_{t' \le a+1} z^{a+1}_{i,k,t'} > 2^k$ then the total volume of class k jobs processed in x^* by time $a+1$ exceeds the total volume of class k jobs scheduled in y till time $a + 1$ by at least 2^k. This implies that there is a job, j, of class k which has been released but not scheduled in Q till time $a + 1$. Since $p_j \leq 2^k$, our procedure would have scheduled j at slot $(i, a + 1)$ and this leads to a contradiction.

Now suppose there are idle machines at time t in our schedule Q. Let i be such a machine of smallest index. Then we claim that $\sum_{i'=i}^{m} \sum_{t' \leq t} z_{i',k,t'}^{t}$ is at most 2^k. Indeed if it were greater than 2^k, then, as above, we can argue that there must be a job j of class k which has been released by time t but not scheduled in Q. Then machine i should not be idle at this time.

Combining these two arguments, we see that $\sum_i \sum_{t' \leq t} z_{i,k,t'}^{t}$ plus "the total unprocessed volume of class k jobs which have been scheduled but not finished by time t" is at most $m_t \cdot 2^k + 2^k$. Thus $V_{k,t}(x^*) - V_{k,t}(y)$ is at most $(m_t + 1)2^k$ which, since $m_t \geq 1$ is at most $m_t \cdot 2^{k+1}$.

Thus we get

$$\Delta'(y) - \Delta'(x^*) \leq \sum_k \sum_t m_t \cdot 2^{k+1}/2^k$$
$$= (2 \cdot \log P) \sum_t m_t$$
$$= 2 \log P \cdot P(y)$$
$$\leq 2 \log P \cdot P(x^*),$$

where the last inequality follows from Lemma 2.

The solution y can be converted into a schedule Q by splitting each slot (i, t) into sub-slots of length proportional to $y_{i,j,t}$ and scheduling job j in the sub-slot corresponding to it.

We are ready to relate the flow-time of the schedule Q to that of OPT.

Lemma 4. *The total flow-time of our schedule Q is at most $2\Delta'(y) + P(y) \cdot \log P$.*

Proof. Fix a job j. Let C_j be its completion time in schedule Q. It is easy to see that $2\Delta_j'(y)$ is at least the flow-time of j minus "the total time for which j is preempted in schedule Q". Since on any machine, there can be at most one job of class k which is preempted, we see that the sum $2 \sum_{j \in J_k} \Delta_j'(y)$ is at least the total flow time of all class k jobs minus "the total processing time of schedule Q". Thus we get that the flow-time of our schedule is at most $2\Delta'(y) + \log P \cdot P(y)$.

We are now ready to state the main theorem.

Theorem 1. *There is a polynomial time $O(\log P)$ approximation algorithm for minimizing flow time on related machines.*

Proof. Combining Lemmas 4, 2 and 3, we see that the flow-time of schedule Q is at most $5P(x^*) \log P + 2\Delta'(x^*)$. Now $P(x^*)$ is at most twice of $\Delta'(x^*)$ (because of the additive $1/2$ term in the objective function). Lemma 1 finally implies the theorem.

4   The On-Line Algorithm

We now give an $O(\log^2 P)$-competitive algorithm for this problem. It is easy to see that a solution to the linear program can be rounded into a non-migratory solution in an on-line manner. However, we do not know how to find an optimum solution to the linear program in an online manner. In this section we first present a modified linear program and argue that its optimum solution can be computed online. We then relate the optimum of this linear program to the flow-time of the optimal schedule.

Let x be a solution to the linear program. We now define the fractional flow time of job j, $\Delta_j''(x)$ as

$$\Delta_j''(x) = \sum_i \sum_t x_{i,j,t} \cdot \frac{t - r_j}{\lceil p_j \rceil \cdot s_i}.$$

Note that this is only less than the fractional flow-time, as defined earlier and so the sum of $\Delta_j''(x)$, which is the objective function, continues to be a lower bound on the optimum.

With this modification we cannot claim any more that $P(x) \le 2\sum_j \Delta_j''(x)$. Bounding the processing time of the fractional solution was crucial to the analysis of the off-line algorithm. Our second modification to the linear program lets us do this. We now require that a job j is scheduled on machine i at least $\lceil p_j \rceil \cdot s_i$ time units after its release. We encode this in the linear program by changing constraint (4) to $x_{i,j,t} = 0$ if $t \le r_j + \lceil p_j \rceil \cdot s_i$. However, now it is no more true that the objective function is a lower bound on the optimum flow time.

The objective function of this modified linear program is equivalent to minimizing $\sum_j \sum_i \sum_t \frac{x_{i,j,t} \cdot t}{\lceil p_j \rceil \cdot s_i}$. This is because of the constraint (3) in the linear program. We claim that the following greedy procedure yields an optimum solution to the linear program : at each time t consider the machines in increasing order of slowness. On machine i schedule the job, j, of smallest class which is available ($t \ge r_j + s_i \cdot \lceil p_j \rceil$) and not completed ($\sum_t \sum_i x_{i,j,t}/s_i < p_j$). Schedule this job to the maximum extent possible (i.e., until the slot is fully occupied or this job finishes) and continue. Note that $x_{i,j,t}$ denotes the fraction of time during $(t, t+1)$ for which j was scheduled on i.

We now argue that the solution obtained by this greedy procedure is optimal.

Lemma 5. *The solution x obtained by the greedy algorithm is an optimal solution to the linear program.*

Proof. Let O' be an optimal solution to the modified LP and let x^* denote the solution corresponding to O'. For a class k of jobs, machine i and time t, define $u_{i,k,t}^*$ as $\sum_{j \in J_k} x_{i,k,t}^*$, i.e., the total processing done on class k during $(t, t+1)$ on i. Define $u_{i,k,t}$ with respect to the greedy solution x similarly. Clearly it is enough to show that $u_{i,k,t} = u_{i,k,t}^*$ for all i, k, t.

So let us suppose for the sake of contradiction that greedy is not optimal. Suppose t is the first time at which the two solution differ in the class of job scheduled, i.e., there is a class k and machine i for which $u^*_{i,k,t} \neq u_{i,k,t}$ (if there are multiple machines i at time t for which this happens, then pick the machine with the smallest index). Then there are jobs j, j' of different class such that greedy processed more of j than j' during $(t, t+1)$ on i while the converse is true for O'. So O' must have processed j at some other time $t' \geq t$ on some other machine i' (the pair (i, t) and (i', t') are different). We reduce $x^*_{i,j',t}$ by an amount ϵ. Then ϵ/s_i amount of job j' has to be accommodated at the slot on machine i' at time t'. So we increase $x^*_{i',j',t'}$ by $\epsilon \cdot s_{i'}/s_i$. To avoid violating the constraint (2) at i', t', we reduce $x^*_{i',j,t'}$ by $\epsilon \cdot s_{i'}/s_i$ and increase $x^*_{i,j,t}$ by ϵ. Clearly we have a valid solution. Let us see the change in the objective function for x^*. The change is equal to

$$\frac{t}{s_i} \cdot \left(\frac{\epsilon}{\lceil p_j \rceil} - \frac{\epsilon}{\lceil p_{j'} \rceil} \right) + \frac{t'}{s_{i'}} \cdot \left(\frac{\epsilon \cdot s_{i'}}{s_i \cdot \lceil p_{j'} \rceil} - \frac{\epsilon \cdot s_{i'}}{s_i \cdot \lceil p_j \rceil} \right) = (t - t') \frac{\epsilon}{s_i} \cdot \left(\frac{1}{\lceil p_j \rceil} - \frac{1}{\lceil p_{j'} \rceil} \right)$$

Since $\lceil p_j \rceil < \lceil p_{j'} \rceil$ and $t \leq t'$ the above quantity is non-positive. Thus we have brought O' closer to our schedule without increasing the cost. By repeatedly doing such operations we can make O' identical to the greedy schedule.

As in the offline case, we convert the greedy solution, x, to obtain a solution y which corresponds to a non-migratory schedule, Q. Define $\Delta''(y) = \sum_j \Delta''_j(y)$ for any feasible solution y. The following claim bounds the processing time of the greedy solution.

Claim. For any feasible solution y, the total processing time, $P(y)$ is at most $\Delta''(y)$.

Proof. $\Delta''(y) = \sum_{j,i,t} \frac{y_{i,j,t}(t-r_j)}{\lceil p_j \rceil \cdot s_i} \geq \sum_{j,i,t} y_{i,j,t} = P(y)$. The last inequality follows from the fact that $y_{i,j,t} > 0$ only if $t \geq r_j + \lceil p_j \rceil \cdot s_i$.

Claim. The total flow time of schedule Q is at most $(4 + 9 \log P) \cdot \Delta''(x)$.

Proof. It is easy to show that if a job gets released at r_j and finishes at t on some machine, its flow-time is at most $4 \cdot \Delta''_j(y)$ plus "the total time for which j was interrupted". Since the interruption intervals for jobs of the same class on a particular machine are disjoint, we can bound the total interruption time for all the jobs by $P(y) \cdot \log P$. So we get that the total flow time is at most $4 \cdot \Delta''(y) + P(y) \log P$. From Lemma 3, we know that $\Delta''(y)$ is at most $\Delta''(x) + 2 \cdot P(y) \cdot \log P$. This implies that the flow time of Q is at most $4 \cdot \Delta''(x) + 9 \cdot P(y) \cdot \log P$.

As in the offline case, we can argue that $P(y) \leq P(x)$. Combining this with the claim above completes the proof.

Finally we bound $\Delta''(x)$ in terms of the flow-time of the optimum schedule OPT (recall that OPT can schedule a job anytime after its release).

Claim. $F(x)$ is at most $O(\log P)$ times the flow-time of OPT.

Proof. Note that $F(x)$ is less than the optimum flow time of the schedule with the restriction that a job, j is scheduled on machine i at least $\lceil p_j \rceil \cdot s_i$ units after its release.

Now consider OPT. Fix a class k and let us only consider the time slots in OPT where we schedule class k jobs. Now we shift each class k job on a machine i by $2^k \cdot s_i$ units (where the shifting is done with respect to these slots only). Note that this shifting does not put any of these jobs into a slot of another class. The shifting process may occupy empty slots at the end of the schedule. If we do this for each k (starting from $k = 1$), we get a schedule which obeys the property that a job j, of class k, is scheduled on a machine i only after $r_j + 2^k \cdot s_i$ time.

Now note that the total increase in the flow time of jobs of class k is at most twice the processing time of the optimum schedule. This implies that the new schedule obtained has flow time at most $(1 + 2 \log P)$ times the flow-time of OPT.

Combining the last two claims we obtain the following theorem.

Theorem 2. *There is an online algorithm for minimizing flow time on related machines which has a competitive ration of $O(\log^2 P)$ where P is the ratio of the maximum to the minimum processing time of a job.*

5   Open Problems

We believe that such a linear programming approach should lead to an offline approximation algorithm for minimizing flow time on unrelated machines. We leave this as an open problem. It would also be interesting to close the gap between the $O(\log^2 P)$ upper bound and the $\Omega(\log P)$ lower bound on the competitive ratio for flow time on related machines.

References

1. Nir Avrahami and Yossi Azar. Minimizing total flow time and total completion time with immediate dispatching. In *SPAA*, pages 11–18, 2003.
2. Awerbuch, Azar, Leonardi, and Regev. Minimizing the flow time without migration. In *STOC: ACM Symposium on Theory of Computing (STOC)*, 1999.
3. Chekuri, Khanna, and Zhu. Algorithms for minimizing weighted flow time. In *STOC: ACM Symposium on Theory of Computing (STOC)*, 2001.
4. Garg and Kumar. Minimizing average flow time on related machines. In *STOC: ACM Symposium on Theory of Computing (STOC)*, 2006.
5. Goel. B.tech thesis. In *Computer Science and Engineering, IIT Delhi*, 2004.
6. Michel X. Goemans. Improved approximation algorthims for scheduling with release dates. In *SODA '97: Proceedings of the eighth annual ACM-SIAM symposium on Discrete algorithms*, pages 591–598, Philadelphia, PA, USA, 1997. Society for Industrial and Applied Mathematics.
7. R. L. Graham, E. L. Lawler, J. K. Lenstra, and A. H. G. Rinnooy Kan. Optimization and approximation in deterministic sequencing and scheduling: A survey. *Ann. Discrete Mathematics*, 5:287–326, 1979.
8. Stefano Leonardi and Danny Raz. Approximating total flow time on parallel machines. In *STOC*, pages 110–119, 1997.

A Push-Relabel Algorithm for Approximating Degree Bounded MSTs

Kamalika Chaudhuri[1], Satish Rao[1], Samantha Riesenfeld[1], and Kunal Talwar[2]

[1] U.C. Berkeley
[2] Microsoft Research Mountain View, CA
{kamalika, satishr, samr}@cs.berkeley.edu,
kunal@microsoft.com

Abstract. Given a graph G and degree bound B on its nodes, the bounded-degree minimum spanning tree (BDMST) problem is to find a minimum cost spanning tree among the spanning trees with maximum degree B. This bi-criteria optimization problem generalizes several combinatorial problems, including the Traveling Salesman Path Problem (TSPP).

An $(\alpha, f(B))$-approximation algorithm for the BDMST problem produces a spanning tree that has maximum degree $f(B)$ and cost within a factor α of the optimal cost. Könemann and Ravi [13,14] give a polynomial-time $(1 + \frac{1}{\beta}, bB(1 + \beta) + \log_b n)$-approximation algorithm for any $b > 1$, $\beta > 0$. In a recent paper [2], Chaudhuri et al. improved these results with a $(1, bB + \sqrt{b}\log_b n)$-approximation for any $b > 1$. In this paper, we present a $(1 + \frac{1}{\beta}, 2B(1 + \beta) + o(B(1 + \beta)))$-approximation polynomial-time algorithm. That is, we give the first algorithm that approximates both degree and cost to within a constant factor of the optimal. These results generalize to the case of non-uniform degree bounds.

The crux of our solution is an approximation algorithm for the related problem of finding a minimum spanning tree (MST) in which the maximum degree of the nodes is minimized, a problem we call the minimum-degree MST (MDMST) problem. Given a graph G for which the degree of the MDMST solution is Δ_{OPT}, our algorithm obtains in polynomial time an MST of G of degree at most $2\Delta_{OPT} + o(\Delta_{OPT})$. This result improves on a previous result of Fischer [4] that finds an MST of G of degree at most $b\Delta_{OPT} + \log_b n$ for any $b > 1$, and on the improved quasipolynomial algorithm of [2].

Our algorithm uses the push-relabel framework developed by Goldberg [7] for the maximum flow problem. To our knowledge, this is the first instance of a push-relabel approximation algorithm for an NP-hard problem, and we believe these techniques may have larger impact. We note that for $B = 2$, our algorithm gives a tree of cost within a $(1 + \epsilon)$-factor of the optimal solution to TSPP and of maximum degree $O(\frac{1}{\epsilon})$ for any $\epsilon > 0$, even on graphs *not* satisfying the triangle inequality.

1 Introduction

Given a graph and upper bounds on the degrees of its nodes, the bounded-degree minimum spanning tree (BDMST) problem is to find a minimum cost spanning

M. Bugliesi et al. (Eds.): ICALP 2006, Part I, LNCS 4051, pp. 191–201, 2006.

tree among the spanning trees that obey the degree bounds. This bi-criteria optimization problem generalizes several combinatorial problems, including the Traveling Salesman Path Problem (TSPP), which corresponds to the case when degrees are restricted to 2 uniformly. Since we do not assume the triangle inquality, approximations for the BDMST problem must relax the degree constraint, unless P equals NP.

Let $c_{opt}(B)$ be the cost of an optimal solution to the BDMST problem, given input graph G and uniform degree bound B. We call a BDMST algorithm an $(\alpha, f(B))$-approximation algorithm if, given graph G and bound B, it produces a spanning tree that has cost at most $\alpha \cdot c_{opt}(B)$ and maximum degree $f(B)$. Könemann and Ravi give, to our knowledge, the first BDMST approximation scheme [13]: a polynomial-time $(1 + \frac{1}{\beta}, bB(1 + \beta) + \log_b n)$-approximation algorithm for any $b > 1$, $\beta > 0$. They illustrate the close relationship between the BDMST problem and the problem of finding an MST in which the maximum degree of the nodes is minimized, a problem we call the minimum-degree MST (MDMST) problem. Using a novel cost-bounding technique based on Lagrangean duality, Könemann and Ravi show that the MDMST problem can essentially be used as a black box in an algorithm for the BDMST problem. In a subsequent paper [14], they use primal dual techniques and give similar results for nonuniform degree bounds.

The BDMST and MDMST problem are different generalizations of the same unweighted problem: given an unweighted graph $G = (V, E)$, find a spanning tree of G of minimum maximum degree. Fürer and Raghavachari [5] give a lovely algorithm for this problem that outputs an MST with degree $\Delta_{OPT} + 1$. Their algorithm finds a sequence of swaps in a laminar family of subtrees of G such that the sequence results in an improvement to the degree of some high-degree node, without creating any new high-degree nodes. The laminar structure relies on the property that an edge $e \in E$ that is not in a spanning tree T can replace *any* tree edge on the induced cycle of $T \cup e$. This property is not maintained in weighted graphs because a non-tree edge can only replace other tree edges of equal cost. The structure of an improving sequence of swaps in a weighted graph can therefore be significantly more complicated.

Könemann and Ravi rely on an MDMST algorithm due to Fischer [4]. Given a graph G for which the MDMST solution is Δ_{OPT}, Fischer's algorithm finds an MST of G of degree at most $b\Delta_{OPT} + \log_b n$ for any $b > 1$. In a recent paper [2], Chaudhuri et al. give an improved MDMST algorithm based on finding augmenting paths of swaps. The algorithm in [2] simultaneously enforces upper *and lower* bounds on degrees, which, by using linear programming duality and techniques of [13,3], is shown to result in an optimal-cost $(1, bB(1 + \beta) + \log_b n)$-approximation BMDST algorithm for any $b > 1$. At the expense of quasipolynomial time, [2] also gives an algorithm that produces an MST with degree at most $\Delta_{OPT} + O(\frac{\log n}{\log \log n})$, leading to a $(1, B + O(\frac{\log n}{\log \log n}))$-approximation for the BDMST problem (in quasipolynomial time).

In this paper, we present a polynomial-time BDMST algorithm which we show to be a $(1 + \frac{1}{\beta}, 2B(1 + \beta) + o(B(1 + \beta)))$-approximation scheme for any $\beta > 0$. That is, we give the first algorithm that approximates both degree and cost to within a constant factor of the optimal.

For example, for $B = 2$, all previous algorithms would produce a tree with near-logarithmic degree and cost within a constant factor of the optimal; our algorithm, in contrast, approximates both the degree and the cost to within a constant factor.

For the sake of a simpler exposition, we describe our BDMST results in the setting of uniform degree bounds. Our techniques imply analogous results even in the case of more general non-uniform degree bounds. Though our BDMST algorithm does not simultaneously enforce upper and lower degree bounds, our techniques here do apply to a version of the BDMST problem in which *lower* bounds on node degrees must be respected, which may be of independent interest.

The crux of our solution is an improved approximation algorithm for the MDMST problem that uses the push-relabel framework invented by Goldberg [7] for the max flow problem (and fully developed by Goldberg and Tarjan [8]). Given a graph G for which the degree of the MDMST solution is Δ_{OPT}, our algorithm obtains in polynomial time an MST of G of degree at most $2\Delta_{\text{OPT}} + o(\Delta_{\text{OPT}})$.

While Fischer's MDMST solution is locally optimal with respect to single edge swaps in the current tree, our algorithm explores a more general set of moves that may consist of long sequences of branching, interdependent changes to the tree. Surprisingly, the push-relabel framework can be delicately adapted to explore these sequences. The basic idea that we borrow from Goldberg [7] is to give each node a label and permit "excess" to flow from a higher labeled node to lower labeled nodes. Nodes are allowed to increase their label when they are unable to get rid of their excess. For max-flow, the excess was a preflow, while in our case, the excess refers to excess degree. To our knowledge, this is the first instance of a push-relabel approximation algorithm for an NP-hard problem and we are intrigued by the possibility that this framework may be extended to search what may appear to be complicated neighborhood structures for other optimization problems.

We note that for $B = 2$, our BDMST algorithm gives a tree of cost within a $(1 + \epsilon)$-factor of the optimal solution to TSPP and of maximum degree $O(\frac{1}{\epsilon})$ for any $\epsilon > 0$. Our work does not assume the triangle inequality; when the triangle inequality holds, Hoogeveen [10] gives a $\frac{3}{2}$-approximation of TSPP based on Christofides' algorithm. The Euclidean version of the BDMST problem has also been widely studied. See, for example, [15,12,1,11].

Independent of our work, Ravi and Singh [16] give an algorithm for the MDMST problem with an additive error of k, where k is the number of distinct weight classes. We note that this bound is incomparable to the one presented here, and does not improve previous results for the BDMST problem. More recently, Goemans [6] has announced an algorithm for the BDMST problem with an additive error of 2.

1.1   Techniques

All known algorithms for the MDMST problem repeatedly swap a non-tree edge $e \in E$ with a tree edge $e' \in T$ of the same weight, where e' is on the induced cycle in $T \cup e$ in the current MST T. Fischer proceeds by executing any swap that improves a degree d node without introducing new degree d nodes, for selected high values of d. He shows that when the tree is locally optimal, the maximum degree of the tree is at most $b\Delta_{OPT} + \log_b n$, for any $b > 1$, where Δ_{OPT} is the degree of the optimal MDMST solution. Moreover, as shown in [2], this analysis is tight.

To illustrate the difficulty of the MDMST problem, we next describe a pathological MST T in a graph G (see Figure 1): the tree T has a long path consisting of $O(n)$ nodes ending in a node u of degree d. The children of u each have degree $(d-1)$; the children of the degree $(d-1)$ nodes have degree $(d-2)$, and so on until we get to the leaves. Each edge on the path has cost ϵ, and an edge from a degree $(d-i+1)$ node to its degree $(d-i)$ child has cost i. In addition, each of the degree $(d-i)$ nodes has a cost-i edge to one of the nodes on the path. For some d with $d = O(\log n / \log \log n)$, the number of nodes in the graph is $O(n)$.

Note that an MST of G with optimal degree consists of the path along with the non-tree edges and has maximum degree three. On the other hand, every cost-neutral swap that improves the degree of a degree-$(d-i)$ node in the current tree increases the degree of a degree-$(d-i-1)$ node. Hence the tree T is locally optimal for the algorithms of [4,13]. Moreover, all the improving edges are incident on a single component of low degree nodes; one can verify that the algorithm of [2] starting with this tree will not be able to improve the maximum degree. In

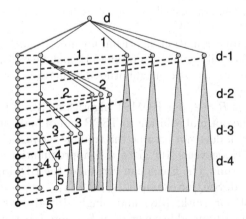

Fig. 1. Graph G and a locally optimal tree. The shaded triangles represent subtrees identical to the corresponding ones shown rooted at the same level. The bold nodes represent a path and the bold dotted edges correspond to a set of edges going to similar nodes in the subtrees denoted by shaded triangles.

fact, a slightly modified instance, G', where several of the non-tree edges are incident on the same node on the path, is not improvable beyond $O(d)$. Previous techniques do not discriminate between different nodes with degree less than $d - 1$ and hence cannot distinguish between G and G'.

On the other hand, our MDMST algorithm, described in Section 2, may perform a swap that improves the degree of a degree d node by creating one or more new degree d nodes. In turn, it attempts to improve the degree of these new degree d nodes. We note that in the process the algorithm may end up undoing the original move. However, the labels ensure that this process cannot continue indefinitely. These two new degree d nodes cannot necessarily be improved independently since they may rely on the same edge or use edges that are incident to the same node. Moreover, this effect snowballs as more and more degree d nodes are created.

As previously mentioned, Goldberg's push-relabel framework helps us tame this beast of a process. A high degree node may only relieve a unit of excess degree using a non-tree edge that is incident to nodes of lower labels. Thus, while two high degree nodes may be created by a swap, at least they are guaranteed to have lower labels than the label of the node initiating the swap.

We define a notion of a *feasible* labeling and prove that our MDMST algorithm maintains one. During the course of the algorithm, there is eventually a label L such that the number of nodes with that label is not much larger than the number of nodes with label $L + 1$. We use feasibility to show that all nodes with labels L and higher must have high average degree in *any* MST, thus obtaining a lower bound on Δ_{OPT}. This degree lower bound also holds for any fractional MST in the graph.

Combining our MDMST algorithm with the cost-bounding techniques of Könemann and Ravi [13] gives us our result for the BDMST problem.

2   Minimum-Degree MSTs

MDMST Problem: Given a weighted graph $G = (V, E, c)$, find an MST T of G such that $\max_{v \in V} \{\deg_T(v)\}$ is minimized.

2.1   The MDMST Algorithm

Our algorithm is based on the *push-relabel* scheme used in an efficient algorithm for the max flow problem [7,9]. Starting with an arbitrary MST of the graph, our algorithm runs in phases. The idea is to reduce the maximum degree in each phase using a push-relabel technique. If we fail to make an improvement at some phase, we get a set of nodes with high labels which serves as a certificate of near-optimality.

More formally, let Δ_i be the maximum degree of any node in the tree T_i at the beginning of phase i, also called the Δ_i-phase. During the Δ_i-phase, we either modify T_i to get T_{i+1} such that the maximum degree in T_{i+1} is less than Δ_i *or* we find a proof that $\Delta_i \leq 2\Delta_{\mathrm{OPT}} + O(\sqrt{\Delta_{\mathrm{OPT}}})$. (The constants hidden in the big-O notation are small—see Section 2.3.)

We now describe a general phase of the algorithm. Let T be the tree at the beginning of the current phase, and let Δ be the maximum degree over all nodes in T. Let $\mathbf{N}$ be the set of nonnegative integers. Given a *labeling* $l : V \to \mathbf{N}$, we extend the labeling to E by defining $l(e) = \max\{l(u), l(v)\}$ for $e = (u, v)$. The label of a node is a measure of its potential. In our algorithm, all nodes are initialized at the beginning of each phase to have label $l(v) = 0$. At any time, let *level i* be defined as the set of nodes that currently have label i.

In addition to being assigned a label, each node is given an initial *excess*. (Excess can also be formally defined as a function from V to $\mathbf{N}$.) At the beginning of the phase, each vertex with degree Δ is initialized to have an excess of 1; all other vertices initially have excess 0. We call a vertex that has positive excess *overloaded*.

We define a *swap* to be a pair of edges (e, e') such that $e \in T$, $e' \notin T$, $c(e) = c(e')$, and e lies on the unique cycle of $T \cup e'$. For a node u and a tree T, let S_u^T denote the set of swaps (e, e') such that e is incident on u and e' is not incident on u. We call a swap in S_u^T *useful* for u because it can be used to decrease the degree of u. We say that a labeling l is *feasible* for a tree T if for all nodes $u \in V$, for every swap $(e, e') \in S_u^T$, $l(e) \leq l(e') + 1$. A swap $(e, e') \in S_u^T$ is called *permissible* for u if $l(u) \geq l(e') + 1$. This notion of feasibility is crucial in establishing a lower bound on the optimal degree (and hence proving an approximation guarantee) when the algorithm terminates.

The current phase proceeds as follows: Let L be the label of the lowest level containing overloaded nodes. If there is an overloaded node u in level L that has a permissible, useful swap $(e, e') \in S_u^T$, modify T by deleting $e = (u, v)$ and adding $e' = (u', v')$. Then decrease the excess on u by one. If u' now has degree Δ or more, add one to its excess; if v' has degree Δ or more, add one to its excess. If no overloaded node in level L has a permissible, useful swap, then *relabel* to $L + 1$ all overloaded nodes in level L. Repeat this loop.

The phase ends either when no node is overloaded, or when there is an overloaded node with label $\log_2 n$. Note that if the phase ends for the former reason, then the tree at the end of the phase has maximum degree at most $\Delta - 1$. See Figure 2 for a formal algorithm.

Since each node is relabeled at most $\log_2 n$ times, the number of iterations in any phase of the algorithm is bounded by $n^2 \log_2 n$. Hence the algorithm runs in polynomial time.

In the next few sections, we build the tools used to argue that when a phase ends with some overloaded node in level $\log_2 n$, the algorithm produces a *witness* to the fact that $\Delta \leq 2\Delta_{\mathrm{OPT}} + O(\sqrt{\Delta_{\mathrm{OPT}}})$.

The proof of the following lemma is omitted from the extended abstract.

Lemma 1. *The algorithm always maintains a feasible labeling.*

Algorithm push_relabel_MDMST

$T \leftarrow$ arbitrary MST of G.
While witness not found **do**
 $\Delta \leftarrow$ maximum degree over nodes in T.
 Initialize labels to zero; put excess of 1 on nodes with degree Δ.
 Repeat
  $L \leftarrow$ lowest level that contains overloaded nodes.
  $U_L \leftarrow$ overloaded nodes with label L.
  If there is a node $u \in U_L$ that has a permissible, useful swap (e, e') where $e = (u, v)$
   $T \leftarrow T \setminus \{e\} \cup \{e'\}$;
   Set excess on u and v to 0;
   If an endpoint of e' has degree Δ or more
    set its excess to 1.
  else
   Relabel all nodes in U_L to $L + 1$.
 until there is an overloaded node with label at least $\log_2 n$
  or there are no more overloaded nodes.
 If some node has label $\log_2 n$
  Pick L such that level L is the highest sparse level.
  Let W be the set of nodes with label strictly higher than L.
  Let W' be the set of nodes with label higher than or equal to L.
  Output tree T and witness (W, W').
endwhile.

Fig. 2. An algorithm for the MDMST problem

2.2 Cascades and Involuntary Losses

For an integer L, let V_L be level L, i.e. the set of nodes with label L, and let U_L be the set of overloaded nodes in V_L. For convenience, we imagine placing flags on nodes when we relabel them. We start with all the pending flags cleared.

In each iteration of the algorithm, we find the lowest L such that U_L is non-empty, i.e. there are some overloaded nodes with label L. If we can find any swap (e, e') that is permissible and useful for a node in U_L, we execute the swap and clear pending flags (if set) on the endpoints of e. Let us call this a label-L swap. If no such swaps exist, we raise the label of all nodes in U_L by one and set their pending flags.

Lemma 2. *During the Δ-phase, no node ever has degree more than Δ.*

Proof. We use induction on the number of swaps. In the beginning of the phase, the maximum degree is Δ. Any swap (e, e') decreases the degree of a node in U_L and adds at most one to the degree of a node with strictly lower label. By choice of L, all nodes with lower labels had degree at most $\Delta - 1$ before the swap. Since a swap adds at most one to any vertex degree, the induction holds. The lemma follows.

Consider a swap (e, e'), where $e = (u, v)$ and $e' = (u', v')$, that is useful for an overloaded node u. If v is *not* overloaded, then we say that the swap (e, e') *causes* v an *involuntary loss* in degree.

We call a swap a *root swap* if it is a useful swap for a node with its pending flag set. Let (e, e') be a non-root swap that occurs in the sequence of swaps made by the algorithm. The swap (e, e') was performed in order to decrease the degree of node u (where $e = (u, v)$). There is a unique swap (f, f') in the sequence that most recently (before the (e, e') swap was done) increased the degree of u (so u is an endpoint of edge f'). We say that swap (e, e') can *blame* swap (f, f'), and we call (f, f') the *parent swap* of (e, e'). Recall that a label-L swap reduces the degree of a node with label L, and note that every non-root swap has a label strictly smaller than its parent. Moreover each swap is the parent of at most two other swaps. This parent relation naturally defines a directed graph on the set of swaps, each component of which is an in-tree rooted at one of the root swaps. We call the set of swaps in a component a *cascade*. In other words, a cascade corresponds to the set of swaps sharing a single root swap as an ancestor. Note that one cascade does not necessarily finish before another begins. The label of the cascade is defined to be the label of the root swap in it.

As noted above, each swap has at most two children and they have a strictly smaller label. Thus it follows that:

Lemma 3. *A label-i cascade contains at most 2^{i-j} label-j swaps.*

We say that the cascade *contains* an involuntary loss if some swap in the cascade causes it. Since each swap causes at most one involuntary loss, the lemma above implies:

Corollary 4. *A label-i cascade contains at most $2^{i-j+1} - 1$ involuntary losses to nodes with labels at least j.*

Proof. An involuntary loss to a label-k node must be caused by a swap with label k or higher.

2.3   Obtaining the Witness

We now show that when the algorithm terminates, we can find a combinatorial structure, that we call the witness, which establishes the near optimality of the final tree. Our witness consists of a partition $\mathcal{C} = \{W, C_1, \ldots, C_k\}$ and a subset $W' \subset V$ such that $W \subseteq W'$. Call an edge e MST-worthy if there is some minimum spanning tree of G that contains e. The witness has the following property: any MST-worthy edge $e = (u, v)$ leaving C_i has at least one endpoint in W', i.e. for any MST-worthy edge $(u, v) : u \in C_i, v \notin C_i, |\{u, v\} \cap W'| \geq 1$. The following lemma, essentially contained in [4] shows that a witness establishes a lower bound on the minimum degree of any MST.

Lemma 5. [4] *Let $\mathcal{W} = \{\mathcal{C}, W'\}$ be a witness defined as above. Then any minimum spanning tree of G has maximum degree at least $(k + |W| - 1)/|W'|$.*

Proof. Consider any MST T of G and let G' be the graph formed by shrinking each of the sets $C_1, \ldots, C_k$ to a single node. T must contain a spanning tree

T' of G' and hence must have at least $k + |W| - 1$ edges from G'. Moreover, each such edge is MST-worthy and hence has at least one endpoint in W'. The average degree of W' is thus at least $(k + |W| - 1)/|W'|$.

Let the Δ-phase be the last phase of the final iteration of the algorithm. It ends with an overloaded node at level $l = \log_2 n$. Let $W_L = \cup_{i \geq L} V_i$ be the set of nodes with label at least L and $s_L = |W_L|$ be the final number of nodes in W_L. Fix a constant $c \geq 2$. We search top-down for the highest level labeled $l - j$, for some j, $0 \leq j < l$, such that $s_{l-(j+1)} < c \cdot s_{l-j}$. We call level $l - (j + 1)$ *sparse*. Since $l \geq \log_c n$, such a sparse level must exist for any $c \geq 2$. Then for every level i such that $i \geq g$, it is true that $s_i \geq c \cdot s_{i+1}$.

We first show that the average degree of W_g is high.

Lemma 6. *The average degree of W_g in the final tree is at least $\Delta - 1 - \frac{2c}{c-2}$.*

Proof. Each node enters the set W_g with degree Δ, after which it may lose at most one degree from a useful swap and it may suffer some involuntary losses. Thus the total degree of W_g is at least $(\Delta - 1)|W_g|$ minus the number of involuntary losses to W_g.

Each involuntary loss to W_g occurs in a cascade and by Corollary 4, the number of involuntary losses to W_g in a label-i cascade is at most 2^{i-g+1}. Recall that the root swap of a cascade is useful to a pending node that moved up one label. The total number of involuntary losses to W_g during the course of the phase is at most

$$\sum_{i \geq g} 2^{i-g+1}|W_i| = \sum_{i \geq g} 2^{i-g+1} s_i$$

$$\leq \sum_{i \geq g} 2^{i-g+1} \frac{s_g}{c^{i-g}}$$

$$\leq 2s_g \sum_{i \geq g} \left(\frac{2}{c}\right)^{i-g}$$

$$\leq 2|W_g|(\frac{c}{c-2})$$

Thus the average degree Δ_{av} of W_g is at least $\Delta - 1 - 2\frac{c}{c-2}$.

Our witness is now constructed as follows: W' is the set W_{g-1} and W is W_g. $C_1, \ldots, C_k$ are the components formed by deleting W_g from the final tree. The feasibility of the labeling implies that this indeed is a witness. Moreover, k is at least $\Delta_{av}|W_g| - 2(|W_g| - 1)$, where Δ_{av} is the average degree of W_g in T. Lemma 5 then implies that the maximum degree of any MST must be at least

$$\left(\Delta - 2 - \frac{2c}{c-2}\right)\frac{|W|}{|W'|} = \left(\Delta - 2 - \frac{2c}{c-2}\right)\left(\frac{|W_g|}{|W_{g-1}|}\right) \geq \frac{\Delta}{c} - \frac{2}{c-2} - 2$$

Rearranging, we get

$$\Delta \leq c(\Delta_{\mathrm{OPT}} + 2 + \frac{2}{c-2})$$

Setting c to be $2 + \frac{2}{\sqrt{\Delta_{\mathrm{OPT}}}}$, we get

$$\Delta \leq 2\Delta_{\mathrm{OPT}} + 4\sqrt{\Delta_{\mathrm{OPT}}} + 6 + \frac{4}{\sqrt{\Delta_{\mathrm{OPT}}}}$$

Theorem 7 summarizes the results of Section 2.

Theorem 7. *Given a graph G, the* push_relabel_MDMST *algorithm obtains in polynomial time an MST of degree Δ, where $\Delta \leq 2\Delta_{\mathrm{OPT}} + O(\sqrt{\Delta_{\mathrm{OPT}}})$.*

This algorithm along with Lagrangean relaxation techniques from [13] gives a bi-criteria approximation for the Bounded-degree MST problem. We omit the proof from this extended abstract.

Theorem 8. *For any $\beta > 0$, there is a polynomial-time algorithm that, given a graph G and degree bound B, computes a spanning tree T with maximum degree at most $2(1 + \beta)B + O(\sqrt{(1 + \beta)B})$ and cost at most $\left(1 + \frac{1}{\beta}\right)\mathrm{OPT}_{LD(B)}$.*

References

1. T. M. Chan. Euclidean bounded-degree spanning tree ratios. In *Proceedings of the nineteenth annual symposium on Computational geometry*, pages 11–19. ACM Press, 2003.
2. K. Chaudhuri, S. Rao, S. Riesenfeld, and K. Talwar. What would edmonds do? augmenting paths and witnesses for bounded degree msts. In *Proceedings of AP-PROX/RANDOM*, 2005.
3. J. Edmonds. Maximum matching and a polyhedron with 0–1 vertices. *Journal of Research National Bureau of Standards*, 69B:125–130, 1965.
4. T. Fischer. Optimizing the degree of minimum weight spanning trees. Technical Report 14853, Dept of Computer Science, Cornell University, Ithaca, NY, 1993.
5. M. Fürer and B. Raghavachari. Approximating the minimum-degree Steiner tree to within one of optimal. *Journal of Algorithms*, 17(3):409–423, Nov. 1994.
6. M. Goemans. Personal communication, 2006.
7. A. V. Golberg. A new max-flow algorithm. Technical Report MIT/LCS/TM-291, Massachussets Institute of Technology, 1985. Technical Report.
8. A. V. Goldberg and R. E. Tarjan. A new approach to the maximum flow problem. In *Proceedings of the eighteenth annual ACM symposium on Theory of computing*, pages 136–146. ACM Press, 1986.
9. A. V. Goldberg and R. E. Tarjan. A new approach to the maximum-flow problem. *J. ACM*, 35(4):921–940, 1988.
10. J. A. Hoogeveen. Analysis of christofides' heuristic: Some paths are more difficult than cycles. *Operation Research Letters*, 10:291– 295, 1991.
11. R. Jothi and B. Raghavachari. Degree-bounded minimum spanning trees. In *Proc. 16th Canadian Conf. on Computational Geometry (CCCG)*, 2004.
12. S. Khuller, B. Raghavachari, and N. Young. Low-degree spanning trees of small weight. *SIAM J. Comput.*, 25(2):355–368, 1996.
13. J. Könemann and R. Ravi. A matter of degree: Improved approximation algorithms for degree-bounded minimum spanning trees. *SIAM Journal on Computing*, 31(6):1783–1793, Dec. 2002.

14. J. Könemann and R. Ravi. Primal-dual meets local search: approximating MST's with nonuniform degree bounds. In ACM, editor, *Proceedings of the Thirty-Fifth ACM Symposium on Theory of Computing, San Diego, CA, USA, June 9–11, 2003*, pages 389–395, New York, NY, USA, 2003. ACM Press.

15. C. H. Papadimitriou and U. Vazirani. On two geometric problems related to the traveling salesman problem. *J. Algorithms*, 5:231–246, 1984.

16. R. Ravi and M. Singh. Delegate and conquer: An LP-based approximation algorithm for minimum degree msts. In *Proceedings of ICALP*, 2006.

Edge Disjoint Paths in Moderately Connected Graphs

Satish Rao[1,*] and Shuheng Zhou[2,**]

[1] University of California, Berkeley, CA 94720, USA
satishr@cs.berkeley.edu
[2] Carnegie Mellon University, Pittsburgh, PA 15213, USA
szhou@ece.cmu.edu

Abstract. We study the Edge Disjoint Paths (EDP) problem in undirected graphs: Given a graph G with n nodes and a set T of pairs of terminals, connect as many terminal pairs as possible using paths that are mutually edge disjoint. This leads to a variety of classic NP-complete problems, for which approximability is not well understood. We show a polylogarithmic approximation algorithm for the undirected EDP problem in general graphs with a moderate restriction on graph connectivity; we require the global minimum cut of G to be $\Omega(\log^5 n)$. Previously, constant or polylogarithmic approximation algorithms were known for trees with parallel edges, expanders, grids and grid-like graphs, and most recently, even-degree planar graphs. These graphs either have special structure (e.g., they exclude minors) or there are large numbers of short disjoint paths. Our algorithm extends previous techniques in that it applies to graphs with high diameters and asymptotically large minors.

1 Introduction

In this paper, we explore approximation for the edge disjoint paths (EDP) problem: Given a graph with n nodes and a set of terminal pairs, connect as many of the specified pairs as possible using paths that are mutually edge disjoint. EDP has a multitude of applications in areas such as VLSI design, routing and admission control in large-scale, high-speed and optical networks. Moreover, EDP and its variants have also been prominent topics in combinatorics and theoretical computer science for decades. For example, the celebrated theory of graph minors of Robertson and Seymour [29] gives a polynomial time algorithm for routing all the pairs given a constant number of pairs. However, varying the number of terminal pairs leads to a variety of classic NP-complete problems, for which approximability is an interesting problem. In a recent breakthrough [3], Andrews and Zhang showed an $\Omega(\log^{\frac{1}{3}-\epsilon} n)$ lower bound on the hardness of approximation for undirected EDP.

---

* Supported in part by NSF Award CCF-0515304.

** This material is based on research sponsored in part by the Army Research Office, under agreement number DAAD19–02–1–0389 and NSF grant CNF–0435382. This work was done while the author was visiting UC Berkeley.

M. Bugliesi et al. (Eds.): ICALP 2006, Part I, LNCS 4051, pp. 202–213, 2006.

In this work, we show a polylogarithmic approximation algorithm for the undirected EDP problem in general graphs with a moderate restriction on graph connectivity; we require that there are $\Omega(\log^5 n)$ edge disjoint paths between every pair of vertices, i.e., the global min cut is of size $\Omega(\log^5 n)$. If this moderately connected case holds, we can route $\Omega(\mathsf{OPT}/\operatorname{polylog} n)$ pairs using disjoint paths with congestion 1, where OPT is the maximum number of pairs that one can route edge disjointly for the given EDP instance. Previously, constant or polylogarithmic approximation algorithms were known for trees with parallel edges, expanders, grids and grid-like graphs, and most recently, even-degree planar graphs [20]. The results rely either on excluding a minor (or other structural properties), or the fact that very short paths exist. Our algorithm extends previous techniques; for example, our graphs can have high diameter and contain very large minors. We are hopeful that this constraint on the global minimum cut can be removed if congestion on each edge is allowed to be $O(\log \log n)$. Formally, we have the following result.

Theorem 1. *There is a polylog n-approximation algorithm for the edge disjoint path problem in a general graph $\mathcal{G}$ with minimum cut and node degree $\Omega(\log^5 n)$.*

1.1 The Approach

We begin with a fractional relaxation of the problem, where each terminal pair can route a real-valued amount of flow between 0 and 1, and this flow can be split fractionally across a set of distinct paths. This can be expressed as an LP and can be solved efficiently. We denote the value of an optimal fractional LP solution as OPT^*. Our algorithm routes a polylogarithmic fraction of this value using integral edge-disjoint paths.

The algorithm proceeds by decomposing the graph into well-connected subgraphs, based on OPT^*, so that a subset of the terminal pairs, that remain within each subgraph are "well-connected", following a decomposition procedure of Chekuri, Khanna, and Shepherd (CKS05) [11]. Then, for each well connected subgraph G, we construct an expander graph that can be embedded into G using its terminal set. We use a result by Khandekar, Rao and Vazirani in [19], where they show that one can build an expander graph H on a set of nodes V by constructing $O(\log^2 n)$ perfect matchings $M_1, \ldots, M_{O(\log^2 n)}$ between $O(\log^2 n)$ sets of equal partitions of V in an iterative manner.

Our contribution along this line is to route each perfect matching $M_t, \forall t$, on one of the $O(\log^2 n)$ (edge-disjoint) subgraphs of G. The "splitting procedure", motivated by Karger's theorem [18], simply assigns edges of G uniformly at random into $O(\log^2 n)$ subgraphs. Using Karger's arguments, we show that all cuts in each subgraph have approximately the correct size with high probability. Here we crucially use the polylogarithmic lower bound on the min-cut. We then route each matching M_t on a unique split subgraph using a max-flow computation with unit capacities. Thus, we can route all $O(\log^2 n)$ matchings edge disjointly in G and embed an expander graph H integrally with congestion 1 on G.

After we construct such an expander graph H for each G, we route terminal pairs in H greedily via short paths. This is effective since there are plenty of short disjoint paths in an expander graph[7,21]. Since a node in H maps to a cluster of nodes in G that is connected by a spanning tree, we put a capacity constraint on $V(H)$: we allow only a single path to go through each node. We greedily connect a pair of terminals from G via a path in H while taking both nodes and edges along the chosen path away from H, until no short paths remain between any unrouted terminal pair. For the pairs we indeed route, we know the congestion is 1 in the original graph G, since we use each edge and node in H only once, and edges and nodes of H correspond to disjoint paths of G. We use a lemma in [15] to show that such a greedy method ensures that we route a sufficiently large number of such pairs; We note that this method was proposed but analyzed somewhat differently by Kleinberg and Rubinfeld [21]. Our analysis is more like that of Obata [27], and yields somewhat stronger bounds. Our approximation factor is $O(\log^{10} n)$. (A breakdown of this factor is described in Theorem 4.)

1.2   Related Work

Much of recent work on EDP has focused on understanding the polynomial-time approximability of the problem. Previously, constant or polylogarithmic approximation algorithms were known for trees with parallel edges [15], expanders [21,26], grids and grid-like graphs [5,6,22,23], and even-degree planar graphs [20]. For general graphs, the best approximation ratio for EDP in directed graphs is $O(\min(n^{2/3}, \sqrt{m}))$ [8,24,25,30,31], where m denotes number of edges in the input graph. This is matched by the $\Omega(m^{\frac{1}{2}-\epsilon})$-hardness of approximation result by Guruswami et al [17]. For undirected and directed acyclic graphs, the upper bound has been improved to $O(\sqrt{n})$ [13]. For even-degree planar graphs, an $O(\log^2 n)$-approximation [20] is obtained recently.

A variant is the EDP with Congestion (EDPwC) problem, where the goal is to route as many terminals as possible, such that at most ω demands can go through any edge in the graph. For EDPwC on planar graphs, for $\omega = 2$ and 4, $O(\log n)$ [10,11] and constant [12] approximations have been obtained respectively. For undirected graphs, the hardness results [1] are $\Omega(\log^{1/2-\epsilon} n)$ for EDP and $\Omega(\log^{(1-\epsilon)/(\omega+1)} n)$ for EDPwC.

A closely related problem is the congestion minimization problem: Given a graph and a set of terminal pairs, connect *all* pairs with integral paths while minimizing the maximum number of paths through any edge. Raghavan and Thompson [28] show that by applying a randomized rounding to a linear relaxation of the problem one obtains an $O(\log n/\log\log n)$ approximation for both directed and undirected graphs. For hardness of approximation, Andrews and Zhang [2] show a result of $\Omega((\log\log^{1-\epsilon} m))$ for undirected and an almost-tight result [4] of $\Omega(\log^{1-\epsilon} m)$ for directed graphs, improving that of $\Omega(\log\log m)$ by Chuzhoy and Naor [14]. Finally, the All-or-Nothing Flow (ANF) problem [9,11] is to choose a subset of terminal pairs such that for each chosen pair, one can fractionally route a unit of flow for all the chosen pairs. The hardness result for ANF and ANF with Congestion is the same as that of EDP and EDPwC [1]. Currently, there exists an $O(\log^2 n)$ [11] approximation for ANF. Indeed, we build on the techniques developed in this approximation algorithm for ANF.

2 Definitions and Preliminaries

We work with graph $G = (V, E)$ with unit-capacity edges, where we allow parallel edges, unless we specify a capacity function for edges explicitly. For a capacitated graph $G = (V, E, c)$, where c is an integer capacity function on edges, one can replace each edge $e \in E$ with $c(e)$ parallel edges. For a cut $(S, \bar{S} = V \setminus S)$ in G, let $\delta_G(S)$, or simply $\delta(S)$ when it is clear, denote the set of edges with exactly one endpoint in S in G. Let $\mathsf{cap}(S, \bar{S}) = |\delta_G(S)|$ denote the total capacity of edges in the cut. The edge expansion of a cut $(S, \bar{S})$, where $|S| \leq |V|/2$, is $\phi(S) = \frac{\mathsf{cap}(S, \bar{S})}{|S|}$. The expansion of a graph G is the minimum expansion over all cuts in G. We call a graph G an expander if its expansion is at least a constant.

An instance of a routing problem consists of a graph $\mathcal{G} = (V, E)$ and a set of terminals pairs $\mathcal{T} = \{(s_1, t_1), (s_2, t_2), \ldots, (s_k, t_k)\}$. Nodes in $\mathcal{T}$ are referred to as terminals. Given an EDP instance $(\mathcal{G}, \mathcal{T})$ with k pairs of terminals, we will use the following LP relaxation as specified in (2.1), to obtain an optimal fractional solution. Let $\mathcal{P}_i, \forall i$, denote the set of paths joining s_i and t_i in $\mathcal{G}$.

$$\max \sum_{i=1}^{k} x_i \ s.t. \tag{2.1}$$

$$x_i - \sum_{p \in \mathcal{P}_i} f(p) = 0, \forall 1 \leq i \leq k \tag{2.2}$$

$$\sum_{p:e \in p} f(p) \leq 1, \forall e \in E \tag{2.3}$$

$$x_i, f(p) \in [0, 1], \forall 1 \leq i \leq k, \forall p \tag{2.4}$$

We let $\mathsf{OPT}^*(\mathcal{G}, \mathcal{T})$ be the value of this linear program for the optimal solution $\bar{f}$ of the LP. In the text, where we always refer to a single instance, we primarily use OPT^*. The following definitions come from [11].

Definition 1. (CKS2005 [11]) *Given a non-negative weight function* $\pi : X \to \mathbb{R}^+$ *on a set of nodes* X *in* G, X *is* π-cut-linked *in* G *if* $\forall S$ *such that* $\pi(S \cap X) = \sum_{x \in S \cap X} \pi(x) \leq \pi(X)/2, |\delta(S)| \geq \pi(S \cap X)$; *We also refer to* (G, X) *as a* π-cut-linked *instance.*

Definition 2. (CKS2005 [11]) *A set of nodes* X *is* well-linked *in* G *if* $\forall S$ *such that* $|S \cap X| \leq |X|/2, |\delta(S)| \geq |S \cap X|$.

3 Decomposition and an Outline of Routing Procedure

In this section, we first present Theorem 2 regarding a preprocessing phase of our algorithm that decomposes and processes $(\mathcal{G}, \mathcal{T})$ into a collection of cut-linked instances with a min-cut $\Omega(\log^3 n)$ in each subgraph. We then state our main theorem with a breakdown of the polylog n approximation factor. Finally, we give an outline on how we route terminal pairs in each cut-linked instance (G, T); Note that we use G to refer to a subgraph that we obtain through Theorem 2 starting from Section 3.1 till the end of the paper, while $\mathcal{G}$ refers to the original input graph. We first specify the following parameters.

- **Parameters related to original EDP instance $(\mathcal{G}, \mathcal{T})$**
  - $\omega \log^2 n$ is the number of matchings as in Figure 1;
  - min-cut $\kappa = \Omega(\log^3 n) = \frac{12(\ln n)(\omega \log^2 n + 1)}{\epsilon^2}$, where $\epsilon < 1$;
  - $\beta(\mathcal{G}) = O(\log n)$: the worst-case mincut-maxflow gap on product commodity flow instances on $\mathcal{G}$;
  - $\lambda(n) = 10\beta(\mathcal{G}) \log \mathsf{OPT}^*(\mathcal{G}, \mathcal{T}) = O(\log^2 n)$: as introduced in [11].

Theorem 2. *There is a polynomial time decomposition algorithm, that given an EDP instance $(\mathcal{G}, \mathcal{T})$, where $\mathcal{G}$ has a min-cut of size $\Omega(\kappa \log^2 n)$, and a solution $\bar{f}$ to the fractional EDP problem, with $x_i, \forall i$, being specified as in (2.1), produces a disjoint set of subgraphs and a weight function $\boldsymbol{\pi} : V(\mathcal{G}) \rightarrow \mathbb{R}^+$ on $V(\mathcal{G})$ where*

1. *there are $\alpha_1, \ldots, \alpha_k$ such that $\forall u$ in a subgraph H, $\boldsymbol{\pi}(u) = \sum_{i:s_i=u,t_i \in H} \alpha_i x_i$, (note that this implies $\forall s_i t_i \in \mathcal{T}$, x_i contributes the same amount of weight to $\boldsymbol{\pi}(s_i)$ and $\boldsymbol{\pi}(t_i)$);*
2. *the set of nodes $V(H)$ in each subgraph H is $\boldsymbol{\pi}$-cut-linked in H;*
3. *each subgraph H has min-cut $\kappa = \Omega(\log^3 n)$;*
4. *$\forall u$ in a subgraph H s.t. $\boldsymbol{\pi}(H) \geq \Omega(\log^3 n)$, $\boldsymbol{\pi}(u) \leq \sum_{i:s_i=u,t_i \in H} \frac{x_i}{\beta(\mathcal{G})\lambda(n)}$;*
5. *and $\boldsymbol{\pi}(\mathcal{G}) = \Omega(\mathsf{OPT}^*/\beta(\mathcal{G})\lambda(n))$.*

The decomposition essentially says that summing across all subgraphs G, a fair fraction of terminal pairs in $\mathcal{T}$ remain (condition 4, 5); indeed, we lose only a constant fraction of the terminal pairs (by assigning a zero weight to those lost terminals) of $\mathcal{T}$. In addition, each subgraph G is well connected with respect to X, the set of induced terminals of $\mathcal{T}$ in G, in the sense of (G, X) being a $\boldsymbol{\pi}$-cut-linked instance. This decomposition is essentially the same as that of Chekuri, Khanna, and Shepherd [11]. We need to do some additional work to ensure that the min-cut condition (condition 3) holds. We prove a dual (flow-based) version of the result in the full version of the paper.

3.1   Overall Routing Algorithm in Each Decomposed Subgraph G

We assume that we have the $\boldsymbol{\pi}$-cut-linked subgraphs given by Theorem 2. We will treat each subgraph and its induced subproblem (G, T) independently. We use $\boldsymbol{\pi}(G)$ to denote $\boldsymbol{\pi}(V(G))$ in the following sections. Let X be the set of terminals of T that is assigned with a positive weight by function $\boldsymbol{\pi}$ in instance G. We further assume that $\boldsymbol{\pi}(G) = \Omega(\log^7 n)$. If not, we just route an arbitrary pair of terminals in T; otherwise, we use PROCEDURE EMBEDANDROUTE$(G, T, \boldsymbol{\pi})$ in Figure 1 to route. We first specify a few more parameters and conditions related to (G, T); We then state Theorem 3, which we prove through the rest of the paper. Combining Theorem 3 and Theorem 2 proves Theorem 4.

- **Parameters and conditions related to an induced subproblem (G, T)**
  - sampling probability $p = 12(\ln n)/\epsilon^2 \kappa = 1/(\omega \log^2 n + 1)$

0. Given graph G with min-cut $\Omega(\log^3 n)$ and a weight function $\pi : V(G) \to \mathbb{R}^+$
1. $\{G^1, \ldots, G^Z\} = \text{SPLIT}(G, Z, \pi)$
2. $\{\mathcal{X}, \mathcal{C}\} = \text{CLUSTERING}(G^Z, \pi)$, where $\mathcal{X} = \{X_1, \ldots, X_r\}$ and $\mathcal{C} = \{C_1, \ldots, C_r\}$
3. Given a set of superterminals $\mathcal{X}$ of size r
4. Let $\mathcal{X}$ map to vertex set $V(H)$ of Expander H
5. For $t = 1$ to $\omega \log^2 n$
6. $(S, \bar{S} = \mathcal{X} \setminus S) = \text{KRV-FINDCUT}(\mathcal{X}, \{M_k : k < t\})$ s. t. $|S| = |\bar{S}| = r/2$
7.   Matching $M_t = \text{FINDMATCH}(S, \bar{S}, G^t)$ s.t. M_t is routable in G^t
8. Combine $M_1, \ldots, M_{\omega \log^2 n}$ to form the edge set F on vertices $V(H)$
9. $\text{EXPANDERROUTE}(H, T, X)$
10. End

Fig. 1. Procedure $\text{EMBEDANDROUTE}(G, T, \pi)$

- number of split subgraphs $Z = 1/p = \omega \log^2 n + 1$
- $W = (\omega \log^2 n + 1)/(1 - \epsilon)$, for some $\epsilon < 1$;
- $r \geq \max\{1, (\pi(G) - (W - 1))/(2W - 1)\}$, such that $\forall i \in [1, \ldots, r], 2W - 1 \geq \pi(X_i) = \sum_{v \in X_i} \pi(v) \geq W$ and $\pi(\mathcal{X}) \geq \pi(G) - (W - 1)$: i.e., at most $W - 1$ unit of weight is not counted in $\mathcal{X}$.

Theorem 3. *Given an induced instance (G, T) with min-cut of G being $\Omega(\log^3 n)$ and a weight function $\pi : V(G) \to \mathbb{R}^+$ such that X is π-cut-linked in G and $\pi(G) = \Omega(\log^7 n)$, EMBEDANDROUTE routes at least $\max\{1, \Omega(\pi(G)/\log^7 n)\}$ pairs of T in G edge disjointly.*

Theorem 4. *Given an EDP instance $(\mathcal{G}, T)$, where $\mathcal{G}$ has a min-cut $\Omega(\lambda(n)\kappa)$, we can route $\Omega(OPT^*(\mathcal{G}, T)/f)$ terminal pairs edge disjointly in $\mathcal{G}$, where the approximation factor f is $O(\lambda(n)\beta(\mathcal{G})W \log^5 n)$.*

4   Obtaining Z Split Subgraphs of G

In this section, we analyze a procedure that splits a graph G, with min-cut $\kappa = \Omega(\log^3 n)$, into Z subgraphs b extending a uniform sampling scheme from Karger [18]. We thus obtain a set of cut-linked instances as in Lemma 1, which immediately follows from Theorem 5.

Procedure Split(G, Z, π): Given a graph $G = (V, E)$ with min-cut $\kappa = \Omega(\log^3 n)$, a weight function $\pi : V(G) \to \mathbb{R}^+$, a set of terminals X in G such that (G, X) is a π-cut-linked instance, and probability $p = 1/Z$.
Output: A set of randomized split subgraphs $G^1, \ldots, G^Z$ of G.
Each split subgraph $G^j, \forall j = 1, \ldots, Z$ inherits the same set of vertices of G; Edges of G are placed independently and uniformly at random into the Z subgraphs; each $e = (u, v) \in E$ is placed between the same endpoints u, v in the chosen subgraph. We retain the same weight function π for all nodes in V in each split subgraph $G^j, \forall j$.

Lemma 1. *With high probability,* X *is* $\frac{(1-\epsilon)\pi}{Z}$*-cut-linked in* $G^j, \forall j$*, for some* $\epsilon < 1$*.*

Theorem 5 says that all cuts can be preserved in all split graphs $G^1, \ldots, G^Z$ of G we thus obtain. Recall for $S \in V$, $|\delta_G(S)|$ denote the size of $(S, V \setminus S)$ in G. For the same cut $(S, V \setminus S)$, we have $\mathbf{E}[|\delta_{G^j}(S)|] = p\,|\delta_G(S)|$ in $G^j, \forall j$, where p is the probability that an edge $e \in E$ is placed in $G^j, \forall j$.

Theorem 5. *Let* $G = (V, E)$ *be any graph with unit-weight edges and min cut* κ*. Let* $\epsilon = \sqrt{3(d+2)(\ln n)/p\kappa}$*. If* $\epsilon \leq 1$*, then with probability* $1 - O(\log^2 n/n^d)$*, every cut* $(S, V \setminus S)$ *in every subgraph* $G^1, G^2, \ldots, G^Z$ *of* G *has value between* $(1 - \epsilon)$ *and* $(1 + \epsilon)$ *times its expected value* $p\,|\delta_G(S)|$*.*

Proof. We sketch a proof, leaving details in the full paper. We first give a definition by Karger [18], regarding a uniform random sampling scheme on an unweighted graph $G = (V, E)$; Lemma 2 immediately follows from this definition. We then state Karger's theorem regarding preserving all cuts of G in a sampled subgraph, under a certain min-cut condition.

Definition 3. **(Karger94 [18])** *A* p*-skeleton of* G *is a random subgraph* $G(p)$ *constructed on the same vertices of* G *by placing each edge* $e \in E$ *in* $G(p)$ *independently with probability* p*.*

Lemma 2. *Every randomized subgraph* $G^j, \forall j$*, is a* p*-skeleton of* G*.*

Theorem 6. **(Karger94 [18])** *Let* G *be a graph with unit-weight edges and min-cut* κ*. Let* $p = 3(d+2)(\ln n)/\epsilon^2\kappa$*. With probability* $1 - O(1/n^d)$*, every cut in a* p*-skeleton of* G *has value between* $(1 - \epsilon)$ *and* $(1 + \epsilon)$ *times its expected value.*

To prove Theorem 6, Karger uses a union bound to show that the sum of probabilities of all *bad* events in a p-skeleton of G is $O(1/n^d)$, where a bad event refers to some cut in a p-skeleton of G diverges from its expected value k by more than ϵk. Given that every random split subgraph $G^j, \forall j$, is a p-skeleton of G by Lemma 2, we apply the essential statement in Karger's proof to all subgraphs G^j with $p = 12(\ln n)/\epsilon^2\kappa$ and $\kappa = 12(\ln n)(\omega \log^2 n + 1)/\epsilon^2$ for a given ϵ. We can then use a union bound to sum up probabilities of bad events across all split subgraphs $G^1, \ldots, G^Z$ of G, which is $O(\log^2 n/n^2)$ for $d = 2$.   ∎

5   Forming Superterminals That Are Well-Linked

The procedure in this section constructs superterminals as follows. It finds connected subgraphs C in G^Z, where $\pi(C) = \Omega(\log^2 n)$, each connecting a subset of terminals. Roughly, the idea is that these clustered terminals are better connected than individual terminals. They are well linked in the sense that any cut that splits off K superterminals as one entity contains at least K edges in $G^j, \forall j$ This allows us to compute congestion-free maximum flows in Section 6.1.

Given split subgraphs $G^1, \ldots, G^Z$ of G, each with the same weight function π on its vertex set $V(G^j) = V, \forall j$, that we obtain through PROCEDURE

SPLIT(G, Z, π), we aim to find a set $\mathcal{X} = \{X_1, \ldots, X_r\}$ of node-disjoint "superterminals", where each superterminal $X_i \in \mathcal{X}$ consists of a subset of terminals in X and each X_i gathers a weight between W and $2W - 1$. In addition, we want to find an edge-disjoint set of clusters $\mathcal{C} = \{C_1, \ldots, C_r\}$, where $C_i = (V_i, E_i)$, such that $X_i \subseteq V_i$ and C_i is a connected component, and hence all nodes in X_i are connected through E_i. W.l.o.g., we pick G^Z for forming such clusters $C_i, \forall i$; note that G^Z is a connected graph with a min-cut of $\Omega(\log n)$, whp, by Theorem 5.

Procedure Clustering(G^Z, π): Given a split subgraph G^Z and a weight function $\pi : V(G^Z) \to \mathbb{R}^+$ and $\pi(V(G^Z)) = \pi(G) \geq W$.
Output: $\mathcal{X} = \{X_1, \ldots, X_r\}$ and $\mathcal{C} = \{C_1, \ldots, C_r\}$ as specified in Lemma 3.
We group subsets of vertices of V in an edge-disjoint manner, following a procedure from [9], by choosing an arbitrary rooted spanning tree of G^Z and greedily partitioning the tree into a set $\mathcal{C}$ of edge-disjoint subgraphs of G^Z.

Lemma 3. (CKS2004 [9]) *Let G^Z be a connected graph with a weight function $\pi : V(G^Z) \to [0, W]$ such that $\pi(V(G^Z)) \geq W$. We can find $r \geq \max\{1, (\pi(G) - (W-1))/(2W-1)\}$ edge-disjoint connected subgraphs, $C_1 = (V_1, E_1), \ldots, C_r = (V_r, E_r)$, such that there exist vertex-disjoint subsets $X_1, \ldots, X_r$ and for each i: (a) $X_i \subseteq V_i$ and (b) $2W - 1 \geq \sum_{v \in X_i} \pi(v) \geq W$.*

Result. To get an intuition of the purpose of forming such clusters, consider a cut $(U, V \setminus U)$ in a split subgraph $G^j, \forall j$. Let U be a subset of $V(G)$ such that $\pi(U) = \sum_{x \in U \cap X} \pi(x) \leq \pi(X)/2$. Let K be the number of superterminals that are contained in U. We have the following lemma, which captures the notion of superterminals being "well-linked", with a hint of Definition 2.

Lemma 4. $\forall$ *split subgraphs $G^1, \ldots, G^Z$, where $Z = 1/(\omega \log^2 n + 1)$, and $\forall U \subset V(G)$ s.t. $\pi(U) \leq \pi(X)/2$, $|\delta_{G^j}(U)| \geq K$, where $K = |\{X_i \in \mathcal{X} : X_i \subseteq U\}|$.*

6   Construct and Embed an Expander H in G

In this section, we use the superterminals from the previous section as nodes in an expander H that we embed in G. The edges of H are defined using a technique in [19] that builds an expander using $O(\log^2 n)$ matchings. We embed this expander in G by routing each matching in one of the split graphs using a maximum flow computation. This allows us to embed H into G with no congestion. The following procedure restates this outline. Theorem 7 is a main technical contribution of this paper.

Procedure EmbedExpander$(G^1, \ldots, G^{\omega \log^2 n}, \mathcal{X})$:
Output: An expander $H = (V', F)$ routable in G s.t. $|V'| = r$ and $\forall i \in V'$, $\pi(i) = \pi(X_i)$ and $\pi(H) = \pi(\mathcal{X})$; F consists of $M_1, \ldots, M_{\omega \log^2 n}$.
We use Step (3) to (8) of PROCEDURE EMBEDANDROUTE in Figure 1, where we substitute PROCEDURE FINDMATCH with Figure 3 while relying on an existing PROCEDURE KRV-FINDCUT [19]. At each round t, we use KRV-FINDCUT

---

0. Given a set of points $V(H)$ of size k
1. for $t = 1$ to $\omega \log^2 n$
2. $(S, \bar{S} = V(H) \setminus S) = $ KRV-FINDCUT$(V(H), \{M_k : k < t\})$ s.t. $|S| = |\bar{S}| = k/2$
3. $M_t = $ FINDMATCH$(S, \bar{S})$ s.t. M_t is a matching between S and $\bar{S}$
4. Combine $M_1, \ldots, M_{\omega \log^2 n}$ to form the edge set F on vertices $V(H)$
5. End

Fig. 2. KRV-Procedure CONSTRUCTING AN α-EXPANDER H

to generate an equal-sized partition $(S, \mathcal{X} \setminus S = \bar{S})$; we then find a matching M_t between S and $\bar{S}$ by computing a single-commodity max-flow using FINDMATCH$(S, \bar{S}, G^t)$ in G^t, that we add to F as edges.

Theorem 7. *(a)* EMBEDEXPANDER *constructs a $1/4$-expander $H = (V', F)$; (b) in addition, H is embedded into G as follows. Each node i of H corresponds to a superterminal X_i in $\mathcal{X}$ in G such that all superterminals are mutually node disjoint and each superterminal is connected by a spanning tree, T_i, in G. Each edge (i, j) in H corresponds to a path, P_{ij} from a node in X_i to a node in X_j. All paths P_{ij} and trees T_i are mutually edge disjoint in G.*

Proof. The expander property (a) follows from a result of Khandekar, Rao and Vazirani [19]; they show the procedure in Figure 2 produces an expander H.

Theorem 8. (KRV2005 [19]) *Given a set of nodes $V(H)$ of size k, $\exists$ a KRV-FINDCUT procedure s.t. given any FINDMATCH procedure, the KRV-PROCEDURE as in Figure 2. produces an α-expander graph H, for $\alpha \geq 1/4$.*

Each edge $e = (i, j)$ in the matching M_t maps to an integral flow path that connects X_i and X_j in G^t; all such flow paths can be simultaneously routed in G^t edge disjointly due to the max-flow computation as we show in Lemma 5. Since each matching M^t is on a unique split subgraph G^t, the entire set of edges in $M_1, \ldots, M_{\omega \log^2 n}$, that comprise the edge set F of H, correspond to edge disjoint paths in $G^1, \ldots, G^{Z-1}$, where $Z = \omega \log^2 n + 1$. Finally, all spanning trees $T_i, \forall i$, are constructed using disjoint set of edges in G^Z as in Lemma 3. ∎

6.1   Finding a Matching Through a Max-Flow Construction

In this section, we show that given an arbitrary equal partition $(S, \bar{S})$ of the set $\mathcal{X} = \{X_1, \ldots, X_r\}$, that we obtain through PROCEDURE CLUSTERING(G^Z, π), we can use the following procedure to route a max-flow of size $r/2$, such that the integral flow paths that we obtain through flow decomposition induce a perfect matching between S and $\bar{S}$. Let $S = \{X_{i_1}, \ldots, X_{i_{r/2}}\}$ and $\bar{S} = \{X_{j_1}, \ldots, X_{j_{r/2}}\}$.

Lemma 5. *In each sampled graph G^t, FINDMATCH produces a perfect matching M_t between an equal partition $(S, \bar{S})$ of $\mathcal{X}$ such that for each edge in $e = (i, j) \in M_t$, there is an integral unit-flow path P_{ij} from a terminal in $X_i \in S$ to a terminal in $X_j \in \bar{S}$. All paths $P_{ij}, s.t.(i, j) \in M_t$ are edge disjoint in G^t.*

0. Given an equal partition $(S, \bar{S})$ of $\mathcal{X}$, we form a flow graph G' from G^t
    by adding auxiliary nodes and directed unit-capacity edges:
1. Add a special source and sink nodes s_0 and t_0;
2.   Add nodes $s_1, \ldots, s_{r/2}$ and an edge from s_0 to $s_k, \forall k = 1, \ldots, r/2$;
3.   Add nodes $t_1, \ldots, t_{r/2}$; from each $t_k, \forall k = 1, \ldots, r/2$, add an edge to t_0
4. From each $s_k, \forall k$, add an edge to each terminal $x \in X_{i_k}$ s.t. $X_{i_k} \in S$
5. To each node t_k, add an edge from each terminal $x \in X_{j_k}$ s.t. $X_{j_k} \in \bar{S}$
6. Route a max-flow from s_0 to t_0
7. Decompose the flow to obtain a matching between S and $\bar{S}$
8. End

Fig. 3. Procedure FINDMATCH$(S, \bar{S}, G^t)$

Lemma 6. *Every $s_0 - t_0$ cut has size at least $r/2$ in the flow graph G'.*

Proof of Lemma 5: By Lemma 6 (proof appears in the full version), and the fact
that there $\exists$ a $s_0 - t_0$ cut of size $r/2$, (e.g., $(\{s_0\}, V(G') \setminus \{s_0\})$) we know the
$s_0 - t_0$ min-cut is $r/2$. Hence by the max-flow min-cut theorem, we know that
there $\exists$ a max-flow of size $r/2$ from s_0 to t_0. We next decompose the max-flow
into $r/2$ integer flow paths, which induce a perfect matching M_t between S and
$\bar{S}$ as follows. Consider an integral flow path $P_k, \forall k = 1, \ldots, r/2$. Let directed
path P_k start with s_0 and go through $s_k, x \in X_{i_k} \in S$ for some x; and let P_k
end with $y \in X_{j_{k'}} \in \bar{S}, t_{k'}, t_0$ for some $k' \in [1, \ldots, r/2]$ and some terminal y.
No other path in the max-flow can go through the same pair of superterminals
$X_{i_k}, X_{j_{k'}}$ due to the capacity constraints on edges (s_0, s_k) and $(t_{k'}, t_0)$. Hence
$M_t = \{(i_k, j_{k'}), \forall k \in [1, \ldots, r/2], \text{ where } k' \in [1, \ldots, r/2]\}$ is a perfect matching
between S and $\bar{S}$.   ∎

7 Routing on an Expander H Node Disjointly

In this section, we show that the following greedy algorithm routes $\Omega(K/\log^5 n)$
pairs of terminals, where $K = |V(H)| = \Omega(\pi(G)/W)$, in H.

Procedure ExpanderRoute(H, T, X): Given an uncapacitated expander H
with at least $512 \log^5 n$ nodes, with node degree $\omega \log^2 n$. While there is a pair
(s, t) in $T \subseteq \mathcal{T}$ whose path length is less than D in $H = (V, E)$, where $D =
a_3 \omega \log^3 n$ and $a_3 = 32$ is a constant; Remove both nodes and edges from H,
along a path through which we connect a pair of terminals in T.

   Since we take away both nodes and edges as we route a path across the
expander H due to the node capacity constraints on $V(H)$, routing the set P
of pairs via integral paths on H induces no congestion in G by Theorem 7. We
now argue that $|P|$ is large to finish our proof. Let H' be the remaining graph of
expander $H = (V, E)$, after we take away nodes and edges along the paths used
to route P. Note that all remaining pairs $T' \subseteq T$ in H' must have distance at
least D. This is the main condition that allows us to prove the following theorem.

Theorem 9. *The procedure above routes $\Omega(K/\log^5 n)$ pairs, node disjointly, in degree-$(\omega \log^2 n)$ expander $H = (V, E)$ with $K \geq 512 \log^5 n$ nodes.*

Proof Sketch: Let us first state the following lemma regarding a multicut in H' which follows from arguments of Garg, Vazirani and Yannakakis [16].

Lemma 7. *If all remaining terminal pairs in $T' \subseteq T$ have distances at least D in H', then there exists a multicut L in $H' = (V', E')$ of size $|E'| \log n/D$ in H' that separates every source and sink pair $s_i t_i \in T'$.*

Lemma 7 implies that there is a multicut of size at most $K\omega \log^3 n/2D = K/2a_3$ given that $|E'| \leq |E| = K\omega \log^2 n/2$ in the remaining graph H'.

We finish, by noting that condition 1 of Theorem 2 implies that any multicut of the terminals in H' ensures that no piece in H' separated by L contains more than half the weight of all terminals in H. We use this fact to show that the multicut L can be rearranged to find a "weight-balanced" cut in H', which corresponds to a node-balanced cut in H. Any node-balanced cut, however, in H must have at least $\Omega(K)$ edges. Using a proper choice of a_3, we force this balanced cut to contain at most half as many edges in H' as in H. Thus, we show $\Omega(K)$ edges have been removed when routing P. Since routing each such pair removes at most $D\omega \log^2 n(O(\log^5 n)$ edges. We conclude $|P|$ must be $\Omega(K/\log^5 n)$.

In more detail, in H', we alter π slightly to generate a new function $\pi'(i), \forall i \in V(H')$, so that only remaining pairs $uv \in T'$ contribute a positive weight to $\pi'(H')$ according to their flow in $\bar{f}$ like that of condition 1 in Theorem 2; hence each connected component in H', separated by multicut L, has a weight of at most $\pi'(H')/2$. We then use L to find a balanced cut $(U', V' \setminus U')$ in H' such that each side has weight at least $\pi'(H')/4$, where $\pi'(H') \geq \pi(G) - (W - 1) - 2(2W - 1)D |P|$. It is straightforward to verify that any partition $(U, V(H) \setminus U)$ in H, such that $U' \subseteq U$ and $(V' \setminus U') \subseteq (V(H) \setminus U)$, is node-balanced in H. The rest of the proof follows the outline in the previous paragraph.   ∎

References

1. M. Andrews, J. Chuzhoy, S. Khanna, and L. Zhang. Hardness of the undirected edge-disjoint paths problem with congestion. In *Proceedings of the 46th IEEE FOCS*, 2005.
2. M. Andrews and L. Zhang. Hardness of the undirected congestion minimization problem. In *Proceedings of the 37th ACM STOC*, 2005.
3. M. Andrews and L. Zhang. Hardness of the undirected edge-disjoint path problem. In *Proceedings of the 37th ACM STOC*, 2005.
4. M. Andrews and L. Zhang. Logarithmic hardness of the directed congestion minimization problem. In *Proceedings of the 38th ACM STOC*, 2006.
5. Y. Aumann and Y. Rabani. Improved bounds for all-optical routing. In *Proceedings of the 6th ACM-SIAM SODA*, pages 567–576, 1995.
6. B. Awerbuch, R. Gawlick, F. T. Leighton, and Y. Rabani. On-line admission control and circuit routing for high performance computing and communication. In *Proceedings of the 35th IEEE FOCS*, pages 412–423, 1994.
7. A. Broder, A. Frieze, and E. Upfal. Existence and construction of edge-disjoint paths on expander graphs. *SIAM Journal of Computing*, 23:976–989, 1994.

8. C. Chekuri and S. Khanna. Edge disjoint paths revisited. In *Proceedings of the 14th ACM-SIAM SODA*, 2003.
9. C. Chekuri, S. Khanna, and F. B. Shepherd. The all-or-nothing multicommodity flow problem. In *Proceedings of the 36th ACM STOC*, 2004.
10. C. Chekuri, S. Khanna, and F. B. Shepherd. Edge-disjoint paths in planar graphs. In *Proceedings of the 45th IEEE FOCS*, 2004.
11. C. Chekuri, S. Khanna, and F. B. Shepherd. Multicommodity flow, well-linked terminals, and routing problems. In *Proceedings of the 37th ACM STOC*, 2005.
12. C. Chekuri, S. Khanna, and F. B. Shepherd. Edge-disjoint paths in planar graphs with constant congestion. In *Proceedings of the 38th ACM STOC*, 2006.
13. C. Chekuri, S. Khanna, and F. B. Shepherd. An $O(\sqrt{n})$ approximation and integrality gap for disjoint paths and unsplittable flow. *Journal of Theory of Computing*, 2:137–146, 2006.
14. J. Chuzhoy and J. Naor. New hardness results for congestion minimization and machine scheduling. In *Proceedings of the 36th ACM STOC*, pages 28–34, 2004.
15. N. Garg, V. V. Vazirani, and M. Yannakakis. Primal-dual approximation algorithms for integral flow and multicut in trees. In *Proc. of the 20th ICALP*, 1993.
16. N. Garg, V. V. Vazirani, and M. Yannakakis. Approximate max-flow min-(multi)cut theorems and their applications. *SIAM J. of Computing*, 25:235–251, 1996.
17. V. Guruswami, S. Khanna, R. Rajaraman, F. B. Shepherd, and M. Yannakakis. Near-optimal hardness results and approximation algorithms for edge-disjoint paths and related problems. In *Proceedings of the 31th ACM STOC*, 1999.
18. D. R. Karger. Random sampling in cut, flow, and network design problems. In *Proceedings of the 26th ACM STOC*, 1994.
19. R. Khandekar, S. Rao, and U. Vazirani. Graph partitioning using single commodity flows. In *Proceedings of the 38th ACM STOC*, 2006.
20. J. Kleinberg. An approximation algorithm for the disjoint paths problem in even-degree planar graphs. In *Proceedings of the 46th IEEE FOCS*, 2005.
21. J. Kleinberg and R. Rubinfeld. Short paths in expander graphs. In *Proceedings of the 37th IEEE FOCS*, 1996.
22. J. Kleinberg and E. Tardos. Approximations for the disjoint paths problem in high-diameter planar networks. In *Proceedings of the 27th ACM STOC*, 1995.
23. J. Kleinberg and E. Tardos. Disjoint paths in densely embedded graphs. In *Proceedings of the 36th IEEE FOCS*, pages 52–61, 1995.
24. J. M. Kleinberg. *Approximation Algorithms for Disjoint Paths Problems*. PhD thesis, MIT, Cambridge, MA, 1996.
25. S. G. Kolliopoulos and C. Stein. Approximating disjoint-path problems using greedy algorithms and packing integer programs. In *Proceedings of IPCO*, 1998.
26. P. Kolman and C. Scheideler. Simple on-line algorithms for the maximum disjoint paths problem. In *Proceedings of the 13th ACM SPAA*, 2001.
27. K. Obata. Approximate max-integral-flow/min-multicut theorems. In *Proceedings of the 36th ACM STOC*, 2004.
28. P. Raghavan and C. D. Thompson. Randomized roundings: a technique for provably good algorithms and algorithms proofs. *Combinatorica*, 7:365–374, 1987.
29. N. Robertson and P. D. Seymour. An outline of a disjoint paths algorithm. *Paths, Flows and VLSI-design, Algorithms and Combinatorics*, 9:267–292, 1990.
30. A. Srinivasan. Improved approximations for edge-disjoint paths, unsplittable flow, and related routing problems. In *Proceedings of the 38th IEEE FOCS*, 1997.
31. K. Varadarajan and G. Venkataraman. Graph decomposition and a greedy algorithm for edge-disjoint paths. In *Proceedings of the ACM-SIAM SODA*, 2004.

A Robust APTAS for the Classical Bin Packing Problem

Leah Epstein[1] and Asaf Levin[2]

[1] Department of Mathematics, University of Haifa, 31905 Haifa, Israel
lea@math.haifa.ac.il
[2] Department of Statistics, The Hebrew University, Jerusalem, Israel
levinas@mscc.huji.ac.il

Abstract. Bin packing is a well studied problem which has many applications. In this paper we design a robust APTAS for the problem. The robust APTAS receives a single input item to be added to the packing at each step. It maintains an approximate solution throughout this process, by slightly adjusting the solution for each new item. At each step, the total size of items which may migrate between bins must be bounded by a constant factor times the size of the new item. We show that such a property cannot be maintained with respect to optimal solutions.

1 Introduction

Consider the classical online bin packing problem where items arrive one by one and are assigned irrevocably to bins. Items have sizes bounded by 1 and are assigned to bins of size 1 so as to minimize the number of bins used. The associated offline problem assumes that the complete input is given in advance.

We follow [12] and allow the "online" algorithm to change the assignment of items to bins whenever a new item arrives, subject to the constraint that the total size of the moved items is bounded by β times the size of the arriving item. The value β is called the *Migration Factor* of the algorithm. We call algorithms that solve an offline problem in the traditional way *static*, whereas algorithms that receive the input items one by one, assign them upon arrival, and can do some amount of re-packing using a constant migration factor are called *dynamic* or *robust*. An example we introduce later shows that an optimal solution for bin packing cannot be maintained using a dynamic (exponential) algorithm. Consequently, we focus on polynomial-time approximation algorithms.

In our point of view, the main advantage in obtaining an APTAS for the classical bin packing problem with a bounded migration factor is that such type of schemes possess a structure. Hence we are able to gain insights into the structure of the solution even though it results from exhaustive enumeration of a large amount of information.

Sanders, Sivadasan and Skutella [12] studied the generalization of the online scheduling problem where jobs that arrive one by one are assigned to identical parallel machines with the objective of minimizing the makespan. In their generalization they allow the current assignment to be changed whenever a new job

M. Bugliesi et al. (Eds.): ICALP 2006, Part I, LNCS 4051, pp. 214–225, 2006.

arrives, subject to the constraint that the total size of moved jobs is bounded by β times the size of the arriving job. They obtained a dynamic polynomial time approximation scheme for this problem extending an earlier polynomial time approximation scheme of Hochbaum and Shmoys [7] for the static problem. They noted that this result is of particular importance if considered in the context of sensitivity analysis. While a newly arriving job may force a complete change of the entire structure of an optimal schedule, only very limited local changes suffice to preserve near-optimal solutions.

For an input X of the bin packing problem we denote by $OPT(X)$ the minimal number of bins needed to pack the items of X, and let $SIZE(X)$ denote the sum of all sizes of items. Clearly, $OPT(X) \geq SIZE(X)$. For an algorithm $\mathcal{B}$ we denote by $\mathcal{B}(X)$ the number of bins used by $\mathcal{B}$.

It is known that no approximation algorithm for the classical bin packing problem can have a cost within a constant factor r of the minimum number of required bins for $r < \frac{3}{2}$ unless $\mathcal{P} = \mathcal{NP}$. This leads to the usage of the standard quality measure for the performance of bin packing algorithms which is the *asymptotic approximation ratio* or *asymptotic performance guarantee*. The asymptotic approximation ratio for an algorithm **A** is defined to be

$$\mathcal{R}(\mathbf{A}) = \limsup_{n \to \infty} \sup_{X} \left\{ \frac{\mathbf{A}(X)}{OPT(X)} \middle| OPT(X) = n \right\}.$$

The natural question, which was whether this measure allows to find an approximation scheme for bin packing, was answered affirmatively by Fernandez de la Vega and Lueker [3]. They designed an algorithm whose output never exceeds $(1 + \varepsilon)OPT(I) + g(\varepsilon)$ bins for an input I and a given $\varepsilon > 0$. The running time was linear in n, but depended exponentially on ε, and such a class of algorithms is considered to be an APTAS (Asymptotic Polynomial Time Approximation Scheme). The function $g(\varepsilon)$ depends only on ε and grows with $\frac{1}{\varepsilon}$.

Two later modifications simplified and improved this seminal result. The first modification allows to replace the function $g(\varepsilon)$ by 1 (i.e. one additional bin instead of some function of ε), see [17], Chapter 9. The second one by Karmarkar and Karp [10] allows to develop an AFPTAS (Asymptotic Fully Polynomial Time Approximation Scheme). This means that using a similar (but much more complex) algorithm, it is possible to achieve a running time which depends on $\frac{1}{\varepsilon}$ polynomially. The dependence on n is much worse than linear, and is not better than $\Theta(n^8)$. In this case the additive term remains $g(\varepsilon)$. Karmarkar and Karp [10] also designed an algorithm which uses at most $OPT(I) + \log^2[OPT(I)]$ bins for an input I.

Related work. The classical online problem was studied in many papers, see the survey papers of [2,1]. It was first introduced and investigated by Ullman [15]. The currently best results are an algorithm of asymptotic performance ratio 1.58889 given by Seiden [14] and a lower bound of 1.5401 [16]. From this lower bound we can deduce that in order to maintain a solution which is very close to optimal, the algorithm cannot be online in the usual sense. Several attempts were made to give a semi-online model which allows a small amount of modifications to the solution produced by the algorithm. We next review these attempts.

Gambosi, Postiglione and Talamo [5,6] introduced a model where a constant number of items (or small items grouped together) can be moved after each arrival of an item. They presented two algorithms. The first moves at most three items on each arrival and has the performance guarantee $\frac{3}{2} = 1.5$. The second algorithm moves at most seven items on each arrival and the performance guarantee $\frac{4}{3} \approx 1.33333$. The running times of these two algorithms are $\Theta(n)$ and $\Theta(n \log n)$ respectively, where n is the number of items.

Ivkovic and Lloyd [9] gave an algorithm which uses $O(\log n)$ re-packing moves (these moves are again of a single item or a set of grouped small items). This algorithm is designed to deal with departures of items as well as arrivals, and has performance guarantee $\frac{5}{4}$. Ivkovic and Lloyd [9,8] considered an amortized analysis as well, and show that for every $\varepsilon > 0$, the performance guarantee $1 + \varepsilon$ can be maintained, with $O(\log n)$ amortized number of re-packing moves if ε is seen as a constant, and with $O(\log^2 n)$ re-packing moves if the running time must be polynomial in $\frac{1}{\varepsilon}$. However, the amortized notion here refers to a situation that for most new items no re-packing at all is done, whereas for some arrivals the whole input is re-packed.

Galambos and Woeginger [4] adapted the notion of bounded space online algorithms (see [11]), where an algorithm may have a constant number of active bins, and bins that are no longer active, cannot be activated. They allow complete re-packing of the active bins. It turned out that the same lower bound as for the original (bounded space) problem holds for this problem as well, and re-packing only allowed to obtain the exact best possible competitive ratio having three active bins, instead of in the limit.

Outline. We review the adaptation of the algorithm of Fernandez de la Vega and Lueker [3], as it appears in [17], in Section 2. We then state some further helpful adaptations that can be made to the static algorithm. In Section 3 we describe our dynamic APTAS and prove its correctness. This algorithm uses many ideas from [3], however the adaptation into a dynamic APTAS requires careful changes to the scheme. We show that the number of bins used by our APTAS never exceeds $(1 + \varepsilon)OPT(X) + 1$ where X is the list of items that has been considered so far. The running time is $O(n \log n)$ where n is the number of items, since the amount of work done upon arrival of an item is a function of ε times $\log n$. In the full version we show an example in which there is no optimal solution that can be maintained with a constant migration factor.

2 Preliminaries

We review a simple version of the very first asymptotic polynomial time approximation scheme. This is the algorithm of Fernandez de la Vega and Lueker [3]. The algorithm is *static*, i.e., it considers the complete set of items in order to compute the approximate solution. Later in this section we adapt it and present another version of it, which is still static. This new version is used in our dynamic APTAS that is presented in the next section.

We are given a value $0 < \varepsilon < 1$ such that the asymptotic performance guarantee should be at most $1 + \varepsilon$, and $\frac{1}{\varepsilon}$ is an integer. Consider an input I for the bin packing problem. We define an item to be *small*, if its size is smaller than $\frac{\varepsilon}{2}$. Other items are *large*.

Algorithm FL [3]

1. Items are partitioned into two sets according to their size. The multiset of large items is denoted L and the multiset of small items is denoted T. We have $I = L \cup T$.

2. A linear grouping is performed to the large items. Let n be the number of large items in the input ($n = |L|$), and let $a_1 \geq \ldots \geq a_n$ be these items. Let $m = \frac{2}{\varepsilon^2}$. We partition the sorted set of large items into m consecutive sequences S_j ($j = 1, \ldots, m$) of $k = \lceil \frac{n}{m} \rceil = \lceil \frac{n\varepsilon^2}{2} \rceil$ items each (to make the last sequence be of the same cardinality, we define $a_i = 0$ for $i > n$). I.e., $S_j = \{a_{(j-1)k+1}, \ldots, a_{(j-1)k+k}\}$ for $j = 1, 2, \ldots, m$. For $j \geq 2$, we define a modified sequence $\hat{S}_j$ which is based on the sequence S_j as follows. $\hat{S}_j$ is a multiset which contains exactly k items of size $a_{(j-1)k+1}$, i.e., all items are rounded up to the size of the largest element of S_j. The set S_1 is not rounded and therefore $\hat{S}_1 = S_1$. Let L' be the union of all multisets $\hat{S}_j$ and let $L'' = \bigcup_{j=2}^{m} \hat{S}_j$.

3. The input L'' is solved optimally.

4. A packing of the complete input is obtained by first replacing the items of $\hat{S}_j$ in the packing by items of S_j (the items of S_j are never larger than the items of $\hat{S}_j$, and so the resulting packing is feasible), and second, using k bins to pack each item of S_1 in a separate bin. Last, the small items are added to the packed bins (with the original items without rounding) using Any Fit Algorithm. Additional bins can be opened for small items if necessary. Step 3 can be executed in polynomial time by solving an integer programming in a fixed dimension.

Lemma 1. *Algorithm FL is a polynomial time algorithm.*

We next analyze the performance guarantee of Algorithm FL. For two multisets A, B, we say that A is *dominated* by B and denote $A \leq B$ if there exists an injection $h : A \rightarrow B$ such that for all $a \in A$, $h(a) \geq a$.

Lemma 2. *If A and B are multisets such that $A \leq B$, then $OPT(A) \leq OPT(B)$.*

Theorem 1. *Algorithm FL is an APTAS.*

Algorithm Revised FL: We now design a new static adaptation Algorithm FL which is later generalized into a dynamic APTAS. We modify only Step 2 as follows:

- The multisets S_j are defined similarly to before, with the following changes. The multisets do not need to have the same size, but their cardinalities need to be monotonically non-increasing. I.e., for all j, $|S_j| \geq |S_{j+1}|$. Moreover, we require that if $n\varepsilon^2 \geq 8$, then $|S_1| \leq \lceil \frac{\varepsilon^2 n}{2} \rceil$ and $|S_m| \geq \frac{|S_1|}{4}$, and otherwise each set has a single element.

- The rounding is done as follows. Given a multiset S_j which consists of elements $c_1 \geq \ldots \geq c_k$, the elements are rounded up into two values. Let $1 < s \leq k$, then all elements $c_1, \ldots, c_{s-1}$ are rounded into c_1, and the elements $c_s, \ldots, c_k$ are rounded into c_t for some $2 \leq t \leq s$.

The proof of Theorem 1 extends easily to this adaptation as well. Specifically, the amount of distinct sizes in the rounded instance is constant (which depends on ε) and so is the number of patterns. The mapping is defined similarly, and the set S_1 still satisfies $|S_1| \leq \lceil \frac{\varepsilon^2 n}{2} \rceil$ so it is small enough to be packed into separate bins. If the small items which are are added using Any Fit cause the usage of additional bins, the situation is exactly the same as before. Therefore, we establish the following theorem.

Theorem 2. *Algorithm Revised FL is an APTAS.*

In the sequel we show how to maintain the input grouped as required by Algorithm Revised FL. We also show that the difference between packing of two subsequent steps is small enough that it can be achieved by using a constant (as a function of ε) migration factor as in [12].

3   APTAS with $f(\frac{1}{\varepsilon})$ Migration Factor

In this Section we describe our dynamic APTAS for bin packing with the additional property that the migration factor of the scheme is $f(\frac{1}{\varepsilon})$ (i.e., it is a function f of the term $\frac{1}{\varepsilon}$), and therefore a constant migration factor for fixed value of ε.

We use the following notations. The number of large items among the first t items is denoted $n(t)$. The size of the ith arriving item is b_i. The value m is defined as in the previous section $m = \frac{2}{\varepsilon^2}$. We denote by $OPT(t)$ the number of bins used by an optimal solution for the first t items. We assume that after the first t items, we have a feasible solution that uses at most $(1+\varepsilon)OPT(t)+1$ bins and show how to maintain such a solution after the arrival of a new item of index $t+1$. Later on we describe several structural properties that our solution satisfies, and show how to maintain these properties as well. These structural properties help us to establish the desired migration factor.

Similarly to Algorithm Revised FL we treat small items and large items differently. Recall that an item is small, if its size is smaller than $\frac{\varepsilon}{2}$. When a small item arrives, we use Any Fit Algorithm to find it a suitable bin.

It remains to consider the case where the $t+1$-th item is a large item, i.e., $b_{t+1} \geq \frac{\varepsilon}{2}$. We keep the following structural properties throughout the algorithm.

1. If $n(t) \geq 4m+1$ then the sorted list of large items which arrived so far is partitioned into $M(t)$ consecutive sequences $S_1(t), \ldots, S_{M(t)}(t)$, where $4m+1 \leq M(t) \leq 8m+1$ such that $|S_1(t)| \geq |S_2(t)| = |S_3(t)| = \cdots = |S_{M(t)}(t)|$.

Otherwise, if $n(t) \leq 4m$ then the sorted list of large items which arrived so far is partitioned into $n(t)$ consecutive sequences each of them has a single large element $S_1(t), \ldots, S_{n(t)}(t)$.

2. Denote $K(t) = |S_{M(t)}(t)|$, then $|S_1(t)| \leq 4K(t)$.
3. There are two special subsets of $S_1(t)$ denoted by $S'_1(t)$ and $S''_1(t)$ (these sets might be empty at some times). Each of these two sets contains at most $K(t)$ items. The sets are special in the sense that we treat them as separate sets while rounding, and in the rounded up instance $S'_1(t)$ and $S''_1(t)$ are rounded. However in the analysis the cost of the solution is bounded allowing each element of $S_1(t)$ to be packed in its own bin.

The time index t can be omitted from the notation of a set or a parameter if the time it belongs to is clear from the context. Therefore, when e.g. we discuss the set S_1 this is the set $S_1(t)$ that is associated with the discussed time. The size of the set S_1 is defined according to algorithm Revised FL. We have $n \geq MK > 4mK$, and therefore $S_1 \leq 4K < \frac{n}{m} = \frac{n\varepsilon^2}{2}$.

Note that for $n \leq 4m$ each element has its own list, and therefore we solve optimally the instance of the large items excluding the largest item.

We turn now to discuss *steps*, where each step is an arrival of a large item. We partition the steps into three types, which are, *regular steps*, *creation steps* and *union steps*. An *insertion of a new large item* operation takes place in all types of steps. A *creation of new sets* operation takes place only in creation steps. A *union of sets* operation takes place only in union steps.

We also maintain the following property. During regular steps or creation steps, no set is rounded to a pair of values but for each set S_j $(j \geq 2)$, the items of S_j are rounded up to the largest size of any element of S_j. However, during union steps, each set is rounded up to a pair of values (as in Algorithm Revised FL).

Lemma 3. *The arrival of a small item and allocating it according to Any Fit Algorithm maintains the structural properties.*

We next define a series of operations on the sequences so the properties are maintained after a new item arrives, and the migration factor of the resulting solution is constant. When a new item arrives, we first apply the *insertion of a new large item* operation. Afterwards if the current step is a creation step, we apply the *creation of new sets* operation, whereas if the current step is a union step, we apply the *union of sets* operation.

When we bound the migration factor, note that changes to the allocation of items into bins are made only when large items arrive. Hence, the size of the arriving item of index $t + 1$ is at least $\frac{\varepsilon}{2}$. Therefore, if we can prove that the allocation is changed only for a set of items that we allocated to a fixed number of bins (their number is a function of ε) then we get a constant migration factor throughout the algorithm.

Insertion of a new large item. When a large item arrives then if $n \leq 4m + 1$ we add a new list that contains the new item as its unique element. Note that in this case the resulting set of lists satisfies the structural properties. Otherwise (i.e., $n \geq 4m + 2$) we first compute the list to which it belongs, and add it there. The list to which it belongs, S_j, is defined as follows. If the new item is larger than an existing item in S_1, then this list is S_1. Otherwise, we find the set S_{j+1} of smallest index such that the new item is larger than all its elements, and the list to which the item belongs is S_j. We *move up* items from S_i to S_{i-1} for all $2 \leq i \leq j$, this operations is defined as follows. We move the largest item of S_i to S_{i-1} and afterwards we change the value which the size of the items of S_i is rounded up to, into the size of the new largest item of S_i. When we consider the effect this operation has on the feasibility of the integer program, we can see that the right hand side does not change (the size of S_r is not affected for all values of r such that $2 \leq r \leq M$), however new patterns arise as the size of the rounded up instance is smaller (and so we can pack more items to a bin in some cases). The additional patterns mean new columns of the feasibility constraint matrix. Note that adding the new large item into its list takes $O(\log n)$ time (as we need to maintain sorted lists of the large elements).

Theorem 3. *[Corollary 17.2a, [13]] Let A be an integral $m \times d$ matrix such that each sub-determinant of A is at most Δ in absolute value, let $\hat{u}$ and u' be column m-vectors, and let v be a row d-vector. Suppose $\max\{vx | Ax \leq \hat{u}; x \text{ is integral}\}$ and $\max\{vx | Ax \leq u'; x \text{ is integral}\}$ are finite. Then, for each optimum solution y of the first maximum there exists an optimum solution y' of the second maximum such that $||y - y'||_\infty \leq d\Delta (||\hat{u} - u'||_\infty + 2)$.*

Lemma 4. *Let A be the constraint matrix of the feasibility integer program. Let d be the number of columns of A and let Δ be the maximum value in absolute value of a sub-determinant of A. Then, throughout the algorithm $d \cdot \Delta$ is bounded by a constant (for a fixed value of ε).*

The proof of the following lemma is similar to the analysis of [12].

Lemma 5. *Assume that before the arrival of the current large item, there is a feasible solution y to the feasibility integer program of the rounded up instance L''. After we apply an insertion of a new large item operation, if the feasibility integer program is feasible then there is a solution to it y' such that it is suffices to re-pack the items that reside in a constant number of bins.*

Proof. Denote by $Ay = u$ the constraints of the feasibility integer program of the rounded up instance. The matrix A is the constraint matrix and u is the right hand side. Note that the columns of A correspond to patterns and the rows correspond to different sizes of items. The constraint that corresponds to an item size a has the following meaning. The amount of the items that we allocate along all possible patterns of items with size a is exactly the number of items in the rounded up instance with size a.

We first assume that $n(t) \geq 4m + 1$. Now consider the change in the constraint matrix A when a new large item of size b_{t+1} arrives. Let A' denote the modified A, which is the feasibility constraint matrix for all the items in L'' and the one extra new element of S_1. First, the cardinality of S_1 increases by 1, (at the end of the move up operation), and this does not change the constraint matrix as we do not have a constraint associated with S_1. Next, we consider the decrease in the size of rounded up items. Note that all the patterns that were feasible in the previous stage clearly remain feasible (given a set of items that can be packed in a single bin, decreasing the size of some of the items still allows to pack them into a single bin). Therefore, the matrix A of the previous stage is a sub-matrix of the matrix after we apply the insertion of a new large item operation. The difference is a possible addition of columns (that correspond to patterns that were infeasible before we decrease the size of some items and before we add the new item, however these patterns are now feasible).

To bound the change in u, denote the new right hand side by u'. The set S_1 is not represented in A, therefore there is no change and $u' = u$. Let $\hat{u} = u$. Then, in the case where $n(t) \geq 4m + 1$ the change in the right hand side is bounded by a constant, i.e., $||u' - \hat{u}||_\infty = 0$.

Otherwise, $n(t) \leq 4m$ and the feasibility integer program has a new row that corresponds to the new large item and whose right hand side value is 1. Note that all the patterns that were feasible in the previous stage clearly remain feasible. Therefore, the matrix A of the previous stage is a sub-matrix of the matrix after we apply the insertion of a new large item operation. The difference is a possible addition of columns that pack the new item as well, and a new row that corresponds to the new item. To bound the change in u, denote the new right hand side by u'. Let $\hat{u}$ be a right hand side equal to u' beside one entry that is in the component that correspond to the new row of A where it equals zero. In this case $||u' - \hat{u}||_\infty = 1$. In the remainder of this proof we do not distinguish between the case where $n \geq 4m + 1$ and the case where $n \leq 4m$.

We can extend y to a vector $\hat{y}$ that is a feasible solution of $A'\hat{y} = \hat{u}$. To do so, we define the entries of y that correspond to the new columns in A' compared to A to be zero. In the other components (whose columns exist in A) the value of $\hat{y}$ is exactly the value of y.

In order to prove the claim it is enough to show that there is a feasible solution y' such that $||\hat{y} - y'||_\infty$ is a constant (then we re-pack the items from the bins that correspond to the difference between $\hat{y}$ and y'). Recall that we assume that $Ay = \hat{u}$ is feasible integer program. Therefore, the assumptions of Theorem 3 are satisfied. Therefore, by Theorem 3, there is a feasible integer solution y' such that $||\hat{y} - y'||_\infty \leq d\Delta (||\hat{u} - u'||_\infty + 2)$. We would like to bound by a constant the right hand side of the last inequality.

By Lemma 4, d and Δ are bounded by a constant. We have already bounded $||\hat{u} - u'||_\infty$ by the constant 1 and this completes the proof.   □

Therefore, the moving up operation causes constant migration. However in order to prove the performance guarantee of the algorithm we need to show how to maintain the structural properties. To do so, note that the only sets whose cardinality increases during the moving up operation is S_1, and therefore we need to show how to deal with cases where S_1 is too large.

Creation of new sets: After S_1 has exactly $3K$ items we start a new operation that we name *creation of new sets* that lasts for K steps (in each such step a new large item arrives and we charge the operation done in the step to this new item). We consider the items $c_1 \geq c_2 \geq \cdots \geq c_{3K}$ of S_1. We create new sets S_1' and S_1'' where eventually $S_1' = \{c_{K+1}, \ldots, c_{2K}\}$ and $S_1'' = \{c_{2K+1}, \ldots, c_{3K}\}$. In each step we will have already rounded i items from S_1' and from S_1'' to its target value and i is increased by 1 each step. So after i steps of this creation of new sets operation, the rounded up instance has i copies of the items c_{K+1} and c_{2K+1}. Then, the resulting instance can be solved in polynomial time where we put each item of S_1 that has not already rounded up to either c_{K+1} or to c_{2K+1} in a separate bin. Rounding up two items at each step results in a constant change in the right hand side of the feasibility integer programming and therefore increase the migration factor within an additive constant factor only as we prove in Lemma 6 below. At the end of K steps we declare the sets S_1' and S_1'' as the new S_2 and S_3 and increasing M by two. Each of the new sets contains exactly K elements and the new S_1 contains exactly $2K$ elements, and therefore if $M(t) \leq 8m - 1$ we are done while keeping a constant migration factor. Otherwise, next time the creation new sets operation takes place we will violate the second structural property, and therefore we currently initiate the *union of sets* operation that lasts for K steps as well. Note that we never apply both the creation of new sets and the union of sets operations at the same step.

The moving up procedure during the insertion of a new large item operation will increase S_1 further, and this case we also apply this procedure to S_1' and S_1'' and thus decreasing the value to which we round up the items that belong to S_1' and S_1''. So in fact during the creation of new sets operation S_1 is partitioned into five sets $S_1^1, S_1', S_1^2, S_1'', S_1^3$ such that the items in S_1^1 are the largest items of S_1, S_1^2 contains the items with size that is smaller than the items in S_1' but they are larger than the items in S_1'', and S_1^3 contains the other items of S_1. Then, in each step we increase the size of S_1^1, S_1' and S_1'' by one item each, whereas the size of S_1^2 and S_1^3 is decreased by one. This means that during the *move up* operation we will move up items also in these collection of the five subsets of S_1.

Lemma 6. *Assume that before we apply the creation of new sets operation and after we finish the insertion of a new large item operation, there is a feasible solution y' to the feasibility integer program of the rounded up instance L''. After we apply the creation new sets operation, if the feasibility integer program is feasible then there is a solution to it y'' such that it is enough to re-pack the items that reside in a constant number of bins.*

Union of sets: When the number of sets reaches $8m + 1$ we start the following operation that lasts for K steps (where again a step means an arrival of a large item). First we declare each pair of consecutive sets as a new set. That is for $2 \leq j \leq 4m + 1$ we let $S_j(t + 1)$ be $S_{2j-1}(t) \cup S_{2j-2}(t)$, but we still do not change the way the rounding is performed. So in the resulting partition into sets, each set has exactly $2K$ items, and we declare this the new value of K, i.e., $K(t+1) = 2K(t)$, and the number of sets now is $M = 4m + 1$. However, each set is rounded to a pair of values, and this is something we will recover in the next steps. While the rounding of each set is to two values, denote by S_j' the items that we round to the largest item of S_j and by $S_j'' = S_j \setminus S_j'$. In each step for all $j \geq 2$ we move the largest item of S_j'' to S_j'. Thus, we round this moved item to the largest element of S_j and do not change the value to which we round up the items of S_j''. Both these changes do not increase the migration factor by much as we prove below in Lemma 7. At the end of this procedure we end up with a collection of subsets each rounded to a common value and finish the union of sets operation.

Lemma 7. *Assume that before we apply the union of sets operation and after we finish the insertion of a new large item operation, there is a feasible solution y' to the feasibility integer program of the rounded up instance L''. After we apply the union of sets operation, if the feasibility integer program is feasible then there is a solution to it y'' such that it is enough to re-pack the items that reside in a constant number of bins.*

We now describe the algorithm we apply each time a new item arrives denoted as **Algorithm Dynamic APTAS**. If the new arriving item is small then we use Any Fit Algorithm to pack it into an existing bin or open a new bin for it if it cannot fit into any other existing bin (when we need to pack a small item, we consider the original sizes of large items that are already packed and not their rounded sizes). In the case where the new small items causes an addition of a new bin, we maintain the solution to the feasibility integer program by adding a bin whose pattern is the *empty pattern* that does not pack any large item.

Otherwise, the item is large and we are allowed (while still getting a constant migration factor) to re-pack the items of a constant number of bins. We consider the optimal rounded-up solution y before the current item has arrived. Note that y is the integer solution after the previous large item was added with the possibility of introducing empty patterns bins in case we *open* new bins to small items.

Then, after we apply the insertion of a new large item operation, we use Lemma 5 and look for a feasible packing of the resulting new rounded-up instance, y' that is close to y (i.e., the norm infinity of their difference is at most a constant that is given by the proof of Lemma 5). The restriction that y' is close to y is given by linear inequalities, and therefore we can solve the resulting feasibility integer program in polynomial time (for fixed value of ε). If the resulting integer program is infeasible, then we must open a new bin for the new rounded up instance, and then we can put the new item in the new bin and the rest of the items as they were packed in the solution before the insertion of a new large item operation occurs. Otherwise, we obtain such an integer solution y' that is close to y.

Similarly, if we need to apply either the creation of new sets operation or the union the sets operation we construct a solution y'' that is close to y' (the norm infinity of their difference is a constant). This restriction is again given by linear inequalities and therefore we can solve the resulting feasibility integer program in polynomial time (for fixed value of ε). If the resulting integer program is infeasible, then we must open a new bin for the new rounded up instance, and then we can put the new item in the new bin and the rest of the items as they were packed in the solution before the insertion of a new large item operation occurs. Otherwise, we obtain such an integer solution y'' that is close to y. If we did not apply the creation of new sets operation or the union the sets operation, then we denote $y'' = y'$.

Note that during creation steps our algorithm packs the items from $S_1' \cup S_1''$ according to their packing in the feasibility integer program. I.e., the integer program has a row for S_1' and a row for S_1''. However, in the analysis of the performance guarantee of the algorithm in Theorem 4 below we bound the cost of the solution by a different solution that packs each item of S_1 in its own bin (this is also with for the items of $S_1' \cup S_1''$). Such a solution is not better than our resulting solution as it corresponds to a solution to the integer program that choose such patterns for its covering of the elements of $S_1' \cup S_1''$.

It remains to show how to construct a solution to the bin packing instance using the vector y''. For each pattern we change the packing of $\max\{y_p - y_p'', 0\}$ bins that were packed according to pattern p. For pattern p we select such bins arbitrarily (from the bins that we pack according to pattern p). We complete the packing of the items to a packing that correspond to y''. This will pack all the large items and all the small items that were not packed in the bins we decided to re-pack. Then, we apply Any Fit Algorithm for the small unpacked items.

This algorithm re-packs a constant number of bins in case a large item arrives and it does not change the packing in case a small item arrives.

Corollary 1. *Algorithm Dynamic APTAS has a constant migration factor for fixed value of ε.*

The proof of the following theorem is based on the analysis of Algorithm Revised FL.

Theorem 4. *Algorithm Dynamic APTAS is a polynomial time algorithm that has a constant migration factor and uses at most $(1 + \varepsilon)OPT(t) + 1$ bins after t items arrives.*

Remark 1. The feasibility integer program to find vectors y' and y'' (in the notations of the algorithm), which are close to y, can be solved by only one integer program of a fixed size.

4 Concluding Remarks

A similar approach allows to prove the following result. Given bins of size $1 + \varepsilon$ instead of size 1, it is possible to design a dynamic algorithm which uses at time t, at most $OPT(t)$ bins to pack the items. The algorithm works as follows. Items are partitioned into large (at least $\frac{\varepsilon}{4}$) and small (all other items). The sizes of large items are rounded up into powers of $1 + \frac{\varepsilon}{4}$. This rounding is permanent and the original size is ignored in all steps of the algorithm. Feasible patterns of large items are defined similarly to [12]. At each arrival of a large item we check whether the previous amount of bins still allows a feasible solution. It is possible to show that in such a case, a limited amount of re-packing is needed. Otherwise the new item is packed in a new bin. Small items are packed greedily using Any Fit Algorithm.

A question that is left open is whether there exists a robust AFPTAS for the classical bin packing problem. It would be interesting to find out which other problems can benefit from the study of robust approximation algorithms.

References

1. E. G. Coffman, M. R. Garey, and D. S. Johnson. Approximation algorithms for bin packing: A survey. In D. Hochbaum, editor, *Approximation algorithms*. PWS Publishing Company, 1997.
2. J. Csirik and G. J. Woeginger. On-line packing and covering problems. In *A. Fiat and G. J. Woeginger, editors,* Online Algorithms: The State of the Art, pages 147–177, 1998.
3. W. Fernandez de la Vega and G. S. Lueker. Bin packing can be solved within $1 + \varepsilon$ in linear time. *Combinatorica*, 1:349–355, 1981.
4. G. Galambos and G. J. Woeginger. Repacking helps in bounded space online bin packing. *Computing*, 49:329–338, 1993.
5. G. Gambosi, A. Postiglione, and M. Talamo. On-line maintenance of an approximate bin-packing solution. *Nordic Journal on Computing*, 4(2):151–166, 1997.
6. G. Gambosi, A. Postiglione, and M. Talamo. Algorithms for the relaxed online bin-packing model. *SIAM Journal on Computing*, 30(5):1532–1551, 2000.
7. D. S. Hochbaum and D. B. Shmoys. Using dual approximation algorithms for scheduling problems: theoretical and practical results. *Journal of the ACM*, 34(1):144–162, 1987.
8. Z. Ivkovic and E. L. Lloyd. Partially dynamic bin packing can be solved within 1 + ε in (amortized) polylogarithmic time. *Information Processing Letters*, 63(1):45–50, 1997.
9. Z. Ivkovic and E. L. Lloyd. Fully dynamic algorithms for bin packing: Being (mostly) myopic helps. *SIAM Journal on Computing*, 28(2):574–611, 1998.
10. N. Karmarkar and R. M. Karp. An efficient approximation scheme for the one-dimensional bin-packing problem. In *Proceedings of the 23rd Annual Symposium on Foundations of Computer Science (FOCS'82*, pages 312–320, 1982.
11. C. C. Lee and D. T. Lee. A simple online bin packing algorithm. *Journal of the ACM*, 32(3):562–572, 1985.
12. P. Sanders, N. Sivadasan, and M. Skutella. Online scheduling with bounded migration. In *Proc. of the 31st International Colloquium on Automata, Languages and Programming (ICALP2004)*, pages 1111–1122, 2004.
13. A. Schrijver. *Theory of Linear and Integer Programming*. John Wiley & Sons, 1986.
14. S. S. Seiden. On the online bin packing problem. *Journal of the ACM*, 49(5):640–671, 2002.
15. J. D. Ullman. The performance of a memory allocation algorithm. Technical Report 100, Princeton University, Princeton, NJ, 1971.
16. A. van Vliet. An improved lower bound for online bin packing algorithms. *Information Processing Letters*, 43(5):277–284, 1992.
17. V. V. Vazirani. *Approximation Algorithms*. Springer-Verlag, 2001.

Better Inapproximability Results for MaxClique, Chromatic Number and Min-3Lin-Deletion

Subhash Khot[*] and Ashok Kumar Ponnuswami

College of Computing
Georgia Institute of Technology
Atlanta GA 30309, USA
{khot, pashok}@cc.gatech.edu

Abstract. We prove an improved hardness of approximation result for two problems, namely, the problem of finding the size of the largest clique in a graph and the problem of finding the chromatic number of a graph. We show that for any constant $\gamma > 0$, there is no polynomial time algorithm that approximates these problems within factor $n/2^{(\log n)^{3/4+\gamma}}$ in an n vertex graph, assuming $\mathrm{NP} \not\subseteq \mathrm{BPTIME}(2^{(\log n)^{O(1)}})$. This improves the hardness factor of $n/2^{(\log n)^{1-\gamma'}}$ for some small (unspecified) constant $\gamma' > 0$ shown by Khot [20]. Our main idea is to show an improved hardness result for the Min-3Lin-Deletion problem.

An instance of Min-3Lin-Deletion is a system of linear equations modulo 2, where each equation is over three variables. The objective is to find the minimum number of equations that need to be deleted so that the remaining system of equations has a satisfying assignment. We show a hardness factor of $2^{\Omega(\sqrt{\log n})}$ for this problem, improving upon the hardness factor of $(\log n)^{\beta}$ shown by Håstad [18], for some small (unspecified) constant $\beta > 0$. The hardness results for clique and chromatic number are then obtained using the reduction from Min-3Lin-Deletion as given in [20].

1 Introduction

A clique in a graph is a subset of vertices such that any pair of vertices in the subset is connected by an edge. MaxClique is the problem of finding the size of the largest clique in a graph. It has been a pivotal problem in the field of inapproximability, leading to the development of many important tools in this field.

The best approximation algorithm for MaxClique was given by Feige [11]. The algorithm achieves an approximation factor of $O(\frac{n(\log \log n)^2}{\log^3 n})$, where n is the number of vertices in the input graph. It was conjectured that the Lovász θ-function might be a $O(\sqrt{n})$ approximation for MaxClique (see [22] for details). Since the Lovász θ-function can be computed to any desired degree of accuracy in polynomial time, the conjecture implies a $O(\sqrt{n})$ approximation algorithm for

---

[*] The research is partly supported by the Microsoft New Faculty Fellowship.

M. Bugliesi et al. (Eds.): ICALP 2006, Part I, LNCS 4051, pp. 226–237, 2006.

MaxClique. For perfect graphs, Lovász θ-function equals the size of the largest clique. For random graphs, the gap between these two values can be as bad as $\Omega(\sqrt{n}/\log n)$. The conjecture says that this may essentially be the worst possible gap. Feige [10] disproved the conjecture by showing that the Lovász θ-function does not approximate MaxClique better than $\frac{n}{2^{\sqrt{c\log n}}}$, where $c > 0$ is a constant.

The first inapproximability result for MaxClique was obtained by Feige et al. [12] who discovered the connection between hardness of approximation and Probabilistically Checkable Proofs(PCPs). We summarize the progress on showing hardness results for MaxClique in Table 1. Let $PCP_{c,s}(r(n), q(n))$ denote the class of languages that have a non-adaptive verifier with the following properties. For an input string of length n, the verifier uses $r(n)$ random bits and queries $q(n)$ from the proof. If the input belongs to the language, there is a correct proof that is accepted with probability c. Otherwise, no proof is accepted with probability more than s. Feige et al. [12] showed that NP $\subseteq$ $PCP_{1,1/2}(O(\log n \log \log n), O(\log n \log \log n))$. Arora and Safra [3] and Arora et al. [2] improved this result to show that NP $\subseteq PCP_{1,1/2}(O(\log n), O(1))$, a result known as the *PCP Theorem*. Since then, many different PCP constructions for languages in NP have led to inapproximability results for several other problems in addition to MaxClique.

Bellare and Sudan [6] defined a parameter called *amortized free bits* for PCPs. They showed that if problems in NP have PCPs that use logarithmic randomness and $\bar{f}$ amortized free bits, then MaxClique is hard to approximate within a factor of $n^{1/(1+\bar{f})-\epsilon}$ unless NP $\subseteq$ ZPP. They constructed PCPs with $3 + \delta$ amortized free bits for arbitrarily small $\delta > 0$. This implies a hardness factor of $n^{1/4-\epsilon}$ for MaxClique. The result was improved by Bellare et al. [4] by constructing PCPs with $2 + \delta$ amortized free bits. Finally, Håstad [16] gave a construction that achieved an amortized free bit complexity of δ for any constant $\delta > 0$, proving $n^{1-\epsilon}$ hardness for MaxClique. Simpler proofs of Håstad's result were given by Samorodnitsky and Trevisan [24] and Håstad and Wigderson [19]. Both these results achieved amortized free bit complexity δ and *amortized query complexity* $1 + \delta$ for any constant $\delta > 0$ (both parameters are optimal).

Khot [20] showed that MaxClique cannot be approximated within a factor of $\frac{n}{2^{(\log n)^{1-\gamma'}}}$ for some small constant $\gamma' > 0$, assuming NP $\not\subseteq$ ZPTIME($2^{(\log n)^{O(1)}}$). We believe that it is an important open problem whether inapproximability of MaxClique can be improved to $\frac{n}{2^{O(\sqrt{\log n})}}$. As mentioned before, Feige [10] showed that the Lovász θ-function can have an approximation ratio as bad as $\frac{n}{2^{c\sqrt{\log n}}}$ for some constant $c > 0$. It would be interesting to prove the same lower bound for *any* polynomial time algorithm. It would also fit in nicely with Trevisan's [27] lower bound of $\frac{d}{2^{O(\sqrt{\log d})}}$ for MaxClique on degree d graphs (d thought of as a large constant). Blum [7] showed that if there exists a factor $\frac{n}{2^{\sqrt{b\log n}}}$ quasi-polynomial time approximation algorithm for MaxClique, then there exists a quasi-polynomial time algorithm to color a 3-colorable graph with n^ϵ colors, where $\epsilon = O(1/b)$. Therefore, strong lower bounds for MaxClique give evidence that the graph coloring problem is hard. Another motivation comes from a result of Feige and Kogan [15] who showed that if the *balanced bipartite clique problem* can be approximated within a constant factor, then there is a $\frac{n}{2^{O(\sqrt{\log n})}}$ approximation for MaxClique. We refer to Srinivasan's paper [26] for several other interesting consequences of proving strong hardness results for MaxClique.

Table 1. Hardness Results for MaxClique

	Hardness Factor	Assumption
Feige et al. [12]	$2^{\log^{1-\epsilon} n}$, for any $\epsilon > 0$	NP $\nsubseteq$ DTIME($2^{(\log n)^{O(1)}}$)
Arora and Safra [3]	$2^{(\log n)^{1/2-\epsilon}}$	P $\neq$ NP
Arora et al. [2]	n^c, for some $c > 0$	P $\neq$ NP
Bellare et al. [5]	$n^{1/30}$	NP $\nsubseteq$ BPP
Bellare et al. [5]	$n^{1/25}$	NEXP $\nsubseteq$ BPEXP
Feige and Kilian [13]	$n^{1/15}$	NP $\nsubseteq$ coRP
Bellare and Sudan [6]	$n^{1/4-\epsilon}$	NP $\nsubseteq$ ZPP
Bellare et al. [4]	$n^{1/3-\epsilon}$	NP $\nsubseteq$ ZPP
Håstad [17]	$n^{1/2-\epsilon}$	NP $\nsubseteq$ coRP
Håstad [16]	$n^{1-\epsilon}$	NP $\nsubseteq$ ZPP
Engebretsen and	$\frac{n}{2^{O(\log n/\sqrt{\log\log n})}}$	NP $\nsubseteq$
Holmerin [9]		ZPTIME($2^{O(\log n(\log\log n)^{3/2})}$)
Khot [20]	$\frac{n}{2^{(\log n)^{1-\gamma'}}}$, for some $\gamma' > 0$	NP $\nsubseteq$ ZPTIME($2^{(\log n)^{O(1)}}$)

The chromatic number of a graph G, denoted by $\chi(G)$, is the minimum number of colors required to color the vertices of G such that for any edge, its end-points receive different colors. Feige and Kilian [14] showed the connection between *randomized PCPs* and inapproximability of chromatic number. Using this result, they prove that it is hard to approximate chromatic number within a factor better than $n^{1-\epsilon}$ for any constant $\epsilon > 0$, assuming NP $\nsubseteq$ ZPP. Khot [20] constructs a more efficient verifier and obtains a hardness factor of $\frac{n}{2^{(\log n)^{1-\gamma'}}}$ for some constant $\gamma' > 0$, assuming NP $\nsubseteq$ ZPTIME($2^{(\log n)^{O(1)}}$). We would like to emphasize that the constant γ' in Khot's hardness results for MaxClique and Chromatic Number is a non-explicit (possibly extremely tiny) constant that depends on the proof of Raz's Parallel Repetition Theorem [23].

2   Our Results and Techniques

We show the following inapproximability results for MaxClique and chromatic number, taking us closer to the goal of $\frac{n}{2^{O(\sqrt{\log n})}}$ (or even $n/\text{polylog}(n)$).

Theorem 1. *Assuming* NP $\nsubseteq$ BPTIME($2^{(\log n)^{O(1)}}$), *for any constant* $\gamma > 0$, *MaxClique on an n vertex graph cannot be approximated within a factor better than* $n/2^{(\log n)^{3/4+\gamma}}$ *by any probabilistic polynomial time algorithm.*

Theorem 2. *Assuming* NP $\nsubseteq$ ZPTIME($2^{(\log n)^{O(1)}}$), *for any constant* $\gamma > 0$, *chromatic number of an n vertex graph cannot be approximated within a factor better than* $n/2^{(\log n)^{3/4+\gamma}}$ *by any probabilistic polynomial time algorithm.*

Our main idea is to show an improved hardness factor for the Min-3Lin-Deletion problem that is defined next.

Definition 1. *Given a system of linear equations modulo 2*

$$\{a_{i0} \oplus (\bigoplus_{j=1}^{m} a_{ij}x_j) = 0\}_{i=1,2,\ldots,l}$$

as an input, Min-Lin-Deletion is the problem of finding the minimum number of equations that need to be deleted so that the remaining system of equations has a satisfying assignment. Min-3Lin-Deletion is the special case where exactly three of the coefficients $a_{i1}, a_{i2}, \ldots, a_{im}$ are non-zero for all i (that is, each equation is over exactly 3 variables). An instance of the Min-Lin-Deletion problem can be specified by a $(l, m + 1)$ matrix

$$A = \begin{bmatrix} a_{10} \; a_{11} \; \ldots \; a_{1m} \\ a_{20} \; a_{21} \; \ldots \; a_{2m} \\ \vdots \quad \vdots \qquad \vdots \\ a_{l0} \; a_{l1} \; \ldots \; a_{lm} \end{bmatrix}$$

In this case, we say that A is a (l, m)-Min-Lin-Deletion instance. We refer to the minimum fraction of equations that need to be deleted to find a satisfying assignment as the optimum of A, denoted by $Opt(A)$. That is, $Opt(A)$ is the minimum possible fraction of 1s in AX, where the minimum is taken over all vectors $X = (1, x_1, \ldots, x_m)$.

Min-Lin-Deletion-(c, s) is the problem of deciding whether the optimum of the input is at most c or at least s (we let c and s depend on the size of the input). The parameters c and s are called the completeness and soundness of this problem.

We say a Min-Lin-Deletion instance is k-restricted if each equation is over at most k variables. We say a Min-Lin-Deletion instance is k-regular if every variable appears in exactly k equations.

All instances of Min-Lin-Deletion considered in this paper have the property that the maximum number of variables in an equation is at most the number of linear equations in that instance. Therefore, for simplicity, we assume that the size of a Min-Lin-Deletion instance is the number of equations in it.

The following theorem was shown by Håstad [18].

Theorem 3. *For any constants $\epsilon, \delta > 0$, there exists a polynomial time algorithm $\mathcal{A}_1$ that when given a 3SAT formula ϕ of size n produces a Min-3Lin-Deletion instance A_1 of size $N_1 = n^{O(1)}$ such that:*

- *(Yes Case:) If ϕ is satisfiable, then there exists an assignment that satisfies all but at most ϵ fraction of the equations. That is, $Opt(A_1) \le \epsilon$.*
- *(No Case:) If ϕ is not satisfiable, then no assignment satisfies more than $1/2 + \delta$ fraction of the equations. That is, $Opt(A_1) \ge 1/2 - \delta$.*

Håstad also showed that the theorem holds with $\epsilon = \delta = (\log N_1)^{-\beta}$ for some (tiny) constant $\beta > 0$ if N_1 and the running time of the reduction are allowed to be slightly super-polynomial in n. In particular Min-3Lin-Deletion-$((\log N_1)^{-\beta}, 0.4)$ is hard. This is the starting point for Khot's [20] hardness results for clique and chromatic number. Our main contribution is the following improved hardness result for Min-3Lin-Deletion. This in turn implies improved hardness results for MaxClique and chromatic number:

Theorem 4. *There exists a $2^{O(\log^2 N_1)}$ time algorithm $\mathcal{A}$ that when given a Min-3Lin-Deletion instance $\boldsymbol{A_1}$ of size N_1 outputs a 7-regular Min-3Lin-Deletion instance $\boldsymbol{A}$ of size $N = 2^{O(\log^2 N_1)}$ such that:*

- *(Yes Case:) If $Opt(\boldsymbol{A_1}) \leq 0.1$, then $Opt(\boldsymbol{A}) \leq 2^{-\Omega(\sqrt{\log N})}$.*
- *(No Case:) If $Opt(\boldsymbol{A_1}) \geq 0.4$, then $Opt(\boldsymbol{A}) \geq \Omega(\log^{-3} N)$.*

2.1   Reduction from Min-3Lin-Deletion to MaxClique

We briefly explain here how the improved hardness result for the Min-3Lin-Deletion problem leads to improved hardness results for MaxClique (and similarly for chromatic number). Khot's [20] reduction from Min-3Lin-Deletion to MaxClique proceeds in two steps. First, the Min-3Lin-Deletion instance is reduced to the so-called *Raz Verifier* and a PCP is built on top of the Raz Verifier. Then, the hardness result for MaxClique follows from the PCP construction using known techniques.

The strength of the hardness result for clique depends directly on the strength of the Raz Verifier. To be precise, one would like to have a Raz Verifier with as low soundness as possible, without losing *much* in completeness. Khot [20] starts with a size N instance of Min-3Lin-Deletion-$((\log N)^{-\beta}, 0.4)$ which is shown to be hard by Håstad [18]. The Raz Verifier is obtained via Parallel Repetition of a certain protocol constructed from the Min-3Lin-Deletion instance. If u is the number of repetitions, then the soundness of the Raz Verifier is $2^{-\Omega(u)}$. Thus the soundness can be lowered by taking u large enough. However, the completeness of the Raz Verifier suffers with parallel repetition. The completeness of the Min-3Lin-Deletion instance is $(\log N)^{-\beta}$ and this limits u to be at most $(\log N)^{\beta}$. Note that $\beta > 0$ is a tiny constant.

On the other hand, we start with the Min-3Lin-Deletion-$(2^{-\Omega(\sqrt{\log N})}, \Omega(\log^{-3} N))$ instance given by Theorem 4. The completeness is good enough so that we may take up to $u = 2^{\Omega(\sqrt{\log N})}$ repetitions (we however take much fewer repetitions since we do not want to blow up the size of the Raz Verifier). For some fixed constant c_0, the soundness of the Raz Verifier is $(1 - (1/\log^3 N)^{c_0})^u$, which is roughly $2^{-u/\log^{3c_0} N}$. We pick $u = (\log N)^{K+3c_0}$ for a large constant K and achieve a Raz Verifier with much lower soundness than earlier.

2.2   Overview of Our Construction

The main steps involved in showing inapproximability of Min-3Lin-Deletion are shown in Fig. 1. We start with the Min-3Lin-Deletion-$(0.1, 0.4)$ problem shown to be NP-hard by Håstad [18]. We repeatedly perform two operations called *tensoring* and *boosting* on this problem. This gives a reduction to a version of Min-Lin-Deletion that has a big gap between completeness and soundness. But the instances of Min-Lin-Deletion produced by the reduction can have equations with large number of variables. We first reduce the number of variables appearing in an equation significantly by using the Sum-Check protocol and the Low-degree Test. We then break each of the linear equations into equations over at most three variables in a trivial way by introducing auxiliary variables. We now describe these steps in more detail and explain the new ideas involved.

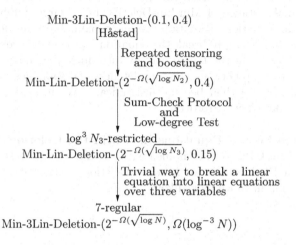

Fig. 1. The main steps in proving an improved factor for Min-3Lin-Deletion

Hardness of Approximation Result for Min-Lin-Deletion. The *tensoring* operation we use on a Min-Lin-Deletion instance is similar to an operation defined by Dumer et al. [8] on linear codes. Tensoring involves taking all possible pairs of linear equations, computing their "product", and then replacing the terms of the form $x_i x_j$ with x_{ij} and x_i with x_{ii} to get back a linear equation. Tensoring converts a Min-Lin-Deletion-(c, s) instance to a Min-Lin-Deletion-(c^2, s^2) instance. Our aim is to bring the completeness close to zero, while keeping the soundness close to $1/2$. Therefore, we cannot use tensoring repeatedly by itself (otherwise the soundness would also tend to zero). We use a *boosting* step after every tensoring operation to work around this problem. Given a Min-Lin-Deletion instance, boosting produces a new Min-Lin-Deletion instance by picking $O(1)$ equations from its input and adding all possible linear combinations of these equations to the output instance. The idea is that even if one the $O(1)$ equations are not satisfied by some assignment, half of the linear combinations of these equations will also not be satisfied by the assignment. To keep the size of the output Min-Lin-Deletion instance small, we pseudo-randomly generate only $O(n)$ of the $n^{O(1)}$ possible ways to pick $O(1)$ equations from n equations. When given a Min-Lin-Deletion-$(c, 0.16)$ instance as an input, boosting produces a Min-Lin-Deletion-$(\sigma c, 0.4)$ instance as the output, for some absolute constant σ. Here, σ is the length of a random walk we need to perform on the expander so that the probablilty of visiting a subset containing 0.16 fraction of the vertices is at least 0.8. By an appropriate choice of the expander, we can assume $\sigma \leq 5$.

After applying tensoring and boosting once to a Min-Lin-Deletion-$(c, 0.4)$ instance, we get a Min-Lin-Deletion-$(\sigma c^2, 0.4)$. If we start with the completeness $c = 1/(2\sigma) = 0.1$, we could apply tensoring and boosting repeatedly. The completeness will decrease each time while the soundness stays at 0.4.

Both tensoring and boosting increase the number of variables appearing in each equation. As a result, even though we start with a Min-3Lin-Deletion instance, the final instance of Min-Lin-Deletion has a large number of variables appearing in an equation. To obtain the inapproximability result for Min-3Lin-Deletion, we cannot simply break the equations into smaller equations with at most three variables in the trivial way by introducing auxiliary variables. The reason is that the gap between the completeness and soundness of the Min-Lin-Deletion problem will then become insignificant. We instead use a technique based on the Sum-Check Protocol.

Reducing the Size of Equations in a Min-Lin-Deletion Instance. We use the Sum-Check Protocol combined with the Low-degree Test (see Arora [1]) to construct a PCP verifier for Min-Lin-Deletion. A typical constraint of the Min-Lin-Deletion instance looks like:

$$x_1 \oplus x_2 \oplus \ldots \oplus x_n = a \tag{1}$$

The verifier tries to verify that this constraint is satisfied using $\log^{O(1)} n$ queries (with access to auxiliary tables as explained below). The test of the verifier is a linear predicate in the $\log^{O(1)} n$ bits read. This gives a reduction to the Min-Lin-Deletion problem in which every equation is over at most $\log^{O(1)} n$ variables. We can then break the equations into smaller equations in the trivial way.

Let $\mathbb{F}$ be a field. Given a r-variate degree d polynomial f and $a \in \mathbb{F}$, the Sum-Check Protocol can be used to verify if the sum of the values of f on the sub-hypercube S^r of $\mathbb{F}^r$ is equal to a, without having to read the value of f at all points on S^r. The prover needs to provide some auxiliary data in the form of a *Partial Sums Table*. The non-adaptive verifier under this protocol randomly reads a few values from the Partial Sums Table and uses the value of f at one point in $\mathbb{F}^r$ and accepts or rejects the proof based on these values. We use this protocol to check if a constraint of the input Min-Lin-Deletion instance (such as (1)) is satisfied. We fix a field $\mathbb{F}$ of characteristic 2 and associate the values of the variables to points in S^r, for some appropriately chosen values of the parameters $|\mathbb{F}|$, $|S|$ and r. There exists a polynomial f of "low" degree that takes these values on S^r. Checking if a linear equation is satisfied is then basically the task of checking if the sum of the values of f on the points in S^r (weighted by the coefficients of the variables in the equation) is a certain target value. We use the Sum-Check Protocol for this purpose. The polynomial f is not known to the verifier (since the values of the variables are not known to the verifier). Hence we expect the prover to also provide a *Points Table*, a table with the value of f at all the points in $\mathbb{F}^r$. For the protocol to work, we need to make sure that the Points Table is in fact "close" to a low degree polynomial. We use the Low-degree test for this purpose.

The Low-degree test expects a *Lines Table* as an auxiliary input. The Lines Table is supposed to contain the restriction of f to every line in $\mathbb{F}^r$. The test picks a random point and a random line through the point. It then checks that the value of the point as reported in the Points Table and the value of the line as reported in the Lines Table are consistent. All the tests performed by the Sum-Check protocol and the Low-degree Test are linear in the field elements. Since the field $\mathbb{F}$ has characteristic 2, these can be replaced by linear tests over boolean values if the field elements are encoded as appropriate bit strings. The number of queries is $\log^{O(1)} n$ and hence we get linear constraints with $\log^{O(1)} n$ size.

3   Hardness for Min-3Lin-Deletion

In this section, we formally state the second step mentioned in Fig. 1. We start by assuming that the first step in Fig. 1 yields the following gap for Min-3Lin-Deletion.

Theorem 5. *There exists a $2^{O(\log^2 n)}$ time algorithm $\mathcal{A}_2$ that when given a Min-3Lin-Deletion instance $\mathbf{A}_1$ of size N_1 outputs a Min-Lin-Deletion instance $\mathbf{A}_2$ of size $N_2 = 2^{O(\log^2 n)}$ such that:*

- *(Yes Case:) If $Opt(\mathbf{A}_1) \leq 0.1$, then $Opt(\mathbf{A}_2) \leq 2^{-\Omega(\sqrt{\log N_2})}$*
- *(No Case:) If $Opt(\mathbf{A}_1) \geq 0.4$, then $Opt(\mathbf{A}_2) \geq 0.4$*

The reduction $\mathcal{A}_2$ works by repeatedly tensoring and boosting $\log\log N_1$ times. The equations in $\mathbf{A}_2$ can be over a large number of variables. We use the Sum-Check Protocol combined with the Low-degree Test (see Arora [1]) to get the number of variables in an equation down to a poly-logarithmic number.

Theorem 6. *There exists a polynomial-time reduction $\mathcal{A}_3$ from a Min-Lin-Deletion instance $\mathbf{A}_2$ of size N_2 to a Min-Lin-Deletion instance $\mathbf{A}_3$ such that:*

- *$\mathbf{A}_3$ has size $N_3 = N_2^{O(1)}$. Also, $\mathbf{A}_3$ is $\log^3 N_3$-restricted. That is, each equation of $\mathbf{A}_3$ is over at most $\log^3 N_3$ variables.*
- *$Opt(\mathbf{A}_3) \leq Opt(\mathbf{A}_2)$. This implies that $Opt(\mathbf{A}_3) \leq 2^{-\Omega(\sqrt{\log N_3})}$ if $Opt(\mathbf{A}_2) \leq 2^{-\Omega(\sqrt{\log N_2})}$.*
- *Assuming N_2 is large enough, $Opt(\mathbf{A}_2) \geq 0.4$ implies $Opt(\mathbf{A}_3) \geq 0.15$.*

Proof. To prove the theorem, we first construct a probabilistic polynomial-time verifier $\mathcal{V}$ for Min-Lin-Deletion with the following properties:

1. The acceptance test is non-adaptive. That is, $\mathcal{V}$ determines the test only based on $\boldsymbol{A}_2$ and its random string.
2. Each test of $\mathcal{V}$ is a logical AND of k smaller tests, where $k = O(\log N_2)$ and each of the smaller tests is a linear equation mod 2 over $O(\log^2 N_2)$ bits of the certificate/proof. We call these the basic tests corresponding to the random string.
3. $\mathcal{A}_3$ uses $O(\log N_2)$ random bits.
4. *(Yes Case:)* There exists a certificate for $\boldsymbol{A}_2$ that is accepted with probability $\geq 1 - Opt(\boldsymbol{A}_2)$.
5. *(No Case:)* If $Opt(\boldsymbol{A}_2) \geq 0.4$, then any certificate for $\boldsymbol{A}_2$ is accepted with probability at most 0.7.

The variables in the output Min-Lin-Deletion instance $\boldsymbol{A}_3$ then correspond to the bits in the certificate. For each possible random string of length $O(\log N_2)$ to $\mathcal{V}$, $\mathcal{A}_3$ adds $2^k = 2^{O(\log N_2)}$ equations, corresponding to all possible linear combinations of the basic tests mentioned in Property 2. Therefore, $\boldsymbol{A}_3$ has $N_3 = N_2^{O(1)}$ linear equations where each equation is over $kO(\log^2 N_2) = O(\log^3 N_2)$ variables. By Property 4, there is an assignment to the variables in $\boldsymbol{A}_3$ such that all the 2^k equations corresponding to $1 - Opt(\boldsymbol{A}_2)$ fraction of the random strings are accepted. If $Opt(\boldsymbol{A}_2) \geq 0.4$, then it follows from Property 5 that for at least a 0.3 fraction of the random strings, one or more of the $O(\log N_2)$ basic tests corresponding to each of these strings must fail. This implies that if $Opt(\boldsymbol{A}_2) \geq 0.4$, then $Opt(\boldsymbol{A}_3) \geq 0.3 \times 1/2 = 0.15$. We next describe the verifier.

Let the number of variables in $\boldsymbol{A}_2$ be n_2. Then, $n_2 \leq N_2^2$ since each equation in $\boldsymbol{A}_2$ is over at most N_2 variables. Define $h = \lceil \log n_2 \rceil$, $m = \lceil \log n_2 / \log \log n_2 \rceil$ and $d = (h - 1)m$. Then, $h^m \geq n_2$. The verifier $\mathcal{V}$ picks a field $(\mathbb{F}, +, \cdot)$ of characteristic 2 with 2^q elements, where $2^q \geq d^3 m$. Let S be any subset of $\mathbb{F}$ of size h. The elements in $\mathbb{F}$ can be represented by bit strings of length q such that for any $\alpha = (\alpha_1, \alpha_2, \ldots, \alpha_q)$ and $\alpha' = (\alpha_1', \alpha_2', \ldots, \alpha_q')$ in $\mathbb{F}$, we have

- $\alpha + \alpha'$ is the bitwise xor of the two strings.
- The kth bit of $\alpha \cdot \alpha'$ is $\oplus_{i,j=1,2,\ldots,q} \alpha_i c_{ijk} \alpha_j'$, where the c_{ijk} only depends on the field.

Let the prover wish to prove that a certain assignment to the variables $x_1, x_2, \ldots, x_{n_2}$ in $\boldsymbol{A}_2$ satisfies a "large" number of linear equations. The prover is expected to provide a certificate consisting of:

- The values of a multivariate polynomial $f(y_1, y_2, \ldots, y_m)$ on the points in $\mathbb{F}^m$, where f has degree $h - 1$ in each of the m variables. This is called the *Points Table*.
- For every line l in $\mathbb{F}^m$, a univariate degree $d = m(h - 1)$ polynomial $g_l(t)$. This is called the *Lines Table*.
- For the ith line in $\mathbb{F}^m$, for every $k = 0, 1, \ldots, m-1$ and every $(\theta_1, \theta_2, \ldots, \theta_k) \in \mathbb{F}^k$, the coefficients of a univariate polynomial $p_{i,\theta_1,\theta_2,\ldots,\theta_k}(y_{k+1})$ with degree $2(h - 1)$. This is called the *Partial Sums Table*. The polynomials for $k = 0$ are denoted as $p_{i,\emptyset}(y_1)$.

All the field elements are expected to be provided as bit-strings with the aforementioned properties. $\mathcal{V}$ associates the variables in A_2 to different points in S^m. The last bit of $f(y_1, y_2, \ldots, y_m)$ is supposed to be the value to be assigned to the variable x_j corresponding to $(y_1, y_2, \ldots, y_m)$. It can be verified that the degree of f is large enough to represent any assignment to the variables x_j. Let $c_i(y_1, y_2, \ldots, y_m)$ be the unique multivariate polynomial of degree $\leq h-1$ in each variable such that $c_i(y_1, y_2, \ldots, y_m)$ is the coefficient of the variable x_j corresponding to $(y_1, y_2, \ldots, y_m)$ in the ith equation of A_2 (If no variable x_j corresponds to $(y_1, y_2, \ldots, y_m) \in S^m$, then $c_i(y_1, y_2, \ldots, y_m)$ is defined to be zero). Note that $c_i(y_1, y_2, \ldots, y_m)$ can be computed from A_2. Then testing that an equation $a_{i0} \oplus (\overset{m}{\underset{j=1}{\oplus}} a_{ij}x_j) = 0$ is satisfied reduces to checking if

$$\sum_{(y_1, y_2, \ldots, y_m) \in S^m} c_i(y_1, y_2, \ldots, y_m) f(y_1, y_2, \ldots, y_m) = a_{i0}$$

Note that $c_i(y_1, y_2, \ldots, y_m) f(y_1, y_2, \ldots, y_m)$ is a degree $2(h-1)m$ polynomial. The polynomial $p_{i,\theta_1,\theta_2,\ldots,\theta_k}(y_{k+1})$ is supposedly the unique degree $2(h-1)$ univariate polynomial such that

$$p_{i,\theta_1,\theta_2,\ldots,\theta_k}(y_{k+1}) =$$
$$\sum_{y_{k+2},\ldots,y_m \in S^{m-k-1}} c_i(\theta_1, \ldots, \theta_k, y_{k+1}, \ldots, y_m) f(\theta_1, \ldots, \theta_k, y_{k+1}, \ldots, y_m)$$

Also, $g_l(t)$ is the supposed restriction of the polynomial f to the line l. That is, if $l(t) = \boldsymbol{\theta} + \boldsymbol{\theta}'t$ is the parametric representation of the line l for some $\boldsymbol{\theta}, \boldsymbol{\theta}' \in \mathbb{F}^m$, then $g_l(t) = f(l(t))$.

The verifier performs the following tests:

1. (The Low-degree Test) Repeat the following $4\delta^{-1}$ times, where $\delta = 10^{-4}/2$. Pick a line l in $\mathbb{F}^m$ and $t \in \mathbb{F}$ uniformly at random. Check that $g_l(t) = f(l(t))$.

2. (The Sum-Check Protocol) Pick $i \in \{1, 2, \ldots, N_2\}$ uniformly at random (i.e. the verifier is trying to verify the ith equation of A_2). Pick $\boldsymbol{\theta} = (\theta_1, \theta_2, \ldots, \theta_m)$ from $\mathbb{F}^m$ uniformly at random.

   (a) Check that $\sum_{y_1 \in S} p_{i,\emptyset}(y_1) = a_{i0}$

   (b) Check that $\forall j : 1 \leq j \leq m-1$,

   $$\sum_{y_{j+1} \in S} p_{i,\theta_1,\ldots,\theta_j}(y_{j+1}) = p_{i,\theta_1,\ldots,\theta_{j-1}}(\theta_j)$$

   (c) Check that $p_{i,\theta_1,\ldots,\theta_{m-1}}(\theta_m) = c_i(\theta_1, \ldots, \theta_m) f(\theta_1, \ldots, \theta_m)$

The verifier accepts if all the above tests succeed. The test is clearly non-adaptive. All the $O(1) + 1 + O(m) + 1 = O(m)$ basic tests mentioned above are linear in the field elements read, and hence can be broken down into $k = O(m \log |\mathbb{F}|)$

$= O(m \log(d^3 m)) = O(\log N_2)$ smaller linear tests over the bits read. Each of these smaller tests is over at most $O(d) \log |\mathbb{F}| = O(\log^2 N_2)$ bits read. The randomness used is $O(\log |\mathbb{F}|) + O(\log N_2) + O(m \log |\mathbb{F}|) = O(\log N_2)$. It is easy to see that if the prover selects the best assignment to A_2 and constructs the proof honestly as expected, it is accepted with probability $\geq 1 - Opt(A_2)$. On the other hand, if $Opt(A_2) \geq 0.4$), then no proof is accepted with probability ≥ 0.7. The analysis of Sum-Check Protocol and Low-degree Test are based on Theorem 4.15 and Lemma 4.12 of Arora [1]. We skip the details due to lack of space. □

4   Conclusion

Recently, Samorodnitsky and Trevisan [25] showed that, assuming the Unique Games Conjecture of Khot [21], it is hard to approximate MaxClique in degree d graphs better than $d/\text{polylog}(d)$. This suggests that MaxClique on general graphs could be hard to approximate within $n/\text{polylog}(n)$. We think it is a challenging (and important) open problem to prove such a hardness result, or even to improve the hardness result to $\frac{n}{2^{O(\sqrt{\log n})}}$.

References

1. S. Arora. *Probabilistic checking of proofs and the hardness of approximation problems.* Ph.D. thesis, UC Berkeley, 1994.
2. S. Arora, C. Lund, R. Motawani, M. Sudan, and M. Szegedy. Proof verification and the hardness of approximation problems. *Journal of the ACM*, 45(3):501–555, 1998.
3. S. Arora and S. Safra. Probabilistic checking of proofs : A new characterization of NP. *Journal of the ACM*, 45(1):70–122, 1998.
4. M. Bellare, O. Goldreich, and M. Sudan. Free bits, PCPs and non-approximability. *Electronic Colloquium on Computational Complexity, Technical Report TR95-024*, 1995.
5. M. Bellare, S. Goldwasser, C. Lund, and A. Russell. Efficient probabilistic checkable proofs and applications to approximation. In *Proc. 25th ACM Symposium on Theory of Computing*, pages 294–304, 1993.
6. M. Bellare and M. Sudan. Improved non-approximability results. In *Proc. 26th ACM Symposium on Theory of Computing*, pages 184–193, 1994.
7. Avrim Louis Blum. *Algorithms for approximate graph coloring.* PhD thesis, Massachusetts Institute of Technology, Cambridge, MA, USA, 1992.
8. I. Dumer, D. Micciancio, and M. Sudan. Hardness of approximating the minimum distance of a linear code. In *Proc. 40th IEEE Symposium on Foundations of Computer Science*, 1999.
9. L. Engebretsen and J. Holmerin. Towards optimal lower bounds for clique and chromatic number. *Electronic Colloquium on Computational Complexity (ECCC)*, (TR01-003), 2001.
10. U. Feige. Randomized graph products, chromatic numbers, and the lovász θ-function. In *Proc. 27th ACM Symposium on Theory of Computing*, pages 635–640, 1995.

11. U. Feige. Approximating maximum clique by removing subgraphs. *SIAM J. Discrete Math.*, 18(2):219–225, 2004.
12. U. Feige, S. Goldwasser, L. Lovász, S. Safra, and M. Szegedy. Interactive proofs and the hardness of approximating cliques. *Journal of the ACM*, 43(2):268–292, 1996.
13. U. Feige and J. Kilian. Two prover protocols: low error at affordable rates. In *Proc. 26th ACM Symposium on Theory of Computing*, pages 172–183, 1994.
14. U. Feige and J. Kilian. Zero knowledge and the chromatic number. In *Proc. 11thIEEE Conference on Computational Complexity*, pages 278–287, 1996.
15. Uriel Feige and Shimon Kogan. Hardness of approximation of the balanced complete bipartite subgraph problem. Technical Report MCS04-04, The Weizmann Institute of Science, 2004.
16. J. Hastad. Clique is hard to approximate within $n^{1-\epsilon}$. In *Proc. 37th IEEE Symposium on Foundations of Computer Science*, pages 627–636, 1996.
17. J. Håstad. Testing of the long code and hardness for clique. In *Proc. 28th ACM Symposium on Theory of Computing*, pages 11–19, 1996.
18. J. Hastad. Some optimal inapproximability results. In *Proc. 29th ACM Symposium on Theory of Computing*, pages 1–10, 1997.
19. J. Hastad and A. Wigderson. Simple analysis of graph tests for linearity and PCP. In *Proc. 16th IEEE Conference on Computational Complexity*, 2001.
20. S. Khot. Improved inapproximability results for maxclique, chromatic number and approximate graph coloring. In *Proc. 42nd IEEE Annual Symposium on Foundations of Computer Science*, 2001.
21. Subhash Khot. On the power of unique 2-prover 1-round games. In *STOC '02: Proceedings of the thiry-fourth annual ACM symposium on Theory of computing*, pages 767–775, New York, NY, USA, 2002. ACM Press.
22. Donald E. Knuth. The sandwich theorem. *Electr. J. Comb.*, 1, 1994.
23. R. Raz. A parallel repetition theorem. *SIAM J. of Computing*, 27(3):763–803, 1998.
24. A. Samorodnitsky and L. Trevisan. A PCP characterization of NP with optimal amortized query complexity. In *Proc. 32nd ACM Symposium on Theory of Computing*, pages 191–199, 2000.
25. A. Samorodnitsky and L. Trevisan. Gowers uniformity, influence of variables, and pcps. *Electronic Colloquium on Computational Complexity (ECCC)*, (TR05-116), 2005.
26. Aravind Srinivasan. The value of strong inapproximability results for clique. In *STOC '00: Proceedings of the thirty-second annual ACM symposium on Theory of computing*, pages 144–152, 2000.
27. L. Trevisan. Non-approximability results for optimization problems on bounded degree instances. In *Proc. 33rd ACM Symposium on Theory of Computing*, pages 453–461, 2001.
28. D. Zuckerman. On unapproximable versions of NP-complete problems. *SIAM J. on Computing*, pages 1293–1304, 1996.

Approximating the Orthogonal Knapsack Problem for Hypercubes

Rolf Harren[*]

Graduate School of Informatics, Kyoto University, Japan and
Fachbereich Informatik, Universität Dortmund, Germany
rolf.harren@ls2.cs.uni-dortmund.de

Abstract. Given a list of d-dimensional cuboid items with associated profits, the *orthogonal knapsack problem* asks for a packing of a selection with maximal profit into the unit cube. We restrict the items to hypercube shapes and derive a $(\frac{5}{4} + \epsilon)$-approximation for the two-dimensional case. In a second step we generalize our result to a $(\frac{2^d+1}{2^d}+\epsilon)$-approximation for d-dimensional packing.

1 Introduction

The *knapsack problem* is one of the most fundamental optimization problems in computer science. The classical one-dimensional variant and its applications are subject to a great number of articles, see [15] and [12] for surveys. Not surprisingly, a geometrical generalization called *d-dimensional orthogonal knapsack problem (OKP-d)* is also popular. It is defined as follows.

Given a list $I = (r_1, \ldots, r_n)$ of cuboid items $r_i = (a_{i,1}, a_{i,2}, \ldots, a_{i,d})$ with associated profit $p_i > 0$ and the unit hypercube $B = [0,1]^d$ as a bin. The objective is to find a *feasible*, i.e., *orthogonal*, *non-rotational* and *non-overlapping* packing of a selection $I' \subset I$ into B such that the overall packed profit is maximized. An *orthogonal* packing requires that the items are packed parallel to the axis of the bin. Items are *non-overlapping* if their interiors are disjoint. For the two-dimensional case, i.e., packing rectangles into a unit square, the best-known general result is a $(2 + \epsilon)$-approximation given by Jansen and Zhang [11]. As the difficulty of the problem is increasing drastically with the dimension, only recently a $(7 + \epsilon)$-approximation for the three-dimensional case was derived [4].

In this paper we restrict the items to *hypercube* shapes (squares instead of rectangles in the two-dimensional case) and investigate how much easier the problem becomes. Note that this restriction is quite popular in the literature ([1], [6], [10]) and yields great potential. Bansal et al. [1] showed for two-dimensional *bin packing*, that even though it is APX-complete in the general case, the restriction to *hypercube bin packing* admits an APTAS. Furthermore, their results hold for higher dimensions as well.

---

[*] This work was partly supported by DAAD (German Academic Exchange Service) and the German National Academic Foundation.

M. Bugliesi et al. (Eds.): ICALP 2006, Part I, LNCS 4051, pp. 238–249, 2006.

Our Contribution. Our main result is an approximation algorithm for *square packing*, i.e., *hypercube OKP-2*, with an approximation ratio of $(\frac{5}{4} + \epsilon)$. Moreover, we show that our result can be extended to d-dimensional packing, deriving an $(\frac{2^d + 1}{2^d} + \epsilon)$-approximation. Note that we improve the known approximations of $(2 + \epsilon)$ and $(7 + \epsilon)$ for general two- and three-dimensional knapsack packing significantly. Furthermore, we reverse the effect of rising approximation ratios for higher dimensions. In fact, our approximation ratio is improving exponentially with the dimension.

Related Problems. Besides the *orthogonal knapsack problem*, there are two other common generalizations of packing problems. The previously mentioned *d-dimensional orthogonal bin packing problem (OBPP-d)* has the objective of minimizing the total number of unit-size bins in order to pack a list I of cuboid items. The *d-dimensional orthogonal strip packing problem (OSPP-d)*, on the other hand, asks to pack into a strip of bounded basis and unlimited height such as to minimize the total height of the packing.

In 1990 Leung et al. [14] proved the NP-hardness in the strong sense for the special case of determining whether a set of squares can be packed into a bigger square or not. Therefore, already a very special two-dimensional case and all generalizations are strongly NP-hard. In spite of that, the NP-hardness of *hypercube* bin, strip and knapsack packing is still an open problem for $d > 2$.

In the *strip packing* setting, *OSPP-2* admits an asymptotic full polynomial time approximation scheme (AFPTAS) for the rotational and non-rotational case, see Jansen and van Stee [9], and Kenyon and Rémila [13]. For *OSPP-3*, Jansen and Solis-Oba [8] gave a $(2 + \epsilon)$-approximation. For general 2-dimensional *bin packing (OBPP-2)*, the best-known result is a $1,691...$-approximation by Caprara et al. [2].

Apart from the general *knapsack packing* results mentioned earlier, *OKP-2* has also been studied in different variants. For the restriction of packing squares into a rectangle in order to maximize the number, Jansen and Zhang gave an AFPTAS [10]. Maximizing the packed area of squares admits a PTAS, as Fishkin et al. showed [6]. In the case that the rectangles are much smaller than the bin, a better approximation is possible. We refer to this case as *packing with large resources*. Fishkin et al. [5] showed that a solution with weight at least $(1 - \epsilon)$ of the optimum can be found if the side length of the bin differs by at least $1/\epsilon^4$.

An application for the two-dimensional *knapsack problem* is job scheduling with a due date, where the jobs have to be assigned to a consecutive line of processors and the overall profit of accepted jobs has to be maximized. Further applications of packing problems include container loading, VLSI design and advertisement placement [7], i.e., placing rectangular ads on a given board.

In order to generalize our *square packing* result to higher dimensions, we derived an APTAS for d-dimensional *hypercube strip packing* and a result similar to [5] for *hypercube knapsack packing with large resources*. Both results are motivated by their two-dimensional equivalents in [13] and [5] and thus also stand for themselves. Due to page limitations we can not give the proofs for these results in this paper.

Presentation of the Paper. We begin with some Preliminaries in Section 2. In Section 3 we describe the $(\frac{5}{4} + \epsilon)$-algorithm for *hypercube OKP-2*. Before the presentation of the generalization in Section 5 we state our results on *hypercube OSPP-d* and on *knapsack packing with large resources* in Section 4. In Section 6 we conclude our presentation and point out future work.

2   Notations and Preliminaries

Since the items are squares (or hypercubes) throughout the paper, we refer to both, the items and their sizes by a_i. Let I be a set of items. We denote the volume of I by $\text{Vol}(I) = \sum_{i \in I} a_i^2$ ($\sum_{i \in I} a_i^d$ in the d-dimensional setting), the profit of I by $p(I) = \sum_{i \in I} p_i$ and the optimal profit by $\text{OPT}(I)$.

Bansal et al. showed in [1] how to check the *feasibility*, i.e., whether a given set of items can be packed into the bin, in constant time when the number of items is bounded by a constant. We refer to this method by *constant packing*.

Coffman et al. [3] analysed the *Next Fit Decreasing Height (NFDH)* heuristic for the two-dimensional case. Their work was generalized by Bansal et al. [1] for d-dimensional packing. We will use *NFDH* for packing small items.

Lemma 1. *NFDH*
Given a set S of small items $a_i \leq \delta$, then

1. *The total wasted, i.e., uncovered, volume of a packing P of S into a cuboid bin $B = (b_1, \ldots, b_d)$ with $b_i \leq 1$ by NFDH is bounded by $\delta \sum_{i=1}^{d} b_i \leq \delta d$.*
2. *If the total volume V of the given space is at least δ and the total wasted volume when packing with NFDH is at most δ^2 then we can pack the small items with profit at least $(1 - 2\delta)\text{OPT}(S)$.*

Note that the bin B is not a unit cube but a cuboid bin and that the given space with volume V can have an arbitrary shape.

Proof. Part 1 is shown in [1]. To see Part 2, we use an instance of the fractional knapsack problem, i.e., one-dimensional knapsack packing where fractions of the items can be packed. Note that the well known greedy algorithm finds an optimal solution with one fractional item at the most. Let $\text{FracKnap}(S, V - 2\delta^2)$ be the fractional knapsack instance with volume bound $V - 2\delta^2$ and the items in S be given by their volume. Let S' be the optimal solution derived by the greedy algorithm, including the possibly fractional item. Since the volume of every item is at most δ^2, we get $\text{Vol}(S') \leq V - \delta^2$. Therefore a packing of S' into the volume is possible. Observe that $p(S') \geq \frac{V - 2\delta^2}{V}\text{OPT}(S) \geq (1 - \frac{2\delta^2}{\delta})\text{OPT}(S) = (1 - 2\delta)\text{OPT}(S)$. $\qquad\square$

To restrict the number of gaps in a packing, Bansal et al. [1] showed

Lemma 2. *Let P be a packing of m hypercubes in $[0, 1]^d$ such that there is a hypercube touching each of the hyperplanes $x_i = 0$ for $i = 1, \ldots, d$. Then, the remaining space $[0, 1]^d \setminus P$ can be divided into at most $(2m)^d$ non-overlapping cuboids.*

For $d = 2$, the number of rectangles is bounded by $3m$.

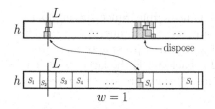

Fig. 1. Freeing a line L

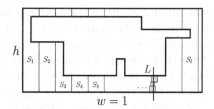

Fig. 2. An irregular shaped R

Note that the *constant packing* method creates suitable packings where each hyperplane $x_i = 0$ for $i = 1, \ldots, d$ is touched by a hypercube.

Finally, we introduce a *shifting technique* that we use several times to free a given line L inside a packing P without losing too much profit. Fishkin et al. used a similar technique in [5].

Lemma 3. *Given a packing P of a list $I = (a_1, \ldots, a_n)$ of small $(a_i \leq \delta)$ squares into a rectangle $R = (w, h)$ with width $w = 1$ and a vertical line L. If $\delta \leq \frac{1}{2}$, we can derive a packing P' of a selection $I' \subseteq I$ into R with profit $p(I') \geq (1 - 4\delta)p(I)$ in polynomial time such that L does not intersect with any item.*

Proof. Let I_L be the set of items that intersect L. Partition R into $l = \lfloor \frac{1}{\delta} \rfloor \geq \frac{1}{\delta} - 1$ rectangular slices $S_1, \ldots, S_l$ of width δ and a possible smaller one by drawing lines with a distance of δ parallel to the bins height as in Figure 1. Find an index i such that the items, that intersect with S_i have minimal profit. Remove all items that intersect with S_i and copy the items I_L left-aligned into S_i. The remaining profit is

$$p(S') \geq p(S) - \frac{2p(S)}{\frac{1}{\delta} - 1} \geq (1 - 4\delta)p(S)$$

since every item intersects with at most two rectangle S_i, S_{i+1}. $\qquad\square$

Note that the proof is also valid for a rectangle R with $h = 1$ and a horizontal line L. Furthermore, it is not necessary that R is a rectangle as long as the cutting line is at the thinnest part of R so that copying the items is possible, see Figure 2 for another possible setting.

3   Square Packing

We now describe our main result for the two-dimensional case. Later we will generalize it for d-dimensional packing. In order to ease the generalization, we split the description into several parts.

Outline. The first step of the algorithm is a separation of the items into sets of large, medium and small items. This yields a gap in size between large and small items and a profit of the medium items that is negligible. Since the number of large items in the bin is bounded by a constant, we can enumerate all possible selections and thus assume the knowledge of an optimal packing of large items. After that, we consider three different cases for packing: 1) the large items leave *enough remaining space* to pack the small items, 2) there are *several large items*, and 3) there is only *one very large item*.

We derive almost optimal solutions for the first and third case and an almost $\frac{k+1}{k}$-optimal solution for the second case, where k is the number of large items. By showing that any packing with $k < 4$ can be reduced to the first or the third case, we derive an overall approximation ratio of $(\frac{5}{4} + \epsilon)$.

Let $0 < \epsilon \le 1/2^{10}$, $\epsilon' = \epsilon/3$. The following *separation technique* divides an optimal solution I_{opt} into sets L_{opt} of large, M_{opt} of medium and S_{opt} of small items such that $p(M_{opt}) \le \epsilon' \mathrm{OPT}(I)$ and thus we can neglect the medium items.

Separation Technique. Let $r = \lceil 1/\epsilon' \rceil$. Consider an optimal solution I_{opt} and the sequence $\alpha_0 = \epsilon'$, $\alpha_{i+1} = \alpha_i^4 \, \epsilon'$ for $i = 0, \ldots, r$. Define the partition of I_{opt} into sets $M_0 = \{s \in I_{opt} : s \ge \alpha_1\}$, $M_i = \{s \in I_{opt} : s \in [\alpha_{i+1}, \alpha_i[\}$ for $1 \le i \le r$ and $M_{r+1} = \{s \in I_{opt} : s < \alpha_{r+1}\}$. Observe, that there is an index $i^* \in \{1, \ldots r\}$ such that $p(M_{i^*}) \le \epsilon' p(I_{opt}) = \epsilon' \mathrm{OPT}(I)$. Let $L_{opt} = M_0 \cup \ldots \cup M_{i^*-1}$ be the set of large, $M_{opt} = M_{i^*}$ the set of medium and $S_{opt} = M_{i^*+1} \cup \ldots \cup M_{r+1}$ the set of small items. Thus $p(L_{opt} \cup S_{opt}) \ge (1 - \epsilon') \mathrm{OPT}(I)$ and it is sufficient to approximate this almost optimal solution. Let $S = \{s \in I : s < \alpha_{i^*+1}\}$, obviously $S_{opt} \subseteq S$ and thus $\mathrm{OPT}(L_{opt} \cup S) \ge (1 - \epsilon')\mathrm{OPT}(I)$.

Since $s \ge \alpha_{i^*}$ for $s \in L_{opt}$, there are at most $1/\alpha_{i^*}^2$ items in L_{opt}. Thus we can enumerate over all $i \in \{1, \ldots r\}$ and L with $|L| \le 1/\alpha_i^2$ and use the *constant packing* method to check the feasibility of L. Hence assume the knowledge of i^* and L_{opt}. Let $P_{L_{opt}}$ be a packing of L_{opt} by the constant packing method.

The gap in size between the large and the small items is needed to obtain an efficient packing of some of the small items in S with *NFDH* into the gaps of $P_{L_{opt}}$. Since $|L_{opt}| \le 1/\alpha_{i^*}^2$, there are at most $3/\alpha_{i^*}^2$ gaps in $P_{L_{opt}}$ - see Lemma 2. Lemma 1 Part 1 bounds the wasted volume for every gap by $2\alpha_{i^*+1}$. Hence we can bound the overall wasted volume of a packing with *NFDH* of the small items in S into the gaps of $P_{L_{opt}}$ by $\frac{3}{\alpha_{i^*}^2} \cdot 2\alpha_{i^*+1} = 6\frac{\alpha_{i^*}^4 \epsilon'}{\alpha_{i^*}^2} = 6\epsilon'\alpha_{i^*}^2 \le \alpha_{i^*}^2$, which is a lower bound for the volume of an item in L_{opt}.

Now let us see how to derive a packing in three different cases: 1) enough remaining space for the small items $(\mathrm{Vol}(L_{opt}) \le 1 - \alpha_{i^*})$, 2) several large items $(|L_{opt}| = k)$, and 3) one very large item $(a_{max} \ge 1 - \epsilon'^4)$, where a_{max} is the biggest item in L_{opt}.

Lemma 4. *Enough Remaining Space*
If $\mathrm{Vol}(L_{opt}) \le 1 - \alpha_{i^*}$, *we can find a selection* $S' \subseteq S$ *of small items in polynomial time such that* L_{opt} *and* S' *can be packed together and* $p(L_{opt} \cup S') \ge (1 - 3\epsilon')\mathrm{OPT}(I)$.

Proof. The remaining space is at least α_{i^*} and the overall wasted volume is at most $\alpha_{i^*}^2$. As all small items have size at most $\alpha_{i^*+1} \le \alpha_{i^*}$ we can apply Lemma 1 Part 2 with $\delta = \alpha_{i^*}$ to find a feasible selection $S' \subseteq S$ with $p(S') \ge$

$(1 - 2\delta)\mathrm{OPT}(S) \geq (1 - 2\epsilon')\mathrm{OPT}(S)$, where $\mathrm{OPT}(S)$ is the optimal profit for packing S into the remaining space.   □

Lemma 5. *Several Large Items*
If $|L_{opt}| = k$, we can find a selection $S' \subseteq S$ of small items in polynomial time such that L_{opt} and S' can be packed together and $p(L_{opt} \cup S') \geq (\frac{k}{k+1} - 2\epsilon')\mathrm{OPT}(I)$.

Proof. Let $\mathrm{Knapsack}(S, V, \epsilon)$ denote to a solution with accuracy ϵ for a one-dimensional knapsack instance with items S and volume bound V. The items are given by their volume. Let $S' = \mathrm{Knapsack}(S, 1 - \mathrm{Vol}(L_{opt}), \epsilon')$. Note that $p(L_{opt} \cup S') \geq (1 - 2\epsilon')\mathrm{OPT}(I)$. Consider the packing $P_{L_{opt}}$ and use *NFDH* to add as much as possible of S' into the gaps. Let the profit be P_1. If S' is completely packed, $P_1 = p(L_{opt} \cup S') \geq (1 - 2\epsilon')\mathrm{OPT}(I)$. Otherwise consider a second packing. Therefore remove the item a^* with smallest profit from L_{opt} and pack the remaining items of L_{opt} together with S' into a bin. This is possible since $\mathrm{Vol}(a^*) \geq \alpha_{i*}^2$ and the total waste is bounded by α_{i*}^2. Let this profit be P_2. We state that $\max(P_1, P_2) \geq \frac{k}{k+1}p(L_{opt} \cup S') \geq (\frac{k}{k+1} - 2\epsilon')\mathrm{OPT}(I)$. Assume $L_{opt} = \{a_1, \ldots a_k\}$ and $a^* = a_k$. Then,

$$P_1 \geq \sum_{i=1}^{k} p_i \geq k\, p_k \qquad \text{and} \qquad P_2 = p(L_{opt} \cup S') - p_k$$

For $p_k \in [0, \frac{p(L_{opt} \cup S')}{k+1}]$, $P_2 \geq p(L_{opt} \cup S') - \frac{p(L_{opt} \cup S')}{k+1} \geq \frac{k}{k+1}p(L_{opt} \cup S')$ and for $p_k \in [\frac{p(L_{opt} \cup S')}{k+1}, \frac{p(L_{opt} \cup S')}{k}]$, $P_1 \geq \frac{k}{k+1}p(L_{opt} \cup S')$. Note that $p_k \leq \frac{p(L_{opt} \cup S')}{k}$ as a_k is the item with smallest profit in L_{opt}.   □

Lemma 6. *One Very Large Item*
If $a_{max} \geq 1 - \epsilon'^4$, we can find a selection $S' \subseteq S$ of small items in polynomial time such that L_{opt} and S' can be packed together and $p(L_{opt} \cup S') \geq (1 - 3\epsilon')\mathrm{OPT}(I)$.

Proof. The proof consists of two parts. First we show that the big item a_{max} can be packed into the lower left corner of the bin. Second we use the result for *packing with large resources* by Fishkin et al. [5] to find an almost optimal packing for the remaining space.

Consider an optimal packing of I_{opt} where a_{max} is not placed in the lower left corner. Notice that the free space to all sides has width at most $1 - a_{max} \leq \epsilon'^4$. Draw three lines S_1, S_2, S_3 as on the left side of Figure 3. As the items might have high profit we cannot dispose them directly, but with the *shifting technique* of Lemma 3 and $\delta = \epsilon'^4$ we obtain a packing without any item intersecting lines S_1, S_2, S_3. Thus replace the packing as in Figure 3 on the right side, such that a_{max} is placed in the lower left corner.

For $\epsilon \leq 1/2^{10}$, Fishkin et al. [5] described an algorithm that finds a packing for a subset S' of a set of rectangles S into a bin $(1, b)$ where $b \geq 1/\epsilon^4$ with profit $p(S') \geq (1 - \epsilon)\mathrm{OPT}(S)$. We can consider the remaining space in the bin as a strip of size $(1 - a_{max}, 1 + a_{max})$ by cutting at S_4 and rotating a part of

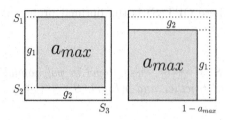

Fig. 3. Almost optimal solution with a_{max} in lower left corner

Fig. 4. Shifting the remaining space

the space as shown in Figure 4. Scaling this strip and all small items by $\frac{1}{1-a_{max}}$ gives a strip of size $(1, b)$ where $b = \frac{1+a_{max}}{1-a_{max}} \geq 1/\epsilon'^4$ (as $a_{max} \geq 1 - \epsilon'^4$). Thus we can find a packing with profit at least $(1 - \epsilon')\mathrm{OPT}(S)$. By cutting again at S_5, the solution can be adopted to the original shape. The rotation is possible since we only have square items. As we have a total of five applications of the *shifting technique*, the loss is bounded by $5 \cdot 4\epsilon'^4\mathrm{OPT}(S) \leq \epsilon'\mathrm{OPT}(S)$.   □

We now give a simple but very important lemma, which takes the full advantage of the square shapes of the items, namely that any packing with $k < 4$ large items can be reduced to either the first or the third case. Our intuition is, that it is impossible to fill a unit-size bin with either two or three equally big squares. This also turns out to be the reason for the improving approximation ratio with higher dimensions, e.g., either one very large or more than seven cubes are needed to fill a cube bin almost completely.

Lemma 7. *If $|L_{opt}| < 4$, then $\mathrm{Vol}(L_{opt}) \leq 1 - \alpha_{i^*}$ or $a_{max} \geq 1 - \epsilon'^4$.*

Proof. Suppose that $|L_{opt}| \in \{1, 2, 3\}$. If $a_{max} \leq 1/2$, then $\mathrm{Vol}(L_{opt}) \leq 3/4 \leq 1 - \alpha_{i^*}$. With $a_{max} > 1/2$ the smaller items in L_{opt} can have a size of at most $1 - a_{max}$ so that L_{opt} is still feasible. As there are at most two more items in L_{opt}, we can bound the total volume by $\mathrm{Vol}(L_{opt}) \leq f(a_{max}) := a_{max}^2 + 2(1 - a_{max})^2$. It is easy to show, that $f(a_{max}) \leq 1 - \alpha_1 = 1 - \epsilon'^5$ for $a_{max} \in [\frac{1}{2}, 1 - \epsilon'^4]$.   □

> **for** *every $i \in \{1, \ldots, r\}$ and feasible $L \subset \{s \in I : s \geq \alpha_i\}$ with $|L| \leq 1/\alpha_i^2$* **do**
>   **case** $\mathrm{Vol}(L) \leq 1 - \alpha_i$: solve almost optimal with Lemma 4
>   **case** $a_{max} \geq 1 - \epsilon'^4$: solve almost optimal with Lemma 6
>   **case** $|L| \geq 4$: solve with Lemma 5
> **end**
> output the packing with the best profit

Algorithm 1. $(\frac{5}{4} + \epsilon)$-algorithm A for square packing

---

The complete algorithm A is summed up in Algorithm 1. The following theorem is immediate since $\frac{1}{\frac{4}{5} - 2\epsilon'} \leq \frac{5}{4} + \epsilon$.

Theorem 1. *There is a polynomial time algorithm for hypercube OKP-2 with performance ratio $(\frac{5}{4} + \epsilon)$.*

4   Useful Tools for Hypercube Packing

In the previous section we used a result on *packing with large resources* to derive the algorithm for the case of one very large item. In order to generalize our algorithm we need a d-dimensional variant of this result. The original two-dimensional algorithm from Fishkin et al. [5] is based on an AFPTAS for *strip packing (OSPP-2)* by Kenyon and Rémila [13]. Similarly, we require an APTAS for *hypercube strip packing* to derive our result on *hypercube knapsack packing with large resources*.

Hypercube Strip Packing is defined as follows. Let $C \geq 1$ be a bound for the size of the basis. Given a list $I = (a_1, \ldots, a_n)$ of hypercubes $a_i \in (0, 1]$ and a $(d-1)$-dimensional cuboid basis of the strip $B = (b_1, b_2, \ldots, b_{d-1})$ with $1 \leq b_i \leq C$. The problem is to find a feasible packing P of I into a strip with basis B and unlimited height such that the total height of the packed items is minimized. Using methods from [13] and [1] we derived an algorithm A_{Strip} that holds the following

Theorem 2. A_{Strip} *is an asymptotic polynomial time approximation scheme (APTAS) for* hypercube OSPP-d *with additive constant* $K_{Strip,\epsilon}$ *for fixed* $\epsilon > 0$ *and* $C \geq 1$.

Hypercube Knapsack Packing with Large Resources is defined as follows. Given a list $I = (a_1, \ldots, a_n)$ of hypercubes $a_i \in (0, 1]$, associated profits $p_i > 0$ and a bin $B = (b_1, b_2, \ldots, b_d)$ with sizes $b_i \geq 1$. The problem is to find a feasible packing P of a selection $I' \subseteq I$ into the bin B with maximal profit. Let $V = \prod_{i=1}^{d} b_i$ be the volume of the bin. Using the algorithm A_{Strip} and ideas from [5], we derived an algorithm A_{LR} that satisfies the following

Theorem 3. *If* $V \geq K_{LR,\epsilon}$ *then algorithm* A_{LR} *finds a feasible packing for a selection* $I' \subseteq I$ *with profit at least* $(1 - \epsilon)\mathrm{OPT}(I)$.

The running time of A_{LR} is polynomial and $K_{LR,\epsilon}$ is constant for fixed $\epsilon > 0$.

5   Hypercube Knapsack Packing

Now we are ready to present the generalization of our main result, a $(\frac{2^d + 1}{2^d} + \epsilon)$-approximation for *hypercube knapsack packing*. In the *square packing* algorithm we considered three different cases, packing with *enough remaining space*, packing with *several large items* and packing with *only one large item*. The latter case was motivated by the observation, that three squares cannot fill a unit bin almost completely unless one of the squares is huge. This observation is generalized to a number of $2^d - 1$ hypercubes in the d-dimensional case.

Outline. First, we give new parameters for the separation step such that the first two cases hold for hypercubes. Second, we show how to handle the third case, applying A_{LR} from the previous section. Finally, we observe that for a number of up to $2^d - 1$ hypercubes, either the remaining space is big enough or there is only one very large item.

Separation. Let $\epsilon' = \epsilon/3$ and $K \geq K_{LR,\epsilon'}$, where $K_{LR,\epsilon'}$ is the constant for algorithm A_{LR} as in the previous section. Let $r = \lceil 1/\epsilon' \rceil$.

Use the sequence $\alpha_0 = \frac{1}{K}, \alpha_{i+1} = \alpha_i^{3d}\epsilon'$ for $i = 0, \ldots, r$ to separate an optimal solution I_{opt} into the sets L_{opt}, M_{opt} and S_{opt} as before. Similar to the *square packing* algorithm, the parameters α_i are chosen such that the overall wasted volume of a packing of small items into the gaps of L with *NFDH* is bounded by α_i^d, the lower bound of the volume of a large item in L - see Lemmas 1 and 2. Again we enumerate over all $i \in \{1, \ldots r\}$ and $|L| \leq 1/\alpha_i^d$ and assume the knowledge of i^* and L_{opt}.

Since the overall wasted volume is bounded by the size of an item in L_{opt}, the first two cases can be handled similarly - see Lemma 4 and Lemma 5.

Now, we show how an almost optimal packing can be derived for $a_{max} \geq 1 - \frac{1}{K}$. First, we show that a special packing structure, similar to packing a_{max} into the lower left corner, does not change the optimal value significantly and second, we use the *shifting technique* and some rotations to apply Theorem 3. Note that the *shifting technique* is similar for d-dimensional hypercubes, as long as one direction of the space R has length 1.

Well-structured Packing. A packing P is called *well-structured* if the biggest item a_{max} is located in the origin $(0, \ldots, 0)$ and the hypercube space of size $1 - a_{max}$ in the opposite corner as well as all hyperplanes, defined by the facets of a_{max} are completely free of items. See Figure 5 for a *well-structured* packing. Similar to the two-dimensional case, we can apply the *shifting technique* to reorder an optimal solution and derive.

Lemma 8. *There is a well-structured packing of a selection* $I' \subseteq I_{opt}$ *with profit* $p(I') \geq (1 - 2\epsilon')\mathrm{OPT}(I)$.

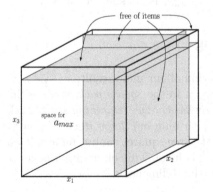
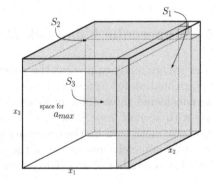

Fig. 5. Free space in a *well-structured* packing for $d = 3$

Fig. 6. Division of the remaining space into $S_1, \ldots, S_d$ for $d = 3$

Applying Algorithm A_{LR} of Theorem 3. We cut and rotate the remaining space of a *well-structured* packing of I' around a_{max} such that it builds a cuboid bin that is much bigger than the remaining items. Then we apply Theorem 3 and by cutting again and reassembling to the original position a valid solution is derived.

Observe, that the remaining space in the bin, with the exception of a hypercube of size $1 - a_{max} \leq \frac{1}{K}$ in the opposite corner of the origin, can be divided into d differently rotated spaces $S_1, \ldots, S_d$ of size $(1 - a_{max}, a_{max}, \ldots, a_{max}, 1)$ - see Figure 6. Note that, since we consider a *well-structured* packing, all items of the near optimal solution I' are completely included in one of these spaces. Rotate all spaces into the same orientation, assemble them to a bin of size $(1 - a_{max}, a_{max}, \ldots, a_{max}, d)$ and scale the bin and all small items with $\frac{1}{1-a_{max}}$. The volume of the bin is bigger than $\frac{1}{1-a_{max}} \geq K$ (since $a_{max} \geq 1 - \frac{1}{K}$). So we can apply Theorem 3 and therefore find a packing for a selection S' of items with profit $p(L \cup S') \geq (1 - 3\epsilon')\mathrm{OPT}(I')$.

Reassembling the strip-like bin requires $d-1$ applications of the *shifting technique* and can thus be done with losing at most another $\epsilon'\mathrm{OPT}(I)$ of the profit. Let S' be the set of small items after the reassembling. We proved

Lemma 9. *If $a_{max} \geq 1 - \frac{1}{K}$, we can find a selection $S' \subseteq S$ of small items in polynomial time such that L and S' can be packed together and $p(L \cup S') \geq (1 - 4\epsilon')\mathrm{OPT}(I)$.*

Now let us see that, if $|L_{opt}| < 2^d$, then $\mathrm{Vol}(L_{opt}) \leq 1 - \alpha_i$ or $a_{max} \geq 1 - \frac{1}{K}$. Similar to the two-dimensional analysis, we get a volume bound of $\mathrm{Vol}(L_{opt}) \leq f_d(a_{max}) = a_{max}^d + (2^d - 2)(1 - a_{max})^d$ for $a_{max} \in [\frac{1}{2}, 1]$, see Figure 7. With the second derivate it is easy to see, that $f_d(a_{max}) \leq 1 - \frac{1}{2^d}$ for $a_{max} \in [\frac{1}{2}, \frac{3}{4}]$ and $f_d(a_{max}) \leq a_{max}$ for $a_{max} \in [\frac{3}{4}, 1]$. Thus $a_{max} \leq 1 - \frac{1}{K}$ implies $\mathrm{Vol}(L_{opt}) \leq 1 - \frac{1}{K}$ for $\frac{1}{K} \leq \frac{1}{2^d}$. Note that $\frac{1}{K} \leq \frac{1}{2^d}$ can be achieved by choosing $K \geq 2^d$. We showed

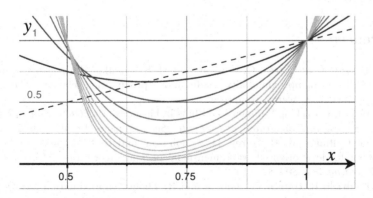

Fig. 7. The volume functions $f_d(a_{max})$ for $d = 2, \ldots, 10$ and $a_{max} \in [\frac{1}{2}, 1]$ (*solid*), and the, on $x \in [\frac{3}{4}, 1]$, dominating function $g(x) = x$ (*slashed*)

Theorem 4. *There is a polynomial time algorithm for* hypercube OKP-d *with performance ratio* $(\frac{2^d+1}{2^d} + \epsilon)$.

6   Conclusion and Future Work

For the special case of packing hypercube items we derived an approximation algorithm for *OKP-d* with performance ratio $(\frac{2^d+1}{2^d} + \epsilon)$ that is, surprisingly, improving with the dimension. Already for the two- and three-dimensional case, we significantly improve upon the best-known general algorithms.

We gave PTAS-like approximations for the cases that either the remaining volume after packing the large items is big enough or there is only one very large item. In the case of *several large items*, the gap structure becomes more complicated with an increasing number of items. Although for *square packing* it seems to be possible to handle the cases $|L_{opt}| = 4$, since the remaining space has the shape of four strip-like bins, and $|L_{opt}| = 5$, since it can be reduced to the case with four large items or the case with enough remaining space, we could not derive a general method to cope with large numbers of items in L_{opt}. Further research should thus be concentrated on the case of several large items in order to solve the question whether or not a PTAS for *hypercube OKP-d* exists.

Acknowledgments

I would like to thank Han Xin for introducing me to the subject and Thomas Hofmeister, Kazuo Iwama and the anonymous referees for many valuable comments and suggestions regarding the presentation of the paper. I am grateful to Ingo Wegener for his encouragement and to Kazuo Iwama for his hospitality. Finally, I thank Annalena for supporting me in all respects.

The research was done while visiting the Graduate School of Informatics at Kyoto University.

References

[1] N. Bansal, J. R. Correa, C. Kenyon, and M. Sviridenko. Bin packing in multiple dimensions - inapproximability results and approximation schemes. *Mathematics of Operations Research*, to appear.

[2] A. Caprara. Packing 2-dimensional bins in harmony. In *FOCS: Proc. 43rd IEEE Symposium on Foundations of Computer Science*, pages 490–499, 2002.

[3] E. G. Coffman, M. R. Garey, D. S. Johnson, and R. E. Tarjan. Performance bounds for level-oriented two-dimensional packing algorithms. *SIAM Journal of Computing*, 9(4):808–826, 1980.

[4] F. Diedrich, R. Harren, K. Jansen, and R. Thöle. Approximation algorithms for a three-dimensional orthogonal knapsack problem. to appear, 2006.

[5] A. V. Fishkin, O. Gerber, and K. Jansen. On weighted rectangle packing with large resources. In *IFIP TCS '04: 18th World Computer Congress, TC1 3rd Int. Conference on Theoretical Computer Science*, pages 237–250, 2004.

[6] A. V. Fishkin, O. Gerber, K. Jansen, and R. Solis-Oba. Packing weighted rect-
     angles into a square. In *MFCS '05: Proc. 30th Int. Symposium on Mathematical
     Foundations of Computer Sience*, pages 352–363, 2005.
[7] A. Freund and J. Naor. Approximating the advertisement placement problem.
     Journal of Scheduling, 7(5):365–374, 2004.
[8] K. Jansen and R. Solis-Oba. An asymptotic approximation algorithm for 3d-strip
     packing. In *SODA '06: Proc. 17th ACM-SIAM symposium on Discrete algorithm*,
     pages 143–152, 2006.
[9] K. Jansen and R. van Stee. On strip packing with rotations. In *STOC '05: Proc.
     37th annual ACM symposium on Theory of Computing*, pages 755–761, 2005.
[10] K. Jansen and G. Zhang. Maximizing the number of packed rectangles. In *SWAT
     '04: Proc. 9th Scandinavian Workshop on Algorithm Theory*, pages 362–371, 2004.
[11] K. Jansen and G. Zhang. On rectangle packing: Maximizing benefits. In *SODA
     '04: Proc. 15th ACM-SIAM Symposium on Discrete Algorithms*, pages 204–213,
     2004.
[12] H. Kellerer, U. Pferschy, and D. Pisinger. *Knapsack Problems*. Springer, Berlin,
     2004.
[13] C. Kenyon and E. Rémila. A near optimal solution to a two-dimensional cutting
     stock problem. *MOR: Mathematics of Operations Research*, 25:645–656, 2000.
[14] J. Y.-T. Leung, T. W. Tam, C. S. Wong, G. H. Young, and F. Y. Chin. Packing
     squares into a square. *Journal of Parallel and Distributed Computing*, 10(3):271–
     275, 1990.
[15] S. Martello and P. Toth. *Knapsack Problems: Algorithms and Computer Imple-
     mentations*. Wiley, 1990.

A Faster Deterministic Algorithm for Minimum Cycle Bases in Directed Graphs

Ramesh Hariharan[1], Telikepalli Kavitha[1,*], and Kurt Mehlhorn[2,**]

[1] Indian Institute of Science, Bangalore, India
{ramesh, kavitha}@csa.iisc.ernet.in
[2] Max-Planck-Institut für Informatik, Saarbrücken, Germany
mehlhorn@mpi-inf.mpg.de

Abstract. We consider the problem of computing a minimum cycle basis in a directed graph. The input to this problem is a directed graph G whose edges have non-negative weights. A cycle in this graph is actually a cycle in the underlying undirected graph with edges traversable in both directions. A $\{-1, 0, 1\}$ edge incidence vector is associated with each cycle: edges traversed by the cycle in the right direction get 1 and edges traversed in the opposite direction get -1. The vector space over $\mathbb{Q}$ generated by these vectors is the cycle space of G. A minimum cycle basis is a set of cycles of minimum weight that span the cycle space of G. The current fastest algorithm for computing a minimum cycle basis in a directed graph with m edges and n vertices runs in $\tilde{O}(m^{\omega+1}n)$ time (where $\omega < 2.376$ is the exponent of matrix multiplication). Here we present an $O(m^3 n + m^2 n^2 \log n)$ algorithm. We also slightly improve the running time of the current fastest randomized algorithm from $O(m^2 n \log n)$ to $O(m^2 n + mn^2 \log n)$.

1 Introduction

Let $G = (V, E)$ be a directed graph with m edges and n vertices. A *cycle* in G is actually a cycle in the underlying undirected graph, i.e., edges are traversable in both directions. Associated with each cycle is a $\{-1, 0, 1\}$ edge incidence vector: edges traversed by the cycle in the right direction get 1, edges traversed in the opposite direction get -1, and edges not in the cycle at all get 0. The vector space over $\mathbb{Q}$ generated by these vectors is the *cycle space* of G. A set of cycles is called a *cycle basis* if it forms a basis for this vector space. When G is connected, the cycle space has dimension $d = m - n + 1$. We assume that there is a weight function $w : E \to \mathbb{R}^{\geq 0}$, i.e., the edges of G have non-negative weights assigned to them. The weight of a cycle basis is the sum of the weights of its cycles. A *minimum cycle basis* of G is a cycle basis of minimum weight. We consider the problem of computing a minimum cycle basis in a given digraph.

---
* This research was partially supported by a "Max Planck-India Fellowship" provided by the Max Planck Society.
** Partially supported by the Future and Emerging Technologies programme of the EU under contract number IST-1999-14186 (ALCOM-FT).

M. Bugliesi et al. (Eds.): ICALP 2006, Part I, LNCS 4051, pp. 250–261, 2006.

A related problem pertains to undirected graphs, where a $\{0, 1\}$ edge incidence vector is associated with each cycle; edges in the cycle get 1 and others get 0. Unlike directed graphs where the cycle space is defined over $\mathbb{Q}$, cycle spaces in undirected graphs are defined as vector spaces over $\mathbb{Z}_2$. Transforming cycles in a directed cycle basis by replacing both -1 and 1 by 1 does not necessarily yield a basis for the underlying undirected graph. In addition, lifting a minimum cycle basis of the underlying undirected graph by putting back directions does not necessarily yield a *minimum* cycle basis for the directed graph. Examples of both phenomena were given in [16]. Thus, one cannot find a minimum cycle basis for a directed graph by simply working with the underlying undirected graph. Books by Deo [6] and Bollobás [3] have an in-depth coverage of cycle bases.

Motivation. Apart from its interest as a natural question, an efficient algorithm for computing a minimum cycle basis has several applications. A minimum cycle basis is primarily used as a preprocessing step in several algorithms. That is, a cycle basis is used as an input for a later algorithm, and using a minimum cycle basis instead of any arbitrary cycle basis reduces the amount of work that has to be done by this later algorithm. Such algorithms span diverse applications like structural engineering [4], cycle analysis of electrical networks [5], and chemical ring perception [7]. The network graphs of interest are frequently directed graphs. Further, specific kinds of cycle bases of directed graphs have been studied in [14,15,8]. One special class is integral cycle bases [14,15], in which the $d \times m$ cycle-edge incidence matrix has the property that all regular $d \times d$ submatrices have determinant ± 1; such cycle bases of minimum length are important in cyclic timetabling. Cycle bases in strongly connected digraphs where cycles are forced to follow the direction of the edges were studied in [8]; such cycle bases are of particular interest in metabolic flux analysis.

Previous Work and Our Contribution. There are several algorithms for computing a minimum cycle basis in an undirected graph [2,5,9,10,13] and the current fastest algorithm runs in $O(m^2 n + mn^2 \log n)$ time [13]. The first polynomial time algorithm for computing a minimum cycle basis in a directed graph had a running time of $\tilde{O}(m^4 n)$ [12]. Liebchen and Rizzi [16] gave an $\tilde{O}(m^{\omega+1} n)$ algorithm for this problem, where $\omega < 2.376$ is the exponent of matrix multiplication; this was the current fastest deterministic algorithm. A faster randomized algorithm of Monte Carlo type with running time $O(m^2 n \log n)$ exists [11].

We present an $O(m^3 n + m^2 n^2 \log n)$ deterministic algorithm for this problem and improve the running time of the randomized algorithm to $O(m^2 n + mn^2 \log n)$. The running time of our deterministic algorithm is m times the running time of the fastest algorithm for computing minimum cycle bases in undirected graphs and we leave it as a challenge to close the gap. The increased complexity seems to stem from the larger base field. Arithmetic in $\mathbb{Z}_2$ suffices for undirected graphs. For directed graphs, the base field is $\mathbb{Q}$ and this seems to necessitate the handling of large numbers. Also, the computation of a shortest cycle that has a non-zero dot product with a given vector seems more difficult in directed graphs than in undirected graphs.

2   Preliminaries

We are given a digraph $G = (V, E)$, where $|V| = n$ and $|E| = m$. Without loss of generality, the underlying undirected graph of G is connected. Then $d = m - n + 1$ is the dimension of the cycle space of G. The minimum cycle basis of G consists of d cycles $C_1, \ldots, C_d$. We describe cycles by their incidence vectors in $\{-1, 0, +1\}^m$. We assume that we have ordered the edges in the edge set $E = \{e_1, \ldots, e_m\}$ so that edges $e_{d+1}, \ldots, e_m$ form the edges of a spanning tree T of the underlying undirected graph. This means that the first d coordinates each of $C_1, \ldots, C_d$ correspond to edges outside the tree T and the last $n - 1$ coordinates are the edges of T. This will be important in our proofs in Section 3.

We can also assume that there are no multiple edges in G. It is easy to see that whenever there are two edges from u to v, the heavier edge (call it a) can be deleted from E and the least weight cycle (call it $C(a)$) that contains the edge a can be added to the minimum cycle basis computed on $(V, E \setminus \{a\})$. The cycle $C(a)$ consists of the edge a and the shortest path between u and v in the underlying undirected graph. All such cycles can be computed by an all-pairs-shortest-paths computation in the underlying undirected graph of G, which takes $\tilde{O}(mn)$ time. Hence we can assume that $m \leq n^2$.

Framework. We begin with a structural characterization of a minimum cycle basis, which is simple to show. This framework was introduced by de Pina [5]; it uses auxiliary rational vectors $N_1, \ldots, N_d$ which serve as a scaffold for proving that $C_1, \ldots, C_d$ form a minimum cycle basis. We use $\langle v_1, v_2 \rangle$ to denote the standard inner product or dot product of the vectors v_1 and v_2.

Theorem 1. *Cycles $C_1, \ldots, C_d$ form a minimum cycle basis if there are vectors $N_1, \ldots, N_d$ in $\mathbb{Q}^m$ such that for all i, $1 \leq i \leq d$:*

1. **Prefix Orthogonality:** N_i *is orthogonal to all previous C_j, i.e., $\langle N_i, C_j \rangle = 0$ for all j, $1 \leq j < i$.*
2. **Non-Orthogonality:** $\langle N_i, C_i \rangle \neq 0$.
3. **Shortness:** C_i *is a shortest cycle with $\langle N_i, C_i \rangle \neq 0$.*

3   A Simple Deterministic Algorithm

We present the simple deterministic algorithm from [12], that computes N_i's and C_i's satisfying the criteria in Theorem 1.

The algorithm Deterministic-MCB:

1. Initialize the vectors $N_1, \ldots, N_d$ of $\mathbb{Q}^m$ to the first d vectors $e_1, \ldots, e_d$ of the standard basis of $\mathbb{Q}^m$. (The vector e_i has 1 in the i-th position and 0's elsewhere.)
2. For $i = 1$ to d do
   - compute C_i to be a shortest cycle such that $\langle C_i, N_i \rangle \neq 0$.

- for $j = i + 1$ to d do

$$\text{update } N_j \text{ as: } N_j = N_j - N_i \frac{\langle C_i, N_j \rangle}{\langle C_i, N_i \rangle}$$

$$\text{normalize } N_j \text{ as: } N_j = N_j \frac{\langle C_i, N_i \rangle}{\langle C_{i-1}, N_{i-1} \rangle}$$

(We take $\langle C_0, N_0 \rangle = 1$.) The above algorithm needs the vector N_i in the i-th iteration to compute the cycle C_i. Instead of computing N_i from scratch in the i-th iteration, it obtains N_i by update and normalization steps through iterations 1 to $i-1$. We describe how to compute a shortest cycle C_i such that $\langle C_i, N_i \rangle \neq 0$ in Section 5. Let us now show that the N_i's obey the prefix orthogonality property. Lemma 1, proved in [12], shows that and more.

Lemma 1. *For any i, at the end of iteration $i - 1$, the vectors $N_i, \ldots, N_d$ are orthogonal to $C_1, \ldots, C_{i-1}$ and moreover, for any j with $i \leq j \leq d$, $N_j = \langle N_{i-1}, C_{i-1} \rangle (x_{j,1}, \ldots, x_{j,i-1}, 0, \ldots, 0, 1, 0, \ldots, 0)$, where 1 occurs in the j-th coordinate and $\mathbf{x} = (x_{j,1}, \ldots, x_{j,i-1})$ is the unique solution to the set of equations:*

$$\begin{pmatrix} \tilde{C}_1 \\ \vdots \\ \tilde{C}_{i-1} \end{pmatrix} \mathbf{x} = \begin{pmatrix} -c_{1j} \\ \vdots \\ -c_{(i-1)j} \end{pmatrix}. \tag{1}$$

Here $\tilde{C}_k$, $1 \leq k < i$, is the restriction of C_k to its first $i - 1$ coordinates and c_{kj} is the j-th coordinate of C_k.

Remark. Note that the i-th coordinate of N_i is non-zero. This readily implies that there is at least one cycle that has non-zero dot product with N_i, namely the fundamental cycle F_{e_i} formed by the edge e_i and the path in the spanning tree T connecting its endpoints. The dot product $\langle F_{e_i}, N_i \rangle$ is equal to the i-th coordinate of N_i, which is non-zero.

We next give an alternative characterization of these N_j's. This characterization helps us in bounding the running time of the algorithm Deterministic-MCB. Let M denote the $(i - 1) \times (i - 1)$ matrix of $\tilde{C}_k$'s in Equation (1) and b_j denote the column vector of $-c_{kj}$'s on the right. We claim that solving $M\mathbf{x} = \det(M) \cdot b_j$ leads to the same vectors N_j for all j with $i \leq j \leq d$. First, note that $\langle N_{i-1}, C_{i-1} \rangle = \det(M)$, it is easy to show this.

By Lemma 1, $(x_{j,1}, \ldots, x_{j,i-1})$ is the unique solution to $M\mathbf{x} = b_j$. Hence $\det(M)(x_{i,1}, \ldots, x_{i,i-1}, 1, 0, \ldots)$, which is N_i in the i-th iteration (that is when it is used in the algorithm to compute the cycle C_i), could have obtained directly in the i-th iteration by solving the set of equations $M\mathbf{x} = \det(M)b_i$ and appending $(\det(M), 0, \ldots, 0)$ to $\mathbf{x}$. However, such an algorithm would be slower - it would take time $\tilde{\Theta}(m^{\omega+2})$, where $\omega < 2.376$ is the exponent of matrix multiplication. The updates and normalizations in the algorithm Deterministic-MCB achieve the same result in a more efficient manner.

Let us now bound the running time of the i-th iteration of Deterministic-MCB. We will show in Section 5 that a shortest cycle C_i such that $\langle C_i, N_i \rangle \neq 0$ can be computed in $O(m^2 n + mn^2 \log n)$ time. Let us look at bounding the time taken for update and normalization steps. We take $O(m)$ arithmetic steps for updating and scaling each N_j since each N_j has m coordinates. Thus the total number of arithmetic operations in the i-th iteration is $O((d-i)m) = O(md)$ over all j, $i+1 \leq j \leq d$. We next estimate the cost of arithmetic. The coordinates of N_j are determined by the system $M\mathbf{x} = \det(M)b_j$ and hence are given by Cramer's rule. Fact 1, which follows from Hadamard's inequality, shows that each entry in N_j is bounded by $d^{d/2}$. Thus we pay $\tilde{O}(d)$ time per arithmetic operation. Thus the running time of the i-th iteration is $\tilde{O}(m^3)$ and hence the running time of Deterministic-MCB is $\tilde{O}(m^4)$.

Fact 1. *Since M is a $\pm 1, 0$ matrix of size $(i-1) \times (i-1)$ and b_j is a $\pm 1, 0$ vector, all determinants used in Cramer's Rule are bounded by $i^{i/2}$. Therefore, the absolute value of each entry in N_j, where $j \geq i$, is bounded by $i^{i/2}$.*

4   A Faster Deterministic Algorithm

The update and normalization steps form the bottleneck in Deterministic-MCB. We will reduce their cost from $\tilde{O}(m^4)$ to $\tilde{O}(m^{\omega+1})$.

- First, we delay updates until after several new cycles C_i have been computed. For instance, we update $N_{\lfloor d/2 \rfloor+1}, \ldots, N_d$ not after each new cycle but in bulk after *all* of $C_1, C_2, \ldots, C_{\lfloor d/2 \rfloor}$ are computed.
- Second, we use a fast matrix multiplication method to do the updates for all of $N_{\lfloor d/2 \rfloor+1}, \ldots, N_d$ together, and not individually as before.

The Scheme. The faster deterministic algorithm starts with the same configuration for the N_i's as before, i.e., N_i is initialized to the i-th unit vector, $1 \leq i \leq d$. It then executes 3 steps. First, it computes $C_1, \ldots, C_{\lfloor d/2 \rfloor}$ and $N_1, \ldots, N_{\lfloor d/2 \rfloor}$ recursively, leaving $N_{\lfloor d/2 \rfloor+1}, \ldots, N_d$ at their initial values. Second, it runs a bulk update step in which $N_{\lfloor d/2 \rfloor+1}, \ldots, N_d$ are modified so that they become orthogonal to $C_1, \ldots, C_{\lfloor d/2 \rfloor}$. And third, $C_{\lfloor d/2 \rfloor+1}, \ldots, C_d$ are computed recursively modifying $N_{\lfloor d/2 \rfloor+1}, \ldots, N_d$ in the process. Such a scheme was used earlier in [13,11].

A crucial point to note about the second recursive call is that it modifies $N_{\lfloor d/2 \rfloor+1}, \ldots, N_d$ while ignoring $C_1, \ldots, C_{\lfloor d/2 \rfloor}$ and $N_1, \ldots, N_{\lfloor d/2 \rfloor}$; how then does it retain the orthogonality of $N_{\lfloor d/2 \rfloor+1}, \ldots, N_d$ with $C_1, \ldots, C_{\lfloor d/2 \rfloor}$ that we achieved in the bulk update step? The trick lies in the fact that whenever we update any $N_j \in \{N_{\lfloor d/2 \rfloor+1}, \ldots, N_d\}$ in the second recursive call, we do it as $N_j = \sum_{k=\lfloor d/2 \rfloor+1}^{d} \alpha_k N_k, \alpha_k \in \mathbb{Q}$. That is, the updated N_j is obtained as a rational linear combination of $N_{\lfloor d/2 \rfloor+1}, \ldots, N_d$. Since the bulk update step prior to the second recursive call ensures that $N_{\lfloor d/2 \rfloor+1}, \ldots, N_d$ are all orthogonal to $C_1, \ldots, C_{\lfloor d/2 \rfloor}$ at the beginning of this step, the updated N_j's remain orthogonal to $C_1, \ldots, C_{\lfloor d/2 \rfloor}$. This property allows the second recursive call to work strictly in the bottom half of the data without looking at the top half.

The base case for the recursion is a subproblem of size 1 (let this subproblem involve C_i, N_i) in which case the algorithm simply retains N_i as it is and computes C_i using the algorithm in Section 5. As regards time complexity, the bulk update step will be shown to take $O(md^{\omega-1})$ arithmetic operations.

4.1   The Bulk Update Procedure

We describe the bulk update procedure in the recursive call that computes the cycles $C_\ell, \ldots, C_h$ for some h and ℓ with $h > \ell$. This recursive call works with the vectors $N_\ell, \ldots, N_h$: all these vectors are already orthogonal to $C_1, \ldots, C_{\ell-1}$. The recursive call runs as follows:

1. compute the cycles $C_\ell, \ldots, C_{mid}$ where $mid = \lceil (\ell + h)/2 \rceil - 1$, using the vectors $N_\ell, \ldots, N_{mid}$, recursively.
2. modify $N_{mid+1}, \ldots, N_h$, which are untouched by the first step, to make them orthogonal to $C_\ell, \ldots, C_{mid}$.
3. compute $C_{mid+1}, \ldots, C_h$ using these $N_{mid+1}, \ldots, N_h$, recursively.

Step 2 is the bulk update step. We wish to update each N_j, $mid + 1 \le j \le h$, to a rational linear combination of $N_\ell, \ldots, N_{mid}$ and N_j as follows[1]:

$$N_j = \frac{\langle N_{mid}, C_{mid} \rangle}{\langle N_{\ell-1}, C_{\ell-1} \rangle} N_j + \sum_{t=\ell}^{mid} \alpha_{tj} N_t$$

where the α_{tj}'s are to be determined in a way which ensures that N_j becomes orthogonal to $C_\ell, \ldots, C_{mid}$. That is, we want for all i, j, where $\ell \le i \le mid$ and $mid + 1 \le j \le h$,

$$\frac{\langle N_{mid}, C_{mid} \rangle}{\langle N_{\ell-1}, C_{\ell-1} \rangle} \langle C_i, N_j \rangle + \sum_{t=\ell}^{mid} \alpha_{tj} \langle C_i, N_t \rangle = 0. \qquad (2)$$

Rewriting this in matrix form, we get $A \cdot \mathcal{N}_d \cdot D = -A \cdot \mathcal{N}_u \cdot X$, where (let $k = mid - \ell + 1$)

 - A is a $k * m$ matrix, the i-th row of which is $C_{\ell+i-1}$,
 - $\mathcal{N}_d$ is an $m * (h - k)$ matrix, the j-th column of which is N_{mid+j},
 - D is an $(h - k) * (h - k)$ diagonal matrix with $\langle N_{mid}, C_{mid} \rangle / \langle N_{\ell-1}, C_{\ell-1} \rangle$ in the diagonal,
 - $\mathcal{N}_u$ is an $m * k$ matrix, the t-th column of which is $N_{\ell+t-1}$,
 - X is the $k * (h - k)$ matrix of variables α_{tj}, with t indexing the rows and j indexing the columns.

---

[1] Note that the coefficient $\langle N_{mid}, C_{mid} \rangle / \langle N_{\ell-1}, C_{\ell-1} \rangle$ for N_j is chosen so that the updated vector N_j here is exactly the same vector N_j that we would have obtained at this stage using the algorithm Deterministic-MCB (Section 3).

To compute the α_{tj}'s, we solve for $X = -(A \cdot \mathcal{N}_u)^{-1} \cdot A \cdot \mathcal{N}_d \cdot D$. Using fast matrix multiplication, we can compute $A \cdot \mathcal{N}_u$ and $A \cdot \mathcal{N}_d$ in $O(mk^{\omega-1})$ time, by splitting the matrices into d/k square blocks and using fast matrix multiplication to multiply the blocks. Multiplying each element of $A \cdot \mathcal{N}_d$ with the scalar $\langle N_{mid}, C_{mid} \rangle / \langle N_{\ell-1}, C_{\ell-1} \rangle$ gives us $A \cdot \mathcal{N}_d \cdot D$. Thus we compute the matrix $A \cdot \mathcal{N}_d \cdot D$ with $O(mk^{\omega-1})$ arithmetic operations. Next, we find the inverse of $A \cdot \mathcal{N}_u$ with $O(k^{\omega})$ arithmetic operations (this inverse exists because $A \cdot \mathcal{N}_u$ is a lower triangular matrix whose diagonal entries are $\langle C_i, N_i \rangle \neq 0$). Then we multiply $(A \cdot \mathcal{N}_u)^{-1}$ with $A \cdot \mathcal{N}_d \cdot D$ with $O(k^{\omega})$ arithmetic operations. Thus we obtain X. Finally, we obtain $N_{mid+1}, \ldots, N_d$ from X using the product $\mathcal{N}_u \cdot X$, which we can compute in $O(mk^{\omega-1})$ arithmetic operations, and adding $\mathcal{N}_d \cdot D$ to $\mathcal{N}_u \cdot X$. The total number of arithmetic operations required for the bulk update step is thus $O(mk^{\omega-1})$.

What is the cost of the arithmetic? In the algorithm presented above, the entries in $(A \cdot \mathcal{N}_u)^{-1}$ could be very large. The elements in $A \cdot \mathcal{N}_u$ have values up to $d^{\Theta(d)}$, which would result in the entries in $(A \cdot \mathcal{N}_u)^{-1}$ being as large as $d^{\Theta(d^2)}$. So each arithmetic operation then costs us up to $\tilde{\Theta}(d^2)$ time and the overall time for the outermost bulk update step would be $\tilde{\Theta}(m^{\omega+2})$ time, which makes this approach slower than the algorithm Deterministic-MCB.

The good news is that the numbers α_{tj}'s are just *intermediate* numbers in our computation. That is, they are the coefficients in

$$\sum_{t=\ell}^{mid} \alpha_{tj} N_t + \frac{\langle N_{mid}, C_{mid} \rangle}{\langle N_{\ell-1}, C_{\ell-1} \rangle} N_j.$$

Our final aim is to determine the updated coordinates of N_j which are at most $d^{d/2}$ (refer Fact 1), since we know $N_j = (y_1, \ldots, y_{mid}, 0, \ldots, \langle N_{mid}, C_{mid} \rangle, \ldots, 0)$, where $\mathbf{y} = (y_1, \ldots, y_{mid})$ is the solution to the linear system: $M\mathbf{y} = \det(M)b_j$; M is the $mid \times mid$ matrix of $C_1, \ldots, C_{mid}$ truncated to their first mid coordinates and b_j is the column vector of negated j coordinates of $C_1, \ldots, C_{mid}$. Since the final coordinates are bounded by $d^{d/2}$ while the intermediate values could be much larger, this suggests the use of modular arithmetic here. We could work over the finite fields $\mathbb{F}_{p_1}, \mathbb{F}_{p_2}, \ldots, \mathbb{F}_{p_s}$ where $p_1, \ldots, p_s$ are small primes (say, in the range d to d^2) and try to retrieve N_j from $N_j \bmod p_1, \ldots, N_j \bmod p_s$, which is possible (by the Chinese Remainder Theorem) if $s \approx d/2$. Arithmetic in $\mathbb{F}_p$ takes $O(1)$ time and we thus spend $O(smk^{\omega-1})$ time for the update step now. However, if it is the case that some p is a divisor of some $\langle N_i, C_i \rangle$ where $\ell \leq i \leq mid$, then we cannot invert $A \cdot \mathcal{N}_u$ in the field $\mathbb{F}_p$. Since each number $\langle N_i, C_i \rangle$ could be as large as $d^{d/2+1}$, it could be a multiple of up to $\Theta(d)$ primes which are in the range $d, \ldots, d^2$. So in order to be able to determine d primes which are relatively prime to each of $\langle N_\ell, C_\ell \rangle, \ldots, \langle N_{mid}, C_{mid} \rangle$, we might in the worst case have to test about $(mid - \ell + 1) \cdot d = kd$ primes. Testing kd primes for divisibility w.r.t. k d-bit numbers costs us $k^2 d^2$ time. We cannot afford so much time per update step.

Another idea is to work over just one finite field $\mathbb{F}_q$ where q is a large prime. If $q > d^{d/2+1}$, then it can never be a divisor of any $\langle N_i, C_i \rangle$, so we can always carry out our arithmetic in $\mathbb{F}_q$ without any problem. Arithmetic in $\mathbb{F}_q$ costs us $\tilde{\Theta}(d)$ time if $q \approx d^d$. Then our update step takes $\tilde{O}(m^2 k^{\omega-1})$ time which will result in a total time of $\tilde{O}(m^{\omega+1})$ for all the update steps, which is our goal. But computing such a large prime q is a difficult problem.

The solution is to work over a suitable ring instead of over a field; note that fast matrix multiplication algorithms work over rings. Let us do the above computation modulo a large integer R, say $R \approx d^d$. Then intermediate numbers do not grow more than R and we can retrieve N_j directly from $N_j \bmod R$, because R is much larger than any coordinate of N_j.

What properties do we need of R? The integer R must be relatively prime to the numbers: $\langle N_\ell, C_\ell \rangle$, $\langle N_{\ell+1}, C_{\ell+1} \rangle$, ..., $\langle N_{mid}, C_{mid} \rangle$ so that that triangular matrix $A \cdot N_d$ which has these elements along the diagonal is invertible in $\mathbb{Z}_R$. And R must also be relatively prime to $\langle N_{\ell-1}, C_{\ell-1} \rangle$ so that the number $\langle N_{mid}, C_{mid} \rangle / \langle N_{\ell-1}, C_{\ell-1} \rangle$ is defined in $\mathbb{Z}_R$. Once we determine such an R, we will work in $\mathbb{Z}_R$. We stress the point that such an R is a number used only in this particular bulk update step - in another bulk update step of another recursive call, we need to compute another such large integer.

It is easy to see that the number R determined below is a large number that is relatively prime to $\langle N_{\ell-1}, C_{\ell-1} \rangle$, $\langle N_\ell, C_\ell \rangle$, $\langle N_{\ell+1}, C_{\ell+1} \rangle$, ..., and $\langle N_{mid}, C_{mid} \rangle$.

1. Right at the beginning of the algorithm, compute d^2 primes $p_1, ..., p_{d^2}$, where each of these primes is at least d. Then form the d products: $P_1 = p_1 \cdots p_d$, $P_2 = p_1 \cdots p_{2d}$, $P_3 = p_1 \cdots p_{3d}$, ..., $P_d = p_1 \cdots p_{d^2}$.
2. Then during our current update step, compute the product: $\mathcal{L} = \langle N_{\ell-1}, C_{\ell-1} \rangle \langle N_\ell, C_\ell \rangle \cdots \langle N_{mid}, C_{mid} \rangle$.
3. By doing a binary search on $P_1, \ldots, P_d$, determine the smallest $s \geq 0$ such that P_{s+1} does not divide $\mathcal{L}$.
4. Determine a $p \in \{p_{sd+1}, \ldots, p_{sd+d}\}$ that does not divide $\mathcal{L}$. Compute $R = p^d$.

Cost of computing R: The value of $\pi(r)$, the number of primes less than r, is given by $r/6 \log r \leq \pi(r) \leq 8r/\log r$ [1]. So all the the primes $p_1, \ldots, p_{d^2}$ are $\tilde{O}(d^2)$, and computing them takes $\tilde{O}(d^2)$ time using a sieving algorithm. The products $P_1, \ldots, P_d$ are computed just once in a preprocessing step. We will always perform arithmetic on large integers using Schönhage-Strassen multiplication, so that it takes $\tilde{O}(d)$ time to multiply two d-bit numbers. Whenever we perform a sequence of multiplications, we will perform it using a tree so that d numbers (each of bit size $\tilde{O}(d)$) can be multiplied in $\tilde{O}(d^2)$ time. So computing $P_1, \ldots, P_d$ takes $\tilde{O}(d^3)$ preprocessing time.

In the update step, we compute $\mathcal{L}$, which takes $\tilde{O}(d^2)$ time. The product $p_{sd+1} \cdots p_{sd+d}$ is found in $\tilde{O}(d^2)$ time by binary search. Determine a $p \in \{p_{sd+1}, \ldots, p_{sd+d}\}$ that does not divide $\mathcal{L}$ by testing which of the two products: $p_{sd+1} \cdots p_{sd+\lfloor d/2 \rfloor}$ or $p_{sd+\lfloor d/2 \rfloor+1} \cdots p_{sd+d}$ does not divide $\mathcal{L}$ and recurse on the product that does not divide $\mathcal{L}$. Thus R can be computed in $\tilde{O}(d^2)$ time.

Computation in $\mathbb{Z}_R$. We need to invert the matrix $A \cdot N_u$ in the ring $\mathbb{Z}_R$. Recall that this matrix is lower triangular. Computing the inverse of a lower triangular matrix is easy. If

$$A \cdot \mathcal{N}_u = \begin{pmatrix} W & 0 \\ Y & Z \end{pmatrix}, \text{ then we have } (A \cdot \mathcal{N}_u)^{-1} = \begin{pmatrix} W^{-1} & 0 \\ -Z^{-1}YW^{-1} & Z^{-1} \end{pmatrix}.$$

Hence to invert $A \cdot \mathcal{N}_u$ in $\mathbb{Z}_R$ we need the multiplicative inverses of only its diagonal elements: $\langle C_\ell, N_\ell \rangle, \ldots, \langle C_{mid}, N_{mid} \rangle$ in $\mathbb{Z}_R$. Using Euclid's gcd algorithm each inverse can be computed in $\tilde{O}(d^2)$ time since each of the numbers involved here and R have bit size $\tilde{O}(d)$. The matrix $A \cdot \mathcal{N}_u$ is inverted via fast matrix multiplication and once we compute $(A \cdot \mathcal{N}_u)^{-1}$, the matrix X, that consists of all the coordinates α_{tj} that we need (refer Equation (2)), can be easily computed in $\mathbb{Z}_R$ as $-(A \cdot \mathcal{N}_u)^{-1} \cdot A \cdot \mathcal{N}_d \cdot D$ by fast matrix multiplication. Then we determine all $N_j \bmod R$ for $mid + 1 \le j \le h$ from $\mathcal{N}_u \cdot X + \mathcal{N}_d \cdot D$. It follows from the discussion presented at the beginning of Section 4.1 that the time required for all these operations is $\tilde{O}(m^2 k^{\omega - 1})$ since each number is now bounded by d^d.

Retrieving the actual N_j. Each entry of N_j can have absolute value at most $d^{d/2}$ (from Fact 1). The number R is much larger than this, $R > d^d$. So if any coordinate, say n_l in $N_j \bmod R$ is larger than $d^{d/2}$, then we can retrieve the original n_l as $n_l - R$. Thus we can retrieve N_j from $N_j \bmod R$ in $O(d^2)$ time. The time complexity for the update step, which includes matrix operations, gcd computations and other arithmetic, is $\tilde{O}(m^2 k^{\omega - 1} + d^2 k)$ or $\tilde{O}(m^2 k^{\omega - 1})$. Thus our recurrence becomes $T(k) = 2T(k/2) + \tilde{O}(m^2 k^{\omega - 1})$ when $k > 1$. We shall show the following lemma in the next section.

Lemma 2. *A shortest cycle C_i such that $\langle C_i, N_i \rangle \neq 0$ can be computed in $O(m^2 n + mn^2 \log n)$ time.*

Thus $T(1) = O(m^2 n + mn^2 \log n)$. Our recurrence solves to $T(k) = O(k(m^2 n + mn^2 \log n) + k^\omega m^2 \cdot \text{poly}(\log m))$ and hence $T(d) = O(m^3 n + m^2 n^2 \log n) + \tilde{O}(m^{\omega+1})$, which is $O(m^3 n + m^2 n^2 \log n)$, because $m \le n^2$ implies $\tilde{O}(m^{\omega+1})$ is always $o(m^3 n)$. We can conclude with the following theorem.

Theorem 2. *A minimum cycle basis in a weighted directed graph with m edges and n vertices and non-negative edge weights can be computed in $O(m^3 n + m^2 n^2 \log n)$ time.*

5   Computing Non-orthogonal Shortest Cycles

Now we come to the second key routine required by our algorithm - given a directed graph G with non-negative edge weights, compute a shortest cycle in G whose dot product with a given vector $N \in \mathbb{Z}^m$ is non-zero. We will first consider the problem of computing a shortest cycle C_p such that $\langle C_p, N \rangle \neq 0 \pmod{p}$ for a number $p = O(d \log d)$. Recall that C_p can traverse edges of G in both forward and reverse directions; the vector representation of C_p has a 1 for every forward edge in the cycle, a -1 for every reverse edge, and a 0 for edges not present at all in the cycle. This vector representation is used for computing dot products with N. The weight of C_p itself is simply the sum of the weights of the edges in the cycle. We show how to compute C_p in $O(mn + n^2 \log n)$ time.

Definitions. To compute shortest paths and cycles, we will work with the undirected version of G. Directions will be used only to compute the *residue class* of a path or cycle, i.e., the dot product between the vector representation of this path or cycle and N modulo p. Let p_{uv} denote a shortest path between vertices u and v and let f_{uv} denote its length and r_{uv} its residue class. Let s_{uv} be the length of a shortest path, if any, between u and v in a residue class distinct from r_{uv}. Observe that the value of s_{uv} is independent of the choice of p_{uv}.

We will show how to compute f_{uv} and s_{uv} for all pairs of vertices u, v in $O(mn + n^2 \log n)$ time. As is standard, we will also compute paths realizing these lengths in addition to computing the lengths themselves. The following claim tells us how these paths can be used to compute a shortest non-orthogonal cycle - simply take each edge uv and combine it with s_{vu} to get a cycle. The shortest of all these cycles having a non-zero residue class is our required cycle.

Lemma 3. *Let $C = u_0 u_1 \ldots u_k u_0$ be a shortest cycle whose residue class is non-zero modulo p and whose shortest edge is $u_0 u_1$. Then the path $u_1 u_2 \ldots u_k u_0$ has a residue class different from the residue class of the edge $u_1 u_0$ and the length of the path $u_1 u_2 \ldots u_k u_0$ equals $s_{u_1 u_0}$ and the length of the edge $u_0 u_1$ equals $f_{u_1 u_0}$.*

Proof. First, we show that the path $u_1 u_2 \ldots u_k u_0$ and the edge $u_1 u_0$ have different residue classes. Let x denote the residue class of the path and y denote the residue class of the edge $u_0 u_1$. Since C is in a non-zero residue class, $x + y \not\equiv 0 \pmod{p}$, so $x \not\equiv -y \pmod{p}$. Since the incidence vector corresponding to $u_1 u_0$ is the negation of the incidence vector corresponding to $u_0 u_1$, the residue class of the edge $u_1 u_0$ is $-y$. Thus the claim follows.

Now, if the length of $u_1 u_0$ is strictly greater than $f_{u_1 u_0}$, then consider any shortest path π between u_1 and u_0 (which, of course, has length $f_{u_1 u_0}$). Combining π with $u_1 u_0$ yields a cycle and combining π with $u_1 u_2 \ldots u_k u_0$ yields another cycle. These cycles are in distinct residue classes and are shorter than C. This contradicts the definition of C. Therefore, the edge $u_1 u_0$ has length $f_{u_1 u_0}$.

Since $u_1 u_2 \ldots u_k u_0$ has a different residue class from the edge $u_1 u_0$, the length of $u_1 u_2 \ldots u_k u_0$ cannot be smaller than $s_{u_1 u_0}$, by the very definition of $s_{u_1 u_0}$. Suppose, for a contradiction that the length $u_1 u_2 \ldots u_k u_0$ is strictly larger than $s_{u_1 u_0}$. Then combining the path between u_1 and u_0 which realizes the length $s_{u_1 u_0}$ along with the edge $u_1 u_0$ yields a cycle which is shorter than C and which has a non-zero residue class modulo p. This contradicts the definition of C. The lemma follows. □

Computing f_{uv} and s_{uv}. We first find any one shortest path (amongst possibly many) between each pair of vertices u and v by Dijkstra's algorithm; this gives us p_{uv}, f_{uv}, and r_{uv}, for each pair u, v. The time taken is $O(mn + n^2 \log n)$. For each pair u, v, we now need to find a shortest path between u, v with residue class distinct from r_{uv}; the length of this path will be s_{uv}. Use q_{uv} to denote any such path. We show how a modified Dijkstra search can compute these paths in $O(mn + n^2 \log n)$ time. The following lemma shows the key prefix property of the q_{uv} paths needed for a Dijkstra-type algorithm.

Lemma 4. *For any u and v, the path q_{uv} can be chosen from the set $\{p_{uw} \circ wv, q_{uw} \circ wv : wv \in E\}$. Here $p \circ e$ denotes the path p extended by the edge e.*

Proof. Consider any path π between u and v realizing the value s_{uv}, i.e., it has length s_{uv} and residue class distinct from r_{uv}. Let w be the penultimate vertex on this path and let π' be the prefix path from u to w. Clearly, π cannot be shorter than $p_{uw} \circ wv$. Hence, if the residue class of $p_{uw} \circ wv$ is distinct from r_{uv}, we are done. So assume that $p_{uw} \circ wv$ has residue class r_{uv}. Then π' must have a residue class distinct from p_{uw} and hence q_{uw} exists. Also, the length of π' must be at least the length of q_{uw} and the residue class of $q_{uw} \circ wv$ is distinct from the residue class of $p_{uw} \circ wv$ and hence distinct from r_{uv}. Thus $q_{uw} \circ wv$ realizes s_{uv}.   □

We now show how to compute the s_{uv}'s for any fixed u in time $O(m+n \log n)$ with a Dijkstra-type algorithm. Repeating this for every source gives the result. The algorithm differs from Dijkstra's shortest path algorithm only in the initialization and update steps, which we describe below. We use the notation key_{uv} to denote the key used to organize the priority heap; key_{uv} will finally equal s_{uv}.

Initialization. We set key_{uv} to the minimal length of any path $p_{uw} \circ wv$ with residue class distinct from r_{uv}. If there is no such path, we set it to ∞.

The Update Step. Suppose we have just removed w from the priority queue. We consider the u to w path of length key_{uw} which was responsible for the current key value of w. For each edge wv incident on w, we extend this path via the edge wv. We update key_{uv} to the length of this path provided its residue class is different from r_{uv}.

Correctness. We need to show that key_{uv} is set to s_{uv} in the course of the algorithm (note that one does not need to worry about the residue class since any path that updates key_{uv} in the course of the algorithm has residue class different from r_{uv}). This follows immediately from Lemma 4. If s_{uv} is realized by the path $p_{uw} \circ wv$ for some neighbor w, then key_{uv} is set to s_{uv} in the initialization step. If s_{uv} is realized by the path $q_{uw} \circ wv$ for some neighbor w, then key_{uv} is set to s_{uv} in the update step. This completes the proof of correctness.

Thus we have given an $O(mn + n^2 \log n)$ algorithm to compute a shortest cycle C_p whose dot product with N is non-zero modulo p. A slower algorithm with running time $O(mn \log n)$, which computes a layered graph, was given in [11] to compute such a cycle C_p. Using the algorithm described here instead of this slower algorithm results in a randomized algorithm with running time $O(m^2n + mn^2 \log n)$ for the minimum cycle basis problem in directed graphs. We state this result as the following theorem.

Theorem 3. *A minimum cycle basis in a directed graph G can be computed with probability at least $3/4$ in $O(m^2n + mn^2 \log n)$ time.*

The original problem. Our original problem here was to compute a shortest cycle C such that $\langle C, N \rangle \neq 0$. Any cycle C which satisfies $\langle C, N \rangle \neq 0$ satisfies

$\langle C, N \rangle \neq 0 \pmod{p}$ for some $p \in \{p_1, \ldots, p_d\}$ where $p_1, \ldots, p_d$ are distinct primes, each of which is at least d. This follows from the isomorphism of $\mathbb{Z}_{\prod p_i}$ and $\mathbb{Z}_{p_1} \times \mathbb{Z}_{p_2} \times \cdots \times \mathbb{Z}_{p_d}$. We have $|\langle C, N \rangle| \leq \|N\|_1 \leq d \cdot d^{d/2} < \prod_{i=1}^d p_i$. So if $\langle C, N \rangle$ is non-zero, then it is a non-zero element in $\mathbb{Z}_{\prod p_i}$ and so it satisfies $\langle C, N \rangle \neq 0 \pmod{p}$ for some p in $\{p_1, \ldots, p_d\}$. Thus a shortest cycle C such that $\langle C, N \rangle \neq 0$ is the shortest among all the cycles C_p, $p \in \{p_1, \ldots, p_d\}$, where C_p is a shortest cycle such that $\langle C_p, N \rangle \neq 0 \pmod{p}$. Hence the time taken to compute C is $O(d \cdot (mn + n^2 \log n))$ or $O(m^2 n + mn^2 \log n)$. This completes the proof of Lemma 2.

References

1. T. M. Apostol. *Introduction to Analytic Number Theory*. Springer-Verlag, 1997.
2. F. Berger, P. Gritzmann, and S. de Vries. Minimum Cycle Bases for Network Graphs. *Algorithmica*, 40(1): 51-62, 2004.
3. B. Bollobás. *Modern Graph Theory*, volume 184 of *Graduate Texts in Mathematics*, Springer, Berlin, 1998.
4. A. C. Cassell, J. C. Henderson and K. Ramachandran. Cycle bases of minimal measure for the structural analysis of skeletal structures by the flexibility method *Proc. Royal Society of London Series A*, 350: 61-70, 1976.
5. J.C. de Pina. *Applications of Shortest Path Methods*. PhD thesis, University of Amsterdam, Netherlands, 1995.
6. N. Deo. *Graph Theory with Applications to Engineering and Computer Science*. Prentice-Hall Series in Automatic Computation. Prentice-Hall, Englewood Cliffs, 1982.
7. P. M. Gleiss. *Short cycles: minimum cycle bases of graphs from chemistry and biochemistry*. PhD thesis, Universität Wien, 2001.
8. P. M. Gleiss, J. Leydold, and P. F. Stadler. Circuit bases of strongly connected digraphs. *Discussiones Math. Graph Th.*, 23: 241-260, 2003.
9. Alexander Golynski and Joseph D. Horton. A polynomial time algorithm to find the minimum cycle basis of a regular matroid. In *8th Scandinavian Workshop on Algorithm Theory*, 2002.
10. J. D. Horton. A polynomial-time algorithm to find a shortest cycle basis of a graph. *SIAM Journal of Computing*, 16:359–366, 1987.
11. T. Kavitha. An $\tilde{O}(m^2 n)$ Randomized Algorithm to compute a Minimum Cycle Basis of a Directed Graph. In *Proc. of ICALP*, LNCS 3580: 273-284, 2005.
12. T. Kavitha and K. Mehlhorn. Algorithms to compute Minimum Cycle Bases in Directed Graphs. Full version to appear in special issue of *TOCS*, preliminary version in *Proc. of STACS*, LNCS 3404: 654-665, 2005.
13. T. Kavitha, K. Mehlhorn, D. Michail, and K. Paluch. A faster algorithm for Minimum Cycle Bases of graphs. In *Proc. of ICALP*, LNCS 3142: 846-857, 2004.
14. Christian Liebchen. Finding Short Integral Cycle Bases for Cyclic Timetabling. In *Proc. of ESA*, LNCS 2832: 715-726, 2003.
15. C. Liebchen and L. Peeters. On Cyclic Timetabling and Cycles in Graphs. Technical Report 761/2002, TU Berlin.
16. C. Liebchen and R. Rizzi. A Greedy Approach to compute a Minimum Cycle Basis of a Directed Graph. Information Processing Letters, 94(3): 107-112, 2005.

Finding the Smallest H-Subgraph in Real Weighted Graphs and Related Problems

Virginia Vassilevska[1], Ryan Williams[1], and Raphael Yuster[2]

[1] Computer Science Department, Carnegie Mellon University, Pittsburgh, PA
{virgi, ryanw}@cs.cmu.edu
[2] Department of Mathematics, University of Haifa, Haifa, Israel
raphy@math.haifa.ac.il

Abstract. Let G be a graph with real weights assigned to the vertices (edges). The weight of a subgraph of G is the sum of the weights of its vertices (edges). The MIN H-SUBGRAPH problem is to find a minimum weight subgraph isomorphic to H, if one exists. Our main results are new algorithms for the MIN H-SUBGRAPH problem. The only operations we allow on real numbers are additions and comparisons. Our algorithms are based, in part, on fast matrix multiplication.

For vertex-weighted graphs with n vertices we obtain the following results. We present an $O(n^{t(\omega,h)})$ time algorithm for MIN H-SUBGRAPH in case H is a fixed graph with h vertices and $\omega < 2.376$ is the exponent of matrix multiplication. The value of $t(\omega, h)$ is determined by solving a small integer program. In particular, the smallest triangle can be found in $O(n^{2+1/(4-\omega)}) \le o(n^{2.616})$ time, the smallest K_4 in $O(n^{\omega+1})$ time, the smallest K_7 in $O(n^{4+3/(4-\omega)})$ time. As h grows, $t(\omega, h)$ converges to $3h/(6 - \omega) < 0.828h$. Interestingly, only for $h = 4, 5, 8$ the running time of our algorithm essentially matches that of the (unweighted) H-subgraph detection problem. Already for triangles, our results improve upon the main result of [VW06]. Using rectangular matrix multiplication, the value of $t(\omega, h)$ can be improved; for example, the runtime for triangles becomes $O(n^{2.575})$. We also present an algorithm whose running time is a function of m, the number of edges. In particular, the smallest triangle can be found in $O(m^{(18-4\omega)/(13-3\omega)}) \le o(m^{1.45})$ time.

For edge-weighted graphs we present an $O(m^{2-1/k} \log n)$ time algorithm that finds the smallest cycle of length $2k$ or $2k - 1$. This running time is identical, up to a logarithmic factor, to the running time of the algorithm of Alon et al. for the unweighted case. Using the color coding method and a recent algorithm of Chan for distance products, we obtain an $O(n^3 / \log n)$ time randomized algorithm for finding the smallest cycle of any fixed length.

1 Introduction

Finding cliques or other types of subgraphs in a larger graph are classical problems in complexity theory and algorithmic combinatorics. Finding a maximum clique is NP-Hard, and also hard to approximate [Ha98]. This problem is also conjectured to be *not* fixed parameter tractable [DF95]. The problem of finding

M. Bugliesi et al. (Eds.): ICALP 2006, Part I, LNCS 4051, pp. 262–273, 2006.
© Springer-Verlag Berlin Heidelberg 2006

(induced) subgraphs on k vertices in an n-vertex graph has been studied extensively (see, e.g., [AYZ95, AYZ97, CN85, EG04, KKM00, NP85, YZ04]). All known algorithms for finding an induced subgraph on k vertices have running time $n^{\Theta(k)}$. Many of these algorithms use fast matrix multiplication to obtain improved exponents.

The main contribution of this paper is a set of improved algorithms for finding an (induced) k-vertex subgraph in a real vertex-weighted or edge-weighted graph. More formally, let G be a graph with real weights assigned to the vertices (edges). The weight of a subgraph of G is the sum of the weights of its vertices (edges). The MIN H-SUBGRAPH problem is to find an H-subgraph of minimum weight, if one exists. Some of our algorithms are based, in part, on *fast* matrix multiplication. In several cases, our algorithms use fast *rectangular* matrix multiplication algorithms. However, for simplicity reasons, we express most of our time bounds in terms of ω, the exponent of fast *square* matrix multiplications. The best bound currently available on ω is $\omega < 2.376$, obtained by Coppersmith and Winograd [CW90]. This is done by reducing each rectangular matrix product into a collection of smaller square matrix products. Slightly improved bounds can be obtained by using the best available rectangular matrix multiplication algorithms of Coppersmith [Cop97] and Huang and Pan [HP98]. In all of our algorithms we assume that the graphs are *undirected*, for simplicity. All of our results are applicable to directed graphs as well. Likewise, all of our results on the MIN-H-SUBGRAPH problem hold for the analogous MAX-H-SUBGRAPH problem. As usual, we use the *addition-comparison* model for handling real numbers. That is, real numbers are only allowed to be compared or added.

Our first algorithm applies to *vertex-weighted* graphs. In order to describe its complexity we need to define a small integer optimization problem. Let $h \geq 3$ be a positive integer. The function $t(\omega, h)$ is defined by the following optimization program.

Definition 1

$$b_1 = \max\{b \in N \; : \; \frac{b}{4-\omega} \leq \lfloor \frac{h-b}{2} \rfloor\}. \tag{1}$$

$$s_1 = h - b_1 + \frac{b_1}{4-\omega}. \tag{2}$$

$$s_2(b) = \max\{h - b + \lfloor \frac{h-b}{2} \rfloor \; , \; h - (3-\omega)\lfloor \frac{h-b}{2} \rfloor\}. \tag{3}$$

$$s_2 = \min\{s_2(b) \; : \; \lfloor \frac{h-b}{2} \rfloor \leq b \leq h-2\}. \tag{4}$$

$$t(\omega, h) = \min\{s_1, s_2\}. \tag{5}$$

By using fast rectangular matrix multiplication, an alternative definition for $t(\omega, h)$, resulting in slightly smaller values, can be obtained (note that if $\omega = 2$, as conjectured by many researchers, fast rectangular matrix multiplication has no advantage over fast square matrix multiplication).

Theorem 1. *Let H be a fixed graph with h vertices. If $G = (V, E)$ is a graph with n vertices, and $w : V \to \Re$ is a weight function, then an induced H-subgraph of G (if exists) of minimum weight can be found in $O(n^{t(\omega, h)})$ time.*

It is easy to establish some small values of $t(\omega, h)$ directly. For $h = 3$ we have $t(\omega, 3) = 2 + 1/(4 - \omega) < 2.616$ by taking $b_1 = 1$ in (1). Using fast rectangular matrix multiplication this can be improved to 2.575. In particular, a triangle of minimum weight can be found in $o(n^{2.575})$ time. This should be compared to the $O(n^\omega) \leq o(n^{2.376})$ algorithm for detecting a triangle in an *unweighted* graph. For $h = 4$ we have $t(\omega, 4) = \omega + 1 < 3.376$ by taking $b = 2$ in (4). Interestingly, the fastest algorithm for detecting a K_4, that uses square matrix multiplication, also runs in $O(n^{\omega+1})$ time [NP85]. The same phenomena also happens for $h = 5$ where $t(\omega, 5) = \omega + 2 < 4.376$ and for $h = 8$ where $t(\omega, 8) = 2\omega + 2 < 6.752$, but in no other cases! We also note that $t(\omega, 6) = 4 + 2/(4-\omega)$, $t(\omega, 7) = 4 + 3/(4-\omega)$, $t(\omega, 9) = 2\omega + 3$ and $t(\omega, 10) = 6 + 4/(4 - \omega)$. However, a closed formula for $t(\omega, h)$ cannot be given. Already for $h = 11$, and for infinitely many values thereafter, $t(\omega, h)$ is only piecewise linear in ω. For example, if $7/3 \leq \omega < 2.376$ then $t(\omega, 11) = 3\omega + 2$, and if $2 \leq \omega \leq 7/3$ then $t(\omega, 11) = 6 + 5/(4 - \omega)$. Finally, it is easy to verify that both s_1 in (2) and s_2 in (4) converge to $3h/(6 - \omega)$ as h increases. Thus, $t(\omega, h)$ converges to $3h/(6 - \omega) < 0.828h$ as h increases.

Prior to a few months ago, the only known algorithm for MIN H-SUBGRAPH in the vertex-weighted case was the naïve $O(n^h)$ algorithm. Very recently, [VW06] gave an $O(n^{h \cdot \frac{\omega+3}{6}}) \leq o(n^{0.896h})$ randomized algorithm, for h divisible by 3. Our algorithms are deterministic, and uniformly improve upon theirs, for all values of h.[1]

A slight modification in the algorithm of Theorem 1, without increasing its running time by more than a logarithmic factor, can also answer the decision problem: "is there an H-subgraph whose weight is in the interval $[w_1, w_2]$ where $w_1 \leq w_2$ are two given reals?" Another feature of Theorem 1 is that it makes a relatively small number of comparisons. For example, the smallest triangle can be found by the algorithm using only $O(m + n \log n)$ comparisons, where m is the number of edges of G.

Since Theorem 1 is stated for induced H-subgraphs, it obviously also applies to not-necessarily induced H-subgraphs. However, the latter problem can, in some cases, be solved faster. For example, we show that the $o(n^{2.616})$ time bound for finding the smallest triangle also holds if one searches for the smallest H-subgraph in case H is the complete bipartite graph $K_{2,k}$.

Several H-subgraph detection algorithms take advantage of the fact that G may be sparse. Improving a result of Itai and Rodeh [IR78], Alon, Yuster and Zwick obtained an algorithm for detecting a triangle, expressed in terms of m [AYZ97]. The running time of their algorithm is $O(m^{2\omega/(\omega+1)}) \leq o(m^{1.41})$. This is faster than the $O(n^\omega)$ algorithm when $m = o(n^{(\omega+1)/2})$. The best known

---

[1] [VW06] also give a deterministic $O(B \cdot n^{(\omega+3)/2}) \leq o(B \cdot n^{2.688})$ algorithm, where B is the number of bits needed to represent the (absolute) maximum weight. Note this algorithm is *not* strongly polynomial.

running times in terms of m for $H = K_k$ when $k \geq 4$ are given in [EG04]. Sparseness can also be used to obtain faster algorithms for the vertex-weighted MIN H-SUBGRAPH problem. The triangle algorithm of [VW06] extends to a randomized $O(m^{1.46})$ algorithm. We prove:

Theorem 2. *If $G = (V, E)$ is a graph with m edges and no isolated vertices, and $w : V \rightarrow \Re$ is a weight function, then a triangle of G with minimum weight (if exists) can be found in $O(m^{(18-4\omega)/(13-3\omega)}) \leq o(m^{1.45})$ time.*

We now turn to edge-weighted graphs. An $O(m^{2-1/\lceil k/2 \rceil})$ time algorithm for detecting the existence of a cycle of length k is given in [AYZ97]. A small improvement was obtained later in [YZ04]. However, the algorithms in both papers fail when applied to edge-weighted graphs. Using the *color coding* method, together with several additional ideas, we obtain a randomized $O(m^{2-1/\lceil k/2 \rceil})$ time algorithm in the edge-weighted case, and an $O(m^{2-1/\lceil k/2 \rceil} \log n)$ deterministic algorithm.

Theorem 3. *Let $k \geq 3$ be a fixed integer. If $G = (V, E)$ is a graph with m edges and no isolated vertices, and $w : E \rightarrow \Re$ is a weight function, then a minimum weight cycle of length k, if exists, can be found with high probability in $O(m^{2-1/\lceil k/2 \rceil})$ time, and deterministically in $O(m^{2-1/\lceil k/2 \rceil} \log n)$ time.*

In a recent result of Chan [Ch05] it is shown that the distance product of two $n \times n$ matrices with real entries can be computed in $O(n^3 / \log n)$ time (again, reals are only allowed to be compared or added). [VW06] showed how to reduce the MIN H-SUBGRAPH problem in edge-weighted graphs to the problem of computing a distance product. (The third author independently proved this as well.)

Theorem 4 ([VW06]). *Let H be a fixed graph with h vertices. If $G = (V, E)$ is a graph with n vertices, and $w : E \rightarrow \Re$ is a weight function, then an induced H-subgraph of G (if exists) of minimum weight can be found in $O(n^h / \log n)$ time.*

We can strengthen the above result considerably, in the case where H is a cycle. For (not-necessarily induced) cycles of fixed length we can combine distance products with the color coding method and obtain:

Theorem 5. *Let k be a fixed positive integer. If $G = (V, E)$ is a graph with n vertices, and $w : E \rightarrow \Re$ is a weight function, a minimum weight cycle with k vertices (if exist) can be found, with high probability, in $O(n^3 / \log n)$ time.*

In fact, the proof of Theorem 5 shows that a minimum weight cycle with $k = o(\log \log n)$ vertices can be found in (randomized) sub-cubic time.

Finally, we consider the related problem of finding a certain chromatic H-subgraph in an edge-colored graph. We consider the two extremal chromatic cases. An H-subgraph of an edge-colored graph is called *rainbow* if all the edges have distinct colors. It is called *monochromatic* if all the edges have the same color. Many combinatorial problems are concerned with the existence of rainbow and/or monochromatic subgraphs.

We obtain a new algorithm that finds a rainbow H-subgraph, if it exists.

Theorem 6. *Let H be a fixed graph with $3k + j$ vertices, $j \in \{0, 1, 2\}$. If $G = (V, E)$ is a graph with n vertices, and $c : E \to C$ is an edge-coloring, then a rainbow H-subgraph of G (if exists) can be found in $O(n^{\omega k+j} \log n)$ time.*

The running time in Theorem 6 matches, up to a logarithmic factor, the running time of the induced H-subgraph detection problem in (uncolored) graphs.

We obtain a new algorithm that finds a monochromatic H-subgraph, if it exists. For fixed H, the running time of our algorithm matches the running time of the (uncolored) H-subgraph detection problem, except for the case $H = K_3$.

Theorem 7. *Let H be a fixed connected graph with $3k+j$ vertices, $j \in \{0, 1, 2\}$. If $G = (V, E)$ is a graph with n vertices, and $c : E \to C$ is an edge-coloring, then a monochromatic H-subgraph of G (if exists) can be found in $O(n^{\omega k+j})$ time, unless $H = K_3$. A monochromatic triangle can be found in $O(n^{(3+\omega)/2}) \leq o(n^{2.688})$ time.*

Due to space limitation, the proofs of Theorems 6 and 7 will appear in the journal version of this paper.

The rest of this paper is organized as follows. In Section 2 we focus on vertex-weighted graphs, describe the algorithms proving Theorems 1 and 2, and some of their consequences. Section 3 considers edge-weighted graphs and contains the algorithms proving Theorems 3, 4 and 5. The final section contains some concluding remarks and open problems.

2   Minimal H-Subgraphs of Real Vertex-Weighted Graphs

In the proof of Theorem 1 it would be convenient to assume that $H = K_h$ is a clique on h vertices. The proof for all other induced subgraphs with h vertices is only slightly more cumbersome, but essentially the same.

Let $G = (V, E)$ be a graph with real vertex weights, and assume $V = \{1, \ldots, n\}$. For two positive integers a, b, the *adjacency system* $A(G, a, b)$ is the 0-1 matrix defined as follows. Let S_x be the set of all $\binom{n}{x}$ x-subsets of vertices. The *weight* $w(U)$ of $U \in S_x$ is the sum of the weights of its elements. We *sort* the elements of S_x according to their weights. This requires $O(n^x \log n)$ time, assuming x is a constant. Thus, $S_x = \{U_{x,1}, \ldots, U_{x,\binom{n}{x}}\}$ where $w(U_{x,i}) \leq w(U_{x,i+1})$. The matrix $A(G, a, b)$ has its rows indexed by S_a. More precisely, the j'th row is indexed by $U_{a,j}$. The columns are indexed by S_b where the j'th column is indexed by $U_{b,j}$. We put $A(G, a, b)[U, U'] = 1$ if and only if $U \cup U'$ induces a K_{a+b} in G (this implies that $U \cap U' = \emptyset$). Otherwise, $A(G, a, b)[U, U'] = 0$. Notice that the construction of $A(G, a, b)$ requires $O(n^{a+b})$ time.

For positive integers a, b, c, so that $a + b + c = h$, consider the Boolean product $A(G, a, b, c) = A(G, a, b) \times A(G, b, c)$. For $U \in S_a$ and $U' \in S_c$ for which $A(G, a, b, c)[U, U'] = 1$, define their *smallest witness* $\delta(U, U')$ to be the smallest element $U'' \in S_b$ for which $A(G, a, b)[U, U''] = 1$ and also $A(G, b, c)[U'', U'] = 1$. For each $U \in S_a$ and $U' \in S_c$ with $A(G, a, b, c)[U, U'] = 1$ and with $U \cup U'$

inducing a K_{a+c}, if $U'' = \delta(U, U')$ then $U \cup U' \cup U''$ induces a K_h in G whose weight is the smallest of all the K_h copies of G that contain $U \cup U'$. This follows from the fact that S_b is sorted. Thus, by computing the smallest witnesses of all plausible pairs $U \in S_a$ and $U' \in S_c$ we can find a K_h in G with minimum weight, if it exists, or else determine that G does not have K_h as a subgraph.

Let $A = A_{n_1 \times n_2}$ and $B = B_{n_2 \times n_3}$ be two 0-1 matrices. The *smallest witness matrix* of AB is the matrix $W = W_{n_1 \times n_3}$ defined as follows. $W[i, j] = 0$ if $(AB)[i, j] = 0$. Otherwise, $W[i, j]$ is the smallest index k so that $A[i, k] = B[k, j] = 1$. Let $f(n_1, n_2, n_3)$ be the time required to compute the smallest witness matrix of the product of an $n_1 \times n_2$ matrix by an $n_2 \times n_3$ matrix. Let $h \geq 3$ be a fixed positive integer. For all possible choices of positive integers a, b, c with $a + b + c = h$ denote

$$f(h, n) = \min_{a+b+c=h} f(n^a, n^b, n^c).$$

Clearly, the time to sort S_b and to construct $A(G, a, b)$ and $A(G, b, c)$ is overwhelmed by $f(n^a, n^b, n^c)$. It follows from the above discussion that:

Lemma 1. *Let $h \geq 3$ be a fixed positive integer and let G be a graph with n vertices, each having a real weight. A K_h-subgraph of G with minimum weight, if exists, can be found in $O(f(h, n))$ time. Furthermore, if $f(n^a, n^b, n^c) = f(h, n)$ then the number of comparisons needed to find a minimum weight K_h is $O(n^b \log n + z(G, a + c))$ where $z(G, a + c)$ is the number of K_{a+c} in G.*

In fact, if $b \geq 2$, the number of comparisons in Lemma 1 can be reduced to only $O(n^b + z(G, a + c))$. Sorting S_b reduces to sorting the sums $X + X + \ldots + X$ (X repeated b times) of an n-element set of reals X. Fredman showed in [Fr76a] that this can be achieved with only $O(n^b)$ comparisons.

A simple randomized algorithm for computing (not necessarily first) witnesses for Boolean matrix multiplication, in essentially the same time required to perform the product, is given by Seidel [Sei95]. His algorithm was derandomized by Alon and Naor [AN96]. However, computing the matrix of first witnesses seems to be a more difficult problem. Improving an earlier algorithm of Bender et al. [BFPSS05], Kowaluk and Lingas [KL05] show that $f(3, n) = O(n^{2+1/(4-\omega)}) \leq o(n^{2.616})$. This already yields the case $h = 3$ in Theorem 1. We will need to extend and generalize the method from [KL05] in order to obtain upper bounds for $f(h, n)$. Our extension will enable us to answer more general queries such as "is there a K_h whose weight is within a given weight interval?"

Proof of Theorem 1. Let $h \geq 3$ be a fixed integer. Suppose a, b, c are three positive integers with $a + b + c = h$ and suppose that $0 < \mu \leq b$ is a real parameter. For two 0-1 matrices $A = A_{n^a \times n^b}$ and $B = B_{n^b \times n^c}$ the *μ-split* of A and B is obtained by splitting the columns of A and the rows of B into consecutive parts of size $\lceil n^\mu \rceil$ or $\lfloor n^\mu \rfloor$ each. In the sequel we ignore floors and ceilings whenever it does not affect the asymptotic nature of our results. This defines a partition of A into $p = n^{b-\mu}$ rectangular matrices $A_1, \ldots, A_p$, each with n^a rows and n^μ columns, and a partition of B into p rectangular matrices $B_1, \ldots, B_p$, each with

n^μ rows and n^c columns. Let $C_i = A_i B_i$ for $i = 1, \ldots, p$. Notice that each element of C_i is a nonnegative integer of value at most n^μ and that $AB = \sum_{i=1}^{p} C_i$. Given the C_i, the smallest witness matrix W of the product AB can be computed as follows. To determine $W[i,j]$ we look for the smallest index r for which $C_r[i,j] \neq 0$. If no such r exists, then $W[i,j] = 0$. Otherwise, having found r, we now look for the smallest index k so that $A_r[i,k] = A_r[k,j] = 1$. Having found k we clearly have $W[i,j] = (r-1)n^\mu + k$.

We now determine a choice of parameters a, b, c, μ so that the time to compute $C_1, \ldots, C_p$ and the time to compute the first witnesses matrix W, is $O(n^{t(\omega,h)})$. By Lemma 1, this suffices in order to prove the theorem. We will only consider $\mu \leq \min\{a,b,c\}$. Taking larger values of μ results in worse running times. The rectangular product C_i can be computed by performing $O(n^{a-\mu}n^{c-\mu})$ products of square matrices of order n^μ. Thus, the time required to compute C_i is

$$O(n^{a-\mu}n^{c-\mu}n^{\omega\mu}) = O(n^{a+c+(\omega-2)\mu}).$$

Since there are p such products, and since each of the n^{a+c} witnesses can be computed in $O(p + n^\mu)$ time, the overall running time is

$$O(pn^{a+c+(\omega-2)\mu} + n^{a+c}(p + n^\mu)) = O(n^{h-(3-\omega)\mu} + n^{h-\mu} + n^{h-b+\mu})$$

$$= O(n^{h-(3-\omega)\mu} + n^{h-b+\mu}). \tag{6}$$

Optimizing on μ we get $\mu = b/(4-\omega)$. Thus, if, indeed, $b/(4-\omega) \leq \min\{a,c\}$ then the time needed to find W is $O(n^{h-b+b/(4-\omega)})$. Of course, we would like to take b as large as possible under these constraints. Let, therefore, b_1 be the largest integer b so that $b/(4-\omega) \leq \lfloor (h-b)/2 \rfloor$. For such a b_1 we can take $a = \lfloor (h-b_1)/2 \rfloor$ and $c = \lceil (h-b_1)/2 \rceil$ and, indeed, $\mu \leq \min\{a,c\}$. Thus, (6) gives that the running time to compute W is

$$O(n^{h-b_1+b_1/(4-\omega)}).$$

This justifies s_1 appearing in (2) in the definition of $t(\omega,h)$. There may be cases where we can do better, whenever $b/(4-\omega) > \min\{a,c\}$. We shall only consider the cases where $a = \mu = \lfloor (h-b)/2 \rfloor \leq b$ (other cases result in worse running times). In this case $c = \lceil (h-b)/2 \rceil$ and, using (6), the running time is

$$O(n^{h-(3-\omega)\lfloor \frac{h-b}{2} \rfloor} + n^{h-b+\lfloor \frac{h-b}{2} \rfloor}).$$

This justifies s_2 appearing in (4) in the definition of $t(\omega,h)$. Since $t(\omega,h) = \min\{s_1, s_2\}$ we have proved that W can be computed in $O(n^{t(\omega,h)})$ time. ∎

As can be seen from Lemma 1 and the remark following it, the number of comparisons that the algorithm performs is relatively small. For example, in the case $h = 3$ we have $a = b = c = 1$ and hence the number of comparisons is $O(n \log n + m)$. In all the three cases $h = 4, 5, 6$ the value $b = 2$ yields $t(\omega,h)$. Hence, the number of comparisons is $O(n^2)$ for $h = 4$, $O(n^2 + mn)$ for $h = 5$ and $O(n^2 + m^2)$ for $h = 6$.

Suppose $w : \{1, \ldots, n^b\} \to \Re$ so that $w(k) \le w(k+1)$. The use of the μ-split in the proof of Theorem 1 enables us to determine, for each i, j and for a real interval $I(i, j)$, whether or not there exists an index k so that $A[i, k] = B[k, j] = 1$ and $w(k) \in I(i, j)$. This is done by performing a binary search within the $p = n^{b-\mu}$ matrices $C_i, \ldots, C_p$. The running time in (6) only increases by a $\log n$ factor. We therefore obtain the following corollary.

Corollary 1. *Let H be a fixed graph with h vertices, and let $I \subset \Re$. If $G = (V, E)$ is a graph with n vertices, and $w : V \to \Re$ is a weight function, then, deciding whether G contains an induced H-subgraph with total weight in I can be done $O(n^{t(\omega,h)} \log n)$ time.*

Proof of Theorem 2. We partition the vertex set V into two parts $V = X \cup Y$ according to a parameter Δ. The vertices in X have degree at most Δ. The vertices in Y have degree larger than Δ. Notice that $|Y| < 2m/\Delta$. In $O(m\Delta)$ time we can scan all triangles that contain a vertex from X. In particular, we can find a smallest triangle containing a vertex from X. By Theorem 1, a smallest triangle induced by Y can be found in $O((m/\Delta)^{t(\omega,3)}) = O((m/\Delta)^{2+1/(4-\omega)})$ time. Therefore, a smallest triangle in G can be found in

$$O\left(m\Delta + \left(\frac{m}{\Delta}\right)^{2+1/(4-\omega)}\right)$$

time. By choosing $\Delta = m^{(5-\omega)/(13-3\omega)}$ the result follows.   ∎

The results in Theorems 1 and 2 are useful not only for real vertex weights, but also when the weights are large integers. Consider, for example, the graph parameter $\beta(G, H)$, the H *edge-covering number* of G. We define $\beta(G, H) = 0$ if G has no H-subgraph. Otherwise, $\beta(G, H)$ is the maximum number of edges incident with an H-subgraph of G. To determine $\beta(G, K_k)$ we assign to each vertex a weight equal to its degree. We now use the algorithm of Theorem 1 to find the *maximum* weighted K_k. If the weight of the maximum weighted K_k is w, then $\beta(G, K_k) = w - \binom{k}{2}$. In particular, $\beta(G, K_k)$ can be computed in $O(n^{t(\omega,k)})$ time.

Finally, we note that Theorems 1 and 2 apply also when the weight of an H-subgraph is not necessarily defined as the sum of the weights of its vertices. Suppose that the weight of a triangle (x, y, z) is defined by a function $f(x, y, z)$ that is monotone in each variable separately. For example, we may consider $f(x, y, z) = xyz$, $f(x, y, z) = xy + xz + yz$ etc. Assuming that $f(x, y, z)$ can be computed in constant time given x, y, z, it is easy to modify Theorems 1 and 2 to find a triangle whose weight is minimal with respect to f in $O(n^{2+1/(4-\omega)})$ time and $O(m^{(18-4\omega)/(13-3\omega)})$ time, respectively.

We conclude this section with the following proposition.

Proposition 1. *If $G = (V, E)$ is a graph with n vertices, and $w : V \to \Re$ is a weight function, then a (not necessarily induced) minimum weight $K_{2,k}$-subgraph can be found in $O(n^{2+1/(4-\omega)})$.*

Proof. To find the smallest $K_{2,k}$ we simply need to find, for any two vertices i, j, the first k smallest weighted vertices $v_1, \ldots, v_k$ so that each v_i is a common neighbor of i and j. As in Lemma 1, this reduces to finding the first k smallest witnesses of a 0-1 matrix product. A simple modification of the algorithm in Theorem 1 achieves this goal in the same running time (recall that k is fixed).

3   Minimal H-Subgraphs of Real Edge-Weighted Graphs

Given a vertex-colored graph G with n vertices, an H-subgraph of G is called *colorful* if each vertex of H has a distinct color. The *color coding* method presented in [AYZ95] is based upon two important facts. The first one is that, in many cases, finding a colorful H-subgraph is easier than finding an H-subgraph in an uncolored graph. The second one is that in a random vertex coloring with k colors, an H-subgraph with k vertices becomes colorful with probability $k!/k^k > e^{-k}$ and, furthermore, there is a derandomization technique that constructs a family of not too many colorings, so that each H-subgraph is colorful in at least one of the colorings. The derandomization technique, described in [AYZ95], constructs a family of colorings of size $O(\log n)$ whenever k is fixed.

By the color coding method, in order to prove Theorem 3, it suffices to prove that, *given* a coloring of the vertices of the graph with k colors, a colorful cycle of length k of minimum weight (if exists) can be found in $O(m^{2-1/\lceil k/2 \rceil})$ time.

Proof of Theorem 3. Assume that the vertices of G are colored with the colors $1, \ldots, k$. We first show that for each vertex u, a minimum weight colorful cycle of length k that passes through u can be found in $O(m)$ time. For a permutation π of $1, \ldots, k$, we show that a minimum weight cycle of the form $u = v_1, v_2, \ldots, v_k$ in which the color of v_i is $\pi(i)$ can be found in $O(m)$ time. Without loss of generality, assume π is the identity. For $j = 2, \ldots, k$ let V_j be the set of vertices whose color is j so that there is a path from u to $v \in V_j$ colored consecutively by the colors $1, \ldots, j$. Let $S(v)$ be the set of vertices of such a path with minimum possible weight. Denote this weight by $w(v)$. Clearly, V_j can be created from V_{j-1} in $O(m)$ time by examining the neighbors of each $v \in V_{j-1}$ colored with j. Now, let $w_u = \min_{v \in V_k} w(v) + w(v, u)$. Thus, w_u is the minimum weight of a cycle passing through u, of the desired form, and a cycle with this weight can be retrieved as well.

We prove the theorem when k is even. The odd case is similar. Let $\Delta = m^{2/k}$. There are at most $2m/\Delta = O(m^{1-2/k})$ vertices with degree at least Δ. For each vertex u with degree at least Δ we find a minimum weight colorful cycle of length k that passes through u. This can be done in $O(m^{2-2/k})$ time. It now suffices to find a minimum weight colorful cycle of length k in the subgraph G' of G induced by the vertices with maximum degree less than Δ. Consider a permutation π of $1, \ldots, k$. For a pair of vertices x, y, let S_1 be the set of all paths of length $k/2$ colored consecutively by $\pi(1), \ldots, \pi(k/2), \pi(k/2+1)$. There are at most $m\Delta^{k/2-1} = m^{2-2/k}$ such paths and they can be found using the greedy algorithm in $O(m^{2-2/k})$ time. Similarly, let S_2 be the set of all paths of length $k/2$ colored consecutively by $\pi(k/2+1), \ldots, \pi(k), \pi(1)$. If u, v are endpoints of at least one path in S_1 then let $f_1(\{u, v\})$ be the minimum weight of such a path. Similarly define $f_2(\{u, v\})$. We can therefore find, in $O(m^{2-2/k})$ a pair u, v (if exists) so that $f_1(\{u, v\}) + f_2(\{u, v\})$ is minimized. By performing this procedure for each permutation, we find a minimum weight colorful cycle of length k in G'. ∎

Let $A = A_{n_1 \times n_2}$ and $B = B_{n_2 \times n_3}$ be two matrices with entries in $\Re \cup \infty$. The *distance product* $C = A \star B$ is an $n_1 \times n_3$ matrix with $C[i, j] = \min_{k=1...,n_2} A[i, k] + B[k, j]$. Clearly, C can be computed in $O(n_1 n_2 n_3)$ time in the addition-comparison model. However, Fredman showed in [Fr76] that the distance product of two square matrices of order n can be performed in $O(n^3 (\log \log n / \log n)^{1/3})$ time. Following a sequence of improvements over Fredman's result, Chan gave an $O(n^3 / \log n)$ time algorithm for distance products. By partitioning the matrices into blocks it is obvious that Chan's algorithm computes the distance product of an $n_1 \times n_2$ matrix and an $n_2 \times n_3$ matrix in $O(n_1 n_2 n_3 / \log \min\{n_1, n_2, n_3\})$ time. Distance products can be used to solve the MIN H-SUBGRAPH problem in edge weighted graphs.

Proof of Theorem 4. We prove the theorem for $H = K_h$. The proof for other induced H-subgraphs is essentially the same. Partition h into a sum of three positive integers $a + b + c = h$. Let S_a be the set of all K_a-subgraphs of G. Notice that $|S_a| < n^a$ and that each $U \in S_a$ is an a-set. Similarly define S_b and S_c. We define A to be the matrix whose rows are indexed by S_a and whose columns are indexed by S_b. The entry $A[U, U']$ is defined to by ∞ if $U \cup U'$ does not induce a K_{a+b}. Otherwise, it is defined to be the sum of the weights of the edges induced by $U \cup U'$. We define B to be the matrix whose rows are indexed by S_b and whose columns are indexed by S_c. The entry $A[U, U']$ is defined to by ∞ if $U \cup U'$ does not induce a K_{b+c}. Otherwise, it is defined to be the sum of the weights of the edges induced by $U \cup U'$ with *at least* one endpoint in U'. Notice the difference in the definitions of A and B. Let $C = A \star B$. The time to compute C using Chan's algorithm is $O(n^h / \log n)$. Now, for each $U \in S_a$ and $U' \in S_c$ so that $U \cup U'$ induces a K_{a+c}, let $w(U, U')$ be the sum of the weights of the edges with one endpoint in U and the other in U' plus the value of $C[U, U']$. If $w(U, U')$ is finite then it is the weight of the smallest K_h that contains $U \cup U'$. Otherwise, no K_h contains $U \cup U'$. ∎

The weighted DENSE k-SUBGRAPH problem (see, e.g., [FKP01]) is to find a k-vertex subgraph with maximum total edge weight. A simple modification of the algorithm of Theorem 4 solves this problem in $O(n^k / \log n)$ time. To our knowledge, this is the first non-trivial algorithm for this problem. Note that the maximum total weight of a k-subgraph can potentially be much larger than a k-clique's total weight.

Proof of Theorem 5. We use the color coding method, and an idea similar to Lemma 3.2 in [AYZ95]. Given a coloring of the vertices with k colors, it suffices to show how to find the smallest colorful path of length $k - 1$ connecting any pair of vertices in $2^{O(k)} n^3 / \log n$ time. It will be convenient to assume that k is a power of two, and use recursion. Let C_1 be a set of $k/2$ distinct colors, and let C_2 be the complementary set of colors. Let V_i be the set of vertices colored by colors from C_i for $i = 1, 2$. Let G_i be the subgraph induced by V_i. Recursively find, for each pair of vertices in G_i, the minimum weight colorful path of length $k/2 - 1$. We record this information in matrices A_1, A_2, where the rows and columns of A_i are indexed by V_i. Let B be the matrix whose rows are indexed by V_1 and whose columns are indexed by V_2 where $B[u, v] = w(u, v)$. The distance product $D_{C_1, C_2} = (A_1 \star B) \star A_2$ gives, for each pair of vertices of G, all shortest paths of length $k - 1$ where the first $k/2$ vertices are colored by colors from C_1 and the last $k/2$ vertices are colored by colors from C_2. By considering all $\binom{k}{k/2} < 2^k$ possible choices for (C_1, C_2), and computing D_{C_1, C_2} for each choice, we can obtain an $n \times n$ matrix D where $D[u, v]$ is the shortest colorful path of length $k - 1$ between u and v. The number of distance products computed using this approach satisfies the recurrence $t(k) \leq 2^k t(k/2)$. Thus, the overall running time is $2^{O(k)} n^3 / \log n$. ∎

The proof of Theorem 5 shows that, as long as $k = o(\log \log n)$, a cycle with k vertices and minimum weight can be found, with high probability, in $o(n^3)$ time. The previous best known algorithm (to our knowledge) for finding a minimum weight cycle of length k, in real weighted graphs, has running time $O(k! n^3 2^k)$ [PV91].

4   Concluding Remarks and Open Problems

We presented several algorithms for MIN H-SUBGRAPH in both real vertex weighted or real edge weighted graphs, and results for the related problem of finding monochromatic or rainbow H-subgraphs in edge-colored graphs. It may be possible to improve upon the running times of some of our algorithms. More specifically, we raise the following open problems.

(i) Can the exponent $t(\omega, 3)$ in Theorem 1 be improved? If so, this would immediately imply an improved algorithm for first witnesses.
(ii) Can the logarithmic factor in Theorem 3 be eliminated? We know from [AYZ97] that this is the case in the unweighted version of the problem. Can the logarithmic factor in Theorem 6 be eliminated?
(iii) Can monochromatic triangles be detected faster than the $O(n^{(3+\omega)/2})$ algorithm of Theorem 7? In particular, can they be detected in $O(n^\omega)$ time?

Acknowledgment

The authors thank Uri Zwick for for some useful comments.

References

[AN96]   N. Alon and M. Naor. Derandomization, witnesses for Boolean matrix multiplication and construction of perfect hash functions. *Algorithmica*, 16:434–449, 1996.

[AYZ95]   N. Alon, R. Yuster, and U. Zwick. Color-coding. *Journal of the ACM*, 42:844–856, 1995.

[AYZ97]   N. Alon, R. Yuster, and U. Zwick. Finding and counting given length
             cycles. *Algorithmica*, 17:209–223, 1997.
[BFPSS05]   M. Bender, M. Farach-Colton, G. Pemmasani, S. Skiena, and P. Sumazin.
             Lowest common ancestors in trees and directed acyclic graphs. *J. Algo-
             rithms*, 57(2):75–94, 2005.
[Ch05]   T.M. Chan. All-Pairs Shortest Paths with Real Weights in $O(n^3/\log n)$
             Time. In *Proc. of the 9th WADS*, Lecture Notes in Computer Science
             3608, Springer (2005), 318–324.
[CN85]   N. Chiba and L. Nishizeki. Arboricity and subgraph listing algorithms.
             SIAM Journal on Computing, 14:210–223, 1985.
[Cop97]   D. Coppersmith. Rectangular matrix multiplication revisited. *Journal of
             Complexity*, 13:42–49, 1997.
[CW90]   D. Coppersmith and S. Winograd. Matrix multiplication via arithmetic
             progressions. *J. Symbol. Comput.*, 9:251–280, 1990.
[DF95]   R.G. Downey and M.R. Fellows. Fixed-parameter tractability and com-
             pleteness II. On completeness for W[1]. *Theoret. Comput. Sci.*, 141(1-
             2):109-131, 1995.
[EG04]   F. Eisenbrand and F. Grandoni. On the complexity of fixed parameter
             clique and dominating set. *Theoret. Comput. Sci.*, 326(1-3):57–67, 2004.
[FKP01]   U. Feige, G. Kortsarz and D. Peleg. The Dense k-Subgraph Problem.
             Algorithmica, 29(3):410–421, 2001.
[Fr76]   M.L. Fredman. New bounds on the complexity of the shortest path prob-
             lem. *SIAM Journal on Computing*, 5:49–60, 1976.
[Fr76a]   M.L. Fredman. How good is the information theory bound in sorting?
             Theoret. Comput. Sci., 1:355–361, 1976.
[Ha98]   J. Håstad. Clique is hard to approximate within $n^{1-\epsilon}$. *Acta Math.*,
             182(1):105-142, 1998.
[HP98]   X. Huang and V.Y. Pan. Fast rectangular matrix multiplications and
             applications. *Journal of Complexity*, 14:257–299, 1998.
[IR78]   A. Itai and M. Rodeh. Finding a minimum circuit in a graph. *SIAM
             Journal on Computing*, 7:413–423, 1978.
[KKM00]   T. Kloks, D. Kratsch, and H. Müller. Finding and counting small induced
             subgraphs efficiently. *Inf. Process. Lett.*, 74(3-4):115–121, 2000.
[KL05]   M. Kowaluk and A. Lingas. LCA Queries in Directed Acyclic Graphs.
             In *Proc. of the 32nd ICALP*, Lecture Notes in Computer Science 3580,
             Springer (2005), 241–248.
[NP85]   J. Nešetřil and S. Poljak. On the complexity of the subgraph problem.
             Comment. Math. Univ. Carolin., 26(2):415–419, 1985.
[PV91]   J. Plehn and B. Voigt. Finding Minimally Weighted Subgraphs. In *Pro-
             ceedings of the 16th International Workshop on Graph-Theoretic Concepts
             in Computer Science (WG)*, Springer-Verlag, 1991.
[Sei95]   R. Seidel. On the All-Pairs-Shortest-Path Problem in Unweighted Undi-
             rected Graphs. *J. Comput. Syst. Sci.*, 51(3):400–403, 1995.
[VW06]   V. Vassilevska and R. Williams. Finding a maximum weight triangle in
             $n^{3-\delta}$ time, with applications. In *Proceedings of the 38th Annual ACM
             Symposium on Theory of Computing (STOC)*, to appear.
[YZ04]   R. Yuster and U. Zwick. Detecting short directed cycles using rectangular
             matrix multiplication and dynamic programming. In *Proc. of the 15th
             ACM-SIAM Symposium on Discrete Algorithms (SODA)*, ACM/SIAM
             (2004), 247–253.

Weighted Bipartite Matching in Matrix Multiplication Time*

[Extended Abstract]

Piotr Sankowski

Institute of Informatics, Warsaw University,
Banacha 2, 02-097, Warsaw, Poland
sank@mimuw.edu.pl

Abstract. In this paper we consider the problem of finding maximum weighted matchings in bipartite graphs with nonnegative integer weights. The presented algorithm for this problem work in $\tilde{O}(Wn^{\omega})^1$ time, where ω is the matrix multiplication exponent, and W is the highest edge weight in the graph. As a consequence of this result we obtain $\tilde{O}(Wn^{\omega})$ time algorithms for computing: minimum weight bipartite vertex cover, single source shortest paths and minimum weight vertex disjoint s-t paths.

1 Introduction

The weighted matching problem is one of the fundamental problems in combinatorial optimization. The first algorithm for this problem in the bipartite case was proposed in the fifties of the last century by Kuhn [14]. His result has been improved several times since then, the known results are summarized in the Table 1. The bold font indicates an asymptotically best bound in the tables. In particular the presented here algorithm is faster than the algorithm of Gabow and Tarjan [8] and the algorithm of Edmonds and Karp [5] in the case of dense graphs with small integer weights. Note, that in this summary there are no algorithms that use matrix multiplication. However, in the papers studying the parallel complexity of the problem [13],[18], such algorithms are implicitly constructed. These results lead to $O(Wn^{\omega+2})$ sequential time algorithms. In this paper we improve the complexity by factor of n^2. The improvement in the exponent by 1 is achieved with use of the very recent results of Storjohann [24], who had shown faster algorithms for computing polynomial matrix determinants. His results are summarized in Section 1.1. Further improvement is achieved by a novel reduction technique, that allows us to reduce the weighted version of the problem to unweighted one. The four steps of the reduction are schematically presented on Figure 1. As a step of the reduction we also compute the bipartite weighted cover of the graph. The unweighted problem is then solved with use of the $O(n^{\omega})$ time algorithms developed two years ago by Mucha and

---

* Research supported by KBN grant 1P03A01830.
[1] $\tilde{O}$ denotes the so-called "soft O" notation, i.e. $f(n) = \tilde{O}(g(n))$ iff $f(n) = O(g(n)\log^k n)$ for some constant k.

M. Bugliesi et al. (Eds.): ICALP 2006, Part I, LNCS 4051, pp. 274–285, 2006.

Table 1. The complexity results for the bipartite weighted matching

Complexity	Author
$O(n^4)$	Khun (1955) [14] and Munkers (1957) [19]
$O(n^2m)$	Iri (1960) [10]
$O(n^3)$	Dinic and Kronrod (1969) [4]
$O(nm)$	Edmonds and Karp (1970) [5]
$O(n^{\frac{3}{4}}m\log W)$	Gabow (1983) [7]
$O(\sqrt{n}m\log(nW))$	Gabow and Tarjan (1989) [8]
$O(\sqrt{n}mW)$	Kao, Lam, Sung and Ting (1999) [12]
$\tilde{O}(n^\omega W)$	this paper

Sankowski [17]. Storjohann's result can also be used to compute the maximum weight of a perfect matching in general graphs. However, the problem of finding such matching remains unsolved.

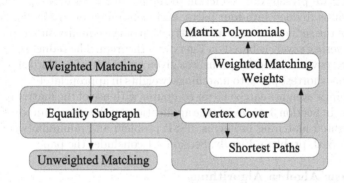

Fig. 1. The scheme of the reduction from weighted to unweighted matchings

The weighted matching problem is not only interesting by itself, but also it can be used to solve many other problems in combinatorial optimization. In particular, the presented algorithm for finding maximum weighted perfect matchings can be used to find minimum weighted perfect matching, as well as, maximum and minimum weighted matchings. Moreover, the minimum weighted perfect matching algorithm can be used for computing the minimum weight of k vertex disjoint s-t paths, whereas the minimum weighted vertex cover can be used to solve the single source shortest paths (SSSP) problem with negative edge weights. The complexity of the algorithms for computing the minimum weight of k vertex disjoint s-t paths, follow exactly the results in Table 1. The author is not aware of any special algorithms for this problem. The complexity results for the SSSP problem with negative edge weights are summarized in Table 2.

Table 2. The complexity results for the SSSP problem with negative weights. The bold font indicates an asymptotically best bound in the table.

Complexity	Author
$O(n^4)$	Shimbel (1955) [23]
$O(n^2 mW)$	Ford (1956) [11]
$\boldsymbol{O(nm)}$	Bellman (1958) [1], Moore (1959) [16]
$O(n^{\frac{3}{4}} m \log W)$	Gabow (1983) [7]
$O(\sqrt{n}m \log(nW))$	Gabow and Tarjan (1989) [8]
$\boldsymbol{O(\sqrt{n}m \log(W))}$	Goldberg (1993) [9]
$\tilde{O}(n^\omega W)$	Yuster and Zwick (2005) [25], Sankowski (2005) [21], this paper

The rest of the paper is organized as follows. In the remainder of this introductory section, we summarize the results in linear algebra algorithms, we show the randomization technique used later in this paper and we recall the result from [17]. In Section 2 we present our reduction technique in the case of bipartite graphs. This section is divided into four parts, each containing one reduction step. The first step of the reduction, i.e., the construction of the equality subgraph from the minimum vertex cover, is based on Egervárys theorem. The reduction from minimum vertex cover to shortest paths was given by Iri [10]. The third step is the reduction from shortest paths to matchings' weights in appropriately defined graph. The matchings' weights can be latter computer with use of the matrix polynomial algorithms. In Section 3 we review the applications of the algorithm to: other kinds of the weighted matching problems, SSSP problem and minimum weight vertex disjoint *s-t* path problem. Finally, Section 4.1 concludes the paper.

1.1   Linear Algebra Algorithms

We denote by ω the exponent of a square matrix multiplication. The best bound on $\omega \leq 2.376$ is due to Coppersmith and Winograd [2]. The interaction of the matrix multiplication and linear algebra is well understood. The best known algorithms for many problems in linear algebra work in matrix multiplication time, i.e., the determinant of an $n \times n$ matrix A, or the solution to the linear system of equations, can be computed in $O(n^\omega)$ arithmetic operations. Very recently Storjohann [24] has shown that for polynomial matrices these problems can be solved with the same exponent.

Theorem 1 (Storjohann '03). *Let $A \in K[x]^{n \times n}$ be a polynomial matrix of degree d and $b \in K[x]^{n \times 1}$ be a polynomial vector of the same degree, then*

- *determinant* $\det(A)$,
- *rational system solution* $A^{-1}b$,

can be computed in $\tilde{O}(n^\omega d)$ operations in K, with high probability.

For our purposes it would be convenient if one could compute the inverse matrix A^{-1} in the above time bounds. This is impossible because the rational functions in the inverse can have degrees as high as nd, so just the output size is $\Omega(n^3 d)$, and no algorithm can work faster. Thus the approach similar to the one in papers [20],[17], where the inverse matrix is used to find edges belonging to a perfect matching, will not lead to $\tilde{O}(Wn^\omega)$ time algorithm.

1.2 Zippel-Schwartz Lemma

In this paper we will reduce a weighted perfect matching problem to testing if some set of polynomials is non-zero. In the simpler case of perfect matchings, in order test if a graph has a perfect matchings we have to test if if appropriately defined adjacency matrix is non-singular [15]. We can verify that the matrix is non-singular by computing its determinant, which is a polynomial of the entries of the matrix. However, this determinant may have exponentially many terms, so it cannot be computed symbolically in polynomial time. The following lemma due to Zippel [26] and Schwartz [22] can be used to overcome this obstacle.

Lemma 2. *If $p(x_1, \ldots, x_m)$ is a non-zero polynomial of degree d with coefficients in a field and S is a subset of the field, then the probability that p evaluates to 0 on a random element $(s_1, s_2, \ldots, s_m) \in S^m$ is at most $d/|S|$. We call such event a false zero.*

Corollary 3. *If a polynomial of degree n is evaluated on random values modulo prime number p of length $(1 + c) \log n$, then the probability of false zero is at most $\frac{1}{n^c}$, for any $c > 0$.*

In the standard RAM model we assume the word size to be $O(\log n)$. Thus the finite field arithmetic modulo p, except division, can be implemented in constant time. The divisions can be realized in $O(\log n)$ time. Nevertheless in our algorithms divisions are not time dominating operations. These assumptions are used to establish the time bounds for our algorithms.

1.3 Unweighted Matchings

As stated in the introduction we show here the reduction from the weighted matching problem to the unweighted one. Next the unweighted perfect matching problem is solved in $O(n^\omega)$ randomized time as stated in the following theorem [17].

Theorem 4 (Mucha and Sankowski '04). *A perfect matching in a graph can be computed in $O(n^\omega)$ time, with high probability.*

For the simplicity in the remainder of this paper we assume that in the considered graphs there is always a perfect matching. When necessary this assumption can be checked with use of this theorem.

2   Weighted Matchings in Bipartite Graphs

A weighted bipartite n-vertex graph G is a tuple $G = (U, V, E, w)$. The vertex sets are given by $U = \{u_1, \ldots, u_n\}$ and $V = \{v_1, \ldots, v_n\}$. $E \subseteq U \times V$ denotes the edge set, and the function $w : E \to \mathcal{Z}_+$ ascribes weights to the edges. We denote by W the maximum weight in w.

In the *maximum weighted bipartite perfect matching problem* we seek a perfect matching M in a weighted bipartite graph G to maximize the total weight $w(M) = \sum_{e \in M} w(e)$.

A *weighted cover* is a choice of labels $y(v_1), \ldots, y(v_n), y(u_1), \ldots, y(u_n)$ such that $y(v) + y(u) \geq w(vu)$ for all $v \in V$ and $u \in U$. The *minimum weighted cover problem* is that of finding a cover of minimum weight. The following theorem states that the vertex cover problem and the maximum weighted matching problem are dual.

Theorem 5 (Egerváry '31). *Let $G = (U, V, E, w)$ be a weighted bipartite graph. The maximum weight of a perfect matching of G is equal to weight of the minimum weighted cover of G.*

In order to find weighted perfect matchings we explore this duality. In the next two subsections we will show how to solve one of these problems when we already have the solution to the other.

2.1   From Equality Subgraph to Vertex Cover

The following lemma is one of the ingredients of the Hungarian method. The *equality subgraph* G_p for a weighted graph G and its vertex cover p is defined as $G_p = (U, V, E')$, where $E' = \{uv : uv \in E \text{ and } p(u) + p(v) = w(uv)\}$. The following lemma is a direct consequence of Egervárys Theorem.

Lemma 6. *Consider a weighted bipartite graph $G = (U, V, E, w)$ and a minimum vertex cover p of G. The matching M is a perfect matching in G_p iff it is a maximum weighted perfect matching in G.*

Thus if we show how to find a minimum weighted bipartite vertex cover p in $\tilde{O}(Wn^\omega)$ time, then we will be able to find the equality subgraph G_p. Next we find the maximum weighted bipartite matching in G by finding the perfect matching in G_p with use of Theorem 4. Similar observation to Lemma 6, but phrased with use of allowed edges, was formulated in [6].

2.2   From Vertex Cover to Shortest Paths

The following technique was given by Iri [10]. Let M be a maximum weighted perfect matching in a weighted bipartite graph $G = (U, V, E, w)$. Construct a directed graph $D = (U \cup V \cup \{r\}, A, w_d)$, and

- for all edges $uv \in E$, with $u \in U$ and $v \in V$, add a directed edge (u, v) to A, with weight $w_d((u, v)) := -w(uv)$,

- for all edges $uv \in M$, with $u \in U$ and $v \in V$, add a directed edge (v, u) to A, with weight $w_d((v, u)) := w(uv)$,
- add zero weight edges (r, v) for each $v \in V$.

Let $\text{dist}_G(u, v)$ be the distance from vertex u to vertex v in the graph G.

Lemma 7 (Iri '60). *Set $y(u) := \text{dist}_D(r, u)$ for $u \in U$ and $y(v) := -\text{dist}_D(r, v)$ for $v \in V$, then y is a minimum weighted vertex cover in G.*

2.3 From Shortest Paths to Matching Weights

In this section we show how to compute the distances in D by computing just some matching weights in G. Consider a weighted bipartite graph $G = (U, V, E, w)$. Add a new vertex $s = u_{n+1}$ to U and a new vertex $t = v_{n+1}$ to V. Connect s with all vertices from V with zero weight edges, and connect the vertex t with a vertex u in U. Let us denote by $G(u)$ the resulting graph and by $M(u)$ the maximum weighted perfect matching in this graph. Note that $G(u)$ has a perfect matching because G has a perfect matching. By $G(*)$ we denote the graph where the vertex t is connected with all vertices in U.

Lemma 8. *Let $G = (U, V, E, w)$ be a weighted bipartite graph and let M be the maximum weighted perfect matching M in G, then $\text{dist}_D(r, u) = w(M) - w(M(u))$, for all $u \in U$.*

Proof. Consider the matchings $M(u)$ and M. Direct all edges in M from V to U and all edges in $M(u)$ from U to V. In this way we obtain a directed path p from s to t and a set C of even length alternating cycles (see Figure 2). The path p

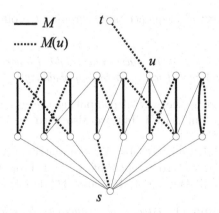

Fig. 2. Graph $G(u)$ with the matching M and $M(u)$. The sum $M \cup M(u)$ gives a path from s to t.

corresponds to the path in graph D from r to u. And the cycles in $\mathcal{C}$ correspond to cycles in D. All these cycles have a zero weight because M and $M(u)$ are maximum. Thus the weight of p is exactly $w(M) - w(M(u))$ and $\text{dist}(r, u) \leq y_u$.

Now note that from each path p from r to u in D we can construct a perfect matching $M'(u)$ in $G(u)$. We simply take $M'(u) = p \oplus M$. Thus $w(M'(u)) + w(M) = w(p)$, and so $\text{dist}(r, u) \geq y_u$. $\square$

Note that if the minimum vertex cover is computed correctly for U then for V it can be determined by the following equation,

$$y(v) := \max_{i \in U}\{w(vi) - y(i)\} \text{ for } v \in V. \tag{1}$$

Thus we can restrict ourselves to computing the vertex cover only for vertices in U, as in Lemma 8.

2.4   From Matching Weights to Matrix Polynomials

In this subsection we show how to compute the weights of the matchings $M(u)$ with use of the matrix polynomials. We start by showing a way of computing the weight of a maximum weighted perfect matching.

For a weighted bipartite graph $G = (U, V, E, w)$, define a $n \times n$ matrix $\tilde{B}(G, x)$ by

$$\tilde{B}(G, x)_{i,j} = \begin{cases} z_{i,j} x^{w(v_i v_j)} & \text{if } v_i v_j \in E, \\ 0 & \text{otherwise,} \end{cases}$$

where $z_{i,j}$ are distinct variables corresponding to edges in G.

Lemma 9 (Karp, Upfal and Wigderson '86). *The degree of x in $\det(\tilde{B}(G, x))$ is the weight of the maximum weighted perfect matching in G.*

The following is a direct consequence of Theorem 1, Corollary 3 and the above lemma.

Corollary 10. *The weight of the maximum weighted bipartite perfect matching can be computed in $\tilde{O}(Wn^\omega)$ time, with high probability.*

The above corollary gives the way of computing the weight of the maximum weighted perfect matching $w(M)$, but we also need to compute the weights $w(M(u))$, for all $u \in U$.

For a given matrix A, we denote by $A^{i,j}$ the matrix A with elements in i-th row and j-th column set to zero except that $A_{i,j} = 1$. From the definition of the adjoint we have that $\text{adj}(A)_{i,j} = \det(A^{j,i})$ and $A^{-1}\det(A) = \text{adj}(A)$.

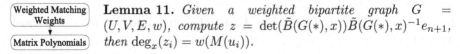

Lemma 11. *Given a weighted bipartite graph $G = (U, V, E, w)$, compute $z = \det(\tilde{B}(G(*), x))\tilde{B}(G(*), x)^{-1}e_{n+1}$, then $\deg_x(z_i) = w(M(u_i))$.*

Proof. We have

$$z_i = \det(\tilde{B}(G(*), x)) \left(\tilde{B}(G(*), x)^{-1} e_{n+1} \right)_i = \left(\text{adj}(\tilde{B}(G(*), x)) e_{n+1} \right)_i =$$

$$= \text{adj}(\tilde{B}(G(*), x))_{n+1,i} = \det(\tilde{B}(G(*), x)^{i,n+1}) = \det(\tilde{B}(G(u_i), x)),$$

and now from Lemma 9 we obtain that

$$\deg_x(z_i) = \deg_x \left(\det(\tilde{B}(G(u_i), x)) \right) = w(M(u_i)). \qquad \square$$

As a a consequence of Theorem 1, Corollary 3 and the above corollary, we obtain that the values $w(M(u))$ can be computed in $\tilde{O}(Wn^\omega)$ time. Joining this observation with Lemma 8, Lemma 7 and Lemma 6 we obtain

Theorem 12. *Given a weighted bipartite graph $G = (U, V, E, w)$, a minimum weighted vertex cover of G and a maximum weighted perfect matching in G, can be computed in $\tilde{O}(Wn^\omega)$ time, with high probability.*

3   Applications of the Matching Algorithm

3.1   Maximum and Minimum Matchings

In the *minimum weighted bipartite perfect matching problem* we seek a perfect matching M in a weighted bipartite graph G to minimize the total weight $w(M)$. The instance of the minimization problem can be turned into the maximization problem by defining a new weight function $w' : E \to \mathcal{Z}_+$ as

$$w'(e) = -w(e) + W.$$

By taking the negative weights we turn the minimization problem into the maximization problem and by adding W we guarantee that edge weights are positive. This does not change the solution because the weight of all perfect matchings is increased by nW.

In the *maximum weighted bipartite matching problem* we want to construct a matching M (not necessary perfect) in a weighted bipartite graph G, that maximizes the total weight $w(M)$. In order to obtain the instance of the perfect matching problem we simply have to add zero weight edges between all not connected vertices in G. The maximum weighted perfect matching in the resulting graph corresponds to the maximum weighted matching in G.

Theorem 13. *Given a weighted bipartite graph $G = (U, V, E, w)$, a minimum weighted perfect matching, as well as, minimum and maximum meighted matching in G, can be computed in $\tilde{O}(Wn^\omega)$ time, with high probability.*

3.2   Single Source Shortest Paths

A *weighted directed graph* G is a tuple $G = (V, E, w)$, where the vertex set is given by $V = \{v_1, \ldots, v_n\}$, $E \subseteq V \times V$ denotes the edge set, and the function $w : E \to \mathcal{Z}$ ascribes the lengths to the edges. We denote by W is the maximum absolute value of the edge length. In the *single source shortest paths problem* want to compute the distances from a given vertex v to all other vertices.

The following reduction was given by Gabow [7]. Define a bipartite graph $G' = (U', V', E', w')$ in the following way:

$$
\begin{aligned}
U' &= \{u'_1, \ldots, u'_n\}, \\
V' &= \{v'_1, \ldots, v'_n\}, \\
E' &= \{u'_i v'_j : (v_i, v_j) \in E\} \cup \{u'_i v'_i : 1 \le i \le n\},
\end{aligned}
$$

$$
w'(u'_i v'_j) = \begin{cases} -w((v_i, v_j)) & \text{if } (v_i, v_j) \in E, \\ 0 & \text{otherwise.} \end{cases}
$$

A perfect matching in G' corresponds to a set of cycles in the graph G. If a maximum weight of a perfect matching in G' is greater than zero, there is a negative weight cycle in G. Therefore, we can detect if G has a negative length cycle in $\tilde{O}(Wn^\omega)$ time. If this is not the case, we compute a minimum vertex cover y' of G'. There is always a zero weight matching in G', so the weight of y' is zero. Note also that $y'(u'_i) = -y'(v'_i)$, because the edges $u'_i v'_i$ are tight. Thus we have for an edge (v_i, v_j)

$$
\begin{aligned}
y'(u'_i) + y'(v'_j) &\ge w'(u'_i v'_j) \\
y'(u'_i) - y'(u'_j) &\ge -w((v_i, v_j)) \\
w((v_i, v_j)) &\ge y'(u'_j) - y'(u'_i).
\end{aligned}
$$

We see that $p : V \to \mathcal{Z}$, defined as $p(v_i) := y'(u'_i)$, is a good potential function in G. Hence we can use it to map the edge weights to positive values conserving the shortest paths (for details see [3]). With the use of the nonnegative weights we can compute the shortest paths with Dijkstra algorithm.

Theorem 14. *Given a weighted directed graph $G = (V, E, w)$ and a vertex $v \in V$, a negative cycle in G can be detected, or the distances from v can be computed, in $\tilde{O}(Wn^\omega)$ time, with high probability.*

3.3   Minimum Weight Disjoint Paths

Let $G = (V, E, w)$ be a weighted directed graph with two distinguished vertices $s, t \in V$. In the *minimum weight vertex disjoint s-t paths* problem we want to find the minimum weight of k vertex disjoint paths in G from s to t. The vertices s and t can be common for the paths. For given G and k, let us define a bipartite graph $G^k = (U^k, V^k, E^k, w^k)$, in the following way

$$U^k = \{u_1^k, \ldots, u_n^k\} \cup \{s_1^k, \ldots, s_k^k\},$$
$$V^k = \{v_1^k, \ldots, v_n^k\} \cup \{t_1^k, \ldots, t_k^k\},$$
$$E^k := \{u_i^k v_j^k : (v_i, v_j) \in E, \ v_i \neq s, \ v_j \neq t\} \cup$$
$$\cup \ \{u_i^k v_i^k : 1 \leq i \leq n\} \cup$$
$$\cup \ \{s_i^k v_j^k : (s, v_j) \in E, \ 1 \leq i \leq k\} \cup$$
$$\cup \ \{u_j^k t_i^k : (v_j, t) \in E, \ 1 \leq i \leq k\}.$$

$$w^k(u_i^k v_j^k) = \begin{cases} w((v_i, v_j)) & \text{if } (v_i, v_j) \in E, \\ 0 & \text{otherwise.} \end{cases}$$

Consider a graph G' obtained from G by adding an edge (t, s). Similarly as above, a perfect matching in G^k corresponds to a set C of cycles in G'. These cycles are, in general, vertex disjoint, but the vertices s and t may be common. The vertices s_i^k and t_i^k cannot be matched in G^k with themselves. Hence C must contain the edge (t, s) exactly k times, because there are k copies of s and t. By removing the edge (t, s) from C we obtain k vertex disjoint paths from s to t. The minimum weight of the s-t paths corresponds to the minimum weight of the perfect matching in G^k. Moreover G^k contains a perfect matching if and only if G contains at least k vertex disjoint s-t paths.

Theorem 15. *Given a weighted directed graph $G = (V, E, w)$, a minimum weight of k vertex disjoint s-t paths can be computed in $\tilde{O}(Wn^\omega)$ time, with high probability.*

4 Conclusions and Open Problems

4.1 Weighted Matchings in General Graphs

Storjohann's theorem can be applied in the general case as well. Let us define for a weighted graph $G = (V, E, w)$ a matrix

$$\tilde{A}(G, x)_{i,j} = \begin{cases} z_{i,j} x^{w(ij)} & \text{if } (i, j) \in E \text{ and } i < j, \\ -z_{j,i} x^{w(ij)} & \text{if } (i, j) \in E \text{ and } i > j, \\ 0 & \text{otherwise.} \end{cases}$$

The following is the generalization of Lemma 9 to general graphs.

Lemma 16 (Karp, Upfal and Widgerson '86). *The degree of x in $\det(\tilde{A}(G, x))$ is twice the weight of the maximum weighted perfect matching in G.*

Similarly as in the bipartite case we obtain the following corollary.

Corollary 17. *The weight of the maximum weighted perfect matching can be computed in $\tilde{O}(Wn^\omega)$ time, with high probability.*

4.2   Conclusions

We have shown that the algebraic approach to finding perfect matching can be used in the case of weighted bipartite graphs as well. We have presented an algorithm solving this problem in $\tilde{O}(Wn^\omega)$ time. For small edge weights this improves over previous fastest algorithms for this problem that work in $\tilde{O}(n^{2.5}\log(nW))$ time [8] and in $\tilde{O}(nm)$ time [5]. We have also shown how to compute the maximum weight of a matching in general graphs in $\tilde{O}(Wn^\omega)$ time. Nevertheless, the finding of this matching remains an open problem. Similar ideas cannot be used for the general case because the approach of Lemma 6 does not work there. It is impossible to construct a subgraph of a graph that will contain all the maximum weight perfect matchings and no other perfect matchings. The following counterexample was given by Eppstein [6]: let $G = K_6$ with edge weights two on two disjoint triangles and unit edge weights on the remaining edges. Each weight one edge is in unique maximum weight matching, so no edges can be removed from G. Thus in general case one have to use a different approach. The algorithm presented here uses the fact, that a dual problem is pretty simple and it can be deduced from the adjoint matrix. In the general case we have to compute so called Edmonds sets, i.e. blossoms. The problem here is that they are not defined unambiguously, so it seems improbable that they could be obtained in a simple way from the adjoint of $\tilde{A}(G, x)$. This is a similar situation as in the case of the NC algorithms for the perfect matchings in planar graphs, where the algorithm for the general case is not known. It seems also that we are facing here exactly the same problem, i.e., how to compute Edmonds sets in polylogarithmic time. However, the most intriguing open problem is whether the dependence of the complexity on W in our algorithm can be reduced. Such reduction would imply also a faster algorithm for the SSSP problem.

References

1. R. Bellman. On a Routing Problem. *Quarterly of Applied Mathematics*, 16(1):87–90, 1958.
2. D. Coppersmith and S. Winograd. Matrix multiplication via arithmetic progressions. In *Proceedings of the nineteenth annual ACM conference on Theory of computing*, pages 1–6. ACM Press, 1987.
3. T.H. Cormen, C.E. Leiserson, and R.L. Rivest. *Introduction to Algorithms*. MIT Press, Cambridge Mass., 1990.
4. E.A. Dinic and M. A. Kronrod. An Algorithm for the Solution of the Assignment Problem. *Soviet Math. Dokl.*, 10:1324–1326, 1969.
5. J. Edmonds and R.M. Karp. Theoretical Improvements in Algorithmic Efficiency for Network Flow Problems. *J. ACM*, 19(2):248–264, 1972.
6. David Eppstein. Representing all Minimum Spanning Trees with Applications to Counting and Generation. Technical Report ICS-TR-95-50, 1995.
7. H.N. Gabow. Scaling Algorithms for Network Problems. *J. Comput. Syst. Sci.*, 31(2):148–168, 1985.
8. H.N. Gabow and R.E. Tarjan. Faster Scaling Algorithms for Network Problems. *SIAM J. Comput.*, 18(5):1013–1036, 1989.

9. Andrew V. Goldberg. Scaling algorithms for the shortest paths problem. In *SODA '93: Proceedings of the fourth annual ACM-SIAM Symposium on Discrete algorithms*, pages 222–231. Society for Industrial and Applied Mathematics, 1993.
10. M. Iri. A new method for solving transportation-network problems. *Journal of the Operations Research Society of Japan*, 3:27–87, 1960.
11. L.R. Ford Jr. Network Flow Theory. Paper P-923, The RAND Corperation, Santa Moncia, California, August 1956.
12. M.-Y. Kao, T. W. Lam, W.-K. Sung, and H.-F. Ting. A decomposition theorem for maximum weight bipartite matchings with applications to evolutionary trees. In *Proceedings of the 7th Annual European Symposium on Algorithms*, pages 438–449, 1999.
13. R. M. Karp, E. Upfal, and A. Wigderson. Constructing a perfect matching is in random nc. *Combinatorica*, 6(1):35–48, 1986.
14. H.W Kuhn. The Hungarian Method for the Assignment Problem. *Naval Research Logistics Quarterly*, 2:83–97, 1955.
15. L. Lovász. On determinants, matchings and random algorithms. In L. Budach, editor, *Fundamentals of Computation Theory*, pages 565–574. Akademie-Verlag, 1979.
16. E. F. Moore. The Shortest Path Through a Maze. In *Proceedings of the International Symposium on the Theory of Switching*, pages 285–292. Harvard University Press, 1959.
17. Marcin Mucha and Piotr Sankowski. Maximum matchings via gaussian elimination. In *Proceedings of the 45th annual IEEE Symposium on Foundations of Computer Science*, pages 248–255, 2004.
18. K. Mulmuley, U.V. Vazirani, and V.V. Vazirani. Matching is as easy as matrix inversion. In *STOC '87: Proceedings of the nineteenth annual ACM conference on Theory of computing*, pages 345–354. ACM Press, 1987.
19. J. Munkres. Algorithms for the Assignment and Transportation Problems. *Journal of SIAM*, 5(1):32–38, 1957.
20. M. O. Rabin and V. V. Vazirani. Maximum matchings in general graphs through randomization. *Journal of Algorithms*, 10:557–567, 1989.
21. P. Sankowski. Shortes paths in matrix multiplication time. In *Proceedings of the 13th Annual European Symposium on Algorithms, LNCS 3669*, pages 770–778, 2005.
22. J. Schwartz. Fast probabilistic algorithms for verification of polynomial identities. *Journal of the ACM*, 27:701–717, 1980.
23. A. Shimbel. Structure in Communication Nets. In *In Proceedings of the Symposium on Information Networks*, pages 199–203. Polytechnic Press of the Polytechnic Institute of Brooklyn, Brooklyn, 1955.
24. Arne Storjohann. High-order lifting and integrality certification. *J. Symb. Comput.*, 36(3-4):613–648, 2003.
25. R. Yuster and U. Zwick. Answering distance queries in directed graphs using fast matrix multiplication. In *In Proceedings of the 46th Annual Symposium on Foundations of Computer Science*, pages 90–100. IEEE, 2005.
26. R. Zippel. Probabilistic algorithms for sparse polynomials. In *International Symposium on Symbolic and Algebraic Computation*, volume 72 of *Lecture Notes in Computer Science*, pages 216–226, Berlin, 1979. Springer-Verlag.

Optimal Resilient Sorting and Searching in the Presence of Memory Faults*

Irene Finocchi[1], Fabrizio Grandoni[1], and Giuseppe F. Italiano[2]

[1] Dipartimento di Informatica, Università di Roma "La Sapienza", Via Salaria 113, 00198 Roma, Italy
{finocchi, grandoni}@di.uniroma1.it
[2] Dipartimento di Informatica, Sistemi e Produzione, Università di Roma "Tor Vergata", Via del Politecnico 1, 00133 Roma, Italy
italiano@disp.uniroma2.it

Abstract. We investigate the problem of reliable computation in the presence of faults that may arbitrarily corrupt memory locations. In this framework, we consider the problems of sorting and searching in optimal time while tolerating the largest possible number of memory faults. In particular, we design an $O(n \log n)$ time sorting algorithm that can optimally tolerate up to $O(\sqrt{n \log n})$ memory faults. In the special case of integer sorting, we present an algorithm with linear expected running time that can tolerate $O(\sqrt{n})$ faults. We also present a randomized searching algorithm that can optimally tolerate up to $O(\log n)$ memory faults in $O(\log n)$ expected time, and an almost optimal deterministic searching algorithm that can tolerate $O((\log n)^{1-\epsilon})$ faults, for any small positive constant ϵ, in $O(\log n)$ worst-case time. All these results improve over previous bounds.

1 Introduction

The need for reliable computations in the presence of memory faults arises in many important applications. In fault-based cryptanalysis, for instance, some recent optical and electromagnetic perturbation attacks [12] work by manipulating the non-volatile memories of cryptographic devices, so as to induce very timing-precise controlled faults on given individual bits: this forces the devices to output wrong ciphertexts that may allow the attacker to determine the secret keys used during the encryption. Applications that make use of large memory capacities at low cost also incur problems of memory faults and reliable computation. Indeed, the unpredictable failures known as *soft memory errors* tend to increase with memory size and speed [8]. Although the number of faults could be reduced by means of error checking and correction circuitry, this imposes non-negligible costs in terms of performance (as much as 33%), size (20% larger areas), and money (10% to 20% more expensive chips). For these reasons, this

---

* This work has been partially supported by the Sixth Framework Programme of the EU under Contract Number 507613 (Network of Excellence EuroNGI) and by MIUR, under Projects WEB MINDS and ALGO-NEXT.

M. Bugliesi et al. (Eds.): ICALP 2006, Part I, LNCS 4051, pp. 286–298, 2006.

is not typically implemented in low-cost memories. Data replication is a natural approach to protect against destructive memory faults. However, it can be very inefficient in highly dynamic contexts or when the objects to be managed are large and complex: copying such objects can indeed be very costly, and in some cases we might not even know how to do this (for instance, when the data is accessed through pointers, which are moved around in memory instead of the data itself, and the algorithm relies on user-defined access functions). In these cases, we cannot assume either the existence of *ad hoc* functions for data replication or the definition of suitable encoding mechanisms to maintain a reasonable storage cost. As an example, consider Web search engines, which need to store and process huge data sets (of the order of Terabytes), including inverted indices which have to be maintained sorted for fast document access: for such large data structures, even a small failure probability can result in bit flips in the index, that may become responsible of erroneous answers to keyword searches [9]. In all these scenarios, it makes sense to assume that it must be the algorithms themselves, rather than specific hardware/software fault detection and correction mechanisms, that are responsible for dealing with memory faults. Informally, we have a *memory fault* when the correct value that should be stored in a memory location gets altered because of a failure, and we say that an algorithm is resilient to memory faults if, despite the corruption of some memory values before or during its execution, the algorithm is nevertheless able to get a correct output (at least) on the set of uncorrupted values.

The problem of computing with unreliable information has been investigated in a variety of different settings, including the *liar model* (see, e.g., [3,6,11]), fault-tolerant sorting networks [1,10], resiliency of pointer-based data structures [2], parallel models of computation with faulty memories [5]. In [7], we introduced a *faulty-memory random access machine*, i.e., a random access machine whose memory locations may suffer from memory faults. In this model, an adversary may corrupt up to δ memory words throughout the execution of an algorithm. The algorithm cannot distinguish corrupted values from correct ones and can exploit only $O(1)$ *safe* memory words, whose content never gets corrupted. Furthermore, whenever it reads some memory location, the read operation will temporarily store its value in the safe memory. The adversary is adaptive, but has no access to information about future random choices of the algorithm: in particular, loading a random memory location into safe memory can be considered an atomic operation.

In this paper we address the problems of resilient sorting and searching in the faulty-memory random access machine. In the *resilient sorting* problem we are given a sequence of n keys that need to be sorted. The value of some keys can be arbitrarily corrupted (either increased or decreased) during the sorting process. The resilient sorting problem is to order correctly the set of uncorrupted keys. This is the best that we can achieve in the presence of memory faults, since we cannot prevent keys corrupted at the very end of the algorithm execution from occupying wrong positions in the output sequence. In the *resilient searching* problem we are given a sequence of n keys on which we wish to perform membership queries. The keys are stored in increasing order, but some keys may be corrupted (at any instant of time) and thus may occupy wrong positions in the sequence. Let x be the key to be searched for. The resilient searching problem is either to find a key equal to x, or to determine that there is no correct key equal to x. Also in this case, this is the best we can hope for, because memory faults can make x appear or disappear in the sequence at any time.

In [7] we contributed a first step in the study of resilient sorting and searching. In particular, we proved that any resilient $O(n \log n)$ comparison-based sorting algorithm can tolerate the corruption of at most $O(\sqrt{n \log n})$ keys and we presented a resilient algorithm that tolerates $O(\sqrt[3]{n \log n})$ memory faults. With respect to searching, we proved that any $O(\log n)$ time deterministic searching algorithm can tolerate at most $O(\log n)$ memory faults and we designed an $O(\log n)$ time searching algorithm that can tolerate up to $O(\sqrt{\log n})$ memory faults. The main contribution of this paper is to close the gaps between upper and lower bounds for resilient sorting and searching. In particular:

- We design a resilient sorting algorithm that takes $O(n \log n + \delta^2)$ worst-case time to run in the presence of δ memory faults. This yields an algorithm that can tolerate up to $O(\sqrt{n \log n})$ faults in $O(n \log n)$ time: as proved in [7], this bound is optimal.
- In the special case of integer sorting, we present a randomized algorithm with expected running time $O(n + \delta^2)$: thus, this algorithm is able to tolerate up to $O(\sqrt{n})$ memory faults in expected linear time.
- We prove an $\Omega(\log n + \delta)$ lower bound on the expected running time of resilient searching algorithms: this extends the lower bound for deterministic algorithms given in [7].
- We present an optimal $O(\log n + \delta)$ time randomized algorithm for resilient searching: thus, this algorithm can tolerate up to $O(\log n)$ memory faults in $O(\log n)$ expected time.
- We design an almost optimal $O(\log n + \delta^{1+\epsilon'})$ time deterministic searching algorithm, for any constant $\epsilon' \in (0, 1]$: this improves over the $O(\log n + \delta^2)$ bound of [7] and yields an algorithm that can tolerate up to $O((\log n)^{1-\epsilon})$ faults, for any small positive constant ϵ.

Notation. We recall that δ is an upper bound on the total number of memory faults. We also denote by α the *actual* number of faults that happen during a specific execution of an algorithm. Note that $\alpha \leq \delta$. We say that a key is *faithful* if its value is never corrupted by any memory fault, and *faulty* otherwise. A sequence is *faithfully ordered* if its faithful keys are sorted, and *k-unordered* if there exist k (faithful) keys whose removal makes the remaining subsequence faithfully ordered. Given a sequence X of length n, we use $X[a\,;b]$, with $1 \leq a \leq b \leq n$, as a shortcut for the subsequence $\{X[a], X[a+1], \ldots, X[b]\}$. Two keys $X[p]$ and $X[q]$, with $p < q$, form an *inversion* in the sequence X if $X[p] > X[q]$: note that, for any two keys forming an inversion in a faithfully ordered sequence, at least one of them must be faulty. A sorting or merging algorithm is called *resilient* if it produces a faithfully ordered sequence.

2   Optimal Resilient Sorting in the Comparison Model

In this section we describe a resilient sorting algorithm that takes $O(n \log n + \delta^2)$ worst-case time to run in the presence of δ memory faults. This yields an $O(n \log n)$ time algorithm that can tolerate up to $O(\sqrt{n \log n}\,)$ faults: as proved in [7], this bound is optimal if we wish to sort in $O(n \log n)$ time, and improves over the best known resilient algorithm, which was able to tolerate only $O(\sqrt[3]{n \log n}\,)$ memory faults [7]. We first present a fast resilient merging algorithm, that may nevertheless fail to insert all the input values in the faithfully ordered output sequence. We next show how to use this algorithm to solve the resilient sorting problem within the claimed $O(n \log n + \delta^2)$ time bound.

The Purifying Merge Algorithm. Let X and Y be the two faithfully ordered sequences of length n to be merged. The merging algorithm that we are going to describe produces a faithfully ordered sequence Z and a disordered fail sequence F in $O(n + \alpha \delta)$ worst-case time. It will be guaranteed that $|F| = O(\alpha)$, i.e., that only $O(\alpha)$ keys can fail to get inserted into Z.

The algorithm, called `PurifyingMerge`, uses two auxiliary input buffers of size $(2\delta + 1)$ each, named $\mathcal{X}$ and $\mathcal{Y}$, and an auxiliary output buffer of size δ, named $\mathcal{Z}$. The input buffers $\mathcal{X}$ and $\mathcal{Y}$ are initially filled with the first $(2\delta + 1)$ values in X and Y, respectively. The merging process is divided into rounds: the algorithm maintains the invariant that, at the beginning of each round, both input buffers are full while the output buffer is empty (we omit here the description of the boundary cases). Each round consists of merging the contents of the input buffers until either the output buffer becomes full or an inconsistency in the input keys is found. In the latter case, we perform a *purifying step*, where two keys are moved to the fail sequence F. We now describe the generic round in more detail.

The algorithm fills buffer $\mathcal{Z}$ by scanning the input buffers $\mathcal{X}$ and $\mathcal{Y}$ sequentially. Let i and j be the running indices on $\mathcal{X}$ and $\mathcal{Y}$: we call $\mathcal{X}[i]$ and $\mathcal{Y}[j]$ the *top keys* of $\mathcal{X}$ and $\mathcal{Y}$, respectively. The running indices i and j, the top keys of $\mathcal{X}$ and $\mathcal{Y}$, and the last key copied to $\mathcal{Z}$ are all stored in $O(1)$ size safe memory. At each step, we compare $\mathcal{X}[i]$ and $\mathcal{Y}[j]$: without loss of generality, assume that $\mathcal{X}[i] \leq \mathcal{Y}[j]$ (the other case being symmetric). We next perform an *inversion check* as follows: if $\mathcal{X}[i] \leq \mathcal{X}[i + 1]$, $\mathcal{X}[i]$ is copied to $\mathcal{Z}$ and index i is advanced by 1 (note that the key copied to $\mathcal{Z}$ is left in $\mathcal{X}$ as well). If the inversion check fails, i.e., $\mathcal{X}[i] > \mathcal{X}[i + 1]$, we perform a purifying step on $X[i]$ and $X[i + 1]$: we move these two keys to the fail sequence F, we append two new keys from X at the end of buffer $\mathcal{X}$, and we restart the merging process of the buffers $\mathcal{X}$ and $\mathcal{Y}$ from scratch by simply resetting all the buffer indices (note that this makes the output buffer $\mathcal{Z}$ empty). Thanks to the comparisons between the top keys and to the inversion checks, the last key appended to $\mathcal{Z}$ is always smaller than or equal to the top keys of $\mathcal{X}$ and $\mathcal{Y}$ (considering their values stored in safe memory): we call this *top invariant*. When $\mathcal{Z}$ becomes full, we check whether all the remaining keys in $\mathcal{X}$ and $\mathcal{Y}$ (i.e., the keys not copied into $\mathcal{Z}$) are larger than or equal to the last key $\mathcal{Z}[\delta]$ copied into $\mathcal{Z}$ (*safety check*). If the safety check fails on $\mathcal{X}$, the top invariant guarantees that there is an inversion between the current top key

$\mathcal{X}[i]$ of $\mathcal{X}$ and another key remaining in $\mathcal{X}$: in that case, we execute a purifying step on those two keys. We do the same if the safety check fails on $\mathcal{Y}$. If all the checks succeed, the content of $\mathcal{Z}$ is flushed to the output sequence Z and the input buffers $\mathcal{X}$ and $\mathcal{Y}$ are refilled with an appropriate number of new keys taken from X and Y, respectively.

Lemma 1. *Algorithm* PurifyingMerge, *given two faithfully ordered sequences of length n, merges the sequences in $O(n + \alpha \delta)$ worst-case time and returns a faithfully ordered sequence Z and a fail sequence F such that $|F| = O(\alpha)$.*

Proof. We first show that the output sequence Z is faithfully ordered. We say that a round is *successful* if it terminates by flushing the output buffer into Z, and *failing* if it terminates by adding keys to the fail sequence F. Since failing rounds do not modify Z, it is sufficient to consider successful rounds only. Let $\mathcal{X}'$ and X' be the remaining keys in $\mathcal{X}$ and X, respectively, at the end of a successful round. The definition of $\mathcal{Y}'$ and Y' is similar. We denote by $\widetilde{\mathcal{Z}}[h]$ the value of the h-th key inserted into $\mathcal{Z}$ at the time of its insertion. The sequence $\widetilde{\mathcal{Z}}$ must be sorted, since otherwise an inversion check would have failed at some point. Since $\widetilde{\mathcal{Z}}[h] = \mathcal{Z}[h]$ for each faithful key $\mathcal{Z}[h]$, it follows that (i) Z is faithfully ordered. Consider now the largest faithful key $z = \widetilde{\mathcal{Z}}[k]$ in $\mathcal{Z}$ and the smallest faithful key x in $\mathcal{X}' \cup X'$. We will show that $z \leq x$ (if one of the two keys does not exist, there is nothing to prove). Note that x must belong to $\mathcal{X}'$. In fact, all the faithful keys in $\mathcal{X}'$ are smaller than or equal to the faithful keys in X'. Moreover, either $\mathcal{X}'$ contains at least $(\delta + 1)$ keys (and thus at least one faithful key), or X' is empty. All the keys in $\mathcal{X}'$ are compared with $\widetilde{\mathcal{Z}}[\delta]$ during the safety check. In particular, $x \geq \widetilde{\mathcal{Z}}[\delta]$ since the safety check was successful. From the order of $\widetilde{\mathcal{Z}}$, we obtain $\widetilde{\mathcal{Z}}[\delta] \geq \widetilde{\mathcal{Z}}[k] = z$, thus implying $x \geq z$. A symmetric argument shows that z is smaller than or equal to the smallest faithful key y in $\mathcal{Y}' \cup Y'$. Hence (ii) all the faithful keys in $\mathcal{Z}$ are smaller than or equal to the faithful keys in $\mathcal{X}' \cup X'$ and $\mathcal{Y}' \cup Y'$. The claim follows from (i) and (ii) by induction on the number of successful rounds. The two values discarded in each failing round form an inversion in one of the input sequences, which are faithfully ordered. Thus, at least one of such discarded values must be corrupted, proving that the number of corrupted values in F at any time is at least $|F|/2$. This implies that $|F|/2 \leq \alpha$ and that the number of failing rounds is bounded above by α. Note that at each round we spend time $\Theta(\delta)$. When the round is successful, this time can be amortized against the time spent to flush δ values to the output sequence. We therefore obtain a total running time of $O(n + \alpha \delta)$.   □

The Sorting Algorithm. We first notice that a naive resilient sorting algorithm can be easily obtained from a bottom-up iterative implementation of MergeSort by taking the minimum among $(\delta+1)$ keys per sequence at each merge step. We call this NaiveSort. The running time of NaiveSort is $O(\delta n \log n)$ and becomes $O(\delta n)$ when $\delta = \Omega(n^\epsilon)$, for some $\epsilon > 0$. In order to obtain a more efficient sorting algorithm, we will use the following merging subroutine, called ResilientMerge. We first merge the input sequences using algorithm

`PurifyingMerge`: this produces a faithfully ordered sequence Z and a disordered fail sequence F of length $O(\alpha)$. We sort F with algorithm `NaiveSort` and produce a faithfully ordered sequence F' in time $O(\alpha \delta)$. We finally merge Z and F' in time $O(|Z| + (|F'| + \alpha)\delta) = O(n + \alpha \delta)$ using the algorithm `UnbalancedMerge` of [7]. Overall, algorithm `ResilientMerge` faithfully merges two faithfully ordered sequences of length n in $O(n + \alpha \delta)$ worst-case time. This implies the following:

Theorem 1. *There is a resilient algorithm that sorts n keys in $O(n \log n + \alpha \delta)$ worst-case time and linear space.*

This yields an $O(n \log n)$ time resilient sorting algorithm that can tolerate up to $O(\sqrt{n \log n})$ memory faults. As shown in [7], no better bound is possible.

3   Resilient Integer Sorting

In this section we consider the problem of faithfully sorting a sequence of n integers in the range $[0, n^c - 1]$, for some constant $c \geq 0$. We will present a randomized algorithm with expected running time $O(n + \delta^2)$: thus, this algorithm is able to tolerate up to $O(\sqrt{n})$ memory faults in expected linear time. Our algorithm is a resilient implementation of (least significant digit) `RadixSort`, which works as follows. Assume that the integers are represented in base b, with $b \geq 2$. At the i-th step, for $1 \leq i \leq \lceil c \log_b n \rceil$, we sort the integers according to their i-th least significant digit using a linear time, stable bucket sorting algorithm (with b buckets).

We can easily implement radix sort in faulty memory whenever the base b is constant: we keep an array of size n for each bucket and store the address of those arrays and their current length (i.e., the current number of items in each bucket) in the $O(1)$-size safe memory. It is not hard to show that this algorithm correctly sorts the faithful elements in $O(n \log n)$ worst-case time and linear space, while tolerating an arbitrary number of memory faults.

Unfortunately, in order to make `RadixSort` run in linear time, we need $b = \Omega(n^\epsilon)$, for some constant $\epsilon \in (0, 1]$. However, if the number of buckets is not constant, we might need more than linear space. More importantly, $O(1)$ safe memory words would not be sufficient to store the initial address and the current length of the b arrays. We will now show how to overcome both problems. We store the b arrays contiguously, so that their initial addresses can be derived from a unique address β (which is stored in safe memory). However, we cannot store in the $O(1)$ safe memory the current length of each array. Hence, in the i-th step of radix sort, with $1 \leq i \leq \lceil c \log_b n \rceil$, we have to solve b instances of the following *bucket-filling* problem. We receive in an online fashion a sequence of $n' \leq n$ integers (faithfully) sorted up to the i-th least significant digit. We have to copy this input sequence into an array $\mathcal{B}_0$ whose current length cannot be stored in safe memory: $\mathcal{B}_0$ must maintain the same faithful order as the order in the input sequence. In the rest of this section we will show how to solve the bucket-filling problem in $O(n' + \alpha \delta)$ expected time and $O(n' + \delta)$ space, where

α is the actual number of memory faults occurring throughout the execution of the bucket-filling algorithm. This will imply the following theorem.

Theorem 2. *There is a randomized algorithm that faithfully sorts n polynomially bounded integers in $O(n + \alpha \delta)$ expected time. The space required is linear when $\delta = O(n^{1-\epsilon})$, for any small positive constant ϵ.*

The Bucket-Filling Problem. We first describe a deterministic bucket-filling algorithm with running time $O(n' + \alpha \delta^{1.5})$. The algorithm exploits the use of buffering techniques. We remark that the input integers are (faithfully) sorted up to the i-th least significant digit and that we cannot store the current length of the buffers in safe memory. In order to circumvent this problem, we will use redundant variables, defined as follows. A *redundant $|p|$-index p* is a set of $|p|$ positive integers. The *value* of p is the majority value in the set (or an arbitrary value if no majority value exists). Assigning a value x to p means assigning x to all its elements: note that both reading and updating p can be done in linear time and constant space (using, e.g., the algorithm in [4]). If $|p| \geq 2\delta + 1$, we say that p is *reliable* (i.e., we can consider its value faithful even if p is stored in faulty memory). A *redundant $|p|$-pointer p* is defined analogously, with positive integers replaced by pointers. Besides using redundant variables, we periodically restore the ordering inside the buffers by means of a (bidirectional) BubbleSort, which works as follows: we compare adjacent pairs of keys, swapping them if necessary, and alternately pass through the sequence from the beginning to the end and from the end to the beginning, until no more swaps are performed. Interestingly enough, BubbleSort is resilient to memory faults and its running time depends only on the disorder of the input sequence and on the actual number of faults occurring during its execution.

Lemma 2. *Given a k-unordered sequence of length n, algorithm BubbleSort faithfully sorts the sequence in $O(n + (k + \alpha) n)$ worst-case time.*

We now give a more detailed description of our bucket-filling algorithm. Besides the output array $\mathcal{B}_0$, we use two buffers to store temporarily the input keys: a buffer $\mathcal{B}_1$ of size $|\mathcal{B}_1| = 2\delta + 1$, and a buffer $\mathcal{B}_2$ of size $|\mathcal{B}_2| = 2\sqrt{\delta} + 1$. All the entries of both buffers are initially set to a value, say $+\infty$, that is not contained in the input sequence. We associate a redundant index p_i to each $\mathcal{B}_i$, where $|p_0| = |\mathcal{B}_1| = 2\delta + 1$, $|p_1| = |\mathcal{B}_2| = 2\sqrt{\delta} + 1$, and $|p_2| = 1$. Note that only p_0 is reliable, while p_1 and p_2 could assume faulty values. Both buffers and indexes are stored in such a way that their address can be derived from the unique address β stored in safe memory. The algorithm works as follows. Each time a new input key is received, it is appended to $\mathcal{B}_2$. Whenever $\mathcal{B}_2$ is full (according to index p_2), we *flush* it as follows: (1) we remove any $+\infty$ from $\mathcal{B}_2$ and sort $\mathcal{B}_2$ with BubbleSort considering the i least significant digits only; (2) we append $\mathcal{B}_2$ to $\mathcal{B}_1$, and we update p_1 accordingly; (3) we reset $\mathcal{B}_2$ and p_2. Whenever $\mathcal{B}_1$ is full, we *flush* it in a similar way, moving its keys to $\mathcal{B}_0$. We flush buffer $\mathcal{B}_j$, $j \in \{1, 2\}$, also whenever we realize that the index p_j points to an entry outside $\mathcal{B}_j$ or to an entry of value different from $+\infty$ (which indicates that a fault happened either in p_j or in $\mathcal{B}_j$ after the last time $\mathcal{B}_j$ was flushed).

Lemma 3. *The algorithm above solves the bucket-filling problem in $O(n' + \alpha \delta^{1.5})$ worst-case time.*

Proof. To show the correctness, we notice that all the faithful keys eventually appear in $\mathcal{B}_0$. All the faithful keys in $\mathcal{B}_j$, $j \in \{1, 2\}$, at a given time precede the faithful keys not yet copied into $\mathcal{B}_j$. Moreover we sort $\mathcal{B}_j$ before flushing it. This guarantees that the faithful keys are moved from $\mathcal{B}_j$ to $\mathcal{B}_{j-1}$ in a first-in-first-out fashion. Consider the cost paid by the algorithm between two consecutive flushes of $\mathcal{B}_1$. Let α' and α'' be the number of faults in $\mathcal{B}_1$ and p_1, respectively, during the phase considered. If no fault happens in either $\mathcal{B}_1$ or p_1 ($\alpha' + \alpha'' = 0$), flushing buffer $\mathcal{B}_1$ costs $O(|\mathcal{B}_1|) = O(\delta)$. If the value of p_1 is faithful ($\alpha'' \leq \sqrt{\delta}$), the sequence is $O(\alpha')$-unordered: in fact, removing the corrupted values from $\mathcal{B}_1$ produces a sorted subsequence. Thus sorting $\mathcal{B}_1$ costs $O((1 + \alpha')\delta)$. Otherwise ($\alpha'' > \sqrt{\delta}$), the sequence $\mathcal{B}_1$ can be $O(\delta)$-unordered and sorting it requires $O((1 + \delta + \alpha')\delta) = O(\delta^2)$ time. Thus, the total cost of flushing buffer $\mathcal{B}_1$ is $O(n' + \alpha/\sqrt{\delta}\,\delta^2 + \alpha\,\delta) = O(n' + \alpha\,\delta^{1.5})$. Using a similar argument, we can show that the total cost of flushing buffer $\mathcal{B}_2$ is $O(n' + \alpha\,\delta)$. The claimed running time immediately follows.   □

The deterministic running time can be improved by choosing more carefully the buffer size and by increasing the number of buffers. Specifically, we can obtain an integer sorting algorithm with $O(n + \alpha\,\delta^{1+\epsilon})$ worst-case running time, for any small positive constant ϵ. The details will be included in the full paper.

A Randomized Approach. We now show how to reduce the (expected) running time of the bucket-filling algorithm to $O(n' + \alpha\,\delta)$, by means of randomization. As we already observed in the proof of Lemma 3, a few corruptions in p_1 can lead to a highly disordered sequence $\mathcal{B}_1$. Consider for instance the following situation: we corrupt p_1 twice, in order to force the algorithm to write first δ faithful keys in the second half of $\mathcal{B}_1$, and then other $(\delta + 1)$ faithful keys in the first half of $\mathcal{B}_1$. In this way, with $2(\sqrt{\delta} + 1)$ corruptions only, one obtains an $O(\delta)$-unordered sequence, whose sorting requires $O(\delta^2)$ time. This can happen $O(\alpha/\sqrt{\delta})$ times, thus leading to the $O(\alpha\,\delta^{1.5})$ term in the running time.

The idea behind the randomized algorithm is to try to avoid such kind of pathological situations. Specifically, we would like to detect early the fact that many values after the last inserted key are different from $+\infty$. In order to do that, whenever we move a key from $\mathcal{B}_2$ to $\mathcal{B}_1$, we select an entry uniformly at random in the portion of $\mathcal{B}_1$ after the last inserted key: if the value of this entry is not $+\infty$, the algorithm flushes $\mathcal{B}_1$ immediately.

Lemma 4. *The randomized algorithm above solves the bucket-filling problem in $O(n' + \alpha\,\delta)$ expected time.*

Proof. Let α' and α'' be the number of faults in $\mathcal{B}_1$ and p_1, respectively, between two consecutive flushes of buffer $\mathcal{B}_1$. Following the proof of Lemma 3 and the discussion above, it is sufficient to show that, when we sort $\mathcal{B}_1$, the sequence to be sorted is $O(\alpha' + \alpha'')$-unordered in expectation. In order to show that, we will describe a procedure which obtains a sorted subsequence from $\mathcal{B}_1$ by removing an expected number of $O(\alpha' + \alpha'')$ keys.

First remove the α' corrupted values in $\mathcal{B}_1$. Now consider what happens either between two consecutive corruptions of p_1 or between a corruption and a reset of p_1. Let $\widetilde{p}_1$ be the value of p_1 at the beginning of the phase considered. By A and B we denote the subset of entries of value different from $+\infty$ after $\mathcal{B}_1[\widetilde{p}_1]$ and the subset of keys added to $\mathcal{B}_1$ in the phase considered, respectively. Note that, when A is large, the expected cardinality of B is small (since it is more likely to select randomly an entry in A). More precisely, the probability of selecting at random an entry of A is at least $|A|/|\mathcal{B}_1|$. Thus the expected cardinality of B is at most $|\mathcal{B}_1|/|A| = O(\delta/|A|)$. The idea behind the proof is to remove A from $\mathcal{B}_1$ if $|A| < \sqrt{\delta}$, and to remove B otherwise. In both cases the expected number of keys removed is $O(\sqrt{\delta})$. At the end of the process, we obtain a sorted subsequence of $\mathcal{B}_1$. Since p_1 can be corrupted at most $O(\alpha''/\sqrt{\delta})$ times, the total expected number of keys removed is $O(\alpha' + \sqrt{\delta}\,\alpha''/\sqrt{\delta}) = O(\alpha' + \alpha'')$. $\square$

The space usage of the bucket-filling algorithm can be easily reduced to $O(n' + \delta)$ via doubling without increasing the asymptotic running time.

4   Resilient Searching Algorithms

In this section we prove upper and lower bounds on the resilient searching problem. Namely, we first prove an $\Omega(\log n + \delta)$ lower bound on the expected running time, and then we present an optimal $O(\log n + \delta)$ expected time randomized algorithm. Finally, we sketch an $O(\log n + \delta^{1+\epsilon'})$ time deterministic algorithm, for any constant $\epsilon' \in (0, 1]$. Both our algorithms improve over the $O(\log n + \delta^2)$ deterministic bound of [7].

A Lower Bound for Randomized Searching. We now show that every searching algorithm which tolerates up to δ memory faults must have expected running time $\Omega(\log n + \delta)$ on sequences of length n, with $n \geq \delta$.

Theorem 3. *Every (randomized) resilient searching algorithm must have expected running time $\Omega(\log n + \delta)$.*

Proof. An $\Omega(\log n)$ lower bound holds even when the entire memory is safe. Thus, it is sufficient to prove that every resilient searching algorithm takes expected time $\Omega(\delta)$ when $\log n = o(\delta)$. Let $\mathcal{A}$ be a resilient searching algorithm. Consider the following (feasible) input sequence I: for an arbitrary value x, the first $(\delta + 1)$ values of the sequence are equal to x and the others are equal to $+\infty$. Let us assume that the adversary arbitrarily corrupts δ of the first $(\delta + 1)$ keys before the beginning of the algorithm. Since a faithful key x is left, $\mathcal{A}$ must be able to find it. Observe that, after the initial corruption, the first $(\delta + 1)$ elements of I form an arbitrary (unordered) sequence. Suppose by contradiction that $\mathcal{A}$ takes $o(\delta)$ expected time. Then we can easily derive from $\mathcal{A}$ an algorithm to find a given element in an unordered sequence of length $\Theta(\delta)$ in sub-linear expected time, which is not possible (even in a safe-memory system). $\square$

Optimal Randomized Searching. Let I be the sorted input sequence and x be the key to be searched for. At each step, the algorithm considers a subsequence $I[\ell; r]$. Initially $I[\ell; r] = I[1; n] = I$. Let $C > 1$ and $0 < c < 1$ be two constants such that $cC > 1$. The algorithm has a different behavior depending on the length of the current interval $I[\ell; r]$. If $r - \ell > C\delta$, the algorithm chooses an element $I[h]$ uniformly at random in the central subsequence of $I[\ell; r]$ of length $(r - \ell)c$, i.e., in $I[\ell'; r'] = I[\ell + (r - \ell)(1 - c)/2; \ell + (r - \ell)(1 + c)/2]$ (for the sake of simplicity, we neglect ceilings and floors). If $I[h] = x$, the algorithm simply returns the index h. Otherwise, it continues searching for x either in $I[\ell; h - 1]$ or in $I[h + 1; r]$, according to the outcome of the comparison between x and $I[h]$. Consider now the case $r - \ell \leq C\delta$. Let us assume that there are at least 2δ values to the left of ℓ and 2δ values to the right of r (otherwise, it is sufficient to assume that $X[i] = -\infty$ for $i < 1$ and $X[i] = +\infty$ for $i > n$). If x is contained in $I[\ell - 2\delta; r + 2\delta]$, the algorithm returns the corresponding index. Else, if both the majority of the elements in $I[\ell - 2\delta; \ell]$ are smaller than x and the majority of the elements in $I[r; r + 2\delta]$ are larger than x, the algorithm returns no. Otherwise, at least one of the randomly selected values $I[h_k]$ must be faulty: in that case the algorithm simply restarts from the beginning. Note that all the variables require total constant space and can be stored in safe memory.

Theorem 4. *The algorithm above performs resilient searching in $O(\log n + \delta)$ expected time.*

Proof. Consider first the correctness of the algorithm. We will later show that the algorithm halts with probability one. If the algorithm returns an index, the answer is trivially correct. Otherwise, let $I[\ell; r]$ be the last interval considered before halting. According to the majority of the elements in $I[\ell - 2\delta; \ell]$, x is either contained in $I[\ell + 1; n]$ or not contained in I. This is true since the mentioned majority contains at least $(\delta + 1)$ elements, and thus at least one of them must be faithful. A similar argument applied to $I[r; r + 2\delta]$ shows that x can only be contained in $I[1; r - 1]$. Since the algorithm did not find x in $I[\ell + 1; n] \cap I[1; r - 1] = I[\ell + 1; r - 1]$, there is no faithful key equal to x in I.

Now consider the time spent in one iteration of the algorithm (starting from the initial interval $I = I[1; n]$). Each time the algorithm selects a random element, either the algorithm halts or the size of the subsequence considered is decreased by at least a factor of $2/(1 + c) > 1$. So the total number of selection steps is $O(\log n)$, where each step requires $O(1)$ time. The final step, where a subsequence of length at most $4\delta + C\delta = O(\delta)$ is considered, requires $O(\delta)$ time. Altogether, the worst-case time for one iteration is $O(\log n + \delta)$.

Thus, it is sufficient to show that in a given iteration the algorithm halts with some positive constant probability $P > 0$, from which it follows that the expected number of iterations is constant. Let $I[h_1], I[h_2] \ldots I[h_t]$ be the sequence of randomly chosen values in a given iteration. If a new iteration starts, this implies that at least one of those values is faulty. Hence, to show that the algorithm halts, it is sufficient to prove that all those values are faithful with positive probability. Let $\overline{P}_k$ denote the probability that $I[h_k]$ is faulty. Consider the last interval $I[\ell; r]$ in which we perform random sampling. The length of this interval is at least $C\delta$. So the value $I[h_t]$ is chosen in a subsequence of length at least $cC\delta > \delta$, from which we obtain $\overline{P}_t \le \delta/(cC\delta) = 1/(cC)$. Consider now the previous interval. The length of this interval is at least $2C\delta/(1+c)$. Thus $\overline{P}_{t-1} \le (1+c)/(2cC)$. More generally, for each $i = 0, 1, \ldots (t-1)$, we have $\overline{P}_{t-i} \le ((1+c)/2)^i/(cC)$. Altogether, the probability P that all the values $I[h_1], I[h_2] \ldots I[h_t]$ are faithful is equal to $\prod_{i=0}^{t-1}(1 - \overline{P}_{t-i})$ and thus

$$
P \ge \prod_{i=0}^{t-1}\left(1 - \frac{1}{cC}\left(\frac{1+c}{2}\right)^i\right) \ge \left(1 - \frac{1}{cC}\right)^{\sum_{i=0}^{t-1}(\frac{1+c}{2})^i} \ge \left(1 - \frac{1}{cC}\right)^{\frac{2}{1-c}} > 0,
$$

where we used the fact that $(1 - xy) \ge (1 - x)^y$ for every x and y in $[0, 1]$. $\qquad\square$

Almost Optimal Deterministic Searching. We now sketch our deterministic algorithm, which we refer to as `DetSearch`. We first introduce the notion of *k-left-test* and *k-right-test* over a position p, for $k \ge 1$ and $1 \le p \le n$. In a k-left-test over p, we consider the neighborhood of p of size k defined as $I[p-k\,;\,p-1]$: the test *fails* if the majority of keys in this neighborhood is larger than the key x to be searched for, and *succeeds* otherwise. A k-right-test over p is defined symmetrically on the neighborhood $I[p+1\,;\,p+k]$. Note that in the randomized searching algorithm described in the previous section we execute a $(2\delta + 1)$-left-test and a $(2\delta + 1)$-right-test at the end of each iteration. The idea behind our improved deterministic algorithm is to design less expensive left and right tests, and to perform them more frequently. More precisely, the basic structure of the algorithm is as in the classical (deterministic) binary search: in each step we consider the current interval $I[\ell; r]$ and we update it as suggested by the central value $I[(\ell + r)/2]$. Every $\sqrt{\delta}$ searching steps, we perform a $\sqrt{\delta}$-left-test over the left boundary ℓ and a $\sqrt{\delta}$-right-test over the right boundary r of the current interval $I[\ell; r]$. If one of the two $\sqrt{\delta}$-tests fails, we revert to the smallest interval $I[\ell'; r']$ suggested by the failed test and by the last $\sqrt{\delta}$-tests previously performed (the boundaries ℓ' and r' can be maintained in safe memory, and are updated each time a $\sqrt{\delta}$-test is performed). Every δ searching steps, we proceed analogously, where $\sqrt{\delta}$-tests are replaced by $(2\delta + 1)$ tests. For lack of space, we defer the low-level details, the description of the boundary cases, and the proof of correctness of algorithm `DetSearch` to the full paper. We now sketch the running time analysis. We say that a boundary p is *misleading* if the value $I[p]$ is faulty and guides the search towards a wrong direction. Similarly, a k-left-test over p is misleading if the majority of the values in $I[p-k\,;\,p-1]$ are misleading.

Theorem 5. *Algorithm* `DetSearch` *performs resilient searching in* $O(\log n + \alpha\sqrt{\delta})$ *worst-case time.*

Proof. (Sketch) Assume that the algorithm takes at some point a wrong search direction (*misled search*). Let us analyze the running time wasted due to a misled search. We first consider a misled search where there is no misleading $\sqrt{\delta}$-test. Without loss of generality, consider the case where the algorithm encounters a misleading left boundary, say p: then, the search erroneously proceeds to the right of p. Consider the time when the next $\sqrt{\delta}$-left-test is performed, and let ℓ be the left boundary involved in the test. Note that it must be $p \leq \ell$ and, since p is misleading, then ℓ must be also a misleading left boundary. Due to the hypothesis that $\sqrt{\delta}$-tests are not misleading, the $\sqrt{\delta}$-left-test over ℓ must have failed, detecting the error on p and recovering the proper search direction: hence, the uncorrect search wasted only $O(\sqrt{\delta})$ time, which can be charged to the faulty value $I[\ell]$. Since $I[\ell]$ is out of the interval on which the search proceeds, each faulty value can be charged at most once and we will have at most α uncorrect searches of this kind. The total running time will thus be $O(\alpha\sqrt{\delta})$. We next analyze the running time for a misled search when there exists at least one misleading $\sqrt{\delta}$-test. In this case, an error due to a misleading $\sqrt{\delta}$-test will be detected at most δ steps later, when the next $(2\delta + 1)$-test is performed. Using similar arguments, we can prove that there must exist $\Theta(\sqrt{\delta})$ faulty values that are eliminated from the interval in which the search proceeds, and we can charge the $O(\delta)$ time spent for the uncorrect search to those values. Thus, we will have at most $O(\alpha/\sqrt{\delta})$ uncorrect searches of this kind, requiring $O(\delta)$ time each. The total running time will be again $O(\alpha\sqrt{\delta})$. Since the time for the correct searches is $O(\log n)$, the claimed bound of $O(\log n + \alpha\sqrt{\delta})$ follows. □

The running time can be reduced to $O(\log n + \alpha\,\delta^{\epsilon'})$, for any constant $\epsilon' \in (0, 1]$, by exploiting the use of $(2\delta^{i\,\epsilon'} + 1)$-tests, with $i = 1, 2, \ldots (1/\epsilon')$. This yields a deterministic resilient searching algorithm that can tolerate up to $O((\log n)^{1-\epsilon})$ memory faults, for any small positive constant ϵ, in $O(\log n)$ worst-case time, thus getting arbitrarily close to the lower bound. For lack of space, we defer the details of the algorithm and of its analysis to the full paper.

References

1. S. Assaf and E. Upfal. Fault-tolerant sorting networks. *SIAM J. Discrete Math.*, 4(4), 472–480, 1991.
2. Y. Aumann and M. A. Bender. Fault-tolerant data structures. *Proc. 37th IEEE Symp. on Foundations of Computer Science*, 580–589, 1996.
3. R. S. Borgstrom and S. Rao Kosaraju. Comparison based search in the presence of errors. *Proc. 25th ACM Symp. on Theory of Computing*, 130–136, 1993.
4. R. Boyer and S. Moore. MJRTY - A fast majority vote algorithm. University of Texas Tech. Report, 1982.
5. B. S. Chlebus, L. Gasieniec and A. Pelc. Deterministic computations on a PRAM with static processor and memory faults. *Fund. Informaticae*, 55, 285–306, 2003.
6. U. Feige, P. Raghavan, D. Peleg, and E. Upfal. Computing with noisy information. *SIAM Journal on Computing*, 23, 1001–1018, 1994.
7. I. Finocchi and G. F. Italiano. Sorting and searching in the presence of memory faults (without redundancy). *Proc. 36th ACM Symp. on Theory of Computing*, 101–110, 2004.
8. S. Hamdioui, Z. Al-Ars, J. Van de Goor, and M. Rodgers. Dynamic faults in Random-Access-Memories: Concept, faults models and tests. *Journal of Electronic Testing: Theory and Applications*, 19, 195–205, 2003.
9. M. Henzinger. The past, present and future of Web Search Engines. Invited talk. *31st Int. Coll. Automata, Languages and Programming*, 12–16 2004.

10. T. Leighton and Y. Ma. Tight bounds on the size of fault-tolerant merging and sorting networks with destructive faults. *SIAM Journal on Computing*, 29(1):258–273, 1999.
11. A. Pelc. Searching games with errors: Fifty years of coping with liars. *Theoretical Computer Science*, 270, 71–109, 2002.
12. S. Skorobogatov and R. Anderson. Optical fault induction attacks. *Proc. 4th Int. Workshop on Cryptographic Hardware and Embedded Systems*, 2–12, 2002.

Reliable and Efficient Computational Geometry Via Controlled Perturbation[*]

Kurt Mehlhorn, Ralf Osbild, and Michael Sagraloff

Max-Planck-Institut für Informatik, Stuhlsatzenhausweg 85, 66123 Saarbrücken, Germany

Abstract. Most algorithms of computational geometry are designed for the Real-RAM and non-degenerate input. We call such algorithms idealistic. Executing an idealistic algorithm with floating point arithmetic may fail. Controlled perturbation replaces an input x by a random nearby $\tilde{x}$ in the δ-neighborhood of x and then runs the floating point version of the idealistic algorithm on $\tilde{x}$. The hope is that this will produce the correct result for $\tilde{x}$ with constant probability provided that δ is small and the precision L of the floating point system is large enough. We turn this hope into a theorem for a large class of geometric algorithms and describe a general methodology for deriving a relation between δ and L. We exemplify the usefulness of the methodology by examples.

1 Introduction

Most algorithms of computational geometry are designed under two simplifying assumptions: the availability of a Real-RAM and non-degeneracy of the input. A Real-RAM computes with real numbers in the sense of mathematics. The notion of degeneracy depends on the problem; examples are collinear or cocircular points or three lines with a common point. We call an algorithm designed under the two simplifying assumptions an *idealistic algorithm*. Implementations have to deal with the precision problem (caused by the Real-RAM assumption) and the degeneracy problem (caused by the non-degeneracy assumption). The *exact computation paradigm* [10,9,3,14,12,13] addresses the precision problem. It proposes to implement a Real-RAM tuned to geometric computations. The degeneracy problem is addressed by reformulating the algorithms so that they can handle all inputs. This may require non-trivial changes. The approach is followed in systems like LEDA and CGAL. Halperin et al. [5,7,6] proposed *controlled perturbation* to overcome both problems. The idea is to solve the problem at hand not on the input given but on a nearby input. The perturbed input is carefully chosen, hence the name *controlled perturbation*, so that it is non-degenerate and can be handled with approximate arithmetic. They applied the idea to three problems (computing polyhedral arrangements, spherical arrangements, and arrangements of circles) and showed that variants of the respective idealistic algorithms can be made to work. Funke et al. [4] extended their work and showed how to use controlled perturbation in the context of randomized algorithms, in particular randomized incremental constructions, and designed specific schemes for planar Delaunay triangulations and convex hulls and Delaunay triangulations in arbitrary dimensions. We extend their work further. We prove

---

[*] Partially supported by the IST Programme of the EU under Contract No IST-006413, Algorithms for Complex Shapes (ACS).

M. Bugliesi et al. (Eds.): ICALP 2006, Part I, LNCS 4051, pp. 299–310, 2006.
© Springer-Verlag Berlin Heidelberg 2006

that controlled perturbation and guarded tests are a general conversion strategy for a wide class of geometric algorithms; the papers cited above hint at this possibility but do not prove it. Moreover, we develop a general methodology for analyzing controlled perturbation, in particular, for deriving quantitative relations between the amount of perturbation and the precision of the approximate arithmetic.

2   Controlled Perturbation (Review from [4])

Geometric algorithms branch on geometric predicates. Typically, geometric predicates can be expressed as the sign of an arithmetic formula E. For example, the *orientation predicate* for $d + 1$ points in $\mathbb{R}^d$ is given by the sign of a $(d + 1) \times (d + 1)$ determinant: the determinant has one row for each point and the row for a point contains the coordinates of the point followed by the entry 1 and evaluates to zero iff the $d + 1$ points lie in a common hyperplane. This is considered a degeneracy.

When evaluating an arithmetic formula E using floating-point arithmetic, round-off error occurs which might result in the wrong sign being reported. If this stays undetected, the program may enter an illegal state and disasters may happen, see [11] for some instructive examples. In order to guard against round-off errors, we postulate the availability of a predicate $\mathscr{G}_E$ with the following *guard property: If $\mathscr{G}_E$ evaluates to true when evaluated with floating point arithmetic, the floating point evaluation (fp-evaluation) of E yields the correct sign.* In an idealistic algorithm A we now guard every sign test by first testing the corresponding guard. If it fails, we abort. We call the resulting algorithm a *guarded algorithm* and use A_g to denote it.

The controlled perturbation version of idealistic algorithm A is as follows: Let δ be a positive real. On input x, we first choose a δ-perturbation $\tilde{x}$ of x and then run the guarded algorithm A_g on $\tilde{x}$. If it succeeds, fine. If not, repeat. What is a δ-perturbation? A δ-perturbation of a point is a random point in the δ-cube (or δ-ball) centered at the point and for a set of points a δ-perturbation is simply a δ-perturbation of each point in the set. For more complex objects, alternative definitions come to mind, e.g., for a a circle one may want to perturb the center or the center and the radius. The goal is now to show experimentally and/or theoretically that A_g has a good chance of working on a δ-perturbation of any input and a small value of δ. More generally, one wants to derive a relation between the precision L of the floating point system (= length of the mantissa), a characteristic of the input set, e.g., the number of points in the set and an upper bound on the maximal coordinate of any point in the input, and δ. Halperin et al. have done so for arrangements of polyhedral surfaces, arrangements of spheres, and arrangements of circles and Funke et al. have done so for Delaunay diagrams and convex hulls in arbitrary dimensions.

We want to stress that a guarded algorithm can be used without any analysis. Suppose we want to use it with a certain δ. We execute it with a certain precision L. If it does not succeed, we double L and repeat. *Our main result states that this simple strategy terminates for a wide class of geometric algorithms. Moreover, it gives a quantitative relation between δ and L and characteristic quantities of the instance.*

Guard predicates must be safe and should be effective, i.e., if a guard does not fire, the approximate sign computation must be correct, and guards should not fire too often

Table 1. Rules for calculating error bounds. $\oplus$, $\ominus$, $\odot$, $\oslash$, and $\sqrt{}$ stand for floating point addition, subtraction, multiplication, division, and square-root, respectively.

E	$\widetilde{E}$	$\widetilde{E_{\text{sup}}}$	ind_E
$c = const$	c	$\lvert c \rvert$	0
$x + y$ or $x - y$	$\widetilde{x} \oplus \widetilde{y}$ or $\widetilde{x} \ominus \widetilde{y}$	$\widetilde{x_{\text{sup}}} \oplus \widetilde{y_{\text{sup}}}$	$1 + \max(\text{ind}_x, \text{ind}_y)$
$x \cdot y$	$\widetilde{x} \odot \widetilde{y}$	$\widetilde{x_{\text{sup}}} \odot \widetilde{y_{\text{sup}}}$	$1 + \text{ind}_x + \text{ind}_y$
$x^{1/2}$	$\sqrt{\widetilde{x}}$	$\begin{cases} (\widetilde{x_{\text{sup}}} \oslash \widetilde{x}) \odot \sqrt{\widetilde{x}} & \text{if } \widetilde{x} > 0 \\ \sqrt{\widetilde{x_{\text{sup}}}} \odot 2^{p/2} & \text{if } \widetilde{x} = 0 \end{cases}$	$1 + \text{ind}_x$

unnecessarily. It is usually difficult to analyze the floating point evaluation of $\mathcal{G}_E$ directly. For the purpose of the analysis, we therefore postulate the existence of a *bound predicate* $\mathcal{B}_E$ with the property: *If $\mathcal{B}_E$ holds, $\mathcal{G}_E$ evaluates to true when evaluated with floating point arithmetic.* We next give some concrete examples for guard and bound predicates.

When E is evaluated by a straight-line program, it is easy to come up with suitable predicates $\mathcal{G}_E$ and $\mathcal{B}_E$ using forward error analysis. For example, the rules in Table 1 ([1]) recursively define two quantities $\overline{E_{\text{sup}}}$ and ind_E for every arithmetic expression E such that $\lvert E - \widetilde{E} \rvert \leq B_E := \overline{E_{\text{sup}}} \cdot \text{ind}_E \cdot 2^{-L}$ where $\widetilde{E}$ denote the value of E computed with floating point arithmetic and L denotes the mantissa length of the floating-point system. (i.e. $L = 52$ for IEEE doubles). We can then use $\mathcal{G}_E \equiv \left(\lvert \widetilde{E} \rvert > B_E \right)$ and $\mathcal{B}_E \equiv (\lvert E \rvert > 2B_E)$, where $\mathcal{B}_E$ is valid since it guarantees that $\lvert \widetilde{E} \rvert = \lvert E \rvert - \lvert E - \widetilde{E} \rvert > 2B_E - B_E = B_E$ by the inverse triangle inequality. For the orientation test of three points in the plane, one obtains $B_{orient} = 24 \cdot M^2 2^{-L}$ and for the incircle test of four points in the plane, one obtains $B_{incircle} = 432 \cdot M^4 2^{-L}$. In both cases, it is assumed that all point coordinates are bounded by M in absolute value.

We assume for this paper that input values are bounded by M in absolute value and that bound predicates are of the form $c_E M^{e_E} 2^{-L}$ where c_E and e_E are constants depending on the predicate expression E. If E is a polynomial, e_E is the degree of the polynomial.

3   The Class of Algorithms

Our result applies to algorithms which can be viewed as decision trees. There is a decision tree T_n for each input size n. We assume that the input consists of a set of n points with coordinates bounded by M in absolute value. Boundedness is essential in some of our arguments and we leave it as a challenge to remove this restriction. The internal nodes of the decision tree are labelled by predicate evaluations $\text{sign} f(x_{i_1}, \ldots, x_{i_k})$ where f stems from a fixed finite set of real-valued functions (for example, orientation of three points or the incircle test of four points) and the x_{i_j} are input points. The tree is ternary and branches according to the sign of f. Observe that predicates can only be applied to input points and not to computed points. This restriction can be relieved somewhat. For example, if the input consists of a set of line segments, each specified by a pair of points, then a predicate applied to an intersection point of two segments is easily reduced to a more complex predicate involving only input points. What predicate functions are allowed? We require that the functions f fulfil the postulates set forth in Section 5.

Many algorithms of computational geometry are within the model, e.g., Delaunay diagram and Voronoi diagram computations, convex hulls, line arrangements, ... It is important to understand the limitations of the model. Gaussian Elimination for $n \times n$ matrices is outside the model since it tests the sign of expressions depending on all n^2 matrix entries. So the number of predicate functions is infinite and their arity is not bounded. Observe however, that Gaussian elimination on $d \times d$ matrices used in an algorithm to compute convex hulls of n points in $\mathbb{R}^d$ is within the model as d does not depend on the input size. Algorithms whose running time depends on actual point coordinates and not just on the number of points are also outside the model. It is the subject of further work to weaken this restriction.

4   The Basic Idea

We concentrate on a single predicate, say $P(x_1,\ldots,x_k) = \mathrm{sign} f(x_1,\ldots,x_k)$, of k points in the plane. The treatment readily generalizes to points in higher dimensions. Forward error analysis gives us an expression B_f which upper bounds the error in the evaluation of f. For this extended abstract, we assume that B_f is a constant as discussed above. We can make B_f arbitrarily small by increasing the precision L of the floating point system.

We want to prove a result of the following form: If each coordinate of any input point is modified by a random number in $[-\delta, +\delta]$ and the program is executed with sufficiently high floating point precision L, the guarded program succeeds with probability at least 1/2. It is clear that such a result is true if f is continuous and the zero set of f is lower-dimensional, because then the set of k-tuples for which $|f| < 2B_f$ is within a small neighborhood of the zero set. In order to obtain a quantitative relation between δ and L, we need to estimate the maximal volume of the set of k-tuples with $|f| < 2B_f$ within an arbitrary axis-oriented 2δ-cube.

We suggest a general approach for deriving such estimates exploiting the fact that functions f underlying geometric predicates have structure. As a first step, we split the arguments of f into $k-1$ points $\mathbf{x}$ and a single point x. We write $f(\mathbf{x},x)$ even if x is not the last argument of f. We consider the points in $\mathbf{x}$ fixed and the point x variable. Geometric predicates can usually be interpreted as follows: $\mathbf{x}$ defines a partition of the plane into regions and P tells the location of x with respect to this partition. P returns zero if x lies on a region boundary, $+1$ if x lies in the positive regions, and -1 is x lies in the negative regions. We use $C_{\mathbf{x}} = \{x : f(\mathbf{x},x) = 0\}$ to denote the zero set of f and call it the *curve of degeneracy*. If $\lambda x.f(\mathbf{x},x)$ is identically zero[1], we call $\mathbf{x}$ *degenerate*. We call it *regular*, otherwise.

Some examples: (1) in the orientation predicate of three points p, q, and r, the first two points ($\mathbf{x}$ comprises p and q) define an oriented line $\ell(p,q)$ and $orient(p,q,r)$ tells the location of r (x corresponds to r) with respect to this line. The curve of degeneracy

---

[1] We use the notation $\lambda x.f(\mathbf{x},x)$ to emphasize that we view f as a function of x and keep $\mathbf{x}$ fixed.

is the line $\ell(p,q)$ if $p \neq q$. The pair (p,q) is degenerate if $p = q$. (2) in the side of circle predicate of four points p, q, r, and s, the first three points define an oriented circle $C(p,q,r)$ and $soc(p,q,r,s)$ tells the location of s with respect to this circle. The curve of degeneracy is $C(p,q,r)$ and the triple (p,q,r) is degenerate if it contains equal points. (3) in the side-of-wedge predicate of four points p, q, r, and s, the first three points define a wedge with boundaries $\ell(p,q)$ and $\ell(p,r)$ and $sow(p,q,r,s)$ tells the location of s with respect to this wedge. The curve of degeneracy is $\ell(p,q) \cup \ell(p,r)$ and the triple (p,q,r) is degenerate if either $q = p$ or $r = p$.

The function $\lambda x.f(\mathbf{x},x)$ is zero on the curve of degeneracy. It will be small near it and larger further away, i.e., $|f(\mathbf{x},x)|$ measures, in some sense, locally the distance of x from the curve of degeneracy.

In our examples this is quite explicit: (1) $orient(p,q,r) = \mathrm{sign} f_o(p,q,r)$ where[2] $|f_o(p,q,r)| = dist(p,q) \cdot dist(r,l(p,q))$, (2) $soc(p,q,r,s) = \mathrm{sign} f_{soc}(p,q,r,s)$ where[3] $|f_{soc}(p,q,r,s)| \geq (1/2)dist(p,q)dist(p,r)dist(q,r)dist(C,s)$ and C denotes the circle or line defined by the first three points, and finally (3) $sow(p,q,r,s) = \mathrm{sign} f_{sow}(p,q,r,s)$ where $|f_{sow}(p,q,r,s)| = |f_o(p,q,s) \cdot f_o(p,r,s)| = dist(p,q) \cdot dist(p,r) \cdot dist(s,\ell(p,q)) \cdot dist(s,\ell(p,r)) \geq dist(p,q) \cdot dist(p,r) \cdot dist(s,\ell(p,q) \cup \ell(p,r))^2$.

Assume we have a function $g(\mathbf{x},d)$ such that $|f(\mathbf{x},x)| \geq g(\mathbf{x},dist(x,C_{\mathbf{x}})) \geq 0$ that is non-zero if $dist(x,C_{\mathbf{x}}) > 0$, i.e., we bound $f(\mathbf{x},x)$ from below by a function in $\mathbf{x}$ and the distance of x from the curve of degeneracy. The requirement $|f(\mathbf{x},x)| \geq 2B_f$ would then translate into the condition $g(\mathbf{x},dist(C_{\mathbf{x}},x)) \geq 2B_f$, i.e., if x lies outside a certain tubular neighborhood of the curve of degeneracy $C_{\mathbf{x}}$, $|f(\mathbf{x},x)|$ is guaranteed to be at least $2B_f$. The width of the tubular region is related to the growth of g and depends on $\mathbf{x}$. What can we say about the growth of g?

Again it is useful to consider our examples. For the orientation predicate, we have $g(p,q,d) = dist(p,q) \cdot d$ and so g grows linearly in d with slope $dist(p,q)$, for the side-of-circle predicate, we have $g(p,q,r,d) \geq dist(p,q)dist(p,r)dist(q,r) \cdot d$ and so g grows at least linearly in d with slope $dist(p,q)dist(p,r)dist(q,r)$, and for the side-of-wedge predicate, we have $g(p,q,r,d) \geq dist(p,q) \cdot dist(p,r) \cdot d^2$ and so g grows at least quadratically in d with factor $dist(p,q) \cdot dist(p,r)$. The slope (factor) is zero for degenerate $\mathbf{x}$ ($p = q$ for the orientation predicate, $|\{p,q,r\}| \leq 2$ for the side-of-circle predicate, and $p \in \{q,r\}$ for the in-wedge-predicate) and grows in the distance of $\mathbf{x}$ from degeneracy. We want to guarantee that the slope (factor) has a certain guaranteed size because this allows us to control the width of the forbidden region for x.

So we proceed as follows. We fix the width of the forbidden region for x at some value γ and then study the function $g(\mathbf{x},\gamma)$. We study the conditions on $\mathbf{x}$ guaranteeing $g(\mathbf{x},\gamma) \geq 2B_f$. Now $g(\mathbf{x},\gamma)$ has one less argument and so continuing in this way k times, we arrive at a trivial case. The non-trivial details are given in the next section.

Let us consider our examples: In all three examples the perturbation must guarantee that points have a certain minimum distance. This will guarantee that $dist(p,q)$, $dist(p,q) \cdot dist(p,r)$, and $dist(p,q)dist(p,r)dist(q,r)$ have certain minimum values.

---

[2] f_o is the value of a 3×3 determinant. The value of the determinant is twice the signed area of the triangle formed by the three points which in turn is the distance of the first two points times the distance of the third point from the line through the first two points.

[3] We are going to prove this in section 6.

5 The General Scheme

We concentrate on a single predicate f of k point variables.

Requirement 1. f *is continuous and* f *is not identically zero.*

In order to apply the scheme, one needs to determine a family of non-negative continuous functions f_s, one for every sequence $s = (s_0, s_1, \ldots, s_{\ell-1})$ with $1 \leq \ell \leq k$ and $s_j \in [1 \ldots k - j]$ for $0 \leq j \leq \ell - 1$; s describes the order in which we eliminated variables, we first eliminated the s_0-th variable of a k-argument function, then the s_1-th variable of a $k - 1$-argument function, and so on. The function f_s depends on $k - \ell$ point variables. For the empty sequence ε, we set $f_\varepsilon = |f|$. One also needs to fix positive constants γ_ℓ. Consider a fixed s with $\ell := |s| < k$; f_s is a function of $k - \ell$ point variables. Let $h \in [1 \ldots k - \ell]$ be arbitrary and let $t = s \circ h$. We use x to denote the h-th variable of f_s and $\mathbf{x}$ to denote the remaining $k - \ell - 1$ variables. We write $f_s(\mathbf{x}, x)$ instead of $f_s(\mathbf{x}', x, \mathbf{x}'')$ where $\mathbf{x}'$ comprises the first $h - 1$ arguments and $\mathbf{x}''$ comprises the last $k - \ell - h$ arguments. For each $\mathbf{x}$, let $C_{\mathbf{x}}^t = \{x : f_s(\mathbf{x}, x) = 0\}$. We call $C_{\mathbf{x}}^t$ the *curve of degeneracy*. If $C_{\mathbf{x}}^t = \emptyset$, we set $C_{\mathbf{x}}^t$ to an arbitrary singleton set for purely technical reasons. We call $\mathbf{x}$ *degenerate* if $\lambda x. f_s(\mathbf{x}, x)$ is identically zero and *regular* otherwise.

We next define the lower bound function. Let $U_d = \{x \in U : dist(x, C_{\mathbf{x}}^t) \geq d\}$ and let d_0 be maximal such that U_d is non-empty. Define $g_t(\mathbf{x}, d) := \min_{x \in U_d} f_s(\mathbf{x}, x)$ for $0 \leq d \leq d_0$ and $g_t(\mathbf{x}, d) := g_t(\mathbf{x}, d_0)$ for $d \geq d_0$.

Lemma 1. *The function* $g_t(\mathbf{x}, d)$ *is non-decreasing in its second argument,* $g_t(\mathbf{x}, d) > 0$ *for* $d > 0$ *if* $\mathbf{x}$ *is regular, and* $g_t(\mathbf{x}, d) = 0$ *for all* d *if* $\mathbf{x}$ *is degenerate.*

Proof. If $\mathbf{x}$ is degenerate, $\lambda x. f_s(\mathbf{x}, x)$ is identically zero and hence $g_t(\mathbf{x}, d) = 0$ for all d. If $\mathbf{x}$ is regular, $C_{\mathbf{x}}$ is a closed proper subset of U and hence $d_0 > 0$. Also U_d is a closed non-empty subset of U for $0 < d \leq d_0$. Since f_s is continuous, $\inf_{x \in U_d} f_s(\mathbf{x}, x)$ is attained for a point $x \in U_d$ and $f_s(\mathbf{x}, x) > 0$. Thus $g_t(\mathbf{x}, d) > 0$. $\quad\blacksquare$

Requirement 2. $f_t(\mathbf{x})$ *is a continuous function with* $0 \leq f_t(\mathbf{x}) \leq g_t(\mathbf{x}, \gamma_\ell)$ *and* $f_t(\mathbf{x}) = 0$ *iff* $\mathbf{x}$ *is degenerate.*

In our applications, $g_t(\mathbf{x}, \gamma_\ell)$ is continuous and we may choose $f_t(\mathbf{x}) = g_t(\mathbf{x}, \gamma_\ell)$. Allowing an inequality, gives additional flexibility. However, there are situations where $g_t(\mathbf{x}, \gamma_\ell)$ is not continuous.

Lemma 2. *If* $|s| = k$, f_s *is a positive constant.*

Proof. Let $(x_1, \ldots, x_k) \in U^k$ be such that $f(x_1, \ldots, x_k) \neq 0$. We may assume without loss of generality that $s = (k, k - 1, \ldots, 1)$, i.e., we remove the arguments from the end. We prove $f_{(k,k-1,\ldots,k-i+1)}(x_1, \ldots, x_{k-i}) \neq 0$ by induction on i. For $i = 0$, there is nothing to prove. So assume $i \geq 1$. We have $f_{(k,k-1,\ldots,k-i+2)}(x_1, \ldots, x_{k-i}, x_{k-i+1}) \neq 0$ by induction hypothesis. So, $\mathbf{x} = (x_1, \ldots, x_{k-i})$ is regular and hence $f_{(k,k-1,\ldots,k-i+1)}(x_1, \ldots, x_{k-i}) > 0$.

The Perturbation: Our input are points $q_1, q_2, \ldots$. We move each q_i to a random point p_i in the δ-cube centered at q_i. Assume that we have already chosen p_1 to p_{n-1} and that the following *perturbation property* (PP) holds true: for every sequence s, $\ell := |s|$, and every tuple of distinct indices j_1 to $j_{k-\ell}$ in $[1..n-1]$: $f_s(p_{j_1}, \ldots, p_{j_{k-\ell}}) \geq 2B_f$. For $n = 1$, the conditions are vacuously true if $\ell < k$. For $\ell = k$, f_s is a positive constant and hence by making the mantissa length L large enough, we can satisfy all conditions.

Requirement 3. *Precision L is large enough so that $2B_f \leq f_s$ for all s with $|s| = k$.*

We now choose p_n.

Lemma 3. *If for all ℓ, $0 \leq \ell < k$, any t with $|t| = \ell + 1$, and any tuple of distinct indices j_1 to $j_{k-\ell-1}$ in $[1..n-1]$ and $\mathbf{p} = (p_{j_1}, \ldots, p_{j_{k-\ell-1}})$, p_n does not lie in the γ_ℓ-neighborhood of $C_{\mathbf{p}}^t$, (PP) holds for n. Moreover, if these neighborhoods together cover at most a fraction $1/(2n)$ of the δ-cube centered at q_n, the precondition fails with probability at most $(1/2n)$.*

Proof. Consider the application of any f_s to a $k - \ell$ tuple of perturbed points. If p_n is not among them, (PP) holds by induction hypothesis. If p_n is among them, assume it is the h-th argument where $1 \leq h \leq k - \ell$. Let $t = s \circ h$ and let $\mathbf{p}$ be the remaining arguments. By induction hypothesis we have $f_t(\mathbf{p}) \geq 2B_f$. Also $dist(p_n, C_{\mathbf{p}}^t) \geq \gamma_\ell$ and hence $f_s(\mathbf{p}, p_n) \geq g_t(\mathbf{p}, dist(p_n, C_{\mathbf{p}}^t)) \geq g_t(\mathbf{p}, \gamma_\ell) = f_t(\mathbf{p}) \geq 2B_f$.

We also need that the local geometry of the curves of degeneracy is simple.

Requirement 4. *There are constants C and δ_0 such that for $0 \leq \delta \leq \delta_0$ and all $\mathbf{p}$ satisfying (PP), the γ_ℓ-neighborhood of $C_{\mathbf{p}}^t$ covers at most an area $C \cdot \gamma_\ell \cdot \delta$ of any δ-cube.*

The requirement excludes space filling curves and sets $C_{\mathbf{p}}^t$ containing an open set. In some cases, the space estimate can be improved. In particular, if $C_{\mathbf{p}}^t$ consists only of a constant number of points, the estimate can be improved to $C\gamma_\ell^2$. Our final requirement relates δ and the γ_l.

Requirement 5. $2Ck! \cdot \sum_{0 \leq \ell < k} n^{k-\ell} \gamma_\ell \leq \delta$.

Theorem 1. *If requirements (1) to (5) hold and the input consists of n points, then with probability $1/2$ the fp-evaluation of f yields the correct sign for any k-tuple of distinct perturbed points.*

Proof. Consider any ℓ with $0 \leq \ell < k$. There are no more than $k!$ sequences t with $|t| = \ell + 1$. Also there are at most $n^{k-\ell-1}$ tuples of $k - \ell - 1$ distinct indices in $[1, n]$. Thus the total area covered by the γ_ℓ-neighborhoods of all $C_{\mathbf{p}}^t$ is at most $k! n^{k-\ell-1} \cdot C \cdot \gamma_\ell \cdot \delta$. The sum over all ℓ of this quantity is at most $\delta^2/(2n)$ by requirement (5). Thus the probability that the choice of p_i for any fixed i with $1 \leq i \leq n$ does not support the induction step is at most $1/(2n)$ and hence the probability that some induction step fails is at most $1/2$. Thus with probability at least $1/2$, we have (PP) for n. Since $f_\varepsilon = |f|$, this implies that the fp-evaluation of f yields the correct sign for any k-tuple of distinct perturbed points.

We give an example. $f(p,q,r) = orient(p,q,r) = dist(p,q) \cdot dist(r, \ell(p,q))$. We compute $f_{(3,2,1)}$. Let $t = (3)$. We have $C_{\mathbf{x}}^t = \ell(p,q)$. Then $g_{(3)}(p,q,d) = dist(p,q) \cdot d$ and hence $f_{(3)}(p,q) = dist(p,q) \cdot \gamma_0$. Let $t = (3,2)$. We have $C_{\mathbf{x}}^t = \{p\}$. Then $g_{(3,2)}(p,d) = d \cdot \gamma_0$ and hence $f_{(3,2)}(p) = \gamma_1 \cdot \gamma_0$. Let $t = (3,2,1)$. We have $C_{\mathbf{x}}^t = \emptyset$ and set it to $C_{\mathbf{x}}^t = \{(0,0)\}$. Then $g_{(3,2,1)}(d) = \gamma_1 \cdot \gamma_0$ and hence $f_{(3,2,1)}(p) = \gamma_1 \cdot \gamma_0$. We may use $C = 4$, need $48M^2 2^{-L} \le \gamma_1 \cdot \gamma_0$ to satisfy requirement 3 and $48 \cdot (n^3 \gamma_0 + n^2 \gamma_1 + n \gamma_2) \le \delta$ to satisfy requirement 5. With $\gamma_2 = 0$, $\gamma_0 = \delta/(96n^3)$, $\gamma_1 = \delta/(96n^2)$, the requirement for L becomes $L \ge 2\log(M/\delta) + 5\log n + O(1)$. This can be improved somewhat by using the fact that requirement 5 can be replaced by $24(n^2 \gamma_0 \delta + n \gamma_1^2) \le \delta^2/(2n)$ since for $t = (3,2)$, the curve of degeneracy consists of a single point. With $\gamma_0 = \delta/(96n^3)$ and $\gamma_1 = \delta/(\sqrt{96}n)$, the requirement for L becomes $L \ge 2\log(M/\delta) + 4\log n + O(1)$.

We summarize: In order to apply the scheme, one first fixes δ to a value suitable for the application. Then one fixes the γ_ℓ to values obeying requirement 5 and determines suitable functions f_t. This might require some ingenuity and is the subject of the discussion below. Finally, one determines C and makes L large enough to guarantee requirement 3. We next specialize and make the general scheme more concrete.

The function f is frequently *symmetric* in its arguments up to change of sign, i.e., permuting the arguments does not change the absolute value. In the case of symmetric functions, the functions f_t only depend on the length of t and not on the actual structure of t. Writing f_ℓ for f_t with $|t| = \ell$ we obtain a sequence of functions f_0, f_1 to f_k where f_ℓ has $k - \ell$ arguments. For simplicity, we restrict most of the further discussion to symmetric functions.

In our examples, the functions g_ℓ are *separable*, i.e., we have $g_{\ell+1}(\mathbf{x}, d) = h_{\ell+1}(\mathbf{x}) \cdot d^{e_\ell} \cdot \prod_{0 \le i < \ell} \gamma_i^{e_i}$ for some function $h_{\ell+1}$ and some integers e_i. More frequently, we can locally bound $g_\ell(\mathbf{x}, d)$ from below by a separable function, i.e., we have a positive constant d_ℓ and a continuous function $h_{\ell+1}(\mathbf{x})$ with $h_{\ell+1}(\mathbf{x}) = 0$ iff $\mathbf{x}$ is degenerate such that $g_{\ell+1}(\mathbf{x}, d) \ge h_{\ell+1}(\mathbf{x}) \cdot d^{e_\ell} \cdot \prod_{0 \le i < \ell} \gamma_i^{e_i}$ for $d \le d_\ell$ and $g_{\ell+1}(\mathbf{x}, d) \ge h_{\ell+1}(\mathbf{x}) \cdot d_\ell^{e_\ell} \cdot \prod_{0 \le i < \ell} \gamma_i^{e_i}$ for $d \ge d_\ell$. With $\gamma_\ell \le d_\ell$ for all ℓ, we obtain $f_{\ell+1}(\mathbf{x}) = h_{\ell+1}(\mathbf{x}) \cdot \prod_{0 \le i \le \ell} \gamma_i^{e_i}$ and hence $f_k = c \cdot \gamma_0^{e_0} \cdots \gamma_{k-1}^{e_{k-1}}$ for some constant $c = h_k$. Thus requirement 3 becomes $2B_f \le c \cdot \gamma_0^{e_0} \cdots \gamma_{k-1}^{e_{k-1}}$.

What is a good choice for the γ_ℓ? The condition $\gamma_\ell \le d_\ell$ makes it difficult to give a general answer. We therefore assume $d_\ell = \infty$ for all ℓ. We want to minimize $\sum_\ell n^{k-\ell} \gamma_\ell$ subject to the constraint $2B_f \le c \cdot \gamma_0^{e_0} \cdots \gamma_{k-1}^{e_{k-1}}$. There is an extremal point where the inequality is an equality. The Kuhn-Tucker conditions tell us that at an extremal point the partial derivatives of the objective function and the constraint with respect to the γ_ℓ must line up, i.e., there is a λ such that $n^{k-\ell} = \lambda 2B_f e_\ell / \gamma_\ell$ for all ℓ. Thus $\gamma_\ell = \lambda 2B_f e_\ell / n^{k-\ell}$ and hence $\lambda = ((2B_f/c)n^S / \prod_\ell e_\ell^{e_\ell})^{1/E} / (2B_f)$ where $E = e_0 + \ldots + e_{k-1}$ and $S = \sum_\ell (k-\ell) e_\ell$. Then $\delta \ge 2Ck! \sum_\ell n^{k-\ell} \gamma_\ell = 2Ck!E ((2B_f/c)n^S / \prod_\ell e_\ell^{e_\ell})^{1/E}$. This becomes $\delta \ge 2Ck!E (2(c_f/c)M^d 2^{-L} n^S / \prod_\ell e_\ell^{e_\ell})^{1/E}$ for f a polynomial of degree d in k point variables and hence $B_f = c_f M^d 2^{-L}$. Thus we need

$$L \ge E\log(1/\delta) + S\log n + d\log M + O(1).$$

The major terms in this lower bound can be explained intuitively. We have to compute with numbers as large as M^d and this requires $d\log M$ bits before the binary point. We want to perturb by as little as δ and hence we need at least $\log(1/\delta)$ after the binary point. This is multiplied by the sum E of exponents. The number of potential predicate evaluations grows like n^k and hence there should be a term $O(k\log n)$ to account for them. We have no intuitive explanation for the term $S\log n$.

A key step in applying our methodology is to find the appropriate functions f_t. We give some guidelines on how to find them.

Consider f and a regular $\mathbf{x}$. For any $x \in U$, let x_0 be the point on $C_{\mathbf{x}}$ closest to x. Define $g_{\mathbf{x}}(d) = f(\mathbf{x}, x_0 + d(x - x_0)/\|x - x_0\|)$ where $d \in \mathbb{R}_{\geq 0}$ and consider the Taylor or Puiseux expansion of $g_{\mathbf{x}}$ at 0. If the Taylor expansion exist, there are $D > 0$, $e \geq 1$, and $c > 0$ depending on $\mathbf{x}$ and x_0 such that $|g_{\mathbf{x}}(d)| \geq c \cdot d^e$ provided that $d \leq D$. We may choose e as the index of the first non-zero coefficient in the Taylor expansion and c as one half of this Taylor coefficient. If e and D can be chosen indecently of $\mathbf{x}$ and x and $c = c(\mathbf{x})$ depends only on $\mathbf{x}$ but not on x, we have a locally valid bound of the desired form: $|f(\mathbf{x}, x)| \geq c(\mathbf{x}) \cdot dist(x, C_{\mathbf{x}})^e$ for $dist(x, C_{\mathbf{x}}) \leq D$.

A frequently occurring case is that $C_{\mathbf{x}}$ has no singularities whenever $\mathbf{x}$ is regular. Let ∇f be the vector of partial derivatives of f with respect to the coordinates of x and let $|\nabla f|(\mathbf{x}, x)$ be the length of the gradient vector at $(\mathbf{x}, x)$. If $C_{\mathbf{x}}$ has no singularities, $|\nabla f|(\mathbf{x}, x_0) > 0$ for all $x_0 \in C_{\mathbf{x}}$ and hence the minimum length of the gradient over all points on $C_{\mathbf{x}}$ is positive. Let $h(\mathbf{x}) = \min_{x_0 \in C_{\mathbf{x}}} |\nabla f|(\mathbf{x}, x_0)$. Then $h(\mathbf{x}) > 0$ if $\mathbf{x}$ is regular and $h(\mathbf{x}) = 0$ if $\mathbf{x}$ is degenerate. Also, $g_{\mathbf{x}}(d) \approx |\nabla f|(\mathbf{x}, x_0) \cdot d \geq (1/2)h(\mathbf{x}) \cdot d$ and we have a separable representation which is linear in the distance and is valid for small d.

Another frequently occurring case is that f is a polynomial in the point coordinates. If $\mathbf{x}$ is regular, the curve $C_{\mathbf{x}}$ has a finite number of singularities. Let S be the set of singularities. For all points x such that x_0 is at least ε away from any singularity, we can proceed as above and obtain a linear estimate in d. Near singularities, we proceed as follows. Let s be a singularity and assume w. l. o. g that s is the origin. Let m be all terms of minimal degree in f. Then $f(\mathbf{x}, x) \approx m(\mathbf{x}, x)$ for x near s. The terms in m have common degree e in the point coordinates of x; the coefficients are polynomials in the point coordinates of the points in $\mathbf{x}$. Over the reals, m factors into a product of linear factors and irreducible quadratic factors. Each linear factor ℓ_i defines a line through the origin whose coefficients are functions in $\mathbf{x}$ and hence $|\ell_i(\mathbf{x}, x)| = c_i(\mathbf{x})dist(x, \ell_i)$. An irreducible quadratic factor q_j contains only a single real point, namely the origin, and the function value of q_j grows quadratically in the distance of $\mathbf{x}$ from s, i.e., $|q_j(\mathbf{x}, x)| \geq c_j(\mathbf{x})dist(x, s)^2$ for some $c_j(\mathbf{x})$. Thus

$$m(\mathbf{x}, x) \geq \prod_i c_i(\mathbf{x}) \cdot dist(x, \ell_i) \cdot \prod_j c_j(\mathbf{x}) \cdot dist(\mathbf{x}, s)^2$$

$$\geq \prod_i c_i(\mathbf{x}) \cdot \prod_j c_j(\mathbf{x}) \cdot dist(x, \cdots \cup \ell_i \cup \cdots)^e \approx \prod_i c_i(\mathbf{x}) \cdot \prod_j c_j(\mathbf{x}) \cdot dist(x, C_{\mathbf{x}})^e ,$$

and there is hope for a separable bound which grows like d^e.

6 Applications

We apply the methodology to the side-of-circle test of four points in the plane. It tells the side of a query point with respect to an oriented circle defined by three points. We have three points $p_i = (x_i, y_i)$, $1 \leq i \leq 3$, and a query point $p = (x, y)$. Let us assume first, that the three points are not collinear. Let R be the radius of the circle C defined by the first three points. We may assume w. l. o. g. that the circle is centered at the origin. Then $x_i^2 + y_i^2 = R^2$ for all i. Let Δ be the signed area of the triangle (p_1, p_2, p_3). The side-of-circle test is given by the sign of the determinant

$$f_0(p_1, p_2, p_3, p) = \begin{vmatrix} 1 & x_1 & y_1 & x_1^2 + y_1^2 \\ 1 & x_2 & y_2 & x_2^2 + y_2^2 \\ 1 & x_3 & y_3 & x_3^2 + y_3^2 \\ 1 & x & y & x^2 + y^2 \end{vmatrix} = - \begin{vmatrix} x_1 & y_1 & R^2 \\ x_2 & y_2 & R^2 \\ x_3 & y_3 & R^2 \end{vmatrix} + (x^2 + y^2) \cdot \begin{vmatrix} 1 & x_1 & y_1 \\ 1 & x_2 & y_2 \\ 1 & x_3 & y_3 \end{vmatrix}$$

$$= -R^2 2\Delta + (x^2 + y^2) 2\Delta = 2\Delta (x^2 + y^2 - R^2) .$$

This predicate was already analyzed in [4] using non-trivial geometric reasoning. The purpose of this section is to show that the same result, in fact, a slightly better result, can be obtained by generic reasoning. The curve of degeneracy is the cycle C and the normal vector at $p_0 = (x_0, y_0) \in C$ is $(4\Delta x_0, 4\Delta y_0)$ and has norm $4\sqrt{x_0^2 + y_0^2}|\Delta| = 4R|\Delta|$. This is independent of p_0. So the first order approximation of f_0's absolute value is $4R|\Delta| \cdot dist(p, C)$. In fact, one half of this is even a global lower bound, namely,

$$|f_0(p_1, p_2, p_3, p)| = 2|\Delta| \cdot |\sqrt{x^2 + y^2} - R| \cdot (\sqrt{x^2 + y^2} + R) \geq 2R|\Delta| \cdot dist(p, C) .$$

Let $a = dist(p_1, p_2)$, $b = dist(p_1, p_3)$, $c = dist(p_2, p_3)$, and let α be the angle at p_3 in the triangle (p_1, p_2, p_3). Then $2R = a/\sin\alpha$ and $|\Delta| = (1/2)bc\sin\alpha$ and hence $2R|\Delta| = 1/2 \cdot abc$. Thus

$$|f_0(p_1, p_2, p_3, p)| \geq \frac{1}{2} dist(p_1, p_2) dist(p_1, p_3) dist(p_2, p_3) dist(C, p)$$

and by continuity of the determinant the latter inequality is also true if the points p_1, p_2, and p_3 are collinear. In this case, C is the line passing through the first three points. The formula also tells us that the triple (p_1, p_2, p_3) is regular iff the points are pairwise distinct.

We next consider $f_1(p_1, p_2, p)/\gamma_0 = (1/2) \cdot dist(p_1, p_2) dist(p_1, p) dist(p_2, p)$. We consider p_1 and p_2 as fixed and treat $p = p_3 = (x, y)$ as a variable. If $p_1 = p_2$, the function is identically zero. If $p_1 \neq p_2$ we have $f_1(p_1, p_2, p) = 0$ iff $p = p_1$ or $p = p_2$. Thus the curve of degeneracy consists of the two isolated points p_1 and p_2 and its tubular neighborhood is two circles. We want to bound f_1 from below. We may assume that p is closer to p_1 than to p_2. Then $dist(p, p_2) \geq dist(p_1, p_2)/2$ and hence $f_1(p_1, p_2, p)/\gamma_0 \geq dist(p_1, p_2)^2/4 \cdot dist(p, \{p_1, p_2\})$. Therefore $f_2(p_1, p)/(\gamma_0 \gamma_1) = dist(p_1, p)^2/4$ and further $f_3(p)/(\gamma_0 \gamma_1 \gamma_2) = 1/4$. In fact, there is no real reason to go down to f_3. We can also argue about f_0 directly. If any two points have a certain minimum distance m, $f_0(p_1, p_2, p_3, p) \geq m^3/2 \cdot dist(p, C)$. In [4], $\Delta^{3/2}$ was considered instead of ΔR and then the curve of degeneracy is the line spanned by p_1 and p_2. The tubular neighborhood is then a strip and the forbidden region is larger.

The computation of Voronoi diagrams of line segments is computationally difficult. The available exact algorithms [2] are slow, the fast algorithm of M. Held [8] is not guaranteed to work for all inputs. The key test in the algorithms is the side-of-circle test: A circle C is specified by three sites (points or lines) and the position of a fourth site (point or line) with respect to C is to be determined, see Figure 6. We discuss the situation where the fourth site is a line segment given by points p and q. Let C have center c and radius R and let p be outside C. The query point is $q = (x,y)$. We want to know whether the line $\ell(p,q)$ intersects, touches, or misses C. There are different ways of realizing this test. For simplicity let us put p at $(0,0)$ and c at $(c_0,0)$.

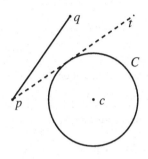

In [2] the test is realized by comparing R and $dist(\ell(p,q),c)$, the distance of c from the line ℓ. The line has equation $y\tilde{x} - x\tilde{y} = 0$ (here x and y are the coefficients and $\tilde{x}$ and $\tilde{y}$ are the variables. The signed distance of c from this line is $yc_0/\sqrt{x^2+y^2}$ and hence the test is realized by the formula $yc_0 - \pm R\sqrt{x^2+y^2}$. For each choice of sign, the curve of degeneracy is one of the tangents t from p at C; the equations for the tangents are $y = \pm Rx/\sqrt{c_0^2 - R^2}$. We leave it to the reader to verify that the norm of the normal vector has the same value for all points on the curve of degeneracy.

Alternatively, we may locate q with respect to the tangents from p at C. We further discuss this method. Let us concentrate on one of the tangents. We refer to it as t.

Then the location of $q = (x,y)$ is given by the sign of $E = \sqrt{c_0^2 - R^2} \cdot y - R \cdot x$. Observe that c_0 is the distance between c and p. Hence the general form for arbitrary p and c is given by $E = \sqrt{dist(p,c)^2 - R^2} \cdot (y - y_p) + R \cdot (x - x_p)$.

The circle C is defined by three sites. We treat the case of three points sites, the other cases are somewhat more involved. Let our three points be $p_i = (x_i, y_i)$, $1 \leq i \leq 3$. The center c has coordinates (it is the intersection of two bisectors)

$$x_c = \frac{\begin{vmatrix} (x_2^2 - x_1^2 + y_2^2 - y_1^2)/2 & y_2 - y_1 \\ (x_3^2 - x_1^2 + y_3^2 - y_1^2)/2 & y_3 - y_1 \end{vmatrix}}{2\Delta} \qquad y_c = \frac{\begin{vmatrix} x_2 - x_1 & (x_2^2 - x_1^2 + y_2^2 - y_1^2)/2 \\ x_3 - x_1 & (x_3^2 - x_1^2 + y_3^2 - y_1^2)/2 \end{vmatrix}}{2\Delta}$$

where Δ is the area of the triangle (p_1, p_2, p_3). Write $x_c = A/(2\Delta)$ and $y_c = B/(2\Delta)$. The radius of the circle is given by $R = \sqrt{(x_1 - x_c)^2 + (y_1 - y_c)^2} = \sqrt{D}/(2|\Delta|)$, where $D = (2x_1\Delta - A)^2 + (2y_1\Delta - B)^2$. Next observe $dist(p,c)^2 = (x_p - x_c)^2 + (y_p - x_c)^2 = ((2x_p\Delta - A)^2 + (2y_p\Delta - B)^2)/(4\Delta^2)$. Plugging into our expression E and multiplying by 2Δ yields the simplified expression

$$E = \sqrt{(2x_p\Delta - A)^2 + (2y_p\Delta - B)^2 - (2x_1\Delta - A)^2 - (2y_1\Delta - B)^2} \cdot (y - y_p)$$
$$+ \sqrt{(2x_1\Delta - A)^2 + (2y_1\Delta - B)^2} \cdot (x - x_p) .$$

Table 1 yields $B_E = c_E M^4 2^{-L}$ for some constant c_E. Next observe that $|E| = |H(y - y_p) + G(x - x_p)| = \sqrt{H^2 + G^2} \cdot dist(t,q)$ because E is a linear function in x and y and hence the first order approximation is exact. Also the norm of the normal vector is

$\sqrt{H^2 + G^2}$. Finally, observe $H^2 + G^2 = (2x_p\Delta - A)^2 + (2y_p\Delta - B)^2 = 4\Delta^2 dist(p,c)^2$ and hence $\sqrt{H^2 + G^2} = 2|\Delta|dist(p,c)$. So the requirement $|E| \geq 2B_E$ boils down to

$$|dist(q,t)| \geq \frac{2B_E}{2|\Delta|dist(p,c)} \geq \frac{B_E}{|\Delta|R} = \frac{c^E M^4 2^{-L}}{|\Delta|R}$$

and the quantity ΔR is familiar to us from the side-of-circle test for points. In order to guarantee a lower bound for it, it suffices to guarantee a minimum distance for the defining points of C.

References

1. C. Burnikel, S. Funke, and M. Seel. Exact arithmetic using cascaded computation. In *SocCG*, pages 175–183, 1998.
2. C. Burnikel, K. Mehlhorn, and S. Schirra. How to compute the Voronoi diagram of line segments: Theoretical and experimental results. In *ESA*, volume 855 of *LNCS*, pages 227–239, 1994.
3. S. Fortune and C. van Wyk. Efficient exact integer arithmetic for computational geometry. In *7th ACM Conference on Computational Geometry*, pages 163–172, 1993.
4. S. Funke, Ch. Klein, K. Mehlhorn, and S. Schmitt. Controlled perturbation for Delaunay triangulations. SODA, pages 1047–1056, 2005.
5. Halperin and Shelton. A perturbation scheme for spherical arrangements with application to molecular modeling. *CGTA*, 10, 1998.
6. D. Halperin and E. Leiserowitz. Controlled perturbation for arrangements of circles. *IJCGA*, 14(4):277–310, 2004.
7. D. Halperin and S. Raab. Controlled perturbation for arrangements of polyhedral surfaces with application to swept volumes. available from Halperin's home page; a preliminary version appeared in SoCG 1999, pages 163–172.
8. M. Held. VRONI: An engineering approach to the reliable and efficient computation of Voronoi diagrams of points and line segments. *Comput. Geom.*, 18(2):95–123, 2001.
9. M. Jünger, G. Reinelt, and D. Zepf. Computing correct Delaunay triangulations. *Computing*, 47:43–49, 1991.
10. M. Karasick, D. Lieber, and L.R. Nackman. Efficient Delaunay triangulation using rational arithmetic. *ACM Transactions on Graphics*, 10(1):71–91, January 1991.
11. L. Kettner, K. Mehlhorn, S. Pion, S. Schirra, and C. Yap. Classroom examples of robustness problems in geometric computations. In *ESA*, volume 3221 of *LNCS*, pages 702–713, 2004.
12. K. Mehlhorn and S. Näher. The implementation of geometric algorithms. In *Proceedings of the 13th IFIP World Computer Congress*, volume 1, pages 223–231. Elsevier, 1994.
13. K. Mehlhorn and S. Näher. *The LEDA Platform for Combinatorial and Geometric Computing*. Cambridge University Press, 1999. 1018 pages.
14. C.K. Yap. Towards exact geometric computation. In *Proceedings of the 5th Canadian Conference on Computational Geometry (CCCG'93)*, pages 405–419, 1993.

Tight Bounds for Selfish and Greedy Load Balancing[*]

Ioannis Caragiannis[1], Michele Flammini[2], Christos Kaklamanis[1],
Panagiotis Kanellopoulos[1], and Luca Moscardelli[2]

[1] Research Academic Computer Technology Institute and
Dept. of Computer Engineering and Informatics
University of Patras, 26500 Rio, Greece
[2] Dipartimento di Informatica, Università di L' Aquila
Via Vetoio, Coppito 67100, L' Aquila, Italy

Abstract. We study the load balancing problem in the context of a
set of clients each wishing to run a job on a server selected among a
subset of permissible servers for the particular client. We consider two
different scenarios. In *selfish load balancing*, each client is selfish in the
sense that it selects to run its job to the server among its permissible
servers having the smallest latency given the assignments of the jobs of
other clients to servers. In *online load balancing*, clients appear online
and, when a client appears, it has to make an irrevocable decision and
assign its job to one of its permissible servers. Here, we assume that the
clients aim to optimize some global criterion but in an online fashion.
A natural local optimization criterion that can be used by each client
when making its decision is to assign its job to that server that gives the
minimum increase of the global objective. This gives rise to *greedy* online
solutions. The aim of this paper is to determine how much the quality
of load balancing is affected by selfishness and greediness.

We characterize almost completely the impact of selfishness and greediness in load balancing by presenting new and improved, tight or almost
tight bounds on the price of anarchy and price of stability of selfish load
balancing as well as on the competitiveness of the greedy algorithm for
online load balancing when the objective is to minimize the total latency
of all clients on servers with linear latency functions.

1 Introduction

We study the load balancing problem in the context of a set of clients each
wishing to run a job on a server selected among a subset of permissible servers
for the particular client. We consider two different scenarios. In the first, called
selfish load balancing (or *load balancing games*), each client is selfish in the sense
that it selects to run its job to the server among its permissible servers having
the smallest latency given the assignments of the jobs of other clients to servers.

---

[*] This work was partially supported by the European Union under IST FET Integrated
Project 015964 AEOLUS and COST Action 293 GRAAL.

M. Bugliesi et al. (Eds.): ICALP 2006, Part I, LNCS 4051, pp. 311–322, 2006.

In the second scenario, called *online load balancing*, clients appear online and, when a client appears, it has to make an irrevocable decision and assign its job to one of its permissible servers. Here, we assume that the clients are not selfish and aim to optimize some global objective but in an online fashion (i.e., without any knowledge of clients that may arrive in the future). A natural local optimization criterion that can be used by each client when making its decision is to assign its job to that server that gives the minimum increase of the global objective. This gives rise to greedy online solutions. The aim of this paper is to answer the question of how much the quality of load balancing is affected by selfishness and greediness.

Load balancing games are special cases of the well-known *congestion games* introduced by Rosenthal [22] and studied in a sequence of papers [4,7,8,11,13,19,23, 24]. In congestion games there is a set E of resources, each having a non-negative and non-decreasing latency function f_e defined over non-negative numbers, and a set of n players. Each player i has a set of strategies $S_i \subseteq 2^E$ (each strategy of player i is a set of resources). An assignment $A = (A_1, ..., A_n)$ is a vector of strategies, one strategy for each player. The cost of a player for an assignment A is defined as $cost(i) = \sum_{e \in A_i} f_e(n_e(A))$, where $n_e(A)$ is the number of players using resource e in A, while the cost of an assignment is the total cost of all players. An assignment is a *pure Nash equilibrium* if no player has any incentive to unilaterally deviate to another strategy, i.e., $cost_i(A) \leq cost_i(A_{-i}, s)$ for any player i and for any $s \in S_i$, where (A_{-i}, s) is the assignment produced if just player i deviates from A_i to s. This inequality is also known as the *Nash condition*. We use the term *social cost* to refer to the cost of a pure Nash equilibrium. In *weighted congestion games*, each player has a weight w_i and the latency of a resource e depends on the total weight of the players that use e. For this case, a natural social cost function is the weighted sum of the costs of all players (or the weighted average of their costs). In *linear congestion games*, the latency function of resource e is of the form $f_e(x) = \alpha_e x + b_e$ with non-negative constants α_e and b_e. Load balancing games are linear congestion games where the strategies of players are singleton sets. In load balancing terminology, we use the terms server and client instead of the terms resource and player. The set of strategies of a client contains the servers that are permissible for the client.

We evaluate the quality of solutions of a load balancing game by comparing the social cost of Nash equilibria to the cost of the optimal assignment (i.e., the minimum cost). We use the notions of *price of anarchy* introduced in a seminal work of Koutsoupias and Papadimitriou [16] (see also [20]) and *price of stability* (or *optimistic price of anarchy*) defined as follows. The price of anarchy/stability of a load balancing game is defined as the ratio of the maximum/minimum social cost over all Nash equilibria over the optimal cost. The price of anarchy/stability for a class of load balancing games is simply the highest price of anarchy/stability among all games belonging to that class.

[10,12,13,14,15,18] study various games which can be thought of as special cases of congestion games with respect to the complexity of computing equilibria of best/worst social cost and the price of anarchy when the social cost is defined as the maximum latency experienced by any player. The social cost of the total latency has been studied in [4,7,17,26]. The authors in [17] study symmetric load balancing games where all servers are permissible for any client and show tight bounds on the price of anarchy of 4/3 for arbitrary servers and 9/8 for identical servers with weighted clients. In two recent papers, Awerbuch et al. [4] and Christodoulou and Koutsoupias [7] prove tight bounds on the price of anarchy of congestion games with linear latency functions. Among other results, they show that the price of anarchy of pure Nash equilibria is 5/2 while for mixed Nash equilibria or pure Nash equilibria of weighted clients it is $\frac{3+\sqrt{5}}{2} \approx 2.618$.

Does the fact that load balancing games are significantly simpler than congestion games in general have any implications for their price of anarchy? We give a negative answer to this question by showing that the 5/2 upper bound (as well as the $\frac{3+\sqrt{5}}{2}$ upper bound for weighted clients) is tight. This is interesting since the upper bounds for congestion games (as well as an earlier upper bound of 5/2 proved specifically for load balancing [26]) are obtained using only the *Nash inequality* (i.e., the inequality obtained by summing up the Nash condition inequalities over all players' strategies) and the definition of the social cost. So, it is somewhat surprising that load balancing games are as general as congestion games in terms of their price of anarchy and that the Nash inequality provides sufficient information to characterize their price of anarchy.

An important special case of load balancing is when servers have identical linear latency functions. Here, better upper bounds on the price of anarchy can be obtained. Note that this is not the case for congestion games since, as it was observed in [7], any congestion game can be transformed to a congestion game on identical resources (and, hence, the lower bounds of [4,7] hold for congestion games with identical resources as well). Suri et al. [26] prove that the price of anarchy of selfish load balancing on identical servers is between $1+2/\sqrt{3} \approx 2.1547$ and 2.012067. Again, the upper bound is obtained by using the Nash inequality and the definition of the social cost. We improve this result by showing that the lower bound is essentially tight. Besides the Nash inequality, our proof also exploits structural properties of the game with the highest price of anarchy. We argue that this game can be represented as a directed graph (called the *game graph*) and, then, structural properties of the game follow as structural properties of this graph. Furthermore, for weighted clients and identical servers, we prove that the price of anarchy is at least 5/2.

The price of stability of congestion games has been recently studied in [8] where it was shown that it is between $1 + 1/\sqrt{3} \approx 1.577$ and 1.6. The technique used to obtain the upper bound is to consider pure Nash equilibria with potential not larger than the potential of the optimal assignment and bound their social cost in terms of the optimal cost using the Nash inequality. Using the same technique but also tightening the analysis, we show that the lower bound is tight. Does the fact that load balancing games are significantly simpler than congestion games have any implications in their price of stability? We give a positive answer to this question by showing that the price of stability of selfish load balancing is 4/3. The proof of the upper bound makes use of completely different arguments since the techniques used for congestion games provably cannot be used to obtain this bound.

From the algorithmic point of view, load balancing has been studied extensively, including papers studying online versions of the problem (e.g., [1,2,3,5,6,9, 21,25,26]). In online load balancing, clients appear in online fashion; when a client appears, it has to make an irrevocable decision and assign its job to a server. In our model, servers have linear latency functions and the objective is to minimize the total latency, i.e., the sum of the latencies experienced by all clients. Clients may also own jobs with non-negative weights; in this case, the objective is to minimize the weighted sum of the latencies experienced by all clients. A natural greedy algorithm proposed in [3] for this problem is to assign each client to that server that yields the minimum increase to the total latency (ties are broken arbitrarily). This results to *greedy assignments*. Given an instance of online load balancing, an assignment of clients to servers is called a greedy assignment if the assignment of a client to a server minimizes the increase in the cost of the instance revealed up to the time of its appearance. Following the standard performance measure in competitive analysis, we evaluate the performance of this algorithm in terms of its *competitiveness* (or *competitive ratio*). The competitiveness of the greedy algorithm on an instance is the maximum ratio of the cost of any greedy assignment over the optimal cost and its competitiveness on a class of load balancing instances is simply the maximum competitiveness over all instances in the particular class.

The performance of greedy load balancing with respect to the total latency has been studied in [3,26]. Awerbuch et al. [3] consider a more general model where each client owns a job with a load vector denoting the impact of the job to each server (i.e., how much the assignment of the job to a server will increase its load) and the objective is to minimize the L_p norm of the load of the servers. In the context similar to the one studied in the current paper, their results imply a $3 + 2\sqrt{2} \approx 5.8284$ upper bound. This result applies also in the case of weighted clients where the objective is to minimize the weighted average latency. Suri et al. [26] consider the same model as ours and show upper bounds of $17/3$ and $2 + \sqrt{5} \approx 4.2361$ for arbitrary servers and identical servers, respectively. In a way similar to the study of the price of anarchy of congestion games, [26] develops a *greedy inequality* which is used to obtain the upper bounds on competitiveness. They also present a lower bound of 3.0833 for the competitiveness of greedy assignments in the case of identical servers.

The main question left open by the work of [26] is whether arbitrary servers do hurt the competitiveness of greedy load balancing. We give a positive answer to this question as well. By a rather counterintuitive construction, we show that the $17/3$ upper bound of [26] is tight. This is interesting since it indicates that the greedy inequality is powerful enough to characterize the competitiveness of greedy load balancing. We also consider the case of identical servers where we almost close the gap between the upper and lower bounds of [26] by showing that the competitiveness of greedy load balancing is between 4 and $\frac{2}{3}\sqrt{21} + 1 \approx 4.05505$. In the proof of the upper bound, we use the greedy inequality but, more importantly, we also use arguments for the structure of greedy and optimal assignments of instances that yield the worst competitiveness. In a similar way to the case of selfish load balancing, we argue that such instances can be represented as directed graphs (called *greedy graphs*) that enjoy particular structural properties. In the case of weighted clients, we present a tight lower bound of $3 + 2\sqrt{2}$ on identical servers matching the upper bound of [3].

The rest of the paper is structured as follows. We present the bounds on the price of stability of linear congestion games and selfish load balancing in Section 2. The bounds on the price of anarchy are presented in Section 3 while the bounds on the competitiveness of greedy load balancing are presented in Section 4. We discuss extensions of the results to selfish and greedy load balancing when clients are weighted and conclude with open problems in Section 5. Due to lack of space, many proofs have been omitted from this extended abstract.

2 Bounds on the Price of Stability

We present a tight upper bound on the price of stability of congestion games. Our proof (omitted) uses the main idea in the proof of [8] and bounds the social cost of any Nash equilibrium having a potential smaller than the potential of the optimal assignment. In the proof we also make use of the Nash inequality which together with the inequality on the potentials yields the upper bound. However, the two inequalities may not be equally important in order to achieve the best possible bound and this is taken into account in our analysis. We obtain the following result. A matching lower bound is presented in [8].

Theorem 1. *The price of stability of congestion games with linear latency functions is at most $1 + 1/\sqrt{3}$.*

In the following we show a tight upper bound of 4/3 on the price of stability of load balancing games. We note that the use of the inequality on the potentials does not suffice since load balancing games may have pure Nash equilibria with potential smaller than the potential of an optimal assignment and with cost strictly larger than 4/3 times the optimal cost. So, in order to prove the 4/3 upper bound on the price of stability of load balancing games, we will use entirely different arguments. Starting from any assignment, we let the clients move (one client moves at each step) until they converge to a pure Nash equilibrium. At each step, the moving client is selected arbitrarily among the clients with current strategy at a server of maximum latency which have an incentive to change their strategy. In our proof, we actually show that the social cost of the pure Nash equilibrium at convergence is no more than 4/3 times the cost of the initial assignment. As a corollary, by starting from an optimal solution, we will obtain that the price of stability is at most 4/3.

Theorem 2. *The price of stability of load balancing games is at most 4/3.*

Proof. Consider a load balancing game, an initial assignment with o_j clients at server j for any j, and the moves as defined above. We denote by n_j the number of clients at server j at the Nash equilibrium. Also, we denote by $f_j(x) = \alpha_j x + b_j$ the latency function of server j.

We define *segments* as follows. For each server j, consider the set of moves $\mu_1, \mu_2, ..., \mu_k$ into server j at steps $t_1, t_2, ..., t_k$ so that $t_1 < t_2 < ... < t_k$, and the set of moves $\mu'_1, \mu'_2, ..., \mu'_{k'}$ out of server j at steps $t'_1, t'_2, ..., t'_{k'}$ so that $t'_1 < t'_2 < ... < t'_{k'}$. For $i = 1, ..., k$, we match move μ_i with the first move (if any) $\mu'_{i'}$ that happens after move μ_i and has not been matched to any of the moves $\mu_1, ..., \mu_{i-1}$. In this way we obtain *passing segments* which are pairs of a move into server j and a move out of server j, *starting segments* which consist of single moves out of server j which were not matched to any incoming move, and *ending segments* which consist of single moves into server j which were not matched to any outgoing move.

We construct *chains* (i.e., sequence of moves) using the segments defined. A chain begins with the move in a starting segment, terminates with a move in an ending segment, while any two consecutive moves in the chain (if any), one into and one out of the same server j, belong to the same passing segment of server j. A chain may consist of a single move if this belongs to both a starting and an ending segment. For each server j, denote by s_j and e_j the number of starting and ending segments defined at server j, respectively. Equivalently, s_j is the number of chains beginning with a move out of server j and e_j is the number of chains terminating with a move into server j.

To obtain the desired bound, we will use the following lemma. The proof is lengthy and hence omitted; it relies on an inductive argument.

Lemma 1. $\sum_j f_j(o_j) s_j \geq \sum_j f_j(n_j) e_j$.

Using Lemma 1, we have

$$\sum_j f_j(o_j) o_j = \sum_j \left(f_j(o_j)(o_j - s_j) + f_j(o_j) s_j \right)$$

$$\geq \sum_j \left(f_j(o_j)(o_j - s_j) + f_j(n_j) e_j \right)$$

$$= \sum_j \left(f_j(o_j)(o_j - s_j) + f_j(n_j)(n_j - o_j + s_j) \right)$$

$$= \sum_j \left(\alpha_j \left(o_j^2 - s_j(o_j - n_j) + n_j^2 - n_j o_j \right) + b_j n_j \right) \qquad (1)$$

We distinguish between two cases to show that $o_j^2 - s_j(o_j - n_j) + n_j^2 - n_j o_j \geq \frac{3}{4} n_j^2$, for any j. If $n_j \leq o_j$, then since $s_j \leq o_j$, it is $o_j^2 - s_j(o_j - n_j) + n_j^2 - n_j o_j \geq o_j^2 - o_j(o_j - n_j) + n_j^2 - n_j o_j = n_j^2$. If $n_j \geq o_j$, it is $o_j^2 - s_j(o_j - n_j) + n_j^2 - n_j o_j \geq o_j^2 + n_j^2 - n_j o_j = (o_j - n_j/2)^2 + \frac{3}{4} n_j^2 \geq \frac{3}{4} n_j^2$. Hence, (1) yields that

$$\sum_j f_j(o_j) o_j \geq \frac{3}{4} \sum_j \left(\alpha_j n_j^2 + b_j n_j \right) = \frac{3}{4} \sum_j f_j(n_j) n_j. \qquad \square$$

To show that the above result is tight, it suffices to consider, for arbitrarily small $\epsilon > 0$, a game with two servers with latency functions $f_1(x) = (2 + \epsilon)x$ and $f_2(x) = x$ and two clients having both servers as strategies.

3   Bounds on the Price of Anarchy

For the study of the price of anarchy, we can consider load balancing games in which each client has at most two strategies. This is clearly sufficient when proving lower bounds. In order to prove upper bounds, we can assume that the highest price of anarchy is obtained by such a game. Consider any load balancing game and let O and N be the optimal assignment and the Nash equilibrium that yields the worst social cost, respectively. The game with the same clients and servers in which each client has its strategies in O and N as strategies also has the same optimal assignment and the same Nash equilibrium (and, consequently the same price of anarchy). We represent such games as directed graphs (called *game graphs*) having a node for each server and a directed edge for each client; the direction of each edge is from the strategy of the client in the optimal assignment to the strategy of the client in the Nash equilibrium. A self-loop indicates that the client has just one strategy.

The next theorem states that the upper bound of $5/2$ presented in [26] (and also implied by the results in [4,7] for congestion games) is tight. This bound was known to be tight for congestion games in general but the constructions in the lower bounds in [4,7] are not load balancing games.

Theorem 3. *For any $\epsilon > 0$, there is a load balancing game with price of anarchy at least $5/2 - \epsilon$.*

Proof. We construct a game graph G consisting of a complete binary tree with $k + 1$ levels and $2^{k+1} - 1$ nodes with a line of $k + 1$ edges and $k + 1$ additional nodes hung at each leaf. So, graph G has $2k + 2$ levels $0, ..., 2k + 1$, with 2^i nodes at level i for $i = 0, ..., k$ and 2^k nodes at levels $k + 1, ..., 2k + 1$. The servers corresponding to nodes of level $i = 0, ..., k - 1$ have latency functions $f_i(x) = (2/3)^i x$, the servers corresponding to nodes of level $i = k, ..., 2k$ have latency functions $f_i(x) = (2/3)^{k-1}(1/2)^{i-k}x$, and the servers corresponding to nodes of level $2k + 1$ have latency functions $f_{2k+1}(x) = (2/3)^{k-1}(1/2)^k x$. The assignment where all clients select servers corresponding to the endpoint of their corresponding edge which is closer to the root of the game graph can be easily verified that it is a Nash equilibrium. Its cost is $\sum_{i=0}^{k-1} 4 \cdot 2^i (2/3)^i + \sum_{i=k}^{2k} 2^k (2/3)^{k-1}(1/2)^{i-k} = 15(4/3)^k - (2/3)^{k-1} - 12$. To compute an upper bound for the cost of the optimal assignment, it suffices to consider the assignment where all clients select the servers corresponding to nodes which are further from the root. We obtain that the cost of the optimal assignment is at most $\sum_{i=1}^{k-1} 2^i (2/3)^i + \sum_{i=k}^{2k} 2^k (2/3)^{k-1}(1/2)^{i-k} + 2^k (2/3)^{k-1}(1/2)^k = 6(4/3)^k - 4$. Hence, for any $\epsilon > 0$ and for sufficiently large k, the price of anarchy of the game is larger than $5/2 - \epsilon$.   □

In the case of identical servers we can show a tight bound on the price of anarchy of approximately 2.012067; a matching lower bound has been presented in [26]. Here, we present the main idea in our analysis to obtain a slightly weaker result; the improved analysis will appear in the final version of the paper.

We will consider the game with the highest price of anarchy and upper-bound the ratio of the social cost of the worst Nash equilibrium to the optimal cost of the particular game. We represent the game by a game graph. We say that server j is of type n_j/o_j meaning that it has n_j clients in the Nash equilibrium and o_j clients in the optimal assignment (equivalently, server j has in-degree n_j and out-degree o_j in the game graph). After observing that each server of type $1/1$ can be associated with a neighboring server of type $0/1$, the idea behind the proof is to account for their contribution in the social cost together. By extending the neighborhood considered together with each server of type $1/1$, we can obtain better and better upper bounds which converge to the lower bound of 2.012067.

In the proof, we make use of the following technical lemma.

Lemma 2. *For any integers x, y, define the functions $g(x,y) = xy + \frac{18+7\sqrt{21}}{30}y - \frac{7\sqrt{21}-12}{30}x$ and $h(x,y) = \frac{6-\sqrt{21}}{10}x^2 + \frac{6+\sqrt{21}}{6}y^2$. For any non-negative integers x, y such that either $x \neq 1$ or $y \neq 1$, it holds that $g(x,y) \leq h(x,y)$. Furthermore, $g(0,1) + g(1,1) = h(0,1) + h(1,1)$.*

Theorem 4. *The price of anarchy of selfish load balancing on identical servers is at most $\frac{2}{3}\sqrt{21} - 1$.*

Proof. Consider a load balancing game on servers with latency function $f(x) = x + b$ and clients having at most two strategies which has the highest price of anarchy. Consider a server j of type $1/1$. If a client c had server j as its only strategy (this corresponds to a self-loop in the corresponding game graph), then we may construct a new game by excluding server j and client c from the original one; it can be easily seen that the new game has worse price of anarchy since both the cost of the optimal assignment and the social cost of the Nash equilibrium are decreased by $1 + b$. So, let j' and j'' be the servers to which server j is connected corresponding to clients c_1 and c_2 selecting servers j' and j in the optimal assignment and servers j and j'' in the Nash assignment, respectively.

Server j' is of type $0/1$. Assume otherwise that it is of type $n_{j'}/o_{j'}$ for $n_{j'} > 0$ or $o_{j'} > 1$. If $n_{j'} > 0$, we can construct a new game by excluding server j and substituting clients c_1 and c_2 by a client selecting server j' in the optimal assignment and server j'' in the Nash assignment. If $o_{j'} > 1$, then we can add a new server j'_1 and change the strategy of client c_1 to $\{j'_1, j\}$. In both cases, we obtain games with higher price of anarchy.

Denote by F the set of servers of type $1/1$ and by S the set of servers of type $0/1$ which are connected through an edge to a server in F in the game graph. Also, for each server j in F we denote by $S(j)$ the server of S from which the client destined for j originates. By the Nash inequality, we obtain that $\sum_j (n_j^2 + bn_j) \leq \sum_j (o_j n_j + (1+b)o_j)$ and, since $\sum_j n_j = \sum_j o_j$, we have that

$$\sum_j n_j^2 \le \sum_j (n_j o_j + o_j) = \sum_j \left(n_j o_j + \frac{18 + 7\sqrt{21}}{30} o_j - \frac{7\sqrt{21} - 12}{30} n_j \right)$$

$$= \sum_{j \notin F \cup S} g(n_j, o_j) + \sum_{j \in F} \left(g(n_{S(j)}, o_{S(j)}) + g(n_j, o_j) \right)$$

$$\le \sum_{j \notin F \cup S} h(n_j, o_j) + \sum_{j \in F} \left(h(n_{S(j)}, o_{S(j)}) + h(n_j, o_j) \right)$$

$$= \frac{6 - \sqrt{21}}{10} \sum_j n_j^2 + \frac{6 + \sqrt{21}}{6} \sum_j o_j^2$$

where the first equality follows since $\sum_j n_j = \sum_j o_j$, the second equality follows by the definition of function g, the second inequality follows by Lemma 2, and the last equality follows by the definition of function h. Hence, we obtain that the price of anarchy is

$$\frac{\sum_j \left(n_j^2 + b n_j \right)}{\sum_j \left(o_j^2 + b o_j \right)} \le \frac{\sum_j n_j^2}{\sum_j o_j^2} \le \frac{2}{3}\sqrt{21} - 1. \qquad \square$$

4   Greedy Load Balancing

Similarly to the case of selfish load balancing, in the study of the competitiveness of greedy load balancing, we consider load balancing instances in which each client has at most two strategies. This is clearly sufficient when proving lower bounds. In order to prove upper bounds, we can assume that the highest competitiveness is obtained by such an instance. Consider any load balancing instance and let O and N be the optimal assignment and the greedy assignment of the highest cost, respectively. The instance with the same clients and servers in which each client has its strategies in O and N as strategies also has the same optimal assignment and the same greedy assignment (and, consequently the same competitiveness). We represent such instances as directed graphs (called *greedy graphs*) having a node for each server and a directed edge with timing information for each client; the direction of each edge is from the strategy of the client in the optimal assignment to the strategy of the client in the greedy assignment and the timing information denotes the time the client appears. We can show that the upper bound of [26] for arbitrary servers is tight.

Theorem 5. *For any $\epsilon > 0$, greedy load balancing has competitiveness at least $17/3 - \epsilon$.*

We also study the case of identical servers with latency function $f(x) = x + b$. By reasoning about the structure of the load balancing instance that yields the worst competitiveness and using the greedy inequality developed in [26], we can prove the following theorem.

Theorem 6. *Greedy load balancing on identical servers has competitiveness at most $\frac{2}{3}\sqrt{21} + 1$.*

We also present an almost matching lower bound.

Theorem 7. *For any $\epsilon > 0$, greedy load balancing on identical servers has competitiveness at least $4 - \epsilon$.*

Proof. We assume that there are m servers $s_1, s_2, ..., s_m$, and k groups of clients $g_1, ..., g_k$, where group g_j has m/j^2 clients c_i^j, $1 \leq i \leq m/j^2$. We assume that m is such that all groups have integer size. Each client c_i^j has $s_1, s_2, ..., s_i$ as permissible servers. The clients appear in non-increasing order according to index i, i.e., $c_m^1, c_{m-1}^1, ..., c_{m/4+1}^1, c_{m/4}^2, c_{m/4}^1, c_{m/4-1}^2, c_{m/4-1}^1, ..., c_{m/9+1}^2, c_{m/9+1}^1, c_{m/9}^3, c_{m/9}^2, c_{m/9}^1, ...,$ etc.

To upper bound the optimal cost opt, it suffices to consider the assignment where each client c_i^j chooses server s_i. We obtain that

$$opt \leq \sum_{i=1}^{k-1} i^2 (|g_i| - |g_{i+1}|) + k^2 |g_k| = m + m \sum_{i=1}^{k-1} i^2 \left(\frac{1}{i^2} - \frac{1}{(i+1)^2} \right)$$

$$= m \left(1 + 2 \sum_{i=1}^{k-1} 1/(i+1) - \sum_{i=1}^{k-1} 1/(i+1)^2 \right) \leq m(2H_k + \zeta_1)$$

for some positive constant ζ_1, where H_k is the k-th Harmonic number.

A greedy assignment is obtained by making each client select the server with the smallest index among its permissible servers having the minimum number of clients. In the analysis we make use of sets of clients called *columns*. A client belongs to column col_i if, when it selects its server, it is the i-th client selecting that server. For example, clients $c_m^1, c_{m-1}^1, ... c_{m/2+1}^1$ select servers $s_1, ..., s_{m/2}$, respectively; each of them is the first client in its server, so they belong to col_1. Then, $c_{m/2}^1, ..., c_{m/4+1}^1$ select servers $s_1, ..., s_{m/4}$; they belong to col_2. We can verify that the set of servers selected by clients in col_{i+1} is subset of the set of servers selected by clients in col_i for $i = 1, ..., 2k - 3$, that columns col_{2i-1} and col_{2i} contain clients of groups $g_1, ..., g_i$, and that $|col_{2i}| = \frac{m}{(i+1)^2}$ and $|col_{2i-1}| = \frac{m}{i(i+1)}$ for any $i = 1, ..., k - 1$. So, for $i = 1, ..., 2k - 3$, the number of servers receiving exactly i clients in the greedy assignment is $|col_i| - |col_{i+1}|$. We compute a lower bound on the cost gr of the greedy assignment by considering only the servers with at most $2k - 4$ clients. We have that

$$gr \geq m \sum_{i=1}^{k-2} \left((2i-1)^2 (|col_{2i-1}| - |col_{2i}|) + (2i)^2 (|col_{2i}| - |col_{2i+1}|) \right)$$

$$= m \sum_{i=1}^{k-2} \left((2i-1)^2 \left(\frac{1}{i(i+1)} - \frac{1}{(i+1)^2} \right) + (2i)^2 \left(\frac{1}{(i+1)^2} - \frac{1}{(i+1)(i+2)} \right) \right)$$

$$\geq m \sum_{i=1}^{k-2} \left(\frac{8}{i+1} - \frac{20}{(i+1)^2} \right) \geq m(8H_k - \zeta_2)$$

for some positive constant ζ_2. We conclude that for any $\epsilon > 0$ and sufficiently large k and m, the competitiveness of the greedy assignment is at least $4 - \epsilon$. $\square$

By slightly modifying the argument in the proof of Theorem 7 we can show that the lower bound holds for any deterministic online algorithm.

5 Extensions and Open Problems

We have also considered clients with non-negative weights. In the case of clients with weights, upper bounds of $\frac{3+\sqrt{5}}{2} \approx 2.618$ and $3 + 2\sqrt{2} \approx 5.8284$ for the price of anarchy of selfish load balancing and the competitiveness of greedy load balancing follow by the analysis of [4,7] for weighted linear congestion games and by the analysis of [3], respectively. We have shown that both bounds are tight. In particular, the second lower bound holds for greedy load balancing on identical servers. For selfish load balancing of weighted clients on identical servers, we can show a lower bound of $5/2$ on the price of anarchy. It is interesting to close the gap between this lower bound and the upper bound of $\frac{3+\sqrt{5}}{2}$ which has been proved for congestion games [4]. We believe that our lower bound is tight. Another interesting open problem is to compute tight bounds for the price of stability of weighted load balancing games. We have considered pure Nash equilibria of load balancing games. Our results hold or can be extended to hold for mixed and correlated equilibria [8] as well. There is also a small gap between 4 and 4.05505 for the competitiveness of greedy load balancing on identical servers. We believe that it can be further narrowed by extending our upper bound technique.

References

1. N. Alon, Y. Azar, G. J. Woeginger and T. Yadid. Approximation schemes for scheduling. In *Proc. of the 8th Annual ACM-SIAM Symposium on Discrete Algorithms (SODA '97)*, pp. 493-500, 1997.
2. A. Avidor, Y. Azar and J. Sgall. Ancient and new algorithms for load balancing in the L_p norm. *Algorithmica*, 29(3): 422-441, 2001.
3. B. Awerbuch, Y. Azar, E. F. Grove, M.-Y. Kao, P. Krishnan, and J. S. Vitter. Load balancing in the L_p norm. In *Proc. of the 36th Annual Symposium on Foundations of Computer Science (FOCS '95)*, pp. 383-391, 1995.
4. B. Awerbuch, Y. Azar, and A. Epstein. The price of routing unsplittable flow. In *Proc. of the 37th Annual ACM Symposium on Theory of Computing (STOC '05)*, pp. 57-66, 2005.
5. Y. Azar and A. Epstein. Convex programming for scheduling unrelated parallel machines. In *Proc. of the 37th Annual ACM Symposium on Theory of Computing (STOC '05)*, pp. 331-337, 2005.
6. A. K. Chandra and C. K. Wong. Worst-case analysis of a placement algorithm related to storage allocation. *SIAM Journal on Computing*, 4(3): 249-263, 1975.
7. G. Christodoulou and E. Koutsoupias. The price of anarchy of finite congestion games. In *Proc. of the 37th Annual ACM Symposium on Theory of Computing (STOC '05)*, pp. 67-73, 2005.
8. G. Christodoulou and E. Koutsoupias. On the price of anarchy and stability of correlated equilibria of linear congestion games. In *Proc. of the 13th Annual European Symposium on Algorithms (ESA '05)*, LNCS 3669, Springer, pp. 59-70, 2005.

9. R. A. Cody and E. G. Coffman. Record allocation for minimizing expected retrieval costs on drum-like storage devices. *Journal of the ACM*, 23(1): 103-115, 1976.

10. A. Czumaj and B. Vöcking. Tight bounds for worst-case equilibria. In *Proc. of the 13th Annual ACM-SIAM Symposium on Discrete Algorithms (SODA '02)*, pp. 413-420, 2002.

11. A. Fabrikant, C. Papadimitriou and K. Talwar. On the complexity of pure equilibria. In *Proc. of the 36th Annual ACM Symposium on Theory of Computing (STOC '04)*, pp. 604-612, 2004.

12. D. Fotakis, S. Kontogiannis, E. Koutsoupias, M. Mavronicolas and P. Spirakis. The structure and complexity of Nash equilibria for a selfish routing game. In *Proc. of the 29th International Colloquium on Automata, Languages and Programming (ICALP '02)*, LNCS 2380, Springer, pp. 123-134, 2002.

13. D. Fotakis, S. Kontogiannis, and P. Spirakis. Selfish unsplittable flows. In *Proc. of the 31st International Colloquium on Automata, Languages, and Programming (ICALP '04)*, LNCS 3142, Springer, pp. 593-605, 2004.

14. M. Gairing, T. Lücking, M. Mavronicolas and B. Monien. Computing Nash equilibria for scheduling on restricted parallel links. In *Proc. of the 36th Annual ACM Symposium on Theory of Computing (STOC '04)*, pp. 613-622, 2004.

15. E. Koutsoupias, M. Mavronicolas and P. Spirakis. Approximate equilibria and ball fusion. *Theory of Computing Systems*, 36(6): 683-693, 2003.

16. E. Koutsoupias and C. Papadimitriou. Worst-case equilibria. In *Proc. of the 16th International Symposium on Theoretical Aspects of Computer Science (STACS '99)*, LNCS 1563, Springer, pp. 404-413, 1999.

17. T. Lücking, M. Mavronicolas, B. Monien, and M. Rode. A new model for selfish routing. In *Proc. of the 21st International Symposium on Theoretical Aspects of Computer Science (STACS '04)*, LNCS 2996, Springer, pp. 547-558, 2004.

18. M. Mavronicolas and P. Spirakis. The price of selfish routing. In *Proc. of the 33rd Annual ACM Symposium on Theory of Computing (STOC '01)*, pp. 510-519, 2001.

19. D. Monderer and L. S. Shapley. Potential games. *Games and Economic Behavior*, 14: 124-143, 1996.

20. C. Papadimitriou. Algorithms, games and the internet. In *Proc. of the 33rd Annual ACM Symposium on Theory of Computing (STOC '01)*, pp. 749-753, 2001.

21. S. Phillips and J. Westbrook. Online load balancing and network flow. In *Proc. of the 25th Annual ACM Symposium on Theory of Computing (STOC '93)*, pp. 402-411, 1993.

22. R. Rosenthal. A class of games possessing pure-strategy Nash equilibria. *International Journal of Game Theory*, 2: 65-67, 1973.

23. T. Roughgarden and E. Tardos. How bad is selfish routing? *Journal of the ACM*, 49(2): 236-259, 2002.

24. T. Roughgarden and E. Tardos. Bounding the inefficiency of equilibria in nonatomic congestion games. *Games and Economic Behavior*, 47(2): 389-403, 2004.

25. D. Shmoys, J. Wein and D. Williamson. Scheduling parallel machines on-line. *SIAM Journal on Computing*, 24(6): 1313-1331, 1995.

26. S. Suri, C. Tóth and Y. Zhou. Selfish load balancing and atomic congestion games. In *Proc. of the 16th Annual ACM Symposium on Parallelism in Algorithms and Architectures (SPAA '04)*, pp. 188-195, 2004.

Lower Bounds of Static Lovász-Schrijver Calculus Proofs for Tseitin Tautologies

Arist Kojevnikov[1,*] and Dmitry Itsykson[2,**]

[1] St. Petersburg Dep. of Steklov Institute of Mathematics, St.Petersburg, Russia
arist@pdmi.ras.ru
[2] St. Petersburg State University, St.Petersburg, Russia
dmitrits@logic.pdmi.ras.ru

Abstract. We prove an exponential lower bound on the size of static Lovász-Schrijver calculus refutations of Tseitin tautologies. We use several techniques, namely, translating static LS_+ proof into *Positivstellensatz* proof of Grigoriev et al., extracting a "good" expander out of a given graph by removing edges and vertices of Alekhnovich et al., and proving linear lower bound on the degree of *Positivstellensatz* proofs for Tseitin tautologies.

1 Introduction

It is a known approach in $\{0, 1\}$-programming to translate a problem to a linear programming problem and repeatedly refining it with new linear inequalities, "cutting planes", that are satisfied only by the $\{0, 1\}$-solutions. The first such method appeared in the works of Gomory [1] and Chvátal [2]. It derives each new inequality as a linear combination and rounding of existing inequalities, using the fact that all variables have values in $\{0, 1\}$. In 1991, Lovász and Schrijver [3] introduced a variety of cutting planes methods that derive new inequalities by first lifting the existing inequalities to higher dimensional space (where a more convenient formulation may give a tighter relaxation) and then linearizing the resulting inequalities using the fact that $x^2 = x$ for $x \in \{0, 1\}$.

We can also use these methods to solve propositional formulas by mapping them into systems of linear inequalities. The obtained proof systems are very strong. They have polynomial-size proofs for tautologies such as the propositional pigeonhole principle that are known to require superpolynomial-size proofs in the resolution proof system and constant-depth Frege systems and no exponential lower bounds are known for Lovász-Schrijver proof systems (though there are exponential bounds for systems of inequalities that are *not* produced from Boolean tautologies; see [4] and references therein).

If we implement a SAT-solver that operates with inequalities and run it on some examples, we would see that it takes a long time on some instances. We

---
* Supported in part by Russian Science Support Foundation, RFBR grants 05-01-00932, 06-01-00502 and INTAS grant 04-83-3836.
** Supported in part by RFBR grant 06-01-00502 and INTAS grant 04-77-7173.

M. Bugliesi et al. (Eds.): ICALP 2006, Part I, LNCS 4051, pp. 323–334, 2006.
© Springer-Verlag Berlin Heidelberg 2006

can explain it in assumption that **NP** $\neq$ co-**NP** or more mild assumption from communication complexity [5]. However, in this paper we prove that Tseitin tautologies require subexponential-size static LS proofs unconditionally.

From SAT-solver point of view, the "static" and "treelike" qualifiers mean that there is no caching and reuse of discovered clauses. This might seem like a strong restriction, and it is. However, even with this constraint, the proved lower bound is important.

The paper is organized as follows. Sect. 2 contains the necessary definitions. The proof of the main result is based on ideas from [4] and is divided into four parts. In Sect. 3 we prove that if a graph G with n vertices is a "good" expander then we can extract a "good" expander out of G by removing $O(n)$ vertices. In order to do this, we use the technique of Alekhnovich et al. [6]. Grigoriev [7] proved the linear lower bound on degree of the *Positivstellensatz* for Tseitin tautologies in binomial form. It can be easily extended to linear lower bound on the Boolean degree. Sect. 4 contains the transformation of the lower bound for *Positivstellensatz* into a Boolean degree lower bound for static LS. Finally, in Sect. 5 we obtain exponential lower bounds for Tseitin tautologies in static and tree-like LS with squares.

2   Preliminaries

A *proof system* [8] for a language L is a polynomial-time computable function mapping words (treated as proof candidates) to L (whose elements are considered as theorems).

A *propositional proof system* is a proof system for the co-**NP**-complete language **TAUT** of all Boolean tautologies in disjunctive normal form (DNF). Since this language is in co-**NP**, any proof system for a co-**NP**-hard language L can be considered as a propositional proof system. However, we need to fix a concrete reduction of **TAUT** to L before compare them.

An *algebraic proof system* is a proof system for the co-**NP**-hard language of unsolvable systems of polynomial equations: we are given several polynomials over a field $\mathbb{F}$ and the question is whether these polynomials have no common roots in $\mathbb{F}$. The polynomials are represented as sums of monomials $c \cdot x_1 \cdots x_s$, where $x_1, \ldots, x_s$ are variables and $c \in \mathbb{F}^* = \mathbb{F} \setminus \{0\}$ is a constant given in some reasonable (e.g., binary) notation. To show that such proof system is a propositional proof system, one translates Boolean tautologies into systems of polynomial equations.

Given a formula F in DNF with n variables and m clauses, we take its negation $\neg F$ in CNF and translate each clause C_i containing variables $x_{i_1}, \ldots, x_{i_l}$ into a polynomial equation of the form

$$(1 - l_{i_1}) \cdot \ldots \cdot (1 - l_{i_l}) = 0 \ , \tag{1}$$

where $l_{i_j} = x_{i_j}$ if the variable x_{i_j} occurs in C_i positively, and $l_{i_j} = (1 - x_{i_j})$ if it occurs negatively. For each variable v_i, $1 \leq i \leq n$, we also add the equation $v_i^2 - v_i = 0$ to this system.

Note that the formula F is a tautology if and only if the obtained system $\mathcal{D}$ of polynomial equations $f_1, \ldots, f_{m+n}$ has no solutions. Therefore, to prove F it suffices to derive a contradiction from the system $\mathcal{D}$.

In *Polynomial Calculus* (**PC**) [9], one starts with the polynomial equation system $\mathcal{D}$ and derives new polynomials using the following two rules:

$$\frac{f = 0 \quad g = 0}{f + g = 0} \quad \text{and} \quad \frac{f = 0}{f \cdot g = 0} \ .$$

A proof in this system is a derivation of $1 = 0$ from $\mathcal{D}$ using these rules.

Positivstellensatz [10] operates on polynomials over a real field. The proof $\mathcal{D}$ consists of polynomials $g_1, \ldots, g_{m+n}$ and $h_1, \ldots, h_l$ such that

$$\sum_{i=1}^{m+n} f_i g_i = 1 + \sum_{j=1}^{l} h_j^2 \ . \tag{2}$$

It is a "static" proof in the sense that it contains only one step. Note that the right-hand side of (2) is the derivation of 0 in **PC**.

A *semialgebraic proof system* operates with language of unsolvable systems of polynomial inequalities. They are much more powerful than algebraic proof systems. No nontrivial complexity lower bounds for some of them are known so far. Moreover, in semialgebraic systems there exist short proofs of many tautologies that are hard for other proof systems [4].

To define a propositional proof system working with inequalities, we translate each formula $\neg F$ with n variables in CNF into a system $\mathcal{D}$ of linear inequalities such that F is a tautology if and only if the system $\mathcal{D}$ has no solution in $\{0,1\}$-variables. For a formula F, we translate each clause C_i of $\neg F$ with variables $x_{j_1}, \ldots, x_{j_t}$, into the inequality

$$l_1 + \ldots + l_t - 1 \geq 0 \ , \tag{3}$$

where $l_i = x_{j_i}$ if the variable x_{j_i} occurs positively in the clause, and $l_i = 1 - x_j$ if x_{j_i} occurs negatively. For every variable x_i, $1 \leq i \leq n$, we also add to the system $\mathcal{D}$ the inequalities

$$0 \leq x_i \leq 1 \ . \tag{4}$$

In the Lovász-Schrijver proof system (LS) [3], one obtains the contradiction $0 \geq 1$ using the rules

$$\frac{f \geq 0 \quad g \geq 0}{\lambda_f f + \lambda_g g \geq 0} \ , \quad \frac{h \geq 0}{hx \geq 0} \ , \quad \frac{h \geq 0}{h(1 - x) \geq 0} \ ,$$

where $\lambda_f, \lambda_g \geq 0$, the polynomial h is linear and x is a variable. Also, the set of axioms (4) is extended by the inequalities

$$x_i^2 - x_i \geq 0 \ , \qquad \text{for every variable } x_i, 1 \leq i \leq n \ . \tag{5}$$

The system LS_+ [3] has the same axioms and derivation rules as LS and the addition axiom, $h^2 \geq 0$, for every linear h. The proof is *tree-like* if the underlying directed acyclic graph, representing the implication structure of the proof, is a tree. That is, every inequality in the proof, except for the initial inequalities, is used at most one as an antecedent of an implication.

A proof of Boolean formula F with n variables in *static LS_+* [4] consists of positive real coefficients $c_{i,l}$ and multisets $U_{i,l}^+, U_{i,l}^-$ determining the polynomials $g_{i,l} = c_{i,l} \cdot \prod_{k \in U_{i,l}^+} x_k \cdot \prod_{k \in U_{i,l}^-} (1 - x_k)$ such that

$$\sum_{i=1}^{M} f_i \sum_l g_{i,l} = -1 \ , \tag{6}$$

where each f_i is either a left-hand part of inequality given by translation (3) from the formula F, or polynomials of the form x_j, $(1 - x_j)$, $(x_j^2 - x_j)$ for some variable x_j, $1 \leq j \leq n$, or a square of linear polynomial h_j^2.

The original definition of the *size* of a refutation in static LS_+ [4] is the length of a reasonable bit representation of all polynomials $g_{i,l}$, f_i and thus is at least the number of $u_{i,l}$'s.

Tseitin tautologies and expanders. Let $G = (V, E)$ be an undirected graph on the set of vertices V, with the set of edges E. To each edge $e \in E$ we attach a $\{0, 1\}$-variable x_e. The negation T_G of Tseitin tautologies with respect to G is a family of formulas meaning that, for each vertex $v \in V' \subset V$, the sum ranging over the edges incident to v is odd and for each vertex $v \in V \setminus V'$, the sum ranging over the edges incident to v is even. If the cardinality of V' is odd then T_G is contradictory.

For subsets I, I_1 of vertices and subset of edges $J \subseteq E$ we define *boundary operation* ∂:

$$N_{V \setminus I, E \setminus J}(I_1) = \{v \in V \setminus I : (v, v') \in E \setminus J \text{ for some } v' \in I_1\} \setminus I_1 \ ,$$
$$\partial_{V \setminus I, E \setminus J}(I_1) = \{(v, v') \in E \setminus J : v \in N_{V \setminus I, E \setminus J}(I_1) \text{ and } v' \in I_1\}.$$

In what follows we will use $\partial_{V,E}(I)$ as short notation for $\partial_{V \setminus \emptyset, E \setminus \emptyset}(I)$. We say that a graph $G = (V, E)$ is an (r, d, c)-*expander* if the maximal degree of a vertex is d, and for every set $X \subseteq V$ of cardinality at most r,

$$|\partial_{V,E}(X)| \geq c \cdot |X| \ .$$

For an (r, d, c)-expander G, Tseitin formula T_G is given by the clauses

$$\bigvee_{e \in S_v \setminus S'_v} x_e \vee \bigvee_{e \in S'_v} (1 - x_e) \ , \tag{7}$$

for each vertex $v \in V'$ and for each subset S'_v of even cardinality of the set S_v of edges incident to v, for each vertex $v \in V \setminus V'$ and for each S'_v of odd cardinality.

In this paper we use three different representations of Boolean formula T_G. One of them is the system of linear inequalities provided by translation (3). It translates every clause (7) into an inequality

$$f^A_{v,S'_v} = \sum_{e \in S_v \setminus S'_v} x_e + \sum_{e \in S'_v} (1 - x_e) - 1 \geq 0 \ . \tag{8}$$

Denote obtained system including axioms (4) and (5) for all variables of the T_G by T^A_G. Our main goal is to prove that every static LS_+ refutation of T^A_G has size $exp(\Omega(n))$.

In our proof we also need the representation of T_G as the equation system obtained by translation (1):

$$f^M_{v,S'_v} = \prod_{e \in S_v \setminus S'_v} (1 - x_e) \cdot \prod_{e \in S'_v} x_e = 0 \ , \tag{9}$$

for every clause of T_G. Denote obtained equation system including axioms $x^2 - x = 0$ for all variables of the T_G by T^M_G.

For lower bounds on *Positivstellensatz* refutations the following *binomial representation of Tseitin formulas* is used [7]. To each edge of the graph G we assign a $\{1, -1\}$-variable y_k. The system T^B_G contains the equations

$$Y(v) = c_v \cdot \prod_{e \ni v} y_e = 1 \tag{10}$$

for each vertex $v \in V'$ with constant $c_v = -1$ and for each vertex $v \in V \setminus V'$ with constant $c_v = 1$ and $y^2_e = 1$ for all $e \in E$.

We use the following notation from [7]. For a monomial $m = x^{i_1}_1 \cdots x^{i_k}_k$, its *multilinearization in $\{0, 1\}$-variables* is the polynomial obtained by the reduction of m modulo $(x^2_i = x_i)$, $1 \leq i \leq k$, i.e. $\overline{m} = x_1 \cdots x_k$. The *multilinearization in $\{1, -1\}$-variables* of the same monomial is $m' = x_{j_1} \cdots x_{j_{k'}}$, its reduction modulo $(x^2_i = 1)$. The corresponding *multilinearization* of a polynomial is a sum of its multilinearized monomials.

We define the *Boolean degree*, $\mathrm{Bdeg}(f)$ of a polynomial f as the degree of its *multilinearization*.

Lemma 1. *The Boolean degree of a polynomial in $\{0, 1\}$-variables is at least its Boolean degree in $\{1, -1\}$-variables.*

Proof. The degree of a multilinearization in $\{0, 1\}$-variables of a monomial $m = x^{i_1}_1 \cdots x^{i_k}_k$ is equal to k. The degree of multilinearization in $\{1, -1\}$-variables of m is equal to $\sum^k_{j=1}(i_j \bmod 2) \leq k$. □

3 Closure Operator on Expanders

In this section we recall the definition of the closure operation of a set of edges w.r.t. graph G (originally defined in [11,6]) and examine properties of graphs under this operation.

For a graph $G = (V, E)$ and a subset of its edges $J \subseteq E$ we define an inference relation $\vdash_J$ on subsets of vertices $I, I_1 \subseteq V$:

$$I \vdash_J I_1 \quad \overset{def}{\Longleftrightarrow} \quad |I_1| \leq r/2 \wedge \left|\partial_{V \setminus I, E \setminus J}(I_1)\right| < c/2|I_1| \ .$$

For a subset of vertices I and a set of edges J we consider the following *cleaning* step:

- If there exists a nonempty $I_1 \subseteq V$, such that $I \vdash_J I_1$ and $I \cap I_1 = \emptyset$, then take the first canonically[1] such I_1 and add it to I.
- Repeat the cleaning step as long as it is applicable.

Let the *closure* $Cl(J)$ of J be the set of all vertices I that can be inferred via $\vdash_J$ from the empty set, $Cl(J) = \{v \mid \emptyset \vdash_J \{v\}\}$.

Lemma 2 ([6], Lemma 3.4). *Assume that graph $G = (V, E)$ is an (r, d, c)-expander and J is a subset of its edges. Let $I' = Cl(J)$ and $J' = \{(v, x) \in E : v \in I' \text{ or } x \in I'\}$. Denote by $G' = (V \setminus I', E \setminus J')$ the graph that results from G by removing the vertices corresponding to I' and edges corresponding to J'. If G' is non-empty then it is an $(r/2, d, c/2)$-expander.*

In the next lemma we show that if we take J of small cardinality, then the graph G' from Lemma 2 is non-empty.

Lemma 3 ([6], Lemma 3.5). *Let a graph $G = (V, E)$ be an (r, d, c)-expander and $|J| < cr/4$. Then $|Cl(J)| < 2c^{-1}|J|$.*

4   Simulation of Static LS_+ in *Positivstellensatz*

In this section we transform the proof in static LS_+ of the system of linear inequalities T_G^A into the *Positivstellensatz* proof of the system of binomial equations T_G^B with constant increase of Boolean degree, then we apply the result of the previous section to obtain a linear lower bound on the Boolean degree of static LS_+ refutations on Tseitin Boolean formulas. The *Boolean degree of a static LS_+ refutation* (6) is the maximum Boolean degree of polynomials $g_{i,l}$ in it. We define *Boolean degree of Positivstellensatz refutation* as the maximum Bdeg of polynomials $f_i g_i$, $1 \le i \le n$, and h_j^2, $1 \le j \le M$ in (2).

Next two lemmas can be applied to a static LS_+ proof P of arbitrary Boolean formula F, they show that P can be transformed into the *Positivstellensatz* proof of F with only constant increase of Boolean degree.

Fix for the time being a Boolean formula F with m clauses and n variables, let F^A be set of linear inequalities provided by translation (3) and F^M be set of equations provided by (1) from formula F.

Lemma 4. *In static LS_+, every proof P of F^A can be transformed into a proof P' of the polynomial equation system F^M. Moreover, if $Bdeg(P) = k$ and the number of variables in every inequality of F^A is at most d, then $Bdeg(P') \le k+d$.*

Proof. The proof P can be represented in the form

$$\sum_{i=1}^{m} f_i^A \sum_l g_{i,l} + \sum_{i=m+1}^{n} f_i \sum_l g_{i,l} = -1 \ , \tag{11}$$

---

[1] It does not matter what particular order to take, we take canonical to exclude ambiguity.

where $g_{i,l} = c_{i,l} \prod_{k \in U_{i,l}^+} x_k \cdot \prod_{k \in U_{i,l}^-} (1 - x_k)$ for appropriate multisets of variables $U_{i,l}^+, U_{i,l}^-$ and a positive real $c_{i,l}$.

We show that the translation of a clause $C_i = (l_1 \vee \ldots \vee l_{d_i})$ into an inequality $f_i^A = \sum_{k=1}^{d_i} l_k - 1 \geq 0$ can be represented as the translation of the clause C_i into an equation $f_i^M = \prod_{k=1}^{d_i} (1 - l_k) = 0$:

$$f_i^A = -f_i^M + \rho(l_1, \ldots, l_{d_i}) , \tag{12}$$

where the second summand $\rho(l_1, \ldots, l_{d_i})$ is nonnegative and equal to sum of literal products. The induction base is $\rho(l_1) = 0 \geq 0$, the induction step is $\rho(l_1, \ldots, l_{d_i}) = \rho(l_1, \ldots, l_{d_i-1})(1 - l_{d_i}) + \sum_{k=1}^{d_i-1} l_k \cdot l_{d_i} \geq 0$.

Let us replace each f_i^A in proof P by (12). As a result, we obtain the proof P':

$$\sum_{i=1}^{m} -f_i^M \sum_l g_{i,l}' + \sum_{i=m+1}^{n'} f_i \sum_l g_{i,l}' = -1 , \tag{13}$$

where $g_{i,l}' = c_{i,l}' \cdot \prod_{k \in U_{i,l}^+} x_k \cdot \prod_{k \in U_{i,l}^-} (1 - x_k)$ for appropriate multisets $U_{i,l}^+, U_{i,l}^-$ and positive real $c_{i,l}'$.

Since the right-hand side of (12) has the Boolean degree at most d, the Boolean degree of the new refutation is at most $k + d$.   □

Lemma 5. *Every static LS_+ proof P of F^M can be transformed into Positivstellensatz proof P' of it. If $Bdeg(P) = k$ and $Bdeg(f_i) \leq d$, then $Bdeg(P') \leq 2k + d$.*

Proof. We use ideas from the proof of Lemma 9.3, [4]. The refutation P can be represented in the form

$$\sum_{i=1}^{n+m} f_i \sum_l g_{i,l} + \sum_{j=1}^{n'} h_{0,j}^2 \cdot g_{m+n+1,j} + \sum_{j=n'+1}^{n''} g_{m+n+1,j} = -1 ,$$

where f_i, $1 \leq i \leq m$ are translations of Boolean clauses, $f_{m+i} = x_i^2 - x_i$, $1 \leq i \leq n$ and $g_{i,l} = c_{i,l} \cdot \prod_{k \in U_{i,l}^+} x_k \cdot \prod_{k \in U_{i,l}^-} (1 - x_k)$ for appropriate multisets of variables $U_{i,l}^+, U_{i,l}^-$, positive real $c_{i,l}$, and linear $h_{0,j}$.

Let us replace each occurrence of x_e in $g_{m+n+1,j}$ by $(x_e - x_e^2) + x_e^2 = -f_{m+e} + x_e^2$ and each occurrence of $1 - x_e$ by $(x_e - x_e^2) + (1 - x_e)^2 = -f_{m+e} + (1 - x_e)^2$, expand the factors obtained, gather all the terms containing at least one of f_i and the products of squares. As a result, we obtain *Positivstellensatz* proof P' of the form

$$\sum_{i=1}^{n+m} f_i g_i + \sum_{j=1}^{n'''} h_j^2 = -1 ,$$

for appropriate polynomials g_i, h_j. The Boolean degrees of all g_i, h_j are at most $2 \cdot Bdeg(g_{i,l})$(was obtained by substitutions of degree 2 polynomials for variables not in squares or f_i) and Boolean degrees of all f_i are at most d, so Boolean degree of P' is at most $2k + d$.   □

Next part of the reductions depends on Tseitin formula $T = T_G$ constructed according to graph $G = (V, E)$ and its representations as linear inequalities, equations and binomial system.

Lemma 6. *Every Positivstellensatz proof P of T_G^M can be transformed into a Positivstellensatz proof P' of T_G^B. The Boolean degree of P' is at most $\mathrm{Bdeg}(P) + d$.*

Proof. Assume the proof P is as follows:

$$\sum_{v, S_v} f_{v, S_v}^M \cdot g_{v, S_v} + \sum_{e \in E} (x_e^2 - x_e) \cdot g_e = 1 + \sum_j h_j^2 \ .$$

First of all, we replace each occurrence of x_e by $(1 - y_e)/2$. Note that the substitution transforms each $x_e^2 - x_e = 0$ into $(y_e^2 - 1)/4 = 0$, and each (9) into

$$\prod_{e \in S_v \setminus S_v'} \frac{1 + y_e}{2} \cdot \prod_{e \in S_v'} \frac{1 - y_e}{2} = 0 \ . \tag{14}$$

Due to Lemma 1 the Boolean degree of the new proof is at most $\mathrm{Bdeg}(P)$.

Next, we multiply (14) and (10) for $v \in V'$:

$$\prod_{e \in S_v \setminus S_v'} \frac{1 + y_e}{2} \prod_{e \in S_v'} \frac{1 - y_e}{2} (\prod_{e \ni v} y_e + 1) =$$

$$\prod_{e \in S_v \setminus S_v'} \frac{y_e + y_e^2}{2} \prod_{e \in S_v'} \frac{y_e - y_e^2}{2} + \prod_{e \in S_v \setminus S_v'} \frac{1 + y_e}{2} \prod_{e \in S_v'} \frac{1 - y_e}{2} =$$

$$\prod_{e \in S_v \setminus S_v'} \frac{y_e + 1}{2} \prod_{e \in S_v'} \frac{y_e - 1}{2} + \prod_{e \in S_v \setminus S_v'} \frac{1 + y_e}{2} \prod_{e \in S_v'} \frac{1 - y_e}{2} =$$

$$2 \cdot \prod_{e \in S_v \setminus S_v'} \frac{1 + y_e}{2} \prod_{e \in S_v'} \frac{1 - y_e}{2} \ .$$

The set S_v' has even cardinality, so $\prod_{e \in S_v'} (y_e - 1) = \prod_{e \in S_v'} (1 - y_e)$. A similar equality holds for $v \in V \setminus V'$.

Now we can write down the transformed proof P':

$$\sum_{v, S_v} (\prod_{e \ni v} y_e + 1) \cdot 2 \cdot f_{v, S_v'}'^M \cdot g_{v, S_v'}' + \sum_{e \in E} 2^{-2} \cdot (y_e^2 - 1) \cdot g_e' = 1 + \sum_j h_j'^2 \ ,$$

where the polynomials $f_{v, S_v'}'^M, g_e', h_j'^2$ are obtained from $f_{v, S_v'}^M, g_e, h_j^2$ by applying the substitution $x_i = (1 - y_e)/2$.

The Boolean degree of each equation (10) is at most d, hence $\mathrm{Bdeg}(P') \leq \mathrm{Bdeg}(P) + d$. $\qquad \square$

Theorem 1. *Every static LS_+ proof of the T_G^A can be transformed into a Positivstellensatz proof T_G^B. We can bound the Boolean degree of the new proof by $2k + 4d$, where k is the Boolean degree of the static LS_+ proof.*

Proof. Fix for the time being a static LS_+ proof P of (8) and apply Lemma 4 to obtain a static LS_+ proof P' of the equation system (9). Next, transform P' into a *Positivstellensatz* proof P'' of (9) by Lemma 5. Finally, due to Lemma 6 we can transform P'' into a *Positivstellensatz* proof P''' of system (10). The Boolean degree of P''' is at most $2k + 4d$. □

The following theorem originally was proved for degree but not Boolean degree. Nevertheless, the similar argument works for Boolean degree. We prove it in extended version of this paper [12].

Theorem 2 ([7], Corollary 1). *The Boolean degree of any Positivstellensatz refutation of the equation system T_G^B is at least $cr/2$.*

Corollary 1. *The Boolean degree of any static LS_+ refutation of Tseitin formula (7) with respect to (r, d, c)-expander G is at least ϵr, where $0 < \epsilon < 1$ depends only on G.*

Proof. Let P be a static LS_+ proof of the formula (7) represented as the system of linear inequalities (8), and Boolean degree of P is k. We apply Theorem 1 and transform it to into a *Positivstellensatz* proof P' of the equation system (10) extended by $y_e^2 - 1 = 0, e \in E$. The Boolean degree of P' is at most $2k + 4d$.

Theorem 2 implies that $2k + 4d \geq cr/2$, hence, there are such $0 < \epsilon < 1$ and $N \in \mathbb{Z}$, that for all $r \geq N$, $k \geq \epsilon r$. □

5   An Exponential Lower Bound on the Size of Static LS_+ Refutation of Tseitin Formulas

In this section we apply the results of previous section to obtain an exponential lower bound on the size of static LS_+ refutation of Tseitin formulas.

Lemma 7 ([4], Lemma 9.2). *Let M denote the number of $g_{i,l}$ in (6) that have Boolean degrees at least k and N denote the number of different variables in (6). Then there is a variable x and a value $a \in \{0, 1\}$ such that the result of substituting $x = a$ in (6) contains at most $M(1 - k/(2N))$ nonzero polynomials $g_{i,l}|_{x=a}$ of Boolean degrees at least k.*

The previous lemma appeared in [9], but as a separate statement was formulated and proved in [4].

By the definition Tseitin formula is a formula corresponding to the graph. The substitution to the formula variables corresponds to the removing of edges in the graph. In the following we will speak about Tseitin formulas in terms of graphs.

In Sect. 3 the operator Cl was defined for sets of edges. We extend it for use with partial assignments: $\overline{Cl}(\rho) = Cl(\{e \mid \rho(e) \text{ is set to 0 or 1}\})$. A substitution ρ is said to be *locally consistent* w.r.t. Tseitin formula T_G if and only if ρ can be extended to an assignment satisfying the subformula $F' = \{C|C$ is a clause of F and contains one of variables $y_{(v,u)}$, where v or u from $\overline{Cl}(\rho)\}$.

In the following theorem we use graphs with a positive expansion constant $c > 2$. For sufficiently large n there are such graphs of degree bounded by a constant (see, e. g. the proof in the Sect. 4 of [13] that for any d-regular graph $G = (V, E)$ and any subset of vertices $A \subseteq V$

$$\frac{|\partial A|}{|A|} \geq (d - \lambda_1)\frac{|V \setminus A|}{|V|} \; ,$$

where λ_1 is the second eigenvalue of G. If we consider only A with $|A| \leq |X|/2$, then $(d - \lambda_1)/2$ is an expander constant for G. Recall that a *Ramanujan graph* is a d-regular graph satisfying $\lambda_1 \leq 2\sqrt{d-1}$ and use the explicit construction of Ramanujan graphs, Sect. 5 of [13] or [14]).

The idea of the proof is following: using Lemma 7 we remove an exponential number of monomials from a static LS_+ proof of the Tseitin formula T_G and after that show that the proof is still nonempty. Unfortunately, using Lemma 7 directly we can obtain the proof of trivially unsatisfiable formula, so we need local consistency of the substitution.

Theorem 3. *Any static LS_+ proof of a Tseitin formula T_G with respect to a connected d-regular $(r = n/2, d, c)$-expander $G = (V, E)$ with n vertices and $c > 2$ has size $exp(\Omega(n))$.*

Proof. Let P be a static LS_+ proof of the T_G. We set $k = \lceil \epsilon n/5 \rceil$, where ϵ is from Corollary 1 for an $(r/2, d, c/2)$-expander.

We apply Lemma 7 repeatedly $\kappa = \lfloor \frac{cr}{13} \rfloor$ times. Before each application we remove some edges by a procedure that will be defined later. We define the sequence of graphs $G_0 = (V, E_0), G_1 = (V, E_1), \ldots, G_\kappa = (V, E_\kappa)$, where $E_0 = E \setminus B_0$, $E_{i+1} \subset E_i$, $E_i \setminus E_{i+1} = \{e_{i+1}\} \cup B_{i+1}, 0 \leq i \leq \kappa - 2$, $E_\kappa = E_{\kappa-1} \setminus \{e_\kappa\}$, edge e_i, $1 \leq i \leq \kappa$ corresponds to variable from the i-th application of Lemma 7 and B_i, $0 \leq i \leq \kappa - 1$ is a set (may be empty) of edges defined by $(i+1)$-th call of our procedure. We also need another sequence of graphs corresponding to all substitutions made by Lemma 7 and our procedure: $G = \widetilde{G}_0 = (V, \widetilde{E}_0), \widetilde{G}_1 = (V, \widetilde{E}_1), \ldots, \widetilde{G}_\kappa = (V, \widetilde{E}_\kappa)$, where graph $\widetilde{G}_i$ $(1 \leq i \leq \kappa)$ is obtained after i-th application of Lemma 7.

We call an edge e in a graph as a *bridge* if the removing of e from the graph split one of a connected component into two disconnected components. Now we describe our procedure $((i+1)$-th call) before the $(i+1)$-th application of Lemma 7 $(0 \leq i \leq \kappa - 1)$. We need a variable for current graph called as Y. The initial value of Y is $\widetilde{G}_i = (V, \widetilde{E}_i)$. Suppose that there is a bridge e (to exclude ambiguity let e be lexicographically first bridge) in the Y. Let the removing of e from Y split one of a connected component H into two disconnected components H_1 and H_2 (assume w.l.o.g. $|H_1| \leq |H_2|$). We substitute value of variable corresponding to e in such a way that a subformula associating with H_1 becomes satisfiable (it can be done since H_1 is connected). After that we remove the sub-formula corresponding to H_1 by a satisfying assignment of it. We also remove all assigned edges from Y. We repeat this procedure for other bridges if any in current graph. Let B_{i+1} be a set of all considered bridges, that we remove from $\widetilde{G}_i$ to obtain $\widetilde{G}_{i+1}$ after removing the edge e_{i+1}.

Informally speaking our procedure removes all edges from the small connected components in the graph $\tilde{G}_i$ but this edges are remained the graph G_i.

Suppose that G_{κ} contains l "satisfied" in a formula corresponding to $\tilde{G}_{\kappa}$ components: $H^{(1)}$, $H^{(2)}$, ..., $H^{(l)}$. Note that l is the number of substitutions to bridges by our procedure. By construction $|H^{(i)}| \leq \frac{n}{2} \leq r$. In the graph G, $|\partial H^{(i)}| \geq \lceil c|H^{(i)}|\rceil \geq 3$. Hence, the number of deleted edges in G_{κ} with respect to G is at least $\frac{3l}{2}$. Therefore $|E| - |E_{\kappa}| = \kappa + l \geq \frac{3l}{2}$, then $l \leq 2\kappa$ and $l + \kappa \leq 3\kappa \leq \frac{3cr}{13} < \frac{cr}{4}$. The size of each $H^{(i)}$ is less than $\frac{r}{2}$, otherwise in G we have $|\partial H^{(i)}| \geq \frac{cr}{2}$, but we deleted $l + \kappa < \frac{cr}{4}$ edges. So we have that $H^{(i)} \subseteq \overline{Cl}(\rho)$ (since $|\partial H^{(i)}| = 0$ in G_{κ}), where ρ is a substitution obtained by applications of Lemma 7 and by substitutions to all bridge variables.

Denote by τ a substitution ρ extended with satisfying assignments for all $H^{(i)}$ and by σ an extension of τ satisfying the subformula $F' = \{C|C$ is a clause of F and contains one of variables $y_{(v,u)}$, where v or u from $\overline{Cl}(\rho)\}$. The substitution σ exists since $\overline{Cl}(\rho)$ contains vertices from $H^{(i)}$, $1 \leq i \leq l$ and all subformulas corresponding to $H^{(i)}$ are satisfied by τ and a formula corresponding to strict subset of vertices of connected component not in $\overline{Cl}(\rho)$ is obviously satisfiable. Denote the result of substitution $P|_{\sigma}$ by P'.

Due to Lemma 2 and Lemma 3 the proof P' is a proof of the Tseitin formula with respect to a $(r/2, d, c/2)$-expander. By Corollary 1, the degree of P' is at least $\epsilon n/4 > k$.

Let M_0 denote the number of polynomials $g_{i,l}$ of degree at least k in P. Let us denote strictly positive constants $(1-\epsilon/(5d))$ by D and $\frac{c}{13}$ by C. By Lemma 7, the refutation P' contains at most $M_0(1-k/(2N))^{\kappa} \leq M_0 \cdot D^{Cn}$ nonzero polynomials $g'_{i,l}$ of degrees at least k. Since there is at least one polynomial $g'_{i,l}$ of such degree, we have $M_0 \cdot D^{Cn} \geq 1$, i.e., $M_0 \geq (1/D)^{Cn}$, which proves the theorem.   □

Corollary 2. *Any tree-like LS_+ refutation of (7) for a connected d-regular ($r = n/2, d, c$)-expander G with n vertices and $c > 2$ has size $exp(\Omega(n))$.*

Proof. We can easily simulate any tree-like LS_+ proof by a static LS_+ proof and apply Theorem 3 afterwards.   □

Acknowledgment

Authors are grateful to Dima Grigoriev, Edward A. Hirsch, Alexander S. Kulikov and Sergey I. Nikolenko for useful discussions and to anonymous referees for numerous comments that improved the quality of this paper.

References

1. Gomory, R.E.: An algorithm for integer solutions of linear programs. In Graves, R.L., Wolfe, P., eds.: Recent Advances in Mathematical Programming. McGraw-Hill (1963) 269–302
2. Chvátal, V.: Edmonds polytopes and a hierarchy of combinatorial problems. Discrete Mathematics 4 (1973) 305–337

3. Lovász, L., Schrijver, A.: Cones of matrices and set-functions and 0-1 optimization. SIAM J. Optimization **1**(2) (1991) 166–190
4. Grigoriev, D., Hirsch, E.A., Pasechnik, D.V.: Complexity of semialgebraic proofs. Moscow Mathematical Journal **2**(4) (2002) 647–679
5. Beame, P., Pitassi, T., Segerlind, N.: Lower Bounds for Lovász-Schrijver Systems and Beyond Follow from Multiparty Communication Complexity. In *ICALP'05* (2005) 1176–1188
6. Alekhnovich, M., Hirsch, E.A., Itsykson, D.: Exponential lower bounds for the running time of DPLL algorithms on satisfiable formulas. Technical Report 04-041, ECCC (2004)
7. Grigoriev, D.: Linear lower bound on degrees of Positivstellensatz Calculus proofs for the Parity. TCS **259** (2001) 613–622
8. Cook, S.A., Reckhow, R.A.: The relative efficiency of propositional proof systems. The Journal of Symbolic Logic **44**(1) (1979) 36–50
9. Clegg, M., Edmonds, J., Impagliazzo, R.: Using the Groebner basis algorithm to find proofs of unsatisfiability. In *STOC'96* (1996) 174–183
10. Grigoriev, D., Vorobjov, N.: Complexity of Null- and Positivstellensatz proofs. APAL **113**(1-3) (2001) 153–160
11. Alekhnovich, M., Razborov, A.: Lower bounds for polynomial calculus: Non-binomial case. In *FOCS'01* (1996) 190–199
12. Kojevnikov, A., Itsykson, D.: Lower Bounds of Static Lovász-Schrijver Calculus Proofs for Tseitin Tautologies. Manuscript in preparation (2006)
13. Murty, R.: Ramanujan graphs. Journal of the Ramanujan Math. Society **18**(1) (2003) 1–20
14. Lubotzky, A., Phillips, R., Sarnak, P.: Ramanujan graphs. Combinatorica **8**(3) (1988) 261–277

Extracting Kolmogorov Complexity with Applications to Dimension Zero-One Laws

Lance Fortnow[1], John M. Hitchcock[2,*], A. Pavan[3,**],
N.V. Vinodchandran[4,***], and Fengming Wang[3,†]

[1] Department of Computer Science, University of Chicago
fortnow@cs.uchicago.edu
[2] Department of Computer Science, University of Wyoming
jhitchco@cs.uwyo.edu
[3] Department of Computer Science, Iowa State University
{pavan, wfengm}@cs.iastate.edu
[4] Department of Computer Science and Engineering, University of Nebraska-Lincoln
vinod@cse.unl.edu

Abstract. We apply recent results on extracting randomness from independent sources to "extract" Kolmogorov complexity. For any $\alpha, \epsilon > 0$, given a string x with $K(x) > \alpha|x|$, we show how to use a constant number of advice bits to efficiently compute another string y, $|y| = \Omega(|x|)$, with $K(y) > (1-\epsilon)|y|$. This result holds for both classical and space-bounded Kolmogorov complexity.

We use the extraction procedure for space-bounded complexity to establish zero-one laws for polynomial-space strong dimension. Our results include:

(i) If $\text{Dim}_{\text{pspace}}(E) > 0$, then $\text{Dim}_{\text{pspace}}(E/O(1)) = 1$.
(ii) $\text{Dim}(E/O(1) \mid \text{ESPACE})$ is either 0 or 1.
(iii) $\text{Dim}(E/\text{poly} \mid \text{ESPACE})$ is either 0 or 1.

In other words, from a dimension standpoint and with respect to a small amount of advice, the exponential-time class E is either minimally complex or maximally complex within ESPACE.

1 Introduction

Kolmogorov complexity quantifies the amount of randomness in an individual string. If a string x has Kolmogorov complexity m, then x is often said to contain m bits of randomness. Given x, is it possible to compute a string of length m that is Kolmogorov-random? In general this is impossible but we do make progress in this direction if we allow a tiny amount of extra information. We give a *polynomial-time computable procedure* which takes x with an additional constant amount of advice and outputs a nearly Kolmogorov-random string whose length

---

* This research was supported in part by NSF grant 0515313.
** This research was supported in part by NSF grant 0430807.
*** This research was supported in part by NSF grant 0430991.
† This research was supported in part by NSF grant 0430807.

is linear in m. Formally, for any $\alpha, \epsilon > 0$, given a string x with $K(x) > \alpha|x|$, we show how to use a constant number of advice bits to compute another string y, $|y| = \Omega(|x|)$, in polynomial-time that satisfies $K(y) > (1 - \epsilon)|y|$. The number of advice bits depends only on α and ϵ, but the content of the advice depends on x. This computation needs only polynomial time, and yet it extracts unbounded Kolmogorov complexity.

Our proofs use a recent construction of extractors using multiple independent sources. Traditional extractor results [13,22,19,12,21,15,16,20,9,18,17,4] show how to take a distribution with high min-entropy and some truly random bits to create a close to uniform distribution. Recently, Barak, Impagliazzo, and Wigderson [2] showed how to eliminate the need for a truly random source when several independent random sources are available. We make use of these extractors for our main result on extracting Kolmogorov complexity. Barak et al. [3] and Raz [14] have further extensions on extracting from independent sources.

To make the connection, consider the uniform distribution on the set of strings x whose Kolmogorov complexity is at most m. This distribution has min-entropy about m and x acts like a random member of this set. We can define a set of strings $x_1, \ldots, x_k$ to be independent if $K(x_1 \ldots x_k) \approx K(x_1) + \cdots + K(x_k)$. By symmetry of information this implies $K(x_i|x_1, \ldots, x_{i-1}, x_{i+1}, \ldots, x_k) \approx K(x_i)$. Combining these ideas we are able to apply the extractor constructions for multiple independent sources to Kolmogorov complexity.

To extract the randomness from a string x, we break x into a number of substrings $x_1, \ldots, x_l$, and view each substring x_i as coming from an independent random source. Of course, these substrings may not be independently random in the Kolmogorov sense. We find it a useful concept to quantify the *dependency within* x as $\sum_{i=1}^{l} K(x_i) - K(x)$. Another technical problem is that the randomness in x may not be nicely distributed among these substrings; for this we need to use a small (constant) number of nonuniform advice bits.

This result about extracting Kolmogorov-randomness also holds for polynomial-space bounded Kolmogorov complexity. We apply this to obtain zero-one laws for the dimensions of certain complexity classes. Polynomial-space dimension [11] and strong dimension [1] have been developed to study the quantitative structure of classes that lie in E and ESPACE. These dimensions are resource-bounded versions of Hausdorff dimension and packing dimension, respectively, the two most important fractal dimensions. Polynomial-space dimension and strong dimension refine pspace-measure [10] and have been shown to be duals of each other in many ways [1]. Additionally, polynomial-space strong dimension is closely related to pspace-category [7]. In this paper we focus on polynomial-space strong dimension which quantifies PSPACE and ESPACE in the following way:

- $\mathrm{Dim}_{\mathrm{pspace}}(\mathrm{PSPACE}) = 0$.
- $\mathrm{Dim}_{\mathrm{pspace}}(\mathrm{ESPACE}) = 1$.

It is interesting to consider the dimension of a complexity class $\mathcal{C}$, where $\mathcal{C}$ is contained in ESPACE. The dimension is always a real number between zero and one inclusive. Can a reasonable complexity class have a fractional dimension? In particular consider the class E. Deciding the polynomial-space dimension of

E would imply a major complexity separation, but perhaps we can show that E must have dimension either zero or one, a "zero-one" law for dimension.

We can show such a zero-one law if we add a small amount of nonuniform advice. An equivalence between space-bounded Kolmogorov complexity rates and strong pspace-dimension allows us to use our Kolmogorov-randomness extraction procedure to show the following results.

(i) If $\text{Dim}_{\text{pspace}}(E) > 0$, then $\text{Dim}_{\text{pspace}}(E/O(1)) = 1$.
(ii) $\text{Dim}(E/O(1) \mid \text{ESPACE})$ is either 0 or 1.
(iii) $\text{Dim}(E/\text{poly} \mid \text{ESPACE})$ is either 0 or 1.

2 Preliminaries

2.1 Kolmogorov Complexity

Let M be a Turing machine. Let $f : \mathbb{N} \to \mathbb{N}$. For any $x \in \{0,1\}^*$, define

$$K_M(x) = \min\{|\pi| \mid M(\pi) \text{ prints } x\}$$

and

$$KS_M^f(x) = \min\{|\pi| \mid M(\pi) \text{ prints } x \text{ using at most } f(|x|) \text{ space}\}.$$

There is a universal machine U such that for every machine M, there is some constant c such that for all x, $K_U(x) \leq K_M(x) + c$ and $KS_U^{cf+c}(x) \leq KS_M^f(x) + c$ [8]. We fix such a machine U and drop the subscript, writing $K(x)$ and $KS^f(x)$, which are called the *(plain) Kolmogorov complexity of* x and f-*bounded (plain) Kolmogorov complexity of* x. While we use plain complexity in this paper, our results also hold for prefix-free complexity.

The following definition quantifies the fraction of randomness in a string.

Definition. For a string x, the *rate* of x is $rate(x) = K(x)/|x|$. For a polynomial g, the g-*rate of* x is $rate^g(x) = KS^g(x)/|x|$.

2.2 Polynomial-Space Dimension

We now review the definitions of polynomial-space dimension [11] and strong dimension [1]. For more background we refer to these papers and the recent survey paper [6].

Let $s > 0$. An s-*gale* is a function $d : \{0,1\}^* \to [0,\infty)$ satisfying $2^s d(w) = d(w0) + d(w1)$ for all $w \in \{0,1\}^*$.

For a language A, we write $A \upharpoonright n$ for the first n bits of A's characteristic sequence (according to the standard enumeration of $\{0,1\}^*$) and $A \upharpoonright [i,j]$ for the subsequence beginning from the ith bit and ending at the jth bit. An s-gale d *succeeds on* a language A if $\limsup_{n \to \infty} d(A \upharpoonright n) = \infty$ and d *succeeds strongly on* A if $\liminf_{n \to \infty} d(A \upharpoonright n) = \infty$. The *success set* of d is $S^\infty[d] = \{A \mid d \text{ succeeds on } S\}$. The *strong success set* of d is $S_{\text{str}}^\infty[d] = \{A \mid d \text{ succeeds strongly on } S\}$.

Definition. Let X be a class of languages.

1. The pspace-*dimension* of X is

$$\dim_{\text{pspace}}(X) = \inf \left\{ s \,\middle|\, \begin{array}{l} \text{there is a polynomial-space computable} \\ s\text{-gale } d \text{ such that } X \subseteq S^{\infty}[d] \end{array} \right\}.$$

2. The *strong* pspace-*dimension* of X is

$$\text{Dim}_{\text{pspace}}(X) = \inf \left\{ s \,\middle|\, \begin{array}{l} \text{there is a polynomial-space computable} \\ s\text{-gale } d \text{ such that } X \subseteq S^{\infty}_{\text{str}}[d] \end{array} \right\}.$$

For every X, $0 \leq \dim_{\text{pspace}}(X) \leq \text{Dim}_{\text{pspace}}(X) \leq 1$. An important fact is that ESPACE has pspace-dimension 1, which suggests the following definitions.

Definition. Let X be a class of languages.

1. The *dimension of X within* ESPACE is

$$\dim(X \mid \text{ESPACE}) = \dim_{\text{pspace}}(X \cap \text{ESPACE}).$$

2. The *strong dimension of X within* ESPACE is

$$\text{Dim}(X \mid \text{ESPACE}) = \text{Dim}_{\text{pspace}}(X \cap \text{ESPACE}).$$

In this paper we will use an equivalent definition of the above dimensions in terms of space-bounded Kolmogorov complexity.

Definition. Given a language L and a polynomial g the *g-rate of L* is

$$rate^g(L) = \liminf_{n \to \infty} rate^g(L \restriction n).$$

strong g-rate of L is

$$Rate^g(L) = \limsup_{n \to \infty} rate^g(L \restriction n).$$

Theorem 2.1. (Hitchcock [5]) *Let* poly *denote all polynomials. For every class X of languages,*

$$\dim_{\text{pspace}}(X) = \inf_{g \in \text{poly}} \sup_{L \in X} rate^g(L).$$

and

$$\text{Dim}_{\text{pspace}}(X) = \inf_{g \in \text{poly}} \sup_{L \in X} Rate^g(L).$$

3   Extracting Kolmogorov Complexity

Barak, Impagliazzo, and Wigderson [2] recently gave an explicit multi-source extractor.

Theorem 3.1. *([2]) For every constant $0 < \sigma < 1$, and $c > 1$ there exist $l = poly(1/\sigma, c)$, a constant r and a computable function $E : \Sigma^{\ell n} \to \Sigma^n$ such that if $H_1, \cdots, H_l$ are independent distributions over Σ^n, each with min entropy at least σn, then $E(H_1, \cdots, H_l)$ is 2^{-cn}-close to U_n, where U_n is the uniform distribution over Σ^n. Moreover, E runs in time n^r.*

We show that the above extractor can be used to produce nearly Kolmogorov-random strings from strings with high enough complexity. The following notion of dependency is useful for quantifying the performance of the extractor.

Definition. Let $x = x_1 x_2 \cdots x_k$, where each x_i is an n-bit string. The *dependency within x*, $dep(x)$, is defined as $\sum_{i=1}^{k} K(x_i) - K(x)$.

Theorem 3.2. *For every $0 < \sigma < 1$ and large enough n, there exist a constant $l > 1$, and a polynomial-time computable function E such that if $x_1, x_2, \cdots x_l$ are n-bit strings with $K(x_i) \geq \sigma n$, $1 \leq i \leq l$, then*

$$K(E(x_1, \cdots, x_l)) \geq n - 10l \log n - dep(x).$$

Proof. Let $0 < \sigma' < \sigma$. By Theorem 3.1, there is a constant l and a polynomial-time computable multi-source extractor E such that if $H_1, \cdots, H_l$ are independent sources each with min-entropy at least $\sigma' n$, then $E(H_1, \cdots, H_l)$ is 2^{-5n} close to U_n.

We show that this extractor also extracts Kolmogorov complexity. We prove by contradiction. Suppose the conclusion is false, i.e,

$$K(E(x_1, \cdots x_l)) < n - 10l \log n - dep(x).$$

Let $K(x_i) = m_i$, $1 \leq i \leq l$. Define the following sets:

$$I_i = \{y \mid y \in \Sigma^n, K(y) \leq m_i\},$$

$$Z = \{z \in \Sigma^n \mid K(z) < n - 10l \log n - dep(x)\},$$

$$Small = \{\langle y_1, \cdots, y_l \rangle \mid y_i \in I_i, \text{ and } E(y_1, \cdots y_l) \in Z\}.$$

By our assumption $\langle x_1, \cdots x_l \rangle$ belongs to *Small*. We use this to arrive at a contradiction regarding the Kolmogorov complexity of $x = x_1 x_2 \cdots x_l$. We first calculate an upper bound on the size of *Small*.

Observe that the set $\{xy \mid x \in \Sigma^{\sigma' n}, y = 0^{n-\sigma' n}\}$ is a subset of each of I_i. Thus the cardinality of each of I_i is at least $2^{\sigma' n}$. Let H_i be the uniform distribution on I_i. Thus the min-entropy of H_i is at least $\sigma' n$.

Since H_i's have min-entropy at least $\sigma' n$, $E(H_1, \cdots, H_l)$ is 2^{-5n}-close to U_n. Then

$$\left| P[E(H_1, \ldots, H_l) \in Z] - P[U_n \in Z] \right| \leq 2^{-5n}. \tag{3.1}$$

Note that the cardinality of I_i is at most 2^{m_i+1}, as there are at most 2^{m_i+1} strings with Kolmogorov complexity at most m_i. Thus H_i places a weight of at least 2^{-m_i-1} on each string from I_i. Thus $H_1 \times \cdots \times H_l$ places a weight of at least $2^{-(m_1+\cdots+m_l+l)}$ on each element of $Small$. Therefore,

$$P[E(H_1,\ldots,H_l) \in Z] = P[(H_1,\ldots,H_l) \in Small] \geq |Small| \cdot 2^{-(m_1+\cdots+m_l+l)},$$

and since $|Z| \leq 2^{n-10l \log n - dep(x)}$, from (3.1) we obtain

$$|Small| < 2^{m_1+1} \times \cdots \times 2^{m_l+1} \times \left(\frac{2^{n-10l \log n - dep(x)}}{2^n} + 2^{-5n} \right)$$

Without loss of generality we can take $dep(x) < n$, otherwise the theorem is trivially true. Thus $2^{-5n} < 2^{-10l \log n - dep(x)}$. Using this and the fact that l is a constant independent of n, we obtain

$$|Small| < 2^{m_1+\cdots+m_l-dep(x)-8l \log n},$$

when n is large enough. Since $K(x) = K(x_1) + \cdots + K(x_l) - dep(x)$,

$$|Small| < 2^{K(x)-8l \log n}.$$

We first observe that there is a program Q that, given the values of m_i's, n, l, and $dep(x)$ as auxiliary inputs, recognizes the set $Small$. This program works as follows: Let $z = z_1 \cdots z_l$, where $|z_i| = n$. For each program P_i of length at most m_i check whether P_i outputs z_i, by running the P_i's in a dovetail fashion. If it is discovered that for each of z_i, $K(z_i) \leq m_i$, then compute $y = E(z_1, \cdots, z_l)$. Now verify that $K(y)$ is at most $n - dep(x) - 10l \log n$. This again can be done by running programs of the length at most $n - dep(x) - 10l \log n$ in a dovetail manner. If it is discovered that $K(y)$ is at most $n - dep(x) - 10l \log n$, then accept z.

So given the values of parameters n, $dep(x)$, l and m_is, there is a program P that enumerates all elements of $Small$. Since by our assumption x belongs to $Small$, x appears in this enumeration. Let i be the position of x in this enumeration. Since $|Small|$ is at most $2^{K(x)-8l \log n}$, i can be described using $K(x) - 8l \log n$ bits.

Thus there is a program P' based on P that outputs x. This program takes i, $dep(x)$, n, $m_1, \cdots, m_l$, and l, as auxiliary inputs. Since the m_i's and $dep(x)$ are bounded by n,

$$K(x) \leq K(x) - 8l \log n + 2 \log n + l \log n + O(1)$$
$$\leq K(x) - 5l \log n + O(1),$$

which is a contradiction. □

If $x_1, \cdots x_l$ are independent strings with $K(x_i) \geq \sigma n$, then $E(x_1, \cdots, x_l)$ is a Kolmogorov random string of length n.

Corollary 3.3. *For every constant $0 < \sigma < 1$, there exists a constant l, and a polynomial-time computable function E such that if $x_1, \cdots x_l$ are n-bit strings such $K(x_i) \geq \sigma n$, and $K(x_1 x_2 \cdots x_l) = \sum K(x_i) - O(\log n)$, then $E(x_1, \cdots, x_l)$ is Kolmogorov random, i.e.,*

$$K(E(x_1, \cdots, x_l)) > n - O(\log n).$$

This theorem says that given $x \in \Sigma^{ln}$, if each piece x_i has high enough complexity and the dependency with x is small, then we can output a string y whose Kolmogorov rate is higher than the Kolmogorov rate of x, i.e, y is relatively more random than x. What if we only knew that x has high enough complexity but knew nothing about the complexity of individual pieces or the dependency within x? Our next theorem states that in this case also there is a procedure producing a string whose rate is higher than the rate of x. However, this procedure needs constant bits of advice.

Theorem 3.4. *For all real numbers $0 < \alpha < \beta < 1$ there exist a constant $0 < \gamma < 1$, constants $c, l, n_0 \geq 1$, and a procedure R such that the following holds. For any string x with $|x| \geq n_0$ and $rate(x) \geq \alpha$, there exists an advice string a_x such that*

$$rate(R(x, a_x)) \geq \min\{rate(x) + \gamma, \beta\}$$

where $|a_x| = c$. Moreover, R runs in polynomial time, and $|R(x, a_x)| = \lfloor |x|/l \rfloor$.
The number c depends only on α, β and is independent of x. However, the contents of a_x depend on x.

Proof. Let $\alpha' < \alpha$ and $\epsilon < \min\{1 - \beta, \alpha'\}$. Let $\sigma = (1 - \epsilon)\alpha'$. Using parameter σ in Theorem 3.2, we obtain a constant $l > 1$ and a polynomial-time computable function E that extracts Kolmogorov complexity.

Let $\beta' = 1 - \frac{\epsilon}{2}$, and $\gamma = \frac{\epsilon^2}{2l}$. Observe that $\gamma \leq \frac{1 - \beta'}{l}$ and $\gamma < \frac{\alpha' - \sigma}{l}$.

Let x have $rate(x) = \nu \geq \alpha$. Let $n, k \geq 0$ such that $|x| = ln + k$ and $k < l$. We strip the last k bits from x and write $x = x_1 \cdots x_l$ where each $|x_i| = n$. Let $\nu' = rate(x)$ after this change. We have $\nu' > \nu - \gamma/2$ and $\nu' > \alpha'$ if $|x|$ is sufficiently large.

We consider three cases.

Case 1. There exists j, $1 \leq j \leq l$ such that $K(x_j) < \sigma n$.
Case 2. Case 1 does not hold and $dep(x) \geq \gamma ln$.
Case 3. Case 1 does not hold and $dep(x) < \gamma ln$.

We have two claims about Cases 1 and 2:

Claim 3.5. *Assume Case 1 holds. There exists i, $1 \leq i \leq l$, such that $rate(x_i) \geq \nu' + \gamma$.*

Proof of Claim 3.5. Suppose not. Then for every $i \neq j$, $1 \leq i \leq l$, $K(x_i) \leq (\nu' + \gamma)n$. We can describe x by describing x_j which takes σn bits, and all the x_i's, $i \neq j$. Thus the total complexity of x would be at most

$$(\nu' + \gamma)(l - 1)n + \sigma n + O(\log n)$$

Since $\gamma < \frac{\alpha'-\sigma}{l}$ and $\alpha' < \nu'$ this quantity is less than $\nu'ln$. Since the rate of x is ν', this is a contradiction. $\square$ *Claim 3.5*

Claim 3.6. Assume Case 2 holds. There exists i, $1 \le i \le l$, $rate(x_i) \ge \nu' + \gamma$.

Proof of Claim 3.6. By definition,

$$K(x) = \sum_{i=1}^{l} K(x_i) - dep(x)$$

Since $dep(x) \ge \gamma ln$ and $K(x) \ge \nu' ln$,

$$\sum_{i=1}^{l} K(x_i) \ge (\nu' + \gamma)ln.$$

Thus there exists i such that $rate(x_i) \ge \nu' + \gamma$. $\square$ *Claim 3.6*

We can now describe the constant number of advice bits. The advice a_x contains the following information: which of the three cases described above holds, and

- If Case 1 holds, then from Claim 3.5 the index i such that $rate(x_i) \ge \nu' + \gamma$.
- If Case 2 holds, then from Claim 3.6 the index i such that $rate(x_i) \ge \nu' + \gamma$.

Since $1 \le i \le l$, the number of advice bits is bounded by $O(\log l)$. We now describe procedure R. When R takes an input x, it first examines the advice a_x. If Case 1 or Case 2 holds, then R simply outputs x_i. Otherwise, Case 3 holds, and R outputs $E(x)$. Since E runs in polynomial time, R runs in polynomial time.

If Case 1 or Case 2 holds, then

$$rate(R(x, a_x)) \ge \nu' + \gamma \ge \nu + \tfrac{\gamma}{2}.$$

If Case 3 holds, we have $R(x, a_x) = E(x)$ and by Theorem 3.2, $K(E(x)) \ge n - 10 \log n - \gamma ln$. Since $\gamma \le \frac{1-\beta'}{l}$, in this case

$$rate(R(x, a_x)) \ge \beta' - \tfrac{10 \log n}{n}.$$

For large enough n, this value is at least β. Therefore in all three cases, the rate increases by at least $\gamma/2$ or reaches β. $\square$

We now prove our main theorem.

Theorem 3.7. *Let α and β be constants with $0 < \alpha < \beta < 1$. There exist a polynomial-time procedure $P(\cdot, \cdot)$ and constants C_1, C_2, n_1 such that for every x with $|x| \ge n_1$ and $rate(x) \ge \alpha$ there exists a string a_x with $|a_x| = C_1$ such that*

$$rate(P(x, a_x)) \ge \beta$$

and $|P(x, a_x)| \ge |x|/C_2$.

Proof. We apply the procedure R from Theorem 3.4 iteratively. Each application of R outputs a string whose rate is at least β or is at least γ more than the rate of the input string. Applying R at most $k = \lceil (\beta - \alpha)/\gamma \rceil$ times, we obtain a string whose rate is at least β.

Note that $R(y, a_y)$ has output length $|R(y, a_y)| = \lfloor |y|/l \rfloor$ and increases the rate of y if $|y| \geq n_0$. If we take $n_1 = (n_0+1)kl$, we ensure that in each application of R we have a string whose length is at least n_0. Each iteration of R requires c bits of advice, so the total number of advice bits needed is $C_1 = kc$. Thus C_1 depends only on α and β. Each application of R decreases the length by a constant fraction, so there is a constant C_2 such that the length of the final outputs string is at least $|x|/C_2$. $\qquad\square$

The proofs in this section also work for space-bounded Kolmogorov complexity. For this we need a space-bounded version of dependency.

Definition. Let $x = x_1 x_2 \cdots x_k$ where each x_i is an n-bit string, let f and g be two space bounds. The (f, g)-*bounded dependency within* x, $dep_g^f(x)$, is defined as $\sum_{i=1}^k KS^g(x_i) - KS^f(x)$.

We obtain the following version of Theorem 3.2.

Theorem 3.8. *For every polynomial g there exists a polynomial f such that for every $0 < \sigma < 1$, there exist a constant $l > 1$, and a polynomial-time computable function E such that if $x_1, \cdots, x_l$ are n-bit strings with $KS^f(x_i) \geq \sigma n$, $1 \leq i \leq l$, then*

$$KS^g(E(x_1, \cdots, x_l)) \geq n - 10l \log n - dep_g^f(x).$$

Similarly we obtain the following extension of Theorem 3.7.

Theorem 3.9. *Let g be a polynomial and let α and β be constants with $0 < \alpha < \beta < 1$. There exist a polynomial f, polynomial-time procedure $R(\cdot, \cdot)$, and constants C_1, C_2, n_1 such that for every x with $|x| \geq n_1$ and $rate^f(x) \geq \alpha$ there exists a string a_x with $|a_x| = C_1$ such that*

$$rate^g(R(x, a_x)) \geq \beta$$

and $|R(x, a_x)| \geq |x|/C_2$.

4   Zero-One Laws

In this section we establish zero-one laws for the dimensions of certain classes within ESPACE. Our most basic result is the following, which says that if E has positive dimension, then the class $E/O(1)$ has maximal dimension.

Theorem 4.1. *If* $\mathrm{Dim}_{\mathrm{pspace}}(\mathrm{E}) > 0$, *then* $\mathrm{Dim}_{\mathrm{pspace}}(\mathrm{E}/O(1)) = 1$.

For the theorem we use the following lemma, which can be proved using Theorem 3.9. We omit the proof due to space constraints.

Lemma 4.2. *Let g be any polynomial and α, θ be rational numbers with $0 < \alpha < \theta < 1$. Then there is a polynomial f such that if there exists $L \in E$ with $Rate^f(L) \geq \alpha$, then there exists $L' \in E/O(1)$ with $Rate^g(L') \geq \theta$.*

Proof of Theorem 4.1. We will show that for every polynomial g, and real number $0 < \theta < 1$, there is a language L' in $E/O(1)$ with $Rate^g(L) \geq \theta$. By Theorem 2.1, this will show that the strong pspace-dimension of $E/O(1)$ is 1.

The assumption states that the strong pspace-dimension of E is greater than 0. If the strong pspace-dimension of E is actually one, then we are done. If not, let α be a positive rational number that is less than $\text{Dim}_{\text{pspace}}(E)$. By Theorem 2.1, for every polynomial f, there exists a language $L \in E$ with $Rate^f(L) \geq \alpha$.

By Lemma 4.2, from such a language L we obtain a language L' in $E/O(1)$ with $Rate^g(L') \geq \theta$. Thus the strong pspace-dimension of $E/O(1)$ is 1.   □

Observe that in the above construction, if the original language L is in $E/O(1)$, then also L' is in $E/O(1)$, and similarly membership in E/poly is preserved. Additionally, if $L \in$ ESPACE, it can be shown that $L' \in$ ESPACE. With these observations, we obtain the following zero-one laws.

Theorem 4.3. *Each of the following is either 0 or 1.*

1. $\text{Dim}_{\text{pspace}}(E/O(1))$.
2. $\text{Dim}_{\text{pspace}}(E/\text{poly})$.
3. $\text{Dim}(E/O(1) \mid \text{ESPACE})$.
4. $\text{Dim}(E/\text{poly} \mid \text{ESPACE})$.

We remark that in Theorems 4.1 and 4.3, if we replace E by EXP, the theorems still hold. The proofs also go through for other classes such as BPEXP, NEXP $\cap$ coNEXP, and NEXP/poly.

Theorems 4.1 and 4.3 concern strong dimension. For dimension, the situation is more complicated. Using similar techniques, we can prove that if $\dim_{\text{pspace}}(E) > 0$, then $\dim_{\text{pspace}}(E/O(1)) \geq 1/2$. Analogously, we can obtain zero-half laws for the pspace-dimension of E/poly, etc.

Acknowledgments

We thank Xiaoyang Gu and Philippe Moser for several helpful discussions.

References

1. K. B. Athreya, J. M. Hitchcock, J. H. Lutz, and E. Mayordomo. Effective strong dimension in algorithmic information and computational complexity. *SIAM Journal on Computing*. To appear.
2. B. Barak, R. Impagliazzo, and A. Wigderson. Extracting randomness using few independent sources. In *Proceedings of the 45th Annual IEEE Symposium on Foundations of Computer Science*, pages 384–393. IEEE Computer Society, 2004.

3. B. Barak, G. Kindler, R. Shaltiel, B. Sudakov, and A. Wigderson. Simulating independence: new constructions of condensers, ramsey graphs, dispersers, and extractors. In *Proceedings of the 37th ACM Symposium on Theory of Computing*, pages 1–10, 2005.

4. B. Chor and O. Goldreich. Unbiased bits from sources of weak randomness and probabilistic communication complexity. In *Proceedings of the 26th Annual IEEE Conference on Foundations of Computer Science*, pages 429–442, 1985.

5. J. M. Hitchcock. *Effective Fractal Dimension: Foundations and Applications*. PhD thesis, Iowa State University, 2003.

6. J. M. Hitchcock, J. H. Lutz, and E. Mayordomo. The fractal geometry of complexity classes. *SIGACT News*, 36(3):24–38, September 2005.

7. J. M. Hitchcock and A. Pavan. Resource-bounded strong dimension versus resource-bounded category. *Information Processing Letters*, 95(3):377–381, 2005.

8. M. Li and P. M. B. Vitányi. *An Introduction to Kolmogorov Complexity and its Applications*. Springer-Verlag, Berlin, 1997. Second Edition.

9. C-J. Lu, O. Reingold, S. Vadhan, and A. Wigderson. Extractors: Optimal up to a constant factor. In *Proceedings of the 35th Annual ACM Symposium on Theory of Computing*, pages 602–611, 2003.

10. J. H. Lutz. Almost everywhere high nonuniform complexity. *Journal of Computer and System Sciences*, 44(2):220–258, 1992.

11. J. H. Lutz. Dimension in complexity classes. *SIAM Journal on Computing*, 32(5):1236–1259, 2003.

12. N. Nisan and A. Ta-Shma. Extracting randomness: A survey and new constructions. *Journal of Computer and System Sciences*, 42(2):149–167, 1999.

13. N. Nisan and D. Zuckerman. Randomness is linear in space. *Journal of Computer and System Sciences*, 52(1):43–52, 1996.

14. R. Raz. Extractors with weak random seeds. In *Proceedings of the 37th ACM Symposium on Theory of Computing*, pages 11–20, 2005.

15. O. Reingold, R. Shaltiel, and A. Wigderson. Extracting randomness via repeated condensing. In *Proceedings of the 41st Annual Conference on Foundations of Computer science*, 2000.

16. O. Reingold, S. Vadhan, and A. Wigderson. Entropy waves, the zig-zag graph product, and new constant-degree expanders and extractors. In *Proceedings of the 41st Annual IEEE Conference on Foundations of Computer Science*, 2000.

17. M. Santha and U. Vazirani. Generating quasi-random sequences from slightly random sources. In *Proceedings of the 25th Annual IEEE Conference on Foundations of Computer Science*, pages 434–440, 1984.

18. R. Shaltiel and C. Umans. Simple extractors for all min-entropies and a new pseudo-random generator. In *Proceedings of the 42nd Annual Conference on Foundations of Computer Science*, 2001.

19. A. Srinivasan and D. Zuckerman. Computing with very weak random sources. *SIAM Journal on Computing*, 28(4):1433–1459, 1999.

20. A. Ta-Shma, D. Zuckerman, and M. Safra. Extractors from reed-muller codes. In *Proceedings of the 42nd Annual Conference on Foundations of Computer Science*, 2001.

21. L. Trevisan. Extractors and pseudorandom generators. *Journal of the ACM*, 48(1):860–879, 2001.

22. D. Zuckerman. Randomness-optimal oblivious sampling. *Random Structures and Algorithms*, 11:345–367, 1997.

The Connectivity of Boolean Satisfiability: Computational and Structural Dichotomies

Parikshit Gopalan[1], Phokion G. Kolaitis[2,*],
Elitza N. Maneva[3], and Christos H. Papadimitriou[3]

[1] Georgia Tech
[2] IBM Almaden Research Center
[3] UC Berkeley

Abstract. Given a Boolean formula, do its solutions form a connected subgraph of the hypercube? This and other related connectivity considerations underlie recent work on random Boolean satisfiability. We study connectivity properties of the space of solutions of Boolean formulas, and establish computational and structural dichotomies. Specifically, we first establish a dichotomy theorem for the complexity of the st-connectivity problem for Boolean formulas in Schaefer's framework. Our result asserts that the tractable side is more generous than the tractable side in Schaefer's dichotomy theorem for satisfiability, while the intractable side is PSPACE-complete. For the connectivity problem, we establish a dichotomy along the same boundary between membership in coNP and PSPACE-completeness. Furthermore, we establish a structural dichotomy theorem for the diameter of the connected components of the solution space: for the PSPACE-complete cases, the diameter can be exponential, but in all other cases it is linear. Thus, small diameter and tractability of the st-connectivity problem are remarkably aligned.

1 Introduction

In 1978, T.J. Schaefer [1] introduced a rich framework for expressing variants of Boolean satisfiability and proved a remarkable *dichotomy theorem*: the satisfiability problem is in P for certain classes of Boolean formulas, while it is NP-complete for all other classes in the framework. In a single stroke, this result pinpoints the computational complexity of all well-known variants of SAT, such as 3-SAT, HORN 3-SAT, NOT-ALL-EQUAL 3-SAT, and 1-IN-3 SAT. Schaefer's work paved the way for a series of investigations establishing dichotomies for several aspects of satisfiability, including optimization [2,3,4], counting [5], inverse satisfiability [6], minimal satisfiability [7], 3-valued satisfiability [8] and propositional abduction [9].

Our aim in this paper is to carry out a comprehensive exploration of a different aspect of Boolean satisfiability, namely, the connectivity properties of the space of solutions of Boolean formulas. The solutions (satisfying assignments) of a given n-variable Boolean formula φ induce a subgraph $G(\varphi)$ of the n-dimensional

---

* On leave from UC Santa Cruz.

M. Bugliesi et al. (Eds.): ICALP 2006, Part I, LNCS 4051, pp. 346–357, 2006.

hypercube. Thus, the following two decision problems, called the *connectivity problem* and the *st-connectivity problem*, arise naturally: (i) Given a Boolean formula φ, is $G(\varphi)$ connected? (ii) Given a Boolean formula φ and two solutions s and t of φ, is there a path from s to t in $G(\varphi)$?

We believe that connectivity properties of Boolean satisfiability merit study in their own right, as they shed light on the structure of the solution space of Boolean formulas. Moreover, in recent years the structure of the space of solutions for random instances has been the main consideration at the basis of both algorithms for and mathematical analysis of the satisfiability problem [10,11,12,13]. It has been conjectured for 3-SAT [12] and proved for 8-SAT [14,15], that the solution space fractures as one approaches the *critical region* from below. This apparently leads to performance deterioration of the standard satisfiability algorithms, such as WalkSAT [16] and DPLL [17]. It is also the main consideration behind the design of the survey propagation algorithm, which has far superior performance on random instances of satisfiability [12]. This body of work has served as a motivation to us for pursuing the investigation reported here. While there has been an intensive study of the structure of the solution space of Boolean satisfiability problems for random instances, our work seems to be the first to explore this issue from a worst-case viewpoint.

Our first main result is a dichotomy theorem for the *st*-connectivity problem. This result reveals that the tractable side is much more generous than the tractable side for satisfiability, while the intractable side is PSPACE-complete. Specifically, Schaefer showed that the satisfiability problem is solvable in polynomial time precisely for formulas built from Boolean relations all of which are bijunctive, or all of which are Horn, or all of which are dual Horn, or all of which are affine. We identify new classes of Boolean relations, called *tight* relations, that properly contain the classes of bijunctive, Horn, dual Horn, and affine relations. We show that *st*-connectivity is solvable in linear time for formulas built from tight relations, and PSPACE-complete in all other cases. Our second main result is a dichotomy theorem for the connectivity problem: it is in coNP for formulas built from tight relations, and PSPACE-complete in all other cases.

In addition to these two complexity-theoretic dichotomies, we establish a structural dichotomy theorem for the diameter of the connected components of the solution space of Boolean formulas. This result asserts that, in the PSPACE-complete cases, the diameter of the connected components can be exponential, but in all other cases it is linear. Thus, small diameter and tractability of the *st*-connectivity problem are remarkably aligned.

To establish these results, we first show that all *tight* relations have "good" structural properties. Specifically, in a tight relation every component has a unique minimum element, or every component has a unique maximum element, or the Hamming distance coincides with the shortest-path distance in the relation. These properties are inherited by every formula built from tight relations, and yield both small diameter and linear algorithms for *st*-connectivity.

Next, the challenge is to show that for non-tight relations, both the connectivity problem and the *st*-connectivity problem are PSPACE-hard. In Schaefer's

Dichotomy Theorem, NP-hardness of satisfiability was a consequence of an *expressibility* theorem, which asserted that every Boolean relation can be obtained as a projection over a formula built from clauses in the "hard" relations. Schaefer's notion of expressibility is inadequate for our problem. Instead, we introduce and work with a delicate and more strict notion of expressibility, which we call *faithful expressibility*. Intuitively, faithful expressibility means that, in addition to definability via a projection, the space of witnesses of the existential quantifiers in the projection has certain strong connectivity properties that allow us to capture the graph structure of the relation that is being defined. It should be noted that Schaefer's Dichotomy Theorem can also be proved using a Galois connection and Post's celebrated classification of the lattice of Boolean clones (see [18]). This method, however, does not appear to apply to connectivity, as the boundaries discovered here cut across Boolean clones. Thus, the use of faithful expressibility or some other refined definability technique seems unavoidable.

The first step towards proving PSPACE-completeness is to show that both connectivity and st-connectivity are hard for 3-CNF formulae; this is proved by a reduction from a generic PSPACE computation. Next, we identify the simplest relations that are not tight: these are ternary relations whose graph is a path of length 4 between assignments at Hamming distance 2. We show that these paths can faithfully express all 3-CNF clauses. The crux of our hardness result is an *expressibility* theorem to the effect that one can faithfully express such a path from any set of relations which is not tight.

Our original hope was that tractability results for connectivity could conceivably inform heuristic algorithms for satisfiability and enhance their effectiveness. In this context, our findings are *prima facie* negative: we show that when satisfiability is intractable, then connectivity is also intractable. But our results do contain a glimmer of hope: there are broad classes of intractable satisfiability problems, those built from tight relations, with polynomial st-connectivity and small diameter. It would be interesting to investigate if these properties make random instances built from tight relations easier for WalkSAT and similar heuristics, and if so, whether such heuristics are amenable to rigorous analysis.

For want of space, some proofs, as well as some additional results, are omitted here; they can be found in the full version available at ECCC.

2   Basic Concepts and Statements of Results

A *logical relation* R is a non-empty subset of $\{0, 1\}^k$, for some $k \geq 1$; k is the *arity* of R. Let S be a finite set of logical relations. A CNF(S)-*formula* over a set of variables $V = \{x_1, \ldots, x_n\}$ is a finite conjunction $C_1 \wedge \ldots \wedge C_n$ of clauses built using relations from S, variables from V, and the constants 0 and 1; this means that each C_i is an expression of the form $R(\xi_1, \ldots, \xi_k)$, where $R \in S$ is a relation of arity k, and each ξ_j is a variable in V or one of the constants 0, 1.

The *satisfiability problem* SAT(S) associated with a finite set S of logical relations asks: given a CNF(S)-formula φ, is it satisfiable? All well known restrictions of Boolean satisfiability, such as 3-SAT, NOT-ALL-EQUAL 3-SAT, and

POSITIVE 1-IN-3 SAT, can be cast as $\mathrm{SAT}(S)$ problems, for a suitable choice of S. For instance, POSITIVE 1-IN-3 SAT is $\mathrm{SAT}(\{R_{1/3}\})$, where $R_{1/3} = \{100, 010, 001\}$. Schaefer [1] identified the complexity of *every* satisfiability problem $\mathrm{SAT}(S)$. To state Schaefer's main result, we need to define some basic concepts.

Definition 1. Let R be a logical relation.

(1) R is *bijunctive* if it is the set of solutions of a 2-CNF formula.

(2) R is *Horn* if it is the set of solutions of a Horn formula, where a Horn formula is a CNF formula such that each conjunct has at most one positive literal.

(3) R is *dual Horn* if it is the set of solutions of a dual Horn formula, where a dual Horn formula is a CNF formula such that each conjunct has at most one negative literal.

(4) R is *affine* if it is the set of solutions of a system of linear equations over $\mathbb{Z}_2$.

Each of these types of logical relations can be characterized in terms of *closure* properties [1]. A relation R is bijunctive if and only if it is closed under the *majority* operation (if $\mathbf{a}, \mathbf{b}, \mathbf{c} \in R$, then $\mathrm{maj}(\mathbf{a}, \mathbf{b}, \mathbf{c}) \in R$, where $\mathrm{maj}(\mathbf{a}, \mathbf{b}, \mathbf{c})$ of $\mathbf{a}, \mathbf{b}, \mathbf{c}$ is the vector whose i-th bit is the majority of a_i, b_i, c_i). A relation R is Horn if and only if it is closed under $\vee$ (if $\mathbf{a}, \mathbf{b} \in R$, then $\mathbf{a} \vee \mathbf{b} \in R$, where, $\mathbf{a} \vee \mathbf{b}$ is the vector whose i-th bit is $a_i \vee b_i$). Similarly, R is dual Horn if and only if it is closed under $\wedge$. Finally, R is affine if and only if it is closed under $\mathbf{a} \oplus \mathbf{b} \oplus \mathbf{c}$.

Definition 2. A set S of logical relations is *Schaefer* if at least one of the following holds: (1) Every relation in S is bijunctive; (2) Every relation in S is Horn; (3) Every relation in S is dual Horn; (4) Every relation in S is affine.

Theorem 1. (Schaefer's Dichotomy Theorem [1]) *If S is Schaefer, then* $\mathrm{SAT}(S)$ *is in* P; *otherwise,* $\mathrm{SAT}(S)$ *is* NP-*complete.*

Note that the closure properties of Schaefer sets yield a cubic algorithm for determining, given a finite set S of relations, whether $\mathrm{SAT}(S)$ is in P or NP-complete (the input size is the sum of the sizes of relations in S).

Here, we are interested in the connectivity properties of the space of solutions of $\mathrm{CNF}(S)$-formulas. If φ is a $\mathrm{CNF}(S)$-formula with n variables, then $G(\varphi)$ denotes the subgraph of the n-dimensional hypercube induced by the solutions of φ. Thus, the vertices of $G(\varphi)$ are the solutions of φ, and there is an edge between two solutions of $G(\varphi)$ precisely when they differ in a single variable. We consider the following two algorithmic problems for $\mathrm{CNF}(S)$-formulas.

(1) The *st-connectivity* problem ST-CONN(S): given a $\mathrm{CNF}(S)$-formula φ and two solutions $\mathbf{s}$ and $\mathbf{t}$ of φ, is there a path from $\mathbf{s}$ to $\mathbf{t}$ in $G(\varphi)$?

(2) The *connectivity* problem CONN(S): given a $\mathrm{CNF}(S)$-formula φ, is $G(\varphi)$ connected?

To pinpoint the computational complexity of ST-CONN(S) and CONN(S), we need to introduce certain new types of relations.

Definition 3. Let $R \subseteq \{0, 1\}^k$ be a logical relation.

(1) R is *componentwise bijunctive* if every connected component of $G(R)$ is bijunctive.

(2) R is OR-*free* if the relation OR $= \{01, 10, 11\}$ cannot be obtained from R by setting $k - 2$ of the coordinates of R to a constant $\mathbf{c} \in \{0, 1\}^{k-2}$. In other words, R is OR-free if $(x_1 \vee x_2)$ is not definable from R by fixing $k - 2$ variables.

(3) R is NAND-*free* if $(\bar{x}_1 \vee \bar{x}_2)$ is not definable from R by fixing $k - 2$ variables.

The next lemma is proved using the closure properties of bijunctive, Horn, and dual Horn relations. (We skip the easy proof).

Lemma 1. *Let R be a logical relation.*
(1) *If R is bijunctive, then R is componentwise bijunctive.*
(2) *If R is Horn, then R is OR-free.*
(3) *If R is dual Horn, then R is NAND-free.*
(4) *If R is affine, then R is componentwise bijunctive, OR-free, and NAND-free.*

These containments are proper. For instance, $R_{1/3} = \{100, 010, 001\}$ is componentwise bijunctive, but not bijunctive as $\mathrm{maj}(100, 010, 001) = 000 \notin R_{1/3}$.

We are now ready to introduce the key concept of a *tight* set of relations.

Definition 4. A set S of logical relations is *tight* if at least one of the following three conditions holds: (1) Every relation in S is componentwise bijunctive; (2) Every relation in S is OR-free; (3) Every relation in S is NAND-free.

In view of Lemma 1, if S is Schaefer, then it is tight. The converse, however, does not hold. It is also easy to see that there is a polynomial-time algorithm for testing whether a given finite set S of logical relations is tight. Our first main result is a dichotomy theorem for the computational complexity of ST-CONN(S).

Theorem 2. *Let S be a finite set of logical relations. If S is tight, then ST-CONN(S) is in P; otherwise, CONN(S) is PSPACE-complete.*

Our second main result asserts that the dichotomy in the computational complexity of ST-CONN(S) is accompanied by a parallel structural dichotomy in the size of the diameter of $G(\varphi)$ (where, for a CNF(S)-formula φ, the *diameter of* $G(\varphi)$ is the maximum of the diameters of the components of $G(\varphi)$).

Theorem 3. *Let S be a finite set of logical relations. If S is tight, then for every CNF(S)-formula φ, the diameter of $G(\varphi)$ is linear in the number of variables of φ; otherwise, there are CNF(S)-formulas φ such that the diameter of $G(\varphi)$ is exponential in the number of variables of φ.*

Our third main result establishes a dichotomy for the complexity of CONN(S).

Theorem 4. *Let S be a finite set of logical relations. If S is tight, then CONN(S) is in coNP; otherwise, it is PSPACE-complete.*

We also show that if S is tight, but not Schaefer, then CONN(S) is coNP-complete. Our results and their comparison to Schaefer's Dichotomy Theorem are summarized in the table below.

S	$\text{SAT}(S)$	$\text{ST-CONN}(S)$	$\text{CONN}(S)$	Diameter
Schaefer	in P	in P	in coNP	$O(n)$
Tight, not Schaefer	NP-compl.	in P	coNP-compl.	$O(n)$
Not tight	NP-compl.	PSPACE-compl.	PSPACE-compl.	$2^{\Omega(\sqrt{n})}$

As an application, the set $S = \{R_{1/3}\}$, where $R_{1/3} = \{100, 010, 001\}$, is tight, but not Schaefer. It follows that $\text{SAT}(S)$ is NP-complete (recall that this problem is POSITIVE 1-IN-3 SAT), $\text{ST-CONN}(S)$ is in P, and $\text{CONN}(S)$ is coNP-complete. Consider also the set $S = \{R_{\text{NAE}}\}$, where $R_{\text{NAE}} = \{0,1\}^3 \setminus \{000, 111\}$. This set is not tight, hence $\text{SAT}(S)$ is NP-complete (this problem is POSITIVE NOT-ALL-EQUAL 3-SAT), while both $\text{ST-CONN}(S)$ and $\text{CONN}(S)$ are PSPACE-complete.

We conjecture that if S is Schaefer, then $\text{CONN}(S)$ is in P. If this conjecture is true, it will follow that the complexity of $\text{CONN}(S)$ exhibits a *trichotomy*: if S is Schaefer, then $\text{CONN}(S)$ is in P; if S is tight, but not Schaefer, then $\text{CONN}(S)$ is coNP-complete; if S is not tight, then $\text{CONN}(S)$ is PSPACE-complete.

3   The Easy Cases of Connectivity

In this section, we determine the complexity of $\text{CONN}(S)$ and $\text{ST-CONN}(S)$ for tight sets S of logical relations, and also show that for such sets, the diameter of $G(\varphi)$ of $\text{CNF}(S)$-formula φ is linear. We prove only the key structural properties of tight relations here, and defer the rest to the full version.

We will use $\mathbf{a}, \mathbf{b}, \ldots$ to denote Boolean vectors, and $\mathbf{x}$ and $\mathbf{y}$ to denote vectors of variables. We write $|\mathbf{a}|$ to denote the Hamming weight (number of 1's) of a Boolean vector $\mathbf{a}$. Given two Boolean vectors $\mathbf{a}$ and $\mathbf{b}$, we write $|\mathbf{a} - \mathbf{b}|$ to denote the Hamming distance between $\mathbf{a}$ and $\mathbf{b}$. Finally, if $\mathbf{a}$ and $\mathbf{b}$ are solutions of a Boolean formula φ and lie in the same component of $G(\varphi)$, then we write $d_\varphi(\mathbf{a}, \mathbf{b})$ to denote the shortest-path distance between $\mathbf{a}$ and $\mathbf{b}$ in $G(\varphi)$.

3.1   The ST-CONN Problem for Tight Sets

Lemma 2. *Let S be a set of componentwise bijunctive relations and φ a $\text{CNF}(S)$-formula. If $\mathbf{a}$ and $\mathbf{b}$ are two solutions of φ that lie in the same component of $G(\varphi)$, then $d_\varphi(\mathbf{a}, \mathbf{b}) = |\mathbf{a} - \mathbf{b}|$.*

Proof. (Sketch) Consider first the special case in which every relation in S is bijunctive. In this case, φ is equivalent to a 2-CNF formula and so the space of solutions of φ is closed under maj. We show that there is a path in $G(\varphi)$ from $\mathbf{a}$ to $\mathbf{b}$, such that along the path only the assignments on variables with indices from the set $D = \{i : a_i \neq b_i\}$ change. This implies that the shortest path is of length $|D|$ by induction on $|D|$. Consider any path $\mathbf{a} \to \mathbf{u}^1 \to \cdots \to \mathbf{u}^r \to \mathbf{b}$ in $G(\varphi)$. We construct another path by replacing $\mathbf{u}^i$ by $\mathbf{v}^i = \text{maj}\,(\mathbf{a}, \mathbf{u}^i, \mathbf{b})$ for $i = 1, \ldots, r$, and removing repetitions. This path has the desired property.

For the general case, it can be shown that every component F of $G(\varphi)$ is the solution space of a 2-CNF formula φ'. If C is a clause of φ involving a relation R in S, then the projection of F on the variables of C is contained in a component of R. Then the formula φ' is obtained from φ as follows: replace each clause C of φ by a 2-CNF formula expressing the component of R that contains the projection of F on the variables of C.   □

Corollary 1. *Let S be a set of componentwise bijunctive relations. Then (1) for every $\varphi \in \mathrm{CNF}(S)$ with n variables, the diameter of each component of $G(\varphi)$ is bounded by n; (2) $\mathrm{ST\text{-}CONN}(S)$ is in P; and (3) $\mathrm{CONN}(S)$ is in coNP.*

Next, we consider sets of OR-free relations (sets of NAND-free relations are handled dually). Define the *coordinate-wise partial order* $\leq$ on Boolean vectors as follows: $\mathbf{a} \leq \mathbf{b}$ if $a_i \leq b_i$, for each i.

Lemma 3. *Let S be a set of OR-free relations and φ a $\mathrm{CNF}(S)$-formula. Every component of $G(\varphi)$ contains a minimum solution with respect to the coordinate-wise order; moreover, every solution is connected to the minimum solution in the same component via a monotone path.*

Proof. Suppose there are two distinct minimal assignments $\mathbf{u}$ and $\mathbf{u}'$ in some component of $G(\varphi)$. Consider the path between them where the maximum Hamming weight of assignments on the path is minimized. If there are many such paths, pick one where the smallest number of assignments have the maximum Hamming weight. Denote this path by $\mathbf{u} = \mathbf{u^1} \to \mathbf{u^2} \cdots \to \mathbf{u^r} = \mathbf{u}'$. Let $\mathbf{u^i}$ be the assignment of largest Hamming weight in the path. Then $\mathbf{u^i} \neq \mathbf{u}$ and $\mathbf{u^i} \neq \mathbf{u}'$, since $\mathbf{u}$ and $\mathbf{u}'$ are minimal. The assignments $\mathbf{u^{i-1}}$ and $\mathbf{u^{i+1}}$ differ in exactly 2 variables, say, in x_1 and x_2. So $\{u_1^{i-1}u_2^{i-1},\ u_1^i u_2^i,\ u_1^{i+1}u_2^{i+1}\} = \{01, 11, 10\}$. Let $\hat{\mathbf{u}}$ be such that $\hat{u}_1 = \hat{u}_2 = 0$, and $\hat{u}_i = u_i$ for $i > 2$. If $\hat{\mathbf{u}}$ is a solution, then the path $\mathbf{u^1} \to \mathbf{u^2} \to \cdots \to \mathbf{u^i} \to \hat{\mathbf{u}} \to \mathbf{u^{i+1}} \to \cdots \to \mathbf{u^r}$ contradicts the way we chose the original path. Therefore, $\hat{\mathbf{u}}$ is not a solution. This means that there is a clause that is violated by it, but is satisfied by $\mathbf{u^{i-1}}$, $\mathbf{u^i}$, and $\mathbf{u^{i+1}}$. So the relation corresponding to that clause is not OR-free, which is a contradiction.

The unique minimal solution in a component is its minimum solution. Furthermore, starting from any assignment $\mathbf{s}$ in the component, and repeatedly flipping variables from 1 to 0 provides a monotone path to the minimum.   □

Corollary 2. *Let S be a set of OR-free relations. Then (1) For every $\varphi \in \mathrm{CNF}(S)$ with n variables, the diameter of each component of $G(\varphi)$ is bounded by $2n$; (2) $\mathrm{ST\text{-}CONN}(S)$ is in P; and (3) $\mathrm{CONN}(S)$ is in coNP.*

4   The PSPACE-Complete Cases of Connectivity

If $k \geq 2$, then a k-*clause* is a disjunction of k variables or negated variables. For $0 \leq i \leq k$, let D_i be the set of all satisfying truth assignments of the k-clause whose first i literals are negated, and let $S_k = \{D_0, D_1, \ldots, D_k\}$. Thus, $\mathrm{CNF}(S_k)$ is the collection of k-CNF formulas.

The starting point of the proof is to show that $\mathrm{CONN}(S_3)$ and $\mathrm{ST\text{-}CONN}(S_3)$ are PSPACE-complete. The proof is fairly intricate, and is via a direct reduction from the computation of a polynomial-space Turing machine. We also show that

3-CNF formulas can have exponential diameter, by inductively constructing a path of length at least $2^{\frac{n}{2}}$ on n variables and then identifying it with the solution space of a 3-CNF formula with $O(n^2)$ clauses.

Lemma 4. ST-CONN(S_3) *and* CONN(S_3) *are* PSPACE-*complete.*

Lemma 5. *For n even, there is a 3-CNF formula φ_n with n variables and $O(n^2)$ clauses, such that $G(\varphi_n)$ is a path of length greater than $2^{\frac{n}{2}}$.*

4.1 Faithful Expressibility

From here onwards, all our hardness results are proved by showing that if S is a non-tight set, then every 3-clause is expressible from S in a certain special way that we describe next. In his dichotomy theorem, Schaefer [1] used the following notion of expressibility: a relation R is *expressible from* a set S of relations if there is a CNF(S)-formula φ so that $R(\mathbf{x}) \equiv \exists \mathbf{y} \; \varphi(\mathbf{x}, \mathbf{y})$. This notion, is not sufficient for our purposes. Instead, we introduce a more delicate notion, which we call *faithful expressibility*. Intuitively, we view the relation R as a subgraph of the hypercube, rather than just a subset, and require that this graph structure be also captured by the formula φ.

Definition 5. *A relation R is* faithfully expressible *from a set of relations S if there is a CNF(S)-formula φ such that:*

(1) $R = \{\mathbf{a} : \exists \mathbf{y} \; \varphi(\mathbf{a}, \mathbf{y})\};$

(2) *For every $\mathbf{a} \in R$, the graph $G(\varphi(\mathbf{a}, \mathbf{y}))$ is connected;*

(3) *For $\mathbf{a}, \mathbf{b} \in R$ with $|\mathbf{a} - \mathbf{b}| = 1$, there exists a $\mathbf{w}$ such that $(\mathbf{a}, \mathbf{w})$ and $(\mathbf{b}, \mathbf{w})$ are solutions of φ.*

For $\mathbf{a} \in R$, the *witnesses* of $\mathbf{a}$ are the $\mathbf{y}$'s such that $\varphi(\mathbf{a}, \mathbf{y})$. The last two conditions say that the witnesses of $\mathbf{a} \in R$ are connected, and that neighboring $\mathbf{a}, \mathbf{b} \in R$ have a common witness. This allows us to simulate an edge $(\mathbf{a}, \mathbf{b})$ in $G(R)$ by a path in $G(\varphi)$, and thus relate the connectivity properties of the solution spaces. There is however, a price to pay: it is much harder to come up with formulas that faithfully express a relation R. An example is when S is the set of all paths of length 4 in $\{0,1\}^3$, a set that plays a crucial role in our proof. While S_3 is easily expressible from S in Schaefer's sense, the CNF(S)-formulas that faithfully express S_3 are fairly complicated and have a large witness space.

Lemma 6. *Let S and S' be sets of relations such that every $R \in S$ is faithfully expressible from S'. Given a CNF(S)-formula $\psi(\mathbf{x})$, one can efficiently construct a CNF(S')-formula $\varphi(\mathbf{x}, \mathbf{y})$ such that:*

(1) $\psi(\mathbf{x}) \equiv \exists \mathbf{y} \; \varphi(\mathbf{x}, \mathbf{y});$

(2) *if $(\mathbf{s}, \mathbf{w^s}), (\mathbf{t}, \mathbf{w^t}) \in \varphi$ are connected in $G(\varphi)$ by a path of length d, then there is a path from $\mathbf{s}$ to $\mathbf{t}$ in $G(\psi)$ of length at most d;*

(3) *If $\mathbf{s}, \mathbf{t} \in \psi$ are connected in $G(\psi)$, then for every witness $\mathbf{w^s}$ of $\mathbf{s}$, and every witness $\mathbf{w^t}$ of $\mathbf{t}$, there is a path from $(\mathbf{s}, \mathbf{w^s})$ to $(\mathbf{t}, \mathbf{w^t})$ in $G(\varphi)$.*

Proof. Suppose ψ is a formula on n variables that consists of m clauses $C_1, \ldots, C_m$. For clause C_j, assume that the set of variables is $V_j \subseteq [n]$, and that it involves relation $R_j \in S$. Thus, $\psi(\mathbf{x})$ is $\wedge_{j=1}^m R_j(\mathbf{x}_{V_j})$. Let φ_j be the faithful expression for R_j from S', so that $R_j(\mathbf{x}_{V_j}) \equiv \exists \mathbf{y}_j\ \varphi_j(\mathbf{x}_{V_j}, \mathbf{y}_j)$. Let $\mathbf{y}$ be the vector $(\mathbf{y}_1, \ldots, \mathbf{y}_m)$ and let $\varphi(\mathbf{x}, \mathbf{y})$ be the formula $\wedge_{j=1}^m \varphi_j(\mathbf{x}_{V_j}, \mathbf{y}_j)$. Then $\psi(\mathbf{x}) \equiv \exists \mathbf{y}\ \varphi(\mathbf{x}, \mathbf{y})$.

Statement (2) follows from (1) by projection of the path on the coordinates of $\mathbf{x}$. For statement (3), consider $\mathbf{s}, \mathbf{t} \in \psi$ that are connected in $G(\psi)$ via a path $\mathbf{s} = \mathbf{u}^0 \to \mathbf{u}^1 \to \cdots \to \mathbf{u}^r = \mathbf{t}$. For every $\mathbf{u}^i, \mathbf{u}^{i+1}$, and clause C_j, there exists an assignment $\mathbf{w}^i{}_j$ to $\mathbf{y}_j$ such that both $(\mathbf{u}^i{}_{V_j}, \mathbf{w}^i{}_j)$ and $(\mathbf{u}^{i+1}{}_{V_j}, \mathbf{w}^i{}_j)$ are solutions of φ_j, by condition (2) of faithful expressibility. Thus $(\mathbf{u}^i, \mathbf{w}^i)$ and $(\mathbf{u}^{i+1}, \mathbf{w}^i)$ are both solutions of φ, where $\mathbf{w}^i = (\mathbf{w}^i{}_1, \ldots \mathbf{w}^i{}_m)$. Further, for every $\mathbf{u}^i$, the space of solutions of $\varphi(\mathbf{u}^i, \mathbf{y})$ is the product space of the solutions of $\varphi_j(\mathbf{u}^i{}_{V_j}, \mathbf{y}_j)$ over $j = 1, \ldots, m$. Since these are all connected by condition (3) of faithful expressibility, $G(\varphi(\mathbf{u}^i, \mathbf{y}))$ is connected. The following describes a path from $(\mathbf{s}, \mathbf{w}^{\mathbf{s}})$ to $(\mathbf{t}, \mathbf{w}^{\mathbf{t}})$ in $G(\varphi)$: $(\mathbf{s}, \mathbf{w}^{\mathbf{s}}) \rightsquigarrow (\mathbf{s}, \mathbf{w}^0) \to (\mathbf{u}^1, \mathbf{w}^0) \rightsquigarrow (\mathbf{u}^1, \mathbf{w}^1) \to \cdots \rightsquigarrow (\mathbf{u}^{r-1}, \mathbf{w}^{r-1}) \to (\mathbf{t}, \mathbf{w}^{r-1}) \rightsquigarrow (\mathbf{t}, \mathbf{w}^{\mathbf{t}})$. Here $\rightsquigarrow$ indicates a path in $G(\varphi(\mathbf{u}^i, \mathbf{y}))$. $\qquad\square$

Corollary 3. *Suppose S and S' are as in Lemma 6.*

(1) *There are polynomial time reductions from* CONN(S) *to* CONN(S'), *and from* ST-CONN(S) *to* ST-CONN(S').
(2) *Given a* CNF(S)-*formula $\psi(\mathbf{x})$ with m clauses, one can efficiently construct a* CNF(S')-*formula $\varphi(\mathbf{x}, \mathbf{y})$ such that the length of $\mathbf{y}$ is $O(m)$ and the diameter of the solution space does not decrease.*

4.2   Expressing 3-Clauses from Non-tight Sets of Relations

In order prove Theorems 2, 3 and 4, it suffices to prove the following Lemma:

Lemma 7. *If set S of relations is non-tight, S_3 is faithfully expressible from S.*

First, observe that all 2-clauses are faithfully expressible from S. There exists $R \in S$ which is not OR-free, so we can express $(x_1 \vee x_2)$ by substituting constants in R. Similarly, we can express $(\bar{x}_1 \vee \bar{x}_2)$ using a relation that is not NAND-free. The last 2-clause $(x_1 \vee \bar{x}_2)$ can be obtained from OR and NAND by a technique that corresponds to reverse resolution. $(x_1 \vee \bar{x}_2) = \exists y\ (x_1 \vee y) \wedge (\bar{y} \vee \bar{x}_2)$. It is easy to see that this gives a faithful expression. From here onwards we assume that S contains all 2-clauses. The proof now proceeds in four steps.

Step 1: *Faithfully expressing a relation in which some distance expands.*
For a relation R, we say that the distance between $\mathbf{a}$ and $\mathbf{b}$ *expands* if $\mathbf{a}$ and $\mathbf{b}$ are connected in $G(R)$, but $d_R(\mathbf{a}, \mathbf{b}) > |\mathbf{a} - \mathbf{b}|$. By Lemma 2 no distance expands in componentwise bijunctive relations. This property also holds for the relation $R_{\text{NAE}} = \{0, 1\}^3 \setminus \{000, 111\}$, which is not componentwise bijunctive. However, we show that if Q is not componentwise bijunctive, then, by adding 2-clauses, we can faithfully express a relation Q' in which some distance expands. For instance, when $Q = R_{\text{NAE}}$, then we can take $Q'(x_1, x_2, x_3) = R_{\text{NAE}}(x_1, x_2, x_3) \wedge (\bar{x}_1 \vee \bar{x}_3)$.

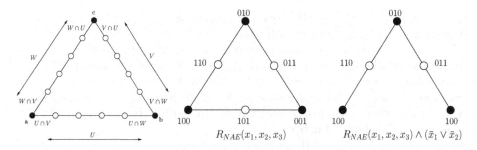

Fig. 1. Proof of Lemma 8

The distance between $\mathbf{a} = 100$ and $\mathbf{b} = 001$ in Q' expands. Similarly, in the general construction, we identify $\mathbf{a}$ and $\mathbf{b}$ on a cycle, and add 2-clauses that eliminate all the vertices along the shorter arc between $\mathbf{a}$ and $\mathbf{b}$.

Step 2: *Isolating a pair of assignments whose distance expands.*
The relation Q' obtained in Step 1 may have several disconnected components. This *cleanup* step isolates a pair of assignments whose distance expands. By adding 2-clauses, we obtain a relation T that consists of a pair of assignments $\mathbf{a}, \mathbf{b}$ of Hamming distance r and a path of length $r + 2$ between them.

Step 3: *Faithfully expressing paths of length 4.*
Let P denote the set of all ternary relations whose graph is a path of length 4 between two assignments at Hamming distance 2. Up to permutations of coordinates, there are 6 such relations. Each of them is the conjunction of a 3-clause and a 2-clause. For instance, the relation $M = \{100, 110, 010, 011, 001\}$ can be written as of $(x_1 \vee x_2 \vee x_3) \wedge (\bar{x}_1 \vee \bar{x}_3)$. These relations are "minimal" examples of relations that are not componentwise bijunctive. By projecting out intermediate variables from the path T obtained in Step 2, we faithfully express one of the relations in P. We faithfully express other relations in P using this relation.

Step 4: *Faithfully expressing S_3.*
We faithfully express $(x_1 \vee x_2 \vee x_3)$ from M using a formula derived from a gadget in [19]. This gadget expresses $(x_1 \vee x_2 \vee x_3)$ in terms of "Protected OR", which corresponds to our relation M. From this, we express the other 3-clauses.

Lemma 8. *There exist a* $\mathrm{CNF}(S)$*-definable relation* Q' *and* $\mathbf{a}, \mathbf{b} \in Q'$ *such that the distance between them expands.*

Proof. Since S is not tight, it contains a relation Q which is not componentwise bijunctive. If Q contains $\mathbf{a}, \mathbf{b}$ where the distance between them expands, we are done. So assume that for all $\mathbf{a}, \mathbf{b} \in G(Q)$, $d_Q(\mathbf{a}, \mathbf{b}) = |\mathbf{a} - \mathbf{b}|$. Since Q is not componentwise bijunctive, there exists a triple of assignments $\mathbf{a}, \mathbf{b}, \mathbf{c}$ lying in the same component such that $\mathrm{maj}(\mathbf{a}, \mathbf{b}, \mathbf{c})$ is not in that component (which also easily implies it is not in Q). Choose the triple such that the sum of pairwise distances $d_Q(\mathbf{a}, \mathbf{b}) + d_Q(\mathbf{b}, \mathbf{c}) + d_Q(\mathbf{c}, \mathbf{a})$ is minimized. Let $U = \{i | a_i \neq b_i\}$, $V = \{i | b_i \neq c_i\}$, and $W = \{i | c_i \neq a_i\}$. Since $d_Q(\mathbf{a}, \mathbf{b}) = |\mathbf{a} - \mathbf{b}|$, a shortest path does not flip variables outside of U, and each variable in U is flipped exactly once. We note some useful properties of the sets U, V, W.

1) *Every index $i \in U \cup V \cup W$ occurs in exactly two of U, V, W.*
Consider going by a shortest path from **a** to **b** to **c** and back to **a**. Every $i \in U \cup V \cup W$ is seen an even number of times along this path since we return to **a**. It is seen at least once, and at most thrice, so in fact it occurs twice.
2) *Every pairwise intersection $U \cap V, V \cap W$ and $W \cap U$ is non-empty.*
Suppose the sets U and V are disjoint. From Property 1, we must have $W = U \cup V$. But then it is easy to see that $\mathrm{maj}(\mathbf{a}, \mathbf{b}, \mathbf{c}) = \mathbf{b}$ which is in Q. This contradicts the choice of $\mathbf{a}, \mathbf{b}, \mathbf{c}$.
3) *The sets $U \cap V$ and $U \cap W$ partition the set U.*
By Property 1, each index of U occurs in one of V and W as well. Also since no index occurs in all three sets U, V, W this is in fact a disjoint partition.
4) *For each index $i \in U \cap W$, it holds that $\mathbf{a} \oplus \mathbf{e}_i \notin Q$.*
Assume for the sake of contradiction that $\mathbf{a}' = \mathbf{a} \oplus \mathbf{e}_i \in R$. Since $i \in U \cap W$ we have simultaneously moved closer to both **b** and **c**. Hence $d_Q(\mathbf{a}', \mathbf{b}) + d_Q(\mathbf{b}, \mathbf{c}) + d_Q(\mathbf{c}, \mathbf{a}') < d_Q(\mathbf{a}, \mathbf{b}) + d_Q(\mathbf{b}, \mathbf{c}) + d_Q(\mathbf{c}, \mathbf{a})$. Also $\mathrm{maj}(\mathbf{a}', \mathbf{b}, \mathbf{c}) = \mathrm{maj}(\mathbf{a}, \mathbf{b}, \mathbf{c}) \notin Q$. But this contradicts our choice of $\mathbf{a}, \mathbf{b}, \mathbf{c}$.

Property 4 implies that the shortest paths to **b** and **c** diverge at **a**, since for any shortest path to **b** the first variable flipped is from $U \cap V$ whereas for a shortest path to **c** it is from $W \cap V$. Similar statements hold for the vertices **b** and **c**. Thus along the shortest path from **a** to **b** the first bit flipped is from $U \cap V$ and the last bit flipped is from $U \cap W$. On the other hand, if we go from **a** to **c** and then to **b**, all the bits from $U \cap W$ are flipped before the bits from $U \cap V$. We use this crucially to define Q'. We will add a set of 2-clauses that enforce the following rule on paths starting at **a**: *Flip variables from $U \cap W$ before variables from $U \cap V$*. This will eliminate all shortest paths from **a** to **b** since they begin by flipping a variable in $U \cap V$ and end with $U \cap W$. The paths from **a** to **b** via **c** survive since they flip $U \cap W$ while going from **a** to **c** and $U \cap V$ while going from **c** to **b**. However all remaining paths have length at least $|\mathbf{a} - \mathbf{b}| + 2$ since they flip twice some variables not in U.

Take all pairs of indices $\{(i, j) | i \in U \cap W, j \in U \cap V\}$. The following conditions hold from the definition of U, V, W: $a_i = \bar{c}_i = \bar{b}_i$ and $a_j = c_j = \bar{b}_j$. Add the 2-clause C_{ij} asserting that the pair of variables $x_i x_j$ must take values in $\{a_i a_j, c_i c_j, b_i b_j\} = \{a_i a_j, \bar{a}_i a_j, \bar{a}_i \bar{a}_j\}$. The new relation is $Q' = Q \wedge_{i,j} C_{ij}$. Note that $Q' \subset Q$. We verify that the distance between **a** and **b** in Q' expands. It is easy to see that for any $j \in U$, the assignment $\mathbf{a} \oplus \mathbf{e}_j \notin Q'$. Hence there are no shortest paths left from **a** to **b**. On the other hand, it is easy to see that **a** and **b** are still connected, since the vertex **c** is still reachable from both.   □

Due to space constraints, all remaining proofs are in the full version.

5   Discussion and Open Problems

In Section 2, we conjectured a trichotomy for CONN(S). We have made progress towards this conjecture; what remains is to pinpoint the complexity of CONN(S) when S is Horn or dual-Horn. We can extend our dichotomy theorem for *st*-connectivity to formulas without constants; the complexity of connectivity for formulas without constants is open. We conjecture that when S is not tight, one can improve the diameter bound from $2^{\Omega(\sqrt{n})}$ to $2^{\Omega(n)}$. Finally, we believe that our techniques can shed light on other connectivity-related problems, such as approximating the diameter and counting the number of components.

References

1. Schaefer, T.: The complexity of satisfiability problems. In: Proc. 10th ACM Symp. Theory of Computing. (1978) 216–226
2. Creignou, N.: A dichotomy theorem for maximum generalized satisfiability problems. Journal of Computer and System Sciences **51** (1995) 511–522
3. Creignou, N., Khanna, S., Sudan, M.: Complexity classification of Boolean constraint satisfaction problems. SIAM Monographs on Disc. Math. Appl. 7 (2001)
4. Khanna, S., Sudan, M., Trevisan, L., Williamson, D.: The approximability of constraint satisfaction problems. SIAM J. Comput., 30(6):1863–1920 (2001)
5. Creignou, N., Hermann, M.: Complexity of generalized satisfiability counting problems. Information and Computation **125**(1) (1996) 1–12
6. Kavvadias, D., Sideri, M.: The inverse satisfiability problem. SIAM J. Comput. **28**(1) (1998) 152–163
7. Kirousis, L., Kolaitis, P.: The complexity of minimal satisfiability problems. Information and Computation **187**(1) (2003) 20–39
8. Bulatov, A.: A dichotomy theorem for constraints on a three-element set. In: Proc. 43rd IEEE Symp. Foundations of Computer Science. (2002) 649–658
9. Creignou, N., Zanuttini, B.: A complete classification of the complexity of propositional abduction. To appear in SIAM Journal on Computing (2006)
10. Achlioptas, D., Naor, A., Peres, Y.: Rigorous location of phase transitions in hard optimization problems. Nature **435** (2005) 759–764
11. Mézard, M., Zecchina, R.: Random k-satisfiability: from an analytic solution to an efficient algorithm. Phys. Rev. E **66** (2002)
12. Mézard, M., Parisi, G., Zecchina, R.: Analytic and algorithmic solution of random satisfiability problems. Science **297, 812** (2002)
13. Maneva, E., Mossel, E., Wainwright, M.J.: A new look at survey propagation and its generalizations. In: Proc. 16th ACM-SIAM Symp. Discrete Algorithms. (2005) 1089–1098
14. Mora, T., Mézard, M., Zecchina, R.: Clustering of solutions in the random satisfiability problem. Phys. Rev. Lett. (2005) In press.
15. Achlioptas, D., Ricci-Tersenghi, F.: On the solution-space geometry of random constraint satisfaction problems. In: 38th ACM Symp. Theory of Computing. (2006)
16. Selman, B., Kautz, H., Cohen, B.: Local search strategies for satisfiability testing. In: Cliques, coloring, and satisfiability : second DIMACS implementation challenge, October 1993, AMS (1996)
17. Achlioptas, D., Beame, P., Molloy, M.: Exponential bounds for DPLL below the satisfiability threshold. In: Proc. 15th ACM-SIAM Symp. Discrete Algorithms. (2004) 132–133
18. Böhler, E., Creignou, N., Reith, S., Vollmer, H.: Playing with Boolean blocks, Part II: constraint satisfaction problems. ACM SIGACT-Newsletter **35**(1) (2004) 22–35
19. Hearne, R., Demaine, E.: The Nondeterministic Constraint Logic model of computation: Reductions and applications. In: 29th Intl. Colloquium on Automata, Languages and Programming. (2002) 401–413

Suffix Trays and Suffix Trists: Structures for Faster Text Indexing

Richard Cole[1], Tsvi Kopelowitz[2], and Moshe Lewenstein[2]

[1] New York University
cole@cs.nyu.edu
[2] Bar-Ilan University
{kopelot, moshe}@cs.biu.ac.il

Abstract. Suffix trees and suffix arrays are two of the most widely used data structures for text indexing. Each uses linear space and can be constructed in linear time [3,5,6,7]. However, when it comes to answering queries, the prior does so in $O(m \log |\Sigma|)$ time, where m is the query size, $|\Sigma|$ is the alphabet size, and the latter does so in $O(m + \log n)$, where n is the text size. We propose a novel way of combining the two into, what we call, a *suffix tray*. The space and construction time remain linear and the query time improves to $O(m + \log |\Sigma|)$.

We also consider the online version of indexing, where the indexing structure continues to update the text online and queries are answered in tandem. Here we suggest a *suffix trist*, a cross between a suffix tree and a suffix list. It supports queries in $O(m + \log |\Sigma|)$. The space and text update time of a suffix trist are the same as for the suffix tree or the suffix list.

1 Introduction

Indexing is one of the most important paradigms in searching. The idea is to preprocess a text and construct a mechanism that will later provide answer to queries of the form "does a pattern P occur in the text" in time proportional to the size of the *pattern* rather than the text. The suffix tree [3,9,10,11] and suffix array [5,6,7,8] have proven to be invaluable data structures for indexing.

Both suffix trees and suffix arrays use $O(n)$ space, where n is the text length. In fact for alphabets from a polynomially sized range, both can be constructed in linear time, see [3,5,6,7].

The query time is slightly different in the two data structures. Namely, in suffix trees queries are answered in $O(m \log |\Sigma| + occ)$, where m is the length of the query, Σ is the alphabet, $|\Sigma|$ is the alphabet size and occ is the number of occurrences of the query. In suffix arrays the time is $O(m + \log n + occ)$. For the rest of this paper we assume that we are only interested in one occurrence of the pattern in the text, and note that we can find all of the occurrences of the pattern with another additive occ cost in the query time.

The differences in the running times follows from the different way queries are answered. In a suffix tree queries are answered by traversing the tree from the root. At each node one needs to know how to continue the traversal and one needs to decide between $|\Sigma|$ options which are sorted, which explains the $O(\log |\Sigma|)$ factor. In

M. Bugliesi et al. (Eds.): ICALP 2006, Part I, LNCS 4051, pp. 358–369, 2006.

suffix arrays one performs a binary search on all suffixes (hence the $\log n$ factor) and uses longest common prefix (LCP) queries to quickly decide whether the pattern needs to be compared to a specific suffix (see [8] for full details).

It is easy to construct a data structure with optimal $O(m)$ query time. This can be done by simply putting a $|\Sigma|$ length array at every node of the suffix tree. Hence, when traversing the suffix tree with the query we will spend constant time at each node. However, the size of this structure is $O(n|\Sigma|)$.

The question of interest here is whether one can construct an $O(n)$ space structure that will answer queries in time faster than the query time of suffix arrays and suffix trees. We indeed propose to do so with the *Suffix Tray*, a new data structure that extracts the advantages of suffix trees and suffix arrays by combining their structures. This yields an $O(m + \log|\Sigma|)$ query time.

We are also concerned with texts that allow online update of the text. In other words, given an indexing structure supporting indexing queries on S, we would also like to support extensions of the text to Sa, where $a \in \Sigma$. We assume that the text is given in reverse, i.e. from the last character towards the beginning. So, an indexing structure of our desire when representing S will also support extensions to aS where $a \in \Sigma$. We call the change of S to aS a *text extension*. The "reverse" assumption that we use is not strict, as most indexing structures can handle online texts that are reversed (e.g. instead of a suffix tree one can construct a prefix tree and answer the queries in reverse. Likewise, a prefix array can be constructed instead of a suffix array).

Online constructions of indexing structures have been suggested previously. McCreight's suffix tree algorithm [9] was the first online construction. It was a reverse construction (in the sense mentioned above). Ukkonen's algorithm [10] was the first online algorithm that was not reversed. In both these algorithms text extensions take $O(1)$ amortized time, but $O(n)$ worst-case time. In [1] an online suffix tree construction (under the reverse assumption) was proposed with $O(\log n)$ worst-case text extensions. In all these constructions a full suffix tree is constructed and hence queries are answered in $O(m \log|\Sigma|)$ time. An on-line variant of suffix arrays was also proposed in [1] with $O(\log n)$ worst-case for text extensions and $O(m + \log n)$ for answering queries. Similar results can be obtained by using the results in [4].

The problem we deal with in the second part of the paper is how to build an indexing structure that supports both text extensions and supports fast(er) indexing. We will show that if there exists an online construction for a linear-space suffix tree such that the cost of adding a character is $O(f(n, |\Sigma|))$ (n is the size of the current text), then we can construct an online linear-space data-structure for indexing that supports indexing queries in time $O(m + \log|\Sigma|)$, where the cost of adding a character is $O(f(n, |\Sigma|) + \log|\Sigma|)$. We will call this data structure the *Suffix Trist*[1].

---

[1] The name Suffix Trist is derived from the combination of suffix trees and suffix lists, the dynamic version of the suffix array. Of course, one may argue that it may be preferable to work with prefix arrays and prefix lists. Then one would receive Prefix Prays and Prefix Priests.

2 Suffix Trees, Suffix Arrays and Suffix Intervals

Consider a text S of length n and let $S^1, \cdots, S^n$ be the suffixes of S. Two classical data structures for indexing are the suffix tree and the suffix array. We assume that the reader is familiar with the suffix tree. Let $S^{i_1}, ..., S^{i_n}$ be the lexicographic ordering of the suffixes. The suffix array of S is defined to be $SA(S) = < i_1, ..., i_n >$, i.e. the indices of the lexicographic ordering of the suffixes. We will sometimes refer to location j of the suffix array as the location of S^{i_j} (instead of the location of i_j).

We let $ST(S)$ and $SA(S)$ denote the suffix tree and suffix array of S, respectively. As with all suffix array constructions to date, we make the assumption that every node in a suffix tree maintains its children in lexicographic order. Therefore, the leaves ordered by an inorder traversal correspond to the suffixes in lexicographic order, which is also the order maintained in the suffix array. Hence, one can view the suffix tree as a tree over the suffix array.

We will now sharpen this connection between suffix arrays and suffix trees. For strings R and R' we say that $R <_L R'$ if R is lexicographically smaller than R'. $leaf(S^i)$ denotes the leaf corresponding to S^i in $ST(S)$, the suffix tree of S. We define $L(v) = SA^{-1}(i)$ if $leaf(S^i)$ is the leftmost leaf of the subtree of v, i.e. $L(v)$ is the location of S^i in the suffix array. Note that since we assume that the children of a node in a suffix tree are maintained in lexicographic order, it follows that for all S^j such that $leaf(S^j)$ is a descendant of v, $S^i \leq_L S^j$. Likewise, we define $R(v) = SA^{-1}(i)$ if $leaf(S^i)$ is the rightmost leaf of the subtree of v. Therefore, for all S^j such that $leaf(S^j)$ is a descendant of v, $S^i \geq_L S^j$. Hence, the interval $[L(v), R(v)]$ is an interval of the suffix array which contains exactly all the suffixes S^j for which $leaf(S^j)$ is a descendant of v.

Moreover, under the assumption that the children of a node in a suffix tree are maintained in lexicographic ordering we can state the following.

Lemma 1. *Let S be a string and $ST(S)$ its suffix tree. Let v be a node in $ST(S)$ and let $v_1, ..., v_r$ be its children. Let $1 \leq i \leq j \leq r$, and let $[L(v_i), R(v_j)]$ be an interval of the suffix array. Then $k \in [L(v_i), R(v_j)]$ if and only if $leaf(S^k)$ is in one of the subtrees rooted at $v_i, ..., v_j$.*

This leads us to the following concept.

Definition 1. *Let S be a string and $\{S^{i_1}, ..., S^{i_n}\}$ be the lexicographic ordering of its suffixes. The interval $[j, k] = \{i_j, ..., i_k\}$, for $j \leq k$, is called a suffix interval.*

Obviously, suffix intervals are intervals of the suffix array. We note that, as mentioned above, for a node v in a suffix tree, $[L(v), R(v)]$ is a suffix interval and we call the interval v's *suffix interval*. Also, by Lemma 1 for v's children $v_1, ..., v_r$ and for any $1 \leq i \leq j \leq r$, $[L(v_i), R(v_j)]$ is a *suffix interval* and we call this interval the (i, j)-*suffix interval*.

3   Faster Indexing on Static Text

3.1   Suffix Trays

We now elaborate on the suffix tray. The suffix tray will use the concept of suffix intervals from the previous section which, as we have seen, is common to both suffix arrays and suffix trees.

For suffix trays we will create special nodes, which correspond to suffix intervals. We call these nodes *suffix interval nodes*. Part of the suffix tray will be a suffix array. Each suffix interval node can be viewed as a node that maintains the endpoints of the interval within the complete suffix array.

Second, we use the idea of the space-inefficient $O(n|\Sigma|)$ suffix tree solution mentioned in the introduction. We maintain $|\Sigma|$ length arrays at a selected subset of nodes, a subset that contains no more that $\frac{n}{|\Sigma|}$ nodes, which maintains the $O(n)$ space bound. To choose this selected subset of nodes we define the following.

Definition 2. *Let S be a string over alphabet Σ. A node u in $ST(S)$ is called a σ-node if the number of leaves in the subtree of $ST(S)$ rooted at u is at least $|\Sigma|$. A σ-node u is called a branching-σ-node, if at least two of u's children in $ST(S)$ are σ-nodes and is called a σ-leaf if all its children in $ST(S)$ are not σ-nodes.*

Note that if a node u is a σ-node, then all of its ancestors are σ-nodes. This also implies that in a σ-leaf's subtree there are no σ-nodes. The following property of branching-σ-nodes is crucial to our result.

Lemma 2. *Let S be a string of size n over an alphabet Σ and let $ST(S)$ be its suffix tree. The number of branching-σ-nodes in $ST(S)$ is $O(\frac{n}{|\Sigma|})$.*

Proof. The number of σ-leaves is at most $\frac{n}{|\Sigma|}$ because (1) they each have at least $|\Sigma|$ leaves in their subtree and (2) their subtrees are disjoint. Let T be the tree induced by the σ-nodes and contracted onto the branching-σ-nodes and σ-leaves only. Then T is a tree with $\frac{n}{|\Sigma|}$ leaves and with every internal node having at least 2 children. Hence, the lemma follows.   □

This means that we can afford to maintain arrays at every branching-σ-node which will be very helpful in answering queries as we shall see in subsection 3.3.

3.2   Suffix Tray Construction

A *suffix tray* is constructed from a suffix tree as follows. The suffix tray contains all the σ-nodes of the suffix tree. We also add some suffix interval nodes to the suffix tray as children of σ-nodes. Here is how each σ-node is converted from the suffix tree to the suffix tray.

- σ-leaf u: u becomes a suffix interval node with suffix interval $[L(u), R(u)]$.
- non-leaf σ-node u: Let $u_1, ..., u_r$ be u's children in the suffix tree and $u_{l_1}, ..., u_{l_x}$ be the subset of children that are σ-nodes. Then u will be in the suffix tray with interleaving suffix interval nodes and σ-nodes, i.e. $(1, l_1 - 1)$-suffix interval node, u_{l_1}, $(l_1 + 1, l_2 - 1)$-suffix interval node, u_{l_2}, ..., u_{l_x}, $(l_x + 1, r)$-suffix interval node.

At each branching-σ-node u in the suffix tray we maintain an array of size $|\Sigma|$, denoted by A_u, that contains the following data. For every child v of u that is a σ-node, location τ in A_u where τ is the first character on the edge (u, v), points to v. The rest of the locations in A_u point to the appropriate suffix-interval node, or to a NIL pointer if no such suffix interval exists.

At each σ-node u which is not a branching-σ-node and not a σ-leaf, i.e. it has exactly one child v which is a σ-node, we store the first character τ on the edge (u, v), which we call the *separating character*.

We now claim that the suffix tray is of linear size.

Lemma 3. *Let S be a string of size n. Then the size of the suffix tray for S is $O(n)$.*

Proof. The suffix array is clearly of size $O(n)$ and the number of suffix interval nodes is bounded by the number of nodes in $ST(S)$. Also, for each non branching-σ-node the auxiliary information is of constant size.

The auxiliary information held in each branching-σ-node is of size $O(|\Sigma|)$. By Lemma 2 there are $\frac{n}{|\Sigma|}$ branching-σ-nodes. Hence, this all is of size $O(n)$. $\square$

Obviously, given a suffix tree and suffix array, a suffix tray can be constructed in linear time (using depth-first searches, and standard techniques). Since both suffix arrays and suffix trees can be constructed in linear time for alphabets from a polynomially sized range [3,5,6,7], so can suffix trays.

3.3   Navigating on Index Queries

We now turn to the important feature of suffix trays, answering index queries.

Upon receiving a query $P = p_1...p_m$ we begin traversing the suffix tray from the root. Assume that we have traversed the suffix tray with $p_1...p_{i-1}$ and need to continue with $p_i...p_m$. At each branching-σ-node u we access location p_i of the array A_u in order to know which suffix tray node to navigate to. Obviously, since this is an array lookup this takes us constant time. For other σ-nodes that are not σ-leaves and not branching-σ-nodes we compare p_i with the separator character τ. Recall that these nodes have only one child v that is a σ-node. Hence, in the suffix tray the children of u are (1) a suffix interval node to the left of v, say u's left interval, (2) v, and (3) a suffix interval node to the right of v, say u's right interval. If $p_i < \tau$ we navigate to u's left interval. If $p_i > \tau$ we navigate to u's right interval. If $p_i = \tau$ we navigate to the only child of u that is a σ-node. If u is a σ-leaf then we are at u's suffix interval.

To search within a suffix interval $[j, k]$ we apply the standard suffix array search beginning with boundaries $[j, k]$. The time to search in this structure is $O(m + \log I)$, where I is the interval size. Hence, the following is important.

Lemma 4. *Every suffix interval in a suffix tray is of size $O(|\Sigma|^2)$.*

Proof. Consider an (i, j)-suffix interval, i.e. the interval $[L(v_i), R(v_j)]$ which stems from a node v with children $v_1, ..., v_r$. Note that by Lemma 1 the (i, j)-suffix interval contains the suffixes which are represented by leaves in the subtrees of

$v_i, ..., v_j$. However, $v_i, ..., v_j$ are not σ-nodes (by suffix tray construction). Hence, each subtree of those nodes contains at most $|\Sigma| - 1$ leaves. Since $j - i + 1 \leq |\Sigma|$ the overall size of the (i, j)-suffix interval is $O(|\Sigma|^2)$.

A suffix interval $[L(v), R(v)]$ is maintained only for σ-leaves. As none of the children of v are σ-nodes this is a special case of the (i, j)-suffix interval.   □

By the discussion above and Lemma 4 the running time for answering an indexing query is $O(m + \log |\Sigma|)$. Summarizing the discussion of the whole section we can claim the following.

Theorem 1. *Let S be a length n string over an alphabet Σ. The suffix tray of S is (1) of size $O(n)$, (2) can be constructed in $O(n + construct_{ST}(n, \Sigma) + construct_{SA}(n, \Sigma))$ time (where $construct_{ST}(n, \Sigma)$ and $construct_{SA}(n, \Sigma)$ are the times to construct the suffix tree and suffix array) and (3) supports indexing queries (of size m) in time $O(m + \log |\Sigma|)$.*

4   The Online Scenario

In this section we deal with the problem of how to build an indexing structure that supports both text extensions and supports fast(er) indexing. We show that if there exists an online construction for a linear-space suffix tree such that the cost of adding a character is $O(f(n, |\Sigma|))$ (n is the size of the current text), then we can construct an online linear-space data-structure for indexing that supports indexing queries in time $O(m + \log |\Sigma|)$, where the cost of adding a character is $O(f(n, |\Sigma|) + \log |\Sigma|)$. During the construction, we will treat the online linear-space suffix-tree construction as a suffix-tree oracle that provides us with the appropriate updates to the suffix tree. Specifically, the best known current construction supports text extensions in $O(\log n)$, see introduction. As already mentioned, the data structure we present is called the Suffix Trist.

The suffix trist imitates the suffix trays. We still use σ-nodes and branching-σ-nodes in the suffix tree, and the method for answering indexing queries is similar. However, new issues arise in the online model:

- Suffix arrays are static data structures and, hence, do not support insertion of new suffixes.
- The status of nodes changes as time progresses (non-σ-nodes become σ-nodes, and σ-nodes become branching-σ-nodes).

4.1   Balanced Indexing Structures and Suffix Trists

In order to solve the first of the two problems we turn to a dynamic variant of suffix arrays which can be viewed as a structure above a suffix list.

The Balanced-Indexing-Structure, BIS for short, was introduced in [1]. BIS is a binary search tree over the suffixes where the ordering is lexicographic. In [1] it was shown how the BIS can be updated in $O(\log n)$ time for every *text extension*, where n is the current text size. Moreover, a BIS supports indexing queries in time $O(m + \log n)$, where m is the query size.

This leads to the following idea for creating a *Suffix Trist* instead of a suffix tray. Take a separate BIS for every suffix interval. Since the suffix intervals are of size $O(|\Sigma|^2)$ the search time in those small BISs will be $O(m + \log |\Sigma|)$.

However, things are not as simple as they seem. Insertion of suffix aS into a BIS for a string S assume that we have all the suffixes of S in the BIS, or more specifically assumes that the suffix S itself is in the BIS. This may not be the case if we limit the BIS to a suffix interval, which contains only part of the suffixes. Nevertheless, in our case there is a way to circumvent this problem. We describe the solution in the next subsection.

Also, we still need to deal with the second problem of nodes changing status. Our solution is a direct deamortized solution and is presented in Section 5.

4.2 Inserting New Nodes into BISs

When we perform a text extension from S to aS, the suffix tree is updated to represent the new text aS (by our suffix-tree oracle). Specifically, a new leaf, corresponding to the new suffix, is added to the suffix tree, and perhaps one internal node is also added. If such an internal node is inserted, then that node is the parent of the new leaf and this happens in the event that the new suffix diverges from an edge (in the suffix tree of S) at a location where no node previously existed. In this case an edge needs to be broken into two and the internal node is added at that point.

Since we assume that the online suffix tree is given to us, what we need to show is how to update the suffix trist using the suffix tree (updated by the oracle). The problem is (1) to find the correct BIS in which to insert the new node and (2) to actually insert it into this BIS. Of course, this may change the status of internal nodes, which we handle in Section 5. We focus on solving (1) and mention that (2) can be solved by BIS tricks in $O(\log |\Sigma|)$ time, which we defer to a full version, but mention that they are similar to what appears in [1] and hence we find it somewhat less interesting here.

The following lemma which we state without proof will be useful and follows from the definition of suffix trists.

Lemma 5. *For a node u in a suffix tree, if u is not a σ-node, then all of the leaves in u's subtree are in the same BIS.*

It will also be handy to maintain a pointer $leaf(u)$ to some leaf in u's subtree for every node u in the suffix tree. This variant can easily be maintained under text extensions using standard techniques.

In order to find the correct BIS in which the new node is to be inserted we consider two cases. First, consider the case where the new leaf u in the suffix tree is inserted as a child of an already existing internal node v. If v is not a σ-node, then from Lemma 5 we know that $leaf(v)$ and u need to be in the same BIS. By traversing up from $leaf(v)$ to the root of the BIS (in $O(\log |\Sigma|)$ time) we can find the root of the BIS which needs to include the new node u. If v is a σ-node, then we can locate the root of the appropriate BIS in constant time: if v is a branching-σ-node, then we can find the BIS in constant time from the array in v (we will guarantee that this holds in the online setting as well). If v is a σ-leaf then there is only one possible BIS. Otherwise, (v is a non-leaf σ-node but not a branching-σ-node) we can the find correct BIS of the two possible BISs by examining the separating character maintained in v.

Next, consider the case where the new leaf u in the suffix tree is inserted as a child of a new internal node v. Let w be v's parent, and let w' be v's other child (not u). We first ignore u completely by treating the new tree as the suffix tree of S where the edge (w, w') is broken into two, creating the new node v. After we show how to update the trist to include v, we can add u as we did in the case that v was already an internal node. In order to determine the status of v, note that v cannot be a branching-σ-node. Moreover, note that the number of leaves in v's subtree is the same as the number of leaves in the subtree rooted at w' (as we are currently ignoring u). So, if w' is not a σ-node, v is not a σ-node, and otherwise, v is a σ-node with a separating character that is the first character of the label of edge (v, w'). Note that the entire process takes $O(\log |\Sigma|)$ time, as required.

5   When a Node Changes Status

Before explaining how to update a node that becomes a σ-node, we must explain how to detect that this event has taken place. This is explained next.

5.1   Detecting a New σ-Node

Let u be a new σ-node and let v be its parent. Just before u becomes a σ-node, (1) v must have already been a σ-node and (2) $u \in \{v_i, ..., v_j\}$ and is associated with an (i, j)-suffix interval represented by a suffix interval node w that is a child of v in the suffix trist. Hence, we will be able to detect when a new σ-node is created if we maintain counters for each of the (suffix tree) nodes $v_i, ..., v_j$ to count the number of leaves in their subtrees (in the suffix tree). These counters are maintained in a binary search tree, which we associate with the BIS, and each counter v_k is indexed by the the first character on the edge (v, v_k).

The update of the counters can be done as follows. When a new leaf is added into a given BIS of a suffix trist at suffix interval node w of the BIS, where u is the parent of w, we need to increase the counter of v_k in the BIS, where v_k is the one node (of the nodes of $v_i, ..., v_j$ of the suffix interval of the BIS) which is an ancestor of the new leaf. The counter of v_k can be found in $O(\log |\Sigma|)$ (as there are at most $|\Sigma|$ nodes in the binary search tree for the counters). The appropriate counter is found by searching with the character that appears on the new suffix at the location immediately after $|label(v)|$; the character can be found in constant time by accessing it from a pointer from node v into the text.

Note that when a new internal node was inserted into the suffix tree as described in the previous section, it is possible that the newly inserted internal node is now one of the nodes $v_i, ..., v_j$ for an (i, j)-suffix interval. In such a case, when the new node is inserted, it copies the number of leaves in its subtree from its only child (as we explained in the previous section we ignore the newly inserted leaf), as that child was previously maintaining the number of leaves in its subtree. Furthermore, from now on we will only update the size of the subtree of the newly inserted node, and not the size of its child subtree.

5.2   Updating the New σ-Node

Let u be the new σ-node (which is, of course, a σ-leaf) and let v be its parent. As discussed in the previous subsection $u \in \{v_i, ..., v_j\}$ where $v_i, ..., v_j$ are the children of v (in the suffix tree) and, as discussed in the previous subsection, just before becoming a σ-node there was a suffix interval node w that was an (i, j)-suffix interval with a BIS representing it.

Updating the new σ-node will require two things. First, we need to split the BIS into 3 parts; two new BISs and the new σ-leaf that separates between them. Second, for the new σ-leaf we will need to add the separating character (easy) and to create a new set of counters for the children of u (more complicated).

The first goal will be to split the BIS that has just been updated into three - the nodes corresponding to suffixes in u's subtree, the nodes corresponding to suffixes that are lexicographically smaller than the suffixes in u's subtree, and the nodes corresponding to suffixes that are lexicographically larger than the suffixes in u's subtree.

As is well-known, for a given a value x, splitting a BST, balanced suffix tree, into two BSTs at value x can be implemented in $O(h)$ time, where h is the height of the BST. The same is true for BISs (although there is a bit more technicalities to handle the auxiliary information). Since the height of BISs is $O(\log |\Sigma|)$ we can split a BIS into two BISs in $O(\log |\Sigma|)$ time and by finding the suffixes (nodes in the BIS) that correspond to the rightmost and leftmost leaves of the subtree of u, we can split the BIS into the three desired parts in $O(\log |\Sigma|)$ time. Fortunately, we can find the two nodes in the BIS in $O(\log |\Sigma|)$ time using $leaf(u)$, the length of $label(u)$, and the auxiliary data in the BIS. We leave the full details for the full version.

We now turn to creating the new counters. Denote the children of u by $u_1, ..., u_k$. We first note that the number of suffixes in a subtree of u_i can be counted in $O(\log |\Sigma|)$ time by a traversal in the BIS using classical tricks of binary search trees. We now show that we have enough time to update all the counters of $u_1, ..., u_k$ before one of them becomes a σ-node, while still maintaining the $O(\log |\Sigma|)$ bound per update.

Specifically, we will update the counters during the first k insertions into the BIS of u (following the event of u becoming a σ-node). At each insertion we update one of the counters. What is critical for this to be done in time for the counters to be useful, i.e. in time to detect a new σ-node occurring in the subtree of u. The following lemma is precisely what is needed.

Lemma 6. *Let u be a node in the suffix tree, and let $u_1, ..., u_k$ be u's children (in the suffix tree). Say u has just become a σ-node. Then at this time, the number of leaves in each of the subtrees of u's children is at most $|\Sigma| - k + 1$.*

Proof. Assume by contradiction that this is not the case. Specifically, assume that child v_i has at least $|\Sigma| - k + 2$ leaves in its subtree at this time. Clearly, the number of leaves in each of the subtrees is at least one. So summing up the number of leaves in all of the subtrees of $u_1, ..., u_k$ is at least $|\Sigma| - k + 2 + k - 1 = |\Sigma| + 1$, contradicting the fact that u just became a σ-node (it should have already been a σ-node). □

5.3 When a σ-Leaf Loses Its Status

The situation where a σ-leaf becomes a non-leaf σ-node is actually a case that we have covered in the previous subsection. Let v be a σ-leaf that is about to change its status to a non-leaf σ-node. This happens because one of its children v_k is about to become a σ-leaf. Note that just before the change v is a suffix interval node. As in the previous subsection we will need to split the BIS representing the suffix interval into three parts, and the details are exactly the same as in the previous subsection. As before this is done in $O(\log|\Sigma|)$ time.

5.4 When a σ-Node Becomes a Branching-σ-Node

Let v be a σ-node that is changing its status to a branching-σ-node. Just before it changes its status it had exactly one child v_j which was a σ-node. The change in status must occur because another child (in the suffix tree), say v_i, has become a σ-leaf (and now that v has two children that are σ-nodes it has become a branching-σ-node).

Just before becoming a branching-σ-node v contained a separating character τ, the first character on the edge (v, v_j), and two suffix interval nodes w and x, corresponding to the left interval of v and the right interval of v, respectively. Now that v_i became a σ-leaf w was split into three parts (as described in subsection 5.2). Assume, without loss of generality, that v_i precedes v_j in the list of v's children. So, in the suffix trist the children of v are (1) a suffix interval node w_L, (2) a σ-leaf v_i, (3) a suffix interval node w_L, (4) a σ-node v_j, and (5) a suffix interval node x. We will denote with B_1, B_2 and B_3 the BISs that represent the suffix interval nodes w_L, w_R and x.

The main problem here is that constructing the array A_v takes too much time, so we must use a different approach and spread the construction over some time. We first give a pseudo-amortized solution and then mention how to (really) deamortize it. The following lemma allows us this time.

Lemma 7. *From the time that w becomes a branching-σ-node, at least $|\Sigma|$ insertions are required into B_1, B_2 or B_3 before any node in the subtree of v (in the suffix tree) that is not in the subtrees of v_i or v_j becomes a branching-σ-node.*

Proof. Clearly, at this time, any node in the subtree of v (in the suffix tree) that is not in the subtrees of v_i or v_j has fewer than $|\Sigma|$ leaves in its subtree. On the other hand, note that any branching-σ-node must have at least $2|\Sigma|$ leaves in its subtree, as it has at least two children that are σ-nodes, each contributing at least $|\Sigma|$ leaves. Thus, in order for a node in the subtree of v (in the suffix tree) that is not in the subtrees of v_i or v_j to become a branching-σ-node, at least $|\Sigma|$ leaves need to be added into its subtree, as required.   □

This yields the pseudo-amortized result, as we can always amortize the A_v construction over its insertions into B_1, B_2 and B_3. The crucial observation that follows from Lemma 7 that on any given search path we charge for at most one branching-σ-nodes construction, even if we go through several branching-σ-nodes.

The reason that we call this a pseudo-amortized result is because the A_v construction charges on future insertions that may not occur. So, we take a lazy approach to solve this problem and this also yields the deamortized result.

We start by using the folklore trick of initializing the array A_v in constant time. Then every time an insertion takes place into one of B_1, B_2 or B_3 we add one more element to the array A_v. Lemma 7 assures us that A_v will be constructed before we begin to handle a branching-σ-node that is a descendant of v but not of v_i or v_j.

This scheme allows us to construct A_v while maintaining the $O(\log |\Sigma|)$ time bound. However, it is still unclear how an indexing query should be answered when encountering v on the traversal of the suffix tree. This is because on the one hand A_v might not be fully constructed, and on the other hand, as time progresses, v might have a non-constant number of children that are σ-nodes. We overcome this issue as follows. We continue to maintain the initial separating character τ of v and another separator τ', the first character on the edge (v, v_i), until A_v is fully constructed. If we are at v during a traversal for a query and the continuation of the traversal is to either v_i or v_j then we can discover this in constant time from tau' or τ. For the rest of the children of v that are σ-nodes, maintain them all in a BST, so that when answering an indexing query, we can discover the appropriate place to continue in $O(log|\Sigma|)$ time (as there are only $|\Sigma|$ children). This does not affect the time it takes to answer an indexing query as we are guaranteed by Lemma 7 that if we need to use the BST of the children that are σ-nodes, then we will not encounter any more branching-σ-nodes afterwards. Thus, we at most add another $O(log|\Sigma|)$ to the query time.

There is one more loose end that we need to deal with. When other children of v (other, as opposed to v_i and v_j) become σ-nodes during the construction of A_v, this can affect many of the locations of A_v. Specifically, updating accordingly could take too much time (or might require too many insertions in order to complete it). In order to solve this problem we define A_v in a slightly different way as opposed to the static case in order to support this. Each entry in A_v will point us to the edge whose label begins with the character of that entry, if such an edge exists. If no such edge exists, we simply put a NIL. This still allows us to spend constant time per branching-σ-node when answering an indexing query. However, when we go on to the edge pointed by the appropriate location (during the process of answering a query), we look at the node v' on the other side of the edge. If v' is a σ-node, we continue to traverse from there. If v' is not a σ-node, then we can find the appropriate BIS by following $leaf(u)$, and traversing up to the root of the BIS in $O(\log|\Sigma|)$ time. Now, when a new node becomes a σ-node, and its parent is already a branching-σ-node, no more changes are required.

Theorem 2. *Let S be a string over an alphabet Σ. The suffix trist of S is (1) of size $O(n)$, (2) supports text extensions in time $O(\log|\Sigma|) + extension_{ST}(n, \Sigma))$ time (where $extension_{ST}(n, \Sigma)$is the time for a text extension in the suffix tree) and (3) supports indexing queries (of size m) in time $O(m + \log|\Sigma|)$.*

References

1. A. Amir, T. Kopelowitz, M. Lewenstein, and N. Lewenstein. Towards Real-Time Suffix Tree Construction. *Proc. of Symp. on String Processing and Information Retrieval (SPIRE)*, 67-78, 2005.
2. T.H. Cormen, C.E. Leiserson, R.L. Rivest, and C. Stein. *Introduction to Algorithms*. MIT Press, second edition, 2001.
3. M. Farach. Optimal suffix tree construction with large alphabets. *Proc. 38th IEEE Symposium on Foundations of Computer Science*, pages 137–143, 1997.
4. R. Grossi and G. F. Italiano. Efficient techniques for maintaining multidimensional keys in linked data structures. In *Proc. 26th Intl. Col. on Automata, Languages and Programming (ICALP)*, LNCS 1644, pages 372–381, 1999.
5. Juha Kärkkäinen and Peter Sanders. Simple linear work suffix array construction. In *Proc. 30th International Colloquium on Automata, Languages and Programming (ICALP 03)*, LNCS 2719, pages 943–955, 2003.
6. D.K. Kim, J.S. Sim, H. Park, and K. Park. Linear-time construction of suffix arrays. *Proc. of 14th Symposium on Combinatorial Pattern Matching*, 186-199, LNCS 2676, 2003.
7. P. Ko and S. Aluru. Space efficient linear time construction of suffix arrays. *Proc. of 14th Symposium on Combinatorial Pattern Matching*, 200-210, LNCS 2676, 2003.
8. U. Manber and E.W. Myers. Suffix arrays: A new method for on-line string searches. *SIAM J. on Computing*, 22(5):935-948, 1993.
9. E. M. McCreight. A space-economical suffix tree construction algorithm. *J. of the ACM*, 23:262–272, 1976.
10. E. Ukkonen. On-line construction of suffix trees. *Algorithmica*, 14:249–260, 1995.
11. P. Weiner. Linear pattern matching algorithm. *Proc. 14th IEEE Symposium on Switching and Automata Theory*, pages 1–11, 1973.

Optimal Lower Bounds for Rank and Select Indexes

Alexander Golynski

David R. Cheriton School of Computer Science, University of Waterloo
agolynski@cs.uwaterloo.ca

Abstract. We develop a new lower bound technique for data structures. We show an optimal $\Omega(n \lg \lg n / \lg n)$ space lower bounds for storing an index that allows to implement rank and select queries on a bit vector B provided that B is stored explicitly. These results improve upon [Miltersen, SODA'05]. We show $\Omega((m/t) \lg t)$ lower bounds for storing rank/select index in the case where B has m 1-bits in it (e.g. low 0-th entropy) and the algorithm is allowed to probe t bits of B. We simplify the select index given in [Raman *et al.*, SODA'02] and show how to implement both rank and select queries with an index of size $(1 + o(1))(n \lg \lg n / \lg n) + O(n / \lg n)$ (i.e. we give an explicit constant for storage) in the RAM model with word size $\lg n$.

1 Introduction

The term *succinct data structure* was first used by Jacobson in [2], where he defined and proposed a solution to the following problem of implementing rank and select queries. We are given a bit vector B of length n. The goal is to represent B in such a way that rank and select queries about B can be answered efficiently. Query $rank_B(i)$ returns the number of 1-bits in B before (and including) the position i, and $select_B(i)$ query returns the position of the i-th occurrence of 1 in B. We require that the representation should be succinct, that is, the amount of space S it occupies is close to the information-theoretic minimum, namely $S = n + o(n)$ in the case of bit vectors of length n. We consider this problem in the RAM model with word size $w = \Theta(\lg n)$. Jacobson proposed a data structure to perform rank queries that uses $n + O(n \lg \lg n / \lg n)$ bits of space and requires only $O(1)$ time to compute the answer. His implementation of the select query requires $O(\lg n)$ bit accesses, but it does not take advantage of word parallelism and runs in time that is more than a constant in RAM model. It was subsequently improved by Clark [1], Munro *et al.* [4,5], and Raman *et al.* [6]. The index proposed by Raman *et al.* [6] occupies $O(n \lg \lg n / \lg n)$ bits, and the select query is implemented in $O(1)$ time. All these data structures belong to a class of *indexing data structures*. An indexing data structure stores data in "raw form" (i.e. B is stored explicitly) plus a small index I to facilitate implementation of queries, such as rank and select. We denote the size of the index by r.

Miltersen [3] showed that any indexing data structure that allows $O(1)$ time implementation of rank (select) queries must use an index of size at least

M. Bugliesi et al. (Eds.): ICALP 2006, Part I, LNCS 4051, pp. 370–381, 2006.

$\Omega(n \lg \lg n / \lg n)$ bits (respectively $\Omega(n / \lg n)$ bits). The purpose of this paper is to develop a new technique for showing lower bounds for indexing data structures. This technique allows to improve lower bounds of Miltersen [3] for both rank and select problems to match the corresponding upper bounds.

For our lower bounds, we use the general *indexing model* which can be described as follows. Our goal is to store and perform a set of queries on a given family of combinatorial objects (e.g. rank/select on bit vectors). Let B be a "raw" representation of a combinatorial object (e.g. a raw bit vector). We assume that B is given to us free of charge, e.g. it is stored in external memory such as a large database; or is provided by outside world, e.g. web graph [7]. Let I be an index that aids performing the set of queries efficiently; presumably it is stored in a relatively fast, expensive and/or limited memory. We are charged 1 unit of space for each bit in I, while access to I is free of charge. An algorithm that performs a query has unlimited computation power, however we are charged 1 unit of time when it accesses (e.g. probes one bit) B.

We show that any algorithm in the indexing model that performs rank (respectively select) queries with the time cost $O(\lg n)$, must have the space cost at least $\Omega(n \lg \lg n / \lg n)$. Note that this setting is general enough; in particular, it subsumes $O(1)$ time RAM algorithms with word size $O(\lg n)$. Hence, (i) for the select index, we improve the lower bound of Miltersen to the optimal; (ii) for the rank index, we show the same bound, but in a more general setting.

We also consider the case where the number of 1-bits in a bit vector B is some given number m (we call it *cardinality*). In this setting, for both rank and select problems, we prove a theorem that any algorithm with the time cost t has the space cost $\Omega((m/t) \lg t)$. In particular this lower bound is optimal for bit vectors of constant 0-th order entropy. This theorem also yields strong lower bounds in the case $m = \Omega(n / \lg \lg n)$.

We also give an implementation of select query that is simpler than the one proposed by Raman *et al.* [6]. We also give an index that allows to implement both rank and select queries in $O(1)$ time and uses space $(1 + o(1))(n \lg \lg n / \lg n) + O(n / \lg n)$. Thus, we give an explicit constant in front of the leading term $n \lg \lg n / \lg n$. This index is simple and space efficient, and it might be of interest to practitioners.

This paper is organized as follows. In the section 2, we give an implementation for rank and select queries. In the section 3, we prove lower bounds for rank and select indexes. In the section 4, we generalize the lower bounds from the section 3 to the case of bit vectors with a given cardinality.

2 Upper Bounds

In this section, we will simplify the result of Raman *et. al* [6] that gives an optimal index for the select query of size $O(n \lg \lg n / \lg n)$. We assume that the word size is $w = \lg n$ for the part of the paper that deals with upper bounds; in contrast, all lower bounds are shown in the indexing model (bit probes).

Then we will construct an optimal index for rank query of size $(1 + o(1))(n \lg \lg n / \lg n) + O(n / \lg n)$. A similar result was obtained by Jacobson [2]; however we implement both the rank and the select indexes simultaneously, such that the space used is just $n + (1 + o(1))(n \lg \lg n / \lg n) + O(n / \lg n)$.

Both of these indexes share a component of size $(1 + o(1))(n \lg \lg n / \lg n)$ that we call a *count index*. The count index is constructed as follows: we split our bit vector B into *chunks* of size $\lg n - 3 \lg \lg n$. Then we store the number of 1-bits in each chunk (we call it *cardinality of a chunk*) in equally spaced fields of size $\lg \lg n$ for a total of $n \lg \lg n / (\lg n - 3 \lg \lg n) = (1 + o(1)) n \lg \lg n / \lg n$ bits.

2.1 Optimal Select Index

In this subsection, we describe a new simplified select index that uses the count index plus an additional $O(n / \lg n)$ bits. Let B be the bit vector of length n. Let $S_1 = (\lg n)^2$. We store the locations of each (iS_1)-th occurrence of 1-bit in B, for each $1 \le i \le n / S_1$. Note that normally the number of 1-bits in B is less than n, so part of the range is unused. This takes $O(n / \lg n)$ bits in total. We call regions from position select(iS_1) to position select$((i+1)S_1) - 1$ *upper blocks*. To perform select$_B(i)$, we first compute $j = \lfloor i / S_1 \rfloor$ the number of the upper block that the i-th bit is in, so that

$$\text{select}_B(i) = \text{select}_B(jS_1) + \text{select}_{UB_i}(i \bmod S_1) \tag{1}$$

where select$_{UB_i}$ denotes the select query with respect to the i-th upper block. We call such an operation *reduction from cardinality n to S_1*. Now we need to implement the select query for upper blocks. We call an upper block *sparse* if its length is at least $(\lg n)^4$. For a sparse block, we can just explicitly write answers for all possible select queries, this will use at most $(\lg n)^3$ bits. Intuitively, this is roughly at most $O(1 / \lg n)$ indexing bits per one bit from B, so that the total space used up by this part of the index sums up to $O(n / \lg n)$ bits. We will repeatedly use this $1 / \lg n$ rule.

Let us consider a non-sparse upper block. It is a bit vector of cardinality S_1 and length at most $(\lg n)^4$. Thus, it takes $O(\lg \lg n)$ bits to encode a pointer within such a block. We perform cardinality reduction from S_1 to $S_2 = \lg n \lg \lg n$. Similarly to upper blocks, we introduce *middle blocks*, each having cardinality S_2. That is, encode every (iS_2)-th occurrence of 1-bit in an upper block. This information occupies $O(\lg \lg n \cdot \lg n / \lg \lg n) = O(\lg n)$ bits for an upper block of length at least $(\lg n)^2$, so that we use $1 / \lg n$ bits for index per one bit from B, for a total of at most $O(n / \lg n)$ bits. We call a middle block *sparse* if it has length more than $(\lg n \lg \lg n)^2$. If a middle block is sparse, then we can explicitly write positions of all occurrences of 1-bits in it, this uses at most $\lg n (\lg \lg n)^2$ bits (at most $O(1 / \lg n)$ indexing bits per one original bit). We call a middle block *dense* if its length is at most $\frac{(\lg n)^2}{4 \lg \lg n}$.

If a middle block is neither sparse nor dense, then use cardinality reduction from S_2 to $S_3 = (\lg \lg n)^3$. Call the resulting blocks of cardinality S_3 *lower blocks*. That is, store every (iS_3)-th occurrence of 1 in a middle block. This uses $\lg n / \lg \lg n$ bits per block of length at least $\frac{(\lg n)^2}{2 \lg \lg n}$, hence $O(1 / \lg n)$ indexing bits per original bit. We say that lower block is *sparse* if it has length at least

$\lg n (\lg \lg n)^4$ and *dense* otherwise. If a lower block is sparse, then we can explicitly encode all 1-bit occurrences in it.

It remains to implement select query for dense middle and lower blocks. Consider, for example, a dense middle block MB and implement $select_{MB}(i)$ on it. We first assume that MB is aligned with chunks, i.e. its starting (ending) position coincide with starting (ending) position of some chunk (chunks are of the size $\lg n - 3 \lg \lg n$). Recall that the length of MB is at most $(\lg n)^2 / 4 \lg \lg n$, so that the part P of the count index that covers block MB (i.e. P encodes cardinality of each chunk inside MB) is of the size at most $(\lg n)/2$. Hence, we can read P in one word and perform a lookup to a table T to compute the number of the chunk where i-th 1-bit of MB is located. Table T is of size at most $\sqrt{n} \lg n \lg \lg n \times \lg \lg n$, and it stores for each possible choice of P and for each $j = O(\lg n \lg \lg n)$ the number of the chunk where j-th occurrence of 1 is located (denote the corresponding chunk by C), and the rank of that occurrence inside C (denote it by p). Now we can compute $select_{MB}(j)$ by reading chunk C and performing $select_C(p)$ using a lookup to a table Q. Table Q is of size at most $O(2^{\lg n - 3 \lg \lg n} \cdot \lg n \times \lg \lg n) = O(n / \lg n)$, and it stores for each possible chunk C and for each position k the result of $select_C(k)$. The case where MB is not aligned with chunks can be resolved by counting number of 1-bits in the first chunk that partially belongs to MB (e.g. a lookup to a table that computes rank within a chunk, we discuss this table later in the next subsection) and adjusting j accordingly. Clearly, select query for the case of dense lower blocks can be implemented in the same way.

2.2 Optimal Rank Index

In this subsection, we show how to design the rank index using the count index and additional $O(n / \lg n)$ bits.

We divide the bit vector B into equally sized *upper blocks* of size $S_1 = (\lg n)^2$ bits each. For each upper block, we write the rank of the position preceding its first position ($rank_B(0) = 0$). This information uses $O(n / \lg n)$ bits total. Now we can compute $rank_B(i)$ as follows: first we compute $j = \lfloor i / S_1 \rfloor$, the number of the upper block that contains i-th bit of B (denote the upper chunk by UC), so that

$$rank_B(i) = \text{rank}_B(jS_1 - 1) + rank_{UC}(i \bmod S_1) \qquad (2)$$

We call such an operation a *length reduction* from n to S_1. Then we perform another length reduction from S_1 to $S_2 = \lg n \lg \lg n$. We call the corresponding blocks of length S_2 *middle blocks*. It takes $\lg n$ bits per an upper block of length $(\lg n)^2$ to describe ranks of the starting positions of middle blocks (each rank uses $\lg \lg n$ bits), so that we use $1 / \lg n$ bits for index per one bit of B. Without loss of generality, we can assume that middle blocks are always aligned with chunks. Let MB be a middle block, we implement $rank_{MB}(i)$ as follows. Let $j = O(\lg \lg n)$ be the number of the chunk (denote it by C) that contains the i-th bit of MB, $j = \lfloor i / S_3 \rfloor$, where $S_3 = \lg n - 3 \lg \lg n$ denotes the length of a chunk. One middle block of size S_2 corresponds to a part P of counting index of size at most $O((\lg \lg n)^2)$ bits, so that we can read it in one word and use one lookup to a table T to compute $\text{rank}_{MB}(jS_3 - 1)$. Table T is of size $(\lg n)^{O(1)} \lg \lg \lg n \times \lg \lg n$, and it stores $\text{rank}_{MB}(jS_3 - 1)$ for each possible part P and chunk number j. Thus,

$$\text{rank}_{MB}(i) = \text{rank}_{MB}(jS_3 - 1) + \text{rank}_C(j \bmod S_3) \tag{3}$$

and the latter rank can be also computed by one lookup to the table Q. Recall that table Q of size $O(n/\lg n)$ stores for each possible chunk C and for each position k the result of $\text{select}_C(k)$.

3   Lower Bounds

In this section, we consider lower bounds for rank and select algorithms in the indexing model with the time cost $O(\lg n)$, we denote the space cost (i.e. index size) by r.

3.1   Rank Index

In this subsection we develop a new combinatorial technique and obtain

$$r = \Omega(n \lg n / \lg \lg n) \tag{4}$$

lower bound for the rank index.

Let us fix the mapping between bit vectors B and indexes I and fix an algorithm that performs the rank query (i.e. it computes $\text{rank}_B(p)$ for a given p). As we mentioned before, an algorithm is allowed to perform unlimited number of bit probes to I and has unlimited computation power; we only limit the number of bit probes it can perform to the bit vector B. Let us fix the number of bit probes $t = f \lg n$ for some constant $f > 0$. We split the bit vector B into p blocks of size $k = t + \lg n$ each. Let n_i be the number of 1-bits in the i-th block, we call n_i the *cardinality* of i-th block. For each block i, $1 \le i \le p$, we simulate the rank query on the last position of the i-th block, $s_i = \text{rank}_B(ik - 1)$, so that $n_i = s_{i+1} - s_i$. Note that we will have at least $n - pt = p \lg n = \Omega(n)$ unprobed bits after the computation is complete. Now we will construct a binary *choices tree*. The first r levels correspond to all possible choices of index. At each node at depth r of the tree constructed so far, we will attach the decision tree of the computation that rank algorithm performed for the query $\text{rank}_B(k - 1)$ when index I is fixed. The nodes are labeled by the positions in the bit vector that algorithm probes and two outgoing edges are labeled 0 or 1 depending on the outcome of the probe; we call the corresponding probe a 0-*probe* or 1-*probe* respectively. At each leaf the previously constructed tree, we attach the decision tree for $\text{rank}_B(2k - 1)$ and so on. Thus, the height of the tree is at most $r + tp$. If the computation probes the same bit twice (even if the previous probe was performed for a different rank query), we do not create a binary node for the second and latter probes; instead we use the result of the first probe to the bit. At the leaves of the tree all block cardinalities n_i are computed. Let us fix a leaf x, we call a bit vector B *compatible* with x iff: (1) the index (i.e. first r

nodes) on the root to the leaf path correspond to the index for B; and (2) the remaining nodes on the root to the leaf path correspond to the choices made by the computation described above.

Let us bound the number $C(x)$ of bit vectors B that are compatible with a given leaf x (in what follows we will use C to denote $C(x)$).

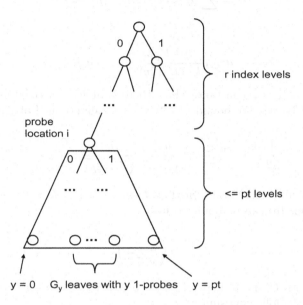

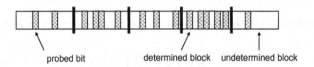

Fig. 1. Choices tree. The leaves could be at different levels. Notation G_y will be defined and used later in Section 4.

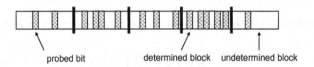

Fig. 2. Bit vector at a leaf

Let u_i be the number of unprobed bits in the block i, so that $u_i \leq k$ and

$$\sum_{i=1}^{p} u_i = U \tag{5}$$

where U is the total number of unprobed bits. At a given leaf, we have computed all n_i's, and hence the sum of all unprobed bits (denote it by v_i) in the block i equals to n_i minus the number of 1-probes in the i-th block. Therefore, we can bound the number of bit vectors compatible with x by

$$\frac{C}{2^U} \leq \frac{\binom{u_1}{v_1} \binom{u_2}{v_2}}{2^{u_1} \, 2^{u_2}} \cdots \frac{\binom{u_p}{v_p}}{2^{u_p}} \tag{6}$$

Let us classify blocks into two categories: determined and undetermined. We call block i *determined* if $u_i \leq U/(2p)$ (intuitively, when it has less than half of the "average" number of unprobed bits) and call it *undetermined* otherwise. Let d be the number of determined blocks. Then

$$U = \sum_i u_i \leq dU/(2p) + (p-d)k \tag{7}$$

hence

$$d \leq p\frac{k - U/p}{k - U/(2p)} \leq (1-a)p \tag{8}$$

where $0 < a < 1$ is a constant. Thus, there is at least a constant fraction of undetermined blocks. We bound $\binom{u_i}{v_i}/2^{u_i} < 1$ for determined blocks, and

$$\frac{\binom{u_i}{v_i}}{2^{u_i}} \leq \frac{\binom{u_i}{u_i/2}}{2^{u_i}} < \frac{b}{\sqrt{u_i}} \leq \frac{b}{\sqrt{U/(2p)}} < \frac{c}{\sqrt{\lg n}} \tag{9}$$

for undetermined blocks using Stirling formula, where $b > 0$ and $c > 0$ are constants. Thus (6) can be bounded by

$$\frac{C}{2^U} \leq \left(\frac{c}{\sqrt{\lg n}}\right)^{ap} \tag{10}$$

Recall that both C and U depend on x, so that $U(x) = n + r - \text{depth}(x)$. We can compute the following sum

$$\sum_{x \text{ is a leaf}} 2^{U(x)} = 2^{n+r} \sum_{x \text{ is a leaf}} 2^{-\text{depth}(x)} = 2^{n+r} \tag{11}$$

The total number of bit vectors B compatible with some leaf is at most

$$\sum_{x \text{ is a leaf}} C(x) \leq 2^{n+r}\left(\frac{c}{\sqrt{\lg n}}\right)^{ap} \tag{12}$$

However, each bit vector has to be compatible with at least one leaf

$$2^n \leq \sum_{x \text{ is a leaf}} C(x) \tag{13}$$

Thus

$$r = \Omega(n \lg \lg n / \lg n) \tag{14}$$

The index presented in the previous section matches this lower bound up to a constant factor.

Note that the techniques given by Miltersen [3] does not allow to obtain the bound (4) in the case where we can perform $O(\lg n)$ bit probes to the bit vector B; although in a more restricted case where only $O(1)$ word probes are allowed his lower bound (4) is optimal. Miltersen [3] showed that the rank index has to be of size r, such that

$$2(2r + \lg(w + 1))tw \geq n \lg(w + 1) \tag{15}$$

where w denotes the word size, t denotes the number of word probes, and r is the size of an index. Miltersen reduces

(i) a set of $\Omega(n/\lg n)$ independent problems; problem i is to compute $n_i \bmod w$, where n_i the number of 1-bits in the region $[2ciw, 2c(i + 1)w]$, where c is a constant to

(ii) the problem of computing rank for positions $2ci \lg n$ for all i, $0 \leq i \leq n/(2c \lg n)$.

In each region of (i), $2c$ numbers $j_1, j_2, \ldots, j_{2c}$, such that $0 \leq j_k \leq w$ are encoded using unary representation $1^{j_k} 0^{w - j_k}$. He shows a lower bound (4) for (i) when $w = \Theta(\lg n)$. For the case $w = 2$, this method only yields $r = \Omega(n/\lg n)$. One can try to generalize Miltersen's approach to allow $O(\lg n)$ bit probes instead of $O(1)$ word probes. The difficulty is that in the bit probe model, a number at most $\lg n$ represented in unary can be recognized using binary search in $\lg \lg n$ bit probes, so that each independent problem of (i) can be solved in $O(\lg \lg n)$ bit probes without using an index. One can also try to "shuffle" bits in unary representation to disallow such binary searches, however it is not clear whether such a proof can be completed. If one tries an approach, where regions of (i) are of the length $\Omega(\lg n)$ and $w = 2$, but then it suffices to store all the $O(n/\lg n)$ answer bits as the index, so that no bit probes are needed to the bit vector.

3.2   Select Index

In this subsection, we apply a similar combinatorial technique to show an optimal lower bound for the select index. Fix the number of probes to the bit vector B to be $t = f \lg n$ (for some constant $f > 0$) that select algorithm uses and let $k = t + \lg n$ as before.

Let us restrict ourselves to bit vectors B of cardinality $n/2$ ($n/2$ bits are 0 and $n/2$ bits are 1). Let us perform the following $p = n/(2k)$ queries: for each $1 \leq i \leq p$ we simulate $\text{select}(ik)$. Similarly, we construct choices tree for these queries. To compute the number of compatible bit vectors for a given leaf, we split each vector B into p blocks, i-th block is from position $\text{select}_B((i-1)k)+1$ to position $\text{select}_B(ik)$ (we define $\text{select}_B(0) = 0$ for convenience). Note that there are exactly k ones in each block. The total number of unprobed bits U is at least $n - pt = n(1 - t/k) = \Omega(n)$. We can count the number of compatible nodes C for each leaf x by applying the same technique as for rank, and obtain (similarly to (10))

$$\frac{C}{2^U} \leq \left(\frac{c}{\sqrt{\lg n}} \right)^{ap} \tag{16}$$

where $0 < a < 1$ and $0 < c$ are positive constants. Next, we can obtain the bound on the total number of bit vectors B that are compatible with at least one node in the choices tree. Similarly to (12), we have

$$2^{n+r} \left(\frac{c}{\sqrt{\lg n}} \right)^{ap} \tag{17}$$

The total number of bit vectors we are considering is $\binom{n}{n/2}$, thus

$$\binom{n}{n/2} \leq 2^{n+r} \left(\frac{c}{\sqrt{\lg n}} \right)^{ap} \tag{18}$$

and hence

$$r = \frac{\lg \binom{n}{n/2}}{n} \Omega \left(\frac{n \lg \lg n}{\lg n} \right) = \Omega \left(\frac{n \lg \lg n}{\lg n} \right) \tag{19}$$

Now we give an argument that techniques from Miltersen [3] cannot be improved from $r = \Omega(n/\lg n)$ to the optimal $r = \Omega(n \lg \lg n / \lg n)$. Miltersen used only bit vectors that only have $O(n/\lg n)$ 1-bits. However, for such vectors, we can construct an index of size $O(n/\lg n)$ that allows $O(1)$ select queries. Let us divide B into p subregions of size $(\lg n)/2$, for each subregion we count number of 1 bits in it (denote it by n_i) and represent it in unary. We construct the following bit vector

$$L = 1^{n_1} 0 1^{n_2} 0 \ldots 0 1^{n_p} \tag{20}$$

of length $p + O(n/\lg n) = O(n/\lg n)$. To perform $\text{select}_B(j)$ on B, we first find $x = \text{select}_L(i)$ and then $i = x - \text{rank}_L(x)$, the number of 0-bits before the position x in L. Hence i gives us the number of the block of B where j-th 1-bit is located (denote the block K). Next, we compute $z = \text{select0}_L(i)$ (where $\text{select0}_L(j)$ gives position of j-th occurrence of 0-bit in L), the starting position of i-th block in L. And then compute $t = x - z$, so that j-th 1-bit of B is t-th 1-bit of K. Finally, $\text{select}_K(j)$ can be done by a lookup to a table of size $\sqrt{n}(\lg \lg n)^2$ bits that stores results of all possible select queries for all possible blocks. Note that rank and select on L requires at most $o(n/\lg n)$ bits in addition to storing L as we discussed in the previous section. Thus, the total space requirement for the index is $O(n/\lg n)$ bits. It follows that for such bit vectors B select indexes of size $O(n/\lg n)$ are optimal.

We state the results for the rank and the select indexes as the following

Theorem 1. *Let B be a bit vector of length n. Assume that there is an algorithm that uses $O(\lg n)$ bit probes to B plus unlimited access to an index of size r and unlimited computation power to answer rank (respectively, select) queries. Then $r = \Omega(\frac{n \lg \lg n}{\lg n})$.*

4   Density-Sensitive Lower Bounds

In this section, we consider the case where the bit vector B contains some fixed number m of 1-bits and express lower bounds for the rank and the select indexes in terms of both parameters m and n, the length of B. We will use techniques similar to the previous section, however, the calculations are slightly more involved in this case. We will prove a lower bound for the rank index and omit the proof for the select index. Throughout this section, we will use the same notation as in the section 3: p denotes the number of simulation of rank (select) queries in the choices tree; k denotes the size of a block; t denotes the running time of one rank (select) query; $U(x)$ (and its short form U) denotes the number of unprobed bits for a given leaf x of the choices tree; r denotes the size of an index.

First, we assume that all leaves in the choices tree are at the same level $pt+r$, i.e. on every root to leaf path the rank algorithm probes exactly pt bits. If some node x is z levels above it, we perform z fake probes, in order to split it into 2^z nodes at the required level, so that $U = n - pt$ for all leaves. We will choose parameter p, such that $pt \leq n/2$, so that at least half of the bits are unprobed at the end. We will partition all the leaves x into m groups depending on the total number of 1-probes on the root to leaf path to x (excluding the first r levels for the index). Let G_y be the group of leaves for which we performed y 1-probes. Clearly, $|G_y| \leq 2^r \binom{pt}{y}$. For each leaf $x \in G_y$ we can bound the number of compatible bit vectors by:

$$\binom{u_1}{v_1}\binom{u_2}{v_2} \cdots \binom{u_p}{v_p} \tag{21}$$

where

$$u_1 + u_2 + \ldots + u_p = U \tag{22}$$
$$v_1 + v_2 + \ldots + v_p = V \tag{23}$$

where $U = n - pt$ is the number of unprobed bits and $V = m - y$. Similar to the previous section, u_i denotes the number of unprobed bits in i-th block and $v_i \leq u_i$ denotes the sum of these bits. Recall that v_i equals to n_i minus number of 1-probes in i-th block, and hence is fixed for a given leaf. We will combine blocks into larger *superblocks* as follows. The 1-st superblock will contain blocks $1, 2, \ldots, z_1$, such that $k \leq u_1 + u_2 + \ldots u_{z_1} \leq 2k$, the i-th superblock will contain blocks $z_{i-1}, \ldots, z_i$ such that $k \leq u_i^s \leq 2k$, where

$$u_i^s = u_{z_{i-1}+1} + u_{z_{i-1}+2} + \ldots + u_{z_i} \tag{24}$$

is the *size* of i-th superblock. Note that this is always possible, since $u_i \leq k$ for all i. Let q be the number of superblocks, clearly $n/(4k) \leq q \leq n/k$ (equivalently $p/4 \leq q \leq p$), since $U \geq n/2$.

For each superblock, we will use the inequality

$$\binom{u_{z_{i-1}+1}}{v_{z_{i-1}+1}}\binom{u_{z_{i-1}+2}}{v_{z_{i-1}+2}} \cdots \binom{u_{z_i}}{v_{z_i}} \leq \binom{u_i^s}{v_i^s} \tag{25}$$

where $v_i^s = v_{z_{i-1}+1} + v_{z_{i-1}+2} + \ldots + v_{z_i}$. So that

$$\binom{u_1}{v_1}\binom{u_2}{v_2} \cdots \binom{u_p}{v_p} \leq \binom{u_1^s}{v_1^s}\binom{u_2^s}{v_2^s} \cdots \binom{u_q^s}{v_q^s} \tag{26}$$

Observe that for any $q_1 < p_1$ and $p_2 < q_2$

$$\frac{\binom{p_1}{q_1+1}\binom{p_2}{q_2}}{\binom{p_1}{q_1}\binom{p_2}{q_2+1}} = \frac{\frac{p_1-q_1}{q_1+1}}{\frac{p_2-q_2}{q_2+1}} \tag{27}$$

That is

$$\binom{p_1}{q_1+1}\binom{p_2}{q_2} > \binom{p_1}{q_1}\binom{p_2}{q_2+1}, \text{ if } \frac{q_1+1}{p_1+1} < \frac{q_2+1}{p_2+1} \tag{28}$$

We can interpret this inequality as follows. Let us maximize the product (26) with fixed values of u_i^s's, subject to the constraint $v_1^s + v_2^s + \ldots + v_q^s = V$. The point $(v_1^s, v_2^s, \ldots, v_q^s)$ is a local maximum if we cannot increase v_i^s by 1 and decrease v_j^s by 1 for some $i \neq j$, so that (26) increases. Intuitively, at a local maximum all fractions v_i^s/u_i^s are roughly equal, or otherwise we can "transfer" 1 from enumerators of larger fractions to enumerators of smaller ones. We can show (proof omitted) that if $V > 2q$ then $|v_i^s/u_i^s - V/U| < 2/k$ (recall that all u_i^s satisfy $k \leq u_i^s \leq 2k$, so that each "transfer" does not change each of v_i^s/u_i^s fractions by more than $1/k$).

We will use the Stirling approximation

$$\binom{u}{v} = \Theta\left(\frac{\sqrt{u}}{\sqrt{v}\sqrt{u-v}} \frac{(u/e)^u}{(v/e)^v((u-v)/e)^{u-v}}\right) \tag{29}$$

$$= \Theta\left(\frac{1}{\sqrt{v}}\left(\frac{1}{\xi}\right)^v\left(\frac{1}{1-\xi}\right)^{u-v}\right) \tag{30}$$

where $\xi = v/u$, here we assumed that v is large enough, so that we can use the Stirling approximation for $v!$, but $v/u \leq 1/2$. Now we can bound

$$\frac{\binom{u_1^s}{v_1^s}\binom{u_2^s}{v_2^s}\cdots\binom{u_q^s}{v_q^s}}{\binom{U}{V}} = 2^{\Theta(q)}\left(\frac{1}{kV/U}\right)^{q/2}\left(\frac{\phi}{\psi}\right)^V\left(\frac{1-\phi}{1-\psi}\right)^{U-V} \tag{31}$$

where $\phi = V/U \leq 1/2$ and ψ is some number, such that $|\psi - \phi| < 2/k$ (again, we assumed $V/U \leq 1/2$). The latter expression is less than

$$2^{\Theta(q)}\left(\frac{1}{kV/U}\right)^{q/2}\left(1 + \frac{4}{kV/U}\right)^V\left(1 + \frac{4}{k}\right)^U \leq \tag{32}$$

$$\leq 2^{\Theta(q)}\left(\frac{1}{kV/U}\right)^{q/2} 2^{\Theta(U/k)} \leq 2^{\Theta(q)}\left(\frac{1}{kV/U}\right)^{q/2} \tag{33}$$

Here we used two facts (i) $(1 + 1/\alpha)^\beta = 2^{\Theta(\beta/\alpha)}$ for large enough α; and (ii) $U/k \leq q/2$.

Note that there are could have at most pt different groups of leaves ($y \leq pt$). For a given group, we bound the total number of compatible bit vectors by (31). Denote $V^* = m - pt$, and note that $V = m - y \geq V^*$. So that the total number of compatible bit vectors for all groups is at most

$$2^r \sum_{y=0}^{pt} \binom{pt}{y}\binom{n-pt}{m-y}\left(\frac{O(1)}{kV^*/U}\right)^{\Theta(p)} = 2^r\binom{n}{m}\left(\frac{O(1)}{kV^*/U}\right)^{\Theta(p)} \tag{34}$$

However, all possible bit vectors of length n with m ones have to be compatible with at least one leaf, so that

$$r = \Omega(p \lg \frac{n(m - pt)}{p(n - pt)}) = \Omega(p \lg(m/p - t)) \tag{35}$$

Choosing $p = m/(2t)$ gives $r = \Omega((m/t) \lg t)$. Essentially the same technique is applicable to the case of select index (proof omitted).

Theorem 2. *Let B be a bit vector of length n with m ones in it. Assume that there is an algorithm that uses t bit probes to B plus unlimited access to an index of size r and unlimited computation power to answer rank (respectively, select) queries. Then $r = \Omega((m/t) \lg t)$.*

Note that this theorem gives an optimal lower bound for the case of constant density bit vectors (i.e. when $m/n < 1/2$ is a constant). For the select index, it also yields a lower bound better than the one given by Miltersen [3] for the case where $m = \Omega(nt/(\lg n \lg t))$, or $m = \Omega(n/ \lg \lg n)$ when $t = \Theta(\lg n)$.

Acknowledgments

We thank Ian Munro, Prabhakar Ragde, and Jeremy Barbay for fruitful discussions and proof reading this paper. We thank S. Srinivasa Rao for bringing the problem to our attention and pointing out the results of Miltersen. We thank Peter Widmayer who helped to improve presentation of the indexing model while author was visiting ETH in Zurich. We also thank anonymous referees for their helpful comments.

References

1. David R. Clark. *Compact Pat Trees.* PhD thesis, University of Waterloo, 1996.
2. Guy Jacobson. *Succinct Static Data Structures.* PhD thesis, Carnegie Mellon University, January 1989.
3. Peter Bro Miltersen. Lower bounds on the size of selection and rank indexes. In *Proceedings of the 16th Annual ACM-SIAM Symposium on Discrete Algorithms*, pages 11–12, 2005.
4. J. Ian Munro, Venkatesh Raman, and S. Srinivasa Rao. Space efficient suffix trees. In *Proceedings of the 18th Conference on the Foundations of Software Technology and Theoretical Computer Science*, volume 1530 of *Lecture Notes in Computer Science*, pages 186–196. Springer, 1998.
5. J. Ian Munro, Venkatesh Raman, and S. Srinivasa Rao. Space efficient suffix trees. *Journal of Algorithms*, 39(2):205–222, 2001.
6. Rajeev Raman, Venkatesh Raman, and S. Srinivasa Rao. Succinct indexable dictionaries with applications to encoding k-ary trees and multisets. In *Proceedings of the 13th Annual ACM-SIAM Symposium on Discrete Algorithms*, pages 233–242, 2002.
7. Peter Widmayer. Personal communication. 2006.

Dynamic Interpolation Search Revisited[*]

Alexis Kaporis[1,2], Christos Makris[1], Spyros Sioutas[1], Athanasios Tsakalidis[1,2],
Kostas Tsichlas[1], and Christos Zaroliagis[1,2]

[1] Dept of Computer Eng and Informatics, University of Patras, 26500 Patras, Greece
[2] Computer Technology Institute, N. Kazantzaki Str, Patras University Campus,
26500 Patras, Greece
{kaporis, makri, sioutas, tsak, tsihlas, zaro}@ceid.upatras.gr

Abstract. A new dynamic Interpolation Search (IS) data structure is
presented that achieves $O(\log \log n)$ search time with high probability
on unknown continuous or even discrete input distributions with measurable probability of key collisions, including power law and Binomial
distributions. No such previous result holds for IS when the probability of key collisions is measurable. Moreover, our data structure exhibits
$O(1)$ expected search time with high probability for a wide class of input distributions that contains all those for which $o(\log \log n)$ expected
search time was previously known.

1 Introduction

The dynamic dictionary search problem is one of the fundamental problems in
computer science. In this problem we have to maintain a set of elements subject
to insertions and deletions such that given a query element y we can retrieve the
largest element in the set smaller or equal to y. Well known search methods use
an arbitrary rule to *select* a splitting element and *split* the stored set into two
subfiles; in binary search, each recursive split selects as splitting element, in a
"blind" manner, the middle (or a close to the middle) element of the current file.
Using this technique, known balanced search trees (e.g., (a, b)-trees [11]) support
search and update operations in $O(\log n)$ time when storing n elements. In the
Pointer Machine (PM) model of computation, the search time cannot be further
reduced, since the lower bound of $\Omega(n \log n)$ for sorting n elements would be
violated. In the RAM model of computation, which we consider in this work, a
lower bound of $\Omega(\sqrt{\frac{\log n}{\log \log n}})$ was proved by Beame and Fich [4]; a data structure
achieving this time bound has been presented by Andersson and Thorup [2].

The aforementioned lower bounds can be surpassed if we take into account the
input distribution of the keys and consider expected complexities; in this case,
the extra knowledge about the probabilistic nature of the keys stored in the file
may lead to better selections of splitting elements. The main representative of

---

[*] This work was partially supported by the FET Unit of EC (IST priority – 6th
FP), under contracts no. IST-2002-001907 (integrated project DELIS) and no. FP6-
021235-2 (project ARRIVAL), and by the Action PYTHAGORAS with matching
funds from the European Social Fund and the Greek Ministry of Education.

M. Bugliesi et al. (Eds.): ICALP 2006, Part I, LNCS 4051, pp. 382–394, 2006.

these techniques is the method of *Interpolation Search* (IS) introduced by Peterson [21], where the splitting element was selected close to the expected location of the target key. Yao and Yao [28] proved a $\Theta(\log \log n)$ average search time for stored elements that are uniformly distributed. In [9,10,18,19,20] several aspects of IS are described and analyzed. Willard [26] proved the same search time for the extended class of *regular* input distributions. The IS method was recently generalized [5] to non-random input data that possess enough "pseudo-randomness" for effective IS to be applied. The study of dynamic insertions of elements with respect to the uniform distribution and random deletions was initiated in [8,12]. In [8] an implicit data structure was presented supporting insertions and deletions in $O(n^\epsilon)$, $\epsilon > 0$, time and IS with expected time $O(\log \log n)$. The structure of [12] has expected insertion time $O(\log n)$, amortized insertion time $O(\log^2 n)$ and it is claimed, without rigorous proof, that it supports IS. Mehlhorn and Tsakalidis [16] demonstrated a novel dynamic version of the IS method, the *Interpolation Search Tree (IST)*, with $O(\log \log n)$ expected search and update time for a larger class than the regular distributions. In particular, they considered μ-random insertions and random deletions[1] by introducing the notion of a (f_1, f_2)-*smooth* probability density μ, in order to control the distribution of the elements in each subinterval dictated by an ID index. Informally, a distribution defined over an interval I is smooth if the probability density over any subinterval of I does not exceed a specific bound, however small this subinterval is (i.e., the distribution does not contain sharp peaks). The class of smooth distributions is a superset of uniform, bounded, and several non-uniform distributions (including the class of regular distributions). The results in [16] hold for $(n^\alpha, \sqrt{n})$-smooth densities, where $1/2 \le \alpha < 1$. Andersson and Mattson [1], generalized and refined the notion of smooth distributions, presenting a variant of the IST called *Augmented Sampled Forest* extending the class of input distributions for which $\Theta(\log \log n)$ search time is expected. In particular, the time complexities of their structure holds for the larger class of $(\frac{n}{(\log \log n)^{1+\epsilon}}, n^\delta)$-smooth densities, where $\delta \in (0,1), \epsilon > 0$. Moreover, their structure exhibited $o(\log \log n)$ expected search time for some classes of input distributions. Finally in [13], a finger search version of these structures was presented.

The analysis of all the aforementioned IS structures was heavily based on the assumption that the conditional distribution on the subinterval dictated by an arbitrary interpolation step remains unaffected. In particular, in [1,13,16] IS is performed on each node of a tree structure under the assumption that all elements in the subtree dictated by the previous interpolation step remain μ-random.

Our first contribution in this work (Section 2) is to show that the above assumption is valid only when the produced elements are *distinct* (as indeed assumed in [1,9,10,13,16,19,20,21,26,28]), i.e., they are produced under some continuous distribution where the probability of collision is zero; otherwise, it *fails*.

---

[1] An insertion is μ-random if the key to be inserted is drawn randomly with density function μ; a deletion is random if every key present in the data structure is equally likely to be deleted.

This means that the probabilistic analyses of previous dynamic interpolation search data structures are inapplicable to sequences of non-distinct elements, produced by discrete probability distributions with measurable (non-zero) probability of key collisions.

This lack of generalization does not have only theoretical, but also serious practical implications. There exist applications where we need to store duplicates, and thus the theoretically used density distribution modelling the input process should *not* produce distinct elements. A classical example is the creation of secondary indices in databases [15]. In a secondary index, duplicate values correspond to different records and they should be stored as distinct entities. There are also specific applications where interpolation search comes into play. For instance, the case of searching tables with alphabetic keys (e.g., names, dictionary entries) [18]. The keys in such tables follow a non-uniform, (unknown) discrete probability distribution and collisions *do* occur. Other useful applications of interpolation search in non-uniform data are discussed in [3,7,18,20,22]. In these papers it has been empirically observed that interpolation search has a very poor performance in such data. To alleviate this problem a series of heuristics have been introduced in [3,7,18,20,22], but no rigorous performance analyses have been provided. In [18,19], it was suggested that such an analysis would be possible if one considers the idea in [10] that translates any continuous input distribution to a uniform one.

In Section 2, we also show that this idea of taking advantage of the cumulative distribution [10,18,19] does *not* apply to discrete distributions with measurable probability of key collisions (a fact that was indeed experimentally verified in [18]). The above pluralism of efforts demonstrates the necessity to handle non-uniform data generated by discrete distributions with measurable probability of key collisions.

One could be tempted to argue that the inapplicability of the previous analyses could be faced by simply storing duplicate elements once; moreover, in these structures the main rebalancing tool is local/global rebuilding, which can be easily modified to produce input sequences with distinct elements. Both arguments are wrong, however, since the new sequences of distinct elements are artificial sequences, different from the initial. Consequently, important statistical properties of the elements are destroyed and the probabilistic analyses fail.

Our second contribution in this paper is a new dynamic interpolation search data structure (Section 3) that overcomes the above problems, and in which the elements stored in each subtree preserve the input distribution, conditioning only on the interval that corresponds to the current subtree. The new structure is quite simple, it exhibits similar expected $O(\log \log n)$ search time as the previous dynamic interpolation structures [1,9,10,16,19,20,21,26,28]), its probabilistic analysis is *always* valid irrespectively of the distinctness or not of the elements in the input sequence (i.e., *regardless* of whether they are produced by a continuous or a discrete distribution), it applies to the same classes of distributions as those in [1,16] and it holds with high probability, while those in [1,9,10,16,19,20,21,26,28]) did not grant such guarantee. Finally, as a by-product of our construction, we get a dynamic search data structure with $O(1)$ expected search time for a wide class of input distributions (Section 3). This result significantly extends the class of input distributions in [1] under which $O(1)$ expected search time was possible. In addition, this search time also holds with high probability, while those in [1] did not grant such property.

Although the class of smooth distributions includes, for appropriate choices of f_1 and f_2, any other probability distribution, the effective range of f_1, f_2 for which $O(\log \log n)$ IS time is achieved excludes distributions of major practical importance; for instance, power law [17], Binomial, etc. We are able to show (Section 3) that a slight modification of our data structure achieves $O(\log \log n)$ time with high probability for power law and Binomial distributions. No previous IS structure achieves such a time bound for these distributions (recall the deterioration of IS that was experimentally observed in [3,7,18,20,22]).

Our data structure is *robust* (as those in [1,13,16,26]), i.e., it remains efficient *without* apriori knowledge of the particular continuous or discrete distribution. Due to space limitations, several details and proofs are omitted and can be found in the full version [14].

2 Probabilistic Analysis of the IS-Tree Revisited

Consider an unknown *continuous* probability distribution over the interval $[a, b]$ with density function $\mu(x) = \mu[a, b](x)$. Given two functions f_1 and f_2, then $\mu(x) = \mu[a, b](x)$ is (f_1, f_2)-*smooth* [1,16] if there exists a constant β, such that for all c_1, c_2, c_3, $a \leq c_1 < c_2 < c_3 \leq b$, and all integers n, it holds that

$$\Pr[X \in [c_2 - \frac{c_3 - c_1}{f_1(n)}, c_2] \mid c_1 \leq X \leq c_3] = \int_{c_2 - \frac{c_3 - c_1}{f_1(n)}}^{c_2} \mu[c_1, c_3](x)dx \leq \frac{\beta f_2(n)}{n} \quad (1)$$

where $\mu[c_1, c_3](x) = 0$ for $x < c_1$ or $x > c_3$, and $\mu[c_1, c_3](x) = \mu(x)/p$ for $c_1 \leq x \leq c_3$ where $p = \int_{c_1}^{c_3} \mu(x)dx$. Similarly, for an unknown *discrete* probability distribution of elements $x_1, \ldots, x_N$ spread over $[a, b]$, with probability function $\mu(x_i) = \mu[a, b](x_i)$ we have

$$\Pr[c_2 - \frac{c_3 - c_1}{f_1(n)} \leq X \leq c_2 \mid c_1 \leq X \leq c_3] = \sum_{c_2 - \frac{c_3 - c_1}{f_1(n)}}^{c_2} \mu[c_1, c_3](x_i) \leq \frac{\beta f_2(n)}{n} \quad (2)$$

where $\mu[c_1, c_3](x_i) = 0$ for $x_i < c_1$ or $x_i > c_3$, and $\mu[c_1, c_3](x_i) = \mu(x_i)/p$ for $x_i \in [c_1, c_3]$ where $p = \sum_{x_i \in [c_1, c_3]} \mu(x_i)$. Intuitively, function f_1 partitions an arbitrary subinterval $[c_1, c_3] \subseteq [a, b]$ into f_1 equal parts, each of length $\frac{c_3 - c_1}{f_1} = O(\frac{1}{f_1})$; that is, f_1 measures how *fine* is the partitioning of an arbitrary subinterval. Function f_2 guarantees that no part, of the f_1 possible, gets more probability mass than $\frac{\beta \cdot f_2}{n}$; that is, f_2 measures the *sparseness* of any subinterval $[c_2 - \frac{c_3 - c_1}{f_1}, c_2] \subseteq [c_1, c_3]$. The class of (f_1, f_2)-smooth distributions (for appropriate choices of f_1 and f_2) is a superset of both regular and uniform classes of distributions, as well as of several non-uniform classes [1,16]. Actually, *any* probability distribution is $(f_1, \Theta(n))$-smooth, for a suitable choice of β.

Consider the random file $S = \{X_1, \ldots, X_n\}$, where each key $X_i \in [a, b] \subset \mathbb{R}$, obeys an unknown (discrete or continuous) distribution μ, $i = 1, \ldots, n$. Let $P = \{X_{(1)}, \ldots, X_{(n)}\}$ be an increasing ordering of file S. The goal is to find the largest key $X_{(j)} \in P$ that precedes a *target* element y. We describe how the *Augmented Sampled Forest (ASF)* [1], which is a generalization of the *Interpolation Search Tree (IST)* [16], can be used to search for this target element y.

Assume that the (discrete or continuous) distribution μ is $(I(n), n/R(n))$-smooth, where $I(n)$, $R(n)$ are two nondecreasing functions. The ASF is a two level data structure; the top level is an *ideal* static IST [16] while the bottom level is a sequence of buckets. The structure is maintained by using the global rebuilding technique and its expected search time is dominated by the expected search time at the top level. At the top level, the root node has $R(n)$ children, and similarly each child node has $R(\frac{n}{R(n)})$ sub-children. The root node corresponds to the ordered file P of size n. Each child corresponds to a part of file P of size $\frac{n}{R(n)}$. That is, these $R(n)$ children partition the ordered file P into $R(n)$ equal subfiles $P_1, \ldots, P_{R(n)}$, of the form $\{X_{(1)}, \ldots, X_{(\frac{n}{R(n)})}\}, \ldots, \{X_{((R(n)-1)\frac{n}{R(n)}+1)}, \ldots, X_{(n)}\}$. Each node of this tree contains a pair of arrays, namely ID and REP, that help to locate the appropriate child eligible to contain the target element y. In the root node the set of indices of the ID array is $[1, \ldots, I(n)]$ and the set of indices of the REP array is $[1, \ldots, R(n)]$. The role of the ID array of the root node is to partition the interval $[a, b]$ into $I(n)$ equal parts, each of length $\frac{b-a}{I(n)}$. When searching for an element y, the first *interpolation* step determines within $O(1)$ time the number j

$$j = \left\lfloor \frac{y - a}{b - a} I(n) \right\rfloor + 1 \tag{3}$$

which denotes the j-th interval I_j of length $\frac{b-a}{I(n)}$ that contains the target y:

$$I_j = \left[a + (j - 1)\frac{b - a}{I(n)}, \ a + j\frac{b - a}{I(n)} \right] \tag{4}$$

The role of the array REP$[1, \ldots, R(n)]$ of the root node is to partition the ordered file P into $R(n)$ equal subfiles, each of size $\frac{n}{R(n)}$. Index REP$[i], i = 1, \ldots, R(n)$, points to the i-th subfile P_i, where $P_i = \{X \in P \mid X_{((i-1)\frac{n}{R(n)})} < X \leq X_{(i\frac{n}{R(n)})}\}$. Alternatively, REP$[i]$ can be seen as the *representative* of the element $X_{(i\frac{n}{R(n)})}$ of P_i. The first interpolation step, provided by Eq. (3), determines within $O(1)$ time the subinterval I_j described in Eq. (4), where the target element y belongs. If in this subinterval correspond $O(1)$ REP indices, then within $O(1)$ time we can determine the unique REP index that corresponds to the subfile that element y may belong. Hence, the search efficiency highly depends on the distribution of the REP indices over each ID subinterval of $[a, b]$. In other words, each ID index that corresponds to a *dense* subinterval of $[a, b]$ causes a great slow-down of the search speed. Most importantly, suppose that the second interpolation step now yields $\mathrm{REP}[s-1] < y \leq \mathrm{REP}[s]$. Then, y must be searched for into the subfile $P_s = \{X_{((s-1)\frac{n}{R(n)})}, \ldots, X_{(s\frac{n}{R(n)})}\}$. A crucial observation is that its endpoints $X_{((s-1)\frac{n}{R(n)})}, X_{(s\frac{n}{R(n)})}$ may in general be neither μ-random nor smooth.

The analyses in [1,13,16] assume that the elements into an arbitrary subfile dictated by an interpolation step *remain μ-randomly distributed* conditioned on the subinterval that all these elements belong, i.e., for a random element $X = \lambda$ in subfile P_v with endpoints $a' = X_{(v-1)\frac{n}{R(n)}}$ and $b' = X_{v\frac{n}{R(n)}}$, its probability density is given by Expression (5). Also, the analyses in [10,18,19] ingeniously apply the cumulative distribution function F on the ordered keys in $P = \{X_{(1)}, \ldots, X_{(n)}\}$, yielding $P_F = \{F(X_{(1)}), \ldots, F(X_{(n)})\}$. Now, each $F(X_{(i)}) \in P_F$ is *uniformly* distributed over $[0,1]$, since $\Pr[F(X_{(i)}) \leq t] = \Pr[X_{(i)} \leq F^{-1}(t)] = F(F^{-1}(t)) = t$ (see [6, pp. 36-37]). Thus, file P_F is very suitable for applying IS on it; i.e., to search for target key y, split P_F on key $F(X_{(j_y)}) \approx \frac{y - F(X_{(1)})}{F(X_{(n)}) - F(X_{(1)})}$, and recursively apply IS to $P_F^- = \{F(X_{(1)}), \ldots, F(X_{(j_y)})\}$, if $y \leq F(X_{(j_y)})n$, otherwise to $P_F^+ = P_F \setminus P_F^-$. However, this approach *also* tacitly assumes that the conditional distribution of the keys in subfiles P_F^-, P_F^+ remains unaffected and obeys Expression (5) with a', b' the corresponding endpoints of the appropriate subfile P_F^- or P_F^+.

In the following, we prove the validity of these assumptions under continuous or discrete distributions with zero probability of element collisions, and we will depict the subtle case of discrete distributions with measurable probability of key-collisions where all the above assumptions *fail*.

Continuous or discrete distributions with zero probability of element collisions. Consider the simple case of three stored elements (random variables) $X_1, X_2, X_3 \in [a, b]$ drawn according to some μ-random smooth distribution (the general case of n variables can be easily deduced from this case by a simple induction argument). These elements are identically and independently distributed and it is assumed that they take *distinct* values. Since the collision probability for continuous distributions is 0, we concentrate our discussion to distinct elements. The *conditional*, on the arbitrary interval with *fixed* endpoints $(a', b'] \subseteq [a, b]$, probability density equals

$$\Pr[X = \lambda \mid a' < X \leq b'] = \frac{\Pr[X = \lambda]}{\Pr[X \leq b'] - \Pr[X \leq a']}. \tag{5}$$

According to definitions (1) and (2), Exp. (5) plays a crucial role in tuning the probability mass in subinterval $(a', b']$ using parameters f_1, f_2. For each $i = 1, 2$ the corresponding $\mathrm{REP}[i]$ is a *new* random variable defined as $\mathrm{REP}[1] \equiv X_{(1)} = \min\{X_1, X_2, X_3\}$, $\mathrm{REP}[2] \equiv X_{(3)} = \max\{X_1, X_2, X_3\}$. We want to show that the random element X that belongs into the subinterval $[\mathrm{REP}[1], \mathrm{REP}[2]]$ is μ-randomly distributed. We have

$$\Pr[X = \lambda \mid \text{REP}[1] = a' < X \le \text{REP}[2] = b'] = \Pr[X = \lambda \mid X_{(1)} = a' \cap X_{(3)} = b']$$

$$= \frac{\Pr[X = \lambda \cap X_{(1)} = a' \cap X_{(3)} = b']}{\Pr[X_{(1)} = a' \cap X_{(3)} = b']}, \tag{6}$$

where $a' < \lambda < b'$. The event $\{X_{(1)} = a' \cap X_{(3)} = b'\}$ occurs if at least one of the following mutually disjoint events occur:

$$\{X_1 = a',\ X_2 = b',\ a' < X_3 < b'\},\quad \{X_2 = a',\ X_1 = b',\ a' < X_3 < b'\},$$
$$\{X_1 = a', X_3 = b',\ a' < X_2 < b'\},\quad \{X_3 = a',\ X_1 = b',\ a' < X_2 < b'\},$$
$$\{X_2 = a',\ X_3 = b', a' < X_1 < b'\},\quad \{X_3 = a',\ X_2 = b',\ a' < X_1 < b'\}. \tag{7}$$

Hence, $\Pr[X_{(1)} = a' \cap X_{(3)} = b'] = 6\Pr[X = a']\Pr[X = b']\Pr[a' < X < b']$ \quad (8)

Similarly, the event $\{X = \lambda \cap X_{(1)} = a' \cap X_{(3)} = b'\}$, with $a' < \lambda < b'$, occurs if one of the following mutually disjoint events occur:

$$\{X_2 = \lambda,\ X_3 = a',\ X_1 = b'\},\quad \{X_1 = \lambda,\ X_2 = a',\ X_3 = b'\},$$
$$\{X_3 = \lambda,\ X_1 = a', X_2 = b'\},\quad \{X_1 = \lambda,\ X_3 = a',\ X_2 = b'\},$$
$$\{X_3 = \lambda,\ X_2 = a',\ X_1 = b'\},\quad \{X_2 = \lambda, X_1 = a',\ X_3 = b'\}. \tag{9}$$

Combining (8) and (9), the conditional probability (6) becomes

$$\Pr[X = \lambda | X_{(1)} = a' \cap X_{(3)} = b'] = \frac{6\Pr[X = \lambda]\Pr[X = a']\Pr[X = b']}{6\Pr[X = a']\Pr[X = b']\Pr[a' < X < b']}$$
$$= \frac{\Pr[X = \lambda]}{\Pr[a' < X < b']} \tag{10}$$

where $a \le a' < \lambda < b' \le b$. This probability equals Exp. (5) and thus is μ-random and consequently smooth (due to definitions (1) and (2)). Hence, we have shown that in the case where the input elements have non measurable probability of collisions, all previous analyses carry over correctly.

Discrete distributions with measurable probability of element collisions. In this case, the event $\{X_{(1)} = a' \cap X_{(3)} = b'\}$ occurs if, besides the events listed in (7), at least one of the following mutually disjoint events occur:

$$\{X_{1,2} = a',\ X_3 = b'\}, \{X_{1,2} = b',\ X_3 = a'\}, \{X_{1,3} = a',\ X_2 = b'\},$$
$$\{X_{1,3} = b', X_2 = a'\}, \{X_{2,3} = a',\ X_1 = b'\}, \{X_{2,3} = b',\ X_1 = a'\} \tag{11}$$

Hence, \quad $\Pr[X_{(1)} = a' \cap X_{(3)} = b'] = 3\Pr[X = a']^2\Pr[X = b'] +$
$$3\Pr[X = a']\Pr[X = b']^2 + 6\Pr[X = a']\Pr[X = b']\Pr[a' < X < b'] \tag{12}$$

If $a \le a' < \lambda < b' \le b$, by combining (11) and (12), now Expression (6) becomes

$$\Pr[X = \lambda | X_{(1)} = a' \cap X_{(3)} = b'] = \frac{\Pr[X = \lambda]}{\frac{\Pr[X=a'] + \Pr[X=b']}{2} + \Pr[a' < X < b']} \tag{13}$$

and if $\lambda = a'$ or $\lambda = b'$, Expression (6) becomes the half of Expression (13). Clearly, (13) is different from (5) and in general may be *neither μ-random nor smooth* (see the full version of the paper [14] for more details). We conclude that, when the probability of collisions is measurable, the net effect of choosing, as endpoints of subintervals, not deterministically obtained values is to *destroy* the smoothness of the distribution of the elements that belong in it.

3   The New IS Data Structure

Consider a dynamic file S containing $O(n)$ elements drawn from the interval $[a, b]$, according to a continuous or discrete distribution μ, which is $(f_1, f_2) = (n^\alpha, n^\delta)$-smooth with *arbitrary* $\alpha, \delta \in (0, 1)$. Our structure consists of LAYERS of bins. The 1st LAYER partitions interval $[a, b]$ into $f_1(n)$ equal-length bins. We define[2] as BIN(j_1), the j_1-th bin in the 1st LAYER of bins, which corresponds to the subinterval $[a + (j_1 - 1)\frac{b-a}{f_1(n)},\ a + j_1\frac{b-a}{f_1(n)}] = [a_{j_1},\ b_{j_1}] \subset [a,\ b], j_1 = 1, \dots, f_1(n)$. Any key $X \in S$ is stored in BIN(j_1), iff X is spread according to μ into the subinterval $[a_{j_1},\ b_{j_1}], j_1 = 1, \dots, f_1(n)$. This subfile $S_{j_1} \subseteq S$ consists of n_{j_1} elements and is stored in BIN(j_1), where $n_1 + \dots + n_{f_1(n)} = |S| = O(n)$, and $j_1 = 1, \dots, f_1(n) = n^\alpha$.

The 2nd LAYER of bins is constructed by recursively partitioning each BIN (j_1) of the 1st LAYER into $f_1(n_{j_1})$ equal-length bins, $j_1 = 1, \dots, f_1(n)$, i.e., BIN(j_1) containing n_{j_1} elements is partitioned into equal-length bins BIN(j_1, j_2), with corresponding indices $j_1 = 1, \dots, f_1(n) = n^\alpha$ and $j_2 = 1, \dots, f_1(n_{j_1}) = (n_{j_1})^\alpha$. Now BIN$(j_1, j_2)$ corresponds to the subinterval $[a_{j_1} + (j_2 - 1)\frac{b_{j_1} - a_{j_1}}{f_1(n_{j_1})}, a_{j_1} + j_2\frac{b_{j_1} - a_{j_1}}{f_1(n_{j_1})}] = [a_{j_1, j_2},\ b_{j_1, j_2}] \subset [a_{j_1},\ b_{j_1}] \subset [a, b]$. An arbitrary element $X \in S$ is stored in BIN(j_1, j_2), iff X is spread according to μ into the subinterval $[a_{j_1, j_2},\ b_{j_1, j_2}], j_2 = 1, \dots, f_1(n_{j_1})$ and $j_1 = 1, \dots, f_1(n)$. The subfile $S_{j_1, j_2} \subseteq S_{j_1}$ consists of n_{j_1, j_2} elements stored in BIN(j_1, j_2), such that $n_{j_1, 1} + \dots + n_{j_1, f_1(n_{j_1})} = |S_{j_1}| = n_{j_1}$.

We proceed recursively for the subsequent LAYERS of bins; however, no bin with less than poly log n keys becomes further partitioned (n is the initial number of keys in the structure), i.e., it becomes a leaf of the structure. Finally, the elements associated with each leaf bin are stored as a q^*-heap. The q^*-*heap* [27] is a search tree data structure having the following useful property: let M be the current number of elements in the q^*-heap and let N be an upper bound on the maximum number of elements ever stored in the q^*-heap. Then, insertion, deletion and search operations are carried out in $O(1 + \log M / \log \log N)$ worst-case time after an $O(N)$ preprocessing overhead. Choosing $M = \text{polylog}(N)$, all operations can be performed in $O(1)$ time. Hence, by setting N to be n, the use of q^*-*heap* at the leaves of the structure permits the manipulation of search and update operations in the leaf bins in worst-case $O(1)$ time.

In the above data structure, we can search for a target element y as follows. Given that a bin containing y at the current LAYER has been located, we perform interpolation search on its offspring of bins to locate the particular bin of the

---

[2] From now on, the subscript i of j_i will denote the i-th LAYER of bins.

next LAYER that y may belong. Since target y may belong in at most one bin of each LAYER, as the LAYERS evolve, this process highly prunes the size of the search space (the occupancy number of the currently scanned bin).

The careful reader should have noticed that the endpoints selected as representatives in each subtree are independent of the particular characteristics of the input distribution μ, thus confronting the weakness of the constructions in all previous approaches. This crucial randomness invariance property of the new data structure is given by Lemma 1 (whose proof is in [14]).

Lemma 1. *Consider an arbitrary bin $BIN(j_1, \ldots, j_i)$ with corresponding subinterval $[a_{j_1,\ldots,j_i}, b_{j_1,\ldots,j_i}]$ of the ith LAYER of bins. Then, the $n_{j_1,\ldots,j_i}$ elements in $BIN(j_1, \ldots, j_i)$ are μ-randomly distributed in the subinterval $[a_{j_1,\ldots,j_i}, b_{j_1,\ldots,j_i}]$.*

Theorem 1 below shows that w.h.p. each IS step prunes drastically the size of the dictated subfile (its proof is in [14]), i.e., a child bin has size at most f_2(elements of father bin).

Theorem 1. *Consider the bin $BIN(j_1, \ldots, j_i)$ of the i-th LAYER of bins and let $n_{j_1,\ldots,j_i}$ be its number of balls at the end of the t-th insertion/deletion operation. These balls are μ-randomly distributed in its subinterval $[a_{j_1,\ldots,j_i}, b_{j_1,\ldots,j_i}]$. Then,*

$$\Pr[\exists \ BIN(j_1, \ldots, j_i, j_{i+1}) : \ n_{j_1,\ldots,j_i,j_{i+1}} = \omega(f_2(n_{j_1,\ldots,j_i}))] \to 0, \ as \ n \to \infty,$$

where $j_{i+1} = 1, \ldots, f_1(n_{j_1,\ldots,j_i})$.

This in turn yields the search time bound in Lemma 2 below (its proof in [14]).

Lemma 2. *For every target element y, the path from its leaf bin to the root of the tree will have length not exceeding $\log \log n$ with high probability.*

Moreover, for every node v of the tree, the subtree of any of its children will have at most half the size of the subtree of v, with high probability. We call a tree with these properties *ideal*; our high probability bound implies that for a given set of μ random elements with cardinality n, such a tree can be found and be built in $O(n)$ expected time. Moreover, by using the arguments in [16, Lemma 2, p. 626], we can straightforwardly show that the space complexity of the described data structure is linear.

Consequently, by embedding the *ideal* version of our new data structure as the top level in the *Augmented Sampled Forest (ASF)* of [1] and by maintaining the leaf bins as a q^*-heap [27], while keeping in parallel a worst-case data structure [2] (in a manner e.g., similar to [13]), we get the following theorem.

Theorem 2. *Consider a file with n (not necessarily distinct) elements that was produced by a sequence of μ-random insertions and random deletions, where μ is a (n^α, n^δ)-smooth density, for any arbitrary $0 < \alpha, \delta < 1$. Then, there exists a dynamic interpolation search tree with $O(\log \log n)$ expected search time with high probability; the space usage of the data structure is $\Theta(n)$, the worst-case update time (position given) is $O(1)$, and the worst-case search time is $O(\sqrt{\log n / \log \log n})$.*

Remark. It is easy to see that every part of our analysis remains valid if we replace the function $f_1(n) = n^\alpha$ with the function $f_1(n) = \frac{n}{(\log \log n)^{1+\epsilon}}$, where $\epsilon > 0$. Hence, our structure can handle within the same time and space complexities, as those mentioned in Theorem 2, the larger class of $(\frac{n}{(\log \log n)^{1+\epsilon}}, n^\delta)$-smooth densities.

The difference of our data structure with those in [1,16] is in the absence of REP arrays. These arrays guarantee that when we move to a child of a node whose subtree contains N nodes, then this child node will be the root of a subtree containing $\sqrt{N}$ nodes. In our case, this is not guaranteed (it is easy to come up with a setting where all elements are in a very small region and thus the height of our tree structure is large). However, assuming that the input elements are generated by a smooth distribution, it is *very unlikely* that this bad scenario will happen, since we prove that the height of our tree structure is doubly logarithmic with high probability. Our data structure is in a sense "similar" to other data structures that partition the space (e.g., quadtrees). Indeed, our structure partitions the universe until each region has a bounded number of elements. On the contrary, the use of REP arrays allows for a partition according to the number of elements (like e.g., in range trees), thus guaranteeing that each partition has geometrically less elements.

$O(1)$ **search time with high probability.** We study a random process of rn insert (or delete) operations on this structure where in each operation, $j = 1, \ldots, rn$, with probability $p = (0, 1]$ a new element $X \in [a, b]$ obeying an unknown $(f_1(n), f_2(n)) = (\frac{n}{g(n)}, \ln^{O(1)} n)$-smooth distribution μ, is inserted, otherwise a random existing key is deleted; here $g(n)$ denotes a function which is either constant or slowly growing with n (i.e., $\ln^* n$). The class $(f_1(n), f_2(n)) = (\frac{n}{g(n)}, \ln^{O(1)} n)$-smooth distributions includes that of bounded $((n, 1)$-smooth) densities, for which $O(1)$ expected search time was known [1], as well as all those for which a $o(\log \log n)$ expected search time could be achieved [1]; for instance, the density $\mu[0, 1](x) = -\ln x$ is $(n/(\log^* n)^{1+\epsilon}, \log^2 n)$-smooth, and an expected search time complexity of $\Theta(\log^* n)$ was given in [1]. Our result implies $O(1)$ search time with high probability for all the aforementioned densities.

The idea is as follows. We can prove (see [14]) that during each step $j = 1, \ldots, rn$, there are $O(n)$ elements stored. Then Theorem 1 establishes that during each step $j = 1, \ldots, rn$, *no bin* of the 1st LAYER gets more than poly $\log n$ elements (balls), with high probability. That is, the whole tree-structure reduces to a single LAYER. Since each BIN(j_1), $j_1 = 1, \ldots, f_1(n)$, is implemented as a q^*-heap, we can search for element y in it within $O(1)$ time. Finally, we can determine within $O(1)$ time the bin BIN(j_1) that y may belong using the Expr. (3).

Power Law Distributions. As shown in Section 2, the efficiency of an arbitrary interpolation step dictating subtree p highly relies on how the total of n_p elements belonging to subtree p are sparsely distributed in its associated subinterval $[a_p, b_p]$. This sparsity fails for power laws, as we show next. Let the *discrete* universe of possible keys be $U = \{1, 2, \ldots, N\}$, with N arbitrarily large, spread over interval $I = [1, b]$ and listed in decreasing frequency. Each random key X is drawn according to the power law distribution $\Pr[X \geq x] = cx^{-\beta}$ for constants $c, \beta > 0$ [17, Sec. 2]. The probability mass accumulated on subinterval $I_1 = [1, n^{\alpha}]$ containing the subset of keys $\{1, \ldots, n^{\alpha}\} \subseteq U$ equals:

$$\Pr[X \in I_1] = 1 - \Pr[X \geq n^{\alpha}] = 1 - \frac{c}{(n^{\alpha})^{\beta}} = \omega(\frac{n^{\delta}}{n}), \ \delta < 1 \qquad (14)$$

which according to definition (2) means that subinterval I_1 is *not* $(f_1(n), f_2(n)) = (n^{\alpha}, n^{\delta})$-smooth for any constant $0 < \alpha < 1$. This rules out any attempt to employ IS on the whole interval $I = [1, b]$. However, $I_2 = I \setminus I_1$ can be arbitrarily sparse, as a function of α, since $\Pr[X \in I_2] = \Pr[X \geq n^{\alpha}] = \frac{c}{(n^{\alpha})^{\beta}}$ and by setting $\alpha = \alpha(\beta) \geq \frac{1}{\beta}$, we get $\Pr[X \in I_2] = O(\frac{1}{n}) = O(\frac{f_2(n)}{n})$, with $f_2(n) = poly \log n$. That is, if we draw a random key $X \in [1, b]$ according to power law $\Pr[X \geq x] = cx^{-\beta}$, it will belong to an *arbitrary* subinterval of I_2 with probability $O(\frac{f_2(n)}{n}) \leq \frac{poly \log n}{n}$. The later implies that the power law distribution with parameters c, β, if restricted to I_2 remains $(f_1(n), poly \log n)$-smooth. Thus, if the target element $y \in I_2$, then Theorem 2 guarantees that IS on I_2 takes $O(1)$ search time with high probability. On the other hand, observe that the discrete subuniverse of U, which is spread in $[1, n^{\alpha(\beta)}]$, has cardinality $|\{1, \ldots, n^{\alpha(\beta)}\}| = O(n^{\alpha(\beta)})$. That is, if the target y belongs to I_1, then it can be searched amongst $n^{\alpha(\beta)}$ possible keys, which is considerably smaller that the universe's cardinality $N = |U|$. Therefore, if $y \in I_1$, we can employ the van Emde Boas structure [24,25], which yields a time complexity $O\left(\log \log \left(|\{1, \ldots, n^{\alpha(\beta)}\}|\right)\right) = O\left(\log \log \left(n^{\alpha(\beta)}\right)\right) = O(\log \log n)$. The splitting key $n^{\alpha(\beta)}$ yielding $\Pr[X \in I_2] = O(\frac{1}{n}) = O(\frac{f_2(n)}{n})$, with $f_2(n) = poly \log n$, can be approximated by a key x^* during the initialization of the structure, without knowledge of the parameters (c, β) (details are given in [14]).

Binomial Distributions. We can identify the dense subinterval $I_1 = [np - \Delta, np + \Delta] \subseteq [a, b]$ around the mean value np for any binomial distribution $B(n, p)$. Notice that $|I_1| = 2\Delta$ and since we can safely set $\Delta = o(n)$, we can similarly apply a van Emde Boas structure on I_1. Taking advantage of the binomial sharp tail bounds, the remaining subinterval $I_2 = [a, b] \setminus I_1$ will remain sparse enough to apply IS. The rest of the details follow similarly to those for power law distributions.

Acknowledgment. We are indebted to Lefteris Kirousis for various helpful discussions.

References

1. A. Andersson and C. Mattson. Dynamic Interpolation Search in $o(\log \log n)$ Time. In *Proc. 20th Coll. on Automata, Languages and Programming* – ICALP'93, LNCS Vol. 700 (Springer 1993), pp. 15-27.
2. A. Anderson and M. Thorup. Tight(er) Worst-case Bounds on Dynamic Searching and Priority Queues. In *Proc. 32nd ACM Symposium on Theory of Computing* – STOC 2001, pp.335-342. ACM, 2000.

3. F.W. Burton and G.N. Lewis. A robust variation of Interpolation Search. *Information Processing Latters*, 10, 198–201, 1980.

4. P. Beame and F. Fich. Optimal bounds for the predecessor problem and related problems. *Journal of Computer and System Sciences*, 65(1):38-72, 2002.

5. E. Demaine, T. Jones, and M. Patrascu. Interpolation Search for Non-Independent Data. In *Proc. 15th ACM-SIAM Symp. on Discrete Algorithms* – SODA 2004, pp. 522-523.

6. W. Feller. *An Introduction to Probability Theory and Its Applications* Vol. II, 2nd Edition, Wiley, New York 1971.

7. K.E. Foster. A statistically based interpolation binary search. TR, Winthrop College, SC.

8. G. Frederickson. Implicit Data Structures for the Dictionary Problem. *Journal of the ACM* 30(1):80-94, 1983.

9. G. Gonnet. Interpolation and Interpolation-Hash Searching. PhD Thesis. Waterloo: University of Waterloo 1977.

10. G. Gonnet, L. Rogers, and J. George. An Algorithmic and Complexity Analysis of Interpolation Search. *Acta Informatica* 13:39-52, 1980.

11. S. Huddleston and K. Mehlhorn. A new data structure for representing sorted lists. *Acta Informatica*, vol. 17 (1982), pp.157–184.

12. A. Itai, A. Konheim, and M. Rodeh. A Sparse Table Implementation of Priority Queues. In *Proc. 8th Coll. on Automata, Languages and Program.* – ICALP'81, LNCS Vol. 115 (Springer 1981), pp. 417-431.

13. A.C. Kaporis, C. Makris, S. Sioutas, A. Tsakalidis, K. Tsichlas and C. Zaroliagis. Improved bounds for finger search on a RAM. In *Algorithms* – ESA 2003, LNCS Vol. 2832 (Springer 2003), pp. 325-336.

14. A.C. Kaporis, C. Makris, S. Sioutas, A. Tsakalidis, K. Tsichlas and C. Zaroliagis. Dynamic Interpolation Search Revisited. Computer Technology Institute Tech. Report TR 2006/04/02, April 2006.

15. Y. Manolopoulos, Y. Theodoridis, V. Tsotras. *Advanced Database Indexing*. Kluwer Academic Publishers, 2000.

16. K. Mehlhorn and A. Tsakalidis. Dynamic Interpolation Search. *Journal of the ACM*, 40(3):621-634, July 1993.

17. M. Mitzenmacher. A Brief History of Generative Models for Power Law and Lognormal Distributions. *Internet Mathematics*, 1(2):226-251, 2004.

18. Y. Perl and L. Gabriel. Arithmetic Interpolation Search for Alphabet Tables. *IEEE Transactions on Computers*, 41(4):493-499, 1992.

19. Y. Perl, A. Itai, and H. Avni. Interpolation Search – A log log N Search. *Communications of the ACM* 21(7):550-554, 1978.

20. Y. Perl, E. M. Reingold. Understanding the Complexity of the Interpolation Search. *Information Processing Letters* 6(6):219-222, December 1977.

21. W.W. Peterson. Addressing for Random Storage. *IBM Journal of Research and Development* 1(4):130-146, 1957.

22. N. Santorno and J.B. Sidney. Interpolation binary search. Information Processing Letters, 20, 179–181, 1985.

23. J. Spencer. Ten Lectures on The Probabilistic Method. Society for Industrial and Applied Mathematics, 2nd Ed. (1994)

24. P. van Emde Boas. Preserving order in a forest in less than logarithmic time and linear space. Information Processing Letters, 6:80-82, 1977.

25. P. van Emde Boas, R. Kaas, E. Zijlstra. Design and implementation of an efficient priority queue. Mathematical Systems Theory, 10:99-127, 1977.

26. D.E. Willard. Searching Unindexed and Nonuniformly Generated Files in log log N Time. *SIAM Journal of Computing* 14(4):1013-1029, 1985.
27. D.E. Willard. Examining Computational Geometry, Van Emde Boas Trees, and Hashing from the Perspective of the Fusion Tree. *SIAM Journal on Computing*, 29(3), 1030-1049, 2000.
28. A.C. Yao and F.F. Yao. The Complexity of Searching an Ordered Random Table. In *Proc. 17th IEEE Symp. on Foundations of Computer Science* – FOCS'76, pp. 173-177, 1976.

Dynamic Matrix Rank

Gudmund Skovbjerg Frandsen[1] and Peter Frands Frandsen[2]

[1] BRICS*, University of Aarhus, Denmark
gudmund@daimi.au.dk
[2] Rambøll Management, Aarhus, Denmark
Peter.Frandsen@r-m.com

Abstract. We consider maintaining information about the rank of a matrix under changes of the entries. For $n \times n$ matrices, we show an upper bound of $O(n^{1.575})$ arithmetic operations and a lower bound of $\Omega(n)$ arithmetic operations per change. The upper bound is valid when changing up to $O(n^{0.575})$ entries in a single column of the matrix. Both bounds appear to be the first non-trivial bounds for the problem. The upper bound is valid for arbitrary fields, whereas the lower bound is valid for algebraically closed fields. The upper bound uses fast rectangular matrix multiplication, and the lower bound involves further development of an earlier technique for proving lower bounds for dynamic computation of rational functions.

1 Introduction

1.1 The Problem

Given a field k the function rank : $k^{n^2} \mapsto \{0, \ldots, n\}$ denotes the rank of an $n \times n$ matrix, i.e. the maximal number of linearly independent columns in the matrix, or, equivalently, the maximal number of linearly independent rows in the matrix. The dynamic matrix rank problem consists in maintaining the rank of an $n \times n$ matrix $M = \{m_{ij}\}$ under the operations change$_{ij}$, $i, j = 1, \ldots, n$, where change$_{ij}(v)$ assigns the value $v \in k$ to m_{ij}.

1.2 Earlier Work

The authors are not aware of earlier nontrivial bounds for dynamic matrix rank, i.e. the best upper bound until now appears to arise from computing the rank from scratch. Off-line rank computation reduces to matrix multiplication via computing the row-echelon form of the matrix using a nice recursive construction due to Schönhage [11] and Keller-Gehrig [8]. There is a selfcontained description in [2, sect. 16.5]. The reduction implies that matrix rank can be computed using $O(n^{2.376})$ arithmetic operations [3].

Sankowski [9] gives several dynamic algorithms for computing matrix inverse, matrix determinant and solving systems of linear equations. The best of these

---

* Basic Research in Computer Science, Centre of the Danish National Research Foundation.

M. Bugliesi et al. (Eds.): ICALP 2006, Part I, LNCS 4051, pp. 395–406, 2006.

algorithms obtains worst case time $O(n^{1.495})$ per update/query. Sankowskis algorithms rely on the matrix staying nonsingular trough all updates, and he states that the time bound gets worse if the matrix is allowed to become singular. Hence, it is not clear that similar upper bounds can be given for dynamic matrix rank.

Frandsen et al. [4] gives $\Omega(n)$ lower bounds for dynamic matrix inverse and matrix determinant in a model based on algebraic computation trees. The lower bound is based on an incompressibility result from algebraic geometry and works only for dynamic evaluation of a set of polynomials or rational functions over a given field. It is not clear that the technique can be adapted to matrix rank that is essentially a constant function except on the algebraic subset of singular matrices.

1.3   Results

We give two dynamic algorithms. They use the techniques from two of Sankowskis algorithms, but as mentioned earlier some modifications are necessary to make the techniques work for rank.

The first algorithm which is quite elementary finds the rank by recomputing a reduced row echelon form of the matrix for every change. This can be done using $O(n^2)$ arithmetic operations per change. This bound is valid also when a change alters arbitrarily many entries in a single column of the matrix.

The second algorithm maintains the rank by using an implicit representation of the reduced row-echelon form. This implicit representation is kept sufficiently compact by using fast rectangular matrix multiplication for global rebuilding, obtaining a worst case complexity of $O(n^{1.575})$ arithmetic operations per change. This bound is still valid when a change alters up to $O(n^{0.575})$ entries in a single column of the matrix.

We show a lower bound $\Omega(n)$ on the worst case time used per change of a matrix entry, when maintaining the rank of a matrix over a field. The lower bound is valid for any algebraically closed field. Our model of computation combines the classical algebraic computation trees used for off-line algebraic computation [2] with the notion of *history dependence* [4] that let us extend the model to dynamic computations. For the computation trees we allow the four arithmetic operations in computation nodes, zero-tests in branching nodes, and leaves are labelled with the rank. The history dependence may be interpreted as a technique for letting the computation trees branch also on any discrete information that was obtained in earlier change operations. Technically, the history dependence works by assigning (infinitely) many computation trees to each change operation, viz. one tree for each history, where a history is every bit of discrete information the system has obtained so far; in particular that includes the result of every branching test made in earlier operations. All our upper bound algorithms can be formulated in this model, and it seems to be a natural model for dynamic algebraic computations that are generic in the sense of being valid for all fields, though it is slightly weaker than the model used in [4].

Our proof technique is a nontrivial adaptation of a technique from [4]. The earlier technique works for dynamic evaluation of rational functions, and it exploits that a rational function is uniquely determined from its values on a small subset (via the Schwartz-Zippel theorem). However, the function of matrix rank is mostly constant and all the interesting behaviour occurs on a lower dimensional subset (the singular matrices). We manage to augment the earlier technique to show lower bounds for dynamic verification of evaluation of a rational function. Then we get the lower bound for dynamic rank using a reduction via dynamic verification of matrix vector multiplication.

1.4   Later Work

Sankowski has recently shown that dynamic matrix rank has an upper bound of $O(n^{1.495})$ over infinite fields when allowing randomization and a small probability of error [10].

1.5   Applications

A mixed matrix is a matrix where some entries are undefined. The maximum rank matrix completion problem consists in assigning values to the undefined entries in a mixed matrix such that the rank of the resulting fully defined matrix is maximized. Geelen [5] has described a simple polynomial time algorithm for maximum rank matrix completion of complexity $O(n^9)$ that uses a data structure for dynamic matrix rank. However, this application has been superseded by newer results. Berdan [1] has introduced improvements reducing the complexity to $O(n^4)$. Harvey et al [6] has an algorithm of complexity $O(n^3 \log n)$ for maximum rank matrix completion using a different technique.

2   Upper Bound on Dynamic Matrix Rank

Our algorithms are inspired by two of Sankowskis algorithms [9]. However, Sankowski considers only updates that preserves nonsingularity. Maintaining rank information means that we must concentrate on the nonsingular case for which reason our algorithms are somewhat different.

2.1   Preliminaries

In the following all matrices will be $n \times n$ matrices over some fixed field k. A vector v is an n-dimensional column vector, and we use the notation v^T for row vectors. Let e_i denote a vector with a 1 in the ith entry and zeros elsewhere. Hence ve_l^T denotes a matrix that has vector v in the lth column and zero's elsewhere. We let A_l denote the lth column of matrix A.

We call an entry in the matrix *leading* if it is the first non-zero entry in its row. Recall that a matrix is in *reduced row-echelon form* when

- the leading entry in any row is 1 (call such entry a leading 1),
- a column containing a leading 1 has zeros in all other entries, and

– rows are sorted according to position of leading 1, i.e. if row i has a leading 1 in position j then any other row $i' < i$ must have a leading 1 in some position $j' < j$. In particular, all zero-rows are at the bottom of the matrix.

Using the notation of [7,9], we let $\omega(1, \epsilon, 1)$ denote the exponent of multiplying $n \times n^\epsilon$ matrices by $n^\epsilon \times n$ matrices, i.e. $O(n^{\omega(1,\epsilon,1)})$ arithmetic operations suffice for this matrix multiplication.

Proposition 1. *(Huang and Pan [7, sect. 8.2]) Let $\omega = \omega(1,1,1) < 2.376$ and let $\alpha = 0.294$. Then*

$$\omega(1, \epsilon, 1) \leq \begin{cases} 2 + o(1), & 0 \leq \epsilon \leq \alpha \\ \frac{2(1-\epsilon)+(\epsilon-\alpha)\omega}{1-\alpha}, & \alpha < \epsilon \leq 1 \end{cases} .$$

2.2 Elementary Dynamic Algorithm

Though the algorithm is elementary, we describe it fairly detailed in Algorithm 1, since it is the basis for the faster algorithm in the next section.

For a matrix A, we maintain matrices U and E under changes of the entries in A such that

$$U \text{ is nonsingular, } E \text{ is in reduced row-echelon form, and } UA = E \quad (1)$$

Clearly, the rank of A is the number of nonzero rows in E.

The initialization for a given matrix A consists in computing a matrix E in reduced row echelon form and corresponding transformation matrix U. This can be done using $O(n^3)$ arithmetic operations by Gaussian elimination or using $O(n^\omega)$ arithmetic operations by an augmentation of the technique for computing echelon form asymptotically fast [2, sect.16.5].

It turns out that our update algorithm works as well for update of an entire column of A as for update of a single entry, so let us assume that an update changes A into $A' = A + ve_l^T$, for some column index l and vector v. We must find U', E' that together with A' satisfies (1). Note that

$$UA' = U(A + ve_l^T) = E + Uve_l^T = E + v'e_l^T,$$

for some vector v'. We need only find some row operations that will bring $D = E + v'e_l^T$ into reduced row-echelon form E' and then apply the same row operations to U to get U'. This may be divided into 3 parts.

First, if column l in D contains a leading entry it may be necessary to clean up the column, i.e. ensure that all entries in the column become zero except for the leading entry which becomes 1. This is handled in lines 3-7 of Algorithm 1 resulting in the matrix D'. The matrix $(I + W')$ where W' has at most a single nonzero column represents the row operations that transform D into D'.

Secondly, we may in the process of computing D (or D') have cancelled a former leading entry in column l of E creating a new leading entry in some column $t > l$, which makes it necessary to clean up column t as well. This is handled in lines 8-11 of Algorithm 1 resulting in matrix D'' with row operations represented by matrix $(I + W'')$, where W'' has at most a single nonzero column. Note that the cleaning up of column t cannot cancel any further leading entries, since E is in *reduced* row echelon form.

Finally, we may have to permute rows. This is done in line 12-13 of the algorithm, resulting in the updated E'. All row operations applied so far to E are also applied to U in line 14.

Algorithm 1. Dynamic Matrix Rank (elementary version)

Memory $n \times n$ matrices A, U, E, where
$\quad U$ is nonsingular, E is in reduced row-echelon form, and $UA = E$

Change Given vector v and column index l
compute new memory A', U', E' such that $A' = A + ve_l^T$:

1: $v' \leftarrow Uv$;
2: $D \leftarrow E + v'e_l^T = \{d_{ij}\}$;
3: If D has a leading entry in column l then
4: $\quad$ select row k such that $d_{k,l}$ is a leading entry and if $d_{k',l}$ is any other leading entry then the number of zeros that immediately follow $d_{k,l}$ in row k is at least as large as the number of zeros that immediately follow $d_{k',l}$ in row k' (equality between number of immediately following zeros only occurs if both rows have all zeros in E).
5: $\quad v_1 \leftarrow \text{cleanColumn}(D_l, k)$; $\quad W' \leftarrow v_1 e_k^T$;
6: else $W' \leftarrow 0$;
7: $D' \leftarrow (I + W')D$;
8: if $E = \{e_{ij}\}$ has leading entry e_{sl} for some s and
$\quad s$'th row of $D' = \{d'_{ij}\}$ has leading entry d'_{st} for some $t > l$ then
9: $\quad v_2 \leftarrow \text{cleanColumn}(D'_t, s)$; $\quad W'' \leftarrow v_2 e_s^T$;
10: else $W'' \leftarrow 0$;
11: $D'' \leftarrow (I + W'')D'$;
12: Select P to be a matrix that permutes the rows of D'' into echelon form
13: $E' \leftarrow PD''$;
14: $U' \leftarrow P(I + W'')(I + W')U$;

where **cleanColumn** given vector v and index r with $v_r \neq 0$, returns vector w such that $(I + we_r^T)$ represents the row operations needed to change any column of a matrix from v to e_r:

$$w_i \leftarrow \begin{cases} -\frac{v_i}{v_r} & \text{for } i \neq r \\ \frac{1}{v_r} - 1 & \text{for } i = r \end{cases}$$

Note that a complete update needs only $O(n^2)$ arithmetic operations. We have shown

Theorem 1. *Dynamic matrix rank over an arbitrary field can be solved using $O(n^2)$ arithmetic operations per change (worst case). This bound is valid when a change alters arbitrarily many entries in a single column. Given an initial matrix the data structure for the dynamic algorithm can be built using $O(n^\omega)$ arithmetic operations.*

2.3   Asymptotically Faster Dynamic Algorithm

To speed up the dynamic algorithm, we only maintain an implicit representation of the reduced row-echelon form of the matrix. Let $\epsilon \in [0, 1]$ be determined later. We maintain matrices $A, T, C, S, R, (I + R)^{-1}$ and an array L such that:

- A is the current matrix.
- R and S have at most n^ϵ nonzero columns.
- $U = (I + R)T$ is invertible
- $E = (I + R)(C + S)$ is in reduced row-echelon form except possibly for a permutation of rows
- $UA = E$
- $L(i)$ contains the column index of the leading 1 in row i of E if it exists, and otherwise (row i is all zeros) it contains 0.

Compared to the simple algorithm in the previous subsection, we don't permute rows in our matrices. In stead the permutation needed to bring E into reduced row-echelon form is represented implicitly by the array L. E itself is represented by the three matrices C, S and R, where C is an old version of E, S represents the changed columns since C was valid, and $(I + R)$ represents the row operations needed to transform $C + S$ into reduced row echelon form. Similarly, U is represented by R and T.

We only allow updates that change at most n^ϵ entries of A, and all changes must be in a single column. When performing an update, we change R to incorporate the additional row-operations needed. In this way, R and S may eventually get more than n^ϵ nonzero columns, and we recompute T and C, while R and S are reset to zero-matrices. This recomputation involves multiplying rectangular matrices and can be done in the background using global rebuilding. It turns out that the existence of asymptotically very fast algorithms for rectangular matrix multiplication suffices to ensure a good worst case bound on the complexity of an update.

In the following we first consider the changes to Algorithm 1 caused by the implicit representation of E and U, and discuss details of the global rebuilding afterwards.

We assume an update changes A into into $A' = A + v e_l^T$, for some column index l and vector v that has at most n^ϵ nonzero entries. If we can compute the vector v' and the matrices W', W'' used in Algorithm 1, then we may compute the updated versions R', S' and $(I + R')^{-1}$ of the matrices in the data structure as follows

- $S' = S + (I + R)^{-1} v' e_l^T$
- $I + R' = (I + W'')(I + W')(I + R)$
- $(I + R')^{-1} = (I + R)^{-1}(I + W')^{-1}(I + W'')^{-1}$

Note that R' has at most 2 more nonzero columns than R, and S' has at most 1 more nonzero column than S.

Note that $(I + R)^{-1} - I$ has the same number of nonzero columns as R namely $O(n^\epsilon)$. Similarly, if $(I + W')^{-1} = (I + \bar{W})$ and $(I + W'')^{-1} = (I + \bar{\bar{W}})$ then each of $\bar{W}$, $\bar{\bar{W}}$, W' and W'' has at most a single nonzero column. Therefore R', S' and $(I+R')^{-1}$ may all be computed using $O(n^{1+\epsilon})$ arithmetic operations, though we still need to argue that v', W', W'' can be computed within the same bound. It is not necessary to compute all entries of matrices D, D', D''. It suffices to know

- $v' = Uv = (I+R)(Tv)$
- the lth column of E, viz. $E_l = (I+R)((C+S)e_l)$
- the lth column of D, viz. $D_l = E_l + v'$
- the tth column of D', where $t \neq l$, viz. $D'_t = (I+W')E_t = (I+W')(I+R)((C+S)e_t)$
- the sth row of D', viz. $e_s^T D' = (e_s^T(I+W')(I+R))(C+S) + e_s^T(I+W')v'e_l^T$

which may all be computed using $O(n^{1+\epsilon})$ arithmetic operations, when parenthesizing as above and recalling that v has at most $O(n^\epsilon)$ nonzero entries, R has at most $O(n^\epsilon)$ nonzero columns and W' has at most a single nonzero column.

In order to determine whether D has a leading entry in column l and select the row k in lines 3-4 of Algorithm 1 it suffices to scan D_l and the list L, since D is identical to E except possibly for column l. Similarly, we can determine whether E has a leading entry in column l (line 8 of Algorithm 1) within time $O(n)$.

Finally L is updated. At line 5, $L(k)$ is changed to l, and at line 8, $L(s)$ is updated to t or 0 as appropriate. Clearly, we can maintain the rank of matrix A with no extra cost, since the rank is the number of nonzero entries in L.

The entire update has so far used $O(n^{1+\epsilon})$ arithmetic operations. To bound the number of nonzero columns in R, we need to recompute C, T and reset R, S to zero. This may be done using two multiplications of $n \times n^\epsilon$ by $n^\epsilon \times n$ matrices taking $O(n^{\omega(1,\epsilon,1)})$ arithmetic operations. This recomputation may be distributed over n^ϵ updates implying that each update uses $O(n^{1+\epsilon} + n^{\omega(1,\epsilon,1)-\epsilon})$ arithmetic operations. Choosing ϵ to balance the two terms, one obtains based on Proposition 1 the bound of $O(n^{1.575})$ arithmetic operations per update. We have shown

Theorem 2. *Dynamic matrix rank over an arbitrary field can be solved using $O(n^{1.575})$ arithmetic operations per change (worst case). This bound is valid when a change alters up to $O(n^{0.575})$ entries in a single column. Given an initial matrix the data structure for the dynamic algorithm can be built using $O(n^\omega)$ arithmetic operations.*

3   Lower Bound on Dynamic Matrix Rank

Our lower bound proof has two steps. We introduce the intermediate problem of dynamic matrix vector multiplication verification (MVMV), where the MVMV problem consists in verifying that $Mx = y$ for square matrix M and column vectors x and y.

In the first step we use the technique of Frandsen et al [4] to show a lower bound for dynamic MVMV. A refinement of the technique is necessary, since the original technique applies to dynamic computation of rational functions, whereas we need a lower bound for dynamically verifying such a computation. Since verification is potentially easier than computation, one might expect it to be harder to prove a lower bound. Though we succeed, our proof works only for algebraically closed fields rather than infinite fields in general, and for a restricted model namely history dependent algebraic computation trees where branching is based on =-comparisons rather than general predicates.

In the second step, we give a reduction that implies the wanted lower bound for dynamic matrix rank.

3.1   Model of Computation

Let k be a field. Recall that an algebraic subset $W \subset k^n$ is an intersection of sets of the form $\{\mathbf{x} \in k^n | p(\mathbf{x}) = 0\}$, where p is a non-trivial multivariate polynomial.

Both the computation of matrix rank and the verification of a matrix vector product may be seen as instances of the following more general problem:

Given a field k and a family $\mathbf{W} = \{W_i\}_{i=0}^{l-1}$ of algebraic subsets of k^n, let the function $f_{\mathbf{W}} : k^n \mapsto \{0, 1, \dots, l\}$ be defined by

$$f_{\mathbf{W}}(\mathbf{x}) = \begin{cases} \min\{i | \mathbf{x} \in W_i\} & \text{if } \mathbf{x} \in \cup_{i=0}^{l-1} W_i \\ l & \text{otherwise} \end{cases}$$

MVMV arises as $f_{W_0} : k^{n^2+n+n} \mapsto \{0, 1\}$, for $W_0 = \{(M, x, y) \in k^{n^2} \times k^n \times k^n | Mx = y\}$, where the function value 0 is interpreted as *true* and 1 as *false*. Similarly, matrix rank arises as $f_{W_0,\dots,W_{n-1}} : k^{n^2} \mapsto \{0, \dots, n\}$ for W_i consisting of those matrices where all $(i+1) \times (i+1)$ minors are zero.

The problem of computing $f_{\mathbf{W}} : k^n \mapsto \{0, 1, \dots, l\}$ dynamically consists in maintaining the value of $f(x_1, \dots, x_n)$, under the operations $\text{change}_i(c)$, that assigns x_i the value $c \in k$, for $i = 1, \dots, n$. We assume that initially $(x_1, \dots, x_n) = (0, \dots, 0)$.

Our change algorithms will be a specific kind of algebraic computation trees. Compared to [4,2] we allow only branching based on =-comparison, and the output is encoded directly in the type of leaf.

We first introduce our version of algebraic computation trees for off-line sequential computation. Given a function $f_{\mathbf{W}}$, we may compute it by an algebraic branching tree, that has 3 types of nodes:

- an internal computation node, labelled with a program line of the form $y_i \leftarrow y_j \circ y_k$, where $\circ \in \{+, -, \cdot, /\}$ and y_j, y_k are variables or inputs. When the computation path goes through this node, variable y_i is assigned the value of $y_j \circ y_k$.
- an internal branching node, labelled with y, a variable or an input. A branching node always has precisely two descendents. The path of computation chooses a branch based on whether the value of y is zero or not.
- a leaf node that is labelled with one of $\{0, 1, \dots, l\}$. When reaching a leaf node the label is the function value.

The complexity of an algebraic branching tree is its depth, i.e. the length of the longest path from the root to a leaf.

For dynamic computation we use *history dependent algebraic computation trees* [4], i.e. for each $change_i(x)$ operation ($i = 1, \ldots, n$), we assign not just one algebraic branching tree, but we assign (infinitely) many trees, namely one tree for each history, where a history is every bit of discrete information the system has obtained so far; namely, the sequence of input variables that were changed, what other variables have been assigned a value, and the result of every branching test made so far during the execution of the operations performed. When we execute a change operation, we find the tree corresponding to the current history and execute that. The complexity of a solution is the depth of its deepest tree.

Note that the dynamic algorithms for matrix rank from the previous section can be interpreted as families of history dependent algebraic computation trees of complexity $O(n^2)$ and $O(n^{1.575})$, respectively. All discrete information is encoded in the history. In particular, the contents of the array L from the asymptotically fast dynamic algorithm is represented as part of the history, so there are different trees for the different possible contents of array L.

3.2   Lower Bound for Dynamic MVMV

Basically we want to modify the lower bound proof for dynamic matrix vector multiplication [4] to be valid for the corresponding verification problem, dynamic MVMV. An essential ingredient of the mentioned proof is to use Schwartz-Zippels theorem to extrapolate correct behaviour (computing a specific polynomial) from a large finite subset to everywhere. However, in the case of a verification problem we compute a $0, 1$-valued function that is constant except on an algebraic subset, so correct behaviour cannot be extrapolated in the same way.

We manage to get around this problem, but have to restrict the computation trees to branch based on =-comparisons only, and the field must be algebraically closed.

We will prove our lower bound specifically for the dynamic MVMV problem, but the technique do apply more generally to verification of polynomial or rational functions (similar to [4, Theorem 2.1].

We need the following incompressibility result.

Proposition 2. *[4, Lemma 2.1] Let k be an algebraically closed field. Let W be an algebraic subset of k^m and let $\phi = (f_1/g_1, \ldots, f_n/g_n) : k^m \setminus W \mapsto k^n$ be a rational map where $f_i, g_i \in k[x_1, \ldots, x_m]$ for $i = 1, \ldots, n$. Assume that there exists $\mathbf{y} \in k^n$ such that $\phi^{-1}(\mathbf{y})$ is non-empty and finite. Then $m \le n$.*

and an additional technical result.

Proposition 3. *[4, Lemma 2.4] Let k be a field. Let $0 \le l \le n$ and let W be a proper algebraic subset of $k^n = k^l \times k^{n-l}$. There exists a proper algebraic subset $W_1 \subset k^{n-l}$ such that for all $\mathbf{a} \in k^{n-l} \setminus W_1$, we can find a proper algebraic subset $W_{\mathbf{a}} \subset k^l$ such that*

$$W \subseteq \{(\mathbf{x}, \mathbf{a}) \in k^l \times k^{n-l} \mid \mathbf{a} \in W_1 \text{ or } \mathbf{x} \in W_{\mathbf{a}}\}.$$

Theorem 3. *Let k be an algebraically closed field. Then any history dependent algebraic computation tree solution for dynamic evaluation of $MVMV(M, x, y)$ where $(M, x, y) \in k^{n^2} \times k^n \times k^n$ has complexity at least $n/4$.*

Proof. Let a family of algebraic computation trees solving dynamic evaluation of $MVMV$ be given, and let the max depth of any computation tree representing a change be d.

If we concatenate several change operations into a composite change, we may compose the associated computation trees into a larger tree by letting the root of a tree replace a leaf in a previous tree. Let in this way $P = P_1; P_2; P_3$ denote the algebraic computation tree for off-line $MVMV(M, x, y)$ that arises by concatenating changes in the following order (with no prior history, all inputs are initially zero) assuming input variables $M = \{m_{ij}\}$, $x = \{x_i\}$, and $y = \{y_j\}$.

$$P_1 : \text{change}_1(m_{11}); \cdots ; \text{change}_{n^2}(m_{nn});$$
$$P_2 : \text{change}_{n^2+1}(x_1); \cdots ; \text{change}_{n^2+n}(x_n);$$
$$P_3 : \text{change}_{n^2+n+1}(y_1); \cdots ; \text{change}_{n^2+2n}(y_n);$$

Define a modified tree $P' = P_1; P_2; P_3'$ where

$$P_3' : \text{change}_{n^2+n+1}((Mx)_1); \cdots ; \text{change}_{n^2+2n}((Mx)_n);$$

Note that P' is essentially P pruned to contain only leaves labelled *true*.

Given specific values for M, x the computation will follow a specific path through P'. Note that among possible computation paths, there will be a unique main path $\pi = \pi_1; \pi_2; \pi_3$ satisfying that there is an algebraic subset $W \in k^{n^2+n}$ such that all $M, x \in k^{n^2+n} \setminus W$ will follow the path π. Here π_1 denotes the portion of the path running through P_1 etc. The path π can also be found in the tree P, since P' is essentially a pruning of P, though π will not be the main path in P.

By proposition 3, there is an algebraic subset W_1 such that for $M \in k^{n^2} \setminus W_1$ there is an algebraic subset W_M such that for $x \in k^n \setminus W_M$ we have $M, x \in k^{n^2+n} \setminus W$, i.e. M, x takes the path π through P'.

Let V be the set of the variables that are written by computation nodes on π_1 and read by computation and branching nodes on $\pi_2; \pi_3$. Let $\mathbf{v} \in k^{|V|}$ denote the contents of V after the execution of π_1 but before the execution of $\pi_2; \pi_3$. Clearly, $\mathbf{v}$ is a rational function of M. Let $g : k^{n^2} \setminus W_1 \mapsto k^{|V|}$ denote that rational function.

We will now argue that g is injective. Assume to the contrary that we can find specific matrices $M_1, M_2 \in k^{n^2} \setminus W_1$ with $M_1 \neq M_2$ and $g(M_1) = g(M_2)$. Let $W_2 = \{x \mid M_1 x = M_2 x\}$, which is an algebraic subset of k^n. Choose an arbitrary $x_1 \in k^n \setminus (W_2 \cup W_{M_1} \cup W_{M_2})$. When the algebraic computation tree P is applied to the input $(M_1, x_1, M_1 x_1)$ it will follow path π and compute *true* as it should. However, when P is applied to input $(M_2, x_1, M_1 x_1)$ it will also follow path π, since $g(M_1) = g(M_2)$, and therefore also answer *true*, which is incorrect. By contradiction, we have shown that g is injective.

Using that $g : k^{n^2} \backslash W_1 \mapsto k^{|V|}$ is injective, Proposition 2 implies that $|V| \geq n^2$. However, since the path $\pi_2; \pi_3$ contains at most $2dn$ computation and branching nodes each of which can read at most 2 variables, it follows that $4dn \geq |V|$, implying that $d \geq n/4$.

3.3   Dynamic MVMV Reduces to Dynamic Matrix Rank

Let k be a field. Given an instance $(M, x, y) \in k^{n^2+n+n}$ of MVMV, create an instance $M' \in k^{(2n)^2}$ of matrix rank, where

$$M' = \begin{bmatrix} I & xe_1^T \\ M & ye_1^T \end{bmatrix}$$

where ze_1^T is the $n \times n$ matrix with vector z in the first column and zero's elsewhere. Clearly, $\text{rank}(M') \in \{n, n+1\}$ and $\text{rank}(M') = n$ if and only if $Mx = y$. Since the change of an input in (M, x, y) corresponds to a single change of M', we have reduced dynamic MVMV to dynamic matrix rank, and Theorem 3 implies

Theorem 4. *Let k be an algebraically closed field. Then any history dependent algebraic computation tree solution for dynamic computation of* $\text{rank}(M)$ *where* $M \in k^{n^2}$ *has complexity at least $n/8$.*

References

1. Mark Berdan. A matrix rank problem. Master's thesis, University of Waterloo, December 2003.
2. Peter Bürgisser, Michael Clausen, and M. Amin Shokrollahi. *Algebraic complexity theory*, volume 315 of *Grundlehren der Mathematischen Wissenschaften [Fundamental Principles of Mathematical Sciences]*. Springer-Verlag, Berlin, 1997. With the collaboration of Thomas Lickteig.
3. Don Coppersmith and Shmuel Winograd. Matrix multiplication via arithmetic progressions. *J. Symbolic Comput.*, 9(3):251–280, 1990.
4. Gudmund Skovbjerg Frandsen, Johan P. Hansen, and Peter Bro Miltersen. Lower bounds for dynamic algebraic problems. *Inform. and Comput.*, 171(2):333–349, 2001.
5. James F. Geelen. Maximum rank matrix completion. *Linear Algebra Appl.*, 288(1-3):211–217, 1999.
6. Nicholas J. A. Harvey, David R. Karger, and Kazuo Murota. Deterministic network coding by matrix completion. In *SODA '05: Proceedings of the sixteenth annual ACM-SIAM symposium on Discrete algorithms*, pages 489–498, Philadelphia, PA, USA, 2005. Society for Industrial and Applied Mathematics.
7. Xiaohan Huang and Victor Y. Pan. Fast rectangular matrix multiplication and applications. *J. Complexity*, 14(2):257–299, 1998.
8. Walter Keller-Gehrig. Fast algorithms for the characteristic polynomial. *Theoret. Comput. Sci.*, 36(2-3):309–317, 1985.

9. Piotr Sankowski. Dynamic transitive closure via dynamic matrix inverse (extended abstract). In *FOCS '04: Proceedings of the 45th Annual IEEE Symposium on Foundations of Computer Science (FOCS'04)*, pages 509–517, Washington, DC, USA, 2004. IEEE Computer Society.
10. Piotr Sankowski. Faster dynamic matchings and vertex connectivity (extended abstract). April 2006.
11. A. Schönhage. Unitäre Transformationen grosser Matrizen. *Numer. Math.*, 20:409–417, 1972/73.

Nearly Optimal Visibility Representations
of Plane Graphs

Xin He[1,*] and Huaming Zhang[2,**]

[1] Department of Computer Science and Engineering
SUNY at Buffalo
Buffalo, NY, 14260, USA
xinhe@cse.buffalo.edu
[2] Computer Science Department
University of Alabama in Huntsville
Huntsville, AL, 35899, USA
hzhang@cs.uah.edu

Abstract. The *visibility representation* (VR for short) is a classical representation of plane graphs. VR has various applications and has been extensively studied in literature. One of the main focuses of the study is to minimize the size of VR. It is known that there exists a plane graph G with n vertices where any VR of G requires a size at least $(\lfloor \frac{2n}{3} \rfloor) \times (\lfloor \frac{4n}{3} \rfloor - 3)$.

In this paper, we prove that every plane graph has a VR with height at most $\frac{2n}{3} + 2\lceil \sqrt{n/2} \rceil$, and a VR with width at most $\frac{4n}{3} + 2\lceil \sqrt{n} \rceil$. These representations are nearly optimal in the sense that they differ from the lower bounds only by a lower order additive term. Both representations can be constructed in linear time. However, the problem of finding VR with optimal height and optimal width simultaneously remains open.

1 Introduction

A *visibility representation* (VR for short) of a plane graph G is a drawing of G, where the vertices of G are represented by non-overlapping horizontal line segments (called *vertex segment*), and each edge of G is represented by a vertical line segment touching only the vertex segments of its end vertices. In this paper, VR refers to weak visibility representation, in which not all mutually visible horizontal segments need to correspond to an edge in the graph.

As in many other graph drawing problems, one of the main concerns in VR research is to minimize the size of the representation. For the lower bounds, it was shown in [9] that there exists a plane graph G with n vertices where any VR of G requires a size at least $(\lfloor \frac{2n}{3} \rfloor) \times (\lfloor \frac{4n}{3} \rfloor - 3)$. For the upper bounds, it is known that every plane graph has a VR with height at most $\lfloor \frac{4n-1}{5} \rfloor$ [11], and a VR with width at most $\lfloor \frac{13n-24}{9} \rfloor$ [10].

In this paper, we prove that every plane graph of n vertices has a VR with height at most $\frac{2n}{3} + 2\lceil \sqrt{n/2} \rceil$, and a VR with width at most $\frac{4n}{3} + 2\lceil \sqrt{n} \rceil$.

* Research supported in part by NSF Grant CCR-0309953.
** Corresponding author.

M. Bugliesi et al. (Eds.): ICALP 2006, Part I, LNCS 4051, pp. 407–418, 2006.
© Springer-Verlag Berlin Heidelberg 2006

These representations are nearly optimal in the sense that they differ from the lower bounds only by a lower order additive term. Both representations can be constructed in linear time.

The present paper is organized as follows. Section 2 introduces preliminaries. Section 3 presents the construction of a VR with nearly optimal height. Section 4 presents the construction of a VR with nearly optimal width.

2 Preliminaries

When discussing VR, we assume the input graph G is a plane triangulation. (If not, we get a plane triangulation G' by adding dummy edges into G. After constructing a VR for G', we can get a VR of G by deleting the vertical line segments for the dummy edges.) We abbreviate the words "counterclockwise" and "clockwise" as **ccw** and **cw** respectively.

A *numbering* $\mathcal{O}$ of a set $S = \{a_1, \ldots, a_k\}$ is a one-to-one mapping between S and the set $\{1, 2, \cdots, k\}$. We write $\mathcal{O} = < a_{i_1}, a_{i_2}, \ldots, a_{i_k} >$ to indicate $\mathcal{O}(a_{i_1}) = 1$, $\mathcal{O}(a_{i_2}) = 2$ ... etc. A set S with a numbering written this way is called an *ordered list*. For two elements a_i and a_j, if a_i is assigned a smaller number than a_j in $\mathcal{O}$, we write $a_i \prec_{\mathcal{O}} a_j$. Let S_1 and S_2 be two sets with empty intersection. If $\mathcal{O}_1$ is a numbering of S_1 and $\mathcal{O}_2$ is a numbering of S_2, their concatenation, written as $\mathcal{O} = < \mathcal{O}_1, \mathcal{O}_2 >$, is the numbering of $S_1 \cup S_2$ where $\mathcal{O}(x) = \mathcal{O}_1(x)$ for all $x \in S_1$ and $\mathcal{O}(y) = \mathcal{O}_2(y) + |S_1|$ for all $y \in S_2$.

G is called a *directed graph* (digraph for short) if each edge of G is assigned a direction. An *orientation* of a (undirected) graph G is a digraph obtained from G by assigning a direction to each edge of G. We will use G to denote both the resulting digraph and the underlying undirected graph unless otherwise specified. (Its meaning will be clear from the context.)

Let $G = (V, E)$ be an undirected graph. A numbering $\mathcal{O}$ of V induces an orientation of G as follows: each edge of G is directed from its lower numbered end vertex to its higher numbered end vertex. The resulting digraph is called the *orientation derived from* $\mathcal{O}$ which, obviously, is an acyclic digraph. We use $\text{length}_G(\mathcal{O})$ (or simply $\text{length}(\mathcal{O})$ if G is clear from the context) to denote the length of the longest path in the orientation of G derived from $\mathcal{O}$. (The length of a path is the number of edges in it.)

For a 2-connected plane graph G and an external edge (s, t), an orientation of G is called an *st-orientation* if the resulting digraph is acyclic with s as the only source and t as the only sink. Such a digraph is also called an *st-graph*. Lempel et. al. [4] showed that for every 2-connected plane graph G and an external edge (s, t), there exists an *st*-orientation. For more properties of *st*-orientation and *st*-graph, we refer readers to [5].

Let G be a 2-connected plane graph and (s, t) an external edge. An *st-numbering* of G is a one-to-one mapping $\xi : V \rightarrow \{1, 2, \cdots, n\}$, such that $\xi(s) = 1$, $\xi(t) = n$, and each vertex $v \neq s, t$ has two neighbors u, w with $\xi(u) < \xi(v) < \xi(w)$, where u (w, resp.) is called a *smaller neighbor* (*bigger neighbor*, resp.) of v. Given an *st*-numbering ξ of G, the orientation of G derived from ξ is obviously an *st*-orientation of G. On the other hand, if $G = (V, E)$

has an st-orientation $\mathcal{O}$, we can define an 1-1 mapping $\xi : V \rightarrow \{1, \cdots, n\}$ by topological sort. It is easy to see that ξ is an st-numbering and the orientation derived from ξ is $\mathcal{O}$. From now on, we will interchangeably use the term "an st-numbering" of G and the term "an st-orientation" of G, where each edge of G is directed accordingly.

Definition 1. Let G be a plane graph with an st-orientation $\mathcal{O}$, where (s, t) is an external edge drawn on the external face of G. The st-*dual* graph G^* of G and the dual orientation $\mathcal{O}^*$ of $\mathcal{O}$ is defined as follows:

- Each face f of G corresponds to a node f^* of G^*. In particular, the unique internal face adjacent to the edge (s, t) corresponds to a node s^* in G^*, the external face corresponds to a node t^* in G^*.
- For each edge $e \neq (s, t)$ of G separating a face f_1 on its left and a face f_2 on its right, there is a dual edge e^* in G^* from f_1^* to f_2^*.
- The dual edge of the external edge (s, t) is directed from s^* to t^*.

It is well known that the st-dual graph G^* defined above is an st-graph with source s^* and sink t^*. The correspondence between an st-orientation $\mathcal{O}$ of G and the dual st-orientation $\mathcal{O}^*$ is a one-to-one correspondence. The following theorem was given in [6,8]:

Theorem 1. *Let G be a 2-connected plane graph with an st-orientation $\mathcal{O}$. Let $\mathcal{O}^*$ be the dual st-orientation of the st-dual graph G^*. A VR of G can be obtained from $\mathcal{O}$ in linear time. The height of the VR is* length($\mathcal{O}$). *The width of the VR is* length($\mathcal{O}^*$).

The following concept was introduced in [7] and is of central importance in our VR construction.

Definition 2. Let G be a plane triangulation of n vertices with three external vertices v_1, v_2, v_n in ccw order. A *realizer* $\mathcal{R} = \{T_1, T_2, T_n\}$ of G is a partition of its internal edges into three sets T_1, T_2, T_n of directed edges such that the following hold:

- For each $i \in \{1, 2, n\}$, the internal edges incident to v_i are in T_i and directed toward v_i.

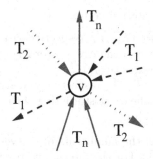

Fig. 1. Edge pattern around an internal vertex v

– For each internal vertex v of G, v has exactly one edge leaving v in each of T_1, T_2, T_n. The ccw order of the edges incident to v is: leaving in T_1, entering in T_n, leaving in T_2, entering in T_1, leaving in T_n, and entering in T_2. Each entering block may be empty. (See Figure 1.)

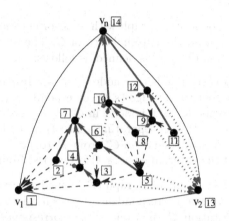

Fig. 2. A plane triangulation G and a realizer $\mathcal{R}$ of G

Figure 2 shows a realizer of a plane triangulation G. The dashed lines (dotted lines and solid lines, respectively) are the edges in T_1 (T_2 and T_n, respectively).

In [7], Schnyder showed that every plane triangulation G has a realizer which can be constructed in linear time. It was also shown that each set T_i of a realizer is a tree rooted at v_i. For each T_i, we denote by $\overline{T}_i$ the tree composed of T_i augmented with the two edges of the external face incident to the root v_i of T_i. Obviously $\overline{T}_i$ is a spanning tree of G. For example, in Figure 2, $\overline{T}_n$ is T_n (the tree in solid lines) augmented with the edges (v_1, v_n) and (v_2, v_n), which is a spanning tree of G.

For each internal vertex v of G and $i \in \{1, 2, n\}$, $p_i(v)$ denotes the path in $\overline{T}_i$ from v to the root v_i of $\overline{T}_i$. It was shown in [7] that $p_1(v), p_2(v)$ and $p_n(v)$ have only the vertex v in common, and for two vertices $u \neq v$ and two indices $i \neq j$, $p_i(u)$ and $p_j(v)$ can have at most one common vertex. The following property was shown in [1,2].

Property 1. Let v be an internal vertex of G.

1. All ancestors of v in $\overline{T}_1$ ($\overline{T}_2$, $\overline{T}_n$ respectively) constitute a nonempty set and they appear before v in the ccw postordering of the vertices of G with respect to $\overline{T}_n$ ($\overline{T}_1$, $\overline{T}_2$ respectively).
2. All ancestors of v in $\overline{T}_2$ ($\overline{T}_n$, $\overline{T}_1$ respectively) constitute a nonempty set and they appear before v in the cw postordering of the vertices of G with respect to $\overline{T}_n$ ($\overline{T}_1$, $\overline{T}_2$ respectively).

For example, in Figure 2, the numbering of the vertices of G is the ccw postordering with respect to $\overline{T}_n$. For the internal vertex 5, its ancestor set in $\overline{T}_1$ is $\{1, 3\}$. It is nonempty and all its elements appear before 5 in this numbering.

3 Visibility Representation of Nearly Optimal Height

Definition 3. Let $T = (V, E)$ be a tree drawn in the plane. The *balanced partition* of T is the partition of V into three ordered subsets A, B, C as follows. Let a_i be the ith vertex of T in ccw postordering and b_i the ith vertex of T in cw postordering. We mark the vertices of T in the order $a_1, b_1, a_2, b_2, \ldots, a_i, b_i \ldots$ Continue this process as long as the next pair of the vertices a_{i+1}, b_{i+1} have not been marked. We stop when either $a_{k+1} = b_{k+1}$ or b_{k+1} is already marked. This vertex a_{k+1} is called the *merge vertex* of T. When the marking process stops, the un-marked vertices of T form a single path from the merge vertex a_{k+1} to the root of T. We call this path the *leftover* path of T. We define $A =< a_1, a_2, \ldots, a_k >$, $B =< b_1, b_2, \ldots, b_k >$ and $C =< c_1(= a_{k+1}), c_2, \ldots, c_p >$ is the leftover path of T ordered from the merge vertex to the root of T.

For example, the balanced partition of the tree $\overline{T}_n$ in Figure 2 is: $A =< 1, 2, 3, 4, 5, 6 >$, $B =< 13, 11, 12, 9, 8, 10 >$, and $C =< 7, 14 >$.

Definition 4. Let $T = (V, E)$ be a tree with the balanced partition (A, B, C). A numbering $\mathcal{O}$ of V is *consistent with respect to* T if the following hold.

1. For any $i < j$, $a_i \prec_\mathcal{O} a_j$, $b_i \prec_\mathcal{O} b_j$, and $c_i \prec_\mathcal{O} c_j$.
2. For any vertices $a_i \in A, b_j \in B$ and $c_l \in C$, $a_i \prec_\mathcal{O} c_l$ and $b_j \prec_\mathcal{O} c_l$.

By the definition of the consistent numbering, it is easy to see that the following property holds:

Property 2. Let T be a tree with balanced partition (A, B, C) and $\mathcal{O}$ a numbering consistent with respect to T. If u is a child of v in T, then $u \prec_\mathcal{O} v$.

Lemma 1. *Let $\mathcal{R} = \{T_1, T_2, T_n\}$ be a realizer of a plane triangulation G. For $i \in \{1, 2, n\}$, let $\mathcal{O}_i$ be a consistent numbering with respect to $\overline{T}_i$. Then $\mathcal{O}_i$ is an st-numbering of G.*

Proof. We only prove the case $i = n$. The other two cases are symmetric. Let (A, B, C) be the balanced partition of $\overline{T}_n$.

For any vertex other than the root of $\overline{T}_n$, its parent is assigned a bigger number in $\mathcal{O}_n$. The root of $\overline{T}_n$ is assigned n.

For any non-leaf vertex of $\overline{T}_n$, their children are assigned smaller numbers in $\mathcal{O}_n$ by Property 2. Consider a leaf $u \neq v_1, v_2$ of $\overline{T}_n$. Clearly, u is an internal vertex of G. If $u \in A$ or $u \in C$ is the merge vertex, then according to Property 1 (1), its ancestor set in $\overline{T}_1$ is nonempty and all its members appear before u in ccw postordering with respect to $\overline{T}_n$. So the parent of u in $\overline{T}_1$ is a smaller neighbor of u. Similarly, by Property 1 (2), if $u \in B$, the parent of u in $\overline{T}_2$ is a smaller neighbor of u. For v_1 and v_2, one of them is assigned 1, and it becomes a smaller neighbor of the other. Thus $\mathcal{O}_n$ is an st-numbering of G.

Definition 5. A *ladder graph* of order k is a plane graph $L = (V_L, E_L)$. The vertex set V_L is partitioned into two ordered lists $A =< a_1, a_2, \ldots, a_k >$ and

$B =< b_1, b_2, \ldots, b_k >$. $E_L = E_A \cup E_B \cup E_{cross}$ where: $E_A = \{(a_i, a_{i+1}) | 1 \leq i < k\}$; $E_B = \{(b_i, b_{i+1}) | 1 \leq i < k\}$; and E_{cross} consists of edges between a vertex $a_i \in A$ and a vertex $b_j \in B$. The edges in E_{cross} are called *cross edges* of L.

For a cross edge (a_i, b_j), we define slope$(a_i, b_j) = i - j$. A cross edge (a_i, b_j) is called a *level* (or *up* or *down*, respectively) edge if slope$(a_i, b_j) = 0$ (or slope$(a_i, b_j) < 0$ or slope$(a_i, b_j) > 0$, respectively).

A numbering $\mathcal{O}$ of the vertices of a ladder graph $L = (A \cup B, E_L)$ is *consistent* with respect to L if for any $i < j$, $a_i \prec_{\mathcal{O}} a_j$ and $b_i \prec_{\mathcal{O}} b_j$.

Lemma 2. *Let* $L = (A \cup B, E_L)$ *be a ladder graph of order* k. *Let* l *be the number of level edges in* L. *Then* L *has a consistent numbering* $\mathcal{O}$ *such that* length$(\mathcal{O}) \leq (k-1) + l$. *This numbering can be constructed in linear time.*

Proof. Let $E_{cross} =< e_1, e_2, \ldots, e_p >$ be the ordered list of cross edges of L ordered from bottom up. We partition E_{cross} into *blocks* $E_1, E_2, \ldots, E_q$. Each E_j is a maximal sublist of E_{cross} that contains only up edges, or level edges, or down edges. E_j is called an up, or a level, or a down block accordingly.

For each block E_j, let x_j be the largest index such that either a_{x_j} or b_{x_j} is incident to an edge in E_j. The *span interval* of E_j (for $1 \leq j \leq q$) is defined to be: $I_1 = [1, x_1]$; $I_j = (x_{j-1}, x_j]$ (for $1 < j < q$); $I_q = (x_{q-1}, k]$. Note that $I_1, \ldots, I_q$ partition the interval $[1, k]$. For each j ($1 \leq j \leq q$), we define a numbering $\mathcal{O}_j$ of the vertex set $\{a_t, b_t \mid t \in I_j\}$ as follows:

If E_j is an up or a level block, let $\mathcal{O}_j =< a_{x_{j-1}+1}, a_{x_{j-1}+2}, \ldots, a_{x_j}, b_{x_{j-1}+1}, b_{x_{j-1}+2}, \ldots, b_{x_j} >$.

If E_j is a down block, let $\mathcal{O}_j =< b_{x_{j-1}+1}, b_{x_{j-1}+2}, \ldots, b_{x_j}, a_{x_{j-1}+1}, a_{x_{j-1}+2}, \ldots, a_{x_j} >$.

Define a numbering of the vertices in L by $\mathcal{O} =< \mathcal{O}_1, \mathcal{O}_2, \ldots, \mathcal{O}_q >$.

It is clear that $\mathcal{O}$ is a consistent numbering of L and the following hold: (i) the path consisting of the edges in E_A and the path consisting of the edges in E_B each has length $k - 1$; (ii) up edges and down edges do not increase the length of the longest path in the orientation of L derived from $\mathcal{O}$, and (iii) each level edge can increase the length of the longest path in the orientation of L derived from $\mathcal{O}$ by at most 1. Thus length$(\mathcal{O}) \leq k - 1 + l$. It is straightforward to construct $\mathcal{O}$ in linear time.

For an example, Figure 3 (1) shows a ladder graph L of order 7. E_{cross} is partitioned into 6 blocks: $E_1 = \{e_1\}$, $E_2 = \{e_2, e_3\}$, $E_3 = \{e_4, e_5\}$, $E_4 = \{e_6\}$, $E_5 = \{e_7, e_8\}$, and $E_6 = \{e_9\}$. $I_1 = [1, 2]$, $I_2 = (2, 3]$, $I_3 = (3, 5]$, $I_4 = (5, 5] = \emptyset$, $I_5 = (5, 7]$, and $I_6 = (7, 7] = \emptyset$. The numbering constructed according to Lemma 2 is $\mathcal{O} =< b_1, b_2, a_1, a_2, a_3, b_3, a_4, a_5, b_4, b_5, a_6, a_7, b_6, b_7 >$, where length$(\mathcal{O}) = 8$. We omit the details of proving the following lemma. Its basic idea is to lower the left side A by t levels, t is from 1 to $sqrt(k)$. Applying Lemma 2 to each of them leads to an st-numbering. we then pick the best st-numbering among them.

Lemma 3. *Let* $L = (A \cup B, E_L)$ *be a ladder graph of order* k. *Then* L *has a consistent numbering* $\mathcal{O}$ *such that* length$(\mathcal{O}) \leq k + 2\lceil \sqrt{k} \rceil - 1$. $\mathcal{O}$ *can be constructed in linear time.*

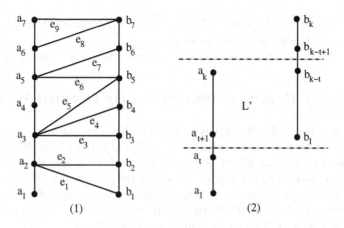

Fig. 3. (1) A ladder graph L; (2) Idea of proving Lemma 3

Lemma 4. *Let $G = (V, E)$ be a plane triangulation with n vertices. Let $\overline{T}$ be a tree obtained from a realizer $\mathcal{R}$ of G. Let (A, B, C) be the balanced partition of $\overline{T}$. Then G has an st-numbering $\mathcal{O}$ such that $\mathrm{length}(\mathcal{O}) \leq n/2 + |C|/2 + 2\lceil\sqrt{n/2}\rceil - 1$. $\mathcal{O}$ can be constructed in linear time.*

Proof. Let $k = (n - |C|)/2$. Then $|A| = |B| = k$. Let $L = (A \cup B, E_L)$ be the ladder graph constructed as follows. The vertex set of L is $A \cup B$. E_L contains all edges of G between A and B, and the edges $\{(a_i, a_{i+1}), (b_i, b_{i+1})|1 \leq i < k\}$. Then L is a ladder graph of order k. By Lemma 3, L has a consistent numbering $\mathcal{O}_L$ with $\mathrm{length}(\mathcal{O}_L) \leq k + 2\lceil\sqrt{k}\rceil - 1$.

Define $\mathcal{O} = < \mathcal{O}_L, C >$. Clearly this is a numbering of G consistent with respect to $\overline{T}$. By Lemma 1, $\mathcal{O}$ is an st-numbering of G. We have $\mathrm{length}(\mathcal{O}) \leq |C| + \mathrm{length}(\mathcal{O}_L) \leq |C| + (n - |C|)/2 + 2\lceil\sqrt{k}\rceil - 1 \leq n/2 + |C|/2 + 2\lceil\sqrt{n/2}\rceil - 1$. It is straightforward to construct $\mathcal{O}$ in linear time.

Lemma 5. *Let G be a plane triangulation of n vertices with a realizer $\mathcal{R} = \{T_1, T_2, T_n\}$. For $i \in \{1, 2, n\}$, let (A_i, B_i, C_i) be the balanced partition of the tree $\overline{T}_i$. Then there exists $i \in \{1, 2, n\}$ such that $|C_i| \leq n/3 + 1$.*

Proof. Each C_i is a path in $\overline{T}_i$. By the property of realizers, if $i \neq j$, C_i and C_j can intersect at most one vertex. Thus there exists $i \in \{1, 2, n\}$ such that $|C_i| \leq (n + 3)/3 = n/3 + 1$.

We now can prove our main theorem in this section.

Theorem 2. *Let G be a plane triangulation with n vertices. Then G has a VR with height $\leq 2n/3 + 2\lceil\sqrt{n/2}\rceil$, which can be constructed in linear time.*

Proof. Let $\mathcal{R} = \{T_1, T_2, T_n\}$ be a realizer of G. For each $i \in \{1, 2, n\}$, let (A_i, B_i, C_i) be the balanced partition of $\overline{T}_i$. By Lemma 5, there exists $i \in \{1, 2, n\}$ such that $|C_i| \leq n/3 + 1$. By Lemma 4, the st-numbering $\mathcal{O}_i$ associated with this $\overline{T}_i$ satisfies $\mathrm{length}(\mathcal{O}_i) \leq n/2 + |C_i|/2 + 2\lceil\sqrt{n/2}\rceil - 1 \leq$

$n/2+n/6+2\lceil\sqrt{n/2}\rceil-1+1 = 2n/3+2\lceil\sqrt{n/2}\rceil$. By Theorem 1, the st-numbering $\mathcal{O}_i$ leads to a VR with height at most $2n/3 + 2\lceil\sqrt{n/2}\rceil$.

By Lemma 4 and Theorem 1, we can construct this VR in linear time.

4   Visibility Representation of Nearly Optimal Width

In order to find a VR with nearly optimal width, our approach is parallel to that in Section 3: start with a realizer $\mathcal{R}^* = \{T_i^*, T_2^*, T_n^*\}$ of the dual graph G^*, we find an st-numbering $\mathcal{O}_i^*$ of G^* for each $i \in \{1, 2, n\}$ and show that one of them has the desired upper bound. However, the dual graph G^* is a 3-regular plane graph, not a plane triangulation. Therefore the concept of realizer defined in Definition 2 must be modified. In [3], the realizer concept is generalized from plane triangulation to 3-connected plane graph as follows.

Definition 6. Let G be a 3-connected plane graph with three external vertices v_1, v_2, v_n in ccw order. A *realizer* of G is a triplet of rooted spanning trees $\{T_1, T_2, T_n\}$ of G with the following properties:

1. For $i \in \{1, 2, n\}$, the root of T_i is v_i, the edges of G are directed from children to parent in T_i.
2. Each edge e of G is contained in at least one and at most two spanning trees. If e is contained in two spanning trees, then it has different directions in the two trees.
3. For each vertex $v \notin \{v_1, v_2, v_n\}$ of G, v has exactly one edge leaving v in each of T_1, T_2, T_n. The ccw order of the edges incident to v is: leaving in T_1, entering in T_n, leaving in T_2, entering in T_1, leaving in T_n, and entering in T_2. Each entering block may be empty. An edge with two opposite directions is considered twice. The first and the last incoming edges are possibly coincident with the outgoing edges. (Figure 4 shows two examples of edge pattern around an internal vertex v. In the second example, the edge leaving v in T_n and an edge entering v in T_2 are the same edge).
4. For $i \in \{1, 2, n\}$, all the edges incident to v_i belong to T_i.

We color the edges in T_1 by blue, T_2 by green, and T_n by red. According to the definition, each edge of G is assigned one or two colors, and is said to be *1-colored* or *2-colored*, respectively.

Consider a realizer $\mathcal{R} = \{T_1, T_2, T_n\}$ (as defined in Definition 2) of a plane triangulation G. The triplet of the three trees $\{\overline{T}_1, \overline{T}_2, \overline{T}_n\}$ (each of the three external edges $(v_1, v_2), (v_2, v_n), (v_n, v_1)$ is in two trees) is a special case of the realizer for 3-connected plane graphs defined here, where the three external edges are the only 2-colored edges.

For each vertex v of G and $i \in \{1, 2, n\}$, $p_i(v)$ denotes the path in T_i from v to the root v_i of T_i. A subpath of $p_i(v)$ between the end vertex v and an ancestor u of v in T_i is denoted by $p_i(v, u)$. The subpath of the external face of G with end vertices v_1 and v_2 and not containing v_n is denoted by $ext(v_1, v_2)$. The subpaths $ext(v_2, v_n)$ and $ext(v_n, v_1)$ are defined similarly.

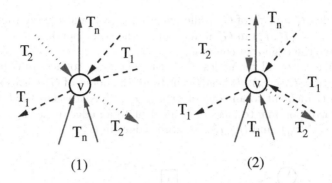

(1)   (2)

Fig. 4. Two examples of edge pattern around an internal vertex v

It was shown in [3] that every 3-connected plane graph has a realizer, which can be computed in linear time. The properties of realizer have been studied extensively in [3] which are summarized in the following lemma.

Lemma 6. *Let* $G = (V, E)$ *be a 3-connected plane graph with* $|V| = n$ *and* $|E| = m$. *Let* $\mathcal{R} = (T_1, T_2, T_n)$ *be a realizer of* G, *where* T_i *is rooted at the vertex* v_i *for* $i \in \{1, 2, n\}$.

1. *The number of 2-colored edges of* G *is* $3n - m - 3$.
2. *For each vertex* v *of* G, $p_1(v)$, $p_2(v)$ *and* $p_n(v)$ *have only the vertex* v *in common.*
3. *For* $i, j \in \{1, 2, n\}$ $(i \neq j)$ *and two vertices* u *and* v, *the intersection of* $p_i(u)$ *and* $p_j(v)$ *is either empty or a common subpath.*
4. *For vertices* v_1, v_2, v_n *the following hold:* $p_1(v_2) = p_2(v_1) = ext(v_1, v_2)$; $p_2(v_n) = p_n(v_2) = ext(v_2, v_n)$; $p_n(v_1) = p_1(v_n) = ext(v_n, v_1)$.

Similar to plane triangulation, we have the following version of Property 1 for 3-connected plane graph G with a realizer $\mathcal{R} = \{T_1, T_2, T_n\}$.

Property 3. Let v be an internal vertex of G.

1. All ancestors of v in T_1 (T_2, T_n respectively) constitute a nonempty set and they appear before v in the ccw postordering of the vertices of G with respect to T_n (T_1, T_2 respectively).
2. All ancestors of v in T_2 (T_n, T_1 respectively) constitute a nonempty set and they appear before v in the cw postordering of the vertices of G with respect to T_n (T_1, T_2 respectively).

Definition 7. [3]. Let $G = (V, E)$ be a 3-connected plane graph, with three external vertices v_1, v_2, v_n in ccw order. The *extended dual* graph G_e^* of G is defined as follows.

1. Each internal face f of G corresponds to a node f^* in G_e^*; the external face of G corresponds to three nodes v_1^*, v_2^*, and v_n^* in G_e^*.
2. Each edge e of G corresponds to an edge e^* in G_e^*.

3. Two nodes f_1^* and f_2^* of G_e^* (different from v_1^*, v_2^*, and v_n^*) are connected by an edge $e^* = (f_1^*, f_2^*)$ in G_e^* if and only if the corresponding faces f_1 and f_2 in G have the edge e in common.

4. v_1^* is adjacent to all the nodes of G_e^* corresponding to faces of G incident to an edge of $ext(v_2, v_n)$; v_2^* is adjacent to all the nodes of G_e^* corresponding to faces of G incident to an edge of $ext(v_n, v_1)$; v_n^* is adjacent to all the nodes of G_e^* corresponding to faces of G incident to an edge of $ext(v_1, v_2)$.

5. (v_1^*, v_2^*), (v_2^*, v_n^*), and (v_n^*, v_1^*) are three edges in G_e^*.

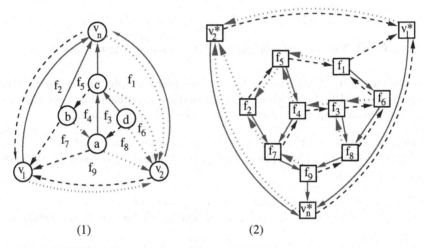

(1)   (2)

Fig. 5. (1) A plane triangulation G with a realizer $\mathcal{R}$; (2) Its extended dual G_e^* with dual realizer $\mathcal{R}^*$

Figure 5 (1) shows a plane triangulation G. Figure 5 (2) shows its extended dual G_e^*. Note that the extended dual G_e^* of a 3-connected plane triangulation G is also a 3-connected plane graph.

Let $G = (V, E)$ be a 3-connected plane graph with a realizer $\mathcal{R} = \{T_1, T_2, T_n\}$. Let G_e^* be the extended dual of G. In [3], it was shown that $\mathcal{R}$ *induces a dual realizer* $\mathcal{R}^* = (T_1^*, T_2^*, T_n^*)$ of G_e^*. In our application, we only need the case where G is a plane triangulation. The following definition of $\mathcal{R}^*$ is restricted to this special case.

Definition 8. Let $G = (V, E)$ be a plane triangulation with a realizer $\mathcal{R} = \{T_1, T_2, T_n\}$. Then $\mathcal{R}$ induces a *dual realizer* $R^* = \{T_1^*, T_2^*, T_n^*\}$ of the extended dual G_e^* as follows.

1. Let $f^* \notin \{v_1^*, v_2^*, v_n^*\}$ be a node of G_e^*. Let f be the face of G corresponding to f^* and e an internal edge of G that is on the boundary of f. We color and direct the dual edge e^* in G_e^* according to the following rules:
   (a) If e is blue (i.e. in T_1), then e^* is colored by green and red (i.e. e^* is in both T_2^* and T_n^*). If e is cw on f, then e^* is an outgoing red and incoming green edge for f^*. If e is ccw on f, then e^* is an outgoing green and incoming red edge for f^*.

(b) If e is green (i.e. in T_2), then e^* is colored by red and blue (i.e. e^* is in both T_n^* and T_1^*). If e is cw on f, then e^* is an outgoing blue and incoming red edge for f^*. If e is ccw on f, then e^* is an outgoing red and incoming blue edge for f^*.

(c) If e is red (i.e. in T_n), then e^* is colored by blue and green (i.e. e^* is in both T_1^* and T_2^*). If e is cw on f, then e^* is an outgoing green and incoming blue edge for f^*. If e is ccw on f, then e^* is an outgoing blue and incoming green edge for f^*.

2. The dual edge of (v_1, v_2) is red (i.e. in T_n^*) and directed to v_n^*. The dual edge of (v_2, v_n) is blue (i.e. in T_1^*) and directed to v_1^*. The dual edge of (v_n, v_1) is green (i.e. in T_2^*) and directed to v_2^*.

3. The edge (v_1^*, v_2^*) is blue and green (i.e. in both T_1^* and T_2^*). The edge (v_2^*, v_n^*) is green and red (i.e. in both T_2^* and T_n^*). The edge (v_n^*, v_1^*) is red and blue (i.e. in both T_n^* and T_1^*). Their directions are shown in Figure 5 (2).

Figure 5 (1) shows a plane triangulation G and a realizer $\mathcal{R}$. Figure 5 (2) shows the dual realizer $\mathcal{R}^*$ of the extended dual graph G_e^*. The extended dual graph G_e^* and the dual realizer $\mathcal{R}^*$ of a plane triangulation G with a realizer $\mathcal{R}$ satisfy the following:

Property 4. Let $\mathcal{R}$ be a realizer of a plane triangulation G with n vertices, and $\mathcal{R}^*$ the dual realizer of $\mathcal{R}$ of the extended dual graph G_e^*.

1. The number N of nodes in G_e^* is the number of faces of G plus two (since the external face of G corresponds to 3 nodes in G_e^*). Thus $N = 2n - 4 + 2 = 2n - 2$.

2. The three edges $(v_1, v_2)^*, (v_2, v_n)^*, (v_n, v_1)^*$ are the only 1-colored edges in $\mathcal{R}^*$.

3. Since $\mathcal{R}^*$ is just a special case of realizers defined for 3-connected plane graphs, Property 3 holds for $\mathcal{R}^*$.

For each tree T_i^* ($i \in \{1, 2, n\}$) of $\mathcal{R}^*$, we can define the balanced partition of T_i^*, and the consistent numbering $\mathcal{O}_i^*$ with respect to T_i^*, exactly the same way as in Section 3.

It turns out that almost all results in Section 3 remain valid and with almost identical proof (with one major exception for Lemma 5), we state its corresponding version in the following lemma.

Lemma 7. *Let G be a plane triangulation of n vertices with a realizer $\mathcal{R} = \{T_1, T_2, T_n\}$. Let G_e^* be the extended dual of G with $N = 2n - 2$ nodes, and $\mathcal{R}^* = \{T_1^*, T_2^*, T_n^*\}$ the dual realizer of $\mathcal{R}$. For each $i \in \{1, 2, n\}$, let (A_i, B_i, C_i) be the balanced partition of T_i^*. Then there exists $i \in \{1, 2, n\}$ such that $|C_i| \leq N/3 + 4$.*

We now state our main theorem in this section, its proof is omitted:

Theorem 3. *Let G be a plane triangulation of n vertices. Then G has a VR with width $\leq 4n/3 + 2\lceil \sqrt{n} \rceil$, which can be constructed in linear time.*

References

1. N. Bonichon, C. Gavoille, and N. Hanusse, An information-theoretic upper bound of planar graphs using triangulation, in *Proc. STACS'03*, pp 499-510, Lectures Notes in Computer Science, Vol. 2607, Springer-Verlag, 2003.
2. Y.-T. Chiang, C.-C. Lin and H.-I. Lu, Orderly spanning trees with applications to graph encoding and graph drawing, in *Proc. of the 12th Annual ACM-SIAM SODA*, pp. 506-515, ACM Press, New York, 2001.
3. G. Di Battista, R. Tamassia and L. Vismara, Output-sensitive Reporting of Disjoint Paths. *Algorithmica* 23, no.4 (1999), 302-340.
4. A. Lempel, S. Even and I. Cederbaum, An algorithm for planarity testing of graphs, in *Theory of Graphs (Proc. of an International Symposium, Rome, July 1966)*, pp. 215-232, Rome, 1967.
5. P. Ossona de Mendez, Orientations bipolaires. PhD thesis, Ecole des Hautes Etudes en Sciences Sociales, Paris, 1994.
6. P. Rosenstiehl and R. E. Tarjan, Rectilinear planar layouts and bipolar orientations of planar graphs. *Discrete Comput. Geom.* 1 (1986), 343-353.
7. W. Schnyder, Planar graphs and poset dimension. *Order* 5 (1989), 323-343.
8. R. Tamassia and I.G.Tollis, An unified approach to visibility representations of planar graphs. *Discrete Comput. Geom.* 1 (1986), 321-341.
9. H. Zhang and X. He, Visibility Representation of Plane Graphs via Canonical Ordering Tree, *Information Processing Letters* 96 (2005), 41-48.
10. H. Zhang and X. He, Improved Visibility Representation of Plane Graphs, *Computational Geometry: Theory and Applications* 30 (2005), 29-29.
11. H. Zhang and X. He, An Application of Well-Orderly Trees in Graph Drawing, in: Proc. GD'2005, Lectures Notes in Computer Science, Vol. 3843, 458-467, Springer-Verlag, 2006.

Planar Crossing Numbers of Genus g Graphs

Hristo Djidjev[1,*] and Imrich Vrt'o[2,**]

[1] Los Alamos National Laboratory
P.O. Box 1663, Los Alamos, NM 87545, USA
[2] Institute of Mathematics, Slovak Academy of Sciences
Dúbravská 9, 841 04 Bratislava, Slovak Republic
djidjev@lanl.gov, vrto@savba.sk

Abstract. Pach and Tóth [14] proved that any n-vertex graph of genus g and maximum degree d has a planar crossing number at most $c^g dn$, for a constant $c > 1$. We improve on this results by decreasing the bound to $O(dgn)$, if $g = o(n)$, and to $O(g^2)$, otherwise, and also prove that our result is tight within a constant factor.

1 Introduction

A *drawing* of a graph G in the plane is an injection of the set of the vertices of G into points of the plane and a mapping of the set of the edges of G into simple continuous curves such that the endpoints of each edge are mapped onto the endpoints of its image curve. Moreover, no curve should contain an image of a vertex in its inside and no three curves should intersect in the same point, unless it is an endpoint. The *planar crossing number* (or simply the *crossing number*) of G, denoted by $\mathrm{cr}(G)$, is the minimum number of edge crossings over all drawings of G in the plane.

The concept of crossing numbers was introduced by Turán [18] more than 50 years ago. Although there have been scores of results and publications since, because of the difficulty of the problem there are only a few infinite classes of graphs with determined exact crossing numbers. For instance, Glebsky and Salazar recently proved that the crossing number of the Cartesian product of two cycles $C_m \times C_n$ is $(m-2)n$ [11]. But the exact crossing numbers for such important graphs as the complete graph K_m and the bipartite graph $K_{m,n}$ are not known.

From algorithmic point of view, crossing numbers have been studied by Leighton [13], who was motivated by their application in VLSI design. In graph drawing, crossing numbers have been used for finding aesthetic drawing of non-planar graphs and graph-like structures [3]. Typically, such graphs are drawn in the plane with a small number of crossings and next each crossing point is

---

* This work has been supported by the Department of Energy under contract W-705-ENG-36 and by the Los Alamos National Laboratory LDRD-DR Grant "Statistical Physics of Infrastructure Networks"

** This research was partially supported by the VEGA grant No. 2/6089/26 and the EPSRC grant GR/S76694/01.

M. Bugliesi et al. (Eds.): ICALP 2006, Part I, LNCS 4051, pp. 419–430, 2006.
© Springer-Verlag Berlin Heidelberg 2006

replaced by a new vertex of degree 4. The resulting planar graph is then drawn in the plane using an existing algorithm for nicely drawing a planar graph, and finally the new vertices are removed and replaced back by edge crossings. The general drawing heuristics are usually based on the divide and conquer approach, using good separators, or using 2-page layouts [4,13].

The problem of finding the crossing number of a given graph was first proved to be NP-hard by Garey and Johnson [9] and, more recently, it was shown to be NP-hard even for cubic graphs [12]. There is only one exact algorithm of practical use [2], but it works for small and sparse graphs only. The best polynomial algorithm approximates the crossing number with a polylogarithmic factor [8].

Another direction of research is to estimate crossing numbers in terms of basic graph parameters, like density and edge separators. There are only a few results of this type [1,13]. And although the crossing number and the genus of the graph are two of the most important measures for nonplanarity, there are only a few results that study the relationship between them. Pach and Tóth [14] showed that any n-vertex d-degree *toroidal* graph G (i.e., graph that can be drawn on the torus with no intersections) has crossing number $O(dn)$. If G is of genus g (i.e., can be drawn on a surface S_g of genus g with no intersections), they proved that $\mathrm{cr}(G) \leq c^g dn$, for some constant $c > 1$. Unfortunately, the constant c is very large and, as a consequence, their result can be useful for very small values of g only. Although their proofs are of a constructive type, Pach and Tóth do not discuss algorithmic issues.

In this paper we show that $\mathrm{cr}(G) = O(dgn)$, if $g = o(n)$, or $\mathrm{cr}(G) = O(g^2)$, otherwise. This result is tight within a constant factor. (The lower bound proof will be included in the full version of the paper.) Our approach allows one to estimate the surface g' crossing numbers of genus g graphs drawn on any surface $S_{g'}$ of genus g', for $g' < g$.

Our result is also interesting because of the fact that it relates the crossing numbers of a given graph on two different surfaces. Specifically, let $\mathrm{cr}_g(G)$ denote the surface g crossing number of G, i.e., the minimum number of edge crossings over all drawings of G in S_g. The above type of results says that if $\mathrm{cr}_g(G) = 0$, then $\mathrm{cr}(G)$ cannot be very large. We further strengthen this result by showing as a corollary of our main result that $\mathrm{cr}(G) = O(\mathrm{cr}_g(G)\, g + gn)$ for bounded degree graphs.

This paper is organized as follows. In Section 2 we give some basic definitions and facts about embeddings and surfaces. In Section 3, we prove our main result and describe the drawing algorithm based on our upper bound proof.

2   Preliminaries

In this paper by G we denote an undirected graph and by $V(G)$ and $E(G)$ we denote the set of the vertices and the set of the edges of G, respectively. The *size* of G is $|G| = |V(G)| + |E(G)|$. For any vertex v, the number of the adjacent vertices to v is called the *degree* of v and is denoted by $\deg(v)$. The maximum degree of any vertex of G is called the *degree* of G. The set of the vertices adjacent to v is called the *neighborhood* of v and is denoted by $N(v)$. For any set of vertices X the *neighborhood* of X is $N(X) = \bigcup_{v \in X} N(v)$.

The *bisection width* of G, denoted by $\mathrm{bw}(G)$, is the smallest number of edges whose removal divides the graph into parts having no more than $2|V(G)|/3$ vertices each.

By a *surface* we mean a closed manifold and by S_g we denote a surface of genus g. A *drawing* of G on S_g is any injection of the vertices of G onto points of S_g and the edges of G onto continuous simple curves of S_g so that the endpoints of any edge are mapped onto the endpoints of its corresponding curve. The drawing is called an *embedding*, if no two curves intersect, except possibly at an endpoint. The *genus* of G, denoted by $g(G)$, is the smallest genus of a surface G can be embedded in. G is *planar*, if the genus of G is zero. Every planar graph can be drawn in the plane without any edge intersections. Throughout this paper we will use *combinatorial representations* of embeddings, where each undirected edge of G is replaced by a pair of opposite directed edges, and the cyclic list of outgoing edges from any vertex v (called *edge-orbit*) specifies the counterclockwise order in which the edges appear around v in the embedding. In a *facial walk*, the successor of any edge (v, w) is the edge after (w, v) in the edge-orbit for w. The *faces* of embedding are all simple closed facial walks and they correspond to the maximal connected regions into which the drawing of G divides the plane. The *outer face* of a planar embedding corresponds to the infinite face of the corresponding drawing. In a combinatorial embedding, any face can be chosen to be the outer face.

In the remainder of this paper, we will use n, m, and d to denote the numbers of vertices, edges, and the degree of G, respectively. We also assume that we are given an embedding $\mu(G)$ of G in S_g as an input. If f denotes the number of the faces of $\mu(G)$, then the *Euler characteristic* $\mathcal{E}(\mu(G))$ of $\mu(G)$, denoted simply by $\mathcal{E}(G)$ when the embedding is clear from the context, is defined as $\mathcal{E}(G) = \mathcal{E}(\mu(G)) = n - m + f$. The relation between the Euler characteristic and the genus g of the embedding is given by the *Euler formula*

$$n - m + f = 2 - 2g. \tag{1}$$

For any subgraph K of G, let $\mu(K)$ denote the embedding of K induced by $\mu(G)$, let $g_\mu(K)$ denote the genus of $\mu(K)$, and let $g(K)$ denote the genus of K. Note that $g_\mu(K)$ and $g(K)$ may not be equal. In order to simplify notations, we denote $g_K = g_\mu(K)$.

3 The Drawing Algorithm

We will start in Section 3.1 by describing a procedure for partitioning G into components with special properties, which we divide into three classes. In Section 3.2 we will outline the rest of the algorithm that draws each component according to its type and then combines all drawings into a drawing of the original graph.

3.1   Dividing the Graph into Components

Without a loss of generality we assume that G is biconnected, since otherwise one can draw the biconnected components separately in the plane and then combine their drawings into a planar drawing of G. Triangulate $\mu(G)$ by inserting a suitable number of additional edges in each face that is not a triangle. Assign weights 1 to all original edges of G and weights 0 to all new edges.

In order to simplify the notations, we will continue to denote by G and $\mu(G)$ the modified graph and embedding, respectively, and will refer to the edges of weights 1 and 0 as *original* and *new* edges of G, respectively. For any set $X \subseteq V(G)$, let $wt(X)$ denote the sum of the weights of all edges of X. Since in our algorithms we will only be interested in intersections between original edges of G, we introduce the term *original crossings* to refer to crossings where both intersecting edges are original.

Select any vertex t and divide the vertices of G into levels according to their distance to t. For a constant r to be determined later, denote by L_j, for $0 \leq j < r$, the set of all edges between level i and level $i + 1$ vertices, for all i satisfying $i \bmod r = j$. Assume that the number of all levels is at least r. Then there exists an $i^* < r$ such that

$$wt(L_{i^*}) \leq \lfloor m/r \rfloor. \tag{2}$$

Replace each edge $e = (v, w) \in L_{i^*}$ by a pair of new edges $s_1 = (v, x_1)$ and $s_2 = (w, x_2)$ called *stubs*, where x_1 and x_2 are new vertices. This has the effect of "cutting" e. The stubs s_1 and s_2 are a *matching pair* of stubs and e is a *parent* of s_1 and s_2. For any stub (v, x), where $v \in V(G)$ and $x \notin V(G)$, vertex v is called *attached* and vertex x is called *unattached*. Our drawing algorithm will eventually join each pair of stubs back into their parent edge.

Compute the connected components of the resulting graph, G'. For any component K of G', let q_K and q'_K denote the number of the bicomponents induced by the vertices on the lowest and on the highest level of K, respectively, that are incident to edges from L_{i^*}. Without loss of generality, we assume that $q_K > 0$ for any component K except the one containing t, since otherwise K will be degenerate (a forest), which is easy to draw without intersections. Denote $L_K = N(V(K)) \cap L_{i^*}$.

Let $\mu(K)$ be the embedding of K induced by the embedding of G, let n_K and m_K be the numbers of the vertices and the edges of K, and let f_K be the number of faces of $\mu(G)$ whose all edges are in K.

Lemma 1. *The Euler characteristic of* $\mu(K)$ *is* $\mathcal{E}(K) = n_K - m_K + f_K + q_K + q'_K$.

Corollary 1. *The genus of* $\mu(K)$ *is* $g_K = 1 - (n_K - m_K + f_K + q_K + q'_K)/2$.

Assign a label (q_K, q'_K) to K. Consider the set M of all components K with $g_K = 0$ (i.e., such that $\mu(K)$ is a planar embedding) and with label $(1, 1)$ or $(1, 0)$. Merge any two components K_1 and K_2 from M that contain a pair of matching stubs by replacing all pairs of matching stubs by their parent edges. Assign a label $(1, 1)$ to the resulting component, $\bar{K}$, and continue until no more merges are possible.

Let $\mathcal{K}(G)$ denote the set of all resulting components of G. We will use the following lemma from [7].

Lemma 2. *Let $q_{G'}$ denote the sum of q_K over all components K of G'. Then $g_\mu(G') \le g(G) - q_{G'} + |\mathcal{K}(G)| - 1$.*

By using Lemma 2 we shows that, after the merges, the number of the resulting components is $O(g)$.

Lemma 3. $|\mathcal{K}(G)| < 2g$.

We will divide all components K of G' into three classes depending on their labels (q_K, q'_K) and on g_K as follows. (i) If $g_K > 0$, then K will be of *non-planar* type (note that K can actually be planar if the genus of K is smaller than the genus of $\mu(K)$). (ii) If $g_K = 0$ and the label of K is $(1, 1)$ or $(1, 0)$, then K will be of *l-planar* (for "long planar") type. (iii) If $g_K = 0$ and the label of K is not $(1, 1)$ or $(1, 0)$, then K will be of *s-planar* (for "short planar") type. Recall that components of non-planar and s-planar types have no more than r levels, but components of l-planar type can have larger number of levels.

3.2   Algorithm Outline

The rest of the algorithm draws each component K of G' in the plane according to its type. The goal is to have the unattached endpoints of all stubs drawn in the outer face and a relatively small number of original crossings between edges of K. After all components are drawn in this way, all pairs of matching stubs are joined into their parent edges. As all stubs are already in the outer face, intersections may occur only between pairs of stubs. Since, by (2), the weight of all stubs is $O(m/r)$, this final step will increase the total number of original crossings by $O((m/r)^2)$.

3.3   Drawing Non-planar Components

If K is non-planar, then we will show that a subgraph of K of relatively small size can be found such that "cutting" the embedding of K along the edges of that subgraph and appropriately pasting a face f along the cut produces a planar surface. Then we will draw K in the plane with f as an outer face and redraw the edges that were destroyed by the cut. Since those edges will be entirely in f, they will not intersect other edges of K. Finally, we will route all stubs to the outer face.

3.3.1   Finding a Planarizing Set for K

Consider a component K such that $g_K > 0$ and let l_K^- and l_K^+ denote the lowest and the highest levels of K. Define a spanning forest F_K of K with q_K trees as follows. For each connected bicomponent Q defined by the vertices of K on level l_K^- define a spanning tree T_Q for Q and add the edges of T_Q to F_K. For each vertex v on level greater than l_K^- choose any vertex w on a lower level adjacent to v and add edge (v, w) to F_K. Finally, for each stub s from K, make the attached endpoint of s parent of its unattached endpoint. Clearly, F_K contains q_K trees, one for each connected bicomponent induced by level l_K^- in K.

We will call an F_K-*cycle* any simple cycle in K that has exactly one non-forest edge. Since K has no-more than r levels, any F_K-cycle will contain no more than $2(r-1)$ vertices of K, excluding the vertices on level l_K^-.

For any non-forest edge e of K incident to two different faces f_1 and f_2 of the embedding, remove e and merge f_1 and f_2 into a single face. Since this operation eliminates one edge and one face, the Euler characteristic does not change. Continue until no such edge e remains. Then any of the remaining non-forest edges should be incident only to f. Clearly, f should be the only face of the resulting embedding, since any face must contains a non-forest edge (otherwise T will contain a cycle). Next, remove any edge that is incident to a degree-1 vertex as well as the degree-1 vertex itself. Since each removal reduces the number of the vertices and the number of the edges by one, this operation preserves the Euler characteristic of $\mu(K)$.

Denote by $Pl(K)$ the resulting graph. We will think of $Pl(K)$ as a "planarizing" graph since, as we will show in Step 4.2, it can be used to transform the embedding of K into a planar embedding. Denote by n_{Pl} and m_{Pl} the number of the vertices and the number of the edges of $Pl(K)$. By (1) we have

$$n_{Pl} - m_{Pl} + 1 = 2 - 2g_K, \qquad (3)$$

and hence $m_{Pl} = (n_{Pl} - 1) + 2g_K$, which implies that the number of remaining non-forest edges is $2g_K$. Therefore, $Pl(K)$ is a union of $2g_K$ F_K-cycles.

We proved the following.

Lemma 4. *The embedding $\mu(Pl(K))$ of $Pl(K)$ has a single face, genus g_K, and no more than $2g_K(r-1)$ vertices whose levels are in the interval $(l_K^-, l_K^+]$.*

We will use $Pl(K)$ in the next subsection to "planarize" $\mu(K)$.

3.3.2 Transforming $\mu(K)$ into a Planar Embedding

Next we transform $\mu(K)$ by modifying $Pl(K)$ so that it is transformed into a new face f bounded by a simple cycle c. See the example on Figure 1 (b). Next we describe more formally the transformation of the different elements of K. We denote by $e = (w_1, w_2)$ an edge of K with at least one endpoint in $Pl(K)$.

1) Vertices of $Pl(K)$. Let v be any vertex of degree k from $Pl(K)$ and let $< e_1, \cdots, e_k >$ be the counterclockwise permutation of the edges of $Pl(K)$ incident to v. Define k new vertices that will replace v and denote them by $\{e_1, e_2\}, \{e_2, e_3\}, \cdots, \{e_k, e_1\}$ (Figure 1 (a)).

2) Edges not in $Pl(K)$. Let $e \notin E(Pl(K))$. We will define an edge (w_1', w_2') to replace e. If w_1 is not from $Pl(K)$ let $w_1' = w_1$. Else denote by $< e_1, \cdots, e_k = e_0 >$ the edge-orbit of w_1 and let e_j be the first edge from $Pl(K)$ in a counterclockwise direction from e and let e_{j-1} be the first edge from $Pl(K)$ in a clockwise direction. Then define $w_1' = \{e_i, e_j\}$. Similarly define a vertex w_2' corresponding to w_2. Replace $e = (w_1, w_2)$ by the edge (w_1', w_2'), which we will denote by $new(w_1, w_2)$.

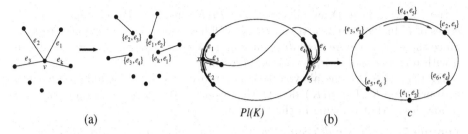

Fig. 1. The transformation of $Pl(K)$. (a) Replacing v by k new vertices. (b) Replacing $Pl(K)$ by a simple cycle c. The arrows show the direction of the face walk.

3) Edges from $Pl(K)$. Let $e \in E(Pl(K))$ and let e' be the first edge from $Pl(K)$ in a clockwise direction from e in the edge-orbit of w_2 and let e'' be the first edge from $Pl(K)$ in a counterclockwise direction of the edge-orbit of w_1. Define a new edge $\overrightarrow{new}(w_2, w_1) = (\{e, e''\}, \{e', e\})$. Similarly, define an edge $\overrightarrow{new}(w_1, w_2)$ by swapping w_1 and w_2. Finally, replace e by the two edges $\overrightarrow{new}(w_2, w_1)$ and $\overrightarrow{new}(w_1, w_2)$ (note that both those new edges are undirected).

4) Updating the edge-orbits. Next update the edge-orbits for the vertices incident to the new edges as follows. Let w be a vertex of $Pl(K)$ and let $< e_1 = (w, v_1), \cdots, e_k = (w, v_k) >$ be the counterclockwise permutation of the edges of K incident to w. For any pair of edges (w, v_i) and (w, v_j), $1 \leq i, j \leq k$, such that (w, v_j) is the first edge from $Pl(K)$ in a counterclockwise direction from (w, v_i) define the edge-orbit of the new vertex $w(e_i, e_j)$ as follows: $< \overrightarrow{new}(w, v_i), new(w, v_{i+1}), \cdots, new(w, v_{j-1}), \overrightarrow{new}(v_j, w) >$.

Denote by K' the resulting component, by $\bar{\mu}(K')$ its embedding, and by c the cycle corresponding to $Pl(K)$..

In order to simplify notations, let $V_{Pl} = V(Pl(K))$, $V_c = V(c)$. By construction, we have the following.

Lemma 5. *The resulting component K', its embedding $\bar{\mu}(K')$, and the cycle c constructed by the transformation of $Pl(K)$ have the following properties:*

(a) $V(K) = V(K') \setminus V_c \cup V_{Pl}$, $E(K) = E(K') \setminus N(V_c) \cup N(V_{Pl})$;
(b) $\{N(v) \mid v \in V_{Pl}\} \setminus V_{Pl} = \{N(v) \mid v \in V_c\} \setminus V_c$;
(c) The number of the faces of $\bar{\mu}(K')$ exceeds the number of the faces of $\mu(K)$ by one.

By computing the Euler characteristic of $\bar{\mu}(K')$ we prove the following lemma.

Lemma 6. *The embedding $\bar{\mu}(K')$ is planar.*

3.3.3 Transforming $\bar{\mu}(K')$ into a Planar Drawing of K with a Small Crossing Number

Recall that the cycle c in K' corresponds to the subgraph $Pl(K)$ of K. Replacing c with $Pl(K)$ will transform K' back into K.

Without loss of generality assume that the face corresponding to c is not the outer face. Remove all vertices from c and all of their incident edges. Denote by

h the resulting face. Draw all vertices of $Pl(K)$ inside h. By Lemma 5 (b), all edges of K incident to a vertex in $Pl(K)$ will have both their endpoints inside h. Since there are no more than dn_{Pl} such original edges, they can be drawn inside h with no more than $(dn_{Pl})^2$ original crossings.

By Lemma 5 (a), the above operation transforms the embedding of K' into a drawing of K. Let $\bar{\mu}(K)$ denote the resulting drawing. We summarize the properties of that drawing in the following lemma.

Lemma 7. $\bar{\mu}(K)$ *is a drawing of K in the plane that has no more than $(dn_{Pl})^2$ original crossings.*

3.3.4   Routing the Stubs of K to the Outer Face

Assign length 0 to all edges joining two vertices on level l_K^- and assign length 1 to all other edges of K. A *length* of a path in K is defined as the sum of the lengths of its edges. We will make use of the following fact.

Lemma 8. *Between any pair of vertices of K there exists a path entirely in K of length no more than $2q_K(r-1) + q_K - 1$.*

Finally, route all stubs of K to the outer face of $\bar{\mu}(K)$ using Lemma 8. More precisely, let s be a stub corresponding to an original edge and let p be the path constructed by the procedure of Lemma 8 for s. Informally, s will be routed along a path "parallel" to p that avoids vertices of p and that, for the portions of the path on level l_K^-, makes a "shortcut" inside the corresponding faces in order to minimize the number of intersections. More formally, remove all edges incident to vertices of p on levels greater than l_K^- (i.e., that have lengths 1) or on level l_K^-, but not on p. This operation creates a new face f that includes all faces defined by the vertices on level l_K^- and that have at least one edge from p. Route s inside f avoiding vertices from p. Then s will intersect no more than $l(s){\cdot}d$ original edges of G plus a number of stubs of K. (We will separately bound the number of *all* intersecting pairs of original stubs in our analysis below.)

Recall that $L_K = N(V(K)) \cap L_{i^*}$, where L_{i^*} was defined in (2). The next lemma summarizes the results of this section regarding the drawing of K.

Lemma 9. *The constructed drawing of K has less than $8(dg_K r)^2 + 2wt(L_K)^2 + 2dq_K r$ original crossings.*

When we apply the above algorithm to all non-planar components, the total number of original crossings is estimated in the following lemma.

Lemma 10. *All non-planar components of G can be drawn in the plane so that the unattached endpoints of all stubs are in the outer face of the drawing and the total number of original crossings is no more than $8(dgr)^2 + 12dgr + 2(m/r)^2$.*

3.4   Drawing s-Planar Components

This case is similar to the case of non-planar components, except that there is no need to planarize. We state the result in the following lemma.

Lemma 11. *All s-planar components of G can be drawn in the plane so that the unattached endpoints of all stubs are in the outer face and the total number of original crossings is no more than $12dgr$.*

3.5   Drawing l-Planar Components

Since l-planar components may have up to $\Omega(n)$ levels, the bound derived from (2) on the number of edges that have to be routed is not sufficient to guarantee a small crossing number. Hence, we have to additionally cut each l-planar component along some small set of edges joining two consecutive levels.

Let K be any l-planar component and let l_K^- and l_K^+ be the lowest and the highest levels of K, respectively. By Lemma 1, the embedding of K^- induced by the embedding of G is planar.

Denote by f^- and f^+ the faces defined by the set of the vertices on levels l_K^- and l_K^+, respectively. We will make use of the following fact.

Lemma 12. *Let v be a vertex on level i, where $l_K^- \leq i \leq l_K^+$. There exists a continuous line in the plane joining v to a point inside f^- (respectively f^+) that contains no other vertices of K except v and with no more than $(i - l_K^- + 1)d$ (respectively $(l_K^+ - i + 1)d$) original crossings with edges from K, excluding the edges joining a vertex on level i and a vertex on level $i + 1$.*

Lemma 13. *There exists a drawing of K in the plane with at most $|V(K)| + wt(L_K)^2$ original crossings such that all stubs have their unattached endpoints drawn in the outer face.*

Proof. Find a level i such that the number of the original edges of K joining a vertex on level i with a vertex on level $i + 1$ is minimum. Replace each edge joining levels i and $i + 1$ by a pair of stubs as we did with G in Section 3.1. This splits K into components K^-, induced by the vertices in K on levels less than or equal to i, and K^+, induced by the vertices in K on levels greater than i.

Convert $\mu(K^-)$ into a drawing with an outer face f^-. For any stub s of K^- incident to a vertex from level i, use Lemma 12 to route s to f^- so that s intersects no more than $(i - l_K^- + 1)d$ edges of K^-. Repeat the procedure for all other stubs of K^- incident to vertices from level i. The total number of original crossings produced in this step is no more than $L(i)(i - l_K^- + 1)d \leq |V(K^-)|d$, where $L(i)$ denotes the number of original edges joining a vertex on level i and a vertex on level $i + 1$.

Similarly, convert $\mu(K^+)$ into a drawing with an outer face f^+ and route all stubs of K^+ incident to vertices from level $i + 1$ to f^+, producing no more than $|V(K^+)|d$ original crossings. Add to the drawing the one of K^- found in the previous step. Finally, merge any pair of stubs incident to vertices from levels i and $i + 1$ into their parent edge of G. Since all stubs were already in the outer face of the drawing, all new intersections are between stubs incident to vertices from levels i and $i + 1$. Hence the number of new original crossings is no more than $(L(i))^2$, resulting in a total number of original crossings no more than $|V(K^-)|d + |V(K^+)|d + (L(i))^2 = |V(K)|d + (L(i))^2 \leq |V(K)|d + wt^2(L_K)$.   □

We summarize the results of this section in the following lemma.

Lemma 14. *All l-planar components can be drawn in the plane so that the unattached endpoints of all stubs are in the outer face of the drawing and the total number of original crossings is no more than $nd + (m/r)^2$.*

3.6 Reconnecting the Embedded Components

After all components of $G - S$ are drawn in the plane by applying the algorithms described in Subsections 3.3, 3.4, and 3.5, all the $wt(L_{i^*})$ original stubs will have their unattached endpoints in the outer face. Joining all pairs of original stubs into their parent edges so that no two stubs intersect more than once will produce at most $wt(L_{i^*})^2$ additional original crossings. This leads to the following result.

Theorem 1. *Any n-vertex graph of maximum degree d embedded in S_g can be drawn in the plane with $O(dgn)$, if $g = o(n)$, or $O(g^2)$, if $g = \Omega(n)$, edge crossings.*

Proof. By Lemmas 10, 11, and 14, the total number of original crossings from drawing individual components is no more than $8(dgr)^2 + 24dgr + nd + 3(m/r)^2 = O((dgr)^2 + nd + 3(m/r)^2$, where m is the number of original edges of G. Choosing $r = \lceil \sqrt{m/(gd)} \rceil$ and adding the number of the original crossings resulting from joining the stubs in the final step, which is bounded by $wt^2(L_{i^*}) = O((m/r)^2)$, the number of all original crossings is $O(dgm + dn) = O(dgm)$.

Without loss of generality we can assume that G is connected and has no vertex of degree two. Then we have the inequalities $m \geq 3/2n$ and $m \geq 3/2f$, which we will use next, where f is the number of the faces of the embedding.

If $g = o(n)$, then from the Euler formula (1) and the above inequalities $m = O(n)$ and hence $O(dgm) = O(dgn)$, which proves the theorem. If $g = \Omega(n)$, then by (1) $m = \Omega(n)$ and $g = \Omega(m)$. Since any straightline drawing of G in the plane has less than $m^2 = O(g^2)$ crossings, it satisfies the theorem. $\square$

Corollary 2. *Let G be any n-vertex bounded degree graph and let $0 < g = o(n)$. Then*
$$\mathrm{cr}(G) = O(\mathrm{cr}_g(G)\,g + gn).$$

3.7 Complete Algorithm and Complexity Analysis

Here we describe the entire algorithm and analyze its complexity.

Algorithm DRAW

Input: An n-vertex, d-degree graph G, an embedding $\mu(G)$ of G in S_g.
Output: A drawing of G with $O(\max\{dgn, g^2\})$ crossings.

1. If $g = \Omega(n)$, construct an arbitrary straightline drawing of G and exit.
2. Triangulate $\mu(G)$ assigning weight 0 to any new edge and weight 1 to any original edge of G.
3. Set $r = \lceil \sqrt{|E(G)|/(gd)} \rceil$. Divide the vertices of G into levels depending on their distances to a chosen vertex. Cut a subset of selected edges joining consecutive levels, as described in Section 3.1, producing components of the following three types: (i) non-planar components, having at most r levels and induced genus greater than zero; (ii) s-planar components, having at most r levels and genus 0; and (iii) l-planar components, whose vertices on their lowest and on their highest levels define single faces and that have genus 0. Each cut edge is replaced by a pair of stubs.

4. For each component K, draw K in the plane applying one of the Steps 5, 6, or 7.
5. If K is non-planar, then
   5.1. Construct a subgraph $Pl(K)$ of K such that (i) $Pl(K)$ contains at most $2g_K(r-1)$ vertices not counting the vertices on the highest and the lowest levels of K; (ii) converting $Pl(K)$ into a simple cycle c that is a face, denoted by f, of the new embedding as described in Section 3.3.2 transforms the embedding of K into an embedding of the updated graph, denoted by K', in S_0. Moreover, $Pl(K)$ and c have the same set, M, of edges joining them to K and K', respectively.
   5.2. Draw K' in the plane with f as an outer face.
   5.3. In order to transform K' back to K, remove c, draw the vertices of $Pl(K)$ in the resulting face, f', and draw all edges of M. Since both endpoints of any edge from M are on or inside f', intersections will occur only between pairs of edges from M.
   5.4. Route any stubs of K to the infinite face of the drawing as described in the proof of Lemma 8 and join matching pairs of stubs into their parent edges.
6. If K is s-planar, then draw K in the plane with one of the cycles determined by the vertices on the lowest level of K as outer face and continue as in Step 5.4.
7. If K is l-planar, choose a pair of adjacent levels of K such that the total weight of the edges between vertices on those levels is minimum. Replace these edges by pairs of stubs and draw the resulting planar components on the plane so that the cycles on the lowest and on the highest level of K are outer faces. Route each stub to the corresponding outer face as in the proof of Lemma 8. Merge all matching pairs of stubs with both endpoints in K into their parent edges.
8. After all components are drawn in the plane, restore G by merging matching pairs of the remaining stubs (all located in the infinite face) into their parent edges.

Theorem 2. *Algorithm DRAW constructs a drawing satisfying Theorem 1 of any n-vertex d-degree graph embedded in S_g in $O(dgn)$ time, if $g = o(n)$, or in $O(g^2)$ time, if $g = \Omega(n)$.*

References

1. M. Ajtai, V. Chvátal, M. M. Newborn, E. Szemerédy. Crossing-free subgraphs. *Theory and Practice of Combinatorics*, North Holland Mathematical Studies 60, Annals of Discrete Mathematics 12, North-Holland, Amsterdam, 1982, 9-12.
2. C. Buchheim, D. Ebner, M. Jünger, G. W. Klau, P. Mutzel, R. Weiskircher. Exact crossing minimization. *13th Intl. Symposium on Graph Drawing*, Lecture Notes in Computer Science, Springer, Berlin, 2006.
3. G. Di Battista, P. Eades, R. Tamassia, I.G. Tollis. Graph Drawing: Algorithms for Visualization of Graphs. Prentice Hall, 1999.

4. R. Cimikowski. Algorithms for the fixed linear crossing number problem. *Discrete Applied Mathematics* 122:93–115, 2002.
5. H. N. Djidjev. A separator theorem. *Compt. rend. Acad. bulg. Sci.*, 34:643–645, 1981.
6. H. N. Djidjev. Partitioning planar graphs with vertex costs: Algorithms and applications. *Algorithmica*, 28(1):51–75, 2000.
7. H. N. Djidjev, S. Venkatesan. Planarization of graphs embedded on surfaces. *21st International Workshop on Graph-Theoretic Concepts in Computer Science*, pages 62–72, 1995.
8. G. Even, S. Guha, B. Schieber. Improved Approximations of Crossings in Graph Drawings and VLSI Layout Areas. *SIAM J. Computing*, 32(1): 231-252, 2002.
9. M. R. Garey, D. S. Johnson. Crossing number is NP-complete *SIAM J. Algeraic and Discrete Methods*, 1983:4, 312–316.
10. J. R. Gilbert, J. P. Hutchinson, R. E. Tarjan. A separator theorem for graphs of bounded genus. *J. Algorithms*, 5:391–407, 1984.
11. L.Y. Glebsky, G. Salazar. The crossing numbert of $cr(C_m \times C_n)$ is as conjectured for $n \geq m(m+1)$. *J. Graph Theory* 47:53–72, 2004.
12. P. Hliněný. Crossing number is hard for cubic graphs. *Mathematical Foundations of Computer Science*, Lecture Notes in Computer Science 3153, Springer, Berlin, 772-782, 2004.
13. F.T. Leighton. Complexity Issues in VLSI. M.I.T. Press, Cambridge, 1983.
14. J. Pach, G. Tóth. Crossing number of toroidal graphs. *13th Intl. Symposium on Graph Drawing*, Lecture Notes in Computer Science 3843, Springer, Berlin, 334–342, 2006.
15. J. Pach, F. Shahrokhi, M. Szegedy. Applications of the crossing number. *Algorithmica*, 16:111-117, 1996.
16. O. Sýkora, I. Vrt'o. Optimal VLSI Layouts of the star graph and related networks. *Integration the VLSI Journal*, 17:83-94, 1994.
17. O. Sýkora, I. Vrt'o. Edge separators for graphs of bounded genus with applications. *Theoretical Computer Science*, 112:419-429, 1993.
18. K. Zarankievicz. On a problem of P. Turán concerning graphs. *Fund. Math.*, 41:137–145, 1954.

How to Trim an MST: A 2-Approximation Algorithm for Minimum Cost Tree Cover

Toshihiro Fujito*

Department of Information & Computer Sciences
Toyohashi University of Technology
Tempaku, Toyohashi 441-8580 Japan
fujito@ics.tut.ac.jp

Abstract. The *minimum cost tree cover* problem is to compute a minimum cost tree T in a given connected graph G with costs on the edges, such that the vertices of T form a vertex cover for G. The problem is supposed to arise in applications of vertex cover and edge dominating set when connectivity is additionally required in solutions. Whereas a linear-time 2-approximation algorithm for the unweighted case has been known for quite a while, the best approximation ratio known for the weighted case is 3. Moreover, the known 3-approximation algorithm for such case is far from practical in its efficiency.

In this paper we present a fast, purely combinatorial 2-approximation algorithm for the minimum cost tree cover problem. It constructs a good approximate solution by trimming some leaves within a minimum spanning tree (MST), and to determine which leaves to trim, it uses both of the primal-dual schema and the local ratio technique in an interlaced fashion.

1 Introduction

In an undirected graph $G = (V, E)$ a set C of vertices is a *vertex cover* if every edge in G has at least one of its end-vertices in C, whereas an edge set D is an *edge dominating set* if every edge not in D is adjacent to some edge in D. A tree $T \subseteq E$ in a connected graph G is called a *tree cover* if it is an edge dominating set for G. Or equivalently, it is a tree such that the set of vertices induced by T is a vertex cover for G. The *minimum cost tree cover* problem is to compute a tree cover of minimum total cost in a given connected graph $G = (V, E)$ with a nonnegative cost l_e on each edge $e \in E$. The problem is clearly NP-hard even in the unweighted case since it then becomes equivalent to the *connected vertex cover* problem, which in fact is known to be as hard (to approximate) as the vertex cover problem [11]. In fact, while it is possible to approximate minimum vertex cover to within a factor slightly better than 2 [3,15,12], doing so within any factor smaller than $10\sqrt{5} - 21 \approx 1.36067$ is NP-hard [6].

* Supported in part by a Grant in Aid for Scientific Research of the Ministry of Education, Science, Sports and Culture of Japan. Also affiliated with Intelligent Sensing System Research Center, Toyohashi Univ. of Tech.

M. Bugliesi et al. (Eds.): ICALP 2006, Part I, LNCS 4051, pp. 431–442, 2006.

The tree cover problem was introduced by Arkin, Halldórsson, and Hassin [1], and they were partially motivated by closely related problems of locating tree-shaped facilities on a graph such that all the vertices are dominated by chosen facilities. They presented a 2-approximation algorithm for the unweighted version, as well as a 3.55-approximation algorithm for the case of general costs. In fact a simpler 2-approximation algorithm appeared earlier for the unweighted case, due to Savage [17], although it was designed for vertex cover and not intended for connected vertex cover. A better approximation algorithm was later developed for minimum weight tree cover by Könemann et al. [14] and independently by Fujito[8], lowering the approximation ratio down to 3, and it is currently the best bound for the problem. Thus, whereas vertex cover, edge dominating set [9,16], and many problems closely related to them are known to be approximable to within a factor of 2, regardless of associated costs, it is not the case for tree cover. Even worse, the algorithms of [14] and [8] are far from practical in their efficiency; either one requires to solve optimally an LP of huge size (see (P) in Sect. 1.1), and to do so, it inevitably resorts to calling the ellipsoid method as their subroutine.

In this paper we present a fast, purely combinatorial 2-approximation algorithm for the minimum cost tree cover problem. All the previous algorithms for general costs [1,14,8] are in the similar style of computing a vertex cover C first, and then connecting all the vertices in C by a Steiner tree. Our algorithm in contrast is designed based on a hunch that a good approximate solution can be always found in the vicinity of a minimum spanning tree (MST)[1].

1.1 Bidirected Formulation

An instance of the minimum cost tree cover problem consists of an undirected graph $G = (V, E)$ and nonnegative costs l_e for all edges $e \in E$. Let $\vec{G} = (V, \vec{E})$ denote the directed graph obtained by replacing every edge $e = \{u, v\}$ of G by two anti-parallel arcs, (u, v) and (v, u), each having the same cost $c(\{u, v\})$ as the original edge e. Pick one vertex in V as the *root*, and suppose $\vec{T'} \subseteq \vec{E}$ is a *branching* (or a directed tree) rooted at r. It is assumed throughout that the arcs in a branching are always directed away from the root to a leaf. (Note: we will often use $\vec{T}$ and T interchangeably, to denote a branching and an undirected tree, respectively, with a root in common). In the *bidirected formulation* of the tree cover problem, one seeks for a minimum cost branching $\vec{T'}$ rooted at r in $\vec{G}$ such that T' is a tree cover rooted at r in G. We call either of such a branching or an undirected tree an *r-tree cover* for $\vec{G}$ (or G).

A set $S \subseteq V - \{r\}$ is called *dependent* if S induces at least one edge in G, and let $\mathcal{D}$ denote the family of such sets. Let $\delta^-(S)$ denote the set of arcs with heads in S and tails out of S (when needed, we use $\delta^-_H()$ to specify that only the arcs of graph H are considered). We call the arc set $\delta^-(S)$ in $\vec{G}$

---

[1] Interestingly, it was already tried by Arkin et al. [1] to use (a modification of) the Prim's or Kruskal's algorithm for MST problem, and either of them was found not to perform well.

an *r-edge cut* if $S \subseteq V - \{r\}$ is a dependent set. By using a max-flow/min-cut argument, one can see that the bidirected formulation of the minimum cost tree cover problem can be modeled by the integer program:

$$\min\{l^T x \mid x \in \{0,1\}^{\vec{E}}, x(\delta^-(S)) \geq 1, \forall S \in \mathcal{D}\},$$

where $x(\vec{F}) = \sum_{a \in \vec{F}} x_a$ for $\vec{F} \subseteq \vec{E}$, as an r-tree cover must pick at least one arc from every r-edge cut. Replacing the integrality constraints by $x \geq 0$, we have the LP relaxation of form:

$$\min \quad \sum_{a \in \vec{E}} l_a x_a$$

(P)   subject to: $\quad x(\delta^-(S)) \geq 1 \qquad \forall S \in \mathcal{D}$

$$x_a \geq 0 \qquad \qquad \forall a \in \vec{E}$$

Unlike the algorithms of [8,14], our algorithm also makes good use of the LP dual of (P):

$$\max \quad \sum_{S \in \mathcal{D}} y_S$$

(D)   subject to: $\quad \sum_{S \in \mathcal{D}: a \in \delta^-(S)} y_S \leq l_a \qquad \forall a \in \vec{E}$

$$y_S \geq 0 \qquad \qquad \forall S \in \mathcal{D}$$

At this point one may notice that the bidirected minimum cost tree cover problem has some similarity with another well-known combinatorial optimization problem, no matter how superficial it might be. In a directed graph $D = (V, A)$ with $r \in V$ an *r-arborescence* $A' \subseteq A$ is a spanning tree of the underlying undirected graph of D such that each vertex of D other than r is entered by exactly one arc of A' (and no arc enters r). An arc set $C \subseteq A$ is called an *r-cut* if $C = \delta^-(U)$ for some nonempty $U \subseteq V - \{r\}$. The *shortest r-arborescence problem* is to, given D, r, and nonnegative costs l_a for all the arcs $a \in A$, compute an r-arborescence of minimum cost.

Suppose now that the set of constraints in (P) concerning all the r-edge cuts is enlarged such that it consists of $x(\delta^-(S)) \geq 1$ for all nonempty $S \subseteq V - \{r\}$; that is, replace $\mathcal{D}$ by $\mathcal{D}' = \{S \subseteq V - \{r\} \mid S \neq \emptyset\}$, and denote it (P'). It was shown by Edmonds that the shortest r-arborescence problem can be formulated *exactly* by (P') [7]. Likewise, replace $\mathcal{D}$ by $\mathcal{D}'$ in (D), and call it (D'). Then, (D'), which is the LP dual of (P'), formulates the problem of maximum (fractional) *r-cut packing*, and it was shown by Fulkerson that (D') has integer optimum solutions if l is integral [10] (thus, there exist an r-arborescence and an integral r-cut packing of the same cost). Recall now that original (P) and (D) are actually based on graphs in bidirected forms of undirected graphs, and not on arbitrary digraphs, and if graphs in (P') are also restricted as such, the problem formulated by (P') reduces to the one on undirected graphs, namely, the *minimum spanning tree problem.* It is this observation that has motivated us to investigate the possibility of whether an MST, as an integer optimal solutions in (P'), could give us a lead when they are cast in (P) (or an r-cut packing when cast in (D)).

1.2   Primal-Dual Schema vs. Local Ratio Technique

Among various methods for design and analysis of approximation algorithms, the *primal-dual schema* and the *local ratio technique* have been popular and applied to a wide range of problems. While it is often possible to interpret algorithms in one framework within the other [3,5,2], and moreover, these two methods have been shown essentially equivalent [4], yet it could be of great use to have both of them at our disposal, as they can provide different lines of approaches to a problem of concern.

Certainly, the primal-dual method could be helpful in approximating the tree cover problem. Consider, for instance, the Savage's 2-approximation algorithm for unweighted tree cover, which simply returns the tree T_{tc} remaining after all the leaves are trimmed from a depth-first-search (DFS) spanning tree T [17]. The directed version $\vec{T}_{tc}$ of T_{tc} rooted at r is clearly feasible to (P). To estimate its cost $|T_{tc}|$, let M be a matching on T such that all the internal nodes of T but r are matched by M (Note: it is easy to find such a matching). Since M is a matching, r-edge cuts, $\delta^-(e)$ and $\delta^-(e')$, are disjoint for any two different edges e and e' of M. Hence, y with $y_e = 1$ for each $e \in M$ and $y_D = 0$ for all the other dependent sets D, is feasible to (D). To show that $|T_{tc}|$ is a factor of at most 2 away from the optimum, we need only to verify that $|\vec{T}_{tc}| \leq 2|M|$, by simple combinatorial arguments, for then, $|\vec{T}_{tc}| \leq 2 \sum y_D \leq 2$(optimal value of (P)).

It does not look so easy, however, to find a way to go from here to the case of arbitrary costs, under guidance of the known primal-dual schema only, and it was not until introducing the local ratio technique on top of it that we could find one. One basic scenario in the paradigm of local ratio technique is to "decompose" a cost function w defined on a problem instance I into many "slices" of cost functions $w_0, w_1, \ldots, w_{k-1}$, such that $w = \sum_i w_i$ and $w_i \geq 0, \forall i$. It is expected that an easily computable solution such as a minimally feasible solution, is a good enough approximation to the optimal one under each of w_i's, and if so, putting all such solutions together would yield a good approximation in the original instance.

A brief overview of our algorithm can be stated now as follows. It first decomposes (G, c) into uniformly costed instances of $(G_0, c_0), (G_1, c_1), \ldots, (G_{k-1}, c_{k-1})$, and it does so according to the costs of edges in an MST T. The algorithm next employs the primal-dual schema on each slice of (G_i, c_i)'s, and sets up a dual solution y^i for each of them. Finally, it determines which leaves of T to be removed using these y^i's. So in our algorithm, both of the primal-dual schema and the local ratio technique are used in an interlaced fashion. As mentioned earlier, quite a number of approximation algorithms have been developed so far using either of these two methods, yet to the best of our knowledge, no algorithm has been designed based on both.

2   Algorithm

Let $S \subseteq V$ be the set of "special" nodes, and M be a matching on a spanning tree T rooted at r. We say M is *dense* if

 - r and all the special nodes are left unmatched by M, and
 - every internal node ($\neq r$) of T with none of its children special is matched by M.

(Note: it does not matter for M to be dense whether any internal node having a special child or any leaf is matched by M or not). A dense matching $M \subseteq T$ can be efficiently computed by a DFS-like procedure (see Fig. 1).

Initialize $M = \emptyset$, and mark root r and all the special nodes "matched".
Call DFS-MATCH(r).

DFS-MATCH(u)
If u is a leaf **then** return
If u is unmatched and has an unmatched child v **then**
    Pick $e = \{u, v\}$ and add it to M by setting $M \leftarrow M \cup \{e\}$.
    Mark both u and v "matched".
For each child v of u **do**
    Call DFS-MATCH(v).

Fig. 1. A DFS-like procedure for computing a dense matching M on tree T

Let T denote any MST in G. Suppose that T consists of edges with k different costs, $w_0, w_1, \ldots, w_{k-1}$, ($k \leq n - 1$) such that $w_0 < w_1 < \cdots < w_{k-1}$. Let $\Delta_0 = w_0$ and $\Delta_i = w_i - w_{i-1}$ for $1 \leq i \leq k - 1$ (so, $\Delta_i > 0, \forall i$). In the following algorithm a sequence of trees, $T_1, T_2, \ldots, T_{k-1}$, and a sequence of graphs, $G_1 = (V_1, E_1), G_2 = (V_2, E_2), \ldots, G_{k-1} = (V_{k-1}, E_{k-1})$, will be generated from $T_0 = T$ and the original graph $G_0 = G$, respectively, in such a way that T_{i+1} (G_{i+1}) is the one obtained from T_i (G_i, resp.) by contracting all the edges of cost w_i in T_i. Such contractions might introduce parallel edges and/or self-loops in G_i's (but not in T_i's), and we may keep all the parallel edges but none of the self-loops.

When tree edges are contracted, the set of vertices connected together by these edges is replaced by a single new vertex (and it becomes a new root labeled r if r is among those merged into one), and such vertices in G_i's are called *s-nodes* (for *special* nodes). Clearly, any s-node u in any G_i corresponds naturally to some set S of vertices, all of them connected together by contracted edges, in original G. Let $D(u) \subseteq V$ denote the set of vertices merged into an s-node u. Then, $\delta_{G_i}(u) = \delta_G(D(u))$ for any s-node $u \in V_i$, and hence, $\delta_{\vec{G}_i}^-(u)$ coincides with the r-edge cut $\delta_{\vec{G}}^-(D(u))$ (for dependent $D(u)$) if $u \neq r$.

Given $G = (V, E)$ and $r \in V$, the algorithm TC computes a tree cover rooted at r (see Fig. 2). Starting with $G_0 = G$ and $T_0 = $ any MST T in G, it computes a sequence of graphs, $G_1, \ldots, G_{k-1}$, and a sequence of trees, $T_1, \ldots, T_{k-1}$, for

---

1. Set $G_0 \leftarrow G, T \leftarrow$ any MST in G, and $T_0 \leftarrow T$. /* initialization */
2. For $i = 0$ to $k - 1$ do
   /* M_i's and S_i's are constructed in this phase */
   2-1. Let S_i be the set of s-nodes in G_i.
        /* set $y_D^i = \Delta_i, \forall D \in \mathcal{D}(S_i)$ */
   2-2. Compute a dense matching M_i on T_i.
        /* set $y_e^i = \Delta_i, \forall e \in M_i$ */
   2-3. Let $T_{i+1}(G_{i+1})$ be the tree (graph) obtained by contracting all the edges
        of cost w_i within T_i.
3. For each leaf edge e of T do
   3-1. Set $\bar{l}_e = l_e - \sum_{i:e \in M_i} \Delta_i$.   /* $= l_e - \sum_{e \in M_i} y_e^i$ */
4. While there exists an edge f between two leaves of T, u and v, with
   $\min\{\bar{l}_{e(u)}, \bar{l}_{e(v)}\} > 0$ do
   4-1. Set $y_f = \min\{\bar{l}_{e(u)}, \bar{l}_{e(v)}\}$.
   4-2. Subtract y_f from each of $\bar{l}_{e(u)}$ and $\bar{l}_{e(v)}$.
5. Let $T_{\text{tc}} \leftarrow$ (T with any of its leaf edges e removed if $\bar{l}_e > 0$), and output T_{tc}.

Fig. 2. Algorithm TC for computing a tree cover T in G

each $0 \leq i \leq k - 2$, by contracting all the edges of cost w_i on T_i (in Step 2). At the same time a set $S_i \subseteq V_i$ of s-nodes and a dense matching $M_i \subseteq T_i$ are constructed for each $0 \leq i \leq k - 1$. Call an edge of a tree T *leaf edge* if it is incident to a leaf u of T, and denote it $e(u)$ (or call an arc of $\vec{T}$ *leaf arc*, and denote it $\vec{e}(u)$). In Step 3 the "residual" cost $\bar{l}_e$ on each leaf edge e of T is set to initial cost l_e less $\sum_{i:e \in M_i} \Delta_i$. Using these residual costs, duals on those edges connecting leaves of T are maximally increased in Step 4; for any f between leaves u and v, y_f is set to a maximal value such that y_f does not exceed either of $\bar{l}_{e(u)}$ and $\bar{l}_{e(v)}$, y_f is next subtracted from each of $\bar{l}_{e(u)}$ and $\bar{l}_{e(v)}$, and repeated by going to any other edge connecting leaves of T, until no longer possible to raise duals on such edges. So after this step, no residual cost remains positive on at least one of leaf edges $e(u)$ and $e(v)$ for any pair of leaves u and v of T connected by an edge. The algorithm outputs an r-tree cover T_{tc} by trimming any leaf edge of T with a positive residual cost still remaining on it.

It is rather easy to see that T_{tc} thus computed is indeed an r-tree cover since 1) the internal structure of T (i.e., the subtree of T obtained by removing all the leaves from T) is completely maintained in T_{tc}, and 2) for any edge connecting two leaves of T, at least one of them is kept in T_{tc} as well.

Now the whole algorithm is to pick any edge $e = \{u, v\}$ in given G, compute both of u- and v-tree covers by calling TC twice, and choose the lighter of them as a tree cover for G.

3 Dual Solution and Its Feasibility

In this section we show how a dual feasible solution y is computed *implicitly* within the algorithm, along with an r-tree cover T_{tc}, and that it is feasible to (D). Let us begin with an easy but very basic observation:

Lemma 1. *For any $e \in E$, $e \notin E_i$ if $l_e < w_i$.*

Proof. For the sake of contradiction, suppose there exists $e \in E_i$ with $l_e < w_i$. Since T_i is the tree resulting from contracting all the edges of cost $< w_i$ within T, e cannot occur within T_i.

So, $e \in E_i - T_i$. Since we always shrink edges of spanning T, every T_i is a spanning tree in G_i. If e not in T_i is lighter than any edge of T_i, a spanning tree T'_e strictly lighter than T_i would arise in G_i, by adding e to T_i and removing some edge from T_i, say e' (recall that e cannot be a self-loop). But then, $(T - \{e\}) \cup \{e'\}$, which is strictly lighter than T, would be a spanning tree in G, and this contradicts the fact that T is an MST in G. □

During the first phase (i.e., within the for-loop of Step 2) of algorithm TC, a dense matching M_i and a set S_i of s-nodes are computed for each i. Recall that $D(u) \subseteq V$ denotes the set of those vertices merged into an s-node u by edge contraction, and let $\mathcal{D}_i = \{D(u) \mid u \in S_i\}$. A dual solution y is set up by letting each of y_e ($e \in M_i$) and y_D ($D \in \mathcal{D}_i$) be given a fixed nonnegative value uniformly for each i as follows; y will then be determined by the component-wise accumulation of them:

1. For each i

$$y_e^i = \begin{cases} \Delta_i & \text{if } e \in M_i \\ 0 & \text{otherwise} \end{cases}$$

and $y_e = \sum_i y_e^i = \sum_{i:e \in M_i} \Delta_i$ for each edge $e \in T$;

2. For each i

$$y_D^i = \begin{cases} \Delta_i & \text{if } D \in \mathcal{D}_i \\ 0 & \text{otherwise} \end{cases}$$

and $y_D = \sum_i y_D^i = \sum_{D \in \mathcal{D}(S_i)} \Delta_i$ for any dependent set $D \subseteq V - \{r\}$.

(Note: Quite possibly, an edge $e \in M_i$ could happen to be identical to $D(u) \in \mathcal{D}_{i'}$ for some s-node $u \in S_{i'}$ if $i \neq i'$. If so, y_e and $y_{D(u)}$ actually correspond to the same component of y. For a clearer presentation, however, they will be distinguished from each other in the sequel.)

Any edge in any M_i is certainly an edge of an MST T, and any D in any $\mathcal{D}(S_i)$ is the vertex set of some subtree of T as all the vertices in D are merged into an s-node by contracting edges of T. We will also need to use an r-edge cut of form $\delta^-(D)$ such that D is *not* a part of T, and its value y_D is to be explicitly determined by the algorithm:

3. $y_e =$ as assigned at step 4 in algorithm TC, for any e joining leaves of T.

From the way the algorithm assigns values to these y_e's (at step 4) and that any leaf edge f is removed at step 5 if $\bar{l}_f > 0$, it is clear that, for any leaf edge $f = e(u)$, $l_f = \sum_{f \in M_i} \Delta_i + \sum(y_e : e \in E$ connects u with another leaf of T) if f remains in T_{tc}.

By setting all the other dual variables to zero, the r-edge cut packing is completed, and this is the dual solution $y \in \mathbb{R}^{\mathcal{D}}$ which in what follows will be paired with the integral primal solution $\vec{T}_{tc} \subseteq \vec{E}$, the directed counterpart of the r-tree cover computed.

Lemma 2. *For all* $\vec{e} = (u, v) \in \vec{E}$, $Y_{\vec{e}} = \sum_{D \in \mathcal{D}: \vec{e} \in \delta^-(D)} y_D \leq l_e$.

Proof. Let T be an MST used in algorithm TC.

Case v is an internal node of T. Notice first that the edges $\in M_i$ and the vertex sets $\in \mathcal{D}_i = \{D(u) \mid u \in S_i\}$ are mutually vertex disjoint in G for each i since so are the edges $\in M_i$ and the nodes $\in S_i$ in G_i. Therefore, at most one among the edges $\in M_i$ and the vertex sets $\in \mathcal{D}_i$ contains v in it, and hence, the contribution to $Y_{\vec{e}}$ from y_f^i's ($f \in M_i$) and y_D^i's ($D \in \mathcal{D}_i$) together is at most Δ_i, for each $0 \leq i \leq k - 1$. All the other dependent sets with positive duals are such edges that are incident to leaves only, and they do not show up within $\sum_{D \in \mathcal{D}: \vec{e} \in \delta^-(D)} y_D$.

By Lemma 1, if $w_j \leq l_e < w_{j+1}$, e does not appear in G_i for $i = j + 1, \ldots, k - 1$. Therefore,

$$
\begin{aligned}
Y_{\vec{e}} &\leq \sum_{v \in e \in M_i} \Delta_i + \sum_{v \in D \in \mathcal{D}} \Delta_i \\
&= \sum_{i=0}^{j} [(\Delta_i : v \in e \in M_i) + (\Delta_i : v \in D \in \mathcal{D})] \\
&\leq \Delta_0 + \Delta_1 + \cdots + \Delta_j \\
&= w_0 + (w_1 - w_0) + \cdots + (w_j - w_{j-1}) = w_j.
\end{aligned}
$$

Case v is a leaf of T. Let f denote $e(v)$, the leaf edge of T incident to v. Suppose $w_j \leq l_e < w_{j+1}$ and $l_f = w_{j'}$. Because T is an MST in G it must be the case that $l_f \leq l_e$, and thus, $j' \leq j$. Since $l_f = \sum_{i=0}^{j'} \Delta_i$, $l_e \geq w_j = \sum_{i=0}^{j} \Delta_i = l_f + \sum_{i=j'+1}^{j} \Delta_i$.

Since v is a leaf of T, among the duals assigned on the edges of dense matchings, only those placed on f can contribute to $Y_{\vec{e}}$.

Certainly, v does not become an s-node before $f = e(v)$ gets contracted at $i = j' + 1$, whereas e disappears at $i = j + 1$ and thereafter. Therefore, if $\vec{e} \in \delta^-(D(w))$ for some s-node w, w can occur only in G_i's for $i = j' + 1, \ldots, j$, and the total contribution of $y_{D(w)}$'s for s-nodes w to $Y_{\vec{e}}$ is at most $\sum_{i=j'+1}^{j} \Delta_i$.

What remains to be accounted for are the duals on those edges joining v with other leaves of T. Whereas actual values placed on those edges are determined within the algorithm, it can be observed, from the way it works, that

$$
\sum (y_g : g \in E \text{ joins } v \text{ with another leaf of } T) \leq \bar{l}_f
$$

where $\bar{l}_f = l_f - \sum (\Delta_i : f \in \text{a dense matching } M_i)$. Therefore,

$$Y_{\vec{e}} \leq \sum (\Delta_i : f \in \text{a dense matching } M_i) + \sum_{i=j'+1}^{j} \Delta_i + \bar{l}_f$$

$$= l_f + \sum_{i=j'+1}^{j} \Delta_i = w_j.$$

$\square$

It follows from this lemma that the dual solution $y \in \mathbb{R}^{\mathcal{D}}$ set up as above is feasible to (D).

4 Approximation Ratio

In approximation algorithms based on the primal-dual schema, the approximation ratios are obtained by relating the value of a computed integral (primal) solution with that of a simultaneously computed dual solution, and these values are usually related by means of complementary slackness conditions, in somehow relaxed forms. In case of (P) and (D), these conditions can be stated as follows, where α and β (with each ≥ 1) denote relaxation factors of the respective conditions:

Primal Complementary Slackness Conditions (PCSC)
    For each $\vec{e} \in \vec{E}$, $x_{\vec{e}} > 0$ implies that $l_e/\alpha \leq \sum_{D \in \mathcal{D}:\vec{e} \in \delta^-(D)} y_D \leq l_e$.
Dual Complementary Slackness Conditions (DCSC)
    For each $D \in \mathcal{D}$, $y_D > 0$ implies that $1 \leq x(\delta^-(D)) \leq \beta$.

It can be shown that, if an algorithm produces x and y satisfying the conditions above, its approximation ratio is at most $\alpha\beta$.

In case of algorithm TC, however, the primal solution $\vec{T}_{\text{tc}}$ and the dual solution y are not related in such a simple manner (even in the unweighted case of $l = \vec{1}$), and the way they satisfy PCSC and DCSC can be seen to be as follows:

- for each $\vec{e} \in \vec{E}$ with $x_{\vec{e}} > 0$ (i.e., $\vec{e} \in \vec{T}_{\text{tc}}$),
    - PCSC may not hold for any α if $\vec{e}$ is not a leaf arc of $\vec{T}$,
    - PCSC is satisfied at $\alpha = 1$ if $\vec{e}$ is a leaf arc of $\vec{T}$, and
- for each $D \in \mathcal{D}$ with $y_D > 0$,
    - DCSC is satisfied at $\beta = 1$ if D is an edge in a dense matching M_i or $D = D(u)$ for some s-node u (Note: in either case D corresponds to a subtree of T),
    - DCSC is satisfied at $\beta = 2$ if D is an edge connecting two leaves of T.

We will use more direct arguments in what follows to show that an r-tree cover $\vec{T}_{\text{tc}}$ computed by the algorithm is of cost no more than twice the cost of the dual feasible solution y computed simultaneously.

The first idea is to "decompose" T into the uniformly weighted trees, $T_0, T_1, \ldots, T_{k-1}$, where every edge of T_i is of cost Δ_i, and then to pay for at least *half* the costs of all the internal (i.e., non-leaf) arcs of $\vec{T}_i$ (and possibly more) by the duals associated with the edges in M_i and the nodes in S_i, for each i. The dual value placed on each $e \in M_i$ and each $u \in S_i$ is Δ_i, and so we may use it to pay for half the costs of two arcs in $\vec{T}_i$. Suppose we use y_e^i to pay for the costs of $\vec{e}$ itself and the arc preceding $\vec{e}$ in $\vec{T}_i$, $\Delta_i/2$ to each, for each $e \in M_i$. Likewise, $y_{D(u)}^i$ is used to pay for the costs of the arc of $\vec{T}_i$ entering to an s-node u and its predecessor in $\vec{T}_i$, again $\Delta_i/2$ to each.

Lemma 3. *Every non-leaf arc of $\vec{T}_i$ gets at least half paid by the duals on M_i and S_i.*

Proof. Let $\vec{e} = (u, v)$ be a non-leaf arc of $\vec{T}_i$. If v is an s-node, $\vec{e}$ gets half paid by $y_{D(v)}$. Since v is not a leaf, it has at least one child, and if any of them is an s-node, say w, then, as $\vec{e}$ is a predecessor of (v, w), $\vec{e}$ gets half paid by $y_{D(w)}$. So, assume that neither of v nor any of its children is an s-node. Then, a matching M_i on T_i must match v if it is dense. If so, $e \in M_i$, or otherwise, $\{v, w\} \in M_i$ for some child w of v, and in either case, $\vec{e}$ gets half paid by either e or $\{v, w\}$. $\quad\square$

Let us now say that an arc $\vec{e}$ in $\vec{T}$ is *half paid* if at least half of its cost, $l_e/2$, is paid in total to e by the duals on M_i's and $\mathcal{D}(S_i)$'s.

Lemma 4. *Every non-leaf arc of $\vec{T}$ gets half paid.*

Proof. By Lemma 3, every non-leaf arc of $\vec{T}$ gets paid for at least $\Delta_i/2$ everytime it occurs in $\vec{T}_i$. Suppose $l_e = w_j$. Then, e appears in $G_0, G_1, \ldots$, up to G_j (but no further). The total amount paid to $\vec{e}$ is hence at least $\Delta_0/2 + \Delta_1/2 + \cdots + \Delta_j/2 = (\Delta_0 + \Delta_1 + \cdots + \Delta_j)/2 = w_j/2 = l_e/2$. $\quad\square$

To account next for the cost of any leaf arc $\vec{e}(u)$ of $\vec{T}$ remaining in $\vec{T}_{tc}$, let us recall that $\bar{l}_e = l_e - \sum_{i:e \in M_i} \Delta_i$ and that $\vec{e}(u)$ remains in $\vec{T}_{tc}$ only if the dual values on the edges joining u with other leaves total to $\bar{l}_e$; that is, $\sum(y_f : f \in E$ joins u with another leaf of $T) = \bar{l}_e$. Thus, if we spend y_f, for any f joining two leaves w and z of T, to pay for the costs of the leaf arcs of $\vec{T}$, $\Delta_i/2$ each to $\vec{e}(w)$ and $\vec{e}(z)$, a leaf arc $\vec{e}$ avoids being got rid of only if it gets half paid. To be precise, one half of $\sum_{i:e \in M_i} \Delta_i$ and one half of $\bar{l}_e$ get paid to $\vec{e}$, totaling to $(\sum_{i:e \in M_i} \Delta_i)/2 + (l_e - \sum_{i:e \in M_i} \Delta_i)/2 = l_e/2$. Therefore,

Lemma 5. *Every leaf edge of $\vec{T}$ remaining in $\vec{T}_{tc}$ also gets half paid.*

It follows immediately from Lemmas 4 and 5 that algorithm TC computes an r-tree cover $\vec{T}_{tc} \subseteq \vec{T}$ and a dual feasible $y \in \mathbb{R}^{\mathcal{D}}$ such that the cost $l(\vec{T}_{tc})$ of $\vec{T}_{tc}$ is no larger than twice the value $\sum_{D \in \mathcal{D}} y_D$ of y. Therefore,

Theorem 6. *The algorithm TC approximates the minimum cost r-tree cover to within a factor of 2; consequently, the approximation ratio of the whole algorithm for approximating the minimum cost tree cover is bounded by 2.*

The integrality gap of (P) is known to be no smaller than 2 [14].

Corollary 7. *The integrality gap of (P) is 2 when the graph is in the bidirected form of an undirected graph.*

It is also clear, from the way the dual solution $y \in \mathbb{R}^{\mathcal{D}}$ is determined as above, that y can be ensured to be integral if l is integral:

Corollary 8. *When l is integral and the graph is in the bidirected form of an undirected graph in (D), there exists an integral r-edge cut packing the cost of which is at least $1/2$ of the cost of an optimal fractional r-edge cut packing. Moreover, such an integral r-edge cut packing is efficiently computable.*

5   Final Remarks

The paper has shown that the minimum cost tree cover can be efficiently approximated to within a factor of 2 of the optimum. As the minimum tree cover problem is as hard to approximate as the minimum vertex cover problem [11], a further improvement on this factor would imply that the minimum vertex cover problem is approximable within a factor better than 2, which has been conjectured by some to be highly unlikely [13].

A natural and equally interesting direction of further research would be in the directed version of the tree cover (DTC) problem; given here is a *directed* graph G, and it is required to compute a directed tree (a branching) T of minimum cost in G such that either head or tail (or both of them) of every arc in G is touched by T. As mentioned in [14], the problem has remained wide open. If G is unweighted, however, it is not hard to find a 2-approximation for it, by extending the approach of the current paper a bit further. Letting V' be the set of vertices reachable from the root vertex r, compute an arborescence T spanning V' entirely (and V' must be a vertex cover for G if it is a feasible instance of DTC). Compute next a dense matching M on T (with no s-nodes), and while there exists an arc connecting two unmatched leaves, add it to M. Finally trim any leaf u from T if u is unmatched *and* there is no arc entering to u from $V - V'$.

Such an approach appears to fall short, however, once arbitrary arc costs are allowed on G. In fact one can come up with an instance in which any feasible solution contained in any spanning arborescence incurs such a cost larger than the optimum by an unbounded factor. Thus, the approximability of minimum cost DTC problem, as well as a related issue of the integrality gap of (P) on arbitrary directed graphs, still remain wide open.

References

1. E.M. Arkin, M.M. Halldórsson and R. Hassin. Approximating the tree and tour covers of a graph. *Inform. Process. Lett.*, 47:275–282, 1993.
2. R. Bar-Yehuda. One for the price of two: A unified approach for approximating covering problems. *Algorithmica*, 27(2):131–144, 2000.
3. R. Bar-Yehuda and S. Even. A local-ratio theorem for approximating the weighted vertex cover problem. *Annals of Discrete Mathematics*, 25:27–46, 1985.
4. R. Bar-Yehuda and D. Rawitz. On the equivalence between the primal-dual schema and the local ratio technique. *SIAM J. Discrete Math.*, 19(3):762–797, 2005.

5. F.A. Chudak, M.X. Goemans, D.S. Hochbaum, and D.P. Williamson. A primal-dual interpretation of recent 2-approximation algorithms for the feedback vertex set problem in undirected graphs. *Oper. Res. Lett.*, 22:111–118, 1998.
6. I. Dinur and S. Safra. The importance of being biased. In *Proc. 34th ACM Symp. Theory of Computing*, pages 33–42, 2002.
7. J. Edmonds. Optimum branchings. *J. Res. Nat. Bur. Standards B*, 71:233–240, 1967.
8. T. Fujito. On approximability of the independent/connected edge dominating set problems. *Inform. Process. Lett.*, 79(6):261–266, 2001.
9. T. Fujito and H. Nagamochi. A 2-approximation algorithm for the minimum weight edge dominating set problem. *Discrete Appl. Math.*, 118:199–207, 2002.
10. D.R. Fulkerson. Packing rooted directed cuts in a weighted directed graph. *Math. Programming*, 6:1–13, 1974.
11. M.R. Garey and D.S. Johnson. The rectilinear Steiner-tree problem is NP-complete. *SIAM J. Appl. Math.*, 32(4):826–834, 1977.
12. E. Halperin. Improved approximation algorithms for the vertex cover problem in graphs and hypergraphs. *SIAM J. Comput.*, 31(5): 1608-1623, 2002.
13. S. Khot and O. Regev. Vertex cover might be hard to approximate to within $2 - \epsilon$. In *Proc. 18th IEEE Conf. Computational Complexity*, pages 379–386, 2003.
14. J. Könemann, G. Konjevod, O. Parekh and A. Sinha. Improved approximations for tour and tree covers. *Algorithmica*, 38(3): 441–449, 2003.
15. B. Monien and E. Speckenmeyer. Ramsey numbers and an approximation algorithm for the vertex cover problem. *Acta Informat.*, 22:115–123, 1985.
16. O. Parekh. Edge dominating and hypomatchable sets. In *Proc. 13th ACM-SIAM Symp. Discrete Algorithms*, pages 287–291, 2002.
17. C. Savage. Depth-first search and the vertex cover problem. *Inform. Process. Lett.*, 14(5):233–235, 1982.

Tight Approximation Algorithm
for Connectivity Augmentation Problems

Guy Kortsarz[1] and Zeev Nutov[2]

[1] Rutgers University, Camden, NJ, USA
[2] The Open University of Israel, Raanana, Israel

Abstract. The *S-connectivity* $\lambda_G^S(u,v)$ of (u,v) in a graph G is the maximum number of uv-paths that no two of them have an edge or a node in $S - \{u,v\}$ in common. The corresponding *Connectivity Augmentation* (CA) problem is: given a graph $G_0 = (V, E_0)$, $S \subseteq V$, and requirements $r(u,v)$ on $V \times V$, find a minimum size set F of new edges (any edge is allowed) so that $\lambda_{G_0+F}^S(u,v) \geq r(u,v)$ for all $u,v \in V$. Extensively studied particular cases are the *edge*-CA (when $S = \emptyset$) and the *node*-CA (when $S = V$). A. Frank gave a polynomial algorithm for *undirected* edge-CA and observed that the directed case even with $r(u,v) \in \{0,1\}$ is at least as hard as the Set-Cover problem. Both directed and undirected node-CA have approximation threshold $\Omega(2^{\log^{1-\varepsilon} n})$. We give an approximation algorithm that matches these approximation thresholds. For both directed and undirected CA with arbitrary requirements our approximation ratio is: $O(\log n)$ for $S \neq V$ arbitrary, and $O(r_{\max} \cdot \log n)$ for $S = V$, where $r_{\max} = \max_{u,v \in V} r(u,v)$.

1 Introduction and Preliminaries

1.1 The Problem and Previous Work

Let $G = (V, E)$ be a graph and let $S \subseteq V$. The *S-connectivity* $\lambda_G^S(u,v)$ of (u,v) in G is the maximum number of uv-paths such that no two of them have an edge or a node in $S - \{u,v\}$ in common. We consider the following problem:

Connectivity Augmentation (CA)
Instance: A directed/undirected graph $G_0 = (V, E_0)$, $S \subseteq V$, and a nonnegative integer *requirement function* $r(u,v)$ on $V \times V$.
Objective: Add a minimum size set F of new edges to G_0 so that for $G = G_0 + F$

$$\lambda_G^S(u,v) \geq r(u,v) \qquad \text{for all } (u,v) \in V \times V. \tag{1}$$

CA is a particular case of the **Generalized Steiner Network** (GSN) problem: given a complete directed/undirected graph $\mathcal{G} = (V, \mathcal{E})$ with edge-costs $\{c_e : e \in \mathcal{E}\}$, a node subset $S \subseteq V$, and a requirement function $r(u,v)$ on $V \times V$, find a minimum cost spanning subgraph G of $\mathcal{G}$ so that (1) holds for G. Clearly, GSN with $\{0,1\}$-costs is the CA problem.

Extensively studied particular choices of S in CA/GSN instances are: $S = \emptyset$ (the *edge*-CA/GSN), $S = V$ (the *node*-CA/GSN), and any S so that $r(u,v) = 0$

M. Bugliesi et al. (Eds.): ICALP 2006, Part I, LNCS 4051, pp. 443–452, 2006.

whenever $u \in S$ or $v \in S$ (the *element-*CA/GSN). Except the general requirements, two special types of requirement functions are studied in the literature. The uniform requirements when $r(u,v) = k$ for all $u, v \in V$, and the rooted (single source/sink) requirements when there is $s \in V$ so that if $r(u,v) > 0$ then: $u = s$ for directed graphs, and $u = s$ or $v = s$ for undirected graphs. Similar variants (edge/node/element cases and general/uniform/rooted requirements) are also extensively studied for other types of GSN costs (e.g., general, $\{1, \infty\}$-costs, and metric costs). Note also that *the Directed Steiner Tree* problem is the special case of directed GSN with rooted $\{0, 1\}$-requirements.

For *undirected* graphs the best known approximation ratios for GSN are as follows. For edge-GSN Jain [19] gave a 2-approximation algorithm. This result was extended to element-GSN in [5,9]. For node-GSN no nontrivial approximation algorithms for arbitrary costs are known. Recently, Cheriyan and Vetta [6] gave an $O(\log n)$-approximation algorithm for the undirected *metric* node-GSN (namely, when $S = V$ and the edge costs satisfy the triangle inequality). For *directed* graphs, nontrivial approximation algorithms are known only for $\{0, 1\}$-requirements (in this case all choices of S are equivalent). Dodis and Khanna [7] showed that even this simple case cannot be approximated within $O(2^{\log^{1-\varepsilon} n})$ for any $\varepsilon > 0$ unless NP $\subseteq$ DTIME $\left(n^{\text{polylog}(n)}\right)$. Charikar et. al [2] gave an $O(p^{2/3} \log^{1/3} p)$-approximation algorithm where $p = |\{(u, v) : r(u, v) = 1\}|$ is the number of pairs that are to be connected. Feldman and Ruhl [8] gave an exact algorithm with running time $O(n^{4p})$. For rooted $\{0, 1\}$-requirements (this is the Directed Steiner Tree problem) [2] gave an $O(n^{\varepsilon}/\varepsilon^3)$-approximation algorithm for any constant $\varepsilon > 0$. See also surveys in [23,27] on various GSN problems.

As CA is a particular case of GSN, these approximation ratios (but not the hardness results) are valid for CA problems as well, except the $O(\log n)$-approximation algorithm for the undirected metric node-GSN of [6]. The result of [6] is not valid for CA since in CA problems the costs are usually *not* metric; furthermore, a polylogarithmic approximation for the node-CA is unlikely, since as shown in [31], the node-CA cannot be approximated within $O(2^{\log^{1-\varepsilon} n})$ for any fixed $\varepsilon > 0$ unless NP $\subseteq$ DTIME$(n^{\text{polylog}(n)})$.

In many cases, for CA better approximation ratios are known than for its generalization GSN. For undirected CA the following results are known. A. Frank [10] gave a polynomial algorithm for undirected edge-CA based on Mader's undirected splitting off theorem for edge-connectivity [29]. The node-CA (and the element-CA) turned to be NP-hard even when the input graph G_0 is connected and $r(u, v) \in \{0, 2\}$ (c.f., [30]). However, while the undirected element-CA admits a 7/4-approximation algorithm [31], the undirected node-CA with $r(u, v) \in \{0, k\}$ cannot be approximated within $O(2^{\log^{1-\varepsilon} n})$ for any fixed $\varepsilon > 0$, unless NP $\subseteq$ DTIME$(n^{\text{polylog}(n)})$, see [31]. For uniform requirements $r(u, v) = k$ for all $u, v \in V$ the complexity status is not known for undirected graphs, but the problem is in P for directed graphs [13]; this implies a 2-approximation algorithm for undirected graphs. For undirected graphs an algorithm that computes a solution of size roughly opt $+ k(k - k_0)/2$ is given in [17], where k_0 is the connectivity of G_0; furthermore, for any fixed k an optimal solution can be computed in polynomial time [18]. For rooted uniform requirements (in undirected graphs) the situation is similar, see [32].

For directed graphs it was already observed by A. Frank [10] that even for rooted $\{0, 1\}$-requirements the edge-CA is at least as hard as the Set-Cover problem. Combined with the result of [33] this implies an $\Omega(\log n)$-approximation threshold for this simple variant (namely, the problem cannot be approximated within $c \ln n$ for some universal constant $c < 1$, unless P=NP). By extending the construction from [10], a similar threshold was shown in [32] for the undirected rooted CA with root s and $S = V - \{s\}$, but for $\{0, k\}$-requirements with $k = \Theta(n)$.

Summarizing, both directed and undirected CA have the following approximation thresholds. An $\Omega(\log n)$-approximation threshold for $S \neq V$ (specifically, for rooted requirements with $S = V - \{s\}$, where s is the "root") [10,32], and for directed graphs this is so even for $\{0, 1\}$-requirements and $S = \emptyset$. An $O(2^{\log^{1-\varepsilon} n})$-approximation threshold for $S = V$ [31] and $\{0, k\}$-requirements with $k = \Theta(n)$.

For more work on CA problems see, e.g., [1,10,13,21,18,30,32,31], and surveys in [10,11,12,34]. The only polylogarithmic approximation algorithm known for CA on directed graphs is for the special case of rooted requirements. Even for this special case the best previously known ratio is $\Theta(\log^2 n)$ [32]. To the best of our knowledge, no nontrivial approximation algorithms were known for the general directed CA even for $S = \emptyset$, nor for undirected CA with S arbitrary.

For work on other types of GSN costs see c.f., [19,9,4,6,14,15,26,25,24], and detailed surveys in [23,27] on known upper and lower bounds with respect to approximation.

1.2 Our Result and Its Significance

Previous work on CA problems that does not follow from results for GSN dealt mainly with algorithm for some special cases, for which were given either polynomial algorithm (c.f., [35,10,13,11]), or constant ratio approximation algorithms (c.f., [21,22,3,17,18,30,28,32,31]). We give a tight approximation algorithm for the most general case of CA:

Theorem. *Both directed and undirected* CA *admit an $O(\log n)$-approximation algorithm except the case $S = V$ for which there exists an $O(r_{\max} \cdot \log n)$-approximation algorithm, where $r_{\max} = \max_{u,v \in V} r(u, v)$ and $n = |V|$.*

The first part of the Theorem extends to GSN, provided there is $s \in V - S$ so that only edges incident to s can be added. As was mentioned, even for undirected graphs our result is the best possible, and it cannot be deduced from the $O(\log n)$-approximation algorithm for the *undirected metric* node-GSN of [6], since for CA problems the costs are usually not metric, and since the node-CA is unlikely to have a polylogarithmic approximation [31].

We elaborate on few more points that should be emphasized. Usually it seems hard to give tight results to meaningful subproblems of the *directed* GSN. The main reason that approximation algorithm for directed GSN are rare is that even for $r(u,v) \in \{0,1\}$ the $\{0,1,\infty\}$-costs case cannot be approximated within $2^{\log^{1-\epsilon} n}$ for any constant $\epsilon > 0$ unless NP $\subseteq$ DTIME($n^{\text{polylog}(n)}$) [7], while the best known approximation ratio for this simplest case is $O(n^{1+\epsilon}/\epsilon^3) = \Omega(n)$ [2]. This hardness result is valid also for the metric costs case, which easily follows by taking metric completion of the construction in [7]. In particular, for *directed* graphs our result is unlikely to be extended to more general cost functions. Even for GSN with rooted $\{0,1\}$-requirements, which is the Directed Steiner Tree problem, there is still a large gap between known approximation ratio and threshold. For the Directed Steiner Tree problem the best known approximation ratio is $O(n^\epsilon/\epsilon^3)$ for any constant ϵ [2], while the known approximation threshold is $\Omega(\log^{2-\epsilon} n)$ [16].

This should be contrasted with the $\{0,1\}$-costs variant studied here; we are able to deal both with the most general type of connectivity – the S-connectivity (bridging between edge- and node-connectivity) and directed graphs to get tight results for (almost) all cases.

Another point is the following irregularity. Our approximation ratio is tight for $S \neq V$ since rooted CA has an $\Omega(\ln n)$-approximation threshold (for directed graphs even for $S = \emptyset$ and $\{0,1\}$-requirements). For $S = V$ our approximation ratio is tight for small requirements, but may seem weak if $r_{\max}$ is large. However, it might be that a much better approximation algorithm does not exist: in [31] it is proved that for $S = V$ and $k = \Theta(n)$, CA with $r(u,v) \in \{0,k\}$ cannot be approximated within $2^{\log^{1-\epsilon} n}$ for any constant $\varepsilon > 0$ unless NP $\subseteq$ DTIME($n^{\text{polylog}(n)}$). Thus there is a large gap in approximability between the case $S = V \setminus \{v\}$ (for any $v \in V$) for which we show an $O(\log n)$-approximation, and the substantially harder case $S = V$.

The techniques used for proving our result for directed CA (the undirected case follows from the directed one) is a combination of some known techniques in addition to some new ones. First, we show a new method to decompose the problem into two subproblems, each one of an "almost" rooted type, and consider the subproblems separately. Second, for each subproblem, we use the well known extension of the set-cover approximation techniques. This is "submodular cover" problems approximation techniques [36] that are based on density considerations (c.f., [20]). Loosely speaking, the density is the "increase in feasibility" or the "decrease in the deficiency" of an added edge set over its size. Our definition of deficiency is different from the commonly used one that is based on "setpair formulation", c.f., [13,9,5]. We define the deficiency of (u,v) as $\max\{r(u,v) - \lambda^S(u,v), 0\}$ and the total deficiency as the sum of the deficiencies of all the node pairs. In order to prove that we can find a subset of appropriate density we use the well known method of uncrossing deficient sets.

1.3   Notation and Preliminaries

An edge from u to v is denoted by uv. A *uv-path* is a path from u to v. For arbitrary two sets A, B of nodes and edges (or graphs) $A - B$ is the set (or graph) obtained by deleting B from A (deletion of a node implies deletion of the edges incident to it); similarly, $A + B$ denotes the set (graph) obtained by adding B to A. Let H be a (possibly directed) graph or an edge set on node set V. For disjoint $X, Y \subseteq V$ we denote by $\delta_H(X, Y)$ the set $\{uv \in E : u \in X, v \in Y\}$ of the edges in H from X to Y and $d_H(X, Y) = |\delta_H(X, Y)|$; for brevity, $\delta_H(X) = \delta_H(X, V - X)$ and $d_H(X) = |\delta_H(X)|$. Let $\Gamma_H(X)$ be the set $\{v \in V - X : uv \in E$ for some $u \in X\}$ of *neighbors* of X in H. We sometimes omit the subscripts if they are clear from the context. We call the new edges that are added to a given graph *links* in order to distinguish them from the existing edges. Let opt denote the optimal solution value of an instance at hand.

2   Proof of the Theorem

We need the following formulation of Menger's theorem for S-connectivity, which can be easily deduced from its original theorem by standard constructions. In this formulation C represents a "mixed" cut, which may include edges and nodes from $S - \{u, v\}$.

Theorem 1 (Menger's Theorem). *Let u, v be two nodes of a (directed or undirected) graph $G = (V, E)$ and let $S \subseteq V$. Then*

$$\lambda_G^S(u, v) = \min\{|C| : C \subseteq E + S - \{u, v\}, G - C \text{ has no } uv\text{-path}\} .$$

We prove the Theorem for the directed case and the statement for the undirected CA follows the following proposition (c.f., [27]), which indicates that undirected CA problems cannot be much harder to approximate than the directed ones.

Proposition 1. *If there is a ρ-approximation algorithm for the directed CA then there is a 2ρ-approximation algorithm for the undirected CA.*

Let F' be an arbitrary solution to an instance G_0, S, r of directed CA. Subdivide every edge in F' by a new node, and then identify all these new nodes into a node s. The obtained graph satisfies the requirements between nodes in V, and the number of links incident to s is $2|F'|$. Now, if $V - S \neq \emptyset$, then by identifying s with some node $v \in V - S$ we get that the new links added form a feasible solution for G_0, S, r. This implies:

Corollary 1. *For any solution F' for directed CA with $S \neq V$ and any $s \in V - S$, there exists a solution F with $|F| \leq 2|F'|$ such that all the links in F are incident to s.*

If $S = V$, we make $r_{\max}$ copies $s_1, \ldots, s_{r_{\max}}$ of s and of the links incident to s, choose arbitrary $r_{\max}$ nodes $\{v_1, \ldots, v_{r_{\max}}\}$, and identify every s_i with v_i. Again, it is easy to see that the new links added form a feasible solution to the CA instance, and that the number of links added is $2|F'|r_{\max}$.

Given an instance G_0, S, r for directed CA, let $H_0 = G_0 + s$ (note that $s \notin S$). We say that a set F of links incident to s is a feasible solution for H_0 if $H_0 + F$ satisfies the S-connectivity requirements defined by r. *The H_0-problem* is to find a feasible solution for H_0 of minimum size. We will give an $O(\log n)$-approximation algorithm for the H_0-problem. This is done by approximating the following two problems. Let H_0^+ be obtained from H_0 by adding r_{max} edges from s to every node in V, and H_0^- is obtained by adding r_{max} edges from every node in V to s. We say that a set F^+ (F^-) of links entering s (leaving s) is a feasible solution for H_0^+ (for H_0^-) if $H_0^+ + F^+$ (if $H_0^- + F^-$) satisfies the S-connectivity requirements defined by r. The H_0^+-problem is to find a feasible solution for H_0^+ of minimum size, and the H_0^- problem is defined similarly. From Corollary 1 it follows that $\text{opt}^+, \text{opt}^- \leq \text{opt}$, where opt^+ and opt^- denote the optimal solution values for H^+ and H^-, respectively, and opt is the optimal solution value for H_0.

We will prove the following two statements:

Lemma 1. *Let F^+ and F^- be a feasible solution for the H_0^+ and for the H_0^- problems, respectively. Then $F = F^+ + F^-$ is a feasible solution for the H_0 problem.*

Lemma 2. *The H_0^+-problem (and the H_0^--problem) admits an $O(\log n)$-approximation algorithm.*

The algorithm for directed CA with $S \neq V$ is as follows.

1. Using the algorithm from Lemma 2 find a solutions F^+ for the H_0^+-problem and F^- for the H_0^--problem, so that $|F^+| = O(\log n) \cdot \text{opt}^+$ and $|F^-| = O(\log n) \cdot \text{opt}^-$.
2. Let $F = F^+ + F^-$, and let $H = H_0 + F$.
   Obtain a graph G from H by identifying s with an arbitrary node in $V - S$.

The algorithm computes a feasible solution, by Corollary 1 and Lemma 1. Since $\text{opt}^+, \text{opt}^- \leq \text{opt}$ the approximation ratio is $O(\log n)$, by Lemma 2.

To finish the proof of the Theorem it remains to prove Lemmas 1 and 2. We need the following statement that stems from Menger's Theorem.

Proposition 2. $\lambda_G^S(u, v) \geq r(u, v)$ *if, and only if, $|Q| + d_G(X, Y) \geq r(u, v)$ for any partition X, Q, Y of V with $u \in X$, $v \in Y$, and $Q \subseteq S$.*

Proof of Lemma 1. Let $H = H_0 + F$. Suppose to the contrary that there are $u, v \in V$ so that $\lambda_H^S(u, v) \leq r(u, v) - 1$. Then by Fact 2 there exists a partition X, Q, Y of $V + s$ with $u \in X$, $v \in Y$, and $Q \subseteq S$ such that $|C| \leq r(u, v) - 1$ for $C = Q + \delta_H(X, Y)$. Note that $s \notin C$, so $s \in X$ or $s \in Y$. If $s \in X$ then $\delta_{H^-}(X, Y) = \delta_H(X, Y)$, so $H^- - C$ has no uv-path. Since $|C| \leq r(u, v) - 1$, we conclude that $\lambda_{H^-}^S(u, v) \leq r(u, v) - 1$, contradicting that F^- is a feasible solution for H_0^-. The proof of the case $s \in Y$ is similar.

In the rest of this section we prove Lemma 2. We use a result due to Wolsey [36] about the performance of the greedy algorithm for a certain type of covering problems. A *covering problem* is defined as follows:

Instance: An integer non-decreasing function p given by an evaluation oracle on subsets of a groundset $\mathcal{E}$.

Objective: Find $F \subseteq \mathcal{E}$ of minimum size so that $p(F) = p(\mathcal{E})$.

Note that the function p may not be given explicitly. The *Greedy Algorithm* starts with $F = \emptyset$ and adds elements to the solution one after the other using the following simple greedy rule. As long as $p(F) < p(\mathcal{E})$ it adds to F an element $e \in \mathcal{E}$ that has maximum $p(F + e) - p(F)$; if this step can be performed in polynomial time, then the algorithm can be implemented to run in polynomial time. Let $\Delta_p = \max_{e \in \mathcal{E}}(p(e) - p(\emptyset))$, and for an integer k let $H(k)$ denote the kth harmonic number.

Theorem 2 ([36]). *Suppose that for an instance of a covering problem*

$$\sum_{e \in F_2} (p(F_1 + e) - p(F_1)) \geq p(F_1 + F_2) - p(F_1) \quad \forall F_1, F_2 \subseteq \mathcal{E}, F_1 \cap F_2 = \emptyset. \quad (2)$$

Then the Greedy Algorithm produces a solution of size at most $H(\Delta_p)$ times the optimal.

We formulate the H_0^+-problem as a covering problem and using Theorem 2 show that it admits an $O(\log n)$-approximation algorithm. The set $\mathcal{E}$ is obtained by taking $r_{\max}$ links from v to s for every $v \in V$. We also need to define a function p on the subsets of $\mathcal{E}$. For $(u,v) \subseteq V \times V$ and $F^+ \subseteq \mathcal{E}$, let $q(F^+, (u,v)) = \max\{r(u,v) - \lambda^S_{H_0^+ + F^+}(u,v), 0\}$ be the deficiency of (u,v) in $H_0^+ + F^+$. Let

$$q(F^+) = \sum_{(u,v) \in V \times V} q(F^+, (u,v)) \quad (3)$$

be the total deficiency of $H_0^+ + F^+$. Then $p(F^+) = q(\emptyset) - q(F^+)$. In other words, $p(F^+)$ is the decrease in the total deficiency as a result of adding F^+ to H_0^+; in the corresponding covering problem, the goal is to find a minimum size $F^+ \subseteq \mathcal{E}$ so that $p(F^+) = p(\mathcal{E})$ (that is, $q(F^+) = 0$). Clearly, p is monotone non-decreasing. The Greedy Algorithm can be implemented in polynomial time, as $p(F^+)$ can be computed in polynomial time for any link set F^+. Clearly, $\Delta_p \leq n^2$. We prove that (2) holds for p, and thus Theorem 2 implies that the Greedy Algorithm produces a solution of size $H(\Delta_p) \cdot \text{opt}^+ \leq H(n^2) \cdot \text{opt}^+ = O(\log n) \cdot \text{opt}^+$.

Let $F_1, F_2 \subseteq \mathcal{E}$ be disjoint link sets. We need to prove that:

$$\sum_{e \in F_2} (p(F_1 + e) - p(F_1)) \geq p(F_1 + F_2) - p(F_1) .$$

To simplify the notation, denote $J = H_0^+ + F_1$, $F = F_2$, and denote by $\Delta(F(u,v))$ the decrease in the deficiency of (u,v) as a result of adding F to J. Namely, $\Delta(F, (u,v))$ is obtained by subtracting the deficiency of (u,v) in $J + F$ from the deficiency of (u,v) in J. Then for our choice of p, the last inequality is equivalent to:

$$\sum_{e \in F} \sum_{(u,v) \in V \times V} \Delta(e, (u,v)) \geq \sum_{(u,v) \in V \times V} \Delta(F, (u,v)) .$$

Consequently, it would be sufficient to show that:

$$\sum_{e \in F} \Delta(e, (u,v)) \geq \Delta(F, (u,v)) \quad \forall (u,v) \in V \times V . \quad (4)$$

Let $u, v \in V$. If $\lambda_J^S(u, v) \geq r(u, v)$, then (4) is valid, since its both sides are zero. Note that $\lambda_{J+F}(u, v) - \lambda_J(u, v) \geq \Delta(F, (u, v))$, while if $\Delta(e, (u, v)) = \lambda_{J+e}^S(u, v) - \lambda_J^S(u, v)$ if $\lambda_J^S(u, v) \leq r(u, v) - 1$. Thus if $\lambda_J^S(u, v) \leq r(u, v) - 1$, it would be sufficient to prove that for any link set F entering s:

$$\sum_{e \in F} \left(\lambda_{J+e}^S(u, v) - \lambda_J^S(u, v) \right) \geq \lambda_{J+F}(u, v) - \lambda_J(u, v) \quad \forall (u, v) \in V \times V .$$

Let us say that $X \subseteq V$ is (u, v)-tight (in J) if there exists a partition X, Q, Y of V with $u \in X$, $v \in Y$, and $Q \subseteq S$ such that $|Q| + d_J(X, Y) = \lambda_J^S(u, v)$. It is well known and easy to show that:

Proposition 3. *The intersection and union of two (u, v)-tight sets are also (u, v)-tight. Thus an inclusion-minimal (u, v)-tight set is unique.*

For $u \in V$ let X_u be the unique minimal (u, v)-tight set in J. By Fact 2 and the definition of J, $\lambda_{J+e}^S(u, v) - \lambda_J^S(u, v) = 1$ if e connects X_u with s. Let $t = \lambda_{J+F}^S(u, v) - \lambda_J^S(u, v)$. Then at least t links in F must connect X_v with s. Thus, each one of these t links contributes 1 to $\sum_{e \in F} \left(\lambda_{J+e}^S(u, v) - \lambda_J^S(u, v) \right)$. This finishes the proof of Lemma 2.

References

1. A. Benczúr and A. Frank. Covering symmetric supermodular functions by graphs. *Math Programming*, 84:483–503, 1999.
2. M. Charikar, C. Chekuri, T. Cheung, Z. Dai, A. Goel, S. Guha, and M. Li. Approximation algorithms for directed Steiner problems. *Journal of Algorithms*, 33:73–91, 1999.
3. J. Cheriyan and R. Thurimella. Fast algorithms for k-shredders and k-node connectivity augmentation. *Journal of Algorithms*, 33:15–50, 1999.
4. J. Cheriyan, S. Vempala, and A. Vetta. An approximation algorithm for the minimum-cost k-vertex connected subgraph. *SIAM Journal on Computing*, 32(4):1050–1055, 2003.
5. J. Cheriyan, S. Vempala, and A. Vetta. Network design via iterative rounding of setpair relaxations. *manuscript*, 2003.
6. J. Cheriyan and A. Vetta. Approximation algorithms for network design with metric costs. In *Symposium on the Theory of Computing (STOC)*, pages 167–175, 2005.
7. Y. Dodis and S. Khanna. Design networks with bounded pairwise distance. In *Symposium on the Theory of Computing (STOC)*, pages 750–759, 1999.
8. J. Feldman and M. Ruhl. The directed steiner network problem is tractable for a constant number of terminals. In *Symposium on the Foundations of Computer Science (FOCS)*, page 299, 1999.
9. L. K. Fleischer, K. Jain, and D. P. Williamson. An iterative rounding 2-approximation algorithm for the element connectivity problem. In *Symposium on the Foundations of Computer Science (FOCS)*, pages 339–347, 2001.
10. A. Frank. Augmenting graphs to meet edge-connectivity requirements. *SIAM Journal on Discrete Math*, 5(1):25–53, 1992.

11. A. Frank. Connectivity augmentation problems in network design. *Mathematical Programming: State of the Art*, pages 34–63, 1995.
12. A. Frank. Edge-connection of graphs, digraphs, and hypergraphs, EGRES TR No 2001-11. 2001.
13. A. Frank and T. Jordán. Minimal edge-coverings of pairs of sets. *J. on Comb. Theory B*, 65:73–110, 1995.
14. A. Frank and E. Tardos. An application of submodular flows. *Linear Algebra and its Applications*, 114/115:329–348, 1989.
15. H. N. Gabow, M. X. Goemans, E. Tardos, and D. P. Williamson. Approximating the smallest k-edge connected spanning subgraph by LP-rounding. In *Symposium on Discrete Algorithms (SODA)*, pages 562–571, 2005.
16. E. Halperin and R. Krauthgamer. Polylogarithmic inapproximability. In *Symposium on the Theory of Computing (STOC)*, pages 585–594, 2003.
17. B. Jackson and T. Jordán. A near optimal algorithm for vertex connectivity augmentation. In *Symposium on Algorithms and Computation (ISAAC)*, pages 313–325, 2000.
18. B. Jackson and T. Jordán. Independence free graphs and vertex connectivity augmentation. In *Integer Programming and Combinatorial Optimization (IPCO)*, pages 264–279, 2001.
19. K. Jain. A factor 2 approximation algorithm for the generalized Steiner network problem. *Combinatorica*, 21(1):39–60, 2001.
20. D. S. Johnson. Approximation algorithms for combinatorial problems. *Joural of Computing and System Sciences*, 9(3):256–278, 1974.
21. T. Jordán. On the optimal vertex-connectivity augmentation. *J. on Comb. Theory B*, 63:8–20, 1995.
22. T. Jordán. A note on the vertex connectivity augmentation. *J. on Comb. Theory B*, 71(2):294–301, 1997.
23. S. Khuller. *Approximation algorithms for for finding highly connected subgraphs*, Chapter 6 in *Approximation Algorithms for NP-hard problems*, D. S. Hochbaum Ed., pages 236-265. PWS, 1995.
24. S. Khuller and B. Raghavachari. Improved approximation algorithms for uniform connectivity problems. *Journal of Algorithms*, 21:434–450, 1996.
25. S. Khuller and U. Vishkin. Biconnectivity approximations and graph carvings. *Journal of the Association for Computing Machinery*, 41(2):214–235, 1994.
26. G. Kortsarz and Z. Nutov. Approximation algorithm for k-node connected subgraphs via critical graphs. In *Symposium on the Theory of Computing (STOC)*, pages 138–145, 2004.
27. G. Kortsarz and Z. Nutov. *Approximating minimum cost connectivity problems*, in *Approximation Algorithms and Metaheuristics*, T. F. Gonzalez ed.,. 2005. To appear.
28. G. Liberman and Z. Nutov. On shredders and vertex-connectivity augmentation. *Manuscript*, 2005.
29. W. Mader. A reduction method for edge-connectivity in graphs. *Annals of discrete Math*, 3:145–164, 1978.
30. H. Nagamochi and T. Ishii. On the minimum local-vertex-connectivity augmentation in graphs. *Discrete Applied Mathematics*, 129(2-3):475–486, 2003.
31. Z. Nutov. Approximating connectivity augmentation problems. In *Symposium on Discrete Algorithms (SODA)*, pages 176–185, 2005.
32. Z. Nutov. Approximating rooted connectivity augmentation problems. *Algorithmica*, 2005. To appear.

33. R. Raz and S. Safra. A sub-constant error-probability low-degree test and a sub-constant error-probability PCP characterization of NP. In *Symposium on the Theory of Computing (STOC)*, pages 475–484, 1997.
34. Z. Szigeti. On edge-connectivity augmentation of graphs and hypergraphs. *manuscript*, 2004.
35. T. Watanabe and A. Nakamura. Edge-connectivity augmentation problems. *Computer and System Sciences*, 35(1):96–144, 1987.
36. L. A. Wolsey. An analysis of the greedy algorithm for the submodular set covering problem. *Combinatorica*, 2:385–393, 1982.

On the Bipartite
Unique Perfect Matching Problem

Thanh Minh Hoang[1,*], Meena Mahajan[2], and Thomas Thierauf[3]

[1] Abt. Theor. Inform., Universität Ulm, 89069 Ulm, Germany
[2] Inst. of Math. Sciences, Chennai 600 113, India
[3] Fak. Elektr. und Inform., Aalen University, 73430 Aalen, Germany

Abstract. In this note, we give tighter bounds on the complexity of the bipartite unique perfect matching problem, bipartite-UPM. We show that the problem is in $\mathbf{C_{=}L}$ and in $\mathbf{NL}^{\oplus \mathbf{L}}$, both subclasses of $\mathbf{NC}^2$.

We also consider the (unary) weighted version of the problem. We show that testing uniqueness of the minimum-weight perfect matching problem for bipartite graphs is in $\mathbf{L}^{\mathbf{C_{=}L}}$ and in $\mathbf{NL}^{\oplus \mathbf{L}}$.

Furthermore, we show that bipartite-UPM is hard for $\mathbf{NL}$.

1 Introduction

The perfect matching problem PM asks whether there exists a perfect matching in a given graph. PM was shown to be in $\mathbf{P}$ by Edmonds [5], but it is still open whether there is an $\mathbf{NC}$-algorithm for PM. In fact, PM remains one of the most prominent open questions in complexity theory regarding parallelizability. It is known to be in randomized $\mathbf{NC}$ ($\mathbf{RNC}$) by Lovász [12]; subsequently Karp, Upfal and Wigderson [9], and then Mulmuley, Vazirani, and Vazirani [13] showed that constructing a perfect matching, if one exists, is in $\mathbf{RNC}^3$ and $\mathbf{RNC}^2$, respectively. Recently, Allender, Reinhardt, and Zhou [3] showed that PM (both decision and construction) is in non-uniform $\mathbf{SPL}$. However, to date no $\mathbf{NC}$ algorithm is known for PM.

In this paper we consider the complexity of the unique perfect matching problem posed by Lovász (see [10]), UPM for short. That is, for a given graph G, one has to decide whether there is precisely one perfect matching in G. Furthermore, we consider the problem of testing if a (unary) weighted graph has a unique minimum-weight perfect matching. The latter problem has applications in computational biology. The unique maximum-weight perfect matching can be used to predict the folding structure of RNA molecules (see [17]).

Gabow, Kaplan, and Tarjan [6] observed that UPM is in $\mathbf{P}$. Kozen, Vazirani, and Vazirani [10,11] showed that UPM for bipartite graphs is in $\mathbf{NC}$. Their techniques don't seem to generalize to arbitrary graphs (see Section 3.2 for more detail)[1].

---

* Supported by DFG grant Scho 302/7-1.
[1] In [10] it is claimed that the technique works also for the general case, but this was later retracted in a personal communication by the authors; see also [11].

M. Bugliesi et al. (Eds.): ICALP 2006, Part I, LNCS 4051, pp. 453–464, 2006.
© Springer-Verlag Berlin Heidelberg 2006

In this paper we give tighter bounds on the complexity of UPM for bipartite graphs (bipartite-UPM, for short). Our bounds place bipartite-UPM into complexity classes lying between logspace $\mathbf{L}$ and $\mathbf{NC}^2$. The classes we consider are non-deterministic logspace ($\mathbf{NL}$), exact counting in logspace ($\mathbf{C_=L}$), and logspace counting modulo 2 ($\oplus\mathbf{L}$). Some known relationships among these classes and their relativized versions are as follows:

$$\mathbf{L} \subseteq \mathbf{NL} \subseteq \mathbf{C_=L} \subseteq \mathbf{L}^{\mathbf{C_=L}} = \mathbf{NL}^{\mathbf{C_=L}} \subseteq \mathbf{NC}^2, \qquad \mathbf{L} \subseteq \oplus\mathbf{L} \subseteq \mathbf{NL}^{\oplus\mathbf{L}} \subseteq \mathbf{NC}^2.$$

These classes are important because they capture, via completeness, the complexities of important naturally defined problems. Reachability in directed graphs is complete for $\mathbf{NL}$, as also 2-CNF-SAT. Testing whether a square matrix over integers is singular is complete for $\mathbf{C_=L}$, and computing the rank of an integer matrix is complete for $\mathbf{L}^{\mathbf{C_=L}}$. A complete problem for $\oplus\mathbf{L}$ is deciding whether the number of perfect matchings in a bipartite graph is odd.

Our results (from Section 3) place bipartite-UPM in $\mathbf{C_=L} \cap \mathbf{NL}^{\oplus\mathbf{L}}$. The first upper bound implies that G is in bipartite-UPM if and only if an associated matrix A, obtainable from G via very simple reductions (projections), is singular. We show in Section 4 that (unary) weighted bipartite-UPM is in $\mathbf{L}^{\mathbf{C_=L}} \cap \mathbf{NL}^{\oplus\mathbf{L}}$. By the preceding upper bounds, it might well be the case that bipartite-UPM is easier than the perfect matching problem. However, we show in Section 5 that bipartite-UPM is hard for $\mathbf{NL}$; thus the best known lower bounds for PM and for UPM coincide. Our results thus place bipartite-UPM between $\mathbf{NL}$ and $\mathbf{C_=L}$. Furthermore, our results provide a new complete problem for $\mathbf{NL}$. This is the problem of testing if a given perfect matching is unique in a bipartite graph.

2   Preliminaries

Complexity Classes: $\mathbf{L}$ and $\mathbf{NL}$ denote languages accepted by deterministic and nondeterministic logspace bounded Turing machines, respectively. For a nondeterministic Turing machine M, we denote the number of accepting and rejecting computation paths on input x by $acc_M(x)$ and by $rej_M(x)$, respectively. The difference of these two quantities is gap_M, i.e., for all x: $gap_M(x) = acc_M(x) - rej_M(x)$. The complexity class $\mathbf{GapL}$ is defined as the set of all functions $gap_M(x)$ where M is a nondeterministic logspace bounded Turing machine. The class $\mathbf{C_=L}$ (*Exact Counting in Logspace*) is the class of sets A for which there exists a function $f \in \mathbf{GapL}$ such that $\forall\, x:\ x \in A \iff f(x) = 0$. $\mathbf{C_=L}$ is closed under union and intersection, but is not known to be closed under complement. $\oplus\mathbf{L}$ is the class of sets A for which there exists a function $f \in \mathbf{GapL}$ such that $\forall\, x:\ x \in A \iff f(x) \equiv 0 \pmod 2$. $\oplus\mathbf{L}$ is closed under Turing reductions. Circuit classes $\mathbf{NC}^k$ are all families of languages or functions that can be computed by polynomial-size circuits of depth $O(\log^k n)$.

Perfect Matchings: Let $G = (V, E)$ be an undirected graph. A *matching* in G is a set $M \subseteq E$ such that no two edges in M have a vertex in common. A matching M is called *perfect* if every vertex from V occurs as the endpoint of some edge in M. By $\#\,\mathrm{pm}(G)$ we denote the number of perfect matchings in G.

The perfect matching problem and the unique perfect matching problem are defined as $\text{PM} = \{\, G \mid \# \operatorname{pm}(G) > 0 \,\}$ and $\text{UPM} = \{\, G \mid \# \operatorname{pm}(G) = 1 \,\}$. Restricted to bipartite graphs, we denote the problem bipartite-UPM. We also consider the problem of testing whether there exists precisely one perfect matching with minimal weight in a weighted graph.

For graph G with n vertices, the (order n) *skew-symmetric adjacency matrix* A is as defined below:

$$a_{i,j} = \begin{cases} 1 & \text{if } (i,j) \in E \text{ and } i < j, \\ -1 & \text{if } (i,j) \in E \text{ and } i > j, \\ 0 & \text{otherwise.} \end{cases}$$

By transforming $a_{i,j} \;\mapsto\; a_{i,j}(x) = a_{i,j} x_{i,j}$, for indeterminate $x_{i,j} = x_{j,i}$, we get a skew-symmetric variable matrix $A(x)$ called the *Tutte's matrix* of G.

Theorem 1 (Tutte 1952). $G \in \text{PM} \iff \det(A(x)) \neq 0$.

Since $\det(A(X))$ is a symbolic multivariate polynomial, it can have exponential length in n, when written as a sum of monomials. However, there are randomized identity tests for polynomials that just need to evaluate a polynomial at a random point [18,15]. Since the determinant of an integer matrix is complete for **GapL**, a subclass of $\mathbf{NC}^2$, Lovász observed that Tutte's Theorem puts PM in $\mathbf{RNC}^2$.

It is well known from linear algebra that, for an $n \times n$ skew-symmetric matrix $(A = -A^T)$, $\det(A) = 0$ if n is odd and $\det(A) = \det(A^T) \geq 0$ if n is even.

The following fact is a consequence of Tutte's Theorem:

Fact 1. *1.* $\# \operatorname{pm}(G) = 0 \implies \det(A) = 0$,
2. $G \in \text{UPM} \implies \det(A) = 1$.

Rabin and Vazirani [14] used Fact 1 for reconstructing the unique perfect matching as follows. Let G be in UPM with the unique perfect matching M. Let $G_{i,j} = G - \{i,j\}$ denote the subgraph of G obtained by deleting vertices i and j, and let $A_{i,j}$ be the skew-symmetric adjacency matrix of $G_{i,j}$. For each edge $(i,j) \in E$, one can decide whether (i,j) belongs to M or not by:

$$(i,j) \in M \implies G_{i,j} \in \text{UPM} \implies \det(A_{i,j}) = 1,$$
$$(i,j) \notin M \implies G_{i,j} \notin \text{PM} \implies \det(A_{i,j}) = 0.$$

Hence, if $G \in \text{UPM}$ we can compute the perfect matching by looking at the values $\det(A_{i,j})$ for all edges (i,j) of G.

3   Testing Unique Perfect Matching

3.1   Bipartite UPM Is in $\mathbf{L}^{\mathbf{C=L}} \cap \mathbf{NL}^{\oplus\mathbf{L}}$

As seen in the last section, if $G \in \text{UPM}$ then the unique perfect matching M in G can be easily computed [14]. Our approach is to assume $G \in \text{UPM}$ and attempt to construct some perfect matching M as above. If this succeeds, then we check whether M is unique.

Note that any perfect matching can be represented as a symmetric permutation matrix. We construct the matrix $B = (b_{i,j})$ of order n, where

$$b_{i,j} = |a_{i,j}| \det(A_{i,j}).$$

Since A and each $A_{i,j}$ are skew-symmetric, B is symmetric non-negative. From the discussion above we have

Lemma 1. $G \in \mathrm{UPM} \implies B$ *is a symmetric permutation matrix.*

The first step of our algorithm for UPM is to check that the symmetric matrix B is indeed a permutation matrix. This is so if and only if every row contains precisely one 1 and all other entries are 0. This is equivalent to

$$\sum_{i=1}^{n} \left(\left(\sum_{j=1}^{n} b_{ij} \right) - 1 \right)^2 = 0. \tag{1}$$

Since all b_{ij}'s can be computed in **GapL**, the expression on the left hand side in equation (1) can be computed in **GapL** too. We conclude

Lemma 2. $\{ G \mid B$ *is a permutation matrix* $\} \in \mathbf{C_=L}$.

Now assume that B is a permutation matrix (if not, already $G \notin \mathrm{UPM}$), and therefore defines a perfect matching M in G. Suppose there is another perfect matching M' in G. Then the graph $(V, M \triangle M')$ is a union of disjoint alternating cycles, defined below.

Definition 1. *Let M be a perfect matching in G. An* alternating cycle *in G with respect to M is an even simple cycle that has alternate edges in M and not in M.*

Lemma 3. *Let M be a perfect matching in G. $G \in \mathrm{UPM}$ if and only if G has no alternating cycle with respect to M.*

ALTERNATING CYCLE(G, B)
1 **guess** $s \in V$
2 $i \leftarrow s$
3 **repeat**
4   **guess** $j \in V$
5   **if** $b_{i,j} = 0$ **then reject**
6   **guess** $k \in V \setminus \{i\}$
7   **if** $a_{j,k} = 0$ **then reject**
8     $i \leftarrow k$
9 **until** $k = s$
10 **accept**

Given graph G and a permutation matrix B that defines some perfect matching M in G, algorithm ALTERNATING CYCLE searches for an alternating cycle in G with respect to M in nondeterministic logspace. It guesses a node s of an alternating cycle in G (line 1). Assume that we are at node i in the moment. Then ALTERNATING CYCLE makes two steps away from i. The first step is to node j such that $(i, j) \in M$ (line 4 and 5), the second step is to a neighbor $k \neq i$ of j such that $(j, k) \notin M$ (line 6 and 7). If $k = s$ in line 9 we closed a cycle of length at least 4 that has edges alternating in M and not in M. Note that the cycle may not be simple. However, if G is bipartite, then all cycles in G have even length, and so we have visited at least one alternating cycle on the way. Since **NL** equals co-**NL** [8,16], we have

Lemma 4. *Given a bipartite graph G and a matching M in G, testing if $G \in$ UPM is in* **NL**.

Consider the setting that ALTERNATING CYCLE has just graph G as input, and matrix B is provided by the oracle set $S = \{ (G, i, j, c) \mid b_{i,j} = c \} \in \mathbf{C_=L}$. Then ALTERNATING CYCLE can be implemented in $\mathbf{NL}^S$. Note that the oracle access is very simple: the queries are just a copy of the input and of some variables. In particular, this fulfills the Ruzzo-Simon-Tompa restrictions for oracle access by space bounded Turing machines. Since $\mathbf{NL}^{\mathbf{C_=L}} = \mathbf{L}^{\mathbf{C_=L}}$ [2], we have

Lemma 5. *Let G be a bipartite graph such that B is a permutation matrix. Then in $\mathbf{L}^{\mathbf{C_=L}}$ we can decide whether G is in* UPM.

Combining the algorithms from Lemma 2 and 5, we obtain the following:

Theorem 2. bipartite-UPM $\in \mathbf{L}^{\mathbf{C_=L}}$.

Unfortunately, ALTERNATING CYCLE works correctly only for bipartite graphs, since these do not have any odd cycles. On input of an non-bipartite graph, AL-TERNATING CYCLE might accept a cycle of alternating edges with respect to some perfect matching which is *not* simple, thereby possibly giving a false answer.

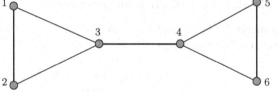

The graph G alongside provides an example. The unique perfect matching is $M = \{(1, 2), (3, 4), (5, 6)\}$, but ALTERNATING CYCLE outputs 'accept'.

```
PERMUTATION(B)

1  for i ← 1 to n do
2      k ← 0; l ← 0
3      for j ← 1 to n do
4          if b_{i,j} ≡ 1 (mod 2)
                  then k ← k + 1
5          if b_{j,i} ≡ 1 (mod 2)
                  then l ← l + 1
6      if k ≠ 1 or l ≠ 1
              then reject
7  accept
```

Interestingly, we can also obtain $\mathbf{NL}^{\oplus \mathbf{L}}$ as an upper bound for bipartite-UPM. Recall the oracle set S we use in the above algorithm. In all the queries we have $c = 0$ or $c = 1$. Suppose we replace S by the set $T \in \oplus \mathbf{L}$,

$$T = \{ (G, i, j, c) \mid b_{i,j} \equiv c \pmod 2 \}.$$

It is easy to design a deterministic logspace algorithm, see PERMUTATION alongside, that, with oracle access to T, checks whether B is a permutation matrix over $\mathbf{Z_2}$.

Since $\mathbf{L}^{\oplus \mathbf{L}} = \oplus \mathbf{L}$, we have

Lemma 6. $\{\,G \mid B \text{ is a permutation matrix}\,\} \in \oplus \mathbf{L}$.

Consider algorithm ALTERNATING CYCLE with oracle T. Although we might get different oracle answers when switching from S to T, it is not hard to check that we anyway get the correct final answer. Again we combine the two steps and get

Corollary 1. bipartite-UPM $\in \mathbf{NL}^{\oplus \mathbf{L}}$.

3.2 Bipartite-UPM Is in $\mathbf{C_{=}L}$

Based on the method in [11], the upper bound $\mathbf{L}^{\mathbf{C_{=}L}}$ for bipartite UPM can be improved to $\mathbf{C_{=}L}$. Note that we do not know whether $\mathbf{L}^{\mathbf{C_{=}L}} = \mathbf{C_{=}L}$.

Let $G = (U, V, E)$ be a bipartite graph with $|U| = |V| = n$. Let A be the bipartite adjacency matrix of G; A is of order n. Then the skew-symmetric adjacency matrix of G is of the form $S = \begin{pmatrix} 0 & A \\ -A^T & 0 \end{pmatrix}$. Since $\det(S) = \det^2(A)$, Fact 1 gives the following for bipartite graph G.

Fact 2. *1.* $\#\operatorname{pm}(G) = 0 \implies \det(A) = 0$,
2. $G \in \text{UPM} \implies \det(A) = \pm 1$.

The following lemma puts the idea of Kozen, Vazirani, and Vazirani [10] in such a way that we get $\mathbf{C_{=}L}$ as an upper bound.

Lemma 7. *For bipartite graph G with $2n$ vertices and bipartite adjacency matrix A, define matrices $B = (b_{i,j})$ and C of order n as follows*

$$b_{i,j} = a_{i,j}\,\det^2(A_{i|j}),\ \text{for } 1 \le i, j \le n,$$
$$C = I - AB^T,$$

where I is the $n \times n$ identity matrix and $A_{i|j}$ is the sub-matrix obtained by deleting the i-th row and the j-th column of A. Then G has a unique perfect matching if and only if

(i) B is a permutation matrix, and
(ii) the characteristic polynomial of C is $\chi_C(x) = x^n$.

We provide some intuition to the lemma. Just as in the general case of Lemma 2, if B is a permutation matrix, then matrix B describes a perfect matching. The product AB^T puts the matching edges on the main diagonal of the matrix. Then $I - AB^T$ takes out the matching edges. Now consider C as the adjacency matrix of a (directed) graph, say H. This can be thought of as identifying vertex i from the left-hand side with vertex i from the right-hand side of the bipartite graph AB^T. (i.e. if G has vertices $U = \{u_1, \ldots, u_n\}$ and $W = \{w_1, \ldots, w_n\}$, then AB^T

matches u_i with w_i, and edge (i, j) in H corresponds to path (u_i, w_i, u_j) in G.) Then any cycle in graph H corresponds to an alternating cycle in G. Hence there should be no cycles in H. Equivalently, all coefficients of the characteristic polynomial of C should be 0.

Another way of seeing this is as follows: $\chi_C(x) = \det(xI - C) = \det((x-1)I + AB^T)$. But $(x-1)I + AB^T$ is the bipartite adjacency matrix of G when vertices are renumbered to get the matching edges (of B) on the main diagonal, and with weights x on these matched edges, weights 1 on other edges. So condition (ii) checks if the determinant of this matrix is x^n.

We consider the complexity of checking the conditions of Lemma 7. Regarding the condition (i), the problem of testing if B is a permutation matrix is essentially the same as in the general case and can be done in $\mathbf{C_=L}$ (Lemma 2). Consider the condition (ii). Here the elements of matrix C are not given as input, they are certain determinants. However, a result in [1] shows that composition of determinants is computable again in $\mathbf{GapL}$, i.e. the coefficients of the characteristic polynomial of C can be computed in $\mathbf{GapL}$ and they can be verified in $\mathbf{C_=L}$ [7]. Therefore condition (ii) can be checked in $\mathbf{C_=L}$. We conclude:

Theorem 3. bipartite-UPM $\in \mathbf{C_=L}$.

The above technique doesn't seem to generalize to non-bipartite graphs. The graph G shown here provides an example where the technique doesn't seem to work. Observe that $G \in$ UPM with the unique perfect matching $M = \{(1,2),(3,4),(5,6)\}$. Permute G and take out the edges of M as described above. This leads to graph H. Since $G \in$ UPM, H should have no directed cycles. But H contains 4 directed cycles, one of them is $(1,4,5,3)$. Another problem comes from the sign of the cycles in H.

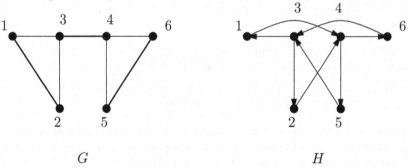

$$G \qquad\qquad\qquad\qquad H$$

Considering the concept of the skew-symmetric matrix of a non-bipartite graph, we ask whether there is an analog to Lemma 7. Namely, let A be the skew-symmetric adjacency matrix of G, and let B be defined as: $b_{i,j} = a_{i,j}\det(A_{i,j|i,j})$. ($A_{i,j|i,j}$ is the skew-symmetric adjacency matrix of the graph $G_{i,j|i,j}$ obtained by deleting vertices i and j, so we delete these rows and columns from A.) Then an analogous test for whether $G \in$ UPM would be: Does B correspond to some perfect matching M, and does weighting the edges of M with x give matrix A_x with determinant x^n? Unfortunately, this is not true.

The graph G shown alongside provides an example (even in the bipartite case). Obviously, G is not in UPM. But the matrix $B = (b_{i,j})$ computed by the expression $b_{i,j} = a_{i,j}\det(A_{i,j|i,j})$ is a symmetric permutation matrix which corresponds to the perfect matching $M = \{(1,2),(3,4),(5,6),(7,8),(9,10),(11,12),$ $(13,14),(15,16),(17,18)\}$. Let B' be the skew-symmetric adjacency matrix corresponds to M. By Maple we get $\chi_C(x) = x^{18}$ where $C = I - AB'^T$.

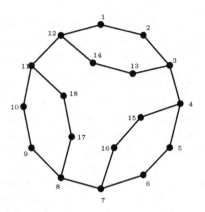

4   Unique Minimum-Weight Perfect Matching

We now consider graphs with positive edge-weights. We assume that the weights are given in unary (or are bounded by a polynomial in the number of vertices). It has been shown by Mulmuley, Vazirani, and Vazirani [13] that there is a $\mathbf{RNC}^2$ algorithm (by using Isolating Lemma) for computing some perfect matching in graph G. This algorithm chooses random weights for the edges of G, then the Isolating Lemma states that with high probability, there is a unique perfect matching of minimal weight. We describe briefly the procedure by [13] for reconstructing that unique minimum-weight perfect matching.

Let $w_{i,j}$ be the weight of edge (i,j) in G. Consider the skew-symmetric adjacency matrix $D = (d_{i,j})$ defined as follows: for $i < j$, $d_{i,j} = 2^{w_{i,j}}$ if (i,j) is an edge and $d_{i,j} = 0$ if (i,j) is not an edge; for $i > j$, $d_{i,j} = -d_{j,i}$. Then the following facts hold:

1. If there is a unique minimum-weight perfect matching M with weight W, then $\det(D) \neq 0$; moreover, the highest power of 2 dividing $\det(D)$ is 2^{2W}.
2. Furthermore, edge (i,j) is in M if and only if $\frac{\det(D_{i,j})2^{w_{i,j}}}{2^{2W}}$ is odd   where $D_{i,j}$ is obtained by deleting rows i,j and columns i,j of D.

In our case the weights belong to the input, so the Isolating Lemma does not help. However, we can still use the above reconstruction procedure to construct some perfect matching which is potentially of minimal weight, and then test the uniqueness separately.

Constructing the unique minimum-weight perfect matching. For the first part of our algorithm (finding a symmetric permutation matrix associated to a perfect matching), unfortunately we cannot argue as elegantly as in the case of unweighted graphs that we treated in the last section, if we follow the reconstruction procedure by [13]. The reason is that we need values of a $\mathbf{GapL}$ function modulo 2^k for some k, and $\mathbf{GapL}$ is not known to be closed under integer division. Thus a $\mathbf{L}^{\mathbf{C=L}}$ upper bound as in the unweighted case may not hold by this way. Instead, we describe another method for computing the unique minimum-weight perfect matching.

Let x be an indeterminate. We relabel all the edges (i, j) of G with $x^{w_{i,j}}$. Let $G(x)$ be the new graph and $A(x)$ its Tutte matrix. Then $\det(A(x))$ is a polynomial, $p(x) = \det(A(x)) = c_N x^N + c_{N-1} x^{N-1} + \cdots + c_0$, $c_N \neq 0$. Note that the degree N of p is bounded by the sum of all edge-weights of G. Thus when all the weights $w_{i,j}$ of G are polynomially bounded, N is also polynomially bounded, and all coefficients of $p(x)$ can be computed in **GapL**.

Assume for a moment that graph G has the unique minimum-weight perfect matching M with weight W. Observe that M corresponds to the lowest term x^{2W} in $p(x)$, moreover we have $c_{2W} = 1$ and $c_i = 0$, for all $0 \leq i < 2W$.

We denote by $G_{i,j}(x)$ the graph obtained from $G(x)$ by deleting the edge (i, j) and by $A_{i,j}(x)$ the Tutte matrix associated with $G_{i,j}(x)$. Furthermore, let $p_{i,j}(x) = \sum_{k \geq 0} c_k^{(i,j)} x^k = \det(A_{i,j}(x))$. Observe that

- If $(i, j) \in M$ then $c_{2W}^{(i,j)} = c_t^{(i,j)} = 0$, for all $0 \leq t \leq 2W$, because M can not be the unique minimum-weight perfect matching in $G - (i, j)$ which is obtained from G by deleting the edge (i, j). Graph $G - (i, j)$ has potentially other perfect matchings with weights bigger than W.
- If $(i, j) \notin M$ then $c_{2W}^{(i,j)} = 1 = c_{2W}$ and $c_t^{(i,j)} = 0$, for all $0 \leq t < 2W$, because M remains as the unique minimum-weight perfect matching in $G - (i, j)$.

Let $A = (a_{i,j})$ be the adjacency matrix of G. Define symmetric matrices $B_t = \left(b_{i,j}^{(t)} \right)$ by

$$b_{i,j}^{(t)} = b_{j,i}^{(t)} = a_{i,j} \sum_{k=0}^{t} \left(c_k - c_k^{(i,j)} \right), \text{ for } 0 \leq t \leq N. \tag{2}$$

It is clear that the elements of B_t are **GapL**-computable. As a consequence of the above observations we have the following fact.

Fact 3. *If G has a perfect matching M with minimal weight W, then B_{2W} is a symmetric permutation matrix and $B_t = 0$, for all $0 \leq t < 2W$.*

Since all elements of matrices B_t are computable in **GapL**, in $\mathbf{C_{=}L}$ we can test if B_t is a permutation matrix or a zero-matrix.

Lemma 8. *If G has unique minimum-weight perfect matching, then there exists $0 \leq t \leq N$ that $c_t = 1$, $c_s = 0$, B_t is a symmetric permutation matrix, and $B_s = 0$, for all $0 \leq s < t$. All these conditions can be tested in $\mathbf{C_{=}L}$.*

Testing uniqueness. The second part of our algorithm is to test if a given perfect matching M is unique with minimal weight in G. For weighted bipartite graphs we can develop an **NL**-algorithm that has B as input, where B is the permutation matrix associated to M. In analogy to the unweighted version of UPM, we don't know whether there is an **NC**-algorithm for weighted non-bipartite UPM.

In the bipartite case we look for an alternating cycle C (with respect to M) where the edges not in M have the same or less total weight than the edges of

M. If such a cycle exists, then $M \triangle C$ gives another matching M' with weight no more than that of M. If no such cycle exists, then M is the unique minimum-weight perfect matching in G. Algorithm ALT-WEIGHTED-CYCLE nondeterministically searches for such cycles.

With oracle access to B, the algorithm is in **NL**. Note that here too it is crucial that weights are given in unary; thus the cumulative weights a_M and $a_{\overline{M}}$ can be stored on a logspace tape.

Lemma 9. *Algorithm* ALT-WEIGHTED-CYCLE *tests correctly if a given perfect matching M is unique with minimal weight in a bipartite graph. It can be implemented in* **NL**.

ALT-WEIGHTED-CYCLE(G, B)

1   $a_M \leftarrow 0$; $a_{\overline{M}} \leftarrow 0$
2   **guess** $s \in V$
3   $i \leftarrow s$
4   **repeat**
5   **guess** $j \in V$
6   **if** $b_{i,j} = 0$ **then reject**
7   **guess** $k \in V \setminus \{i\}$
8   **if** $a_{j,k} = 0$ **then reject**
9       $a_M \leftarrow a_M + w_{i,j}$
10      $a_{\overline{M}} \leftarrow a_{\overline{M}} + w_{j,k}$
11      $i \leftarrow k$
12 **until** $k = s$
13 **if** $a_{\overline{M}} \leq a_M$ **then reject**
14 **accept**

By combining Lemma 8 and 9 we can test if a bipartite graph G has unique minimum-weight perfect matching. Namely, the algorithm computes all matrices B_t, then it searches the potential perfect matching M by Lemma 8. Thereafter G and M are the inputs for algorithm ALT-WEIGHTED-CYCLE. By this way we can show that the weighted-bipartite UPM is in $\mathbf{NL}^{\mathbf{C=L}}$ which is equal to $\mathbf{L}^{\mathbf{C=L}}$.

In analogy to the unweighted case we can modify the computation of the perfect matching M by $B'_t = B_t$ (mod 2). The matrices B'_t are computed in $\oplus\mathbf{L}$. The rest of the algorithm is the same, giving an upper bound of $\mathbf{NL}^{\oplus\mathbf{L}}$.

Theorem 4. Weighted-bipartite UPM *with polynomially bounded weights is in* $\mathbf{L}^{\mathbf{C=L}} \cap \mathbf{NL}^{\oplus\mathbf{L}}$.

5   Unique Perfect Matching Is Hard for NL

Chandra, Stockmeyer, and Vishkin [4] have shown that the perfect matching problem is hard for **NL**. We modify their reduction to show that UPM is hard for **NL**. Recall that UPM might be an easier problem than the general perfect matching problem.

Let $G = (V, E)$ be a directed acyclic graph, and let $s, t \in V$ be two vertices. By $\#\,\mathrm{path}(G, s, t)$ we denote the number of paths in G from s to t. The connectivity problem asks whether $\#\,\mathrm{path}(G, s, t) > 0$ and it is complete for **NL**. Since **NL** is closed under complement [8,16], asking whether $\#\,\mathrm{path}(G, s, t) = 0$ is also complete for **NL**.

In [4], the following undirected graph H is constructed: $H = (V_H, E_H)$,
$V_H = \{s_{out}, t_{in}\} \cup \{u_{in}, u_{out} \mid u \in V \setminus \{s, t\}\}$
$E_H = \{(u_{in}, u_{out}) \mid u \in V \setminus \{s, t\}\} \cup \{(u_{out}, v_{in}) \mid (u, v) \in E\}$
It is easy to see that $\#\,\text{path}(G, s, t) = \#\,\text{pm}(H)$. Therefore s_{out} is connected to t_{in} if and only if $H \in \text{PM}$.

Now obtain H' from H by adding the edge (s_{out}, t_{in}). Observe that H' has at least one perfect matching, namely $M_H = \{(s_{out}, t_{in})\} \cup \{(u_{in}, u_{out}) \mid u \in V - \{s, t\}\}$. Other than this, H' and H have the same perfect matchings. We conclude that $\#\,\text{pm}(H) + 1 = \#\,\text{pm}(H')$. In summary, $\#\,\text{path}(G, s, t) = 0 \iff \#\,\text{pm}(H') = 1$. Note that each edge in H' is of the form (u_{in}, v_{out}); thus the partition $S, V_H \setminus S$ where $S = \{u_{in} \mid u_{in} \in V_H\}$ witnesses that H' is bipartite.

Theorem 5. UPM *is hard for* **NL**, *even when restricted to bipartite graphs.*

As a consequence of the hardness of UPM, we consider the problem of testing if a given perfect matching M is unique in a graph G. The problem for bipartite graphs can be solved in **NL** by Lemma 4. For non-bipartite graphs we don't know whether the considered problem is in **NC** (note that if this problem for non-bipartite graph is in **NC**, then UPM for non-bipartite graphs is also in **NC**, because the unique perfect matching can always be computed in **NC**). Furthermore, the problem is hard for **NL** because in the above construction, $\#\,\text{path}(G, s, t) = 0$ if and only if the perfect matching M_H is unique in the constructed graph H'. Thus we have

Corollary 2. *The problem of testing if a given perfect matching is unique in a bipartite graph is complete for* **NL**.

Summary and Open Problems

We showed in the paper that the unique perfect matching problem for bipartite graphs for both cases weighted or unweighted is in **NC**. We have placed bipartite UPM between **NL** and $\mathbf{C_=L} \cap \mathbf{NL}^{\oplus \mathbf{L}}$ and the unique minimum-weight perfect matching problem between **NL** and $\mathbf{L}^{\mathbf{C_=L}} \cap \mathbf{NL}^{\oplus \mathbf{L}}$. Some questions remain open: 1) *Is non-bipartite UPM in* **NC**? 2) *Can we improve the lower bound* **NL** *for UPM?* A possible improvement seems to be important because if UPM is hard for $\mathbf{C_=L}$, we could conclude that $\mathbf{C_=L} \subseteq \mathbf{NL}^{\oplus \mathbf{L}}$ (which is an open question), if UPM is hard for $\oplus \mathbf{L}$, we could conclude that $\oplus \mathbf{L} \subseteq \mathbf{L}^{\mathbf{C_=L}}$ (which is open too).

The same question about the upper bound can be asked for weighted non-bipartite graphs. Also, we restricted the weights to be polynomially bounded. It is not clear how to handle exponential weights. The current technique to determine one perfect matching would then lead to double exponential numbers. This is no longer in $\mathbf{NC}^2$. Note however, that the weight of an alternating cycle requires summing up at most n weights, which can be done in $\mathbf{NC}^2$.

Note that the results involving $\oplus \mathbf{L}$, namely Lemma 6, Corollary 1 and Theorem 4, carry over to $\mathbf{Mod_pL}$ for any p as well. The open questions listed above concerning $\oplus \mathbf{L}$ are open for all these classes as well.

Clearly, the most important open problem is: *Is the perfect matching problem in* **NC**?

Acknowledgments

We thank Jochen Messner and Ilan Newman for very interesting discussions. Eric Allender gave very helpful comments on an earlier version of the paper. We also thank the anonymous referees for pointing out the generalization to Mod_pL, and for comments which improved the presentation.

References

1. E. Allender, V. Arvind, and M. Mahajan. Arithmetic complexity, Kleene closure, and formal power series. *Theory Comput. Syst.*, 36(4):303–328, 2003.
2. E. Allender, R. Beals, and M. Ogihara. The complexity of matrix rank and feasible systems of linear equations. *Computational Complexity*, 8(2):99–126, 1999.
3. E. Allender, K. Reinhardt, and S. Zhou. Isolation, matching and counting: uniform and nonuniform upper bounds. *Journal of Computer and System Sciences*, 59:164–181, 1999.
4. A. Chandra, L. Stockmeyer, and U. Vishkin. Constant depth reducibility. *SIAM Journal on Computing*, 13(2):423–439, 1984.
5. J. Edmonds. Maximum matching and a polyhedron with 0-1 vertices. *Journal of Research National Bureau of Standards*, 69:125–130, 1965.
6. H. N. Gabow, H. Kaplan, and R. E. Tarjan. Unique maximum matching algorithms. In *31st Symposium on Theory of Computing (STOC)*, pages 70–78. ACM Press, 1999.
7. T. M. Hoang and T. Thierauf. The complexity of the characteristic and the minimal polynomial. *Theoretical Computer Science*, 295:205–222, 2003.
8. N. Immerman. Nondeterministic space is closed under complementation. *SIAM Journal on Computing*, 17(5):935–938, 1988.
9. R. M. Karp, E. Upfal, and A. Wigderson. Constructing a perfect matching is in random NC. *Combinatorica*, 6:35–48, 1986.
10. D. Kozen, U. Vazirani, and V. Vazirani. NC algorithms for comparability graphs, interval graphs, and testing for unique perfect matching. In *Proceedings of FST&TCS Conference, LNCS Volume 206*, pages 496–503. Springer-Verlag, 1985.
11. D. Kozen, U. Vazirani, and V. Vazirani. NC algorithms for comparability graphs, interval graphs, and testing for unique perfect matching. Technical Report TR86-799, Cornell University, 1986.
12. L. Lovasz. On determinants, matchings and random algorithms. In L. Budach, editor, *Proceedings of Conference on Fundamentals of Computing Theory*, pages 565–574. Akademia-Verlag, 1979.
13. K. Mulmuley, U. Vazirani, and V. Vazirani. Matching is as easy as matrix inversion. *Combinatorica*, 7(1):105–131, 1987.
14. M. Rabin and V. Vazirani. Maximum matchings in general graphs through randomization. *Journal of Algorithms*, 10(4):557–567, 1989.
15. J. Schwartz. Fast probabilistic algorithms for verification of polynomial identities. *Journal of the ACM*, 27:701–717, 1980.
16. R. Szelepcsényi. The method of forced enumeration for nondeterministic automata. *Acta Informatica*, 26(3):279–284, 1988.
17. J. E. Tabaska, R. B. Cary, H. N. Gabow, , and G. D. Stormo. An RNA folding method capable of identifying pseudoknots and base triples. *Bioinformatics*, 14(8):691–699, 1998.
18. R. Zippel. Probabilistic algorithms for sparse polynomials. In *International Symposium on Symbolic and Algebraic Computation, LNCS 72*, pages 216–226, 1979.

Comparing Reductions to NP-Complete Sets

John M. Hitchcock[1,*] and A. Pavan[2,**]

[1] Department of Computer Science, University of Wyoming
[2] Department of Computer Science, Iowa State University

Abstract. Under the assumption that NP does not have p-measure 0, we investigate reductions to NP-complete sets and prove the following:

1. Adaptive reductions are more powerful than nonadaptive reductions: there is a problem that is Turing-complete for NP but not truth-table-complete.
2. Strong nondeterministic reductions are more powerful than deterministic reductions: there is a problem that is SNP-complete for NP but not Turing-complete.
3. Every problem that is many-one complete for NP is complete under length-increasing reductions that are computed by polynomial-size circuits.

The first item solves one of Lutz and Mayordomo's "Twelve Problems in Resource-Bounded Measure" (1999). We also show that every problem that is complete for NE is complete under one-to-one, length-increasing reductions that are computed by polynomial-size circuits.

1 Introduction

A language $L \in$ NP is NP-complete if every language in NP is *reducible* to L. There are several possible interpretations of the word "reducible." Polynomial-time many-one reducible is the most typical meaning, but there are many other reducibilities, each providing a potentially different NP-completeness notion. Are there languages that are NP-complete using one type of reduction but not complete under another type of reduction? Are there two apparently different notions of reductions for which the corresponding completeness notions coincide? We study these questions for several types of reductions.

1.1 Adaptive Versus Nonadaptive Reductions

A many-one reduction ($\leq_m^P$) from A to B converts a question about membership in A to an equivalent question about membership in B. Formally, there is a function f such that $x \in A$ if and only if $f(x) \in B$. A variation on this theme is to allow the use of B as an oracle to solve A. Here there is an algorithm M that takes as input an instance x of A and may ask multiple queries about instances of B before outputting its decision for x. There are two basic forms of this type

---

* This research was supported in part by NSF grant 0515313.
** This research was supported in part by NSF grant 0430807.

M. Bugliesi et al. (Eds.): ICALP 2006, Part I, LNCS 4051, pp. 465–476, 2006.

of reduction: adaptive and nonadaptive. In an adaptive reduction (also called a Turing reduction, $\leq_T^p$) M receives the answer for each query before asking its next query – subsequent queries may depend on the answers to previous queries. In a nonadaptive reduction (also called a truth-table reduction, $\leq_{tt}^p$) M asks all of its queries before receiving any answers.

Lutz and Mayordomo [23] showed that if NP does not have p-measure zero (written $\mu_p(\mathrm{NP}) \neq 0$), then adaptive completeness for NP is different from many-one completeness. In fact, they showed this hypothesis yields a problem that is complete for NP under adaptive reductions that make only two queries, but is not complete under many-one reductions. In the conclusion of their paper, Lutz and Mayordomo conjectured that the measure hypothesis would yield separations of other completeness notions between $\leq_m^p$ and $\leq_T^p$ for NP, similar to what is known unconditionally for E and NE [29,9].

Since then there have been several results in this direction. Ambos-Spies and Bentzien [5] used a genericity hypothesis on NP, an assumption which is implied by the measure hypothesis, to separate essentially all bounded-query completeness notions for NP. It is also known that some of these separations can be obtained under bi-immunity hypotheses [27,18], which are even weaker assumptions. For a survey of these results see [25].

However, so far a separation of adaptive completeness from nonadaptive completeness for NP has been elusive. This question has been asked in several survey papers [11,22,12,24], most prominently as one of Lutz and Mayordomo's "Twelve Problems in Resource-Bounded Measure," Problem 9:

Does $\mu_p(\mathrm{NP}) \neq 0$ imply the existence of a problem that is $\leq_T^p$-complete, but not $\leq_{tt}^p$-complete, for NP?

The only partial result on this problem was by Pavan and Selman [26] who used a strong hypothesis about UP to separate these two completeness notions. We affirmatively answer the above question. Our proof combines the connection between the measure of NP and the NP-machine hypothesis [17] with results about nonadaptive reductions to P-selective sets [10,28].

1.2   Nondetermistic Versus Deterministic Reductions

Adleman and Manders [1] observed that while most problems can be shown to be NP-complete using polynomial-time reductions, some problems resist this approach. To classify such problems, they proposed what are now called *strong nondeterministic many-one reductions*. (Adleman and Manders called these reductions γ-reductions.) If a language that is NP-complete under strong nondeterministic reductions admits an efficient algorithm, then NP = coNP. Therefore, if we believe NP $\neq$ coNP, strong nondeterministic completeness can also be taken as evidence that the problem in hand is intractable.

Adleman and Manders showed that some number-theoretic problems are NP-complete under strong nondeterministic many-one reductions. Chung and Ravikumar [13] showed that certain questions regarding comparator networks are also NP-complete under these reductions. It is not known whether these problems remain complete if we use polynomial-time reductions.

This situation raises the following question: are there languages that are complete under strong nondeterministic reductions, but not complete under polynomial-time reductions? We show that if $\mu_p(\text{NP}) \neq 0$, then the answer to this question is yes, even if we consider polynomial-time adaptive reductions.

1.3 Length-Increasing Reductions

It has been observed that many NP-completeness results hold under very restrictive reductions. For example, SAT is complete under polynomial-time reductions that are one-to-one and length-increasing. In fact, all known many-one complete problems for NP are complete under this type of reduction [8]. This raises the following question: are there languages that are complete under polynomial-time many-one reductions but not complete under polynomial-time, one-to-one, length-increasing reductions?

Berman [7] showed that every many-one complete set for E is complete under one-to-one, length-increasing reductions. Thus for E, these two completeness notions coincide. A weaker result is known for NE. Ganesan and Homer [15] showed that all NE-complete sets are complete via one-to-one reductions that are exponentially honest.

For NP, until very recently there had not been any progress on this question. Agrawal [3] showed that if one-way permutations exist, then all NP-complete sets are complete via one-to-one, length-increasing, p/poly-reductions. Agrawal's result also holds for the NE-complete sets under the same hypothesis.

In this paper, we show that if $\mu_p(\text{NP}) \neq 0$, then all NP-complete sets are complete via length-increasing, p/poly-reductions. We note that the measure hypothesis on NP is apparently incomparable with Agrawal's hypothesis that one-way permutations exist. Regarding NE-completeness, we show that Agrawal's result can be made unconditional. That is, we unconditionally show that all NE-complete sets are complete via one-to-one, length-increasing, p/poly-reductions.

2 Preliminaries

We assume that the readers are familiar with polynomial-time many-one reductions. A language A is *polynomial-time Turing reducible* to B ($A \leq_T^p B$) if there is a polynomial-time oracle Turing M such that $A = L(M^B)$. A language A is *polynomial-time truth-table reducible* to a language B ($A \leq_{tt}^p B$) if there exist polynomial-time computable functions g and h such that on input x, $g(x) = \{q_1, \cdots, q_m\}$, and $x \in A$ if and only if $h(x, B(q_1), \cdots, B(q_m)) = 1$. Given a reducibility $\leq_r^\alpha$, a set S in NP is $\leq_r^\alpha$-complete for NP if every set in NP is $\leq_r^\alpha$-reducible to S.

2.1 Resource-Bounded Measure

Lutz [21] introduced resource-bounded measure to study the quantitative structure of complexity classes.

A *martingale* is a function $d : \Sigma^* \rightarrow \mathbb{Q}$ with the property that for every $w \in \Sigma^*$, $2d(w) = d(w0) + d(w1)$. A martingale d *succeeds* on a language A if

$$\limsup_{n \rightarrow \infty} d(A|n) = \infty,$$

where $A|n$ is the length n prefix of A's characteristic sequence.

Given a time bound $t(n)$, a language L is $t(n)$-*random* [6] if no $O(t(n))$-time computable martingale succeeds on L. A class of languages X has p-measure zero, written $\mu_p(X) = 0$, if there exists a polynomial t such that every language in X in not $t(n)$-random.

Lutz suggested studying the structure of the class NP under the hypothesis "NP does not have p-measure 0," which is written $\mu_p(\text{NP}) \neq 0$. Since then several believable consequences of this hypothesis have been obtained. For a survey of these results see [22,24].

2.2   NP-Machine Hypothesis

Our proofs crucially make use of the following hypothesis. Several variants of this hypothesis have been studied earlier [14,16].

NP-Machine Hypothesis. There exists an NP-machine M and $\epsilon > 0$ such that M accepts 0^* and no 2^{n^ϵ}-time-bounded Turing machine computes infinitely many accepting computations of M.

It is known that the measure hypothesis implies the NP-machine hypothesis.

Theorem 2.1. (Hitchcock and Pavan [17]) *If $\mu_p(\text{NP}) \neq 0$, then the* NP-*machine hypothesis holds.*

Observation 2.2. *Assume that the* NP-*machine hypothesis is true and let p be any polynomial. Then there exists an* NP-*machine N that accepts 0^*, and no $2^{p(n)}$-time-bounded machine computes infinitely many accepting computations of N.*

2.3   Reductions to P-Selective Sets

A set S is p-selective if there exists a polynomial-time computable function $f : \Sigma^* \times \Sigma^* \rightarrow \Sigma^*$ such that for all x, y, $f(x,y) \in \{x,y\}$, and if at least one of x and y belongs to S, then $f(x,y)$ belongs to S.

Let P-sel denote the class of p-selective sets. For a reduction $\leq_r^\alpha$ and a class $\mathcal{C}$, let

$$R_r^\alpha(\mathcal{C}) = \{A \mid (\exists B \in \mathcal{C}) A \leq_r^\alpha B\}.$$

Theorem 2.3. (Buhrman and Longpré [10], Wang [28]) $R_{tt}^p(\text{P-sel})$ *has p-measure 0.*

Let $\leq_{tt}^{t(n)\restriction p}$ denote a truth-table reduction that is computable in $t(n)$ time, but where the number and length of the queries is bounded by a polynomial. It is straightforward to extend the arguments in [10] or [28] to show that Theorem 2.3 extends to these reductions when $t(n)$ is linear-exponential.

Theorem 2.4. *For every $c \in \mathbb{N}$, the class $R_{tt}^{2^{cn}\restriction p}(\text{P-sel})$ has p-measure 0.*

3   Adaptive Versus Nonadaptive Reductions

We now present our solution to Problem 9 of Lutz and Mayordomo [24].

Theorem 3.1. *If $\mu_p(NP) \neq 0$, then there is a problem that is $\leq^P_T$-complete for NP but not $\leq^P_{tt}$-complete.*

Proof. Assume that $\mu_p(NP) \neq 0$. From Theorem 2.1 and Observation 2.2 we obtain an NP-machine M that accepts 0^* such that no 2^{n^2}-time machine can compute infinitely many of its accepting computations.

For each n, let a_n be the lexicographically maximum accepting computation of $M(0^n)$. Let a be the infinite sequence $a = a_0 a_1 a_2 \ldots$. Let

$$A = \{\langle x, w \rangle \mid x \in \text{SAT and } w \text{ is an accepting computation of } M(0^{|x|})\},$$

$$B = L(a) = \{x \mid x < a\},$$

where $<$ is the standard dictionary order. Let

$$C = 0A \cup 1B.$$

Then C is $\leq^P_T$-complete for NP: to decide whether $x \in \text{SAT}$, we can adaptively query B to find $a_{|x|}$ and then ask if $\langle x, a_{|x|} \rangle \in A$.

Suppose that C is $\leq^P_{tt}$-complete for NP. Then for every $L \in \text{NP}$, $L \leq^P_{tt} C$ via some reduction (g, h).

Claim 3.2. *For all but finitely many x, all queries of $g(x)$ to strings of the form $0\langle y, w \rangle$ must satisfy $|y| \leq |x|$.*

Proof of Claim 3.2. Consider the following algorithm.

> input 0^n;
> for all $x \in \{0,1\}^{<n}$:
>   compute $g(x)$;
>   for all queries in $g(x)$ that are of the form $0\langle y, w \rangle$, where $|y| = n$:
>   if w is an accepting computation of $M(0^n)$
>   output w and halt;

This algorithm runs in $O(2^n \cdot \text{poly}(n))$ time, and would compute infinitely many accepting computations of M if the claim is false.   □ *Claim 3.2*

Claim 3.3. $L \leq^{2^n \upharpoonright p}_{tt} B$.

Proof of Claim 3.3. Given x, compute $g(x)$. By Claim 3.2, all queries of $g(x)$ to strings of the form $0\langle y, w \rangle$ must satisfy $|y| \leq |x|$. We can decide whether these queries are in A in 2^n time by checking if $y \in \text{SAT}$ in exponential time and whether w is an accepting computation of $M(0^{|y|})$ in polynomial time. Our reduction to B simply solves the queries to A directly.   □ *Claim 3.3*

Since B is a left-cut, it is p-selective, so it follows from Claim 3.3 that NP $\subseteq R^{2^n \upharpoonright p}_{tt}(\text{P-sel})$. By Theorem 2.4, this implies $\mu_p(\text{NP}) = 0$, a contradiction.   □

4 Nondeterministic Versus Deterministic Reductions

Definition. [1,20] A language A is *strong nondeterministic many-one reducible* to a language B, written $A \leq_m^{SNP} B$, if there is a nondeterministic polynomial-time machine M such that the following conditions hold.

- On an input x, every path of M either outputs a string y or outputs the special symbol "?". At least one path outputs a string.
- If x belongs to A, then every output y belongs to B, and if x does not belong to A, then every output y does not belong to B.

Adleman and Manders [1] also called this γ-reducibility and denoted it $\leq_\gamma$.
 Long [20] showed that the following are equivalent:

- for all A, B, $A \leq_m^{SNP} B$ implies $A \leq_T^P B$
- every NPMV total function has a polynomial-time refinement.

The latter has been called Proposition Q in [14]. To separate $\leq_m^{SNP}$-completeness from $\leq_T^P$-completeness for NP, we clearly need a hypothesis that at least implies Q is false. The NP-machine hypothesis fits the bill:

Theorem 4.1. *If the* NP*-machine hypothesis holds, then there is a problem that is* $\leq_m^{SNP}$*-complete for* NP *but not* $\leq_T^P$*-complete.*

Proof. By Observation 2.2, there exists an NP machine M that accepts 0^* for which no 2^{3n}-time bounded machine can compute infinitely many accepting computations. Consider the following language L

$$L = \{\langle x, a \rangle \mid x \in \text{SAT and } a \text{ is an accepting computation of } M(0^{|x|})\}.$$

Then $L \in$ NP, and we claim that L is strong nondeterministic many-one complete. Consider a nondeterministic machine N that on input x guesses a string a, and if a is an accepting computation of $M(0^{|x|})$, then it outputs $\langle x, a \rangle$. If a is not an accepting computation of $M(0^{|x|})$, then N outputs ?. Then N is a strong nondeterministic many-one reduction from SAT to L. It follows that L is strong nondeterministic many-one complete for NP.

We will show that L is not Turing complete for NP. Suppose to the contrary that it is Turing complete. Consider the following language S.

$$S = \{\langle 0^n, w \rangle \mid w \text{ is a prefix of an accepting computation of } M(0^n)\}.$$

Since S in in NP, there is a polynomial-time oracle Turing machine R such that $S = R^L$. Consider the following procedure $\mathcal{A}$ that tries to compute accepting computations of M.

1. Input 0^n.
2. Set $y = \epsilon$.
3. Run $R(\langle 0^n, y0 \rangle)$. When R generates a query $q = \langle x, z \rangle$, $|x| = t$ do the following:

   (a) If z is not an accepting computation of $M(0^t)$, then continue simulation of R with answer "No".

   (b) Else, z is an accepting computation of $M(0^t)$.

   (c) If $t \geq n$, then Output "Unsuccessful", print z and halt.

   (d) Otherwise, decide whether $\langle x, z \rangle \in L$ by checking whether $x \in$ SAT. Since $t < n$ this takes at most 2^n time. Use this answer to continue the simulation.

4. If R accepts $\langle 0^n, y0 \rangle$, then set $y = y0$. Else set $y = y1$.

5. If y is an accepting computation of $M(0^n)$, then output y and halt. Else, GoTo Step 3.

Observe that the most expensive step in the above computation is Step 3d. This takes 2^n time. Since this step is repeated at most polynomial number of steps, the above algorithm halts in 2^{2n} steps.

Next we make two claims about the behavior of the algorithm $\mathcal{A}$.

Claim 4.2. If $\mathcal{A}(0^n)$ outputs "Unsuccessful" for infinitely many n, then there is a 2^{3n}-time algorithm that outputs infinitely many accepting computations of $M(0^n)$.

Proof of Claim 4.2. Observe that if $\mathcal{A}(0^n)$ outputs "Unsuccessful", then there exists a $t \geq n$ and $\mathcal{A}(0^n)$ outputs an accepting computation of $M(0^t)$. Thus if there exist infinitely many n for which $\mathcal{A}(0^n)$ outputs "Unsuccessful", then there exists infinitely many t for which there exists $n \leq t$, and $\mathcal{A}(0^n)$ outputs an accepting computation of $M(0^t)$. Now consider the following algorithm: On input 0^t, run $\mathcal{A}(0^j)$, $1 \leq j \leq t$. If any of the runs of $\mathcal{A}$ outputs an accepting computation of $M(0^t)$, then output that accepting computation.

This algorithm outputs an accepting computation of $\mathcal{A}(0^t)$ for infinitely many t. The running time of the algorithm is bounded by $\sum_{j=1}^{t} 2^{2j} \leq 2^{3t}$. This establishes the claim.   □ *Claim 4.2*

Claim 4.3. If $\mathcal{A}(0^n)$ does not output "Unsuccessful", then it outputs an accepting computation of $M(0^n)$ in time 2^{2n}.

Proof of Claim 4.3. Observe that $\mathcal{A}(0^n)$ is trying to compute an accepting computation of $M(0^n)$ by doing a prefix search. This is accomplished by running the Turing reduction R, and whenever the reduction generates a query it is attempting to find the answer to the query without actually making the query. Thus if all the queries are answered correctly, it will compute an accepting computation of $M(0^n)$. We argue that $\mathcal{A}(0^n)$ computes all query answers correctly. Let $q = \langle x, y \rangle$ be a query that is generated.

If y is not an accepting computation of M, then q does not belong to L. Thus $\mathcal{A}$ answers the query correctly in 3a. So assume y is an accepting computation of $M(0^t)$. Since $\mathcal{A}(0^n)$ does not output "Unsuccessful", $t < n$. Thus the algorithm reaches Step 3d. In this step, it decides whether $x \in$ SAT by a running a deterministic algorithm for SAT. Thus the query answer is computed correctly in this step.

Thus $\mathcal{A}(0^n)$ computes all query answers correctly. Thus $\mathcal{A}(0^n)$ outputs an accepting computation of $M(0^n)$. Recall that the running time of $\mathcal{A}$ is bounded by 2^{2n}. □ *Claim 4.3*

Now, if $\mathcal{A}(0^n)$ outputs "Unsuccessful" for infinitely many n, then, by Claim 4.2, there is a 2^{3n}-time algorithm that computes infinitely many accepting computations of $M(0^n)$. This contradicts the NP-machine hypothesis. Thus for all but finitely many n, $\mathcal{A}(0^n)$ does not output "Unsuccessful". Thus, by Claim 4.3, for all but finitely many n, $\mathcal{A}(0^n)$ outputs an accepting computation of $M(0^n)$ in time 2^{2n}. This again contradicts the NP-machine hypothesis.

Thus there is no Turing reduction from S to L. Thus L is not Turing complete for NP. □

By Theorem 2.1, we immediately have the following.

Corollary 4.4. *If $\mu_p(\text{NP}) \neq 0$, there is a problem that is $\leq_m^{\text{SNP}}$-complete for NP but not $\leq_T^p$-complete.*

5 Length-Increasing Reductions and Polynomial-Size Circuits

In this section we study one-to-one, length-increasing reductions. (All reductions in this section are many-one reductions. We say that a many-one reduction f is *length-increasing* if $|f(x)| > |x|$ for all strings x and that f is *one-to-one* if for all strings $x \neq y$, $f(x) \neq f(y)$.)

Berman proved [7] that every $\leq_m^p$-complete set for E is also is complete under one-to-one, length-increasing reductions. This proof makes essential use of the fact that E is closed under complementation, so it does not go through for nondeterministic classes. As a partial result, Ganesan and Homer [15] showed that every $\leq_m^p$-complete set for NE is complete under one-to-one, exponentially-honest reductions. See also the survey paper [19] by Homer.

Agrawal [3] showed that if one-way permutations exist, then many-one complete sets for NP and NE are complete via one-to-one, length-increasing, p/poly reductions. (A p/poly reduction is computed by a nonuniform family of polynomial-size circuits, one for each input length.) A well-known fact is that coNE ⊆ NE/poly: to determine if a string x is not in an NE language, we can give as advice the number of strings in the language at x's length. Then an NE machine can guess all of these strings and determine whether or not x is in the language. We make use of this idea and Berman's technique to prove the following theorem.

Theorem 5.1. *Every set that is $\leq_m^p$-complete for NE is complete under one-to-one, length-increasing, p/poly reductions.*

Next we will show that if NP does not have p-measure zero, then all NP-complete sets are complete via length-increasing p/poly reductions. In the proof we will consider whether a language R has the following property.

Property 5.2. There is a 2^{cn}-time computable function f such that for every n, $f(0^n)$ either outputs $\bot$ or outputs a tuple $\langle a, b, u, v \rangle$. Whenever $f(0^n) = \langle a, b, u, v \rangle$, the following hold.

- $|a| = |b| = n$.
- $R(a)R(b) \neq uv$, and uv is either 00 or 11.

And for infinitely many n, $f(0^n) \neq \bot$.

Informally, f either finds two strings such that at least one of them is in R, or finds two strings such that at least one of them does not belong to R.

Lemma 5.3. *If R has Property 5.2, then R is not n^c-random.*

Proof. We describe a martingale d that can win an infinite amount of money while betting on R. Let $d(n)$ denote the amount of money that the martingale has before it starts betting on strings of length n. Before starting betting on strings of length n, the martingale runs $f(0^n)$. If $f(0^n) = \bot$, then d does not bet on any string of length n. Suppose $f(0^n) = \langle a, b, u, v \rangle$. Without loss of generality we can assume $a < b$. Consider the case $uv = 00$. In this case at least one of a and b must be in R. The martingale bets 1/3rd of its amount on $a \in R$. If a really belongs to R, then d does not bet on any other string of length n. So if $a \in R$, then $d(n+1) = 4d(n)/3$. However, if $a \notin R$, then d is left with capital $2d(n)/3$. However, since at least one of a and b must be in R, b must belong to R. Now d bets all its money on $b \in R$. Thus in this case also $d(n+1) = 4d(n)/3$. The case $uv = 11$ is handled via a symmetric argument.

Since $f(0^n) \neq \bot$ for infinitely many n, for infinitely many n, $d(n+1) \geq 4d(n)/3$. Thus $d(n)$ approaches infinity as n tends to ∞. Since f runs in 2^{cn}-time, d runs in time $O(n^c)$. Thus R is not n^c-random. □

Now we are ready to prove the theorem regarding complete sets for NP.

Theorem 5.4. *If $\mu_p(NP) \neq 0$, then every NP-complete language is complete under length-increasing, p/poly reductions.*

Proof. Let L be any NP-complete language. We show that there is a p/poly, length-increasing reduction from SAT to L. We first define an intermediate language S such that SAT is p/poly, length increasing reducible to S, and S is honest polynomial-time reducible to L. Combing these two reductions we obtain the desired reduction from SAT to L. Let $L \in \text{DTIME}(2^{n^k})$.

If NP does not have p-measure 0, then there is an n^4-random language R in NP. The randomness of R implies that both R and $\overline{R}$ have at least one string at each length. Let

$$S = \{\langle x, y, z \rangle \mid |x| = |y| = |z| \text{ and } \texttt{MAJ}\{x \in R, y \in \text{SAT}, z \in R\} = 1\}.$$

Here $\texttt{MAJ}\{\phi, \psi, \tau\} = 1$ if a majority of ϕ, ψ, and τ are true.

It is clear that S is NP. For every n, fix two strings a_n and b_n of length n such that $a_n \in R$ and $b_n \notin R$. Consider the following reduction from SAT to S: Given an input y of length n the reduction outputs $\langle a_n, y, b_n \rangle$. Now $y \in$ SAT $\Leftrightarrow \langle a_n, y, b_n \rangle \in S$. The reduction takes a_n and b_n as advice. It is clear that this reduction is length increasing. Therefore we have established that SAT is p/poly, length-increasing reducible to S.

Since S is in NP and L is NP-complete, there is a many-one reduction f from S to L. We now argue that f must be a honest reduction on strings of form $\langle x, y, z \rangle$ where $|x| = |y| = |z|$.

Claim 5.5. Let $T = \{\langle x, y, z \rangle \mid |x| = |y| = |z|\}$. For all but finitely many strings w from T, $|f(w)| \geq n^{1/k}$.

Proof of Claim 5.5. Consider the following set

$$U = \{w = \langle x, y, z \rangle \in T \mid |x| = n, |f(w)| < n^{1/k}\}.$$

We show that if U is infinite, then R has Property 5.2.

Recall that L can be decided in time 2^{n^k}. Thus if a string w belongs to U, then the membership of $f(w)$ in L can be decided in time $2^{|f(w)|^k} < 2^n$. Since f is a many-one reduction from S to L, for every string w in U, its membership in S can be computed in time 2^n.

Define a function f as follows. In input 0^n, cycle through all tuples $w = \langle x, y, z \rangle$, $|x| = |y| = |z| = n$, and check if $w \in U$ by computing $f(w)$. If none of the w's are in U, then output $\perp$. Else, let $w = \langle x, y, z \rangle$ be the first string that belongs to U.

Compute the membership of w in S. We first consider the case $w \in S$. In this case,

$$\text{MAJ}\{x \in R, y \in \text{SAT}, z \in R\} = 1.$$

Thus it can not be the case that both x and z are out of R. So f outputs $\langle x, z, 0, 0 \rangle$. Similarly, if $w \notin S$, then it cannot be the case that both x and z are in S. So f outputs $\langle x, z, 1, 1, \rangle$.

Observe that the running time of f is bounded by $O(2^{3n})$. If U is infinite, then for infinitely many n, $f(0^n) \neq \perp$. So, if U is infinite, then R has Property 5.2, and by Lemma 5.3, R is not n^3-random. Since R is n^4-random, U is finite.

Thus for all but finitely many strings from T, $|f(w)| \geq n^{1/k}$. $\Box$ *Claim 5.5*

Now consider the following reduction g from SAT to L: On input y of length n, output $f(\langle a_n, y, b_n \rangle)$. By Claim 5.5, $|f(\langle a_n, y, b_n \rangle)| \geq n^{1/k}$. Thus g is an honest, p/poly-reduction from SAT to L. Since SAT is paddable, there exists a length-increasing p/poly-reduction from SAT to L. Thus L is complete via length-increasing, p/poly reductions. $\Box$

6   Conclusion

We now know that the measure hypothesis separates nearly all polynomial-time completeness notions for NP. It would be interesting to separate completeness notions for NP under weaker hypotheses such as "NP is hard on average".

Theorem 4.1 gives evidence that when we give more resources to the reductions, we obtain a richer class of complete sets. What happens when we decrease the resource bound of the reductions? Agrawal et al [4,2] showed that NC^0-completeness and AC^0-completeness for NP coincide whereas AC^0-completeness and $AC^0[mod2]$-completeness for NP differ. It would be interesting to extend these results other resource bounds.

Results of Agrawal [3] and results in Section 5 indicate that complete sets for NP and NE are complete under one-to-one, length-increasing reductions. However these reductions need polynomial advice. Can we eliminate the advice?

Acknowledgments. We thank the anonymous reviewers for helpful comments and corrections.

References

1. L. Adleman and K. Manders. Reducibility, randomness, and intractability. In *Proceedings of the 9th ACM Symposium on Theory of Computing*, pages 151–163, 1977.
2. A. Agrawal, E. Allender, R. Impagliazzo, T. Pitassi, and S. Rudich. Reducing the complexity of reductions. *Computational Complexity*, 10:117–138, 2001.
3. M. Agrawal. Pseudo-random generators and structure of complete degrees. In *17th Annual IEEE Conference on Computational Complexity*, pages 139–145, 2002.
4. M. Agrawal, E. Allender, and S. Rudich. Reductions in circuit complexity: An isomorphism theorem and a gap theorem. *Journal of Computer and System Sciences*, 57(2):127–143, 1998.
5. K. Ambos-Spies and L. Bentzien. Separating NP-completeness notions under strong hypotheses. *Journal of Computer and System Sciences*, 61(3):335–361, 2000.
6. K. Ambos-Spies, S. A. Terwijn, and X. Zheng. Resource bounded randomness and weakly complete problems. *Theoretical Computer Science*, 172(1–2):195–207, 1997.
7. L. Berman. *Polynomial Reducibilities and Complete Sets*. PhD thesis, Cornell University, 1977.
8. L. Berman and J. Hartmanis. On isomorphism and density of NP and other complete sets. *SIAM Journal on Computing*, 6(2):305–322, 1977.
9. H. Buhrman, S. Homer, and L. Torenvliet. Completeness notions for nondeterministic complexity classes. *Mathematical Systems Theory*, 24:179–200, 1991.
10. H. Buhrman and L. Longpré. Compressibility and resource bounded measure. *SIAM Journal on Computing*, 31(3):876–886, 2002.
11. H. Buhrman and L. Torenvliet. On the structure of complete sets. In *Proceedings of the Ninth Annual Structure in Complexity Theory Conference*, pages 118–133. IEEE Computer Society, 1994.
12. H. Buhrman and L. Torenvliet. Complete sets and structure in subrecursive classes. In *Logic Colloquium '96*, volume 12 of *Lecture Notes in Logic*, pages 45–78. Association for Symbolic Logic, 1998.
13. M. J. Chung and B. Ravikumar. Strong nondeterministic Turing reduction—a technique for proving intractability. *Journal of Computer and System Sciences*, 39:2–20, 1989.
14. S. Fenner, L. Fortnow, A. Naik, and J. Rogers. Inverting onto functions. *Information and Computation*, 186(1):90–103, 2003.

15. K. Ganesan and S. Homer. Complete problems and srong polynomial reducibilities. *SIAM Journal on Computing*, 21(4), 1991.
16. L. Hemaspaandra, J. Rothe, and G. Wechsung. Easy sets and hard certificate schemes. *Acta Informatica*, 34(11):859–879, 1997.
17. J. M. Hitchcock and A. Pavan. Hardness hypotheses, derandomization, and circuit complexity. In *Proceedings of the 24th Conference on Foundations of Software Technology and Theoretical Computer Science*, pages 336–347. Springer-Verlag, 2004.
18. J. M. Hitchcock, A. Pavan, and N. V. Vinodchandran. Partial bi-immunity, scaled dimension, and NP-completeness. *Theory of Computing Systems*. To appear.
19. S. Homer. Structural properties of complete sets for exponential time. In L. A. Hemaspaandra and A. L. Selman, editors, *Complexity Theory Retrospective II*, pages 135–153. Springer-Verlag, 1997.
20. T. J. Long. On γ-reducibility versus polynomial time many-one reducibility. *Theoretical Computer Science*, 14:91–101, 1981.
21. J. H. Lutz. Almost everywhere high nonuniform complexity. *Journal of Computer and System Sciences*, 44:220–258, 1992.
22. J. H. Lutz. The quantitative structure of exponential time. In L. A. Hemaspaandra and A. L. Selman, editors, *Complexity Theory Retrospective II*, pages 225–254. Springer-Verlag, 1997.
23. J. H. Lutz and E. Mayordomo. Cook versus Karp-Levin: Separating completeness notions if NP is not small. *Theoretical Computer Science*, 164:141–163, 1996.
24. J. H. Lutz and E. Mayordomo. Twelve problems in resource-bounded measure. *Bulletin of the European Association for Theoretical Computer Science*, 68:64–80, 1999. Also in *Current Trends in Theoretical Computer Science: Entering the 21st Century*, pages 83–101, World Scientific Publishing, 2001.
25. A. Pavan. Comparison of reductions and completeness notions. *SIGACT News*, 34(2):27–41, June 2003.
26. A. Pavan and A. Selman. Separation of NP-completeness notions. *SIAM Journal on Computing*, 31(3):906–918, 2002.
27. A. Pavan and A. Selman. Bi-immunity separates strong NP-completeness notions. *Information and Computation*, 188:116–126, 2004.
28. Y. Wang. NP-hard sets are superterse unless NP is small. *Information Processing Letters*, 61(1):1–6, 1997.
29. O. Watanabe. A comparison of polynomial time completeness notions. *Theoretical Computer Science*, 54:249–265, 1987.

Design Is as Easy as Optimization

Deeparnab Chakrabarty[1], Aranyak Mehta[2], and Vijay V. Vazirani[1,*]

[1] Georgia Institute of Technology, Atlanta, USA
[2] IBM Almaden Research Center, San Jose, USA

Abstract. We identify a new genre of algorithmic problems – design problems – and study them from an algorithmic and complexity-theoretic view point. We use the learning techniques of Freund-Schapire [FS99] and its generalizations to show that for a large class of problems, the design version is as easy as the optimization version.

1 Introduction

Over the last four decades, theoreticians have identified several fundamental genres of algorithmic problems and have studied their computational complexity and the inter-relationships among them. These include decision, search, optimization, counting, enumeration, random generation, and approximate counting problems. In this paper, we define and study the complexity of design problems.

This new genre of algorithmic problems should come as no surprise. In the past, several researchers have studied natural design problems – we provide some prominent examples below. Moreover, practitioners have always been faced with such problems and have sought intelligent solutions to them. However, to the best of our knowledge, this genre has not been formally defined before and subjected to a systematic complexity-theoretic study.

Every optimization problem leads to a natural design problem. This process is formally defined in Section 2. Let us illustrate it in the context of the sparsest cut problem. We are given an undirected graph $G(V, E)$ and a bound B on the total weight. The problem is to find a way to distribute weight B on the edges of G so that the weight of the sparsest cut is maximized. Note that this design problem is a maxmin problem.

Three examples of natural design problems considered in the past are: Boyd, Diaconis and Xiao [BDX04] study the design of the Fastest Mixing Markov Chain on a graph with a budget constraint on the weights of the edges of a fixed graph. Elson, Karp, Papadimitriou and Shenker [EKPS04] study the Synchronization Design Problem in sensornets. Baiou and Barahona [BB05] and Frederickson and Solis-Oba [FSO99] study a cost-based design version of maximizing the minimum weight spanning tree. A closely related problem of budgeted optimization was studied by Juttner [JÖ3]

The main result in this paper is that for a large class of optimization problems, the design version of a problem is as easy as the optimization version. We provide several different techniques to show this:

---

* Work supported by NSF grants 0311541, 0220343 and 0515186.

M. Bugliesi et al. (Eds.): ICALP 2006, Part I, LNCS 4051, pp. 477–488, 2006.

- In Section 3.1, we observe that if the objective functions in the minimization (maximization) problem Π are concave (convex), then the maxmin (minmax) design problem $D(\Pi)$ can be set up as a convex optimization problem. Moreover, Π itself appears as the separation oracle required in the ellipsoid method. Thus $D(\Pi)$ is no harder than Π in terms of complexity. Further, we show using techniques of [JMS03] that if Π has an α-factor approximation algorithm, then $D(\Pi)$ also has an α-factor approximation.
- Since the ellipsoid method takes a long time in practice, we seek more efficient methods. In Section 3.2 we observe that if the optimization problem Π can be set up as a linear program, then the design problem $D(\Pi)$ can be set up as another linear program. If Π has an LP-relaxation which gives a factor of α, then $D(\Pi)$ also has an LP-based solution which gives factor α. In Section 3.3 we show that if the optimization problem possesses certain structural (packing) properties, then we can use these to solve the design problem more efficiently. We give examples to illustrate these specific methods.
- In Section 4, we give what is perhaps the central algorithmic result of this paper – we provide a second general method for solving the design problem. This method is much more efficient than the ellipsoid method of Section 3.1. We set up the design problem as a two player zero-sum game and show that the design problem seeks the minmax value of the game. We apply the techniques of Freund-Schapire [FS99] in the additive case and that of Zinkevich [Zin03] and Flaxman et.al [FKM05] in the convex/concave cases to solve the game. This technique also requires an (approximation) algorithm for the optimization problem. If this algorithm has a worst case factor of α, then we will be able to solve the design problem upto a factor of α with an additional additive error of an arbitrarily small ϵ.
- In Section 5 we ask how hard is the design version of a problem if the optimization version is NP-hard. We provide an example in which the design version is in P and another in which the design version is NP-hard. In Section 3.1 we have established that if the optimization version is in P then so is the design version.
- In Section 6, we observe the close relationship between maxmin design problems and fractional packing of the corresponding combinatorial structures. We use this to prove some results about fractional packings of spanning and Steiner trees.

2   Problem Definition

We present a general framework to define the design version of optimization problems. An **optimization problem** Π consists of a set of *valid instances* $\mathcal{I}_\Pi$. Each instance I is a triple $(E_I, \mathcal{S}_I, \mathbf{w}_I)$. E_I is a universe of *elements*, and each element $e \in E_I$ has an associated weight w_e, a rational number, giving the vector $\mathbf{w}_I$. Each instance also has a set of *feasible solutions* $\mathcal{S}_I$, where each $S \in \mathcal{S}_I$ is a subset of E. The number of feasible solutions may be exponential in $|E_I|$. For an instance $I = (E_I, \mathcal{S}_I, \mathbf{w}_I)$, and a feasible solution $S \in \mathcal{S}_I$, the objective function value for S is given as some function of the solution and the weight vector $obj(S) = f_S(\mathbf{w}_I)$. In most optimization problems like the Travelling Salesman problem, Sparsest Cut problem, etc., the function f_S is just the sum of weights of elements in S. These class of problems are called *additive* optimization problems. A more general class of problems is the one in which the functions f_S are a convex or concave function of the weight vector. In a minimization (maximization) problem one wishes to find a feasible solution of minimum (maximum) objective function value.

The ***maxmin design version*** $D(\Pi)$ of a minimization problem Π is defined as follows: For every collection of valid instances of Π of the form $I = (E_I, \mathcal{S}_I, \cdot)$, there is one valid instance of $D(\Pi)$: $J = (E_J, \mathcal{S}_J, B_J)$, where $E_J = E_I$, $\mathcal{S}_J = \mathcal{S}_I$, and B_J is a rational number, called the weight budget. A feasible solution to J is a weight vector $\mathbf{w} = (w_e)_{e \in E_J}$, which satisfies the *budget constraint* $\sum_{e \in E_J} w_e \leq B_J$. Every feasible solution $\mathbf{w}$ to J leads to an instance $I = (E_I, \mathcal{S}_I, \mathbf{w})$ of the optimization problem Π.

The goal of the maxmin design problem is to find a feasible solution $\mathbf{w}$ so that the minimum objective function value of the resulting instance of the minimization problem is as large as possible. That is,

$$OPT_{D(\Pi)}\left((E, \mathcal{S}, B)\right) = \max_{\mathbf{w}: \sum_e w_e \leq B} OPT_{\Pi}((E, \mathcal{S}, \mathbf{w}))$$

The minmax design version of a maximization problem is defined similarly.

3   Solving Design Problems

3.1   A General Technique Based on the Ellipsoid Method

Let the design problem at hand be a maxmin design problem. The analysis for minmax design problems is similar. Let $(E, \mathcal{S}, B)$ be an instance of the design problem, and let $f_S(.)$ be the function giving the objective value for the solution $S \in \mathcal{S}$. In this section, we assume $f_S(.)$ to be a concave function in the weights[1]. Consider the following program

$$\max\{\ \lambda \quad \text{s.t.} \quad f_S(\mathbf{w}) \geq \lambda \ \ \forall S \in \mathcal{S}; \quad \sum_{e \in E} w_e \leq B\} \qquad (1)$$

Firstly note that the feasible region in the above program is convex, and thus program 1 is a convex program. This is because if $(\lambda, \mathbf{w})$ and $(\lambda', \mathbf{w}')$ are feasible solutions, then so is their convex combination: For any $0 \leq \mu \leq 1$, $\forall\ S \in \mathcal{S}$,

$$f_S(\mu\mathbf{w} + (1 - \mu)\mathbf{w}') \geq \mu f_S(\mathbf{w}) + (1 - \mu)f_S(\mathbf{w}') \geq \mu\lambda + (1 - \mu)\lambda'$$

where the first inequality uses the fact that f_S is concave.

---

[1] $f_S(.)$ can be a concave function of the weights of all elements in E, not just the elements in S, which is used here only for indexing. Recall that we defined the special case of additive functions to have $f_S(\mathbf{w})$ as the sum of weights of elements in S.

Therefore we can use the ellipsoid method to solve the convex program. Given a candidate point $(\lambda, \mathbf{w})$, the separation oracle needs to check whether it is feasible or return a set S as a certificate of infeasibility. Note that solving the optimization problem $(E, \mathcal{S}, \mathbf{w})$ suffices: if the minimum is greater than λ then the solution is feasible, otherwise the set with the minimum objective value is the certificate of infeasibility[2]. Thus we have the following theorem:

Theorem 1. *If we have an algorithm which solves the optimization problem Π in polynomial time, then for any $\epsilon > 0$, we can solve the corresponding design problem $D(\Pi)$ up to an additive error of ϵ in time polynomial in n and $\log \frac{1}{\epsilon}$.*

Suppose we can not solve the optimization problem exactly but only have an α-approximation for it, for some $\alpha \geq 1$. That is, we have a polytime algorithm which, given $(E, \mathcal{S}, \mathbf{w})$, returns a set S with objective function value guaranteed to be at most α-factor away from the actual optimum: $f_S(\mathbf{w}) \leq \alpha \min_{T \in \mathcal{S}} f_T(\mathbf{w})$. Then we can use the methods of [JMS03] to obtain an α approximation to the convex program 1.

Theorem 2. *If we have a polynomial time algorithm returning an α-approximation to the optimization problem Π, then we can find, for any $\epsilon > 0$, an approximation algorithm for the design problem $D(\Pi)$, with a multiplicative factor of α and an additive error of ϵ.*

Note that in both Theorems 1 and 2, if the problems are additive, then we do not need the additive error of ϵ. The ellipsoid method may need to take a number of steps equal to a large polynomial. In each step we need to solve an instance of the optimization problem Π. The ellipsoid method also takes a huge time in practice. This motivates us to look for faster algorithms for the design problem. Below, we provide two techniques which are much faster and which apply if the given problem has a special structure. In Section 4, we will provide a general method which also works much faster.

3.2   A Technique Based on LP-Relaxation

Suppose we have a linear programming relaxation for the minimization problem Π, which yields an α-approximation algorithm, for some $\alpha \geq 1$. That is, corresponding to $(E, \mathcal{S}, \mathbf{w})$ there is a linear program:

$$\min\{ \ \mathbf{w} \cdot \mathbf{x} \quad \text{s.t} \quad A\mathbf{x} \geq \mathbf{b}; \ \mathbf{x} \geq \mathbf{0} \ \} \tag{2}$$

with the property that the optimum value of the LP, call it L, is a lower bound on the optimum of the given instance - $\forall \ T \in \mathcal{S}, \ L \leq f_T(\mathbf{w})$. Moreover, there is a guarantee that for any weight vector $\mathbf{w}$, given an optimum solution to LP 2, one can produce in polynomial time a set S such that $f_S(\mathbf{w}) \leq \alpha L$.

---

[2] Here, and throughout, we will say that an (approximation) algorithm solves a optimization problem if it gives the (approximately) optimum value as well as a set S which achieves this (approximately) optimum value.

To solve the design problem, we look at the dual of LP 2.

$$\max\{ \ \mathbf{b} \cdot \mathbf{y} \quad \text{s.t} \quad \mathbf{y}^T A \leq \mathbf{w}; \ \mathbf{y} \geq \mathbf{0} \ \} \tag{3}$$

We note that the weight vector $\mathbf{w}$ is no longer in the objective function but appears in the constraints. Parametrizing the program on $\mathbf{w}$, let the optimum solution to LP 3 be $D(\mathbf{w})$. From the previous supposition, we know there is an algorithm giving a set S with the guarantee, $D(\mathbf{w}) \leq f_S(\mathbf{w}) \leq \alpha D(\mathbf{w})$ for all weight vectors $\mathbf{w}$.

To solve the design problem, we consider $\mathbf{w}$ as a variable in LP 3, and add the constraint that the total weight is bounded by B. Thus we solve the following LP

$$\max\{ \ \mathbf{b} \cdot \mathbf{y} \quad \text{s.t} \quad \mathbf{y}^T A - \mathbf{w} \leq 0; \ \mathbf{w} \cdot \mathbf{1} \leq B; \ \mathbf{y}, \mathbf{w} \geq \mathbf{0} \ \} \tag{4}$$

Let the optimum solution to LP 4 be D^*. Let $\mathbf{w}'$ be the optimum vector returned in the solution of LP 4. Note that for any weight vector $\mathbf{w}$ satisfying $\mathbf{w} \cdot \mathbf{1} \leq B$, we have $D(\mathbf{w}) \leq D^*$ with equality at $\mathbf{w}'$. Solve LP 2 with $\mathbf{w}'$ and obtain a set T with the guarantee $D^* \leq f_T(\mathbf{w}') \leq \alpha D^*$.

We now claim that $T, \mathbf{w}'$ gives an α approximation to the design problem. To see this, suppose $\mathbf{w}^*$ was the weight vector acheiving the maxmin design. Moreover, suppose S was the set that minimized its objective value given $\mathbf{w}^*$. We need to show $\alpha f_T(\mathbf{w}') \geq f_S(\mathbf{w}^*)$. To see this note $f_S(\mathbf{w}^*) \leq \alpha D(\mathbf{w}^*) \leq \alpha D^* \leq \alpha f_T(\mathbf{w}')$. Thus we have:

Theorem 3. *If we have an LP relaxation for the optimization problem Π, and a polynomial time algorithm producing a solution within $\alpha \geq 1$ times the LP optimum, then we can produce an α approximation algorithm for the corresponding design problem $D(\Pi)$ which requires solving an LP having one constraint more than that of the LP relaxation.*

As a corollary we get a $\log n$ approximation to maximum min-multicut, a $\log n$ approximation to the maximum sparsity cut, a 2-approximation to the maximum min weighted vertex cover and many such problems which have approximation algorithms via LP-relaxations.

3.3 A Technique Based on Integral Packing

Suppose we have an instance of the additive minimization problem $(E, \mathcal{S}, \mathbf{w})$ with the following structure: There exist solutions $S_1, S_2, \cdots, S_k$ which are disjoint. In this case, we see that B/k is an upper bound on the optimum of the maxmin design problem $(E, \mathcal{S}, B)$. This is because no matter how we distribute the weight vector $\mathbf{w}$, one of the sets S_i will have $\sum_{e \in S_i} w_e \leq B/k$, since these sets are disjoint. If we can demonstrate a solution of value B/k, then this is optimal.

As an example, consider the maxmin $s-t$ cut problem. If l is the length of the shortest path from s to t, we can pack l edge disjoint $s-t$ cuts, e.g. the level cuts of the BFS tree from s to t. By the above argument, we have an upper bound of B/l on the maxmin $s-t$ cut. Now take any shortest path and distribute the weight B equally on all the edges in the path. Since any $s - t$ cut contains at least one of these edges, we see that this solution has value B/l, hence optimal.

A second example is the design version of minimum weight spanning tree in a graph - find a weight distribution to maximize the weight of an MST. Here the upper bound comes from the Nash-Williams and Tutte Theorem on packing of edge disjoint spanning trees [NW61, Tut61] and can be achieved via giving weights to the cross edges in the optimal partition. In fact, in Section 6 we shall see a close relation between maxmin design problems and fractional packing of solutions, and how the maxmin design framework can be used to prove results about fractional packing.

4   Faster Algorithms for Design Problems

In this section we provide a general method to solve design problems which works much faster than the method in Section 3.1. In Section 4.1 we solve the additive case, before solving the more general concave/convex cases in Section 4.2.

4.1   Additive Design Problems, Zero-Sum Games and Multiplicative Update

In the additive case, the maxmin design problem $(E, \mathcal{S}, B = 1)$ can be formulated as a two-player zero-sum game $G(E, \mathcal{S})$: The row player (the maxminimizer) has $|E|$ rows, corresponding to the elements, and the column player has $|\mathcal{S}|$ columns, corresponding to the solutions. The $|E| \times |\mathcal{S}|$ matrix has 0 or 1 entries, with the entry $(e, S) = 1 \iff e \in S$. This is the amount that the column player pays the row player.

A probability distribution on the pure row strategies corresponds to a distribution of the weight on the elements. Now the column player's best responses (in pure strategies) correspond to sets $S \in \mathcal{S}$ with mimimum weight with respect to the given distribution of weight.

Proposition 1. *The set of optimal weight distributions for the maxmin design problem $(E, \mathcal{S}, B)$ is equal to the set of maxmin strategies for the row player in the game $G(E, \mathcal{S})$, scaled by B.*

Thus the goal of the maxmin assignment problem is precisely to find a maxmin strategy for the row player. Since $|\mathcal{S}|$ may be very large, one cannot just solve the game by traditional means, say, using linear programming. We use the technique developed in [FS99] to approximate zero-sum games to approximate design problems. The algorithms and proofs remaining section mimic [FS99] in our setting.

Assume $B = 1$ for notational ease. The algorithm proceeds in rounds. In each round we define a new weight function $\mathbf{w}_t$. We assume that we have a polynomial time oracle which given any weight vector, is guaranteed to return a solution of cost within α times the minimum cost set. That is, at each round we get a solution S_t such that $f_{S_t}(\mathbf{w}_t) \leq \alpha \min_S f_S(\mathbf{w}_t)$. Note that in this additive case,

$f_S(\mathbf{w}) = \sum_{e \in S} \mathbf{w}(e)$.

We then apply the multiplicative update rule:

- Initialize $\forall\, e:\ z_1(e) = 1$. Let $\mathbf{w}_1(e) = z_1(e)/\sum_e z_1(e)$.
- **Multiplicative update:** Suppose the oracle on input $\mathbf{w}_t$ returns solution S_t. Then the new weights are found as follows:
  $z_{t+1}(e) = z_t(e)\beta^{M(e,S_t)}$, $\quad \mathbf{w}_{t+1}(e) = z_{t+1}(e)/\sum_e z_{t+1}(e)$
  where $M(e, S_t) = 1$ if S_t contains e, 0 otherwise.

We run this algorithm for T steps. Define the regret after T steps as

$$R_T := \max_{\mathbf{w}:\sum_e \mathbf{w}(e)=1} \sum_{t=1}^{T} f_{S_t}(\mathbf{w}) - \sum_{t=1}^{T} f_{S_t}(\mathbf{w}_t)$$

The following theorem was proved in [FS99].

Theorem FS: $R_T \le \sqrt{T}O(\sqrt{\ln n})$

Run the algorithm for T rounds, and take the average of all the weight vectors over the T rounds: $\overline{\mathbf{w}} := \frac{1}{T}\sum_{t=1}^{T} \mathbf{w}_t$. We prove in the next lemma that $\overline{\mathbf{w}}$ is an α approximation with additive error to the maxmin design problem. In particular we shall show

Lemma 1. $\min_S f_S(\overline{\mathbf{w}}) \ge \frac{1}{\alpha} \max_{\mathbf{w}:\sum_e \mathbf{w}(e)=1} \min_S f_S(\mathbf{w}) - O(\frac{1}{\alpha}\sqrt{\frac{\ln n}{T}})$

Proof. We follow the proof as in [FS99]. In this when we use subscript $\mathbf{w}$ we assume that sum of weights is equal to 1 and not explicitly mention it. We have

$$\min_S f_S(\overline{\mathbf{w}}) = \min_S \frac{1}{T}\sum_{t=1}^{T} f_S(\mathbf{w}_t) \qquad (\text{ by linearity of } f_S)$$

$$\ge \frac{1}{T}\sum_{t=1}^{T} \min_S f_S(\mathbf{w}_t)$$

$$\ge \frac{1}{T}\sum_{t=1}^{T} \frac{1}{\alpha} f_{S_t}(\mathbf{w}_t) \qquad (\text{ oracle is } \alpha \text{ approximate})$$

$$\ge \frac{1}{\alpha} \max_{\mathbf{w}} \frac{1}{T}\sum_{t=1}^{T} f_{S_t}(\mathbf{w}) - O(\frac{1}{\alpha}\sqrt{\frac{\ln n}{T}}) \qquad (\text{ by Theorem FS})$$

$$\ge \frac{1}{\alpha} \max_{\mathbf{w}} \min_S f_S(\mathbf{w}) - O(\frac{1}{\alpha}\sqrt{\frac{\ln n}{T}}) \qquad (\text{minimum} \le \text{average})$$

Thus if we run for $T = \frac{\ln n}{\epsilon}$ rounds, we get an ϵ additive error. Thus we get the following theorem

Theorem 4. *Given a maxmin design problem (E, S, B), suppose we have (as a black box) an approximation algorithm which solves the corresponding minimization problem upto a factor $\alpha \ge 1$. Then we can design an algorithm which can solve the maxmin design problem upto a factor of α.*

4.2   Extending the Framework to Concave Utility Functions and Convex Cost Functions

In this section, we extend the technique described in Section 4.1 to solve maxmin (minmax) design problems with concave utility functions (convex cost functions). We use the following online optimization setting, defined in [Zin03] and modified in [FKM05]: There is an unknown collection $\mathcal{C}$ of concave utility functions over a convex feasible region F. The optimization proceeds in rounds. In round t, the algorithm has to choose a vector $\mathbf{w}_t \in F$, and then the adversary will provide a utility function $c_t \in \mathcal{C}$. The algorithm will suffer a cost of $c_t(\mathbf{w}_t)$. Zinkevich [Zin03] considered the case when the function c_t is revealed, while Flaxman et.al [FKM05] considered the bandit setting: only the value $\mathbf{c}_t(\mathbf{w}_t)$ is revealed. The regret of the algorithm after T rounds is defined as

$$R_T := \max_{\mathbf{w} \in F} \sum_{t=1}^{T} c_t(\mathbf{w}) - \sum_{t=1}^{T} c_t(\mathbf{w}_t)$$

Flaxman et.al. [FKM05] provide an algorithm called Bandit Gradient Descent (BGD), with the following guarantee:

Theorem (Flaxman et al.). If all the functions c_t defined on a set S are bounded in an interval $[-C, C]$, the regret of the BGD algorithm is

$$R_T \leq 6nCT^{5/6} \tag{5}$$

where n is the dimension of the convex set.

For our application of this setting, we will take $\mathcal{C}$ to be the collection of functions f_S of the instance $(E, \mathcal{S}, B)$ of the design problem. Suppose there exists a poly-time algorithm A which given a weight vector $\mathbf{w}$ returns a solution S such that $f_S(\mathbf{w}) \leq \alpha \min_{T \in \mathcal{S}} f_T(\mathbf{w})$. For our application, we will choose the adversary to play $c_t = f_{S_t}$ where S_t is the solution returned by the algorithm A on input $\mathbf{w}_t$. We shall also assume that all functions f_S are bounded by a polynomial $C(n)$, where n is the number of elements. The BGD algorithm also requires that the convex set S has a membership oracle. For our application, the convex set will be the n-simplex corresponding to the weight distribution over the n elements. We get the following theorem whose proof is similar to the previous subsection.

Theorem 5. *Suppose we are given an approximation algorithm for the concave minimization problem Π, then we can obtain an α-approximation algorithm for the maxmin problem $D(\Pi)$.*

Example: Designing graphs to minimize commute time and cover time. As an application of the framework for convex functions, we show how to design the transition probabilities on a graph to minimize the maximum commute time on a graph.

Suppose we are given a budget B on the total weight and we have to assign weight on each edge. These weights determine the transition probabilities of a random walk: the probability of moving from a vertex u to a vertex v is $p_{uv} = \frac{w_{(uv)}}{\sum_{e \sim u} w_e}$. The goal is to place weights in such a manner so that the maximum commute time among all pairs of vertices is minimized. We note that in a related result, Boyd et.al [BDX04] investigate a similar problem of assigning transition probabilities to the edges of a graph such that the *mixing time* is minimized.

We note that the commute time can be found in polynomial time (see for example [MR95]). It is also known that the commute time is a convex function of the edge weights (see [EKPS04], also [GBS06]) [3]. Thus this problem falls in the framework and thus we can apply the BGD algorithm to obtain the minmax commute time. Moreover, the Matthews bound states that the cover time is within $\log n$ of the maximum commute time. Thus we have

Theorem 6. *Given a graph and a budget B on the total edge weights, one can find (upto additive error) a weight distribution on the edges so that the maximum commute time between two vertices is minimized over all possible weight distributions. Moreover, the same distribution also gives a $\log n$ approximation to the minimum cover time over all possible distributions.*

5   The Complexity of Design Problems

In this section we study the relationship of the complexity of design problems and the complexity of the corresponding optimization problems.

The main result of this paper as described in Sections 3 and 4 is that solving a design problem $D(\Pi)$ is as easy as solving the corresponding optimization problem Π, for the class of concave (convex) minimization (maximization) problems (upto arbitrarily small additive errors). This is proved via two different general techniques to give Theorem 2 and Theorem 5.

A natural question is if the converse also holds, i.e. whether the complexity of the optimization and design version of a problem are the same. The following shows that this is not the case:

Theorem 7. *There exists an additive minimization problem Π such that finding the value of the minimum is **NP**-complete, but its design version $D(\Pi)$ can be solved in polynomial time.*

Proof. Call a graph a bridged clique if it consists of two cliques K_1 and K_2, and two edges $(u, u'), (v, v')$ with $u, v \in K_1$ and $u', v' \in K_2$. Consider the problem of finding (the value of) the cheapest tour on a weighted bridged clique. This problem is **NP**-hard as it involves finding the cheapest hamiltonian paths between u, v and u', v' respectively. Now consider the design version of the problem. We

---

[3] Note the problem is a *minmax* problem and hence we require convex objective functions.

have to find a distribution of the weight budget on a bridged clique so that the cost of the minimum weight tour is maximized. Since any tour will have to pick both edges of the bridge, the optimal strategy is to divide the weights only on the bridge edges. Thus the design version of this problem can be solved trivally in polynomial time. This construction extends to any **NP**-complete problem.

We have seen that all design problems are as easy as their optimization versions, and that some are polynomial time sovable even though the optimization versions are **NP**-hard. To complete the picture we show below that not all design problems are easy:

Theorem 8. *There exists an* **NP***-complete additive minimization problem such that the corresponding design problem is also* **NP***-complete.*

Proof. Consider the problem of finding the minimum weight Steiner tree in a weighted graph. We prove in Section 6 (Theorem 9) that the value of the maxmin Steiner tree is exactly the reciprocal of the maximum number of Steiner trees that can be fractionally packed in the weighted graph. However, the fractional packing number of Steiner trees is known to be **NP**-hard, as proved by Jain et al. [JMS03].

We mention here a related result of Fortnow et al. [FIKU05], in which they study the complexity of solving a succinctly represented zero-sum game. Our setting is different in that the number of row strategies is part of the input size, and we have access to an (approximately) best-response oracle.

6   MaxMin Design Problems and Packing Problems

For lack of space we defer most of the definitions and proofs of this section to the full version of the paper, while providing a sketch of the main results.

6.1   Fractional Packing and Maxmin Design

Consider an instance $(E, \mathcal{S}, 1)$ of a design problem with a budget of 1. A collection of sets $S_1, S_2, \cdots, S_k \in \mathcal{S}$ are said to pack fractionally with weights $\lambda_1, \cdots, \lambda_k$, if for each element e, $\sum_{S_i : e \in S_i} \lambda_{S_i} \leq 1$. The value of the packing is $\sum \lambda_i$.

Theorem 9. *Given a set of elements E and a collection of subsets $\mathcal{S}$ of E, the maximum number of sets that can be packed fractionally is exactly equal to the reciprocal of the maxmin design of the additive instance $(E, \mathcal{S}, 1)$.*

Proof. (Sketch) The LPs for fractional packing and maxmin design are duals of each other upto taking reciprocals.

6.2   Packing Steiner Trees Fractionally

In this subsection we look at the special case of Steiner trees. Given a graph $G(V, E)$ with a set of required nodes R and Steiner nodes $S = V \setminus R$, a Steiner tree is a subtree of G containing all the nodes in R. Let τ denote the set of all Steiner trees in G. Let k_f denote the maximum number of Steiner trees that can be packed fractionally. Thus

$$k_f = \max\{\sum_{T \in \tau} \lambda_T \quad \text{s.t.} \quad \forall e \in E : \sum_{T : e \in T} \lambda_T \le 1\}$$

We shall call this the fractional packing number for Steiner trees. In this section, we use the LP framework developed in Section 3.2 to relate the fractional packing number of Steiner trees to a quantity called the strength of a graph via the well-known bidirected LP relaxation for minimum weight Steiner tree.

Given a partition P of vertices with a required vertex in each partition, the strength of a partition $\gamma(P)$ is defined as the ratio of the number of cross-edges and the size of the partition minus 1. The strength of a graph, γ is defined as the minimum over all partitions. The bidirected-cut relaxation is an LP-relaxation for the minimum Steiner tree problem (see e.g. [Vaz00]). Evaluating the integrality gap α of this relaxation is a major open problem and currently it is known that $8/7 \le \alpha \le 2$. We prove the following result.

Theorem 10. *Fractional Packing number of Steiner trees is within 2α of the strength, that is, $\frac{\gamma}{2\alpha} \le k_f \le \gamma$.*

The proof proceeds by proving that the *maxmin Steiner tree* is within 2α of the reciprocal of the strength and by Theorem 9 we are done. This is proved by giving feasible solutions to the LP relaxations obtained from the bidirected-cut relaxation, as in Section 3.2.

For the special case of spanning trees, we prove a stronger result.

Theorem 11. *The fractional packing number of spanning tree is exactly the strength of the graph.*

The proof uses the same techniques as the last proof and the fact that the spanning tree can be found via a greedy algorithm. All these proofs can be found in the full version of the paper and have been omitted here for sake of brevity.

Remark. We note that Jain et.al [JMS03] proved that evaluating the fractional packing number is **NP**-hard, and an α-approximation to the minimum Steiner tree problem implies existence of an α-approximation to the fractional packing problem. We note that they do not show any relation to the strength of the graph while Theorem 10 wishes to investigate the relationship with strength. Also, Theorem 11 can be directly inferred from the Nash-Williams and Tutte theorems [NW61, Tut61], but our proof techniques do not use these theorems.

We find it an interesting question as to whether the relationship between packing and maxmin design problems can be used to produce other intersting packing theorems in other combinatorial settings. Regarding the Steiner tree setting, it follows from a conjecture of Kriesell [Kri03], that the maximum number of Steiner trees that can be packed *integrally* is within 2 times the strength of the graph. Recently, Lap Chi Lau [Lau04] has proved this within a factor of 26. Improving this factor and settling Kriesell's conjecture seems to be a challenging problem.

References

[BB05]   Francisco Barahona and Mourad Baiou. A linear programming approach to increasing the weight of all minimum spanning trees. *INFORMS*, 2005.

[BDX04]   S. Boyd, P. Diaconis, and L. Xiao. The fastest mixing markov chain on a graph. *SIAM Review*, 2004.

[EKPS04]   J. Elson, R. Karp, C. Papadimitriou, and S. Shenker. Global synchronization in sensornets. *LATIN*, 2004.

[FIKU05]   L. Fortnow, R. Impagliazzo, V. Kabanets, and C. Umans. On the complexity of succinct zero-sum games. *IEEE Conference on Computational Complexity*, pages 323–332, 2005.

[FKM05]   Abraham Flaxman, Adam Tauman Kalai, and H. Brendan McMahan. Online convex optimization in the bandit setting: gradient descent without a gradient. In *SODA*, 2005.

[FS99]   Y. Freund and R. Schapire. Adaptive game playing using multiplicative weights. *Games and Economic Behavior*, 29:79–103, 1999.

[FSO99]   G. Frederickson and R. Solis-Oba. Increasing the weight of minimum spanning trees. *J. Algorithms*, 1999.

[GBS06]   A. Ghosh, S. Boyd, and A. Saberi. Minimizing effective resistance of a graph. *Manuscript*, 2006.

[Jö3]   A. Jüttner. On budgeted optimization problems. *Proc. 3rd Hungarian-Japanese Symposium on Discrete Mathematics and Its Applications*, pages 194–203, 2003.

[JMS03]   K. Jain, M. Mahdian, and M. Salavatipour. Packing steiner trees. In *SODA*, pages 266–274, 2003.

[Kri03]   M. Kriesell. Edge-disjoint trees containing some given vertices in a graph. *J. Comb. Theory, Ser. B 88(1)*, pages 53–65, 2003.

[Lau04]   L.C. Lau. An approximate max-steiner-tree-packing min-steiner-cut theorem. In *FOCS*, pages 61–70, 2004.

[MR95]   R. Motwani and P. Raghavan. **Randomized Algorithms**. Cambridge University Press, 1995.

[NW61]   C. St. J. A. Nash-Williams. Edge disjoint spanning trees of finite graphs. *J. Lond. Math. Soc.*, 1961.

[Tut61]   W. T. Tutte. On the problem of decomposing a graph into n connected factors. *J. Lond. Math. Soc.*, 1961.

[Vaz00]   Vijay V. Vazirani. **Approximation Algorithms**. Springer, 2000.

[Zin03]   M. Zinkevich. Online convex programming and generalized infinitesimal gradient ascent. In *ICML*, pages 928–936, 2003.

On the Complexity of 2D Discrete
Fixed Point Problem
(Extended Abstract)

Xi Chen[1] and Xiaotie Deng[2,*]

[1] Department of Computer Science, Tsinghua University
`xichen00@mails.tsinghua.edu.cn`
[2] Department of Computer Science, City University of Hong Kong
`deng@cs.cityu.edu.hk`

Abstract. While the 3-dimensional analogue of Sperner's problem in the plane was known to be complete in class **PPAD**, the complexity of **2D-SPERNER** itself is not known to be **PPAD**-complete or not. In this paper, we settle this open problem proposed by Papadimitriou [9] fifteen years ago. The result also allows us to derive the computational complexity characterization of a discrete version of the 2-dimensional Brouwer fixed point problem, improving a recent result of Daskalakis, Goldberg and Papadimitriou [4]. Those hardness results for the simplest version of those problems provide very useful tools to the study of other important problems in the **PPAD** class.

1   Introduction

The classical lemma of Sperner [11], which is the combinatorial characterization behind Brouwer's fixed point theorem, states that any admissible 3-coloring of any triangulation of a triangle has a trichromatic triangle. Naturally, it defines a search problem **2D-SPERNER** of finding such a triangle in an admissible 3-coloring for an exponential size triangulation, typical of problems in **PPAD**, a complexity class introduced by Papadimitriou to characterize mathematical structures with the path-following proof technique [10]. Many important problems, such as the Brouwer fixed point, the search versions of Smith's theorem, as well as the Borsuk-Ulam theorem, belong to this class [10]. The computational complexity issue for those problems is of interest only when the search space is exponential in the input parameter.

For problem **2D-SPERNER** as an example, with an input parameter n, we consider a right angled triangle with a side length $N = 2^n$. Its triangulation is into right angled triangles of side length one. There is a (3-coloring) function which, given any vertex in the triangulation, outputs its color in the coloring. The color function is guaranteed to be admissible and is given by a polynomial-time Turing machine. The problem is to find a triangle that has all three colors. Its 3-dimensional analogue **3D-SPERNER** is the first natural problem proved

---

* Work supported by an SRG grant (No. 7001838) of City University of Hong Kong.

M. Bugliesi et al. (Eds.): ICALP 2006, Part I, LNCS 4051, pp. 489–500, 2006.

to be **PPAD**-complete [10]. Whether the 2-dimensional case is complete or not was left as an open problem. Since then, progress has been made toward the solution of this problem: In [7], Grigni defined a non-oriented version of **3D-SPERNER** and proved that it is **PPA**-complete. Friedl, Ivanyos, Santha and Verhoevenproved showed [5,6] that the locally 2-dimensional case of Sperner's problem is complete in **PPAD**. Despite those efforts, the original 2-dimensional Sperner's problem remains elusive.

In this article, we prove that **2D-SPERNER** is **PPAD**-complete and thus settle the open problem proposed by Papadimitriou [9] fifteen years ago. Furthermore, this result also allows us to derive the **PPAD**-completeness proof of a discrete version of the 2D fixed point problem (**2D-BROUWER**). Our study is motivated by the complexity results in [1] and [8] for finding a discrete Brouwer fixed point in d-dimensional space with a function oracle. The combinatorial structure there is similar to the one here. It was proved that, for any $d \geq 2$, the fixed point problem for the oracle model unconditionally requires an exponential number (in consistency with d) of queries. Although the computational models in these two problems are different, we moved into the direction of a hardness proof expecting that the complexity hierarchy in Sperner's problem may have a similar structure with respect to the dimension.

The class **PPAD** is the set of problems that are polynomial-time reducible to the problem called **LEAFD** [10]. It considers a directed graph of an exponential number, in the input parameter n, of vertices, numbered from 0 to $N - 1$ where $N = 2^n$. Each vertex has at most one incoming edge and at most one outgoing edge. There is a distinguished vertex, 0, which has no incoming edge and has one outgoing edge. The required output is another vertex for which the sum of its incoming degree and outgoing degree is one. To access the directed graph, we have a polynomial-time Turing machine which, given any vertex as an input, outputs its possible predecessor and successor. In examination into the **PPAD**-completeness proof of problem **3D-SPERNER**, we found that the main idea is to embed complete graphs in 3-dimensional search spaces [10]. Such an embedding, obviously impossible in the plane, would allow us to transform any Turing machine which generates a directed graph in **LEAFD** to a Turing machine which produces an admissible coloring on a 3-dimensional search space of **3D-SPERNER**.

We take a different approach for the proof which can be clearly divided into two steps. First, we define a new search problem called **RLEAFD** (restricted-**LEAFD**). While the input graph has the same property as those in problem **LEAFD** (that is, both the incoming degree and outgoing degree of every vertex are at most one), it is guaranteed to be a sub-graph of some predefined planar grid graph. The interesting result obtained is that, even with such a strong restriction, the problem is still complete in **PPAD**. In the second step, we reduced **RLEAFD** to **2D-SPERNER** and proved that the latter is also complete. The main idea represents an improved understanding of **PPAD** reductions and may be of general applicability in related problems.

The completeness result of **2D-SPERNER** allows us to deduce that a discrete version of the two dimensional Brouwer fixed point search problem is also **PPAD**-complete. The discrete version considers a function g on a 2D grid such that, for every point **p** in the grid, $g(\mathbf{p})$ is equal to **p** plus an incremental vector with only three possible values: $(1,0)$, $(0,1)$ and $(-1,-1)$. A fixed point is a set of four corners of an orthogonal unit square such that incremental vectors at those point include all the three possibilities, an analogue to that of the three dimensional case introduced in [4]. Such a definition of a fixed point, which is different from the original Brouwer fixed point but is related to Sperner's lemma, has a natural connection with approximation [8], and is consistent in spirit with the recent algorithmic studies on discrete fixed points [1]. On a first careful look at the new definition, its natural link to the Sperner's fully colored triangle is only in one direction. We overcome the difficulty in the other direction to show the reduction is indeed complete.

The **PPAD**-completeness of both **2D-SPERNER** and **2D-BROUWER**, in their simplicities, can serve better benchmarks as well as provide the much needed intuition to derive completeness proofs for complicated problems, such as in the subsequent result of non-approximability (and also smoothed complexity) of the bimatrix game Nash Equilibrium [3]. In particular, an important key lemma in the non-approximability result is a **PPAD**-completeness proof of a discrete fixed point problem on high-dimensional hypergrids with a constant side length, which can be most conveniently derived from our hardness result on the 2D discrete fixed point problem.

2   Preliminaries

2.1   TFNP and PPAD

Definition 1 (TFNP). *Let $R \subset \{0,1\}^* \times \{0,1\}^*$ be a polynomial-time computable, polynomially balanced relation (that is, there exists a polynomial $p(n)$ such that for every pair $(x,y) \in R$, $|y| \leq p(|x|)$). The **NP** search problem Q_R specified by R is this: given an input $x \in \{0,1\}^*$, return a string $y \in \{0,1\}^*$ such that $(x,y) \in R$, if such a y exists, and return the string "no" otherwise.*

*An **NP** search problem Q_R is said to be total if for every $x \in \{0,1\}^*$, there exists a $y \in \{0,1\}^*$ such that $(x,y) \in R$. We use **TFNP** to denote the class of total **NP** search problems.*

An **NP** search problem $Q_{R_1} \in$ **TFNP** is *polynomial-time reducible* to problem $Q_{R_2} \in$ **TFNP** if there exists a pair of polynomial-time computable functions (f,g) such that, for every input x of Q_{R_1}, if y satisfies $(f(x),y) \in R_2$, then $(x,g(y)) \in R_1$. We now define a total **NP** search problem called **LEAFD** [10].

Definition 2 (LEAFD). *The input of the problem is a pair $(M, 0^k)$ where M is the description of a polynomial-time Turing machine which satisfies: **1**). for any $v \in \{0,1\}^k$, $M(v)$ is an ordered pair (u_1, u_2) where $u_1, u_2 \in \{0,1\}^k \cup \{no\}$; **2**). $M(0^k) = \{no, 1^k\}$ and the first component of $M(1^k)$ is 0^k. M generates a*

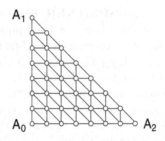

Fig. 1. The standard 7×7 triangulation of a triangle

directed graph $G = (V, E)$ where $V = \{0, 1\}^k$. An edge uv appears in E iff v is the second component of $M(u)$ and u is the first component of $M(v)$.

The output is a directed leaf (with in-degree + out-degree = 1) other than 0^k.

PPAD [9] is the set of total **NP** search problems that are polynomial-time reducible to **LEAFD**. From its definition, **LEAFD** is complete for **PPAD**.

2.2   2D-SPERNER

One of the most interesting problems in **PPAD** is **2D-SPERNER** whose totality is based on Sperner's Lemma [11]: any admissible 3-coloring of any triangulation of a triangle has a trichromatic triangle.

In problem **2D-SPERNER**, we consider the standard $n \times n$ triangulation of a triangle which is illustrated in Figure 1. Every vertex in the triangulation corresponds to a point in $\mathbb{Z}^2$. Here $A_0 = (0, 0)$, $A_1 = (0, n)$ and $A_2 = (n, 0)$ are the three vertices of the original triangle. The vertex set T_n of the $n \times n$ triangulation is defined as $T_n = \{\, \mathbf{p} \in \mathbb{Z}^2 \mid p_1 \geq 0, \ p_2 \geq 0, \ p_1 + p_2 \leq n \,\}$. A 3-coloring of the $n \times n$ triangulation is a function f from T_n to $\{0, 1, 2\}$. It is said to be admissible if **1)** $f(A_i) = i$, for all $0 \leq i \leq 2$; **2)** for every point $\mathbf{p}$ on segment $A_i A_j$, $f(\mathbf{p}) \neq 3 - i - j$.

A unit size well-oriented triangle is a triple $\Delta = (\mathbf{p}^0, \mathbf{p}^1, \mathbf{p}^2)$ where $\mathbf{p}^i \in \mathbb{Z}^d$ for all $0 \leq i \leq 2$. It satisfies either $\mathbf{p}^1 = \mathbf{p}^0 + \mathbf{e}_1$, $\mathbf{p}^2 = \mathbf{p}^0 + \mathbf{e}_2$ or $\mathbf{p}^1 = \mathbf{p}^0 - \mathbf{e}_1$, $\mathbf{p}^1 = \mathbf{p}^0 - \mathbf{e}_2$. In other words, the triangle has a northwest oriented hypotenuse. We use S to denote the set of all such triangles.

From Sperner's Lemma, we define problem **2D-SPERNER** as follows.

Definition 3 (2D-SPERNER [9]). *The input instance is a pair $(F, 0^k)$ where F is a polynomial-time Turing machine which produces an admissible 3-coloring f on T_{2^k}. Here $f(\mathbf{p}) = F(\mathbf{p}) \in \{0, 1, 2\}$, for every vertex $\mathbf{p} \in T_{2^k}$.*

The output is a trichromatic triangle $\Delta \in S$ of coloring f.

In [9], it was shown that **2D-SPERNER** is in **PPAD**. They also defined a 3-dimensional analogue **3D-SPERNER** of **2D-SPERNER** and proved that it is **PPAD**-complete. The completeness of the 2-dimensional case was left as an open problem.

3   Definition of Search Problem RLEAFD

Before the definition of problem **RLEAFD**, we describe a class of planar grid graphs $\{G_i\}_{i\geq 1}$, where $G_n = (V_n, E_n)$ and vertex set

$$V_n = \left\{\, \mathbf{u} \in \mathbb{Z}^2 \;\middle|\; 0 \leq u_1 \leq 3(n^2 - 2), 0 \leq u_2 \leq 3(2n - 1) \,\right\}.$$

Informally speaking, G_n is a planar embedding of the complete graph K_n with vertex set $\{0, 1 \ldots n - 1\}$. For every $0 \leq i < n$, vertex i of K_n corresponds to the vertex $(0, 6i)$ of G_n. For every edge $ij \in K_n$, we define a path E_{ij} from vertex $(0, 6i)$ to $(0, 6j)$. To obtain the edge set E_n of G_n, we start from an empty graph $(V_n, \emptyset)$, and then add all the paths E_{ij}. There are $O(n^2)$ vertices in V_n, which are at the intersection of two paths added previously. Since K_n is not a planar graph when $n \geq 5$, there is no embedding which can avoid those crossing points. For each of those crossing points, we add four more edges into E_n.

We define E_n formally as follows. E_n can be divided into two parts: E_n^1 and E_n^2 such that $E_n = E_n^1 \cup E_n^2$ and $E_n^1 \cap E_n^2 = \emptyset$. The first part $E_n^1 = \cup_{ij \in K_n} E_{ij}$ and path E_{ij} is defined as follows.

Definition 4. *Let $\mathbf{p}^1, \mathbf{p}^2 \in \mathbb{Z}^2$ be two points with the same x-coordinate or the same y-coordinate. Let $\mathbf{u}^1, \mathbf{u}^2 \ldots \mathbf{u}^m \in \mathbb{Z}^2$ be all the integral points on segment $\mathbf{p}^1\mathbf{p}^2$ which are labeled along the direction of $\mathbf{p}^1\mathbf{p}^2$. We use $E(\mathbf{p}^1\mathbf{p}^2)$ to denote the path which consists of $m - 1$ directed edges: $\mathbf{u}^1\mathbf{u}^2$, $\mathbf{u}^2\mathbf{u}^3$, $\ldots$ $\mathbf{u}^{m-1}\mathbf{u}^m$.*

Definition 5. *For every edge $ij \in K_n$ where $0 \leq i \neq j < n$, we define a path E_{ij} as $E(\mathbf{p}^1\mathbf{p}^2) \cup E(\mathbf{p}^2\mathbf{p}^3) \cup E(\mathbf{p}^3\mathbf{p}^4) \cup E(\mathbf{p}^4\mathbf{p}^5)$, where $\mathbf{p}^1 = (0, 6i)$, $\mathbf{p}^2 = (3(ni + j), 6i)$, $\mathbf{p}^3 = (3(ni + j), 6j + 3)$, $\mathbf{p}^4 = (0, 6j + 3)$ and $\mathbf{p}^5 = (0, 6j)$.*

One can show that, every vertex in V_n has at most 4 edges (including both incoming and outgoing edges) in E_n^1. Moreover, if $\mathbf{u}$ has 4 edges, then $3 \mid u_1$ and $3 \mid u_2$. We now use $\{\mathbf{u}^i\}_{1 \leq i \leq 8}$ to denote the eight vertices around $\mathbf{u}$. For each $1 \leq i \leq 8$, $\mathbf{u}^i = \mathbf{u} + \mathbf{x}^i$ where $\mathbf{x}^1 = (-1, 1)$, $\mathbf{x}^2 = (0, 1)$, $\mathbf{x}^3 = (1, 1)$, $\mathbf{x}^4 = (1, 0)$, $\mathbf{x}^5 = (1, -1)$, $\mathbf{x}^6 = (0, -1)$, $\mathbf{x}^7 = (-1, -1)$ and $\mathbf{x}^8 = (-1, 0)$. If $\mathbf{u} \in V_n$ has 4 edges in E_n^1, then it must satisfy the following two properties:

1. either edges $\mathbf{u}^4\mathbf{u}, \mathbf{u}\mathbf{u}^8 \in E_n^1$ or $\mathbf{u}^8\mathbf{u}, \mathbf{u}\mathbf{u}^4 \in E_n^1$;
2. either edges $\mathbf{u}^2\mathbf{u}, \mathbf{u}\mathbf{u}^6 \in E_n^1$ or $\mathbf{u}^6\mathbf{u}, \mathbf{u}\mathbf{u}^2 \in E_n^1$.

Now for every vertex $\mathbf{u} \in V_n$ which has 4 edges in E_n^1, we add four more edges into E_n. For example, if $\mathbf{u}^4\mathbf{u}, \mathbf{u}\mathbf{u}^8, \mathbf{u}^2\mathbf{u}, \mathbf{u}\mathbf{u}^6 \in E_n^1$ (that is, the last case in Figure 2), then $\mathbf{u}^4\mathbf{u}^5, \mathbf{u}^5\mathbf{u}^6, \mathbf{u}^2\mathbf{u}^1, \mathbf{u}^1\mathbf{u}^8 \in E_n^2$. All the four possible cases are summarized in Figure 2.

An example (graph G_3) is showed in Figure 3. We can draw it in two steps. In the first step, for each $ij \in K_3$, we add path E_{ij} into the empty graph. In the second step, we search for vertices of degree four. For each of them, 4 edges are added according to Figure 2. One can prove the following property of G_n.

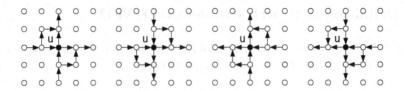

Fig. 2. Summary of cases in the construction of E_n^2

Lemma 1. *Every vertex in G_n has at most 4 edges. There is a polynomial-time Turing machine M^* such that, for every input instance $(n, \mathbf{u})$ where $\mathbf{u} \in V_n$, it outputs all the predecessors and successors of vertex $\mathbf{u}$ in graph G_n.*

We use C_n to denote the set of graphs $G = (V_n, E)$ such that $E \subset E_n$ and for every $\mathbf{u} \in V_n$, both of its in-degree and out-degree are no more than one.

The new problem **RLEAFD** is similar to **LEAFD**. The only difference is that, in **RLEAFD**, the directed graph G generated by the input pair $(K, 0^k)$ always belongs to C_{2^k}. By Lemma 1, one can prove that **RLEAFD** $\in$ **PPAD**.

Definition 6 (RLEAFD). *The input instance is a pair $(K, 0^k)$ where K is the description of a polynomial-time Turing machine which satisfies: **1)**. for every vertex $\mathbf{u} \in V_{2^k}$, $K(\mathbf{u})$ is an ordered pair $(\mathbf{u}^1, \mathbf{u}^2)$ where $\mathbf{u}^1, \mathbf{u}^2 \in V_{2^k} \cup \{no\}$; **2)**. $K(0,0) = (no, (1,0))$ and the first component of $K(1,0)$ is $(0,0)$. K generates a directed graph $G = (V_{2^k}, E) \in C_{2^k}$. An edge $\mathbf{uv}$ appears in E iff $\mathbf{v}$ is the second component of $K(\mathbf{u})$, $\mathbf{u}$ is the first component of $K(\mathbf{v})$ and edge $\mathbf{uv} \in E_{2^k}$.*

The output of the problem is a directed leaf other than $(0,0)$.

4   RLEAFD Is PPAD-Complete

In this section, we will describe a polynomial-time reduction from **LEAFD** to **RLEAFD** and prove that **RLEAFD** is also complete in **PPAD**.

Let G be a directed graph with vertex set $\{0, 1...n - 1\}$ which satisfies that the in-degree and out-degree of every vertex are at most one. We now build the graph $C(G) \in C_n$ in two steps. An important observation here is that $C(G)$ is not a planar embedding of G, as the structure of G is mutated dramatically in $C(G)$. However, it preserves the leaf nodes of G and does not create any new leaf node. Graph $C(G)$ is constructed as follows.

1. Starting from an empty graph $(V_n, \emptyset)$, for every $ij \in G$, add path E_{ij};
2. For every $\mathbf{u} \in V_n$ of degree 4, remove all the four edges which have $\mathbf{u}$ as an endpoint and add four edges around $\mathbf{u}$ using Figure 2.

One can check that, for each vertex in graph $C(G)$, both of its in-degree and out-degree are no more than one, and thus, we have $C(G) \in C_n$. For example, Figure 4 shows $C(G)$ where $G = (\{0, 1, 2\}, \{02, 21\})$. The following lemma is easy to check.

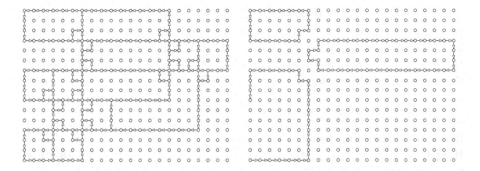

Fig. 3. The planar grid graph G_3 **Fig. 4.** Graph $C(G) \in C_3$

Lemma 2. *Let G be a directed graph with vertex set $\{0, ...n-1\}$ which satisfies that the in-degree and out-degree of every vertex are at most one. For every vertex $0 \leq k \leq n-1$ of G, it is a directed leaf of G iff $\mathbf{u} = (0, 6k) \in V_n$ is a directed leaf of $C(G)$. On the other hand, if $\mathbf{u} \in V_n$ is a directed leaf of $C(G)$, then $u_1 = 0$ and $6 \mid u_2$.*

Lemma 3. *Search problem* **RLEAFD** *is* **PPAD**-*complete.*

Proof. Let $(M, 0^k)$ be an input instance of search problem **LEAFD**, and G be the directed graph specified by M. We can construct a TM K which satisfies the conditions in Definition 6. The construction is described in the full version [2]. It's tedious, but not hard to check that, pair $(K, 0^k)$, as an input of **RLEAFD**, generates graph $C(G) \in C_{2^k}$. On the other hand, Lemma 2 shows that, given a directed leaf of $C(G)$, we can locate a directed leaf of G easily.

5   2D-SPERNER Is PPAD-Complete

In this section, we will present a polynomial-time reduction from **RLEAFD** to **2D-SPERNER** and finish the completeness proof of **2D-SPERNER**.

Let $(K, 0^k)$ be an input instance of **RLEAFD** and $G \in C_{2^k}$ be the directed graph generated by K. We will build a polynomial-time Turing machine F that defines an admissible 3-coloring on $T_{2^{2k+5}}$. Given a trichromatic triangle $\Delta \in S$, a directed leaf of G can be found easily. To clarify the presentation here, we use $\mathbf{u}$, $\mathbf{v}$, $\mathbf{w}$ to denote vertices in V_{2^k}, and $\mathbf{p}$, $\mathbf{q}$, $\mathbf{r}$ to denote vertices in $T_{2^{2k+5}}$.

To construct F, we first define a mapping $\mathcal{F}$ from V_{2^k} to $T_{2^{2k+5}}$. Since $G \in C_{2^k}$, its edge set can be uniquely decomposed into a collection of paths and cycles $P_1, P_2, ... P_m$. By using $\mathcal{F}$, every P_i is mapped to a set $I(P_i) \subset T_{2^{2k+5}}$. Only vertices in $I(P_i)$ have color 0 (with several exceptions around A_0). All the other vertices in $T_{2^{2k+5}}$ are colored carefully with either 1 or 2. Let $\Delta \in S$ be a trichromatic triangle of F and $\mathbf{p}$ be the point in Δ with color 0, then the construction of F guarantees that $\mathcal{F}^{-1}(\mathbf{p}^i) \in V_{2^k}$ is a directed leaf of G, which is different from $(0,0)$.

Firstly, the mapping $\mathcal{F}$ from V_{2^k} to $T_{2^{2k+5}}$ is defined as $\mathcal{F}(\mathbf{u}) = \mathbf{p}$ where $p_1 = 3u_1 + 3$ and $p_2 = 3u_2 + 3$. For each $\mathbf{uv} \in E_{2^k}$, we use $I(\mathbf{uv})$ to denote the set of four vertices in $T_{2^{2k+5}}$ which lie on the segment between $\mathcal{F}(\mathbf{u})$ and $\mathcal{F}(\mathbf{v})$. Let $P = \mathbf{u}^1 ... \mathbf{u}^t$ be a simple path or cycle in G_{2^k} where $t > 1$ (if P is a cycle, then $\mathbf{u}^1 = \mathbf{u}^t$), then we define $I(P) = \cup_{i=1}^{t-1} I(\mathbf{u}^i \mathbf{u}^{i+1})$ and $O(P) \subset T_{2^{2k+5}}$ as

$$O(P) = \left\{ \mathbf{p} \in T_{2^{2k+5}} \text{ and } \mathbf{p} \notin I(P) \mid \exists\, \mathbf{p}' \in I(P),\ \|\mathbf{p} - \mathbf{p}'\|_\infty = 1 \right\}.$$

If P is a simple path, then we decompose $O(P)$ into $\{\mathbf{s}_P, \mathbf{e}_P\} \cup L(P) \cup R(P)$. Here $\mathbf{s}_P = \mathcal{F}(\mathbf{u}^1) + (\mathbf{u}^1 - \mathbf{u}^2)$ and $\mathbf{e}_P = \mathcal{F}(\mathbf{u}^t) + (\mathbf{u}^t - \mathbf{u}^{t-1})$. Starting from $\mathbf{s}_P$, we enumerate vertices in $O(P)$ clockwise as $\mathbf{s}_P, \mathbf{q}^1 ... \mathbf{q}^{n_1}, \mathbf{e}_P, \mathbf{r}^1 ... \mathbf{r}^{n_2}$, then

$$L(P) = \left\{ \mathbf{q}^1, \mathbf{q}^2 ... \mathbf{q}^{n_1} \right\} \quad \text{and} \quad R(P) = \left\{ \mathbf{r}^1, \mathbf{r}^2 ... \mathbf{r}^{n_2} \right\}.$$

If P is a simple cycle, then we decompose $O(P)$ into $L(P) \cup R(P)$ where $L(P)$ contains all the vertices on the left side of the cycle and $R(P)$ contains all the vertices on the right side of the cycle.

As the graph G specified by $(K, 0^k)$ belongs to C_{2^k}, we can uniquely decompose its edge set into $P_1, ... P_m$. For every $1 \leq i \leq m$, P_i is either a maximal path (that is, no path in G contains P_i), or a cycle in graph G. For both cases, the length of P_i is at least 1. One can prove the following two lemmas.

Lemma 4. *For every $1 \leq i \neq j \leq m$, $(I(P_i) \cup O(P_i)) \cap (I(P_j) \cup O(P_j)) = \emptyset$.*

Lemma 5. *Let $(K, 0^k)$ be an input instance of problem **RLEAFD** and $G \in C_{2^k}$ be the directed graph specified, we can construct a polynomial-time TM M_K in polynomial time. Given any vertex $\mathbf{p} \in T_{2^{2k+5}}$, it outputs an integer t: $0 \leq t \leq 5$. Let the unique decomposition of graph G be $P_1, P_2 ... P_m$, then: if $\exists\, i$, $\mathbf{p} \in I(P_i)$, then $t = 1$; if $\exists\, i$, $\mathbf{p} \in L(P_i)$, then $t = 2$; if $\exists\, i$, $\mathbf{p} \in R(P_i)$, then $t = 3$; if $\exists\, i$, $\mathbf{p} = \mathbf{s}_{P_i}$, then $t = 4$; if $\exists\, i$, $\mathbf{p} = \mathbf{e}_{P_i}$, then $t = 5$; otherwise, $t = 0$.*

Turing machine F is described by the algorithm in Figure 5. For example, let $G \in C_2$ be the directed graph generated by pair $(K, 0^1)$, which is illustrated in Figure 6, then Figure 7 shows the 3-coloring F on T_{128}. As T_{128} contains so many vertices, not all of them are drawn in Figure 7. For every omitted vertex $\mathbf{p} \in T_{128}$, if $p_1 = 0$, then $F(\mathbf{p}) = 1$, otherwise, $F(\mathbf{p}) = 2$.

One can prove the following two properties of TM F: **1)**. the 3-coloring f specified by F is admissible; **2)**. let $\Delta \in S$ be a trichromatic triangle and $\mathbf{p}$ be the vertex in Δ with color 0, then $\mathbf{u} = \mathcal{F}^{-1}(\mathbf{p})$ is a directed leaf of G, which is different from $(0,0)$. By these two properties, we get the following theorem.

Theorem 1. *Search problem **2D-SPERNER** is **PPAD**-complete.*

6   2D-BROUWER Is PPAD-Complete

Recently, Daskalakis, Goldberg and Papadimitriou [4] proved that the problem of computing Nash equilibria in games with four players is **PPAD**-complete. In

---

Turing Machine F with input $\mathbf{p} = (p_1, p_2) \in T_{2^{2k+5}}$

1: **if** $p_1 = 0$ **then**
2:     case $p_2 \leq 3$, output 0 ; case $p_2 > 3$, output 1
3: **else if** $p_1 = 1$ **then**
4:     case $p_2 = 3$, output 0 ; case $p_2 = 4$, output 1 ; otherwise, output 2
5: **else if** $p_1 = 2$ and $p_2 = 3$ **then**
6:     output 0
7: let $t = M_K(\mathbf{p})$. case $t = 1$, output 0 ; case $t = 2$, output 1 ; otherwise, output 2

Fig. 5. Behavior of Turing Machine F

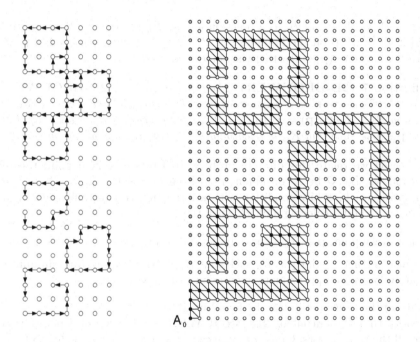

Fig. 6. Graph G_2 and $G \in C_2$ **Fig. 7.** F: black $-$ 0, gray $-$ 1, white $-$ 2

the proof, they define a 3-dimensional Brouwer fixed point problem and proved it is **PPAD**-complete. By reducing it to 4-NASH, they show that the latter one is also complete in **PPAD**.

In this section, we first define a new problem **2D-BROUWER** which is a 2-dimensional analogue of the 3-dimensional problem in [4]. By reducing **2D-SPERNER** to **2D-BROUWER**, we prove the latter is **PPAD**-complete.

For every $n > 1$, we let

$$B_n = \{\, \mathbf{p} = (p_1, p_2) \in \mathbb{Z}^2 \;\big|\; 0 \leq p_1 < n - 1 \text{ and } 0 \leq p_2 < n - 1 \,\}.$$

The boundary of B_n is the set of points $\mathbf{p} \in B_n$ with $p_i \in \{0, n-1\}$ for some $i \in \{1, 2\}$. For every $\mathbf{p} \in \mathbb{Z}^2$, we let $K_{\mathbf{p}} = \{\, \mathbf{q} \in \mathbb{Z}^2 \mid q_i = p_i \text{ or } p_i + 1, \, \forall \, i \in$

Turing Machine F' with input $\mathbf{p} = (p_1, p_2) \in B_{3n}$

1: let $p_1 = 3l + i$ and $p_2 = 3k + j$, where $0 \le i, j \le 2$
2: if $(i, j) = (0, 0), (1, 0)$ or $(0, 1)$ then
3:     $F'(\mathbf{p}) = F(\mathbf{q})$ where $q_1 = l$ and $q_2 = k$
4: else if $(i, j) = (1, 1), (2, 0)$ or $(2, 1)$ then
5:     $F'(\mathbf{p}) = F(\mathbf{q})$ where $q_1 = l + 1$ and $q_2 = k$
6: else [when $j = 2$]
7:     $F'(\mathbf{p}) = F(\mathbf{q})$ where $q_1 = l$ and $q_2 = k + 1$

Fig. 8. The construction of Turing machine F'

$\{1, 2\}$ }. A 3-coloring of B_n is a function g from B_n to $\{0, 1, 2\}$. It is said to be valid if for every $\mathbf{p}$ on the boundary of B_n: if $p_2 = 0$, then $g(\mathbf{p}) = 2$; if $p_2 \ne 0$ and $p_1 = 0$, then $g(\mathbf{p}) = 0$; otherwise, $g(\mathbf{p}) = 1$.

Definition 7 (2D-BROUWER). *The input instance of* **2D-BROUWER** *is a pair* $(F, 0^k)$ *where* F *is a polynomial-time TM which produces a valid 3-coloring* g *on* B_{2^k}. *Here* $g(\mathbf{p}) = F(\mathbf{p}) \in \{0, 1, 2\}$ *for every* $\mathbf{p} \in B_{2^k}$. *The output is a point* $\mathbf{p} \in B_{2^k}$ *such that* $K_{\mathbf{p}}$ *is trichromatic, that is,* $K_{\mathbf{p}}$ *has all the three colors.*

The reason we relate this discrete problem to Brouwer's fixed point theorem is as follows. Let $\mathcal{G}$ be a continuous map from $[0, n-1] \times [0, n-1]$ to itself. If $\mathcal{G}$ satisfies a Lipschitz condition with a large enough constant, then we can construct a valid 3-coloring g on B_n such that:

1. For every point $\mathbf{p} \in B_n$, $g(\mathbf{p})$ only depends on $\mathcal{G}(\mathbf{p})$;

2. Once getting a point $\mathbf{p} \in B_n$ such that $K_{\mathbf{p}}$ is trichromatic, one can immediately locate an approximate fixed point of map $\mathcal{G}$.

Details of the construction can be found in [1].

Notice that the output of **2D-BROUWER** is a set $K_{\mathbf{p}}$ of 4 points which have all the three colors. Of course, one can pick three vertices in $K_{\mathbf{p}}$ to form a trichromatic triangle Δ, but it's possible that $\Delta \notin S$. Recall that every triangle in S has a northwest oriented hypotenuse. In other words, the hypotenuse of the trichromatic triangle in $K_{\mathbf{p}}$ might be northeast oriented. As a result, **2D-BROUWER** could be easier than **2D-SPERNER**.

Motivated by the discussion above, we define a problem **2D-BROUWER*** whose output is similar to **2D-SPERNER**. One can reduce **2D-SPERNER** to **2D-BROUWER*** easily and prove the latter is complete in **PPAD**.

Definition 8 (2D-BROUWER*). *The input instance is a pair* $(F, 0^k)$ *where* F *is a polynomial-time Turing machine which generates a valid 3-coloring* g *on* B_{2^k}. *Here* $g(\mathbf{p}) = F(\mathbf{p}) \in \{0, 1, 2\}$ *for every* $\mathbf{p} \in B_{2^k}$.
The output is a trichromatic triangle $\Delta \in S$ *which has all the three colors.*

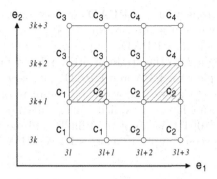

Fig. 9. F': $c_1 = F(l, k)$, $c_2 = F(l+1, k)$, $c_3 = F(l, k+1)$ and $c_4 = F(l+1, k+1)$

We now give a reduction from **2D-BROUWER*** to **2D-BROUWER**.

Let $(F, 0^k)$ be an input pair of problem **2D-BROUWER***, and $n = 2^k$. In Figure 8, we describe a new Turing machine F' which generates a 3-coloring on B_{3n}. For integers $0 \leq l, k < n$, Figure 9 shows the 3-coloring produced by F' on $\{3l, 3l+1, 3l+2, 3l+3\} \times \{3k, 3k+1, 3k+2, 3k+3\} \subset B_{3n}$. Clearly, F' is also a polynomial-time TM, which can be computed from F in polynomial time. Besides, F' generates a valid 3-coloring on B_{3n}. We prove that, for every $\mathbf{p} \in B_{3n}$ such that set $K_{\mathbf{p}}$ is trichromatic in F', one can recover a trichromatic triangle $\Delta \in S$ in F easily.

Let $p_1 = 3l + i$ and $p_2 = 3k + j$, where $0 \leq i, j \leq 2$. By examining Figure 9, we know that either $(i, j) = (0, 1)$ or $(i, j) = (2, 1)$. Furthermore,

1. if $(i, j) = (0, 1)$, then $\Delta = (\mathbf{p}^0, \mathbf{p}^1, \mathbf{p}^2) \in S$ is a trichromatic triangle in F, where $\mathbf{p}^0 = (k, l)$, $\mathbf{p}^1 = \mathbf{p}^0 + \mathbf{e}_1$ and $\mathbf{p}^2 = \mathbf{p}^0 + \mathbf{e}_2$;

2. if $(i, j) = (2, 1)$, then $\Delta = (\mathbf{p}^0, \mathbf{p}^1, \mathbf{p}^2) \in S$ is a trichromatic triangle in F, where $\mathbf{p}^0 = (k+1, l+1)$, $\mathbf{p}^1 = \mathbf{p}^0 - \mathbf{e}_1$ and $\mathbf{p}^2 = \mathbf{p}^0 - \mathbf{e}_2$.

Finally, we get an important corollary of Theorem 1.

Theorem 2. *Search problem* **2D-BROUWER** *is* **PPAD***-complete.*

7   Concluding Remarks

All the **PPAD**-completeness proofs of Sperner's problems before rely heavily on embeddings of complete graphs in the standard subdivisions. That is, edges in the complete graph correspond to independent paths which are composed of neighboring triangles or tetrahedrons in the standard subdivision. Such an embedding is obviously impossible in the plane, as complete graphs with order no less than 5 are not planar. We overcome this difficulty by placing a carefully designed gadget (which looks like a switch with two states) at each intersection of two paths. While the structure of the graph is mutated dramatically (e.g. Figure 4), the property of a vertex being a leaf is well maintained.

An important corollary of the **PPAD**-completeness of **2D-SPERNER** is that, the computation of discrete Brouwer fixed points in 2-dimensional spaces (**2D-BROUWER**) is also **PPAD**-complete. Our new proof techniques may provide helpful insight into the study of other related problems: Can we show more problems complete for **PPA** and **PPAD**? For example, is **2D-TUCKER** [10] **PPAD**-complete? Can we find a natural complete problem for either **PPA** or **PPAD** that doesn't have an explicit Turing machine in the input? For example, is **SMITH** [10] **PPA**-complete? Finally and most importantly, what is the relationship between complexity classes **PPA**, **PPAD** and **PPADS**?

References

1. X. Chen and X. Deng. On Algorithms for Discrete and Approximate Brouwer Fixed Points. In *STOC 2005*, pages 323–330.
2. X. Chen and X. Deng. On the Complexity of 2D Discrete Fixed Point Problem. *ECCC, TR06-037*, 2006.
3. X. Chen, X. Deng, and S.-H. Teng. Computing Nash equilibria: approximation and smoothed complexity.. *ECCC, TR06-023*, 2006.
4. C. Daskalakis, P.W. Goldberg, and C.H. Papadimitriou. The Complexity of Computing a Nash Equilibrium. *STOC*, 2006.
5. K. Friedl, G. Ivanyos, M. Santha, and Y. Verhoeven. Locally 2-dimensional Sperner problems complete for the Polynomial Parity Argument classes. *submitted*.
6. K. Friedl, G. Ivanyos, M. Santha, and Y. Verhoeven. On the complexity of Sperner's Lemma. *Research report NI05002-QIS at the Isaac Newton Institute for Mathematical studies*.
7. M. Grigni. A Sperner lemma complete for PPA. *Inform. Process. Lett.*, 77(5-6):255–259, 2001.
8. M.D. Hirsch, C. Papadimitriou and S. Vavasis. Exponential lower bounds for finding Brouwer fixed points. *J.Complexity*, 5:379–416, 1989.
9. C.H. Papadimitriou. On graph-theoretic lemmata and complexity classes. In *In Proceedings 31st Annual Symposium on Foundations of Computer Science*, pages 794–801, 1990.
10. C.H. Papadimitriou. On the complexity of the parity argument and other inefficient proofs of existence. *JCSS*, pages 498–532, 1994.
11. E. Sperner. Neuer Beweis fur die Invarianz der Dimensionszahl und des Gebietes. *Abhandlungen aus dem Mathematischen Seminar Universitat Hamburg*, 6:265–272, 1928.

Routing (Un-) Splittable Flow in Games with Player-Specific Linear Latency Functions[*]

Martin Gairing, Burkhard Monien, and Karsten Tiemann[**]

Faculty of Computer Science, Electrical Engineering and Mathematics,
University of Paderborn, Fürstenallee 11, 33102 Paderborn, Germany
{gairing, bm, tiemann}@uni-paderborn.de

Abstract. In this work we study weighted *network congestion games* with *player-specific* latency functions where selfish *players* wish to route their *traffic* through a shared *network*. We consider both the case of *splittable* and *unsplittable* traffic. Our main findings are as follows:

- For routing games on parallel links with linear latency functions without a constant term we introduce two new potential functions for unsplittable and for splittable traffic respectively. We use these functions to derive results on the convergence to pure Nash equilibria and the computation of equilibria. We also show for several generalizations of these routing games that such potential functions do not exist.
- We prove upper and lower bounds on the price of anarchy for games with linear latency functions. For the case of unsplittable traffic the upper and lower bound are asymptotically tight.

1 Introduction

Motivation and Framework. Large scale communication networks, like e.g. the Internet, often lack a central regulation for several reasons. For instance the size of the network may be too large, or the users may be free to act according to their private interests. Such an environment – where users neither obey some central control instance nor cooperate with each other – can be modeled as a *non-cooperative game*. The concept of *Nash equilibria* [21] has become an important mathematical tool for analyzing non-cooperative games. A Nash equilibrium is a state in which no player can improve his private objective by unilaterally changing his strategy.

For a special class of non-cooperative games, now widely known as *congestion games*, Rosenthal [23] showed the existence of pure Nash equilibria with the help of a certain potential function. In a congestion game, the strategy set of each player is a subset of the power set of given resources and the private cost function of a player is defined as the sum (over the chosen resources) of functions in the number of players sharing this resource. An extension to congestion games in which the players have

---

[*] This work has been partially supported by the DFG-SFB 376 and by the European Union within the 6th Framework Programme under contract 001907 (DELIS). Parts of this work were done, while the second author visited the University of Cyprus and the University of Texas in Dallas.

[**] International Graduate School of Dynamic Intelligent Systems

M. Bugliesi et al. (Eds.): ICALP 2006, Part I, LNCS 4051, pp. 501–512, 2006.

weights and thus different influence on the congestion of the resources are *weighted congestion games*. Weighted congestion games provide us with a general framework for modeling any kind of non-cooperative resource sharing problem. A typical resource sharing problem is that of routing. In a routing game the strategy sets of the players correspond to paths in a network. Routing games where the demand of the players cannot be split among multiple paths are also called *(weighted) network congestion games*.

Another model for selfish routing where traffic flows can be split arbitrarily – the *Wardrop model* – was already studied in the 1950's (see e.g. [4,26]) in the context of road traffic systems. In a *Wardrop equilibrium* each player assigns its traffic in such a way that the latency experienced on all used paths is the same and minimum among all possible paths for the player. The Wardrop model can be understood as a special network congestion game with infinitely many players each carrying a negligible demand.

In order to measure the degradation of social welfare due to the selfish behavior, Koutsoupias and Papadimitriou [17] used a global objective function, usually termed as *social cost*. They defined the *price of anarchy* as the worst-case ratio between the value of social cost in a Nash equilibrium and that of some social optimum. Thus, the price of anarchy measures the extent to which non-cooperation approximates cooperation.

In weighted network congestion games, as well as in the Wardrop model, players have complete information about the system. However in many cases users of a routing network only have incomplete information about the system. Harsanyi [15] defined the Harsanyi transformation that transforms strategic games with incomplete information into *Bayesian* games where the players uncertainty is expressed in a probability distribution. A Bayesian routing game where players have incomplete information about each others traffic was introduced and studied by Gairing et al. [13]. Georgiou et al. [14] introduced a routing game where the players only have incomplete information about the vector that contains all edge latency functions. Each user's uncertainty about the latency functions is modelled with a probability distribution over a set of different possible latency function vectors. Georgiou et al. [14] showed, that such an incomplete information routing game can be transformed into a complete information routing game where the latency functions are *player-specific*. The resulting games with player-specific latency functions were earlier studied by Milchtaich [19]. Monderer [20] showed that games with player-specific latency functions are of particular importance since each game in strategic form is isomorphic to a congestion game with player-specific latency functions. In this paper we study routing games with player-specific latency functions for both splittable and unsplittable traffic.

Related Work. Routing Games: The class of *weighted congestion games* has been extensively studied (see [12] for a survey). Fotakis et al. [10] proved that a pure Nash equilibrium always exists if the latency functions are linear. For non-linear latency functions they showed, that a pure Nash equilibrium might not exist, even if there are only 2 players (this was also observed earlier by Libman and Orda [18]). For the class of weighted congestion games on parallel links a pure Nash equilibrium always exists, if all edge latency functions are non-decreasing. The *price of anarchy* was studied for congestion games with social cost defined as the total latency. For linear latency functions, it is exactly $\frac{5}{2}$ for unweighted [7] and $\frac{3+\sqrt{5}}{2}$ for weighted congestion games [2]. The exact price of anarchy is also known for polynomials with non-negative coefficients [1].

Inspired by the arisen interest in the price of anarchy Roughgarden and Tardos [25] re-investigated the *Wardrop model* and used the total latency as a social cost measure. In this context the price of anarchy was shown to be $\frac{4}{3}$ for linear latency functions [25] and $\Theta(\frac{d}{\ln d})$ for polynomials of degree at most d with non-negative coefficients [24]. If all latency functions are linear and do not include a constant, then every Wardrop equilibrium has optimum social cost [25]. Since a Wardrop equilibrium is a solution to a convex program it can be computed in polynomial time using the ellipsoid method of Khachyan [16]. This results also implies that the total latency is the same for all Wardrop equilibria. There are several papers (see e.g. [6,8,22]) studying games with a finite number of atomic players where each player can split its traffic over the available paths with the objective to minimize its latency. In this setting the price of anarchy is at most $\frac{3}{2}$ for linear latency functions [8].

Routing Games with Player-Specific Latency Functions: Weighted congestion games on parallel links with player-specific latency functions were studied by Milchtaich [19]. For the case of unweighted players and non-decreasing latency functions, Milchtaich showed that such games do in general not posses the finite improvement property but always admit a pure Nash equilibrium. In case of weighted players a pure Nash equilibrium might not exist, even for a game with 3 players and 3 edges (links) [19]. This is a tight result since such games possess the finite best-reply property in case of 2 players and the finite improvement property in case of 2 edges [19]. Georgiou et al. [14] studied the same class of games as Milchtaich but they only allowed linear latency functions without a constant term. They were able to prove upper bounds on the price of anarchy for both social cost defined as the maximum private cost of a player and social cost defined as the sum over the private cost of all players. Furthermore they presented a polynomial time algorithm to compute a pure Nash equilibrium in case of two edges.

Orda et al. [22] studied a splittable flow routing game with certain player-specific latency functions and a finite number of players each minimizing its latency. They showed that there is a unique Nash equilibrium for each game on parallel links. They also described a game on a more complex graph possessing two different Nash equilibria.

Contribution. In this work we generalize weighted network congestion games and the Wardrop model, to accommodate player-specific latency functions. Our main contributions are the definition of new potential functions and the extension of the techniques from [2,7] to prove upper bounds on the price of anarchy also for games with player-specific latency functions. More specifically, we prove:

- For routing games on parallel links with linear latency functions without a constant term we introduce two new potential functions for unsplittable and for splittable traffic respectively.
  - In the case of unsplittable traffic we use our potential function to show that games with unweighted players possess the *finite improvement property*. We also show that games with weighted players do not possess the finite improvement property even if $n = 3$.
  - In the case of splittable traffic we show that our other convex potential function is minimized if and only if the corresponding assignment is an equilibrium. This result implies that an equilibrium can be computed in polynomial time.

We also show for several generalizations of the above games that such potential functions do not exist.

- We prove upper and lower bounds on the price of anarchy for games with linear latency functions. For the case of unsplittable traffic the upper and lower bound are asymptotically tight.

Road Map. In Sect. 2 we define the games we consider. We present our results for unsplittable traffic in Sect. 3 and for splittable traffic in Sect. 4. Due to lack of space we have to omit many proofs.

2   Notation

For all $k \in \mathbb{N}$ denote $[k] = \{1, \ldots, k\}$. For a vector $\mathbf{v} = (v_1, \ldots, v_n)$ let $\mathbf{v}_{-i} = (v_1, \ldots, v_{i-1}, v_{i+1}, \ldots, v_n)$ and $(\mathbf{v}_{-i}, v_i') = (v_1, \ldots, v_{i-1}, v_i', v_{i+1}, \ldots, v_n)$.

Routing with Splittable Traffic. A *Wardrop game with player-specific latency functions* is a tuple $\Upsilon = (n, G, \mathbf{w}, Z, \mathbf{f})$. Here, n is the number of *players* and $G = (V, E)$ is an undirected *(multi)graph*. The vector $\mathbf{w} = (w_1, \ldots, w_n)$ defines for every player $i \in [n]$ its *traffic* $w_i \in \mathbb{R}^+$. For each player $i \in [n]$ the set $Z_i \subset 2^E$ consists of all possible routing paths in $G = (V, E)$ from some node $s_i \in V$ to some other node $t_i \in V$. Denote $Z = Z_1 \times \ldots \times Z_n$. Edge latency functions $\mathbf{f} = (f_{ie})_{i \in [n], e \in E}$ are player-specific and $f_{ie} : \mathbb{R}_0^+ \to \mathbb{R}_0^+$ is the non-negative, non-decreasing and continuous *player-specific latency function* that player $i \in [n]$ assigns to edge $e \in E$. Notice that we are in the setting of the regular Wardrop game if $f_{ie} = f_{ke}$ for all $i, k \in [n]$, $e \in E$. In the majority of cases we consider *linear* player-specific latency functions $f_{ie}(u) = a_{ie} \cdot u + b_{ie}$ with $a_{ie}, b_{ie} \geq 0$. For a Wardrop game Υ with linear latency functions denote $\Delta(\Upsilon) = \max_{e \in E; i, k \in [n]} \{a_{ie}/a_{ke}; a_{ie} < \infty, a_{ke} < \infty\}$ with the understanding that $\frac{0}{0} = 1$ and $\frac{c}{0} = \infty$ if $c > 0$. $\Delta(\Upsilon)$ describes the maximum factor by which the slopes of the player-specific linear latency functions deviate. Note that $\Delta(\Upsilon)$ does not depend on the constants b_{ie} of the latency functions. We will use the term Wardrop game with *player-specific capacities* to denote a game where all latency functions are of the form $f_{ie}(u) = a_{ie} \cdot u, a_{ie} > 0$. In this case, we write $\mathbf{a}$ instead of $\mathbf{f}$ to denote the vector $\mathbf{a} = (a_{ie})_{i \in [n], e \in E}$. We will often consider games on a parallel link multi-graph $G = (V, E)$ that has two nodes $V = \{s, t\}$, $s = s_1 = \ldots s_n$, $t = t_1 = \ldots t_n$, and $|E|$ edges connecting these two nodes.

Strategies and Strategy Profiles. A player $i \in [n]$ can split its traffic w_i over the paths in Z_i. A (pure) *strategy* for player $i \in [n]$ is a tuple $\mathbf{x}_i = (x_{iR_i})_{R_i \in Z_i}$ with $\sum_{R_i \in Z_i} x_{iR_i} = w_i$ and $x_{iR_i} \geq 0$ for all $R_i \in Z_i$. Denote by $\mathcal{X}_i = \{\mathbf{x}_i \mid \mathbf{x}_i$ is a strategy for player $i\}$ the set of all strategies for player i. Note, that $\mathcal{X}_i$ is an infinite, compact and convex set. A *strategy profile* $\mathbf{x} = (\mathbf{x}_1, \ldots, \mathbf{x}_n)$ is an n-tuple of strategies for the players. Define $\mathcal{X} = \mathcal{X}_1 \times \ldots \times \mathcal{X}_n$ as the set of all possible strategy profiles.

Wardrop Equilibria. For a strategy profile $\mathbf{x}$ the load $\delta_e(\mathbf{x})$ on an edge $e \in E$ is given by $\delta_e(\mathbf{x}) = \sum_{i \in [n]} \sum_{R_i \in Z_i, R_i \ni e} x_{iR_i}$. A strategy profile $\mathbf{x}$ is a *Wardrop equilibrium*, if for every player $i \in [n]$, and every $R_i, R_i' \in Z_i$ with $x_{iR_i} > 0$ it holds that

$$\sum_{e \in R_i} f_{ie}(\delta_e(\mathbf{x})) \leq \sum_{e \in R_i'} f_{ie}(\delta_e(\mathbf{x})).$$

Observe that in a Wardrop equilibrium all flow paths of a player have equal latency. We can regard each player $i \in [n]$ as a service provider who has many clients each handling a negligible small amount of traffic. In a Wardrop equilibrium each service provider satisfies all his clients because none of them can improve its experienced latency.

Social Cost and Price of Anarchy. Associated with a game and a strategy profile $\mathbf{x}$ is the *social cost* $\mathsf{SC}(\mathbf{x})$ as a measure of social welfare:

$$\mathsf{SC}(\mathbf{x}) = \sum_{i \in [n]} \sum_{R_i \in Z_i} x_{iR_i} \sum_{e \in R_i} f_{ie}(\delta_e(\mathbf{x})).$$

This social cost is motivated by the interpretation as a game with infinitely many players with negligible demand and models the sum of the players latencies. The *optimum* associated with a game is defined by $\mathsf{OPT} = \min_{\mathbf{x} \in \mathcal{X}} \mathsf{SC}(\mathbf{x})$. The *price of anarchy*, also called *coordination ratio* and denoted PoA, is the maximum value, over all instances and Wardrop equilibria $\mathbf{x}$, of the ratio $\frac{\mathsf{SC}(\mathbf{x})}{\mathsf{OPT}}$.

Routing with Unsplittable Traffic. We also consider the case where players have to assign their traffic integrally to a single path. Denote such a *weighted network congestion game with player-specific latency functions* by $\Gamma = (n, G, \mathbf{w}, Z, \mathbf{f})$. The players are *unweighted* if they are all of traffic 1, i.e. $w_1 = \ldots = w_n = 1$. In this case we write $\mathbf{1}$ instead of $\mathbf{w}$. A pure strategy x_i for player $i \in [n]$ is a tuple $\mathsf{x}_i = (x_{iR_i})_{R_i \in Z_i}$ with $\sum_{R_i \in Z_i} x_{iR_i} = w_i$ and $x_{iR_i} \in \{0, w_i\}$ for all $R_i \in Z_i$. Alternatively, with a slight abuse of notation, we write $\mathbf{R} = (R_1, \ldots, R_n)$ where $R_i \in Z_i, 1 \leq i \leq n$, to denote a strategy profile such that $x_{iR_i} = w_i$ for all $i \in [n]$. In this setting, $Z = Z_1 \times \ldots \times Z_n$ is the set of all pure strategy profiles. We define the *private cost* of player $i \in [n]$ as the sum over the player-specific latencies of all used edges: $\mathsf{PC}_i(\mathbf{R}) = \sum_{e \in R_i} f_{ie}(\delta_e(\mathbf{R}))$.

Given a pure strategy profile $\mathbf{R} = (R_1, \ldots, R_n)$ a *selfish step* of a player $i \in [n]$ is a deviation to strategy profile $(\mathbf{R}_{-i}, R_i')$ where $\mathsf{PC}_i(\mathbf{R}_{-i}, R_i') < \mathsf{PC}_i(\mathbf{R})$ and $R_i' \in Z_i$. Such a selfish step is a *greedy selfish step* if there is for player i no strategy $R_i'' \in Z_i$ such that $\mathsf{PC}_i(\mathbf{R}_{-i}, R_i'') < \mathsf{PC}_i(\mathbf{R}_{-i}, R_i')$.

A game Γ possesses the *finite best-reply property* if any sequence of greedy selfish steps is finite. If even any sequence of selfish steps is finite it possesses in addition the *finite improvement property*. Note, that the finite improvement property implies the finite best-reply property which again implies the existence of a pure Nash equilibrium.

We also consider *mixed strategies* P_i for the players. Then, $\mathsf{P}_i = (p(i, R_i))_{R_i \in Z_i}$ is a probability distribution over Z_i and $p(i, R_i)$ denotes the probability that player i chooses path R_i. A *mixed strategy profile* $\mathbf{P} = (\mathsf{P}_1, \ldots, \mathsf{P}_n)$ is represented by an n tuple of mixed strategies. For a mixed strategy profile $\mathbf{P}$ denote $p(\mathbf{R}) = \prod_{i \in [n]} p(i, R_i)$ as the probability that the players choose the pure strategy profile $\mathbf{R} = (R_1, \ldots, R_n)$.

Nash Equilibrium, Social Cost, and Price of Anarchy. For a mixed strategy profile $\mathbf{P}$ the private cost of player $i \in [n]$ is $\mathsf{PC}_i(\mathbf{P}) = \sum_{\mathbf{R} \in Z} p(\mathbf{R}) \cdot \mathsf{PC}_i(\mathbf{R})$. For a pure strategy profile $\mathbf{R}$ the social cost $\mathsf{SC}(\mathbf{R})$ is defined as before whereas for a mixed strategy profile $\mathbf{P}$ the social cost is given by $\mathsf{SC}(\mathbf{P}) = \sum_{\mathbf{R} \in Z} p(\mathbf{R}) \cdot \mathsf{SC}(\mathbf{R})$. A strategy profile $\mathbf{P}$ is a Nash equilibrium if no player $i \in [n]$ can decrease its private cost PC_i if the other players stick to their strategies. More formally, $\mathbf{P} = (\mathsf{P}_1, \ldots, \mathsf{P}_n)$ is a Nash equilibrium if $\mathsf{PC}_i(\mathbf{P}) \leq \mathsf{PC}_i(\mathbf{P}_{-i}, \mathsf{P}_i')$ for all probability distributions P_i' over Z_i and for all $i \in [n]$. In the unsplittable setting the price of anarchy PoA is the worst-case ratio between the social cost of a mixed Nash equilibrium and that of some social optimum.

3   Results for Unsplittable Traffic

3.1   Unweighted Players: Finite Improvement Property

Milchtaich [19] showed that network congestion games on parallel links with player-specific latency functions and unweighted players do not possess the finite improvement property in general. In Theorem 1 we show that we achieve the finite improvement property if we restrict to player-specific capacities. In Theorem 2 we give counterexamples to show that a slight deviation from this model yields a loss of the finite improvement property. For the positive result in Theorem 1 we define for every strategy profile $\mathbf{R} = (R_1, \ldots, R_n)$ the following potential function:

$$\Phi(\mathbf{R}) = \prod_{i \in [n]} a_{i\,R_i} \cdot \prod_{e \in E} \delta_e(\mathbf{R})!$$

In contrast to all other potential functions we know, Φ does not contain any summation.

Theorem 1. *Every network congestion game on parallel links with unweighted players and player-specific capacities possesses the finite improvement property.*

Proof. Consider a selfish step $\mathbf{R} \to \mathbf{R}'$ of a player $i \in [n]$ from edge $j \in E$ to edge $k \in E$, i.e. $\mathbf{R} = (R_1, \ldots, R_{i-1}, j, R_{i+1}, \ldots, R_n)$ and $\mathbf{R}' = (R_1, \ldots, R_{i-1}, k, R_{i+1}, \ldots, R_n)$. If A denotes the common part of the expressions $\Phi(\mathbf{R})$ and $\Phi(\mathbf{R}')$ they can be written as $\Phi(\mathbf{R}') = A \cdot a_{i\,k} \cdot (\delta_k(\mathbf{R}) + 1)$ and $\Phi(\mathbf{R}) = A \cdot a_{i\,j} \cdot \delta_j(\mathbf{R})$. Since $\mathbf{R} \to \mathbf{R}'$ is a selfish step we have that $\mathsf{PC}_i(\mathbf{R}') = a_{i\,k} \cdot (\delta_k(\mathbf{R}) + 1) < a_{i\,j} \cdot \delta_j(\mathbf{R}) = \mathsf{PC}_i(\mathbf{R})$. Thus $\Phi(\mathbf{R}') = A \cdot \mathsf{PC}_i(\mathbf{R}') < A \cdot \mathsf{PC}_i(\mathbf{R}) = \Phi(\mathbf{R})$. The claim follows since the number of strategy profiles is finite.   □

Theorem 2. *Network congestion games on a graph G with unweighted players and player-specific latency functions do (in general) not possess*

(a) *the finite best-reply property if the game has 3 players, linear latency functions, and G is a parallel links graph.*

(b) *the finite improvement property if the game has 2 players, player-specific capacities, and G is a concatenation of 2 parallel link graphs connected in series.*

(c) *a pure Nash equilibrium if the game has 3 players, player-specific capacities, and all paths in G are of length at most 2.*

3.2   Weighted Players: Finite Improvement Property

For weighted congestion games on parallel links with player-specific capacities Georgiou et al. [14] showed that a Nash equilibrium always exists in the case of 3 players. For arbitrary many players it is an open problem whether such a game still admits a pure Nash equilibrium or not. Theorem 3 implies that the finite improvement property can not be used to solve the open problem even if there are only 3 players. We would like to note that for the case of 2 players we can give a potential function showing that the finite improvement property is fulfilled.

Theorem 3. *There is a weighted congestion game on parallel links with 3 players and player-specific capacities that does not possess the finite improvement property.*

Proof. The 3 players of the game are of traffic $w_1 = 1$, $w_2 = 2$, and $w_3 = 79$. The player-specific capacities of the 11 edges are listed in this table ($\epsilon_1, \epsilon_2, \epsilon_3 > 0$ are small numbers we will discuss later):

j	1	2	3	4	5	6	7	8	9	10	11
a_{1j}	$\frac{3^8}{80} - \epsilon_3$	∞	1	$3 - \epsilon_1$	$(3-\epsilon_1)^2$	$(3-\epsilon_1)^3$	$(3-\epsilon_1)^4$	$(3-\epsilon_1)^5$	$(3-\epsilon_1)^6$	$(3-\epsilon_1)^7$	$(3-\epsilon_1)^8$
a_{2j}	∞	1	$\frac{2}{3} - \epsilon_1$	$\frac{2^2}{3^2} - \epsilon_1$	$\frac{2^3}{3^3} - \epsilon_1$	$\frac{2^4}{3^4} - \epsilon_1$	$\frac{2^5}{3^5} - \epsilon_1$	$\frac{2^6}{3^6} - \epsilon_1$	$\frac{2^7}{3^7} - \epsilon_1$	$\frac{2^8}{3^8} - \epsilon_1$	∞
a_{3j}	1	∞	$\frac{80}{79} - \epsilon_1$	∞	∞	∞	∞	∞	∞	$\frac{80^2}{79^2} \cdot \frac{79}{81} - \epsilon_2$	$\left(\frac{80}{79} - \epsilon_1\right)^2$

Our cycle of selfish steps starts in the initial strategy profile $(3, 2, 1)$. We now perform 8 double-steps (A). A double-step (A) consists of a first step that moves player 2 from an edge j to the edge k player 1 is assigned to and a second step to an empty edge l that player 1 does. Both steps of a double-step (A) are selfish iff $a_{2k}/a_{2j} < \frac{2}{3}$ and $a_{1l}/a_{1k} < 3$. In each step of our 8 double-steps (A) the deviating player moves from edge t to edge $t+1$. After the double-steps (A) the strategy profile $(11, 10, 1)$ is reached. Notice that all 16 steps performed up to now are selfish since $\epsilon_1 > 0$.

The cycle continues with double-steps (B). A double-step (B) starts with a move of player 1 from edge j to the edge k used by player 3 followed by a step of player 3 to an empty edge l. Observe that (B) is a pair of selfish steps iff $a_{1k}/a_{1j} < \frac{1}{80}$ and $a_{3l}/a_{3k} < \frac{80}{79}$. We conduct 2 double-steps (B): $(11,10,1) \to (1,10,1) \to (1,10,3) \to (3,10,3) \to (3,10,11)$. These steps are selfish if:

$$1 \cdot (3 - \epsilon_1)^8 > 80 \cdot \left(\frac{3^8}{80} - \epsilon_3\right) \text{ i.e. } \epsilon_3 > \frac{3^8}{80} - \frac{(3-\epsilon_1)^8}{80} \text{ and} \tag{1}$$

$$1 \cdot \left(\frac{3^8}{80} - \epsilon_3\right) > 80 \cdot 1 \text{ i.e. } \epsilon_3 < \frac{3^8}{80} - 80. \tag{2}$$

Starting from the strategy profile $(3,10,11)$ we proceed with a double-step (C) that moves player 3 to the edge 10 player 2 is assigned to and continues with a step of player 2 to the empty edge 2. This double-step consists of selfish steps iff $a_{3\,10}/a_{3\,11} < \frac{79}{81}$ and $a_{2\,2}/a_{2\,10} < \frac{81}{2}$. The double-step (C) is selfish if:

$$79 \cdot \left(\frac{80}{79} - \epsilon_1\right)^2 > 81 \cdot \left(\frac{80^2 \cdot 79}{79^2 \cdot 81} - \epsilon_2\right) \text{ i.e. } \epsilon_2 > \frac{80^2}{79 \cdot 81} - \frac{79}{81}\left(\frac{80}{79} - \epsilon_1\right)^2 \text{ and} \tag{3}$$

$$81 \cdot \left(\frac{2^8}{3^8} - \epsilon_1\right) > 2 \cdot 1 \text{ i.e. } \epsilon_1 < \frac{2^8}{3^8} - \frac{2}{81}. \tag{4}$$

The 11 double-steps explained up to now are followed by a final step that moves player 3 back to edge 1: $(3,2,10) \to (3,2,1)$. It is selfish if:

$$79 \cdot \left(\frac{80^2}{79^2} \cdot \frac{79}{81} - \epsilon_2\right) > 79 \cdot 1 \text{ i.e. } \epsilon_2 < \frac{80^2}{79 \cdot 81} - 1. \tag{5}$$

It is possible to select $\epsilon_1, \epsilon_2, \epsilon_3 > 0$ fulfilling $(1) - (5)$. Thus the claim follows. $\square$

3.3 General Networks and Linear Latency Functions: Price of Anarchy

In this section we study the price of anarchy for weighted congestion games with linear player-specific latency functions. To prove our upper bound we use similar techniques as Christodoulou and Koutsoupias [7] and Awerbuch et al. [2]. The proof is also based on the following technical lemma.

Lemma 1. *For all* $u, v \in \mathbb{R}_0^+$ *and* $c \in \mathbb{R}^+$ *we have* $v(u + v) \leq c \cdot u^2 + \left(1 + \frac{1}{4c}\right) \cdot v^2$.

Theorem 4. *Let* Γ *be a weighted network congestion game with player-specific linear latency functions. Then,* $\mathrm{PoA} \leq \frac{1}{2} \cdot [\Delta(\Gamma) + 2 + \sqrt{\Delta(\Gamma)(\Delta(\Gamma) + 4)}]$.

Proof. Let $\mathbf{P} = (P_1, \ldots, P_n)$ be a mixed Nash equilibrium and let $\mathbf{Q}$ be a pure strategy profile with optimum social cost. Since $\mathbf{P}$ is a Nash equilibrium, player i cannot improve by switching from strategy P_i to Q_i. Thus,

$$
\mathrm{PC}_i(\mathbf{P}) \leq \mathrm{PC}_i(\mathbf{P}_{-i}, Q_i) = \sum_{\mathbf{R} \in Z} p(\mathbf{R}) \left[\sum_{e \in Q_i \cap R_i} f_{ie}(\delta_e(\mathbf{R})) + \sum_{e \in Q_i \setminus R_i} f_{ie}(\delta_e(\mathbf{R}) + w_i) \right]
$$
$$
\leq \sum_{\mathbf{R} \in Z} p(\mathbf{R}) \sum_{e \in Q_i} f_{ie}(\delta_e(\mathbf{R}) + \delta_e(\mathbf{Q})).
$$

It follows that

$$
\mathrm{SC}(\mathbf{P}) = \sum_{\mathbf{R} \in Z} p(\mathbf{R}) \sum_{i \in [n]} w_i \sum_{e \in R_i} f_{ie}(\delta_e(\mathbf{R})) = \sum_{i \in [n]} w_i \cdot \mathrm{PC}_i(\mathbf{P})
$$
$$
\leq \sum_{i \in [n]} \sum_{\mathbf{R} \in Z} p(\mathbf{R}) \sum_{e \in Q_i} w_i \cdot f_{ie}(\delta_e(\mathbf{R}) + \delta_e(\mathbf{Q}))
$$
$$
= \sum_{\mathbf{R} \in Z} p(\mathbf{R}) \sum_{e \in E} \sum_{i, Q_i \ni e} w_i \cdot [a_{ie}(\delta_e(\mathbf{R}) + \delta_e(\mathbf{Q})) + b_{ie}]
$$
$$
= \sum_{\mathbf{R} \in Z} p(\mathbf{R}) \sum_{\substack{e \in E, \delta_e(\mathbf{Q}) > 0, \\ \delta_e(\mathbf{R}) > 0}} \frac{\sum_{i, Q_i \ni e} a_{ie} w_i}{\delta_e(\mathbf{Q})} \cdot \delta_e(\mathbf{Q}) \cdot (\delta_e(\mathbf{R}) + \delta_e(\mathbf{Q}))
$$
$$
+ \sum_{\mathbf{R} \in Z} p(\mathbf{R}) \sum_{\substack{e \in E, \delta_e(\mathbf{Q}) > 0, \\ \delta_e(\mathbf{R}) = 0}} \delta_e(\mathbf{Q}) \cdot \sum_{i, Q_i \ni e} a_{ie} w_i + \sum_{\mathbf{R} \in Z} p(\mathbf{R}) \sum_{e \in E} \sum_{i, Q_i \ni e} w_i b_{ie}.
$$

By Lemma 1 we get for $c \in \mathbb{R}^+$,

$$
\mathrm{SC}(\mathbf{P}) \leq \sum_{\mathbf{R} \in Z} p(\mathbf{R}) \sum_{\substack{e \in E, \delta_e(\mathbf{Q}) > 0, \\ \delta_e(\mathbf{R}) > 0}} \frac{\sum_{i, Q_i \ni e} a_{ie} w_i}{\delta_e(\mathbf{Q})} \cdot \left[\left(1 + \frac{1}{4c}\right) \cdot \delta_e(\mathbf{Q})^2 + c \cdot \delta_e(\mathbf{R})^2 \right]
$$
$$
+ \sum_{\mathbf{R} \in Z} p(\mathbf{R}) \sum_{\substack{e \in E, \delta_e(\mathbf{Q}) > 0, \\ \delta_e(\mathbf{R}) = 0}} \delta_e(\mathbf{Q}) \cdot \sum_{i, Q_i \ni e} a_{ie} w_i + \sum_{\mathbf{R} \in Z} p(\mathbf{R}) \sum_{e \in E} \sum_{i, Q_i \ni e} w_i b_{ie}
$$
$$
\leq \left(1 + \frac{1}{4c}\right) \sum_{\mathbf{R} \in Z} p(\mathbf{R}) \sum_{e \in E} \left(\sum_{i, Q_i \ni e} a_{ie} w_i \right) \delta_e(\mathbf{Q}) + \sum_{\mathbf{R} \in Z} p(\mathbf{R}) \sum_{e \in E} \sum_{i, Q_i \ni e} w_i \cdot b_{ie}
$$
$$
+ c \cdot \sum_{\mathbf{R} \in Z} p(\mathbf{R}) \sum_{\substack{e \in E, \delta_e(\mathbf{Q}) > 0, \\ \delta_e(\mathbf{R}) > 0}} \frac{\sum_{i, Q_i \ni e} a_{ie} w_i}{\delta_e(\mathbf{Q})} \cdot \delta_e(\mathbf{R})^2
$$

$$\leq \left(1+\frac{1}{4c}\right) \cdot \mathsf{SC}(\mathbf{Q}) + c \cdot \sum_{\mathbf{R} \in Z} p(\mathbf{R}) \sum_{\substack{e \in E, \delta_e(\mathbf{Q})>0, \\ \delta_e(\mathbf{R})>0}} \frac{\sum_{i,Q_i \ni e} a_{ie} w_i}{\delta_e(\mathbf{Q})} \cdot \delta_e(\mathbf{R})^2.$$

Observe that $\frac{1}{\delta_e(\mathbf{Q})} \cdot \sum_{i,Q_i \ni e} a_{ie} w_i$ is a weighted average slope of latency functions for edge $e \in E$. With $\frac{a_{ie}}{a_{ke}} \leq \Delta(\Gamma)$ for all $i,k \in [n]$ with $a_{ie}, a_{ke} < \infty$ it follows that $\frac{1}{\delta_e(\mathbf{Q})} \cdot \sum_{i,Q_i \ni e} a_{ie} w_i \leq \Delta(\Gamma) \cdot \frac{1}{\delta_e(\mathbf{R})} \cdot \sum_{i,R_i \ni e} a_{ie} w_i$. We get,

$$\mathsf{SC}(\mathbf{P}) \leq \left(1+\frac{1}{4c}\right) \cdot \mathsf{SC}(\mathbf{Q}) + c \cdot \sum_{\mathbf{R} \in Z} p(\mathbf{R}) \sum_{\substack{e \in E, \delta_e(\mathbf{Q})>0, \\ \delta_e(\mathbf{R})>0}} \Delta(\Gamma) \cdot \frac{\sum_{i,R_i \ni e} a_{ie} w_i}{\delta_e(\mathbf{R})} \cdot \delta_e(\mathbf{R})^2$$

$$\leq \left(1+\frac{1}{4c}\right) \cdot \mathsf{SC}(\mathbf{Q}) + c \cdot \sum_{\mathbf{R} \in Z} p(\mathbf{R}) \sum_{e \in E} \Delta(\Gamma) \cdot \delta_e(\mathbf{R}) \cdot \sum_{i,R_i \ni e} a_{ie} w_i$$

$$\leq \left(1+\frac{1}{4c}\right) \cdot \mathsf{SC}(\mathbf{Q}) + c \cdot \Delta(\Gamma) \cdot \mathsf{SC}(\mathbf{P}).$$

Thus choosing $c = \frac{-\Delta(\Gamma)+\sqrt{\Delta(\Gamma)(\Delta(\Gamma)+4)}}{4\Delta(\Gamma)}$ yields

$$\frac{\mathsf{SC}(\mathbf{P})}{\mathsf{SC}(\mathbf{Q})} \leq \frac{4c+1}{4c(1-c\Delta(\Gamma))} = \frac{\Delta(\Gamma)+2+\sqrt{\Delta(\Gamma)(\Delta(\Gamma)+4)}}{2}.$$

Since $\mathbf{P}$ is an arbitrary (mixed) Nash equilibrium the claim follows. □

Interestingly, we get with Theorem 4 an upper bound of $\frac{1}{2} \cdot (3 + \sqrt{5})$ in the case of $\Delta(\Gamma) = 1$ which matches the exact price of anarchy for weighted congestion games [2] even though our model still allows for player-specific constants $b_{ie} \neq b_{ke}$. We proceed with a lower bound on the price of anarchy that is asymptotically tight. Variations of the games used in the proof of the lower bound were also used in some recent papers to show lower bounds on the price of anarchy in different settings (see e.g. [3,9,11]).

Theorem 5. *For each $l \in \mathbb{N}$ and for each $\epsilon > 0$ there is a congestion game Γ on parallel links with unweighted players and player-specific capacities that possesses a pure Nash equilibrium $\mathbf{R}$ such that $\Delta(\Gamma) \geq l$ and $\mathsf{SC}(\mathbf{R})/\mathsf{OPT} \geq (1 - \epsilon) \cdot \Delta(\Gamma)$.*

The construction in the proof of Theorem 5 uses a large number of players. However, the price of anarchy is unbounded even for 2 player games.

Theorem 6. *For every $k \geq 1$ there is a weighted congestion game on parallel links with 2 players and player-specific capacities that possesses a pure Nash equilibrium $\mathbf{R}$ such that $\mathsf{SC}(\mathbf{R})/\mathsf{OPT} > k$.*

4 Results for Splittable Traffic

4.1 Parallel Links and Player-Specific Capacities: Existence of and Convergence to a Wardrop Equilibrium

In this section we consider Wardrop games on parallel links with player-specific capacities. For such a game and a strategy profile $\mathbf{x}$ define the following function:

$$\Psi(\mathbf{x}) = \sum_{i \in [n]} \sum_{e \in E} x_{ie} \cdot \ln(a_{ie}) + \sum_{\substack{e \in E, \\ \delta_e(\mathbf{x})>0}} \delta_e(\mathbf{x}) \cdot \ln(\delta_e(\mathbf{x})).$$

Note, that $e^{\Psi(\mathbf{x})}$ has a similar form as the potential function Φ in Sect. 3. The next theorem shows that Ψ plays a similar role as the potential function Φ.

Theorem 7. *Let Υ be a Wardrop game on parallel links with player-specific capacities. Moreover let $\mathbf{x}$ be a strategy profile for Υ so that there exists a player $k \in [n]$, two edges $p, q \in E$, and some Λ, $0 < \Lambda \le x_{kp}$ such that: $a_{kp} \cdot (\delta_p(\mathbf{x}) - \Lambda) \ge a_{kq} \cdot (\delta_q(\mathbf{x}) + \Lambda)$. Define a new strategy profile $\mathbf{y}$ by:*

$$y_{ij} = \begin{cases} x_{kp} - \Lambda & \text{if } i = k, j = p, \\ x_{kq} + \Lambda & \text{if } i = k, j = q, \\ x_{ij} & \text{otherwise.} \end{cases}$$

Then $\Psi(\mathbf{y}) < \Psi(\mathbf{x})$.

We now show that $\Psi(\mathbf{x})$ is minimized iff $\mathbf{x}$ is a Wardrop equilibrium.

Theorem 8. *Let Υ be a Wardrop game on parallel links with player-specific capacities. Moreover let $\mathbf{y}$ be a strategy profile for Υ. Then the following two conditions are equivalent:*

(a) $\Psi(\mathbf{y}) = \min_{\mathbf{x} \in \mathcal{X}} \Psi(\mathbf{x})$,
(b) $\mathbf{y}$ is a Wardrop equilibrium.

Proof. (a) $\Rightarrow$ (b) follows immediately with Theorem 7. It is possible to show (b) $\Rightarrow$ (a) with an argumentation based on the Karush-Kuhn-Tucker theorem (see [5]).   □

Since Ψ is a convex function it follows with Theorem 8 that the ellipsoid method of Khachyan [16] can be used to compute a Wardrop equilibrium in time polynomial in the size of the instance and the number of bits of precision required.

4.2 Does There Exist a Convex Potential Function for a More General Setting?

If a game can be described by a convex potential function then the set of Nash equilibria forms a convex set. In this section we show that no such convex function exists for general graphs with player-specific capacities (Theorem 9) whereas the existence remains an open problem for parallel links with strictly increasing player-specific latency functions (Theorem 10).

Theorem 9. *There is a Wardrop game Υ with player-specific capacities that possesses two Wardrop equilibria $\mathbf{x}$ and $\mathbf{y}$ where*

(a) $\delta_j(\mathbf{x}) \ne \delta_j(\mathbf{y})$ for an edge $j \in E$ and $SC(\mathbf{x}) \ne SC(\mathbf{y})$,
(b) the set of Wardrop equilibria for Υ does not form a convex set.

Theorem 10. *Let Υ be a Wardrop game on parallel links with strictly increasing player-specific latency functions. Let $\mathbf{x}$ and $\mathbf{y}$ be Wardrop equilibria for Υ. Then,*

(a) $\delta_j(\mathbf{x}) = \delta_j(\mathbf{y})$ for all $j \in E$ and $SC(\mathbf{x}) = SC(\mathbf{y})$,
(b) the set of Wardrop equilibria for Υ forms a convex set.

4.3 General Networks and Player-Specific Latency Functions: Existence of Wardrop Equilibria

Each Wardrop game possesses a Wardrop equilibrium (see [4]). It is possible to use Brouwer's fixed point theorem to prove the existence of equilibria for our more general class of games.

Theorem 11. *Every Wardrop game Υ with strictly increasing player-specific latency functions possesses a Wardrop equilibrium.*

4.4 General Networks and Linear Latency Functions: Price of Anarchy

In this section we give bounds on the price of anarchy. The proof of the upper bound uses the same technique as the proof of Theorem 4.

Theorem 12. *Let Υ be a Wardrop game with player-specific linear latency functions. Then,*

$$\text{PoA} \leq \begin{cases} \frac{4}{4-\Delta(\Upsilon)} & \text{if } \Delta(\Upsilon) \leq 2, \\ \Delta(\Upsilon) & \text{otherwise.} \end{cases}$$

Theorem 13. *For each $n \in \mathbb{N}$ there is a Wardrop game Υ on 2 parallel links with n unweighted players and player-specific capacities that possesses a Wardrop equilibrium $\mathbf{x}$ such that $\Delta(\Upsilon) = n^2$ and $\text{SC}(\mathbf{x})/\text{OPT} \geq \frac{1}{4} \cdot \sqrt{\Delta(\Upsilon)}$.*

For the Wardrop model with linear latency functions, Roughgarden and Tardos [25] showed that the price of anarchy is exactly $\frac{4}{3}$. Theorem 12 with $\Delta(\Upsilon) = 1$ implies that the price of anarchy does not change even if the linear latency functions of the players have player-specific constants $b_{ie} \neq b_{ke}$. Although our upper bound is tight for $\Delta(\Upsilon) = 1$ there is for large $\Delta(\Upsilon)$ still a gap between the upper bound of $\Delta(\Upsilon)$ and the lower bound.

Acknowledgment. We would like to thank Chryssis Georgiou, Marios Mavronicolas, and Thomas Sauerwald for many fruitful discussions and helpful comments.

References

1. S. Aland, D. Dumrauf, M. Gairing, B. Monien, and F. Schoppmann. Exact Price of Anarchy for Polynomial Congestion Games. In *Proc. of the 23rd International Symposium on Theoretical Aspects of Computer Science*, LNCS Vol. 3884, Springer Verlag, pages 218–229, 2006.
2. B. Awerbuch, Y. Azar, and A. Epstein. The Price of Routing Unsplittable Flow. In *Proc. of the 37th ACM Symposium on Theory of Computing*, pages 57–66, 2005.
3. B. Awerbuch, Y. Azar, Y. Richter, and D. Tsur. Tradeoffs in Worst-Case Equilibria. In *Proc. of the 1st International Workshop on Approximation and Online Algorithms*, LNCS Vol. 2909, Springer Verlag, pages 41–52, 2003.
4. M. Beckmann, C. B. McGuire, and C. B. Winsten. *Studies in the Economics of Transportation.* Yale University Press, 1956.
5. S. Boyd and L. Vandenberghe. *Convex Optimization.* Cambridge University Press, 2004.

6. S. Catoni and S. Pallottino. Traffic Equilibrium Paradoxes. *Transportation Science*, 25(3):240–244, 1991.
7. G. Christodoulou and E. Koutsoupias. The Price of Anarchy of Finite Congestion Games. In *Proc. of the 37th ACM Symposium on Theory of Computing*, pages 67–73, 2005.
8. R. Cominetti, J. R. Correa, and N. E. Stier-Moses. Network Games With Atomic Players. In *Proc. of the 33rd International Colloquium on Automata, Languages, and Programming*, to appear 2006.
9. A. Czumaj and B. Vöcking. Tight Bounds for Worst-Case Equilibria. In *Proc. of the 13th ACM-SIAM Symposium on Discrete Algorithms*, pages 413–420, 2002. Also accepted to *Journal of Algorithms* as Special Issue of SODA'02.
10. D. Fotakis, S. Kontogiannis, and P. Spirakis. Selfish Unsplittable Flows. In *Proc. of the 31st International Colloquium on Automata, Languages, and Programming*, LNCS Vol. 3142, Springer Verlag, pages 593–605, 2004.
11. M. Gairing, T. Lücking, M. Mavronicolas, and B. Monien. Computing Nash Equilibria for Scheduling on Restricted Parallel Links. In *Proc. of the 36th ACM Symposium on Theory of Computing*, pages 613–622, 2004.
12. M. Gairing, T. Lücking, B. Monien, and K. Tiemann. Nash Equilibria, the Price of Anarchy and the Fully Mixed Nash Equilibrium Conjecture. In *Proc. of the 32nd International Colloquium on Automata, Languages, and Programming*, LNCS Vol. 3580, Springer Verlag, pages 51–65, 2005.
13. M. Gairing, B. Monien, and K. Tiemann. Selfish Routing with Incomplete Information. In *Proc. of the 17th ACM Symposium on Parallel Algorithms and Architectures*, pages 203–212, 2005.
14. C. Georgiou, T. Pavlides, and A. Philippou. Network Uncertainty in Selfish Routing. In *Proc. of the 20th IEEE International Parallel & Distributed Processing Symposium*, 2006.
15. J. C. Harsanyi. Games with Incomplete Information Played by Bayesian Players, I, II, III. *Management Science*, 14:159–182, 320–332, 468–502, 1967.
16. L. G. Khachiyan. A Polynomial Time Algorithm in Linear Programming. *Soviet Mathematics Doklady*, 20(1):191–194, 1979.
17. E. Koutsoupias and C. H. Papadimitriou. Worst-Case Equilibria. In *Proc. of the 16th International Symposium on Theoretical Aspects of Computer Science*, LNCS Vol. 1563, Springer Verlag, pages 404–413, 1999.
18. L. Libman and A. Orda. Atomic Resource Sharing in Noncooperative Networks. *Telecommunication Systems*, 17(4):385–409, 2001.
19. I. Milchtaich. Congestion Games with Player-Specific Payoff Functions. *Games and Economic Behavior*, 13(1):111–124, 1996.
20. D. Monderer. *Multipotential Games*. Unpublished manuscript, available at http://ie.technion.ac.il/ dov/multipotential_games.pdf, 2005.
21. J. F. Nash. Non-Cooperative Games. *Annals of Mathematics*, 54(2):286–295, 1951.
22. A. Orda, R. Rom, and N. Shimkin. Competitive Routing in Multiuser Communication Networks. *IEEE/ACM Transactions on Networking*, 1(5):510–521, 1993.
23. R. W. Rosenthal. A Class of Games Possessing Pure-Strategy Nash Equilibria. *International Journal of Game Theory*, 2:65–67, 1973.
24. T. Roughgarden. *Selfish Routing and the Price of Anarchy*. MIT Press, 2005.
25. T. Roughgarden and É. Tardos. How Bad Is Selfish Routing? *Journal of the ACM*, 49(2):236–259, 2002.
26. J. G. Wardrop. Some Theoretical Aspects of Road Traffic Research. In *Proc. of the Institute of Civil Engineers, Pt. II, Vol. 1*, pages 325–378, 1952.

The Game World Is Flat:
The Complexity of Nash Equilibria in Succinct Games

Constantinos Daskalakis*, Alex Fabrikant**, and Christos H. Papadimitriou***

UC Berkeley, Computer Science Division
costis@cs.berkeley.edu
alexf@cs.berkeley.edu
christos@cs.berkeley.edu

Abstract. A recent sequence of results established that computing Nash equilibria in normal form games is a PPAD-complete problem even in the case of two players [11,6,4]. By extending these techniques we prove a general theorem, showing that, for a far more general class of families of succinctly representable multiplayer games, the Nash equilibrium problem can also be reduced to the two-player case. In view of empirically successful algorithms available for this problem, this is in essence a positive result — even though, due to the complexity of the reductions, it is of no immediate practical significance. We further extend this conclusion to extensive form games and network congestion games, two classes which do not fall into the same succinct representation framework, and for which no positive algorithmic result had been known.

1 Introduction

Nash proved in 1951 that every game has a mixed Nash equilibrium [15]. However, the complexity of the computational problem of finding such an equilibrium had remained open for more than half century, attacked with increased intensity over the past decades. This question was resolved recently, when it was established that the problem is PPAD-complete [6] (the appropriate complexity level, defined in [18]) and thus presumably intractable, for the case of 4 players; this was subsequently improved to three players [5,3] and, most remarkably, two players [4].

In particular, the combined results of [11,6,4] establish that the general Nash equilibrium problem for normal form games (the standard and most explicit representation) and for graphical agames (an important succinct representation, see the next paragraph) can all be reduced to 2-player games. 2-player games in turn can be solved by several techniques such as the Lemke-Howson algorithm [14,20], a simplex-like technique that is known empirically to behave well even though exponential counterexamples do exist [19]. *In this paper we extend these results to essentially all known kinds of succinct representations of games, as well as to more sophisticated concepts of equilibrium.*

Besides this significant increase in our understanding of complexity issues, computational considerations also led to much interest in *succinct representations of games.*

---

* Supported by NSF ITR Grant CCR-0121555.
** Supported by the Fannie and John Hertz Foundation.
*** Supported by NSF ITR CCR-0121555 grant and a Microsoft Research grant.

M. Bugliesi et al. (Eds.): ICALP 2006, Part I, LNCS 4051, pp. 513–524, 2006.

Computer scientists became interested in games because they help model networks and auctions; thus we should mainly focus on games with many players. However, multi-player games in normal form require in order to be described an amount of data that is exponential in the number of players. When the number of players is large, the resulting computational problems are hardly legitimate, and complexity issues are hopelessly distorted. This has led the community to consider broad classes of *succinctly representable games*, some of which had been studied by traditional game theory for decades, while others (like the graphical games [13]) were invented by computer scientists spurred by the motivations outline above. (We formally define succinct games in the next section, but also deal in this paper with two cases, network congestion games and extensive form games, that do not fit within this definition).

The first general positive algorithmic result for succinct games was obtained only recently [17]: a polynomial-time algorithm for finding a correlated equilibrium (an important generalization of the Nash equilibrium due to Aumann [1]). The main result in [17] states that a family of succinct games has a polynomial-time algorithm for correlated equilibria provided that there is a polynomial time oracle which, given a strategy profile, computes the expected utility of each player.

In this paper, using completely different techniques inspired from [11], we show a general result (Theorem 2) that is remarkably parallel to that of [17]: The Nash equilibrium problem of a family of succinct games can be reduced to the 2-player case provided that a (slightly constrained) *polynomial-length straight-line arithmetic program* exists which computes, again, the expected utility of a given strategy profile (notice the extra algebraic requirement here, necessitated by the algebraic nature of our techniques). We proceed to point out that *for all major known* families of succinct games such a straight-line program exists (Corollary 1).

We also extend these techniques to two other game classes, Network congestion games [7] and extensive form games, which do not fit into our succinctness framework, because the number of strategies is exponential in the input, and for which the result of [17] does not apply, Theorems 3 and 4, respectively).

2   Definitions and Background

In a *game in normal form* we have $r \geq 2$ players (and for each player $p \leq r$ a finite set S_p of pure strategies. We denote the Cartesian product of the S_p's by S (the set of *pure strategy profiles*) and the Cartesian product of the pure strategy sets of players other than p by S_{-p}. Finally, for each $p \leq r$ and $s \in S$ we have a *payoff* u_s^p.

A *mixed strategy* for player p is a distribution on S_p, that is, $|S_p|$ nonnegative real numbers adding to 1. Call a set of r mixed strategies $x_j^p, p = 1, \ldots, r, j \in S_p$ a *Nash equilibrium* if, for each p, its expected payoff, $\sum_{s \in S} u_s^p \prod_{q=1}^r x_{s_q}^q$ is maximized over all mixed strategies of p. That is, a Nash equilibrium is a set of mixed strategies from which no player has an incentive to deviate. For $s \in S_{-p}$, let $x_s = \prod_{q \neq p} x_{s_q}^q$. It is well-known (see, e.g., [16]) that the following is an equivalent condition for a set of mixed strategies to be a Nash equilibrium:

$$\forall p, j \sum_{s \in S_{-p}} u_{js}^p x_s > \sum_{s \in S_{-p}} u_{j's}^p x_s \implies x_j^p = 0. \tag{1}$$

Also, a set of mixed strategies is an ε-*Nash equilibrium* for some $\varepsilon > 0$ if the following holds:

$$\sum_{s \in S_{-p}} u^p_{js} x_s > \sum_{s \in S_{-p}} u^p_{j's} x_s + \varepsilon \implies x^p_{j'} = 0. \tag{2}$$

We next define the complexity class PPAD. An *FNP search problem* $\mathcal{P}$ is a set of inputs $I_{\mathcal{P}} \subseteq \Sigma^*$ such that for each $x \in I_{\mathcal{P}}$ there is an associated set of solutions $\mathcal{P}_x \subseteq \Sigma^{|x|^k}$ for some integer k, such that for each $x \in I_{\mathcal{P}}$ and $y \in \Sigma^{|x|^k}$ whether $y \in \mathcal{P}_x$ is decidable in polynomial time (notice that this is precisely NP with an added emphasis on finding a witness). For example, r-NASH is the search problem $\mathcal{P}$ in which each $x \in I_{\mathcal{P}}$ is an r-player game in normal form together with a binary integer A (the *accuracy specification*), and $\mathcal{P}_x$ is the set of $\frac{1}{A}$-Nash equilibria of the game.

A search problem is *total* if $\mathcal{P}_x \neq \emptyset$ for all $x \in I_{\mathcal{P}}$. For example, Nash's 1951 theorem [15] implies that r-NASH is total. The set of all total FNP search problems is denoted TFNP. TFNP seems to have no generic complete problem, and so we study its subclasses: PLS [12], PPP, PPA and PPAD [18]. In particular, PPAD is the class of all total search problems reducible to the following:

END OF THE LINE: Given two circuits S and P with n input bits and n output bits, such that $P(0^n) = 0^n \neq S(0^n)$, find an input $x \in \{0,1\}^n$ such that $P(S(x)) \neq x$ or $S(P(x)) \neq x \neq 0^n$.

Intuitively, END OF THE LINE creates a directed graph with vertex set $\{0,1\}^n$ and an edge from x to y whenever $P(y) = x$ and $S(x) = y$ (S and P stand for "successor candidate" and "predecessor candidate"). This graph has indegree and outdegree at most one, and at least one source, namely 0^n, so it must have a sink. We seek either a sink, or a source other than 0^n. Thus, PPAD is the class of all total functions whose totality is proven via the simple combinatorial argument outlined above.

A polynomially computable function f is a *polynomial-time reduction* from total search problem $\mathcal{P}$ to total search problem $\mathcal{Q}$ if, for every input x of $\mathcal{P}$, $f(x)$ is an input of $\mathcal{Q}$, and furthermore there is another polynomially computable function g such that for every $y \in \mathcal{Q}_{f(x)}$, $g(y) \in \mathcal{P}_x$. A search problem $\mathcal{P}$ in PPAD is called *PPAD-complete* if all problems in PPAD reduce to it. Obviously, END OF THE LINE is PPAD-complete; we now know that 2-NASH is PPAD-complete [6,4].

In this paper we are interested in *succinct games*. A succinct game [17] $G = (I, T, U)$ is a set of inputs $I \in$ P, and two polynomial algorithms T and U. For each $z \in I$, $T(z)$ returns a *type*, that is, the number of players $r \leq |z|$ and an r-tuple $(t_1, \ldots, t_r)$ where $|S_p| = t_p$. We say that G is of *polynomial type* if all t_p's are bounded by a polynomial in $|z|$. In this paper we are interested in games of both polynomial (Section 3) and non-polynomial type (Sections 5 and 4). Finally, for any r-tuple of positive integers $s = (s_1, \ldots, s_r)$, where $s_p \leq t_p$, and $p \leq r$, $U(z, p, s)$ returns an integer standing for the utility u^p_s. The game in normal form thus encoded by $z \in I$ is denoted $G(z)$.

Examples of succinct games (due to space constraints we omit the formal definitions, see [17] for more details) are:

- *graphical games* [13], where players are nodes on a graph, and the utility of a player depends only on the strategies of the players in its neighborhood.

- *congestion games* [7], where strategies are sets of *resources*, and the utility of a player is the sum of the delays of the resources in the set it chose, where the delay is a resource-specific function of the number of players who chose this resource.
- *network congestion games*, where the strategies of each player are given implicitly as paths from a source to a sink in a graph; since the number of strategies is potentially exponential, this representation is not of polynomial type; we treat network congestion games in Section 4.
- *multimatrix games* where each player plays a different 2-person game with each other player, and the utilities are added.
- *semi-anonymous games* (a generalization of symmetric games not considered in [17]) in which all players have the same set of strategies, and each player has a utility function that depends solely on the *number* of other players who choose each strategy (and not the identities of these players).
- several other classes such as *local effect games, scheduling games, hypergraphical games, network design games, facility location games,* etc., as catalogued in [17].

Our main result, shown in the next section, implies that the problem finding a Nash equilibrium in all of these classes of games can be reduced to 2-player games (equivalently, belongs to the class PPAD).

Lastly, we define a *bounded (division-free) straight-line program* to be an arithmetic binary circuit with nodes performing addition, subtraction, or multiplication on their inputs, or evaluating to pre-set constants, with the additional constraint that the values of all the nodes remain in $[0, 1]$. This restriction is not severe, as it can be shown that an arithmetic circuit of size n with intermediate nodes bounded in absolute value by $2^{\text{poly}(n)}$ can be transformed in polynomial time to fit the above constraint (with the output scaled down by a factor dependent only on the bound).

3   The Main Result

Given a succinct game, the following problem, called EXPECTED UTILITY, is of interest: Given a mixed strategy profile $x^1, \ldots, x^r$, compute the expected utility of player p. Notice that the result sought is a polynomial in the input variables. It was shown in [17] that a polynomial-time algorithm for EXPECTED UTILITY (for succinct games of polynomial type) implies a polynomial-time algorithm for computing correlated equilibria for the succinct game. Here we show a result of a similar flavor.

3.1   Mapping Succinct Games to Graphical Games

Theorem 1. *If for a succinct game G of polynomial type there is a bounded division-free straight-line program of polynomial length for computing EXPECTED UTILITY, then G can be mapped in polynomial time to a graphical game $\mathcal{G}$ so that there is a polynomially computable surjective mapping from the set of Nash equilibria of $\mathcal{G}$ to the set of Nash equilibria of G.*

Proof. Let G be a succinct game for which there is a bounded straight-line program for computing EXPECTED UTILITY. In time polynomial in $|G|$, we will construct a graphical game $\mathcal{G}$ so that the statement of the theorem holds. Suppose that G has r players, $1, \ldots, r$, with strategy sets $S_p = \{1, \ldots, t_p\}, \forall p \leq r$. The players of game $\mathcal{G}$, which we shall call *nodes* in the following discussion to distinguish them from the players of G, will have two strategies each, strategy 0 and strategy 1. We will interpret the probability with which a node x of $\mathcal{G}$ chooses strategy 1 as a real number in $[0, 1]$, which we will denote, for convenience, by the same symbol x that we use for the node.

Below we describe the nodes of $\mathcal{G}$ as well as the role of every node in the construction. We will describe $\mathcal{G}$ as a directed network with vertices representing the nodes (players) of $\mathcal{G}$ and directed edges denoting directed flow of information as in [11,6].

1. For every player $p = 1, \ldots, r$ of G and for every pure strategy $j \in S_p$, game $\mathcal{G}$ has a node x_j^p. Value x_j^p should be interpreted as the probability with which player p plays strategy j; in fact, we will establish later that, given a Nash equilibrium of $\mathcal{G}$, this interpretation yields a Nash equilibrium of G. As we will see in Item 4 below, our construction will ensure that, at any Nash equilibrium, $\sum_{j=1}^{t_p} x_j^p = 1, \forall p \leq r$. Therefore, it is legitimate to interpret the set of values $\{x_j^p\}_j$ as a mixed strategy for player p in G.

2. For every player $p = 1, \ldots, r$ of G and for every pure strategy $j \in S_p$, game $\mathcal{G}$ has nodes U_j^p and $U_{\leq j}^p$. The construction of $\mathcal{G}$ will ensure that, at a Nash equilibrium, value U_j^p equals the utility of player p for playing pure strategy j if every other player $q \neq p$ plays the mixed strategy specified by the distribution $\{x_j^q\}_j$. Also, the construction will ensure that $U_{\leq j}^p = \max_{j' \leq j} U_{j'}^p$. Without loss of generality, we assume that all utilities in G are scaled down to lie in $[0, 1]$.

3. For every node of type U_j^p there is a set of nodes in $\mathcal{G}$ that simulate the intermediate variables used by the straight-line program computing the expected utility of player p for playing pure strategy j when the other players play according to the mixed strategies specified by $\{\{x_j^q\}_j\}_{q \neq p}$. This is possible due to our constraint on the straight-line program.

4. For every player p of G, there is a set of nodes Ψ_p defining a component $\mathcal{G}_p$ of $\mathcal{G}$ whose purpose is to guarantee the following at any Nash equilibrium of $\mathcal{G}$:
   (a) $\sum_{j=1}^{t_p} x_j^p = 1$
   (b) $U_j^p > U_{j'}^p \implies x_{j'}^p = 0$
   The structure and the functionality of $\mathcal{G}_p$ are described in section 3 of [11], so its details will be omitted here. Note that the nodes of set Ψ_p interact only with the nodes $\{U_j^p\}_j$, $\{U_{\leq j}^p\}_j$ and $\{x_j^p\}_j$. The nodes of types U_j^p and $U_{\leq j}^p$ are not affected by the nodes in Ψ_p and should be interpreted as "input" to $\mathcal{G}_p$, whereas the nodes of type x_j^p are only affected by $\mathcal{G}_p$ and not by the rest of the game and are the "output" of $\mathcal{G}_p$. The construction of $\mathcal{G}_p$ ensures that they satisfy Properties 4a and 4b.

Having borrowed the construction of the components $\mathcal{G}_p, p \leq r$, from [11], the only components of $\mathcal{G}$ that remain to be specified are those that compute expected utilities. With the bound on intermediate variable values, the construction of these components can be easily done using the games $\mathcal{G}_=, \mathcal{G}_\zeta, \mathcal{G}_+, \mathcal{G}_-, \mathcal{G}_*$ for assignment, assignment of a constant ζ, addition, subtraction and multiplication that were defined in [11]. Finally, the components of $\mathcal{G}$ that give values to nodes of type $U_{\leq j}^p$ can be easily constructed using games $\mathcal{G}_{\max}$ from [11]. It remains to argue that, given a Nash equilibrium of $\mathcal{G}$, we can find in polynomial time a Nash equilibrium of G and moreover that this mapping is onto. The first claim follows from the following lemma and the second is easy to verify.

Lemma 1. *At a Nash equilibrium of game $\mathcal{G}$, values $\{\{x_j^p\}_j\}_p$ constitute a Nash equilibrium of game G.*

Proof. From the correctness of games $\mathcal{G}_p, p \leq r$, it follows that, at any Nash equilibrium of game $\mathcal{G}$, $\sum_{j=1}^{t_p} x_j^p = 1, \forall p$. Moreover, from the correctness of games $\mathcal{G}_=, \mathcal{G}_\zeta$, $\mathcal{G}_+, \mathcal{G}_-, \mathcal{G}_*$, it follows that, at any Nash equilibrium of game $\mathcal{G}$, U_j^p will be equal to the utility of player p for playing pure strategy j when every other player $q \neq p$ plays as specified by the values $\{x_j^q\}_j$. From the correctness of $\mathcal{G}_{\max}$ it follows that, at any Nash equilibrium of game $\mathcal{G}$, $U_{\leq j}^p = \max_{j' \leq j} U_{j'}^p, \forall p, j$. Finally, from the correctness of games $\mathcal{G}_p, p \leq r$, it follows that, at any Nash equilibrium of game $\mathcal{G}$, for every $p \leq r$ and for every $j, j' \in S_p, j \neq j' : U_j^p > U_{j'}^p \implies x_{j'}^p = 0$. By combining the above it follows that $\{\{x_j^p\}_j\}_p$ constitute a Nash equilibrium of game G.   □

3.2   Succinct Games in PPAD

We now explore how the mapping described in Theorem 1 can be used in deriving complexity results for the problem of computing a Nash equilibrium in succinct games.

Theorem 2. *If for a succinct game G of polynomial type there is a bounded division-free straight-line program of polynomial length for computing* EXPECTED UTILITY, *then the problem of computing a Nash equilibrium in the succinct game polynomially reduces to the problem of computing a Nash equilibrium of a 2-player game.*

Proof. We will describe a reduction from the problem of computing a Nash equilibrium in a succinct game to the problem of computing a Nash equilibrium in a graphical game. This is sufficient since the latter can be reduced to the problem of computing a Nash equilibrium in a 2-player game [6,4]. Note that the reduction sought does not follow trivially from Theorem 1; the mapping there makes sure that the exact equilibrium points of the graphical game can be efficiently mapped to exact equilibrium points of the succinct game. Here we seek something stronger; we want every approximate Nash equilibrium of the former to be efficiently mapped to an approximate Nash equilibrium of the latter. This requirement turns out to be more delicate than the previous one.

Formally, let G be a succinct game for which there is a straight line program for computing EXPECTED UTILITY and let ε be an accuracy specification. Suppose that G has r players, $1, \ldots, r$, with strategy sets $S_p = \{1, \ldots, t_p\}, \forall p \leq r$. In time polynomial in $|G| + |1/\varepsilon|$, we will specify a graphical game $\mathcal{G}$ and an accuracy ε' with the property that, given an ε'-Nash equilibrium of $\mathcal{G}$, one can recover in polynomial time an ε-Nash equilibrium of G. In our reduction, the graphical game $\mathcal{G}$ will be the same as the one described in the proof of Theorem 1, while the accuracy specification will be of the form $\varepsilon' = \varepsilon/2^{p(n)}$, where $p(n)$ is a polynomial in $n = |G|$ that will be be specified later. Using the same notation for the nodes of game $\mathcal{G}$ as we did in Theorem 1, let us consider if the equivalent of Lemma 1 holds for approximate Nash equilibria.

Observation 1. *For any $\varepsilon' > 0$, there exist ε'-Nash equilibria of game $\mathcal{G}$ in which the values $\{\{x_j^p\}_j\}_p$ **do not** constitute an ε-Nash equilibrium of game G.*

Proof. A careful analysis of the mechanics of gadgets $\mathcal{G}_p$, $p \leq r$, shows that property (2) which is the defining property of an approximate Nash equilibrium is not guaranteed to hold. In fact, there are ε'-equilibria of $\mathcal{G}$ in which $\sum_{s \in S_{-p}} u_{js}^p x_s > \sum_{s \in S_{-p}} u_{j's}^p x_s + \varepsilon'$ for some $p \leq r$, j and j', and, yet, $x_{j'}^p$ is any value in $[0, t_p \cdot \varepsilon']$. The details are omitted.   □

Moreover, the values $\{x_j^p\}_j$ do not necessarily constitute a distribution as specified by the following observation.

Observation 2. *For any $\varepsilon' > 0$, for any $p \leq r$, at an ε'-Nash equilibrium of game $\mathcal{G}$, $\sum_j x_j^p$ is not necessarily equal to 1.*

Proof. Again by carefully analyzing the behavior of gadgets $\mathcal{G}_p$, $p \leq r$, at an ε'-Nash equilibrium of game $\mathcal{G}$, it can be shown that there are equilibria in which $\sum_j x_j^p$ can be any value in $1 \pm 2t_p \varepsilon'$. The details are omitted.   □

Therefore, the extraction of an ε-Nash equilibrium of game G from an ε'-Nash equilibrium of game $\mathcal{G}$ cannot be done by just interpreting the values $\{x_j^p\}$ as the probability distribution of player p. What we show next is that, for the right choice of ε', a *trim and renormalize* strategy succeeds in deriving an ε-Nash equilibrium of game G from an ε'-Nash equilibrium of game $\mathcal{G}$. For any $p \leq r$, suppose that $\{\hat{x}_j^p\}_j$ are the values derived from $\{x_j^p\}_j$ as follows: make all values smaller than $t_p \varepsilon'$ equal to zero (trim) and renormalize the resulting values so that $\sum_j \hat{x}_j^p = 1$. The argument will rely on the tightness of the bounds mentioned above, also obtained from the gadgets' properties:

Observation 3. *In an ε'-Nash equilibrium of game $\mathcal{G}$, $|\sum_j x_j^p - 1| \leq 2t_p \varepsilon'$, and, if $\sum_{s \in S_{-p}} u_{js}^p x_s > \sum_{s \in S_{-p}} u_{j's}^p x_s + \varepsilon'$, then $x_{j'}^p \in [0, t_p \cdot \varepsilon']$.*

Lemma 2. *There exists a polynomial $p(n)$ such that, if $\varepsilon' = \varepsilon/2^{p(n)}$, then, at an ε'-Nash equilibrium of game $\mathcal{G}$, the values $\{\{\hat{x}_j^p\}_j\}_p$ constitute an ε-Nash equilibrium of game G.*

Proof. We will denote by $\mathcal{U}_j^p(\cdot)$ the function defined by the straight-line program that computes the utility of player p for choosing pure strategy j. We need to compare the values $\mathcal{U}_j^p(\hat{x})$ with the values of the nodes U_j^p of the graphical game $\mathcal{G}$ at an ε'-Nash equilibrium. For convenience, let $\hat{U}_j^p \triangleq \mathcal{U}_j^p(\hat{x})$ be the expected utility of player p for playing pure strategy j when the other players play according to $\{\{\hat{x}_j^q\}_j\}_{q \neq p}$. Our ultimate goal is to show that, at an ε'-Nash equilibrium of game $\mathcal{G}$, for all $p \leq r, j \leq t_p$

$$\hat{U}_j^p > \hat{U}_{j'}^p + \varepsilon \implies \hat{x}_{j'}^p = 0 \tag{3}$$

Let us take $c(n)$ to be the polynomial bound on $2t_p$. Using Observation 3, we get that, for all p, j,

$$\hat{x}_j^p(1 - c(n)\varepsilon') \leq x_j^p \leq \max\{c(n)\varepsilon', \hat{x}_j^p(1 + c(n)\varepsilon')\}$$
$$\Rightarrow \quad \hat{x}_j^p - c(n)\varepsilon' \leq x_j^p \leq \hat{x}_j^p + c(n)\varepsilon' \tag{4}$$

To carry on the analysis, note that, although $\hat{U}_j^p$ is the output of function $\mathcal{U}_j^p(\cdot)$ on input $\{\hat{x}_j^p\}_{j,p}$, U_j^p is not the correct output of $\mathcal{U}_j^p(\cdot)$ on input $\{x_j^p\}_{j,p}$. This is, because, at an ε'-Nash equilibrium of game $\mathcal{G}$, the games that simulate the gates of the arithmetic circuit introduce an additive error of absolute value up to ε' per operation. So, to compare U_j^p with $\hat{U}_j^p$, we shall compare the "erroneous" evaluation of the arithmetical circuit on input $\{x_j^p\}_{j,p}$ carried inside $\mathcal{G}$ against the ideal evaluation of the circuit on input $\{\hat{x}_j^p\}_{j,p}$. Let us assign a nonnegative "level" to every wire of the arithmetical circuit in the natural way: the wires to which the input is provided are at level 0 and a wire out of a gate is at level one plus the maximum level of the gate's input wires. Since the arithmetical circuits that compute expected utilities are assumed to be of polynomial length the maximum level that a wire can be assigned to is $q(n)$, $q(\cdot)$ being some polynomial. The "erroneous" and the "ideal" evaluations of the circuit on inputs $\{x_j^p\}_{j,p}$ and $\{\hat{x}_j^p\}_{j,p}$ respectively satisfy the following property which can be shown by induction:

Lemma 3. *Let* v, $\hat{v}$ *be the values of a wire at level* i *of the circuit in the erroneous and the ideal evaluation respectively. Then*

$$\hat{v} - g(i)\varepsilon' \le v \le \hat{v} + g(i)\varepsilon'$$

where $g(i) = 3^i \cdot (c(n) + \frac{1}{2}) - \frac{1}{2}$.

By this lemma, the outputs of the two evaluations will satisfy

$$\hat{U}_j^p - (2^{q(n)} \cdot (c(n) + 1) - 1)\varepsilon' \le U_j^p \le \hat{U}_j^p + (2^{q(n)} \cdot (c(n) + 1) - 1)\varepsilon'$$

Thus, setting $\varepsilon' = \frac{\varepsilon}{8c(n)3^{q(n)}}$ yields $|U_j^p - \hat{U}_j^p| \le \varepsilon/4$. After applying the same argument to $U_{j'}^p$ and $\hat{U}_{j'}^p$, we have that $\hat{U}_j^p > \hat{U}_{j'}^p + \varepsilon$ implies $U_j^p + \varepsilon/4 \ge \hat{U}_j^p > \hat{U}_{j'}^p + \varepsilon \ge U_{j'}^p + 3\varepsilon/4$, and thus $U_j^p > U_{j'}^p + \varepsilon/2 > U_{j'}^p + \varepsilon'$. Then, from Observation 3, it follows that $x_{j'}^p < t_p\varepsilon'$ and, from the definition of our trimming process, that $\hat{x}_{j'}^p = 0$. So (3) is satisfied, therefore making $\{\{\hat{x}_j^p\}_j\}_p$ an ε-Nash equilibrium. $\qquad\square$

In Section 3.4 we point out that the EXPECTED UTILITY problem in typical succinct games of polynomial type is very hard. However, in all well known succinct games in the literature, it turns out that there is a straight-line program of polynomial length that computes EXPECTED UTILITY:

Corollary 1. *The problem of computing a Nash equilibrium in the following families of succinct games can be polynomially reduced to the same problem for 2-player games: graphical games, congestion games, multimatrix games, semi-anonymous games, local effect games, scheduling games, hypergraphical games, network design games, and facility location games.*

Proof. It turns out that, for all these families, there is indeed a straight-line program as specified in Theorem 2. For graphical games, for example, the program computes explicitly the utility expectation of a player with respect to its neighbors; the other mixed strategies do not matter. For multimatrix games, the program computes one quadratic form per constituent game, and adds the expectations (by linearity). For hypergraphical games, the program combines the previous two ideas. For the remaining kinds, the program combines results of several instances of the following problem (and possibly the two previous ideas, linearity of expectation and explicit expectation calculation): Given n Bernoulli variables $x_1, \ldots, x_n$ with $\Pr[x_i = 1] = p_i$, calculate $q_j = \Pr[\sum_{i=1}^{n} x_i = j]$ for $j = 0, \ldots, n$. This can be done by dynamic programming, letting $q_j^k = \Pr[\sum_{i=1}^{k} x_i = j]$ (and omitting initializations): $q_{j+1}^k = (1 - p_i)q_j^{k-1} + p_i q_{j-1}^{k-1}$, obviously a polynomial division-free straight-line program. □

3.3 An Alternative Proof

We had been looking for some time for an alternative proof of this result, not relying on the machinery of [11]. This proof would start by reducing the Nash equilibrium problem to Brouwer by the reduction of [10]. The Brouwer function in [10] maps each mixed strategy profile $x = (x_1, \ldots, x_n)$ to another $(y_1, \ldots, y_n)$, where $y_i = \arg\max\left(E_{(x_{-i}, y_i)}[U_i] - ||y_i - x_i||^2\right)$. That is, y_i optimizes a trade-off between utility and distance from x_i. It should be possible, by symbolic differentiation of the straight-line program, to approximate this optimum and thus the Brouwer function. There are, though, difficulties in proceeding, because the next step (reduction to Sperner's Lemma) seems to require precision incompatible with guarantees obtained this way.

3.4 Intractability

Let us briefly explore the limits of the upper bound in this section.

Proposition 1. *There are succinct games of polynomial type for which* EXPECTED UTILITY *is #P-hard.*

Proof. Consider the case in which each player has two strategies, `true` and `false`, and the utility of player 1 is 1 if the chosen strategies satisfy a given Boolean formula. Then the expected utility, when all players play each strategy with probability $\frac{1}{2}$ is the number of satisfying truth assignments divided by 2^n, a #P-hard problem.   □

Thus, the sufficient condition of our Theorem is nontrivial, and there are games of polynomial type that do not satisfy it. *Are there games of polynomial type for which computing Nash equilibria is intractable beyond PPAD?* This is an important open question. Naturally, computing a Nash equilibrium of a general succinct game is EXP-hard (recall that it is so even for 2-person zero-sum games [8,9], and the nonzero version can be easily seen to be complete for the exponential counterpart of PPAD).

Finally, it is interesting to ask whether our sufficient condition (polynomial computability of EXPECTED UTILITY by a bounded division-free straight-line program) is strictly weaker than the condition in [17] for correlated equilibria (polynomial computability of EXPECTED UTILITY by Turing machines). It turns out[1] that it is, unless ⊕P is in nonuniform polynomial time [2]. Determining the precise complexity nature of this condition is another interesting open problem.

---

[1] Many thanks to Peter Bürgisser for pointing this out to us.

4   Network Congestion Games

A network congestion game [7] is specified by a network with delay functions, that is, a directed graph (V, E) with a pair of nodes (a_p, b_p) for each player p, and also, for each edge $e \in E$, a *delay function* d_e mapping $[n]$ to the positive integers; for each possible number of players "using" edge e, d_e assigns a delay. The set of strategies for player p is the set of all paths from a_p to b_p. Finally, the payoffs are determined as follows: If $s = (s_1, \ldots, s_n)$ is a pure strategy profile, define $c_e(s) = |\{p : e \in s_p\}|$ (here we consider paths as sets of edges); then the utility of player p under s is simply $-\sum_{e \in s_p} d_e(c_e(s))$, the negation of the total delay on the edges in p's strategy. It was shown in [7] that a *pure* Nash equilibrium of a network congestion game (known to always exist) can be found in polynomial time when the game is *symmetric* ($a_p = a_1$ and $b_p = b_1$ for all p), and PLS-complete in the general case. There is no known polynomial-time algorithm for finding Nash equilibria (or *any* kind of equilibria, such as correlated [17]) in general network congestion games. We prove:

Theorem 3. *The problem of computing a Nash equilibrium of a network congestion game polynomially reduces to the problem of computing a Nash equilibrium of a 2-player game.*

Proof. (Sketch.) We will map a network congestion game to a graphical game $\mathcal{G}$. To finish the proof one needs to use techniques parallel to Section 3.2. To simulate network congestion games by graphical games we use a nonstandard representation of mixed strategy: We consider a mixed strategy for player p to be a *unit flow* from a_p to b_p, that is, an assignment of nonnegative values $f_p(e)$ to the edges of the network such that all nodes are balanced except for a_p who has a deficit of 1 and b_p who has a gain of 1. Intuitively, $f_p(e)$ corresponds to the sum of the probabilities of all paths that use e.

It turns out that such flow can be set up in the simulating graphical game by a gadget similar to the one that sets up the mixed strategy of each player. In particular, for every player p and for every edge e of the network there will be a player in the graphical game whose value will represent $f_p(e)$. Moreover, for every node $v \neq a_p, b_p$ of the network, there will be a player S_v^p in the graphical game whose value will be equal to the sum of the flows of player p on the edges entering node v; there will also be a gadget $\mathcal{G}_v^p$ similar to the one used in proof of Theorem 1, whose purpose will be to distribute the flow of player p entering v, i.e. value S_v^p, to the edges leaving node v, therefore guaranteeing that Kirchhoff's first law holds. The distribution of the value S_v^p on the edges leaving v will be determined by finding the net delays between their endpoints and node b_p as specified by the next paragraphs. Finally, note that the gadgets for nodes a_p and b_p are similar but will inject a gain of 1 at a_p and a deficit of 1 at b_p. Some scaling will be needed to make sure that all computed values are in $[0, 1]$.

The rest of the construction is based on the following Lemma, whose simple proof we omit. Fix a player p and a set of unit flows f_q for the other players. These induce an expected delay on each edge e, $\mathrm{E}[d_e(c_e(k))]$ where k is 1 (for player p) plus the sum of $n - 1$ variables that are 1 with probability $f_q(e)$ and else 0. Call this quantity $D_p(e)$.

Lemma 4. *A set of unit flows $f_p(e), p = 1, \ldots, n$ is an ε-Nash equilibrium if and only if $f_p(e) > 0$ implies that e lies on a path whose length (defined as net delay under D_p), is at most ε above the length of the shortest path from a_p to b_p.*

We shall show that these conditions can be calculated by a straight-line program in polynomial time; this implies the Theorem. This is done as follows: First we compute the distances $D_p(e)$ for all edges and players by dynamic programming, as in the proof of Corollary 1. Then, for each player p and edge (u,v) we calculate the shortest path distances, under D_p, (a) from a_p to b_p; (b) from a_p to u and (c) from v to b_p. This is done by the Bellman-Ford algorithm, which is a straight line program with the additional use of the min operator (see [11] for gadget). The condition then requires that the sum of the latter two and $D_p(u,v)$ be at most the former plus ε. This completes the proof. $\square$

5   Extensive form Games

An r-player *extensive form game* (see, e.g., [16]) is represented by a game tree with each non-leaf vertex v assigned to a player $p(v)$, who "plays" by choosing one of the outgoing labeled edges, and with a vector of payoffs u_x^p at each leaf x (let X be the set of leaves). All edges have labels, with the constraint that $l(v,v') \neq l(v,v'')$. The vertex set is partitioned into *information sets* $I \in \mathcal{I}$, with all $v \in I$ owned by the same player $p(I)$, and having identical sets of outgoing edge labels L_I. We also define $\mathcal{I}_p = \{I \in \mathcal{I} | p(I) = p\}$. Information sets represent a player's knowledge of the game state. A *behavioral strategy* σ^p for player p is an assignment of distributions $\{\sigma_j^{p,I}\}_{j \in L_I}$ over the outgoing edge labels of each $I \in \mathcal{I}_p$. A behavioral strategy profile $\sigma = (\sigma^1, \ldots, \sigma^r)$ induces a distribution over the leaves of the game tree, and hence expected utilities. A *behavioral Nash equilibrium* is the natural equivalent of the normal form's mixed Nash equilibrium: a σ such that no player p can change σ^p and increase his expected payoff.

Theorem 4. *The problem of computing a behavioral Nash equilibrium (and, in fact, a subgame perfect equilibrium [16]) in an extensive form game Γ is polynomially reducible to computing a mixed Nash equilibrium in a 2-player normal form game.*

Proof. (Sketch.) As in Section 4, we will map an extensive form congestion game to a graphical game, and omit the rest of the argument, which is also akin to Section 3.2. The graphical game construction is similar to that in Section 3.1. Using nodes with strategy sets $\{0, 1\}$,

1. For every information set I with $p(I) = p$ and an outgoing edge label $j \in L_I$, make a node $\sigma_j^{p,I}$, to represent the probability of picking j.
2. For every information set I and every $j \in L_I$ make a node $\mathcal{U}_j^I$; the value of $\mathcal{U}_j^I$ will represent the utility of player $p(I)$ resulting from the optimal choice of distributions player $p(I)$ can make in the part of the tree below information set I given that the player arrived at information set I and chose j and assuming that the other players play as prescribed by the values $\{\sigma_j^{q,I'}\}_{q=p(I') \neq p}$; the weighting of the vertices of I when computing $\mathcal{U}_j^I$ is defined by the probabilities of the other players on the edges that connect I to the closest information set of $p(I)$ above I. Let $\mathcal{U}^I$ be the maximum over $\mathcal{U}_j^I$. Assuming the values $\mathcal{U}^{I'}$ for the information sets I' below I are computed, value $\mathcal{U}_j^I$ can be found by arithmetic operations.
3. Finally, for every information set, take a gadget $\mathcal{G}_I$ similar to $\mathcal{G}_p$ above that guarantees that (i) $\sum_{j \in L_I} \sigma_j^{p(I),I} = 1$, and (ii) $\mathcal{U}_j^I > \mathcal{U}_{j'}^I \implies \sigma_{j'}^{p(I),I} = 0$.

Further details are omitted. The construction works by arguments parallel to the proof of Theorem 1.   □

References

1. R. J. Aumann. "Subjectivity and Correlation in Randomized Strategies," *Journal of Mathematical Economics*, 1 , pp. 67-95, 1974.
2. P. Bürgisser. "On the structure of Valiant's complexity classes," *Discr. Math. Theoret. Comp. Sci.*, **3**, pp. 73–94, 1999.
3. X. Chen and X. Deng. "3-NASH is PPAD-Complete," *ECCC*, TR05-134, 2005.
4. X. Chen and X. Deng. "Settling the Complexity of 2-Player Nash-Equilibrium," *ECCC*, TR05-140, 2005.
5. C. Daskalakis and C. H. Papadimitriou. "Three-Player Games Are Hard," *ECCC*, TR05-139, 2005.
6. C. Daskalakis, P. W. Goldberg and C. H. Papadimitriou. "The Complexity of Computing a Nash Equilibrium," *Proceedings of 38th STOC*, 2006.
7. A. Fabrikant, C.H. Papadimitriou and K. Talwar. "The Complexity of Pure Nash Equilibria," *Proceedings of 36th STOC*, 2004.
8. J. Feigenbaum, D. Koller and P. Shor. "A game-theoretic classification of interactive complexity classes," *IEEE Conference on Structure in Complexity Theory*, 1995.
9. L. Fortnow, R. Impagliazzo, V. Kabanets and C. Umans. "On the Complexity of Succinct Zero-Sum Games," *IEEE Conference on Computational Complexity*, 2005.
10. J. Geanakoplos. "Nash and Walras Equilibrium via Brouwer," *Economic Theory*, *21*, 2003.
11. P. W. Goldberg and C. H. Papadimitriou. "Reducibility Among Equilibrium Problems," *Proceedings of 38th STOC*, 2006.
12. D. S. Johnson, C. H. Papadimitriou and M. Yannakakis, "How Easy is Local Search?," *J. Comput. Syst. Sci. 37*, 1, pp. 79–100,1988.
13. M. Kearns, M. Littman and S. Singh. "Graphical Models for Game Theory," *In UAI*, 2001.
14. C. E. Lemke and J. T. Howson Jr.. "Equilibrium points of bimatrix games," *Journal of the Society for Industrial and Applied Mathematics*, 12, 413-423, 1964.
15. J. Nash. "Noncooperative Games," *Annals of Mathematics*, 54, 289-295, 1951.
16. M.J. Osborne and A. Rubinstein. *A Course in Game Theory*, MIT Press, 1994.
17. C. H. Papadimitriou. "Computing Correlated Equilibria in Multiplayer Games," *STOC*, 2005.
18. C. H. Papadimitriou. "On the Complexity of the Parity Argument and Other Inefficient Proofs of Existence," *J. Comput. Syst. Sci.* 48, 3, pp. 498–532, 1994.
19. R. Savani and B. von Stengel. "Exponentially many steps for finding a Nash equilibrium in a Bimatrix Game". *Proceedings of 45th FOCS*, 2004.
20. B. von Stengel, Computing equilibria for two-person games. In R.J. Aumann and S. Hart, editors, *Handbook of Game Theory, Vol. 3*, pp. 1723–1759. North-Holland, Amsterdam, 2002.

Network Games with Atomic Players

Roberto Cominetti[1], José R. Correa[2], and Nicolás E. Stier-Moses[3]

[1] Departamento de Ingeniería Matemática, Universidad de Chile, Santiago, Chile
rcominet@dim.uchile.cl
[2] School of Business, Universidad Adolfo Ibáñez, Santiago, Chile
correa@uai.cl
[3] Graduate School of Business, Columbia University, New York, USA
stier@gsb.columbia.edu

Abstract. We study network and congestion games with atomic players that can split their flow. This type of games readily applies to competition among freight companies, telecommunication network service providers, intelligent transportation systems and manufacturing with flexible machines. We analyze the worst-case inefficiency of Nash equilibria in those games and conclude that although self-interested agents will not in general achieve a fully efficient solution, the loss is not too large. We show how to compute several bounds for the worst-case inefficiency, which depend on the characteristics of cost functions and the market structure in the game. In addition, we show examples in which market aggregation can adversely impact the aggregated competitors, even though their market power increases. When the market structure is simple enough, this counter-intuitive phenomenon does not arise.

1 Introduction

In this paper, we study network games with atomic players that can split flow among multiple routes. This type of games readily applies to competition among freight companies, telecommunication network service providers, intelligent transportation systems and, by considering the generalization to congestion games, it also applies to manufacturing with flexible machines. This class of network games was first discussed by [17]. More recently, [25, 24, 8] considered a similar model from the perspective of the price of anarchy, which is the framework of this paper.

Consider a directed network $G = (V, A)$ and players that wish to route flow between origin-destination (OD) pairs. We denote the set of all players by $[K] = \{1, \ldots, K\}$. Each player $k \in [K]$ has to choose a flow $x^k \in \mathbb{R}_+^A$ that routes d_k units of flow from s_k to t_k. Note that players can divide their flows among many paths. We refer collectively to the flows for all players by $\vec{x} := (x^1, \ldots, x^K)$. In addition, to simplify notation we henceforth let $x := \sum_{k \in [K]} x^k$ be the aggregate flow induced by all K players.

As arcs are subject to adverse congestion effects, we associate a cost function $c_a(\cdot) : \mathbb{R}_+ \to \mathbb{R}_+$ to every arc. These functions map the total flow on the arc x_a to its per-unit cost $c_a(x_a)$. Cost functions are assumed nondecreasing, differentiable and convex although for some of our results the convexity assumption can

M. Bugliesi et al. (Eds.): ICALP 2006, Part I, LNCS 4051, pp. 525–536, 2006.

be slightly relaxed. In addition, we only consider separable cost functions, i.e., the cost in one arc only depends on the flow in the same arc. The goal of each competitor is to send its demand minimizing its total cost $C^k(\vec{x}) := \sum_{a \in A} x_a^k c_a(x_a)$. To understand the dependence between the inefficiency of equilibria and the cost functions, we let $\mathcal{C}$ be an arbitrary but fixed set that contains the allowable cost functions. Typical choices include polynomials of degree at most r with r fixed, and the delay functions of M/M/1 queues.

A strategy distribution $\vec{x}$ is a Nash equilibrium when no player has an incentive to unilaterally change her strategy. In other words, the best reply strategy for player k is the flow x^k that solves the following optimization problem in which flows x^i are fixed for $i \neq k$. For ease of notation, we introduce a reverse arc with zero cost between t^k and s^k.

$$(\text{NE}^k) \qquad \min \qquad C^k(\vec{x})$$
$$\sum_{(u,v) \in A} x_{(u,v)}^k - \sum_{(v,w) \in A} x_{(v,w)}^k = 0 \qquad \text{for all } v \in V$$
$$x_{(t^k, s^k)}^k = d_k$$
$$x_a^k \geq 0 \qquad \text{for all } a \in A.$$

Note that our assumptions guarantee that these optimization problems are convex, which implies that an equilibrium always exists [22]. The uniqueness of equilibria is a longstanding open question, except for some particular cases [17, 1]. Using the convexity of $C^k(\vec{x})$ and the first order optimality conditions of (NE^k), we can characterize equilibria with a variational inequality. Indeed, $\vec{x}$ is at equilibrium if and only if, for all $k \in [K]$, x^k solves

$$\sum_{a \in A} c_a^k(\vec{x}_a)(y_a^k - x_a^k) \geq 0 \text{ for any feasible flow } y^k \text{ for player } k. \qquad (1)$$

Here, the modified cost function $c_a^k(\vec{x}_a) := c_a(x_a) + x_a^k c_a'(x_a)$ is the derivative with respect to x_a^k of the term $x_a^k c_a(x_a)$ in $C^k(\vec{x})$. Intuitively, the second term accounts for player k's ability to set prices in the arc.

Following [12], we also consider situations in which some OD pairs are controlled by individual players while others are controlled by infinitely many of them, each in charge of an arbitrary small portion of demand. These games can be viewed as limits of games in which the number of players tends to infinity but some of them retain market power to set prices while others are relegated to be price-takers, and their equilibria can be characterized by a variational inequality similar to (1). The extreme case in which all users are price-takers was first considered by Wardrop [28]. In this case, the game becomes nonatomic and the corresponding solution concept is usually called a Wardrop equilibrium. In other words Wardrop equilibria are those for which all flow-carrying paths are of minimum cost (among all paths serving the same OD pair). Except were otherwise stated, all results in this paper are valid for the three classes of equilibria that we just mentioned because we work with arbitrary market powers.

To measure the quality of equilibria, we need to introduce a social cost function. The most common measure for network models is the sum of the costs among all players. This is easily computed as $C(\vec{x}) = C(x) := \sum_{k \in [K]} C^k(\vec{x}) = \sum_{a \in A} x_a c_a(x_a)$. (Notice that the total cost does not depend on $\vec{x}$ directly, but rather on the total flow x.) A socially optimal flow is a solution $\vec{x}^{\text{OPT}}$ that minimizes $C(x)$ among all feasible solutions $\vec{x}$. It is well known that a system optimum can have a strictly lower social cost than an equilibrium [20, 10]. Moreover, system optima may even be better for all users compared to an equilibrium [4]; the problem is that users may still have an incentive to deviate from it so it may not be stable.

The main objective of this article is to study how much efficiency is lost when competition arises. To that extent, [16] proposes to use the worst-case inefficiency (in terms of ratio) of the social cost of equilibria with respect to that of social optima as a way to quantify the impact of not being able to coordinate players in a game. This ratio became known as the *price of anarchy* [18]. Roughgarden and Tardos initiated the study of the price of anarchy in network games by proving that the inefficiency loss because of selfish behavior in nonatomic network games with affine cost functions is at most 33% [25]. Following their work, a series of papers generalized these results to nonatomic network and congestion games, under less and less restrictive assumptions [23, 27, 7, 5, 19]. Finally, [26, 8] generalize the results to arbitrary nonatomic congestion games by noticing that for the characterization of equilibria with variational inequalities, the network structure is unnecessary.

In addition to nonatomic games, [25, 11, 2, 6] consider the atomic case with unsplittable flows, meaning that competitors have to choose a single path to route all the demand from their (single) origins to their (single) destinations.

Our Results. Section 2 shows an upper bound on the price of anarchy for arbitrary networks when costs belong to a given set of cost functions. In addition, we provide a lower bound that arises from a particular instance. Both the lower and upper bounds are strictly higher than the price of anarchy in the nonatomic case implying that price-setting behavior can hurt the system. These are the first bounds of this type for atomic games. Section 3 concentrates in games with a single OD pair. First, we provide bounds on the price of anarchy that depend on the variability of the market power of the different players. To the best of our knowledge, this is the first bound that depends on the market concentration. It measures the price to pay—in the worst case—when going from monopolies to oligopolies to markets with numerous similar players. Then, in Section 3.2, we study the case in which players are symmetric, i.e., they all share the same OD pair and they all control the same amount of flow. In this setting we are able to give a potential function that characterizes equilibria and use this to show that equilibria with atomic players are at least as efficient as the Wardrop equilibrium. All results in this paper generalize to the more general atomic congestion games with divisible demands.

Due to space limitations, several proofs in this extended abstract are omitted or only sketched. More details and further results can be found in the full version of this article available at the authors' website.

2 Atomic Games with General Players

In this section we study the price of anarchy for atomic games with arbitrary networks and arbitrary configuration of OD pairs. Players with arbitrarily large market power may coexist with price-taking players. As a warm-up exercise and before considering arbitrary sets $\mathcal{C}$, we derive a bound on the price of anarchy for the case in which cost functions are affine functions. To this end, we define an optimization problem whose first order optimality conditions correspond to the equilibrium conditions. In particular, this optimization problem implies that the equilibrium is essentially unique (although, as [3] points out, this is implied by [22]).

Consider an affine cost function of the form $c(x) = qx + r$. Let us define a modified cost function $\hat{c} : \mathbb{R}_+^K \to \mathbb{R}_+$ by $\hat{c}(\vec{x}) := \frac{q}{2}(\sum_{k \in [K]} x^k)^2 + \frac{q}{2} \sum_{k \in [K]} (x^k)^2 + r \sum_{k \in [K]} x^k$. It is easy to see that $\hat{c}(\vec{x})$ is convex, or strictly convex when $q > 0$. We define Problem (NLP-NE) as the minimization of the potential function $\hat{C}(\vec{x}) := \sum_{a \in A} \hat{c}_a(\vec{x}_a)$ among all feasible flows $\vec{x}$. Strict convexity implies that there is a single solution to the previous problem. As its first-order optimality conditions coincide with the conditions that characterize a Nash equilibrium, the latter has to be unique. In addition, (NLP-NE) can be used to approximate a Nash equilibrium up to a fixed additive term in polynomial time [21]. One cannot expect to do better than an additive approximation because an equilibrium may require irrational numbers. We remark that this approach can be easily extended to the setting of games with a mix of atomic and nonatomic players introduced by [12].

As in other settings [25, 15], this potential function can be used to derive bounds on the price of anarchy but those bounds turn out to be loose. Using the variational inequality displayed in (1), we can prove a stronger upper bound on the price of anarchy for atomic congestion games. The upper bound we provide below originates in [24] using ideas from [7]. Let us define, for $c \in \mathcal{C}$,

$$\alpha^K(c) := \sup_{\vec{x}, \vec{y} \in \mathbb{R}_+^K} \frac{xc(x)}{yc(y) + \sum_{k \in [K]} (x^k - y^k)c^k(\vec{x})}. \tag{2}$$

We remind the reader that we have defined $\vec{x} := (x^1, \ldots, x^K)$ and $x := \sum_{k \in [K]} x^k$, and similarly for $\vec{y}$. For this definition and the ones below to work, we shall assume that $0/0 = 0$. Roughgarden proved that $\alpha^K(\mathcal{C}) := \sup_{c \in \mathcal{C}} \alpha^K(c)$ is an upper bound on the price of anarchy of atomic games [24]. To slightly simplify the calculations, we define

$$\beta^K(c) := \sup_{\vec{x}, \vec{y} \in \mathbb{R}_+^K} \frac{\sum_{k \in [K]} \{(c^k(\vec{x}) - c(y))y^k + (c(x) - c^k(\vec{x}))x^k\}}{xc(x)},$$

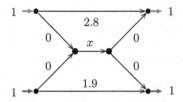

Fig. 1. Example with price of anarchy larger than 4/3

and $\beta^K(\mathcal{C}) := \sup_{c \in \mathcal{C}} \beta^K(c)$. It is straightforward to see that $\beta^K(\mathcal{C}) \geq 0$ and that $\alpha^K(\mathcal{C}) = (1 - \beta^K(\mathcal{C}))^{-1}$ when $\beta^K(\mathcal{C}) < 1$. To simplify notation we will not explicitly distinguish the case of $\beta^K(\mathcal{C}) \geq 1$ and assume that $(1 - \beta^K(\mathcal{C}))^{-1} = +\infty$ in such a case. We now give a bound on the price of anarchy that depends on $\beta^K(\mathcal{C})$. Note that this bound on the price of anarchy is also valid for the mixed atomic and nonatomic games.

Proposition 2.1 ([24]). *Consider an atomic congestion game with K players and with separable cost functions drawn from $\mathcal{C}$. Let $\vec{x}^{\text{NE}}$ be a Nash equilibrium and $\vec{x}^{\text{OPT}}$ be a social optimum. Then, $C(x^{\text{NE}}) \leq (1 - \beta^K(\mathcal{C}))^{-1} C(x^{\text{OPT}})$.*

Proof. Using (1) and the definition of $\beta^K(\mathcal{C})$ in order, we get that

$$C(x^{\text{NE}}) = \sum_{a \in A} \sum_{k \in [K]} \{ (c_a(x_a^{\text{NE}}) - c_a^k(\vec{x}_a^{\text{NE}})) x_a^{\text{NE},k} + c_a^k(\vec{x}_a^{\text{NE}}) x_a^{\text{NE},k} \}$$

$$\leq \sum_{a \in A} \sum_{k \in [K]} \{ (c_a(x_a^{\text{NE}}) - c_a^k(\vec{x}_a^{\text{NE}})) x_a^{\text{NE},k} + c_a^k(\vec{x}_a^{\text{NE}}) x_a^{\text{OPT},k} \} \leq \beta^K(\mathcal{C}) C(x^{\text{NE}}) + C(\vec{x}^{\text{OPT}}).$$

$\square$

Although [24, 8] independently claimed (by providing different proofs) that the price of anarchy in the atomic case cannot exceed that of the nonatomic case, in Fig. 1 we present an instance with affine cost functions that has a price of anarchy larger than α(affine functions) = 4/3. The top OD pair is controlled by a single player while the bottom one is nonatomic. At Nash equilibrium, the common arc has 0.9 and 1 units of demand coming from the atomic and nonatomic OD pairs, respectively, and the total cost is 3.89. Under the social optimum, the common arc has 1 and 0 units of demand and the total cost is 2.9. Dividing, we get a price of anarchy of approximately 1.341. Moreover, optimizing over the parameters, we can get an instance with a price of anarchy of approximately 1.343. Notice that the nonatomic OD pair is not necessary to be worse than in nonatomic games. We could construct a similar example with a finite number of players. That would require replacing the nonatomic OD pair by $K - 1$ atomic players, each controlling $1/(K - 1)$ units of demand. If K is large, both equilibria are similar by continuity (e.g., [13] proves that equilibria in atomic games converge to those in nonatomic games when players lose market power).

We now provide a concrete expression for the price of anarchy under specific sets of cost functions. The key is to first obtain a simpler expression for $\beta^K(c)$.

Theorem 2.2. *Assume that $xc(x)$ is a convex function. Defining $\beta^\infty(c) :=$ $\sup_{0 \leq y \leq x} \frac{y(c(x)-c(y)+c'(x)y/4)}{xc(x)}$, we have that $\beta^K(c) \leq \beta^\infty(c)$.*

Proof. Starting from the definition of $\beta^K(c)$, we get

$$\beta^K(c) = \sup_{\vec{x},\vec{y} \in \mathbb{R}_+^K} \frac{xc(x) - yc(y) + \sum_{k \in [K]} c^k(\vec{x})(y^k - x^k)}{xc(x)}$$

$$= \sup_{\vec{x},\vec{y} \in \mathbb{R}_+^K} \frac{yc(x) - yc(y) + c'(x)\left(\sum_{k \in [K]} y^k x^k - \sum_{k \in [K]}(x^k)^2\right)}{xc(x)} \qquad (3)$$

As c is nondecreasing, $c'(x) \geq 0$. Thus, assuming w.l.o.g. that $x^1 \geq x^k$ for all $k \in [K]$, to make (3) as big as possible we have to set $(y^1, \dots, y^K)$ to $(y, 0, \dots, 0)$. It follows that

$$\beta^K(c) = \sup_{\vec{x} \in \mathbb{R}_+^K; x^1 = \max(\vec{x}); y \in \mathbb{R}_+} \frac{y(c(x) - c(y)) + c'(x)\left(x^1 y - \sum_{k \in [K]}(x^k)^2\right)}{xc(x)} \qquad (4)$$

To find the best choice of $\vec{x}$, it is enough to solve $\max\{x^1 y - \sum_{k \in [K]}(x^k)^2 : \vec{x} \in \mathbb{R}_+^K, x^1 = \max(\vec{x})\}$. By symmetry, an optimal solution to this problem satisfies $x^2 = \cdots = x^K$. Therefore, we replace x_1 by u and the rest of the x_k by v, and solve

$$\max_{u \geq v \geq 0; \, u+(K-1)v=x} uy - u^2 - (K-1)v^2. \qquad (5)$$

The optimal solution satisfies that $u = \min\{x/K + y(K-1)/2K, x\}$. Plugging in $x^1 = \min\{x/K + y(K-1)/2K, x\}$ and $x^k = \max\{x/K - y/2K, 0\}$ for $k = 2, \dots, K$ in (4), we have that

$$\beta^K(c) \leq \sup_{x,y \in \mathbb{R}_+^2} \frac{yc(x) - yc(y) + c'(x)\left(\frac{y^2}{4} - \frac{(x-y/2)^2}{K}\right)}{xc(x)}.$$

Under the convexity assumption, a calculation shows that the optimal solution for the RHS is achieved at $y \leq x$, from where the result follows. $\square$

The definition $\beta^\infty(C)$ is very similar to that of $\beta(C) := \sup_{0 \leq y \leq x} \frac{y(c(x)-c(y))}{xc(x)}$, which provides a bound on the price of anarchy for nonatomic games [7]. The only difference between the two expressions is the last term in the numerator of $\beta^\infty(C)$, which penalizes equilibria in the case of atomic players. Extending the arguments of [7] to $\beta^\infty(C)$, we can prove the following result. In particular, that allows us to conclude that the price of anarchy is at most 3/2, 2.464 and 7.826, for affine, quadratic and cubic cost functions, respectively.

Proposition 2.3. *If C only contains polynomials of degree at most r, the price of anarchy is at most $(1 - \max_{0 \leq u \leq 1} u(1 - u^r + ru/4))^{-1}$.*

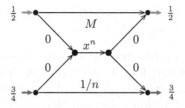

Fig. 2. Example with pseudo-approximation guarantee larger than 2

2.1   Pseudo-approximations

We now concentrate on pseudo-approximation results (also known as bicreteria results) which compare the Nash equilibrium to a social optimum in an instance with expanded demands. Roughgarden and Tardos proved that the social cost of a Wardrop equilibrium is bounded by that of a social optimum of a game with demands doubled [25]. They extended the pseudo-approximation bound to atomic games, which was based on a characterization of equilibria of atomic congestion games. Unfortunately, this characterization is not correct. Figure 2 presents an example for which the Nash equilibrium is more costly than the system optimum with demands doubled. The top OD pair is atomic and the bottom one is nonatomic. Consider $M := (1 - \varepsilon)^n + n(1/4 - \varepsilon)(1 - \varepsilon)^{n-1}$, where ε is such that $(1 - \varepsilon)^n < 1/n$. The parameters M and ε are chosen so that the Nash equilibrium is the flow in which the nonatomic demand routes all its $3/4$ units of flow in the middle arc and the atomic player splits its flow in $1/4 - \varepsilon$ along the middle arc and the rest in the other. The social cost of the equilibrium equals $(1 - \varepsilon)^{n+1} + (1/4 + \varepsilon)M$. Consider the flow routing twice the demand in which ε units of flow take the top arc, $1 - \varepsilon$ units take the middle arc, and $3/2$ units take the bottom arc. Therefore, the social cost of the system optimum is at most $\varepsilon M + (1 - \varepsilon)^{n+1} + 3/(2n)$. Comparing the two costs, we conclude that in order to find a counterexample we need to find n and ε such that $n(1 - \varepsilon)^n < 1$ and $Mn/6 > 1$. This is achieved by taking $\varepsilon = 0.1$ and $n = 34$. Modifying the example slightly, we can obtain a counterexample with polynomials of degree 26. On the other hand, if we allow polynomials of arbitrary degree, it can be seen that the cost of the Nash equilibrium can be made arbitrarily higher than that of the system optimum with demands doubled.

In addition, one cannot expect to prove a theorem of this type with a constant expansion factor if arbitrary cost functions are allowed. To see this, consider the same example as in Fig. 2 and a parameter $0 < \delta < 1$. The nonatomic demand is $1 - \delta$, the demand of the atomic player is 2δ, and the cost functions, from top to bottom, are 2, a step function that is 0 for $x \leq 1$ and 1 otherwise, and 0. It can be seen that there is one equilibrium with total cost equal to 2δ while the system optimum when the demand is amplified by $1/(2\delta)$ has zero cost. The example can be worked out for polynomial cost functions (of arbitrary high degree). The previous discussion leads us to the following result.

Proposition 2.4. *Let $\vec{x}^{\,\mathrm{NE}}$ be a Nash equilibrium and, for an arbitrary $\alpha > 1$, let $\vec{x}^{\,\mathrm{OPT}}$ be a social optimum of the game when demands are multiplied by α. Then,* *there exists an instance of the atomic network game with convex and increasing cost functions such that $C(x^{\text{NE}}) > C(x^{\text{OPT}})$.*

In view of the previous negative results, we now prove a pseudo-approximation result for atomic games that hinges on ideas of [8]. The following proposition provides a bound that depends on the allowable cost functions $\mathcal{C}$. For example, in the case of affine cost functions, the expansion factor for which the social cost of equilibria is bounded by that of the expanded system optimum is 4/3.

Proposition 2.5. *Let $\vec{x}^{\text{NE}}$ be a Nash equilibrium of an atomic congestion game with K players and with separable cost functions drawn from $\mathcal{C}$. If $\vec{x}^{\text{OPT}}$ denotes a social optimum of the game with demands multiplied by $1+\beta^K(\mathcal{C})$, then $C(x^{\text{NE}}) \le C(x^{\text{OPT}})$.*

Proof. Consider the flow $\vec{y} = \vec{x}^{\text{OPT}}$ that optimally routes $(1 + \beta^K(\mathcal{C}))d_k$ units of demand from s_k to t_k for $k \in [K]$. Then,

$$C(x^{\text{NE}}) = (1+\beta^K(\mathcal{C}))\sum_{a \in A}\sum_{k \in [K]}\{(c_a(x_a^{\text{NE}})-c_a^k(\vec{x}_a^{\text{NE}}))x_a^{\text{NE},k}+c_a^k(\vec{x}_a^{\text{NE}})x_a^{\text{NE},k}\}-\beta^K(\mathcal{C})C(x^{\text{NE}})$$

$$\le (1+\beta^K(\mathcal{C}))\sum_{a \in A}\sum_{k \in [K]}\left\{(c_a(x_a^{\text{NE}})-c_a^k(\vec{x}_a^{\text{NE}}))x_a^{\text{NE},k}+\frac{c_a^k(\vec{x}_a^{\text{NE}})\,y_a^k}{1+\beta^K(\mathcal{C})}\right\}-\beta^K(\mathcal{C})C(x^{\text{NE}}),$$

where the inequality follows using (1) with $y_a^k/(1+\beta^K(\mathcal{C}))$. As $c_a(x_a^{\text{NE}})-c_a^k(\vec{x}_a^{\text{NE}})\le 0$,

$$C(x^{\text{NE}}) \le \sum_{a \in A}\sum_{k \in [K]}\{(c_a(x_a^{\text{NE}}) - c_a^k(\vec{x}_a^{\text{NE}}))x_a^{\text{NE},k} + c_a^k(\vec{x}_a^{\text{NE}})y_a^k\} - \beta^K(\mathcal{C})C(x^{\text{NE}}),$$

$$\le \beta^K(\mathcal{C})C(x^{\text{NE}}) + C(y) - \beta^K(\mathcal{C})C(x^{\text{NE}}) = C(y). \qquad \square$$

Remark 2.6. Note that the example above shows that for general cost functions (continuous and convex), $\beta^\infty(\mathcal{C})$ is unbounded.

3   Atomic Games with a Single OD Pair

In this section we concentrate on games played on networks with arbitrary topology in which all K players share the same source s and sink t. Single-source single-sink instances are easier to analyze because the total flow can be decomposed by player in an arc-by-arc fashion. This decomposition will allow us to provide improved results compared to the general case. The presentation is divided into two sections: In the first we consider the case in which different players control different amounts of demand, resulting in different market shares. In the second part, we consider the case of symmetric players in which all players have the same demand to route through the network. These two alternatives have previously been considered by [17], although it is assumed that the network only consists of parallel links.

3.1   Variable Market Power

We consider the case in which different players have different market power as they control different amounts of demand. To that extent, we define the Herfindahl index by $H := \sum_{k \in [K]} (d_k/D)^2$, where $D := \sum_{k \in [K]} d_k$ is the total demand. This index is a number between $1/K$ and 1. A higher index means that the market is less competitive, and the case of $H = 1$ corresponds to a monopoly. The case in which $H = 1/K$ corresponds to symmetric players (see next section). The following proposition reinterprets the definition of $\beta^K(c)$ to improve the bound given by Theorem 2.2. The proof can be found in the full paper.

Proposition 3.1. *If we only consider instances with a single OD pair, the constant $\beta^K(c)$ is at most* $\sup_{0 \le y \le x} \frac{y(c(x) - c(y) + c'(x)yH/4)}{xc(x)}$.

The difference compared to the expression provided by Theorem 2.2 is the factor H in the last term of the numerator. Observe that as $H \le 1$, this result can only reduce the price of anarchy. Moreover, if each player controls at most a fraction $\phi(K)$ of the demand such that $\phi(K) \to 0$ when $K \to \infty$, the price of anarchy is asymptotically equal to that in the nonatomic game. Indeed, the worst case for the market power variability is that there are $1/\phi(K)$ players, each controlling a fraction $\phi(K)$ of the demand, while the rest of the players control an infinitesimal. In that case $H \le (1/\phi(K))\phi(K)^2 = \phi(K) \to 0$. For example, in an oligopoly with K players that control a total demand equal to K, but in which $K/\ln K$ players control $\ln K$ units of demand each and the rest of the players do not have market power, the analysis above shows that this oligopoly approaches the nonatomic game when K grows.

For the case of affine cost functions, the price of anarchy can be bounded by $(4 - H)/(3 - H)$. This generalizes that the price of anarchy is equal to $4/3$ for nonatomic games ($H = 0$) and at most $3/2$ in general (arbitrary H). Nevertheless, we know that when $H = 1$, the price of anarchy equals 1. By perturbing the monopolistic case we can show that the price of anarchy for the case of a single OD pair is strictly less than $3/2$. However, this analysis is quite technical and it is unlikely to provide a bound that is tight.

Dafermos and Sparrow [9] proved that nonatomic games with general networks and cost functions of the form $c_a(x_a) = q_a x_a^p$ for non-negative constants q_a and p, have fully efficient Nash equilibria. The situation for atomic games is totally different: a carefully constructed instance with linear cost functions (constant times flow) implies that the price of anarchy under linear costs is at least 1.17. For games with a single OD pair, [1] proves that the flow that divides a system optimum proportionally to the market power of different players is a Nash equilibrium, implying that the price of anarchy is 1.

Providing bounds that depend on the market concentration for multiple OD pairs is an interesting question that our work leaves open. Our techniques do not easily extend to multiple OD pairs because it is not clear how to create a feasible flow arc by arc. Nevertheless, as (1) holds when competitors have to route from multiple origins to multiple destinations, these results can presented with more general assumptions.

3.2   Symmetric Players

When all players have the same demand d to route through the network, [17] shows that there is a unique Nash equilibrium. Our first contribution in this section is to provide a convex optimization problem whose optimum is the unique equilibrium. This implies that the game with symmetric players is a potential game. To facilitate notation, we add a reverse arc between t and s with zero cost.

$$\text{(SNE)} \qquad \min \quad \sum_{a \in A} x_a c_a(x_a) + (K-1) \sum_{a \in A} \int_0^{x_a} c_a(\tau) d\tau$$

$$\sum_{(u,v) \in A} x_{(u,v)} - \sum_{(v,w) \in A} x_{(v,w)} = 0 \qquad \text{for all } v \in V$$

$$x_{(t,s)} = dK$$

$$x_a \geq 0 \qquad \text{for all } a \in A.$$

Interestingly, (SNE) consists in finding a feasible flow that minimizes a convex combination between the objective functions of the problems used to compute a system optimum and a Nash equilibrium of a nonatomic game. When there is a single player the second part vanishes leaving the social cost only. Instead, when there are many players the second part is dominant and the social cost becomes negligible. It turns out that a solution is optimal for (SNE) if and only if it is a Nash equilibrium. Therefore, if the cost functions are strictly increasing, there is exactly one Nash equilibrium. By comparing the KKT conditions of (SNE) and (NE^k), we get the following result.

Theorem 3.2. *If x solves (SNE), then $\vec{x} = (x/K, \ldots, x/K)$ is a Nash equilibrium of the symmetric game with atomic players.*

We make use of the potential function to derive results on the efficiency of equilibria and the monotonicity of the cost when the number of players increase.

Proposition 3.3. *Let $\vec{x} \in \mathbb{R}_+^K$ be a Nash equilibrium in an atomic game with K players who control d units of flow each; and let $\vec{y} \in \mathbb{R}_+^{\tilde{K}}$ be a Nash equilibrium in an atomic game with $\tilde{K} < K$ players who control $dK/\tilde{K}$ units of flow each. Then, $C(y) \leq C(x)$.*

Proof. Using the optimality of x and y in their respective problems,

$$\sum_{a \in A} x_a c_a(x_a) + (K-1) \sum_{a \in A} \int_0^{x_a} c_a(\tau) d\tau \leq \sum_{a \in A} y_a c_a(y_a) + (K-1) \sum_{a \in A} \int_0^{y_a} c_a(\tau) d\tau$$

$$\leq \sum_{a \in A} x_a c_a(x_a) + (\tilde{K}-1) \sum_{a \in A} \int_0^{x_a} c_a(\tau) d\tau + (K - \tilde{K}) \sum_{a \in A} \int_0^{y_a} c_a(\tau) d\tau.$$

Thus, $\sum_{a \in A} \int_0^{x_a} c_a(\tau) d\tau \leq \sum_{a \in A} \int_0^{y_a} c_a(\tau) d\tau$, from where

$$\sum_{a \in A} y_a c_a(y_a) + (\tilde{K}-1) \sum_{a \in A} \int_0^{y_a} c_a(\tau) d\tau \leq \sum_{a \in A} x_a c_a(x_a) + (\tilde{K}-1) \sum_{a \in A} \int_0^{x_a} c_a(\tau) d\tau$$

$$< \sum_{a \in A} x_a c_a(x_a) + (\tilde{K}-1) \sum_{a \in A} \int_0^{y_a} c_a(\tau) d\tau.$$

$\square$

The previous propositions imply that the price of anarchy in symmetric games with K players increases as the number of players increases. Furthermore, it approaches the price of anarchy in the nonatomic case when the number of players goes to infinity. In particular, we can show that, for affine cost functions the price of anarchy is exactly $(4K^2)/(K+1)(3K-1)$. The conclusion is that for symmetric games non-atomicity does not degrade the quality of equilibria. This stands in clear contrast to the case of atomic asymmetric games whose price of anarchy is larger than that of nonatomic games. Independently of this work, [14] have studied the effect of collusion in network games. In particular, their results imply that, for an atomic game in a parallel link network and divisible demands, the price of anarchy is at most that of the corresponding nonatomic game. The results we just presented have a similar flavor: we have more restrictive assumptions on the players, but our results are valid for arbitrary networks. We believe that a more general result actually holds. Namely, we conjecture that for atomic networks games with splittable flows and a single OD pair, the price of anarchy is at most that of the corresponding nonatomic game.

The results for symmetric players can be generalized to the asymmetric case (but still users that share a single OD pair) if we assume that all players have a positive flow on all arcs ([17] referred to this assumption by "all-positive flows," and proved that in this case there is a unique Nash equilibrium). For those extensions, we just need to decompose the flow in each arc proportionally to the demand of each player (as we have done in Section 3.1). The assumption guarantees that the decomposition is feasible for all players.

Acknowledgements. We thank N. Figueroa, A. Schulz, and G. Weintraub for useful comments that helped us improve the paper. The first author was partially supported by FONDAP in Applied Mathematics, CONICYT-Chile; and the second by Anillo en Redes ACT08 and FONDEYT 1060035, CONICYT-Chile.

References

[1] E. Altman, T. Başar, T. Jimenez, and N. Shimkin. Competitive routing in networks with polynomial costs. *IEEE Transactions on Automatic Control*, 47:92–96, 2002.

[2] B. Awerbuch, Y. Azar, and A. Epstein. The price of routing unsplittable flow. STOC, 57–66, 2005. NY.

[3] T. Boulogne. *Nonatomic Strategic Games and Applications to Networks*. PhD thesis, Université Paris 6, Paris, France, December 2004.

[4] D. Braess. Über ein Paradoxon aus der Verkehrsplanung. *Unternehmensforschung*, 12:258–268, 1968. An English translation appears in Transportation Science, 39:4, Nov 2005, pp. 446–450.

[5] C. K. Chau and K. M. Sim. The price of anarchy for non-atomic congestion games with symmetric cost maps and elastic demands. *Operations Research Letters*, 31:327–334, 2003.

[6] G. Christodoulou and E. Koutsoupias. The price of anarchy of finite congestion games. STOC, 67–73, 2005. NY.

[7] J. R. Correa, A. S. Schulz, and N. E. Stier-Moses. Selfish routing in capacitated networks. *Mathematics of Operations Research*, 29:961–976, 2004.

[8] J. R. Correa, A. S. Schulz, and N. E. Stier-Moses. On the inefficiency of equilibria in congestion games. IPCO, 167–181, 2005.

[9] S. C. Dafermos and F. T. Sparrow. The traffic assignment problem for a general network. *Journal of Research of the U.S. National Bureau of Standards*, 73B:91–118, 1969.

[10] P. Dubey. Inefficiency of Nash equilibria. *Mathematics of Operations Research*, 11:1–8, 1986.

[11] D. Fotakis, S. C. Kontogiannis, and P. G. Spirakis. Selfish unsplittable flows. ICALP, 593–605, 2004.

[12] P. T. Harker. Multiple equilibrium behaviors of networks. *Transportation Science*, 22:39–46, 1988.

[13] A. Haurie and P. Marcotte. On the relationship between Nash-Cournot and Wardrop equilibria. *Networks*, 15:295–308, 1985.

[14] A. Hayrapetyan, É. Tardos, and T. Wexler. Collution in nonatomic games. STOC, 2006. Forthcoming.

[15] R. Johari and J. N. Tsitsiklis. Network resource allocation and a congestion game. *Mathematics of Operations Research*, 29:407–435, 2004.

[16] E. Koutsoupias and C. H. Papadimitriou. Worst-case equilibria. STACS, 404–413, 1999.

[17] A. Orda, R. Rom, and N. Shimkin. Competitive routing in multiuser communication networks. *IEEE/ACM Transactions on Networking*, 1:510–521, 1993.

[18] C. H. Papadimitriou. Algorithms, games, and the Internet. STOC, 749–753, 2001.

[19] G. Perakis. The "price of anarchy" under nonlinear and asymmetric costs. IPCO, 46–58, 2004.

[20] A. C. Pigou. *The Economics of Welfare*. Macmillan, London, 1920.

[21] F. Potra and Y. Ye. A quadratically convergent polynomial algorithm for solving entropy optimization problems. *SIAM Journal on Optimization*, 3:843–860, 1993.

[22] J. B. Rosen. Existence and uniqueness of equilibrium points for concave N-person games. *Econometrica*, 33:520–534, 1965.

[23] T. Roughgarden. The price of anarchy is independent of the network topology. *Journal of Computer and System Sciences*, 67:341–364, 2003.

[24] T. Roughgarden. Selfish routing with atomic players. SODA, 973–974, 2005.

[25] T. Roughgarden and É. Tardos. How bad is selfish routing? *Journal of the ACM*, 49:236–259, 2002.

[26] T. Roughgarden and É. Tardos. Bounding the inefficiency of equilibria in nonatomic congestion games. *Games and Economic Behavior*, 47:389–403, 2004.

[27] A. S. Schulz and N. E. Stier-Moses. On the performance of user equilibria in traffic networks. SODA, 86–87, 2003.

[28] J. G. Wardrop. Some theoretical aspects of road traffic research. *Proceedings of the Institution of Civil Engineers, Part II, Vol. 1*, 325–378, 1952.

Finite-State Dimension and Real Arithmetic

David Doty[*], Jack H. Lutz[**], and Satyadev Nandakumar[***]

Department of Computer Science, Iowa State University, Ames, IA 50011, USA
{ddoty, lutz, satyadev}@cs.iastate.edu

Abstract. We use entropy rates and Schur concavity to prove that, for every integer $k \geq 2$, every nonzero rational number q, and every real number α, the base-k expansions of $\alpha, q + \alpha$, and $q\alpha$ all have the same finite-state dimension and the same finite-state strong dimension. This extends, and gives a new proof of, Wall's 1949 theorem stating that the sum or product of a nonzero rational number and a Borel normal number is always Borel normal.

1 Introduction

The finite-state dimension of a sequence S over a finite alphabet Σ is an asymptotic measure of the density of information in S as perceived by finite-state automata. This quantity, denoted $\dim_{FS}(S)$, is a finite-state effectivization of classical Hausdorff dimension [15,12] introduced by Dai, Lathrop, Lutz, and Mayordomo [9]. A dual quantity, the finite-state *strong* dimension of S, denoted $\mathrm{Dim}_{FS}(S)$, is a finite-state effectivization of classical packing dimension [30,29,12] introduced by Athreya, Hitchcock, Lutz, and Mayordomo [2]. (Explicit definitions of $\dim_{FS}(S)$ and $\mathrm{Dim}_{FS}(S)$ appear in section 2.) In fact both $\dim_{FS}(S)$ and $\mathrm{Dim}_{FS}(S)$ are asymptotic measures of the density of finite-state information in S, with $0 \leq \dim_{FS}(S) \leq \mathrm{Dim}_{FS}(S) \leq 1$ holding in general. The identity $\dim_{FS}(S) = \mathrm{Dim}_{FS}(S)$ holds when S is sufficiently "regular", but, for *any* two real numbers $0 \leq \alpha \leq \beta \leq 1$, there exists a sequence S with $\dim_{FS}(S) = \alpha$ and $\mathrm{Dim}_{FS}(S) = \beta$ [13].

Although finite-state dimension and finite-state strong dimension were originally defined in terms of finite-state gamblers [9,2] (following the gambling approach used in the first effectivizations of classical fractal dimension [21,22]), they have also been shown to admit equivalent definitions in terms of information-lossless finite-state compressors [9,2], finite-state predictors in the log-loss model [16,2], and block-entropy rates [6]. In each case, the definitions of $\dim_{FS}(S)$ and $\mathrm{Dim}_{FS}(S)$ are exactly dual, differing only that a limit inferior appears in one

---

[*] This research was supported in part by National Science Foundation Grant 9972653 as part of their Integrative Graduate Education and Research Traineeship (IGERT) program.
[**] This research was supported in part by National Science Foundation Grant 0344187.
[***] This research was supported in part by National Science Foundation Grant 0344187.

definition where a limit superior appears in the other. These two finite-state dimensions are thus, like their counterparts in fractal geometry, robust quantities and not artifacts of a particular definition.

The sequences S satisfying $\dim_{FS}(S) = 1$ are precisely the *(Borel) normal* sequences, i.e., those sequences in which each nonempty string $w \in \Sigma^*$ appears with limiting frequency $|\Sigma|^{-|w|}$. (This fact was implicit in the work of Schnorr and Stimm [27] and pointed out explicitly in [6].) The normal sequences, introduced by Borel in 1909 [4], were extensively investigated in the twentieth century [25,19,32,10,14]. Intuitively, the normal sequences are those sequences that are random relative to finite-state automata. This statement may seem objectionable when one first learns that the Champernowne sequence

$$01000110110000010100111100\ldots,$$

obtained by concatenating all binary strings in standard order, is normal [8], but it should be noted that a finite-state automaton scanning this sequence will spend nearly all its time in the middle of long strings that are random in the (stronger) sense of Kolmogorov complexity [20] and, having only finite memory, will have no way of "knowing" where such strings begin or end. This perspective is especially appropriate when modeling situations in which a data stream is truly massive relative to the computational resources of the entity processing it.

An informative line of research on normal sequences concerns operations that preserve normality. For example, in his 1949 Ph.D. thesis under D.H. Lehmer, Wall [31] proved that every subsequence that is selected from a normal sequence by taking all symbols at positions occurring in a given arithmetical progression is itself normal. Agafonov [1] extended this by showing that every subsequence of a normal sequence that is selected using a regular language is itself normal; Kamae [17] and Kamae and Weiss [18] proved related results; and Merkle and Reimann [24] proved that a subsequence selected from a normal sequence using a context-free language need not be normal (in fact, can be constant, even if selected by a one-counter language). For another example, again in his thesis, Wall [31] (see also [19,5]) proved that, for every integer $k \geq 2$, every nonzero rational number q, and every real number α that is normal base k (i.e., has a base-k expansion that is a normal sequence), the sum $q + \alpha$ and the product $q\alpha$ are also normal base k. (It should be noted that a real number α may be normal in one base but not in another [7,26].)

This paper initiates the study of operations that preserve finite-state dimension and finite-state strong dimension. This study is related to, but distinct from, the study of operations that preserve normality. It is clear that every operation that preserves finite-state dimension must also preserve normality, but the converse does not hold. For example, a subsequence selected from a sequence according an arithmetical progression need not have the same finite-state dimension as the original sequence. This is because a sequence with finite-state dimension less than 1 may have its information content distributed heterogeneously. Specifically, given a normal sequence S over the alphabet $\{0,1\}$, define a sequence T whose n^{th} bit is the $\frac{n}{2}^{\text{th}}$ bit of S if n is even and 0 otherwise. Then the sequence S and the constant sequence 0^∞ are both selected from T

according to arithmetic progressions, but it is easy to verify that $\dim_{\mathrm{FS}}(T) = \mathrm{Dim}_{\mathrm{FS}}(T) = \frac{1}{2}, \dim_{\mathrm{FS}}(0^\infty) = \mathrm{Dim}_{\mathrm{FS}}(0^\infty) = 0$, and $\dim_{\mathrm{FS}}(S) = \mathrm{Dim}_{\mathrm{FS}}(S) = 1$. Hence, Wall's first above-mentioned theorem does not extend to the preservation of finite-state dimension. Of course, this holds *a fortiori* for the stronger results by Agafonov, Kamae, and Weiss.

Our main theorem states that Wall's second above-mentioned theorem, unlike the first one, does extend to the preservation of finite-state dimension. That is, we prove that, for every integer $k \geq 2$, every nonzero rational number q, and every real number α, the base-k expansions of $\alpha, q + \alpha$, and $q\alpha$ all have the same finite-state dimension and the same finite-state strong dimension.

The proof of our main theorem does not, and probably cannot, resemble Wall's uniform distribution argument. Instead we use Bourke, Hitchcock, and Vinodchandran's block-entropy rate characterizations of $\dim_{\mathrm{FS}}$ and $\mathrm{Dim}_{\mathrm{FS}}$ [6], coupled with the Schur concavity of the entropy function [28,23,3], to prove that finite-state dimension and finite-state strong dimension are contractive functions with respect to a certain "logarithmic block dispersion" pseudometric that we define on the set of all infinite k-ary sequences. (A function is contractive if the distance between its values at sequences S and T is no more than the pseudodistance between S and T.) This gives a general method for bounding the difference between the finite-state dimensions, and the finite-state strong dimensions, of two sequences. We then use this method to prove our main theorem. In particular, this gives a new proof of Wall's theorem on the sums and products of rational numbers with normal numbers.

In summary, our main result is a fundamental theorem on finite-state dimension that is a quantitative extension of a classical theorem on normal numbers but requires a different, more powerful proof technique than the classical theorem.

2   Preliminaries

Throughout this paper, $\Sigma = \{0, 1, \ldots, k - 1\}$, where $k \geq 2$ is an integer. All *strings* are elements of Σ^*, and all *sequences* are elements of Σ^∞. If x is a string or sequence and i, j are integers, $x[i \mathinner{.\,.} j]$ denotes the string consisting of the ith through j^{th} symbols in x, provided that these symbols exist. We write $x[i] = x[i \mathinner{.\,.} i]$ for the ith symbol in x, noting that $x[0]$ is the leftmost symbol in x. If w is a string and x is a string or sequence, we write $w \sqsubseteq x$ to indicate that $w = x[0 \mathinner{.\,.} n - 1]$ for some nonnegative integer n.

A *base-k expansion* of a real number $\alpha \in [0, 1]$ is a sequence $S \in \Sigma^\infty$ such that

$$\alpha = \sum_{n=0}^{\infty} S[n] k^{-(n+1)}.$$

A sequence $S \in \Sigma^\infty$ is *(Borel) normal* if, for every nonempty string $w \in \Sigma^+$

$$\lim_{n \to \infty} \frac{1}{n} \left| \left\{ u \in \Sigma^{<n} \,\middle|\, uw \sqsubseteq S \right\} \right| = |\Sigma|^{-|w|},$$

i.e., if each string w appears with asymptotic frequency $k^{-|w|}$ in S.

If Ω is a nonempty finite set, we write $\Delta(\Omega)$ for the set of all (discrete) probability measures on Ω, i.e., all functions $\pi : \Omega \to [0,1]$ satisfying $\sum_{w \in \Omega} \pi(w)$ $= 1$. We write $\Delta_n = \Delta(\{1, \dots, n\})$.

All logarithms in this paper are base 2. The *Shannon entropy* of a probability measure $\pi \in \Delta(\Omega)$ is

$$H(\pi) = \sum_{w \in \Omega} \pi(w) \log \frac{1}{\pi(w)},$$

where $0 \log \frac{1}{0} = 0$.

We briefly define finite-state dimension and finite-state strong dimension. As noted in the introduction, several equivalent definitions of these dimensions are now known. In this paper, it is most convenient to use the definitions in terms of block-entropy rates, keeping in mind that Bourke, Hitchcock, and Vinodchandran [6] proved that these definitions are equivalent to earlier ones.

For nonempty strings $w, x \in \Sigma^+$, we write

$$\#_\square(w, x) = \left| \left\{ m \le \frac{|x|}{|w|} - 1 \;\middle|\; w = x[m|w| \dots (m+1)|w| - 1] \right\} \right|$$

for the number of *block occurrences* of w in x. Note that $0 \le \#_\square(w, x) \le \frac{|x|}{|w|}$.

For each sequence $S \in \Sigma^\infty$, positive integer n, and string $w \in \Sigma^{<n}$, the nth *block frequency* of w in S is

$$\pi_{S,n}(w) = \frac{\#_\square(w, S[0 \dots n|w| - 1])}{n}.$$

Note that, for all $S \in \Sigma^\infty$ and $0 < l < n$,

$$\sum_{w \in \Sigma^l} \pi_{S,n}(w) = 1,$$

i.e., $\pi_{S,n}^{(l)} \in \Delta(\Sigma^l)$, where we write $\pi_{S,n}^{(l)}$ for the restriction of $\pi_{S,n}$ to Σ^l.

For each sequence $S \in \Sigma^\infty$ and positive integer l, the lth *normalized lower and upper block entropy rates* of S are

$$H_l^-(S) = \frac{1}{l \log k} \liminf_{n \to \infty} H\left(\pi_{S,n}^{(l)} \right)$$

and

$$H_l^+(S) = \frac{1}{l \log k} \limsup_{n \to \infty} H\left(\pi_{S,n}^{(l)} \right),$$

respectively.

Definition 1. *Let $S \in \Sigma^\infty$.*

1. *The* finite-state dimension *of S is*

$$\dim_{\text{FS}}(S) = \inf_{l \in \mathbb{Z}^+} H_l^-(S).$$

2. *The* finite-state strong dimension *of S is*

$$\text{Dim}_{\text{FS}}(S) = \inf_{l \in \mathbb{Z}^+} H_l^+(S).$$

More discussion and properties of these dimensions appear in the references cited in the introduction, but this material is not needed to follow the technical arguments in the present paper.

3   Logarithmic Dispersion and Finite-State Dimension

In this section we prove a general theorem stating that the difference between two sequences' finite-state dimensions (or finite-state strong dimensions) is bounded by a certain "pseudodistance" between the sequences. Recall that $\Delta_n = \Delta(\{1,\ldots,n\})$ is the set of all probability measures on $\{1,\ldots,n\}$.

Definition 2. *Let n be a positive integer. The* logarithmic dispersion *(briefly, the* log-dispersion*) between two probability measures $\pi, \mu \in \Delta_n$ is*

$$\delta(\pi,\mu) = \log m,$$

where m is the least positive integer for which there is an $n \times n$ nonnegative real matrix $A = (a_{ij})$ with the following three properties.

  (i) A is stochastic: *each column of A sums to 1, i.e., $\sum_{i=1}^n a_{ij} = 1$ holds for all $1 \le j \le n$.*
  (ii) $A\pi = \mu$, i.e., $\sum_{j=1}^n a_{ij}\pi(j) = \mu(i)$ holds for all $1 \le i \le n$.
  (iii) No row or column of A contains more than m nonzero entries.

It is clear that $\delta : \Delta_n \times \Delta_n \to [0, \log n]$. We now extend δ to a normalized function $\delta^+ : \Sigma^\infty \times \Sigma^\infty \to [0, 1]$. Recall the block-frequency functions $\pi_{S,n}^{(l)}$ defined in section 2.

Definition 3. *The* normalized upper logarithmic block dispersion *between two sequences $S, T \in \Sigma^\infty$ is*

$$\delta^+(S,T) = \limsup_{l \to \infty} \frac{1}{l \log k} \limsup_{n \to \infty} \delta\left(\pi_{S,n}^{(l)}, \pi_{T,n}^{(l)}\right).$$

    Recall that a *pseudometric* on a set X is a function $d : X \times X \to \mathbb{R}$ satisfying the following three conditions for all $x, y, z \in X$.

  (i) $d(x,y) \ge 0$, with equality if $x = y$.   (nonnegativity)
  (ii) $d(x,y) = d(y,x)$.   (symmetry)
  (iii) $d(x,z) \le d(x,y) + d(y,z)$.   (triangle inequality)

(A pseudometric is a *metric*, or *distance function*, on X if it satisfies (i) with "if" replaced by "if and only if".) The following fact must be known, but we do not know a reference at the time of this writing.

Lemma 3.1. *For each positive integer n, the log-dispersion function δ is a pseudometric on Δ_n.*

It is easy to see that S is not a metric on Δ_n for any $n \geq 2$. For example, if π is any nonuniform probability measure on $\{1, \ldots, n\}$ and μ obtained from π by permuting the values of π nontrivially, then $\pi \neq \mu$ but $\delta(\pi, \mu) = 0$.

Lemma 3.1 has the following immediate consequence.

Corollary 3.2. *The normalized upper log-block dispersion function δ^+ is a pseudometric on Σ^∞.*

If d is a pseudometric on a set X, then a function $f : X \to \mathbb{R}$ is d-*contractive* if, for all $x, y \in X$,
$$|f(x) - f(y)| \leq d(x, y),$$
i.e., the distance between $f(x)$ and $f(y)$ does not exceed the pseudodistance between x and y.

Lemma 3.3. *For each positive integer n, the Shannon entropy function $H : \Delta_n \to [0, \log n]$ is δ-contractive.*

The following useful fact follows easily from Lemma 3.3.

Theorem 3.4. *Finite-state dimension and finite-state strong dimension are δ^+-contractive. That is, for all $S, T \in \Sigma^\infty$,*
$$|\dim_{FS}(S) - \dim_{FS}(T)| \leq \delta^+(S, T)$$
and
$$|\mathrm{Dim}_{FS}(S) - \mathrm{Dim}_{FS}(T)| \leq \delta^+(S, T).$$

In this paper, we only use the following special case of Theorem 3.4.

Corollary 3.5. *Let $S, T \in \Sigma^\infty$. If*
$$\limsup_{n \to \infty} \delta\left(\pi_{S,n}^{(l)}, \pi_{T,n}^{(l)}\right) = o(l)$$
as $l \to \infty$, then
$$\dim_{FS}(S) = \dim_{FS}(T)$$
and
$$\mathrm{Dim}_{FS}(S) = \mathrm{Dim}_{FS}(T).$$

4   Finite-State Dimension and Real Arithmetic

Our main theorem concerns real numbers rather than sequences, so the following notation is convenient. For each real number α and each integer $k \geq 2$, write
$$\dim_{FS}^{(k)}(\alpha) = \dim_{FS}(S)$$

and
$$\mathrm{Dim}_{\mathrm{FS}}^{(k)}(\alpha) = \mathrm{Dim}_{\mathrm{FS}}(S),$$

where S is a base-k expansion of $\alpha - \lfloor \alpha \rfloor$. Note that this notation is well-defined, because a real number α has two base-k expansions if and only if it is a k-adic rational, in which case both expansions are eventually periodic and hence have finite-state strong dimension 0. It is routine to verify the following.

Observation 4.1. *For every integer $k \geq 2$, every positive integer m, and every real number α,*

$$\mathrm{dim}_{\mathrm{FS}}^{(k)}(m + \alpha) = \mathrm{dim}_{\mathrm{FS}}^{(k)}(-\alpha) = \mathrm{dim}_{\mathrm{FS}}^{(k)}(\alpha)$$

and

$$\mathrm{Dim}_{\mathrm{FS}}^{(k)}(m + \alpha) = \mathrm{Dim}_{\mathrm{FS}}^{(k)}(-\alpha) = \mathrm{Dim}_{\mathrm{FS}}^{(k)}(\alpha).$$

The following lemma contains most of the technical content of our main theorem.

Lemma 4.2 (main lemma). *For every integer $k \geq 2$, every positive integer m, and every real number $\alpha \geq 0$,*

$$\mathrm{dim}_{\mathrm{FS}}^{(k)}(m\alpha) = \mathrm{dim}_{\mathrm{FS}}^{(k)}(\alpha)$$

and

$$\mathrm{Dim}_{\mathrm{FS}}^{(k)}(m\alpha) = \mathrm{Dim}_{\mathrm{FS}}^{(k)}(\alpha).$$

Proof. Let k, m, and α be as given, let $S, T \in \Sigma^\infty$ be the base-k expansions of $\alpha - \lfloor \alpha \rfloor$, $m\alpha - \lfloor m\alpha \rfloor$, respectively, and write

$$\pi_{\alpha,n}^{(l)} = \pi_{S,n}^{(l)} \quad , \quad \pi_{m\alpha,n}^{(l)} = \pi_{T,n}^{(l)}$$

for each $l, n \in \mathbb{Z}^+$. By Corollary 3.5, it suffices to show that

$$\limsup_{n \to \infty} \delta\left(\pi_{\alpha,n}^{(l)}, \pi_{m\alpha,n}^{(l)}\right) = o(l) \tag{4.1}$$

as $l \to \infty$.

Let $r = \lfloor \log_k m \rfloor$, let

$$m = \sum_{i=0}^{r} m_i k^i$$

be the base-k expansion of m, and let

$$s = \sum_{i=0}^{r} m_i.$$

The first thing to note is that, in base k, $m\alpha - \lfloor m\alpha \rfloor$ is the sum, modulo 1, of s copies of $\alpha - \lfloor \alpha \rfloor$, with m_i of these copies shifted i symbols to the left, for each $0 \leq i \leq r$.

For each $l \in \mathbb{Z}^+$ and $j \in \mathbb{N}$, let

$$u_j^{(l)} = S[jl \mathinner{..} (j+1)l - 1],$$
$$v_j^{(l)} = T[jl \mathinner{..} (j+1)l - 1]$$

be the jth l-symbol blocks of $\alpha - \lfloor \alpha \rfloor$, $m\alpha - \lfloor m\alpha \rfloor$, respectively. If we let

$$\tau_j^{(l)} = \sum_{i=0}^{r} m_i \sum_{t=(j+1)l}^{\infty} S[t + i]k^{-(t+1)}$$

be the sum of the tails of the above-mentioned s copies of $\alpha - \lfloor \alpha \rfloor$ lying to the right of the jth l-symbol block, then the block $v_j^{(l)}$ of $m\alpha - \lfloor m\alpha \rfloor$ is completely determined by $u_j^{(l)}$, the "carry"

$$c_j^{(l)} = \left\lfloor k^{(j+1)l} \tau_j^{(l)} \right\rfloor,$$

and the longest string of symbols shifted from the right, which is the string $u_{j+1}^{(l)}[0 \mathinner{..} r - 1]$. To be more explicit, note that

$$0 \le c_j^{(l)} \le k^{(j+1)l} \tau_j^{(l)} \le k^{(j+1)l} \sum_{i=0}^{r} m_i \sum_{t=(j+1)l}^{\infty} (k-1)k^{-(t+1)} = s;$$

define the "advice"

$$h_j^{(l)} = \left(c_j^{(l)}, u_j^{(l)}[0 \mathinner{..} r - 1] \right) \in \{0, \dots, s\} \times \Sigma^r;$$

and define the function

$$f^{(l)} : \Sigma^l \times \{0, \dots, s\} \times \Sigma^r \to \Sigma^l$$

by letting $f^{(l)}(x, c, z)$ be the base-k expansion of the integer

$$mn_x^{(k)} + c + \sum_{i=0}^{r} m_i \sum_{t=0}^{i-1} z[t]k^t \bmod k^l,$$

where $n_x^{(k)}$ is the nonnegative integer of which x is a base-k expansion, possibly with leading 0's. (Intuitively, the three terms here are the "block product", the "carry", and the "shift", respectively.) Then, for all integers $l > 0$ and $j \ge 0$,

$$v_j^{(l)} = f^{(l)}(u_j^{(l)}, h_j^{(l)}).$$

For positive integers l and n, define the $k^l \times k^l$ matrix $A^{(l,n)} = \left(a_{y,x}^{(l,n)} \right)$ by

$$a_{y,x}^{(l,n)} = \begin{cases} \dfrac{\left| \left\{ j < n \mid u_j^{(l)} = x \text{ and } f^{(l)}\left(x, h_j^{(l)}(j)\right) = y \right\} \right|}{n\pi_{\alpha,n}^{(l)}(x)} & \text{if } \pi_{\alpha,n}^{(l)}(x) > 0 \\[2mm] 1 & \text{if } \pi_{\alpha,n}^{(l)}(x) = 0 \text{ and } x = y \\[2mm] 0 & \text{otherwise} \end{cases}$$

for all $x, y \in \Sigma^l$. It is routine to verify that

$$\sum_{y \in \Sigma^l} a_{y,x}^{(l,n)} = 1$$

for all $x \in \Sigma^l$, i.e., $A^{(l,n)}$ is stochastic, and that

$$\sum_{x \in \Sigma^l} a_{y,x}^{(l,n)} \pi_{\alpha,n}^{(l)}(x) = \pi_{m\alpha,n}^{(l)}(y)$$

for all $y \in \Sigma^l$, i.e., $A^{(l,n)} \pi_{\alpha,n}^{(l)} = \pi_{m\alpha,n}^{(l)}$. We complete the proof by bounding the number of nonzero entries in each row and column of $A^{(l,n)}$.

Fix a column x of $A^{(l,n)}$. If $\pi_{\alpha,n}^{(l)}(x) = 0$, then there is exactly one nonzero entry in column x of $A^{(l,n)}$. If $\pi_{\alpha,n}^{(l)}(x) > 0$, then the number of nonzero entries in column x is bounded by

$$|\{0, \dots, s\} \times \Sigma^r| = (s+1)k^r \le (s+1)m.$$

Hence there are at most $(s+1)m$ nonzero entries in column x of $A^{(l,n)}$.

Fix a row y of $A^{(l,n)}$. Let g be the greatest common divisor of m and k^l. Note that, for all $n_1, n_2 \in \mathbb{Z}^+$,

$$mn_1 \equiv mn_2 \mod k^l \implies k^l \mid m(n_2 - n_1)$$
$$\implies \frac{k^l}{g} \,\Big|\, \frac{m}{g}(n_2 - n_1)$$
$$\implies \frac{k^l}{g} \,\Big|\, n_2 - n_1$$
$$\implies n_1 \equiv n_2 \mod \frac{k^l}{g}.$$

This implies that each string $y \in \Sigma^l$ has at most g preimages x under the mapping that takes x to the base-k expansion of $mn_x^{(l)} \mod k^l$. This, in turn, implies that there are at most $g|\{0, \dots, s\} \times \Sigma^r| \le g(s+1)m$ nonzero entries in row y of $A^{(l,n)}$.

We have shown that, for each $l, n \in \mathbb{Z}^+$, the matrix $A^{(l,n)}$ testifies that

$$\delta\left(\pi_{\alpha,n}^{(l)}, \pi_{m\alpha,n}^{(l)}\right) \le \log(g(s+1)m) \le \log(m^2(s+1)).$$

Since this bound does not depend on l or n, this proves (4.1). □

We now prove that addition and multiplication by nonzero rationals preserve finite-state dimension and finite-state strong dimension.

Theorem 4.3 (main theorem). *For every integer $k \ge 2$, every nonzero rational number q, and every real number α,*

$$\dim_{\mathrm{FS}}^{(k)}(q + \alpha) = \dim_{\mathrm{FS}}^{(k)}(q\alpha) = \dim_{\mathrm{FS}}^{(k)}(\alpha)$$

and

$$\mathrm{Dim}_{\mathrm{FS}}^{(k)}(q + \alpha) = \mathrm{Dim}_{\mathrm{FS}}^{(k)}(q\alpha) = \mathrm{Dim}_{\mathrm{FS}}^{(k)}(\alpha).$$

Proof. Let k, q, and α be as given, and write $q = \frac{a}{b}$, where a and b are integers with $a \neq 0$ and $b > 0$. By Observation 4.1 and Lemma 4.2,

$$
\begin{aligned}
\dim_{\mathrm{FS}}^{(k)}(q\alpha) &= \dim_{\mathrm{FS}}^{(k)}\left(\frac{|a|}{b}\alpha\right) = \dim_{\mathrm{FS}}^{(k)}\left(b\frac{|a|}{b}\alpha\right) \\
&= \dim_{\mathrm{FS}}^{(k)}(|a|\alpha) = \dim_{\mathrm{FS}}^{(k)}(\alpha),
\end{aligned}
$$

and

$$
\begin{aligned}
\dim_{\mathrm{FS}}^{(k)}(q + \alpha) &= \dim_{\mathrm{FS}}^{(k)}\left(\frac{a}{b} + \alpha\right) = \dim_{\mathrm{FS}}^{(k)}\left(\frac{a + b\alpha}{b}\right) \\
&= \dim_{\mathrm{FS}}^{(k)}\left(b\frac{a + b\alpha}{b}\right) = \dim_{\mathrm{FS}}^{(k)}(a + b\alpha) \\
&= \dim_{\mathrm{FS}}^{(k)}(b\alpha) = \dim_{\mathrm{FS}}^{(k)}(\alpha).
\end{aligned}
$$

Similarly, $\mathrm{Dim}_{\mathrm{FS}}^{(k)}(q\alpha) = \mathrm{Dim}_{\mathrm{FS}}^{(k)}(\alpha)$, and $\mathrm{Dim}_{\mathrm{FS}}^{(k)}(q + \alpha) = \mathrm{Dim}_{\mathrm{FS}}^{(k)}(\alpha)$.   □

Finally, we note that Theorem 4.3 gives a new proof of the following classical theorem.

Corollary 4.4. *(Wall [31]) Let $k \geq 2$. For every nonzero rational number q and every real number α that is normal base k, the sum $q + \alpha$ and the product $q\alpha$ are also normal base k.*

Acknowledgments. The authors thank Philippe Moser and Arindam Chatterjee for useful discussions, and we thank two anonymous reviewers for useful suggestions.

References

1. V. N. Agafonov. Normal sequences and finite automata. *Soviet Mathematics Doklady*, 9:324–325, 1968.
2. K. B. Athreya, J. M. Hitchcock, J. H. Lutz, and E. Mayordomo. Effective strong dimension, algorithmic information, and computational complexity. *SIAM Journal on Computing*. To appear.
3. R. Bhatia. *Matrix Analysis*. Springer, 1997.
4. E. Borel. Sur les probabilités dénombrables et leurs applications arithmétiques. *Rendiconti del Circolo Matematico di Palermo*, 27:247–271, 1909.
5. J. Borwein and D. Bailey. *Mathematics by Experiment: Plausible Reasoning in the 21st Century*. A. K. Peters, Ltd., Natick, MA, 2004.
6. C. Bourke, J. M. Hitchcock, and N. V. Vinodchandran. Entropy rates and finite-state dimension. *Theoretical Computer Science*, 349:392–406, 2005.
7. J. W. S. Cassels. On a problem of Steinhaus about normal numbers. *Colloquium Mathematicum*, 7:95–101, 1959.
8. D. G. Champernowne. Construction of decimals normal in the scale of ten. *J. London Math. Soc.*, 2(8):254–260, 1933.
9. J. J. Dai, J. I. Lathrop, J. H. Lutz, and E. Mayordomo. Finite-state dimension. *Theoretical Computer Science*, 310:1–33, 2004.

10. K. Dajani and C. Kraaikamp. *Ergodic Theory of Numbers.* The Mathematical Association of America, 2002.
11. G. A. Edgar. *Classics on Fractals.* Westview Press, Oxford, U.K., 2004.
12. K. Falconer. *Fractal Geometry: Mathematical Foundations and Applications.* John Wiley & Sons, 1990.
13. X. Gu, J. H. Lutz, and P. Moser. Dimensions of Copeland-Erdös sequences. *Information and Computation,* 2005.
14. G. Harman. One hundred years of normal numbers. In M. A. Bennett, B. C. Berndt, N. Boston, H. G. Diamond, A. J. Hildebrand, and W. Philip (eds.), *Surveys in Number Theory: Papers from the Millennial Conference on Number Theory,* pages 57–74, 2003.
15. F. Hausdorff. Dimension und äusseres Mass. *Mathematische Annalen,* 79:157–179, 1919. English version appears in [11], pp. 75-99.
16. J. M. Hitchcock. Fractal dimension and logarithmic loss unpredictability. *Theoretical Computer Science,* 304(1–3):431–441, 2003.
17. T. Kamae. Subsequences of normal sequences. *Israel Journal of Mathematics,* 16:121–149, 1973.
18. T. Kamae and B. Weiss. Normal numbers and selection rules. *Israel Journal of Mathematics,* 21:101–110, 1975.
19. L. Kuipers and H. Niederreiter. *Uniform Distribution of Sequences.* Wiley-Interscience, 1974.
20. M. Li and P. M. B. Vitányi. *An Introduction to Kolmogorov Complexity and its Applications.* Springer-Verlag, Berlin, 1997. Second Edition.
21. J. H. Lutz. Dimension in complexity classes. *SIAM Journal on Computing,* 32:1236–1259, 2003.
22. J. H. Lutz. The dimensions of individual strings and sequences. *Information and Computation,* 187:49–79, 2003.
23. A. W. Marshall and I. Olkin. *Inequalities: Theory of Majorization and Its Applications.* Academic Press, New York, 1979.
24. W. Merkle and J. Reimann. On selection functions that do not preserve normality. In B. Rovan and P. Vojtás, editors, *MFCS,* volume 2747 of *Lecture Notes in Computer Science,* pages 602–611, Bratislava, Slovakia, 2003. Springer.
25. I. Niven. *Irrational Numbers.* Wiley, 1956.
26. W. Schmidt. On normal numbers. *Pacific Journal of Mathematics,* 10:661–672, 1960.
27. C. P. Schnorr and H. Stimm. Endliche Automaten und Zufallsfolgen. *Acta Informatica,* 1:345–359, 1972.
28. I. Schur. Über eine Klasse von Mittelbildungen mit Anwendungen auf die Determinantentheorie. *Sitzungsberichte der Berliner Mathematischen Gesellschaft,* 22:9–20, 1923.
29. D. Sullivan. Entropy, Hausdorff measures old and new, and limit sets of geometrically finite Kleinian groups. *Acta Mathematica,* 153:259–277, 1984.
30. C. Tricot. Two definitions of fractional dimension. *Mathematical Proceedings of the Cambridge Philosophical Society,* 91:57–74, 1982.
31. D. D. Wall. *Normal Numbers.* PhD thesis, University of California, Berkeley, California, USA, 1949.
32. B. Weiss. *Single Orbit Dynamics.* American Mathematical Society, Providence, RI, 2000.

Exact Algorithms for Exact Satisfiability
and Number of Perfect Matchings

Andreas Björklund and Thore Husfeldt

Department of Computer Science, Lund University, Box 118, 221 00 Lund, Sweden
andreas.bjorklund@anoto.com, thore.husfeldt@cs.lu.se

Abstract. We present exact algorithms with exponential running times for variants of n-element set cover problems, based on divide-and-conquer and on inclusion–exclusion characterisations.

We show that the Exact Satisfiability problem of size l with m clauses can be solved in time $2^m l^{O(1)}$ and polynomial space. The same bounds hold for counting the number of solutions. As a special case, we can count the number of perfect matchings in an n-vertex graph in time $2^n n^{O(1)}$ and polynomial space. We also show how to count the number of perfect matchings in time $O(1.732^n)$ and exponential space.

Using the same techniques we show how to compute Chromatic Number of an n-vertex graph in time $O(2.4423^n)$ and polynomial space, or time $O(2.3236^n)$ and exponential space.

1 Introduction

We present exact algorithms with exponential running times using two simple techniques for a number of related NP-hard problems, including exact satisfiability, disjoint set covering, counting the number of perfect matchings, graph colouring, and TSP.

Our algorithms run in polynomial space and have running times of the form $O(c^n)$, where n is a natural parameter of the instance, such as 'number of vertices' or 'number of clauses,' but smaller than the instance size. For some NP problems, this can be achieved by exhaustive search: a maximum clique can be found by inspecting all $2^{|V|}$ vertex subsets. But in general, this fails: exhaustive search for a Hamiltonian cycle would require checking $(|V| - 1)!$ permutations. For the problems in this paper, finding such algorithms was an open problem.

Techniques

The first idea is divide-and-conquer, where we divide the instance into an *exponential number* of sub-instances, halving n at each step. This leads to running times of the form

$$T(n) = 2^n n^{O(1)} T\left(\tfrac{1}{2}n\right),$$

which is $O(c^n)$, and the space is polynomial in n.

The second idea is inclusion–exclusion, in which we express the problem in the form

$$\sum_{S \subseteq \{1,\dots,n\}} (-1)^{|S|} f(S),$$

M. Bugliesi et al. (Eds.): ICALP 2006, Part I, LNCS 4051, pp. 548–559, 2006.

where $f(S)$ is some easier-to-calculate predicate depending on the problem. The running time is 2^n times the time used to calculate $f(S)$, and the space is dominated by the algorithm for $f(S)$.

Both ideas obviously give rise to polynomial-space algorithms, but we will also consider exponential-space algorithms that reduce the running times. For the problems studied in this paper, the inclusion–exclusion approach always gives the better worst-case time bounds. However, the divide-and-conquer approach is more versatile and can be applied to natural variant problems, such as weighted versions; it may also run faster in practice.

Both ideas also turn out to be old. Gurevich and Shelah used exponential divide-and-conquer as a building block for an expected linear time algorithm [13], and Feige and Kilian use it to compute the bandwidth in an unpublished manuscript [12]. The idea of using inclusion-exclusion was used by Bax [2] to find Hamiltonian Cycles. Before that, a much older example of relevance to the present paper is the Ryser formula for the permanent [20], which counts the number of matchings in a bipartite graph.

Exact Satisfiability

Most of the results and techniques of this paper can be presented in terms of the Exact Satisfiability problem. It is equivalent to a set covering problem and generalises perfect matching in graphs, see Sec. 2 for the precise relationships.

Given a formula in disjunctive normal form with m clauses in n variables and total size $l = O(mn)$, the Exact Satisfiability problem asks for the existence of an assignment that satisfies exactly one literal of each clause. This can be solved easily in polynomial space and $2^n l^{O(1)}$ time by considering all assignments, and many papers have improved the 2^n factor.

Here, we analyse the problem in terms of the number m of clauses. For ordinary (non-exact) satisfiability, the seminal result is the DPLL procedure from 1962 [9], which gives a polynomial space algorithm with running time $2^m n^{O(1)}$. Again, many papers have since improved this result.

However, for *Exact* Satisfiability, no such result was known; the best bounds are exponential time and space $2^m l^{O(1)}$ using dynamic programming, or polynomial space and time $m! l^{O(1)}$ [17]. Prior to the current paper, no polynomial space algorithm with running time $c^m l^{O(1)}$ was known for any constant c, an open problem observed in [5,17].

Proposition 1 provides such an algorithm for many choices of m; the running time is bounded by $c^m l^{O(1)}$ except when $\log n < m \le \log n \log \log n$. The algorithm works for the weighted case as well. Our inclusion–exclusion based algorithm in Proposition 2 is even simpler and faster in the worst case, achieving the desired running time of $2^m l^{O(1)}$. Both algorithms use space polynomial in l.

Number of perfect matchings

The inclusion–exclusion based algorithm actually counts the *number* of exactly satisfying assignments and therefore solves the $\#P$-complete counting variant of XSAT. A well-studied special case of this is *perfect matching*: Given a graph $G = (V, E)$ on n vertices, find a subset $M \subseteq E$ of disjoint edges that cover V.

Although finding a perfect matching if there is one can be done in polynomial time, counting the perfect matchings is #P-complete [22]. For bipartite graphs, the best exact counting algorithm for perfect matchings is to apply the Ryser formula for the permanent [20], which runs in time $O(1.414^n)$. More recently, Jerrum, Sinclair, and Vigoda [15] discovered a fully polynomial time randomized approximation scheme for counting perfect matchings in *bipartite* graphs. For general graphs, Chien's approximation algorithm [6] runs in expected exponential time $O(1.3161^n)$ and polynomial space. The dependency on the approximation ratio has the form ϵ^{-2}, so his algorithm guarantees a $O(c^n)$ running time (for some constant c) as long as the approximation guarantee is not better than d^{-n} (for some constant d).

We solve the exact problem, albeit slower. A $2^n n^{O(1)}$ time, polynomial space, algorithm is immediate from our Prop. 2, and in exponential space we can improve the time bound to $O(1.732^n)$ using Coppersmith–Winograd matrix multiplication [8] and a construction inspired by Williams [23].

TSP and colouring

The techniques used in the present paper suffice to address some open questions about other covering-related graph problems.

First, we observe that the divide-and-conquer algorithm for Hamiltonian Path [13] can be used for TSP without significant modification and obtains running time $O(4^n n^{\log n})$ while retaining the polynomial space guarantee. Such a result was solicited in [17] and [25].

A graph's *chromatic number* is the smallest integer $\chi \leq n$ such that there is a mapping $V \rightarrow \{1, \ldots, \chi\}$ that gives different values ('colours') to neighbouring vertices. This is a well-studied problem with a rich history of exponential-time algorithms. For polynomial space, the 1971 algorithm of Christofides [7] runs in time $n! n^{O(1)}$. Feder and Motwani [11] give a randomised linear space algorithm with running time $O((\chi/e)^n)$ (with high probability), and the running time of a recent algorithm by Angelsmark and Thapper [1] can be given as $O((2+\log \chi)^n)$, an asymptotic improvement over Christofides' result for all values of χ. Prior to the current paper, no polynomial space algorithm with running time $O(c^n)$ was known for any constant c, an open problem observed in [5,17]. We provide two such algorithms, based on divide-and-conquer in time $O(8.33^n)$, and based on inclusion–exclusion in time $O(2.4423^n)$.

Such time bounds were known for exponential space since 1976, with Lawler's algorithm [16] that uses time $O(2.4423^n)$ and space $O(2^n \log n)$. This was improved by Eppstein [10] to time $O(2.4151^n)$, and further by Byskov [4] to time $O(2.4023^n)$ and $O(2^n)$ space, the fastest previous algorithm. We present an algorithm with running time $O(2.3236^n)$ and space $2^n n^{O(1)}$, which addresses Open Problem 3.5 in [24].

Like many previous colouring algorithms, our running times all rely on the Moon–Moser bound on the number of maximal independent subsets of a graph [18]. Using a different approach we have recently improved the colouring results [3].

2   Preliminaries

We write $[n] = \{1, \ldots, n\}$. In a graph, $e(U)$ denotes the number of edges between vertices in U, and $e(U, W)$ denotes the edges between U and W.

Disjoint Set Cover. It is well-known that one can use resolution to reduce any m-clauses XSAT instance to one where no variable appears negated [19]. This 'monotone' problem can be understood as the Disjoint Set Cover problem: Given a family of subsets $A_1, \ldots, A_n \subseteq [m]$, find a disjoint cover, i.e., a subfamily indexed by $I \subseteq [n]$ such that

$$\bigcup_{i \in I} A_i = [m], \qquad \text{and } A_i \cap A_j = \varnothing \text{ for all } i \neq j \in I. \tag{1}$$

To see that this problem is equivalent to (monotone) XSAT let $[m]$ represent the clauses, and for $1 \leq i \leq n$ let A_i contain j if the variable x_i appears in the jth clause. We will adopt whichever formulation is most convenient.

Perfect Matchings. Given a graph $G = (V, E)$, construct for each vertex $v \in V$ the set of its incident edges E_v. We can view this as an instance of XSAT where the sets E_v are the clauses and the graph's edges E are the variables. A satisfying assignment $M \subseteq E$ then corresponds to picking exactly one edge from each E_v; in graph-theoretic terms a *perfect matching*.

3   Exact Satisfiability

3.1   Divide-and-Conquer

Proposition 1. *Exact Satisfiability with n variables and m clauses can be solved in polynomial space and time $4^m l^{O(\log m)}$.*

Proof. After removing all negations according to [19] we may view every clause as a subset c of the variables V.

Fix an ordering $<$ of the variables. To motivate the algorithm consider a specific exactly satisfying assignment $A \subseteq V$, viewed as a subset of variables.

We claim that there is some $y \in A$ for which the clauses can be partitioned into three subsets C_y, C_1, C_2 with $|C_1|, |C_2| \leq \frac{1}{2}m$ such that y appears only in the clauses of C_y and such that clauses in C_1 are satisfied by variables $x < y$ and clauses in C_2 are satisfied by variables $x > y$.

To see that such a partition exists first partition the clauses into sets S_x for $x \in A$ according to which variable in A makes them true (by definition there is exactly one for each clause). Let $y \in A$ be the unique variable for which $\sum_{x < y} |S_x| < \frac{1}{2}m$ but $\sum_{x \leq y} |S_x| \geq \frac{1}{2}m$. Then, $C_y = S_y$, $C_1 = \bigcup_{x < y} S_x$, and $C_2 = \bigcup_{x > y} S_x$ is a partition of the claimed kind.

This suggests the following algorithm: For every variable $y \in V$ and for every partition of the clauses into three sets C_y, C_1, C_2 with $|C_1|, |C_2| \leq \frac{1}{2}m$. We construct two instances of Exact Satisfiability. Instance 1 contains the clauses of C_1, and all variables $x < y$ not part of any clause in $C_y \cup C_2$. Similarly, Instance 2 consists of the clauses of C_2 with all variables $x > y$ not part of any clause in $C_y \cup C_1$. The algorithm checks recursively if both instances admit an exactly satisfying assignment and returns the answer.

For correctness, we already observed that if an exactly satisfying assignment exists, then a partition with the desired properties exists. On the other hand, if the algorithm finds exactly satisfying assignments for Instance 1 and Instance 2, then the original instance can be satisfied by setting y to true, thereby satisfying all clauses. Moreover, this assignment is exact because the construction ensures that the variables appearing in both instances, and in any clause of C_y, are disjoint.

For the time and space bounds, note that the number of clauses is halved at each step of the recursion. Every recursive step checks n variables and fewer than 2^m partitions, each in $l^{O(1)}$ time and space. The stated time bound follows. □

The algorithm works with minor modifications also for weighted versions of XSAT. Either the objective function is to maximise the number (or total weight) of clauses that can be exactly satisfied, or the variables have weights and a minimum (or maximum) total weight satisfying assignment is sought.

3.2   Inclusion–Exclusion

Lemma 1. *Consider a family $\mathcal{A}$ of subsets of $[m]$ and for each $S \subseteq [m]$ define*

$$\mathcal{A}(S) = \{ A \in \mathcal{A} \colon A \cap S = \varnothing \},$$

the subfamily of subsets that avoid S. Let $g(S)$ be the number of multisets of size m obtained as the union of sets in $\mathcal{A}(S)$.

Then the number of disjoint set covers, i.e., subfamilies satisfying (1), is

$$\sum_{S \subseteq [m]} (-1)^{|S|} g(S). \tag{2}$$

Proof. A disjoint set cover of $[m]$ is an m-element sum of sets from $\mathcal{A}$ that avoids no elements from $[m]$, and thus these solutions are counted in $g(\varnothing)$ only. Each other m-element multiset obtained as a sum of sets from $\mathcal{A}$ avoids some elements $T \subseteq [m]$ and therefore is counted in $g(T)$. It is also counted in $g(S)$ for all $S \subseteq T$, but not in any other $g(S)$. Thus the multiset contributes once to the term $(-1)^{|S|} g(S)$ for every $S \subseteq T$. However, these contributions cancel, because every nonempty set T has the same number of odd sized and even sized subsets. □

Proposition 2.   *1. Disjoint Set Cover for n subsets $A_i \subseteq [m]$ of total size l can be solved in polynomial space and time $2^m l^{O(1)}$.*

2. Exact Satisfiability with n variables and m clauses and total size l can be solved in polynomial space and time $2^m l^{O(1)}$.

3. *The number of perfect matchings in a graph on n vertices can be found in polynomial space and time $2^n n^{O(1)}$.*

Proof. For Disjoint Set Cover, the algorithm evaluates the sum (2) and checks if the result is nonzero. For each of the 2^m terms, we calculate the size of $\mathcal{A}(S)$, which can be accomplished in time $O(l)$ by inspecting every A_i. Then $g(S)$ can be obtained through dynamic programming over the sizes of the sets in $\mathcal{A}(S)$, in time $l^{O(1)}$.

The other two claims are equivalent.   □

4   Number of Perfect Matchings

In the special case of perfect matchings the clause structure is given by a graph; every edge being incident to exactly two vertices means that every 'variable' appears in exactly two 'clauses.' We can exploit this structure with an exponential space algorithm to significantly reduce the time bound of Prop. 2.

Proposition 3. *The number of perfect matchings in a graph on n vertices can be found in time $O(1.732^n)$.*

The remainder of this section establishes the above result. First, we need a variation of (2):

Lemma 2. *The number of perfect matchings in a graph with n vertices V is*

$$\frac{1}{(n/2)!} \sum_{X \subseteq V} (-1)^{|X|} \big(e(V - X)\big)^{n/2}. \tag{3}$$

Proof. The term $\big(e(V - X)\big)^{n/2}$ counts the number of ways to pick $n/2$ ordered edges between vertices in $V - X$ with replacement. A perfect matching is such a collection of $n/2$ edges, and is only counted in $\big(e(V - \varnothing)\big)^{n/2}$ in the above formula, because every vertex is covered by some edge. Every other collection of $n/2$ edges misses some vertices W, and therefore is counted once in $\big(e(V-U)\big)^{n/2}$ for every $U \subseteq W$, but in no other terms. As before, these contributions cancel.

This approach counts every perfect matching $(n/2)!$ times, once for every way of ordering its edges.   □

Let G be a graph with n vertices V and m edges. Let $a_0(k)$ count the induced subgraphs of G containing k edges and an *even* number of vertices, and let $a_1(k)$ count those with an *odd* number. Rearranging (3) by summing over the possible values of $e(V - X)$ we can express the number of perfect matchings in G as

$$\frac{1}{(n/2)!} \sum_{k=1}^{m} \big(a_0(k) - a_1(k)\big) k^{n/2}. \tag{4}$$

As we shall see, we can calculate $a_p(k)$ faster than the obvious 2^n.

We follow Williams [23] and divide the vertices of our graph in three sets V_0, V_1, and V_2 of equal size, assuming 3 divides n for readability. Next however, the construction needs to be tailor-made to fit our application. For every triple $n_0, n_1, n_2 \in \{0, \ldots, \frac{1}{3}n\}$ and every triple $e_{01}, e_{12}, e_{20} \in \{0, \ldots, m\}$ we build a tripartite graph called $H(n_0, n_1, n_2, e_{01}, e_{12}, e_{20})$: The graph contains a vertex for every size n_i-subset of V_i ($i = 0, 1, 2$). An edge joins the vertices corresponding to $X_i \subseteq V_i$ and $X_j \subseteq V_j$ for $j = i + 1$ (mod 3) whenever $e(X_i, X_i \cup X_j) = e_{ij}$. For $p \in \{0, 1\}$ and $k \in \{0, \ldots, m\}$ let $\mathcal{H}_{p,k}$ be the family of all graphs $H(n_0, n_1, n_2, e_{01}, e_{12}, e_{20})$ satisfying

$$n_0 + n_1 + n_2 = p \pmod{2} \quad \text{and} \quad e_{01} + e_{12} + e_{20} = k. \tag{5}$$

Lemma 3. *The total number of triangles in $\mathcal{H}_{p,k}$ is $a_p(k)$.*

Proof. We will in fact argue something stronger: the triangles in $\mathcal{H}_{p,k}$ correspond one-to-one to the induced subgraphs of size of parity p containing exactly k edges. Every triangle (X_0, X_1, X_2) in a graph $H(n_0, n_1, n_2, e_{01}, e_{12}, e_{20})$ satisfying (5) corresponds to an induced subgraph in G containing k edges whose intersection with V_i is X_i, of size n_i. In particular, two *different* triangles (X_0, X_1, X_2) and (Y_0, Y_1, Y_2) represent different subgraphs. This is because either they were picked from different graphs in $\mathcal{H}_{p,k}$ and thus $|X_i| \neq |Y_i|$ for some $i = 0, 1, 2$, or they were selected from the same graph in which case $X_i \neq Y_i$ for some $i = 0, 1, 2$.

In the other direction, we note that any subgraph induced by the vertices $U \subseteq V$ containing k edges, is a triangle in the graph

$$H\big(|U_0|, |U_1|, |U_2|, e(U_0, U_0 \cup U_1), e(U_1, U_1 \cup U_2), e(U_2, U_2 \cup U_0)\big),$$

where $U_i = U \cap V_i$ for $i = 0, 1, 2$.   □

Lemma 4. *The number of triangles in a graph on n vertices can be found in time $O(n^\omega)$ where $\omega = 2.376$.*

Proof. Let A be the adjacency matrix of the graph and compute A^2. The sum of all entries in A^2 whose corresponding entry in A is one is six times the number of triangles. This is because the entry at row r and column c in A^2 counts the number of paths of length two from vertex r to vertex c. Since each edge occurs twice in A, the triangles are counted six times.

The matrix product is computed in the given time bound using the algorithm of Coppersmith and Winograd [8].   □

The graphs in $\mathcal{H}_{p,k}$ have at most $3 \cdot 2^{n/3}$ vertices and $3 \cdot 2^{2n/3}$ edges and can be built within a polynomial factor of their size. Combining lemma 4 and lemma 3 for all choices of $p \in \{0, 1\}$ and $k \in \{0, \ldots, m\}$ after noting that $|\mathcal{H}_{p,k}|$ is $n^{O(1)}$, we conclude that we can calculate a table containing all the values $a_p(k)$ for $k \in [m]$ in time $O\big((2^{n/3})^\omega\big) = O(1.732^n)$. With such a table we can evaluate (4), the incurred polynomial factor is absorbed by the asymptotic time bound. This completes the proof of Prop. 3.

5   Applications for TSP and Colouring

5.1   TSP

We observe that the algorithm of [13] can be applied to TSP with minimal modifications.

Proposition 4. *Traveling Salesman can be solved in polynomial space and running time* $O(4^n n^{\log n})$.

Proof. [Gurevich and Shelah] We consider the following variant: given n cities with positive symmetric distances, find the shortest $c_1 c_2$-path that visits every city exactly once.

First observe that if such a path exists then the cities can be partitioned into subsets $C_1 \ni c_1$, $C_2 \ni c_2$ of roughly equal size and a 'middle' city $\{m\}$, such that the path first exhausts C_1, then passes through m, and sub-sequentially remains in C_2. In other words, the path in C_i is a Hamiltonian $c_i m$-path in the subproblem induced by $C_i \cup \{m\}$, and it is the shortest of these paths.

This suggests the following algorithm: For all partitions $\{m\} \cup C_1 \cup C_2$ of the cities with $|C_1| = \lfloor \frac{1}{2} n \rfloor$, $|C_2| = \lfloor \frac{1}{2}(n-1) \rfloor$, recursively find the shortest Hamiltonian $c_i m$-paths P_i in the two instances induced by $C_i \cup \{m\}$ $(i = 1, 2)$. The combined path $P_1 P_2$ is a Hamiltonian $c_1 c_2$-path in the original instance. Return the shortest of these paths.

At each level of the recursion, the algorithm considers $(n-2) \binom{n-2}{\lceil (n-2)/2 \rceil}$ partitions and recurses on two instances with fewer than $\frac{1}{2} n + 1$ cities. The time bound follows.

The space required on each recursion level to enumerate all partitionings is polynomial. Since the recursion depth is logarithmic in n, the polynomial space bound is easily met.   □

5.2   Colouring

In a colouring χ, every colour class $\chi^{-1}(k)$ is an independent set in the underlying graph. Moreover, one of these classes C (for example, the largest) can be assumed to be a *maximal* independent set (m.i.s.), i.e., not a proper subset of another independent set. Otherwise, just change the colours of those vertices to C's colour and observe that this leads to no conflicts.

Proposition 5. *The Chromatic Number of a graph with n vertices can be found in polynomial space and time* $O(8.33^n)$.

Proof. The colour classes correspond to a partition of the graph's vertices into independent sets $I_1 \cup I_2 \cup \cdots \cup I_k$, where I_1 the largest of the sets and maximal independent. The smaller $k-1$ sets can be collected into two families C_1 and C_2, each containing at most $\frac{1}{2} n$ vertices in total. To see this, assume the I_i are ordered according to size, largest first. Find the unique index r where $|I_1| + \cdots + |I_r| \le \frac{1}{2} n$ but $|I_1| + \cdots + |I_{r+1}| > \frac{1}{2} n$. The family C_1 consists of $I_2, \ldots, I_{r+1}$ and the family

C_2 consists of $I_{r+2}, \ldots, I_k$. The two graphs induced by C_1 and C_2 can be viewed as independent colouring problems.

This suggests the following algorithm: Consider all partitions of the vertices into $I \cup C_1 \cup C_2$ with I maximal independent and both $|C_i|$ at most $\frac{1}{2}n$, and recursively find the chromatic numbers χ_i of both C_i. Return the smallest resulting $1 + \chi_1 + \chi_2$ among these partitions.

For the running time, all m.i.s. can be enumerated in time $O(\sqrt[3]{3}^n n^3)$ and space $O(n^3)$ [21]. For each such choice we need to consider fewer than 2^n partitions of the remaining vertices into two sets. The instance size is halved at each level, so the total running time amounts to $(2\sqrt[3]{3})^{2n} n^{O(1)} = O(8.33^n)$. At each level of the recursion, the algorithm uses only polynomial space.   □

We provide an algorithm with a better worst-case running time in Prop. 6. However, the technique presented here is potentially faster for some instances after some modifications. A simple improvement is to add a test for bipartiteness on the graph both before and after the removal of a m.i.s. on each level of recursion.

The algorithm works with slight modifications also for *Chromatic Sum* (sometimes called Minimum Colour Sum), the problem of finding a colouring that minimises $\sum_{v \in V} \chi(v)$.

5.3   Inclusion–Exclusion Algorithm for Colouring

Let $\mathcal{M}$ denote the family of maximal independent sets of a graph $G = (V, E)$ and let $c_k(G)$ denote the number of ways to cover G with k distinct, possibly overlapping, maximal independent sets, i.e., the number of ways to chose $M_1, \ldots, M_k \in \mathcal{M}$ (without replacement) such that $M_1 \cup \cdots \cup M_k = V$. We can use this value to determine if the graph can be k-coloured:

Lemma 5. $\chi(G) = \min\{\, k : c_k(G) > 0 \,\}$.

Proof. First we note that whenever $c_k(G) > 0$ then G can be k-coloured: If $I_1, \ldots, I_k$ is a set of independent subsets (disjoint or not) that cover G then $\chi(v) = \min\{\, r : v \in I_r \,\}$ is a legal colouring.

Now assume $\chi(G) = k$; we will show $c_k(G) > 0$. If G can be k-coloured then the colour classes provide a covering with k independent subsets, each of which can be extended to a maximal independent one. Moreover, if the colouring is optimal, these m.i.s. are all distinct. To see this, assume that $C_1 = \chi^{-1}(1)$ and $C_2 = \chi^{-1}(2)$ are subsets of the same independent set. Then we could recolour $\chi(C_2) = 1$, using only $k - 1$ colours in total.   □

This is useful because the number c_k can be expressed by an inclusion–exclusion formula.

Lemma 6. *For every vertex subset $S \subseteq V$ let $a(S)$ denote the number of $M \in \mathcal{M}$ that do not intersect S. Then*

$$c_k(G) = \sum_{S \subseteq V} (-1)^{|S|} \binom{a(S)}{k}.$$

Proof. A covering of G with k m.i.s. avoids no vertices. and thus contributes to the term corresponding to $S = \varnothing$. On the other hand, every non-covering family of k m.i.s. avoids some vertices $T \subset V$ and thus contributes once to the terms corresponding to every subset $S \subseteq T$ but no other terms. As before, these contributions cancel. □

Proposition 6. *The chromatic number of a graph can be found in time* $O(2.4423^n)$ *and polynomial space.*

Proof. The algorithm evaluates $c_k(G)$ using the above lemma for $k = 1, 2, ..., n$.

For every S, to calculate $a(S)$ we enumerate all m.i.s. in $G[V - S]$ and check for each of them if it is maximal also in G. The number of maximal independent sets in a graph on r vertices is at most $3^{r/3}$ [18], and they can be listed in polynomial space and polynomial overhead [21]. Furthermore, evaluating the binomial coefficients $\binom{a(S)}{k}$ is also a polynomial time task.

The total running time for an n vertex graph is

$$n^{O(1)} \sum_{r=0}^{n} \binom{n}{r} 3^{r/3} = O(2.4423^n).$$

Note that each $\binom{a(S)}{k}$ is $O(2^{n^2})$, and thus polynomial space suffices to hold all calculations. □

The above polynomial space algorithm computes $a(S)$ anew for each S. We can improve the time bound using exponential space to store such computations, relying on fast matrix multiplication [8] as we did in Prop. 3.

Proposition 7. *The chromatic number of a graph can be found in time* $O(2.3236^n)$ *and space* $2^n n^{O(1)}$.

Proof. Partition V into two equal size sets, U and U' (assume n is even for simplicity). Construct two $(0, 1)$-matrices A and B as follows. The $2^{n/2} \times |\mathcal{M}|$ matrix A is indexed by subsets $T \subseteq U$ and m.i.s. $M \in \mathcal{M}$ (in G) such that the corresponding entry is 1 if and only if M avoids T:

$$A_{T,M} = \begin{cases} 1, \text{ if } M \cap T = \varnothing, \\ 0, \text{ otherwise.} \end{cases}$$

Similarly, the $|\mathcal{M}| \times 2^{n/2}$ matrix B is indexed by $M \in \mathcal{M}$ and $T' \subseteq U'$. In the product AB, the entry indexed by T, T' counts the number of m.i.s. (in G) that avoid $T \cup T'$. Thus, after computing the matrix product, we can find the value of $a(S)$ at the entry indexed by $S \cap U, S \cap U'$ in constant time for each $S \subseteq V$.

The running time is dominated by the calculation of AB, which yields the stated time bound. However, *storing* the two matrices A and B requires space $2^{n/2}3^{n/3}n^{O(1)}$ at worst, which is larger than claimed.

We can save space by not storing all of A and B at the same time. Decompose the matrices into $r = \lceil |\mathcal{M}|/2^{n/2} \rceil$ square matrices of dimension $2^{n/2}$,

$A = (A_1, A_2, \cdots, A_r)$ and $B = (B_1, B_2, \cdots, B_r)^t$, possibly padding the last ones with 0s. Then $AB = A_1B_1 + A_2B_2 + \cdots + A_rB_r$. The algorithm computes the partial sums $A_1B_1 + \cdots + A_iB_i$ for $i = 1, 2, \ldots, r$, storing only the ith partial sum. The next term $A_{i+1}B_{i+1}$ can be computed in space $2^n n^{O(1)}$ and time $2^{\omega n/2}$. The total running time is dominated by the $r2^{\omega n/2} \leq |\mathcal{M}|2^{-n/2}2^{\omega n/2} \leq 3^{n/3}2^{(\omega-1)n/2}$ steps to build the matrix (plus the time to perform the calculations of lemma 6), which evaluates to the stated bound, absorbing polynomial factors.   □

Remark. These bounds depend on the number of maximal independent sets in the input instance; graphs with fewer m.i.s. have better guarantees. For example, triangle-free graphs have at most $2^{n/2}$ maximal independent sets [14], so for this class of graphs, the above algorithm runs in time $O(2^{\omega n/2}) = O(2.2784^n)$.

References

1. O. Angelsmark and J. Thapper. Partitioning based algorithms for some colouring problems. In *Recent Advances in Constraints*, volume 3978 of *LNAI*, pages 44–58. Springer Verlag, Berlin, 2005.
2. E. T. Bax. Inclusion and exclusion algorithm for the Hamiltonian Path problem. *Inf. Process. Lett.*, 47(4):203–207, 1993.
3. A. Björklund and T. Husfeldt. Inclusion–exclusion based algorithms for graph colouring. Technical Report TR06-44, Elect. Coll. Comput. Compl., 2006.
4. J. M. Byskov. Enumerating maximal independent sets with applications to graph colouring. *Operations Research Letters*, 32:547–556, 2004.
5. J. M. Byskov. *Exact algorithms for graph colouring and exact satisfiability.* PhD thesis, University of Aarhus, 2004.
6. S. Chien. A determinant-based algorithm for counting perfect matchings in a general graph. In *Proc. 15th SODA*, pages 728–735, 2004.
7. N. Christofides. An algorithm for the chromatic number of a graph. *Computer J.*, 14:38–39, 1971.
8. D. Coppersmith and S. Winograd. Matrix multiplication via arithmetic progressions. *J. Symb. Comput.*, 9(3):251–280, 1990.
9. M. Davis, G. Logemann, and D. Loveland. A machine program for theorem proving. *Communications of the ACM*, 5(7):394–397, 1962.
10. D. Eppstein. Small maximal independent sets and faster exact graph coloring. *J. Graph Algorithms and Applications*, 7(2):131–140, 2003.
11. T. Feder and R. Motwani. Worst-case time bounds for coloring and satisfiability problems. *J. Algorithms*, 45(2):192–201, 2002.
12. U. Feige and J. Kilian. Exponential time algorithms for computing the bandwidth of a graph. Manuscript, cited in [24].
13. Y. Gurevich and S. Shelah. Expected computation time for Hamiltonian path problem. *SIAM Journal on Computing*, 16(3):486–502, 1987.
14. M. Hujter and Z. Tuza. On the number of maximal independent sets in triangle-free graphs. *SIAM J. Discrete Mathematics*, 6(2):284–288, 1993.
15. M. Jerrum, A. Sinclair, and E. Vigoda. A polynomial-time approximation algorithm for the permanent of a matrix with non-negative entries. In *Proc. 33rd STOC*, pages 712–721, 2001.
16. E. L. Lawler. A note on the complexity of the chromatic number problem. *Information Processing Letters*, 5(3):66–67, 1976.

17. B. A. Madsen. An algorithm for exact satisfiability analysed with the number of clauses as parameter. *Information Processing Letters*, 97(1):28–30, 2006.
18. J. W. Moon and L. Moser. On cliques in graphs. *Israel J. Math.*, 1965.
19. S. Porschen. On some weighted satisfiability and graph problems. In *Proc. 31st SOFSEM*, volume 3381 of *LNCS*, pages 278–287, 2005.
20. H. J. Ryser. *Combinatorial Mathematics*. Number 14 in Carus Math. Monographs. Math. Assoc. America, 1963.
21. S. Tsukiyama, M. Ide, H. Ariyoshi, and I. Shirakawa. A new algorithm for generating all the maximal independent sets. *SIAM J. Comput.*, 6(3):505–517, 1977.
22. L. G. Valiant. The complexity of computing the permanent. *Theoretical Computer Science*, 8:189–201, 1979.
23. R. Williams. A new algorithm for optimal constraint satisfaction and its implications. In *Proc. 31st ICALP*, pages 1227–1237, 2004.
24. G. J. Woeginger. Exact algorithms for NP-hard problems: a survey. In *Combinatorial optimization: Eureka, you shrink!*, pages 185–207. Springer, 2003.
25. G. J. Woeginger. Space and time complexity of exact algorithms: Some open problems. In *Proc. 1st IWPEC*, volume LNCS 3162, pages 281–290, 2004.

The Myriad Virtues of Wavelet Trees

Paolo Ferragina[1,*], Raffaele Giancarlo[2,**], and Giovanni Manzini[3,***]

[1] Dipartimento di Informatica, Università di Pisa, Italy
ferragina@di.unipi.it
[2] Dipartimento di Matematica ed Applicazioni, Università di Palermo, Italy
raffaele@math.unipa.it
[3] Dipartimento di Informatica, Università del Piemonte Orientale, Italy
manzini@mfn.unipmn.it

Abstract. Wavelet Trees have been introduced in [Grossi, Gupta and Vitter, SODA '03] and have been rapidly recognized as a very flexible tool for the design of compressed full-text indexes and data compressors. Although several papers have investigated the beauty and usefulness of this data structure in the full-text indexing scenario, its impact on data compression has not been fully explored. In this paper we provide a complete theoretical analysis of a wide class of compression algorithms based on Wavelet Trees. We also show how to improve their asymptotic performance by introducing a novel framework, called *Generalized Wavelet Trees*, that aims for the best combination of binary compressors (like, Run-Length encoders) versus non-binary compressors (like, Huffman and Arithmetic encoders) and Wavelet Trees of properly-designed shapes. As a corollary, we prove high-order entropy bounds for the challenging combination of Burrows-Wheeler Transform and Wavelet Trees.

1 Introduction

The Burrows-Wheeler Transform [3] (bwt for short) has changed the way in which fundamental tasks for string processing and data retrieval, such as compression and indexing, are designed and engineered (see e.g. [4,5,7,9,10,11]). The transform reduces the problem of high-order entropy compression to the apparently simpler task of designing and engineering good order-zero (or memoryless) compressors. This point has lead to the paradigm of compression boosting presented in [4]. However, despite nearly 60 years of investigation in the design of

---

* Partially supported by Italian MIUR grants Italy-Israel FIRB "Pattern Discovery Algorithms in Discrete Structures, with Applications to Bioinformatics", and PRIN "Algorithms for the Next Generation Internet and Web" (ALGO-NEXT); and by Yahoo! Research grant on "Data compression and indexing in hierarchical memories".

** Partially supported by Italian MIUR grants PRIN "Metodi Combinatori ed Algoritmici per la Scoperta di Patterns in Biosequenze" and FIRB "Bioinformatica per la Genomica e La Proteomica" and Italy-Israel FIRB Project "Pattern Discovery Algorithms in Discrete Structures, with Applications to Bioinformatics".

*** Partially supported by Italian MIUR Italy-Israel FIRB Project "Pattern Discovery Algorithms in Discrete Structures, with Applications to Bioinformatics".

M. Bugliesi et al. (Eds.): ICALP 2006, Part I, LNCS 4051, pp. 560–571, 2006.
© Springer-Verlag Berlin Heidelberg 2006

good memoryless compressors, no general theory for the design of order-zero compressors suited for the bwt is available, since it poses special challenges. Indeed, bwt is a string in which symbols following the same context (substring) are grouped together, giving raise to clusters of nearly identical symbols. A good order-zero compressor must both adapt fast to those rapidly changing contexts and compress efficiently the runs of identical symbols. By now it is understood that one needs a clever combination of classic order-zero compressors and run length encoding techniques. However, such a design problem is mostly open. Recently Grossi et al. [7,8] proposed an elegant and effective solution to the posed design problem: the Wavelet Tree. It is a binary tree data structure that reduces the compression of a string over a finite alphabet to the compression of a set of binary strings. The latter problem is then solved via Run Length Encoding or Gap Encoding techniques. A formal definition is given in Section 2.1.

Wavelet Trees are remarkably natural since they use a well known decomposition of entropy in terms of binary entropy and, in this respect, it is surprising that it took so long to define and put them to good use. The mentioned groundbreaking work by Grossi et al. highlights the beauty and usefulness of this data structure mainly in the context of full-text indexing, and investigates a few of its virtues both theoretically and experimentally in the data compression setting [6,8]. Yet, it is still open the fundamental question of whether Wavelet Trees can provide a data structural paradigm based on which one can design good order-zero compression algorithms for the Burrows-Wheeler Transform.

Our main contribution is to answer this question in the affirmative by providing a general paradigm, and associated analytic tools, for the design of good order-zero compressors for the bwt. It is also rather fortunate that a part of our theoretical results either strengthen the ones by Grossi et al. or fully support the experimental evidence presented by those researchers and cleverly used in their engineering choices. The remaining part of our results highlight new virtues of Wavelet Trees. More specifically, in this paper:

(**A**) We provide a complete theoretical analysis of Wavelet Trees as *stand-alone*, general purpose, order-zero compressors for an arbitrary string σ. We consider both the case in which binary strings associated to the tree are compressed using Run Length Encoding (RLE), and refer to it as RLE Wavelet Tree, and the case in which Gap Encoding (GE) is used, and refer to it as GE Wavelet Tree. In both cases, a generic prefix-free encoding of the integers is used as a subroutine, thus dealing with the typical scenario of use for Wavelet Trees (see [6,7,8]). Our analysis is done in terms of the features of these prefix-free encodings and $H_0^*(\sigma)$, the modified order-zero entropy of the string σ, defined in Section 2. As a notable Corollary, we also obtain an analysis of Inversion Frequencies coding [2] offering a theoretical justification of the better compression observed in practice by this technique with respect to Move-to-Front coding.

(**B**) We study the use of Wavelet Trees to compress the output of the bwt. We show that RLE Wavelet Trees achieve a compression bound in terms of $H_k^*(\sigma)$. The technical results are in Section 4 and they improve Theorem 3.2 in [8] in that our final bound does not contain the additive term $|\sigma|$. We also show that GE Wavelet Trees *cannot achieve* analogous bounds. A striking consequence of our analytic results is to give full theoretic support to the engineering choices made in [6,8] where, based on a punctual experimental analysis of the data, the former method is preferred to the latter to compress the output of the bwt.

(**C**) We define *Generalized Wavelet Trees* that add to this class of data structures in several ways. In order to present our results here, we need to mention some facts about Wavelet Trees, when they are used as stand-alone order-zero compressors. The same considerations apply when they are used in (B). Wavelet Trees reduce the problem of compressing a string to that of compressing a set of binary strings. That set is uniquely identified by: (C.1) the shape (or topology) of the binary tree underlying the Wavelet Tree; (C.2) an assignment of alphabet symbols to the leaves of the tree. How to choose the best Wavelet Tree, in terms of number of bits produced for compression, is open. Grossi et al. establish worst-case bounds that hold for the entire family of Wavelet Trees and therefore they do not depend on (C.1) and (C.2). They also bring some experimental evidence that choosing the "best" Wavelet Tree may be difficult [8, Sect. 3.1]. It is possible to exhibit an infinite family of strings over an alphabet Σ for which changing the Wavelet Tree shape (C.1) influences the coding cost by a factor $\Theta(\log|\Sigma|)$, and changing the assignment of symbols to leaves (C.2) influences the coding cost by a factor $\Theta(|\Sigma|)$. So, the choice of the best tree cannot be neglected and remains open. Moreover, (C.3) Wavelet Trees commit to binary compressors, loosing the potential advantage that might come from a mixed strategy in which only some strings are binary and the others are defined on an arbitrary alphabet (and compressed via general purpose order-zero compressors, such as Arithmetic and Huffman coding). Again, it is possible to exhibit an infinite family of strings for which a mixed strategy yields a constant multiplicative factor improvement over standard Wavelet Trees. So, (C.3) is relevant and open.

We introduce the new paradigm of Generalized Wavelet Trees that allows us to reduce the compression of a string σ to the identification of a set of strings, for which only a part may be binary, that are compressed via the mixed strategy sketched above. We develop a combinatorial optimization framework so that one can address points (C.1)-(C.3) simultaneously. Moreover, we provide a *polynomial-time* algorithm for finding the *optimal* mixed strategy for a Generalized Wavelet Tree of *fixed shape* (Theorem 5). In addition, we provide a *polynomial-time* algorithm for selecting the *optimal tree-shape* for Generalized Wavelet Trees when the size of the alphabet is constant and the assignment of symbols to the leaves of the tree is fixed (Theorem 6). Apart from their intrinsic interest, being Wavelet Trees a special case, those two results shed some light on a problem implicitly posed in [8], where it is reported that a closer inspection of the data did not yield any insights as to how to generate a space-optimizing tree, even with the use of heuristics.

Due to space limitations some proofs will be either omitted or simply sketched.

2   Background and Notation

Let s be a string over the alphabet $\Sigma = \{a_1, \ldots, a_h\}$ and, for each $a_i \in \Sigma$, let n_i be the number of occurrences of a_i in s. Throughout this paper we assume that all logarithms are taken to the base 2 and $0 \log 0 = 0$. The 0-*th order empirical entropy* of the string s is defined as $H_0(s) = -\sum_{i=1}^{h}(n_i/|s|)\log(n_i/|s|)$. It is well known that H_0 is the maximum compression we can achieve using a fixed codeword for each alphabet symbol. We can achieve a greater compression if the codeword we use for each symbol depends on the k symbols preceding it, since the maximum compression is now bounded by the k-th order entropy $H_k(s)$ (see [11] for the formal definition).

For highly compressible strings, $|s|\, H_k(s)$ fails to provide a reasonable bound to the performance of compression algorithms (see discussion in [4,11]). For that reason, [11] introduced the notion of 0-*th order modified empirical* entropy:

$$H_0^*(s) = \begin{cases} 0 & \text{if } |s| = 0 \\ (1 + \lfloor \log |s| \rfloor)/|s| & \text{if } |s| \neq 0 \text{ and } H_0(s) = 0 \\ H_0(s) & \text{otherwise.} \end{cases} \quad (1)$$

Note that for a non-empty string s, $|s|H_0^*(s)$ is at least equal to the number of bits needed to write down the length of s in binary. The k-*th order modified empirical entropy* H_k^* is then defined in terms of H_0^* as the maximum compression we can achieve by looking at *no more than* k symbols preceding the one to be compressed.

2.1   Wavelet Trees

For ease of exposition we use a slightly more verbose notation than the one in [7]. Let T_Σ be a complete binary tree with $|\Sigma|$ leaves. We associate one-to-one the symbols in Σ to the leaves of T_Σ and refer to it as an *alphabetic tree*. Given a string s over Σ the *full Wavelet Tree* $W_f(s)$ is the labeled tree returned by the procedure TreeLabel of Fig. 1 (see also Fig. 2). Note that to each internal node $u \in W_f(s)$ we associate two strings of equal length. The first one, assigned in Step 1, is a string over Σ and we denote it by $s(u)$. The second one, assigned in Step 3, is a binary string and we denote it by $s^{01}(u)$. Note that the length of these strings is equal to the number of occurrences in s of the symbols associated to the leaves of the subtree rooted at u.

In this paper we use $\Sigma^{(s)}$ to denote the set of symbols that appear in s. If $\Sigma^{(s)} = \Sigma$, the Wavelet Tree $W_f(s)$ has the same shape as T_Σ and is therefore a complete binary tree. If $\Sigma^{(s)} \subset \Sigma$, $W_f(s)$ is not necessarily a complete binary tree since it may contain unary paths. By contracting all unary paths we obtain a *pruned Wavelet Tree* $W_p(s)$ which is a complete binary tree with $|\Sigma^{(s)}|$ leaves and $|\Sigma^{(s)}| - 1$ internal nodes (see Fig. 2).

As observed in [7], we can always retrieve s given the binary strings $s^{01}(u)$ associated to the internal nodes of a Wavelet Tree and the mapping between leaves and alphabetic symbols. Hence, Wavelet Trees are a tool for *encoding arbitrary strings using only an encoder for binary strings*. Let $\mathcal{C}$ denote any

---

Procedure TreeLabel(u, s)

1. Assign string s to node u. If u has no children return.
2. Let u_L (resp. u_R) denote the left (resp. right) child of u. Let $\Sigma^{(u_L)}$ (reps. $\Sigma^{(u_R)}$) be the set of symbols associated to the leaves of the subtree rooted at u_L (resp. u_R).
3. Assign to node u the binary string obtained from s replacing the symbols in $\Sigma^{(u_L)}$ with 0, and the symbols in $\Sigma^{(u_R)}$ with 1.
4. Let s_L denote the string obtained from s removing the symbols in $\Sigma^{(u_R)}$. If $|s_L| > 0$, TreeLabel(u_L, s_L).
5. Let s_R denote the string obtained from s removing the symbols in $\Sigma^{(u_L)}$. If $|s_R| > 0$, TreeLabel(u_R, s_R).

Fig. 1. Procedure TreeLabel for building the full Wavelet Tree $W_f(s)$ given the alphabetic tree T_Σ and the string s. The procedure is called with $u = root(T)$.

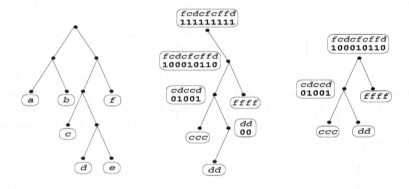

Fig. 2. An alphabetic tree (left) for the alphabet $\Sigma = \{\texttt{a,b,c,d,e,f}\}$. Given the string $s = \texttt{fcdcfcffd}$, we show its full (center) and pruned (right) Wavelet Trees.

algorithm for encoding binary strings. For any internal node u we denote by $C^*(u)$ the length of the encoding of $s^{01}(u)$ via $\mathcal{C}$, that is, $C^*(u) = |\mathcal{C}(s^{01}(u))|$. With a little abuse of notation we write $C^*(W_p(s))$ to denote the total cost of encoding the Wavelet Tree $W_p(s)$. Namely, $C^*(W_p(s)) = \sum_{u \in W_p(s)} C^*(u)$, where the sum is done over the internal nodes only. $C^*(W_f(s))$ is defined similarly. The following fundamental property of pruned Wavelet Trees was established in [7] and shows that there is essentially no loss in compression performance when we compress an arbitrary string s using Wavelet Trees and a binary encoder.

Theorem 1 (Grossi et al., ACM Soda 2003). *Let $\mathcal{C}$ be a binary encoder such that for any binary string z the bound $|\mathcal{C}(z)| \leq \lambda|z|H_0(z) + \eta|z| + \mu$ holds with constant λ, η, μ. Then, for a string s drawn from any alphabet $\Sigma^{(s)}$, we have $C^*(W_p(s)) = \sum_{u \in W_p(s)} |\mathcal{C}(s^{01}(u))| \leq \lambda|s|H_0(s) + \eta|s| + (|\Sigma^{(s)}| - 1)\mu$. The bound holds regardless of the shape of $W_p(s)$. The same result holds when the entropy H_0 is replaced by the modified entropy H_0^*.* □

3 Achieving 0-th Order Entropy with Wavelet Trees

This section contains a technical outline of the results claimed in (**A**) of the Introduction, where Wavelet Trees are used as stand alone, general purpose, order-zero compressors. In particular, we analyze the performance of RLE Wavelet Trees (Section 3.1) and GE Wavelet Trees (Section 3.2) showing that GE is superior to RLE as an order-zero compressor over Wavelet Trees. Nevertheless, we will show in Section 4 that GE Wavelet Trees, unlike RLE Wavelet Trees, are unable to achieve the k-th order entropy when used to compress the output of the Burrows-Wheeler Transform. This provides a theoretical ground to the practical choices and experimentation made in [6,8]. Moreover, a remarkable corollary of this section is a theoretical analysis of Inversion Frequencies coding [2].

Let $\mathcal{C}_{PF}$ denote a prefix-free encoding of the integers having logarithmic cost, namely $|\mathcal{C}_{PF}(n)| \le a \log n + b$, for $n \ge 1$. Note that since $|\mathcal{C}_{PF}(1)| \le b$ we must have $b \ge 1$. Also note that for γ codes we have $a = 2$ and $b = 1$. This means that it is worthwhile to investigate only prefix codes with $a \le 2$. Indeed, a code with $a > 2$ (and necessarily $b \ge 1$) would be worse than γ codes for any n and therefore not interesting. Hence in the following we assume $a \le 2$, $b \ge 1$ and thus $a \le 2b$ and $a \le b + 1$.

3.1 Analysis of RLE Wavelet Trees

For any binary string $s = a_1^{\ell_1} a_2^{\ell_2} \cdots a_k^{\ell_k}$, with $a_i \in \{0, 1\}$ and $a_i \ne a_{i+1}$, we define $\mathcal{C}_{RLE}(s) = a_1 \mathcal{C}_{PF}(\ell_1) \mathcal{C}_{PF}(\ell_2) \cdots \mathcal{C}_{PF}(\ell_k)$. Note that we need to store explicitly the bit a_1 since the values $\ell_1, \ldots, \ell_k$ alone are not sufficient to retrieve s.

Lemma 1. *For any binary string* $s = a_1^{\ell_1} a_2^{\ell_2} \cdots a_k^{\ell_k}$, *with* $a_i \in \{0, 1\}$ *and* $a_i \ne a_{i+1}$, *we have* $|\mathcal{C}_{RLE}(s)| = 1 + \sum_{i=1,k} |\mathcal{C}_{PF}(\ell_i)| \le 2 \max(a, b)|s|H_0^*(s) + b + 1$. □

Combining the above Lemma with Theorem 1 we immediately get:

Corollary 1. *For any string* s *over the alphabet* $\Sigma^{(s)}$, *if the internal nodes of the Wavelet Tree* $W_p(s)$ *are encoded using* RLE *we have*

$$C^*(W_p(s)) \le 2 \max(a, b)|s|H_0^*(s) + (|\Sigma^{(s)}| - 1)(b + 1). \qquad \square$$

Consider now the algorithm rle_wt defined as follows. We first encode $|s|$ using $|\mathcal{C}_{PF}(|s|)| \le a \log |s| + b$ bits. Then we encode the internal nodes of the Wavelet Tree using RLE. The internal nodes are encoded in a predetermined order—for example heap order—such that the encoding of a node u always precedes the encoding of its children (if any).[1] This ensures that from the output of rle_wt we can always retrieve s. To see this, we observe that when we start the decoding of the string $s^{01}(u)$ we already know its length $|s^{01}(u)|$ and therefore no additional bits are needed to mark the end of the run-length encoding. Since the output of rle_wt consists of $|\mathcal{C}_{PF}(|s|)| + C^*(W_p(s))$ bits, by Corollary 1 we get:

Theorem 2. *For any string* s *over the alphabet* $\Sigma^{(s)}$ *we have*

$$|\text{rle\_wt}(s)| \le 2 \max(a, b)|s|H_0^*(s) + (b + 1)|\Sigma^{(s)}| + a \log |s| - 1. \qquad \square$$

---

[1] We are assuming that the Wavelet Tree shape is hard-coded in the (de)compressor.

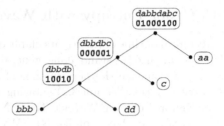

Fig. 3. The skewed Wavelet Tree for the string $s =$ **dabbdabc**. Symbol **b** is the most frequent one and is therefore associated to the leftmost leaf.

3.2   Analysis of GE Wavelet Trees

For any binary string s with exactly r 1's, let $p_1, p_2, \ldots, p_r$ denote their positions in s, and let $g_1, \ldots, g_r$ be defined by $g_1 = p_1$, $g_i = p_i - p_{i-1}$ for $i = 2, \ldots, r$. We denote by $\mathcal{C}_{Gap}(s)$ the concatenation $\mathcal{C}_{PF}(g_1)\mathcal{C}_{PF}(g_2)\cdots\mathcal{C}_{PF}(g_r)$.

Lemma 2. *Let s be a binary string with r 1's. If $1 \le r \le |s|/2$, we have*

$$|\mathcal{C}_{Gap}(s)| = |\mathcal{C}_{PF}(g_1)| + \cdots + |\mathcal{C}_{PF}(g_r)| \le \max(a,b)|s|H_0(s). \qquad \square$$

Let s be a string over the alphabet $\Sigma^{(s)}$. Consider the following algorithm called ge_wt. First we encode the length of s using $a \log |s| + b$ bits and the number of occurrences of each symbol using a total of $|\Sigma^{(s)}| \lceil \log |s| \rceil$ bits. Then, we build a Wavelet Tree completely skewed to the left such that the most frequent symbol is associated to the leftmost leaf. The other symbols are associated to the leaves in reverse alphabetic order (see Fig. 3). Finally, we use GE to encode the strings $s^{01}(u_1), \ldots, s^{01}(u_{|\Sigma^{(s)}|-1})$ associated to the internal nodes of such Wavelet Tree. Note that this information is sufficient to reconstruct the input string s. The crucial point is that the decoding starts with the retrieval of the number of occurrences of each symbol. Hence, we can immediately determine the association between leaves and symbols and when we later decode a string $s^{01}(u_i)$ we already know its length and the number of 1's in it.

Theorem 3. *For any string s, it is*

$$|\mathbf{ge\_wt}(s)| \le \max(a,b)|s|\,H_0(s) + |\Sigma^{(s)}|\,\lceil \log |s| \rceil + a \log |s| + b.$$

Proof. We only need to show that $\sum_i |\mathcal{C}_{Gap}(s^{01}(u_i))| \le \max(a,b)|s|\,H_0(s)$. To this end we observe that assigning the most frequent symbol to the leftmost leaf ensures that each $s^{01}(u_i)$ contains more 0's than 1's. The thesis follows by Lemma 2 and Theorem 1. $\qquad \square$

Let us give a closer look to the ge_wt algorithm when $\Sigma = \{a_1, a_2, \ldots, a_h\}$ and a_h is the most frequent symbol. In this case, when we encode $s^{01}(u_i)$ we are encoding the positions of the symbol a_i in the string s with the symbols $a_1, \ldots, a_{i-1}$ removed. In other words, we are encoding the number of occurrences of $a_{i+1}, \ldots, a_h$ between two consecutive occurrences of a_i. This strategy is known as *Inversion Frequencies* (IF) and was first suggested in [2] as an alternative to Move-to-Front (MTF) encoding. We have therefore the following result.

Corollary 2. *The variant of* IF*-coding in which the most frequent symbol is processed last produces a sequence of integers that we can encode with C_{PF} in at most* $\max(a, b)|s| H_0(s)$ *bits.* $\Box$

Standard analysis of MTF says that combining C_{PF} with MTF outputs at most $a|s| H_0(s) + b|s|$ bits. Hence, the above corollary is the first theoretical justification of the fact, observed by practitioners, that IF-coding is superior to MTF [1,2]. Corollary 2 also provides a theoretical justification for the strategy, suggested in [1], of processing the symbols in order of increasing frequency.

4   Achieving H_k^* with RLE Wavelet Trees and bwt

This section provides the technical details about the results claimed in (**B**) of the Introduction. In particular, we show that by using RLE Wavelet Trees as a post-processor of the bwt one can achieve higher order entropy compression (cfr [10]). We also show that the same result cannot hold for GE Wavelet Trees.

We need to recall a key property of the Burrows-Wheeler Transform of a string σ [11]: If $s = \mathsf{bwt}(\sigma)$ then for any $k \geq 0$ there exists a partition $s = s_1 s_2 \cdots s_t$ such that[2] $t \leq |\Sigma|^k$ and $|\sigma| H_k^*(\sigma) = \sum_{i=1}^{t} |s_i| H_0^*(s_i)$. In other words, the bwt is a tool for achieving the k-th order entropy H_k^* provided that we can achieve the entropy H_0^* on each s_i. An analogous result holds for H_k as well.

The proof idea is to show that compressing the whole s via one RLE Wavelet Tree is not much worse than compressing each string s_i separately. In order to prove such a result, some care is needed. We can assume without loss of generality that $\Sigma^{(s)} = \Sigma$. However, $\Sigma^{(s_i)}$ will not, in general, be equal to $\Sigma^{(s)}$ and this creates some technical difficulties and forces us to consider both full and pruned Wavelet Trees. Indeed, if we "slice" the Wavelet Tree $W_p(s)$ according to the partition $s = s_1 \cdots s_t$ we get *full* Wavelet Trees for the strings s_i's.

Our first lemma states that, for full RLE Wavelet Trees, partitioning a string does not improve compression.

Lemma 3. *Let $\alpha = \alpha_1 \alpha_2$ be a string over the alphabet Σ. We have $C^*(W_f(\alpha)) \leq C^*(W_f(\alpha_1)) + C^*(W_f(\alpha_2))$.* $\Box$

Since Theorem 1 bounds the cost of *pruned* Wavelet Trees, in order to use Lemma 3 we need to bound $C^*(W_f(\alpha_i))$ in terms of $C^*(W_p(\alpha_i))$, for $i = 1, 2$.

Lemma 4. *Let β be a string over the alphabet Σ. We have*

$$C^*(W_f(\beta)) \leq C^*(W_p(\beta)) + (|\Sigma| - 1)(a \log |\beta| + b + 1).$$ $\Box$

---

[2] For simplicity we ignore the end-of-file symbol and the first k symbols of σ that do not belong to any s_i. We will take care of these details in the full version.

We are now able to bound the size of a RLE Wavelet Tree over the string $s = \text{bwt}(\sigma)$ in terms of the k-th order entropy of σ.

Theorem 4. *Let σ denote a string over the alphabet $\Sigma = \Sigma^{(s)}$, and let $s = \text{bwt}(\sigma)$. For any $k \geq 0$ we have*

$$C^*(W_p(s)) \leq 2\max(a,b)|\sigma|\, H_k^*(\sigma) + |\Sigma|^{k+1}(2b + 2 + a\log(|\sigma|)). \qquad (2)$$

In addition, if $|\Sigma| = O(\text{polylog}(|\sigma|))$, for all $k \leq \alpha \log_{|\Sigma|} |\sigma|$, constant $0 < \alpha < 1$, we have

$$C^*(W_p(s)) \leq 2\max(a,b)|\sigma|\, H_k^*(\sigma) + o(|\sigma|). \qquad (3)$$

Proof. Let $s = s_1 \cdots s_t$ denote the partition of s such that $|\sigma|\, H_k^*(\sigma) = \sum_{i=1}^{t} |s_i|\, H_0^*(s_i)$. By Lemma 3, and the fact that $\Sigma = \Sigma^{(s)}$, we have that $C^*(W_p(s)) = C^*(W_f(s)) \leq \sum_{i=1}^{t} C^*(W_f(s_i))$. By Lemma 4, we get

$$C^*(W_p(s)) \leq \sum_{i=1}^{t} C^*(W_p(s_i)) + (|\Sigma| - 1)\sum_{i=1}^{t}(a\log|s_i| + b + 1)$$

$$= \sum_{i=1}^{t} C^*(W_p(s_i)) + (|\Sigma| - 1)\sum_{i=1}^{t} a\log|s_i| + t(|\Sigma| - 1)(b + 1)$$

Since $\sum_{i=1}^{t} \log|s_i| \leq t\log(|s|/t)$ and $t \leq |\Sigma|^k$, using Corollary 1 we get

$$C^*(W_p(s)) \leq 2\max(a,b)\Big(\sum_{i=1}^{t} |s_i|\, H_0^*(s_i)\Big)$$
$$+ t(|\Sigma| - 1)a\log(|s|/t) + 2t(|\Sigma| - 1)(b + 1) \qquad (4)$$
$$\leq 2\max(a,b)|\sigma|\, H_k^*(\sigma) + t(|\Sigma| - 1)(2b + 2 + a\log(|s|/t))$$

which implies (2) since $|s| = |\sigma|$. To prove (3) we start from (4) and note that $|\Sigma|$'s size and the inequality $t \leq |\Sigma|^k$ imply $t|\Sigma|\log(|s|/t) = o(|s|) = o(|\sigma|)$. $\qquad \square$

Theorem 4 shows that RLE Wavelet Trees achieve the k-th order entropy with the same multiplicative constant $2\max(a,b)$ that RLE achieves with respect to H_0^* (Lemma 1). Thus, Wavelet Trees are a sort of *booster* for RLE (cfr. [4]). It is possible to prove (details in the full paper) that if we apply the Compression Boosting algorithm [4] to RLE we get slightly better bounds than the ones of Theorem 4, the improvement being in the term not containing H_k^*. However, Compression Boosting makes use of a non trivial (even if linear time) partitioning algorithm. It is therefore not obvious which approach is preferable in practice.

In proving Theorem 4 we have used some rather coarse upper bounds and we believe that the result can be significantly improved. However, there are some limits to the possible improvements. The following example shows that, even for constant size alphabets, the $o(|\sigma|)$ term in (3) cannot be reduced to $\Theta(1)$.

Example 1. Let $\Sigma = \{1, 2, \ldots, m\}$, and let $\sigma = (123 \cdots m)^n$. We have $|\sigma| H_1^*(\sigma) \approx m\log n$ and $s = \text{bwt}(\sigma) = m^n 1^n 2^n \cdots (m-1)^n$. Consider a balanced Wavelet Tree of height $\lceil \log m \rceil$. It is easy to see that there exists an alphabet ordering such that the internal nodes of the Wavelet Tree all consist of alternate sequences of 0^n and 1^n. Even encoding these sequences with $\log n$ bits each would yield a total cost of about $(m \log m) \log n \approx (\log m)|\sigma|H_1^*(\sigma)$ bits. $\square$

Finally, it is natural to ask whether we can repeat the above analysis and prove a bound for GE Wavelet Trees in terms of the k-th order entropy. Unfortunately the answer is no! The problem is that when we encode s with GE we have to make some global choices—e.g., the shape of the tree in ge_wt, the role of zeros or ones in each internal node in the algorithm of [8]—and these are not necessarily good choices for every substring s_i. Hence, we can still split $W_f(s)$ into $W_f(s_1), \ldots, W_f(s_t)$, but it is not always true that $W_f(s_i) \leq \lambda |s_i| H_0(s_i) + o(|s_i|)$. As a more concrete example, consider the string $\sigma = (01)^n$. We have $|\sigma| H_1^*(\sigma) = \Theta(\log n)$ and $s = \mathsf{bwt}(\sigma) = 1^n 0^n$. $W_p(s)$ consists only of the root with associated string $1^n 0^n$, that can encode the gaps between either 1's or 0's. In both cases the output will be $\Theta(n)$ bits, thus exponentially larger than $|\sigma| H_1^*(\sigma)$.

5   Generalized Wavelet Trees

In point (**C**) of the Introduction we discussed the impact on the cost of a Wavelet Tree of: (C.1) its (binary) shape, (C.2) the assignment of alphabet symbols to its leaves, (C.3) the possible use of non-binary compressors to encode the strings associated to its internal nodes. Those examples motivate us to introduce and discuss Generalized Wavelet Trees, a new paradigm for the design of effective order-zero compressors. Let $\mathcal{C}_{01}$ and $\mathcal{C}_\Sigma$ be two compressors such that $\mathcal{C}_{01}$ is specialized to binary strings while $\mathcal{C}_\Sigma$ is a generic compressor. We assume that $\mathcal{C}_{01}$ and $\mathcal{C}_\Sigma$ satisfy the following property, which holds—for example—when $\mathcal{C}_{01}$ is RLE (with γ codes or order-2 Fibonacci codes used for the coding of integers, see e.g. Lemma 1) and $\mathcal{C}_\Sigma$ is Arithmetic or Huffman coding.

Property 1. (a) For any binary string x, $|\mathcal{C}_{01}(x)| \leq \alpha |x| H_0^*(x) + \beta$ bits, where α and β are constants; (b) For any string y, $|\mathcal{C}_\Sigma(y)| \leq |y| H_0(y) + \eta |y| + \mu$ bits, where η and μ are constants; (c) the running time of $\mathcal{C}_{01}$ and $\mathcal{C}_\Sigma$ is a convex function (say T_{01} and T_Σ) and their working space is a non decreasing function (say S_{01} and S_Σ). $\square$

Given the Wavelet Tree $W_p(s)$, a subset $\mathcal{L}$ of its nodes is a *leaf cover* if every leaf of $W_p(s)$ has a *unique* ancestor in $\mathcal{L}$ (see [4, Sect. 4]). Let $\mathcal{L}$ be a leaf cover of $W_p(s)$ and let $W_p^{\mathcal{L}}(s)$ be the tree obtained by removing all nodes in $W_p(s)$ descending from nodes in $\mathcal{L}$. We assign colors to nodes of $W_p^{\mathcal{L}}(s)$ as follows: all leaves are *black* and the remaining nodes *red*. We use $\mathcal{C}_{01}$ to compress all binary strings $s^{01}(u)$, $u \in W_p^{\mathcal{L}}(s)$ and *red*, while we use $\mathcal{C}_\Sigma$ to compress all strings $s(u)$, $u \in W_p^{\mathcal{L}}(s)$ and *black*. Nodes that are leaves of $W_p(s)$ are ignored (as usual). It is a simple exercise to work out the details on how to make this encoding decodable.[3] The *cost* $C^*(W_p^{\mathcal{L}}(s))$ is the total number of bits produced

---

[3] Note that we need to encode which compressor is used at each node and (possibly) the tree shape. For simplicity in the following we ignore this $\Theta(|\Sigma|)$ bits overhead.

---

(1) If r is the only node, let $C_{opt}(r) \leftarrow |\mathcal{C}_{01}(s)|$ and $\mathcal{L}(r) \leftarrow \{r\}$.
(2) Else, visit $W_p(s)$ in post-order. Let u be the currently visited node.
    (2.1) If u is a leaf, let $Z(u) \leftarrow 0$ and $\mathcal{L}(u) \leftarrow \{u\}$. Return.
    (2.2) Compute $Z(u) \leftarrow \min\{|\mathcal{C}_{\Sigma}(s(u))|, |\mathcal{C}_{01}(s^{01}(u))| + Z(u_L) + Z(u_R)\}$.
    (2.3) If $Z(u) = |\mathcal{C}_{\Sigma}(s(u))|$ then $\mathcal{L}(u) \leftarrow \{u\}$, else $\mathcal{L}(u) \leftarrow \mathcal{L}(u_L) \cup \mathcal{L}(u_R)$.
(3) Set $\mathcal{L}_{\min} \leftarrow \mathcal{L}(root(T))$.

Fig. 4. The pseudocode for the linear-time computation of an optimal leaf cover $\mathcal{L}_{\min}$ for a given decomposition tree T_s

by the encoding process just described. In particular, a *red* node u contributes $|\mathcal{C}_{01}(s^{01}(u))|$ bits, while a *black* node contributes $|\mathcal{C}_{\Sigma}(s(u))|$ bits.

Example 2. When $\mathcal{L} = root(W_p(s))$ we compress s using $\mathcal{C}_{\Sigma}$ only. By Property 1(b) we have $C^*(W_p^{\mathcal{L}}(s)) \leq |s|H_0(s) + \eta|s| + \mu$. The other extreme case is when $\mathcal{L}$ consists of all the leaves of $W_p(s)$. In this case we never use $\mathcal{C}_{\Sigma}$ and we have $C^*(W_p^{\mathcal{L}}(s)) \leq \alpha|s|H_0^*(s) + \beta(|\Sigma| - 1)$ by Property 1(a) and Theorem 1. $\square$

We note that when the algorithms $\mathcal{C}_{01}$ and $\mathcal{C}_{\Sigma}$ are fixed, the cost $C^*(W_p^{\mathcal{L}}(s))$ depends on two factors: the shape of the alphabetic tree T_{Σ}, and the leaf cover $\mathcal{L}$. The former determines the shape of the Wavelet Tree, the latter determines the assignment of $\mathcal{C}_{01}$ and $\mathcal{C}_{\Sigma}$ to the nodes of $W_p(s)$. It is natural to consider the following two optimization problems.

Problem 1. Given a string s and a Wavelet Tree $W_p(s)$, find the optimal leaf cover $\mathcal{L}_{\min}$ that minimizes the cost function $C^*(W_p^{\mathcal{L}}(s))$. Let $C_{opt}^*(W_p(s))$ be the corresponding optimal cost.

Problem 2. Given a string s, find an alphabetic tree T_{Σ} and a leaf cover $\mathcal{L}_{\min}$ for that tree giving the minimum of the function $C_{opt}^*(W_p(s))$. That is, we are interested in finding both a shape of the Wavelet Tree, and an assignment of $\mathcal{C}_{01}$ and $\mathcal{C}_{\Sigma}$ to the Wavelet Tree nodes, so that the resulting compressed string is the shortest possible.

Problem 2 is a global optimization problem, while Problem 1 is a much more constrained local optimization problem. Note that by Example 2 we have $C_{opt}^*(W_p(s)) \leq \min(|s|H_0^*(s) + \eta|s| + \mu, \alpha|s|H_0^*(s) + \beta(|\Sigma| - 1))$.

5.1 Optimization Algorithms (Sketch)

The first algorithm we sketch is an efficient algorithm for the solution of Problem 1. The pseudo-code is given in Figure 4. We have

Theorem 5. *Given two compressors satisfying Property 1 and a Wavelet Tree $W_p(s)$, the algorithm in Figure 4 solves Problem 1 in $O(|\Sigma|(T_{01}(|s|) + T_{\Sigma}(|s|)))$ time and $O(|s|\log|s| + \max(S_{01}(|s|), S_{\Sigma}(|s|)))$ bits of space.*

Proof. (Sketch). The correctness of the algorithm hinges on a *decomposability property* of the cost functions associated to $\mathcal{L}_{\min}$ with respect to the subtrees of

$W_p(s)$. Such a property is essentially the same used in [4, Sect. 4.5], here exploited to devise an optimal Generalized Wavelet Tree. As for the time analysis, it is based on the convexity of the functions $T_{01}(\cdot)$ and $T_\Sigma(\cdot)$ which implies that on any Wavelet Tree level we spend $O(T_{01}(|s|) + T_\Sigma(|s|))$ time.   □

Since we are assuming that the alphabet is of constant size, the algorithm of Figure 4 can be turned into an exhaustive search procedure for the solution of Problem 2. The time complexity would be polynomial in $|s|$ but at least exponential in $|\Sigma|$. Although we are not able to provide algorithms for the global optima with time complexity polynomial both in $|\Sigma|$ and $|s|$, we are able to settle the important special case in which the ordering of the alphabet is assigned. Using Dynamic Programming techniques, we can show:

Theorem 6. *Consider a string s and fix an ordering $\prec$ of the alphabet symbols appearing in the string. Then, one can solve Problem 2 constrained to that ordering of Σ, in $O(|\Sigma|^4(T_{01}(|s|) + T_{gen}(|s|))$ time.*   □

References

1. J. Abel. Improvements to the Burrows-Wheeler compression algorithm: After BWT stages. http://citeseer.ist.psu.edu/abel03improvements.html.
2. Z. Arnavut and S. Magliveras. Block sorting and compression. In *DCC: Data Compression Conference*, pages 181–190. IEEE Computer Society TCC, 1997.
3. M. Burrows and D. Wheeler. A block sorting lossless data compression algorithm. Technical Report 124, Digital Equipment Corporation, 1994.
4. P. Ferragina, R. Giancarlo, G. Manzini, and M. Sciortino. Boosting textual compression in optimal linear time. *Journal of the ACM*, 52:688–713, 2005.
5. P. Ferragina and G. Manzini. Indexing compressed text. *Journal of the ACM*, 52(4):552–581, 2005.
6. L. Foschini, R. Grossi, A. Gupta, and J. Vitter. Fast compression with a static model in high order entropy. In *DCC: Data Compression Conference*, pages 62–71. IEEE Computer Society TCC, 2004.
7. R. Grossi, A. Gupta, and J. Vitter. High-order entropy-compressed text indexes. In *Proc. 14th Annual ACM-SIAM Symp. on Discrete Algorithms (SODA '03)*, pages 841–850, 2003.
8. R. Grossi, A. Gupta, and J. Vitter. When indexing equals compression: Experiments on compressing suffix arrays and applications. In *Proc. 15th Annual ACM-SIAM Symp. on Discrete Algorithms (SODA '04)*, pages 636–645, 2004.
9. R. Grossi and J. Vitter. Compressed suffix arrays and suffix trees with applications to text indexing and string matching. *SIAM Journal on Computing*, 35:378–407, 2005.
10. V. Mäkinen and G. Navarro. Succinct suffix arrays based on rul-length encoding. *Nordic Journal of Computing*, 12(1):40–66, 2005.
11. G. Manzini. An analysis of the Burrows-Wheeler transform. *Journal of the ACM*, 48(3):407–430, 2001.

Atomic Congestion Games Among Coalitions[*]

Dimitris Fotakis[1], Spyros Kontogiannis[2,3], and Paul Spirakis[3]

[1] Dept. of Information and Communication Systems Engineering,
University of the Aegean, 83200 Samos, Greece
`fotakis@aegean.gr`
[2] Computer Science Department, University of Ioannina,
45110 Ioannina, Greece
`kontog@cs.uoi.gr`
[3] Research Academic Computer Technology Institute,
P.O. Box 1382, N. Kazantzaki Str., 26500 Rio–Patra, Greece
{`kontog, spirakis`}`@cti.gr`

Abstract. We consider algorithmic questions concerning the existence, tractability and quality of atomic congestion games, among users that are considered to participate in (static) selfish coalitions. We carefully define a coalitional congestion model among atomic players.

Our findings in this model are quite interesting, in the sense that we demonstrate many similarities with the non–cooperative case. For example, there exist potentials proving the existence of Pure Nash Equilibria (PNE) in the (even unrelated) parallel links setting; the Finite Improvement Property collapses as soon as we depart from linear delays, but there is an exact potential (and thus PNE) for the case of linear delays, in the network setting; the Price of Anarchy on identical parallel links demonstrates a quite surprising threshold behavior: it persists on being asymptotically equal to that in the case of the non–cooperative KP-model, unless we enforce a sublogarithmic number of coalitions.

We also show crucial differences, mainly concerning the hardness of algorithmic problems that are solved efficiently in the non–cooperative case. Although we demonstrate convergence to robust PNE, we also prove the hardness of computing them. On the other hand, we can easily construct a generalized fully mixed Nash Equilibrium. Finally, we propose a new improvement policy that converges to PNE that are robust against (even dynamically forming) coalitions of small size, in pseudo–polynomial time.

Keywords: Game Theory, Atomic Congestion Games, Coalitions, Convergence to Equilibria, Price of Anarchy.

1 Introduction

The new research field of algorithmic game theory, till now, had its focus mostly on *non–cooperative* strategic games. Indeed, the cases where dynamic game elements are studied are rather rare. In addition, the main concept of algorithmic

[*] Partially supported by EU / 6th Framework Programme, contract 001907 (DELIS).

M. Bugliesi et al. (Eds.): ICALP 2006, Part I, LNCS 4051, pp. 572–583, 2006.

game theory, namely, the *price of anarchy* (PoA), has been applied mostly to non–cooperative selfish players with a variety of pure strategy sets and payoff functions. Most of the examples are motivated from network traffic and congestion problems. But real life examples justify the necessity for the consideration of selfish coalitions, since in most cases there is some sort of hierarchy that needs to be taken into account. For example, in a communication network, the service providers are interested in minimizing their own cost for assuring a promised quality of service (eg, bandwidth) to their users, but they are not actually interested in utilizing their users' individual delays, so long as the maximum delay is small. In fact, in some cases they have to sacrifice the utilization of some of their users dictatorially, for the sake of their own (still private and selfish wrt other coalitions) objective. In this scenario, each service provider can be seen as a static coalition of users that tries to minimize the maximum cost that any of its users has to pay. Alternatively, we may actually have altruistic (wrt their participating users) coalitions that try to minimize the cumulative cost that all their users have to pay, by performing joint decisions for all their members.

All these scenarios are captured by (static) coalitional congestion games, in which the selfish players are actually the coalitions that may handle more than one users at will. In this work we are not interested in how this coalitional cost is shared among the members of the coalition, neither for the dynamics that lead to coalition formation. Our main goal is to study the existence, convergence, and quality of produced Nash Equilibria among given (static) coalitions, that are robust not only against unilateral moves of the users, but also against joint moves of subsets of any coalition.

1.1   Related Work

In this work we focus on coalitions of players in atomic congestion games. This implies that each coalition selfishly governs a unique subset of atomic players, whose traffic demands have to be routed via single paths (ie, unsplittably). Nevertheless, a coalition is allowed to choose different routes for different players that it handles. Our concern is the existence and construction of Pure Nash Equilibria in these games, as well as the effect of coalition formation to the quality of the game.

Existence of PNE and convergence issues have been extensively studied for the non–cooperative KP Model (eg, [7,8,10,11,19,23] and references therein) and networks (eg, [12,13,9,19,23]). It is well known that there is always a PNE in the non–cooperative KP Model, even for the case of unrelated parallel links. It is quite impressive that convergence to PNE in the non–cooperative KP Model may vary dramatically, depending on the families of improvement paths we consider. The quality of atomic non–cooperative congestion games has also been a hot topic in the recent literature, both for arbitrary Nash Equilibria (eg, [6,11,12,13,18,20] and references therein) and only among pure Nash Equilibria (eg, [1,2]). For an overview of recent developments on algorithmic questions on non–cooperative atomic congestion games, the interested reader is referred to [16].

A concept similar to atomic coalitional congestion games was recently discussed by Roughgarden [24] and Correa et al. [5] in a different framework (called *atomic splittable* setting): Each coalition has its own atomic weight which it routes selfishly from a source to a destination node in the network, but is allowed to do it in a non–atomic fashion: this weight can be split to infinitesimally small pieces that are then routed to the common destination. Indeed, Cominetti et al. [3] have proved that the proofs in [5,24] have a basic flow and many of the provided bounds on the price of anarchy are actually incorrect. They also provided the correct (albeit slightly weaker) bounds in some cases. Very recently (and independently), Hayrapetyan et al. [15] have dealt with the effect of coalition formation in congestion games. They consider static coalitions as well, but they focus on the TOTAL COST objective. They consider coalitions among splittable flows (which is closely related to the atomic splittable congestion games studied in [5,24]), as well as coalitions among atomic players (ie, whose traffic demands cannot be split at all). But they only consider the case where each atomic player has exactly unit traffic demand. They prove that coalition formation is not necessarily beneficial to the quality of the game (compared to the coalition–free game), unless we have an atomic splittable congestion game in a network of parallel links with convex link delay functions.

In our work, we mainly focus on a network of parallel links and coalitions among atomic players with arbitrary (integer) traffic demands, that have to assure a certain quality of service for the players, but are otherwise completely selfish. Therefore, we consider the MAX COST objective. In our case, it is easy to show that it is always beneficial (or at least, not worse) for the quality of the game if the players formed coalitions. We also provide some preliminary results for the TOTAL COST objective for the coalitions of unit size users in general networks.

1.2 Our Contribution and Roadmap

In section 2 we define the coalitional congestion models that we study and provide the necessary definitions and notation. For the case of parallel links (aka the KP Model [18]), in almost all cases (unless otherwise stated) we consider identical parallel links. Moreover, we consider the MAX COST (ie, the ∞–norm) measure for the coalitions: Each coalition has to pay for the maximum delay that any of its own users suffers. In section 3 we prove the existence of PNE in the coalitional KP Model, even for dynamically forming coalitions and unrelated parallel links.

In section 4 we prove pseudo–polynomial convergence time to 2–robust PNE for a wide family of improvement paths (we call it $SCF(2)$), that combine selfish ($\leqslant 2$–moves of even of dynamically forming coalitions, for identical parallel links. This family of improvement paths always gives priority to selfish moves of smaller coalitions that also involve the smallest number of links possible. We conjecture that similar results hold for k–robust PNE, for any constant $k \geqslant 2$. For efficient constructions of Nash Equilibria we have both bad and good news: Our bad news is that it is **NP**–complete to compute PNE for coalitional congestion games on parallel identical links, even if the number of coalitions is $1 + n(1 - \delta)$, for any constant $\delta \in (0, 1]$. On the positive side, we show pseudo–polynomial convergence time for $SCF(2)$, and also define a generalized fully mixed strategies profile for the coalitions that we then prove to be a mixed Nash Equilibrium.

In section 5 we prove asymptotically tight bounds for the Price of Anarchy of the Coalitional KP Model (with identical links). Indeed, we demonstrate a *threshold behavior* of the Price of Anarchy wrt the number of coalitions: It is as bad as in the case of the traditional (coalition–free) KP Model, so long as there are $\Omega\left(\frac{\log m}{\log\log m}\right)$ coalitions. From that point on, the Price of Anarchy drops linearly with the number of coalitions.

Finally, in section 6 we deal with network congestion games that allow static coalitions of unit-size users. Our results in this case are for the TOTAL COST (ie, the $1-$norm) measure for the coalitions: Each coalition pays for the sum of delays that its own users have to suffer. We prove that even very simple single–commodity network congestion games with linear or $2-$wise linear (ie, max of two linear functions) delay functions that allow static coalitions of players, may not possess the Finite Improvement Property, although a PNE may exist. Therefore, we cannot hope for potential–based arguments in these settings. On the other hand, if we restrict delay functions of the network to be linear, then every multi–commodity network congestion game that allows coalitions of players, possesses an exact potential, which is actually identical to that of the non–cooperative case [12].

Due to lack of space, the proofs are given in the full version of this work [14].

2   The Model of Static Coalitional Congestion Games

2.1   Coalitional KP Model

Consider a collection $[m]$[1] of identical parallel links and a collection $[n]$ of tasks. Each task must be uniquely allocated to any of the m available links. Each task $j \in [n]$ has an integer *demand* $w_j \in \mathbb{N}_+$ (eg, the number of elementary operations for the execution of task j). Let $\widetilde{W} = \{w_j\}_{j\in[n]}$ be the multiset of the tasks' demands. A set of $k \geqslant 1$ (static) coalitions $C_1, \ldots, C_k$ is a partition of $\widetilde{W}$ into k nonempty multisets. Hence: (i) the union (as multisets) of these coalitions is exactly $\widetilde{W}$, (ii) $C_j \neq \emptyset$, $\forall j \in [k]$, and (iii) $C_i \cap C_j = \emptyset$, $\forall i, j \in [k] : i \neq j$. For $j \in [k]$ let $C_j = \{w_j^1, \ldots, w_j^{n_j}\}$, so that $\sum_{j=1}^{k} n_j = n$. Denote by $W_j = \sum_{i=1}^{n_j} w_j^i$ the cumulative demand required by coalition C_j, while $W_{tot} = \sum_{j\in[k]} W_j$ is the overall demand required by the system. Wlog assume that $w_j^1 \geqslant \cdots \geqslant w_j^{n_j}$, $\forall j \in [k]$.

Strategies and Profiles. A *pure strategy* $\sigma_j = (\sigma_j^i)_{i\in[n_j]}$ for coalition C_j defines the deterministic selection of a link $\sigma_j^i \in [m]$ for each $w_j^i \in C_j$. Denote by Σ_j the set of all pure strategies available to coalition C_j. Clearly, $\Sigma_j = M^{n_j}$. A *mixed strategy* for coalition C_j is a probability distribution $\mathbf{p_j}$ on the set Σ_j of its pure strategies (ie, a point of the simplex $\Delta(\Sigma_j) \equiv \{\mathbf{q} \in \mathbb{R}^{n_j} : \mathbf{q} \geqslant \mathbf{0}; \ \mathbf{1}^T\mathbf{q} = 1\}$). In order to indicate the probability of pure strategy σ_j being chosen by C_j when

---

[1] For any integer $k \geqslant 1$, $[k] \equiv \{1, \ldots, k\}$.

p_j has been adopted, we use (for sake of simplicity) the functional notation $p_j(\sigma_j)$, rather than the coordinate of the vector $\mathbf{p_j}$ corresponding to σ_j.

A *pure strategies profile* or *configuration* is a collection $\sigma = (\sigma_j)_{j \in [k]}$ of pure strategies, one per coalition. $\Sigma \equiv \times_{j \in [k]} \Sigma_j$ is the set of all the possible configurations of the game (called the *configuration space*). (σ_{-j}, α_j) denotes the configuration resulting from a configuration σ when coalition C_j unilaterally changes its pure strategy from σ_j to α_j. The simplotope $\Delta(\Sigma) \equiv \times_{j \in [k]} \Delta(\Sigma_j)$ is the *mixed strategies space* of the coalitional game. A *mixed strategies profile* $\mathbf{p} = (\mathbf{p_j})_{j \in [k]} \in \Delta(\Sigma)$ is a collection of mixed strategies, one per coalition. The *support* of coalition C_j in the mixed profile $\mathbf{p}$ is the set $S_j(\mathbf{p}) = \{\sigma_j \in \Sigma_j : p_j(\sigma_j) > 0\}$; thus $S_j(\mathbf{p})$ is the set of pure strategies that C_j chooses with non-zero probability. If $S_j(\mathbf{p}) = \Sigma_j$ for all $j \in [k]$ then $\mathbf{p}$ is a *fully mixed* profile.

A special case of particular interest is when the coalitions are enforced to eventually choose consecutive links for their own tasks. In this case we shall refer to the *Coalitional Chains* model.

Selfish Costs. Fix a configuration $\sigma = (\sigma_j)_{j \in [k]}$. The load on link $\ell \in M$ due to coalition C_j is $\theta_\ell(\sigma_j) \equiv \sum_{i \in [n_j]:\sigma_j^i = \ell} w_j^i$. The total load on link $\ell \in [m]$ is the total demand on link ℓ with respect to σ, ie, $\theta_\ell(\sigma) = \sum_{j=1}^k \theta_\ell(\sigma_j)$. The load induced on link $\ell \in [m]$ by all the coalitions, except for coalition C_j is $\theta_\ell(\sigma_{-j}) = \sum_{j' \in [k] \setminus \{j\}} \sum_{i \in [n_{j'}]:\sigma_{j'}^i = \ell} w_{j'}^i$. The *selfish cost* $\lambda_j(\sigma)$ of coalition C_j is the maximum load over the set of links it uses: $\lambda_j(\sigma) = \max_{i \in [n_j]} \{\theta_{\sigma_j^i}(\sigma)\}$.

For a mixed profile $\mathbf{p}$, the load on each link $\ell \in [m]$ becomes a random variable induced by the probability distributions $\mathbf{p_j}$ for all $j \in [k]$. More specifically, we define the *expected load* on link $\ell \in [m]$ as the expectation of the load on link ℓ according to $\mathbf{p}$: $\theta_\ell(\mathbf{p}) = \sum_{\sigma \in \Sigma} \left[\left(\prod_{j \in [k]} p_j(s_j) \right) \cdot \theta_\ell(\sigma) \right]$. We can also determine the expected load that all the coalitions except for coalition C_j induce to some link ℓ: $\theta_\ell(\mathbf{p_{-j}}) = \sum_{\sigma_{-j} \in \Sigma_{-j}} \left[\left(\prod_{j' \in [k] \setminus \{j\}} p_{j'}(s_{j'}) \right) \cdot \theta_\ell(\sigma_{-j}) \right]$. The *conditional expected selfish cost* of coalition j adopting the pure strategy $\sigma_j \in \Sigma_j$, given that the other players follow the strategies indicated by $\mathbf{p}$, is defined as follows: $\lambda_j(\mathbf{p}, \sigma_j) = \max_{\ell \in S_j(\sigma_j)} \left\{ \left(\sum_{i \in C_j:\sigma_j^i = \ell} w_i \right) + \theta_\ell(\mathbf{p_{-j}}) \right\}$. Ie, coalition C_j pays for the the conditional expectation of its selfish cost, had it adopted the pure strategy $\sigma_j \in \Sigma_j$. This is because coalition C_j has to encounter all the possible alternatives for its own tasks prior to the other coalitions' determination of their actual action, knowing only their probability distributions. The *expected selfish cost* of coalition C_j is defined as the expectation of coalition C_j's conditional expected cost, over all possible actions that can be taken: $\lambda_j(\mathbf{p}) = \sum_{\sigma_j \in \Sigma_j} [p_j(\sigma_j) \cdot \lambda_j(\mathbf{p}, \sigma_j)]$. Observe that each coalition pays for the *expected maximum load* that it would cause as if it was on its own, plus the *expected loads* caused by the other coalitions to each of the links.

Nash Equilibria. The definition of expected selfish costs completes the definition of the finite non-cooperative game involving the κ coalitions of tasks that are to be assigned to the m links: $\Gamma = \langle [k], (\Sigma_j)_{j \in [k]}, (\lambda_j)_{j \in [k]} \rangle$. We are interested in the induced Nash Equilibria [22] of Γ. Informally, a Nash Equilibrium is a (pure or mixed) profile such that no coalition can reduce its selfish cost by unilaterally changing its strategy. Formally: A pure strategies profile $\sigma = (\sigma_j)_{j \in [k]}$ is a *Pure Nash Equilibrium (PNE)* for Γ if, $\forall j \in [k]$, $\forall \alpha_j \in \Sigma_j$, $\lambda_j(\sigma) \leqslant \lambda_j(\sigma_{-j}, \alpha_j)$. A mixed strategies profile $\mathbf{p}$ is a *Nash Equilibrium (NE)* if, $\forall j \in [k]$, $\forall \sigma_j \in \Sigma_j$, it holds that $p_j(\sigma_j) > 0 \Rightarrow \sigma_j \in \arg\min_{\alpha_j \in \Sigma_j} \{\lambda_j(\mathbf{p}, \alpha_j)\}$.

Social Cost, Social Optimum and Price of Anarchy. For any configuration $\sigma = (\sigma_j)_{j \in [k]}$, the *social cost*, denoted $\mathsf{SC}(\sigma)$, is the maximum load over the set of links M with respect to σ, ie, $\mathsf{SC}(\sigma) = \max_{\ell \in M}\{\theta_\ell(\sigma)\} = \max_{j \in [k]}\{\lambda_j(\sigma)\}$. For any mixed profile $\mathbf{p}$ the social cost is defined as the expectation, over all random choices of the coalitions, of the maximum load over the set of links: $\mathsf{SC}(\mathbf{p}) = \sum_{\sigma \in \Sigma} \left(\prod_{j=1}^{k} p_j(\sigma_j)\right) \cdot \max_{\ell \in [m]} \{\theta_\ell(\sigma)\}$. Now let σ^* be a configuration that minimizes the social cost function, ie, $\sigma^* \in \arg\min_\sigma \{\mathsf{SC}(\sigma)\}$. Thus σ^* is an optimal configuration of the set of loads $\widetilde{W}$ to the set of links $[m]$. We denote this value by $\mathsf{OPT} = \mathsf{SC}(\sigma^*)$. The *Price of Anarchy* (also referred to as *Coordination Ratio*) [18], is the worst–case ratio of the social cost paid at any NE over the value of the social optimum of the game: $\mathsf{R} = \max_{\mathbf{p} \text{ is NE}} \left\{\frac{\mathsf{SC}(\mathbf{p})}{\mathsf{OPT}}\right\}$.

Improvement Paths. When we discuss convergence issues, we shall frequently refer to the notion of improvement paths: These are sequences of configurations (ie, points in Σ), such that any two consecutive configurations differ only in the pure strategy of *exactly one* coalition, and additionally the cost of this unique coalition is *strictly less* in the latter configuration than in the former one.

2.2 Coalitional Players in Networks

Let $G = (V, E)$ be a directed network with a non-decreasing delay function $d_e(x)$ for each $e \in E$. Consider also a multiset of users of *identical traffic demands*, willing to be routed between unique source–destination pairs of nodes in the network. A network congestion game with coalitions is defined as follows: The set of players is the set of coalitions $\{C_1, \ldots, C_k\}$. Every coalition C_j consists of n_j users routing their traffic from s_j to t_j. Let $\mathcal{P}_j$ be the set of $s_j - t_j$ paths in G. The set of pure strategies of coalition C_j is $\mathcal{P}_j^{n_j}$. The load of any edge $e \in E$ due to the users of C_j in σ_j is $\theta_e(\sigma_j) = |\{i \in C_j : e \in \sigma_j^i\}|$. Let $\sigma = (\sigma)_{j \in [k]}$ be arbitrary configuration. For every edge $e \in E$, the load of e in σ is $\theta_e(\sigma) = \sum_{j=1}^{k} \theta_e(\sigma_j)$. For every path $\pi \in \bigcup_{j=1}^{k} \mathcal{P}_j$, the delay along π in σ is $d_\pi(\sigma) = \sum_{e \in \pi} d_e(\theta_e(\sigma))$. There are (at least) two natural notions of selfish cost of coalition C_j in σ. The first is the *maximum delay* over all paths used by C_j, denoted $\lambda_j(\sigma)$. Formally, $\lambda_j(\sigma) = \max_{i \in C_j}\{d_{\sigma^i}(\sigma)\}$. The second is the *total delay* of coalition C_j, denoted $\tau_j(\sigma)$. Formally, $\tau_j(\sigma) = \sum_{i \in C_j} d_{\sigma_j^i}(\sigma) = \sum_{e \in E} \theta_e(\sigma_j) d_e(\theta_e(\sigma))$. Both maximum delay and total delay generalize the notion selfish cost used in congestion games in a natural way. Total delay has been used before in the non-atomic setting (see eg [23,24]). The definitions of Nash Equilibria, social costs and price of anarchy extend in a straightforward manner to the case of network congestion games.

3   Tractability of NE in the Coalitional KP Model

It is not hard to prove the existence of PNEs in coalitional congestion games, even if we assume unrelated parallel links and we allow the players to form *dynamic coalitions* (ie, consider the case of arbitrary combinations of players that attempt a joint selfish move). This is done by showing the existence of a generalized ordinal potential function which assures the convergence to a PNE in a finite number of steps. For the case of unrelated parallel links (ie, $\forall j \in [n], \forall \ell \in [m], w_j(\ell) > 0$ indicates the additional load that task j enforces to link ℓ due to its allocation to it), we prove that any improvement path that combines selfish movements of (even dynamically forming) coalitions, has length at most $2^{W_{\text{tot}}}$ where here (for the more general case of unrelated parallel links) $W_{\text{tot}} = \sum_{j \in [n]} \max_{\ell \in [m]} \{w_j(\ell)\}$:

Theorem 1. *For the case of unrelated parallel links and selfish tasks with integer weights, any improvement path that combines arbitrary selfish movements of coalitions of players of size at most k (even if these are formed dynamically) has length at most $\frac{(2k)^{W_{\text{tot}}}}{2k-1}$.*

Proof. See the proof in the full version of the paper.   □

The above argument works not only for integer demands, but also (after some trivial modification) for arbitrary demands. The only difference is that rather than having a minimum difference of 1 when comparing loads, this will have to be substituted with some (sufficiently small) positive quantity and the convergence rate implied becomes much worse. Thus, the exponential potential used captures the power of the *lexicographic ordering* arguments for convergence.

Despite the existence of PNE, it is actually hard to compute one, even if we demand a very large number of coalitions of players, so long as we allow one coalition have large cardinality. This is shown in the next theorem:

Theorem 2. *For arbitrary static coalitional congestion games over identical parallel links among players with integer weights, it is **NP**−complete to find a PNE, even if we enforce a number of coalitions $k = 1 + n(1 - \delta)$, for any constant $\delta \in (0, 1]$.*

Proof. See the proof in the full version of the paper.   □

Observe that (by Theorem 1) any improvement path of maximal length that allows arbitrary selfish $(\leqslant k)$−moves of players converges monotonically to a PNE that is also robust against any (even dynamically forming) coalition of at most k players. We call such a PNE a k−*robust* PNE. In the next section we explore better than the previous bounds on arbitrary improvement paths that end up with k−robust PNE, at least in the restricted case of identical links.

4   Convergence Time to Robust Equilibria

In this section we initiate the study on the speed of convergence to PNE that are robust to arbitrary (even dynamically forming) selfish k−moves. Towards this direction, we seek for upper bounds of the convergence rate to 2−robust PNE. The rate of convergence assured by the exponential potential we use for the existence of PNEs, is rather poor. Our seek now is for a possibly better rate of convergence, at least in the case of identical parallel links and tasks with integer weights that we study in this paper. We already know that rather than using the exponential potential of the unrelated links, in the case of selfish 1−moves in a system of identical parallel links and tasks with integer weights there is a much better potential that assures pseudo−polynomial convergence to a PNE. Indeed, for specific strategies one can be based on combinatorial arguments to show either linear (for the MAX WEIGHT PRIORITY−BEST RESPONSE strategy) or quadratic (for the FIFO PRIORITY−BEST RESPONSE strategy) convergence to an arbitrary PNE (for more details see [7]).But we are also interested

in the convergence rate of an arbitrary improvement path with selfish 1−moves; one can use as a (weighted) potential the square of the loads of the links: $F(t) = \sum_{\ell \in [m]} (L_\ell(t))^2$, in order to prove pseudo−polynomial convergence time. Unfortunately, $F(t)$ is no longer a potential when we allow joint (selfish) moves of players, even if we consider only static coalitions of at most 2 players each. This is shown by the exam-

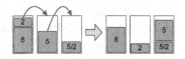

Fig. 1. An example of a selfish 2−move for which difference in the value of F is $-1 < 0$

ple of figure 1. Nevertheless, we shall consider a special family of improvement paths with selfish ($\leq k$)−moves, for which $F(t)$ is actually a potential. This family consists of improvement paths that always give priority to selfish moves of smaller coalitions (therefore we shall call the strategy that creates these paths SMALLER COALITIONS FIRST strategy, $SCF(k)$ in short). In particular, we start by allowing (in arbitrary order) selfish 1−moves, until no such move exists (ie, we are already at some 1−robust PNE). Consequently we check whether there is a selfish 2−move. If this is the case, then we make such a move, and then we perform again a maximal number of selfish 1−moves. If not, we have already reached a 2−robust PNE, and check whether there is a selfish 3−move. If there are selfish 3−moves, we allow one of them arbitrarily and consequently we perform (first) a maximal number of selfish 1−moves and (from 1-robust PNE) selfish 2−moves, before allowing another selfish 3−move. We proceed in this manner, until we reach a k−robust PNE, in which case we stop.

In order to demonstrate the significance of this family of improvement paths, we shall show that indeed for $SCF(2)$ (which allows up to selfish 2−moves and ends up with a 2−robust PNE), $F(t)$ is indeed a generalized ordinal potential (ie, the difference $F(t) - F(t + 1)$ is positive for selfish (≤ 2)−moves). This assures a pseudo−polynomial length for arbitrary improvement paths that mix selfish 1−moves and 2−moves arbitrarily, but always give priority to selfish 1−moves.

Whenever a 2−coalition of tasks (i, j) selfishly defects from their current hosts (ℓ_i, ℓ_j) respectively, towards two new links (ℓ'_i, ℓ'_j), we denote this by $(i, j) \rhd (\ell_i, \ell_j) \mapsto (\ell'_i, \ell'_j)$. For $\ell_i \neq \ell_j$, the 2−move $(i, j) \rhd (\ell_i, \ell_j) \mapsto (\ell_j, \ell_i)$ is called a 2−flip, while the 2−move $(i, j) \rhd (\ell_i, \ell_j) \mapsto (\ell_j, \ell'_j)$ such that $\ell'_j \notin \{\ell_i, \ell_j\}$ is called a $2-chain$. We first explain why we may focus only on selfish $2-$flips and $2-$chains for the paths of $SCF(2)$:

Lemma 1. *The only possible selfish $2-$moves of any element of $SCF(2)$ are either $2-$flips or $2-$chains.*

Proof. See the proof in the full version of the paper. □

The following lemma implies that if we restrict our attention to members of $SCF(2)$ that only allow selfish $2-$flips (and no $2-$chains), then we are fair to the coalitional players, in the sense that no coalition can gain more by doing a $2-$chain rather than its corresponding $2-$flip. This is quite interesting, since a selfish $2-$flip seems much more "selfishly motivated" than a selfish $2-$chain: a selfish $2-$flip is essentially equivalent to (actually, slightly stronger than) the selfish $1-$move of the difference of two weights from the link with the heavier load to the link with the lesser load. In contrast, the $2-$chain cannot be decomposed into a sequence of selfish $1-$moves of weights, since one of the two transfers of weights may not be actually selfish.

Lemma 2. *At an arbitrary PNE, the existence of a selfish $2-$chain implies the existence of a selfish $2-$flip that assures at least the same improvement for the coalitional player that performs the move.*

Proof. See the proof in the full version of the paper. □

The following theorem demonstrates the pseudo–polynomial convergence rate of any element of $SCF(2)$ that only allows selfish $2-$flips:

Theorem 3. *For an arbitrary system of identical parallel links and selfish users with integer weights, the function $F(t) = \sum_{\ell \in [m]} (L_\ell(t))^2$ is a weighted potential for all the members of $SCF(2)$ that only allow selfish $1-$moves and $2-$flips, and thus it assures their convergence to a $2-$robust PNE in at most $\frac{W_{\text{tot}}^2}{2}$ steps.*

Proof. See the proof in the full version of the paper. □

Consequently we shall demonstrate the inexistence of Fully Mixed NE for coalitional congestion games with coalitions of cardinality at least 2. We shall also prove the existence of the so-called generalized fully mixed NE, for any coalitional congestion game on identical parallel links:

Lemma 3. *For the system of parallel identical links and users with arbitrary weights, there exists a fully mixed NE if and only if $n_j = 1$ for all the coalitions $j \in [k]$. If on the other hand there are coalitions of more than one users, then a generalized fully mixed NE that assures no conflict among players of the same coalition, always exists.*

Proof. See the proof in the full version of the paper. □

Remark 1. If each coalition chooses uniformly at random from all the possible optimal allocations (that are also lexicographically minimum wrt the load vector of the links) of its own set of weights *alone* in the m links, then the produced mixed profile is also a generalized fully mixed NE. This is a simple extension of GFMNE to the cases of coalitions with more than m tasks.

5   Price of Anarchy in the Coalitional KP Model

Assume that $k = 1$, ie, there is a single coalition $C_1 = \widetilde{W} = \{w_1, \ldots, w_n\}$. Then, any NE must be an optimum assignment of the set of loads C_1 to the set of links $[m]$ and vice versa, hence in any NE σ, $\mathsf{SC}(\sigma) = \mathsf{OPT}$ and thus $\mathsf{R} = 1$. For $k = n$, ie, $C_j = \{w_j\}$ for all $j \in [n]$, this case reduces to the standard KP model [18] for which it is well known [17] that $\mathsf{R} = \Theta\left(\frac{\log m}{\log \log m}\right)$.

In this section, we prove that the price of anarchy is $\Theta\left(\min\left\{k, \frac{\log m}{\log \log m}\right\}\right)$, where m denotes the number of links and k denotes the number of coalitions. The lower bound (Theorem 4) holds even for identical tasks and coalitions of equal cardinality. The upper bound (Theorem 5) holds for n integer weights $w_1 \geqslant w_2 \geqslant \cdots \geqslant w_n > 0$ and arbitrary coalitions.

Theorem 4. *The price of anarchy is* $\Omega\left(\min\left\{k, \frac{\log m}{\log \log m}\right\}\right)$ *even for identical tasks and coalitions of equal cardinality.*

Proof. We consider m identical parallel links and m unit size tasks partitioned into $k \geqslant 2$ coalitions each with $r \equiv m/k$ tasks (wlog we assume that m/k is an integer). We first prove a lower bound for the Coalitional Chains Model:

Lemma 4. *In the Coalitional Chains Model, when the number of coalitions is* $\kappa = m^\varepsilon$ *for arbitrary constant* $\varepsilon \in (0,1]$, *the price of anarchy is* $\mathsf{R} = \Omega\left(\frac{\log m}{\log \log m}\right)$.

Proof. See the proof in the full version of the paper.   □

For the general case where the coalitions may adopt any possible pure strategy, we prove that the social cost of the GFMNE that we already proved to exist in Lemma 3, is $\Omega\left(\min\left\{k, \frac{\log m}{\log \log m}\right\}\right)$. Since the social optimum of the specific instance is 1, we get the desired lower bound. See the complete proof in the full version of the paper.   □

Theorem 5. *For every NE* $\mathbf{p}$, $\mathsf{SC}(\mathbf{p}) \leqslant \mathcal{O}\left(\min\left\{k, \frac{\log m}{\log \log m}\right\}\right) \mathsf{OPT}$, *where* m *denotes the number of (identical) links.*

Proof. See the proof in the full version of the paper.   □

6   NE in Coalitional Network Games

[23] shows that in a single–commodity network congestion game with coalitions (in the splittable setting) and the selfish cost being the total coalition delay, NE may not be unique even for two coalitions of different cardinalities and very simple networks (see Figure 3.a in full paper). On the other hand, the uniqueness of NE is established for the special case of two coalitions and edge delays $d_e(x) = a_e x^c$, for every $c \in \{1, \ldots, 7\}$.

Computing a coalition's best response for total delay in a single–commodity network congestion game can be performed by first applying a transformation similar to that in [9, Theorem 2] and then computing a min-cost flow. On the other hand, computing a coalition's best response for maximum delay is **NP**-hard even for single–commodity network congestion games with linear delays and a coalition of size 2 (see eg [4, Theorem 3]).

In the following, we focus on the second notion of selfish cost (total delay of coalition). We prove that a congestion game with coalitions is an exact potential game if the edge delays are linear (cf. Theorem 6). On the other hand, we give a simple example of a single–commodity network congestion game with coalitions of equal cardinality and 2-wise linear edge delays that does not have the *Finite Improvement Property* (FIP) (cf. Lemma 5). By [21, Lemma 2.5], this game does not admit any kind of potential function, even a generalized ordinal one.

Lemma 5. *There exist instances of single–commodity network congestion games with coalitions of equal cardinality and edge delays being either linear of 2-wise linear which do not have the Finite Improvement Property.*

Proof. See the proof in the full version of the paper.   □

Theorem 6. *Every multi–commodity (network) congestion game with coalitions and linear edge delays is an exact potential game.*

Proof. See the proof in the full version of the paper.   □

References

1. Awerbuch B., Azar Y., Epstein A. The price of routing unsplittable flow. In *Proc. of the 37th ACM Symp. on Th. of Comp. (STOC '05)*, pages 57–66, 2005.
2. G. Christodoulou and E. Koutsoupias. The Price of Anarchy of Finite Congestion Games. In *Proc. of the 37th ACM Symp. on Th. of Comp. (STOC '05)*, pages 67–73, 2005.
3. Cominetti R., Correa J.R., Stier Moses N.E. Network games with atomic players. In *Proc. of the 33rd Int. Col. on Automata, Lang. and Progr. (ICALP '06)*. Springer-Verlag, 2006. (to appear).
4. Correa J.R., Schulz A.S., Stier Moses N.E. Computational complexity, fairness, and the price of anarchy of the maximum latency problem. In *Proc. of the 10th Conf. on Int. Progr. and Comb. Opt. (IPCO '04) LNCS 3064*, pages 59–73, 2004.
5. Correa J.R., Schulz A.S., Stier Moses N.E. On the inefficiency of equilibria in congestion games. In *Proc. of the 11th Conf. on Int. Progr. and Comb. Opt. (IPCO '05)*, pages 167–181, 2005.
6. Czumaj A., Vöcking B. Tight bounds for worst-case equilibria. In *Proc. of the 13th ACM-SIAM Symp. on Discr. Alg. (SODA '02)*, pages 413–420, 2002.
7. Even-Dar E., Kesselman A., Mansour Y. Convergence time to nash equilibria. In *Proc. of the 30th Int. Col. on Automata, Lang. and Progr. (ICALP '03)*, pages 502–513. Springer-Verlag, 2003.
8. Even-Dar E., Mansour Y. Fast convergence of selfish rerouting. In *Proc. of the 16th ACM-SIAM Symp. on Discr. Alg. (SODA '05)*, pages 772–781. SIAM, 2005.

9. Fabrikant A., Papadimitriou C., Talwar K. The complexity of pure nash equilibria. In *Proc. of the 36th ACM Symp. on Th. of Comp. (STOC '04)*, 2004.

10. Feldmann R., Gairing M., Luecking T., Monien B., Rode M. Nashification and the coordination ratio for a selfish routing game. In *Proc. of the 30th Int. Col. on Automata, Lang. and Progr. (ICALP '03)*, pages 514–526. Springer-Verlag, 2003.

11. Fotakis D., Kontogiannis S., Koutsoupias E., Mavronicolas M., Spirakis P. The structure and complexity of nash equilibria for a selfish routing game. In *Proc. of the 29th Int. Col. on Automata, Lang. and Progr. (ICALP '02)*, pages 123–134. Springer-Verlag, 2002.

12. Fotakis D., Kontogiannis S., Spirakis P. Selfish unsplittable flows. *Theoretical Computer Science*, 348((2-3)):226–239, December 2005. Special Issue dedicated to ICALP 2004 (TRACK-A).

13. Fotakis D., Kontogiannis S., Spirakis P. Symmetry in network congestion games: Pure equilibria and anarchy cost. In *Proc. of the 3rd W. on Approx. and Online Algorithms (WAOA '05)*. Springer-Verlag, 2005.

14. Fotakis D., Kontogiannis S., Spirakis P. Atomic congestion games among coalitions. Technical report, DELIS, 2006. Accessible via http://delis.upb.de/docs/.

15. Hayrapetyan A., Tardos E., Wexler T. The effect of collusion in congestion games. In *Proc. of the 38th ACM Symp. on Th. of Comp. (STOC '06)*, 2006. (to appear).

16. Kontogiannis S., Spirakis P. Atomic selfish routing in networks: A survey. In *Proc. of the 1st W. on Internet and Net. Econ. (WINE '05)*, LNCS 3828, pages 989–1002. Springer, 2005.

17. Koutsoupias E., Mavronicolas M., Spirakis P. Approximate equilibria and ball fusion. *Theory of Computing Systems*, 36(6):683–693, 2003. Special Issue devoted to SIROCCO'02.

18. Koutsoupias E., Papadimitriou C. Worst-case equilibria. In *Proc. of the 16th Annual Symp. on Theor. Aspects of Comp. Sci. (STACS '99)*, pages 404–413. Springer-Verlag, 1999.

19. Libman L., Orda A. Atomic resource sharing in noncooperative networks. *Telecommunication Systems*, 17(4):385–409, 2001.

20. Mavronicolas M., Spirakis P. The price of selfish routing. In *Proc. of the 33rd ACM Symp. on Th. of Comp. (STOC '01)*, pages 510–519, 2001.

21. Monderer D., Shapley L. Potential games. *Games & Econ. Behavior*, 14:124–143, 1996.

22. Nash J. F. Noncooperative games. *Annals of Mathematics*, 54:289–295, 1951.

23. Orda A., Rom R., Shimkin N. Competitive routing in multi-user communication networks. *IEEE/ACM Trans. on Networking*, 1(5):510–521, 1993.

24. Roughgarden T. Selfish routing with atomic players. In *Proc. of the 16th ACM-SIAM Symp. on Discr. Alg. (SODA '05)*, pages 973–974, 2005.

Computing Equilibrium Prices in Exchange Economies with Tax Distortions

Bruno Codenotti[1], Luis Rademacher[2], and Kasturi Varadarajan[3]

[1] IIT-CNR, Pisa, Italy
[2] Department of Mathematics, MIT
[3] Department of Computer Science, The University of Iowa

Abstract. We consider the computation of equilibrium prices in market settings where purchases of goods are subject to taxation. While this scenario is a standard one in applied computational work, so far it has not been an object of study in theoretical computer science. Taxes introduce significant distortions: equilibria are no longer Pareto optimal, sufficient conditions for uniqueness do not continue to guarantee it, existence itself must be revisited. We analyze the effects of these distortions on scenarios which, in the absence of taxes, admit polynomial time algorithms. In spite of the loss of certain structural properties (including uniqueness), we are able to obtain polynomial time algorithms or approximation schemes in several instances where the model without taxes admitted them.

1 Introduction

The equilibrium problem for a pure exchange economy amounts to finding a set of prices and allocations of goods to economic agents such that each agent maximizes her utility, subject to her budget constraints, and the market clears. The equilibrium depends only on the agents' utility functions and initial endowments of goods.

If one aims at analyzing equilibrium problems arising from real world applications, the scenario outlined above has often to be extended. Indeed one needs to take into account the presence of suitable distortions, which might be, depending on the specific application, transaction costs, transportation costs, tariffs, and/or taxes.

In these frameworks, which are standard ones for applied computational work, one has to deal with equilibrium conditions influenced by additional parameters which often change the mathematical properties of the problem. For instance, in models with taxes, (i) the equilibrium allocations might lose their Pareto optimality; (ii) restrictions which imply, in the absence of taxes, the uniqueness of equilibrium prices might become compatible with multiple disconnected equilibria.

In this paper, we consider exchange economies with either uniform or differentiated *ad valorem* taxes (see Section 2 for appropriate definitions). We explore the effects of such tax distortions on models which admit - in the absence of taxes - polynomial time algorithms. In spite of the loss of certain structural properties (including uniqueness), we are able to obtain polynomial time algorithms

M. Bugliesi et al. (Eds.): ICALP 2006, Part I, LNCS 4051, pp. 584–595, 2006.

or approximation schemes in several instances where the model without taxes admitted them.

Background. We now describe the model of an exchange economy, and provide some basic definitions. Let us consider m economic agents which represent traders of n goods. Let $\mathbf{R}_+^n$ denote the subset of $\mathbf{R}^n$ with all nonnegative coordinates. The j-th coordinate in $\mathbf{R}^n$ will stand for good j. Each trader i has a concave utility function $u_i : \mathbf{R}_+^n \to \mathbf{R}_+$, which represents her preferences for the different bundles of goods, and an initial endowment of goods $w_i = (w_{i1}, \ldots, w_{in}) \in \mathbf{R}_+^n$. At given prices $\pi \in \mathbf{R}_+^n$, trader i will demand a bundle of goods $x_i = (x_{i1}, \ldots, x_{in}) \in \mathbf{R}_+^n$ which maximizes $u_i(x)$ subject to the budget constraint $\pi \cdot x \leq \pi \cdot w_i$. Let $W_j = \sum_i w_{ij}$ denote the total amount of good j in the market.

An equilibrium is a vector of prices $\pi = (\pi_1, \ldots, \pi_n) \in \mathbf{R}_+^n$ at which, for each trader i, there is a bundle $\bar{x}_i = (\bar{x}_{i1}, \ldots, \bar{x}_{in}) \in \mathbf{R}_+^n$ of goods such that the following two conditions hold: (i) for each trader i, the vector $\bar{x}_i$ maximizes $u_i(x)$ subject to the constraints $\pi \cdot x \leq \pi \cdot w_i$ and $x \in \mathbf{R}_+^n$; (ii) for each good j, $\sum_i \bar{x}_{ij} \leq W_j$. The celebrated result of Arrow and Debreu [1] states that, under quite mild assumptions, such an equilibrium exists.

For any price vector π, a vector $x_i(\pi)$ that maximizes $u_i(x)$ subject to the constraints $\pi \cdot x \leq \pi \cdot w_i$ and $x \in \mathbf{R}_+^n$ is called the *demand* of the i-th trader. By adding up the traders' demands, one gets the *market demand*.

We now give the definition of approximate equilibrium. A bundle $x_i \in \mathbf{R}_+^n$ is an ε-*approximate demand*, for $0 < \varepsilon < 1$, of trader i at prices π if $u_i(x_i) \geq (1-\varepsilon)u^*$ and $\pi \cdot x_i \leq (1+\varepsilon)\pi \cdot w_i$, where $u^* = \max\{u_i(x)|x \in \mathbf{R}_+^n, \pi \cdot x \leq \pi \cdot w_i\}$. A price vector $\pi \in \mathbf{R}_+^n$ is an ε-approximate equilibrium if there is a bundle x_i for each i such that (1) for each trader i, x_i is an ε-approximate demand of trader i at prices π, and (2) $\sum_i x_{ij} \leq (1+\varepsilon)\sum_i w_{ij}$ for each good j.

An important special case of an exchange economy is the *distributional economy*, where the initial endowments are all *collinear*, i.e., $w_i = \delta_i w$, $\delta_i > 0$, so that the relative incomes of the traders are independent of the prices. This special case is equivalent to *Fisher's model*, which is a market of n goods desired by m utility maximizing buyers with fixed incomes.

A utility function $u(\cdot)$ is *homogeneous* of degree one if it satisfies $u(\alpha x) = \alpha u(x)$, for all $\alpha > 0$, while it is *log-homogeneous* if it satisfies $u(\alpha x) = u(x) + \log \alpha$, for all $\alpha > 0$.

A linear utility function has the form $u_i(x) = \sum_j a_{ij} x_{ij}$. A CES (constant elasticity of substitution) utility function has the form $u(x_i) = (\sum_j (a_{ij} x_{ij})^\rho)^{1/\rho}$, where $-\infty < \rho < 1$, $\rho \neq 0$. The Cobb-Douglas utility function has the form $u_i(x) = \prod_j (x_{ij})^{a_{ij}}$, where $a_{ij} \geq 0$ and $\sum_j a_{ij} = 1$.

Related Work. Substantial work has been done on extending equilibrium models to handle scenarios where good purchases are subject to taxation [11,12,16,17,18]. Such efforts have provided existential results [16,17], evidence of tax-induced multiplicity [18] and of the loss of Pareto-optimality of equilibria [11,12]. Building upon this body of results, applied models have been designed to explicitly take into account tax distortions (see for instance the popular GAMS-MPSGE programming environment).

Previous work within theoretical computer science, which was initiated by [5], has identified several restrictions under which the market equilibrium problem, in its version without taxes, can be solved in polynomial time. These restrictions include (i) distributional economies (the Fisher setting) where the traders have homogeneous utility functions, (ii) exchange economies which satisfy *weak gross substitutability*, and (iii) exchange economies with some families of CES and nested CES utility functions. For detailed references, see the survey [3] and [10].

Our Results. We prove that a distributional economy (which is equivalent to Fisher's model) with uniform ad valorem taxes and homogeneous consumers can be efficiently transformed into an equivalent two-trader exchange economy without taxes (Section 3.1). We then develop a polynomial-time algorithm for approximating the equilibrium (Section 3.2). To analyze some parameters related to the accuracy of the algorithm, we use the tool of implicit differentiation. Note that a two-trader exchange economy with homogeneous consumers admits multiple disconnected equilibria [9]. The example in [9] can be modified to model an exchange economy that is equivalent to a distributional economy with uniform ad valorem taxes. Therefore our algorithm provides the first significant example of polynomial time computation of equilibria in a setting with multiple disconnected equilibria.

We then show that an n-good exchange economy with differentiated ad valorem taxes and m Cobb-Douglas consumers can be efficiently transformed into an equivalent $(n + 1)$-good exchange economy without taxes, with m Cobb-Douglas consumers (Section 2.2). Since the equilibrium for a Cobb-Douglas exchange economy can be computed in polynomial time [6], our reduction shows that the same is possible for the model with non-uniform taxation.

We finally show that the approximation algorithm of Garg and Kapoor [8] for linear utilities can be adapted to handle the differentiated tax scenario (Section 4).

All the proofs are omitted from this extended abstract. The interested reader can obtain from the web a version with proofs and additional results [4].

2   Exchange Economies with Taxes on Consumption

We describe the model of an exchange economy with *ad valorem* taxes as presented by Kehoe ([11], pp. 2127-2128), and distinguish between the uniform case, where the tax rate is uniform across consumers, and the non-uniform case, where the tax rate is differentiated among consumers.

2.1   Uniform Ad Valorem Taxes

Consider a trader i with utility function $u_i(x_i)$ and initial endowment w_i.

Let $\tau_j \geq 0$ be the uniform *ad valorem* tax associated with the consumption of good j. This means that if consumer i purchases x_{ij} units of good j at price π_j, she will spend on this good the amount $\pi_j x_{ij} + \tau_j \pi_j x_{ij}$.

We postulate the presence of a special actor, the *government*, which will rebate the tax revenues to consumers. Let $\theta_i \geq 0$, with $\sum_i \theta_i = 1$, be the share of total tax revenues rebated to consumer i as a lump sum.

Then the classical consumer's maximization problem gets modified as follows:

$$\max u_i(x_i) \tag{1}$$

$$\text{s.t.} \sum_j \pi_j(1 + \tau_j)x_{ij} \leq \sum_j \pi_j w_{ij} + \theta_i R, \tag{2}$$

where $x_i \in \mathbf{R}_+^n$, and R is the total amount of revenues distributed by the government.

In this context, the market equilibrium problem consists of finding $(\bar{\pi}, \bar{x}_i, \bar{R})$ such that

- at prices $\bar{\pi}$, $\bar{x}_i$ solves (1) (2), $\forall i$ (optimality and budget constraint are satisfied for all consumers);
- $\sum_i \bar{x}_{ij} = \sum_i w_{ij}$, $\forall j$ (the market clears all the goods);
- $\bar{R} = \sum_j \bar{\pi}_j \tau_j \sum_i \bar{x}_{ij}$ (the amount of taxes distributed is equal to the amount of taxes collected).

We now show that the equilibria for such an economy are in a one-to-one correspondence with the equilibria of an exchange economy without taxes, where the traders have a different set of initial endowments, obtained by a suitable redistribution of the original ones. This correspondence has been established in [2] (see also Appendix A of [4]), where models of taxation were one of the targets of some experimental work.

Whenever the equilibria of two economies are in a one-to-one correspondence, and can be immediately computed one from the other, we say that the two economies are equivalent.

Proposition 1. *[2] Let $E = E(u_i(\cdot), w_i, \tau, \theta)$ be an exchange economy with uniform ad valorem taxes $\tau = (\tau_1, \ldots, \tau_n)$, and tax shares $\theta = (\theta_1, \ldots, \theta_m)$. E is equivalent to an exchange economy without taxes $E' = E'(u_i(\cdot), w_i')$ where $w_{ij}' = \frac{w_{ij}}{1+\tau_j} + \theta_i \frac{\tau_j}{1+\tau_j} \sum_i w_{ij}$.*

2.2 Differentiated Taxes

We now consider a more general model in which the taxes on purchases are different for each trader. We call this scheme *specific taxation* or *differentiated ad valorem taxation*.

In an exchange economy with specific taxes, $\tau_{ij} \geq 0$ is the tax rate imputed to trader i on purchase of good j. The setting for an exchange economy with specific taxes differs from that with uniform taxes in the budget constraint, which is now given by

$$\sum_j \pi_j(1 + \tau_{ij})x_{ij} \leq \pi \cdot w_i + \theta_i R. \tag{3}$$

At equilibrium, we must have $R = \sum_{ij} \pi_j \tau_{ij} x_{ij}$.

The lack of uniformity of this model, which differentiates between consumers, prevents the possibility of a direct reduction to a pure exchange economy, obtained by redistributing the individual endowments, as in Proposition 1. Nevertheless, this model can be made *similar* to a pure exchange economy with an extra good. Indeed, note that the budget constraint can be rewritten as

$$\pi \cdot x_i + R x_{i,n+1} \leq \pi \cdot w_i + R\theta_i \qquad (4)$$

where $x_{i,n+1} = \frac{\sum_j \pi_j \tau_{ij} x_{ij}}{R}$.

If we interpret R as the price of an additional (fictitious) good, that we call the "tax good", and θ_i and $x_{i,n+1}$ as the i-th trader's initial endowment and demand of such good, then inequality (4) corresponds to the budget constraint of a consumer in a pure exchange economy with an extra good.

To get a reduction to an exchange economy without taxes, one would now need to exhibit a utility function, defined on $n + 1$ goods, which, combined with the budget constraint (4), gives a demand of $x_{ij}(\pi, R)$ for the first n goods, and $x_{i,n+1}(\pi, R) = \frac{\sum_j \pi_j \tau_{ij} x_{ij}}{R}$ for the "tax good".

We do not know if such a reduction is possible in general. However we show below (Proposition 2) that it can be done in the case of exchange economies where the traders have Cobb-Douglas utility functions. See Appendix B of [4] for the proof.

Proposition 2. *Let n and m be the number of goods and the number of traders, respectively. Let $u_i(\cdot)$, $i = 1, \ldots, m$, be Cobb-Douglas utility functions. We denote by $E_{n,m} = E_{n,m}(u_i(\cdot), w_i, \tau_i, \theta)$ a Cobb-Douglas exchange economy with differentiated ad valorem taxes $\tau_i = (\tau_{i1}, \ldots, \tau_{in})$, and tax shares $\theta = (\theta_1, \ldots, \theta_m)$. $E_{n,m}$ is equivalent to an exchange economy without taxes $E'_{n+1,m} = E'_{n+1,m}(v_i(\cdot), w'_i)$ where the $v_i(\cdot)$'s are Cobb-Douglas utility functions, and $w'_i = (w_{i1}, \ldots, w_{in}, \theta_i)$.*

3   Collinear Endowments Distorted by Uniform Taxation

The general reduction of Proposition 1 shows that uniform taxation does not affect algorithms which compute equilibrium prices for exchange economies without exploiting any particular property of the initial endowments. Therefore, several results for pure exchange economies extend to the model with uniform taxation. One interesting case where the redistribution of initial endowments potentially carries negative computational consequences is that of exchange economies with collinear endowments and homogeneous utilities. In this setting an equilibrium (without taxes) can be computed in polynomial time by convex programming, based on certain aggregation properties of the economy which imply the existence of a *representative consumer* [7]. The redistribution of endowments associated with taxation clearly destroys the collinearity of endowments, and thus the collapse to a single consumer's problem. We show that in an exchange economy with collinear endowments and homogeneous utilities, the model with uniform taxation is equivalent to an exchange economy with *two representative consumers*. Building upon this property, we then show how to compute an approximate equilibrium in polynomial time for a wide family of problems.

3.1   Reduction to Two Representative Consumers

Recall that a *distributional economy* is an exchange economy where the initial endowments of the traders are collinear. In other words, the k-th trader has endowments of the form $w_k = \gamma_k w$, for $k = 1, \ldots, m$, where $w = (w_1, \ldots, w_n)$ describes the overall amount of each good in the market, and γ_k is a positive constant less then one. We have $\sum_k \gamma_k = 1$. In this scenario, the relative incomes of the traders are constants, so that the model is equivalent to Fisher's model.

If we specialize the model of Section 2.1 to an economy with collinear initial endowments, then the redistribution described by Proposition 1, gives $w'_{ij} = \gamma_i \frac{w_j}{1+\tau_j} + \theta_i \frac{\tau_j}{1+\tau_j} w_j$.

Notice that the matrix whose columns represent the new initial endowments of the traders has rank at most two. Indeed all the columns are linear combinations of the vectors z and s, whose j-th entries are $z_j = \frac{w_j}{1+\tau_j}$, and $s_j = \frac{\tau_j}{1+\tau_j} w_j$, respectively. Note that $w'_i = \gamma_i z + \theta_i s$. (Note also that $w = z + s$ and verify that $\sum_i w'_i = w$.) Thus the effect of uniform taxation on economies with proportional endowments amounts to an increase of the rank of the endowment matrix from one to two.

Whenever the consumers are homogeneous (or log-homogeneous), exchange economies with a rank two endowment matrix can be reduced to a two-trader economy, according to the following scheme:

1. Let z be the n-vector whose j-th component is $\frac{w_j}{1+\tau_j}$, and s be the n-vector whose j-th component is $\frac{w_j \tau_j}{1+\tau_j}$.
2. For all k, split the k-th trader into two traders, which have the same utility function of the original trader and initial endowments $\gamma_k z$, and $\theta_k s$, respectively. This procedure produces two groups of m traders each, where the traders in each group have proportional endowments.
3. Based on the properties in [7], aggregate all the consumers from each group into one representative consumer, with endowment given by the sum of their endowments and utility function obtained by aggregating the utility functions as in [7]. This gives a two-trader economy.

These arguments lead to the following result.

Proposition 3. *Let $u_i(\cdot)$, $i = 1, \ldots, m$, be log-homogeneous utility functions. Let $E_m = E(u_i(\cdot), w, \tau, \theta, \gamma)$ be an m-trader distributional economy with uniform ad valorem taxes $\tau = (\tau_1, \ldots, \tau_n)$, tax shares $\theta = (\theta_1, \ldots, \theta_m)$, and income shares $\gamma = (\gamma_1, \ldots, \gamma_m)$. E_m is equivalent to a two-traders exchange economy without taxes $E_2 = E_2'(v_1(\cdot), v_2(\cdot), z, s)$, where v_1 and v_2 are the log-homogeneous utility functions of the two consumers, and z and s are their initial endowment vectors. Here, $v_1(x)$ (resp. $v_2(x)$) is defined to be the maximum of $\sum_i \gamma_i u_i(x_i)$ (resp. $\sum_i \theta_i u_i(x_i)$) over all $x_1, \ldots, x_m \in \mathbf{R}^n_+$ such that $\sum_i x_i = x$.*

3.2   The Algorithm

The reduction summarized in Proposition 3, combined with some results of Mantel [13] on two-trader economies, suggest the following algorithm, which we call RSR (Reduce-Solve-Reconstruct), for the computation of an approximate equilibrium. For the analysis, we assume that $\tau_j > 0$ for each j. Let $\tau_{min} = \min_j \tau_j$ and $\tau_{max} = \max_j \tau_j$.

Algorithm RSR

1. The input is given in terms of m log-homogeneous utility functions u_i, $i = 1,\ldots,m$, and vectors w, γ, θ, τ.
2. Apply the transformation of Proposition 3, which returns an economy with two log-homegeneous consumers (with utility functions v_1 and v_2, and initial endowments z and s) and n goods.
3. Consider the following constrained maximization problem:

$$\max \quad \alpha v_1(x_1) + (1 - \alpha)v_2(x_2)$$
$$\text{s.t.} \quad x_1 + x_2 = w$$
$$x_1, x_2 \geq 0$$

For a given $0 \leq \alpha \leq 1$, let $x_1(\alpha)$ and $x_2(\alpha)$ be maximizing allocations, and let $\pi(\alpha)$ be the vector of shadow prices (Lagrange multipliers). It can be shown that $\pi(\alpha) \cdot x_1(\alpha) = \alpha$, $\pi(\alpha) \cdot x_2(\alpha) = 1 - \alpha$, and thus $\pi(\alpha) \cdot w = 1$. Moreover, $x_1(\alpha)$ and $x_2(\alpha)$ have the "right shape" – they are proportional to the optimal bundles demanded by the two traders at the price $\pi(\alpha)$. Let $B_1(\alpha) = \pi(\alpha) \cdot (z - \bar{x}_1(\alpha))$ and $B_2(\alpha) = \pi(\alpha) \cdot (s - \bar{x}_2(\alpha))$ be the functions expressing the (positive or negative) *savings* of consumer 1 and 2, respectively. Note that $B_1(\alpha) + B_2(\alpha) = 0$, since $z + s = w$. Thus, if $B_1(\alpha) = 0$, then we have $\pi(\alpha) \cdot x_1(\alpha) = \pi(\alpha) \cdot z$, and $\pi(\alpha) \cdot x_2(\alpha) = \pi(\alpha) \cdot s$. In this case, $x_1(\alpha)$ and $x_2(\alpha)$ are not merely proportional to the optimal bundles, but they *are* the optimal bundles of the two traders at price $\pi(\alpha)$; thus $\pi(\alpha)$ is an equilibrium for the two trader economy [14].

We therefore find an approximate equilibrium for the two-trader economy by finding a value of α such that $B_1(\alpha)$ and $B_2(\alpha)$ are sufficiently close to zero. In such a case, $\pi(\alpha)$, $x_1(\alpha)$ and $x_2(\alpha)$ form an approximate equilibrium, provided that the functions $B_i(\alpha)$ are smooth enough (see the extensive discussion below). The search for an appropriate value of α can be done by the bisection method, i.e., binary search, guided by the value of $B_i(\alpha)$, computed from the values $x_i(\alpha)$ and $\pi(\alpha)$ returned by the solution of the maximization problem above. The applicability of bisection method builds upon some results by Mantel [13] on the global convergence of the welfare adjustment process when applied to two-trader economies.

4. From the solution to the two-trader problem, reconstruct the solution to the $2m$-trader problem, i.e., the corresponding allocations, and then to the m-trader problem without taxes.
5. Compute approximate equilibrium prices $\tilde{\pi}_j$, $j = 1, \ldots,$ for the original economy with taxes, by scaling prices $\pi_j(\alpha)$, i.e., $\tilde{\pi}_j = \frac{\pi_j(\alpha)}{1+\tau_j}$.

3.3   Analysis of Algorithm RSR

For any price vector π, it is easy to see that $\frac{\pi \cdot z}{\pi \cdot w}$ lies in the interval $[\alpha_{min} = \frac{1}{1+\tau_{max}}, \alpha_{max} = \frac{1}{1+\tau_{min}}]$. Thus if $B_1(\bar{\alpha}) = 0$, then $\bar{\alpha} = \frac{\pi(\bar{\alpha}) \cdot x_1(\bar{\alpha})}{\pi(\bar{\alpha}) \cdot w} = \frac{\pi(\bar{\alpha}) \cdot z}{\pi(\bar{\alpha}) \cdot w}$ lies in the range $[\alpha_{min}, \alpha_{max}]$. It is also easy to verify that $B_1(\alpha_{min}) \geq 0$ and $B_1(\alpha_{max}) \leq 0$. So we perform our binary search in the interval $[\alpha_{min}, \alpha_{max}]$.

The binary search is described and analyzed in Appendix C of [4], where we show that Algorithm RSR computes an ε-approximate equilibrium in time of the order of $T(n)(\log M + \log \frac{1}{\varepsilon} + \log \frac{1}{\alpha_{min}} + \log \frac{1}{(1-\alpha_{max})})$, where $T(n)$ is the polynomial bound on the time required to solve the convex program, and M is an upper bound on the absolute value of the derivative of $B_1(\alpha)$ in the interval $[\alpha_{min}, \alpha_{max}]$. The next section takes a close look at the parameter M that influences the running time.

3.4   Sensitivity of the Welfare Maximization Problem

The running time of algorithm RSR depends on the logarithm of M, where M is an upper bound on $|B'_1(\alpha)|$ for α in $[\alpha_{min}, \alpha_{max}]$. We now show how to estimate M for a wide family of utility functions. For lack of space the details of the analysis are omitted from this extended abstract; the interested reader can find them in [4].

The two-trader maximization problem occurring in step 3 of Algorithm RSR, when written in its expanded form, becomes the $2m$-trader maximization problem:

$$\max \quad \alpha \sum_i \gamma_i u_{1i}(x_i^1) + (1 - \alpha) \sum_i \theta_i u_{2i}(x_i^2)$$
$$\text{s.t.} \quad x_1^1 + \ldots + x_m^1 + x_1^2 + \ldots + x_m^2 = w$$
$$x_i^1, x_i^2 \geq 0$$

where $u_{1i}() = u_{2i}() = u_i()$. Let $x_i^\ell(\alpha)$ denote the solution of this problem, and $\pi(\alpha)$ the corresponding shadow prices. We will simply denote $x_i^\ell(\alpha)$ by x_i^ℓ and $\pi(\alpha)$ by p. Since $B_1(\alpha) = p \cdot z - \alpha$, we can upper bound $|B'_1(\alpha)|$ by bounding the elements of the vector $\frac{\partial p}{\partial \alpha}$.

Let us consider the first order optimality conditions for the problem above, which we will denote by

$$H(\overbrace{x, p}^{y}, \alpha) = 0.$$

These equations can be explicitly written as:

$$\begin{cases} \alpha \gamma_i \nabla u_{1i}(x_i^1) - p = 0; \\ (1 - \alpha) \theta_i \nabla u_{2i}(x_i^2) - p = 0; \\ x_1^1 + \ldots + x_m^1 + x_1^2 + \ldots + x_m^2 - w = 0. \end{cases}$$

By *Implicit Differentiation*, from the first order conditions we obtain the following equation:

$$\nabla_y H(x,p,\alpha)\frac{\partial}{\partial\alpha}y(\alpha) + \frac{\partial}{\partial\alpha}H(x,p,\alpha) = 0. \tag{5}$$

Let A_i and B_i denote the Hessian $\alpha\gamma_i\nabla^2 u_{1i}(x_i^1)$, and $(1-\alpha)\theta_i\nabla^2 u_{2i}(x_i^2)$, respectively. Equation 5 takes the form:

$$\begin{bmatrix} A_1 & & & & & -I \\ & \ddots & & & & \vdots \\ & & A_m & & & -I \\ & & & B_1 & & -I \\ & & & & \ddots & \vdots \\ & & & & B_m & -I \\ I & \dots & I & I & \dots & I & 0 \end{bmatrix} \begin{bmatrix} \frac{\partial x_1^1}{\partial\alpha} \\ \vdots \\ \frac{\partial x_m^1}{\partial\alpha} \\ \frac{\partial x_1^2}{\partial\alpha} \\ \vdots \\ \frac{\partial x_m^2}{\partial\alpha} \\ \frac{\partial p}{\partial\alpha} \end{bmatrix} + \begin{bmatrix} \gamma_1\nabla u_{11}(x_1^1) \\ \vdots \\ \gamma_m\nabla u_{1m}(x_m^1) \\ -\theta_1\nabla u_{21}(x_1^2) \\ \vdots \\ -\theta_m\nabla u_{2m}(x_m^2) \\ 0 \end{bmatrix} = \begin{bmatrix} 0 \\ 0 \\ 0 \\ \vdots \\ 0 \\ 0 \\ 0 \end{bmatrix}$$

Assume that the A_i's and B_i's be nonsingular, i.e., that they are negative definite. In this case the matrix of the linear system above has an inverse, which is

$$\begin{bmatrix} A_1^{-1} - A_1^{-1}KA_1^{-1} & -A_1^{-1}KA_2^{-1} & \cdots & & -A_1^{-1}KB_m^{-1} & A_1^{-1}K \\ \vdots & & \ddots & & \vdots & \vdots \\ & & & \ddots & \vdots & \vdots \\ -B_m^{-1}KA_1^{-1} & \cdots & & \cdots & B_m^{-1} - B_m^{-1}KB_m^{-1} & B_m^{-1}K \\ -KA_1^{-1} & \cdots & & \cdots & -KB_m^{-1} & K \end{bmatrix},$$

where $K = \left(A_1^{-1} + \dots + A_m^{-1} + B_1^{-1} + \dots + B_m^{-1}\right)^{-1}$.

Let now $d_{1i} = \gamma_i\nabla u_{1i}(x_i^1)$, and $d_{2i} = \theta_i\nabla u_{2i}(x_i^2)$. We obtain the expression

$$\frac{\partial p}{\partial\alpha} = -K\left(\sum_i A_i^{-1}d_{1i} - \sum_i B_i^{-1}d_{2i}\right), \tag{6}$$

from which we can upper bound the absolute value of any element of $\frac{\partial p}{\partial\alpha}$ in terms of $\|A_i^{-1}\|$, $\|B_i^{-1}\|$, $\|K\|$, and $\nabla u_{\ell i}(x_i^\ell)$, where $\|\cdot\|$ denotes the spectral norm of a matrix, which, in the case of semi-definite matrices, coincides with the spectral radius.

Using a classical result from linear algebra (see [15], p. 192), we can then bound the spectral norm of K in terms of those of A_i and B_i, .

This allows us to bound $|B_1'(\alpha)|$ in terms of the Hessians of the utility functions $u_{\ell i}$ evaluated at x_i^ℓ.

For an interesting class of utility functions, one can proceed even further, and obtain bounds in terms of the input data. This requires some calculations, which are shown in [4].

4   Linear Exchange Economies with Differentiated Taxes

We consider an economy with m traders and n goods where each trader has a linear utility function. Let $u_i = \sum_j a_{ij}x_j$ denote the utility function of the i-th trader. Each trader has the initial endowment $w_i \in \mathbf{R}_+^n$. Let $\tau_{ij} \geq 0$ denote the tax rate of the i-th trader for the consumption of the j-th good. And let θ_i denote the share of the i-th trader in the overall tax collected. We have $\sum_i \theta_i = 1$. An equilibrium is a price vector $\pi = (\pi_1, \ldots, \pi_n)$ and a number $R \geq 0$ at which there are bundles $x_i \in \mathbf{R}_+^n$ for each trader i such that (1) x_i maximizes $u_i(x)$ over all $x \in \mathbf{R}_+^n$ such that the cost $\sum_j \pi_j (1 + \tau_{ij}) x_j$ of bundle x is at most the income $\theta_i R + \sum_j \pi_j w_{ij}$; (2) $\sum_i x_{ij} \leq \sum_i w_{ij}$ for each good j; and (3) $\sum_i T_i(x_i, \pi) = R$, where $T_i(x, \pi)$ is defined to be $\sum_j \tau_{ij} \pi_j x_j$, the tax that i has to pay to consume x at price π.

For $\varepsilon > 0$, we define an ε-approximate equilibrium to be a price vector $\pi = (\pi_1, \ldots, \pi_n)$ and a number $R \geq 0$ at which there are bundles $x_i \in \mathbf{R}_+^n$ for each trader i such that (1) $\pi \cdot x_i + T(x_i, \pi) \leq (1 + \varepsilon)(\theta_i R + \sum_j \pi_j w_{ij})$, and $u_i(x_i) \geq (1 - \varepsilon) v_i(\pi, R)$, where $v_i(\pi, R)$ is the maximum value of $u_i(x)$ over all $x \in \mathbf{R}_+^n$ such that $\sum_j \pi_j (1 + \tau_{ij}) x_j \leq \theta_i R + \sum_j \pi_j w_{ij}$; (2) $\sum_i x_{ij} \leq (1 + \varepsilon) \sum_i w_{ij}$ for each good j; and (3) $(1 - \varepsilon) R \leq \sum_i T_i(x_i, \pi) \leq (1 + \varepsilon) R$.

We now describe our algorithm, an adaptation of the auction based algorithm of Garg and Kapoor [8], for computing an approximate equilibrium of the model. The analysis of the algorithm assumes that $a_{ij} > 0$ for each i and j. Let τ_{max} denote $\max_{i,j} \tau_{ij}$. For simplifying some expressions, we also assume, without loss of generality, that $\sum_i w_{ij} = 1$ for each j. Let $w_{min} = \min_{i,j} w_{ij}$. Our analysis also assumes that $w_{min} > 0$. Let $\kappa = 1/w_{min}$.

Let $\delta = \frac{\varepsilon}{90n(1+\tau_{max})\kappa}$. The algorithm has variables π_j for the prices, and a variable R that stands for the tax money that is distributed to the buyers. The algorithm starts with all prices set to 1. From time to time, it increases the price of some good by a multiplicative factor of $1 + \delta$. It has variables y_{ij} and h_{ij} corresponding to the amounts of good j allocated to i at the current price π_j and the previous price $\pi_j/(1 + \delta)$, respectively. Let $x_{ij} = y_{ij} + h_{ij}$. Let $T_i'(y_i, h_i, \pi) = \sum_j \tau_{ij}(\pi_j y_{ij} + \frac{\pi_j}{1+\delta} h_{ij})$, the tax that i pays in consuming y_i and h_i. Let

$$D_i(\pi) = \{j | \frac{a_{ij}}{(1 + \tau_{ij})\pi_j} \geq \frac{a_{ik}}{(1 + \tau_{ik})\pi_k} \text{ for } 1 \leq k \leq n\}.$$

Initialize. Let $\pi_j = 1$ for $1 \leq j \leq n$, $R = 1$, $y_{ij} = 0$ and $h_{ij} = 0$ for each i and j.

Phase 1

1. We make a call to the procedure Allocatemore(), described below.
2. If $\sum_i T_i'(y_i, h_i, \pi) = R$, let $R \leftarrow R(1 + \delta)$ and go to Step 1 of Phase 1.
3. If for each i, we have $\sum_j(\pi_j y_{ij} + \frac{\pi_j}{1+\delta} h_{ij}) + T_i'(y_i, h_i, \pi) = \theta_i R + \sum_j \pi_j w_{ij}$, the algorithm ends.
4. If for some i, we have $\sum_j(\pi_j y_{ij} + \frac{\pi_j}{1+\delta} h_{ij}) + T_i'(y_i, h_i, \pi) < \theta_i R + \sum_j \pi_j w_{ij}$, consider any $j \in D_i(\pi)$. An inspection of Allocatemore() tells us that we must have $h_{i'j} = 0$ for every trader i' and $\sum_{i'} y_{i'j} = 1$. We call Raiseprice(j). If $\pi_k > 0$ for every good k, we jump to Step 1 of Phase 2. Otherwise, return to Step 1 of Phase 1.

Phase 2

1. If $R \leq \delta \sum_j \pi_j w_{ij}$ for each i, let $R \leftarrow \sum_i T'(y_i, h_i, \pi)$; the algorithm ends.
2. Make a call to Allocatemore().
3. If $\sum_i T'_i(y_i, h_i, \pi) = R$, the algorithm ends.
4. If for each i, we have $\sum_j(\pi_j y_{ij} + \frac{\pi_j}{1+\delta} h_{ij}) + T'_i(y_i, h_i, \pi) = \theta_i R + \sum_j \pi_j w_{ij}$, the algorithm ends.
5. If for some i, we have $\sum_j(\pi_j y_{ij} + \frac{\pi_j}{1+\delta} h_{ij}) + T'_i(y_i, h_i, \pi) < \theta_i R + \sum_j \pi_j w_{ij}$, consider any $j \in D_i(\pi)$. We must have $h_{i'j} = 0$ for every trader i' and $\sum_{i'} y_{i'j} = 1$. We call Raiseprice(j) and return to Step 1 of Phase 2.

To complete the description of the algorithm, we need to specify the two procedures Allocatemore and Raiseprice.

The procedure Allocatemore. In this procedure, we first solve a linear program, which uses as data the current values of the variables π, y, h, and R. The linear program has non-negative variables y'_{ij} and h'_{ij} for each trader i and good j. The linear program is:

$$\text{Maximize} \quad \sum_{i,j} y'_{ij}$$

$$\text{Subject to}$$

$$\sum_i (y'_{ij} + h'_{ij}) \leq 1 \text{ for each } j$$

$$\sum_i (y'_{ij} + h'_{ij}) = 1 \text{ for each } j \text{ such that } \pi_j > 1$$

$$\sum_i T'_i(y'_i, h'_i, \pi) \leq R$$

$$\sum_j \pi_j y'_{ij} + \frac{\pi_j}{1+\delta} h'_{ij} + T'_i(y'_i, h'_i, \pi) \leq \theta_i R + \sum_j \pi_j w_{ij} \text{ for each } i$$

$$y'_{ij} = 0 \text{ for each } i \text{ and } j \notin D_i(\pi).$$
$$h'_{ij} = 0 \text{ for each } i \text{ and } j \text{ such that } h_{ij} = 0.$$

As we argue below, this linear program will always be feasible. Thus the maximization is well defined. After solving the linear program, we set $y_{ij} = y'_{ij}$ and $h_{ij} = h'_{ij}$ for each i and j. This completes the description of the procedure Allocatemore.

The procedure Raiseprice(j). In this procedure, we set $\pi_j \leftarrow \pi_j(1 + \delta)$, $h_{ij} \leftarrow y_{ij}$ for each i, and $y_{ij} \leftarrow 0$ for each i. This completes the description of Raiseprice.

Phases 1 and 2 of our algorithm are similar to the basic algorithm of Garg and Kapoor [8] - a call to Allocatemore() replaces the sequence of steps in their algorithm occurring between two price raises. Our algorithm needs to track the relation of $\sum_i T'(y_i, h_i, \pi)$ – the tax paid as a consequence of consumption – to R, the tax that is distributed as income. The reason we have Phase 2 is that at the end of Phase 1, $\sum_i T'(y_i, h_i, \pi)$ can be significantly smaller than R.

Theorem 1. *For any $\varepsilon > 0$, our algorithm computes an ε-approximate equilibrium for the model in time that is polynomial in the input size, $1/\varepsilon$, and κ.*

References

1. K.J. Arrow and G. Debreu, Existence of an Equilibrium for a Competitive Economy, Econometrica 22 (3), pp. 265–290 (1954).
2. B.Codenotti, B. McCune and K. Varadarajan, Computing Equilibrium Prices: Does Theory Meet Practice? Proc. ESA 2005.
3. B.Codenotti, S. Pemmaraju and K. Varadarajan, Algorithms Column: The Computation of Market Equilibria, ACM SIGACT News 35(4) December 2004.
4. B. Codenotti, L. Rademacher, K. Varadarajan, Online version of this extended astract, available at http://www.imc.pi.cnr.it/$\sim$codenotti/taxes.pdf.
5. X. Deng, C. Papadimitriou, and S. Safra. On the complexity of equilibria. Proc. STOC 2002.
6. B. C. Eaves, Finite Solution of Pure Trade Markets with Cobb-Douglas Utilities, Mathematical Programming Study 23, pp. 226-239 (1985).
7. E. Eisenberg, Aggregation of Utility Functions. Management Sciences, Vol. 7 (4), 337–350 (1961).
8. R. Garg and S. Kapoor, Auction Algorithms for Market Equilibrium. In *Proc. STOC*, 2004.
9. S. Gjerstad, Multiple Equilibria in Exchange Economies with Homothetic, Nearly Identical Preference, University of Minnesota, Center for Economic Research, Discussion Paper 288, 1996.
10. K. Jain and K. Varadarajan. Equilibria for Economies with Production: Constant-Returns Technologies and Production Planning Constraints. SODA 2006.
11. T.J. Kehoe, Computation and Multiplicity of Equilibria, in Handbook of Mathematical Economics Vol IV, pp. 2049-2144, Noth Holland (1991).
12. T.J. Kehoe, The Comparative Statics Properties of Tax Models, The Canadian Journal of Economics 18(2), 314-334 (1985).
13. R. R. Mantel, The welfare adjustment process: its stability properties. International Economic Review 12, 415-430 (1971).
14. T. Negishi, Welfare Economics and Existence of an Equilibrium for a Competitive Economy, Metroeconomica 12, 92-97 (1960).
15. B.N. Parlett, The Symmetric Eigenvalue Problem, Prentice Hall (1980).
16. J.B. Shoven, J. Whalley, Applied General Equilibrium Models of Taxation and International Trade, Journal of Economic Literature 22, 1007-1051 (1984).
17. J.B. Shoven, J. Whalley, Applying General Equilibrium, Cambridge University Press (1992).
18. J. Whalley, S. Zhang, Tax induced multiple equilibria. Working paper, University of Western Ontario (2002).

New Constructions of Mechanisms with Verification

Vincenzo Auletta, Roberto De Prisco, Paolo Penna,
Giuseppe Persiano, and Carmine Ventre

Dipartimento di Informatica ed Applicazioni, Università di Salerno, Italy
Research funded by the EU through IP AEOLUS

Abstract. A social choice function A is implementable with verification if there exists a payment scheme P such that (A, P) is a truthful mechanism for verifiable agents [Nisan and Ronen, STOC 99]. We give a simple sufficient condition for a social choice function to be implementable with verification for *comparable* types. Comparable types are a generalization of the well-studied one-parameter agents. Based on this characterization, we show that a large class of objective functions μ admit social choice functions that are implementable with verification and minimize (or maximize) μ. We then focus on the well-studied case of one-parameter agents. We give a general technique for constructing *efficiently computable* social choice functions that minimize or approximately minimize objective functions that are *non-increasing* and *neutral* (these are functions that do not depend on the valuations of agents that have no work assigned to them). As a corollary we obtain efficient online and offline mechanisms with verification for some hard scheduling problems on related machines.

1 Introduction

Computations over the Internet often involve self-interested parties (*selfish agents*) which may manipulate the system by misreporting a fundamental piece of information they hold (their own *type* or *valuation*). The system runs some algorithm which, because of the misreported information, is no longer guaranteed to return a "globally optimal" solution (optimality is naturally expressed as a function of agents' types) [1]. Since agents can manipulate the algorithm by misreporting their types, one has to carefully design payment functions which make disadvantageous for an agent to do so. A *mechanism* $M = (A, P)$ consists of a *social choice function* A which, on input the reported types, chooses an outcome, and a payment function P which, on input the reported types, associate a payment to every agent. Payments should guarantee that it is in the agent's interest to report his type correctly. Social choice functions A for which there exists a payment P that guarantees that the *utility* that an agent derives from the chosen outcome and from the payment he receives is maximum when this agent reports his type correctly are called *implementable* (see Sect. 1 for a formal definition of these concepts). In this case the mechanism $M = (A, P)$ is called *truthful*. The main difficulty in designing truthful mechanisms stems from

M. Bugliesi et al. (Eds.): ICALP 2006, Part I, LNCS 4051, pp. 596–607, 2006.

the fact that the utility itself depends on the type of the agent: for instance, payments designed to "compensate" certain costs of the agents should make impossible for an agent to speculate. It is well-known that certain social choice functions cannot be implemented. This poses severe limitations on the class of optimization problems involving selfish agents that one can optimally solve (see e.g. [1,2]).

Notation. The following notations will be useful. For a vector $\mathbf{x} = (x_1, \ldots, x_m)$, we let $\mathbf{x}_{-i}$ denote the vector $(x_1, \ldots, x_{i-1}, x_{i+1}, \ldots, x_m)$ and $(y, \mathbf{x}_{-i})$ the vector $(x_1, \ldots, x_{i-1}, y, x_{i+1}, \ldots, x_m)$. For sets $D_1, \ldots, D_m$, we let D denote the Cartesian product $D_1 \times \cdots \times D_m$ and, for $1 \leq i \leq m$, we let D_{-i} denote the Cartesian product $D_1 \times \cdots \times D_{i-1} \times D_{i+1} \times \cdots \times D_m$.

Implementation with verification. In this paper we focus on so called mechanisms with *verification* as introduced in [1] and studied in [3]. These mechanisms award payments *after* the selected outcome has been "implemented" and this implementation allows some limited "verification" on the agents' reported types. We have a finite set $\mathcal{O}$ of possible outcomes and m selfish rational agents. Agent i has a *valuation* (or *type*) v_i taken from a finite set D_i called the *domain* of agent i. A valuation v_i is a function $v_i : \mathcal{O} \to \Re$; $v_i(X)$ represents how much agent i likes outcome $X \in \mathcal{O}$ (higher valuations correspond to preferred outcomes). The valuation v_i is known to agent i only. A *social choice function* $A : D \to \mathcal{O}$ maps the agents' valuations into a particular outcome $A(v_1, \ldots, v_m)$. A *mechanism* $M = (A, P)$ is a social choice function A coupled with a *payment scheme* $P = (P_1, \ldots, P_m)$, where each P_i is a function $P_i : D \to \Re$. The mechanism elicits from each agent his valuation and we denote by $b_i \in D_i$ the *reported* valuation of agent i. On input the vector $\mathbf{b} = (b_1, \ldots, b_m)$ of *reported* valuations, the mechanism selects outcome X as $X = A(\mathbf{b})$ and assigns agent i payment $P_i(\mathbf{b})$. We assume that agents have quasi-linear utilities; more specifically, the *utility* $u_i^M(\mathbf{b}|v_i)$ of agent i when $\mathbf{b}$ is the vector of reported valuations and v_i is the type of agent i is $u_i^M(\mathbf{b}|v_i) = P_i(\mathbf{b}) + v_i(A(\mathbf{b}))$. Agents are selfish and rational in the sense that each of them will report b_i which maximizes the corresponding utility. We stress that both the outcome and the payments depend on the *reported* valuations $\mathbf{b} = (b_1, \ldots, b_m)$. In particular, for a fixed $\mathbf{b}_{-i}$, the outcome $A(\mathbf{b}_{-i}, b_i)$ is a function $A_{\mathbf{b}_{-i}}(b_i)$ of the reported valuation b_i of agent i.

The classical notion of a mechanism assumes that there is no way of verifying whether an agent reported his type truthfully (that is, whether $b_i = v_i$). Therefore, a selfish rational agent can declare any type that will maximize his utility. In some cases, though, it is reasonable to assume that the mechanism has some limited way of verifying the reported types of the agents. In this paper, we consider *mechanisms with verification* which can detect whether $b_i \neq v_i$ if and only if $v_i(A_{\mathbf{b}_{-i}}(b_i)) < b_i(A_{\mathbf{b}_{-i}}(b_i))$; in this case, agent i will *not* receive any payment.

A scenario that is often considered when dealing with selfish rational agents consists of a social choice function that has to share some work-load among the agents. In this scenario, an outcome X specifies for each agent the task that the agent has to complete. It is thus natural to assume that the valuation $v_i(X)$ of agent i reflects how much time it takes agent i to complete the task assigned to him. For example, one could have $v_i(X) = -T_i(X)$ where $T_i(X)$ is the time needed by agent i to complete the task assigned to him by X; thus higher valuations correspond to outcomes X that assign to agent i tasks that can be completed faster. In this scenario, it is natural to assume that an agent can report to be slower than he actually is and delay the completion of the task assigned to him without being caught by the mechanism (this corresponds to the case in which agent i declares b_i such that $b_i(A_{\mathbf{b}_{-i}}(b_i)) \leq v_i(A_{\mathbf{b}_{-i}}(b_i))$. On the other hand, if agent i declares to be faster that he actually is (that is, he declares b_i such that $b_i(A_{\mathbf{b}_{-i}}(b_i)) > v_i(A_{\mathbf{b}_{-i}}(b_i))$) then agent i will complete his task at time $-v_i(A_{\mathbf{b}_{-i}}(b_i))$ instead of time $-b_i(A_{\mathbf{b}_{-i}}(b_i))$ as expected by the mechanism, given his declared valuation b_i. The mechanism will thus punish agent i by not giving him any payment. The well-studied class of one-parameter agents [4,2] corresponds to the special case in which the task assigned to an agent is described by a weight and the time needed to complete a task is proportional to its weight. In this case, the type of the agent is determined by the time it takes the agent to complete a task of unitary weight. Let us now proceed more formally.

Definition 1 ([1]). *A social choice function A is* implementable with verification *if there exists $P = (P_1, \ldots, P_m)$ such that for all i, all $v_i \in D_i$, all $\mathbf{b}_{-i} \in D_{-i}$, utility $u_i^{(A,P)}(\mathbf{b}|v_i)$ of agent i is maximized by setting $b_i = v_i$.*

In this case, $M = (A, P)$ is called a *truthful mechanism with verification*. It is easy to see that, if A is implementable with verification then there exists $P = (P_1, \ldots, P_m)$ such that, for all $v_i, b_i \in D_i$ and $\mathbf{b}_{-i} \in D_{-i}$, the following inequalities hold:

$$v_i(A_{\mathbf{b}_{-i}}(v_i)) + P_i(v_i, \mathbf{b}_{-i}) \geq v_i(A_{\mathbf{b}_{-i}}(b_i)) \qquad \text{if } v_i(A_{\mathbf{b}_{-i}}(b_i)) < b_i(A_{\mathbf{b}_{-i}}(b_i)) \quad (1)$$
$$v_i(A_{\mathbf{b}_{-i}}(v_i)) + P_i(v_i, \mathbf{b}_{-i}) \geq v_i(A_{\mathbf{b}_{-i}}(b_i)) + P_i(b_i, \mathbf{b}_{-i}) \text{ if } v_i(A_{\mathbf{b}_{-i}}(b_i)) \geq b_i(A_{\mathbf{b}_{-i}}(b_i)) \quad (2)$$

We are interested in social choice functions A which are implementable with verification and that optimize some *objective function* $\mu(\cdot)$ which depends on the agents' valuations $\mathbf{v} = (v_1, \ldots, v_m)$. For maximization (resp., minimization) functions, we let $\text{OPT}_\mu(\mathbf{v})$ be $\max_{X \in \mathcal{O}} \mu(X, \mathbf{v})$ (resp., $\min_{X \in \mathcal{O}} \mu(X, \mathbf{v})$). An outcome $X \in \mathcal{O}$ is an α-approximation of μ for $\mathbf{v} \in D$ if the ratio betweeen $\mu(X, \mathbf{v})$ and the optimum is at most α. A social choice function A is α-*approximate* for μ if, for every $\mathbf{v} \in D$, $A(\mathbf{v})$ is an α-approximation of μ for $\mathbf{v}$. In particular, we say that social choice function A maximizes function μ if, for all $\mathbf{v}$, $A(\mathbf{v}) = \arg\max_{X \in \mathcal{O}} \mu(X, \mathbf{v})$.

Our results. We start by studying a generalization of one-parameter agents which we call *comparable* types. We give a simple sufficient condition for social choice function to be implementable with verification for comparable types and, based on this characterization, we show that a large class of objective functions μ admit social choice functions that are implementable with verification and minimize (or maximize) μ. In particular, we consider maximization (respectively, minimization) functions of the form $\mu(v_1(X), \ldots, v_m(X))$ which are monotone non-decreasing (respectively, non-increasing) in each agent valuation $v_i(X)$. Observe that VCG mechanisms [5,6,7] can only deal with particular functions of this form called *affine maximizers* and the $Q||C_{\max}$ scheduling problem is an example of an optimization problem involving a monotone non-decreasing function (thus our result applies to $Q||C_{\max}$) that is not an affine maximizer. We remark that agents with comparable types are more general than one-parameter agents. In the full version we shows a simple class of latencies (corresponding to comparable types) for which optimization is not implementable if verification is not allowed. We also *characterize* social choice functions implementable with verification for agents with *strongly comparable types*, a reach subclass of comparable types which has the well-studied one-parameter agents as a special instance. In Section 3, the focus is on *efficiently* computable social choice functions and one-parameter agents. We give a general transformation for turning any polynomial-time α-approximate algorithm A for the optimization problem with objective function μ into an $\alpha(1 + \varepsilon)$-approximate social choice function $A^\star$ that is implementable with verification. If the number of agents is constant, $A^\star$ can be computed in polynomial-time and this gives immediate applications to NP-hard scheduling problems (see Section 4). Most of the proofs are omitted fromthis paper but can be found in the full version available from the authors' web pages.

Related work. Mechanisms for *one-parameter* agents have been characterized in [4,2]. Lavi, Mu'alem and Nisan [8] showed that a *weak monotonicity* condition (W-MON) characterizes *order-based* domains with range constraints and this result was extended, in a sequence of papers [9,10], to *convex domains*. These results concern mechanisms which do *not* use verification and cannot be applied to our case. We show that the "counterpart" of W-MON for mechanisms with verification (which we term WMonVer) is not always sufficient, unlike the cases considered in [8,9,10]. This gives evidence that the results about W-MON cannot be imported in mechanisms with verification. The study of social choice functions implementable with verification starts with the work of Nisan and Ronen [1], who gave a truthful $(1 + \varepsilon)$-approximate mechanism for minimizing scheduling on a constant number of unrelated machines. Similar results have been obtained by Auletta *et al.* [3] for scheduling on any number of related machines (see also [11] for the online case). Also the works of [12,13] give mechanisms for agents which are verifiable.

2   Agents with Comparable Types

In this section we consider *comparable types*. The main result of this section (see Theorem 4) shows that, for any monotone non-decreasing function μ, there exists a social choice function A that maximizes μ and that is implementable with verification. In Theorem 5, we give a necessary and sufficient condition for a social choice function to be implementable with verification with respect to a subclass of comparable types which includes one-parameter agents.

Definition 2. *Let a and b be valuations. We say that a is smaller or equal to b, in symbols $a \le b$, if, for all $X \in \mathcal{O}$, $a(X) \le b(X)$. Domain D is comparable if for any $a, b \in D$ either $a \le b$ or $b \le a$.*

In this section we assume that for all i, the domain D_i of agent i is comparable. We also assume domains to have finite cardinality (even though this assumptions can be relaxed in some cases, e.g., for one-parameter agents). For fixed i and $\mathbf{b}_{-i}$, inequalities (1-2) give a system of linear inequalities with unknowns $P^x :=$ $P_i(x, \mathbf{b}_{-i})$, for $x \in D_i$. For $a, b \in D_i$ with $a \le b$, Inequalities (1-2) are equivalent to the following two inequalities

$$P^a - P^b \ge a(A_{\mathbf{b}_{-i}}(b)) - a(A_{\mathbf{b}_{-i}}(a)) \quad \text{if } a(A_{\mathbf{b}_{-i}}(b)) = b(A_{\mathbf{b}_{-i}}(b)), \quad (3)$$

$$P^b - P^a \ge b(A_{\mathbf{b}_{-i}}(a)) - b(A_{\mathbf{b}_{-i}}(b)). \quad (4)$$

As before, for fixed i and $\mathbf{b}_{-i}$ the two inequalities above give rise to a system of inequalities as a and b with $a \le b$ range over D_i. This system of inequalities is compactly encoded by the following graph that is a modification of the graph introduced in [9] to study the case in which verification is not allowed.

Definition 3 (verification-graph). *Let A be a social choice function. For every i and $\mathbf{b}_{-i} \in D_{-i}$, the verification-graph $\mathcal{V}(\mathbf{b}_{-i})$ has a node for each type in D_i. The set of edges of $\mathcal{V}(\mathbf{b}_{-i})$ is defined as follows. For every $a \le b$, add a directed edge (b, a) of weight $\delta_{b,a} := b(A_{\mathbf{b}_{-i}}(b)) - b(A_{\mathbf{b}_{-i}}(a))$ (encoding Inequality (4)). If $a(A_{\mathbf{b}_{-i}}(b)) = b(A_{\mathbf{b}_{-i}}(b))$, then also add directed edge (a, b) of weight $\delta_{a,b} := a(A_{\mathbf{b}_{-i}}(a)) - a(A_{\mathbf{b}_{-i}}(b))$ (encoding Inequality (3)).*

Theorem 1. *A social choice function A is implementable with verification if and only if, for all i and $\mathbf{b}_{-i} \in D_{-i}$, the graph $\mathcal{V}(\mathbf{b}_{-i})$ does not have negative weight cycles.*

The theorem follows from the observation that the system of linear inequalities involving the payment functions is the linear programming dual of the shortest path problem on the verification-graph. Therefore, a simple application of Farkas lemma shows that the system of linear inequalities has solution if and only if the verification-graph has no negative weight cycle. The same argument has been used for the case in which verification is not allowed albeit on a different graph (see [15] and [9]).

We next show that there exists an interesting class of social choice functions whose verification graphs have no cycle with negative weights. As we shall prove below, these functions can be used to design optimal truthful mechanisms with verification.

Definition 4 (stable social choice function). *A social choice function A is stable if, for all i, for all $\mathbf{b}_{-i} \in D_{-i}$, and for all $a, b \in D_i$, with $a \le b$, if $a(A_{\mathbf{b}_{-i}}(b)) = b(A_{\mathbf{b}_{-i}}(b))$, then we have that $A_{\mathbf{b}_{-i}}(a) = A_{\mathbf{b}_{-i}}(b)$.*

The following result is based on the fact that stable social choice functions guarantee that, if $\mathcal{V}(\mathbf{b}_{-i})$ contains a cycle, then all edges in that cycle have zero weight (see full version for a proof).

Theorem 2. *Every stable social choice function A is implementable with verification.*

We use the above result to show that it is possible to implement social choice functions which select the best outcome out of a fixed subset of possible outcomes:

Theorem 3. *For any $X_1, \ldots, X_\ell \in \mathcal{O}$, let $A = MAX_\mu(X_1, \ldots, X_\ell)$ be the social choice function that, on input $(b_1, \ldots, b_m) \in D$, returns the solution X_j of minimum index that maximizes the value*

$$\mu(b_1(X_j), \ldots, b_m(X_j)).$$

If $\mu(\cdot)$ is monotone non-decreasing in each of its arguments then A is stable.

Proof. Fix an agent i and the reported types $\mathbf{b}_{-i} \in D_{-i}$ of all the other agents. Let $a, b \in D_i$ with $a \leq b$, and denote $X_{i_a} := A_{\mathbf{b}_{-i}}(a)$ and $X_{i_b} := A_{\mathbf{b}_{-i}}(b)$. To prove that A is stable we have to show that, if $a(A_{\mathbf{b}_{-i}}(b)) = b(A_{\mathbf{b}_{-i}}(b))$, then $X_{i_a} = X_{i_b}$. Observe that

$$
\begin{aligned}
\mu(b_1(X_{i_b}), \ldots, b_{i-1}(X_{i_b}), b(X_{i_b}), \ldots, b_m(X_{i_b})) = & \quad \text{(by } a(X_{i_b}) = b(X_{i_b})) & (5) \\
\mu(b_1(X_{i_b}), \ldots, b_{i-1}(X_{i_b}), a(X_{i_b}), \ldots, b_m(X_{i_b})) \leq & \quad \text{(definition of } A \text{ and } X_{i_a}) & (6) \\
\mu(b_1(X_{i_a}), \ldots, b_{i-1}(X_{i_a}), a(X_{i_a}), \ldots, b_m(X_{i_a})) \leq & \quad (a \leq b \text{ and } \mu \text{ non decr.)} & (7) \\
\mu(b_1(X_{i_a}), \ldots, b_{i-1}(X_{i_a}), b(X_{i_a}), \ldots, b_m(X_{i_a})) \leq & \quad \text{(definition of } A \text{ and } X_{i_b}) & (8) \\
\mu(b_1(X_{i_b}), \ldots, b_{i-1}(X_{i_b}), b(X_{i_b}), \ldots, b_m(X_{i_b})). & & (9)
\end{aligned}
$$

This implies that all inequalities above hold with "=". Since A chooses the optimal solution of minimal index, equality between (5) and (8) yields $i_b \leq i_a$. Similarly, the equality between (7) and (6) yields $i_a \leq i_b$, thus implying $X_{i_a} = X_{i_b}$.

Combining Theorem 3 and Theorem 2 we obtain the main result of this section.

Theorem 4. *Let $\mu(\cdot)$ be any function monotone non-decreasing in its arguments $b_1(X), \ldots, b_m(X)$, with $X \in \mathcal{O}$ and $b_i \in D_i$. Then, there exists a social choice function OPT_μ which maximizes $\mu(\cdot)$ and is implementable with verification.*

In the full version we exhibit a social choice function which satisfies the hypothesis of Theorem 3 (and thus is implementable with verification) but is not implementable if verification is not allowed. This shows that, for comparable types, verification does help.

If the set $\mathcal{O}$ of outcomes is very large, then social choice function A could not be efficiently computable. Our next result can be used to derive efficiently-computable social choice functions which approximate the objective function by restricting the search to a suitable subset of the possible outcomes.

Definition 5 (approximation preserving). *A set $\mathcal{O}' \subseteq \mathcal{O}$ is α-approximation preserving for μ if, for every $\mathbf{b} \in D$, the set $\mathcal{O}'$ contains a solution X' which is an α-approximation of μ for $\mathbf{b}$.*

Theorem 3 implies the following.

Corollary 1. *Let $\mu(\cdot)$ be any optimization function monotone non-decreasing in its arguments $b_1(X), \ldots, b_m(X)$, with $X \in \mathcal{O}$ and $b_i \in D_i$. For any α-approximation preserving set $\mathcal{O}' \subseteq \mathcal{O}$ the social choice function $APX_\mu :=$ $MAX_{X \in \mathcal{O}'}\{X\}$ is an α-approximation for μ and is implementable with verification. Moreover, social choice function $APX_\mu(\mathbf{b})$ can be computed in time proportional to the time needed for computing values $\mu(X, \mathbf{b})$, for $X \in \mathcal{O}'$.*

Characterization. The following definition is adapted to the verification setting from the W-MON condition (see [8]) which has been proved necessary and sufficient for implementation without verification for convex domains.

Definition 6 (WMonVer). *A social choice function A is WMonVer for domains $D_1, \ldots, D_m$, if, for all i, for all $\mathbf{b}_{-i} \in D_{-i}$, the graph $\mathcal{V}(\mathbf{b}_{-i})$ does not contain 2-cycles of negative weight.*

Obviously, condition WMonVer is necessary for A to be implementable with verification. Next we prove that for strongly comparable types (a restriction of comparable types that includes one-parameter types) WMonVer is a necessary and sufficient condition for a social choice-function A to be implementable with verification. In the full version, we give an example of a WMonVer social choice function that is not implementable with verification for comparable types.

Definition 7 (strongly comparable types). *A domain with comparable types D_i is with strongly comparable types if there exists $\overline{v}_i \in \mathfrak{R}$ such that, for all $X \in \mathcal{O}$: (i) $a(X) \leq \overline{v}_i$, for all $a \in D_i$, and (ii) for all $a, b \in D_i$, $a(X) = b(X)$ implies $a(X) = \overline{v}_i$.*

Theorem 5. *For domains with strongly comparable types, social choice function A is implementable with verification if and only if A is WMonVer.*

3   One-Parameter Agents

In this section we present our results about *one-parameter* agents. One-parameter agents are a special case of agents with strongly comparable types, and thus Theorem 5 gives us a necessary and sufficient condition for a social choice function to be implementable with verification. In this section, the focus is on *efficiently* computable social choice functions (which will also be referred to as *algorithms*). The main result of this section (see Theorem 9) shows, for a large class of optimization functions μ (see Definitions 10 and 11), how to transform a polynomial-time α-approximate algorithm for μ into an efficiently computable social choice function that is implementable with verification for one-parameter agents and $\alpha(1 + \varepsilon)$-approximates μ. The class of function μ to which our transformation applies include several classical scheduling problems (see Section 4). In Theorem 11, we give a similar result for *online* settings.

Definition 8. *[2] The valuation v_i of a one-parameter agent can be written as $v_i(X) = -w_i(X) \cdot t_i$, for some publicly known non-negative function $w_i(\cdot)$ and some real number $t_i \geq 0$ that is privately known to agent i.*

Observe that the valuation of a one-parameter agent is non-positive. We assume that when asked to report his type, an agents replies with a real number r_i, implying that he reports his valuation to be $b_i(X) = -w_i(X) \cdot r_i$. We consider optimization functions $\mu(X, b_1, \ldots, b_m)$ (as opposed to functions of the form $\mu(b_1(X), \ldots, b_m(X))$ of the previous section) that are non-decreasing in each valuation b_i and thus, equivalently, non-increasing in each reported type r_i.

In the rest of this section, we will show how to design social choice functions for one-parameter agents that are implementable with verification and that can be computed in polynomial time. By virtue of Theorem 5, it suffices to focus on social choice functions that are WMonVer for one-parameter agents. We first observe the following:

Fact 1 *For one-parameter agents, a social choice function A is WMonVer if and only if, for all i, $\mathbf{r}_{-i}$ there exists a critical value $\theta_i \in (\Re^+ \cup \infty)$ such that (i) $w_i(r_i, \mathbf{r}_{-i}) = 0$ for $r_i > \theta_i$, and (ii) $w_i(r_i, \mathbf{r}_{-i}) > 0$ for $r_i < \theta_i$.*

Notice that with a slight abuse of notation we have denoted the critical value with θ_i even though it depends on i and $\mathbf{r}_{-i}$. The above property is called *weak monotonicity* in [3], and Theorem 5 implies one of the main results in that work. *The MAX operator.* We are given a function μ and want to design a social choice function A that is implementable with verification (i.e., WMonVer) and, for a given vector $\mathbf{b}$ of declared types, returns an outcome X such that $\mu(X, \mathbf{b})$ is close to the maximum of μ over all choices of $X \in \mathcal{O}$. Moreover, we want A to be efficiently computable. A natural approach is to start from simple social choice functions and combine them together. Mu'Alem and Nisan [14] consider the following "MAX" operator:

$\mathrm{MAX}_\mu(A_1, A_2)$ operator

- compute $X_1 = A_1(\mathbf{b})$ and $X_2 = A_2(\mathbf{b})$;
- if $\mu(X_1, \mathbf{b}) \geq \mu(X_2, \mathbf{b})$ then return X_1 else return X_2.

For minimization problems, one can simply consider a 'MIN' operator defined as $\mathrm{MIN}_\mu(A_1, A_2) := \mathrm{MAX}_{-\mu}(A_1, A_2)$. Notice the slight abuse of notation in using MAX_μ both with social choice functions (as in the description of the MAX_μ operator) and outcomes (as in Theorem 3) as arguments. In general, the fact that A_1 and A_2 are WMonVer does not guarantee that $\mathrm{MAX}_\mu(A_1, A_2)$ is also WMonVer. We borrow (and adapt) the following definition.

Definition 9 ([14]). *A social choice function A is bitonic w.r.t. $\mu(\cdot)$ if it is WMonVer and, for every i and $\mathbf{r}_{-i}$, one of the following two conditions holds for the function $g(x) := \mu(A(x, \mathbf{r}_{-i}), (x, \mathbf{r}_{-i}))$: (i) $g(x)$ is non-increasing for $x < \theta_i$ and non-decreasing for $x \geq \theta_i$; or (ii) $g(x)$ is non-increasing for $x \leq \theta_i$ and non-decreasing for $x > \theta_i$, where θ_i is the critical value.*

The following is the main technical contribution of this sections and will be used to prove Theorem 9.

Theorem 6. *If each A_i is bitonic w.r.t. $\mu(\cdot)$ then social choice function MAX_μ $(A_1, A_2, \ldots, A_k):=MAX_\mu(MAX_\mu(A_1, \ldots, A_{k-1}), A_k)$ is bitonic w.r.t. $\mu(\cdot)$ and WMonVer for one-parameter agents.*

The same results hold for the 'MIN' operator if each A_i is bitonic w.r.t. $-\mu(\cdot)$. Theorem 6 is proved by showing a connection between WMonVer social choice functions and monotone social choice functions for known single minded bidders (a special type of agents for combinatorial auctions studied in [14]).

Efficient WMonVer social choice functions. Theorem 6 provides a powerful tool for efficiently building social choice functions starting from simpler ones. In particular, we will use this result to extend Theorem 3 to a wider class of optimization functions of the form $\mu(X, b_1, \ldots, b_m)$. This allows us to deal with certain scheduling problems where the measure depends on the scheduling policy internal to the machines and therefore cannot be expressed as the machines completion times (i.e., as a function of $w_i(X) \cdot t_i$). We start by defining the notion of neutral functions.

Definition 10. *A function $\mu(\cdot)$ is neutral if, for every X such that $w_i(X) = 0$, it holds that $\mu(X, (b_i, \mathbf{b}_{-i})) = \mu(X, (b_i', \mathbf{b}_{-i}))$, for every b_i, b_i' and every $\mathbf{b}_{-i}$.*

We have the following technical lemma.

Lemma 1. *Let $\mu(X, b_1, \ldots, b_m)$ be neutral and non-decreasing in each b_i, for every $X \in \mathcal{O}$. Then any algorithm returning a fixed outcome X is bitonic w.r.t. $\mu(\cdot)$.*

Theorem 7. *Let $\mu(X, b_1, \ldots, b_m)$ be neutral and non-decreasing in each b_i, for every $X \in \mathcal{O}$. Then, for any $X_1, \ldots, X_\ell \in \mathcal{O}$, the social choice function $A = MAX_\mu(X_1, \ldots, X_\ell)$ is bitonic w.r.t. $\mu(\cdot)$. Hence A is implementable with verification.*

PROOF SKETCH. The proof is based on the observation that a fixed outcome X_j can be seen as an algorithm returning X_j for all inputs. We show that such an algorithm is bitonic and then apply Theorem 6.   □

Theorem 7 above has two important consequences. First of all, we can obtain a result (similar to Theorem 4) that shows that optimization of neutral monotone functions $\mu(X, b_1, \ldots, b_m)$ for one parameter agents can be implemented with verification.

Theorem 8. *For one-parameter agents and for any function $\mu(X, b_1, \ldots, b_m)$ which is neutral and non-decreasing in each b_i, there exists a social choice function OPT_μ that maximizes $\mu(\cdot)$ and is implementable with verification.*

Another consequence is that, if we have an α-approximation preserving set of outcomes $\mathcal{O}'$ for μ, we can apply the above theorem to all outcomes $X \in \mathcal{O}'$.

This gives us a social choice function A which is implementable with verification, α-approximates μ and can be computed in time polynomial in $|\mathcal{O}'|$.

We next introduce the class of smooth functions, for which there exists a small α-approximation preserving set of outcomes.

Definition 11. *Fix $\varepsilon > 0$ and $\gamma > 1$. A function μ is (γ, ε)–smooth if, for any pair of declarations $\mathbf{r}$ and $\tilde{\mathbf{r}}$ such that $r_i \leq \tilde{r}_i \leq \gamma r_i$ for $i = 1, 2, \ldots, m$, and for all possible outcomes X, it holds that $\mu(X, \mathbf{r}) \leq \mu(X, \tilde{\mathbf{r}}) \leq (1 + \varepsilon) \cdot \mu(X, \mathbf{r})$.*

For smooth, neutral functions μ we can transform any α-approximate polynomial-time algorithm A (which is not necessarily implementable with verification) into a social choice function for a constant number of agents which is computable in polynomial-time, implementable with verification and $\alpha(1 + \varepsilon)$-approximates μ.

Theorem 9. *Let A be a polynomial-time α-approximate algorithm for a neutral, non-decreasing (in each b_i) (γ, ε)-smooth objective function $\mu(\cdot)$. Then, for any $\varepsilon > 0$, there exists an $\alpha(1 + \varepsilon)$-approximate social choice function $A^\star$ implementable with verification. If the number of agents is constant, $A^\star$ can be computed in polynomial time.*

PROOF SKETCH. Let $\mathcal{O}'$ be the set of outcomes returned by A when run on bid vectors whose components are powers of γ. For m agents, $|\mathcal{O}'|$ is $O(\max_i\{\log_\gamma |D_i|\}^m)$ which is polynomial for fixed m, and, since μ is (γ, ε)-smooth, $\mathcal{O}'$ is an $\alpha(1 + \varepsilon)$–approximation preserving set for μ. Consider social choice function $A^\star$ that on input $\mathbf{r}$ outputs the outcome $X \in \mathcal{O}'$ that maximizes $\mu(X, \mathbf{r})$. By Theorem 7, $A^\star$ is WMonVer and $\alpha(1 + \varepsilon)$-approximates μ. Moreover, for constant m, $A^\star$ is polynomial-time computable. □

Online mechanism. A natural way of designing an *online* algorithm for scheduling problems is to iterate a "basic-step" algorithm B which, given the current assignment X, the processing requirement of the new job J and the reported types $b_1, \ldots, b_m$ (that is, the reported speed of machine i is $1/r_i$) outputs the index $B(X, J, \mathbf{b})$ of the machine to which the job must be assigned. For algorithm B, the set of outcomes $\mathcal{O}$ consists of all allocations that can be obtained from X by allocating job J to one of the m machines.

Algorithm B-iterated(b)
- $X := \emptyset$;
- while a new job J arrives do
-     assign job J to machine of index $B(X, J, \mathbf{b})$ and modify X accordingly.

Observe that a basic-step algorithm B that is implementable with verification does not necessarily remains implementable with verification when iterated and we need the stronger property of stability.

Theorem 10. *If B is stable then algorithm B-iterated is stable as well.*

Therefore, by Theorem 2, B-iterated is implementable with verification. For example, Graham's [16] online greedy algorithm for $Q||C_{\max}$ can be seen as the iterated version of a simple basic stable step and thus it is implementable with verification. This property holds more in general. Consider the *greedy* algorithm which, at every step, assigns a newly arrived job to the machine that, given the current assignment of previous jobs, maximizes the increase of the objective function $\mu(\cdot)$; ties are broken in a fixed manner, and $-\mu(\cdot)$ is typically a cost function that one wishes to minimize (e.g., the L_p norm defined as $\sqrt[p]{\sum_i (w_i(X) \cdot t_i)^p}$). Then next Theorem says that greedy is stable and thus if the greedy algorithm is α-approximating for $\mu(\cdot)$, then one has a α-approximating algorithm implementable with verification.

Theorem 11. *The greedy algorithm is stable for cost functions $\mu(X, b_1, \ldots, b_m)$ that are neutral and non-decreasing in each b_i, for every $X \in \mathcal{O}$.*

4   Applications

We consider scheduling problems on related machines owned by selfish agents as in [2]. We are given a set of m *related machines* and a set of n jobs. Each job has a weight and a job can be assigned to any machine. Assigning a job to machine i makes the work w_i of that machine to increase by an amount equal to the job weight. Each machine i has a *speed* s_i, and the completion time of machine i is w_i/s_i, where w_i is the *work* assigned to machine i. In the *online* setting, jobs arrive one-by-one, the k-th job must be scheduled before next one arrives, and jobs cannot be reallocated. For an assignment X, we let $w_i(X)$ be the work that this solution assigns to machine i. Each machine i corresponds to a *selfish agent* whose valuation is $-w_i(X)/s_i = -w_i(X) \cdot t_i$ for $t_i = 1/s_i$. The speed of machine i is known to agent i only (everything else is known to the mechanism) and her valuation is the opposite of the completion time of her machine. An agent can thus misreport her speed (i.e., declare $r_i \neq t_i$). Mechanisms with verification compute, for each agent i, an associated payment and award agent i her payment if and only if all jobs assigned to machine i have been released by time $w_i(X) \cdot r_i$, where X is the outcome selected by the mechanism [1,3]. Machine i can misreport her speed and still receive her associated payment if one of the following happens: (i) the declared speed is worse (i.e., $r_i < t_i$) and jobs are released accordingly by adding some delay; (ii) the declared speed is better (i.e., $r_i < t_i$) but this makes the allocation algorithm A to compute an allocation X which does not assign any job to machine i (i.e., $w_i(X) = 0$, in which case no verification is possible).

We consider several variants of this problem depending on the optimization function adopted. All of these problems are *minimization problems* for which it is NP-hard to compute exact solutions, even for $m = 2$. The table below summarizes some of the applications of our techniques to scheduling problems. For the first three problems, no mechanism without verification can attain an approximation factor better than $2/\sqrt{3} > 1$, for all $m \geq 2$ [2]. Our upper bounds (in bold) are the first bounds on these problems which are all NP-hard to solve exactly; bounds for $Q||\sum_j w_j C_j$ and $Q|r_j|\sum_j w_j C_j$ break the $2/\sqrt{3}$ lower bound in [2], which holds also for exponential-time mechanisms; upper bound for the L_p norm is obtained via online mechanisms based on the greedy algorithm (for $p = 2$, the bound is $1 + \sqrt{2}$). Our techniques can be used also to obtain mechanisms without verification for some graph problems (see full version).

Problem version	Upper Bound	
	Exp Time (any m)	Polytime (m constant)
$Q||\sum_j w_j C_j$	OPT [3]	$(1+\varepsilon)$-approximate [Thm. 9 & [17]]
$Q|r_j|\sum_j w_j C_j$	OPT [Thm. 8]	$(1+\varepsilon)$-approximate [Thm. 9 & [17]]
$Q|prec,r_j|\sum_j w_j C_j$	OPT [Thm. 8]	$O(\log m)$-approximate [Thm. 9 & [18]]
L_p norm	OPT [Thm. 8]	$O(p)$-competitive [Thm. 11]

References

1. Nisan, N., Ronen, A.: Algorithmic Mechanism Design. In: Proc. of the STOC. (1999) 129–140
2. Archer, A., Tardos, E.: Truthful mechanisms for one-parameter agents. In: Proc. of FOCS. (2001) 482–491
3. Auletta, V., De Prisco, R., Penna, P., Persiano, G.: The power of verification for one-parameter agents. In: Proc. of ICALP. Volume 3142 of LNCS. (2004) 171–182
4. Myerson, R.: Optimal auction design. Mathematics of Operations Research **6** (1981) 58–73
5. Vickrey, W.: Counterspeculation, Auctions and Competitive Sealed Tenders. Journal of Finance (1961) 8–37
6. Clarke, E.: Multipart Pricing of Public Goods. Public Choice (1971) 17–33
7. Groves, T.: Incentive in Teams. Econometrica **41** (1973) 617–631
8. Lavi, R., Mu'Alem, A., Nisan, N.: Towards a characterization of truthful combinatorial auctions. In Proc. of FOCS. (2003)
9. Gui, H., Muller, R., Vohra, R.V.: Dominant strategy mechanisms with multidimensional types. Technical report (2004)
10. Saks, M., Yu, L.: Weak monotonicity suffices for truthfulness on convex domains. In Proc. of EC 2005. 286–293
11. Auletta, V., De Prisco, R., Penna, P., Persiano, G.: On designing truthful mechanisms for online scheduling. In Proc. of SIROCCO 2005. 3–17.
12. Hajiaghayi, M.T., Kleinberg, R.D., Mahdian, M., Parkes, D.C.: Online auctions with re-usable goods. In Proc. of EC 2005. 165–174.
13. Porter, R.: Mechanism design for online real-time scheduling. In Proc. of EC 2004. 61–70
14. Mu'Alem, A., Nisan, N.: Truthful approximation mechanisms for restricted combinatorial auctions. In: Proc. of 18th AAAI. (2002) 379–384
15. Rochet, J.C.: A condition for rationalizability in a quasi-linear context. Journal of Mathematical Economics **16** (1987) 191–200
16. Graham, R.L.: Bounds for certain multiprocessing anomalies. Bell System Technical Journal **45** (1966) 1563–1581
17. Chekuri, C., Khanna, S.: A PTAS for minimizing weighted completion time on uniformly related machines. In Proc. of ICALP 2001. Volume 2076 of LNCS.
18. Chudak, F., Shmoys, D.: Approximation algorithms for precedence-constrained scheduling problems on parallel machines that run at different speeds. Journal of Algorithms **30**(2) (1999) 323–343

On the Price of Stability for Designing Undirected Networks with Fair Cost Allocations

Amos Fiat, Haim Kaplan, Meital Levy,
Svetlana Olonetsky, and Ronen Shabo

School of Computer Science, Tel-Aviv University, Israel

Abstract. In this paper we address the open problem of bounding the price of stability for network design with fair cost allocation for undirected graphs posed in [1]. We consider the case where there is an agent in every vertex. We show that the price of stability is $O(\log \log n)$. We prove this by defining a particular improving dynamics in a related graph. This proof technique may have other applications and is of independent interest.

1 Introduction

The *price of stability* [1] of a noncooperative game is the ratio between the cost of the least expensive Nash equilibria and the cost of the social optimum. The price of stability for network design games is motivated by the scenario where one may have some centralized control for a limited time when the network is set-up. But, once the network is up and running, it should be stable without central control. Of course, the price of stability is not larger than the *price of anarchy* [6] which is the ratio of the cost of the most expensive Nash Equilibrium and the cost of the social optimum.

We consider the game of network design with fair cost allocation introduced in [1]. In this game, agent i has to choose a path (strategy) from source node s_i to destination node t_i. The cost of an edge e, $c(e)$, is shared equally by all agents i whose chosen path $p_i = s_i, ..., t_i$ includes e.

It follows from the potential function arguments of [7,8] that pure strategy Nash equilibria always exist for general congestion games, and in particular for the network design game that we consider here (both directed and undirected versions)[1]. In the following, we consider the price of stability for this network design game with respect to pure strategies.

The social optimum for this game is a minimum Steiner network connecting all source-destination pairs. Anshelevich *et al.* [1] show that the price of stability of this game is at most $H(n) = 1 + 1/2 + \cdots + 1/n$, where n is the number of agents. They also exhibit a directed network where this bound is tight.

For undirected graphs the upper bound of $H(n)$ on the price of stability still holds but the lower bound does not. Furthermore, for the case of two players and an undirected graph with a single source Anshelevich *et al.* [1] prove a tight

---

[1] Some weighted congestion games do not have Nash equilibria in pure strategies.

M. Bugliesi et al. (Eds.): ICALP 2006, Part I, LNCS 4051, pp. 608–618, 2006.

bound on the price of stability of $4/3$ which is less than $H(2) = 3/2$. Thus, [1] left open the question of whether there is a tighter bound for undirected graphs.

Our Results. We prove that for undirected graphs with an agent in every vertex and a distinguished source vertex r to which all agents must connect, the price of stability of the network design game of [1] is $O(\log \log n)$ where n is the number of agents. In contrast, in directed graphs even when there is a single source and an agent in every vertex the price of stability is still $\Theta(\log n)$. This follows by a slight modification of the lower bound example of [1].

Related Work on Network Games. Much of the work on network games has focused on congestion games [7,8]. In particular, latency minimization and some network construction/design games can be modeled as congestion games or weighted congestion games.

Most of the previous work has been focused on bounding the price of anarchy. The main focus was latency minimization for linear and polynomial latency functions [3,5,9]. The price of stability for linear latency functions has been studied by Christodoulou and Koutsoupias [4].

As most of previously considered games the game that we consider here is also a congestion game where players are source-destination pairs and a strategy of a player is a single path from the source to the destination. The difference is that the cost that a player pays for each edge e on its path is $c(e)/x_e$ where x_e is the number of players using the edge. The price of anarchy for this game can be high as shown in [1]. But we are interested in the price of stability. The price of stability of a different connection game was also considered by Anshelevitz et al. [2].

2   Preliminaries

Our input is an undirected graph $G = (V, E)$, along with a distinguished source vertex $r \in V$, and a cost function $c : E \mapsto R^+$. We will refer to $c(e)$, $e \in E$, as the *cost* of the edge e.

Associated with every vertex $v \in V$ is a selfish player. The network design game defines a strategy of a player v, to be a simple path in G connecting v to the source r. Let S_v denote the strategy chosen by player v, we define the *state* S to be the set of all paths S_v, for all players v. We define $E(S)$ to be the set of edges that appear in one or more of the paths in state S. [2]

It follows that the graph $(V, E(S))$ is a subgraph of G. In state S, let $x_s(e)$ be the number of players whose strategy contains edge $e \in E$. We define the cost of player v in state S, $C_S(v)$, to be $\sum_{e \in S_v} c(e)/x_s(e)$. A state S is in a Nash equilibrium if no player can lower her cost by unilaterally changing her path to the source r.

We shall use the standard potential function Φ, see e.g. [1,7], that maps every state S into a numeric value: $\Phi(S) = \sum_{e \in E} c(e) H(x_s(e))$, where $H(n) = 1 +$

---

[2] Note that if one allow non simple paths as strategies then for every non simple strategy there is always a simple one which is strictly better.

$1/2 + 1/3 + \cdots + 1/n$ is the n'th Harmonic number. If a single player v changes her strategy then the difference between the potential of the new state and the potential of the original state is exactly the change in the cost of player v. This implies that the improving response dynamics converges to a Nash equilibrium in pure strategies.

Notice that the sum of the costs of all players in state S is exactly the sum of the costs of the edges of $E(S)$. It follows that if the social cost function is the sum of the costs of all players then the social optimum of this game is a minimum spanning tree of the graph. We denote by OPT an arbitrary but fixed minimum spanning tree. Let p be the path from vertex u to vertex v in OPT. We define the *distance between u and v in OPT*, denoted by $d_{opt}(u, v)$, to be the sum of the costs of the edges between vertex u and vertex v along p.

Let S be a state and let $e = (x, y) \in S_u$. We say that u *uses e in the direction $x \to y$* if y is closer than x to the r on S_u. Similarly, we say that u *uses e in the direction $y \to x$* if x is closer than y to r on S_u. We say that e *appears in S in the direction $x \to y$* (or simply $x \to y$ appears in S) if there is a player u such that e appears in S_u in the direction $x \to y$.

In the following definitions assume that v is the only player making the change, and we denote the new state by S' which is identical to S except that we replace S_v by S'_v. We say that a player v makes an *improvement move* when the player chooses a new strategy S'_v such that $C_{S'}(v) < C_S(v)$. We limit player v to choose strategies S'_v of the following three types.

EE (Existing Edges) – An improvement move such that $E(S') \subseteq E(S)$. Furthermore, if S'_v uses an edge $e = (x, y)$ in the direction $x \to y$ then $x \to y$ appears in S.

OPT – An improvement move such that $E(S') \subseteq E(S) \cup OPT$, but $E(S') \not\subseteq E(S)$. Furthermore, if S'_v uses an edge $e = (x, y) \notin OPT$ in the direction $x \to y$ then $x \to y$ appears in S.

$\overline{\textbf{OPT}}$ – The first edge $e = (v, w)$ on S'_v is not in $E(S) \cup OPT$, and $E(S') - \{e\} \subseteq E(S)$. Furthermore, if S'_v uses an edge $e' = (x, y)$, $e' \neq e$ in the direction $x \to y$ then $x \to y$ appears in S.

Remark 1. Note that if we start from OPT and perform only EE, OPT, and $\overline{OPT}$ moves then in the state that we reach, no edge $(x, y) \notin OPT$ appears in both directions, $x \to y$ and $y \to x$. It appears in the same direction determined by the $\overline{OPT}$ move that added (x, y).

Overview. In Section 3 we prove that if no player has an improvement move of type EE, OPT, or $\overline{OPT}$ then the state is a Nash equilibrium. We single out a specific Nash equilibrium, denoted by N, that we reach by carefully scheduling EE, OPT, and $\overline{OPT}$ moves. We then prove that the cost of N is larger than the cost of OPT by a factor of at most $O(\log \log n)$.

After an $\overline{OPT}$ move of a player u that adds the edge (u, v) into the current state, we make further OPT and EE moves so that more players use (u, v). We traverse players in increasing distance from u in OPT. Each player that improves her strategy by using the path to u in OPT following by the strategy of u makes the corresponding improvement move.

Let $c(u, v) = z$. This scheduling has two effects which our proof exploits.

1. If there are $O(\log n)$ players whose distance to u in OPT is no larger than $z/4$ then the potential decreases by $O(z \log n)$. Therefore, the total cost introduced into N by such edges is $O(OPT)$.
2. Edges in $N \setminus OPT$ cannot be too close to each other in the metric defined by OPT. This allows us to relate the cost of all other edges in $N \setminus OPT$ to the cost of OPT.

Our scheduling algorithm is described in Section 4. In Section 5 we prove the bound on the price of stability of the Nash Equilibrium obtained by the scheduler. Due to the space limit some of the proofs are omitted.

3   Improvement Moves Result in Nash Equilibria

We now show that if no player has an improvement move of type EE, OPT, or $\overline{OPT}$ then the current set of strategies is a Nash equilibrium.

Lemma 1. *Let S be a state such that no player has an improving move of type EE. Then $(V, E(S))$ is a tree.*

Proof. Assume that $(V, E(S))$ is not a tree. Since our strategies are simple paths there must be some vertex w from which one can follow two paths to r; one path is the strategy S_w of w, and the other path, denoted by $\hat{S}_w$, is a suffix of some path S_u of a vertex u that goes through w. If $\sum_{e \in \hat{S}_w} c(e)/x_s(e) \leq \sum_{e \in S_w} c(e)/x_s(e)$ then w has an improving EE move in which she replaces her path by $\hat{S}_w$ which is a contradiction. On the other hand, if $\sum_{e \in S_w} c(e)/x_s(e) \leq \sum_{e \in \hat{S}_w} c(e)/x_s(e)$ then u has an improving EE move in which she replaces the suffix $\hat{S}_w$ of S_u by S_w. □

Lemma 2. *Let S be a state in which no player can make an OPT, $\overline{OPT}$, or EE improvement move. Then S is in a Nash equilibrium.*

4   Scheduling $\overline{OPT}$, OPT, and EE Improvement Moves

For technical reasons that we will elaborate on later, instead of considering the stability problem on the graph G, we switch to a related multigraph, $\overline{G}$. It would be clear from the definition of $\overline{G}$ that every minimum spanning tree in $\overline{G}$ corresponds to a minimum spanning tree in G with the same cost and vice versa. We also argue that a Nash equilibrium in the multigraph gives us a Nash equilibrium in the original graph with the same cost.

We define $\overline{G}$ as follows. Associate with every edge $e \in G$, not in OPT, an identical edge $e' \in \overline{G}$. Replace an edge $e \in G$ that is in OPT by parallel edges e^1 and e^2 in $\overline{G}$, each of weight $c(e)$. We say that e^1 and e^2 are *associated with* e and vice versa. We can show that:

Lemma 3. *For every Nash equilibrium in* $\overline{G}$ *there is a Nash equilibrium in* G *of the same cost.*

We define EE, OPT, or $\overline{\text{OPT}}$ moves in $\overline{G}$ the same as we defined them in Section 2 where by edges of OPT in $\overline{G}$ we refer to both copies of each edge of OPT in G.

The scheduler: We start the scheduler on $\overline{G}$ from an initial state isomorphic to OPT. We define the initial state S to consist of all edges $e^1 \in \overline{G}$ associated with some $e \in OPT$. The scheduler halts and the process converges when no EE, OPT, or $\overline{\text{OPT}}$ moves are possible. The scheduler works in phases where in each phase we make a single $\overline{\text{OPT}}$ move.

Let S be some state, that includes strategy S_v for player v and S_w for player w. Given that w is a vertex on S_v, we define **Follow**(S, v, w) as a possible alternative strategy for vertex v. Strategy **Follow**(S, v, w) consists of the prefix of S_v up to and including vertex w, followed by S_w.

As an aid to the exposition, we use colors red and blue to label the parallel edges of $\overline{G}$. Initially, for every $e \in OPT$ we assign the edge e^1 the color red and the edge e^2 the color blue. In the beginning of a phase we may change the assignment of the red/blue colors to the parallel edges.

OptFollow(S, v, w) is a strategy for player v that is defined if there is an edge (v, w) that is a copy of an edge in OPT colored blue. The strategy **OptFollow**(S, v, w) consists of the single edge (v, w) followed by S_w.

A phase of the scheduler: Let S be the state at the beginning of a phase. We maintain the invariant that in S no player can make an improving OPT or EE move, and thereby S is a tree according to Lemma 1. Before the phase starts we make a *Recoloring step*. In this step we recolor red each edge in S which is a copy of an edge in OPT, and we color blue the other copy of the edge which not in S.

$\overline{\text{OPT}}$**-move:** The phase starts with some player u changing her strategy by an improving $\overline{\text{OPT}}$ move. We denote by S' the state after this $\overline{\text{OPT}}$ move of u at the beginning of the phase.

OPT-loop: Following this $\overline{\text{OPT}}$ move we start a breadth first search of OPT from u and for each player v in increasing order of $d_{opt}(u, v)$ we do the following. Let $CurS$ be the state right before we process v, and let $p(v)$ be the parent of v in the breadth first search tree. We check if OptFollow$(CurS, v, p(v))$ is an improving strategy for v. If it is improving then v changes her strategy to OptFollow$(CurS, v, p(v))$. If it is not improving then we truncate the breadth first search at v. Note that all these OptFollow moves are defined since we started the phase with a recoloring step. We call this part of the phase of the scheduler the OPT-loop since all improvement moves made in this part are OPT moves. We denote by D the set of players that includes u and players who performed an OPT move in the OPT-loop.

EE-loop: For each player $w \in D$ let M_w be the subset of descendants of w in the tree S rooted at r, such that $v \in M_w$ if and only if $v \notin D$ and w is the first player in D along the path from v to r in S. In the second part of the phase we traverse the vertices in $\bigcup_{w \in D} M_w$. For each player $v \in M_w$, let $CurS$

be the state right after we process w, if the strategy Follow($CurS, v, w$) is an improving strategy for v, then v changes her strategy to Follow($CurS, v, w$). We call this part of the phase of the scheduler the EE-loop since all improvement moves made in this part are EE moves.

In the last part of the scheduler we perform any improving OPT or EE moves until no such improving move exists. Then the phase ends, and we start the next one if there is an improving $\overline{OPT}$ move, or we stop if there isn't.

5   The Price of Stability

In this section we bound the cost of the Nash equilibrium reached by the scheduler.

We introduce the following definitions. Let S be the state which is a tree. Assume we root the tree at r. Let $P_S(v, w)$ be the path from vertex v to w in state S and let $LCA_S(v, w)$ be the lowest common ancestor of v and w in state S (when we root the tree at r). We remove the subscript S when it is clear from the context.

Let $P_w^v = P(w, LCA(v, w))$ and define $C_S^v(w) = \sum_{e \in P_w^v} \frac{c(e)}{x_s(e)+1} + \sum_{e \in S_w - P_w^v} \frac{c(e)}{x_s(e)}$, where S_w is the strategy of w in state S. In other words, we take into account an additional player on the path from w to $LCA(v, w)$ in S. One can think of $C_S^v(w)$ as the cost of w after v changes her strategy to a strategy in which she takes some path to w and then continues to the source according to S_w. It is clear that $C_S^v(w) \leq C_S(w)$ since the share of w in the cost of each edge on P_w^v in $C_S^v(w)$ is smaller than in $C_S(w)$.

Lemma 4. *Assume that no improving OPT moves, and no improving EE moves are possible in a state S. Then for every pair of players v and w the inequality $C_S(v) \leq C_S^v(w) + d_{opt}(v, w)$ holds.*

Proof. Suppose that $C_S(v) > C_S^v(w) + d_{opt}(v, w)$. Consider the strategy S_v' that consists of the path of OPT edges from v to w followed by the strategy of w. The strategy S_v' has cost $C_{S'}(v) \leq C_S^v(w) + d_{opt}(v, w)$, so it is an improving OPT move and we get a contradiction.   □

Let S' be the state after player u performs an $\overline{OPT}$ move during the execution of the scheduler and let S be the state preceding this move. Let the cost of the newly used edge $e' = (u, v)$ be $c(e') = z$. In the following lemma we show that for every player w for which $d_{opt}(u, w) \leq \frac{z}{4}$, w would pay less if she takes the path in OPT to u and then continues as u in S'. The intuition of why this holds is as follows: From Lemma 4 we know that when no OPT moves are possible the cost of u in S could not be much larger than the cost of w. The difference is about $d_{opt}(u, w) \leq \frac{z}{4}$. So if we make w go through u in S her cost may increase by at most $z/2$. It increases by at most $z/4$ for the path to get to u and by at most $z/4$ since the cost of u may be larger by at most $z/4$ from the cost of w. In S' however w will split the cost of the edge (u, v) with u, paying only $z/2$ to go through it and thereby recovering the extra cost to get to u.

Lemma 5. *Let S be a state where no OPT moves and no EE moves which are improving are possible. Let S' be the new state after player u makes an improving $\overline{OPT}$ move defined by the edge $e' = (u, v)$. Let the cost of $c(e')$ be z. Then for every player w for which $d_{opt}(u, w) \leq \frac{z}{4}$, $C_{S'}(w) > C_{S'}(v) + \frac{z}{2} + d_{opt}(u, w)$.*

Proof. The strategy of player u in S' is the edge (u, v) followed by the strategy of player v, S_v, that is $C_{S'}(u) = C_{S'}(v) + z$. Since u performed an improving $\overline{OPT}$ move, $C_{S'}(u) < C_S(u)$, and thus

$$C_{S'}(v) + z < C_S(u) . \tag{1}$$

Since in S there are no improving OPT moves and no improving EE moves, then, by Lemma 4,

$$C_S(u) \leq C_S^u(w) + d_{opt}(u, w) . \tag{2}$$

We claim that $C_S^u(w) \leq C_{S'}(w)$. First note that the strategy S_w is equal to the strategy S'_w, since only the strategy of u is different in S and S'. The cost of w however may be different in S and S'. Split S_w into two pieces. One piece, denoted by P_1, from w to $LCA_S(u, w)$, and the other piece, denoted by P_2, from $LCA_S(u, w)$ to the source (see Figure 1). In S, player w shares with player u the cost of the edges in P_2, but this may not be true in S', so for $e \in P_2$, $x_s(e) \geq x_{s'}(e)$. Consider P_1. In S player w does not share with player u the cost of the edges on P_1, but she may share this cost with u in S'. So for $e \in P_1$ we have $x_s(e) + 1 \geq x_{s'}(e)$. In contrast $C_S^u(w)$ is the tentative cost of w assuming that she shares with u the cost for every edge of her strategy. Therefore,

$$C_S^u(w) = \sum_{e \in P_1} \frac{c(e)}{x_s(e) + 1} + \sum_{e \in P_2} \frac{c(e)}{x_s(e)} \leq \sum_{e \in S'_w} \frac{c(e)}{x_{s'}(e)} = C_{S'}(w) , \tag{3}$$

as we claimed. From inequalities (2) and (3) we obtain

$$C_S(u) \leq C_{S'}(w) + d_{opt}(u, w) . \tag{4}$$

Considering inequalities (1) and (4) we get $C_{S'}(w) + d_{opt}(u, w) > C_{S'}(v) + z$, and therefore

$$C_{S'}(w) > C_{S'}(v) + z - d_{opt}(u, w) .$$

For player w for which $d_{opt}(u, w) \leq \frac{z}{4}$,

$$C_{S'}(w) > C_{S'}(v) + z - d_{opt}(u, w) \geq C_{S'}(v) + \frac{3z}{4} \geq C_{S'}(v) + \frac{z}{2} + d_{opt}(u, w) . \qquad \square$$

Let S' be the state after player u performs an $\overline{OPT}$ move during the execution of the scheduler, defined by the edge $e_u = (u, v)$ whose cost is z. Let $w_0, w_1, w_2, \ldots, w_m$ be the vertices with $d_{opt}(u, w_i) \leq \frac{z}{4}$. Assume that $d_{opt}(u, w_i) \leq d_{opt}(u, w_{i+1})$. In particular $w_0 = u$, and the vertex w_1 is adjacent to u in OPT. Lemma 5 implies that the strategy OptFollow(S, w_1, u) is improving for w_1. But what happens after w_1 changes her strategy? Can w_2 still make an OPT move using some edge which is not in S and lower her cost? The following lemma shows that indeed this is the case.

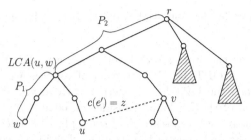

Fig. 1. Player u makes an $\overline{OPT}$-move and buys edge $e' = (u, v)$ of cost z. We assume that $d_{opt}(u, w) \leq \frac{z}{4}$.

Lemma 6. *Let w_k be the vertex following w_i on the path from w_i to u in OPT (that is, w_k is the parent of w_i in the BFS tree traversed by the OPT-loop). Let S^i be the state just before the scheduler processes w_i in its OPT-loop. Then $C_{S^i}(w_i) > C_{S^i}(v) + \frac{z}{2} + d_{opt}(u, w_i)$, and therefore OptFollow(S^i, w_i, w_k) is an improvement move for w_i and the scheduler changes the state of w_i to this strategy.*

Remark 2. To make Lemma 6 work we had to introduce $\overline{G}$. With one set of OPT edges it is possible that when w_i changes her strategy she uses OPT edges that can be part of the strategy of w_ℓ for some $\ell > i$. If these edges are not in S_v, and are not on the path between w_ℓ and u in OPT then this may lower the cost of S_{w_ℓ} such that when the scheduler gets to w_ℓ in the OPT-loop, her alternative OptFollow move is not improving.

The following lemma gives a lower bound on the decrease in the potential during a phase of the scheduler.

Lemma 7. *Let u be the player making the $\overline{OPT}$ move at the beginning of a phase. Let $e' = (u, v)$ be the first edge in the new strategy of player u, and let $z = c(e')$. Let m be the number of players at distance at most $\frac{z}{4}$ from player u in OPT (other than u itself). If $m \geq 2$ then the potential of the state at the end of the phase is smaller by $\Omega(zm)$ from the potential of the state at the beginning of the phase.*

Proof. Let $w_1, \ldots, w_m$ be the players such that $d_{opt}(u, w_i) \leq \frac{z}{4}$. Assume that $d_{opt}(u, w_i) \leq d_{opt}(u, w_{i+1})$. Let S^i be the state right before the scheduler processes w_i in its OPT-loop.

By Lemma 6, when the scheduler processes player w_i we have that $C_{S^i}(w_i) > C_{S^i}(v) + \frac{z}{2} + d_{opt}(u, w_i)$. Also according to Lemma 6 players $w_1, \ldots, w_{i-1}$ already use the edge (u, v) in their strategy in S^i. Therefore the cost of the new strategy OptFollow(S^i, w_i, w_k) for w_i is at most $C_{S^i}(v) + \frac{z}{i+1} + d_{opt}(u, w_i)$. (Here w_k is the vertex adjacent to w_i on the path in OPT from w_i to u.) It follows that player w_i decreases her cost by at least $\frac{z}{2} - \frac{z}{i+1}$. Summing up the decrease in the cost of all m players $w_1, \ldots, w_m$, we get $\sum_{i=1}^m \frac{z}{2} - \frac{z}{i+1} = z(\frac{m}{2} - (H(m+1) - 1)) = \Theta(zm)$. This is also the decrease in the potential since when a single player changes her strategy the change in the potential is equal to the change in the cost of the player.   □

As before, let S' be the state after player u performs an $\overline{\text{OPT}}$ move and uses an edge $e' = (u, v) \notin OPT$. Let D be the set of vertices accumulated while the scheduler performed the OPT-loop, together with u, and let S'' be the state after the execution of the EE-loop. Consider an edge $e \notin OPT$ which was the first edge in the strategy S_w in state S, of some player $w \in D$. By the definition of the scheduler, the first edge in the strategy of w in S'' would be an edge in OPT (or e' for u) and not e. However, it could be that some descendant of w still uses e in her strategy. We want to show that this could not be the case. That is, while performing the EE-loop all these descendants take an alternative strategy that does not use e.

Lemma 8. *Consider a phase of the scheduler. Let S be the starting state, and let D be the set of players that includes player u and the players that change their strategy in the OPT-loop. Let $e \notin OPT$ be the first edge in a strategy S_w, for some $w \in D$. Let S'' be the state after the execution of the EE-loop. Then $e \notin S''$.*

The total cost of the edges in $N \cap OPT$ is no larger than the cost of OPT. We associate each edge $(u, v) \in N \setminus OPT$ with player u that actually improved her strategy by the $\overline{\text{OPT}}$ move that added the edge (u, v) to N. We further partition the edges $e = (u, v)$ in $N \setminus OPT$ according to the number of vertices in OPT in a neighborhood of size $c(e)/4$ around the associated player. Specifically, let $e = (u, v) \in N \setminus OPT$ be associated with player u. We say that e is *crowded* if $|\{w \mid d_{opt}(u, w) \leq \frac{c(e)}{4}\}| \geq \log n$, and we say that e is *light* otherwise.

Lemma 9. *The total cost of all crowded edges is $O(OPT)$.*

Proof. Let e be a crowded edge in $N \setminus OPT$. By Lemma 7, in the phase that started with the $\overline{\text{OPT}}$ move that put e into N, the potential dropped by $\Omega(c(e) \log n)$. Since initially the potential is at most $OPT \cdot \log n$, and is always decreasing, the lemma follows.   □

Lemma 10. *The total cost of all light edges in N is $O(OPT \cdot \log \log n)$.*

Proof. Let U be the set of players assigned to light edges. For a player $v \in U$ we denote the associated light edge by e_v. We define *the cost of v* to be the cost of e_v and denote it by z_v.

We choose a subset $F \subseteq U$ as follows. Start with $T = U$ and $F = \emptyset$. Let $v \in T$ be a player of maximum cost in T. Let $U_v = \{w \in U \mid d_{opt}(v, w) \leq z_v/4, z_w \leq z_v/\log n\}$. Add v to F and continue with $T = T \setminus (\{v\} \cup U_v)$ until T is empty.

Since every vertex $v \in F$ is a light vertex, the total cost of all vertices in U_v is at most z_v, so its enough to prove that the total cost of all vertices in F is $O(OPT \cdot \log \log n)$.

For $v \in F$, consider a ball, B_v, of radius $z_v/12$ around v in OPT. According to Lemma 4, $z_v < d_{opt}(v, r)$, so the ball B_v contains at least one path of length at least $z_v/12$. We prove that every point $\xi \in OPT$ is contained in at most $\log\log n$ balls B_v for $v \in F$. Therefore the total cost of all vertices in F is $O(OPT \cdot \log\log n)$.

Let $e \in OPT$ and let ξ be some point on edge e. Let A_ξ be the set of vertices whose balls contain ξ. We show that $|A_\xi| \leq \log\log n$. Let $v_1, \ldots, v_m$ be the vertices of A_ξ in the order that their light edges $e_{v_1}, \ldots, e_{v_m}$ were added to N (if some edge was added more than once, we consider the last time it was added). Let $1 \leq i < j \leq m$. By Remark 1, when v_j makes the $\overline{OPT}$ move that adds e_{v_j}, v_i was using e_{v_i} in her strategy. Since $e_{v_i} \in N$, that is v_i did not change her strategy in the OPT-loop of the phase where v_j added e_{v_j}, according to Lemma 8, we have

$$d_{opt}(v_i, v_j) > \frac{z_{v_j}}{4} . \tag{5}$$

Since $d_{opt}(v_i, \xi) \leq z_{v_i}/12$ and $d_{opt}(v_j, \xi) \leq z_{v_j}/12$, we obtain

$$d_{opt}(v_i, v_j) \leq \frac{z_{v_i}}{12} + \frac{z_{v_j}}{12} . \tag{6}$$

Substituting $j = i + 1$ and combining the Inequalities (5) and (6), we get $z_{v_{i+1}} < z_{v_i}/2$ and, by induction, $z_{v_{i+1}} < \frac{z_{v_1}}{2^i}$. In particular, for every i we have $z_{v_{i+1}} < z_{v_1}$, so by applying Equation 6 to v_{i+1} and v_1 we get $d_{opt}(v_{i+1}, v_1) \leq z_{v_1}/6$. Therefore, by the definition of F, it must be that $z_{v_{i+1}} > z_{v_1}/\log n$. Since $\frac{z_{v_1}}{\log n} < z_{v_{i+1}} \leq \frac{z_{v_1}}{2^i}$, we get that $i < \log\log n$, and therefore $|A_\xi| \leq \log\log n$. $\square$

The following theorem follows from Lemmas 9 and 10 and is the main result of this work.

Theorem 1. *For a graph with a source vertex and a player in every vertex the price of stability is $O(\log\log n)$.*

Acknowledgement

We would like to thank Micha Sharir for helpful discussions.

References

1. E. Anshelevich, A. Dasgupta, J. Kleinberg, E. Tardos, T. Wexler, and T. Rough-garden. The price of stability for network design with fair cost allocation. In *Proc. of 45th FOCS*, pages 295–304, 2004.
2. E. Anshelevich, A. Dasgupta, E. Tardos, and T. Wexler. Near-optimal network design with selfish agents. In *Proc. of 35th STOC*, pages 511–520, 2003.
3. B. Awerbuch, Y. Azar, and A. Epstein. Large the price of routing unsplittable flow. In *Proc. of 37th STOC*, pages 57–66, 2005.
4. G. Christodoulou and E. Koutsoupias. On the price of anarchy and stability of correlated equilibria of linear congestion games. In *Proc. of 13th ESA*, pages 59–70, 2005.

5. G. Christodoulou and E. Koutsoupias. The price of anarchy of finite congestion games. In *Proc. of 37th STOC*, pages 67–73, 2005.
6. E. Koutsoupias and C. Papadimitriou. Worst-case equilibria. In *Proc. of 16th STACS*, pages 404–413, 1999.
7. D. Monderer and L. Shapley. Potential games. *Games and Economic Behavior*, pages 14:124–143, 1996.
8. R. Rosenthal. A class of games possessing pure-strategy nash equilibria. *International Journal of Game Theory*, pages 2:65–67, 1972.
9. T. Roughgarden and E. Tardos. How bad is selfish routing? *Journal of the ACM*, pages 49(2):236–259, 2002.

Dynamic Routing Schemes for General Graphs

(Extended Abstract)

Amos Korman[1,*] and David Peleg[2,**]

[1] Information Systems Group, Faculty of IE&M, The Technion, Haifa, Israel
`pandit@tx.technion.ac.il`
[2] Department of Computer Science, Weizmann Institute of Science, Rehovot, Israel
`david.peleg@weizmann.ac.il`

Abstract. This paper studies approximate distributed routing schemes on dynamic communication networks. The paper focuses on dynamic weighted general graphs where the vertices of the graph are fixed but the weights of the edges may change. Our main contribution concerns bounding the cost of adapting to dynamic changes. The update efficiency of a routing scheme is measured by the number of messages that need to be sent, following a weight change, in order to update the scheme. Our results indicate that the graph theoretic parameter governing the amortized message complexity of these updates is the local density D of the underlying graph, and specifically, this complexity is $\tilde{\Theta}(D)$. The paper also establishes upper and lower bounds on the size of the databases required by the scheme at each site.

1 Introduction

Motivation: The basic function of a communication network, namely, message delivery, is performed by its *routing scheme*. Subsequently, the performance of the network as a whole may be dominated by the quality of the routing scheme. Thus, constructing an efficient routing scheme is one of the most important tasks when dealing with communication network design.

We distinguish between *static* and *dynamic* routing schemes. In a static routing scheme the databases of the processors are tailored to the particular network topology. However, in most communication networks, the typical setting is highly dynamic, namely, even when the physical infrastructure is relatively stable, the network traffic load patterns undergo repeated changes. Therefore, for a routing scheme to be useful in practice, it should be capable of reflecting up-to-date load information in a dynamic setting, which may require occasional updates to the databases. Ideally, upon a topological change, only a *limited* number of messages are sent in order to update the databases. We rank the update efficiency of dynamic schemes by their *message complexity*, i.e., the amortized number of messages sent per topological change. Note that the message complexity also

---

* Supported in part by the Aly Kaufman fellowship.
** Supported in part by a grant from the Israel Science Foundation.

M. Bugliesi et al. (Eds.): ICALP 2006, Part I, LNCS 4051, pp. 619–630, 2006.
© Springer-Verlag Berlin Heidelberg 2006

bounds from above the amortized number of graph vertices whose database needs to be modified per update operation, hence lower message complexity implies also lower accounting efforts and fewer interruptions to the vertices.

The efficiency of a dynamic scheme is measured not only by its message complexity but also by the quality of the routes it provides and by the memory complexities associated with it. Route quality is measured by the *stretch factor* of the scheme, i.e., the maximum ratio, over all pairs of nodes in the network, between the length of the route provided for them by the routing scheme, and the actual (weighted) distance between them in the network. We focus on β-*approximate* routing schemes, namely, ones that produce a route whose weighted length is perhaps not the shortest possible, but approximates it by a factor of at most β, for some constant $\beta > 1$.

Another consideration is the amount of information stored at each vertex. We distinguish between the *internal database Data(v)* used by each node v to deduce the required information in response to online queries, and the additional external storage $Memory(v)$ at each node v, used during (offline) updates and maintenance operations. For certain applications, the internal database $Data(v)$ is often kept in the router itself, whereas the additional storage $Memory(v)$ is kept on some external storage device. Subsequently, the size of $Data(v)$ is a more critical consideration than the total amount of storage needed for the information maintenance.

The current paper investigates schemes on dynamic settings involving changing link weights. The model studied considers a network whose underlying topology is a fixed graph, i.e., the vertices and edges of the network are fixed but the (positive integer) weights of the edges may change. At each time the weight of one of the edges can increase or decrease by a fixed quanta (which for notational convenience is set to be 1), as long as the weight remains a positive integer. (Our algorithms and bounds apply also for larger weight changes, as clearly, a weight change of $\Delta > 1$ can be handled, albeit naively, by simulating it as Δ individual weight changes of 1. As our focus is on establishing the complexity bounds for the problem, no attempt was made to optimize the performance of our algorithms in case of large weight changes.)

This paper introduces dynamic β-approximate routing schemes that are efficient in terms of their message complexity. We also give lower bounds regarding the complexities of dynamic routing schemes. Our results may indicate that the graph theoretic parameter governing the message complexity is the local density of the underlying graph, defined as follows. For a graph G and integer $r \geq 1$, let $N(v,r)$ denote the set of vertices at distance at most r from the node v. Then the *local density* of G is $D = \max_{v,r}\{|N(v,r)|/2r\}$.

Related work: Many routing schemes and lower bounds for the resources required for routing were presented in the past (cf. [16]). The first studies attempting to characterize and bound the resource tradeoffs involved in routing schemes for general networks were presented in [17,3,4] and were followed by a number of improved constructions (cf. [5,10,11,12,19]). These studies focused on routing schemes with compact routing tables and low stretch factors.

Unfortunately, most known algorithms in this field apply only for static networks, and only a few papers consider dynamic networks. In [18] a partial solution is presented for limited cases of topology changes that keep the network in a tree topology. The following dynamic routing schemes on trees assume that the designer of the routing scheme has the freedom to choose the identities of the vertices. In [2] a routing scheme is presented for the restricted case of dynamic growing trees using identities of size $O(\log^2 n)$, database size $O(\Delta \log^3 n)$ (for graphs of maximum degree Δ) and message complexity $O(\log n)$ where n is an upper bound on the number of vertices in the growing tree. When an upper bound n on the number of vertices in the growing tree is known in advance, a routing scheme with message complexity $O(\frac{\log n}{\log \log n})$ and polylogarithmic database size is given in [14]. In the more general case where in each step, a leaf can either be added to or removed from the tree, a routing scheme with $O(\log^2 n)$ message complexity and $O(\log^2 n)$ database size is presented in [14] and a routing scheme with $\Theta(\log n)$ database size and sublinear message complexity is presented in [13]. All the above mentioned dynamic routing schemes deal only with tree networks. For general graphs there are even fewer results. A lower bound of $\Omega(n)$ is established in [2] for the message complexity of constant approximation routing schemes on general dynamic graphs (where edges may be added or deleted). The routing scheme of [6] for dynamic graphs applies to a somewhat different networking model based on virtual circuits, where the route quality is measured in terms of the number of "super-hops" required for a route, hence those results cannot be directly compared with ours. Also, the analysis therein does not consider the length of the routes produced by the routing scheme, and in fact the scheme may incur a linear stretch factor in some dynamic scenarios.

The maximum database size of β-approximate point-to-point routing schemes (defined below) on general graphs was previously investigated in the static scenario. A lower bound of $\Omega(n^{\frac{1}{2\beta+4}})$ for $\beta \geq 2$ was shown in [17] and a lower bound of $\Omega(n \log n)$ for $1 \leq \beta < 2$ was shown in [8].

In the *sequential* (non-distributed) model, dynamic data structures have been studied extensively. For surveys on dynamic graph algorithms see [7,9].

Model: In this paper the underlying network topologies considered are general graphs. Throughout the paper, denote by n the number of vertices in the network. Let W be the maximum weight assigned to an edge in the network.

This paper studies two types of routing schemes. *Source-directed* routing schemes are routing schemes in which the message originator computes the entire route to the destination and attaches it to the message header. In contrast, *point-to-point* routing schemes route messages on a hop by hop basis, with each intermediate node along the route determining the next edge to be used.

Formally, we make the following definitions. A *point-to-point β-approximate routing scheme* is composed of an *update protocol* for assigning each vertex v of the graph with a local database $Data(v)$, coupled with a *router* algorithm whose inputs are the header of a message M, $Data(v)$ and the identity of a vertex u. If a vertex x wishes to send a message M to vertex y, it first prepares a header for M and attaches it to M. Then x's router algorithm outputs a port of x

on which the message is delivered to the next vertex, until it reaches y. The requirement is that the weighted length of the resulting path connecting x and y is a β-approximation of the weighted distance between x and y at the time the route starts. A *source directed β-approximate routing scheme* is composed of an *update protocol* algorithm for assigning each vertex v of a graph with a local database $Data(v)$, coupled with a *router* algorithm that using *only* the information in $Data(v)$ and the identity of a vertex u, outputs a sequence of port numbers representing a path connecting v and u whose weighted length is a β-approximation of the distance between u and v.

Contributions: Our main contribution focuses on the message complexity of dynamic routing schemes. We use the local density parameter in an attempt at capturing the graph theoretic parameter governing the message complexity of the problem.

In section 3 we present our β-approximate source directed scheme for the class $\mathcal{F}(n, D)$ of n-vertex graphs with local density at most D. In section 4 we present our β-approximate point-to-point routing scheme for $\mathcal{F}(n, D)$. Both schemes incur an amortized message complexity $O(D \log^2 n)$ per weight change. We show that any β-approximate source directed routing scheme on $\mathcal{F}(n, D)$ (for $D > 1$) must have amortized message complexity $\Omega(D)$ per weight change. In the cases where $1 \leq \beta < 3$ we show that any β-approximate point-to-point routing scheme on $\mathcal{F}(n, D)$ (for $D > 1$) must have message complexity $\Omega(D)$ per weight change.

In terms of the database size, we show an upper bound of $O(n \log n)$ and a lower bound of $\Omega(n)$ for the database size of β-approximate source directed routing scheme on $\mathcal{F}(n, D)$ (for $D \geq 1.5$). Our point-to-point β-approximate routing scheme uses database size $O(n \log^2 n)$. Both schemes use $O(|E| \cdot (\log W + \log n))$ bits of memory per vertex. The point-to-point routing scheme uses a header size of $O(\log n)$ bits.

2   Preliminaries

We assume that the vertices and edges of the network are fixed and that the edges of the network are assigned positive integer weights. For two vertices u and v in a weighted graph G with edge weight function ω and for a time t, denote by $d_G^\omega(u, v, t)$ the weighted distance between u and v at time t. Denote by $d_G(u, v)$ the unweighted distance between u and v in G. For $q > 0$, the q-neighborhood of $v \in G$, denoted $\Gamma(v, q)$, is the subgraph of G induced by $\{w \mid d_G(v, w) \leq q\}$. We omit the subscript G when the graph G is clear from the context. It is assumed that the nodes of a given n-vertex graph have distinct identities in the range $1, \cdots, n$. The identity of a node v is denoted by $id(v)$.

The dynamic model: In the dynamic network model considered in this paper, it is assumed that the weights of the network links may change from time to time. For example, the weights may represent the current loads on the links which may change due to queue buildups and nonuniform arrival rates.

The following events may occur:

1. Positive weight change: An edge increases its weight by one.
2. Negative weight change: An edge of weight ≥ 2 decreases its weight by one.

Subsequent to an event on an edge $e = (u, v)$, its endpoints u and v are informed of this event.

In this abstract we deal with scenarios in which at most n topological events occur; in the full version of the paper we show how to deal with the general case.

Routing schemes: We start with the static setting. For fixed $\beta > 1$, a static β-*approximate source directed routing scheme* $\pi_{SD} = \langle \mathcal{P}_{SD}, \mathcal{R}_{SD} \rangle$ for a family of graphs $\mathcal{F}$ is composed of the following components:

1. A *preprocessing* algorithm $\mathcal{P}_{SD}$ that given a graph $G \in \mathcal{F}$, assigns a local database $Data(v)$ to each vertex $v \in G$.
2. A polynomial time *router* algorithm $\mathcal{R}_{SD}$ that given the database $Data(u)$ and $id(v)$ for some vertices u and v in some graph $G \in \mathcal{F}$, outputs a sequence of port numbers representing a path P connecting u and v.

A *static β-approximate point-to-point routing scheme* $\pi_{PTP} = \langle \mathcal{P}_{PTP}, \mathcal{R}_{PTP} \rangle$ for a family of graphs $\mathcal{F}$ is composed of the following components:

1. A *preprocessing* algorithm $\mathcal{P}_{PTP}$ that given a graph $G \in \mathcal{F}$, assigns a local database $Data(v)$ to each vertex $v \in G$.
2. A polynomial time *router* algorithm $\mathcal{R}_{PTP}$ that given the database $Data(u)$ and $id(v)$ for some vertices u and v in some graph $G \in \mathcal{F}$ and a header H of a message M, outputs a port number of u and a new header for M.

Routing a message using a point-to-point routing scheme π_{PTP} is done as follows. If a vertex x wants to send a message M to the vertex y in G, it first prepares a header H for M and attaches it to M. Then the message is delivered via the port $\mathcal{R}_{PTP}(Data(x), id(y), h)$ to x', a neighbor of x. The vertex x' repeats the process, using its own data base $Data(x')$. The message M is thus delivered on the port $\mathcal{R}_{PTP}(Data(x'), id(y), h)$ of x' to the next vertex and so forth. In contrast, when using a source-directed routing scheme π_{SD}, if a vertex x wants to send a message M to the vertex y in G, x computes a path P connecting x and y and attaches it to the message header. Each vertex on the route delivers M to the next vertex on the path P until M reaches its destination y. For the routing scheme (either source-directed or point-to-point) to be a β-approximate routing scheme, the requirement is that the weighted length of resulting path connecting x and y is a β-approximation of $d_G^\omega(x, y)$.

Let us now turn to dynamic routing schemes. In the asynchronous dynamic network model the preprocessing algorithm $\mathcal{P}$ changes into an update protocol $\mathcal{U}$ (denoted $\mathcal{U}_{SD}$ for source directed routing and $\mathcal{U}_{PTP}$ for point-to-point routing) that initially assigns a local database $Data(v)$ to each vertex $v \in G$. After the initial setup, $\mathcal{U}$ is activated after every topological change in order to maintain all local databases so that the corresponding router algorithms work correctly. Observe that in the context of distributed networks, the update algorithms must be implemented as *distributed update protocols*. In particular, the messages sent by $\mathcal{U}$ in order to maintain the databases are sent over the edges of the underlying graph.

It is easier to analyze our protocols assuming that the topological changes occur sequentially and are sufficiently spaced so that the update protocol has enough time to complete its operation in response to a given topological change before the occurrence of the next change. However, our schemes can operate also under weaker assumptions. Specifically, it is allowed for topological changes to occur in rapid succession or even concurrently. Our statements concerning the correctness of our source directed routing scheme (a scheme is correct if every message sent will eventually reach its destination) still hold. Our point-to-point scheme is also correct provided that the topological changes quiet down at some later time for a sufficiently long time period. The quality of our schemes, however, is affected by such behavior of the system as follows. We say that a time t is *quiet* if all updates (of the relevant update protocol) concerning the previous topological changes have occurred by time t. At a quiet time t, the system is said to be quiet. Our demand from a dynamic β-approximate routing scheme (either source directed or point-to-point) is that if a route from u to v starts at some quiet time t and the system remains quiet throughout the rest of the route then the weighted length of resulting route is a β-approximation to $d^\omega(u, y, t)$. The above demand is reasonable, as it can easily be shown that for any routing scheme, if a route from u to v starts at a non-quiet time t then we cannot expect the resulting route to be a β-approximation to $d^\omega(u, v, t)$ for any $\beta > 1$.

For a dynamic β-approximate routing scheme $\pi(\beta) = \langle \mathcal{U}(\beta), \mathcal{R} \rangle$ on the family $\mathcal{F}$, we are interested in the following complexity measures.

Maximum Database Size, $Data(\pi(\beta))$: the maximum number of bits in $Data(v)$ taken over all vertices v and all scenarios on all n-vertex graphs $G \in \mathcal{F}$.
Maximum Memory Complexity, $\mathcal{MMC}(\pi(\beta))$: the maximum number of bits in $Memory(v)$ taken over all vertices v and all scenarios on all n-vertex $G \in \mathcal{F}$.
Message Complexity, $\mathcal{MC}(\pi(\beta))$: the maximum amortized number of messages sent by $\mathcal{U}(\beta)$ per topological change, taken over all scenarios on all $G \in \mathcal{F}$.
For the point-to-point routing scheme we are also interested in the following:
Header Size, $\mathcal{HD}(\pi_{PTP})$: the maximum number of bits attached to the message header by the router protocol π_{PTP} at any step along the route.

3   General Intuition

Our schemes are based on the following ideas. After every change in a weight of an edge e, one of e's end nodes creates a 'report' of this event encoded on a token that is sent to some vertices in the graph. A simple routing scheme would require the update protocol to send each such token to all the vertices. For such a scheme, on the one hand, all nodes have an up-to-date view of the graph and the routings can be made over the shortest weighted paths, but on the other hand, the message complexity is high, namely, $\Omega(n)$.

In order to reduce the message complexity, both our schemes are based on the principle that updates are made in a gradual manner: the tokens are disseminated only to a limited distance and are then stored in intermediate bins of various sizes. Nodes outside this range are thus unaware of the changes represented by these tokens. Whenever sufficiently many tokens accumulate at a bin, they are disseminated further, to a distance proportionate to their number. The analysis of the approximation is based on bounding the possible overall error made in the path selected by the router protocol. This bound is based on the maximal distortion in the way the relevant vertices view the weights of the edges in some subgraph in comparison to the reality. This distortion corresponds to the number of delayed tokens 'stuck' in the various bins of this subgraph.

Algorithms based on the idea of gradual token passing appeared in [1,15,14]. However, a direct application of the method presented in the above papers would not yield a routing scheme with the desired properties. Informally, the reason is that the algorithms used in the above papers were designed for trees. Moreover, using techniques similar to those in the above papers, one can only guarantee that each node x knows an approximation of the weighted length of any path *in some spanning tree* containing x, while for our purposes we are interested in approximations of all paths *in the graph* that pass through x. In order to achieve this, we extend the techniques of the above papers by separating the updates, which are done on the graph, from the token passing, which is carried out on the spanning tree. The token passing implicitly monitors the updates. Each time a bin b becomes full, it is emptied and its contents are used to update vertices on the graph at distance $d(b)$ proportionate to the size of b. In addition, the contents of b are transferred to b', a bin on the spanning tree, at distance $d'(b)$ which is also proportionate to the size of b. Our performance bounds then rely on the fact that $d(b) = c \cdot d'(b)$ for some constant $c > 1$.

When routing a message from x to y in our source directed routing scheme x outputs $\mathcal{R}_{SD}(Data(x), y)$, the shortest path from x to y (according to x's knowledge). It will follow from our analysis that $\mathcal{R}_{SD}(Data(x), y)$ is a good approximation to $d^\omega(x, y)$. The problem becomes more difficult in the point-to-point routing scheme. A natural approach for constructing the point-to-point scheme would be to use the same data structure as in the source directed scheme, except that whenever v receives a message addressed to y, v delivers the message to the next node (i.e., its neighbor) on the path $\mathcal{R}_{SD}(Data(v), y)$. Unfortunately, this may cause the routing process to end up with a message caught in an infinite loop. For example, since $Data(v)$ and $Data(w)$ are not identical, v may think that w is the next node on the shortest path from v to y, and w may think that v is the next node on the shortest path from w to y.

The main technical contribution of this paper is based on the following idea which is used by our point-to-point routing scheme in order to prevent the above undesirable phenomenon from happening. When routing a message from x to y, x first estimates $d^\omega(x, y)$ as in the source directed scheme and uses this estimate to define some value $q = \Theta(d^\omega(x, y))$ that will be attached to the message header. When getting a message destinated at y, the intermediate node v along the route creates a collection $\tilde{\Gamma}_v(y, q)$ of estimates for the weights of all the edges in the q-neighborhood of y, $\Gamma(y, q)$. As established later, the estimations $\tilde{\Gamma}_v(y, q)$ have two important properties. The first property of these estimates is that they are the same for all vertices v on the *route*. This property allows each intermediate node v along the route to mimic the shortest path computation carried by x in order to decide to which node it should pass the message, yielding consistency between the decisions of the nodes on the route. The second property is that although these estimates are potentially weaker than the corresponding estimates obtained by the source directed routing scheme, they are still good enough to ensure that the route is a β-approximation to $d^\omega(x, y)$.

4 The Source Directed Routing Scheme $\pi_{SD}(\beta)$

This section introduces our β-approximate source directed routing scheme $\pi_{SD}(\beta)$ for the family $\mathcal{F}(n, D)$. As mentioned before, in this extended abstract we assume that at most n topological changes occur in each scenario. This assumption only affects the memory size of the scheme. In the full version we describe how to descard this assumption without affecting our complexity bounds.

General structure: Let $T(G)$ be a spanning tree of some graph $G \in \mathcal{F}(n, D)$, rooted at some vertex r. Let $T'(G)$ be the tree obtained from $T(G)$ by extending it with an imaginary n-node path P_r attached to r. Let r' be the end node of P_r not attached to $T(G)$. We view r' as the root of $T'(G)$. In the current section the vertices of G are considered as vertices of $T'(G)$ and not of G.

The token passing mechanism (which we use in order to monitor the message delivery mechanism of the update protocol) is inspired by [1] and [15]. Each node v maintains two bins, a "local" bin b_l and a "global" bin b_g, storing a varying number of tokens throughout the execution. Each token contains information about some weight change in one of the edges. Specifically, a token is of the form $\phi = \langle id(e), \omega(e), c \rangle$, indicating that the c'th weight change on the edge e sets its weight to $\omega(e)$. In the following discussion, unless it might cause confusion, we do not distinguish between a bin and the node holding that bin. Let $H(v)$ denote v's unweighted (hop) distance from r'. For every node v of $T'(G)$, apart from r', the bins b_l and b_g are assigned a *level*, defined as follows: $Level(b_g) = \max\{i \mid 2^i \text{ divides } H(v)\}$, and $Level(b_l) = -1$.

Note that the level of the bin determines whether it is of type b_l or b_g. Hence hereafter,, we omit subscripts g and l unless confusion might arise. For each bin b at node v, the closest ancestor bin in $T'(G)$ (including possibly in v itself), b', satisfying $Level(b') = Level(b) + 1$ is set to be the *supervisor* of b. If there is no such bin, then the global bin of r' is set to be the supervisor of b. Let $sup(b)$ denote the supervisor bin of b. Note that the supervisor bin of a local bin is either the global bin of the same node, or the global bin of its parent in $T'(G)$. This defines a bin hierarchy with the following easy to prove properties.

1. The highest level of the bin hierarchy is at most $\log n + 2$.
2. If $Level(b) = l$ then the path from b to $sup(b)$ have at most $3 \cdot 2^l$ nodes.
3. On any path of length p, the number of level l bins that are supervisors to other bins in that path is at most $\frac{p}{2^{l-1}}$.

$$\text{For } \beta > 1, \text{ let } \quad \alpha = \frac{1 + \sqrt{2\beta - 1}}{2}, \quad \delta = \min\left\{\sqrt{\beta} - 1, \frac{\alpha - 1}{\beta\alpha}\right\}.$$

The number of tokens stored at each bin b at a given time is denoted $\tau(b)$. The *capacity* of each bin depends on its level. Specifically, a bin b on $Level(b) = l$ may store $0 \leq \tau(b) \leq Cap(l)$ tokens, where $Cap(l) = \max\{2^{\lfloor \log \frac{\delta \cdot 2^l}{3D \cdot (\log n + 2)} \rfloor}, 1\}$. In fact, it will follow from the algorithm description that at any given moment, a bin is either empty or half-full, namely, $\tau(b) \in \{0, Cap(l)/2\}$. In particular, a bin of capacity 1 is always empty and therefore, the number of tokens stored in an l-level bin is at most $\frac{\delta \cdot 2^{l-1}}{3D(\log n + 2)}$.

Update protocol $\mathcal{U}_{SD}(\beta)$: The memory structure $Memory(v)$ of each vertex v contains the adjacency matrix $A(v)$ of the entire initial graph. For each edge e, the counter $c(e)$ counts the number of changes in e's weight and is initially set to be 0. Each entry e in $A(v)$ contains two fields, denoted by $\omega(e, v)$ and $c(e, v)$. If the latest change in e's weight that v heard about was the c_0'th change, which has led to the values $c(e) = c_0$ and $\omega(e) = \omega_0$, then e's entry in $A(v)$ is set to $\langle \omega(e, v), c(e, v)\rangle = \langle \omega_0, c_0 \rangle$.

For each edge e, one of its endpoints (say, the one with the smaller id) is said to be *responsible* for e. A token $\phi = \langle id(e), \omega(e), c\rangle$ is said to be *fresh* w.r.t. the matrix $A(v)$ if c is larger than $c(e, v)$. Intuitively, such a token can be used to update the entry corresponding to e in $A(v)$.

Let $b_g(v)$ denote the global bin of a vertex $v \in T'(G)$. Let $T^*(G)$ be the same as $T(G)$, with the same bin hierarchy, except that r has a number of additional global bins. Specifically, the set of added global bins of r is $\{b_g(w) \mid w \in P_r\}$. We next describe the imaginary Protocol BIN on $T'(G)$. In practice we run Protocol BIN* on $T^*(G)$, which is the same as Protocol BIN except that r uses its multiple global bins to simulate the behavior of the imaginary path P_r in Protocol BIN. For a level l bin b, let $Q(b)$ denote the set of all nodes at unweighted distance at most $2^5 \cdot 2^l$ from b (in G). Note that $sup(b) \in Q(b)$.

The update protocol $\mathcal{U}_{SD}$ uses Protocol BIN described below to maintain the databases of the vertices.

Protocol BIN

1. Initially all bins are empty.
2. For an edge e under u's responsibility, each time u learns that the weight $\omega(e)$ has increased or decreased by one, it adds $+1$ to $c(e, u)$ and adds a token to u's local bin, making it full. This token is a triplet $\langle id(e), \omega(e), c(e)\rangle$ where $id(e)$ is the identity of the edge e, $\omega(e)$ is the new weight of e, and $c(e)$ is the value of the counter $c(e, u)$.
3. Whenever a bin b on level l gets filled with tokens, the following happens.
   (a) The bin b is emptied and its content is broadcast to all nodes in $Q(b)$.
   (b) If the $sup(b) \neq r'$ then $sup(b)$ adds the content of b to itself. (Note that if $sup(b) = r'$ then it does not keep these tokens.)

(c) For each node $z \in Q(b)$ and each token $\phi = \langle id(e), \omega(e), c \rangle$ in b: If ϕ is fresh w.r.t. the adjacency matrix $A(z)$, then z updates e's entry in $A(z)$ to be $(\omega(e), c)$.

Data structure $Data(v)$ **and router protocol** $\mathcal{R}_{SD}(\beta)$**:** Let $A'(v)$ be the graph obtained by $A(v)$ where the weight of an edge in $A'(v)$ is $\omega(e, v)$, the first field in the e's entry of $A(v)$. Using an algorithm similar to the Dijkstra or the Bellman-ford algorithms on the graph $A'(u)$, let $Data(v)$ be the BFS tree of $A'(v)$ rooted at v. We therefore obtain the following lemma.

Lemma 1. $Data(\pi_{SD}(\beta)) = O(n \log n)$.

Given $Data(u)$ and $id(v)$, $\mathcal{R}_{SD}(\beta)$ outputs the simple path P connecting u and v in the tree $Data(u)$. The analysis is deferred to the full paper. We have the following theorem.

Theorem 1. *The scheme* $\pi_{SD} = \langle \mathcal{U}_{SD}, \mathcal{R}_{SD} \rangle$ *is a* β-*approximate source directed routing scheme for the family* $\mathcal{F}(n, D)$ *with the following complexities.*

1. $\mathcal{MC}(\pi_{SD}(\beta)) = O(D \log^2 n)$,
2. $Data(\pi_{SD}(\beta)) = O(n \cdot \log n)$,
3. $\mathcal{MMC}(\pi_{SD}(\beta)) = O(|E| \cdot (\log W + \log n))$.

Lower bounds: We now establish lower bounds for the maximum database size and message complexity of source directed routing schemes. The proofs of the following lemmas are deferred to the full paper. Note that any graph with maximum degree greater than 2 has local density greater than 1.5, therefore the following two lemmas apply for all families $\mathcal{F}(n, D)$ except for $\mathcal{F}(n, 1)$.

Lemma 2. *Any* β-*approximate source directed routing scheme* $\pi_{SD} = (\mathcal{U}_{SD}, \mathcal{R}_{SD})$ *for the family* $\mathcal{F}(n, D)$ *for any* $D \geq 1.5$ *has* $\mathcal{MMC}(\pi_{SD}) = \Omega(n)$.

Lemma 3. *Any* exact *source directed routing scheme* $\pi_{SD} = (\mathcal{U}_{SD}, \mathcal{R}_{SD})$ *for the family* $\mathcal{F}(n, D)$ *for any* $D \geq 1.5$ *incurs an message complexity of* $\mathcal{MC}(\pi_{SD}) = \Omega(n)$ *in some scenario.*

Lemma 4. *For constant* β *and* $D > 1$, $\mathcal{MC}(\pi_{SD}(\beta)) = \Omega(D)$ *on* $\mathcal{F}(n, D)$.

5   The Point-to-Point Routing Scheme $\pi_{PTP}(\beta)$

In this section we introduce our β-approximate point-to-point routing scheme $\pi_{PTP}(\beta)$ for the family $\mathcal{F}(n, D)$.

Update protocol $\mathcal{U}_{PTP}(\beta)$**:** The memory structure $Memory(v)$ of each vertex v is $\langle A(v), A_0, L(v) \rangle$ where $A(v)$ is the memory structure given to v by the update protocol $\mathcal{U}_{SD}$ of the source directed scheme, A_0 is the initial matrix $A(v)$ (which corresponds to the initial graph) and $L(v)$ is a table containing $O(n)$ elements. An element of $L(v)$ is a tuple $\langle \phi, b \rangle$ where b is a bin and ϕ is a token. For each token ϕ and vertex z, we maintain the invariant that at most one bin satisfies $\langle\phi, b\rangle \in L(z)$.

The update protocol $\mathcal{U}_{PTP}$ uses Protocol BIN$'$ instead of Protocol BIN. Protocol BIN$'$ is the same as Protocol BIN except that in step 3 we add one more substep 3.(d), described below.

Substep 3.(d) of BIN$'$: Each node $z \in Q(b)$ updates $L(z)$ as follows. Let ϕ be a token in b. If $(\phi, b') \in L(z)$ for some bin b' and b is affected by b' then z extract (ϕ, b') from $L(z)$ and adds (ϕ, b) instead. If $(\phi, b') \notin L(z)$ for any bin b' then z adds (ϕ, b) to $L(z)$.

Router protocol $\mathcal{R}_{PTP}(\beta)$ **and data structure** $Data(u)$: Let M be a message originated at x and destined to y. The sender x initially calculates $h = d^{\omega}_{A'(x)}(x, y)$, the weighted distance between x and y in $A'(x)$. Let $q = \min\{\beta\alpha h, n\}$ and let $l = \lceil \log q \rceil$ (i.e., $2^{l-1} < q \le 2^l$). The sender then attaches the header $H = (id(y), l)$ to the message M. Since H can be encoded with at most $2\log n$ bits, we obtain that the header size $\mathcal{HD}(\pi_{PTP})$ is $O(\log n)$.

Denote the subgraph of G induced by all vertices whose unweighted distance to y is at most 2^l by $Ball(H) = \Gamma(y, 2^l)$. Consider the vertices of $Ball(H)$ as vertices of $T(G)$ and for every node $u_i \in Ball(H)$, let R_i be the path from u_i to r' and let I_i be the subpath of R_i of length 2^l containing u_i. Let $I(Ball(H)) = \bigcup_i I_i$. A bin b is $I(Ball(H))$-*universal* if $sup(b)$ is outside of $I(Ball(H))$.

In an intermediate node u along the route (including the sender x), algorithm $\mathcal{R}_{PTP}(\beta)$ operates as follows. Upon receiving H and $Memory(u)$ as input, $\mathcal{R}_{PTP}(\beta)$ creates the following edge-weight matrix $C(u, H)$ for $Ball(H)$. Initially $C(u, H)$ is A_0 restricted to $Ball(H)$. Now u updates $C(u, H)$ by inspecting its table $L(u)$. Let $\phi = \langle id(e), \omega(e), c \rangle$ be a token in some element of $(\phi, b) \in L(u)$. If b is an $I(Ball(H))$-universal bin, ϕ corresponds to an event happening in $Ball(H)$ and ϕ is fresh w.r.t $C(u, H)$, then $C(u, H)$ updates e's entry to be $(\omega(e), c)$.

After calculating $C(u, H)$, using an algorithm similar to the Dijkstra or the Bellman-ford algorithms on the graph $C(u, H)$, $\mathcal{R}_{PTP}(\beta)$ efficiently calculates and outputs a port number connecting u to the next vertex on a simple path P (contained in $Ball(H)$) connecting u and y which satisfies $\tilde{P} = d^{\omega}_{C(u,H)}(u, y)$.

The database size can be reduced further. For an intermediate node u along the route from x to y, the port on which the message is to be delivered depends only on $Memory(u)$, l and the destination y. We let $Data(u)$ contain a table D_l for each integer $l \le \log n$. Each such table contains n entries corresponding to the vertices of G. The y's entry in D_l contains $\mathcal{R}_{PTP}(Memory(u), id(y), 2^l)$. Given $Data(u)$ and $H = (id(y), l)$, $\mathcal{R}_{PTP}$ outputs the port in the y's entry in D_l. We therefore obtain the following lemma.

Lemma 5. $Data(\pi_{PTP}(\beta)) = O(n \log^2 n)$.

The analysis is deferred to the full paper. We have the following theorem.

Theorem 2. *The scheme* $\pi_{PTP}(\beta) = \langle \mathcal{U}_{PTP}, \mathcal{R}_{PTP} \rangle$ *is a* β-*approximate point-to-point routing scheme for the family* $\mathcal{F}(n, D)$ *with the following complexities.*

1. $\mathcal{MC}(\pi_{PTP}(\beta)) = O(D \log^2 n)$,
2. $Data(\pi_{PTP}(\beta)) = O(n \cdot \log^2 n)$,
3. $\mathcal{MMC}(\pi_{PTP}(\beta)) = O(|E| \cdot (\log W + \log n))$,
4. $\mathcal{HD}(\pi_{PTP}) = O(\log n)$.

A lower bound on communication: The proof of the following lemma is deferred to the full paper.

Lemma 6. *For constant $1 \le \beta < 3$ and $D > 1$, $\mathcal{MC}(\pi_{PTP}(\beta)) = \Omega(D)$ on $\mathcal{F}(n, D)$.*

References

1. Y. Afek, B. Awerbuch, S.A. Plotkin and M. Saks. Local management of a global resource in a communication network. *J. ACM* **43**, (1996), 1–19.
2. Y. Afek, E. Gafni and M. Ricklin. Upper and lower bounds for routing schemes in dynamic networks. In *Proc. 30th Symp. on Foundations of Computer Science*, 1989, 370–375.
3. B. Awerbuch, A. Bar-Noy, N. Linial and D. Peleg. Improved routing strategies with succinct tables. *J. Algorithms* **11**, (1990), 307–341.
4. B. Awerbuch and D. Peleg. Routing with polynomial communication-space trade-off. *SIAM J. Discrete Math.* **5**, (1992), 307–341.
5. L. Cowen. Compact routing with minimum stretch. *J. Algorithms* **38**, (2001), 170–183.
6. S. Dolev, E. Kranakis, D. Krizanc and D. Peleg. Bubbles: Adaptive routing scheme for high-speed dynamic networks. *SIAM J. on Comput.* **29**, (1999), 804–833.
7. D. Eppstein, Z. Galil and G. F. Italiano. Dynamic Graph Algorithms. In *Algorithms and Theoretical Computing Handbook*, M.J. Atallah, Ed., CRC Press, 1999, Ch. 8.
8. P. Fraigniaud and C. Gavoille. Universal routing schemes. *Distributed Computing* **10**, (1997), 65–78.
9. J. Feigenbaum and S. Kannan. Dynamic Graph Algorithms. In *Handbook of Discrete and Combinatorial Mathematics*, CRC Press, 2000.
10. C. Gavoille. Routing in distributed networks: Overview and open problems. *ACM SIGACT News-Distributed Computing Column* **32**, (2001), 36–52.
11. K. Iwama and A. Kawachi. Compact routing with stretch factor less than three. In *Proc. 19th ACM Symp. on Principles of Distributed Computing*, 2000, 337.
12. K. Iwama and M. Okita. Compact Routing for Flat Networks. In *Proc. 17th International Simposium on Distributed Computing*, Oct. 2003.
13. A. Korman. General Compact Labeling Schemes for Dynamic Trees. In *Proc. 19th International Simposium on Distributed Computing*, Sep. 2005.
14. A. Korman, D. Peleg and Y. Rodeh. Labeling schemes for dynamic tree networks. *Theory of Computing Systems* **37**, (2004), 49–75.
15. A. Korman and D. Peleg. Labeling schemes for weighted dynamic trees. In *Proc. 30th Int. Colloq. on Automata, Languages and Prog.*, July 2003, 369–383.
16. D. Peleg. *Distributed computing: A Locality-Sensitive Approach*. SIAM, 2000.
17. D. Peleg and E. Upfal. A tradeoff between size and efficiency for routing tables. *J. ACM* **36**, (1989), 510–530.
18. N. Santoro and R. Khatib. Labeling and implicit routing in networks. *The Computer Journal* **28**, (1985), 5–8.
19. M. Thorup and U. Zwick. Compact routing schemes. In *Proc. 13th ACM Symp. on Parallel Algorithms and Architectures*, July 2001, 1–10.

Energy Complexity and Entropy of Threshold Circuits

Kei Uchizawa[1], Rodney Douglas[2], and Wolfgang Maass[3]

[1] Graduate School of Information Sciences
Tohoku University
[2] Institute of Neuroinformatics
University and ETH Zurich
[3] Institute for Theoretical Computer Science
Technische Universitaet Graz

Abstract. Circuits composed of threshold gates (McCulloch-Pitts neurons, or perceptrons) are simplified models of neural circuits with the advantage that they are theoretically more tractable than their biological counterparts. However, when such threshold circuits are designed to perform a specific computational task they usually differ in one important respect from computations in the brain: they require very high activity. On average every second threshold gate fires (sets a "1" as output) during a computation. By contrast, the activity of neurons in the brain is much more sparse, with only about 1% of neurons firing. This mismatch between threshold and neuronal circuits is due to the particular complexity measures (circuit size and circuit depth) that have been minimized in previous threshold circuit constructions. In this article we investigate a new complexity measure for threshold circuits, *energy complexity*, whose minimization yields computations with sparse activity. We prove that all computations by threshold circuits of polynomial size with entropy $O(\log n)$ can be restructured so that their energy complexity is reduced to a level near the *entropy of circuit states*. This entropy of circuit states is a novel circuit complexity measure, which is of interest not only in the context of threshold circuits, but for circuit complexity in general. As an example of how this measure can be applied we show that any polynomial size threshold circuit with entropy $O(\log n)$ can be simulated by a polynomial size threshold circuit of depth 3.

1 Introduction

The active outputs of neurons are stereotypical electrical pulses (action potentials, or "spikes"). The stereotypical form of these spikes suggests that the output of neurons is analogous to the "1" of a threshold gate. In fact, historically and even currently, threshold circuits are commonly viewed as abstract computational models for circuits of biological neurons. Nevertheless, it has long been recognized by neuroscientists that neurons are generally silent, and that information processing in the brain is usually achieved with a sparse distribution of

M. Bugliesi et al. (Eds.): ICALP 2006, Part I, LNCS 4051, pp. 631–642, 2006.

neural firing [1]. One reason for this sparse activation may be metabolic cost. For example, a recent biological study on the energy cost of cortical computation [6] concludes that "The cost of a single spike is high, and this limits, possibly to fewer than 1 %, the number of neurons that can be substantially active concurrently". The metabolic cost of the active ('1") state of a neuron is very asymmetric. The production of a spike consumes a substantial amount of energy (about 2.4×10^9 molecules of ATP according to [6]), whereas the energy cost of the no-spike rest state, is substantially less. In contrast to neuronal circuits, computations in feedforward threshold circuits (and many other circuit models for digital computation) have the property that a large portion, usually around 50%, of gates in the circuit output a "1" during any computation. Common abstract measures for the energy consumption of electronic circuits treat the cost of the two output states 0 and 1 of a gate symmetrically, and focus instead on the required number of switchings between these two states (see [5] and its references, as well as [11]). An exception are [14,4,1], which provide Shannon-type results for the number of gates that output a "1" in Boolean circuits consisting of gates with bounded fan-in. Circuits of threshold gates (= linear threshold gates = McCulloch-Pitts neurons) are an important class of circuits that are frequently used as simplified models for computations in neural circuits [8,12,10,13]. In this paper we consider how investigations of such abstract threshold circuits can be reconciled with actual activity characteristics of biological neural networks.

In section 2 we give a precise definition of threshold circuits, and also define their *energy complexity*, whose minimization yields threshold circuits that carry out computations with sparse activity: on average few gates output a "1" during a computation. In section 2 we also introduce another novel complexity measure, the *entropy of a computation*. This measure is interesting for many types of circuits, beyond the threshold circuits discussed in this paper. It measures the total number of different patterns of gate states that arise during computations on different circuits inputs. We show in section 3 that the entropy of circuit states defines a coarse lower bound for its energy complexity. This result is relevant for any attempt to simulate a given threshold circuit by another threshold circuit with lower energy complexity, since the entropy of a circuit is directly linked to the algorithm that it implements. Therefore, it is unlikely that there exists a general method permitting any given circuit to be simulated by one with smaller entropy. In this sense the entropy of a circuit defines a hard lower bound for any general method that aims to simulate any given threshold circuit using a circuit with lower energy complexity. However, we will prove in section 3 that there exists a general method that reduces – if this entropy is $O(\log n)$ – the energy complexity of a circuit to a level near the entropy of the circuit. Since the entropy of a circuit is a complexity measure that is interesting in its own right, we also offer in section 4 a first result on the computational power of threshold circuits with low entropy. Some open problems related to the new concepts introduced in this article are listed in section 5.

---

[1] According to recent data [7] from whole cell recordings in awake animals the spontaneous firing rates are on average below 1 Hz.

2   Definitions

A threshold gate g (with weights $w_1, \ldots, w_n \in \mathbb{R}$ and threshold $t \in \mathbb{R}$) outputs 1 for any input $X = (x_1, \ldots, x_n) \in \mathbb{R}^n$ if $\sum_{i=1}^{n} w_i x_i \geq t$, otherwise 0. We write $g(X) = \text{sign}(\sum_{i=1}^{n} w_i x_i - t)$ where $sign(z) = 1$ if $z \geq 0$ and $sign(z) = 0$ if $z < 0$. As usual we assume that threshold gates operate in discrete time, with unit delays between gates.

For a threshold gate g_i within a feedforward circuit C that receives $X = (x_1, \ldots, x_n)$ as *circuit input*, we write $g_i(X)$ for the output that the gate g_i gives for this circuit input X (although the actual input to gate g_i during this computation will in general consist of just some variables x_i from X, and in addition, or even exclusively, of outputs of other gates in the circuit C).

We define the *energy complexity* of a circuit C consisting of threshold gates $g_1, \ldots, g_m$ to be the expected number of 1's that occur in a computation, for some given distribution Q of circuit inputs X, i. e.

$$ EC_Q(C) := E[\sum_{i=1}^{m} g_i(X)], $$

where the expectation is evaluated with regard to the distribution Q over $X \in \mathbb{R}^n$ (or $X \in \{0,1\}^n$). Thus, for the case where Q is the uniform distribution over $\{0,1\}^n$, we have $EC_{uniform} := \frac{1}{2^n} \sum_{X \in \{0,1\}^n} \sum_{i=1}^{m} g_i(X)$.

In some cases it is also interesting to consider the *maximal* energy consumption of a circuit for any input X, defined by

$$ EC_{\max}(C) := \max(\sum_{i=1}^{m} g_i(X) : X \in \mathbb{R}^n). $$

We define the *entropy* of a (feedforward) circuit C to be

$$ H_Q(C) := - \sum_{A \in \{0,1\}^m} P_C(A) \cdot \log P_C(A), $$

where $P_C(A)$ is the probability that the internal gates $g_1, \ldots, g_m$ of the circuit C assume the state $A \in \{0,1\}^m$ during a computation of circuit C (for some given distribution Q of circuit inputs $X \in \mathbb{R}^n$). We often write $H_{\max}(C)$ for the largest possible value that $H_Q(C)$ can assume for any distribution on a given set of circuit states A. If $MAX(C)$ is defined as the total number of different circuit states that circuit C assumes for different inputs $X \in \mathbb{R}^n$, then one has $H_Q(C) = H_{\max}(C)$ if Q is such that these $MAX(C)$ circuit states all occur with the same probability, and $H_{\max}(C)$ is then equal to $\log_2 MAX(C)$. Thus $2^{H_{\max}(C)}$ is the maximal number of circuit states that a circuit C assumes for arbitrary inputs X.

We write $size(C)$ for the number m of gates in a circuit C, and $depth(C)$ for the length of the longest path in C from an input to its output node (which is always assumed to be the node g_m).

3   Construction of Threshold Circuits with Sparse Activity

Obviously, the number of 1's in a computation limits the number of states that the circuit can assume:

$$H_Q(C) \leq \log(\# \text{ of circuit states } A \text{ that } C \text{ assumes})$$

$$\leq \log \sum_{j=0}^{EC_{\max}(C)} \binom{\text{size}(C)}{j}$$

$$\leq \log(\text{size}(C)^{EC_{\max}(C)}) = EC_{\max}(C) \cdot \log \text{size}(C)$$

(for sufficiently large values of $EC_{\max}(C)$ and $size(C)$; log always stands for $\log_2$ in this paper). Hence

$$EC_{\max}(C) \geq H_Q(C)/\log \text{size}(C) \ . \tag{1}$$

In fact, this argument shows that

$$EC_{\max}(C) \geq H_{\max}(C)/\log \text{size}(C) \ . \tag{2}$$

Every Boolean function $f : \{0,1\}^n \to \{0,1\}$ can be computed by a threshold circuit C of depth 2 that represents its disjunctive normal form, in such a way that for every circuit input X at most a single gate on level 1 outputs a 1. This circuit C has the property that $EC_{\max}(C) = 2$ and $H_Q(C) = \log(\text{size}(C) - 1)$ for a suitable distribution Q of circuit inputs. Hence it is in some cases possible to achieve $EC_{\max}(C) < H_Q(C)$, and the factor $\log \text{size}(C)$ in (1) and (2) cannot be eliminated or significantly reduced.

Threshold circuits that represent a Boolean function f in its disjunctive normal form allow us to compute any Boolean function with a circuit C that achieves $EC_{\max}(C) = 2$. However these circuits C have in general exponential size in n. Therefore, the key question is whether one can also construct polynomial size circuits C with small EC_Q or $EC_{\max}$. Because of the a-priori bounds (1) and (2), this is only possible for those functions f that can be computed with a low entropy of circuit states. The following results show that, on the other hand, the existence of a circuit C that computes f with $H_{\max}(C) = O(\log n)$ is sufficient to guarantee the existence of a circuit that computes f with low energy complexity.

Theorem 1. *Assume that a Boolean function $f : \{0,1\}^n \to \{0,1\}$ can be computed by some polynomial size threshold circuit C with $H_{\max}(C) = O(\log n)$. Then f can also be computed by some polynomial size threshold circuit C' with*

$$EC_{\max}(C') \leq H_{\max}(C) + 1 = O(\log n). \tag{3}$$

Furthermore, if Q is any distribution of inputs $X \in \{0,1\}^n$, then it is possible to construct a polynomial size threshold circuit C'' with

$$EC_Q(C'') \leq \frac{H_Q(C)}{2} + 1 = O(\log n). \tag{4}$$

Remark 1. The proof below shows that the following more general statements hold for any function f and any distribution Q:

If f can be computed by some arbitrary (feedforward) threshold circuit C, then f can also be computed by a threshold circuit C' with $size(C') \leq 2^{H_{\max}(C)}$, $depth(C') \leq size(C)+1$, $H_{\max}(C') \leq H_{\max}(C)$, and $EC_{\max}(C') \leq H_{\max}(C)+1$.

Furthermore, f can also be computed by a threshold circuit C'' with $size(C'') \leq 2^{H_{\max}(C)}$, $depth(C'') \leq size(C) + 1$, $H_Q(C'') \leq H_Q(C)$, and $EC_Q(C'') \leq \frac{H_Q(C)}{2} + 1$.

Remark 2. The assumption $H_{\max}(C) = O(\log n)$ is satisfied by standard constructions of threshold circuits for many commonly considered functions f. Examples are all symmetric functions (hence in particular PARITY of n bits), COMPARISON of binary numbers, and BINARY ADDRESSING (routing) where the first k input bits represent an address for one of the 2^k subsequent input bits (thus $n = k + 2^k$). In fact, to the best of our knowledge there is no function known which can be computed by polynomial size threshold circuits, but not by polynomial size threshold circuits C with $H_{\max}(C) = O(\log n)$.

Proof of Theorem 1. The proof is split up into a number of Lemmata (Lemma 1 – 6). The idea is first to simulate in Lemma 1 the given circuit C by a threshold decision tree (i.e., by a decision tree T with threshold gates at its nodes, see Definition 1) that has at most $2^{H_{\max}(C)}$ leaves. Then this threshold decision tree is restructured in Lemma 3 in such a manner that every path in the tree from the root to a leaf takes at most $\log(\# \text{ of leaves})$ times, hence in this case at most $H_{\max}(C)$ times, the right branch at an internal node. Obviously such an asymmetric cost measure is of interest when one wants to minimize an asymmetric complexity measure such as EC, which assigns different costs to gate outputs 0 and 1. Finally, we show in Lemma 5 that the computations of the resulting threshold decision tree can be simulated by a threshold circuit where some gate outputs a "1" whenever the simulated path in the decision tree moves into the right subtree at an internal node of the tree. The proof of this Lemma 5 has to take into account that the control structures of decision trees and circuits are quite different: A gate in a decision tree is activated only when the computation path happens to arrive at the corresponding node of the decision tree, but a gate in a threshold circuit is activated in *any* computation of that circuit. Hence a threshold decision tree with few threshold gates that output "1" does not automatically yield a threshold circuit with low energy complexity. However, we show that all gates in the simulating threshold circuit that do not correspond to a node in the decision tree where the right branch is chosen, receive an additional input with a strongly negative weight (see Lemma 4), so that they output a "0" when they get activated.

Finally, we show in Lemma 6 that the threshold decision tree can be restructure alternatively, so that the *average* number of times when a computation path takes the right subtree at a node remains small (instead of the maximal number of taking the right subtree). This manouvre yields the proof of the second part of the claim of Theorem 1.

Definition 1. *A threshold decision tree(called a linear decision tree in [2]) T is a binary tree in which each internal node has two children, a left and a right one, and is labeled by a threshold gate that is applied to the input $X \in \{0,1\}^n$ for the tree. All the leaves of threshold decision trees are labeled by 0 or 1. To compute the output of a threshold decision tree T on an input X we apply the following procedure from the root until reaching a leaf: we go left if the gate at a node outputs 0, otherwise we go right. If we reach a leaf labeled by $l \in \{0,1\}$, then l is the output of T for input X.*

Note that the threshold gates in a threshold decision tree are only applied to input variables from the external input $X \in \{0,1\}^n$, not to outputs of preceding threshold gates. Hence it is obvious that computations in threshold decision trees have a quite different structure from computations in threshold circuits, although both models use the same type of computational operation at each node.

The depth of a threshold decision tree is the maximum number of nodes from the root to a leaf. We assign binary strings to nodes of T in the usual manner:

- $\hat{g}_\varepsilon$ denotes the root of the tree (where ε is the empty string)
- For a binary string s, let $\hat{g}_{s \circ 0}$ and $\hat{g}_{s \circ 1}$ be the left and right child of the node with label $\hat{g}_s$, where $\circ$ denotes concatenation of strings.

For example, the ancestors of a node $\hat{g}_{1011}$ are $\hat{g}_\varepsilon$, $\hat{g}_1$, $\hat{g}_{10}$ and $\hat{g}_{101}$. Let S_T be the set of all binary strings s that occur as indices of nodes $\hat{g}_s$ in a threshold decision tree T. Then all the descendants of node $\hat{g}_s$ in T can be represented as $\hat{g}_{s \circ *}$ for $s \circ * \in S_T$.

The given threshold circuit C can be simulated in the following way by a threshold decision tree:

Lemma 1. *Let C be a threshold circuit computing a function $f : \{0,1\}^n \to \{0,1\}$ with m gates. Then one can construct a threshold decision tree T with at most $2^{H_{\max}(C)}$ leaves and $depth(T) \le m$ which computes the same function f.*

Proof. Assume that C consists of m gates. We number the gates $g_1, \ldots, g_m$ of C in topological order. Since g_i receives the circuit input X and the outputs of g_j only for $j < i$ as its inputs, we can express the output $g_i(X)$ of g_i for circuit input $X = <x_1, \ldots, x_n>$ as $g_i(X) = sign(\sum_{j=1}^n w_j^i x_j + \sum_{j=1}^{i-1} w_{g_j}^i g_j(X) + t_i)$, where $w_{g_j}^i$ is the weight which g_i applies to the output of g_j in circuit C.

Let S be the set of all binary strings of length up to $m - 1$. We define threshold gates $\hat{g}_s : X \to \{0,1\}$ for $s \in S$ by $\hat{g}_s(X) = sign(\sum_{j=1}^n w_j^{|s|+1} x_j + t_s)$ with $t_s = \sum_{j=1}^{|s|} w_{g_j}^{|s|+1} s_j + t_{|s|+1}$, where s_j is the j-th bit of string s and $|s|$ is the length of s. Obviously these gates $\hat{g}_s$ are variations of gate g_i with different built-in assumptions s about the outputs of preceding gates.

Let T be the threshold decision tree consisting of gates $\hat{g}_s$ for $s \in S$. That is, gate $\hat{g}_\varepsilon = g_1$ is placed at the root of T. We let the left child of $\hat{g}_s$ be $\hat{g}_{s \circ 0}$ and the right child of $\hat{g}_s$ be $\hat{g}_{s \circ 1}$. We let each $\hat{g}_s$ with $|s| = m - 1$ have a leaf labeled by 0 as left child and a leaf labeled 1 as right child. Since $\hat{g}_s$ computes the same function as $g_{|s|+1}$ if the preceding gates g_i output s_i for $1 \leq i \leq |s|$, T computes the same function f as C. We then remove all leaves from T for which the associated paths correspond to circuit states $A \in \{0,1\}^m$ that do not occur in C for any circuit input $X \in \{0,1\}^n$. This reduces the number of leaves in T to $2^{H_{\max}(C)}$. Finally, we iteratively remove all nodes without children, and replace all nodes below which there exists just a single leaf by a leaf. In this way we arrive again at a binary tree. □

We now introduce a cost measure $cost(T)$ for trees T, that like the energy complexity for circuits, measures for threshold decision trees how often a threshold gate outputs a 1 during a computation:

Definition 2. *We denote by $cost(T)$ the maximum number of times where a path from the root to a leaf in a binary tree T goes to the right. If T is a leaf, then $cost(T) = 0$.*

We will show later, in Lemma 5, that one can simulate any threshold decision tree T' by a threshold circuit $C_{T'}$ with $EC_{\max}(C_{T'}) \leq cost(T') + 1$. Hence it suffices for the proof of Theorem 1 to simulate the threshold decision tree T resulting from Lemma 1 by another threshold decision tree T' for which $cost(T')$ is small. This is done in Lemma 4, where we will construct a tree T' that reduces $cost(T')$ down to another cost measure $rank(T)$. This measure $rank(T)$ always has a value $\leq \log(\# \text{ of leaves of } T)$ according to Lemma 2, hence $rank(T) \leq H_{\max}(C)$ for the tree T constructed in Lemma 1.

Definition 3. *The rank of a binary tree T is defined inductively as follows:*

- *If T is a leaf then $rank(T) = 0$.*
- *If T has subtrees T_l and T_r then*

$$rank(T) = \begin{cases} rank(T_l), & if\ rank(T_l) > rank(T_r) \\ rank(T_r) + 1, & if\ rank(T_l) = rank(T_r) \\ rank(T_r), & if\ rank(T_l) < rank(T_r) . \end{cases}$$

Lemma 2. *Let T be any binary tree. Then $rank(T) \leq \log(\# \text{ of leaves of } T)$.* □

Lemma 3. *Let T be a threshold decision tree computing a function $f : \{0,1\}^n \to \{0,1\}$. Then f can also be computed by a threshold decision tree T' which has the same depth and the same number of leaves as T, and which satisfies $cost(T') = rank(t)$.*

Proof. Let T consist of gates g_s for $s \in S_T$. We define T^s as the subtree of T whose root is g_s. Let T_l^s(respectively, T_r^s) denote the left(right) subtree below the root of T^s. We modify T inductively by the following procedure, starting at the nodes g_s of largest depth. If $cost(T_l^s) < cost(T_r^s)$, we replace g_s by its complement, and swap the left subtree and the right subtree. The complement of g_s is here another threshold gate g that outputs 1 if and only if g_s outputs 0. Such

gate g exists since $\sum_{i=1}^{n} w_i x_i < t \Leftrightarrow \sum_{i=1}^{n} (-w_i) x_i > -t \Leftrightarrow \sum_{i=1}^{n} (-w_i) x_i \geq t'$
for another threshold t' (which always exists if the x_i assume only finitely many
values). Let $\hat{T}^s$ be the threshold decision tree which is produced from T^s by this
procedure. By construction it has the following properties:

- If the children of g_s both are both leaves, then we have $cost(\hat{T}^s) = 1$.
- Otherwise,

$$cost(\hat{T}^s) = \begin{cases} cost(\hat{T}_l^s), & \text{if } cost(\hat{T}_l^s) > cost(\hat{T}_r^s) \\ cost(\hat{T}_r^s) + 1, & \text{if } cost(\hat{T}_l^s) = cost(\hat{T}_r^s) \\ cost(\hat{T}_r^s), & \text{if } cost(\hat{T}_l^s) < cost(\hat{T}_r^s) , \end{cases}$$

where $\hat{T}^s$ has subtrees $\hat{T}_l^s$ and $\hat{T}_r^s$.

Since this definition coincides with the definition of the rank, we have constructed
a tree T' with $cost(T') = rank(T)$. This procedure preserves the function that
is computed, the depth of the tree, and the number of leaves.   □

We now show that the threshold decision tree that was constructed in Lemma 3
can be simulated by a threshold circuit with low energy complexity. As a preparation we first observe in Lemma 4 that one can "veto" any threshold gate g
through some extra input. This will be used in Lemma 5 in order to avoid the
event that gates in the simulating circuit that correspond to gates in an inactive
path of the simulated threshold decision tree increase the energy complexity of
the resulting circuit.

Lemma 4. *Let $g(x_1, \ldots, x_n) = sign(\sum_{i=1}^{n} w_i x_i - t)$ be a threshold gate. Then
one can construct a threshold gate g' using an additional input x_{n+1} which has
the following property:*

$$g'(x_1, \ldots, x_n, x_{n+1}) = \begin{cases} 0, & \text{if } x_{n+1} = 1 \\ g(x_1, \ldots, x_n), & \text{if } x_{n+1} = 0 . \end{cases}$$

Proof.   We set $w_{n+1} := -(\sum_{i=1}^{n} |w_i| + |t| + 1)$. Apart from that g' uses the
same weights and threshold as g. It is obvious that the resulting gate g' has the
desired property.   □

Lemma 5. *Let T be a threshold decision tree which consists of k internal nodes
and which computes a function f. Then one can construct a threshold circuit C_T
with $EC_{\max}(C_T) \leq cost(T) + 1$ that computes the same function f. In addition
C_T satisfies $depth(C_T) \leq depth(T) + 1$ and $size(C_T) \leq k + 1$.*

Proof.   We can assume without loss of generality that every leaf with label 1 in T
is the right child of its parent (if this is not the case, swap this leaf with the right
subtree of the parent, and replace the threshold gate at the parent node like in
the proof of Lemma 3 by another threshold gate that always outputs the negation
of the former gate; this procedure does not increase the cost of the tree, nor its
depth or number of internal nodes). Let now $g_s(X) = sign(\sum_{j=1}^{n} w_j^s x_j - t_s)$ be

the threshold gate in T at the node with label $s \in S_T$. Let w_{n+1}^s be the weight constructed in Lemma 4 for an additional input which can force gate g_s to output 0. Set $W := \max\{|w_{n+1}^s| : s \in S_T\}$.

The threshold circuit C_T that simulates T has a gate g_s' for every gate g_s in T, and in addition an OR-gate which receives inputs from all gates g_s' so that g_s has a leaf with label 1 (according to our preceding remark this leaf is reached whenever the gate g_s at node $s \in S_T$ gets activated and g_s outputs a 1). We make sure that any gate g_s' in C_T outputs 1 for a circuit input X if and only if the gate g_s in T gets activated for this input X, and outputs 1. This implies that only gates g_s' in C_T can output 1 that correspond to gates g_s in T with output 1 that lie on the single path of T that gets activated for the present circuit input X. Hence this construction automatically ensures that $EC_{\max}(CT) \leq cost(T) + 1$ (where the "+1"arises from the additional OR-gate in C_T).

In order to achieve this objective, g_s' gets additional inputs from all gates $g_{\tilde{s}}'$ in C_T so that $\tilde{s}$ is a proper prefix of s. The weight for the additional input from $g_{\tilde{s}}'$ is $-W$ if $\tilde{s} \circ 0$ is a prefix of s, and W otherwise. In addition the threshold of g_s' is increased by $l_s \cdot W$, where l_s is the number of $1's$ in the binary string s. In this way g_s' can output 1 if and only if g_s *outputs 1 for the present circuit input* X, *and all gates* $g_{\tilde{s}}$ *of* T *for which* g_s *lies in the right subtree below* $g_{\tilde{s}}$ *output 1, and all gates* $\hat{g}_{\tilde{s}}$ *of* T *for which* g_s *lies in the left subtree below* $g_{\tilde{s}}$ *output 0*. Thus g_s' outputs 1 if and only if the path leading to gate g_s gets activated in T and g_s outputs 1. $\square$

The *proof of the first claim of Theorem 1* follows now immediately from the Lemmata 1–5. Note that the number k of internal nodes in a binary tree is equal to (# of leaves)-1, hence $k \leq 2^{H_{\max}(C)} - 1$ in the case of the decision tree T resulting from applications of Lemma 1 and Lemma 3. This yields $size(C_T) \leq 2^{H_{\max}(C)}$ for the circuit C_T that is constructed in Lemma 5 for this tree T.

The *proof of the second claim of Theorem 1* follows by applying the subsequent Lemma 6 instead of Lemma 3 to the threshold decision tree T resulting from Lemma 1.

Lemma 6. *Let T be a threshold decision tree computing $f: \{0,1\}^n \to \{0,1\}$. Then for any given distribution Q of circuit inputs, there exists a threshold decision tree T' computing f such that the expected number of 1's with regard to Q is at most $H_Q(C)/2$.*

Proof.   Let $P(s)$ be the probability (with regard to Q) that gate g_s outputs 1. We construct T' by modifying T inductively (starting at the nodes of the largest depth m in T) through the following procedure: If $P(s) > 1/2$, replace g_s by a threshold gate which computes its negation and swap the left and right subtree below this node.

Let $cost_Q(s)$ be the expected number of times where one goes to the right in the subtree of T' whose root is the node labeled by s. By construction we have $P(s) \leq 1/2$ for every gate g_s in T'. Furthermore we have:

- If $|s| = m - 1$ then $cost_Q(s) = P(s)$.
- If $0 \leq |s| < m - 1$, then $P(s) \leq 1/2$ and
  $cost_Q(s) = P(s) + P(s)cost_Q(s \circ 1) + (1 - P(s))cost_Q(s \circ 0)$.

One can prove by induction on $|s|$ that $cost_Q(s) \leq H_Q(s)/2$ for all $s \in S_{T'}$, where $H_Q(s)$ is the entropy of states of the ensemble of gates of T' in the subtree below gate g_s.

For the induction step one uses the convexity of the log-function, which implies that $P(s) = -P(s) \cdot (-1) = -P(s) \cdot \log \frac{P(s)+(1-P(s))}{2} \leq -P(s) \left(\frac{\log(P(s))+\log(1-P(s))}{2} \right)$, and the fact that $P(s) \leq 1 - P(s)$ to show that

$$cost_Q(s) \leq P(s) + P(s) \cdot \frac{H_Q(s \circ 1)}{2} + (1 - P(s)) \cdot \frac{H_Q(s \circ 0)}{2}$$

$$\leq -P(s) \cdot \left(\frac{\log P(s) + \log(1 - P(s))}{2} \right) +$$

$$P(s)\frac{H_Q(s \circ 1)}{2} + (1 - P(s)) \cdot \frac{H_Q(s \circ 0)}{2}$$

$$\leq -\frac{P(s)}{2} \log P(s) - \frac{(1 - P(s))}{2} \log(1 - P(s)) + P(s)\frac{H_Q(s \circ 1)}{2}$$

$$+(1 - P(s))\frac{H_Q(s \circ 0)}{2} \leq \frac{H_Q(s)}{2} .$$

□

Remark 3. The results of this section can also be applied to circuits that compute arbitrary functions $f : D \to \{0,1\}$ for some arbitrary finite set $D \subseteq \mathbb{R}^n$ (instead of $\{0,1\}^n$). For domains $D \subseteq \mathbb{R}^n$ of *infinite* size a different proof would be needed, since then one can no longer replace any given threshold gate by another threshold gate that computes its negation (as used in the proofs of Lemma 3, Lemma 5, and Lemma 6).

4   On the Computational Power of Circuits with Low Entropy

The concepts discussed in this article raise the question which functions $f : \{0,1\}^n \to \{0,1\}$ can be computed by polynomial size threshold circuits C with $H_{\max}(C) = O(\log n)$. There is currently no function f in P (or even in NP) known for which this is provably false. But the following result shows that if all functions that can be computed by polynomial size threshold circuits of bounded depth can be computed by a circuit C of the same type which satisfies in addition $H_{\max}(C) = O(\log n)$, then this implies a collapse of the depth hierarchy for polynomial size threshold circuits.

Theorem 2. *Assume that a function $f : \{0,1\}^n \to \{0,1\}$ (or $f : \mathbb{R}^n \to \{0,1\}$) can be computed by a threshold circuit C with polynomially in n many gates and*

$H_{\max}(C) = O(\log n)$. *Then one can compute f with a polynomial size threshold circuit C' of depth 3.*

Proof. According to Lemma 1 there exists a threshold decision tree T with polynomially in n many leaves and $depth(T) \leq size(C)$. Design (similarly as in [2]) for each path p from the root to a leaf with output 1 in T a threshold gate g_p on layer 2 of C' that outputs 1 if and only if this path p becomes active in T. The output gate on layer 3 of C' is simply an OR of all these gates g_p.   □

5   Discussion

In this article we introduced an energy complexity measure for threshold circuits that reflects the biological fact that the firing of a neuron consumes more energy than its non-firing. We also have provided methods for restructuring a given threshold circuit with high energy consumption by a threshold circuit that computes the same function, but with brain-like sparse activity. Theorem 1 implies that the computational power of such circuits is quite large. The resulting circuits with sparse activity may help us to elucidate the way in which circuits of neurons are designed in biological systems. In fact, the structure of computations in the threshold circuits with sparse activity that were constructed in the proof of Theorem 1 is reminiscent of biological results on the structure of computations in cortical circuits of neurons, where there is concern for the selection of different pathways ("dynamic routing") in dependence of the stimulus [9]. In addition our constructions provide first steps towards the design of algorithms for future extremely dense VLSI implementations of neurally inspired circuits, where energy consumption and heat dissipation become critical factors.

The new concepts and results of this article suggest a number of interesting open problems in computational complexity theory. At the beginning of section 3 we showed that the energy complexity of a threshold circuit that computes some functions f cannot be less than the a-priori bound given by the minimal circuit entropy required for computing such a function. This result suggests that the entropy of circuit states required for various practically relevant functions should be investigated. Another interesting open problem is the tradeoff between energy complexity and computation speed in threshold circuits, both in general and for concrete computational problems. Finally, we consider that both the energy complexity and the entropy of threshold circuits are concepts that are of interest in their own right. They give rise to interesting complexity classes that have not been considered previously in computational complexity theory. In particular, it may be possible to develop new lower bound methods for circuits with low entropy, thereby enlarging the reservoir of lower bound techniques in circuit complexity theory.

Acknowledgments

We would like to thank Michael Pfeiffer, Pavel Pudlak and Robert Legenstein for helpful discussions, Kazuyuki Amano and Eiji Takimoto for their advice, and Akira Maruoka for making this collaboration possible. This work was partially supported by the Austrian Science Fund FWF, project # P15386, project # S9102-N04, and projects # FP6-015879 (FACETS) and FP6-2005-015803 (DAISY) of the European Union.

References

1. O. V. Cheremisin. (2003). On the activity of cell circuits realising the system of all conjunctions. *Discrete Mathematics and Applications*, 13(2):209–219.
2. H. D. Gröger and G. Turán. (1991). On linear decision trees computing Boolean functions. *Lecture Notes in Computer Science*, 510:707–718.
3. A. Hajnal, W. Maass, P. Pudlak, M. Szegedy, and G. Turan.(1993) Threshold circuits of bounded depth. *Journal of Computer and System Sciences*, 46:129–154.
4. 0. M. Kasim-Zade. (1992). On a measure of the activeness of circuits made of functional elements (Russian). *Mathematical problems in cybernetics*, 4:218–228, "Nauka", Moscow, see Math. Reviews MR1217502 (94c:94019).
5. G. Kissin. (1991). Upper and lower bounds on switching energy in VLSI. *J. of Assoc. for Comp. Mach.*, 38:222–254.
6. P. Lennie. (2003). The cost of cortical computation. *Current Biology*, 13:493–497.
7. T. W. Margrie, M. Brecht, and B. Sakmann. (2002). In vivo, low-resistance, whole-cell recordings from neurons in the anaesthetized and awake mammalian brain. *Pflugers Arch.*, 444(4):491–498.
8. M. Minsky. and S. Papert. (1988). *Perceptrons: An Introduction to Computational Geometry*. MIT Press, Cambridge, MA.
9. B. A. Olshausen, C. H. Anderson, and D. C. V. Essen. (1995). A multiscale dynamic routing circuit for forming size- and position-invariant object representations. *J. Comput. Neurosci.*, 2(1):45–62.
10. I. Parberry. (1994). *Circuit Complexity and Neural Networks*. MIT Press.
11. J. H. Reif and A. Tyagi. (1990) Energy complexity of optical computations. In *Proceedings of the Second IEEE Symposium on Parallel and Distributed Processing (December 1990)*, 14–21.
12. V. P. Roychowdhury, K. Y. Siu, and A. Orlitsky. (1994). *Theoretical Advances in Neural Computation and Learning*. Kluwer Academic, Boston.
13. K. Y. Siu, V. Roychowdhury, and T. Kailath. (1995). *Discrete Neural Computation; A Theoretical Foundation*. Information and System Sciences Series. Prentice-Hall.
14. M. N. Weinzweig. (1961). On the power of networks of functional elements, (Dokl.Akad.Nauk SSSR 139(1961),320-323 (Russian). *in English: Sov.Phys.Dokl*, 6:545–547, see Math. Reviews MR0134413 (24 #B466).

New Algorithms for Regular Expression Matching

Philip Bille

The IT University of Copenhagen, Rued Langgaards Vej 7,
2300 Copenhagen S, Denmark
beetle@itu.dk

Abstract. In this paper we revisit the classical regular expression matching problem, namely, given a regular expression R and a string Q consisting of m and n symbols, respectively, decide if Q matches one of the strings specified by R. We present new algorithms designed for a standard unit-cost RAM with word length $w \geq \log n$. We improve the best known time bounds for algorithms that use $O(m)$ space, and whenever $w \geq \log^2 n$, we obtain the fastest known algorithms, regardless of how much space is used.

1 Introduction

Regular expressions are a powerful and simple way to describe a set of strings. For this reason, they are often chosen as the input language for text processing applications. For instance, in the lexical analysis phase of compilers, regular expressions are often used to specify and distinguish tokens to be passed to the syntax analysis phase. Utilities such as Grep, the programming language Perl, and most modern text editors provide mechanisms for handling regular expressions. These applications all need to solve the classical REGULAR EXPRESSION MATCHING problem, namely, given a regular expression R and a string Q, decide if Q matches one of the strings specified by R.

The standard textbook solution, proposed by Thompson [8] in 1968, constructs a *non-deterministic finite automaton* (NFA) accepting all strings matching R. Subsequently, a state-set simulation checks if the NFA accepts Q. This leads to a simple $O(nm)$ time and $O(m)$ space algorithm, where m and n are the number of symbols in R and Q, respectively. The full details are reviewed later in Sec. 2 and can found in most textbooks on compilers (e.g. Aho et. al. [1]). Despite the importance of the problem, it took 24 years before the $O(nm)$ time bound was improved by Myers [6] in 1992, who achieved $O(\frac{nm}{\log n} + (n+m)\log n)$ time and $O(\frac{nm}{\log n})$ space. For most values of m and n this improves the $O(nm)$ algorithm by a $O(\log n)$ factor. Currently, this is the fastest known algorithm. Recently, Bille and Farach-Colton [3] showed how to reduce the space of Myers' solution to $O(n)$. Alternatively, they showed how to achieve a speedup of $O(\log m)$ while using $O(m)$ space, as in Thompson's algorithm. These results are all valid on a unit-cost RAM with w-bit words and a standard instruction set including addition, bitwise boolean operations, shifts, and multiplication. Each

M. Bugliesi et al. (Eds.): ICALP 2006, Part I, LNCS 4051, pp. 643–654, 2006.

word is capable of holding a character of Q and hence $w \geq \log n$. The space complexities refer to the number of words used by the algorithm, not counting the input which is assumed to be read-only. All results presented here assume the same model. In this paper we present new algorithms achieving the following complexities:

Theorem 1. *Given a regular expression R and a string Q of lengths m and n, respectively,* REGULAR EXPRESSION MATCHING *can be solved using $O(m)$ space with the following running times:*

$$\begin{cases} O(n\frac{m \log w}{w} + m \log w) & \text{if } m > w \\ O(n \log m + m \log m) & \text{if } \sqrt{w} < m \leq w \\ O(\min(n + m^2, n \log m + m \log m) & \text{if } m \leq \sqrt{w}. \end{cases}$$

This represents the best known time bound among algorithms using $O(m)$ space. To compare these with previous results, consider a conservative word length of $w = \log n$. When the regular expression is "large", e.g., $m > \log n$, we achieve an $O(\frac{\log n}{\log \log n})$ speedup over Thompson's algorithm using $O(m)$ space. Hence, we simultaneously match the best known time and space bounds for the problem, with the exception of an $O(\log \log n)$ factor in time. More interestingly, consider the case when the regular expression is "small", e.g., $m = O(\log n)$. This is usually the case in most applications. To beat the $O(n \log n)$ time of Thompson's algorithm, the fast algorithms [6,3] essentially convert the NFA mentioned above into a *deterministic finite automaton* (DFA) and then simulate this instead. Constructing and storing the DFA incurs an additional exponential time and space cost in m, i.e., $O(2^m) = O(n)$. However, the DFA can now be simulated in $O(n)$ time, leading to an $O(n)$ time and space algorithm. Surprisingly, our result shows that this exponential blow-up in m can be avoided with very little loss of efficiency. More precisely, we get an algorithm using $O(n \log \log n)$ time and $O(\log n)$ space. Hence, the space is improved exponentially at the cost of an $O(\log \log n)$ factor in time. In the case of an even smaller regular expression, e.g., $m = O(\sqrt{\log n})$, the slowdown can be eliminated and we achieve optimal $O(n)$ time. For larger word lengths our time bounds improve. In particular, when $w > \log n \log \log n$ the bound is better in all cases, except for $\sqrt{w} \leq m \leq w$, and when $w > \log^2 n$ it improves all known time bounds regardless of how much space is used.

The key to obtain our results is to avoid explicitly converting small NFAs into DFAs. Instead we show how to effectively simulate them directly using the parallelism available at the word-level of the machine model. The kind of idea is not new and has been applied to many other string matching problems, most famously, the Shift-Or algorithm [2], and the approximate string matching algorithm by Myers [7]. However, none of these algorithms can be easily extended to REGULAR EXPRESSION MATCHING. The main problem is the complicated dependencies between states in an NFA. Intuitively, a state may have long paths of ϵ-transitions to a large number of other states, all of which have to be traversed in parallel in the state-set simulation. To overcome this problem we develop several new techniques ultimately leading to Theorem 1. For instance, we introduce a new hierarchical decomposition of NFAs suitable for a parallel state-set simulation. We also show how state-set simulations of large NFAs efficiently reduces to simulating small NFAs.

The results presented in this paper are primarily of theoretical interest. However, we believe that most of the ideas are useful in practice. The previous algorithms require large tables for storing DFAs, and perform a long series of lookups in these tables. As the tables become large we can expect a high number of cache-misses during the lookups, thus limiting the speedup in practice. Since we avoid these tables, our algorithms do not suffer from this defect.

The paper is organized as follows. In Sec. 2 we review Thompson's NFA construction, and in Sec. 3 we present the above mentioned reduction. In Sec. 4 we present our first simple algorithm for the problem which is then improved in Sec. 5. Combining these algorithms with our reduction leads to Theorem 1.

2   Regular Expressions and Finite Automata

In this section we briefly review Thompson's construction and the standard state-set simulation. The set of *regular expressions* over an alphabet Σ are defined recursively as follows: A character $\alpha \in \Sigma$ is a regular expression, and if S and T are regular expressions, then so is the *catenation*, $S \cdot T$, the *union*, $S|T$, and the *star*, S^* (we often remove the $\cdot$ when writing regular expressions). The *language* $L(R)$ generated by R is the set of all strings matching R. The *parse tree* $T(R)$ of R is the binary rooted tree representing the hiearchical structure of R. Each leaf is labeled by a character in Σ and each internal node is labeled either $\cdot$, $|$, or $*$. A *finite automaton* is a tuple $A = (V, E, \delta, \theta, \phi)$, where V is a set of nodes called *states*, E is set of directed edges between states called *transitions*, $\delta : E \rightarrow \Sigma \cup \{\epsilon\}$ is a function assigning labels to transitions, and $\theta, \phi \in V$ are distinguished states called the *start state* and *accepting state*, respectively. [1] Intuitively, A is an edge-labeled directed graph with special start and accepting nodes. A is a *deterministic finite automaton* (DFA) if A does not contain any ϵ-transitions, and all outgoing transitions of any state have different labels. Otherwise, A is a *non-deterministic automaton* (NFA). We say that A *accepts* a string Q if there is a path from θ to ϕ such that the concatenation of labels on the path spells out Q. Thompson [8] showed how to recursively construct a NFA $N(R)$ accepting all strings in $L(R)$. The rules are shown in Fig. 1.

Readers familiar with Thompson's construction will notice that $N(ST)$ is slightly different from the usual construction. This is done to simplify our later presentation and does not affect the worst case complexity of the problem. Any automaton produced by these rules we call a *Thompson-NFA* (TNFA). By construction, $N(R)$ has a single start and accepting state, denoted θ and ϕ, respectively. θ has no incoming transitions and ϕ has no outgoing transitions. The total number of states is $2m$ and since each state has at most 2 outgoing transitions that the total number of transitions is at most $4m$. Furthermore, all

---

[1] Sometimes NFAs are allowed a *set* of accepting states, but this is not necessary for our purposes.

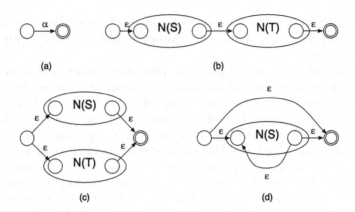

Fig. 1. Thompson's NFA construction. The regular expression for a character $\alpha \in \Sigma$ correspond to NFA (a). If S and T are regular expression then $N(ST)$, $N(S|T)$, and $N(S^*)$ correspond to NFAs (b), (c), and (d), respectively. Accepting nodes are marked with a double circle.

incoming transitions have the same label, and we denote a state with incoming α-transitions an α-*state*. Note that the star construction in Fig. 1(d) introduces a transition from the accepting state of $N(S)$ to the start state of $N(S)$. All such transitions are called *back transitions* and all other transitions are *forward transitions*. We need the following property.

Lemma 1 (Myers [6]). *Any cycle-free path in a TNFA contains at most one back transition.*

For a string Q of length n the standard state-set simulation of $N(R)$ on Q produces a sequence of state-sets $S_0, \ldots, S_n$. The ith set S_i, $0 \leq i \leq n$, consists of all states in $N(R)$ for which there is a path from θ that spells out the ith prefix of Q. The simulation can be implemented with the following simple operations. For a state-set S and a character $\alpha \in \Sigma$, define

Move(S, α): Return the set of states reachable from S via a single α-transition.
Close(S):   Return the set of states reachable from S via 0 or more ϵ-transitions.

Since the number of states and transitions in $N(R)$ is $O(m)$, both operations can be easily implemented in $O(m)$ time. The Close operation is often called an ϵ-*closure*. The simulation proceeds as follows: Initially, $S_0 := \text{Close}(\{\theta\})$. If $Q[j] = \alpha$, $1 \leq j \leq n$, then $S_j := \text{Close}(\text{Move}(S_{j-1}, \alpha))$. Finally, $Q \in L(R)$ iff $\phi \in S_n$. Since each state-set S_j only depends on S_{j-1} this algorithm uses $O(mn)$ time and $O(m)$ space.

3   From Large to Small TNFAs

In this section we show how to simulate $N(R)$ by simulating a number of smaller TNFAs. We will use this to achieve our bounds when R is large.

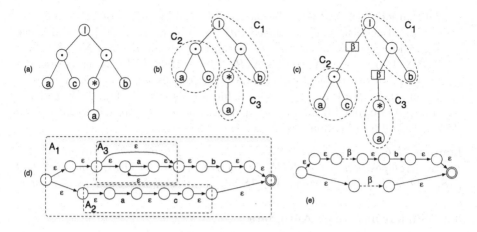

Fig. 2. (a) The parse tree for the regular expression $ac|a^*b$. (b) A clustering of (a) into node-disjoint connected subtrees C_1, C_2, and C_3, each with at most 3 nodes. (c) The clustering from (b) extended with pseudo-nodes. (d) The nested decomposition of $N(ac|a^*b)$. (e) The TNFA corresponding to C_1.

3.1 Clustering Parse Trees and Decomposing TNFAs

Let R be a regular expression of length m. We first show how to decompose $N(R)$ into smaller TNFAs. This decomposition is based on a simple clustering of the parse tree $T(R)$. A *cluster* C is a connected subgraph of $T(R)$ and a *cluster partition* CS is a partition of the nodes of $T(R)$ into node-disjoint clusters. Since $T(R)$ is a binary tree with $O(m)$ nodes, a simple top-down procedure provides the following result:

Lemma 2. *Given a regular expression R of length m and a parameter x, a cluster partition CS of $T(R)$ can be constructed in $O(m)$ time such that $|CS| = O(\lceil m/x \rceil)$, and for any $C \in CS$, the number of nodes in C is at most x.*

For a cluster partition CS, edges adjacent to two clusters are *external edges* and all other edges are *internal edges*. Contracting all internal edges in CS induces a *macro tree*, where each cluster is represented by a single *macro node*. Let C_v and C_w be two clusters with corresponding macro nodes v and w. We say that C_v is the *parent cluster* (resp. *child cluster*) of C_w if v is the parent (resp. child) of w in the macro tree. The *root cluster* and *leaf clusters* are the clusters corresponding to the root and the leaves of the macro tree. An example clustering of a parse tree is shown in Fig. 2(b). Given a cluster partition CS of $T(R)$ we show how to divide $N(R)$ into a set of small nested TNFAs. Each cluster $C \in CS$ will correspond to a TNFA A, and we use the terms child, parent, root, and leaf for the TNFAs in the same way we do with clusters. For a cluster $C \in CS$ with children $C_1, \ldots, C_l$, insert a special *pseudo-node* p_i, $1 \leq i \leq l$, in the middle of the external edge connecting C with C_i. We label each pseudo-node by a special character $\beta \notin \Sigma$. Let T_C be the tree induced by the set of nodes in C and $\{p_1, \ldots, p_l\}$. Each leaf in T_C is labeled with a character from $\Sigma \cup \{\beta\}$, and hence T_C is a well-formed parse tree for some regular expression R_C over $\Sigma \cup \{\beta\}$. Now, the TNFA A corresponding to C is $N(R_C)$. In A, child TNFA A_i is represented by its start and accepting state θ_{A_i} and ϕ_{A_i} and a *pseudo-transition* labeled β connecting them. An example of these definitions is given in Fig. 2. We call any set of TNFAs obtained from a cluster partition as above a *nested decomposition* AS of $N(R)$. From Lemma 2 we have:

Lemma 3. *Given a regular expression R of length m and a parameter x, a nested decomposition AS of $N(R)$ can be constructed in $O(m)$ time such that $|AS| = O(\lceil m/x \rceil)$, and for any $A \in AS$, the number of states in A is at most x.*

3.2   Simulating Large Automata

We now show how $N(R)$ can be simulated using the TNFAs in a nested decomposition. For this purpose we define a simple data structure to dynamically maintain the TNFAs. Let AS be a nested decomposition of $N(R)$ according to Lemma 3, for some parameter x. Let $A \in AS$ be a TNFA, let S_A be a state-set of A, let s be a state in A, and let $\alpha \in \Sigma$. A *simulation data structure* supports the five operations: $\mathsf{Move}_A(S_A, \alpha)$, $\mathsf{Close}_A(S_A)$, $\mathsf{Member}_A(S_A, s)$, and $\mathsf{Insert}_A(S_A, s)$. Here, the operations Move_A and Close_A are defined exactly as in Sec. 2, with the modification that they only work on A and not $N(R)$. The operation $\mathsf{Member}_A(S_A, s)$ return yes if $s \in S_A$ and no otherwise and $\mathsf{Insert}_A(S_A, s)$ returns the set $S_A \cup \{s\}$.

In the following sections we consider various efficient implementations of simulation data structures. For now assume that we have a black-box data structure for each $A \in AS$. To simulate $N(R)$ we proceed as follows. First, fix an ordering of the TNFAs in the nested decomposition AS, e.g., by a preorder traversal of the tree represented given by the parent/child relationship of the TNFAs. The collection of state-sets for each TNFA in AS are represented in a *state-set array* X of length $|AS|$. The state-set array is indexed by the above numbering, that is, $X[i]$ is the state-set of the ith TNFA in AS. For notational convenience we write $X[A]$ to denote the entry in X corresponding to A. Note that a parent TNFA share two states with each child, and therefore a state may be represented more than once in X. To avoid complications we will always assure that X is *consistent*, meaning that if a state s is in included in the state-set of some TNFA, then it is also included in the state-sets of all other TNFAs that share s. If $S = \cup_{A \in AS} X[A]$ we say that X *models* the state-set S and write $S \equiv X$.

Next we show how to do a state-set simulation of $N(R)$ using the operations Move_{AS} and Close_{AS}, which we define below. These operations recursively update a state-set array using the simulation data structures. For any $A \in AS$, state-set array X, and $\alpha \in \Sigma$ define

$\mathsf{Move}_{AS}(A, X, \alpha)$: 1. $X[A] := \mathsf{Move}_A(X[A], \alpha)$
                             2. For each child A_i of A do
                                (a) $X := \mathsf{Move}_{AS}(A_i, X, \alpha)$

$\qquad$ (b) If $\phi_{A_i} \in X[A_i]$ then $X[A] := \mathsf{Insert}_A(X[A], \phi_{A_i})$

$\qquad$ 3. Return X

$\mathsf{Close}_{AS}(A, X)$: $\quad$ 1. $X[A] := \mathsf{Close}_A(X[A])$

$\qquad$ 2. For each child A_i of A in topological order do

$\qquad$ (a) If $\theta_{A_i} \in X[A]$ then $X[A_i] := \mathsf{Insert}_{A_i}(X[A_i], \theta_{A_i})$

$\qquad$ (b) $X := \mathsf{Close}_{AS}(A_i, X)$

$\qquad$ (c) If $\phi_{A_i} \in X[A_i]$ then $X[A] := \mathsf{Insert}_A(X[A], \phi_{A_i})$

$\qquad$ (d) $X[A] := \mathsf{Close}_A(X[A])$

$\qquad$ 3. Return X

The Move_{AS} and Close_{AS} operations recursively traverses the nested decomposition top-down processing the children in topological order. At each child the shared start and accepting states are propagated in the state-set array. For simplicity, we have written Member_A using the symbol $\in$.

The state-set simulation of $N(R)$ on a string Q of length n produces the sequence of state-set arrays $X_0, \ldots, X_n$ as follows: Let A_r be the root automaton and let X be an empty state-set array (all entries in X are $\emptyset$). Initially, set $X[A_r] := \mathsf{Insert}_{A_r}(X[A_r], \theta_{A_r})$ and compute $X_0 := \mathsf{Close}_{AS}(A_r, \mathsf{Close}_{AS}(A_r, X))$. For $i > 0$ we compute X_i from X_{i-1} as follows:

$$X_i := \mathsf{Close}_{AS}(A_r, \mathsf{Close}_{AS}(A_r, \mathsf{Move}_{AS}(A_r, X_{i-1}, Q[i])))$$

Finally, we output $Q \in L(R)$ iff $\phi_{A_r} \in X_n[A_r]$. To see that this algorithm correctly solves REGULAR EXPRESSION MATCHING it suffices to show that for any $i, 0 \leq i \leq n$, X_i correctly models the ith state-set S_i in the standard state-set simulation. We need the following lemma.

Lemma 4. *Let X be a state-set array and let A_r be the root TNFA in a nested decomposition AS. If S is the state-set modeled by X, then*

- $\mathsf{Move}(S, \alpha) \equiv \mathsf{Move}_{AS}(A_r, X, \alpha)$ *and*
- $\mathsf{Close}(S) \equiv \mathsf{Close}_{AS}(A_r, \mathsf{Close}_{AS}(A_r, X))$.

The proof is left for the full version of the paper. Intuitively, the 2 calls to Close_{AS} produce the set of states reachable via a path of forward ϵ-transitions, and the set of states reachable via a path of forward ϵ-transitions and at most 1 back transition, respectively. By Lemma 1 it follows that this is the correct set.

By Lemma 4 the state-set simulation can be done using the Close_{AS} and Move_{AS} operations and the complexity now directly depends on the complexities of the simulation data structure. Putting it all together the following reduction easily follows:

Lemma 5. *Let R be a regular expression of length m over alphabet Σ and let Q a string of length n. Given a simulation data structure for TNFAs with $x < m$ states over alphabet $\Sigma \cup \{\beta\}$, where $\beta \notin \Sigma$, that supports all operations in $O(t(x))$ time, using $O(s(x))$ space, and $O(p(x))$ preprocessing time, REGULAR EXPRESSION MATCHING for R and Q can be solved in $O(\frac{nm \cdot t(x)}{x} + \frac{m \cdot p(x)}{x})$ time using $O(\frac{m \cdot s(x)}{x})$ space.*

The idea of decomposing TNFAs is also present in Myers' paper [6], though he does not give a "black-box" reduction as in Lemma 5. Essentially, he provides a simulation data structure supporting all operations in $O(1)$ time using $O(x \cdot 2^x)$ preprocessing time and space. For $x \leq \log(n/\log n)$ this achieves the result mentioned in the introduction. The result of Bille and Farach [3] does not use Lemma 5. Instead they efficiently encode *all* possible simulation data structures in total $O(2^x + m)$ time and space.

4   A Simple Algorithm

In this section we present a simple simulation data structure for TNFAs, and develop some of the ideas for the improved result of the next section. Let A be a TNFA with $m = O(\sqrt{w})$ states. We will show how to support all operations in $O(1)$ time using $O(m)$ space and $O(m^2)$ preprocessing time.

To build our simulation data structure for A, first sort all states in A in topological ignoring the back transitions. We require that the endpoints of an α-transition are consecutive in this order. This is automatically guaranteed using a standard $O(m)$ time algorithm for topological sorting (see e.g. [4]). We will refer to states in A by their rank in this order. A the state-set of A is represented using a bitstring $S = s_1 s_2 \ldots s_m$ defined such that $s_i = 1$ iff node i is in the state-set. The simulation data structure consists of the following bitstrings:

- For each $\alpha \in \Sigma$, a string $D_\alpha = d_1, \ldots, d_m$ such that $d_i = 1$ iff i is an α-state.
- A string $E = 0e_{1,1}e_{1,2} \ldots e_{1,m} 0 e_{2,1} e_{2,2} \ldots e_{2,m} 0 \ldots 0 e_{m,1} e_{m,2} \ldots e_{m,m}$, where $e_{i,j} = 1$ iff i is ϵ-reachable from j. The zeros are *test bits* needed for the algorithm.
- Three constants $I = (10^m)^m$, $X = 1(0^m 1)^{m-1}$, and $C = 1(0^{m-1}1)^{m-1}$. Note that I has a 1 in each test bit position. [2]

The strings E, I, X, and C are easily computed in $O(m^2)$ time and use $O(m^2)$ bits. Since $m = O(\sqrt{w})$ only $O(1)$ space is needed to store these strings. We store D_α in a hashtable indexed by α. Since the total number of different characters in A can be at most m, the hashtable contains at most m entries. Using perfect hashing D_α can be represented in $O(m)$ space with $O(1)$ worst-case lookup time. The preprocessing time is expected $O(m)$ w.h.p.. To get a worst-case bound we use the deterministic dictionary of Hagerup et. al. [5] with $O(m \log m)$ worst-case preprocessing time. In total the data structure requires $O(m)$ space and $O(m^2)$ preprocessing time.

Next we show how to support each of the operations on A. Suppose $S = s_1 \ldots s_m$ is a bitstring representing a state-set of A and $\alpha \in \Sigma$. The result of $\text{Move}_A(S, \alpha)$ is given by

$$S' := (S >> 1) \ \& \ D_\alpha.$$

This should be understood as C notation, where the right-shift is unsigned. Readers familiar with the Shift-Or algorithm [2] will notice the similarity. To see

---

[2] We use exponentiation to denote repetition, i.e., $1^3 0 = 1110$.

the correctness, observe that state i is put in S' iff state $(i-1)$ is in S and the ith state is an α-state. Since the endpoints of α-transitions are consecutive in the topological order it follows that S' is correct. Here, state $(i-1)$ can only influence state i, and this makes the operation easy to implement in parallel. However, this is not the case for Close_A. Here, any state can potentially affect a large number of states reachable through long ϵ-paths. To deal with this we use the following steps.

$$Y := (S \times X) \ \& \ E$$
$$Z := ((Y \mid I) - (I >> m)) \ \& \ I$$
$$S' := ((Z \times C) << w - m(m+1)) >> w - m$$

We describe in detail why this, at first glance somewhat cryptic sequence, correctly computes S' as the result of $\mathsf{Close}_A(S)$. The variables Y and Z are simply temporary variables inserted to increase the readability of the computation. Let $S = s_1 \ldots s_m$. Initially, $S \times X$ concatenates m copies of S with a zero bit between each copy, that is, $S \times X = s_1 \ldots s_m \times 1(0^m 1)^{m-1} = (0s_1 \ldots s_m)^m$. The bitwise $\&$ with E gives $Y = 0y_{1,1}y_{1,2} \ldots y_{1,m}0y_{2,1}y_{2,2} \ldots y_{2,m}0 \ldots 0y_{m,1}y_{m,2} \ldots y_{m,m}$, where $y_{i,j} = 1$ iff state j is in S and state i is ϵ-reachable from j. In other words, the substring $Y_i = y_{i,1} \ldots y_{i,m}$ indicates the set of states in S that have a path of ϵ-transitions to i. Hence, state i should be included in $\mathsf{Close}_A(S)$ precisely if at least one of the bits in Y_i is 1. This is determined next. First $(Y \mid I) - (I >> m)$ sets all test bits to 1 and subtracts the test bits shifted right by m positions. This ensures that if all positions in Y_i are 0, the ith test bit in the result is 0 and otherwise 1. The test bits are then extracted with a bitwise $\&$ with I, producing the string $Z = z_1 0^m z_2 0^m \ldots z_m 0^m$. This is almost what we want since $z_i = 1$ iff state i is in $\mathsf{Close}_A(S)$. It is easy to check that $Z \times C$ produces a string, where positions $m(m-1)+1$ through m^2 (from the left) contain the test bits compressed into a string of length m. The two shifts zero all other bits and moves this substring to the rightmost position in the word, producing the final result. Since $m = O(\sqrt{w})$ all of the above operations can be done in constant time. Finally, observe that Insert_A and Member_A are trivially implemented in constant time. Thus,

Lemma 6. *For any TNFA with $m = O(\sqrt{w})$ states there is a simulation data structure using $O(m)$ space and $O(m^2)$ preprocessing time which supports all operations in $O(1)$ time.*

The main bottleneck in the above data structure is the string E that represents all ϵ-paths. On a TNFA with m states E requires at least m^2 bits and hence this approach only works for $m = O(\sqrt{w})$. In this next section we show how to use the structure of TNFAs to do better.

5   Overcoming the ϵ-Closure Bottleneck

In this section we show how to compute an ϵ-closure on a TNFA with $m = O(w)$ states in $O(\log m)$ time. Compared with the result of the previous section we quadratically increase the size of the TNFA at the expense of using logarithmic time. The algorithm is easily extended to an efficient simulation data structure. The key idea is a new hierarchical decomposition of TNFAs described below.

5.1 Partial-TNFAs and Separator Trees

First we need some definitions. Let A be a TNFA with parse tree T. Each node v in T uniquely correspond to two states in the A, namely, the start and accepting states $\theta_{A'}$ and $\phi_{A'}$ of the TNFA A' with the parse tree consisting of v and all descendants of v. We say v associates the states $S(v) = \{\theta_{A'}, \phi_{A'}\}$. In general, if C is a cluster of T, i.e., any connected subgraph of T, we say C associates the set of states $S(C) = \cup_{v \in C} S(v)$. We define the partial-TNFA (pTNFA) for C, as the directed, labeled subgraph of A induced by the set of states $S(C)$. In particular, A is a pTNFA since it is induced by $S(T)$. The two states associated by the root node of C are defined to be the start and accepting state of the corresponding pTNFA. We need the following result.

Lemma 7. For any pTNFA P with $m > 2$ states there exists a partitioning of P into two subgraphs P_O and P_I such that

(i) P_O and P_I are pTNFAs with at most $2/3m + 2$ states each,
(ii) any transition from P_O to P_I ends in θ_{P_I} and any transition from P_I to P_O starts in ϕ_{P_I}, and
(iii) the partitioning can be computed in $O(m)$ time.

The proof is left for the full version of the paper. Intuitively, if we draw P, P_I is "surrounded" by P_O, and therefore we will often refer to P_I and P_O as the inner pTNFA and the outer pTNFA, respectively. Applying Lemma 7 recursively gives the following essential data structure. Let P be a pTNFA with m states. The separator tree for P is a binary, rooted tree B defined as follows: If $m = 2$, i.e., P is a trivial pTNFA consisting of two states θ_P and ϕ_P, then B is a single leaf node v that stores the set $X(v) = \{\theta_P, \phi_P\}$. Otherwise $(m > 2)$, compute P_O and P_I according to Lemma 7. The root v of B stores the set $X(v) = \{\theta_{P_I}, \phi_{P_I}\}$, and the children of v are roots of separator trees for P_O and P_I, respectively.

With the above construction each node in the separator tree naturally correspond to a pTNFA, e.g., the root corresponds to P, the children to P_I and P_O, and so on. We denote the pTNFA corresponding to node v in B by $P(v)$. A simple induction combined with Lemma 7(i) shows that if v is a node of depth k then $P(v)$ contains at most $(\frac{2}{3})^k m + 6$ states. Hence, the depth of B is at most $d = \log_{3/2} m + O(1)$. By Lemma 7(iii) each level of B can be computed in $O(m)$ time and thus B can be computed in $O(m \log m)$ total time.

5.2 A Recursive ϵ-Closure Algorithm

We now present a simple ϵ-closure algorithm for a pTNFA, which recursively traverses the separator tree B. We first give the high level idea and then show how it can be implemented in $O(1)$ time for each level of B. Since the depth of B is $O(\log m)$ this leads to the desired result. For a pTNFA P with m states, a separator tree B for P, and a node v in B define

$\mathsf{Close}_{P(v)}(S)$:   1. Compute the set $Z \subseteq X(v)$ of states in $X(v)$ that are ϵ-reachable from S in $P(v)$.

2. If v is a leaf return $S' := Z$, else let u and w be the children of v, respectively:
   (a) Compute the set $G \subseteq V(P(v))$ of states in $P(v)$ that are ϵ-reachable from Z.
   (b) Return $S' := \mathsf{Close}_{P(u)}((S\cup G)\cap V(P(u)))\cup\mathsf{Close}_{P(w)}((S\cup G) \cap V(P(w)))$.

A simple case analysis shows the correctness of $\mathsf{Close}_{P(v)}(S)$. Next we show how to efficiently implement the above algorithm in parallel. The key ingredient is a compact mapping of states into positions in bitstrings. Suppose B is the separator tree of depth d for a pTNFA P with m states. The *separator mapping* M maps the states of P into an interval of integers $[1, l]$, where $l = 3 \cdot 2^d$. The mapping is defined recursively according to the separator tree. Let v be the root of B. If v is a leaf node the interval is $[1, 3]$. The two states of P, θ_P and ϕ_P, are mapped to positions 2 and 3, respectively, while position 1 is left intentionally unmapped. Otherwise, let u and w be the children of v. Recursively, map $P(u)$ to the interval $[1, l/2]$ and $P(w)$ to the interval $[l/2 + 1, l]$. Since the separator tree contains at most 2^d leaves and each contribute 3 positions the mapping is well-defined. The size of the interval for P is $l = 3 \cdot 2^{\log_{3/2} m + O(1)} = O(m)$. We will use the unmapped positions as test bits in our algorithm.

The separator mapping compactly maps all pTNFAs represented in B into small intervals. Specifically, if v is a node at depth k in B, then $P(v)$ is mapped to an interval of size $l/2^k$ of the form $[(i-1) \cdot \frac{l}{2^k} + 1, i \cdot \frac{l}{2^k}]$, for some $1 \leq i \leq 2^k$. The intervals that correspond to a pTNFA $P(v)$ are *mapped* and all other intervals are *unmapped*. We will refer to a state s of P by its mapped position $M(s)$. A state-set of P is represented by a bitstring S such that, for all mapped positions i, $S[i] = 1$ iff the i is in the state-set. Since $m = O(w)$, state-sets are represented in a constant number of words.

To implement the algorithm we define a simple data structure consisting of four length l bitstrings X_k^θ, X_k^ϕ, E_k^θ, and E_k^ϕ for each level k of the separator tree. For notational convenience, we will consider the strings at level k as two-dimensional arrays consisting of 2^k intervals of length $l/2^k$, i.e., $X_k^\theta[i, j]$ is position j in the ith interval of X_k^θ. If the ith interval at level k is unmapped then all positions in this interval are 0 in all four strings. Otherwise, suppose that the interval corresponds to a pTNFA $P(v)$ and let $X(v) = \{\theta_v, \phi_v\}$. The strings are defined as follows:

$$X_k^\theta[i, j] = 1 \text{ iff } \theta_v \text{ is } \epsilon\text{-reachable in } P(v) \text{ from state } j,$$

$$E_k^\theta[i, j] = 1 \text{ iff state } j \text{ is } \epsilon\text{-reachable in } P(v) \text{ from } \theta_v,$$

$$X_k^\phi[i, j] = 1 \text{ iff } \phi_v \text{ is } \epsilon\text{-reachable in } P(v) \text{ from state } j,$$

$$E_k^\phi[i, j] = 1 \text{ iff state } j \text{ is } \epsilon\text{-reachable in } P(v) \text{ from } \phi_v.$$

In addtion to these, we also store a string I_k containing a test bit for each interval, that is, $I_k[i, j] = 1$ iff $j = 1$. Since the depth of B is $O(\log m)$ the strings use $O(\log m)$ words. With a simple depth-first search they can all be computed in $O(m \log m)$ time. It is now a relatively simple matter to simulate the recursive algorithm using techniques similar to those in Sec. 4. Due to lack of space we leave the details for the full version of the paper.

Next we show how to get a full simulation data structure. First, note that in the separator mapping the endpoints of the α-transitions are consecutive (as in Sec. 4). It follows that we can use the same algorithm as in the previous section to compute Move_A in $O(1)$ time. This requires a dictionary of bitstrings, D_α, using additional $O(m)$ space and $O(m \log m)$ preprocessing time. The Insert_A and Member_A operations are trivially implemented in $O(1)$. Putting it all together we have:

Lemma 8. *For a TNFA with $m = O(w)$ states there is a simulation data structure using $O(m)$ space and $O(m \log m)$ preprocessing time which supports all operations in $O(\log m)$ time.*

Combining the simulation data structures from Lemmas 6 and 8 with the reduction from Lemma 5 and taking the best result gives Theorem 1. Note that the simple simulation data structure is the fastest when $m = O(\sqrt{w})$ and n is sufficiently large compared to m.

Acknowledgments. The author wishes to thank Rasmus Pagh and Inge Li Gørtz for many interesting discussions and the anonymous reviewers for many insightful comments.

References

1. A. V. Aho, R. Sethi, and J. D. Ullman. *Compilers: principles, techniques, and tools.* Addison-Wesley Longman Publishing Co., Inc., Boston, MA, USA, 1986.
2. R. Baeza-Yates and G. H. Gonnet. A new approach to text searching. *Commun. ACM*, 35(10):74–82, 1992.
3. P. Bille and M. Farach-Colton. Fast and compact regular expression matching, 2005. Submitted to a journal. Preprint availiable at `arxiv.org/cs/0509069`.
4. T. H. Cormen, C. E. Leiserson, R. L. Rivest, and C. Stein. *Introduction to Algorithms, second edition.* MIT Press, 2001.
5. T. Hagerup, P. B. Miltersen, and R. Pagh. Deterministic dictionaries. *J. of Algorithms*, 41(1):69–85, 2001.
6. E. W. Myers. A four-russian algorithm for regular expression pattern matching. *J. of the ACM*, 39(2):430–448, 1992.
7. G. Myers. A fast bit-vector algorithm for approximate string matching based on dynamic programming. *J. ACM*, 46(3):395–415, 1999.
8. K. Thompson. Regular expression search algorithm. *Comm. of the ACM*, 11:419–422, 1968.

A Parameterized View on Matroid Optimization Problems

Dániel Marx

Institut für Informatik,
Humboldt-Universität zu Berlin,
Unter den Linden 6, 10099
Berlin, Germany
dmarx@informatik.hu-berlin.de

Abstract. Matroid theory gives us powerful techniques for understanding combinatorial optimization problems and for designing polynomial-time algorithms. However, several natural matroid problems, such as 3-matroid intersection, are NP-hard. Here we investigate these problems from the parameterized complexity point of view: instead of the trivial $O(n^k)$ time brute force algorithm for finding a k-element solution, we try to give algorithms with uniformly polynomial (i.e., $f(k) \cdot n^{O(1)}$) running time. The main result is that if the ground set of a represented matroid is partitioned into blocks of size ℓ, then we can determine in $f(k, \ell) \cdot n^{O(1)}$ randomized time whether there is an independent set that is the union of k blocks. As consequence, algorithms with similar running time are obtained for other problems such as finding a k-set in the intersection of ℓ matroids, or finding k terminals in a network such that each of them can be connected simultaneously to the source by ℓ disjoint paths.

1 Introduction

Many of the classical combinatorial optimization problems can be studied in the framework of matroid theory. The polynomial-time solvability of finding minimum weight spanning trees, finding perfect matchings, and certain connectivity problems all follow from the general algorithmic results on matroids.

Deciding whether there is an independent set of size k in the intersection of two matroids can be done in polynomial time, but the problem becomes NP-hard if we have to find a k-element set in the intersection of three matroids. Of course, the problem can be solved in $n^{O(k)}$ time by brute force, hence it is polynomial-time solvable for every fixed value of k. However, the running time is prohibitively large, even for small values of k (e.g., $k = 10$) and moderate values of n (e.g., $n = 1000$). The aim of parameterized complexity is to identify problems that can be solved in *uniformly polynomial time* for every fixed value of the problem parameter k, that is, the running time is of the form $f(k) \cdot n^{O(1)}$. A problem that can be solved in such time is called *fixed-parameter tractable*. Notice the huge qualitative difference between running times such as $O(2^k \cdot n^2)$ and n^k: the former can be efficient even for, say, $k = 15$, while the latter has no chance of working. For more background on parameterized complexity, see [1].

M. Bugliesi et al. (Eds.): ICALP 2006, Part I, LNCS 4051, pp. 655–666, 2006.

The question that we investigate in this paper is whether the NP-hard matroid optimization problems can be solved in uniformly polynomial time, if the parameter is the size of the object that we are looking for. The most general result is the following:

Theorem 1 (Main). *Let $M(E, \mathcal{I})$ be a matroid where the ground set is partitioned into blocks of size ℓ. Given a representation A of M, it can be determined in $f(k, \ell) \cdot \|A\|^{O(1)}$ randomized time whether there is an independent set that is the union of k blocks. ($\|A\|$ denotes the length of A in the input.)*

For $\ell = 2$, this problem is exactly the matroid parity problem, which is polynomial-time solvable for represented matroids [4]. For $\ell \geq 3$, the problem is NP-hard.

As applications of the main result, we show that the following problems are also solvable in $f(k, \ell) \cdot n^{O(1)}$ randomized time:

1. Given a family of subsets each of size at most ℓ, find k of them that are pairwise disjoint.
2. Given a graph G, find k (edge) disjoint triangles in G.
3. Given ℓ matroids over the same ground set, find a set of size k that is independent in each matroid.
4. FEEDBACK EDGE SET WITH BUDGET VECTORS: given a graph with ℓ-dimensional cost vectors on the edges, find a feedback edge set of size at most k such that the total cost does not exceed a given vector C (see Section 5.3 for the precise definition).
5. RELIABLE TERMINALS: select k terminals and connect each of them to the source with ℓ paths such that these $k \cdot \ell$ paths are pairwise disjoint.

The fixed-parameter tractability of the first two problems is well-known: they can be solved either with color coding or using representative systems. However, it is interesting to see that randomized fixed-parameter tractability can be obtained as a straightforward corollary of our results on matroids. We are not aware of any parameterized investigations of the last three problems.

The algorithm behind the main result is inspired by the technique of representative systems introduced by Monien [6] (see also [8,5] and [1, Section 8.2]). Iteratively for $i = 1, 2, \ldots, \ell$, we construct a collection $\mathcal{S}_i$ that contains independent sets arising as the union of i blocks (if there are such independent sets). The crucial observation is that we can ensure that the size of each $\mathcal{S}_i$ is at most a constant depending only on k and ℓ. In [5], this bound is obtained using Bollobás' Inequality. In our case, the bound can be obtained using a linear-algebraic generalization of Bollobás' Inequality due to Lovász [3, Theorem 4.8] (see also [2, Chapter 31, Lemma 3.2]). However, we need an algorithmic way of bounding the size of the $\mathcal{S}_i$'s, hence we do not state and use these inequalities here, but rather reproduce the proof of Lovász in an algorithmic form (Lemma 12). The proof of this lemma is a simple application of multilinear algebra.

The algorithms that we obtain are randomized in the sense that they use random numbers and there is a small probability of not finding a solution even if it exists. The randomized nature of the algorithm comes from the fact that we rely on the Zippel-Schwartz Lemma in some of the operations involving matroid representations. Additionally, when working with representations over finite fields, then some of the algebraic operations are most conveniently done randomized. As the main result is randomized, we do not discuss whether these miscellaneous algebraic operations can be derandomized.

Section 2 summarizes the most important notions of matroid theory. Section 3 discusses how certain operations can be performed on the representations of matroids. Most of these constructions are either easy or folklore. The reason why we discuss them in detail is that we need these results in algorithmic form. The main result is presented in Section 4. In Section 5, the randomized fixed-parameter tractability of certain problems are deduced as corollaries.

2 Preliminaries

A *matroid* $M(E, \mathcal{I})$ is defined by a *ground set* E and a collection $\mathcal{I} \subseteq 2^E$ of *independent sets* satisfying the following three properties:

(I1) $\emptyset \in \mathcal{I}$
(I2) If $X \subseteq Y$ and $Y \in \mathcal{I}$, then $X \in \mathcal{I}$.
(I3) If $X, Y \in \mathcal{I}$ and $|X| < |Y|$, then $\exists e \in Y \setminus X$ such that $X \cup \{e\} \in \mathcal{I}$.

An inclusionwise maximal set of $\mathcal{I}$ is called a *basis* of the matroid. It can be shown that the bases of a matroid all have the same size. This size is called the *rank* of the matroid M, and is denoted by $r(M)$. The rank $r(S)$ of a subset S is the size of the largest independent set in S.

The definition of matroids was motivated by two classical examples. Let $G(V, E)$ be a graph, and let a subset $X \subseteq E$ of edges be independent if X does not contain any cycles. This results in a matroid, which is called the *cycle matroid* of G. The second example comes from linear algebra. Let A be a matrix over an arbitrary field F. Let E be the set of columns of A, and let $X \subseteq E$ be independent if these columns are linearly independent. The matroids that can be defined by such a construction are called *linear matroids,* and if a matroid can be defined by a matrix A over a field F, then we say that the matroid is *representable over* F. In this paper we consider only representable matroids, hence matroids are given by a matrix A over a field F. To avoid complications involving the representations of the elements in the matrix, we assume that F is either a finite field or the rationals. We denote by $\|A\|$ the size of the representation A: the total number of bits required to describe all elements of the matrix.

We say that an algorithm is randomized polynomial time if the running time can be bounded by a polynomial of the input size and the error parameter P, and it produces incorrect answer with probability at most 2^{-P}. Most of the randomized algorithms in this paper are based on the following lemma:

Lemma 2 (Zippel-Schwartz [12,10]). *Let* $p(x_1, \ldots, x_n)$ *be a nonzero polynomial of degree* d *over some field* F, *and let* S *be an* N *element subset of* F. *If each* x_i *is independently assigned a value from* S *with uniform probability, then* $p(x_1, \ldots, x_n) = 0$ *with probability at most* d/N.

3 Representation Issues

The algorithm in Section 4 is based on algebraic manipulations, hence it requires that the matroid is given by a linear representation in the input. Therefore, in the proof of the main result and in its applications, we need algorithmic results on how to find representations for certain matroids, and if some operation is performed on a matroid, then how to obtain a representation of the result.

3.1 Dimension

The rank of a matroid represented by an $m \times n$ matrix is a most m: if the columns are m-dimensional vectors, then more than m of them cannot be independent. Conversely, every linear matroid of rank r has a representation with r rows:

Proposition 3. *Given a matroid M of rank r with a representation A over F, we can find in polynomial time a representation A' over F having r rows.*   □

3.2 Increasing the Size of the Field

The applications of Lemma 2 requires N to be large, so the probability of accidentally finding a root is small. However, N can be large only if the field F contains a sufficient number of elements. Therefore, if a matroid representation is given over some small field F, then we need a method of transforming this representation to a representation over a field F' having at least N elements.

Let $|F| = q$ and let $n = \lceil \log_q N \rceil$. We construct a field F' having $q^n \geq N$ elements. In order to do this, an irreducible polynomial $p(x)$ of degree n over F is required. Such a polynomial $p(x)$ can be found for example by the randomized algorithm of Shoup [11] in time polynomial in n and $\log q$. Now the ring of degree n polynomials over F modulo $p(x)$ is a field F' of size q^n. If a representation over F is given, then each element can be replaced by the corresponding degree 0 polynomial from F', which yields a representation over F'.

Proposition 4. *Let A be the representation of a matroid M over some field F. For every N, it is possible to construct a representation A' of M over some field F' with $|F'| \geq N$ in $(\|A\| \cdot \log N)^{O(1)}$ randomized time.*   □

3.3 Direct Sum

Let $M_1(E_1, \mathcal{I}_1)$ and $M_2(E_2, \mathcal{I}_2)$ be two matroids with $E_1 \cap E_2 = \emptyset$. The *direct sum* $M_1 \oplus M_2$ is a matroid over $E := E_1 \cup E_2$ such that $X \subseteq E$ is independent if and only if $X \cap E_1 \in \mathcal{I}_1$ and $X \cap E_2 \in \mathcal{I}_2$. The notion can be generalized for the sum of more than two matroids.

Proposition 5. *Given representations of matroids $M_1, \ldots, M_k$ over the same field F, a representation of their direct sum can be found in polynomial time.*   □

3.4 Uniform and Partition Matroids

The *uniform matroid* $U_{n,k}$ has an n-element ground set E, and a set $X \subseteq E$ is independent if and only if $|X| \leq k$. Every uniform matroid is linear and can be represented over the rationals by a $k \times n$ matrix where the element in the i-th column of j-th row is $i^{(j-1)}$. Clearly, no set of size larger than k can be independent in this representation, and every set of k columns is independent, as they form a Vandermonde matrix.

A *partition matroid* is given by a ground set E partitioned into k blocks E_1, ..., E_k, and by k integers $a_1, \ldots, a_k$. A set $X \subseteq E$ is independent if and only if $|X \cap E_i| \leq a_i$ holds for every $i = 1, \ldots, k$. As this partition matroid is the direct sum of uniform matroids $U_{|E_1|, a_1}, \ldots, U_{|E_k|, a_k}$, we have

Proposition 6. *A representation over the rationals of a partition matroid can be constructed in polynomial time.* □

3.5 Dual

The *dual* of a matroid $M(E, \mathcal{I})$ is a matroid $M^*(E, \mathcal{I}^*)$ over the same ground set where a set $B \subseteq E$ is a basis of M^* if and only if $E \setminus B$ is a basis of M.

Proposition 7. *Given a representation A of a matroid M, a representation of the dual matroid M^* can be found in polynomial time.*

Proof. Let r be the rank of the matroid M. By Prop. 3, it can be assumed that A is of the form $(I_{r \times r} \ B)$, where $I_{r \times r}$ is the unit matrix of size $n \times n$, and B is a matrix of size $r \times (n - r)$. Now the matrix $A^* = (B^\top \ I_{(n-r) \times (n-r)})$ represents the dual matroid M^*, see any text on matroid theory (e.g., [9]). □

3.6 Truncation

The *k-truncation* of a matroid $M(E, \mathcal{I})$ is a matroid $M'(E, \mathcal{I}')$ such that $S \subseteq E$ is independent in M' if and only if $|S| \leq k$ and S is independent in M.

Proposition 8. *Given a matroid M with a representation A over a finite field F and an integer k, a representation of the k-truncation M' can be found in randomized polynomial time.*

Proof. By Prop. 3 and 4, it can be assumed that A is of size $r \times n$ and the size of F is at least $N := 2^P \cdot k n^k$. Let R be a random matrix of size $k \times r$, where each element is taken from F with uniform distribution. We claim that with high probability, RA is a representation of the k-truncation. Since RA cannot have more than k independent columns, all we have to show is that a k-element set is independent in M' if and only if it is independent in M. Let S be a set of size k, let A_0 be the $r \times k$ submatrix of A formed by the corresponding k columns, and let $B_0 = RA_0$ be the corresponding k columns in RA. If S is not independent in M, then the columns of B_0 are not independent either. This means that S is not independent in the matroid M' represented by RA. Assume now that S is independent in M. The columns of A_0 are independent, thus $\det RA_0 \neq 0$ with positive probability (e.g., there is a matrix R such that RA_0 is the unit matrix). We use Lemma 2 to show that this probability is at least $1 - 2^{-P}/n^k$. The value $\det RA_0$ can be considered as a polynomial, with the kr elements of the matrix

R being the variables. Since $\det RA_0$ is not always zero, the polynomial is not identically zero. As the degree of this polynomial is k, Lemma 2 ensures that $\det RA_0 = 0$ with probability at most $k/N = 2^{-P}/n^k$. Thus the probability that a particular k-element independent set of M is not independent in M' is at most $2^{-P}/n^k$. As M has not more than n^k independent set of size k, the probability that M' is not the k-truncation of M is at most 2^{-P}. $\qquad\square$

3.7 Cycle Matroids

The cycle matroid of $G(V, E)$ can be represented over the 2-element field: consider the $|V| \times |E|$ incidence matrix of G, where the i-th element of the j-row is 1 if and only if the i-th vertex is an endpoint of the j-th edge.

Proposition 9. *Given a graph, a representation of the cycle matroid over the two element field can be constructed in polynomial time.* $\qquad\square$

3.8 Transversal Matroids

Let $G(A, B; E)$ be a bipartite graph. The *transversal matroid* M of G has A as its ground set, and a subset $X \subseteq A$ is independent in M if and only if there is a matching that covers X. That is, X is independent if and only if there is an injective mapping $\phi : X \to B$ such that $\phi(v)$ is a neighbor of v for every $v \in X$.

Proposition 10. *Given a bipartite graph $G(A, B; E)$, a representation of its transversal matroid can be constructed in randomized polynomial time.*

Proof. Let R be a $|B| \times |A|$ matrix, where the i-th element in the j-th row is

- a random integer between 1 and $N := 2^P \cdot |A| \cdot 2^{|A|}$ if the i-th element of A and the j-th element of B are adjacent, and
- 0 otherwise.

We claim that with high probability, R represents the transversal matroid of M. Assume that a subset X of columns is independent. These columns have a $|X| \times |X|$ submatrix with nonzero determinant, hence there is at least one nonzero term in the expansion of this determinant. The nonzero term is a product of $|X|$ nonzero cells, and these cells define a matching covering X.

Assume now that $X \subseteq A$ is independent in the transversal matroid: it can be matched with elements $Y \subseteq B$. This means that the determinant of the $|Y| \times |X|$ submatrix R_0 of R corresponding to X and Y has a term that is the product of nonzero elements. The determinant of R_0 can be considered as a polynomial of degree at most $|A|$, where the variables are the random elements of R_0. The existence of the matching and the corresponding nonzero term in the determinant shows that this polynomial is not identically zero. By Lemma 2, the probability that the determinant of R_0 is zero is at most $2^{-P}/2^{|A|}$, implying that the columns X are independent with high probability. There are at most $2^{|A|}$ independent sets in M, thus the probability that not all of them are independent in the matroid represented by R is at most 2^{-P}. $\qquad\square$

4   The Main Result

In this section we give a randomized fixed-parameter tractable algorithm for determining whether there are k blocks whose union is independent, if a matroid is given with a partition of the ground set into blocks of size ℓ. The idea is to construct for $i = 1, \ldots, k$ the set $\mathcal{S}_i$ of all independent sets that arise as the union of i blocks. A solution exists if and only if $\mathcal{S}_k$ is not empty. The set $\mathcal{S}_i$ is easy to construct if $\mathcal{S}_{i-1}$ is already known. The problem is that the size of $\mathcal{S}_i$ can be as large as $n^{\Omega(i)}$, hence we cannot handle sets of this size in uniformly polynomial time. The crucial idea is that we retain only a constant size subset of each $\mathcal{S}_i$ in such a way that we do not throw away any sets essential for the solution. The property that this reduced collection has to satisfy is the following:

Definition 11. *Given a matroid $M(E, \mathcal{I})$ and a collection $\mathcal{S}$ of subsets of E, we say that a subsystem $\mathcal{S}^* \subseteq \mathcal{S}$ is r-representative for $\mathcal{S}$ if the following holds: for every set $Y \subseteq E$ of size at most r, if there is a set $X \in \mathcal{S}$ disjoint from Y with $X \cup Y \in \mathcal{I}$, then there is a set $X^* \in \mathcal{S}^*$ disjoint from Y with $X^* \cup Y \in \mathcal{I}$.*

That is, if an independent set in $\mathcal{S}$ can be extended to an independent set by r new elements, then there is a set in $\mathcal{S}^*$ that can be extended by the same r elements. 0-representative means that $\mathcal{S}^*$ is not empty if $\mathcal{S}$ is not empty. We use the following lemma to obtain a representative subcollection of constant size:

Lemma 12. *Let M be a linear matroid of rank $r + s$, and let $\mathcal{S} = \{S_1, \ldots, S_m\}$ be a collection of independent sets, each of size s. If $|\mathcal{S}| > \binom{r+s}{s}$, then there is a set $S \in \mathcal{S}$ such that $\mathcal{S} \setminus \{S\}$ is r-representative for $\mathcal{S}$. Furthermore, given a representation A of M, we can find such a set S in $f(r, s) \cdot (\|A\| m)^{O(1)}$ time.*

Proof. Assume that M is represented by an $(r + s) \times n$ matrix A over some field F. Let E be the ground set of the matroid M, and for each element $e \in E$, let x_e be the corresponding $(r+s)$-dimensional column vector of A. Let $w_i = \bigwedge_{e \in S_i} x_e$, a vector in the exterior algebra of the linear space F^{r+s}. As every w_i is the wedge product of s vectors, the w_i's span a space of dimension at most $\binom{r+s}{s}$. Therefore, if $|\mathcal{S}| > \binom{r+s}{s}$, then the w_i's are not independent. Thus it can be assumed that some vector w_k can be expressed as the linear combination of the other vectors.

We claim that if S_k is removed from $\mathcal{S}$, then the resulting subsystem is r-representative for $\mathcal{S}$. Assume that, on the contrary, there is a set Y of size at most r such that $S_k \cap Y = \emptyset$ and $S_k \cup Y$ is independent, but this does not hold for any other S_i with $i \neq k$. Let $y = \bigwedge_{e \in Y} x_e$. A crucial property of the wedge product is that the product of some vectors in F^{r+s} is zero if and only if they are not independent. Therefore, $w_k \wedge y \neq 0$, but $w_i \wedge y = 0$ for every $i \neq k$. However, w_k is the linear combination of the other w_i's, thus, by the multilinearity of the wedge product, $w_k \wedge y \neq 0$ is a linear combination of the values $w_i \wedge y = 0$ for $i \neq k$, which is a contradiction.

It is straightforward to make this proof algorithmic. First we determine the vectors w_i, then a vector w_k that is spanned by the other vectors can be found by standard techniques of linear algebra. Let us fix a basis of F^{r+s}, and express the vectors x_e as the linear combination of the basis vectors. The vector w_i is the wedge product of s vectors, hence, using the multilinearity of the wedge product, each w_i can be expressed as the sum of $(r + s)^s$ terms. Each term is the wedge product of basis vectors of F^{r+s}; therefore, the antisymmetry property can be used to reduce each term to 0 or a basis vector of the exterior algebra. Thus we obtain each w_i as a linear combination of basis vectors. Now Gaussian elimination can be used to determine the rank of the subspace spanned by the w_i's, and to check whether the rank remains the same if one of the vectors is removed. If so, then the set corresponding to this vector can be removed from $\mathcal{S}$, and the resulting subsystem $\mathcal{S}^*$ is representative for $\mathcal{S}$. The running time of the algorithm can be bounded by a polynomial of the number of vectors n, the number of terms in the expression of a w_i (i.e., $(r + s)^s$), the dimension of the subspace spanned by the w_i's (i.e., $\binom{r+s}{s}$), and the size of the representation of M. Therefore, the algorithm is polynomial-time for every fixed value of r and s. $\qquad\square$

Now we are ready to prove the main result:

Proof (of Theorem 1). First we obtain a representation A' for the $k\ell$-truncation of the matroid. By Prop 8, this can be done in time polynomial in $\|A\|$. Using A' instead of A does not change the answer to the problem, as we consider the independence of the union of at most k blocks. However, when invoking Lemma 12, it will be important that the elements are represented as $k\ell$-dimensional vectors.

For $i = 1, \ldots, k$, let $\mathcal{S}_i$ be the set system containing those independent sets that arise as the union of i blocks. Clearly, the task is to determine whether $\mathcal{S}_k$ is empty or not. For each i, we construct a subsystem $\mathcal{S}_i^* \subseteq \mathcal{S}_i$ that is $(k-i)\ell$-representative for $\mathcal{S}_i$. As $\mathcal{S}_k^*$ is 0-representative for $\mathcal{S}_k$, the emptiness of $\mathcal{S}_k$ can be checked by checking whether $\mathcal{S}_k^*$ is empty.

The set system $\mathcal{S}_1$ is easy to construct, hence we can take $\mathcal{S}_1^* = \mathcal{S}_1$. Assume now that we have a set system $\mathcal{S}_i^*$ as above. The set system $\mathcal{S}_{i+1}^*$ can be constructed as follows. First, if $|\mathcal{S}_i^*| > \binom{i\ell + (k-i)\ell}{i\ell} = \binom{k\ell}{i\ell}$, then by Lemma 12, we can throw away an element of $\mathcal{S}_i^*$ in such a way that $\mathcal{S}_i^*$ remains $(k-i)\ell$-representative for $\mathcal{S}_i$. Therefore, it can be assumed that $|\mathcal{S}_i^*| \leq \binom{k\ell}{i\ell}$. To obtain $\mathcal{S}_{i+1}^*$, we enumerate every set S in $\mathcal{S}_i^*$ and every block B, and if S and B are disjoint and $S \cup B$ is independent, then $S \cup B$ is put into $\mathcal{S}_{i+1}^*$. We claim that the resulting system is $(k-i-1)\ell$-representative for $\mathcal{S}_{i+1}$ provided that $\mathcal{S}_i^*$ is $(k-i)\ell$-representative for $\mathcal{S}_i$. Assume that there is a set $X \in \mathcal{S}_{i+1}$ and a set Y of size $(k-i-1)\ell$ such that $X \cap Y = \emptyset$ and $X \cup Y$ is independent. By definition, X is the union of $i+1$ blocks; let B be an arbitrary block of X. Let $X_0 = X \setminus B$ and $Y_0 = Y \cup B$. Now X_0 is in $\mathcal{S}_i$, and we have $X_0 \cap Y_0 = \emptyset$ and $X_0 \cup Y_0 = X \cup Y$ is independent. Therefore, there is a set $X_0^* \in \mathcal{S}_i^*$ with $X_0^* \cap Y_0 = \emptyset$ and $X_0^* \cup Y_0$ independent. This means that the independent set $X^* := X_0^* \cup B$ is put into $\mathcal{S}_{i+1}^*$, and it satisfies $X^* \cap Y = \emptyset$ and $X^* \cup Y$ independent.

When constructing the set system $\mathcal{S}_{i+1}^*$, the amount of work to be done is polynomial in $\|A'\|$ for each member S of $\mathcal{S}_i^*$. As discussed above, the size of each $\mathcal{S}_i^*$ can be bounded by $\binom{k\ell}{i\ell}$, thus the running time is $f(k, \ell) \cdot \|A'\|^{O(1)}$. $\qquad\square$

5   Applications

In this section we derive some consequences of the main result: we list problems that can be solved using the algorithm of Theorem 1.

5.1   Matroid Intersection

Given matroids $M_1(E, \mathcal{I}_1)$, ..., $M_\ell(E, \mathcal{I}_\ell)$ over a common ground set, their *intersection* is the set system $\mathcal{I}_1 \cap \cdots \cap \mathcal{I}_\ell$. In general, the resulting set system is not a matroid, even for $k = 2$. Deciding whether there is a k-element set in the intersection of two matroids is polynomial-time solvable (cf. [9]), but NP-hard for more than two matroids. Here we show that the problem is randomized fixed-parameter tractable for a fixed number of represented matroids:

Theorem 13. *Let M_1, ..., M_ℓ be matroids over the same set, given by their representations A_1, ..., A_ℓ over F. We can decide in $f(k, \ell) \cdot (\sum_{i=1}^{\ell} \|A_i\|)^{O(1)}$ randomized time if there is a k-element set that is independent in every M_i.*

Proof. Let $E = \{e_1, \ldots, e_n\}$. We rename the elements of the matroids to make the ground sets pairwise disjoint: let $e_j^{(i)}$ be the copy of e_j in M_i. By Prop. 5, a representation of $M := M_1 \oplus \cdots \oplus M_\ell$ can be obtained. Partition the ground set of M into blocks of size ℓ: for $1 \leq j \leq n$, block B_j is $\{e_j^{(1)}, \ldots, e_j^{(\ell)}\}$. If M has an independent set that is the union of k blocks, then the corresponding k elements of E is independent in each of $M_1, \ldots, M_\ell$. Conversely, if $X \subseteq E$ is independent in every matroid, then the union of the corresponding blocks is independent in M. Therefore, the algorithm of Theorem 1 answers the question.   □

5.2   Disjoint Sets

Packing problems form a well-studied class of combinatorial optimization problems. Here we study the case when the objects to be packed are small:

Theorem 14. *Let $\mathcal{S} = \{S_1, \ldots, S_n\}$ be a collection of subsets of E, each of size at most ℓ. There is an $f(k, \ell) \cdot n^{O(1)}$ time randomized algorithm for deciding whether it is possible to select k pairwise disjoint subsets from $\mathcal{S}$.*

Proof. By adding dummy elements, it can be assumed that each S_i is of size exactly ℓ. Let $V = \{v_{i,j} : 1 \leq i \leq n, 1 \leq j \leq \ell\}$. We define a partition matroid over V as follows. For every element $e \in E$, let $V_e \subseteq V$ contain $v_{i,j}$ if and only if the j-th element of S_i is e. Clearly, the V_e's form a partition of V. Consider the partition matroid M where a set is independent if and only if it contains at most 1 element from each class of the partition. Let block B_i be $\{v_{i,1}, \ldots, v_{i,\ell}\}$. If k disjoint sets can be selected from $\mathcal{S}$, then the union of the corresponding k blocks is independent in M as every element is contained in at most one of the selected sets. The converse is also true: if the union of k blocks is independent, then the corresponding k sets are disjoint, hence the result follows from Theorem 1.   □

Theorem 14 immediately implies the existence of randomized fixed-parameter tractable algorithms for two well-know problems: DISJOINT TRIANGLES and

EDGE DISJOINT TRIANGLES. In these problems the task is to find, given a graph G and an integer k, a collection of k triangles that are pairwise (edge) disjoint. If E is the set of vertices (edges) of G, and the sets in $\mathcal{S}$ are the triangles of G, then it is clear that the algorithm of Theorem 14 solves the problem.

5.3 Feedback Edge Set with Budget Vectors

Given a graph $G(V, E)$, a *feedback edge set* is a subset X of edges such that $G(V, E \setminus X)$ is acyclic. If the edges of the graph are weighted, then finding a minimum weight feedback edge set is the same as finding a maximum weight spanning forest, hence it is polynomial time solvable. Here we study a generalization of the problem, where each edge has a vector of integer weights:

FEEDBACK EDGE SET WITH BUDGET VECTORS

Input: A graph $G(V, E)$, a vector $\mathbf{x}_e \in [0, 1, \ldots, m]^\ell$ for each $e \in E$, a vector $C \in \mathbb{Z}_+^\ell$, and an integer k.

Parameter: k, ℓ, m

Question: Find a feedback edge set X of $\leq k$ edges such that $\sum_{e \in X} \mathbf{x}_e \leq C$.

Theorem 15. FEEDBACK EDGE SET WITH BUDGET VECTORS *can be solved in* $f(k, \ell, m) \cdot n^{O(1)}$ *randomized time.*

Proof. It can be assumed that $k = |E| - |V| + c(G)$ (where $c(G)$ is the number of components of G): if k is smaller, then there is no solution; if k is larger, then it can be decreased without changing the problem. Let $M_0(E, \mathcal{I}_0)$ be the dual of the cycle matroid of G. The rank of M_0 is k, and a set X of k edges is a basis of M if and only if the complement of X is a spanning forest.

Let $C = [c_1, \ldots, c_\ell]$ and $n = |E|$. For $i = 1, \ldots, \ell$, let $M_i(E_i, \mathcal{I}_i)$ be the uniform matroid U_{nm, c_i}. By Props. 9, 4, 7, 6, and 5, a representation of the direct sum $M = M_0 \oplus M_1 \oplus \cdots \oplus M_k$ can be constructed in polynomial time. For each $e \in E$, let B_e be a block containing $e \in E$ and $x_e^{(i)}$ arbitrary elements of E_i for every $i = 1, \ldots, \ell$ (where $x_e^{(i)} \leq m$ denotes the i-th component of $\mathbf{x}_e$). The set E_i contains nm elements, which is sufficiently large to make the blocks B_i disjoint. The size of each block is at most $\ell' := 1 + m\ell$, hence the algorithm of Theorem 1 can be used to determine in $f(k, \ell') \cdot n^{O(1)}$ randomized time whether there is an independent set that is the union of k blocks. It is clear that every such independent set corresponds to a feedback edge set such that the total weight of the edges does not exceed C at any component.   □

5.4 Reliable Terminals

In this section we give a randomized fixed-parameter tractable algorithm for a combinatorial problem motivated by network design applications.

RELIABLE TERMINALS

> *Input:* A directed graph $D(V, A)$, a source vertex $s \in V$, a set $T \subseteq V \setminus \{s\}$ of possible terminals.

Parameter: k, ℓ

> *Question:* Select k terminals $t_1, \ldots, t_k \in T$ and $k \cdot \ell$ internally vertex disjoint paths $P_{i,j}$ ($1 \leq i \leq k$, $1 \leq j \leq \ell$) such that path $P_{i,j}$ goes from s to t_i.

The problem models the situation when k terminals have to be selected that receive k different data streams (hence the paths going to different terminals should be disjoint due to capacity constraints) and each data stream is protected from $\ell-1$ node failures (hence the ℓ paths of each data stream should be disjoint).

Let $D(V, A)$ be a directed graph, and let $S \subseteq A$ be a subset of vertices. We say that a subset $X \subseteq S$ is *linked to* S if there are $|X|$ vertex disjoint paths going from S to X. (Note that here we require that the paths are disjoint, not only internally disjoint. Furthermore, zero-length paths are also allowed if $X \cap S \neq \emptyset$.) A result due to Perfect shows that the set of linked vertices form a matroid:

Theorem 16 (Perfect [7]). *Let $D(V, A)$ be a directed graph, and let $S \subseteq A$ be a subset of vertices. The subsets that are linked to S form the independent sets of a matroid over V. Furthermore, a representation of this matroid can be obtained in randomized polynomial time.*

Proof. Let $V = \{v_1, \ldots, v_n\}$ and assume that no arc enters S. Let $G(U, W; E)$ be a bipartite graph where a vertex $u_i \in U$ corresponds to each vertex $v_i \in V$, and a vertex $w_i \in W$ corresponds to each vertex $v_i \in V \setminus S$. For each $v_i \in V$, there is an edge $w_i u_i \in E$, and for each $\overrightarrow{v_i v_j} \in A$, there is an edge $u_i w_j \in E$.

The size of a maximum matching in G is at most $|W| = n - |S|$. Furthermore, a matching of size $n - |S|$ can be obtained by taking the edges $u_i w_i$ for every $v_i \notin S$. Let $V_0 \subseteq V$ be a subset of size $|S|$, and let U_0 be the corresponding subset of U. We claim that V_0 is linked to S if and only G has a matching covering $U \setminus U_0$. Assume first that there are $|S|$ disjoint paths going from S to V_0. Consider the matching where $w_i \in W$ is matched to u_j if one of the paths enters v_i from v_j, and w_i is matched to u_i otherwise. This means that u_i is matched if one of the paths reaches v_i and continues further on, or if none of the paths reaches v_i. Thus the unmatched u_i's corresponds to the end points of the paths, as required.

To see the other direction, consider a matching covering $U \setminus U_0$. As $|U \setminus U_0| = n - |S|$, this is only possible if the matching fully covers W. Let v_{i_1} be a vertex of S. Let u_{i_2} be the pair of w_{i_1} in the matching, let u_{i_3} be the pair of w_{i_2}, etc. We can continue this until a vertex u_{i_k} is found that is not covered in the matching. Now $v_{i_1}, v_{i_2}, \ldots, v_{i_k}$ is a path going from S to $v_{i_k} \in V_0$. If this procedure is repeated for every vertex of S, then we obtain $|S|$ paths that are pairwise disjoint, and each of them ends in a vertex of V_0.

If X is linked to S, then X can be extended to a linked set of size exactly $|S|$ by adding vertices of S to it (as they are connected to S by zero-length paths). The observation above shows that linked sets of size $|S|$ are exactly the bases of the dual of the transversal matroid of G, which means that the linked sets are exactly the independent sets of this matroid. By Props. 10 and 7, a representation of this matroid can be constructed in randomized polynomial time.   □

Theorem 17. RELIABLE TERMINALS *is solvable in* $f(k, \ell) \cdot n^{O(1)}$ *randomized time.*

Proof. Let us replace the vertex s with $k \cdot \ell$ independent vertices $S = \{s_1, \ldots, s_{k\ell}\}$ such that each new vertex has the same neighborhood as s. Similarly, each $t \in T$ is replaced with ℓ vertices $t^{(1)}, \ldots, t^{(\ell)}$, but now we remove every outgoing edge from $t^{(2)}, \ldots, t^{(\ell)}$. Denote by D' the new graph. It is easy to see that a set of terminals $t_1, \ldots, t_k$ form a solution for the RELIABLE TERMINALS problem if and only if the set $\{t_i^{(j)} : 1 \leq i \leq k, 1 \leq j \leq \ell\}$ is linked to S. Using Theorem 16, we can construct a representation of the matroid whose independent sets are exactly the sets linked to S in D'. Delete the columns that do not correspond to vertices in T, hence the ground set of the matroid has $\ell|T|$ elements. Partition the ground set into blocks of size ℓ: for every $t \in T$, there is a block $B_t = \{t^1, \ldots, t^\ell\}$. Clearly, the RELIABLE TERMINALS problem has a solution if and only if the matroid has an independent set that is the union of k blocks. Therefore, Theorem 1 can be used to solve the problem.   □

References

1. R. G. Downey and M. R. Fellows. *Parameterized complexity.* Monographs in Computer Science. Springer-Verlag, New York, 1999.
2. R. L. Graham, M. Grötschel, and L. Lovász, editors. *Handbook of combinatorics. Vol. 1, 2.* Elsevier Science B.V., Amsterdam, 1995.
3. L. Lovász. Flats in matroids and geometric graphs. In *Combinatorial surveys (Proc. Sixth British Combinatorial Conf., Royal Holloway Coll., Egham, 1977),* pages 45–86. Academic Press, London, 1977.
4. L. Lovász. Matroid matching and some applications. *J. Combin. Theory Ser. B,* 28(2):208–236, 1980.
5. D. Marx. Parameterized coloring problems on chordal graphs. *Theoret. Comput. Sci.,* 351(3):407–424, 2006.
6. B. Monien. How to find long paths efficiently. In *Analysis and design of algorithms for combinatorial problems (Udine, 1982),* volume 109 of *North-Holland Math. Stud.,* pages 239–254. North-Holland, Amsterdam, 1985.
7. H. Perfect. Applications of Menger's graph theorem. *J. Math. Anal. Appl.,* 22:96–111, 1968.
8. J. Plehn and B. Voigt. Finding minimally weighted subgraphs. In *Graph-theoretic concepts in computer science (WG '90),* LNCS 484, 18–29. Springer, Berlin, 1991.
9. A. Recski. *Matroid theory and its applications in electric network theory and statics,* Springer-Verlag, Berlin, New York and Akadémiai Kiadó, Budapest, 1989.
10. J. T. Schwartz. Fast probabilistic algorithms for verification of polynomial identities. *J. Assoc. Comput. Mach.,* 27(4):701–717, 1980.
11. V. Shoup. Fast construction of irreducible polynomials over finite fields. *J. Symbolic Comput.,* 17(5):371–391, 1994.
12. R. Zippel. Probabilistic algorithms for sparse polynomials. In *Symbolic and algebraic computation (EUROSAM '79),* LNCS 72, 216–226. Springer, Berlin, 1979.

Fixed Parameter Tractability of Binary Near-Perfect Phylogenetic Tree Reconstruction[*]

Guy E. Blelloch[1], Kedar Dhamdhere[2], Eran Halperin[3], R. Ravi[4],
Russell Schwartz[5], and Srinath Sridhar[1]

[1] Computer Science Department, CMU
{guyb, srinath}@cs.cmu.edu
[2] Google Inc, Mountain View, CA
kedar.dhamdhere@gmail.com
[3] ICSI, 1947 Center St, Berkeley, CA
heran@icsi.berkeley.edu
[4] Tepper School of Business, CMU
ravi@cmu.edu
[5] Dept Biological Sciences, CMU
russells@andrew.cmu.edu

Abstract. We consider the problem of finding a Steiner minimum tree in a hypercube. Specifically, given n terminal vertices in an m dimensional cube and a parameter q, we compute the Steiner minimum tree in time $O(72^q + 8^q nm^2)$, under the assumption that the length of the minimum Steiner tree is at most $m + q$.

This problem has extensive applications in taxonomy and biology. The Steiner tree problem in hypercubes is equivalent to the phylogeny (evolutionary tree) reconstruction problem under the maximum parsimony criterion, when each taxon is defined over binary states. The taxa, character set and mutation of a phylogeny correspond to terminal vertices, dimensions and traversal of a dimension in a Steiner tree. Phylogenetic trees that mutate each character exactly once are called *perfect phylogenies* and their size is bounded by the number of characters. When a perfect phylogeny consistent with the data set exists it can be constructed in linear time. However, real data sets often do not admit perfect phylogenies. In this paper, we consider the problem of reconstructing near-perfect phylogenetic trees (referred to as BNPP). A near-perfect phylogeny relaxes the perfect phylogeny assumption by allowing at most q additional mutations. We show for the first time that the BNPP problem is fixed parameter tractable (FPT) and significantly improve the previous asymptotic bounds.

1 Introduction

One of the core areas of computational biology is phylogenetics, the reconstruction of evolutionary trees [13,21]. This problem is often phrased in terms of a

---

[*] Supported in part by NSF grant CCF-043075 and ITR grant CCR-0122581(The ALADDIN project).

parsimony objective, in which one seeks the simplest possible tree to explain a set of observed taxa. Parsimony is a particularly appropriate objective for trees representing short time scales, such as those for inferring evolutionary relationships among individuals within a single species or a few closely related species. Such phylogeny problems have become especially important since we now have identified millions of single nucleotide polymorphisms (SNPs) [16,17], sites at which a single DNA base takes on two common variants. Simply stated, if we examine any specific SNP site on the human genome, then all the individuals in the data sets can be classified into two classes. Therefore an individual's A, C, G, T string can be represented as a binary string with no loss of information.

Consider a $n \times m$ input matrix I, where each row represents an input taxon and is a string over states Σ. The columns of I are called *characters*. A phylogeny is a tree where vertices represent taxa and edges mutations. A phylogeny T for I is a tree that contains all the taxa in I and its length is the sum of the Hamming distances of adjacent vertices. Minimizing the length of a phylogeny is the problem of finding the most parsimonious tree, a well known NP-complete problem, even when $|\Sigma| = 2$ [10]. Researchers have thus focused on either sophisticated heuristics (e.g. [4,11]) or solving optimally for special cases (e.g. [1,18]).

In this work, we focus on the case when the set of states is binary, $|\Sigma| = 2$. The taxa can therefore be viewed as vertices of an m-cube, and the problem is equivalent to finding the Steiner minimum tree in an m-cube. In this setting, a phylogeny for I is called *perfect* if its length equals m. Gusfield showed that such phylogenies can be reconstructed in linear time [12]. If there exists no perfect phylogeny for input I, then one option is to slightly modify I so that a perfect phylogeny can be constructed for the resulting input. Upper bounds and negative results have been established for such problems. For instance, Day and Sankoff [6], showed that finding the maximum subset of characters containing a perfect phylogeny is NP-complete while Damaschke [7] showed fixed parameter tractability for the same problem. The problem of reconstructing the most parsimonious tree without modifying the input I seems significantly harder.

In the general case when $|\Sigma| = s$, a phylogeny for I is called perfect if the length is $m(s - 1)$. In this setting, Bodlaender et al. [3] proved a number of crucial negative results, among them that finding the perfect phylogeny when the number of characters is a parameter is $W[t]$-hard for all t. A problem is fixed parameter tractable on parameter k if there exists an algorithm that runs in time $O(f(k)\mathrm{poly}(|I|))$ where $|I|$ is the input size. Since $FPT \subseteq W[1]$, this shows in particular that the problem is not fixed parameter tractable (unless the complexity classes collapse).

Fernandez-Baca and Lagergren considered the problem of reconstructing optimum near-perfect phylogenies [9]. A phylogeny is q-near-perfect if its length is $m(s-1)+q$. They find the optimum phylogeny in time $nm^{O(q)}2^{O(q^2 s^2)}$, assuming a q-near-perfect phylogeny exists. This bound may be impractical for sizes of m to be expected from SNP data (binary states), even for moderate q. Given the importance of SNP data, it would therefore be valuable to develop methods able to handle large m for the special case of $s = 2$, when all taxa are represented by binary strings. This problem is called Binary Near-Perfect Phylogenetic tree reconstruction (BNPP).

In a prior work, Sridhar et al. [20] solve the BNPP problem in time $O(\binom{m}{q}72^q nm + nm^2)$. The main contribution of the prior work is two-fold: they simplify the previous algorithm [9] which results in the reduction of the exponent in the run-time to q and demonstrate the first empirical results on near-perfect phylogenies. For real data sets that were solved, the range of values for n, m and q were: 15-150, 49-1510 and 1-7 respectively. However, many instances were unsolvable because of the high running time.

Our Work: Here, we present a new algorithm for the BNPP problem that runs in time $O(72^q+8^q nm^2)$. This result significantly improves the prior running time. Fernandez-Baca and Lagergren [9] in concluding remarks state that the most important open problem in the area is to develop a parameterized algorithm or prove $W[t]$ hardness for the near-perfect phylogeny problem. We make progress on this open problem by showing for the first time that BNPP is fixed parameter tractable (FPT). To achieve this, we use a divide and conquer algorithm. Each divide step involves performing a 'guess' (or enumeration) with cost exponential in q. Finding the Steiner minimum tree on a q-cube dominates the run-time when the algorithm bottoms out.

2   Preliminaries

In defining formal models for parsimony-based phylogeny construction, we borrow definitions and notations from a couple of previous works [9,21]. The input to a phylogeny problem is an $n \times m$ binary matrix I where rows $R(I)$ represent *input taxa* and are binary strings. The column numbers $C = \{1, \cdots, m\}$ are referred to as *characters*. In a *phylogenetic tree*, or *phylogeny*, each vertex v corresponds to a taxon (not necessarily in the input) and has an associated label $l(v) \in \{0,1\}^m$.

Definition 1. *A* phylogeny *for matrix* I *is a tree* $T(V, E)$ *with the following properties:* $R(I) \subseteq l(V(T))$ *and* $l(\{v \in V(T)|degree(v) \leq 2\}) \subseteq R(I)$. *That is, every input taxon appears in* T *and every leaf or degree-2 vertex is an input taxon.*

Definition 2. *A vertex* v *of phylogeny* T *is terminal if* $l(v) \in R(I)$ *and Steiner otherwise.*

Definition 3. *For a phylogeny* T, $\texttt{length}(T) = \sum_{(u,v)\in E(T)} d(l(u), l(v))$, *where* d *is the Hamming distance.*

A phylogeny is called an optimum phylogeny if its length is minimized. We will assume that both states $0, 1$ are present in all characters. Therefore the length of an optimum phylogeny is at least m. This leads to the following definition.

Definition 4. *For a phylogeny* T *on input* I, $\texttt{penalty}(T) = \texttt{length}(T) - m$; $\texttt{penalty}(I) = \texttt{penalty}(T^{opt})$, *where* T^{opt} *is any optimum phylogeny on* I.

Definition 5. *A phylogeny T is called* q-near-perfect *if* penalty$(T) = q$ *and* perfect *if* penalty$(T) = 0$.

Note that in an optimum phylogeny, no two vertices share the same label. Therefore, we can equivalently define an edge of a phylogeny as (t_1, t_2) where $t_i \in \{0,1\}^m$. Since we will always be dealing with optimum phylogenies, we will drop the label function $l(v)$ and use v to refer to both a vertex and the taxon it represents in a phylogeny.

The BNPP problem: Given an integer q and an $n \times m$ binary input matrix I, if penalty$(I) \leq q$, then return an optimum phylogeny T, else declare NIL. The problem is equivalent to finding the minimum Steiner tree on an m-cube if the optimum tree is at most q larger than the number of dimensions m or declaring NIL otherwise. An optimum Steiner tree can easily be converted to an optimum phylogeny by removing degree-two Steiner vertices. The problem is fundamental and therefore expected to have diverse applications besides phylogenies.

Definition 6. *We define the following notations.*

- $r[i] \in \{0,1\}$: *the state in character i of taxon r*
- $\mu(e) : E(T) \rightarrow 2^C$: *the set of all characters corresponding to edge $e = (u,v)$ with the property for any $i \in \mu(e)$, $u[i] \neq v[i]$*
- *for a set of taxa M, we use T_M^* to denote an optimum phylogeny on M*

We say that an edge e *mutates* character i if $i \in \mu(e)$. We will use the following well known definition and lemma on phylogenies.

Definition 7. *Given matrix I, the* set of gametes $G_{i,j}$ *for characters i,j is defined as: $G_{i,j} = \{(r[i], r[j]) | r \in R(I)\}$. Two characters i,j* share t gametes *in I i.f.f. $|G_{i,j}| = t$.*

In other words, the set of gametes $G_{i,j}$ is a projection on the i,j dimensions.

Lemma 1. [12] *An optimum phylogeny for input I is not perfect i.f.f. there exists two characters i,j that share (all) four gametes in I.*

Definition 8. (Conflict Graph [15]): *A conflict graph G for matrix I with character set C is defined as follows. Every vertex v of G corresponds to unique character $c(v) \in C$. An edge (u,v) is added to G i.f.f. $c(u), c(v)$ share all four gametes in I. Such a pair of characters are defined to be in* conflict. *Notice that if G contains no edges, then a perfect phylogeny can be constructed for I.*

Simplifications: We assume that the all zeros taxon is present in the input. If not, using our freedom of labeling, we convert the data into an equivalent input containing the all zeros taxon (see section 2.2 of Eskin et al [8] for details). We now remove any character that contains only one state. Such characters do not mutate in the whole phylogeny and are therefore useless in any phylogeny reconstruction. The BNPP problem asks for the reconstruction of an unrooted tree. For the sake of analysis, we will however assume that all the phylogenies are rooted at the all zeros taxon.

3 Algorithm

This section deals with the complete description and analysis of our algorithm for the BNPP problem. For ease of exposition, we first describe a randomized algorithm for the BNPP problem that runs in time $O(18^q + qnm^2)$ and returns an optimum phylogeny with probability at least 8^{-q}. We later show how to derandomize it. In sub-section 3.1, we first provide the complete pseudo-code and describe it. In sub-section 3.2 we prove the correctness of the algorithm. Finally, in sub-section 3.3 we upper bound the running time for the randomized and derandomized algorithms and the probability that the randomized algorithm returns an optimum phylogeny.

3.1 Description

We begin with a high level description of our randomized algorithm. The algorithm iteratively finds a set of edges E that decomposes an optimum phylogeny T_I^* into at most q components. An optimum phylogeny for each component is then constructed using a simple method and returned along with edges E as an optimum phylogeny for I.

We can alternatively think of the algorithm as a recursive, divide and conquer procedure. Each recursive call to the algorithm attempts to reconstruct an optimum phylogeny for an input matrix M. The algorithm identifies a character c s.t. there exists an optimum phylogeny T_M^* in which c mutates exactly once. Therefore, there is exactly one edge $e \in T_M^*$ for which $c \in \mu(e)$. The algorithm, then guesses the vertices that are adjacent to e as r, p. The matrix M can now be partitioned into matrices $M0$ and $M1$ based on the state at character c. Clearly all the taxa in $M1$ reside on one side of e and all the taxa in $M0$ reside on the other side. The algorithm adds r to $M1$, p to $M0$ and recursively computes the optimum phylogeny for $M0$ and $M1$. An optimum phylogeny for M can be reconstructed as the union of *any* optimum phylogeny for $M0$ and $M1$ along with the edge (r, p). We require at most q recursive calls. When the recursion bottoms out, we use a simple method to solve for the optimum phylogeny.

We describe and analyze the iterative method which flattens the above recursion. This makes the analysis easier. For the sake of simplicity we define the following notations.

- For the set of taxa M, $M(i, s)$ refers to the subset of taxa that contains state s at character i.
- For a phylogeny T and character i that mutates exactly once in T, $T(i, s)$ refers to the maximal subtree of T that contains state s on character i.

The pseudo-code for the above described algorithm is provided in Figure 1. The algorithm performs 'guesses' at Steps 2a and 2c. If all the guesses performed by the algorithm are 'correct' then it returns an optimum phylogeny. Guess at Step 2a is correct i.f.f. there exists $T_{M_j}^*$ where $c(v)$ mutates exactly once. Guess at Step 2c is correct i.f.f. there exists $T_{M_j}^*$ where $c(v)$ mutates exactly once and edge

---

buildNPP(input matrix I)

1. let $L := \{I\}, E := \emptyset$
2. while $|\cup_{M_i \in L} N(M_i)| > q$
   (a) **guess** vertex v from $\cup_{M_i \in L} N(M_i)$, let $v \in N(M_j)$
   (b) let $M0 := M_j(c(v), 0)$ and $M1 := M_j(c(v), 1)$
   (c) **guess** taxa r and p
   (d) add r to $M1$, p to $M0$ and (r, p) to E
   (e) remove M_j from L, add $M0$ and $M1$ to L
3. for each $M_i \in L$ compute an optimum phylogeny T_i
4. return $E \cup (\cup_i T_i)$

Fig. 1. Pseudo-code to solve the BNPP problem. For all $M_i \in L$, $N(M_i)$ is the set of non-isolated vertices in the conflict graph of M_i. Guess at Step 2a is correct i.f.f. there exists $T^*_{M_j}$ where $c(v)$ mutates exactly once. Guess at Step 2c is correct i.f.f. there exists $T^*_{M_j}$ where $c(v)$ mutates exactly once and edge $(r, p) \in T^*_{M_j}$ with $r[c(v)] = 1, p[c(v)] = 0$. Implementation details for Steps 2a, 2c and 3 are provided in Section 3.3.

$(r, p) \in T^*_{M_j}$ with $r[c(v)] = 1, p[c(v)] = 0$. Implementation details for Steps 2a, 2c and 3 are provided in Section 3.3. An example illustrating the reconstruction is provided in Figure 2.

3.2 Correctness

We will now prove the correctness of the pseudo-code under the assumption that all the guesses performed by our algorithm are correct. Specifically, we will show that if $\texttt{penalty}(I) \leq q$ then function $\texttt{buildNPP}$ returns an optimum phylogeny. The following lemma proves the correctness of our algorithm.

Lemma 2. *At any point in execution of the algorithm, an optimum phylogeny for I can be constructed as $E \cup (\cup_i T_i)$, where T_i is any optimum phylogeny for $M_i \in L$.*

Proof. We prove the lemma using induction. The lemma is clearly true at the beginning of the routine when $L = \{I\}, E = \emptyset$. As inductive hypothesis, assume that the above property is true right before an execution of Step 2e. Consider any optimum phylogeny $T^*_{M_j}$ where $c(v)$ mutates exactly once and on the edge (r, p). Phylogeny $T^*_{M_j}$ can be decomposed into $T^*_{M_j}(c(v), 0) \cup T^*_{M_j}(c(v), 1) \cup (r, p)$ with length $l = \texttt{length}(T^*_{M_j}(c(v), 0)) + \texttt{length}(T^*_{M_j}(c(v), 1)) + d(r, p)$. Again, since $c(v)$ mutates exactly once in $T^*_{M_j}$, all the taxa in $M0$ and $M1$ are also in $T^*_{M_j}(c(v), 0)$ and $T^*_{M_j}(c(v), 1)$ respectively. Let T', T'' be *arbitrary* optimum phylogenies for $M0$ and $M1$ respectively. Since $p \in M0$ and $r \in M1$ we know that $T' \cup T'' \cup (r, p)$ is a phylogeny for M_j with cost $\texttt{length}(T') + \texttt{length}(T'') + d(r, p) \leq l$. By the inductive hypothesis we know that an optimum phylogeny for I can be constructed using any optimum phylogeny for M_j. We have now shown that using any optimum phylogeny for $M0$ and $M1$ and adding edge (r, p) we can construct an optimum phylogeny for M_j. Therefore the proof follows by induction.   □

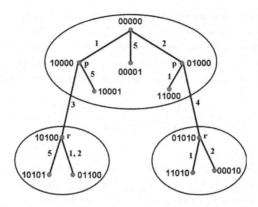

Fig. 2. Example illustrating the reconstruction. Underlying phylogeny is T_I^*; taxa r and p (both could be Steiner) are guessed to create $E = \{(10000, 10100), (01000, 01010)\}$; E induces three components in T_I^*. When all taxa in T_I^* are considered, character 3 conflicts with $1, 2$ and 5 and character 4 conflicts with 1 and 2; two components are perfect (penalty 0) and one has penalty 2; $\texttt{penalty}(I) =_{def} \texttt{penalty}(T_I^*) = 7$.

3.3 Bounds

In this sub-section we bound the probability of correct guesses, analyze the running time and show how to derandomize the algorithm. We perform two guesses at Steps 2a and 2c. Lemmas 3 and 5 bound the probability that all the guesses performed at these Steps are correct throughout the execution of the algorithm.

Lemma 3. *The probability that all guesses performed at Step 2a are correct is at least* 4^{-q}.

Proof. Implementation: The guess at Step 2a is implemented by selecting v uniformly at random from $\cup_i N(M_i)$.

To prove the lemma, we first show that the number of iterations of the while loop (step 2) is at most q. Consider any one iteration of the while loop. Since v is a non-isolated vertex of the conflict graph, $c(v)$ shares all four gametes with some other character c' in some M_j. Therefore, in every optimum phylogeny $T_{M_j}^*$ that mutates $c(v)$ exactly once, there exists a path P starting with edge e_1 and ending with e_3 both mutating c', and containing edge e_2 mutating $c(v)$. Furthermore, the path P contains no other mutations of $c(v)$ or c'. At the end of the current iteration, M_j is replaced with $M0$ and $M1$. Both subtrees of $T_{M_j}^*$ containing $M0$ and $M1$ contain (at least) one mutation of c' each. Therefore, $\texttt{penalty}(M0) + \texttt{penalty}(M1) < \texttt{penalty}(M_j)$. Since $\texttt{penalty}(I) \le q$, there can be at most q iterations of the while loop.

We now bound the probability. Intuitively, if $|\cup_i N(M_i)|$ is very large, then the probability of a correct guess is large, since at most q out of $|\cup_i N(M_i)|$ characters can mutate multiple times in $T_{M_j}^*$. On the other hand if $|\cup_i N(M_i)| = q$ then we terminate the loop. Formally, at each iteration $|\cup_i N(M_i)|$ reduces by at least 1 (guessed vertex v is no longer in $\cup_i N(M_i)$). Therefore, in the worst case (to minimize the probability of correct guesses), we can have q iterations of the loop, with $q+1$ non-isolated vertices in the last iteration and $2q$ in the first iteration. The probability in such a case that all guesses are correct is at least

$$(\frac{q}{2q}) \times (\frac{q-1}{2q-1}) \times \ldots \times (\frac{1}{q+1}) = \frac{1}{\binom{2q}{q}} \geq 2^{-2q}. \qquad \square$$

Buneman Graphs. We now show that r, p can be found efficiently. To prove this we need some tools from the theory of Buneman graphs [21].

Let M be a set of taxa defined by character set C of size m. A Buneman graph F for M is a vertex induced subgraph of the m-cube. Graph F contains vertices v i.f.f. for every pair of characters $i, j \in C$, $(v[i], v[j]) \in G_{i,j}$. Recall that $G_{i,j}$ is the set of gametes (or projection of M on dimensions i, j). Each edge of the Buneman graph is labeled with the character at which the adjacent vertices differ.

Buneman graphs have been defined in previous works on matrices M in which no two characters share exactly two gametes. The definition can be extended to allow such characters while preserving the following lemmas (see expanded version for details). We say that a subgraph F' of F is the same as an edge labeled tree T if F' is a tree and T can be obtained from F' by suppressing degree-two vertices. A phylogeny T is contained in a graph F if there exists an edge-labeled subgraph F' that is the same as the edge labeled (by function μ) phylogeny T. A Buneman graph F for input M has the property that every optimum phylogeny for M is contained in F [21]. From the definition of the Buneman graph F, we know that there exists no vertex $v \in F$ for which $(v[i], v[j]) \notin G_{i,j}$. Therefore, using the above property, we have:

Lemma 4. *In every optimum phylogeny T_M^*, the conflict graph on the set of taxa in T_M^* (Steiner vertices included) is the same as the conflict graph on M.*

Lemma 5. *The probability that all guesses performed at Step 2c are correct is at least 2^{-q}.*

Proof. Implementation: We first show how to perform the guess efficiently. For every character i, we perform the following steps in order.

1. if all taxa in $M0$ contain the same state s in i, then fix $r[i] = s$
2. if all taxa in $M1$ contain the same state s in i, then fix $r[i] = s$
3. if $r[i]$ is unfixed then guess $r[i]$ uniformly at random from $\{0,1\}$

Assuming that the guess at Step 2a (Figure 1) is correct, we know that there exists an optimum phylogeny $T_{M_j}^*$ on M_j where $c(v)$ mutates exactly once. Let $e \in T_{M_j}^*$ s.t. $c(v) \in \mu(e)$. Let r' be an end point of e s.t. $r'[c(v)] = 1$ and p' be the other end point. If the first two conditions hold with the same state s, then character i does not mutate in M_j. In such a case we know that $r'[i] = s$, since $T^*_{M_j}$ is optimal and the above method ensures that $r[i] = s$. Notice that if both conditions are satisfied simultaneously with different values of s then i and $c(v)$ share exactly two gametes in M_j and therefore $i, c(v) \in \mu(e)$. Hence, $r'[i] = r[i]$. We now consider the remaining cases when exactly one of the above conditions hold. We show that if $r[i]$ is fixed to s then $r'[i] = s$. Note that in such a case at least one of $M0, M1$ contain both the states on i and $i, c(v)$ share at least 3 gametes in M_j. The proof can be split into two symmetric cases based on whether r is fixed on condition 1 or 2. One case is presented below:

Taxon $r[i]$ is fixed based on condition 1: In this case, all the taxa in $M0$ contain the same state s on i. Therefore, the taxa in $M1$ should contain both states on i. Hence i mutates in $T^*_{M_j}(c(v), 1)$. For the sake of contradiction, assume that $r'[i] \neq s$. If $i \notin \mu(e)$ then $p'[i] \neq s$. However all the taxa in $M0$ contain state s. This implies that i mutates in $T^*_{M_j}(c(v), 0)$ as well. Therefore i and $c(v)$ share all four gametes on $T^*_{M_j}$. However i and $c(v)$ share at most 3 gametes in M_j - one in $M0$ and at most two in $M1$. This leads to a contradiction to Lemma 4. Once r is guessed correctly, p can be computed since it is is identical to r in all characters except $c(v)$ and those that share two gametes with $c(v)$ in M_j. We make a note here that we are assuming that e does not mutate any character that does not share two gametes with $c(v)$ in M_j. This creates a small problem that although the length of the tree constructed is optimal, r and p could be degree-two Steiner vertices. If after constructing the optimum phylogenies for $M0$ and $M1$, we realize that this is the case, then we simply add the mutation adjacent to r and p to the edge (r, p) and return the resulting phylogeny where both r and p are not degree-two Steiner vertices.

The above implementation therefore requires only guessing states corresponding to the remaining unfixed characters of r. If a character i violates the first two conditions, then i mutates once in $T^*_{M_j}(i, 0)$ and once in $T^*_{M_j}(i, 1)$. If $r[i]$ has not been fixed, then we can associate a pair of mutations of the same character i with it. At the end of the current iteration M_j is replaced with $M0$ and $M1$ and each contains exactly one of the two associated mutations. Therefore if q' characters are unfixed then $\texttt{penalty}(M0) + \texttt{penalty}(M1) \leq \texttt{penalty}(M_j) - q'$. Since $\texttt{penalty}(I) \leq q$, throughout the execution of the algorithm there are q unfixed states. Therefore the probability of all the guesses being correct is 2^{-q}. □

This completes our analysis for upper bounding the probability that the algorithm returns an optimum phylogeny. We now analyze the running time. We use the following lemma to show that we can efficiently construct optimum phylogenies at Step 3 in the pseudo-code.

Lemma 6. *For a set of taxa M, if the number of non-isolated vertices of the associated conflict graph is t, then an optimum phylogeny T^*_M can be constructed in time $O(3^s 6^t + nm^2)$, where $s = \texttt{penalty}(M)$.*

Proof. We use the approach described by Gusfield and Bansal (see Section 7 of [14]) that relies on the Decomposition Optimality Theorem for recurrent mutations. We first construct the conflict graph and identify the non-trivial connected components of it in time $O(nm^2)$. Let κ_i be the set of characters associated with component i. We compute the Steiner minimum tree T_i for character set κ_i. The remaining conflict-free characters in $C \setminus \cup_i \kappa_i$ can be added by contracting each T_i to vertices and solving the perfect phylogeny problem using Gusfield's linear time algorithm [12].

Since $\texttt{penalty}(M) = s$, there are at most $s + t + 1$ distinct bit strings defined over character set $\cup_i \kappa_i$. The Steiner space is bounded by 2^t, since $|\cup_i \kappa_i| = t$. Using the Dreyfus-Wagner recursion [19] the total run-time for solving all Steiner tree instances is $O(3^{s+t}2^t)$.   □

Lemma 7. *The algorithm described solves the BNPP problem in time $O(18^q + qnm^2)$ with probability at least 8^{-q}.*

Proof. For a set of taxa $M_i \in L$ (Step 3, Figure 1), using Lemma 6 an optimum phylogeny can be constructed in time $O(3^{s_i}6^{t_i} + nm^2)$ where $s_i = \texttt{penalty}(M_i)$ and t_i is the number of non-isolated vertices in the conflict graph of M_i. We know that $\sum_i s_i \leq q$ (since $\texttt{penalty}(I) \leq q$) and $\sum_i t_i \leq q$ (stopping condition of the while loop). Therefore, the total time to reconstruct optimum phylogenies for all $M_i \in L$ is bounded by $O(18^q + qnm^2)$. The running time for the while loop is bounded by $O(qnm^2)$. Therefore the total running time of the algorithm is $O(18^q + qnm^2)$. Combining Lemmas 3 and 5, the total probability that all guesses performed by the algorithm is correct is at least 8^{-q}.   □

Lemma 8. *The algorithm described above can be derandomized to run in time $O(72^q + 8^q nm^2)$.*

Proof. It is easy to see that Step 2c can be derandomized by exploring all possible states for the unfixed characters. Since there are at most q unfixed characters throughout the execution, there are 2^q possibilities for the states.

However, Step 2a cannot be derandomized naively. We use the technique of bounded search tree [5] to derandomize it efficiently. We select an arbitrary vertex v from $\cup_i N(M_i)$. We explore both the possibilities on whether v mutates once or multiple times. We can associate a search (binary) tree with the execution of the algorithm, where each node of the tree represents a selection v from $\cup_i N(M_i)$. One child edge represents the execution of the algorithm assuming v mutates once and the other assuming v mutates multiple times. In the execution where v mutates multiple times, we select a different vertex from $\cup_i N(M_i)$ and again explore both paths. The height of this search tree can be bounded by $2q$ because at most q characters can mutate multiple times. The path of height $2q$ in the search tree is an interleaving of q characters that mutate once and q characters that mutate multiple times. Therefore, the size of the search tree is bounded by 4^q.

Combining the two results, the algorithm can be derandomized by solving at most 8^q different instances of Step 3 while traversing the while loop 8^q times for a total running time of $O(144^q + 8^q nm^2)$. This is, however, an over-estimate. Consider any iteration of the while loop when M_j is replaced with $M0$ and $M1$. If a state in character c is unfixed and therefore guessed, we know that there are two associated mutations of character c in both $M0$ and $M1$. Therefore at iteration i, if q_i' states are unfixed, then $\texttt{penalty}(M0) + \texttt{penalty}(M1) \leq \texttt{penalty}(M_j) - q_i'$. At the end of the iteration we can reduce the value of q used in Step 2 by q_i', since the penalty has reduced by q_i'. Intuitively this implies that if we perform a total of q' guesses (or enumerations) at Step 2c, then at Step 3 we only need to solve Steiner trees on $q - q'$ characters. The additional cost $2^{q'}$ that we incur results in reducing the running time of Step 3 to $O(18^{q-q'} + qnm^2)$. Therefore the total running time is $O(72^q + 8^q nm^2)$. □

4   Discussion and Conclusions

Discussion: If all Steiner tree problem instances on the q-cube are solved in a pre-processing step, then our running time just depends on the number of iterations of the while loop, which is $O(8^q nm^2)$. Such pre-processing would be impossible to perform with previous methods. Alternate algorithms for solving Steiner trees may be faster in practice as well.

In Lemma 8, we showed that the guesses performed at Step 2c do not affect the overall running time. We can also establish a trade-off along similar lines for Step 2a that can reduce the theoretical run-time bounds. Details of such trade-offs will be analyzed in the expanded version.

Conclusions: We have presented an algorithm to solve the BNPP problem that is theoretically superior to existing methods. In an empirical evaluation [20], the prior algorithm reconstructed optimum phylogenies for values of q up to 7. Our algorithm should solve for larger values of q since it is clearly expected to out-perform prior methods and its own worst case guarantees. The algorithm is intuitive and simple and hence is one of the few theoretically sound phylogenetic tree reconstruction algorithms that is also expected to be practical.

References

1. R. Agarwala and D. Fernandez-Baca. A Polynomial-Time Algorithm for the Perfect Phylogeny Problem when the Number of Character States is Fixed. In *SIAM Journal on Computing*, 23 (1994).
2. H. Bodlaender, M. Fellows and T. Warnow. Two Strikes Against Perfect Phylogeny. In proc *International Colloquium on Automata, Languages and Programming* (1992).
3. H. Bodlaender, M. Fellows, M. Hallett, H. Wareham and T. Warnow. The Hardness of Perfect Phylogeny, Feasible Register Assignment and Other Problems on Thin Colored Graphs. In *Theoretical Computer Science* (2000).
4. M. Bonet, M. Steel, T. Warnow and S. Yooseph. Better Methods for Solving Parsimony and Compatibility. In *Journal of Computational Biology*, 5(3) (1992).

5. R.G. Downey and M. R. Fellows. Parameterized Complexity. *Monographs in Computer Science* (1999).
6. W. H. Day and D. Sankoff. Computational Complexity of Inferring Phylogenies by Compatibility. *Systematic Zoology* (1986).
7. P. Damaschke. Parameterized Enumeration, Transversals, and Imperfect Phylogeny Reconstruction. In proc *International Workshop on Parameterized and Exact Computation* (2004).
8. E. Eskin, E. Halperin and R. M. Karp. Efficient Reconstruction of Haplotype Structure via Perfect Phylogeny. In *Journal of Bioinformatics and Computational Biology* (2003).
9. D. Fernandez-Baca and J. Lagergren. A Polynomial-Time Algorithm for Near-Perfect Phylogeny. In *SIAM Journal on Computing*, 32 (2003).
10. L. R. Foulds and R. L. Graham. The Steiner problem in Phylogeny is NP-complete. In *Advances in Applied Mathematics* (3) (1982).
11. G. Ganapathy, V. Ramachandran and T. Warnow. Better Hill-Climbing Searches for Parsimony. In *Workshop on Algorithms in Bioinformatics* (2003).
12. D. Gusfield. Efficient Algorithms for Inferring Evolutionary Trees. In *Networks*, 21 (1991).
13. D. Gusfield. Algorithms on Strings, Trees and Sequences. *Cambridge University Press* (1999).
14. D. Gusfield and V. Bansal. A Fundamental Decomposition Theory for Phylogenetic Networks and Incompatible Characters. In proc *Research in Computational Molecular Biology* (2005).
15. D. Gusfield, S. Eddhu and C. Langley. Efficient Reconstruction of Phylogenetic Networks with Constrained Recombination. In Proc *IEEE Computer Society Bioinformatics Conference* (2003).
16. D. A. Hinds, L. L. Stuve, G. B. Nilsen, E. Halperin, E. Eskin, D. G. Ballinger, K. A. Frazer, D. R. Cox. Whole Genome Patterns of Common DNA Variation in Three Human Populations. www.perlegen.com. In *Science* (2005).
17. The International HapMap Consortium. The International HapMap Project. www.hapmap.org. *Nature* 426 (2003).
18. S. Kannan and T. Warnow. A Fast Algorithm for the Computation and Enumeration of Perfect Phylogenies. In *SIAM Journal on Computing*, 26 (1997).
19. H. J. Promel and A. Steger. The Steiner Tree Problem: A Tour Through Graphs Algorithms and Complexity. *Vieweg Verlag* (2002).
20. S. Sridhar, K. Dhamdhere, G. E. Blelloch, E. Halperin, R. Ravi and R. Schwartz. Simple Reconstruction of Binary Near-Perfect Phylogenetic Trees. In proc *International Workshop on Bioinformatics Research and Applications* (2006).
21. C. Semple and M. Steel. Phylogenetics. *Oxford University Press* (2003).
22. M. A. Steel. The Complexity of Reconstructing Trees from Qualitative Characters and Subtrees. In *J. Classification*, 9 (1992).

Length-Bounded Cuts and Flows[*]

Georg Baier[1,**], Thomas Erlebach[2], Alexander Hall[3], Ekkehard Köhler[1],
Heiko Schilling[1], and Martin Skutella[4]

[1] Institute of Mathematics, TU Berlin, Germany
[2] Department of Computer Science, U Leicester, England
[3] Institute of TCS, ETH Zurich, Switzerland
[4] Department of Mathematics, U Dortmund, Germany

Abstract. An L-length-bounded cut in a graph G with source s, and
sink t is a cut that destroys all s-t-paths of length at most L. An L-
length-bounded flow is a flow in which only flow paths of length at most
L are used. We show that the minimum length-bounded cut problem
in graphs with unit edge lengths is $\mathcal{NP}$-hard to approximate within a
factor of at least 1.1377 for $L \geq 5$ in the case of node-cuts and for $L \geq 4$
in the case of edge-cuts. We also give approximation algorithms of ratio
$\min\{L, n/L\}$ in the node case and $\min\{L, n^2/L^2, \sqrt{m}\}$ in the edge case,
where n denotes the number of nodes and m denotes the number of edges.
We discuss the integrality gaps of the LP relaxations of length-bounded
flow and cut problems, analyze the structure of optimal solutions, and
present further complexity results for special cases.

1 Introduction

In a classical article Menger [1], shows a strong relation between cuts and systems
of disjoint paths: let G be a graph and s, t two nodes of G, then the maximum
number of edge-/node-disjoint s-t-paths equals the minimum size of an s-t-edge-
/node-cut (Menger's Theorem); see also Dantzig and Fulkerson [2] and Kotzig [3].
Ford and Fulkerson [4] and Elias, Feinstein, and Shannon [5] generalized the
theorem of Menger to flows in graphs with capacities on the arcs and provided
algorithms to find an s-t-flow and an s-t-cut of the same value.

Lovász, Neumann Lara, and Plummer [6] consider the maximum length-
bounded node-disjoint s-t-paths problem. For length-bounds 2, 3, and 4 a relation holds that is analogous to Menger's theorem, but with a new suitable cut
definition. For length-bounds greater than 4, they give upper and lower bounds
for the gap between the maximum number of length-bounded node-disjoint paths
and the minimum cardinality of a cut. Furthermore, they provide examples showing that some of the bounds are tight. The results were extended independently
to edge-disjoint paths by Exoo [7] and Niepel and Safaríková [8].

[*] This work was partly supported by the Federal Ministry of Education and Re-
search (BMBF grant 03-MOM4B1), by the European Commission - Fet Open project
DELIS IST-001907 (SBF grant 03.0378-1), and by the German Research Foundation
(DFG grants MO 446/5-2 and SK 58/5-3).
[**] Since 12/04 at Corporate Technology, Information & Communication, Siemens AG.

M. Bugliesi et al. (Eds.): ICALP 2006, Part I, LNCS 4051, pp. 679–690, 2006.
© Springer-Verlag Berlin Heidelberg 2006

According to Bondy and Murty [9], Lovász conjectured that there is a constant C such that the size of a minimum L-length-bounded s-t-node-cut, i.e., a minimum node-set disjoint to $\{s, t\}$ which hits each L-length-bounded s-t-path, is at most a factor of $C \cdot \sqrt{L}$ larger than the cardinality of a maximum system of node-disjoint s-t-paths of length at most L. Exoo and Boyles [10] disprove this conjecture. They construct for each length-bound $L > 0$ a graph and a node pair s, t, such that the minimum L-length-bounded s-t-node-cut has size greater than $C \cdot L$ times the maximum number of node-disjoint s-t-paths of length at most L; the constant C is roughly $1/4$.

Itai, Perl, and Shiloach [11] give efficient algorithms to find the maximum number of node-/edge-disjoint s-t-paths with at most 2 or 3 edges; the node-disjoint case is also solved for length-bound 4. On the complexity side they show that the node- and edge-disjoint length-bounded s-t-paths problem is $\mathcal{NP}$-complete for length-bounds greater than 4. Instead of fixing the path length, one can fix the number of paths and look for the minimal value bounding all path lengths. Again both the node- and edge-disjoint version is $\mathcal{NP}$-complete for two paths already.

Guruswami et al. [12] show that the edge-disjoint length-bounded s-t-paths problem is MAX SNP-hard even in undirected networks, and they give an $\mathcal{O}(\sqrt{m})$-approximation algorithm for it. For directed networks, they can show that the problem is hard to approximate within a factor $n^{\frac{1}{2}-\epsilon}$, for any $\epsilon > 0$.

For fractional length-bounded multi-commodity flows in graphs with edge-capacities and edge lengths Baier [13] gives a fully polynomial time approximation scheme (FPTAS). This FPTAS also yields a polynomial time algorithm for fractional length-bounded multi-commodity flows and fractional length-bounded edge-(multi-)cuts in unit-length graphs.

Mahjoub and McCormick [14] present a polynomial algorithm for the 3-length-bounded edge-cut in undirected graphs. Furthermore, they show that the fractional versions of the length-bounded flow- and cut problem are polynomial even if L is part of the input, but that the integral versions are strongly $\mathcal{NP}$-hard even if L is fixed.

Length-bounded path problems arise naturally in a variety of real world optimization problems and therefore many heuristics for finding large systems of length-bounded paths have been developed, see e.g. [15,16,17,18].

Our Contribution. We present various results concerning the complexity and approximability of length-bounded cut and flow problems. After the preliminaries, in Section 3, we show that the minimum length-bounded cut problem in graphs with unit edge-lengths is $\mathcal{NP}$-hard to approximate within a factor of at least 1.1377 for $L \geq 5$ in the case of node-cuts and for $L \geq 4$ in the case of edge-cuts; see Table 1 for an overview of known and new complexity results. We also give approximation algorithms of ratio $\min\{L, n/L\}$ in the node case and $\min\{L, n^2/L^2, \sqrt{m}\}$ in the edge case. For classes of graphs such as constant degree expanders, hypercubes, and butterflies, we state an $\mathcal{O}(\log n)$-approximation algorithm pointed out by [19]. Furthermore, we give instances for which the integrality gap of the LP relaxation is $\Omega(\sqrt{n})$. Section 4 discusses the maximum

Table 1. Known and new (bold type) complexity results; $\varepsilon \in \mathbb{R}^+$ and $c \in \mathbb{N}$ are constants, ε can be arbitrarily small

L	node-cut	edge-cut
1	—	poly.
2	poly.	poly.
3	poly.	poly. [14] (undirected)
4	poly. [6] (undirected)	**inapprox. within 1.1377** (directed & undirected)
$5 \ldots \lfloor n^{1-\varepsilon} \rfloor$	**inapprox. within 1.1377** (directed & undirected)	**inapprox. within 1.1377** (directed & undirected)
$n - c$	**poly.** (directed & undirected)	

length-bounded flow problem. For series-parallel graphs with unit edge lengths and unit edge-capacities, we proof a lower bound of $\Omega(\sqrt{n})$ on the integrality gap of the LP formulation. Furthermore, we show that edge- and path-flows are not polynomially equivalent for length-bounded flows: there is no polynomial algorithm to transform an edge-flow which is known to correspond to a length-bounded path-flow into a length-bounded path-flow. We analyze the structure of optimal solutions and give instances where each maximum flow ships a large percentage of the flow along paths with an arbitrarily small flow value.

2 Preliminaries

We consider (directed or undirected) graphs $G = (V, E)$ with node set $V = V(G)$ and edge set $E = E(G)$. The number of nodes are denoted by n and the number of edges are denoted by m. A graph may contain *multi-edges*, i.e. parallel edges, in which case the graph will be called a *multi-graph*. Sometimes, we call an edge *simple* to distinguish it clearly from multi-edges. The graph G possesses two independent weights, an edge-capacity function $u : E \to \mathbb{Q}_{>0}$ and an edge-length function $d : E \to \mathbb{Q}_{\geq 0}$. If not stated otherwise, we assume unit-lengths and capacities.

Length-Bounded Cuts. Let $s, t \in V$ be two distinct nodes. We call a subset of edges C_e of G an *s-t-edge-cut*, if no path remains from s to t in $G \setminus C_e$. The *value* (or *capacity*) of C_e is the number of edges in C_e (or the total capacity of edges in C_e, if edge-capacities are not unit). Similarly, a node set C_n of G which separates s and t (and contains neither s nor t) is defined as an *s-t-node-cut*; its *value* is the number of nodes in C_n.

Let $\mathcal{P}_{s,t}(L)$ denote the set of all s-t-paths with length at most L. We call a subset of edges C_e^L of G an *L-length-bounded s-t-edge-cut*, if the nodes s and t have a distance greater than L in $G \setminus C_e^L$. This means that C_e^L must hit every path in $\mathcal{P}_{s,t}(L)$. Similarly, a subset C_n^L of the node set of G is called *L-length-bounded s-t-node-cut* if it destroys all paths in $\mathcal{P}_{s,t}(L)$. All of our cuts are s-t-cuts and therefore we will often omit the s-t-prefix. If the type of a cut is clear from the context, we will also omit the superscript L of C as well as the indices e and n of C. The *value* (or *capacity*) of a length-bounded cut is defined as in the standard cut case. In the *Minimum Length-Bounded Cut* problem we are looking for a length-bounded cut of minimum value.

In the linear programming relaxation of the minimum length-bounded edge-cut problem one has to assign to each edge $e \in E$ a dual length ℓ_e such that the dual length of a shortest s-t-path from $\mathcal{P}_{s,t}(L)$ is at least 1 (the LP relaxation for node-cuts is analogous):

$$\min \sum_{e \in E} u_e \ell_e \quad \text{s.t.} \quad \sum_{e \in P} \ell_e \geq 1 \quad (P \in \mathcal{P}_{s,t}(L)), \quad \ell_e \geq 0 \quad (e \in E). \quad (1)$$

An integral solution to this linear program corresponds to a length-bounded s-t-cut, and vice versa. In particular, the minimum length-bounded s-t-cut value and the value of a minimum integral solution are equal. We will refer to feasible solutions of (1) as *fractional cuts* since only a fraction of an edge may contribute to the cut.

Length-Bounded Flows. Length-bounded flows are flows along paths such that the length of every path is bounded. More precisely, an *L-length-bounded s-t-flow* is a function $f : \mathcal{P}_{s,t}(L) \to \mathbb{R}_{\geq 0}$ assigning a flow value f_P to each s-t-path P in G of length at most L. The sum $\sum_{P \in \mathcal{P}_{s,t}(L)} f_P$ is called the *s-t-flow value* of f. The flow f is *feasible*, if edge-capacities are obeyed, i.e., for each edge $e \in E$ the sum of the flow values of paths containing this edge must be bounded by its capacity u_e. Since all our flows are s-t-flows, we will often omit the s-t-prefix.

A natural optimization objective is to find a feasible length-bounded s-t-flow of maximum value. We can formulate this problem as a linear program:

$$\max \sum_{P \in \mathcal{P}_{s,t}(L)} f_P \quad \text{s.t.} \quad \sum_{P : e \in P} f_P \leq u_e \quad (e \in E), \quad f_P \geq 0 \quad (P \in \mathcal{P}_{s,t}(L)). \quad (2)$$

We will refer to feasible solutions of this linear program as *path-flows*. Note that the dual of (2) is the linear program (1) for the minimum length-bounded cut problem. One way to prove the maximum-flow minimum-cut equality for standard flows is to apply duality theory of linear programming. In the case of multiple commodities, a source- and sink-node pair (s_i, t_i) and a length-bound $L_i \geq 0$ is given for each commodity $i = 1, \ldots, k$. An $(L_1, \ldots, L_k)$-*length-bounded multicommodity flow* f is a set of L_i-length-bounded s_i-t_i-flows f_i, for $i = 1, \ldots, k$.

3   Length-Bounded Cuts

It follows from linear programming duality that the maximum fractional length-bounded flow value equals the minimum fractional length-bounded cut value. In the case of standard flows, this equality holds for (integral) cuts as well. In the presence of a length-bound, the maximum flow value and the minimum cut value may be different. This is an immediate consequence of the integrality gap that we state in the following theorem.

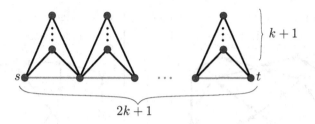

Fig. 1. Example of a large integrality gap of the linear program (1) of the minimum length-bounded cut. The straight s-t-path (*in gray*) contains $2k+1$ edges. Each of these edges is accompanied by $k+1$ parallel paths of length 2.

Theorem 1. *For (un-)directed series-parallel graphs the ratio of the minimum integral length-bounded edge-/node-cut value to the minimum fractional one can be of order $\Omega(\sqrt{n})$. In particular, the ratio of the minimum length-bounded edge-/node-cut size to the maximum number of length-bounded edge-disjoint paths can be of order $\Omega(\sqrt{n})$.*

Proof (sketch). The class of graphs depicted in Fig. 1 for $L = 3k + 1$ have a fractional length-bounded edge-cut value less than 2 but an integral length-bounded edge-cut value $k + 1 \in \theta(\sqrt{n})$. The result for node-cuts follows by considering the corresponding line graph. □

3.1 Complexity and Polynomially Solvable Cases

We present a simple polynomial time algorithm for length-bounded node-cuts with $L = n - c$, where $c \in \mathbb{N}$ is an arbitrary constant.

Theorem 2. *If $c \in \mathbb{N}$ is constant and $L = n-c$, then a minimum length-bounded node-cut can be computed in polynomial time in (un-)directed graphs.*

Proof. Enumerate all $V' \subseteq V$ with $|V'| \leq c$ and return the smallest V' which is a length-bounded node-cut, if there is any. Otherwise, any length-bounded node-cut V' contains at least $c + 1$ nodes so that the longest remaining s-t-path has length at most $n - c - 1$ and therefore V' actually cuts all s-t-paths. Thus, returning a standard minimum node-cut suffices. □

Note that Theorem 2 does not carry over to the edge version of the problem, since by removing c edges one cannot guarantee that a standard cut suffices.

Theorem 3. *For any $\varepsilon > 0$ and $L \in \{5, \ldots, \lfloor n^{1-\varepsilon} \rfloor\}$, it is $\mathcal{NP}$-hard to approximate the minimum length-bounded node-cut in (un-)directed graphs within a factor of 1.1377.*

Proof. We first look at the case $L = 5$ and give a reduction from the well known Vertex Cover problem which has been shown to be $\mathcal{NP}$-hard to approximate within a factor ≈ 1.3606 [20]. Given a Vertex Cover instance G_{vc} with

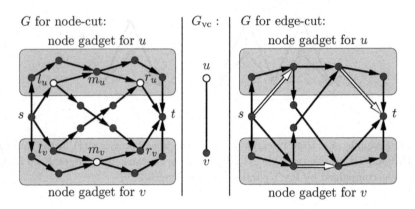

Fig. 2. Gadgets for the reduction of VERTEX COVER to length-bounded node-cut (*left*) and length-bounded edge-cut (*right*), respectively. Both correspond to two connected nodes u, v of the given VERTEX COVER instance, shown in the *middle*. The *highlighted nodes (edges)* are in the cut / vertex cover.

$n_{\mathrm{vc}} = |V_{\mathrm{vc}}|$ nodes, we construct a length-bounded node-cut instance $G = (V, E)$ as follows: start with $V = \{s, t\}$ and no edges. For each node $v \in V_{\mathrm{vc}}$ we add a *node gadget* to G consisting of seven nodes which are interconnected with s, t and themselves as shown in Fig. 2 (left) – the nodes in the bottom half surrounded by a gray box. For each edge $\{u, v\} \in E_{\mathrm{vc}}$ we add an *edge gadget* consisting of four nodes and six edges connecting them to the node gadgets corresponding to u and v as shown in Fig. 2 (left).

Lemma 1. *From a vertex cover V'_{vc} in G_{vc} of size x one can always construct a node-cut V' in G of size $n_{\mathrm{vc}} + x$ and vice versa, for $x < n_{\mathrm{vc}}$.*

We only deal with the easy direction "$\Rightarrow$" in Lemma 1 and omit all further details due to space limitations. Let $V'_{\mathrm{vc}} \subseteq V_{\mathrm{vc}}$ be a vertex cover with $|V'_{\mathrm{vc}}| = x$. For each node $v \in V'_{\mathrm{vc}}$ we add l_v and r_v to our cut $V' \subseteq V$ and for each node $u \in V_{\mathrm{vc}} \setminus V'_{\mathrm{vc}}$ we add m_u to V' (see Fig. 2 for an example). Note that $|V'| = n_{\mathrm{vc}} + x$ and that no path of length at most 5 remains after removing V' from G.

The proof of Theorem 1.1 in [20] gives the following gap. There are graphs G_{vc} for which it is $\mathcal{NP}$-hard to distinguish between two cases: the case where a vertex cover of size $n_{\mathrm{vc}} \cdot (1 - p + \varepsilon')$ exists, and the case where any vertex cover has size at least $n_{\mathrm{vc}} \cdot (1 - 4p^3 + 3p^4 - \varepsilon')$, for any $\varepsilon' \in \mathbb{R}^+$ and $p = (3 - \sqrt{5})/2$. If we plug this into the result of Lemma 1, we have shown that the length-bounded node-cut is hard to approximate within a factor (there is an $\varepsilon' \in \mathbb{R}^+$ for which the inequality holds): $(n_{\mathrm{vc}} + n_{\mathrm{vc}} \cdot (1 - 4p^3 + 3p^4 - \varepsilon'))/(n_{\mathrm{vc}} + n_{\mathrm{vc}} \cdot (1 - p + \varepsilon')) > 1.1377$.

For other values of $L \in \{5, \dots, \lfloor n^{1-\varepsilon} \rfloor\}$, we modify the construction of G as follows: (1) Add a path of length $L - 5$ from a new source node s' to s. Let s' be our new source. (2) Stepwise replace each node on this path after s' and until s (inclusive) by a group of $c \cdot n_{\mathrm{vc}}$ nodes, for some constant c. For each of these groups connect all new nodes with all neighbors of the replaced node. We omit

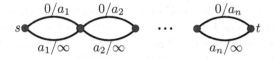

Fig. 3. Reduction of 2-PARTITION to the length-bounded cut problem. The *labels* denote length/capacity.

all further details. To see that the reduction also works for undirected graphs, observe that by removing the edge directions in the gadgets, no new undirected paths of length less than L are introduced. □

The proof of the following theorem is similar to the proof of Theorem 3 with the difference that the adapted gadgets given in Fig. 2 (right) are used, which already work for length-bound $L = 4$.

Theorem 4. *For any $\varepsilon > 0$ and $L \in \{4, \ldots, \lfloor n^{1-\varepsilon} \rfloor\}$, it is $\mathcal{NP}$-hard to approximate the length-bounded edge-cut in (un-)directed (simple) graphs within a factor of 1.1377.*

Lemma 2. *For a series-parallel and outer-planar (un-)directed graph with edge-capacities and lengths it is $\mathcal{NP}$-hard to decide whether there is a length-bounded edge-cut of size less than a given value.*

Proof. We give a reduction from 2-PARTITION. Take an arbitrary 2-PARTITION instance $a_1, \ldots, a_k \in \mathbb{N}$ and consider the graph in Fig. 3. Let the length-bound be $L = B - 1$. It is not difficult to see that there is an edge-cut of size at most B if and only if the instance of 2-PARTITION is a yes-instance. □

We will show in Theorem 8 that it is $\mathcal{NP}$-hard to decide whether a fractional length-bounded flow of given flow value exists even if the graph is outer-planar. Since the primal and dual programs have identical optimal objective function values, the same holds for the fractional length-bounded edge-cut problem.

3.2 Approximation Algorithms

If the length-bound L is so large that the system of L-length-bounded s-t-paths contains the set of all s-t-paths, then length-bounded cuts and flows reduce to standard cuts and flows. The maximum-flow minimum-cut equality holds and there are many efficient algorithms to compute minimum cuts and maximum flows exactly. Another extreme case is if the length-bound equals the distance between s and t, denoted by $\mathrm{dist}(s,t)$. Lovász, Neumann Lara, and Plummer [6] show a special version of the following theorem in the context of length-bounded node-disjoint paths.

Theorem 5. *In (un-)directed multi-graphs with edge-capacities and lengths, for $L = \mathrm{dist}(s,t)$ the minimum length-bounded edge-/node-cut and the maximum length-bounded flow problem can be solved efficiently. In particular, the max flow value and the min cut value coincide if $L = \mathrm{dist}(s,t)$.*

The proof of this theorem is based on considering the sub-graph induced by all edges which are contained in at least one shortest s-t-path. For suitable length functions, like unit-edge-lengths, Theorem 5 yields the following approximation result for the minimum length-bounded cut problem.

Corollary 1. *In (un-)directed multi-graphs one can find an $(L+1-\operatorname{dist}(s,t))$-approximation to the minimum L-length-bounded cut.*

Proof (sketch). Repeatedly compute and remove a minimum $\operatorname{dist}(s,t)$-length-bounded cut from the graph until $\operatorname{dist}(s,t) > L$.   □

It can be shown that the given performance ratio bound is tight for the sketched algorithm. The next theorem establishes bounds on the absolute difference between the sizes of standard minimum cuts and length-bounded minimum cuts.

Theorem 6. *Let $G = (V, E)$ be a (un-)directed multi-graph. A minimum node-cut in G is larger than a minimum length-bounded node-cut by at most $\frac{n}{L}$. If G is a simple graph, a minimum edge-cut is larger than a minimum length-bounded edge-cut by at most $\mathcal{O}(\frac{n^2}{L^2})$.*

Proof. The size of a minimum node-cut is equal to the maximum number of node-disjoint s-t-paths by Menger's theorem. Let C_1 be an optimal length-bounded node-cut. We construct a node-cut C of size at most $|C_1| + \frac{n}{L}$. In $G \setminus C_1$, all s-t-paths have length at least $L+1$. Thus, the number of node-disjoint s-t-paths in $G \setminus C_1$ is at most $(n-2)/L \leq n/L$. Therefore, a minimum node-cut C_2 in $G \setminus C_1$ has cardinality at most n/L. Then $C = C_1 \cup C_2$ is a node-cut in G of the desired cardinality. The proof for edge-cuts is similar. It applies a helpful lemma from [21] which states that if a (directed or undirected) simple graph contains k edge-disjoint s-t-paths, the shortest of these has length $\mathcal{O}(n/\sqrt{k})$.   □

One can show that the bound of $\frac{n}{L}$ on the gap between standard and length-bounded node-cuts given in Theorem 6 is tight with a graph consisting of parallel paths of length $L + 1$ except for one of them having length L. Theorem 6 leads to the following corollary.

Corollary 2. *For (un-)directed multi-graphs there exists an $\mathcal{O}(\frac{n}{L})$-approximation algorithm for the minimum length-bounded node-cut problem. For simple graphs (directed or undirected) there exists an $\mathcal{O}(\frac{n^2}{L^2})$-approximation algorithm for the minimum length-bounded edge-cut problem.*

Now we show that there are approximation algorithms with ratio $\mathcal{O}(\sqrt{n})$ for length-bounded node-cuts and with ratio $\mathcal{O}(\sqrt{m})$ for length-bounded edge-cuts.

Theorem 7. *For (un-)directed graphs there exists an $\mathcal{O}(\min\{L, n/L, \sqrt{n}\})$-approximation algorithm for the minimum length-bounded node-cut problem and an $\mathcal{O}(\min\{L, n^2/L^2, \sqrt{m}\})$-approximation algorithm for the minimum length-bounded edge-cut problem.*

Proof. The upper bounds of $\min\{L, n/L\}$ in the node case and $\min\{L, n^2/L^2\}$ in the edge case follow from Corollaries 1 and 2. Furthermore, we have $\min\{L, n/L\} \leq \sqrt{n}$, so the claimed ratio for length-bounded node cuts follows directly. It remains to show that ratio $\mathcal{O}(\sqrt{m})$ can be achieved for length-bounded edge-cuts.

Let OPT denote the size of a smallest length-bounded edge-cut. If $L \leq \sqrt{m}$, we simply apply the algorithm from Corollary 1. If $L > \sqrt{m}$, we repeatedly find an s-t-path of length at most $\lceil \sqrt{m} \rceil$, add all its edges to the cut, and remove these edges from the graph. Let C_1 denote the set of edges added to the cut in this process. Note that $|C_1| \leq \lceil \sqrt{m} \rceil \cdot \text{OPT}$.

If $G \setminus C_1$ does not contain an s-t-path of length at most L, we output C_1. Otherwise, we compute a minimum edge-cut C_2 in $G \setminus C_1$ and output $C_1 \cup C_2$. It suffices to show that $|C_2| \leq \sqrt{m}$. Let V_i denote the set of nodes at distance i from s in $G \setminus C_1$. Note that the distance from s to t in $G \setminus C_1$ is at least $\lceil \sqrt{m} \rceil + 1$. Let E_i be the set of edges in $G \setminus C_1$ with tail in V_i and head in V_{i+1}. Note that E_i is an edge-cut. Let j be such that E_j has minimum cardinality among the sets E_i for $0 \leq i \leq \lceil \sqrt{m} \rceil - 1$. Observe that $|E_j| \leq m/\lceil \sqrt{m} \rceil \leq \sqrt{m}$. □

For a large class of graphs a better approximation ratio is possible [19]: let F be the *flow number* of G, as defined in [22]. By the Shortening Lemma [22] it follows that if L is at least $4 \cdot F$, a standard minimum-cut is an $\mathcal{O}(1)$-approximation for the L-length-bounded cut. By Corollary 1 this gives an $\mathcal{O}(F)$-approximation for arbitrary L. Since $F = \mathcal{O}(\Delta \alpha^{-1} \log n)$, where Δ is the maximum degree and α the expansion, (cf. [22]) we obtain $\mathcal{O}(\log n)$-approximations for classes of graphs such as constant degree expanders, hypercubes, and butterflies.

4 Length-Bounded Flows

4.1 Edge-Based vs. Path-Based Flows: Complexity

When looking at a given length-bounded flow, we can infer from linear programming theory the existence of a corresponding path-decomposition of small size, where all paths fulfil the length-bound.

Proposition 1. *Given a length-bounded (multi-commodity) path-flow in a graph with edge-capacities and lengths, and m edges. There exists a length-bounded (multi-commodity) path-flow with the same length bound and the same flow value per edge and commodity that uses at most m paths for each commodity.*

The proof of Proposition 1 follows from the fact, that the linear program in (2) has only m linear constraints. Thus, the theory of linear programming can be used to show that there is always a path-flow of maximum flow value which has a small size. Nevertheless, linear programming cannot be used to find maximum fractional length-bounded flows efficiently, unless $\mathcal{P} = \mathcal{NP}$.

Theorem 8. *For a single-commodity length-bounded flow problem in an (un-) directed outer-planar graph with edge-capacities and lengths it is $\mathcal{NP}$-complete to decide whether there is a fractional length-bounded flow of given flow value.*

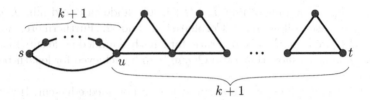

Fig. 4. Graph G_k in which the unique maximum length-bounded flow sends more than one half of the flow along paths with small flow values

Proof (sketch.). The 2-PARTITION problem can be reduced to the integral length-bounded flow problem for a flow of value 2 (similar to the proof of Lemma 2; see also Fig. 3). In a second step one shows that a fractional flow of value 2 in this special graph induces an integral flow of value 2.   □

Finding a maximum length-bounded flow is computationally more difficult than finding a standard maximum flow. Standard flows are usually modeled as edge-flows. Each flow in a path formulation can easily be transformed into an edge-flow. For standard flows the reverse transformation is also possible. If length-bounds are present, one may try to use an edge-flow formulation, too. However, as the following theorem shows, edge- and path-flows are not polynomially equivalent for length-bounded flows. The following result is an immediate consequence of the proof of Theorem 8 and has been shown independently by Correa et al. [23, Corollary 3.4].

Corollary 3. *Unless $\mathcal{P} = \mathcal{NP}$, there is no polynomial algorithm to transform an edge-flow which is known to correspond to a length-bounded path-flow into a length-bounded path-flow, even if the graph is outer-planar.*

4.2   Structure of Optimal Solutions and Integrality Gap

For standard single-commodity flows with integral capacities there is always an integral maximum flow. The situation is completely different in the presence of length constraints. We will not only show that there need not exist an integral maximum flow, but also that there are instances where each maximum flow ships a large percentage of the flow along paths with very small flow values.

Theorem 9. *There are unit-capacity outer-planar graphs of order n such that every maximum length-bounded flow ships more than one half of the total flow along paths with flow values $\mathcal{O}(1/n)$.*

Proof (sketch). Consider the family of unit-capacity and unit-length graphs depicted in Fig. 4. One can show that the unique maximum $(2k+2)$-length-bounded flow contains $k + 1$ paths each with flow value $\frac{1}{k+1}$ and one path with flow value $\frac{k}{k+1}$ (using the sub-path of length $k + 1$ between nodes s and u).   □

For integral length-bounded flows there is a surprising structural difference between path- and edge-flows which is stated in Theorem 10.

Fig. 5. A unit-length graph with an integral edge-flow of value 4 that corresponds to a maximum fractional 6-length-bounded path-flow but which has no integral 6-length-bounded path-decomposition: edge vt has capacity 3, all other edges have unit capacity

Theorem 10. *An integral (maximum) edge-flow corresponding to a (fractional) length-bounded flow in an (un-)directed graph with unit-edge-lengths does not need to have an integral length-bounded path decomposition.*

Proof (sketch). The graph in Fig. 5 has a unique maximum 6-length-bounded flow of value 4. The flow on each edge equals its capacity and is thus integral. But the unique length-bounded path decomposition is half-integral.   □

In [13] it was shown that the length-bounded flow problem can be approximated within arbitrary precision. Having this in mind, it is interesting how far the value of such a fractional solution is away from the maximum integral solution.

Theorem 11. *For unit-capacity graphs with n nodes, the integrality gap of the integer program in (2) can be of order $\Omega(\sqrt{n})$ even for unit-edge-lengths and planar graphs. The length-bound used is of order $\Theta(\sqrt{n})$.*

The proof is based on a unit-capacity graph with n nodes, a maximum integral length-bounded flow of value 1, and a maximum half-integral flow of value $\Omega(\sqrt{n})$. The structure of this graph is a refinement of half a k by k grid. The construction is inspired by Guruswami et al. [12]. We omit all further details in this extended abstract.

The big integrality gap in Theorem 11 is tied to the unit-capacities of the graph used in the proof. Raising the edge-capacities in this graph to 2 brings the integrality gap down to 2. Indeed, the integrality gap is constant for high capacity graphs. The following result is a consequence of the randomized rounding technique of Raghavan and Thompson [24].

Theorem 12. *Consider a graph with minimal edge-capacity at least $c \log n$, for a suitable constant c. Using randomized rounding one can convert any fractional length-bounded flow into an integral length-bounded flow whose value is at most a constant factor smaller (with high probability). In particular, the integrality gap is constant for high capacity graphs.*

Acknowledgement. The authors residing in the lowlands would like to thank the author residing in the highlands for his sedulous commitment and readiness to make sacrifices for the sake of this work.

References

1. Menger, K.: Zur allgemeinen Kurventheorie. Fund. Mathematicae (1927) 96–115
2. Dantzig, G.B., Fulkerson, D.R.: On the max flow min cut theorem of networks. In Kuhn, H.W., Tucker, A.W., eds.: Linear Inequalities and Related Systems. Volume 38 of Annals of Math. Studies. Princeton University Press (1956) 215–221
3. Kotzig, A.: Connectivity and Regular Connectivity of Finite Graphs. PhD thesis, Vysoká Škola Ekonomická, Bratislava, Slovakia (1956)
4. Ford, L.R., Fulkerson, D.R.: Maximal flow through a network. Canadian Journal of Mathematics 9 (1956) 399–404
5. Elias, P., Feinstein, A., Shannon, C.E.: A note on the maximum flow through a network. IRE Transactions on Information Theory 2 (1956)
6. Lovász, L., Neumann Lara, V., Plummer, M.D.: Mengerian theorems for paths of bounded length. Periodica Mathematica Hungarica 9 (1978) 269–276
7. Exoo, G.: On line disjoint paths of bounded length. Discrete Mathematics 44 (1983) 317–318
8. Niepel, L., Safaríková, D.: On a generalization of Menger's Theorem. Acta Mathematica Universitatis Comenianae 42 (1983) 275–284
9. Bondy, J.A., Murty, U.: Graph Theory with Applications. North Holland (1976)
10. Boyles, S.M., Exoo, G.: A counterexample to a conjecture on paths of bounded length. Journal of Graph Theory 6 (1982) 205–209
11. Itai, A., Perl, Y., Shiloach, Y.: The complexity of finding maximum disjoint paths with length constraints. Networks 12 (1982) 277–286
12. Guruswami, V., Khanna, S., Rajaraman, R., Shepherd, B., Yannakakis, M.: Near-optimal hardness results and approximation algorithms for edge-disjoint paths and related problems. In: Proceedings of the Symposium on Theory of Computing, ACM (1999) 19–28
13. Baier, G.: Flows with Path Restrictions. PhD thesis, TU Berlin, Germany (2003)
14. Mahjoub, A.R., McCormick, S.T.: The complexity of max flow and min cut with bounded-length paths. unpublished Manuscript (2003)
15. Perl, Y., Ronen, D.: Heuristics for finding a maximum number of disjoint bounded paths. Networks 14 (1984) 531–544
16. Brandes, U., Neyer, G., Wagner, D.: Edge-disjoint paths in planar graphs with short total length. Technical Report 19, Universität Konstanz (1996)
17. Wagner, D., Weihe, K.: A linear-time algorithm for edge-disjoint paths in planar graphs. Combinatorica 15(1) (1995) 135–150
18. Hsu, D.: On container width and length in graphs, groups, and networks. IEICE Transactions on Fundamentals of Electronics, Communications and Computer Sciences E77-A(4) (1994)
19. Kolman, P.: Personal communication (2006)
20. Dinur, I., Safra, S.: On the hardness of approximating minimum vertex cover. Annals of Mathematics 162(1) (2005) 439–486
21. Galil, Z., Yu, X.: Short length versions of Menger's Theorem. In: Proceedings of the Symposium on Theory of Computing, ACM (1995) 499–508
22. Kolman, P., Scheideler, C.: Improved bounds for the unsplittable flow problem. In: Proceedings of the Symposium on Discrete Algorithms, ACM (2002) 184–193
23. Correa, J.R., Schulz, A.S., Stier Moses, N.E.: Fast, fair, and efficient flows in networks. In: Operations Research, INFORMS (2006) to appear.
24. Raghavan, P., Thompson, C.: Randomized rounding: A technique for provably good algorithms and algorithmic proofs. Combinatorica 7 (1987) 365–374

An Adaptive Spectral Heuristic for Partitioning Random Graphs

Amin Coja-Oghlan

Humboldt-Universität zu Berlin, Institut für Informatik,
Unter den Linden 6, 10099 Berlin, Germany
coja@informatik.hu-berlin.de

Abstract. We study random instances of a general graph partitioning problem: the vertex set of the random input graph G consists of k classes $V_1, \ldots, V_k$, and V_i-V_j-edges are present with probabilities p_{ij} independently. The main result is that with high probability a partition $S_1, \ldots, S_k$ of G that coincides with $V_1, \ldots, V_k$ on a huge subgraph $\mathrm{core}(G)$ can be computed in polynomial time via spectral techniques. The result covers the case of sparse graphs (average degree $O(1)$) as well as the massive case (average degree $\#V(G) - O(1)$). Furthermore, the spectral algorithm is *adaptive* in the sense that it does not require any information about the desired partition beyond the number k of classes.

1 Introduction and Results

1.1 Spectral Techniques for Graph Partitioning

To solve various types of graph partitioning problems, *spectral heuristics* are in common use. Such heuristics represent a given graph by a matrix and compute its eigenvalues and -vectors to solve the combinatorial problem in question. Spectral techniques are used either to deal with "classical" NP-hard problems such as GRAPH COLORING or MAX CUT, or to solve less well defined problems such as recovering a "latent" clustering of the graph. In the present paper we mainly deal with the latter.

Despite their popularity in applications, for most of the known spectral heuristics there are counterexamples known showing that these algorithms perform badly in the worst case. Therefore, to provide a better understanding of such heuristics, quite a few authors have contributed rigorous analyses of their performance on suitable models of *random graphs*. For example, Alon and Kahale [1] invented and analyzed a spectral technique for GRAPH COLORING, Alon, Krivelevich, Sudakov [2] dealt with the MAXIMUM CLIQUE problem, and Boppana [3] and Coja-Oghlan [5] studied MINIMUM BISECTION.

Though the basic ideas of [1,2,3,5] are related, these heuristics are really tailored for the concrete problems (and random graph models) studied in the respective articles. Therefore, McSherry's work [10] on a *general* spectral partitioning heuristic is quite remarkable. McSherry studied a random graph model $G_{n,k}(\psi, \boldsymbol{p})$ that encompasses all the models in [1,2,3,5]. The parameters are the following: n, k are positive integers, and ψ is a map $\{1, \ldots, n\} \rightarrow \{1, \ldots, k\}$. Moreover, $\boldsymbol{p} = (p_{ij})_{1 \leq i,j \leq k}$ is a symmetric $k \times k$ matrix with entries in $[0, 1]$. Now, the vertex set of $G_{n,k}(\psi, \boldsymbol{p})$ is $V = \{1, \ldots, n\}$, and for any two vertices $v, w \in V$ the edge $\{v, w\}$ is present in $G_{n,k}(\psi, \boldsymbol{p})$ with probability $p_{\psi(v)\,\psi(w)}$ independently. Thus, $G_{n,k}(\psi, \boldsymbol{p})$ has a *planted partition* $V_1 = \psi^{-1}(1), \ldots, V_k = \psi^{-1}(k)$. We say that $G_{n,k}(\psi, \boldsymbol{p})$ has a certain property $\mathcal{P}$ *with high probability (w.h.p.)* if the probability that $\mathcal{P}$ holds tends to one as $n \rightarrow \infty$.

McSherry's spectral heuristic recovers the planted partition $V_1, \ldots, V_k$ of a given graph $G = G_{n,k}(\psi, \boldsymbol{p})$ w.h.p. if the parameters $n, k, \boldsymbol{p}, \psi$ satisfy the following condition. Let $\mathcal{E} = (e_{vw})_{v,w \in V}$ be the matrix with entries $e_{vw} = p_{\psi(v)\,\psi(w)}$, let $\mathcal{E}_v = (e_{vw})_{w \in V}$ denote its v-column, and let $\sigma_{\max}^2 = \max_{1 \leq i,j \leq k} p_{ij}(1 - p_{ij})$. Then the assumption is that for all $u, v \in V$ such that $\psi(u) \neq \psi(v)$ we have

$$\|\mathcal{E}_u - \mathcal{E}_v\|^2 \geq c_0 k \cdot \max\left\{\sigma_{\max}^2, \frac{\ln^6 n}{n}\right\} \cdot \left[\frac{n}{\min_{1 \leq i \leq k} \#V_i} + \ln n\right], \qquad (1)$$

where $c_0 > 0$ is a certain constant. In addition to the graph $G = G_{n,k}(\psi, \boldsymbol{p})$, the heuristic requires (suitable lower bounds on) $\min_{u,v:\psi(u) \neq \psi(v)} \|\mathcal{E}_u - \mathcal{E}_v\|$ and $\min_{1 \leq i \leq k} \#V_i$ at the input.

The occurrence of the $\ln n$-terms in (1) implies that the average degree of $G_{n,k}(\psi, \boldsymbol{p})$ must at least be $\ln^3 n$ and at most $n - \ln^3 n$, which is indeed instrumental for the arguments in [10] to go through. The contribution of the present paper is a general partitioning algorithm that also applies to both *sparse graphs* with average degree $O(1)$ and *massive graphs* with average degree $n - O(1)$, or mixtures of both. Indeed, graph partitioning seems to be considerably more difficult in the sparse (or massive) case than for graphs with a "moderate" average degree between polylog(n) and $n -$ polylog(n) (cf. the comments in [1,5]). Also in the present setting the sparse (or massive) case requires new algorithmic ideas, because, as we shall see, there occur considerable fluctuations of the vertex degrees, which cause the spectral methods from [10] to fail. A further novel aspect is that the new algorithm is *adaptive* in the sense that its input *only* consists of the graph $G = G_{n,k}(\psi, \boldsymbol{p})$ and the number k, but no further parameters.

1.2   The Main Result

In this paper we show that a planted partition can be recovered w.h.p. under a different assumption than (1). The new assumption allows for graphs with bounded average degree.

Theorem 1. *There is a polynomial time algorithm* Partition *that satisfies the following. Let k be a number independent of n. Suppose that*

1. *$\mu^* = \max_{1 \leq j \leq k} \sum_{i=1}^{k} \#V_i p_{ij}(1 - p_{ij}) \geq \ln^2(n/\min_j \#V_j)$,*
2. *$\min_{1 \leq i \leq k} \#V_i \geq \ln^{10} n$, and*
3. *for all $u, v \in V$ such that $\psi(u) \neq \psi(v)$ the inequality*

$$\|\mathcal{E}_u - \mathcal{E}_v\|^2 \geq \rho^2 = c_0 \left[\frac{k^3 \mu^*}{n_{\min}} + \ln\left(\mu^* + \frac{n}{n_{\min}} \right) \max_i \sum_{j=1}^{k} p_{ij}(1 - p_{ij}) \right] \quad (2)$$

holds, where $n_{\min} = \min_i \#V_i$ and c_0 is a large enough constant.

Then w.h.p. $G = G_{n,k}(\psi, \boldsymbol{p})$ has an induced subgraph $\mathrm{core}(G)$ on $\geq n - \mu^{*-10}$. $\min_i \#V_i$ vertices that enjoys the following property. On input (G, k) the algorithm outputs a partition $S_1, \ldots, S_k$ of G such that $S_i \cap \mathrm{core}(G) = V_{\tau(i)} \cap \mathrm{core}(G)$ for all $i = 1, \ldots, k$, where τ is a permutation of the indices $1, \ldots, k$.

In the remainder of this section we discuss Theorem 1. First, we explain the conditions (1), (2) and how they relate. Then, we discuss why in the sparse (or massive) case it is *impossible* to recover the planted partition perfectly w.h.p., and explain the subgraph $\mathrm{core}(G)$. Finally, we illustrate Theorem 1 with the examples of MINIMUM BISECTION and GRAPH COLORING.

To explain (2), we let $e(v, V_j)$ signify the number of v-V_j-edges. Moreover, for each $v \in V$ we define a vector $d_v = (d_{vw})_{w \in V}$ by letting $d_{vw} = e(v, V_j)/\#V_j$ for all $w \in V_j$ and $1 \leq j \leq k$. Thus, d_{vw} is the actual edge density between v and the class of w. Then the entries e_{vw} of the matrix $\mathcal{E}$ are just the expectations $e_{vw} = \mathrm{E}(d_{vw})$, so that $\mathrm{E}(\|\mathcal{E}_v - d_v\|^2)$ quantifies the "variance" of $d(v)$. For $v \in V_i$ we can bound this by

$$\mathrm{E}(\|\mathcal{E}_v - d_v\|^2) = \sum_{j=1}^{k} \sum_{w \in V_j} \mathrm{E}\left[(e_{vw} - d_{vw})^2 \right] \leq \frac{\mu^*}{n_{\min}}.$$

Thus, $\mu^*/n_{\min}$ bounds the influence of "random noise" on the vector d_v. Furthermore, if $\psi(v) \neq \psi(w)$, then $\|\mathcal{E}_v - \mathcal{E}_w\|^2$ quantifies how much the planted partition influences $d_v - d_w$. Hence, (2) basically says that Partition can (almost) recover the planted classes $V_1, \ldots, V_k$ if the influence $\|\mathcal{E}_v - \mathcal{E}_w\|^2$ exceeds the bound $\mu^*/n_{\min}$ on the "random noise" by a certain amount.

The two conditions (1) and (2) compare as follows. Since $n\sigma_{\max}^2 \geq \mu^*$, (1) also relates the influence $\|\mathcal{E}_v - \mathcal{E}_w\|^2$ of the planted partition with a bound on the effect of random noise. However, due to the $\ln n$-terms occurring in (1), the condition implies that for each i there is a j such that $\#V_j p_{ij} \geq \ln^3 n$. Thus, if (1) holds, then $G_{n,k}(\psi, \boldsymbol{p})$ has average degree at least $\ln^3 n$ (and $\leq n - \ln^3 n$). By contrast, Theorem 1 also comprises the following three types of graphs.

Sparse graphs. Suppose that $n_{\min} = \Omega(n)$. Then condition (2) allows that the mean $\#V_j e_{vw}$ of $e(v, V_j)$ is $O(1)$ as $n \to \infty$ for all $v \in V_i$, $w \in V_j$, and all $1 \leq i, j \leq k$. In this case the average degree of $G_{n,k}(\psi, \boldsymbol{p})$ is $O(1)$.

Massive graphs. Similarly, it is possible that $e(v, V_j)$ has mean $\#V_j - O(1)$ for all v, j. Then $G_{n,k}(\psi, \boldsymbol{p})$ is a massive graph, i.e., the average degree is $n - O(1)$.

Mixtures of both. The most difficult case algorithmically is a "mixture" of the above two cases: $e(v, V_j)$ has mean either $O(1)$ or $\#V_j - O(1)$ for all v, j; i.e., some of the subgraphs induced on two sets V_i, V_j are sparse, while others are massive.

Furthermore, $n\sigma_{max}^2$ may exceed μ^* significantly if $G_{n,k}(\psi, \boldsymbol{p})$ features a "small" part of medium density (say, $\frac{1}{2}$). In this case (2) can be considerably weaker than (1). Nevertheless, (2) does not strictly improve (1), because if $n\sigma_{max}^2 = O(\mu^*)$, then (2) may be bigger than (1) by a factor of k^2.

Why can't Partition recover the entire planted solution $V_1, \ldots, V_k$ in general? Let us assume that $\#V_j p_{ij} = \Theta(1)$ for all i, j. Then the average degree of $G = G_{n,k}(\psi, \boldsymbol{p})$ is $\Theta(1)$. Moreover, for all v and all i the number $e(v, V_i)$ of v-V_i-edges is asymptotically Poisson with a bounded mean. Therefore, for any constant ζ the probability that $e(v, V_i) = \zeta$ is bounded away from zero. In effect, there occur a linear number of vertices such that d_v deviates significantly from its mean $\mathcal{E}_v$. Indeed, for $\Omega(n)$ vertices $v \in V_i$ the vector d_v will be closer to $\mathcal{E}_w$ for some $w \in V_j \neq V_i$ than to $\mathcal{E}_v$. Consequently, v "looks more like" a vertex in V_j than like a vertex in V_i. Hence, it is simply impossible to recognize that $v \in V_i$, and thus *no* algorithm can recover the partition $V_1, \ldots, V_k$ perfectly w.h.p.

Nonetheless, Theorem 1 states that for on a huge share core(G) we can actually compute the planted solution correctly. The subgraph core(G) basically consists of those vertices for which d_v is close to its mean $\mathcal{E}_v$; core(G) is actually a "canonically" defined subgraph, and not an artefact produced by the algorithm (cf. Section 2 for a precise definition).

In summary, Theorem 1 extends [10] in the following respects.

- The most important point is that Partition can cope with the three types of graphs described above (sparse, massive, and mixed).
- The new algorithm requires only the graph G and the number k of classes at the input, but does not need *any* further information about the desired partition. Hence, the algorithm is "adaptive" in the sense that it finds out "on the fly" what kind of partition it is actually looking for. By comparison, the algorithm as it is described in [10] requires some further information (e.g., a lower bound on $\|\mathcal{E}_v - \mathcal{E}_w\|$ for v, w in distinct classes).
- Partition is deterministic, while the algorithm in [10] is randomized.

Example 2. The following model for MINIMUM BISECTION was considered in [3,5]. Let $k = 2$, and choose $\psi : V \to \{1, 2\}$ such that $\#\psi^{-1}(1) = \#\psi^{-1}(2) = \frac{n}{2}$ at random. Let $\boldsymbol{p} = (p_{ij})_{i,j=1,2}$, where $p_{11} = p_{22} = p' > p_{12} = p_{21} = p$. Then $G_{n,k}(\psi, \boldsymbol{p})$ is a random graph with a planted bisection V_1, V_2. In this case the conditions (1) and (2) can be rephrased in terms of n, p, p' as follows:

- (1) requires that $n(p' - p) \geq c\sqrt{np' \ln n}$ for some constant $c > 0$; this is exactly the regime addressed in [3].
- (2) reads $n(p' - p) \geq c\sqrt{np' \ln(np')}$; this is precisely the assumption needed in [5].

Thus, in the first case we must have $np' \geq c^2 \ln n$, while in the second case it is possible that $np' = O(1)$ as $n \to \infty$. Since for each vertex $v \in V_i$ we have $\mathrm{E}(e(v, V_i) - e(v, V_{3-i})) \sim \frac{n}{2}(p' - p) > 0$, the planted partition V_1, V_2 should be a "good" bisection of G. However, if $np' = O(1)$, then w.h.p. there are $\Omega(n)$ "exceptional" vertices $v \in V_i$ such that $e(v, V_i) < e(v, V_{3-i})$ w.h.p. Hence, on the one hand (V_1, V_2) is no *optimal* solution, and on the other hand `Partition` (or any other algorithm) cannot recognize that $v \in V_i$. Nonetheless, using dynamic programming one can w.h.p. extend the partition S_1, S_2 that `Partition`(G) produces to an *optimal* bisection in polynomial time (cf. [5]) – hence, even though `Partition` does not quite yield the planted solution, its output is rather useful. By contrast, if $np' \geq c^2 \ln n$, then matters are simpler: w.h.p. we have $e(v, V_i) > e(v, V_{3-i})$ for all i, v, and hence (V_1, V_2) is in fact an optimal bisection that can be computed by applying `Partition` directly w.h.p.

Example 3. Alon and Kahale [1] studied the following model of k-colorable graphs. Let $\psi : V \to \{1, \ldots, k\}$ be a random mapping, and let $p_{ij} = p$ if $i \neq j$ and $p_{ii} = 0$ for all i, j. Then $V_1, \ldots, V_k$ is a planted k-coloring of $G = G_{n,k}(\psi, \boldsymbol{p})$. To satisfy (1), we need that $p \geq c(\ln^3 n)/n$ for a constant $c = c(k) > 0$. By comparison, (2) only requires that $p \geq c'/n$ for a constant $c' = c'(k) > 0$. This is the same regime as addressed in [1], where the authors actually point out that the case $np = O(1)$ is more challenging algorithmically than the dense case $np \gg \ln n$. While in the latter situation all vertices $v \in V_i$ have $\sim np/k$ neighbors in all other classes V_j, $j \neq i$ w.h.p., in the case $np = O(1)$ some vertices, e.g., $v \in V_1$ have no neighbors in, e.g., V_2 at all w.h.p. Thus, there is no way to tell whether $v \in V_1$ or $v \in V_2$. Nevertheless, the arguments in [1] show that w.h.p. the partition $S_1, \ldots, S_k$ of core(G) produced by `Partition` can be extended to a k-coloring of the entire graph G in polynomial time.

1.3 Further Related Work

`Partition` generalizes the spectral methods in [1,2,3,5], and in fact the results on GRAPH COLORING, CLIQUE, and MINIMUM BISECTION in these papers can be derived from Theorem 1 by adding just a few problem specific details. In addition, the techniques behind Theorem 1 can also be adapted to rederive the results of Flaxman [8] for random 3-SAT, and Chen and Frieze [4] for hypergraph 2-coloring.

Dasgupta, Hopcroft, and McSherry [6] suggested an even more general model than $G_{n,k}(\psi, \boldsymbol{p})$, namely a random graph with a planted partition featuring a "skewed" degree distribution. This model is very interesting, because it covers, e.g., random "power law" graphs. The main result is that the planted partition can be recovered also in this case w.h.p. under a similar assumption as (1). Thus, also in [6] it is assumed that the graphs are dense enough (average degree $\geq \text{polylog}(n)$).

1.4 Techniques and Outline

The algorithm `Partition` for Theorem 1 generalizes/extends the methods developed by Alon and Kahale [1] and McSherry [10]. The approach in [10] is basically the following. Given $G = G_{n,k}(\psi, \boldsymbol{p})$, the algorithm tries to compute an approximation $\hat{A}$ of the matrix $\mathcal{E}$ using the eigenvalues and eigenvectors of the adjacency matrix $A = A(G)$ of G. This matrix $\hat{A}$ is close to $\mathcal{E}$ in the sense that $\hat{A}_v$ is close to $\mathcal{E}_v$ for most v. Hence, as (1) requires that for vertices v, w

such that $\psi(v) \neq \psi(w)$ the distance $\|\mathcal{E}_v - \mathcal{E}_w\|$ is large, partitioning the vertices according to the columns $\hat{A}_v$ should get us close to $V_1, \ldots, V_k$; in fact, McSherry shows that this yields exactly the planted partition w.h.p. if (1) holds. More precisely, the matrix $\hat{A}$ is a *rank k approximation* of A: let P the projection onto a space generated by k eigenvectors of A corresponding to the k largest eigenvalues in absolute value. Then $\hat{A} = PAP$.

By comparison, `Partition` also approximates $\mathcal{E}$ by a certain matrix $\hat{A}$, but given just the assumption (2) instead of (1) it is no longer feasible to just let $\hat{A}$ be a rank k approximation of A. For example, in the case $\mu^* = O(1)$ fluctuations of the vertex degrees "jumble up" the spectrum of $A(G)$ so that the k largest eigenvalues just correspond to the tails of the degree distribution but not to the planted partition. Therefore, `Partition` needs to "filter" the spectrum. While a related approach was already used in [1], the present general situation is more delicate, because the algorithm *does not know* the expected degrees in advance.

Once `Partition` has obtained a suitable matrix $\hat{A}$, the algorithm tries to produce classes $S_1, \ldots, S_k$ such that $\|\hat{A}_v - \hat{A}_w\|$ is "small" if $v, w \in S_i$ and "large" if $v \in S_i$ and $w \in S_j$ ($i \neq j$). Finally, the algorithm starts a local improvement procedure from $S_1, \ldots, S_k$ that converges to the planted solution $V_1, \ldots, V_k$ on core(G) w.h.p. This procedure is a generalized version of the greedy heuristics used in [1,5].

In the next section we give the precise definition of the subgraph core(G) mentioned in Theorem 1. Then, in Section 3 we present `Partition` in detail.

1.5 Notation

We let $V = \{1, \ldots, n\}$. If $G = (V, E)$ is a graph, then $A(G)$ signifies the adjacency matrix. Moreover, for $X, Y \subset V$ we let $e(X, Y) = e_G(X, Y)$ denote the number of X-Y-edges, and we set $e(X) = e(X, X)$. Further, $\mu(X, Y)$ signifies the *expected* number of X-Y-edges in $G_{n,k}(\psi, \boldsymbol{p})$. In addition, by $d_G(v)$ we denote the degree of v in G.

If $M = (m_{vw})_{v,w \in V}$ is a matrix and $v \in V$, then $M_v = (m_{vw})_{w \in V}$ is the v-column of M. We let $\|M\| = \max_{\xi: \|\xi\|=1} \|M\xi\|$ denote the norm and $\|M\|_F = \sum_{v \in V} \|M_v\|^2$ the Frobenius norm of M.

2 The Core

The subgraph core(G) basically consists of those vertices $v \in V$ for which the numbers $e(v, V_i)$ of v-V_i-edges do not deviate from their means "too much". More precisely, for any two vertices $v, w \in V$ we set $d(v, w) = e(v, V_{\psi(w)})/\#V_{\psi(w)}$. Then, we define a vector $d(v) = (d(v, w))_{w \in V} \in \mathbf{R}^V$ that contains the actual numbers of $e(v, V_i)$-edges. By comparison, the vector $\mathcal{E}_v$ (i.e., the v-column of the matrix $\mathcal{E}$) represents the *expected* numbers of v-V_i-edges.

CR1. Initially, remove all vertices v such that $\|d(v) - \mathcal{E}_v\| > \varepsilon\rho^2$ from G; that is, set $H = G - \{v \in V : \|d(v) - \mathcal{E}_v\| > \varepsilon\rho^2\}$. (Here ρ^2 is the r.h.s. of (2), and $\varepsilon > 0$ is some small enough constant.)

However, in general the result H of CR1 will *not* be such that $e(v, V_i \cap H)$ approximates $\mu(v, V_i)$ well for all $v \in H$. The reason is that there might occur a few vertices v such that "many" neighbors of v get removed. To deal with this, we decompose G into a "sparse" part G_1 and a "dense" part, which are defined as follows. Let $\Phi = (\Phi_{vw})_{v,w \in V}$ be the matrix with entries

$$\Phi_{vw} = 1 \text{ if } p_{\psi(v)\,\psi(w)} > \tfrac{1}{2}, \text{ and } \Phi_{vw} = 0 \text{ otherwise.} \tag{3}$$

Then $G_1 = (V, \{\{v, w\} \in E : \Phi_{vw} = 0\})$. Moreover, $G_2 = (V, \{\{v, w\} \notin E : \Phi_{vw} = 1\})$. Now, as the *expected* degrees $d_{G_i}(v)$ for $i = 1, 2$ and all $v \in V$ are $\leq 2\mu^*$, step CR2 first removes all vertices v such that $d_{G_i}(v)$ is atypically large. Then, we iteratively remove all vertices that have plenty of neighbors that were removed earlier.

CR2. Remove all vertices $v \in H$ such that $\max_{i=1,2} d_{G_i}(v) > 10\mu^*$ from H. Then, while there is a vertex $v \in H$ such that $\max_{i=1,2} e_{G_i}(v, G - H) > 100$, remove v from H.

The final outcome of the process CR1–CR2 is $\mathrm{core}(G) = H$; specialized versions of the core also played an instrumental role in [1,4,5,8].

Proposition 4. *Suppose that (2) holds. Then w.h.p.* $\mathrm{core}(G_{n,k}(\psi, \boldsymbol{p}))$ *contains at least* $n - n_{\min}\mu^{*-10}$ *vertices.*

3   The Algorithm Partition

3.1   Outline

Throughout, we let $G = G_{n,k}(\psi, \boldsymbol{p})$ *be a random graph whose parameters satisfy (2).* Partition *proceeds in three phases. In the first phase (Steps 1–2), the objective is to compute a matrix* $\hat{A}$ *of rank* $\leq 2k$ *that approximates the expectation* $\mathcal{E}$ *of* $A(G_{n,k}(\psi, \boldsymbol{p}))$ *well; more precisely,* $\|\hat{A} - \mathcal{E}\| \leq Ck\sqrt{\mu^*}$ *for a constant* $C > 0$. *Then, since we are assuming that for vertices* $v, w \in V$ *such that* $\psi(v) \neq \psi(w)$ *the corresponding columns* $\mathcal{E}_v, \mathcal{E}_w$ *are "far apart", at least for most such* v, w *the vectors* $\hat{A}_v, \hat{A}_w$ *will have a large distance as well. Therefore, the second phase uses the columns* $\hat{A}_v$ *to obtain an initial partition* $S_1, \ldots, S_k$ *of* G. *Finally, in the third phase we improve this partition* $S_1, \ldots, S_k$ *locally.*

Algorithm 5. Partition(G, k)
Input: A graph $G = (V, E)$ and an integer k.
Output: A partition $S_1, \ldots, S_k$ of G.

1. Run the procedure Identify(G).
2. If Identify fails, then let $\hat{A}$ be a rank k approximation of A; otherwise let $\varphi = (\varphi_{vw})_{v,w \in V}$ be the output of Identify, and let $\hat{A} = $ Approx(G, φ).
3. Let $(S_1, \ldots, S_k, \xi_1, \ldots, \xi_k) = $ Initial$(\hat{A}, k)$.
4. Let $(T_1, \ldots, T_k) = $ Improve$(G, S_1, \ldots, S_k, \xi_1, \ldots, \xi_k)$.

With respect to the first phase, it is fairly easy to obtain a matrix $\hat{A}$ that approximates $\mathcal{E}$ well if the parameter μ^* is not too small – say, $\mu^* \gg \ln n$: in this case we just let $\hat{A}$ be any rank k approximation of the adjacency matrix. Then $\|\hat{A} - \mathcal{E}\| \leq (2 + o(1))k\sqrt{\mu^*}$ w.h.p.

By contrast, if μ^* is small (say, $\mu^* = O(1)$ as $n \to \infty$) then G consists of "extremely sparse" and/or "extremely dense" parts. Indeed, for any two indices i, j and each vertex $v \in V_i$ the expected number $\mu(v, V_j)$ of v-V_j-edges is either as small as μ^* or as large as $\#V_j - \mu^*$. To obtain a good approximation $\hat{A}$ of $\mathcal{E}$, it is instrumental to determine which parts of the graph are sparse and which are dense. This is the aim of the procedure Identify, whose analysis is summarized in the following proposition.

Proposition 6. *W.h.p. the output of* Identify *satisfies the following.*

1. *Either* Identify *outputs "fail" or its output is the matrix Φ defined in (3).*
2. *If $\mu^* \leq \log^3 n$, then the output is Φ.*

Identify essentially performs a "coarse" spectral partitioning of G.

Using the result of Identify, Partition computes $\hat{A}$. If Identify fails, then Partition assumes that μ^* is "large", and thus $\hat{A}$ is just a rank k approximation of $A(G)$. Otherwise the output φ of Identify is handed on to the subroutine Approx, which w.h.p. yields a matrix $\hat{A}$ as desired, cf. Section 3.2.

Proposition 7. *The output $\hat{A}$ of* Approx *is a matrix of rank $\leq 2k$ such that w.h.p. $\|\hat{A} - \mathcal{E}\| \leq Ck\sqrt{\mu^*}$ for a constant $C > 0$.*

Thus, let us assume that Step 2 has successfully computed a matrix $\hat{A}$ of rank $\leq 2k$ such that $\|\mathcal{E} - \hat{A}\| \leq Ck\sqrt{\mu^*}$. Then it turns out that for "most" vertices v the distance $\|\hat{A}_v - \mathcal{E}_v\|$ is "small". Therefore, Initial partitions the vertices $v \in V$ according to the vectors $\hat{A}_v$. More precisely, Initial computes k "centers" $\xi_1, \ldots, \xi_k \in \mathbf{R}^V$ and a partition $S_1, \ldots, S_k$ of V such that essentially S_i consists of those vertices v that are close to ξ_i, cf. Section 3.3.

Proposition 8. *After a suitable permutation of the indices the result of* Initial *satisfies $\|\xi_i - \mathcal{E}_i\|^2 \leq 0.001\rho^2$ for all i and $\sum_{i=1}^{k} \#S_i \triangle V_i < 0.001 n_{\min}$ w.h.p.*

While the initial partition $S_1, \ldots, S_k$ is solely determined by the matrix $\hat{A}$, the subroutine Improve actually investigates combinatorial properties of G. Improve performs iteratively a local improvement of the initial partition $S_1, \ldots, S_k$ that restricted to the subgraph core(G) converges to the planted partition $V_1, \ldots, V_k$.

Proposition 9. *W.h.p. there is a permutation τ such that the output $T_1, \ldots, T_k$ of* Improve *satisfies $T_i \cap \text{core}(G) = V_{\tau(i)} \cap \text{core}(G)$ for all $i = 1, \ldots, k$.*

A detailed description of Improve can be found in Section 3.4. Finally, since all procedures run in polynomial time, Theorem 1 is an immediate consequence of Propositions 4 and 6–9.

3.2 Approximating the Expected Densities

Algorithm 10. Approx(G, φ)
Input: A graph $G = (V, E)$ and a matrix $\varphi = (\varphi_{vw})_{v,w \in V}$. *Output:* A matrix $\hat{A}$.

1. Let $\hat{G}_1 = (V, E \cap \varphi^{-1}(0))$, $\hat{G}_2 = \overline{(V, E \cap \varphi^{-1}(1))}$. Let $\Delta = n$. Set $R_0 = \emptyset$ and $A_0 = (a_{0,vw})_{v,w \in V} = A(G)$.
2. For $t = 1, \dots, \log_2 \Delta$ do
3.   Let $\Delta_t = 2^{-t}\Delta$ and $R_t = \{v \in V : \max_{i=1,2} d_{G_i}(v) > \Delta_t\}$.
        Let $A_t = (a_{t,vw})_{v,w \in V}$ be the matrix with entries $a_{t,vw} = \varphi_{vw}$ if $(v, w) \in R_t \times V \cup V \times R_t$, and $a_{t,vw} = a_{t-1,vw}$ otherwise.
        If there is an $0 \le s < t$ such that $\|A_s - A_t\| > 2C''k\Delta_s^{1/2}$, then abort the for-loop; here C'' is a certain constant (cf. Lemma 11).
4. Let $\hat{t} = \max\{0, t - 1\}$ and return a rank k approximation of $A_{\hat{t}}$.

The aim of Approx is to compute a low rank matrix $\hat{A}$ that approximates $\mathcal{E}$. To this end, Approx analyses the spectrum of A, which is a "noisy" version of the spectrum of $\mathcal{E}$. If $\mu^* \ge \ln^2 n$ is sufficiently large, then it is actually very simple to recover the spectrum of $\mathcal{E}$ (approximately) from the spectrum of A: in this case we just let $\hat{A}$ be a rank k approximation of A (cf. Step 2 of Partition).

On the other hand, if μ^* is "small", say $\mu^* = O(1)$, then the "relevant" eigenvalues of A do not stand out anymore but are actually hidden among "noise" that is due to fluctuations of the vertex degrees. More precisely, remember the decomposition of G into the "sparse" part G_1 and the "dense" part G_2 (cf. Section 2). Then both G_1, G_2 are sparse (random) graphs with average degree $\le (1 + o(1))\mu^*$. However, as $\mu^* = O(1)$ as $n \to \infty$, the degree distributions of G_1, G_2 w.h.p. feature heavy upper tails. Now, vertices of degree $d \gg \mu^*$ in G_1 or G_2 induce eigenvalues of $A(G)$ which are as large as $\sqrt{d}$ in absolute value. By comparison, the "relevant" eigenvalues corresponding to the spectrum of $\mathcal{E}$ are in general $O(k\mu^{*\,1/2}) \ll \sqrt{d}$. Hence, the eigenvalues induced by the high degree vertices "hide" the relevant information.

Of course, if μ^* were known to the algorithm, then we could just delete all vertices v such that $\max\{d_{G_1}(v), d_{G_2}(v)\} > 4\mu^*$, say, from G and compute a low rank approximation $\hat{A}$ of the remaining graph's adjacency matrix. We do, however, *not* assume that μ^* is given at the input. Now, one might object that we could just try all possible values of μ^*. The problem is that possibly we could not tell from the resulting partition which value of μ^* was correct. Indeed, for a wrong value of μ^* the algorithm may easily miss some small planted class V_i but instead split some other big class V_j into two pieces erroneously.

Therefore, Approx pursues the following "adaptive" approach. The algorithm is given the graph G and the matrix φ produced by Identify, which equals the matrix Φ defined in (3) w.h.p. (by Proposition 6). Thus, the two graphs $\hat{G}_1, \hat{G}_2$ set up in Step 1 coincide with the graphs G_1, G_2 from Section 2 w.h.p. Proceeding in $\le \log_2 \Delta$ steps $t = 1, \dots, \log_2 \Delta$, Step 2 of Approx computes sets R_t of vertices of degree $\max\{d_{G_1}(v), d_{G_2}(v)\} \ge \Delta_t = 2^{-t}\Delta$ and matrices A_t. The A_t's are obtained from $A(G)$ by replacing all entries indexed by $V \times R_t \cup R_t \times V$ by the corresponding entries of φ. As soon as for some $s < t$ the matrices A_t

and A_s differ "significantly", the loop is aborted and the output is constructed from the matrix A_{t-1} obtained in the step before.

Why does this procedure yield a good approximation $\hat{A}$ of $\mathcal{E}$ w.h.p.? Suppose that $\Delta_t > 2.1\mu^*$, say. Since the expectations of $d_{G_i}(v)$ are $\leq 2\mu^*$ for all $v \in V$ and $i = 1, 2$, the set R_t consists just of $\leq n_{\min}/\mu^{*\,10}$ vertices of atypically high degree either in G_1 or in G_2 w.h.p. (by Chernoff bounds). Thus, deleting the vertices R_t removes the eigenvalues caused by the fluctuations of the vertex degrees $> \Delta_t$ while leaving the planted partition essentially intact. Therefore, using [7,9], we can estimate $\|A_t - \mathcal{E}\|$ as follows.

Lemma 11. *Suppose that $\Delta_t \geq 2.1\mu^*$. Then $\|A_t - \mathcal{E}\| \leq Ck(\Delta_t + \mu^*)^{1/2}$ w.h.p., where $C'' > 0$ is a constant.*

Let s be such that $2.1\mu^* \leq \Delta_s \leq 4.2\mu^*$. Then by Lemma 11 $\|A_s - \mathcal{E}\| \leq 3C''k\mu^{*\,1/2}$ is "small" w.h.p. However, being not aware of μ^*, Approx does not "notice" this, and may proceed with higher values of $t > s$. Nonetheless, the exit condition in Step 3 ensures that Approx will cancel the *for*-loop as soon as $\|A_t - A_s\|$ exceeds $2C''k\sqrt{\mu^*}$. In effect, the final matrix $A_{\hat{t}}$ obtained in Step 4 satisfies $\|A_{\hat{t}} - \mathcal{E}\| \leq 3C''k\sqrt{\mu^*}$, as desired.

3.3   Computing an Initial Partition

Algorithm 12. Initial$(\hat{A}, k)$
Input: A matrix $\hat{A}$ and the parameter k.
Output: A partition $S_1, \ldots, S_k$ of V and vectors $\xi_1, \ldots, \xi_k \in \mathbf{R}^V$.

1.   For $j = 1, \ldots, 2\log n$ do
2.   Let $\rho_j = n2^{-j}$ and compute $S^{(j)}(v) = \{w \in V : \|\hat{A}_w - \hat{A}_v\| \leq 10\varepsilon\rho_j^2\}$ for all $v \in V$, where $\varepsilon > 0$ is some small constant. Then, determine sets $S_1^{(j)}, \ldots, S_k^{(j)}$ as follows: for $i = 1, \ldots, k$ do
            Pick a vertex $v \in V \setminus \bigcup_{l=1}^{i-1} S_i^{(j)}$ such that $\#S^{(j)}(v) \setminus \bigcup_{l=1}^{i-1} S_i^{(j)}$ is maximum. Set $S_i^{(j)} = S^{(j)}(v) \setminus \bigcup_{l=1}^{i-1} S_i^{(j)}$ and let $\xi_i^{(j)} = \sum_{w \in S_i^{(j)}} \hat{A}_w / \#S_i^{(j)}$.
            Extend $S_1^{(j)}, \ldots, S_k^{(j)}$ to a partition of the entire set V by adding each vertex $v \in V \setminus \bigcup_{l=1}^k S_l^{(j)}$ to a set $S_i^{(j)}$ such that $\|\hat{A}_v - \xi_i^{(j)}\|$ is minimum.
            Set $r_j = \sum_{i=1}^k \sum_{v \in S_i^{(j)}} \|\hat{A}_v - \xi_i^{(j)}\|^2$.
3.   Let J be such that $r^* = r_J$ is minimum.
      Return $S_1^{(J)}, \ldots, S_k^{(J)}$ and $\xi_1^{(J)}, \ldots, \xi_k^{(J)}$.

Initial is given the approximation $\hat{A}$ of $\mathcal{E}$ and the parameter k, and its goal is to compute a partition of V that is "close" to the planted partition $V_1, \ldots, V_k$. By Proposition 7 we may assume that $\hat{A}$ has rank $\leq 2k$ and $\|\hat{A} - \mathcal{E}\| \leq O(k\sqrt{\mu^*})$. Hence, with respect to the Frobenius norm we obtain

$$\|\hat{A} - \mathcal{E}\|_F^2 \leq k^3\mu^* < \varepsilon^2\rho^2 n_{\min} \quad \text{(where } \rho^2 \text{ is the r.h.s. of (2)).} \tag{4}$$

If in addition we were given the parameter ρ, then we could partition G as follows. Since $z = \#\{v \in V : \|\hat{A}_v - \mathcal{E}_v\|^2 > \varepsilon\rho^2\}$ satisfies $\varepsilon\rho^2 z \leq \|\hat{A} - \mathcal{E}\|_F^2$, (2) and (4) yield $z \leq \varepsilon n_{\min}$. Now, consider a $v \in V_i$ such that $\|\hat{A}_v - \mathcal{E}_v\|^2 \leq \varepsilon\rho^2$, and define $S(v) = \{w \in V : \|\hat{A}_v - \hat{A}_w\|^2 \leq 10\varepsilon\rho^2\}$ for $v \in V$. Since $\mathcal{E}_w = \mathcal{E}_v$ for all $w \in V_i$, we have $\#S(v) \cap V_i \geq \#V_i - z \geq (1 - \varepsilon)\#V_i$. Moreover, $\#S(v) \setminus V_i \leq z$, because we assume that for all $w \in V \setminus V_i$ we have $\|\mathcal{E}_v - \mathcal{E}_w\| \geq \rho^2$. Thus, $S(v)$ "almost" coincides with V_i. Hence, we could obtain a good approximation of $V_1, \ldots, V_k$ by just picking iteratively k vertices $v_1, \ldots, v_k$ such that $v_{i+1} \in V \setminus \bigcup_{j=1}^{i-1} S(v_j)$ and $S(v_i)$ has maximum cardinality. A very similar procedure is used in [10], and this is also what Step 2 of `Initial` does.

However, since we do *not* assume that ρ is known to the algorithm, `Initial` has to estimate ρ. To this end, `Initial` applies the clustering procedure described in the previous paragraph for various candidate values $\rho_j = n2^{-j}$, $1 \leq j \leq 2\log_2 n$. Thus, for each j `Initial` obtains a collection $S_1^{(j)}, \ldots, S_k^{(j)}$ of subsets of V and corresponding vectors $\xi_i^{(j)}$. The idea is that $\xi_i^{(j)}$ should approximate $\mathcal{E}_v$ for $v \in V_i$ well if $S_i^{(j)}$ is a good approximation of V_i. Hence, $r_j = \sum_{i=1}^k \sum_{v \in S_i^{(j)}} \|\hat{A}_v - \xi_i\|^2$ should be small if $S_1^{(j)}, \ldots, S_k^{(j)}$ is close to the planted partition $V_1, \ldots, V_k$. Therefore, the output of `Initial` is just the partition $S_1^{(j)}, \ldots, S_k^{(j)}$ with minmal r_j.

Lemma 13. *There is a constant $C > 0$ such that in the case $\rho \leq \rho_j \leq 2\rho$ we have $r_j \leq Ck^3\mu^*$ w.h.p.*

Finally, Proposition 8 is a direct consequence of the following lemma.

Lemma 14. *If $r^* \leq Ck^3\mu^*$, then after a suitable permutation of the indices the result of `Initial` satisfies $\|\xi_i - \mathcal{E}_i\|^2 \leq 0.001\rho^2$ for all $i = 1, \ldots, k$, and $\sum_{i=1}^k \#S_i^*\triangle V_i < 0.001 n_{\min}$.*

3.4 Local Improvement

Algorithm 15. `Improve`$(G, S_1, \ldots, S_k, \xi_1, \ldots, \xi_k)$
Input: The graph $G = (V, E)$, a partition $S_1, \ldots, S_k$ of V, and vectors $\xi_1, \ldots, \xi_k$.
Output: A partition of G.

1. Repeat the following $\log n$ times:
2.   For all $v \in V$, all $l = 1, \ldots, k$, and all $w \in S_l$ compute the numbers $\delta(v, w) = e(v, S_l)/\#S_l$. Let $\delta(v) = (\delta(v, w))_{w \in V} \in \mathbf{R}^V$.
        For all $v \in V$ pick $1 \leq \gamma(v) \leq k$ such that $\|\delta(v) - \xi_{\gamma(v)}\| = \min_{1 \leq i \leq k} \|\delta(v) - \xi_i\|$ (ties are broken arbitrarily). Then, update $S_i = \gamma^{-1}(i)$ for $i = 1, \ldots, k$.
3. Return the partition $S_1, \ldots, S_k$.

Having computed the initial partition $S_1, \ldots, S_k$ with the "centers" $\xi_1, \ldots, \xi_k$, finally `Partition` calls the procedure `Improve` to home in on the planted partition $V_1, \ldots, V_k$ on the subgraph $\text{core}(G)$. In contrast to the previous steps of

Partition, Improve does not rely on spectral methods anymore but just performs a "local" combinatorial procedure.

By Lemma 14 we expect that $\xi_1, \ldots, \xi_k$ approximate the expected densities given by $\mathcal{E}$ well (after a suitable permutation of the indices). The basic idea behind Improve is to compare for each vertex v the actual values $e(v, S_i)$ with the expected values $\mu(v, V_i)$, where in turn the latter are approximated by the entries of ξ_i. More precisely, for each vertex v Improve sets up the vector $\delta(v)$ that encodes the densities $e(v, S_i)/\#S_i$. Then, Improve updates the partition $S_1, \ldots, S_k$ by putting each vertex v into that class S_j such that $\|\delta(v) - \xi_j\|$ is minimum. A detailed analysis of this process yields Proposition 9.

References

1. Alon, N., Kahale, N.: A spectral technique for coloring random 3-colorable graphs. SIAM J. Comput. **26** (1997) 1733–1748
2. Alon, N, Krivelevich, M., Sudakov, B.: Finding a large hidden clique in a random graph. Random Structures and Algorithms **13** (1998) 457–466
3. Boppana, R.: Eigenvalues and graph bisection: an average-case analysis. Proc. 28th FOCS (1987) 280–285
4. Chen, H., Frieze, A.: Coloring bipartite hypergraphs. Proc. 5th IPCO (1996) 345–358
5. Coja-Oghlan, A.: A spectral heuristic for bisecting random graphs. Proc. 16th SODA (2005) 850–859
6. Dasgupta, A., Hopcroft, J.E., McSherry, F.: Spectral Partitioning of Random Graphs. Proc. 45th FOCS (2004) 529–537
7. Feige, U., Ofek, E.: Spectral techniques applied to sparse random graphs. Random Structures and Algorithms **27** (2005) 251–275
8. Flaxman, A.: A spectral technique for random satisfiable 3CNF formulas. Proc. 14th SODA (2003) 357–363
9. Füredi, Z., Komlós, J.: The eigenvalues of random symmetric matrices. Combinatorica **1** (1981) 233–241
10. McSherry, F.: Spectral Partitioning of Random Graphs. Proc. 42nd FOCS (2001) 529–537
11. Pothen, A., Simon, H.D., Kang-Pu, L.: Partitioning sparse matrices with eigenvectors of graphs. SIAM J. Matrix Anal. Appl. **11** (1990) 430–452

Some Results on Matchgates and Holographic Algorithms

Jin-Yi Cai* and Vinay Choudhary

Computer Sciences Department, University of Wisconsin, Madison,
WI 53706. USA. Tsinghua University, Beijing, China
{jyc, vinchr}@cs.wisc.edu

Abstract. We establish a 1-1 correspondence between Valiant's *character* theory of matchgate/matchcircuit [14] and his *signature* theory of planar-matchgate/matchgrid [16], thus unifying the two theories in expressibility. In [3], we had established a complete characterization of general matchgates, in terms of a set of *useful* Grassmann-Plücker identities. With this correspondence, we give a corresponding set of identities which completely characterizes planar-matchgates and their signatures. Applying this characterization we prove some negative results for *holographic algorithms*. On the positive side, we also give a polynomial time algorithm for a simultaneous node-edge deletion problem, using holographic algorithms. Finally we give characterizations of symmetric signatures realizable in the Hadamard basis.

1 Introduction

Recently Valiant has introduced a novel methodology in algorithm design. In a ground breaking paper [14], Valiant initiated a new theory of matchgate/ matchcircuit computations. Subsequently, in [16], he further proposed the theory of holographic algorithms, based on planar matchgates and matchgrids. Underlying both theories are the beautiful ideas of (a) using perfect matchings to encode and organize computations, and (b) applying the algebraic construct called the Pfaffian.

A basic component in both theories is a matchgate. A matchgate is essentially a finite graph with certain nodes designated as inputs or outputs. In the matchcircuit theory, each matchgate defines a *character matrix*, with entries defined in terms of the Pfaffian. In the theory of holographic algorithms, only planar matchgates are considered, and each planar matchgate defines a *signature matrix*, which directly captures the properties of the matchgate under the consideration of (perfect) matchings when certain input and/or output nodes are retained or removed.

These matchgates are combined to form matchcircuits or matchgrids. For a matchcircuit, some of its global properties can be interpreted as realizing certain computations which would seem to take exponential time in the size of the

---

* Supported by NSF CCR-0208013 and CCR-0511679.

M. Bugliesi et al. (Eds.): ICALP 2006, Part I, LNCS 4051, pp. 703–714, 2006.

circuit. However, due to the way the matchcircuits are constructed and the algebraic properties of Pfaffians defining the character matrices of the constituent matchgates, these properties can in fact be computed in polynomial time. For holographic algorithms, a new crucial ingredient was added—a choice of a set of linear basis vectors, in terms of which the computation can be expressed and interpreted. They are called holographic, because the algorithm introduces an exponential number of solution fragments in a pattern of interference, analogous to quantum computing. However, because of the planarity condition, the computation by matchgrids can be expressed via the elegant Fisher-Kasteleyn-Temperley (FKT) method [8,9,12] for planar perfect matchings, and therefore computable in P. Valiant [14] used matchcircuits to show that a non-trivial fragment of quantum circuits can be simulated classically in polynomial time. With holographic algorithms, he was able to devise polynomial time algorithms for several problems, which were not known to be in P, and certain minor variations of which are NP-hard (or even #P-hard). It is not clear what are the ultimate computational capabilities of either theories.

In a paper currently in submission [3], the present authors investigated a number of interesting properties of matchgate computations. In particular, we gave a necessary and sufficient condition, in terms of a set of *useful* Grassmann-Plücker identities, which completely captures the realizability of matchgates with given characters. The study of matchgate identities was already initiated by Valiant in [15]. It was shown in [3] that the matchgates form an algebraic variety, and a certain group action underlies the symmetry present in the character matrices.

In this paper, we first unite the two theories: matchcircuit computation on the one hand and matchgrid computation on the other. We show that, the planarity restriction not withstanding, any matchcircuit computation can be simulated by a matchgrid, and vice versa. In fact we will give an interpretation between the characters and signatures in a one-to-one fashion. Thus, all important theorems in [14] (e.g., its Matchcircuit Theorem and its Main Theorem) can be stated in terms of planar matchgrids. Conversely, to design holographic algorithms, one can ignore the planar restriction on the matchgates. For the proof of this equivalence theorem, in one direction we use a cross-over gadget designed by Valiant [16]; in the other direction we use the FKT method [8,9,12].

As part of this proof, we also define a notion of a *naked character*. Based on our previous work reported in [3], we can derive a corresponding set of matchgate identities, which are necessary and sufficient for naked characters. Then we prove that a matrix is a naked character matrix iff it is a signature matrix. This gives us a complete characterization on the realizability of planar matchgates in terms of their signatures.

Such a characterization provides for the first time the possibility of proving negative results for holographic algorithms. We note that, by definition, even with a fixed number of input and output nodes, a matchgate may consist of an arbitrarily large number of internal nodes. Thus one can prove the existence of a matchgate fulfilling certain computational requirements by construction. But one cannot prove in this way the non-existence of such a matchgate. Our characterization makes this possible. Indeed, we define *holographic templates* to capture a restricted but natural subclass of holographic algorithms, and prove certain non-existence theorems. In particular, we prove that certain natural generalizations of some of the problems solved by P-time algorithms in [16] do not have P-time algorithms by holographic templates by linearly independent basis of dimension 2. In many of the problems in [16], a particular basis **b2** (which can be called the Hadamard basis) was particularly useful. We characterize the representable matchgate signatures that are based on cardinality alone over this basis. This uses the properties of Krawtchouk polynomials. We also give a positive result by deriving a polynomial time algorithm for a problem using holographic templates. It is a simultaneous node-edge deletion problem for a graph to become bipartite, for planar graphs with maximal degree 3. This generalizes both the edge deletion problem, and the node deletion problem which was considered in [16] for such graphs. We note that the edge deletion problem is the same as the MAX-CUT problem [1].

The most intriguing question is whether this new theory leads to any collapse of complexity classes. The kinds of algorithms that are obtained by this theory are quite unlike anything before and almost exotic. If our belief in NP $\neq$ P is based on the sense and experience that the usual algorithmic paradigms are insufficient for NP-hard problems (certainly it is not due to strong lower bounds), then we feel our erstwhile experience does not apply to these new algorithms. Of course it is quite possible that the theory of matchcircuit and holographic algorithms do not in the end lead to any collapse of complexity classes. But even in this eventuality, as Valiant suggested in [16], "any proof of P $\neq$ NP may need to explain, and not only to imply, the unsolvability" of NP-hard problems using this approach. Regardless of its final outcome, this paper is an attempt towards such a fundamental understanding.

The rest of the paper is organized as follows: In Section 2, we give a brief account of the background. In Section 3, we give the equivalence theorem of the two theories. We also discuss matchgate identities for naked characters and signatures. In Section 4, we give a positive result on the simultaneous node-edge deletion problem. In Section 5 we define holographic templates, and give some impossibility results. In Section 6, we characterize symmetric signatures for **b2**. More proof details can be found in [4].

2   Background

Due to space limitations, most details are left out. The readers are referred to [14,15,16,2,3].

Let $G = (V, E, W)$ be a weighted undirected graph. We represent the graph by the *skew-symmetric adjacency* matrix M, where $M(i, j) = w(k_i, k_j)$ if $i < j$, $M(i, j) = -w(k_i, k_j)$ if $i > j$, and $M(i, i) = 0$. The Pfaffian of an $2k \times 2k$ skew-symmetric matrix M is defined to be

---

[1] MAX-CUT for planar graphs is known to be in P [5].

$$\text{Pf}(M) = \sum_{\pi} \epsilon_{\pi} w(i_1, i_2) w(i_3, i_4) \ldots w(i_{2k-1}, i_{2k}),$$

where the sum is over all permutations π, where $i_1 < i_2, i_3 < i_4, \ldots, i_{2k-1} < i_{2k}$ and $i_1 < i_3 < \ldots < i_{2k-1}$, and $\epsilon_{\pi} \in \{-1, 1\}$. The Pfaffian is computable in polynomial time. In particular $(\text{Pf}(M))^2 = \det(M)$. We refer to [14,15] for definitions of Pfaffian Sum $\text{PfS}(M)$, Matchgates and Matchcircuits. Very briefly, each matchgate is assigned a *character matrix*, where an entry of the matrix is $\mu(\Gamma, Z)\text{PfS}(G-Z)$, where $\mu(\Gamma, Z) \in \{-1, 1\}$ is a *modifier* and Z is a set of deleted vertices. A *matchcircuit* is a way of combining matchgates. The character of a matchcircuit is defined in the same way as the character of a matchgate except that there is no modifier μ.

Character matrices of matchgates satisfy a rich set of algebraic constraints called *matchgate identities*. Valiant already derived a number of these identities in [15]. They are derived from the Grassmann-Plücker identities [11].

Theorem 1. *For any $n \times n$ skew-symmetric matrix M, and any $I = \{i_1, \ldots, i_K\} \subseteq [n]$ and $J = \{j_1, \ldots, j_L\} \subseteq [n]$,*

$$\sum_{l=1}^{L} (-1)^l \text{Pf}(j_l, i_1, \ldots, i_K) \text{Pf}(j_1, \ldots, \hat{j_l}, \ldots, j_L) +$$

$$\sum_{k=1}^{K} (-1)^k \text{Pf}(i_1, \ldots, \hat{i_k}, \ldots, i_K) \text{Pf}(i_k, j_1, \ldots, j_L) = 0$$

In our paper [3] we derived a complete set of algebraic identities using the so-called useful Grassmann-Plücker identities. Due to space limitation we will only describe these for 4×4 character matrices B.

Denote by $D(ij, kl) = \begin{vmatrix} B_{ik} & B_{il} \\ B_{jk} & B_{jl} \end{vmatrix}$, the 2×2 minor of B consisting of rows i and j, and columns k and l. Let S denote the set of $\binom{4}{2}$ unordered pairs of $\{1, 2, 3, 4\}$, $S = \{\{1, 2\}, \{1, 3\}, \{1, 4\}, \{2, 3\}, \{2, 4\}, \{3, 4\}\}$. Define an involution σ on S which exchanges the pair $\{1, 4\}$ and $\{2, 3\}$, and leaves everything else fixed. Then it is proved in [3] that B is a character matrix iff the following set of identities hold:

$$D(p, q) = D(\sigma(p), \sigma(q)),$$

for any $(p, q) \in S \times S$. E.g., $B_{11}B_{44} - B_{14}B_{41} = B_{22}B_{33} - B_{23}B_{32}$ and $B_{12}B_{43} - B_{13}B_{42} = B_{21}B_{34} - B_{24}B_{31}$, etc.

Theorem 2. *[3] Let B be a 4×4 matrix over a field F. It satisfies the above set of matchgate identities (there are ten non-trivial identities) iff there exists a matchgate Γ such that $\chi(\Gamma) = B$.*

Theorem 3. *[3] There is an effectively constructible set of matchgate identities which completely characterizes any k input l output matchgate.*

The matchgate identities have far reaching implications. On the positive side, the proof in [3] indicates that whenever B is a $2^k \times 2^l$ character matrix there is a matchgate Γ of size $O(k + l)$ realizing it, thus it can be found in a bounded search. On the negative side, the complete characterization provides us with the tools to prove non-existence for general k and l.

When the weighted graph $G = (V, E, W)$ is planar, we have a *planar matchgate* $\Gamma = (G, X, Y)$ where X is the set of inputs and Y is the set of outputs and we can associate a notion of a *signature matrix* with it. Planar matchgates can be combined to form planar matchgrids. The central definition of the theory of holographic algorithms is the Holant of a matchgrid.

The following Holant Theorem says that the Holant can be efficiently computed using the FKT technique [8,9].

Theorem 4 (Valiant). *For any matchgrid Ω over any basis β, let G be its underlying weighted graph, then* $\mathrm{Holant}(\Omega) = \mathrm{PerfMatch}(G)$.

For the details of the definitions and the proof, we refer the reader to [16] and [2].

3   An Equivalence Between Matchcircuits and Planar Matchgrids

3.1   Naked Characters

In [3], we showed that the set of *useful* Grassmann-Plücker identities gives a complete characterization of matchgate characters, i.e., every character matrix satisfies these equations and any matrix satisfying these is the character of some matchgate. A *useful* Grassmann-Plücker identity is derived from a Grassmann-Plücker identity on (I, J), where I and $J \subseteq V$ are subsets of nodes of the matchgate containing all internal nodes. We refer to [3] for details. For convenience of proof, we also define a *naked character* as a character without the modifiers. Thus, the entries of the *naked character* of a k-input, l-output matchgate is simply $\mathrm{PfS}(G - Z)$ where Z varies over subsets of $X \cup Y$. Since the modifier $\mu(Z)$ does not depend on the internal nodes, the useful Grassmann-Plücker identities can be considered as identities over the entries of the naked character matrix. These identities completely characterize the naked character matrices of matchgates.

3.2   Equivalence of Matchgates and Planar Matchgates

In this subsection, we prove a surprising equivalence between matchgates (which are generally not planar) and planar matchgates. Specifically, we can show that the set of naked character matrices of k-input, l-output matchgates is the same as the set of signature matrices of k-input, l-output planar matchgates. This theorem has remarkable implications. In particular, it implies that the set of matchgate identities (for naked characters) also characterize all signature matrices. With this we obtain a complete algebraic characterization of planar matchgates. This will enable us to prove some impossibility results.

Lemma 1. *Given a matchgate Γ with naked character matrix B, there exists a planar matchgate Γ' with signature B.*

Proof. Recall that the vertices of Γ are numbered 1 through n with the first k being inputs and the last l being outputs. Now arrange the vertices (with their edges) on a *strictly convex* curve, e.g., a upper semicircle, such that as we move clockwise from vertex 1, we encounter all the vertices in increasing order (see Figure 1). By doing this, we have achieved the following: Any two edges (i, j) and (k, l) overlap (i.e. $i < k < j < l$ or $k < i < l < j$) *iff* they physically cross each other as two straight line segments. If any such pair of overlapping edges is present in a matching, it introduces a negative sign to the Pfaffian. Now we can convert this graph into a planar graph by using the gadget given in Figure 8 in [16]. We will replace any physical crossing by a *local* copy of the gadget. We then use the properties of this gadget proved in Proposition 6.3 of [16]. We omit the details, but it can be shown that the MatchSum polynomial of the new graph is the same as the Pfaffian Sum of the original graph. It follows that the signature of this planar graph is the same as the character of Γ except that the signature doesn't consider any external edges and hence, it doesn't have any modifiers. This means that the signature is actually equal to the naked character B. Note that, in this construction, if omittable nodes are present, they are now all on the outer face (in fact all the original nodes of Γ are now on the outer face).

Fig. 1. An example of converting a 2-input, 2-output matchgate to a planar matchgate

Lemma 2. *Given a planar matchgate Γ with signature u, there is a matchgate Γ' with naked character equal to u.*

Proof. The underlying graph of Γ' is the same as that of Γ but we'll change the weights suitably. For that, we have to consider the orientation given to edges by the FKT algorithm to count the number of perfect matchings as described in [9]. For any edge (i, j) where $i < j$, if the direction assigned to it is i to j, then we keep the weight as is, otherwise we multiply the weight by a -1. The matrix whose Pfaffian we evaluate to count the number of perfect matchings in Γ is exactly the same as the (skew-symmetric) adjacency matrix of the new graph. That means that its character, after dropping the modifiers μ, is the same as the signature of Γ.

If omittable nodes are present, we need to evaluate MatchSum. Since the omittable nodes are all on the outer face, one single consistent orientation can be chosen for all edges, as the result of FKT algorithm, simultaneously for all terms of MatchSum. This reduces to a Pfaffian Sum.

Lemmas 1 and 2 prove the following theorem.

Theorem 5. *The set of signature matrices of planar matchgates is the same as the set of naked character matrices of matchgates.*

Using the same technique as in the proof above, we can show that matchcircuits and planar matchgrids are computationally equivalent. In other words, we are able to show that the same fragment of quantum computation that was simulated by matchcircuits in [14] can be simulated by matchgrids. The exact statement of the theorem and its proof can be found in [4].

4   Simultaneous Node-Edge Deletion

In this section we give a holographic algorithm for a simultaneous node-edge deletion problem. This is the first poly-time algorithm for this problem. The problem is a generalization of the PL-NODE-BIPARTITION problem for which the first polynomial time algorithm was given by Valiant [16]. It also generalizes the planar edge deletion problem, which is the same as MAX-CUT. Planar MAX-CUT is known to be in P [5]. We note that the closely related problem of Planar-Max-Bisection (where a bisection is a cut with two equal parts) was a long standing open problem till Jerrum proved it NP-hard (see [6]). There has also been important progress on its approximability [6]. We also note that the status of Planar-Min-Bisection remains open.

PL-NODE-EDGE-BIPARTITION

Input: A planar graph $G = (V, E)$ of maximum degree 3. A non-negative integer $k \leq |V|$. **Output:** The minimal l such that deletion of at most k nodes (including all of their incident edges) and l more edges results in a bipartite graph.

Theorem 6. *There is a polynomial time algorithm for PL-NODE-EDGE-BIPARTITION.*

Proof. We will use the method of holographic algorithms [16]. Let the given input graph be G. First, note that we can simply delete any node of degree 1. We will replace each remaining nodes by recognizers with symmetric signature $[1, x, x, 1]$ or $[1, x, 1]$ depending on their degree. The edges will be replaced by generators with symmetric signature $[y, 1, y]$. This forms a matchgrid Ω. It is known that the above symmetric signatures are realizable in the Hadamard basis $\mathbf{b2} = [n, p] = [(1, 1), (1, -1)]$. (See [16] and also Section 6.) Every term in the Holant corresponds to an assignment of n or p to each end of every connecting edge in Ω. This induces an assignment on the vertices of G. We consider vertices in G that get nnn or nn (depending on the degree) are colored white and those that get ppp or pp are colored black. The remaining vertices are not colored. Now, every colored node contributes 1 to the Holant and every uncolored node contributes x. Any edge that is assigned nn or pp contributes y and any edge that is assigned np or pn contributes 1. It is clear that we can obtain a bipartite graph by deleting the uncolored vertices and the edges that are assigned nn or pp. We define

$\ell(k) = \min\{\ l'\ |\ \text{The coefficient of } x^k y^{l'} \text{ in Holant is non-zero.}\}$

$l(k) = \min\{\ l'\ |\ \exists \text{ a subset } S \subseteq V \text{ of size } k \text{ and some } l' \text{ edges in } G - S \text{ such that}$
$\qquad\qquad \text{removal of the } l' \text{ edges from } G - S \text{ gives a bipartite graph.}\}$

From the discussion earlier, we can see that $l(k) \leq \ell(k)$.

Claim. $l(k)$ is a strictly monotonic decreasing in k, until $l(k) = 0$.

Proof. Let $k' < k$. We show that if $l(k') > 0$, then $l(k') > l(k)$. Let $S \subseteq V$ be a subset of size k', such that the deletion of S and some $l(k')$ edges from $G - S$ results in a bipartite graph. Then, if we are allowed to delete $k > k'$ vertices, we can choose to delete S and some of the vertices to which the other $l(k')$ edges are incident. Then, clearly, $l(k) < l(k')$.

Claim. $\ell(k) \leq l(k)$

Proof. Let $S \subseteq V$ be a subset of size k such that the deletion of S and some $l(k)$ other edges results in a bipartite graph. Assign nnn or nn to the vertices on the left and ppp or pp to those on the right. This means that for a connecting edge incident to a recognizer for a node on the left, we assign n to its end which is incident to it. Similarly for a connecting edge incident to a recognizer for a node on the right we assign p there. The generator corresponding to any edge present in the bipartite graph gets np or pn and the $l(k)$ deleted edges get nn or pp, due to the minimality of $l(k)$. The remaining edges can be of three types: having one end point on the left side and one in S, having one end point on the right and one in S, or having both end points in S. The generators corresponding to all these edges are given np or pn in such a way that any output adjacent to a recognizer on the left gets n and any output adjacent to a recognizer on the right gets p. This can be done, since the remaining edges have at least on end point in S, we have at least one *free* output.

It is easy to see that the degree of y in this term of the Holant is exactly $l(k)$. Note that all the vertices not in S are assigned nnn or ppp (or nn or pp) and contribute 1 to the Holant. We further claim that no vertex in S gets nnn or ppp (or nn or pp). Hence the coefficient of $x^k y^{l(k)}$ is positive, and therefore $\ell(k) \leq l(k)$. If some vertex in S were to get nnn or ppp (or nn or pp), we can add those vertices and their incident edges to the bipartite graph, and we will still have a bipartite graph, since all the edges incident to any vertex in S are assigned either np or pn. This means that for some $k' < k$, $l(k') \leq l(k)$ which is impossible, by Claim 4.

The proof of the theorem is now easy. The required value is $l = l(k)$. As $l(k) = \ell(k)$, we can find this by computing the Holant, which is a polynomial in x and y of degree at most $|V| + |E|$. This is done by evaluating Holant at several values of x and y, and then by polynomial interpolation. Note that, since every term in the Holant contributes either a one or a zero to the coefficient of at most one term in the polynomial, the coefficients are bounded by $2^{2|E|}$, i.e., $O(|E|)$ bits.

Valiant's PL-NODE-BIPARTITION [16] asks for the minimal k such that $l(k) = 0$, while PL-EDGE-BIPARTITION (Planar MAX-CUT for degree ≤ 3) [5] asks for $l(0)$. This problem generalizes both.

5 Limitations of Holographic Algorithms

There is not yet any formal definition of what is computable by holographic algorithms. In this section, we try to define the most basic kinds of holographic algorithms and call these *holographic templates*. The aim is to capture essentially what is computable by using only the Holant and nearly no other meaningful polynomial time computation. Then we look at some generalizations of two of the problems solved by Valiant using holographic algorithms in [16]. We show that there are no holographic templates for these generalizations. To make the impossibility results more meaningful, we will also need a formal definition of the types of problems to which holographic templates can possibly be applied. The formal definition of such problems and of holographic templates are omitted due to space limitations (see [4] for details). The definition captures the notion that local solution fragments of a counting problem are mapped to the non-zero entries in the signature of planar matchgates in such a way that the Holant of the matchcircuit is equal to the answer of the counting problem. All of the holographic algorithms presented by Valiant can essentially be realized in this notion of holographic templates.

By our definition, if there is holographic template for a problem then the answer produced by it, i.e. the Holant of the holographic template, is the answer of the counting problem. Below we will show the non-existence of holographic template algorithms for some problems. Our negative results will only apply to basis of dimension 2. We will only consider holographic templates using planar matchgates and matchgrids without omittable nodes. The impossibility results will be achieved by showing that there are no bases of dimension 2 in which there are recognizers and generators having some required signatures. Suppose we need to find a basis and some generators/recognizers with given signatures w.r.t. the basis. We first translate the signature into standard signatures. The entries of the standard signature will be in terms of the basis vectors. We will then use our algebraic equations that characterize the signature matrices of planar matchgates. These include the parity constraints and the matchgate identities. By parity constraints, we mean the constraint that for any standard signature, either all terms corresponding to deletion of an odd number of nodes are zero or all terms corresponding to deletion of even number of nodes are zero. This is a consequence of perfect matchings. For a number of problems, we will be able to show that there are no bases for which the standard signature satisfies all these constraints, thus concluding that these problems cannot be solved by this method.

Before moving on, we note that if a basis β consists of only two linearly dependent two-dimensional vectors, then the span of any higher tensor $\beta^{\otimes f}$ will also be one-dimensional and thus ruling out any interesting signatures from being in its span. So for the problems we consider, we will only look for linearly independent bases without explicitly proving that any linearly dependent basis of two vectors doesn't work.

#X-Matchings

One problem solved by Valiant by a holographic algorithm [16] is called #X-Matchings. This is motivated by its proximity to counting the number of (not necessarily perfect) matchings in a planar graph, which was proved to be #P-complete by Jerrum [7]. Vadhan [13] subsequently proved that it remains #P-complete for planar bipartite graphs of degree 6. For degree two the problem can be easily solved. For Valiant's #X-Matchings, a planar bipartite graphs $G = (V, E, W)$ is given with bipartition $V = V_1 \cup V_2$, where nodes in V_1 have degree 2 and nodes in V_2 have arbitrary degrees. The problem is to compute $\sum_M m(M)$, where M runs through all (not necessarily perfect) matchings, and the mass $m(M)$ is the product of (1) weights of $e \in M$ and (2) the quantity $-(w_1 + \ldots + w_k)$ for each unmatched node in V_2, where w_i are the weights of edges incident to that node. One can use this to compute the total number of matchings mod 5, if all vertices in V_2 have degree 4.

Still, the quantity $-(w_1 + \ldots + w_k)$ seems artificial. If one were to be able to replace $-(w_1 + \ldots + w_k)$ by 1, then one would be able to count all (not necessarily perfect) matchings in such planar bipartite graphs. However, we prove that this is impossible using holographic templates.

Theorem 7. *There is no holographic template using any basis of two linearly independent vectors to solve the counting problem for all (not necessarily perfect) matchings for such graphs, which is the same as the above problem with $-(w_1 + \ldots + w_k)$ replaced by 1.*

The proof uses our characterizations of realizability of matchgates and the equivalence theorems on characters and signatures.

Several other problems solved by Valiant in [16] use a matchgate with a symmetric signature which is logically a Not-All-Equal gate. This is typified by the following problem:

#PL-3-NAE-ICE

Input: A planar graph $G = (V, E)$ of maximum degree 3.
Output: The number of orientations such that no node has all edges directed towards it or away from it.

If one were to relax the degree bound $k = 3$, some of his problems [16] are known to be NP-hard. We prove that for any $k > 3$, one can not realize a Not-All-Equal by a symmetric signature.

Theorem 8. *There is no holographic template using any basis of two linearly independent vectors to solve the above ICE problem if we replace the degree bound by any $k > 3$.*

Again the proof uses our characterizations including matchgate identities, and is omitted (see [4].) As the proof deals with the non-existence of certain matchgates of prescribed signatures, this is applicable to other problems in addition to #PL-3-NAE-ICE.

6 Symmetric Signatures in b2

The most versatile basis in the design of holographic algorithms so far has been the Hadamard basis **b2**, namely $[n, p] = [(1, 1), (1, -1)]$. In [16], most often, it is used to realize a symmetric signature that has a clear Boolean logical meaning, such as the Not-All-Equal function. In this section, we give a complete characterization of all the symmetric signatures that can be realized by some generators or recognizers (having no omittable nodes) in this basis.

Let T denote the matrix $\begin{pmatrix} 1 & 1 \\ 1 & -1 \end{pmatrix}$. T is symmetric and non-singular, and therefore $T^{\otimes n}$ is a symmetric non-singular $2^n \times 2^n$ matrix. It follows that for **b2**, realizability for a recognizer is the same as for a generator.

The Hamming weight of a row or column index to $T^{\otimes n}$, which is a 0-1 vector in binary representation, is the number of 1's in it. Suppose we have a generator having standard signature u and signature u_{b2} under **b2**. We claim that u_{b2} is a symmetric signature iff u is. Since $T^{-1} = \frac{1}{2}T$, we only need to show this in one direction.

Row vectors u and u_{b2} are related by $u = u_{b2} T^{\otimes n}$. Suppose u_{b2} is a symmetric signature. We sum the rows of equal Hamming weight in $T^{\otimes n}$ to obtain an $(n + 1) \times 2^n$ matrix M. It is clear that M has a full row rank because any linear combination of rows of M is a linear combination of rows of $T^{\otimes n}$, which is non-singular. It can be seen that any two columns of M having indices of the same Hamming weight are equal. So M has at most $n + 1$ distinct columns. Thus u is also symmetric. And since the rank of M is $n + 1$, there must be exactly $n + 1$ distinct columns, and they are linearly independent. Consider the $(n + 1) \times (n + 1)$ matrix $A = [a_{ij}]$ obtained by taking the distinct columns from M. A is non-singular.

In fact, these a_{ij} can be expressed by the Krawtchouk polynomials [10], the properties of which can then be used to derive the theorem below.

Lemma 3. $a_{ij} = \sum_{k=0}^{i}(-1)^k \binom{j}{k}\binom{n-j}{i-k}$. *In particular,* $a_{ij} = (-1)^i a_{i,n-j} = (-1)^j a_{n-i,j}$.

Theorem 9. *A symmetric signature $[x_0, x_1, \ldots, x_n]$ is realizable under the Hadamard basis b2 iff it takes the following form: For all $0 \le i < n/2$,*

$$x_i = \lambda \left(s^{n-i}t^i - t^{n-i}s^i \right),$$

where λ, s and t are arbitrary constants. In addition, for all $0 \le i < n/2$, either all $x_{n-i} = x_i$ or all $x_{n-i} = -x_i$. Finally, for n even, $x_{\frac{n}{2}} = 0$.

Acknowledgments

We would like to thank Eric Bach, Rakesh Kumar, Anand Kumar Sinha, Leslie Valiant, and Andrew Yao and his group of students in Tsinghua University for many interesting discussions.

References

1. A. C. Aitken. Determinants and Matrices, Oliver and Boyd, London, 1951.
2. Jin-Yi Cai, V. Choudhary. Valiant's Holant Theorem and Matchgate Tensors. To appear in the Proceedings of *Theory and Applications of Models of Computation*, TAMC 2006. Lecture Notes in Computer Science. Also at ECCC TR05-118.
3. Jin-Yi Cai, V. Choudhary. On the Theory of Matchgate Computations. *Submitted*. Also available at ECCC TR06-018.
4. Jin-Yi Cai, V. Choudhary. Some Results on Matchgates and Holographic Algorithms. ECCC TR06-048.
5. F. Hadlock. Finding a Maximum Cut of a Planar Graph in Polynomial Time. *SIAM Journal on Computing*, 4 (1975): 221-225.
6. K. Jansen, M. Karpinski, A. Lingas, and E. Seidel. Polynomial Time Approximation Schemes for MAX-BISECTION on Planar and Geometric Graphs. *Proceedings of 18th Symposium on Theoretical Aspects in Computer Science (STACS) 2001*, LNCS 2010: 365-375, Springer (2001).
7. M. R. Jerrum. Two-dimensional Monomer-Dimer Systems are Computationally Intractable. *Journal of Statistical Physics*, 48, 1/2: 121-134 (1987). (Also 59, 3/4: 1087-1088 (1990)).
8. P. W. Kasteleyn. The statistics of dimers on a lattice. *Physica*, 27: 1209-1225 (1961).
9. P. W. Kasteleyn. Graph Theory and Crystal Physics. In *Graph Theory and Theoretical Physics*, (F. Harary, ed.), Academic Press, London, 43-110 (1967).
10. F. MacWilliams and N. Sloane, The Theory of Error-Correcting Codes, NorthHolland, Amsterdam, p. 309, 1977.
11. K. Murota. Matrices and Matroids for Systems Analysis, Springer, Berlin, 2000.
12. H. N. V. Temperley and M. E. Fisher. Dimer problem in statistical mechanics – an exact result. *Philosophical Magazine* 6: 1061– 1063 (1961).
13. S. P. Vadhan. The Complexity of Counting in Sparse, Regular and Planar Graphs. *SIAM Journal on Computing*, 8(1): 398-427 (2001).
14. L. G. Valiant. Quantum circuits that can be simulated classically in polynomial time. *SIAM Journal on Computing*, 31(4): 1229-1254 (2002).
15. L. G. Valiant. Expressiveness of Matchgates. *Theoretical Computer Science*, 281(1): 457-471 (2002). See also 299: 795 (2003).
16. L. G. Valiant. Holographic Algorithms (Extended Abstract). In *Proc. 45th IEEE Symposium on Foundations of Computer Science*, 2004, 306–315. A more detailed version appeared in Electronic Colloquium on Computational Complexity Report TR05-099.
17. L. G. Valiant. Holographic circuits. In *Proc. 32nd International Colloquium on Automata, Languages and Programming*, 1-15, 2005.
18. L. G. Valiant. Completeness for parity problems. In *Proc. 11th International Computing and Combinatorics Conference*, 2005.

Weighted Popular Matchings[*]

Julián Mestre

Department of Computer Science
University of Maryland, College Park, MD 20742
jmestre@cs.umd.edu

Abstract. We study the problem of assigning applicants to jobs. Each applicant has a weight and provides a *preference list*, which may contain ties, ranking a subset of the jobs. An applicant x may prefer one matching over the other (or be indifferent between them, in case of a tie) based on the jobs x gets in the two matchings and x's personal preference. A matching M is *popular* if there is no other matching M' such that the weight of the applicants who prefer M' over M exceeds the weight of those who prefer M over M'.

We present two algorithms to find a popular matching, or in case none exists, to establish so. For the case of strict preferences we develop an $O(n + m)$ time algorithm. When ties are allowed a more involved algorithm solves the problem in $O(\min(k\sqrt{n}, n)m)$ time, where k is the number of distinct weights the applicants are given.

1 Introduction

Consider the problem of assigning applicants to jobs where every applicant provides a *preference list*, which may contain ties, ranking a subset of the jobs. Formally, an instance consists of a bipartite graph $G = (A, J, E)$ with n vertices and m edges between a set of applicants A and a set of jobs J. An edge $(x, p) \in E$ denotes that job p belongs to x's list. Moreover, every edge (x, p) is assigned a rank i, which means p is the ith choice on x's list. An applicant x prefers job p over q if the edge (x, p) is ranked higher than (x, q); if (x, p) and (x, q) have the same rank we have a tie, and thus x is indifferent between them. Likewise, we say x prefers one matching over the other or is indifferent between them based on the two jobs x is assigned by the two matchings. Our ultimate goal is to produce a matching in this graph.

While simple, this framework captures many real-world problems, for instance the assignment of government-subsidized houses to families [9], and graduates to training position [6]. The issue of what constitutes a fair or good assignment has been studied in the Economics literature [1,9,10]. The least restrictive definition of optimality is that of a *Pareto optimal* matching [2,1]. A matching M is Pareto optimal if there is no matching M' such that at least one person prefers M' over M and nobody prefers M over M'. In this paper we study a stronger definition of optimality, that of *popular matchings*. We say M_1 is *more popular than* M_2

---

[*] Research supported by NSF Awards CCR 0113192 and CCF 0430650.

M. Bugliesi et al. (Eds.): ICALP 2006, Part I, LNCS 4051, pp. 715–726, 2006.

if the applicants who prefer M_1 over M_2 outnumber those who prefer M_2 over M_1; a matching M is *popular* if there is no matching more popular than M.

Popular matchings were first considered by Gardenfors [5] who showed that not every instance allows a popular matching. Abraham *et al.* [3] gave the first polynomial time algorithms to determine if a popular matching exists and if so, to produce one. The first algorithm runs in $O(n + m)$ time and works for the special case of strict preference lists; a second $O(\sqrt{n}m)$ time algorithm solves the problem for the general case where ties are allowed. They noted that maximum cardinality matching can be reduced to finding a popular matching in an instance with ties (by letting every edge be of rank 1) thus a linear time algorithm for the general case is unlikely.

Observe that the above definition of popular matching does not make any distinction between individuals—the opinion of every applicant is valued equally. But what if we had some preferred set of applicants we would like to give priority over the rest? This option becomes particularly interesting when jobs are scarce or there is a lot of contention for a few good jobs.

To answer this question we propose a new definition for the more popular than relation under which every applicant x is given a positive weight $w(x)$. The *satisfaction* of M_1 with respect to M_2 is defined as the weight of the applicants that prefer M_1 over M_2 minus the weight of those who prefer M_2 over M_1. Then M_1 is more popular than M_2 if the satisfaction of M_1 w.r.t. M_2 is positive. We believe this is an interesting generalization of popular matchings that addresses the natural need to assign priorities (weights) to the applicants while retaining the one-sided preferences of the original setup.

In this paper we develop algorithms to determine if in a given instance a weighted popular matching exists, and if so, to produce one. For the case of strict preference lists we give an $O(n + m)$ time algorithm. When ties are allowed the problem becomes more involved; a second algorithm solves the general case in $O(\min(k\sqrt{n}, n)m)$ time, where k is the number of distinct weights the applicants are given.

Our approach consists in developing an alternative characterization for popular matchings that will naturally lead to an efficient algorithm. First, we introduce the notion of *well-formed matching*, and show that every popular matching is well-formed. While in the unweighted case one can show [3] that every well-formed matching is popular, this is not always the case when weights are introduced. To weed out those well-formed matchings that are not popular we develop a procedure that identifies and prunes certain bad edges which cannot be part of any popular matching. Finally, we prove that every well-formed matching in the pruned graph is indeed popular.

A related, but not equivalent, problem is that of computing a *rank-maximal matching*. Here we want to maximize the number of rank 1 edges, and subject to this, maximize the number of rank 2 edges, and so on. Irving *et al.* [7] showed how to solve this problem in $O(\min(C\sqrt{n}), n)m)$ time where C is the rank of the lowest ranked edge in a rank-maximal matching. Recent developments on popular matchings are the papers by Abraham *et al.* [4] on *voting paths* and by Mahdian [8] on *random popular matching*.

2 Strict Preference Lists

We first study the case where the preference lists provided by the applicants are strict but need not be complete. Let us partition A into categories $C_1, C_2, \ldots C_k$, such that the weight of applicants in category C_i is w_i, and $w_1 > w_2 > \ldots > w_k > 0$.

In order to ease the analysis we first modify the given instance slightly. For every applicant x, we create a last resort job $l(x)$ and place it at the end of x's preference list. This does not affect whether the instance allows a popular matching or not, but it does force every popular matching to be applicant complete.

The plan is to develop an alternative characterization for popular matchings that will allow us to efficiently test if a given instance admits a popular matching, and if so to produce one. We define the notion of first and second jobs. For every applicant $x \in C_1$, let $f(x)$ be the first job on x's preference list, such a job is said to be an f_1-job; also let $s(x)$ be the first non-f_1-job on x's list. For $x \in C_i$ we recursively define $f(x)$ as the first non-$f_{<i}$-job on x's list, this will be an f_i-job; $s(x)$ will be the first non-$f_{j \leq i}$-job. Notice that $s(x)$ is ill defined when $f(x) = l(x)$. This, however, is not a problem since, as we will shortly see, the second job is used only when there is contention for the first job, which by definition never happens for $l(x)$.

The following properties about first and second jobs are easy to check:

- The set of f_i-jobs is disjoint from the set of f_j-jobs for $i \neq j$.
- The set of f_i-jobs is disjoint from the set of s_j-jobs for $i \leq j$, but may not be for $i > j$.

Our alternative characterization for popular matchings will be based on *well-formed matchings*.

Definition 1. *A matching is well-formed if it has the following two properties: every f_i-job p is matched to $x \in C_i$ where $f(x) = p$, and every applicant x is matched either to $f(x)$ or $s(x)$.*

Realize that when $k = 1$ our definition of well-formed matching becomes the characterization proposed by Abraham *et al.* [3]. For the unweighted case ($k = 1$) they show that a matching is popular if and only if it is well-formed. Unfortunately, for $k > 1$, not every well-formed matching is popular. To exemplify this consider the instance in Fig. 1. There are only two well-formed matchings: $M_1 = \{(x_1, A), (x_2, C), (x_3, D), (x_4, E)\}$ and $M_2 = \{(x_1, A), (x_2, C), (x_3, E), (x_4, D)\}$. However, M_1 is not popular because $\{(x_2, A), (x_3, C), (x_4, D)\}$ is more popular than M_1. On the other hand, as we will see, M_2 is popular.

Nevertheless, we can still prove one direction of the implication.

Theorem 1. *Let M be a popular matching, then M is well-formed.*

$$w(x_1) = 7 \qquad x_1 \begin{array}{|l} A\ B\ C \end{array}$$
$$w(x_2) = 4 \qquad x_2 \begin{array}{|l} A\ C\ D \end{array}$$
$$w(x_3) = 2 \qquad x_3 \begin{array}{|l} C\ A\ D\ E \end{array}$$
$$w(x_4) = 2 \qquad x_4 \begin{array}{|l} A\ D\ E \end{array}$$

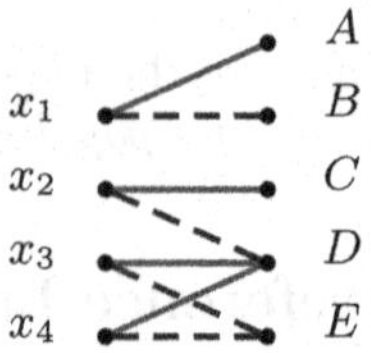

Fig. 1. Instance with strict preferences. The graph on the right shows the first jobs (solid lines) and second jobs (dashed lines).

One could be tempted to discard the current definition of well-formed matching and seek a stronger one that will let us replace the *if then* of Theorem 1 with an *if and only if*. As we will see, in proving the theorem we only use the fact that $w_i > w_{i+1}$. Armed with this sole fact Theorem 1 is the best we can hope for because if the weights are sufficiently spread apart ($w_i \geq 2w_{i+1}$), one can show that every well-formed matching is in fact popular.

We proceed to prove Theorem 1 by breaking it down into Lemmas 1 and 2.

Lemma 1. *Let M be a popular matching, then every f_i-job p is matched to an applicant $x \in C_i$ where $f(x) = p$*

Proof. By induction on i. For the base case let $x \in C_1$ and $f(x) = p$. For the sake of contradiction assume p is matched to y and $f(y) \neq p$. If $y \in C_{s>1}$, then promote x to p and demote y to $l(y)$. The swap improves the satisfaction by $w_1 - w_s > 0$, but this cannot be since M is popular. If $y \in C_1$ then promote x to p and y to $f(y)$, and demote applicant $z = M(f(y))$. The improvement in satisfaction is now $w_1 + w_1 - w(z) > 0$.

For the inductive case let $x \in C_i$. Assume like before that $f(x) = p$ but $M(p) = y$ and $f(y) \neq p$. If $y \in C_{s>i}$, then promote x and demote y to get a change in satisfaction of $w_i - w_s > 0$. If $y \in C_{s<i}$ then by induction $f(y)$ is matched to $z \in C_s$, promoting x to p and y to $f(y)$ while demoting z changes the satisfaction by $w_i + w_s - w_s > 0$. Finally, suppose $y \in C_i$. Let $z = M(f(y))$, if $z \in C_i$ the usual promotions change the satisfaction by $w_i > 0$. Note that if $z \in C_{s \neq i}$ then $f(z) \neq f(y)$, letting y play the role of x above handles the case. In every case we reach the contradiction that M is not popular, therefore the lemma follows. $\qquad\square$

Lemma 2. *Let M be a popular matching, then every $x \in A$ is matched either to $f(x)$ or $s(x)$.*

Proof. As a corollary of Lemma 1 no applicant x can be matched to a job which is strictly better than $f(x)$ or in between $f(x)$ and $s(x)$. Hence we just need to show that x cannot be matched to a job which is strictly worse than $s(x)$. For the sake of contradiction let us assume this is the case.

Let $x \in C_i$ and $p = s(x)$. Note that p must be matched to some applicant y, otherwise we get an immediate improvement by promoting x to p. If $y \in C_{s>i}$ then promoting x and demoting y gives us a more popular matching because $w_i - w_s > 0$. Otherwise y belongs to $C_{s \leq i}$ in which case $f(y) \neq p$. Lemma 1 tells us that there exists $z \in C_s$ matched to $f(y)$. Promoting x to $s(x)$ and y to $f(y)$ while demoting z improves the satisfaction by $w_i + w_s - w_s > 0$. A contradiction.   □

Let G' be a subgraph of G having only those edges between applicants and their first and second jobs. See the graph on the right of Fig. 1. Theorem 1 tells us that every popular matching must be contained in G'. Ideally we would like every well-formed matching in G' to be popular, unfortunately this is not always the case. To fix this situation, we will prune some edges from G' that cannot be part of any popular matching. Then we will argue that every well-formed matching in the pruned graph is popular. In order to understand the intuition behind the pruning algorithm we need the notion of *promotion path*.

Definition 2. *A promotion path w.r.t. a well-formed matching M is a sequence $p_0, x_0, \ldots, p_s, x_s$, such that $p_i = f(x_i)$, $(x_i, p_i) \in M$, and for all $i < s$, applicant x_i prefers p_{i+1} over p_i.*

Such a path can be used to free p_0 by promoting x_i to p_{i+1}, for all $i < s$, and leaving x_s jobless. We say the cost (in terms of satisfaction) of the path is $w(x_s) - w(x_0) - \ldots - w(x_{s-1})$, as everyone gets a better job except x_s. To illustrate this consider the instance in Fig. 1, and the well-formed matching $\{(x_1, A), (x_2, C), (x_3, D), (x_4, E)\}$. The sequence D, x_3, C, x_2, A, x_1 is a promotion path with cost $w(x_1) - w(x_2) - w(x_3) = 1$ that can be used to free D.

To see how promotion paths come into play, let M be a well-formed matching and M' be any other matching. Suppose y prefers M' over M, we will construct a promotion path starting at $p_0 = M'(y)$. Note that p_0 is an f-job and must be matched in M to x_0 such that $f(x_0) = p_0$. Thus, our path starts with p_0, x_0. To extend the path from x_i, check if x_i prefers M' over M, if that is the case, $p_{i+1} = M'(x_i)$ and $x_{i+1} = M(p_{i+1})$, otherwise the path ends at x_i. Notice that if $x_i \in C_s$ then $x_{i+1} \in C_{<s}$; this is important as it implies that the path must eventually end because when we reach $x_i \in C_1$, M' cannot improve on $M(x_i)$. Coming back to y, the applicant who induced the path, note that if $w(y)$ is greater than the cost of the path, then M is not popular because using the promotion path and promoting y to p_0 gives us a more popular matching. On the other hand it is easy to see that if for every applicant y, the cost of the path induced by y is at least $w(y)$, then M' is not more popular than M.

The pruning procedure keeps a label $\lambda(p)$ for every f_i-job p. Based on these labels we will decide which edges to prune. The following invariant states the meaning these labels carry.

Invariant 1. *Let p be an f_i-job and M be any well-formed matching contained in the pruned graph. A minimum cost promotion path out of p w.r.t. M has cost exactly $\lambda(p)$.*

We now describe the pruning procedure whose pseudo-code is given in Fig. 2. The algorithm works in iterations. In the ith iteration we do two things. First, we prune some edges incident to C_i, making sure that these edges do not belong

```
PRUNE-STRICT(G)
  1   All f₁-jobs get a label of w₁.
  2   for i = 2 to k
  3       for x ∈ Cᵢ
  4           if λmin(x, f(x)) < wᵢ
  5               then return "no popular matching exists"
  6       for p ∈ fᵢ-job
  7           let S be the set {x ∈ A|f(x) = p}
  8           if S = {x}
  9               then λ(p) = min(wᵢ, λmin(x, f(x)) − wᵢ)
 10               else  λ(p) = wᵢ
 11                     for x ∈ S such that λmin(x, p) < 2wᵢ
 12                         prune the edge (x, p)
 13   for x ∈ A such that λmin(x, s(x)) < w(x)
 14       prune the edge (x, s(x))
```

Fig. 2. Pruning the graph

to any popular matching. Second, we label all the f_i-jobs such that Invariant 1 holds for them. Note that later pruning cannot falsify the invariant for f_i-jobs as promotion paths out of these jobs only use edges incident to applicants in $C_{\leq i}$.

In the first iteration we do not prune any edges. Notice that a promotion path out of an f_1-job must end in its C_1 mate, therefore line 1 sets the label of all f_1-jobs to w_1.

At the beginning of the ith iteration we know the invariant holds for all $f_{<i}$-jobs. Consider an applicant $x \in C_i$. Let q be a job x prefers over $f(x)$. Note that q must be an $f_{<i}$-job, therefore, in any well-formed matching included in the pruned graph the min cost promotion out q has cost $\lambda(q)$. We can use the path to free q and then promote x to it, the total change in satisfaction is $w_i - \lambda(q)$. Therefore, if $\lambda(q) < w_i$ no popular matching exists. Lines 3–5 check for this, the expression $\lambda_{\min}(x, r)$ is a shorthand notation for $\min_q \lambda(q)$ where q is a job x prefers over r.

Let p be an f_i-job and S be the set of applicants in C_i whose first job is p, also let M be a well-formed matching contained in the pruned graph. Suppose S consists of just one applicant x, then (x, p) must belong to M. A promotion path out of p either ends at x or continues with another job which x prefers over p. Therefore $\lambda(p) = \min(w_i, \lambda_{\min}(x, f(x)) - w_i)$, which must be positive. On the other hand, if $|S| > 1$, only one of these applicants will be matched to p while the rest will get their second job. Suppose $M(p) = x \in S$. Invariant 1 tells us that there exists a promotion path w.r.t. M out of p with cost $\lambda_{\min}(x, p) - w_i$ that can be used to free p, which in turn allows us promote one of the other applicants in $S - x$ to p. Therefore if $\lambda_{\min}(x, p) < 2w_i$, M is not popular, which means the edge (x, p) cannot belong to any popular matching and can safely be pruned. We set $\lambda(p) = w_i$ because in the pruned graph p can only be matched to $x \in S$ such that $\lambda_{\min}(x, p) \geq 2w_i$. Lines 6–12 capture exactly this.

Finally, lines 13–14 prune edges $(x, s(x))$ that cannot be part of any popular matching because of promotion paths out of jobs between $f(x)$ and $s(x)$ on x's list, with cost $\lambda_{\min}(x, s(x)) < w_i$.

Running the pruning algorithm on the example in Fig. 1, the jobs are labeled $\lambda(A) = 7$, $\lambda(C) = 3$, and $\lambda(D) = 2$. The only edge pruned is (x_3, D) because both x_3 and x_4 have D as their first job and $\lambda_{\min}(x_3, D) = 2 < 4 = 2w(x_3)$.

We have argued that no pruned edge can be present in any popular matching, let us now show that every well-formed matching M in the pruned graph is indeed popular. Let M' be any other matching, our goal is to show that M' is not more popular than M. Suppose x prefers M' over M, this induces a promotion path at $M'(x)$ with respect to M. If x gets his first job in M then the cost of such a path is at least $\lambda_{\min}(x, f(x)) \geq w_i$. Otherwise, $M(x) = s(x)$ and lines 13-14 make sure the cost at the promotion path is at least w_i. Since this holds for every applicant x, M' cannot be more popular than M.

It is entirely possible that the pruned graph does not contain any well-formed matching. In this case we know that no popular matching exists.

Theorem 2. *In the case of strict preferences lists, we can find a weighted popular matching, or determine that none exists, in $O(n + m)$ time.*

Let G' be the graph with edge set $\{(x, f(x)), (x, s(x)) \,|\, x \in A\}$. Assuming the applicants are already partitioned into categories C_i, we can compute G' and prune it in $O(n + m)$ time. Thus, finding a popular matching reduces to finding a well-formed matching in G'. Abraham *et al.* [3] showed how to build a well-formed matching in G', if one exists, within the same time bounds.

Recall that at the beginning we modified the instance by adding a dummy last resort job at the end of everyone's list. A natural objective would be to find a popular matching that minimizes the number of applicants getting a dummy job. The cited work shows how to do this in $O(n + m)$ time, and thus it carries over to our problem.

3   Preference Lists with Ties

Needless to say, if ties are allowed in the preference lists, the solution from the previous section does not work anymore. We will work out an alternative definition for first and second jobs which will lead to a new definition of well-formed matchings. Like in the case without ties all popular matchings are well-formed, but the other way around does not always hold. We will show how to prune some edges that cannot be part of any popular matching to arrive at the goal that every well-formed matching in this pruned graph is popular.

Let us start by revising the notion of first job. For $x \in C_1$, let $f(x)$ be the set of jobs on x's list with the highest rank. Let G_1 be the graph with edges between applicants in C_1 and their first jobs. We say a job/applicant is *critical* in G_1 if it is always matched in every maximum matching of G_1. For $x \in C_i$, we inductively define $f(x)$ as the highest ranked jobs on x's list which are not critical in G_{i-1}. The graph G_i includes G_{i-1} and edges between C_i and their first jobs. A job/applicant is critical in G_i if it is always matched in every maximum matching of G_i or was previously declared critical in $G_{<i}$.

If $x \in C_i$ is non-critical in G_i we define $s(x)$ as the highest ranked set of jobs on x's list which are not critical in G_i. If x is critical then $s(x)$ is the empty set. When x is not critical we can show that all the jobs in $f(x)$ are critical, therefore $f(x)$ and $s(x)$ are always disjoint.

Definition 3. *A matching M is* well-formed *if, for all $1 \leq i \leq k$, the matching $M_i = M \cap E[G_i]$ is maximum in G_i, and every applicant x is matched within $f(x) \cup s(x)$.*

Note that when there are no ties all these definitions are identical to the ones given in the previous section. Before proceeding to prove Theorem 1 in the context of ties we review some basic notions of matching theory.

The following definitions are all with respect to a given matching M. An *alternating path* is one that alternates between matched and free edges. An *augmenting path* is an alternating path that starts in a free vertex and ends either in a free vertex or a matched edge[1]. Augmenting along an alternating path P results in the matching $M \oplus P$, the symmetric difference of M and P.

In our proofs we will make use of the following property of non-critical nodes, which is part of the Gallai-Edmonds decomposition theorem. Let H be a bipartite graph and v be a vertex such that there exists a maximum matching of H that leaves v unmatched. Then, in every maximum matching M of H there exists an alternating path starting at v and ending with a free vertex in the same side of the bipartition as v. To see why this is true, consider another maximum matching O which does not use v, then in $O \oplus M$ there must be an alternating path w.r.t. M of even length that starts at v and ends a free (w.r.t. M) vertex.

Lemma 3. *Let M be a popular matching. Then, for all i, $M_i = M \cap E[G_i]$ is maximum in G_i.*

Proof. As usual, we use induction to prove the lemma. For the base case, suppose M_1 is not maximum, then there must be an augmenting path w.r.t. M_1 starting at $x \in C_1$ and ending at p. If p is free in M then we augment along the path to improve the satisfaction by w_1, so let us assume there exists $y = M(p)$. If $y \in C_{s>1}$ then augmenting and demoting y gives us an improvement in satisfaction of $w_1 - w_s > 0$. Suppose then that $y \in C_1$, and let q be a job in $f(y)$. Since $(y, p) \notin M_1$, applicant y must prefer q over p. If the job after x in the augmenting path is q then we can replace x by y and augment, otherwise we can augment along the old path and then promote y to q in which case the satisfaction improves by $w_1 + w_1 - w(M(q)) > 0$. In every case we reach the contraction that M is not popular, thus M_1 must be maximum in G_1.

For the inductive step, if M_i is not maximum we can find like before an augmenting path starting at $x \in C_i$ and ending at a job p. Assume p is matched

---

[1] Observe that this is slightly different from usual definition that requieres the alternating path to start and end in a free vertex.

in M to $y \in C_{s \leq i}$ (the other cases are similar to the base case). Since $(y, p) \notin M_i$, we know by inductive hypothesis that p must be strictly worse than $f(y)$. We augment along the path to get M', in doing so we leave y unmatched. By inductive hypothesis M_s is maximum in $G_{s < i}$, therefore so is M'_s.

Let q be a job in $f(y)$, there are three cases to consider. First, if $y \in C_1$ then we can promote y to q and demote whoever is matched to q, the total change in satisfaction is at least $w_i + w_1 - w(M'(q)) > 0$. Secondly, suppose $y \in C_{1 < s < i}$. By definition of $f(y)$, q is non-critical in G_{s-1}. Thus we can find an alternating path in M'_{s-1} starting at q and ending at a free job r. Note that r cannot be free in M'_s, otherwise it would not be maximum in G_s, therefore r must be matched in M' to $z \in C_s$. Promoting y to p and augmenting along the path, while demoting z gives us a change in satisfaction of $w_i + w_s - w_s > 0$. Finally, we need to consider the case where $y \in C_i$; we proceed to find z as before, except now z need not belong to C_i. A similar case analysis as in the proof of Lemma 1 finishes the argument. □

Armed with Lemma 3, it is easy to show the second part of the characterization, which is stated without a proof due to lack of space.

Lemma 4. *Let M be a popular matching, then every applicant x is matched within $f(x) \cup s(x)$.*

This finishes the proof of Theorem 1 under the new definition of well-formed matching. Thus every popular matching is contained in G', the graph consisting only of those edges between applicants and their first and second jobs. Since the new definition generalizes the one from the previous section we again encounter the problem that not every well-formed matching is popular. We proceed as before, pruning certain edges which are not part of any popular matching. Finally, we show that every well-formed matching in the pruned graph is popular.

It is time to revise the definition of promotion path. Let M be a well-formed matching. Our promotion path starts at p_0, a job critical in G_{i_0}, but non-critical in $G_{i_0 - 1}$. We find an alternating path in G_{i_0} w.r.t. M_{i_0} from p_0 to x_0 which starts and ends with a matched edge; we augment along the path to get M'. Let p_1 be a job which according to x_0 is better than $f(x_0)$ (or as good, but not in $f(x_0)$), moreover let p_1 be critical in G_{i_1}, but not in $G_{i_1 - 1}$. Since $x_0 \in C_{>i_1}$, the matching M'_j is still maximum in $G_{j \leq i_1}$. Find a similar alternating path in G_{i_1} w.r.t. M'_{i_1} from p_1 to x_1, update M', and so on. Finally, every applicant x_i is assigned to p_{i+1}, except for x_s, the last applicant in the path, who is left jobless. The cost of the path is defined as the satisfaction of M with respect to M', or equivalently, $w(x_s)$ minus the weight of those applicants $x_{i<s}$ who like p_{i+1} better than $f(x_i)$, recall that p_{i+1} may be as good as, but not in, $f(x_i)$. This is the price to pay, in terms of satisfaction, to free p_0 using the path.

To see why this is the right definition, let M be a well-formed matching and M' be any other matching. Suppose y prefers M' over M, we will construct a promotion path starting at $p_0 = M'(y)$. Since M is well-formed, p_0 must be critical; let i_0 be the smallest i such that p_0 is critical in G_i. Taking $M_{i_0} \oplus M'_{i_0}$ we can find an alternating path that starts with $(p_0, M(p_0))$ and ends at x_0 which is free in M'_{i_0}—the path cannot end in a job that is free in M_{i_0} because p_0 is critical. Either x_0 gets a worse job under M', in which case the promotion path ends, or gets a job p_1 which is better than $f(x_0)$, or just as good but does not belong to $f(x_0)$. We continue growing the path until we run into an applicant x_s who prefers M over M', notice that since $i_j > i_{j+1}$ we are bound to find such an applicant. Now, if the cost of the path is less than $w(y)$ then we know the well-formed matching M is not popular after all. On the other hand, if the cost of the path induced by y is at least $w(y)$, for all such y, we can claim that M' is not more popular than M.

We are ready to discuss the algorithm for pruning the graph in the presence of ties, which is given in Fig. 3. In the ith iteration we prune some edges incident to applicants in C_i making sure these edges do not belong to any popular matching, and label those jobs that became critical in G_i such that Invariant 2 holds for them.

Invariant 2. *Let p be a critical job in G_i, and M be a matching in the pruned graph, maximum in all $G_{\leq i}$, i.e., $M_j = M \cap E[G_j]$ is maximum in G_j for all $j \leq i$. A minimum cost promotion path out of p w.r.t. M has cost exactly $\lambda(p)$.*

In the first iteration we do not prune any edges from G_1. Let p be a critical job in G_1, and M be a maximum matching in G_1. Every promotion path w.r.t. M out of p must end in some C_1 applicant, therefore, in line 1 we set $\lambda(p) = w_1$.

For the ith iteration we assume the invariant holds for those jobs critical in G_{i-1}. Suppose there exists an applicant $x \in C_i$ such that $\lambda_{\min}(x, f(x)) < w_i$. Then in every well-formed matching in the pruned graph we can use a promotion path to free a job which x prefers over $f(x)$, and then promote x to that job. This improves the satisfaction by $w_i - \lambda_{\min}(x, f(x)) > 0$. Therefore, no popular matching exists. Lines 3–5 check for this.

Pick an arbitrary matching M in the pruned graph, maximum in all $G_{\leq i}$. Let $x \in C_{j \leq i}$ be such that there exists an alternating path w.r.t. M from x to a free applicant $y \in C_i$. Furthermore, suppose $\lambda_{\min}(x, f(x)) < w_j + w_i$ or $\min_q \lambda(q) < w_i$ where q is a job not in $f(x)$, but as good as $f(x)$. Now let O be a matching in the pruned graph, also maximum in all $G_{\leq i}$. Suppose $O(x) \in f(x)$. We know there is an augmenting path w.r.t. O from x to some applicant y', note that $w(y') \geq w_i$. Augment along the path to get O'. While the matching O' may not be maximum in G_j (if $w(x) < w(y')$), it is still maximum in all $G_{<j}$. Invariant 2 tells us we can find a promotion path to free a certain job that x will be promoted to. Theses changes improve the satisfaction of the matching, therefore O cannot be popular, thus x cannot be matched within $f(x)$ in any popular matching. If $x \in C_i$ then the edges from x to $f(x)$ can be safely pruned. On the other hand, if $x \in C_{j<i}$, then x must be critical (otherwise we would have pruned the edges $(x, f(x))$ in the jth iteration) in which case there is no hope of finding a popular matching. Lines 6–10 check for this.

Finally, we must compute $\lambda(p)$ for jobs p that are critical in G_i but not in G_{i-1}. A promotion path out of p must begin with an alternating path starting and ending with a matched edge, going from p to some applicant x. It is not hard to see that there is always such an alternating path to somebody in C_i,

PRUNE-TIES(G)
1   All critical jobs in G_1 get a label of w_1.
2   **for** $i = 2$ **to** k
3   **for** $x \in C_i$
4   **if** $\lambda_{\min}(x, f(x)) < w_i$
5   **then return** "no popular matching exists"
6   **for** $x \in C_{j \leq i}$ having an alternating path to a free applicant $y \in C_i$
7   and $\lambda_{\min}(x, f(x)) < w_j + w_i$ or $\min_{\substack{q \notin f(x) \\ \text{as good as } f(x)}} \lambda(q) < w_i$
8   **if** $x \in C_i$
9   **then** prune edges between x and $f(x)$
10   **else return** "no popular matching exists"
11   **for** p critical in G_i, but non-critical in G_{i-1}
12   let $S = \{x \mid \exists \text{ alternating path from } x \text{ to } p\}$
13         $\lambda(p) = \min_{x \in S} \left\{ w_i, \lambda_{\min}(x, f(x)) - w_i, \min_{\substack{q \notin f(x) \\ \text{as good as } f(x)}} \lambda(q) \right\}$
14   **for** $x \in A$ such that $\lambda_{\min}(x, s(x)) < w(x)$
15   prune the edges between x and $s(x)$

Fig. 3. Pruning the graph

thus $\lambda(p) \leq w_i$. Note that if $x \in C_j$ is non-critical then $\lambda_{\min}(x, f(x)) \geq w_j + w_i$ and $\min_{\substack{q \notin f(x) \\ \text{as good as } f(x)}} \lambda(q) \geq w_i$, otherwise the edges $(x, f(x))$ would have beeen pruned earlier. We shall only explore alternating paths to critical applicants in some arbitrary matching M included in the pruned graph which is maximum in all $G_{\leq i}$. In fact we only care about reaching a critical applicant $x \in C_j$ with $\lambda_{\min}(x, f(x)) < w_j + w_i$ or $\min_{\substack{q \notin f(x) \\ \text{as good as } f(x)}} \lambda(q) < w_i$, as w_i is already given for $\lambda(p)$. Since M is an arbitrary matching, we would like to claim that a similar path can always be found in any other matching O included in the pruned graph, maximum in all $G_{\leq i}$. To show this, augment along the path to get M'_i, the resulting matching is not maximum in G_i any more. Take $M'_i \oplus O$, and consider the alternating path out of p. This path must end at an applicant y, matched in O_i, but free in M'_i—otherwise, if it ends in a job free in O_i, the job p is not critical. For the sake of contradiction suppose that $y \neq x$. Since x is critical, there must be a path in $M'_i \oplus O_i$ from x to y', free in O_i. In order for O to be maximal in all $G_{\leq i}$ and x critical we must have $w(y') < w(x)$. But since $w_i \leq w(y')$, this kind of path should have been found before in lines 6–9, we have reached a contradiction. Thus we set $\lambda(p)$ to the minimum of $\lambda_{\min}(x, f(x)) - w_i$ or $\min_{\substack{q \notin f(x) \\ \text{as good as } f(x)}} \lambda(q)$, for those applicants x that can be reached from p with an alternating path. Lines 11–13 do this.

The last thing to consider are non-critical applicants x who may get their second job. We can promote them to a job p strictly better than $s(x)$ and start a promotion path from there. If such exchange improves the satisfaction then the edges $(x, s(s))$ must be pruned. This is done in lines 14–15.

Theorem 3. *In the presence of ties we can find a weighted popular matching or determine that none exists in $O(\min(k\sqrt{n}, n)m)$ time.*

Due to lack of space the implementation details and the complexity analysis of the algorithm are deferred to the journal version of this paper.

4   Conclusion

We have developed efficient algorithms for finding weighted popular matchings, a natural generalization of popular matchings. It would be interesting to study other definitions of the *more popular than* relation. For example, define the satisfaction of M over R to be the sum (or any linear combination) of the differences of the ranks of the jobs each applicant gets in M and R. Finding a popular matching under this new definition can be reduced to maximum weight matching, and vice versa. Defining the satisfaction to be a linear combination of the sign of the differences we get weighted popular matchings. We leave as an open problem to study other definitions that use a function "in between" these two extremes. Ideally, we would like to have efficient algorithms that can handle any odd step function.

Acknowledgment. Many thanks to Samir Khuller for suggesting the notion of weighted popular matchings and providing comments on earlier drafts.

References

1. A. Abdulkadiroğlu and T. Sönmez. Random serial dictatorship and the core from random endowments in house allocation problems. *Econ.*, 66(3):689–701, 1998.
2. D. J. Abraham, K. Cechlarova, D. F. Manlove, and K. Mehlhorn. Pareto optimality in house allocation problems. In *Proc. of ISAAC*, pages 3–15, 2004.
3. D. J. Abraham, R. W. Irving, T. Kavitha, and K. Mehlhorn. Popular matchings. In *Proc. of SODA*, pages 424–432, 2005.
4. D. J. Abraham and T. Kavitha. Dynamic matching markets and voting paths. To appear in SWAT, 2006.
5. P. Gardenfors. Match making: assignments based on bilateral preferences. *Behavioural Sciences*, 20:166–173, 1975.
6. A. Hylland and R. Zeeckhauser. The efficient allocation of individuals to positions. *Journal of Political Economy*, 87(2):293–314, 1979.
7. R. W. Irving, T. Kavitha, K. Mehlhorn, D. Michail, and K. Paluch. Rank-maximal matchings. In *Proc. of SODA*, pages 68–75, 2004.
8. M. Mahdian. Random popular matchings. To appear in EC, 2006.
9. Y. Yuan. Residence exchange wanted: a stable residence exchange problem. *European Journal of Operational Research*, 90:536–546, 1996.
10. L. Zhou. On a conjecture by Gale about one-sided matching problems. *Journal of Economic Theory*, 52(1):123–135, 1990.

Author Index

Lecture Notes in Computer Science

For information about Vols. 1–3976

please contact your bookseller or Springer